Trigonometric Identities

Basic Identities

$$\sec \theta = \frac{1}{\cos \theta}$$

$$\cot \theta = \frac{\cos \theta}{\sin \theta}$$

$$\csc \theta = \frac{1}{\sin \theta}$$

$$\sin^2 \theta + \cos^2 \theta = 1$$

$$\cot \theta = \frac{1}{\tan \theta}$$

$$1 + \tan^2 \theta = \sec^2 \theta$$

$$\tan \theta = \frac{\sin \theta}{\cos \theta}$$

$$1 + \cot^2 \theta = \csc^2 \theta$$

Sum and Difference Formulas

$$\sin (\theta \pm \phi) = \sin \theta \cos \phi \pm \cos \theta \sin \phi$$

$$\cos (\theta \pm \phi) = \cos \theta \cos \phi \mp \sin \theta \sin \phi$$

Double-Angle Formulas

$$\sin 2\theta = 2 \sin \theta \cos \theta$$

$$\cos 2\theta = \cos^2 \theta - \sin^2 \theta = 1 - 2 \sin^2 \theta = 2 \cos^2 \theta - 1$$

Half-Angle Formulas

$$\sin \frac{\theta}{2} = \pm \sqrt{\frac{1 - \cos \theta}{2}}$$

$$\cos \frac{\theta}{2} = \pm \sqrt{\frac{1 + \cos \theta}{2}}$$

Alternate Half-Angle Formulas

$$\sin^2 x = \frac{1}{2} (1 - \cos 2x)$$

$$\cos^2 x = \frac{1}{2} (1 + \cos 2x)$$

Law of Sines

$$\frac{a}{\sin A} = \frac{b}{\sin B} = \frac{c}{\sin C}$$

Pythagorean Theorem

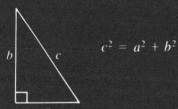

$$c^2 = a^2 + b^2$$

Law of Cosines

$$a^2 = b^2 + c^2 - 2bc \cos A$$

Second Edition

Basic Technical Mathematics with Calculus

Books in the Kuhfittig series in technical mathematics

Basic Technical Mathematics, Second Edition
Basic Technical Mathematics with Calculus, Second Edition
Technical Calculus with Analytic Geometry, Second Edition

Second Edition

Basic Technical Mathematics with Calculus

Peter Kuhfittig
Milwaukee School of Engineering

Brooks/Cole Publishing Company
Pacific Grove, California

Brooks/Cole Publishing Company
A Division of Wadsworth, Inc.

Printed in the United States of America

10 9 8 7 6 5 4 3 2 1

Library of Congress Cataloging in Publication Data

Kuhfittig, Peter K. F.
    Basic technical mathematics with calculus/Peter Kuhfittig. —
2nd ed.
      p. cm.
    Includes index.
    ISBN 0-534-10062-7:
    1. Mathematics—1961–   2. Calculus.   I. Title.
QA37.2.K84 1988
510—dc 19
                                                                   88-23223
                                                                            CIP

Sponsoring Editors: *Jeremy Hayhurst, Sue Ewing*
Editorial Assistant: *Virge Kelmser, Heidi Wieland*
Production Editor: *Joan Marsh*
Manuscript Editor: *Carol Dondrea*
Permissions Editor: *Carline Haga*
Interior and Cover Design: *Lisa Thompson*
Cover Photo: *Fred Otnes*
Art Coordinator: *Sue C. Howard*
Interior Illustration: *Carl Brown*
Print Buyer: *Vena M. Dyer*
Typesetting: *Syntax International*
Cover Printing: *Phoenix Color Corporation, Long Island City, New York*
Printing and Binding: *R.R. Donnelley & Sons, Harrisonburg, Virginia*

P R E F A C E

Purpose of This Book

The main purpose of the second edition of *Basic Technical Mathematics with Calculus* is to enhance the student-oriented features that have contributed to the success of the first edition. Based on classroom experience and on suggestions from users of the first edition, a number of changes have been made:

New Features

1. Certain portions of the book were completely rewritten: Chapters 1, 2, and 9, and Sections 19.3, 20.3, 22.6, 22.7, 23.5, 24.1, 26.6, 28.1, and 28.4. Many other sections have been extensively revised.
2. Several new sections have been introduced: scientific calculator operations (1.4), BASIC programming (Appendix D), arithmetic progressions (18.1), subsections on the order of operations (1.2) and addition of vectors by components (9.3), a separate section on formulas (2.3), and a review of geometry (Appendix E).
3. A new section on integration of rational functions has been introduced in Chapter 25 (Section 25.8). To increase the book's flexibility, however, the discussion of partial fraction expansions in Chapter 29 has been retained. Thus Chapter 29 is independent of Section 25.8.
4. The student-oriented features of the first edition were enhanced through additional diagrams and marginal annotations, and by providing greater detail in the presentation.

Some additional new features are:

5. More space is given to calculator operations throughout the book.
6. BASIC notations are introduced in several places in the text.
7. Most of the exercise sets have been expanded and, in some cases, rearranged. All exercise sets containing algebra word problems have been completely revised.
8. Greater use is made of decimal degrees.

9. The polar form of a complex number is now denoted by $r\angle\theta$.
10. The number of examples and figures has been greatly increased, particularly in the calculus part of the book. More figures have been added to the answer section.
11. All answers to the review and cumulative review exercises are given in the answer section.
12. More technical applications have been included.

Continuing Features

The presentation in this book is exceptionally student-oriented and fills the needs of today's technology student in many ways:

1. The approach is concrete and intuitive.
2. Drill exercises make use of notations commonly encountered in technical areas.
3. Most sections contain exercises that illustrate how mathematics is applied to technical problems.
4. Color ink is not used in a merely decorative way. Its main function is to help explain difficult steps.
5. The most important concepts are boxed and labeled for easy reference; other concepts are identified by marginal labels.
6. Important procedures are summarized; step-by-step procedures are provided whenever appropriate.
7. Calculator operations are discussed throughout the book.
8. Examples are worked out in great detail. Marginal notes are used to help explain the steps.
9. Common pitfalls are pointed out in special segments called *Common Errors.*
10. A large number of drill exercises is included to help reinforce basic concepts.

The use of realistic notations, even in drill exercises, is particularly important, since students tend to have great difficulty in transferring skills to technical problems that use a different notation. No prior knowledge of the different technical areas is assumed, however.

Some other features are:

11. Throughout the book, the use of graphs has been given particular attention.
12. Both metric and English units are employed, with SI notation generally used for the former.
13. Every chapter ends with a set of review exercises. Cumulative Review Exercises are given at the end of every third chapter.
14. A background in algebra and geometry is assumed for most of the material in this book. However, the discussion begins at a sufficiently low level to help overcome deficiencies in some areas.

Flexibility

While this book allows considerable flexibility in the order of presentation of topics, careful attention has been paid to the fact that most technical curricula require certain topics at definite times. For example, although scientific notation is introduced in Chapter 1 and basic trigonometry in Chapter 4, a discussion of these topics may be postponed. On the other hand, trigonometric identities (Chapter 16) can be taken up after Chapter 10. Since determinants up to the third order are discussed in Chapter 3, higher-order determinants and matrices, which are first discussed in Chapter 15, can be introduced earlier or omitted altogether.

The basic plan of the second part of the book is to introduce the differential and integral calculus in Chapters 20–23, with two-year technology programs in mind. The later chapters treat various topics from more advanced areas and may be used according to the requirements of individual programs.

Coverage and Scope

Chapter 1 discusses basic algebra. The topics covered range from signed numbers and order of operations to operations with polynomials. Zero and negative exponents are discussed briefly in case the teacher wants to cover scientific notation (Section 1.9) early in the course, but these topics may be postponed to Chapter 10. Because of the large number of applied problems in this book, calculator operations are also introduced early (Section 1.4). (Calculations with approximate numbers and conversion of units are presented in Appendices A and B, respectively.)

Chapter 2 begins with the solution of linear equations and formulas, followed by a section on ratio and proportion and a section on variation. These topics are a natural continuation of the discussion of equations and formulas and provide a concrete basis for the definition of a function in Section 2.8.

Chapter 3 deals with systems of linear equations and their applications. The methods of solution discussed are graphing, addition or subtraction, and substitution. Determinants up to the third order are also introduced. (Higher-order determinants are covered in Chapter 15.)

Chapter 4 is an introduction to right-triangle trigonometry. The topics covered are basic definitions, values of trigonometric functions, and right-triangle applications.

Chapter 5 on factoring and fractions is organized to allow a gradual mastery of factoring: Basic special products, introduced in Section 5.1, are followed by the corresponding factoring cases in Section 5.2. Section 5.3 introduces more special products, and Section 5.4 the corresponding factoring cases. Factoring by grouping is then discussed in Section 5.5. The rest of the chapter is devoted to operations with fractions, complex fractions, and fractional equations.

Chapter 6 covers quadratic equations and their applications. The methods of solution discussed are factoring, completing the square, and the use of the quadratic formula.

Chapter 7 expands the discussion of trigonometric functions begun in Chapter 4. The topics covered are the functions of angles in any quadrant, functions of special angles, radian measure, and applications of radian measure, including linear and angular velocity.

Chapter 8 covers the graphs of trigonometric functions, as well as graphing by addition of coordinates. Considerable space is given to applications of sinusoidal functions.

Vectors and applications of vectors are treated in Chapter 9. The sections on the sine and cosine laws contain additional vector applications.

Chapter 10 comprises a detailed treatment of exponents and radicals: zero, negative, and fractional exponents, and fundamental operations with radicals.

Chapter 11 covers complex numbers: rectangular, polar, and exponential forms of complex numbers and powers and roots by De Moivre's theorem. The chapter ends with a brief discussion of phasors.

Chapter 12 contains a thorough treatment of logarithmic and exponential functions with special emphasis on applications. Also included are discussions of natural logarithms, properties of logarithms, exponential and logarithmic equations, and graphing on logarithmic paper. As in the case of trigonometric functions, the emphasis is on the use of calculators rather than of tables. (Computations with logarithms are introduced only briefly.)

Chapter 13 begins with a brief discussion of conic sections to provide a basis for solving systems of two quadratic equations by means of graphs. Algebraic methods are discussed next, followed by equations in quadratic form and fractional equations.

Chapter 14 contains a detailed treatment of higher-order equations. Approximation of irrational roots by linear interpolation is also discussed in detail.

Higher-order determinants and a detailed treatment of matrices can be found in Chapter 15.

Chapter 16 is devoted to trigonometric identities, trigonometric equations, and inverse trigonometric functions.

Chapter 17 on inequalities covers both graphical and algebraic methods of solution. The chapter ends with an optional section on linear programming.

Arithmetic progressions, geometric series, the binomial theorem, and binomial series are treated in Chapter 18.

Chapter 19 covers the traditional topics of analytic geometry. Although intended mainly for use in calculus, a number of applications of conic sections are also discussed. The derivative is introduced in Chapter 20 and applications of the derivative in Chapter 21. Chapters 22 and 23 cover basic integration, including the trapezoidal rule and Simpson's rule. Although based on Riemann sums, the emphasis in setting up integrals is on a shortcut using a singly typical element, referred to as a "sloppy Riemann sum." This shortcut is used extensively in Chapter 23 on applications of integration. The purpose is to enable the student to gain the necessary insight for setting up integrals in many different situations, rather than relying on memorized formulas.

Chapter 24 covers transcendental functions and their applications to various technical fields. Also included is a brief discussion of L'Hospital's rule. Chapter 25 develops different integration techniques. These techniques are covered in considerable detail and may be selected according to individual needs. For example, if time constraints do not allow a detailed discussion of trigonometric substitution, integrals of this form can be obtained by use of tables, discussed in Section 25.9. As noted earlier, the discussion of partial fractions (Section 25.8) can be postponed to Chapter 29.

Chapter 26 emphasizes power-series expansions, but it contains an optional section on tests of convergence. A section on Fourier series is included for use in electrical technology curricula.

Chapters 27, 28, and 29 are all devoted to differential equations since these provide particularly interesting and powerful applications to numerous technical fields. Chapter 29 on Laplace transforms includes a section on partial fractions. Although time-consuming, partial fractions are essential to the Laplace transform technique.

A brief discussion of Newton's method is given in Chapter 30. This method of solving equations is particularly useful for technology students.

Appendices A–D discuss, respectively, approximation and measurement, reduction and conversion of units, BASIC programming, and review of geometry.

Supplements

The answers to the odd-numbered exercises and all answers to the review and cumulative review exercises are given in the answer section; the answers to the remaining exercises are published in a separate answer book. A Student Solutions Manual, with approximately every fourth problem worked in detail, is also available.

Acknowledgments

I am indebted to the students at the Milwaukee School of Engineering for having provided me with the stimulus to write this book. I am especially grateful to my colleagues at the Milwaukee School of Engineering for checking all the answers to the exercises and providing numerous valuable suggestions and constructive criticisms. In particular, I wish to thank Professors George L. Edenharder, Stanley J. Guberud, Richard B. Hernday, Dorothy J. Johnson, Ralph J. Jondle, Janet G. Klein, Robert P. Schilleman, and Andrew B. Schmirler.

I should also like to thank the staff of Brooks/Cole Publishing Company and the following reviewers for their cooperation and help in the preparation of the manuscript: Glenn Boston, Catawba Valley Technical College; William Brower, New Jersey Institute of Technology; Gladys Crates, Chattanooga State Technical Community College; Zsolt Domotorffy, Cuyahoga Community College; Molly Fails, Terra Technical College; Harold Hackett, Jr., State University of New York A & T College at Alfred; Wendell Johnson, University of Akron Community and Technical College; John Monroe, University of Akron; Catherine Murphy, Purdue University—Calumet Campus; Joseph Murray, State University of New York, Stony Brook; James Schmeidler; Lary Skane, Catonsville Community College; Al Swimmer, Arizona State University; Jan Wynn, Brigham Young University; and Karen Ann Yoho, Fairmont State College.

Peter Kuhfittig

BRIEF CONTENTS

CONTENTS

Chapter 3

Systems of Linear Equations and Introduction to Determinants   **90**

Chapter 4

Introduction to Trigonometry   **127**

Chapter 5

Factoring and Fractions   **155**

Chapter 11   Complex Numbers   335

Chapter 12   Logarithmic and Exponential Functions   359

Chapter 13   More on Equations and Systems of Equations   393

Chapter 14   Higher-Order Equations in One Variable   413

Chapter **15**   **Determinants and Matrices**   **439**

Chapter **16**   **Additional Topics in Trigonometry**   **479**

Chapter **17**   **Inequalities**   **520**

Appendix E: Tables   988

Appendix F: Answers to Selected Exercises   1001

Index   1070

Second Edition

Basic Technical Mathematics with Calculus

C H A P T E R **1**

Fundamental Operations

1.1 Numbers

Rational number

Irrational number

While the nature of mathematics has changed over the centuries, it is generally agreed that mathematics has always been based on the concepts of number and form. We will begin our study with a brief review of the number concept.

First recall that a number is called **rational** if it can be written as a ratio of two integers. For example, the numbers

$$\frac{2}{3}, \quad \frac{7}{11}, \quad \frac{19}{6}, \quad 3 \text{ or } \frac{3}{1}, \quad 10 \text{ or } \frac{10}{1}$$

are rational. A number that cannot be written as a ratio of two integers is called **irrational**. Examples of irrational numbers are

$$\sqrt{2}, \quad \sqrt{5}, \quad \sqrt[3]{4}, \quad \pi$$

(Thus $\frac{22}{7}$ is not equal to π; it is only an approximation.)

Rational numbers have a unique repeating decimal expansion. For example,

$$\frac{1}{11} = 0.090909\ldots, \text{ or } 0.0909\overline{09}$$

the three dots indicating an omission of the terms that complete the expression. (This equality can be checked by direct division.) Note that the cycle 09 repeats indefinitely. Similarly,

$$\frac{1}{5} = 0.200\bar{0}$$

In this case the zero repeats indefinitely.

Irrational numbers do not have repeating decimal expansions. For example, in

$$\sqrt{2} = 1.4142136\ldots$$

there are no cycles of repeating integers.

Although we cannot pursue this idea any further at this point, it should be noted that irrational numbers fall into two categories: algebraic and transcendental. The former is a root of an algebraic equation, which will be defined later. (Thus rational numbers are by definition algebraic.) The number π and most trigonometric ratios, on the other hand, are transcendental.

Another important type of number that we need to consider is a **negative number**, which is most easily defined geometrically. Let us assume that the set of all positive numbers can be placed on a line so that each positive number corresponds to a unique point. To accomplish this, we select an arbitrary point, called the **origin**, to be associated with the number 0. The positive integers are then placed at equal intervals to the right of the origin (Figure 1.1). The other numbers are assigned to appropriate points between the integers. The number $\frac{3}{2}$, for example, is located halfway between 1 and 2.

Negative numbers are now defined as follows: A positive number a has a "mirror image" on the left side of the origin. This mirror image is denoted by $-a$. Also, $a + (-a) = 0$.

Negative number

Origin

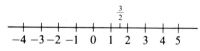

Figure 1.1

Negative Number: For every number a there exists a number $-a$ such that $a + (-a) = 0$. (The number 0 itself is neither positive nor negative.)

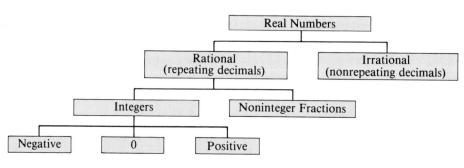

Figure 1.2

Greater than and
less than

Recall that the symbol $>$ means **greater than** and the symbol $<$ means **less than**. These relations carry over to negative numbers. For example, since -5 is to the left of -4, we have $-5 < -4$. For the same reason, $-2 < 3$, while $1 > -3$. Also, the symbol $\geq$ means **greater than or equal to** and $\leq$ means **less than or equal to**.

Real numbers

The numbers considered so far are collectively known as the set of **real numbers**, which are distinguished from **complex numbers**. Although not part of the real number system, complex numbers will play an important role in our later work.

The composition of the real number system is shown in Figure 1.2.

1.2 The Four Fundamental Operations; Order of Operations

Properties of Real Numbers

In performing the fundamental operations of addition, subtraction, multiplication, and division, we follow certain rules that we know and use intuitively. These rules are called the commutative, associative, and distributive laws.

For example,

$$2 + 3 = 3 + 2 \qquad \text{and} \qquad 4 \times 3 = 3 \times 4$$

In other words, addition and multiplication are **commutative**—they are independent of order.

Nor does the sum or product of any two numbers depend on the order of grouping. For example,

$$2 + (4 + 7) = (2 + 4) + 7$$

and

$$(3 \times 6) \times 5 = 3 \times (6 \times 5)$$

(The parentheses indicate that the quantity inside the parentheses is to be evaluated first.) These properties are called the **associative laws** of addition and multiplication, respectively.

Multiplication and addition are also connected by the **distributive law**, which states that the product of one number and the sum of two or more other numbers is equal to the sum of the products of the first number and each of the numbers in the sum. For example,

$$\mathbf{2} \times (\mathbf{7} + \mathbf{9}) = (\mathbf{2} \times \mathbf{7}) + (\mathbf{2} \times \mathbf{9})$$

These properties, together with the identity and inverse properties, constitute the properties of real numbers. They are summarized in the accompanying table.

The Properties of Real Numbers:

Addition	*Multiplication*

Commutative law

$$a + b = b + a \qquad\qquad ab = ba$$

Associative law

$$(a + b) + c = a + (b + c) \qquad\qquad (ab)c = a(bc)$$

Distributive law

$$a(b + c) = ab + ac$$
$$(b + c)a = ba + ca$$

Identity

There exists a unique real number 0 such that $a + 0 = a$ for every a.

There exists a unique real number 1 such that $a \cdot 1 = a$ for every a.

Inverse

For every real number a there exists a unique real number $-a$ such that $a + (-a) = 0$.

For every nonzero real number a there exists a unique real number $1/a$ such that $a \cdot 1/a = 1$.

Multiplication by 0

$a \cdot 0 = 0$ for every real number a.

Although first defined for positive numbers, *the fundamental operations are also valid for negative numbers.* In fact, the rules for signed numbers were designed to preserve the rules for positive numbers. To state these rules, we need the concept of **absolute value**, which is defined next.

Absolute Value: The absolute value of a positive number is the number itself. The absolute value of a negative number is the corresponding positive number. The absolute value of 0 is 0.

For example, the absolute value of 3 is 3, and the absolute value of -5 is 5. The absolute value is denoted by the symbol $| \quad |$, placed around the number. Thus

$$|3| = 3 \qquad \text{and} \qquad |-5| = 5$$

Similarly,

$$|4| = 4, \qquad |-8| = 8, \qquad \text{and} \qquad |-20| = 20$$

Stated informally, the absolute value of a number is the number without the negative sign. A formal definition of absolute value is the following:

$$|x| = \begin{cases} x, & \text{if } x \geq 0 \\ -x, & \text{if } x < 0 \end{cases}$$

For example, we know that $|-2| = 2$. By the formal definition, since $-2 < 0$, we have

$$|-2| = -(-2) = 2 \qquad x = -2$$

Operations with Signed Numbers

The concept of absolute value is used in the rules for operating with signed numbers, which are summarized next.

Rules for Operating with Signed Numbers:

Rule 1. To add two real numbers with like signs, find the sum of their absolute values and affix their common sign to the result.

Rule 2. To add two real numbers with unlike signs, find the difference of their absolute values and affix the sign of the number with the larger absolute value.

Rule 3. To subtract two real numbers, change the sign of the subtrahend (the number to be subtracted) and proceed as in addition.

Rule 4. The product (or quotient) of two real numbers with like signs is the product (or quotient) of their absolute values. The product (or quotient) of two real numbers with unlike signs is the negative of the product (or quotient) of their absolute values.

The following examples illustrate the rules for operating with signed numbers:

E X A M P L E **1** Perform the following operations:

a. $(-3) + (-9)$ **b.** $5 + (-8)$

Solution.

a. Since -3 and -9 have *like* signs, we use Rule 1 and add the absolute values ($3 + 9 = 12$) and then affix the common negative sign to the result:

$$(-3) + (-9) = -12$$

b. The numbers 5 and -8 have *unlike* signs. By Rule 2 we subtract 5 from 8 ($8 - 5 = 3$) and affix a negative sign to the result (since -8 has the larger absolute value):

$$5 + (-8) = -3$$

◀

E　X　A　M　P　L　E　**2**　Perform the indicated operations:

　　　a. $-5 - 10$　　**b.** $8 - 13$

Solution. By the rule for subtraction (Rule 3), we change the sign of the subtrahend and proceed as in addition:

　　　a. $-5 - 10 = -5 - (+10) = -5 + (-10) = -15$
　　　b. $8 - 13 = 8 + (-13) = -5$　　　　　　　　　　　　◀

E　X　A　M　P　L　E　**3**　Carry out the indicated operations:

　　　a. $(-7)(-8)$　　**b.** $\dfrac{-24}{6}$

Solution.

　　a. Note that -7 and -8 have *like* signs. So by Rule 4, the product is positive:

　　　　$(-7)(-8) = 56$

　　b. Here -24 and 6 have *unlike* signs. So by Rule 4, the quotient is negative:

　　　　$\dfrac{-24}{6} = -4$　　　　　　　　　　　　　　◀

E　X　A　M　P　L　E　**4**　**a.** $(-4)(-5)(-8) = (-4)(-5)(-8)$　　like signs
　　　　　　　　　　　　　$= (20)(-8)$　　unlike signs
　　　　　　　　　　　　　$= -160$

　　b. $\dfrac{(-6)(-5)}{-12} = \dfrac{(-6)(-5)}{-12}$　　like signs

　　　　　　$= \dfrac{30}{-12}$　　unlike signs

　　　　　　$= -\dfrac{30}{12}$

　　　　　　$= -\dfrac{5}{2}$　　　　　　　　　　　◀

　　　If a number is multiplied by itself repeatedly, the product can be written by use of *exponents*. For example,

　　$2 \times 2 \times 2$　is written　2^3

The number 2 is called the **base** and the number 3 the **exponent**. Similarly,

　　$4^5 = 4 \times 4 \times 4 \times 4 \times 4$

Here 4 is the base and 5 the exponent. In general,

$$a \cdot a = a^2 \text{ is read "}a \text{ squared" or "}a \text{ to the second power"}$$

$$a \cdot a \cdot a = a^3 \text{ is read "}a \text{ cubed" or "}a \text{ to the third power"}$$

$$a \cdot a \cdot a \cdot a = a^4 \text{ is read "}a \text{ to the fourth power" or simply "}a \text{ to the fourth," and so on}$$

The *n*th Power:

$$a^n = a \cdot a \cdot a \cdot \cdots \cdot a \, (n \text{ times})$$

is read "*a* to the *n*th power"; *a* is called the **base** and *n* the **exponent**.

E X A M P L E **5**   Evaluate 3^3 and $(-4)^4$.

Solution. $3^3 = 3 \times 3 \times 3 = 27$

$$(-4)^4 = (-4) \times (-4) \times (-4) \times (-4)$$

$$= 256 \qquad \blacktriangleleft$$

Order of Operations

If several operations are involved in a calculation, then multiplication and division are performed first. For example,

$$3 + 2 \times 7 = 3 + 14 = 17$$

If a different order is intended, we may use parentheses to indicate that the quantity inside the parentheses is to be evaluated first. For example,

$$(3 + 2) \times 7 = 5 \times 7 = 35$$

The order in which the different operations must be performed is summarized next.

Order of Operations:

  1. Evaluate quantities enclosed in parentheses (or other symbols of grouping).
  2. Evaluate powers.
  3. Perform all multiplications and divisions in the order in which they occur (from left to right).
  4. Perform all additions and subtractions in the order in which they occur (from left to right).

E X A M P L E **6**   **a.** $3 + 4(-5)$

$$= 3 + \mathbf{4(-5)}$$
$$= 3 + \mathbf{(-20)} \qquad \text{multiplication performed first}$$
$$= -17$$

b. $4 - \dfrac{-28}{7}$

$$= 4 - \dfrac{-28}{7}$$
$$= 4 - \mathbf{(-4)} \qquad \text{division performed first}$$
$$= 4 + 4 = 8 \qquad\qquad\qquad\qquad\qquad \blacktriangleleft$$

E X A M P L E **7**   Evaluate:

$$(-3)^2(7 - 9)$$

Solution. $(-3)^2(7 - 9)$

$$= (-3)^2(-2) \qquad \text{evaluating the expression enclosed in parentheses}$$
$$= 9(-2) \qquad\quad \text{evaluating the power}$$
$$= -18 \qquad\qquad\qquad\qquad\qquad\qquad\qquad\qquad \blacktriangleleft$$

E X A M P L E **8**   Evaluate:

$$2 - \frac{(-4)^2}{32} + 5$$

Solution. $2 - \dfrac{(-4)^2}{32} + 5$

$$= 2 - \frac{16}{32} + 5 \qquad \text{evaluating the power}$$
$$= 2 - \frac{1}{2} + 5 \qquad \text{division}$$
$$= \frac{13}{2} \qquad\qquad\quad \text{addition and subtraction} \qquad\qquad \blacktriangleleft$$

E X A M P L E **9**   Evaluate:

$$3(-1 + 6^2) + 7$$

Solution. $3(-1 + 6^2) + 7$

$$= 3(-1 + 36) + 7 \qquad \text{6}^2 \text{ = 36}$$
$$= 3(35) + 7 \qquad \text{-1 + 36 = 35}$$
$$= 105 + 7 \qquad \text{multiplication}$$
$$= 112 \qquad \text{addition} \qquad \blacktriangleleft$$

The fraction bar may also be a symbol of grouping, as shown in the next example.

E X A M P L E **10** Evaluate:

$$\frac{-4 - (-10)}{6} - \frac{|-8|}{4}$$

Solution. $\dfrac{-4 - (-10)}{6} - \dfrac{|-8|}{4}$

$$= \frac{-4 + 10}{6} - \frac{|-8|}{4}$$

$$= \frac{6}{6} - \frac{|-8|}{4}$$

$$= 1 - \frac{8}{4} \qquad \textstyle\frac{6}{6}\text{ = 1, }|-8|\text{ = 8}$$

$$= 1 - 2 = -1 \qquad \blacktriangleleft$$

 E X A M P L E **11** The formula relating Celsius and Fahrenheit temperatures is $C = \frac{5}{9}(F - 32)$. What is $-13°F$ in degrees Celsius?

Solution. If we substitute -13 for F in the above formula, we get

$$C = \frac{5}{9}(-13 - 32) = \frac{5}{9}(-45) = -25°C \qquad \blacktriangleleft$$

1.3 Operations with Zero

For any number a, the following operations with zero hold:

$$a + 0 = a$$
$$a - 0 = a$$
$$a \cdot 0 = 0$$
$$\frac{0}{a} = 0$$

Only division by zero ($a/0$) is undefined. Let's consider the reason. We know that $12 \div 4 = 3$ because $4 \times 3 = 12$. In general,

$$\frac{a}{b} = c \quad \text{if and only if} \quad a = bc$$

In particular, if $\frac{1}{0}$ is defined, then $\frac{1}{0} = x$ for some number x. But then

$$\frac{1}{0} = x \quad \text{is equivalent to} \quad 1 = 0 \cdot x = 0$$

or $1 = 0$, which is incorrect. So $\frac{1}{0}$ is not equal to any number whatsoever. On the other hand,

$$\frac{0}{0} = x \quad \text{is equivalent to} \quad 0 = 0 \cdot x = 0$$

This time we do not get a contradiction, but we are forced to draw another absurd conclusion:

$$\frac{0}{0} = x \text{ for any number } x$$

For this reason, since no specific value can be determined, $\frac{0}{0}$ is called *indeterminate*.

To summarize, **never divide by zero**.

E X E R C I S E S / S E C T I O N   **1.3**

In Exercises 1–10, determine the given absolute values.

1. $|3|$
2. $|-5|$
3. $|-7|$
4. $|10|$
5. $|-\pi|$
6. $|-\sqrt{3}|$
7. $|-(-\sqrt{3})|$
8. $|\sqrt[3]{-8}|$
9. $|\sqrt[3]{-27}|$
10. $|-\sqrt[3]{27}|$

In Exercises 11–20, state whether the given numbers are rational or irrational.

11. $\frac{2}{3}$
12. $\frac{1}{4}$
13. $7\frac{1}{2}$
14. $4\frac{1}{3}$
15. $\sqrt{10}$
16. $\sqrt{7}$
17. $\sqrt{9}$
18. $\sqrt{25}$
19. $\sqrt[3]{9}$
20. $\sqrt[3]{25}$

In Exercises 21–78, carry out the indicated operations.

21. $|-2| - 5$
22. $5 - |-4|$
23. $|1 - 9|$
24. $|-1 - (-3) + (-20)|$
25. $2 + |3 - 12|$
26. $-4 - 7 - 10$
27. $(-5) + (-8) + (-1) + (-3)$
28. $-5 - 8 - 1 - 3$
29. $3 + 9 - 20 - (-3)$
30. $-2 - 6 + 0 - 3$
31. $3 - 4 + (-4) - 6 + 0$
32. $-20 - 1 + 15 - (-6) - (-9)$
33. $15 + (-7) + (-12) - (-20)$
34. $18 - (-30) + 7$
35. $3.2 - (-4.5) + 0.7$
36. $1.74 + (-3.6) - (-2.13)$
37. $-\frac{5}{7} - \left(-\frac{3}{14}\right)$
38. $\frac{1}{15} + \left(-\frac{1}{6}\right)$
39. $-3(7 + 19)$
40. $-4(-25 + 20)$
41. $8(-5 - 10 + 3)$
42. $(6 - 7 + 0) \cdot 6$
43. $(8)(-3)(-4)$
44. $(-1)(-6)(-10)$
45. $(1.3)(-0.7)(-1)$
46. $(7.3)(0)(-8.96)(3.2)$

47. $(-3.22)(-8.7)$

48. $\dfrac{-3}{-9}$

49. $\dfrac{(-6)(15)}{(-3)(-18)}$

50. $\dfrac{(-14)(-1)}{(-2)(7)}$

51. $\dfrac{(3)(-9)}{(-3)(0)}$

52. $\dfrac{\left(\frac{1}{2}\right)\left(\frac{1}{3}\right)}{\left(-\frac{2}{3}\right)\left(-\frac{1}{4}\right)}$

53. $\dfrac{\left(-\frac{2}{7}\right)\left(-\frac{7}{15}\right)}{\left(\frac{15}{16}\right)(-4)}$

54. $\dfrac{\left(-\frac{1}{2}\right)\left(-\frac{1}{10}\right)}{(0)\left(-\frac{1}{2}\right)}$

55. $\dfrac{(3)(0)(-6)}{(-21)(27)}$

56. $\dfrac{(-2)(0)(-6)}{(-8)(7)(0)}$

57. $\dfrac{(2)(-10)-(3)(-5)}{3-8}$

58. $\dfrac{6+8-3(-7)}{-2-2(-5)}$

59. $-2+(3)(-6)$

60. $7-(4)(5)+2$

61. $8+\dfrac{-9}{3}-(2)(4)$

62. $10-\dfrac{20}{5}-(6)(-2)$

63. $2-5+(-3)(-4)-\dfrac{30}{5}$

64. $11-(-3)-(6)(-4)$

65. $7-(-7)-\dfrac{36}{-18}+4$

66. $(4)(-5)-(-3)(6)$

67. $(-3)(-8)-(-4)(-7)$

68. $1-(2)(-4)-(-7)(-8)$

69. $(3+4)\cdot(-3+6)$

70. $(-7-4)\cdot(2+4)$

71. $(-4)^2(6-10)$

72. $(7+2)(-2)^2$

73. $\dfrac{4-7^2}{-9}$

74. $\dfrac{3-(-6)^2}{-5-6}$

75. $\dfrac{2^2-(-3)^2}{5^3}$

76. $\dfrac{(-7)^2-5^2}{(-3)^3}$

77. $\dfrac{2(7-8)}{2^4}$

78. $\dfrac{2^3}{3(6-8)}$

79. If the temperature outside is $-12°C$ and it increases by $5°C$, what is the final temperature?

80. The temperature inside a freezer is $-2.5°F$ and it decreases by $1.6°F$. Determine the new temperature.

81. The voltage across an element with respect to the ground is -18.4 volts initially and then it changes to -15.7 volts. Determine the absolute value of the change in the voltage.

82. The current in a circuit is -1.86 amperes. The value is three times the current measured earlier. What was the current earlier?

83. The highest point in the United States is Mt. McKinley at 20,320 ft, and the lowest point is Death Valley at -282 ft. What is the difference in altitude between these two points?

84. A technician in a liquid oxygen manufacturing plant checks the temperature of incoming air hourly during the first cooling stage. During his eight-hour shift he obtains the following readings (in degrees Celsius): -2.2, -1.3, 0.5, -1.5, 0.6, 1.3, 2.1, and -0.3. Determine the average temperature, which is the algebraic sum of all the readings divided by the number of readings.

85. The overnight low temperatures (in degrees Fahrenheit) in Walhalla, North Dakota, during the first week in January were -10, -12, -1, 2, 4, -8, and -3. Determine the average low temperature during that week.

86. The current in a certain circuit changes from -4.0 to 2.5 amperes. What is the change in the current?

87. For a short time interval, the velocity v (in feet per second) of a certain particle moving along a line is $v = 2.0 - 1.0t - 1.0t^2$, where t is measured in seconds. Determine the velocity a second ago (at $t = -1$).

1.4 The Scientific Calculator

The purpose of this section is to discuss the basic scientific calculator operations. Since many types of calculators are available, we cannot discuss all of them here. So we will confine ourselves to those types used by most students. This includes most scientific calculators using *algebraic logic*.

General Remarks

A calculator cannot accept numbers that are too large or too small for the display. Also, if a number resulting from a calculation is too large or too small, the calculator will give an error indication (such as an E).

Other operations resulting in an error designation are division by zero, taking the square root of a negative number, and (for many calculators) raising a negative number to a power.

The Control Keys

Certain keys, often referred to as *control keys*, are used to "erase" data from the calculator. Most calculators have two types: The clear key $\boxed{C}$ is used to clear the displayed value and all information being calculated. The clear key *will not* erase information stored in the memory.

Another key, designated by $\boxed{CE}$ or $\boxed{CD}$, will erase only the last entry. Its main function is to erase a number entered by mistake.

E X A M P L E **1**   Suppose we wish to enter 4 + 7, but enter 4 + 8 by mistake. The $\boxed{CD}$ key can be used to erase the 8, so that 7 can be entered instead:

$\boxed{4}\boxed{+}\boxed{8}$ **8 entered by mistake**

$\boxed{CD}$ **clears last entry (8)**

$\boxed{7}$ **7 replaces 8**   ◄

The equal sign key $\boxed{=}$ will also clear all information being calculated. However, the display and the information stored in the memory will not be erased.

Data Entry Keys

The data entry keys are used to read in numbers in order to perform calculations with them. The data entry keys are the digits $\boxed{0}$, $\boxed{1}$, $\boxed{2}$, . . . , $\boxed{9}$; the decimal point $\boxed{\cdot}$; the $\boxed{\pi}$ key; the change of sign key $\boxed{+/-}$, which is used to change the sign of the number displayed; and the key $\boxed{EE}$, used to enter numbers written in scientific notation. (Scientific notation is discussed in Section 1.9.)

E X A M P L E **2**   To enter the number 602, use the sequence

$\boxed{6}\boxed{0}\boxed{2}$

The display will now read **602**.   ◄

E X A M P L E **3**   To enter the number 42.87, use the sequence

$\boxed{4}\boxed{2}\boxed{\cdot}\boxed{8}\boxed{7}$

The display will now read **42.87**.   ◄

E X A M P L E **4** When pressing the pi key $\boxed{\pi}$, the display will read

3.1415927

The displayed number is an approximation for π. (The number of digits displayed depends on the type of calculator used.) ◀

E X A M P L E **5** The sequence $\boxed{3}\boxed{+/-}$ will cause −**3** to be displayed.

Arithmetic Functions

The basic arithmetic functions are used to perform addition, subtraction, multiplication, and division with positive and negative numbers. The keys are

$\boxed{+}$, $\boxed{-}$, $\boxed{\times}$, and $\boxed{\div}$

The key $\boxed{=}$ is used to tell the calculator to complete the calculation with the data entered to that point. This key is used both for intermediate and final results. For example, to multiply 236 × 125, enter 236, press $\boxed{\times}$, enter 125, and press $\boxed{=}$. The result is 29,500.

The calculator sequence will be abbreviated as follows:

236 $\boxed{\times}$ 125 $\boxed{=}$ → 29500

(From now on we will omit the boxes around the individual digits.) ◀

E X A M P L E **6** Perform the following addition:

3.76 + 9.82

Solution. The sequence is

3 $\boxed{\cdot}$ 76 $\boxed{+}$ 9 $\boxed{\cdot}$ 82 $\boxed{=}$ → 13.58 ◀

E X A M P L E **7** Perform the following subtraction:

12.0765 − π

Solution. The sequence is

12 $\boxed{\cdot}$ 0765 $\boxed{-}$ $\boxed{\pi}$ $\boxed{=}$ → 8.9349073 ◀

From now on we will omit the decimal point key in describing a sequence.

E X A M P L E **8**   Multiply: 69.4×25.7.

Solution. The sequence is

$$69.4 \boxed{\times} 25.7 \boxed{=} \rightarrow 1783.58$$

Since 69.4 and 25.7 each have three significant digits, the answer is rounded off to three significant digits to become 1780. (For a discussion of approximation and significant digits, see Appendix A.) ◀

Rounding off. To round off a number to a specified number of digits, examine the next digit to the right. If this digit is 5 or more, increase the last digit by 1. This digit becomes the last significant digit. If the next digit is less than 5, leave the last digit as it is.

For example, 35.748 rounded off to three significant digits is 35.7 since the digit immediately to the right of 7 is 4. Similarly,

$$5.62\mathbf{4}9 = 5.62 \qquad \text{to three significant digits}$$

$$183.3\mathbf{5}26 = 183.4 \qquad \text{to four significant digits}$$

$$6.1\mathbf{8} = 6.2 \qquad \text{to two significant digits}$$

Common error

Rounding off a number one digit at a time:

$$8.0347 = 8.035 = 8.04 \qquad \text{incorrect}$$

Instead,

$$8.03\mathbf{4}7 = 8.03 \qquad \text{correct}$$

E X A M P L E **9**   Divide: $743 \div (-14.1)$.

Solution. The sequence is

$$743 \boxed{\div} 14.1 \boxed{+/-} \boxed{=} \rightarrow -52.695035$$

So the answer is -52.7, rounded off to three significant digits. ◀

Order of Operations

Whenever several operations are involved in a calculation, the following order of operations should be followed:

1. Evaluate any expression enclosed in parentheses (or other symbols of grouping).
2. Evaluate powers.
3. Perform all multiplications and divisions from left to right.
4. Perform all additions and subtractions from left to right.

Most scientific calculators are programmed to perform these operations in the correct order.

E X A M P L E **10**   Evaluate:

$$-16.3 + (15.2)(7.06)$$

Solution. The correct order of operations is to do the multiplication first and then the addition of -16.3. Since scientific calculators perform the operations in the correct order automatically, a proper sequence is

$$16.3 \boxed{+/-} \boxed{+} 15.2 \boxed{\times} 7.06 \boxed{=} \rightarrow 91.012$$

Rounding off to three significant digits, we get 91.0.   ◀

For a longer string of calculations the memory feature may be helpful. The **memory** works like an electronic scratch pad and saves you the trouble of recording intermediate results. Any value in the register can be stored by using the **storage key** $\boxed{STO}$ or $\boxed{x \rightarrow M}$. The stored number can be recovered and used directly in a calculation by pressing the **recall key** $\boxed{RCL}$ or $\boxed{MR}$.

E X A M P L E **11**   Evaluate:

$$(12.461)(-8.073) - (4.713)(0.83107)$$

Solution. For scientific calculators that perform the operations in the correct order automatically a correct sequence is

$$12.461 \boxed{\times} 8.073 \boxed{+/-} \boxed{-} 4.713 \boxed{\times} 0.83107 \boxed{=} \rightarrow -104.51449$$

Rounding off to the least accurate numbers (four significant digits), the result is -104.5.

The calculation can also be performed by using the memory feature: obtain the product $(4.713)(0.83107)$ and store the value in the memory. This value can be subtracted directly from the first product by pressing $\boxed{-}\boxed{MR}$. The sequence is

$$4.713 \boxed{\times} 0.83107 \boxed{=} \boxed{STO} \; 12.461 \boxed{\times} 8.073 \boxed{+/-} \boxed{-} \boxed{MR} \boxed{=}$$
$$\rightarrow -104.51449 \qquad ◀$$

Problems arise when the usual order of operations is changed. For example, if 3 is to be multiplied by the sum of 4 and 5, we need to use parentheses to indicate that the addition is to be performed first:

$$3 \times (4 + 5) = 3 \times 9 = 27$$

Without parentheses we have

$$3 \times 4 + 5 = 12 + 5 = 17$$

which is a completely different result.

Expressions containing parentheses can be evaluated with a calculator in two ways:

1. Using parentheses
2. Using the memory feature

Consider the next example.

E X A M P L E **12**   Evaluate:

$$(4.6 + 7.06)(10.4) - (2.14 + 7.3)(-4.2)$$

Solution.

  a. *Using parentheses.* Since most scientific calculators perform the operations in the correct order, the numbers and parentheses can be entered in these calculators in the order in which they occur:

$$\boxed{(}\, 4.6 \,\boxed{+}\, 7.06 \,\boxed{)}\,\boxed{\times}\, 10.4 \,\boxed{-}\,\boxed{(}\, 2.14 \,\boxed{+}\, 7.3 \,\boxed{)}$$
$$\boxed{\times}\, 4.2 \,\boxed{+/-}\,\boxed{=}\, \rightarrow 160.912$$

    Two significant digits: 160.

  b. *Using the memory feature.* The use of the memory feature is a good alternative to the above sequence (especially for nonscientific calculators that do not have parentheses): We find the value of the second quantity, $(2.14 + 7.3)(-4.2)$, store this value in the memory, and then use $\boxed{-}\,\boxed{\text{MR}}$. The complete sequence is

$$2.14 \,\boxed{+}\, 7.3 \,\boxed{=}\,\boxed{\times}\, 4.2 \,\boxed{+/-}\,\boxed{=}\,\boxed{\text{STO}}\, 4.6 \,\boxed{+}\, 7.06 \,\boxed{=}$$
$$\boxed{\times}\, 10.4 \,\boxed{-}\,\boxed{\text{MR}}\,\boxed{=}\, \rightarrow 160.912 \qquad \blacktriangleleft$$

Special Function Keys

Every scientific calculator has a number of special function keys. These special functions are particularly useful because they perform operations on the quantity displayed without your having to press the $\boxed{=}$ key. As a result, the special function keys can be used in a chain of operations.

  The special functions operate only on the quantity displayed.

In this section we will consider only three special functions: the reciprocal, power, and square root keys.

The **reciprocal key** $\boxed{1/x}$ is used to find the reciprocal of a number. For example, to find the reciprocal of 50, $\frac{1}{50}$, we may use the sequence

$$50 \boxed{1/x} \rightarrow 0.02$$

There is no need to press $\boxed{=}$. (Note that the same result can be obtained by using the sequence $1 \boxed{\div} 50 \boxed{=}$.)

E X A M P L E   **13**   To find the reciprocal of 1.25, use the sequence

$$1.25 \boxed{1/x} \rightarrow 0.8$$ ◄

As already noted, the special function keys are particularly useful when several operations are involved, as shown in the next example.

E X A M P L E   **14**   Evaluate:

$$\left(1 + \frac{1}{3.4}\right) \times \frac{1}{7.6}$$

Solution. Using the reciprocal key $\boxed{1/x}$, we get the following sequence:

$$\boxed{(}\, 1 \boxed{+} 3.4 \boxed{1/x} \boxed{)} \boxed{\times} 7.6 \boxed{1/x} \boxed{=} \rightarrow 0.1702786$$

So

$$\left(1 + \frac{1}{3.4}\right) \times \frac{1}{7.6} \approx 0.17$$

The symbol $\approx$ means "approximately equal to."   ◄

The **square key** $\boxed{x^2}$ is used to find the square of a number displayed. For example, to evaluate 16^2, we use the sequence

$$16 \boxed{x^2} \rightarrow 256$$

As with all special function keys, there is no need to press $\boxed{=}$. Also, the square key is particularly useful in a chain of operations.

E X A M P L E   **15**   Evaluate: $(2 + 17^2)^2 + (-15)^2$.

Solution. The sequence is

$$\boxed{(}\, 2 \boxed{+} 17 \boxed{x^2} \boxed{)} \boxed{x^2} \boxed{+} 15 \boxed{+/-} \boxed{x^2} \boxed{=} \rightarrow 84906$$ ◄

The Square Root

Our final operation in this section is evaluating the square root of a given number, defined next.

> **Square Root:** The **square root** of a given number is the number which, when multiplied by itself, is equal to the given number.

For example, we know that $5^2 = 25$. It follows that 5 is the square root of 25. The symbol for square root is $\sqrt{}$. So

$$\sqrt{25} = 5$$

E X A M P L E **16**
$\sqrt{36} = 6$ since $6^2 = 36$

$\sqrt{121} = 11$ since $11^2 = 121$ ◄

A positive integer whose square root is also a positive integer is called a *perfect square*. For example, 49 is a perfect square since $\sqrt{49} = 7$. The square root of a number that is not a perfect square can only be approximated. There exists a method somewhat similar to long division for finding square roots. However, with calculators so readily available, it is no longer necessary to do such a calculation by hand.

To find the square root of a given number with a calculator, we use the special **square root key** $\boxed{\sqrt{}}$. As with the other special function keys, there is no need to press $\boxed{=}$.

E X A M P L E **17**
Use a calculator to find an approximation of $\sqrt{50}$.

Solution. The sequence is

$$50 \; \boxed{\sqrt{}} \; \rightarrow 7.0710678$$

Thus $\sqrt{50} \approx 7.07$, to two decimal places.   ◄

In the next example the square root key is used in a chain of operations.

E X A M P L E **18**
Evaluate:

$$\sqrt{\sqrt{361} - \sqrt{126}}$$

Solution. The sequence is

$$361 \; \boxed{\sqrt{}} \; \boxed{-} \; 126 \; \boxed{\sqrt{}} \; \boxed{=} \; \boxed{\sqrt{}} \; \rightarrow 2.7883737 \qquad ◄$$

EXERCISES / SECTION 1.4

In Exercises 1–20, perform the indicated operations with a calculator. The numbers are assumed to be exact. (For a discussion of exact and approximate numbers, see Appendix A.)

1. $284 + 768$

2. $963 - 254$

3. $24(36 - 17)$

4. $15(83 + 32)$

5. $(14)(-32)(2)$

6. $(17)(-6)(-9)$

7. $40 - (52)(7)$

8. $18 - (-3)(24)$

9. $(37 - 63)(28 - 11)$

10. $(47 - 21)(23 - 49)$

11. $\dfrac{1}{0.25}$ (use $\boxed{1/x}$)

12. $\dfrac{1}{0.125}$ (use $\boxed{1/x}$)

13. $(26)^2$

14. $(38)^2$

15. $(-19)^2$

16. $(-45)^2$

17. $\sqrt{729}$

18. $\sqrt{961}$

19. $\sqrt{1,225}$

20. $\sqrt{2,116}$

In Exercises 21–24, round off the answers to four decimal places.

21. $(-4)^2 - 2\pi$

22. $5\pi - (2)(8)$

23. $3\pi^2 - 25$

24. $33 - 2\pi^2$

In Exercises 25–44, the numbers are assumed to be approximate. (For a discussion of approximation and significant digits, see Appendix A.)

25. $7.4(8.3 - 1.7) - (3.6)(8.4)$

26. $1.2(3.6 - 1.2) - (2.5)(-3.8)$

27. $2.83 - \dfrac{16.7}{2.64}$

28. $-1.89 - \dfrac{4.36}{12.40}$

29. $12.84 - (4.068)^2$

30. $16.64 - (5.104)^2$

31. $\dfrac{3.07 - 10.70}{1.19 - 5.041}$

32. $\dfrac{12.4 - 17.68}{(-5.46)(6.875)}$

33. $4.7\left(\dfrac{1}{6.30} - \dfrac{1}{2.1}\right)$

34. $-3.6\left(\dfrac{1}{15.0} - \dfrac{1}{7.6}\right)$

35. $(-1.46)(7.947) - (1.007)^2$

36. $(2.607)(4.8321) - (-5.409)^2$

37. $\left(-7.46 - \dfrac{1}{8.6}\right)^2$

38. $\left(4.43 - \dfrac{1}{3.8}\right)^2$

39. $(1.2)(-1.3) - (2.3)^2 + \dfrac{7.61}{-8.60}$

40. $\dfrac{5.8}{13.0} - (3.4)^2 - (-2.3)(1.9)$

41. $\dfrac{1}{8.46} - \sqrt{10.62}$

42. $\sqrt{3.494} - (2.86)^2$

43. $\dfrac{2.04 - (0.364)^2}{4.83 - (1.78)^2}$

44. $\dfrac{10.64 - (2.874)^2}{(-3.840)(7.061)}$

1.5 Polynomials

Literal symbols
Variable
Constant

In algebra the letters of the alphabet are customarily used to represent numbers. The use of letters requires some additional definitions and conventions. The basic terms will be discussed in this section.

The letters used to represent numbers are called **literal symbols**. If a literal symbol can take on various values, it is called a **variable**. A number or literal symbol that does not vary is called a **constant**. For example, suppose R (in ohms) is the resistance of a variable resistor. If the current through the resistor is 2 amperes, then the voltage V (in volts) is $V = 2R$. Here 2 is a constant (since the current does not change) and R and V are variables (since the resistance and voltage vary).

Literal symbols often occur in formulas. A *formula* expresses a relationship among variables. For example, the formula for the area A of a circle is $A = \pi r^2$. Here A and r are variables and π is a constant.

Power

Product

Factors

If two literal symbols a and b are multiplied, then ab is called a **product** and a and b are called the **factors**.

If a factor repeats, the product can be written by use of exponents. Let us recall the definition of power from Section 1.2.

Definition of Power

$$a^n = a \cdot a \cdot a \cdots \cdot a \quad (n \text{ factors})$$

is read "a to the nth power"; a is called the **base** and n the **exponent**.

Remark. If a number a occurs only once in a product, then a is raised to the first power, or a^1. However, since $a^1 = a$, the exponent 1 is not usually written.

Coefficients

Coefficient

Sometimes we have a combination of numbers and literal symbols. If one of the factors is a number, that number is called a **numerical coefficient**. For example, 3 is the numerical coefficient of $3x$.

E X A M P L E **1**

$4x$ has numerical coefficient 4

$-3xy$ has numerical coefficient -3

$2.8a^2b$ has numerical coefficient 2.8   ◄

Coefficients do not have to be numerical. Consider, for example, the product xyz^2. Here x is the coefficient of yz^2 and xy is the coefficient of z^2. When we use the word *coefficient*, we usually mean numerical coefficient.

If no numerical coefficient is written, it is understood to be 1. Since $1 \cdot x = x$,

$x = 1x$

we have $x = 1 \cdot x = 1x$, so the numerical coefficient of x is 1.

A combination of numbers and literal symbols connected by the four fundamental operations is called an **expression**. For example,

Expression

$$3y^2, \quad \frac{2a}{b^2}, \quad x + 5, \quad \text{and} \quad V - P$$

are expressions. (Single numbers and literal symbols are also expressions. For example, y is an expression because $y = 1y$.)

When an algebraic expression consists of several parts connected by plus and minus signs, then each part, together with the sign that precedes it, is called a **term**. For example, the expression $4x - 3y$ has two terms ($4x$ and $-3y$).

Term

Polynomials

By our definition of power, an exponent is a positive integer. Later we will extend this definition to zero, negative, and fractional exponents. However, except for a brief discussion of zero and negative exponents in Section 1.8, we will not be dealing with such exponents until Chapter 10. As a result, the algebraic expressions in this and the next several chapters contain only positive exponents.

An expression containing only a real number or the product of a real number and literal factors with positive integral exponents is called a **monomial**.

Monomial

E X A M P L E **2**   Examples of monomials are

$$-9, \quad 3x^3, \quad -4y^2, \quad 5ab^2, \quad \text{and} \quad -2y^2z^3 \qquad \blacktriangleleft$$

Binomial
Trinomial

The algebraic sum of exactly two monomials is called a **binomial**. The algebraic sum of exactly three monomials is called a **trinomial**.

E X A M P L E **3**
$$x + y \text{ is a binomial}$$
$$3x^2 - 4 \text{ is a binomial}$$
$$2x^2 + 3x - 1 \text{ is a trinomial}$$
$$5a^2 + 4ab - b^2 \text{ is a trinomial} \qquad \blacktriangleleft$$

Polynomial

The sum of a finite number of monomials (including a single monomial) is called a **polynomial**. By this definition, monomials, binomials, and trinomials are also polynomials.

E X A M P L E **4**   Examples of polynomials are

$$3w^2, \quad 4x + 1, \quad x^2 + 2x + 2, \quad y^3 - 3y^2 + 2y + 5, \quad \text{and}$$
$$a^4 + 2a^3b - 3a^2b^2 + ab^3 - b^4 \qquad \blacktriangleleft$$

Multinomial

As noted earlier, not every algebraic expression is a polynomial. An algebraic expression consisting of *two or more* terms is called a **multinomial**. For example, $x + \sqrt{y}$ is a multinomial. However, because of the term $\sqrt{y}$, the expression $x + \sqrt{y}$ is not a polynomial. (We will see in Chapter 10 that $\sqrt{y} = y^{1/2}$, which is not an integral power.)

E X A M P L E **5**   The following algebraic expressions are multinomials, but not polynomials:

$$2 + \sqrt{x}, \quad y + \frac{2}{z}, \quad \text{and} \quad \frac{\sqrt{A} + 2}{B} + 1 \qquad \blacktriangleleft$$

Even though multinomials are not necessarily polynomials, some of the terminology used with polynomials is also used with multinomials. For example, a multinomial consisting of two terms is called a *binomial*, and a multinomial consisting of three terms is called a *trinomial*. However, while a polynomial may consist of a single monomial, such an expression is not a multinomial because a multinomial must have two or more terms.

E X A M P L E 6   **a.** $2\sqrt{xy}$ is a monomial, but not a multinomial.

b. $2\sqrt{x} + \dfrac{C}{V}$ is a binomial and therefore a multinomial.

c. $\dfrac{a}{b} + \sqrt{b} + bc$ is a trinomial and therefore a multinomial.

Note that none of these expressions are polynomials.   ◄

Historical note. Much of the material discussed in this chapter, together with equations of first and second degree, had already been developed by the early Middle Ages. Some of these developments took place after the expansion of the Arab state, when Baghdad became a new center of learning, about the middle of the eighth century. Of the many scholars in Baghdad, the best-remembered is al-Khowarizmi (who died before 850 A.D.). His books, particularly the *Al-jabr w'al muqabala*, were to have a profound influence on the development of algebra. In fact, our word *algebra* comes from the name of this book, and the word *algorithm* comes from the author's name.

Our symbols for the digits 0 and 1 through 9 also date from this period. Most of our other algebraic notations, however, were introduced only gradually over many centuries.

E X E R C I S E S / S E C T I O N   **1.5**

In Exercises 1–6, identify the numerical coefficients and the factors.

1. $4xy$

2. $2ab$

3. $-2a^2b$

4. $-4LM^2$

5. $12s^2t$

6. $-7mn^3$

In Exercises 7–20, state in each case whether the given polynomial is a monomial, binomial, or trinomial.

7. $4xy^3$

8. $9ab$

9. $1 - 10cd$

10. $-6xw^2 + 7$

11. $5xz - 10Q$

12. $-18PV + 3P^2$

13. $2w^2$

14. $-11p^2q$

15. $3.2x^2 - 1.5x + 6.3$

16. $4.3 + 7.0y - 9.6y^2$

17. $7Z^2 + 2Z$

18. $9L^2 + 5L - 1$

19. $4s^2 + 5st - 2t^2$

20. $2.4m^2 - 3.0n$

1.6   Addition and Subtraction of Polynomials

In this section we will study two of the fundamental operations involving polynomials, addition and subtraction. To state the rules for addition and subtraction, we need the definition of *like terms*.

> **Like Terms:** Two or more terms are said to be **like terms** if the literal (nonnumerical) parts are the same.

For example,

$$5x, \quad 10x, \quad -3x, \quad \text{and} \quad -7x$$

are like terms, but

    $5x$ and $10y$

are not like terms;

    $7xy^2,$ $-2xy^2,$ and $3xy^2$

are also like terms, but

    $7xy^2$ and $-2xy$

are not because the y factors have different exponents.

    Addition and subtraction of polynomials are carried out in accordance with the following rule.

Addition and Subtraction of Polynomials: When adding (subtracting) polynomials, add (subtract) the coefficients of like terms.

For example, the sum $2x + 5x = 7x$ is obtained by adding the coefficients. This addition can be justified by means of the distributive law: $(b + c)a = ba + ca.$

Using the distributive law in reverse, we have

    $2x + 5x = (2 + 5)x = 7x$

Another way to see this is by noting that $2x = x + x$, since multiplication by a positive integer is actually repeated addition. Thus,

    $2x + 5x = (x + x) + (x + x + x + x + x) = 7x$

On the other hand,

    $2x + 5y$

cannot be combined, since neither x nor y is common to both terms. This can also be seen from

    $2x + 5y = x + x + y + y + y + y + y$

which cannot be written as a single term.

When adding algebraic expressions, only like terms can be combined to form a new term.

E X A M P L E **1**   Add the polynomials

$$-3x^2y + 4xy^2 \qquad \text{and} \qquad -7x^2y - 2xy^2$$

Solution. Since the first terms and the last terms in the expressions are like terms, the polynomials may be arranged in columns. Now add the coefficients in each column:

$$
\begin{array}{l}
-3x^2y + 4xy^2 \\
-7x^2y - 2xy^2 \\
\hline
-10x^2y + 2xy^2
\end{array}
\qquad
\begin{array}{l}
(-3) + (-7) = -10 \\
4 + (-2) = 2
\end{array}
$$

Note that only the like terms are added. The two terms in the sum cannot be combined, since x^2y and xy^2 are not the same.   ◄

E X A M P L E **2**   Add the polynomials

$$10x - 8y + 5z \qquad \text{and} \qquad -9x - 2y + w$$

Solution. Arranging the expressions so that the like terms are in the same column, we get

$$
\begin{array}{l}
10x - 8y + 5z \\
-9x - 2y + w \\
\hline
1x - 10y + 5z + w
\end{array}
\qquad \text{adding coefficients}
$$

Note that $5z$ and w are unlike terms and do not appear in the same column. Also, since $1 \cdot x = x$, the sum should be written

$$x - 10y + 5z + w \qquad\qquad\qquad ◄$$

When adding polynomials, we use the fact that $x = 1x$ and $-x = -1x$, as shown in the next example.

E X A M P L E **3**   Add $-x$ and $6x$.

Solution. Since $-x = -1x$, we get

$$-x + 6x = -1x + 6x = (-1 + 6)x = 5x \qquad\qquad ◄$$

As we saw in Section 1.2

$$a - b = a + (-b)$$

Subtraction

So whenever we subtract one term from another, we simply change the sign of the subtrahend and add algebraically.

E X A M P L E **4** Subtract $-5x - y + z$ from $-6x + 2y + 2z$.

Solution. If we arrange our work in columns and obey the rules regarding the sign changes, we obtain

$$
\begin{array}{l}
-6x + 2y + 2z \\
\underline{5x + \ y - \ z} \qquad -(-5x - y + z) \\
-1x + 3y + 1z \qquad \text{adding coefficients}
\end{array}
$$

or $-x + 3y + z$. ◀

 E X A M P L E **5** The voltage drops (in volts) across two resistors are given by $V_1 = 2.1R - 0.03$ and $V_2 = 1.5R + 0.19$, respectively. Find $V_1 + V_2$, which is the voltage drop across the two resistors connected in series. (The subscripts indicate that V_1 and V_2 are different quantities.)

Solution. $V_1 + V_2 = (2.1R - 0.03) + (1.5R + 0.19)$
$$= 3.6R + 0.16 \text{ (in volts)}$$ ◀

SI units

From now on we are going to use the following SI (International System of Units) notations: Ω for ohm, V for volt, A for ampere, F for farad, and C for coulomb. (See the table on the inside front cover. Also, SI units, reductions, and conversions are discussed in Appendix B.)

E X E R C I S E S / S E C T I O N **1.6**

In Exercises 1–8, simplify the given algebraic expressions.

1. $2x - 5y - 8x$
2. $3x - 4y + 2x$
3. $-7x - 3y + 5x - y$
4. $x - 5y - 2y + x$
5. $2a^2 - 3a - 8a^2 + a$
6. $-3a^2b + 10ab^2 - 2a^2b + 20ab^2$
7. $9xy^2 - 9x^2y - 8xy^2 + 6x^2y$
8. $-19xyz + 5xy - xyz$

In Exercises 9–20, find the sum of each of the given polynomials.

9.
$$
\begin{array}{r}
2x - 5y - 3z \\
-4x + \ y - 2z \\
x - \ y + 6z
\end{array}
$$

10.
$$
\begin{array}{r}
5ab - \ 9b^2 \\
-4ab - \ 9b^2 \\
ab + 18b^2
\end{array}
$$

11.
$$
\begin{array}{r}
x^2 - \ y^2 + 2z^2 \\
3x^2 \qquad + \ z^2 \\
- 5y^2 - 6z^2
\end{array}
$$

12.
$$
\begin{array}{r}
9a - 2b - 9x^2 \\
a - 5b \\
6a \qquad - 8x^2
\end{array}
$$

13. $3a - 5b - c, \ -7a + 2b + 2c, \ 5a + 3b - c$
14. $5a^2 - 10b^2 + 2c, \ -6a^2 - 9b^2 + c, \ -a^2 + 20b^2 - 4c$

15. $2x^2 - 3xy + y^2, \ 8x^2 + 5xy + 2y^2, \ -10x^2 - 3xy - y^2$
16. $2a^2 - 5x, \ a^2 - 5x + 2y, \ -x + y, \ 3a^2 + 3y$
17. $3x - 2y + w, \ -5x + z - 3w, \ -4x - 3y + 2w$
18. $5a^3 - 3c^2 + x, \ a^3 - 6d + 2x, \ -6a^3 + 5c^2 - 4x$
19. $a - b, \ b + 2c, \ 2a - 2b - 3c, \ -5a - 6c$
20. $2x^2 + y, \ -3x^2 - y + z, \ -x^2 - 6z, \ y + z$

In Exercises 21–28, subtract the second expression from the first.

21. $5a - 7b + c, \ -6a - 6b - 6c$
22. $x^2 \div y^2 + a^2, \ -x^2 - y^2 - 2a^2$
23. $6a + 5b - 3c, \ 5a - b - 4c$
24. $w + z + 2a, \ -w + z + 2a$
25. $-3a^2 + 5c^2 - d^2, \ 4a^2 - 5c^2 - e$
26. $2x^2, \ -3x^2 + 5b - d$
27. $-3a + 3b^2 + c, \ -3b^2$
28. $7x^2 - 2xy + 6y^2, \ -2x^2 - 6y^2$

In Exercises 29–32, subtract the third expression from the sum of the first two.

29. $a - 10b - 11c$, $a + b - c$, $2a - 8b - 8c$

30. $-x^2 + 2xy + 2y^2$, $-2x^2 + 4xy - 5y^2$, $-4x^2 + 5xy - 4y^2$

31. $20a^2 + b^2 + c^2$, $a^2 + 10b^2 - 2c^2$, $-20a^2 - b^2 - c^2$

32. $-5ab + 6b^2 - 3ac$, $-4ab + b^2 + 2ac$, $-ab - ac$

33. Write the polynomial representing the perimeter of the foundation wall in Figure 1.3.

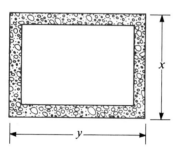

Figure 1.3

34. A shipment contains x gears costing $3 each and y gears costing $4 each. What does the polynomial $3x + 4y$ represent?

35. A box contains x bolts weighing 2 oz each and y bolts weighing 4 oz each. What does the polynomial $2x + 4y$ represent?

36. The capacitance in microfarads (μF) of two capacitors is $C_1 = 0.013C + 3$ and $C_2 = 0.198C - 2$, respectively. Find $C_1 + C_2$, the combined capacitance if the two capacitors are connected in parallel. (C_1 and C_2 denote different quantities.)

37. In determining the currents I_1 and I_2 in a certain two-loop circuit, the following expressions have to be added: $-I_1 + 3I_2$ and $6I_1 - 7I_2$. Perform the indicated operation.

38. One solar panel is $(x + 4a)$ ft long and another is $(3x - a)$ ft long. Find an expression for the total length if the panels are placed end to end.

39. The cost C in dollars of one computer is $6C + a$. The cost of another, less expensive, computer is $4C - 3a$. Find an expression for the difference in the cost.

1.7   Symbols of Grouping

The purpose of this section is to discuss the simplification of algebraic expressions containing symbols of grouping.

In algebra, terms are frequently grouped to indicate that they are to be considered as a single quantity or number. For example, the parentheses in the expression $3(2 + 6)$ indicate that the addition operation is to be carried out before multiplication. Thus, $3(2 + 6) = 3 \times 8 = 24$. Without parentheses we would have

$$3 \cdot 2 + 6 = 6 + 6 = 12 \neq 24$$

since multiplication always precedes addition.

Symbols of grouping

The most common symbols of grouping are parentheses ( ), brackets [ ], and braces { }; these symbols sometimes occur in combination. When simplifying certain algebraic expressions, symbols of grouping must often be removed. This is usually accomplished by means of the **distributive law**. Consider, for example, the expression

$$3[a + 2(a + b)]$$

If we first use the distributive law on the terms inside the parentheses, we get

$$3[a + 2a + 2b] = 3[3a + 2b] = 9a + 6b$$

If the expression to be simplified contains several sets of grouping symbols, we use the distributive law to remove the **innermost** set of grouping symbols and then combine similar terms, if possible. We then repeat the procedure for the resulting expression.

> **Removing Symbols of Grouping:** When removing symbols of grouping, work from the inside out.

E X A M P L E **1**

$$-2[3(x - 2y) - y]$$ original expression
$$= -2[3x - 6y - y]$$ removing inner parentheses
$$= -2[3x - 7y]$$ simplifying
$$= -6x + 14y$$ removing brackets ◀

An important special case involves a set of parentheses preceded by a negative sign, as in the expression

$$a - (a + b)$$

Since the quantity $a + b$ is subtracted from a, we can apply the rule from the previous section and conclude that the signs of both terms have to be changed. Thus

$$a - (a + b) = a - a - b = -b$$

A simple alternative is to write

> $$-(a + b) = -1(a + b) \tag{1.1}$$

and then apply the distributive law to get $-a - b$. If the set of parentheses is preceded by a plus sign (or no sign), then no sign change is required and the parentheses may simply be dropped: $+ (a + b) = a + b$.

E X A M P L E **2** Remove the symbols of grouping and simplify:

$$2[s - (2s - t)]$$

Solution.

$$2[s - (2s - t)]$$ given expression
$$= 2[s - 1(2s - t)]$$ $-x = -1x$
$$= 2[s - 2s + t]$$ removing inner parentheses
$$= 2[-s + t]$$ simplifying
$$= -2s + 2t$$ distributive law ◀

E X A M P L E **3**   Simplify:

$$-[a + (a + 2b)]$$

Solution.

$-[a + (a + 2b)]$	**given expression**
$= -[a + a + 2b]$	$+(a + 2b) = a + 2b$
$= -[2a + 2b]$	**simplifying**
$= -2a - 2b$	**removing brackets** ◀

E X A M P L E **4**   Simplify:

$$-\{x^2 + 2[x - (-x + 3)]\}$$

Solution. Since several sets of grouping symbols have to be removed, it is necessary to work from the inside out.

$-\{x^2 + 2[x - (-x + 3)]\}$	**original expression**
$= -\{x^2 + 2[x + x - 3]\}$	**removing inner parentheses**
$= -\{x^2 + 2[2x - 3]\}$	**simplifying**
$= -\{x^2 + 4x - 6\}$	**distributive law**
$= -x^2 - 4x + 6$	**removing braces** ◀

Inserting parentheses

It is occasionally necessary to reverse the above procedure and *insert* parentheses. Consider the next example.

E X A M P L E **5**   Enclose the last two terms in parentheses preceded by a minus sign:

$$2x^2y - 3 - ab + c$$

Solution. We know from the procedure for removing parentheses that the signs of the last two terms have to be changed. We therefore get

$$2x^2y - 3 - (ab - c)$$

which can be checked by removing the parentheses.   ◀

Common errors

1. Leaving out parentheses in $a(b + c)$ and incorrectly writing $ab + c$. The two expressions are not the same.
2. Changing the sign of only the first term when removing parentheses. Thus, $-(-a - b) = a + b$, not $a - b$.

 E X A M P L E **6**   If two forces $F_1 = 4f_1 - 2f_2$ and $F_2 = -f_1 + 6f_2$ are acting on the same object in opposite directions, then the net force F_n is given by $F_n = F_1 - F_2$. Find the expression for F_n.

Solution. $F_n = F_1 - F_2 = (4f_1 - 2f_2) - (-f_1 + 6f_2)$

$$= 4f_1 - 2f_2 + f_1 - 6f_2$$
$$= 5f_1 - 8f_2$$

◀

E X E R C I S E S / S E C T I O N 1.7

In Exercises 1–28, remove symbols of grouping and simplify.

1. $3(-4 + x)$

2. $2(a - 4)$

3. $2(x + 2y) - 3(x - 2y)$

4. $3(-a - b) + 4(a + 2b)$

5. $-(x - y) - (-2x + 3y)$

6. $x^2 - y^2 - (x^2 - y^2)$

7. $(x^2 - xy) - (3x^2 + 3xy)$

8. $-(a + b) - (2a - 2b) - 3a$

9. $(2a - 3b) - (-a + b)$

10. $(4z - 7w) - (3z - 4w)$

11. $-(-2p - 3q) + (2p - 5q) - (p - q)$

12. $(5s + 2t) - (-s + t) - (3s - 4t)$

13. $-[-V - (2V - C)]$

14. $-[n - (4n - 4m)]$

15. $-[-(2s - 3t) - (s - 4t)]$

16. $-[2 - (3c + 6) + (2c - 6) + 4]$

17. $2 + [-(7w + 4z - 5) - (-7w + 4z)]$

18. $R + [3C + (3R + 2C) - (7R - 12C) + (R - 4C)]$

19. $(x - 2y) - x - [2x - (-6x + 4y)]$

20. $x - (-4x + 2y) - [-3x - (7x + 3y)]$

21. $2[a - (2a - b)] - (a + b)$

22. $-[x^2 - x(2x + y)] + (x^2 - xy)$

23. $-(x - xy) - [3 - (x + y) - xy]$

24. $2[x - 3(x - y) + y]$

25. $-\{x - [(y - x) - y]\}$

26. $-\{2a - 3[-b - (-2b + 4a)]\}$

27. $4\{2a - 2[2a - 3(2a - 3) + 4a] + 3\}$

28. $-3\{x + 2(y - 2z) + 3[x - 4(y + 2z)]\}$

In Exercises 29–42, enclose the last two terms in parentheses preceded by (a) a plus sign and (b) a minus sign.

29. $a + b - c + 2d$

30. $2 + 3x + 2a - 5$

31. $x - 2y - 3x^2 - 2y^2$

32. $5 - 3x + 2y$

33. $7 + 2a - 5b$

34. $3x - 7y - 3 + c$

35. $6w + 5z - 3x + 5y$

36. $5r^2 - 7s^2 + 2a - 1$

37. $7x^2y + 2xy^2 - 6xz + 5yz$

38. $2pq^2 - 3q^2p - 9pq + 3$

39. $-3x^2 + 2y^2 - 7z^2 - 8w^2$

40. $-9 - x^2y - xy^2$

41. $-5x - 7a - 7b$

42. $-6y + 2a + 5b$

43. Simplify the following expression, which comes from a problem in electrical circuits: $6R_1 + 3R_2 - (4R_1 - R_2)$.

44. Simplify the following expression, which arose in a problem involving forces: $-(F_1 - 2F_2) - (3F_1 + 5F_2)$.

45. Write the following expression for the voltage in a circuit with a minus sign in front: $-R_1I_1 + R_2I_2 - R_3I_3$.

46. Repeat Exercise 45 for the expression $-3R_1I_1 - 2R_2I_2 + 5R_3I_3$.

47. The velocity v (in meters per second) of an object falling from level y_1 to level y_2 measured from a point above y_1 is

$$v = \sqrt{19.6(y_2 - y_1)}$$

Write this formula without parentheses.

1.8   Exponents and Radicals

The purpose of this section is to introduce the basic operations with exponents and radicals. A more detailed study will be undertaken in Chapter 10.

Positive Integral Exponents

Recall the definition of exponent from Section 1.5:

$$a^n = a \cdot a \cdot a \cdots a \quad (n \text{ factors})$$

This definition leads at once to the laws of exponents for multiplication and division. For example,

$$a^2 \cdot a^4 = (a \cdot a)(a \cdot a \cdot a \cdot a) = a^{2+4} = a^6$$

Similarly,

$$\frac{a^6}{a^3} = \frac{\overset{1}{\cancel{a}} \cdot \overset{1}{\cancel{a}} \cdot \overset{1}{\cancel{a}} \cdot a \cdot a \cdot a}{\underset{1}{\cancel{a}} \cdot \underset{1}{\cancel{a}} \cdot \underset{1}{\cancel{a}}} = a^{6-3} = a^3$$

Finally,

$$(a^2)^4 = a^2 \cdot a^2 \cdot a^2 \cdot a^2 = a^{4 \cdot 2} = a^{2 \cdot 4} = a^8$$

The basic laws of exponents are summarized next.

Laws of Exponents: If m and n are positive integers,

$$a^m a^n = a^{m+n} \tag{1.2}$$

$$\frac{a^m}{a^n} = a^{m-n} \quad (m > n, \quad a \neq 0) \tag{1.3}$$

$$(a^m)^n = a^{mn} \tag{1.4}$$

$$(ab)^n = a^n b^n \tag{1.5}$$

$$\left(\frac{a}{b}\right)^n = \frac{a^n}{b^n} \quad (b \neq 0) \tag{1.6}$$

These rules can be used to simplify certain algebraic expressions. For example, the expression $2x^3 x^4$ can be written in simpler form by adding exponents: $2x^3 x^4 = 2x^{3+4} = 2x^7$. Consider another example.

E X A M P L E **1**

a. $a^2 a^8 = a^{2+8} = a^{10}$ adding exponents

b. $\dfrac{2x^6}{3x^4} = \dfrac{2}{3} x^{6-4} = \dfrac{2}{3} x^2$ subtracting exponents

c. $(a^3)^5 = a^{3 \cdot 5} = a^{15}$ multiplying exponents

d. $(-2x)^3 = (-2)^3 x^3$ since $(ab)^n = a^n b^n$
$\qquad\quad = -8x^3$ $(-2)^3 = -8$

e. $\left(\dfrac{3}{x}\right)^3 = \dfrac{3^3}{x^3} = \dfrac{27}{x^3}$ since $\left(\dfrac{a}{b}\right)^n = \dfrac{a^n}{b^n}$ ◄

E X A M P L E 2 Simplify the following expressions:

a. $\left(\dfrac{2x}{y}\right)^5$ **b.** $\dfrac{-3(x^3y^2)^3}{6(xy^2)^2}$

Solution.

a. $\left(\dfrac{2x}{y}\right)^5 = \dfrac{(2x)^5}{y^5}$ since $\left(\dfrac{a}{b}\right)^n = \dfrac{a^n}{b^n}$

$= \dfrac{2^5x^5}{y^5} = \dfrac{32x^5}{y^5}$ since $(ab)^n = a^nb^n$

b. $\dfrac{-3(x^3y^2)^3}{6(xy^2)^2} = \dfrac{-3(x^3)^3(y^2)^3}{6x^2(y^2)^2}$ since $(ab)^n = a^nb^n$

$= \dfrac{-3x^9y^6}{6x^2y^4}$ multiplying exponents

$= \dfrac{-x^9y^6}{2x^2y^4}$

$= \dfrac{-x^7y^2}{2}$ subtracting exponents

$= -\dfrac{x^7y^2}{2}$ $\dfrac{-a}{b} = \dfrac{a}{-b} = -\dfrac{a}{b}$ ◄

Common error Failing to distinguish between $(-ax)^2$ and $-(ax)^2$. Note that

$$(-ax)^2 = (-ax)(-ax) = a^2x^2$$

but

$$-(ax)^2 = -1(ax)^2 = -a^2x^2$$

So $(-ax)^2 \neq -(ax)^2$.

E X A M P L E 3 $(-x^2)^7 = -x^{14}$

but

$$(-x^2)^6 = x^{12}$$

In other words, an even power of a negative number is positive and an odd power of a negative number is negative. ◄

E X A M P L E 4
a. $(-3a)^2 = 9a^2$, but $(-3a)^3 = -27a^3$
b. $(-ab^2)^4 = a^4b^8$, but $(-ab^2)^5 = -a^5b^{10}$
c. $(-a^2b^3)^4 = a^8b^{12}$, but $-(a^2b^3)^4 = -1(a^2b^3)^4 = -a^8b^{12}$ ◄

Radical

Square root

*n*th root

Principal square root

Radicals

The other important concept to be discussed briefly here is that of **radical** or **root**. Consider the problem of what number multiplied by itself equals 16. This number, namely 4, is called the **square root** of 16, and is written

$$\sqrt{16} = 4$$

Similarly, $\sqrt[3]{8}$, read "cube root of 8," is the number whose cube is equal to 8. The **nth root** of a is denoted by $\sqrt[n]{a}$; n is called the *index* of the radical and a is called the *radicand*. If $n = 2$, we omit the index and write $\sqrt{a}$.

Remark. The square root of 16 could quite reasonably be given as 4 or -4, and there are times when both roots are important. However, whenever the symbol $\sqrt{\ }$ is employed, by convention only the positive root is intended. Thus

$$\sqrt{16} = 4, \qquad \sqrt{16} \neq -4$$

The positive square root is known as the **principal square root**. Similar conventions apply to the *n*th root. Thus $\sqrt[4]{16} = 2$, not -2, but $\sqrt[5]{-32} = -2$, since $(-2)^5 = -32$, while $2^5 = 32$.

As in the case of exponents, there exist several laws of operations with radicals, some of which will be introduced in this section. However, the justification for these laws must be postponed until Chapter 10, which contains a more elaborate discussion of exponents and radicals.

Laws of Radicals:

$$\sqrt[n]{ab} = \sqrt[n]{a} \cdot \sqrt[n]{b} \tag{1.7}$$

$$\sqrt[n]{\frac{a}{b}} = \frac{\sqrt[n]{a}}{\sqrt[n]{b}} \qquad (b \neq 0) \tag{1.8}$$

$$(\sqrt[n]{x})^n = \sqrt[n]{x^n} = x \tag{1.9}$$

At this point we use these laws only to simplify certain algebraic expressions. For example, since $50 = 25 \cdot 2$ and $\sqrt{25} = 5$, we have by rule (1.7)

$$\sqrt{50} = \sqrt{25 \cdot 2} = \sqrt{25}\sqrt{2} = 5\sqrt{2}$$

E X A M P L E **5**   Simplify the following radicals:

a. $\sqrt{48}$ **b.** $\sqrt{125}$ **c.** $\dfrac{1}{\sqrt{40}}$

Solution.

a. Since $48 = 16 \cdot 3$ and $\sqrt{16} = 4$, we get $\sqrt{48} = \sqrt{16 \cdot 3} = \sqrt{16}\sqrt{3} = 4\sqrt{3}$

b. $\sqrt{125} = \sqrt{25 \cdot 5} = \sqrt{25}\sqrt{5} = 5\sqrt{5}$

c. $\dfrac{1}{\sqrt{40}} = \dfrac{1}{\sqrt{4 \cdot 10}} = \dfrac{1}{2\sqrt{10}}$ ◀

Rationalizing the denominator

We can use law (1.9) to perform an operation called **rationalizing the denominator**: A fraction of the form

$$\frac{1}{\sqrt{x}}$$

can be written without a radical in the denominator. To eliminate the radical, we **multiply numerator and denominator by** $\sqrt{x}$. (Multiplying both numerator and denominator of a fraction by the same number does not affect the value of the fraction.) Observe that

$$\frac{1}{\sqrt{x}} = \frac{1 \cdot \sqrt{x}}{\sqrt{x} \cdot \sqrt{x}} = \frac{\sqrt{x}}{(\sqrt{x})^2} = \frac{\sqrt{x}}{x}$$

since $(\sqrt[n]{x})^n = x$. For example,

$$\frac{1}{\sqrt{2}} = \frac{1}{\sqrt{2}} \cdot \frac{\sqrt{2}}{\sqrt{2}} = \frac{\sqrt{2}}{(\sqrt{2})^2} = \frac{\sqrt{2}}{2}$$

E X A M P L E **6** Rationalize the denominator in each of the following fractions:

a. $\dfrac{1}{\sqrt{10}}$ **b.** $\dfrac{1}{2\sqrt{x}}$ **c.** $\dfrac{3}{2\sqrt{5}}$ **d.** $\dfrac{2}{\sqrt{2}}$

Solution.

a. $\dfrac{1}{\sqrt{10}} = \dfrac{1 \cdot \sqrt{10}}{\sqrt{10}\sqrt{10}} = \dfrac{\sqrt{10}}{(\sqrt{10})^2} = \dfrac{\sqrt{10}}{10}$

b. $\dfrac{1}{2\sqrt{x}} = \dfrac{1 \cdot \sqrt{x}}{2\sqrt{x}\sqrt{x}} = \dfrac{\sqrt{x}}{2(\sqrt{x})^2} = \dfrac{\sqrt{x}}{2x}$

c. $\dfrac{3}{2\sqrt{5}} = \dfrac{3\sqrt{5}}{2\sqrt{5}\sqrt{5}} = \dfrac{3\sqrt{5}}{2(\sqrt{5})^2} = \dfrac{3\sqrt{5}}{2(5)} = \dfrac{3\sqrt{5}}{10}$

d. $\dfrac{2}{\sqrt{2}} = \dfrac{2\sqrt{2}}{\sqrt{2}\sqrt{2}} = \dfrac{2\sqrt{2}}{(\sqrt{2})^2} = \dfrac{2\sqrt{2}}{2} = \sqrt{2}$ ◀

Caution. When writing $\sqrt{a}$, we assume that a is positive or zero, since we cannot find the square root of a negative number. For example, there is no (real) number equal to $\sqrt{-4}$. Thus $\sqrt{x - 3}$ is valid only for $x \geq 3$, and $\sqrt{1 - x}$ is valid only for $x \leq 1$.

Complex number

Numbers of the form $a + b\sqrt{-1}$ are called *complex numbers* and will be studied in detail in Chapter 11. The square root of a negative number such as $\sqrt{-4}$ is called an *imaginary number*.

 E X A M P L E   **7**

The frequency of a series resonance circuit is given by $f = 1/(2\pi\sqrt{LC})$. Write the expression for f without a radical in the denominator.

Solution. $f = \dfrac{1}{2\pi\sqrt{LC}} \cdot \dfrac{\sqrt{LC}}{\sqrt{LC}} = \dfrac{\sqrt{LC}}{2\pi(\sqrt{LC})^2} = \dfrac{\sqrt{LC}}{2\pi LC}$ ◀

Zero and Negative Exponents*

If the laws of exponents are applied to a^4/a^4, we get

$$\frac{a^4}{a^4} = a^{4-4} = a^0$$

On the other hand,

$$\frac{a^4}{a^4} = 1$$

since any nonzero number divided by itself is 1. These results suggest the following definition:

Zero Exponent:

$$a^0 = 1, \quad a \neq 0 \tag{1.10}$$

Similarly,

$$\frac{a^2}{a^5} = a^{2-5} = a^{-3} \qquad \text{and} \qquad \frac{a^2}{a^5} = \frac{1}{a^3}$$

This suggests that $a^{-3} = 1/a^3$. The general case is given next.

Negative Exponent:

$$a^{-n} = \frac{1}{a^n}, \quad a \neq 0 \tag{1.11}$$

---

* Zero and negative exponents are needed at this time only for the discussion of scientific notation in the next section. Consideration of these topics may be postponed until Chapter 10.

The rules for multiplication and division hold for zero and negative exponents. Consider the following example:

E X A M P L E **8**

a. $x^0 x^4 = x^{0+4} = x^4$ adding exponents

b. $\dfrac{a^5}{a^0} = a^{5-0} = a^5$ subtracting exponents

c. $a^5 a^{-3} = a^{5+(-3)} = a^2$ adding exponents

d. $\dfrac{a^5}{a^{-3}} = a^{5-(-3)} = a^8$ subtracting exponents

e. $\dfrac{b^{-4} b^3}{b^{-5}} = b^{-4+3-(-5)} = b^4$

Other operations with negative exponents will be discussed in Chapter 10. ◀

Notation. So far we have used mostly x, y, and z as variables. In technology, however, it is customary to use letters that suggest the meaning of physical quantities directly. For example, the formula for the power delivered to a resistor is given by $P = RI^2$. Here P stands for the power in watts, I for the current in amperes, and R for the resistance in ohms. The use of these letters makes the formula easier to comprehend and to remember. *From now on many of the exercises will contain notations that are frequently encountered in technical fields.*

E X E R C I S E S / S E C T I O N **1.8**

In Exercises 1–48, simplify each of the given expressions.

1. $(-2)^3$

2. $\left(-\dfrac{1}{4}\right)^4$

3. $\sqrt[4]{16}$

4. $\sqrt[5]{32}$

5. $x^2 x^6$

6. $2x(x^3)$

7. $-2(xy)^3$

8. $(-x)^4$

9. $\dfrac{x^{10}}{x^5}$

10. $\dfrac{4y^5}{2y^2}$

11. $\dfrac{-4z^5}{8z}$

12. $\dfrac{-3x^{12}}{-2x^{12}}$

13. $\dfrac{-(2x)^3}{(-2x)^3}$

14. $\dfrac{(-3x)^2}{-(3x)^2}$

15. $\dfrac{(-x)^3 y^6}{-2x^3 y^6}$

16. $\dfrac{-4x^6 y^6}{-2x^6 y^5}$

17. $\dfrac{(-2x)^2 y^3}{(-2x)^3 y^2}$

18. $\dfrac{-2(x^2 y^3)^3}{3(xy^2)^2}$

19. $\dfrac{x^4}{x^5}$

20. $\dfrac{2x^6}{-x^{10}}$

21. $\dfrac{-10x^6 y^3}{-5x^8 y}$

22. $\dfrac{-(xy)(-2x^2 y^2)}{2x^3 y^3}$

23. $\dfrac{(-5ab)(-10ab)}{25a^5 b^4}$

24. $\dfrac{(-2x)(-b)^3}{3x^2 b}$

25. $\sqrt{8}$

26. $\sqrt{24}$

27. $\sqrt{54}$

28. $\sqrt{18}$

29. $\sqrt{75}$

30. $\sqrt{72}$

31. $\sqrt[3]{16}$

32. $\sqrt[4]{32}$

33. $\sqrt[5]{64}$

34. $\sqrt[3]{32}$

35. $\sqrt{32}$

36. $\sqrt{128}$

37. $\dfrac{(-\omega^2)\alpha^3}{-5\omega^2(-\alpha)^4}$

38. $\dfrac{-14L_1{}^2(-L_2)^3}{7L_1 L_2{}^2}$

39. $\dfrac{(4A_1)^2(-A_2)^4}{(-2A_1 A_2)^2}$

40. $\dfrac{(-I_1)^5}{(-2I_1)^3 I_2}$

41. $\dfrac{(\sqrt{3v_1})^2(-v_2)^2}{3v_1v_2{}^2}$

42. $\dfrac{(-2xy^2)(-2x^2y^3)^3}{(x^3y)^4}$

43. $\dfrac{(-a^3b^4)^2(-2a^4b^2c)^3}{-(a^2bc)^3}$

44. $\dfrac{(-xyz)^2(-xyz)^3}{-3x^2(yz)^2}$

45. $\dfrac{(\sqrt{2xy^2})^2(\sqrt{3}xy)}{\sqrt{3}x^2y^2}$

46. $\dfrac{(2x^3y^6)^4(x^2y)^2}{(-x^2y^2)^3}$

47. $\dfrac{(v_1v_2)^2(-2v_1v_2)}{(-v_1v_2)^2}$

48. $\dfrac{(-i_1i_2)^2(-i_1i_2)}{3i_1{}^2i_2{}^3}$

In Exercises 49–60, simplify the given expressions by rationalizing denominators.

49. $\dfrac{1}{\sqrt{5}}$

50. $\dfrac{2}{\sqrt{7}}$

51. $\dfrac{3}{\sqrt{3}}$

52. $\dfrac{1}{2\sqrt{3}}$

53. $\dfrac{a}{\sqrt{b}}$

54. $\dfrac{a}{b\sqrt{c}}$

55. $\dfrac{2}{3\sqrt{5}}$

56. $\dfrac{1}{\sqrt{8}}$

57. $\dfrac{2}{\sqrt{32}}$

58. $\dfrac{4}{\sqrt{20}}$

59. $\dfrac{3}{\sqrt{12}}$

60. $\dfrac{2}{\sqrt{50}}$

61. A box in the shape of a cube has a volume of 729 cubic centimeters. Find the length of one side.

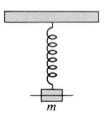

Figure 1.4

62. The frequency, or number of vibrations per unit time, of a mass hanging on a spring (Figure 1.4) is given by

$$f = \frac{1}{2\pi}\sqrt{\frac{k}{m}}$$

where m is the mass and k is a constant. Simplify this expression by rationalizing the denominator.

63. In determining the velocity of a water wave, the expression $\sqrt{4\pi/ad}$ has to be simplified. Rationalize the denominator and simplify.

64. Simplify the following expression from a problem in mechanics:

$$\frac{(3m^2v)^3}{(4m^2v)^2}$$

65. The area of a triangle with sides of length a, b, and c is

$$A = \sqrt{s(s-a)(s-b)(s-c)}$$

where $a + b + c = 2s$. Find the area of a triangle with sides of length 7, 9, and 12 centimeters.

In Exercises 66–83, simplify the expression (see Example 8).

66. a^0a^5

67. $\dfrac{a^6}{a^0}$

68. a^0x^2

69. $a^{-2}a^6$

70. $x^{-4}x^7$

71. $a^{-4}a^5$

72. $\dfrac{x^{-2}}{x^{-8}}$

73. $\dfrac{x^2}{x^{-4}}$

74. $\dfrac{3x^4}{x^{-1}}$

75. $\dfrac{4a^0x^{-2}}{x^{-5}}$

76. $\dfrac{5x^0y^{-1}}{y^{-2}}$

77. $\dfrac{6a^{-2}y^0}{a^{-4}}$

78. $\dfrac{2x^2y^3}{x^{-2}y^{-2}}$

79. $\dfrac{7p^{-1}q^{-1}}{p^{-5}q^{-3}}$

80. $\dfrac{3R_1{}^{-2}R_2}{R_1{}^{-2}R_2{}^{-1}}$

81. $\dfrac{9V_1{}^{-4}V_2}{V_1{}^{-4}V_2{}^{-2}}$

82. $\dfrac{L^0L^3M^{-2}}{L^{-1}M^{-4}}$

83. $\dfrac{4C_1{}^{-2}C_2{}^3C_3{}^0}{8C_1{}^{-6}C_2{}^{-5}}$

1.9   Scientific Notation

Many scientific and technical applications involve very large or very small numbers. A convenient way to write such numbers is by means of scientific notation, which is defined next.

> **Scientific Notation:** A number in **scientific notation** has the form
>
> $$n \times 10^k, \quad 1 \le n < 10 \tag{1.12}$$
>
> where k is an integer.

Before examining scientific notation in detail, let us consider the following powers of 10:

$$10 = 10^1 \qquad \frac{1}{10} = 10^{-1}$$

$$100 = 10^2 \qquad \frac{1}{100} = 10^{-2}$$

$$1,000 = 10^3 \qquad \frac{1}{1,000} = 10^{-3}$$

$$10,000 = 10^4 \qquad \frac{1}{10,000} = 10^{-4}$$

To write the number 30,000 in scientific notation, observe that $30,000 = (3)(10,000) = 3 \times 10^4$. Similarly, $31,000 = (3.1)(10,000) = 3.1 \times 10^4$.

Now consider a small number such as 0.0054. Written as an ordinary fraction, we get

$$0.0054 = \frac{54}{10,000} = \frac{54}{10^4} = \frac{5.4}{10^3} = 5.4 \times 10^{-3}$$

To write a number in **scientific notation**

$$n \times 10^k, \quad 1 \leq n < 10$$

move the decimal point to the right of the first nonzero digit. The number of places that the decimal point is moved is equal to the exponent k.

1. If the decimal point is moved to the **left**, k is **positive**.
2. If the decimal point is moved to the **right**, k is **negative**.

E X A M P L E **1** Write 4,350,000 in scientific notation.

Solution. According to the rule, the decimal point is placed to the right of 4 (to obtain 4.35). To place the decimal point in this position, we must move the original decimal point six places to the *left*; so $k = +6$. The result is

$$4350000. = 4.35 \times 10^6$$

6 places   ◄

E X A M P L E **2** Write 0.0000000832 in scientific notation.

Solution. The decimal point is moved to the right of the first nonzero digit to obtain 8.32. To place the decimal point in this position, we have to move the original decimal point eight places to the *right*. So $k = -8$. The result is

$$0.0000000832 = 8.32 \times 10^{-8}$$

$\underbrace{\qquad\qquad}$
8 places

◄

E X A M P L E **3**

$246{,}000 = 2.46 \times 10^5$ (decimal point moved five places to the left)

$0.00000129 = 1.29 \times 10^{-6}$ (decimal point moved six places to the right)

◄

To change a number from scientific to regular decimal notation, the above procedure is reversed.

E X A M P L E **4**

$9.3 \times 10^4 = 93{,}000$ (decimal point moved four places)

$2.6 \times 10^{-3} = 0.0026$ (decimal point moved three places)

◄

Scientific notation is particularly useful for indicating significant digits, especially when ambiguous final zeros are involved. (For a discussion of significant digits, see Appendix A.)

> When a number is written in scientific notation $n \times 10^k$, n indicates the number of significant digits.

For example, **8.30** $\times 10^9$ has three significant digits.

E X A M P L E **5**   Since

$$3{,}140{,}000 = 3.14 \times 10^6$$

$n = 3.14$ and indicates three significant digits.

◄

Scientific notation deals with final zeros in a simple way. Consider the next example.

E X A M P L E **6**   The number 3,140,000 in Example 5 is written

a. 3.140×10^6 to indicate four significant digits.
b. 3.1400×10^6 to indicate five significant digits.

◄

Multiplication and division can be carried out directly with numbers written in scientific notation, but the final result must be expressed in the proper form.

E X A M P L E **7** Carry out the following multiplication and express the answer in scientific notation:

$$(5.26 \times 10^{10}) \times (9.35 \times 10^{-4})$$

Solution. $(5.26 \times 10^{10}) \times (9.35 \times 10^{-4}) = (5.26)(9.35) \times (10^{10} \times 10^{-4})$
$$= (5.26)(9.35) \times 10^{6}$$
$$= 49.181 \times 10^{6}$$
$$= 49.2 \times 10^{6}$$

to three significant digits. Since the coefficient of 10^{6} has to be between 1 and 10, we write

$$\mathbf{49.2 \times 10^{6} = (4.92 \times 10)} \times 10^{6} = 4.92 \times 10^{7} \qquad \blacktriangleleft$$

E X A M P L E **8** Carry out the following division:

$$\frac{4.360 \times 10^{3}}{8.290 \times 10^{12}}$$

Solution. $\dfrac{4.360 \times 10^{3}}{8.290 \times 10^{12}} = \dfrac{4.360}{8.290} \times 10^{-9} = 0.5259 \times 10^{-9}$

to four significant digits. Since the coefficient has to be between 1 and 10, we write 0.5259 as 5.259/10 to get

$$\mathbf{0.5259 \times 10^{-9}} = \frac{\mathbf{5.259}}{\mathbf{10}} \times 10^{-9}$$
$$= \mathbf{(5.259 \times 10^{-1})} \times 10^{-9}$$
$$= 5.259 \times 10^{-10} \qquad \blacktriangleleft$$

E X A M P L E **9** Express the given numbers in scientific notation before multiplying:

$$(29,600)(0.00000000816)$$

Solution. $(29,600)(0.00000000816) = (2.96 \times 10^{4})(8.16 \times 10^{-9})$
$$= 24.1536 \times 10^{-5}$$
$$= 2.41536 \times 10^{-4}$$
$$= 2.42 \times 10^{-4}$$

to three significant digits. $\qquad \blacktriangleleft$

The next example illustrates the addition of numbers expressed in scientific notation.

E X A M P L E **10**　Add

$$2.78 \times 10^{-5} \quad \text{and} \quad 3.12 \times 10^{-5}$$

Solution.

Numbers	Algebraic comparison	
2.78×10^{-5}	$2.78x$	
$+3.12 \times 10^{-5}$	$+3.12x$	
5.90×10^{-5}	$5.90x$	**adding coefficients**

Note that to add numbers expressed in scientific notation, the powers of 10 must be the same. ◄

CALCULATOR COMMENT

Scientific calculators are programmed to express numbers in scientific notation. In fact, large and small numbers resulting from a calculation are automatically displayed in this form. Moreover, very large and very small numbers must be converted to scientific notation before they can be entered.

To enter numbers expressed in scientific notation, use the EE, EEX, EXP, or similar key to enter the exponent. For example, to enter 9.316×10^{-20}, use the following sequence:

$$9.316 \boxed{\text{EE}} \, 20 \boxed{+/-} \rightarrow 9.316 \quad -20$$

(Note the space between 9.316 and -20 in the display.)

E X A M P L E **11**　Use a calculator to multiply

$$(2.75 \times 10^{8})(4.372 \times 10^{-25})$$

Solution. The sequence is

$$2.75 \boxed{\text{EE}} \, 8 \boxed{\times} \, 4.372 \boxed{\text{EE}} \, 25 \boxed{+/-} \boxed{=} \rightarrow 1.2023 \quad -16$$

The result to three significant digits is 1.20×10^{-16}. ◄

 E X A M P L E **12**　Light travels at the rate of 6.70×10^{8} mi/h. The average distance from the earth to the sun is 9.29×10^{7} mi. How long does it take for the sun's light to reach the earth?

Solution. The time is found by dividing the distance to the sun by the given rate:

$$t = \frac{9.29 \times 10^{7} \text{ mi}}{6.70 \times 10^{8} \text{ mi/h}}$$

The sequence is

$$9.29 \boxed{\text{EE}} \, 7 \boxed{\div} \, 6.70 \boxed{\text{EE}} \, 8 \boxed{=} \; \rightarrow \; 0.1386567$$

So the time is about 0.139 h = 8.34 min.   ◀

E X E R C I S E S / S E C T I O N   **1.9**

In Exercises 1–8, express each number in scientific notation.

1. 26,000

2. 9,230,000

3. 379,200,000

4. 43,000,000

5. 0.00013

6. 0.00000126

7. 0.00008927

8. 0.0000000347

In Exercises 9–16, express each number in decimal notation.

9. 1.2×10^6

10. 3.98×10^{-4}

11. 6.273×10^{-3}

12. 7.8×10^8

13. 2.7×10^{-7}

14. 5.17×10^5

15. 9.56×10^5

16. 3.743×10^{-7}

In the statements of Exercises 17–23, change the given numbers from scientific to decimal notation or from decimal to scientific notation. (For the units employed, see the table on the inside cover.)

17. The distance from the earth to the moon is 240,000 mi.

18. The mean distance from the earth to the sun is 93 million mi.

19. The gravitational constant G is $G = 6.670 \times 10^{-11}$ $N \cdot m^2/kg^2$.

20. The capacitance of a certain capacitor is 4.1×10^{-6} F (farad).

21. 1 kilowatt $\cdot$ hour $= 3.6 \times 10^6$ J (joules).

22. A typical steel wire will break if pulled with a force of 60,000 lb/in.^{2}.

23. A pressure of 1 atmosphere $= 1.013 \times 10^5$ newtons per square meter.

In Exercises 24–41, express the given numbers in scientific notation and carry out the indicated operations with a calculator. Round off the answers to the appropriate number of significant digits. (See Examples 9 and 11.)

24. (34,000)(9,630,000)

25. (823,000)(4,760,000)

26. (2,760,000)(0.00052)

27. (0.0024)(0.00096)(0.00082)

28. (31,600)(0.0000126)(0.0009116)

29. $\dfrac{0.00127}{54,300}$

30. $\dfrac{723,000}{0.003864}$

31. $\dfrac{(2,340)(52,300)}{0.001763}$

32. $\dfrac{(0.00148)(72,800)}{0.0000264}$

33. $\dfrac{(0.00026)(0.00092)}{2,008}$

34. (0.0000007132)(0.000002516)

35. (0.000003651)(0.00000081)

36. $\dfrac{856,000,000}{36,000,000,000}$

37. $\dfrac{792,500,000}{0.00000039}$

38. $\dfrac{0.0000000093}{0.00000000000084}$

39. $\dfrac{234,000,000}{(63,400,000)(0.000000000906)}$

40. $\dfrac{(86,430,000)(0.000000036)}{0.00000000946}$

41. $\dfrac{(0.00000000000064)(0.00000000967)}{0.0000000000000841}$

In Exercises 42–44, perform the indicated operations. (See Example 10.)

42. $(3.785 \times 10^{-5}) + (1.720 \times 10^{-5})$
$(3.785 \times 10^{-5}) - (1.720 \times 10^{-5})$

43. $(8.13 \times 10^{-6}) + (1.04 \times 10^{-6})$
$(8.13 \times 10^{-6}) - (1.04 \times 10^{-6})$

44. $(1.35 \times 10^7) + (2.78 \times 10^7)$
$(1.35 \times 10^7) - (2.78 \times 10^7)$

1.10   Multiplication of Polynomials

Earlier we considered the addition and subtraction of polynomials. We will complete our study of the fundamental operations by taking up multiplication in this section and division in the next.

The basic rules of multiplication and division of single terms can be applied to polynomials by means of the distributive law $a(b + c) = ab + ac$. In particular, to multiply a polynomial by a monomial, we multiply *each* term of the polynomial by the monomial. For example,

Multiplication by a monomial

$$4(2x + y) = 4(2x) + 4(y) = 8x + 4y$$

Consider the following example.

E X A M P L E 1   Multiply $-2x(x^2 - 5)$.

Solution. Since $a(b + c) = ab + ac$, we have

$$-2x(x^2 - 5) = (-2x)(x^2) + (-2x)(-5)$$
$$= -2x^3 + 10x$$

(Always place either a plus or a minus sign between terms.)

Remark. It is probably simpler to skip the intermediate step and to multiply directly, since direct multiplication avoids possible confusion on the signs. In other words, it is easy to see that $(-2x)(x^2) = -2x^3$ and $(-2x)(-5) = 10x$. So it follows in just one step that

$$-2x(x^2 - 5) = -2x^3 + 10x \qquad \blacktriangleleft$$

Polynomials can be multiplied by the rule given next.

Multiplication of Polynomials: To multiply two polynomials, multiply each term of one polynomial by each term of the other and add the results.

As an example, to multiply

$$(x - 2y)(2x + y)$$

the expression $x - 2y$ may be treated as a single number. In other words, in the distributive law $a(b + c) = ab + ac$, let

$$a = x - 2y, \quad b = 2x, \quad \text{and} \quad c = y$$

Then

$$(x - 2y)(2x + y) = (x - 2y)(2x) + (x - 2y)(y)$$
$$= (x - 2y)(2x) + (x - 2y)(y)$$

Now apply the distributive law again to obtain

$$(x - 2y)(2x) + (x - 2y)(y)$$
$$= (x)(2x) + (-2y)(2x) + (x)(y) + (-2y)(y)$$
$$= 2x^2 - 4xy + xy - 2y^2$$

So

$$(x - 2y)(2x + y) = 2x^2 - 3xy - 2y^2$$

To make this procedure more systematic, let us arrange our work so that it resembles ordinary multiplication in arithmetic. First, let us write the polynomials in two rows:

$$\begin{array}{r} x - 2y \\ 2x + y \\ \hline \end{array}$$

Now, working from left to right, multiply $2x$ (in the second row) by both terms in the first row and write the result in the first row below the line:

$$\begin{array}{r} x - 2y \\ \mathbf{2x} + y \\ \hline 2x^2 - 4xy \end{array}$$

Next, multiply $+y$ (in the second row) by both terms in the first row and write the result in the next row under the line, taking care to place like terms in the same column to facilitate the addition step:

$$\begin{array}{r} x - 2y \\ 2x + y \\ \hline 2x^2 - 4xy \\ xy - 2y^2 \\ \hline 2x^2 - 3xy - 2y^2 \end{array} \qquad xy = 1xy$$

after drawing another line and adding.

E X A M P L E **2** Multiply the following polynomials:

$$\begin{array}{r} 3x^2 - 2xy + y^2 \\ 2x - 3y \\ \hline 6x^3 - 4x^2y + 2xy^2 \\ - 9x^2y + 6xy^2 - 3y^3 \\ \hline 6x^3 - 13x^2y + 8xy^2 - 3y^3 \end{array} \qquad \begin{array}{l} 2x(3x^2 - 2xy + y^2) \\ -3y(3x^2 - 2xy + y^2) \\ \text{adding} \end{array}$$ ◄

The characteristic pattern (in particular, the indented row of intermediate products) is always obtained if the terms in the polynomials appear in *descending* powers of x and in *ascending* powers of y. For example,

$$2x^3 - 3x^2y + 2xy^2 - 5y^3$$

has x^3 in the first term, x^2 in the second, x in the third, and no x in the fourth. The powers of y follow the opposite order. If the polynomials are not in this form, then the pattern becomes disrupted.

E X A M P L E **3**

$$
\begin{array}{l}
2x^2 + xy + z \\
\underline{x - y} \\
2x^3 + x^2y + xz \\
 - 2x^2y - xy^2 - yz \\
\hline
2x^3 - x^2y + xz - xy^2 - yz
\end{array}
$$

$x(2x^2 + xy + z)$
$-y(2x^2 + xy + z)$
adding

Note that unlike terms are placed in separate columns. ◀

E X A M P L E **4** Multiply $(3x^2 - xy + 2y^2)(-2x^2 + 4xy - y^2)$.

Solution. After writing the polynomials in two rows, we work again from left to right, starting with $-2x^2$. Note that the intermediate products take up three rows.

$$
\begin{array}{l}
3x^2 - xy + 2y^2 \\
\underline{-2x^2 + 4xy - y^2} \\
-6x^4 + 2x^3y - 4x^2y^2 \\
 12x^3y - 4x^2y^2 + 8xy^3 \\
 3x^2y^2 + xy^3 - 2y^4 \\
\hline
-6x^4 + 14x^3y - 11x^2y^2 + 9xy^3 - 2y^4
\end{array}
$$

$3x^2 - xy + 2y^2$ **multiplied by:**
$-2x^2$
$+4xy$
$-y^2$

◀

 E X A M P L E **5** A circuit has a variable resistance $R = 2.5t + 1.6$ (in ohms), where t is measured in seconds. If the current in the circuit is $I = 0.01t - 0.02$ (in amperes), find the voltage $V = IR$ across the resistor.

Solution. From $V = IR$, we get

$$
\begin{aligned}
V &= (0.01t - 0.02)(2.5t + 1.6) \\
&= (0.01)(2.5)t^2 - (0.02)(2.5)t + (0.01)(1.6)t - (0.02)(1.6) \\
&= 0.025t^2 - 0.05t + 0.016t - 0.032 \\
&= 0.025t^2 - 0.034t - 0.032 \text{ (in volts)}
\end{aligned}
$$

◀

E X E R C I S E S / S E C T I O N   **1.10**

In Exercises 1–60, multiply the given expressions.

1. $(-3a^2b^3)(-4ab)$

2. $(-x^2y^2)(-xy)(3xy^3)$

3. $(2ab)(-3a^3b^4)(-4x)$

4. $(abc)(abc)(-abc)$

5. $xy(3x + 2xy)$

6. $2y(-3x^2 + 10xy)$

7. $-5x^2y^2(6xy^2 - 4x^2y)$

8. $-10ab^2(2ab + 5a^3b)$

9. $ab(2a^2b + 3ab - 4ab^2)$

10. $-5xyz(3x^2y^2z^2 - 15x^3y^3z^3)$

11. $-2a^2x(6a^2b - 5a^3x + 2ab)$

12. $axy(axy - 2a^2x^2y^2 - 3ax)$

13. $2xy(3xy - 5x^2y^2) - 5xy(-2xy + 10x^2y^2)$
(Combine like terms after multiplying.)

14. $ab(-ab + ab^2) - ab(2ab - 5ab^2)$

15. $a^2b(2a^3b - 3ab^2) - ab^2(-5a^4 + 2a^2b)$

16. $-2xy(5x^2y - 2xy^2) - x^2y^2(x + 2y)$

17. $(x - 1)(x - 2)$ **18.** $(x + 2)(x - 5)$

19. $(2x - 1)(x - 3)$ **20.** $(3x - 7)(2x + 2)$

21. $(2x + 2)(5x - 7)$ **22.** $(x - 5)(2x + 2)$

23. $(x - y)(x + y)$ **24.** $(x - y)^2$

25. $(2x - 1)(2x + 1)$ **26.** $(2x - y)(2x + y)$

27. $(3x - 2y)(3x + 2y)$ **28.** $(4x + y)(4x - y)$

29. $(3x - 1)^2$ **30.** $(3x + y)^2$

31. $(2x - 2y)^2$ **32.** $(5x + 1)^2$

33. $(2x + y)^2$ **34.** $(x + 2b)(x - 2b)$

35. $(x^2 - y^2)(x^2 + y^2)$ **36.** $(2x^2 + y^2)(2x^2 - y^2)$

37. $(x - 2b)(5x - b)$ **38.** $(2x - z)(3x + 2z)$

39. $(x - 1)(x^2 - 2x - 3)$ **40.** $(2x + 2)(x^2 + 5x - 4)$

41. $(a - 3b)(a^2 - ab + b^2)$ **42.** $(c - 2d)(c^2 - 3cd + 5d^2)$

43. $(5x - 3y)^2$ **44.** $(3a^2 + 4b^2)^2$

45. $(5x - 2y - 3z)(x - y + z)$

46. $(3a^2 - 2ab - 5b^2)(a^2 - ab + b^2)$

47. $(4x^4 - 2x^2 + 5)(2x^4 - x^2 - 1)$

48. $(2a^4 - 3a^2 - 1)(a^4 - 6a^2 + 3)$

49. $(2x^3y - 4x^2y^2 - xy^3)(x^2y + 2xy^2 - 2y^3)$

50. $(2ab + 3ac - 4bc)(3ab - ac - 2bc)$

51. $(x^4 - x^3 + x^2 - 1)(x + 1)$

52. $(x^3 - x^2y + 2xy^2 + 2y^3)(x^2 + xy - y^2)$

53. $(x + y + z)^2$

54. $(a - b - c)^2$

55. $(2a - 3b + 2c)^2$

56. $(-a - 4c - 6d)^2$

57. $3(l - 2w)(l + 3w)$

58. $(I_1 + I_2)^2 - 2(I_1 + I_2)^2$
(Combine like terms after multiplying.)

59. $3(R_1 + R_2)(2R_1 + 4R_2)$

60. $(2i_1 + 3i_2 + i_3)^2$

61. Find an expression for the area of a square measuring $(x + 2)$ m on the side.

62. Find an expression for the area of a rectangular plate $(3a + 2b)$ ft long and $(a + b)$ ft wide.

63. Find an expression for the total cost of $2a - b$ calculators costing \$25 each and $3a + 2b$ calculators costing \$30 each.

64. A shipment contains $3x + 4$ gears weighing 3 lb each and $5x + 6$ gears weighing 4 lb each. Find an expression for the total weight of the shipment.

65. Find an expression for the area inside the concrete form shown in Figure 1.5.

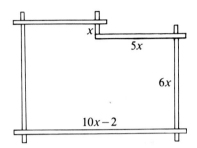

Figure 1.5

66. Find an expression for the volume illustrated in Figure 1.6.

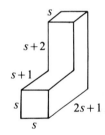

Figure 1.6

67. The voltage (in volts) across a resistor is given by $V = IR$, where I is measured in amperes and R in ohms. If $I = 0.1t - 2.0$ and $R = 0.4t + 9.0$ (t measured in seconds), find V.

68. The power P (in watts) delivered to a resistor is given by $P = RI^2$, where I is in amperes and R in ohms. If $R = 10$ ohms and $I = 2.0t^2 - 4.0t + 1.0$, find P.

69. The energy E radiated by a blackbody can be written $E = k(T^2 + T_0^2)(T - T_0)(T + T_0)$, where T is the temperature of the body, T_0 the temperature of the surrounding medium, and k a constant. Simplify the expression for E.

70. Simplify the following expression for the deflection of a beam: $d = (L - ax/2)(L - ax^2)$.

71. In computing the center of mass of a plate of uniform density, the expression $\frac{1}{2}(y_2 - y_1)(y_2 + y_1)$ has to be simplified. Carry out the simplification.

1.11   Division of Polynomials

Division of polynomials can be carried out by a scheme analogous to that for multiplication: We set up the procedure so that it resembles long division in arithmetic. But first we need to discuss the simpler case of division by a monomial. Consider, for example,

$$(4x^3 - 8x^2) \div 2x, \quad \text{or} \quad \frac{4x^3 - 8x^2}{2x}$$

By the distributive law, this problem is equivalent to

$$\frac{1}{2x}(4x^3 - 8x^2) = \frac{4x^3}{2x} - \frac{8x^2}{2x} = 2x^2 - 4x$$

Division by a monomial

In summary, the quotient of a polynomial and a monomial is found by dividing each term of the polynomial by the monomial and adding the results. The division should be carried out in one step to avoid sign errors.

E X A M P L E   **1**

$$\frac{15a^3b^4 - 20a^6b^7}{-5a^2b} = -3ab^3 + 4a^4b^6$$

Note that

$$\frac{15a^3b^4}{-5a^2b} = -3ab^3 \quad \text{and} \quad \frac{-20a^6b^7}{-5a^2b} = 4a^4b^6 \qquad \blacktriangleleft$$

To carry out the division

$$(x^3 - 4x^2 + x + 2) \div (x - 1)$$

we set up the division problem as follows:

$$x - 1\,\overline{)\,x^3 - 4x^2 + x + 2}$$

(Both polynomials should be written in descending powers of x.) Next, we divide x in the divisor into the first term, x^3, of the dividend and place the resulting quotient, x^2, on top.

$$
\begin{array}{r}
x^2 \\
x - 1\,\overline{)\,x^3 - 4x^2 + x + 2}
\end{array}
\qquad \frac{x^3}{x} = x^2
$$

Now we multiply x^2 (in the quotient) by the *entire* divisor and place the product in the row below:

$$
\begin{array}{r}
x^2 \\
x - 1 \overline{)\; x^3 - 4x^2 + x + 2} \\
\underline{x^3 - x^2}
\end{array}
$$

$x^2(x - 1) = x^3 - x^2$

As in long division in arithmetic, we now subtract:

$$
\begin{array}{r}
x^2 \\
x - 1 \overline{)\; x^3 - 4x^2 + x + 2} \\
\underline{x^3 - x^2} \\
0 \; - 3x^2
\end{array}
$$

$x^3 - x^3 = 0; \; -4x^2 - (-x^2) = -3x^2$

(Remember to change the sign of each term in the subtrahend.) After subtracting, we bring down the next term:

$$
\begin{array}{r}
x^2 \\
x - 1 \overline{)\; x^3 - 4x^2 + x + 2} \\
\underline{x^3 - x^2} \\
- 3x^2 + x
\end{array}
$$

The procedure is now repeated: We divide x into the first term of the last row to obtain $-3x$, the next term in the quotient:

$$
\begin{array}{r}
x^2 - 3x \\
x - 1 \overline{)\; x^3 - 4x^2 + x + 2} \\
\underline{x^3 - x^2} \\
- 3x^2 + x
\end{array}
$$

$\dfrac{-3x^2}{x} = -3x$

After multiplying $-3x$ by the entire divisor and subtracting, we have

$$
\begin{array}{r}
x^2 - 3x \\
x - 1 \overline{)\; x^3 - 4x^2 + x + 2} \\
\underline{x^3 - x^2} \\
- 3x^2 + x \\
\underline{- 3x^2 + 3x} \\
- 2x
\end{array}
$$

$-3x(x - 1) = -3x^2 + 3x$

subtracting

After subtracting, we bring down the next term:

$$
\begin{array}{r}
x^2 - 3x \\
x - 1 \overline{)\; x^3 - 4x^2 + x + 2} \\
\underline{x^3 - x^2} \\
- 3x^2 + x \\
\underline{- 3x^2 + 3x} \\
- 2x + 2
\end{array}
$$

Finally, we divide x into $-2x$ to obtain -2, the last term of the quotient.

Division procedure

$$
\begin{array}{r}
x^2 - 3x \;- 2 \\
x - 1 \overline{)\, x^3 - 4x^2 + \;x + 2} \\
\underline{x^3 - \;x^2} \\
-3x^2 + \;x \\
\underline{-3x^2 + 3x} \\
-2x + 2 \\
\underline{-2x + 2} \qquad -2(x-1) = -2x+2 \\
0 \qquad\quad \text{subtracting}
\end{array}
$$

Since the remainder is 0, we conclude that

$$(x^3 - 4x^2 + x + 2) \div (x - 1) = x^2 - 3x - 2$$

E X A M P L E **2**   Carry out the following division:

$$(2x^3 + 2y^3 + 2xy^2 - 6x^2y) \div (2x - 2y)$$

Solution. Arranging the terms in descending powers of x, we get

$$
\begin{array}{r}
x^2 - 2xy \;- y^2 \\
2x - 2y \overline{)\, 2x^3 - 6x^2y + 2xy^2 + 2y^3} \\
\underline{2x^3 - 2x^2y} \qquad\qquad x^2(2x - 2y) \\
-4x^2y + 2xy^2 \\
\underline{-4x^2y + 4xy^2} \qquad -2xy(2x - 2y) \\
-2xy^2 + 2y^3 \\
\underline{-2xy^2 + 2y^3} \qquad -y^2(2x - 2y) \\
0
\end{array}
$$

◀

If one of the powers in the dividend is missing, we need to leave a space, as shown in the next example. (A missing term merely indicates a coefficient of 0.)

E X A M P L E **3**   Divide: $(4x^3 - 5x + 2) \div (2x - 1)$.

Solution. Note that the x^2 term is missing.

$$
\begin{array}{r}
2x^2 + \;x \;- 2 \\
2x - 1 \overline{)\, 4x^3 \qquad\; - 5x + 2} \\
\underline{4x^3 - 2x^2} \\
2x^2 - 5x \\
\underline{2x^2 - \;x} \\
-4x + 2 \\
\underline{-4x + 2} \\
0
\end{array}
$$

◀

Recall from arithmetic that whenever the remainder is not 0, we divide the remainder by the divisor and add the resulting fraction to the quotient.

E X A M P L E **4**

$$
\begin{array}{r}
2x^2 - \ xy \ - \ y^2 \\
x^2 - 3xy - 2y^2 \ ) \overline{\ 2x^4 - 7x^3y - 2x^2y^2 + 5xy^3 + \ y^4} \\
2x^4 - 6x^3y - 4x^2y^2 \\
\hline
- x^3y + 2x^2y^2 + 5xy^3 \\
- x^3y + 3x^2y^2 + 2xy^3 \\
\hline
- x^2y^2 + 3xy^3 + \ y^4 \\
- x^2y^2 + 3xy^3 + 2y^4 \\
\hline
- \ y^4
\end{array}
$$

The resulting expression is

$$
2x^2 - xy - y^2 + \frac{-y^4}{x^2 - 3xy - 2y^2}
$$

or

$$
2x^2 - xy - y^2 - \frac{y^4}{x^2 - 3xy - 2y^2}
$$ ◀

E X E R C I S E S / S E C T I O N 1.11

In Exercises 1–12, find each of the quotients.

1. $\dfrac{a^4}{a^2}$

2. $\dfrac{3a^6b^5}{ab}$

3. $\dfrac{8x^4y^2}{-4x^2y}$

4. $\dfrac{-10a^6x^5y^4}{-5ax^2}$

5. $(-8x^6 - 4x^4) \div 2x^2$

6. $(24a^2b^2 - 12a^3b^2) \div a^2b^2$

7. $(-12abc + 9a^2bc^2) \div (-3ab)$

8. $(-36ab - 27a^2b^2 + 18ab^2) \div (9ab)$

9. $\dfrac{-17x^2y^3 + 34x^3y^5 - 51x^5y^5}{-17x^2y^3}$

10. $\dfrac{25x^2y^2z^2 - 15x^3y^3z^3}{-5x^2z}$

11. $\dfrac{12x^7y^6a^5 + 6x^6y^4a^3 - 9x^5y^3a^7}{3x^2y^3a^3}$

12. $\dfrac{5a^2y^2z - 10a^3yz^2 + 12a^4y^3z}{a^2yz}$

In Exercises 13–52, divide the first expression by the second.

13. $6x^2 - x - 2,\quad 2x + 1$

14. $3x^2 + 10x - 8,\quad x + 4$

15. $2x^2 - 5x - 3,\quad x - 3$

16. $8x^2 + 2x - 6,\quad 2x + 2$

17. $2x^2 - 8,\quad x - 2$

18. $3a^2 - 27,\quad a - 3$

19. $4x^2 - 4,\quad x + 1$

20. $2x^2 - 18,\quad 2x + 6$

21. $3x^2 - 5x - 12,\quad x - 3$

22. $2x^2 + xy - 6y^2,\quad x + 2y$

23. $12x^2 + 18xy - 12y^2,\quad 4x - 2y$

24. $6x^2 - 11xy + 4y^2,\quad 2x - y$

25. $4a^2 + ab - 3b^2,\quad 4a - 3b$

26. $8s^2 - 2st - 15t^2,\quad 2s - 3t$

27. $6w^2 - 5wz - 4z^2,\quad 2w + z$

28. $2m^2 + mn - 3n^2,\quad m - n$

29. $x^3 - x^2y - xy^2 + y^3,\quad x + y$

30. $a^3 - a^2b - 3ab^2 - 9b^3,\quad a - 3b$

31. $3a^3 + 5a^2b - 17ab^2 + 5b^3,\quad 3a - b$

32. $2x^3 + 3x^2y - xy^2 + 2y^3,\quad x + 2y$

33. $3v^3 + 10v^2 - 4v + 16,\quad 3v^2 - 2v + 4$

34. $3R^3 - 13R^2 + 14R - 6,\quad 3R^2 - 4R + 2$

35. $8V^3 - 16V^2 + 4V + 3,\quad 4V^2 - 2V - 1$

36. $4a^3 + 12a^2 + 7a - 3,\quad 2a^2 + 3a - 1$

37. $6s^3 - 4st^2 - 2t^3$, $2s - 2t$

38. $8R^3 - 8RS^2 - 3S^3$, $2R + S$

39. $4p^3 - 11pq^2 - 3q^3$, $2p + 3q$

40. $9v^3 - 10vt^2 + 4t^3$, $3v - 2t$

41. $4x^3 + xy^2 + y^3$, $2x + y$

42. $9x^3 - 13xy^2 - 6y^3$, $3x + 2y$

43. $x^3 + y^3$, $x + y$

44. $x^3 - y^3$, $x - y$

45. $2x^4 + 2x + 2x^2 - 5x^3 - 1$, $x^2 - x - 1$

46. $4x + 1 + 2x^4 - 5x^3$, $2x^2 - 3x - 1$

47. $6x^3 + 7xy^2 - 11x^2y - 6y^3$, $3x^2 + 2y^2 - xy$

48. $x^3y + x^4 + 14xy^3 - 8x^2y^2 - 8y^4$, $x^2 + 3xy - 4y^2$

49. $y^3 - 3xy^2 + x^3$, $x + 2y$

50. $x^3 - 7x^2y - 2y^3$, $x - 3y$

51. $x^4 - 3x^3 - 3x^2 - 6x + 3$, $x - 4$

52. $4x^3 - 3x^2y + 4xy^2 - y^3$, $x + 2y$

53. If C is the combined capacitance of two capacitors, C_1 and C_2, in series, then

$$\frac{1}{C} = \frac{C_2 + C_1}{C_2 C_1}$$

Rewrite this formula by performing the indicated division on the right side.

54. The area A (in square inches) of a rectangle is $A = 2x^3 + 7x^2 + 7x + 2$ and the length (in inches) is $x + 2$. Find an expression for the width (width = area/length).

55. If you can drive $(3x^3 + 8x^2 + 10x + 4)$ miles on $(3x + 2)$ gallons of gasoline, how many miles per gallon do you get?

1.12   Algebraic Expressions in BASIC (Optional)

In this section we discuss algebraic expressions written in BASIC. (For a brief discussion of the BASIC programming language, see Appendix C.)

In most versions of BASIC, a legal name for a variable can be any single capital letter or a single letter followed by a single digit. Examples of proper variable names are

A, B, C, X, Y, and Z

or

A1, B3, Y9, and Z0

However, in some versions of BASIC, variable names can be longer than two characters.

The BASIC symbols for some of the algebraic operations are given in the following table:

Operation	Meaning	Example	Algebraic notation
$+$	addition	A + B	$A + B$
$-$	subtraction	A $-$ B	$A - B$
$*$	multiplication	A $*$ B	AB or $A \cdot B$
/	division	A/B	$\dfrac{A}{B}$ or A/B
$\uparrow$	power	A $\uparrow$ 2	A^2
SQR(X)	square root	SQR(X $-$ Y)	$\sqrt{X - Y}$
ABS(X)	absolute value	ABS(-2)	$\lvert -2 \rvert$

In BASIC, operations are performed in the usual order:

1. Evaluation of quantities enclosed in parentheses.
2. Evaluation of powers.
3. Multiplication and division from left to right.
4. Addition and subtraction from left to right.

E X A M P L E **1**

	Algebraic notation	BASIC
a.	$\dfrac{x + 3}{6}$	$(X + 3)/6$
b.	$3(x - y)^3$	$3 * (X - Y) \uparrow 3$
c.	$\dfrac{1}{R_1} + \dfrac{1}{R_2}$	$1/R1 + 1/R2$
d.	$2\sqrt{b}$	$2 * \text{SQR(B)}$
e.	$2\sqrt{3 - t_1}$	$2 * \text{SQR}(3 - \text{T1})$ ◄

E X E R C I S E S / S E C T I O N   **1.12**

In Exercises 1–16, write each algebraic expression in BASIC.

1. $-2a^3$

2. $-3v^2$

3. $3x^3 - 4w$

4. $2m^2 - n$

5. $\dfrac{a - 2b}{c}$

6. $\dfrac{v + 3w}{a}$

7. $\sqrt{2 - 6p}$

8. $\sqrt{7 - 2n^2}$

9. $\sqrt{s_1 - 2s_2}$

10. $\sqrt{2p_1 + 3p_2}$

11. $\dfrac{a - b}{\sqrt{t_1}}$

12. $\dfrac{(s + t)^2}{s^2}$

13. $\dfrac{i^2 + k^2}{\sqrt{m}}$

14. $\dfrac{2 + \sqrt{v}}{s_1}$

15. $\dfrac{\sqrt{C_1 - C_2}}{5T_1}$

16. $\dfrac{2T_1}{\sqrt{R_1 - R_2}}$

R E V I E W   E X E R C I S E S / C H A P T E R   **1**

In Exercises 1–12, carry out the indicated operations.

1. $\dfrac{12}{|-4|} - 3$

2. $|-3| - |-5|$

3. $(-5) + (-10) - 1$

4. $-5 - 10 - 1$

5. $(-4) + (-3) + (2) - 5$

6. $(-4)(-3) - (-2)^2$

7. $\dfrac{(-28)(-3)}{(-5)(-14)}$

8. $\dfrac{(-7)(9)(-3)}{(2)(0)(-6)}$

9. $(-2)^2(7 - 12)$

10. $\dfrac{6 - 14}{(-2)^3}$

11. $\dfrac{8 - 6^2}{2 - (-3)^2}$

12. $\dfrac{-2 + 4^2}{2(3 - 7)}$

In Exercises 13–18, remove symbols of grouping and simplify.

13. $(2x - y) + (5x - 7y)$

14. $(3x - 2y) - (6x + 5y)$

15. $(2x^2 - 4xy) - (-3x^2 + 4xy)$

16. $(-a + 2b) + (5a - b) - (7a + 2b)$

17. $-[a - (a + b)]$

18. $-\{x - [y - (2x + y)]\}$

In Exercises 19 and 20, enclose the last three terms in parentheses preceded by (a) a plus sign and (b) a minus sign.

19. $2x + a - x^2 + y - 4w$ **20.** $a - 3b - 2a^2 - b^2 + c^2$

In Exercises 21–28, simplify each of the given expressions.

21. $\dfrac{(-5s^2 t)(3s^3 t^3)}{-25s^{10} t^4}$

22. $\dfrac{(-6R_1{}^2 R_2 R_3)(2R_1{}^3 R_2{}^2 R_3)}{12R_1{}^5 R_2{}^4 R_3}$

23. $\sqrt{50}$ **24.** $\sqrt{48}$ **25.** $\sqrt[3]{81}$

26. $\dfrac{2}{\sqrt{A}}$ **27.** $\dfrac{1}{\sqrt{5}}$ **28.** $\dfrac{5}{\sqrt{50}}$

In each of Exercises 29–32, state whether the given polynomial is a monomial, binomial, or trinomial.

29. $-3x^2 - y^4$ **30.** $2P_1{}^2 P_2{}^3$

31. $4x^2 - 6xy + y^2$ **32.** $6a^2 - 5b^2$

In Exercises 33–44, carry out the indicated operations.

33. $x(x^2 - 2y)$ **34.** $\dfrac{6v^3 - 12v^2}{3v^2}$

35. $(2x - 6y)(-x - 2y)$ **36.** $(3a^2 + 2a + 1)(a - 2)$

37. $(4s^2 - 2st - t^2)(s + 3t)$ **38.** $(x - 2y)(3x - a)$

39. $(x^2 + 3xy - 10y^2) \div (x + 5y)$

40. $(x^2 - xy - 12y^2) \div (x - 4y)$

41. $\dfrac{5a^2 + 2ab - b^2}{a + b}$

42. $\dfrac{4ab^2 + 2a^3 - 5a^2b - b^3}{a - b}$

43. $(a^3 - 8b^3) \div (a - 2b)$

44. $(2x + y)(4x^2 - 2xy + y^2)$

45. Find an expression for the area in Figure 1.7.

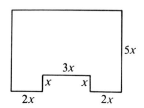

Figure 1.7

46. The voltage across a resistor is the product of the resistance and current. If the current i (in amperes) and the resistance R (in ohms) vary with time t (in seconds) according to the relations $i = 2 - 3t^2$ and $R = 5 + 3t$, find an expression for the voltage.

47. Simplify the following expression for the deflection of a beam: $d = (L - x)(2L - 2x^2)$.

48. The area A of a circular ring with inner radius r_1 and outer radius r_2 is $A = \pi(r_2 - r_1)(r_2 + r_1)$. Simplify this expression.

In Exercises 49–56, carry out the indicated operations with a calculator. Round off the answers to the appropriate number of significant digits.

49. $0.138 + 2.13 - 0.72$ **50.** $10.19 + 3.471 - 2.98$

51. $(1.23)(0.72)(3.980)(0.11)$ **52.** $(10.10)(20.3)(0.123)$

53. $\dfrac{(7.21)(1.32)}{1,001}$ **54.** $(3,200)(5.3)$

55. $(2.70)(0.37)$ **56.** $(4,900)(0.57)$

In Exercises 57–62, change the given numbers from scientific to decimal notation or from decimal notation to scientific notation.

57. The mass of the earth is 5.98×10^{24} kg.

58. The diameter of the sun is 864,000 mi.

59. The rest mass of a lithium nucleus is 0.00000000000000000000001164 g.

60. The diameter of an oxygen molecule (O_2) is 3×10^{-8} cm.

61. The charge of a single electron is 1.60×10^{-19} C (coulomb).

62. The speed of light is approximately 186,000 mi/s.

63. Multiply: $(0.000000792)(0.000000035)$.

64. Multiply: $(0.00000085)(0.000000760)$.

Equations, Ratio and Proportion, Variation, and Functions

2.1 Equations

Most of this chapter is devoted to the study of various types of equations. While important in mathematics, equations play an equally prominent role in science and technology, as we will see later in this chapter.

An **equation** is an equality between two expressions. Examples of equations are

1. $\dfrac{1}{2} = \dfrac{2}{4}$

2. $x = 1$

3. $2y + 3y = 5y$

4. $3x + 1 = 4$

Identity
Conditional equation

Note that equation 3 is valid for all numbers y and for that reason is called an **identity**. Equation 4, on the other hand, is valid only for $x = 1$ and is therefore called a **conditional equation**. The word *equation* is commonly restricted to refer to conditional equations. Let us now consider some of the basic terms.

> The literal symbol in a conditional equation is called the **variable** (which may represent some unknown quantity). The value of the unknown for which the equality holds is called the **solution** or **root** and is said to **satisfy** the equation. Finding the solution is referred to as **solving** the equation.

The most direct way to solve an equation is to make use of the fact that both sides of the equation are numbers. Consequently, the left side of the equation must be equal to the right side. This equality can be preserved at all times if we perform certain algebraic operations on both sides. More precisely:

> The equality can be preserved if we add the same number to both sides of the equation, subtract the same number from both sides, multiply both sides by the same number, or divide both sides by the same nonzero constant.

To illustrate these operations, consider the equation

$$x - 3 = 5$$

Even if we cannot see offhand what the solution is, we still know that $x - 3$ (the left side) must be equal to 5 (the right side). So adding 3 to both sides will not destroy this equality. Note that adding 3 to both sides eliminates the -3 on the left side:

$$x - 3 + \mathbf{3} = 5 + \mathbf{3}$$

or

$$x = 8$$

So $x = 8$ is the root of the equation.
   Similarly, to solve the equation

$$2x = 7$$

we divide both sides of the equation by 2. Dividing by 2 eliminates the coefficient 2 on the left side:

$$\frac{2x}{\mathbf{2}} = \frac{7}{\mathbf{2}}$$

$$x = \frac{7}{2}$$

So $x = \frac{7}{2}$ is the solution.
   The equation

$$4x + 5 = -13 \tag{2.1}$$

requires more than one step. First subtract 5 from both sides (or add -5) to eliminate the $+5$ on the left side:

$$4x + 5 - \mathbf{5} = -13 - \mathbf{5}$$

$$4x = -18$$

Now divide both sides of the equation by **4** to yield the solution:

$$\frac{4x}{4} = \frac{-18}{4} = -\frac{9}{2} \quad \text{or} \quad x = -\frac{9}{2}$$

The solution found can always be checked by substituting the root into the original equation. Letting $x = -\frac{9}{2}$ in the left side of equation (2.1), we get

$$4\left(-\frac{9}{2}\right) + 5 = -18 + 5 = -13$$

which is indeed equal to the right side.

Summary   In summary, we solve an equation by writing all terms containing the variable on one side of the equation and all constants on the opposite side. After combining the terms, we obtain the form $ax = b$. The resulting solution is $x = b/a$.

E X A M P L E **1**   Solve the equation $3x - 3 = x - 6$.

Solution. In this equation the unknown appears on both sides of the equal sign. One of these terms should be eliminated so that the terms containing the unknown are on one side only:

$3x - 3 = x - 6$	given equation
$3x - x - 3 = x - x - 6$	subtracting x from both sides
$2x - 3 = -6$	combining terms
$2x - 3 + 3 = -6 + 3$	adding 3 to both sides
$2x = -3$	combining terms
$\dfrac{2x}{2} = \dfrac{-3}{2}$	dividing by 2
$x = -\dfrac{3}{2}$	reducing the fraction on the left side

To check the solution, we substitute $x = -\frac{3}{2}$ into the given equation.

Left Side	Right Side
$3\left(-\dfrac{3}{2}\right) - 3$	$-\dfrac{3}{2} - 6$
$= -\dfrac{9}{2} - \dfrac{6}{2}$	$= -\dfrac{3}{2} - \dfrac{12}{2}$
$= -\dfrac{15}{2}$	$= -\dfrac{15}{2}$

The solution checks.   ◀

E X A M P L E **2**   Solve the equation $2x + 5 = 5x - 7$.

Solution. In this equation, it is helpful to eliminate the $2x$ from the left side to avoid negative coefficients:

$2x + 5 = 5x - 7$	given equation
$2x - 2x + 5 = 5x - 2x - 7$	subtracting 2x from both sides
$5 = 3x - 7$	combining terms
$5 + 7 = 3x - 7 + 7$	adding 7 to both sides
$12 = 3x$	combining terms
$\dfrac{12}{3} = \dfrac{3x}{3}$	dividing by 3
$4 = x$	reducing
$x = 4$	switching sides

Check:

Left Side	Right Side
$2(4) + 5 = 13$	$5(4) - 7 = 13$

The solution checks. ◀

E X E R C I S E S / S E C T I O N   **2.1**

Solve the following equations for x and check.

1. $x - 6 = 1$
2. $2x = 8$
3. $2x + 3 = -9$
4. $3x - 2 = 8$
5. $2x + 5 = x + 3$
6. $2x - 7 = 3x - 5$
7. $2x - 2 = 4x + 6$
8. $7x - 3 = 1 - 4x$
9. $x + 2 = 5x - 3$
10. $5x - 1 = -x - 2$
11. $1 - 2x = 3x - 4$
12. $2 - 3x = 7 - 5x$
13. $1 - 2x = 6 - 3x$
14. $2x - 1 = 2 - 6x$
15. $2x - 3 = 4 + 6x$

2.2   Transposition

The procedure for solving equations that was studied in the last section can be shortened considerably. To see how, consider the equation

$$5x + 5 = 1 - x$$

Adding x to both sides, we get

$$5x + 5 + x = 1$$

Comparing this equation with the original, it is apparent that moving $-x$ to the other side of the equal sign and changing the sign has the same effect as adding x to both sides. The resulting equation is $6x + 5 = 1$.

Next, subtract 5 from both sides of

$$6x + 5 = 1$$

to obtain

$$6x = 1 - 5$$

Again, moving $+5$ to the other side of the equal sign and changing the sign has the same effect as subtraction. Finally, from $6x = -4$, we get $x = -\frac{2}{3}$.

The procedure just described is called **transposition**. Note that the division step is still carried out last.

> **Transposition:** To **transpose** a term, move the term to the other side of the equal sign and change the sign.

E X A M P L E **1** Solve the equation $3 - x = 2x + 5$.

Solution.

$3 - x = 2x + 5$	given equation
$3 = 2x + x + 5$	transposing $-x$
$3 = 3x + 5$	simplifying
$3 = 3x + \mathbf{5}$	
$3 - \mathbf{5} = 3x$	transposing $+5$
$-2 = 3x$	simplifying
$\dfrac{-2}{3} = x$	dividing by 3
$x = -\dfrac{2}{3}$	switching sides

Check:

Left Side	Right Side
$3 - \left(-\dfrac{2}{3}\right) = \dfrac{11}{3}$	$2\left(-\dfrac{2}{3}\right) + 5 = \dfrac{11}{3}$ ✓

◄

If more than one term must be transposed, all transpositions can be carried out in one step, as shown in the next example.

E X A M P L E **2**   Solve the equation $x - 2(x - 3) = 5x - 12$.

Solution. The first step is to remove the parentheses:

$$x - 2(x - 3) = 5x - 12 \qquad \text{original equation}$$

$$x - 2x + 6 = 5x - 12 \qquad \text{removing parentheses}$$

$$-x + 6 = 5x - 12 \qquad \text{simplifying}$$

$$\mathbf{-}x + 6 = 5x \mathbf{- 12}$$

$$6 + 12 = 5x + x \qquad \text{transposing } -x \text{ and } -12$$

$$18 = 6x \qquad \text{simplifying}$$

$$3 = x \qquad \text{dividing by 6}$$

$$x = 3 \qquad \text{switching sides}$$

Check:

$$3 - 2(3 - 3) = 3, \qquad 5(3) - 12 = 3 \quad \sqrt{} \qquad \blacktriangleleft$$

The equations we have considered so far are called **first-degree equations**, since the variable x is raised to the first power. The procedure for solving first-degree equations to this point is summarized next.

Procedure for Solving First-degree Equations:

1. Remove symbols of grouping.
2. Use transposition to collect all terms containing x on one side of the equation and all constants on the other side. The resulting equation has the form $ax = b$.
3. Divide both sides by the coefficient of x to obtain the solution $x = b/a$.
4. Check the solution by substituting the value of x into the original equation.

E X A M P L E **3**   Solve for x: $2(1 - x) = 3(2 + x)$.

Solution.

$$2(1 - x) = 3(2 + x) \qquad \text{given equation}$$

Step 1. $2 - 2x = 6 + 3x \qquad \text{removing parentheses}$

Step 2. $2 - \mathbf{2}x = \mathbf{6} + 3x$

$$2 - 6 = 3x + 2x \qquad \text{transposing } -2x \text{ and } 6$$

$$-4 = 5x \qquad \text{combining terms}$$

Step 3. $-\dfrac{4}{5} = x$ **dividing by 5**

Step 4. Check: $x = -\dfrac{4}{5}$

Left Side   Right Side

$$2\left[1 - \left(-\frac{4}{5}\right)\right] \qquad 3\left(2 - \frac{4}{5}\right)$$

$$= 2\left(1 + \frac{4}{5}\right) \qquad = 3\left(\frac{6}{5}\right)$$

$$= 2\left(\frac{9}{5}\right) = \frac{18}{5} \qquad = \frac{18}{5} \ \checkmark$$

◀

Equations with Literal Coefficients

Sometimes equations have to be solved in terms of letters, as shown in the next example.

E   X   A   M   P   L   E   **4**   Solve the following equation for x, given that a is a constant:

$$2a(x - 3) = ax + 5$$

Solution.

$$2a(x - 3) = ax + 5 \qquad \text{given equation}$$

Step 1. $2ax - 6a = ax + 5$ **removing parentheses**

Step 2. $2ax - ax = 6a + 5$ **transposing**

$$ax = 6a + 5 \qquad \text{combining terms}$$

Step 3. $x = \dfrac{6a + 5}{a}$ **dividing by a**

Step 4. Check:

Left Side   Right Side

$$2a\left(\frac{6a + 5}{a} - 3\right) \qquad a\left(\frac{6a + 5}{a}\right) + 5$$

$$= 2(6a + 5) - 6a \qquad = 6a + 5 + 5$$

$$= 12a + 10 - 6a \qquad = 6a + 10$$

$$= 6a + 10 \ \checkmark$$

◀

Equations with Fractions

If an equation contains fractions, it is best to simplify the equation first. This can be done by *multiplying both sides of the equation by the lowest common denominator*, abbreviated LCD. This procedure is called **clearing fractions**.

Clearing fractions

E X A M P L E **5** Solve the equation

$$\frac{1}{5}x + 1 = \frac{1}{15}x - \frac{1}{5}$$

Solution. We multiply both sides of the equation by 15, which is the lowest common denominator. Remember to multiply all the terms!

$$15 \cdot \frac{1}{5}x + 15 \cdot 1 = 15 \cdot \frac{1}{15}x - 15 \cdot \frac{1}{5} \qquad \text{multiplying by 15}$$

$$3x + 15 = x - 3 \qquad \text{simplifying}$$

$$3x - x = -3 - 15 \qquad \text{transposing}$$

$$2x = -18$$

$$x = -9 \qquad \blacktriangleleft$$

The unknown can also occur in the denominator. Such equations are important enough to warrant a separate section in Chapter 5. However, the simplest cases can be adequately handled with our present techniques.

E X A M P L E **6** Solve the equation

$$\frac{1}{x} + 2 = 1 - \frac{1}{2x}$$

Solution. All the fractions can be cleared by multiplying both sides by $2x$.

$$\frac{1}{x} + 2 = 1 - \frac{1}{2x} \qquad \text{original equation}$$

$$2x\left(\frac{1}{x}\right) + 2x(2) = 2x(1) - 2x\left(\frac{1}{2x}\right) \qquad \text{clearing fractions (LCD = 2x)}$$

$$2 + 4x = 2x - 1 \qquad \text{simplifying}$$

$$4x - 2x = -1 - 2 \qquad \text{transposing}$$

$$2x = -3$$

$$x = -\frac{3}{2} \qquad \blacktriangleleft$$

Some equations have no solutions at all. For example, the equation

$$x + 1 = x + 2$$

leads to

$$x - x = 2 - 1 \qquad \text{or} \qquad 0 = 1$$

which is impossible. In other words, this equation is not satisfied by any value of x.

E X E R C I S E S / S E C T I O N 2.2

In Exercises 1–40, solve each equation for x.

1. $x - 5 = 4$

2. $x + 5 = -5$

3. $3x = 12$

4. $5x = 16$

5. $2x - 3 = 5$

6. $5x + 2 = 12$

7. $4x - 5 = 7$

8. $3x - 1 = 5$

9. $3 - x = 7$

10. $1 - 5x = 5$

11. $2x - 3 = 5x + 6$

12. $1 - x = -5 + 6x$

13. $-2x - 3 = 5x + 1$

14. $2(x + 5) = -3(x + 2)$

15. $3(2 - x) = 5(2x + 1)$

16. $7(3 + x) = 8(4 - x)$

17. $6(2 - 2x) = 3(2 - 4x)$

18. $6(2 - 3x) = -(x + 2)$

19. $-(x - 1) = -(x + 3)$

20. $2(x + 3) = 2(x + 1)$

21. $\dfrac{1}{2}x + 1 = x + 3$

22. $\dfrac{1}{3}x - \dfrac{1}{6} = \dfrac{1}{6}x - \dfrac{1}{3}$

23. $\dfrac{1}{4}x - \dfrac{1}{12} = \dfrac{1}{6}x - \dfrac{1}{3}$

24. $\dfrac{1}{2}x - \dfrac{1}{3}x = \dfrac{1}{6}x - \dfrac{1}{6}$

25. $\dfrac{1}{5}x - 7 = x - 3$

26. $\dfrac{1}{3}x - \dfrac{1}{5} = \dfrac{1}{7}$

27. $\dfrac{1}{4}(x - 2) = \dfrac{1}{8}(2x + 1)$

28. $\dfrac{1}{3}(2x - 7) = \dfrac{1}{4}(1 - 3x)$

29. $ax = b$

30. $x - 3 = a$

31. $2ax - 1 = 3ax + 2$

32. $5ax + 2 = 8ax - 7$

33. $3ax + b = ax - b$

34. $b - ax = 3b - 5ax$

35. $bx + a = a - 2bx$

36. $2(ax + b) = 3(ax - 5b)$

37. $a(2x - 3) = a - 5ax$

38. $b(ax + 1) = b(3 - 2ax)$

39. $2(bx + 3) = 4(bx - 1)$

40. $b(1 - x) = b(2 - 2x)$

In Exercises 41–48, solve each equation for x.

41. $\dfrac{2}{x} + 1 = \dfrac{1}{x}$

42. $\dfrac{1}{x} - 1 = 1 - \dfrac{2}{x}$

43. $\dfrac{1}{2x} - \dfrac{1}{2} = \dfrac{1}{2x}$

44. $\dfrac{1}{2x} - \dfrac{1}{4} = \dfrac{1}{2} - \dfrac{1}{4x}$

45. $\dfrac{1}{5x} - \dfrac{1}{x} = \dfrac{1}{15}$

46. $\dfrac{1}{6x} - \dfrac{1}{8x} = \dfrac{1}{24}$

47. $\dfrac{1}{2x} - \dfrac{1}{4} = \dfrac{1}{x}$

48. $\dfrac{1}{3x} - \dfrac{1}{x} = \dfrac{1}{6}$

2.3 Formulas

Formula

In the last section we learned to solve equations containing literal terms. In this section we will study literal equations called **formulas**, which are solved for one letter in terms of the rest.

A formula is a relationship between variables that often expresses a geometric property or a physical law. For example, the voltage V across a resistor is equal to the current I times the resistance R, or

$$V = IR$$

In what sense can this formula be solved? Suppose we divide both sides of the equation by I. Then

$$\frac{V}{I} = \frac{IR}{I} \qquad \text{dividing by } I$$

$$\frac{V}{I} = \frac{\cancel{I}R}{\cancel{I}} \qquad \text{canceling}$$

$$R = \frac{V}{I} \qquad \text{switching sides}$$

This "solution" is actually a new formula. This formula says that the resistance R is equal to the voltage V across the resistor divided by the current I. This example shows that there may be good reasons for solving a formula. The term *formula rearrangement* is sometimes used instead of *solution* of a formula. Consider another example.

E X A M P L E **1**   The formula for converting degrees Fahrenheit to degrees Celsius is given by

$$C = \frac{5}{9}(F - 32)$$

Solve the formula for F in terms of C.

Solution.

$$C = \frac{5}{9}(F - 32) \qquad \text{given formula}$$

$$\frac{9}{5}C = \frac{9}{5}\left[\frac{5}{9}(F - 32)\right] \qquad \text{multiplying by } \frac{9}{5}$$

$$\frac{9}{5}C = F - 32 \qquad \frac{9}{5} \cdot \frac{5}{9} = 1$$

$$\frac{9}{5}C + 32 = F \qquad \text{transposing}$$

$$F = \frac{9}{5}C + 32 \qquad \text{switching sides}$$

This is the formula for converting degrees Celsius to degrees Fahrenheit. ◀

The formulas in this section can be solved for the indicated letter by executing the following steps:

Summary

1. Remove symbols of grouping.
2. Clear fractions if necessary.
3. Write the terms containing the letter to be solved for on one side of the equation and collect the remaining terms on the other side.
4. Divide both sides by the coefficient of the letter to be solved for.

E X A M P L E **2**   Solve the formula $P = a(S/2 + b)$ for S.

Solution.

$$P = a\left(\frac{S}{2} + b\right) \qquad \text{given formula}$$

Step 1.

$$P = \frac{aS}{2} + ab \qquad \text{removing symbols of grouping}$$

Step 2.	$2P = aS + 2ab$	multiplying both sides by 2
Step 3.	$2P - 2ab = aS$	transposing 2*ab*
	$aS = 2P - 2ab$	switching sides
Step. 4.	$\dfrac{aS}{a} = \dfrac{2P - 2ab}{a}$	dividing both sides by *a*
	$S = \dfrac{2P - 2ab}{a}$	reducing ◀

E X A M P L E **3** The volume V of a gas under constant pressure varies with the temperature T according to the formula

$$V = V_0[1 + b(T - T_0)]$$

Solve this formula for T.

Solution. Using the four steps listed above, we get

	$V = V_0[1 + b(T - T_0)]$	given formula
Step 1.	$V = V_0[1 + bT - bT_0]$	eliminating symbols of grouping
	$V = V_0 + V_0bT - V_0bT_0$	
Step 2.	Not necessary in this formula	
Step 3.	$V - V_0 + V_0bT_0 = V_0bT$	transposing
	$V_0bT = V - V_0 + V_0bT_0$	switching sides
Step 4.	$\dfrac{V_0bT}{V_0b} = \dfrac{V - V_0 + V_0bT_0}{V_0b}$	dividing by V_0b
	$T = \dfrac{V - V_0 + bT_0V_0}{bV_0}$	canceling V_0b ◀

The steps listed in the summary are not always followed in the order given, as shown in the next example.

 E X A M P L E **4** The number of teeth N in a gear is related to the pitch diameter D of the gear and the outside diameter D_0 by the formula

$$N = \frac{2D}{D_0 - D}$$

Solve this formula for D_0.

Solution. This formula does not contain parentheses. So we first clear fractions by multiplying both sides by the denominator of the right side, $D_0 - D$. This operation introduces $N(D_0 - D)$ on the left side and is followed by removing symbols of grouping.

$$N = \frac{2D}{D_0 - D} \qquad \text{given formula}$$

$$N(D_0 - D) = 2D \qquad \text{multiplying by } D_0 - D$$

$$ND_0 - ND = 2D \qquad \text{distributive law}$$

$$ND_0 - ND + ND = 2D + ND \qquad \text{adding } ND$$

$$ND_0 = 2D + ND \qquad -ND + ND = 0$$

$$\frac{ND_0}{N} = \frac{2D + ND}{N} \qquad \text{dividing by } N$$

$$D_0 = \frac{2D + ND}{N} \qquad \blacktriangleleft$$

E X E R C I S E S / S E C T I O N   **2.3**

In Exercises 1–30, solve each formula for the indicated letter.

1. $2a = b$; a

2. $3P_1 = P_2$; P_1

3. $pq + 3 = s$; q

4. $2 - LD = a$; D

5. $\dfrac{T_1}{T_2} = ab$; T_2

6. $\dfrac{C_1}{C_2} = a + b$; C_2

7. $PR - Q = S$; R

8. $MN - P = Q$; M

9. $\dfrac{1}{2} ab + c = 2$; a

10. $st - 3 = r$; s

11. $2(a + b) = d$; b

12. $2(s + 3t) = a$; t

13. $S = \dfrac{a - b}{P}$; b

14. $T = \dfrac{2c - d}{4}$; d

15. $L = \dfrac{3n + 4}{c}$; n

16. $P = \dfrac{3s - t}{b}$; s

17. $T_0 = \dfrac{2(3T_1 - T_2)}{b}$; T_2

18. $R = \dfrac{2(C_1 - 2C_2)}{a}$; C_2

19. $B = S_0[a + d(D - D_0)]$; D

20. $N = a[R_1 - 2(R_2 - R_3)]$; R_1

21. $M = \dfrac{s - 3}{a - b}$; b

22. $S = \dfrac{3r}{2v - w}$; v

23. $C_0 = \dfrac{t(C_1 + C_2)}{C_2}$; C_1

24. $V = \dfrac{a(b + 2c)}{d - 2}$; c

25. $P = \dfrac{Vr}{a} - b$; V

26. $C = p + \dfrac{Vs}{p}$; V

27. $\dfrac{P_1 V_1}{T_1} = \dfrac{P_2 V_2}{T_2}$; P_2

28. $S = \dfrac{1}{2} at^2 + v_0 t$; a

29. $A = \dfrac{1}{2} h(b_1 + b_2)$; b_2

30. $B = \dfrac{1}{3} s(2n_1 - n_2)$; n_1

31. An investment of P dollars accumulates to A dollars in t years according to the formula

$$A = P + Prt$$

where r is the interest rate. Solve this formula for t.

32. In the study of the photoelectric effect, the formula

$$T = k(v - v_0) \qquad (v = \text{nu})$$

arises. Solve this formula for v.

33. The following formula arises in the study of atomic spectra:

$$E_d = \frac{(K + 1)H^2}{I}$$

Solve this formula for K.

34. The amount of heat conducted through a wall is found from the formula

$$Q = \frac{kA(t_1 - t_2)}{l}$$

Solve this formula for t_1.

35. The power loss in a transmission line is calculated from

$$L = k_1 A + \frac{k_2}{A}$$

Solve this formula for k_2.

2.4 Writing Formulas in BASIC (Optional)

Equations and formulas can be written in BASIC by using the BASIC notation introduced in Section 1.12. It should be noted, however, that in BASIC the symbol = actually means "replaces." For example, the program step

$$3\emptyset \quad X = X + 1$$

means "replace the value of X in a specified storage location by X + 1."

E X A M P L E **1**

Algebraic Notation	BASIC
$y = 2x^3 - 3$	Y = 2 * X ↑ 3 − 3
$C = \dfrac{5}{9}(F - 32)$	C = (5/9) * (F − 32)

◄

E X E R C I S E S / S E C T I O N **2.4**

Write the following equations in BASIC.

1. $3(x - 5) = x$

2. $x = 2(1 - x)$

3. $4(2x - 3) = 0$

4. $-(x - 8) = \dfrac{x}{2}$

5. $V = IR$

6. $E = mc^2$

7. $P = \dfrac{k}{V}$

8. $d = rt$

9. $\dfrac{1}{R} = \dfrac{1}{R_1} + \dfrac{1}{R_2}$

10. $\dfrac{1}{C} = \dfrac{1}{C_1} + \dfrac{1}{C_2}$

11. $s = \dfrac{1}{2}at^2 + vt$

12. $A = P + Prt$

13. $L = k_1 A + \dfrac{k_2}{A}$

14. $Q = \dfrac{kA(t_1 - t_2)}{l}$

2.5 Applications of Equations

Our discussion of formulas in the previous section demonstrated the importance of first-degree equations in science and technology. Equations are also used to solve "word problems." These problems are solved by translating the given statements from ordinary language into algebraic language. For example, if we know that the length of a rectangle is one unit more than the width, we write $l = w + 1$.

The translation from verbal statements to mathematical equations can be difficult. Although no general rule can cover all cases, here are some guidelines.

Guidelines for Solving Word Problems:

1. Read the problem carefully; make sure you understand the situation described. Drawing a figure may help.
2. Identify the known and unknown quantities. Assign a letter to one of the unknown quantities and express the others, if any, in terms of this quantity.
3. Look for information that tells you which quantity or quantities are equal. Use this information to write the equation.
4. Solve the equation.
5. Check the result in the original problem.

E X A M P L E **1**   Two less than twice a given number is 10. What is the number?

Solution.

Step 1. Is the statement of the problem clear? We are looking for a number. If this number is doubled, and if 2 is subtracted from the result, we get 10.

Step 2. The number we are looking for is unknown. Let us denote this number by x. (The known quantities are 2 and 10.)

Step 3. What quantities are equal? Note first that twice the unknown number is $2x$ and 2 less than this is $2x - 2$. So the statement "Two less than twice a given number is 10" translates to

$$2x - 2 = 10$$

Step 4. We solve the equation:

$$2x - 2 = 10$$
$$2x = 10 + 2 \qquad \text{transposing}$$
$$x = 6 \qquad \text{dividing by 2}$$

Step 5. To check the solution, we always return to the given problem, since the equation we found in Step 3 may already be wrong: The number we found is 6. Since $2 \times 6 = 12$ and $12 - 2 = 10$, the number 6 satisfies the conditions in the given problem.   ◀

E X A M P L E **2**   The manager of a machine shop has an order for a rectangular metal plate meeting the following specifications: The length of the plate is 2.00 in. more than the width and the perimeter is 19.2 in. What must the dimensions be?

Solution.

  Step 1. Is the statement of the problem clear? The plate is 2.00 in. longer than it is wide. The perimeter, which is 19.2 in., is the distance around the rim of the plate.

  Step 2. We know neither the length nor the width. However, if we denote the width by x, as in Figure 2.1, then the length (in inches) is $x + 2.00$. (The length is 2.00 in. more than the width.)

  We now have an expression for both unknown quantities in terms of x.

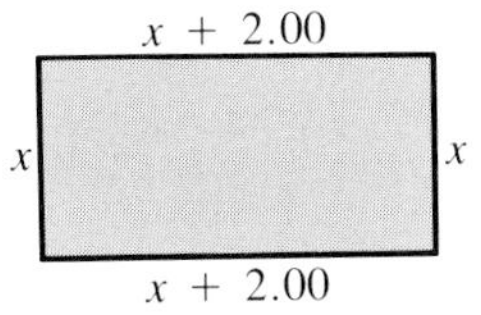

  Step 3. What quantities are equal? Since the perimeter is equal to 19.2 in., we get from Figure 2.1

$$(x + 2.00) + x + (x + 2.00) + x = 19.2$$

  Step 4. To solve this equation, we first combine the terms on the left side:

$$(x + 2.00) + x + (x + 2.00) + x = 19.2$$

$4x + 4.00 = 19.2$	**combining terms**
$4x = 19.2 - 4.00$	**subtracting 4.00**
$4x = 15.2$	
$x = 3.80$ in.	**width**
$x + 2.00 = 5.80$ in.	**length**

  Step 5. As a check, the perimeter is

$$2(3.80 \text{ in.}) + 2(5.80 \text{ in.}) = 19.2 \text{ in.} \quad \checkmark$$ ◀

x + 2.00

x x

x + 2.00

Figure 2.1

E X A M P L E **3**   Three resistors are connected in series. The resistance (in ohms) of the second is $2\frac{1}{2}$ times that of the first, and the resistance of the third is 1.2 Ω more than that of the first. The total resistance is 19.2 Ω. Find the resistance of the first resistor.

Solution.

  Step 1. Since the resistors are connected in series, we make use of the following information: The total resistance R_T of two or more resistors connected in series is equal to the sum of the individual resistances. (See Figure 2.2.)

  Step 2. We are looking for the resistance of the first resistor. Let us denote this unknown quantity by R. (The known quantities are 1.2 Ω and 19.2 Ω.) Since R is the resistance of the first resistor, we have

$$2.5R = \text{resistance of second resistor}$$

and

$$R + 1.2 = \text{resistance of third resistor}$$

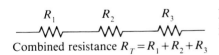

R_1 R_2 R_3

Combined resistance $R_T = R_1 + R_2 + R_3$

Figure 2.2

Step 3. From Step 1 and Figure 2.2:

sum of the three resistances = total resistance

$$R + 2.5R + (R + 1.2) = 19.2 \ \Omega$$

Step 4.

$4.5R + 1.2 = 19.2$	**combining terms**
$4.5R = 19.2 - 1.2$	**subtracting 1.2**
$4.5R = 18$	
$R = 4.0 \ \Omega \cdot$	**dividing by 4.5**

Step 5. To check the solution, we always return to the given problem rather than the equation: If $4.0 \ \Omega$ is the resistance of the first resistor, then $(4.0)(2.5) = 10 \ \Omega$ and $4.0 + 1.2 = 5.2 \ \Omega$ are the respective resistances of the other two. The total is $19.2 \ \Omega$, as required. ◀

E X A M P L E **4**   One brine solution contains 15% salt by volume, and another solution 25%. How many milliliters of each must be mixed to produce 40 mL of brine containing 18% salt?

Solution. Let

$$x = \text{number of milliliters of 15\% solution}$$

Then

$$40 - x = \text{number of milliliters of 25\% solution}$$

The best way to obtain an equation is to work directly with the quantities involved—in this case, the amount of salt. For example, in the 18% solution the fractional part by volume consisting of salt is $(0.18)(40) = 7.20$ mL. (See Figure 2.3.)

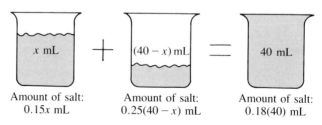

Amount of salt:	Amount of salt:	Amount of salt:
$0.15x$ mL	$0.25(40 - x)$ mL	$0.18(40)$ mL

Figure 2.3

amount of salt (before mixing) = amount of salt (after mixing)

$$0.15x + 0.25(40 - x) = 0.18(40) \qquad \text{see Figure 2.3}$$

$$0.15x + 10 - 0.25x = 7.20 \qquad \text{removing parentheses}$$

$$15x + 1000 - 25x = 720 \qquad \text{multiplying by 100}$$

$$-10x = -280 \qquad \text{simplifying}$$

$$x = 28 \text{ mL} \qquad \text{dividing by } -10$$

$$40 - x = 12 \text{ mL}$$

Hence 28 mL of the 15% solution must be mixed with 12 mL of the 25% solution.   ◄

E X A M P L E **5**   A young woman can row at the rate of 4 mi/h in still water. Rowing downstream in a river, she can travel 3 times as far in 1 h as she can rowing upstream. What is the rate of flow of the river?

Solution. The basic relationship in this kind of problem is distance = rate × time, or

$$d = rt$$

Let x = the rate of flow of the river (in miles per hour). The rate downstream is $4 + x$ (in miles per hour) and the rate upstream $4 - x$ (in miles per hour). In 1 h the respective distances are $(4 + x)\dfrac{\text{mi}}{\cancel{\text{h}}} \cdot 1\,\cancel{\text{h}} = (4 + x)$ mi and $(4 - x)\dfrac{\text{mi}}{\cancel{\text{h}}} \cdot 1\,\cancel{\text{h}} = (4 - x)$ miles. From the given information:

$$\text{distance (downstream)} = 3 \times \text{distance (upstream)}$$

$$(4 + x) \cdot 1 = 3(4 - x) \cdot 1$$

$$4 + x = 12 - 3x$$

$$x + 3x = 12 - 4$$

$$4x = 8$$

$$x = 2$$

So the rate of flow of the river is 2 mi/h.   ◄

Word problems in algebra have often been called artificial. No doubt many of them are. For example, probably no one could row at the same rate for a whole hour regardless of physical condition. Yet this problem does not differ substantially from the problem of detecting the speed of the earth through the "ether" (or the apparent rate of flow of the ether past the earth) if the speed of light is substituted for the speed of the boat. This was the famous Michelson–Morley experiment. Its outcome was negative: The speed of light proved to be the same in the direction of the earth's motion as in the lateral direction, thus opening the door to the discovery of the special theory of relativity by Albert Einstein in 1905.

On a more modest scale, skill in solving algebra problems is often valuable in setting up more interesting problems in calculus. Of course, many problems in algebra are of interest in their own right. Consider the following example.

E X A M P L E **6**   A storage tank can be filled in 18 h and drained in 6 h. If the tank is initially full and both the drain and the intake valve are open, how long will it take to drain the tank?

Solution. The best procedure in this kind of problem is to examine the situation after one time unit. So if

$$x = \text{time taken to drain tank}$$

then $1/x$ is the fractional part drained after 1 h. Similarly, $\frac{1}{6}$ of the tank is drained after 1 h and $\frac{1}{18}$ is filled. Thus

$$\frac{1}{x} = \frac{1}{6} - \frac{1}{18}$$

$$\frac{1}{x} = \frac{3}{18} - \frac{1}{18} = \frac{2}{18} = \frac{1}{9}$$

$$x = 9 \text{ h}$$

Alternatively, we can clear fractions by multiplying by $18x$. Then

$$18x\left(\frac{1}{x}\right) = 18x\left(\frac{1}{6} - \frac{1}{18}\right)$$

$$18 = 3x - x$$

$$18 = 2x$$

$$x = 9$$ ◀

EXERCISES / SECTION **2.5**

1. The sum of two numbers is 30. If one number is 3 less than the other, what are the numbers? (See Example 1.)

2. The sum of two numbers is 54. If one number is twice the other, what are the numbers?

3. The cost of one tool is $2.60 more than that of the other and the total cost is $23.60. Find the cost of each.

4. One machine part costs twice as much as the other. If the total price is $9.63, what is the cost of each?

5. The sum of two currents is 4.3 A. If one of the currents is 1.9 A more than the other, find each of the currents.

6. The sum of two resistances is 97.2 Ω. If the resistance of one is 3 times that of the other, find each of the resistances.

7. The perimeter of a rectangular plate is 44.0 cm. Given that the plate is 3.2 cm longer than it is wide, what are the dimensions? (See Example 2.)

8. The perimeter of a rectangle is 21.0 ft. Find the dimensions, given that the length is 2.5 ft more than the width.

9. A piece of wire 38 in. long is bent into the shape of a rectangle that is twice as long as it is wide. Find its dimensions.

10. A gutter is to be made from a long piece of metal 52 cm wide by turning up the sides so that the cross-sectional area is a rectangle. (See Figure 2.4.) Determine the dimensions of the cross-section if the gutter is to be twice as wide as it is deep.

Figure 2.4

11. Two resistors are connected in series. The resistance of the first resistor is 5.6 Ω more than that of the second. Given that the combined resistance is 22.4 Ω, find the resistance of each. (See Example 3.)

12. Two resistors are connected in series. Find the resistance of each, given that the resistance of the first is twice that of the second and the combined resistance is 95.4 Ω.

13. Three resistors connected in series have a combined resistance of 120.8 Ω. The resistance of the second is twice that of the first, and the resistance of the third is 14.3 Ω more than that of the second. Find the resistance of each.

14. The total cost of three machine parts is $73.10. The cost of the second machine part is $5.70 more than that of the first, and the cost of the third is $6.20 more than that of the second. Find the cost of each.

15. How many liters of a 14% alcohol solution (by volume) must be added to a 20% solution (by volume) to produce 12 L of a 16% solution? (See Example 4.)

16. How many pounds of an alloy containing 20% copper (by weight) must be added to 50 lb of an alloy containing 30% copper (by weight) to produce an alloy containing 24% copper?

17. One alloy contains 10% brass and another 15% brass (by weight). How many pounds of each must be combined to form 100 lb of an alloy containing 12% brass?

18. One brine solution contains 10% salt by volume, and another contains 18%. How many gallons of each must be mixed to produce 16 gal containing 15% salt?

19. How many gallons of a 20% salt solution must be added to 5 gal of a 10% salt solution to produce a 16% salt solution?

20. How much water must be added to 20 gal of a 60% alcohol solution to produce a 45% alcohol solution?

21. A 1-L bottle contains a 20% alcohol solution. How much of the solution must be drained off and replaced by an 80% solution to produce a 50% alcohol solution?

22. A chemist ordered a 5% sulfuric acid solution from the supply room but received an 8% solution by mistake. How much must be drawn off from a 1-L bottle and replaced by distilled water to produce the right concentration?

23. The River Queen, a ship for sightseers, averages 15 mi/h in still water. Traveling downstream for 2 h, she can go twice as far as traveling upstream for 2 h. How fast is the river flowing? (See Example 5.)

24. John can row at the rate of 5 mi/h in still water. Rowing downstream in a certain river, he can travel 4 times as fast as upstream. Find the rate at which the river is flowing.

25. The velocity of a small plane is 140 mi/h in still air. With a tail wind it flies from city A to city B in 3 h, but the return trip (against the wind) requires 4 h. Determine the velocity of the wind.

26. A boat averages 20 mi/h in still water. A trip that takes 3 h traveling downstream requires 5 h to return to the starting point. Find the velocity of the river.

27. A river flows at the rate of 3 mi/h. A motorboat traveling downstream for 2 h requires 6 h to get back to the starting point. Find the speed of the boat.

28. A car traveling at 40 mi/h leaves a certain intersection 45 min before a second car traveling at 52 mi/h. How long will it take for the second car to overtake the first?

29. Paul usually takes his bicycle to school, which is only 5 mi from his house. One morning he is in a hurry and asks his mother to drive him. If the car goes 3 times as fast as the bike and the trip takes 20 min less, what is the speed of the bike?

30. On Saturday Jane drives home from school, a distance of 120 mi. Due to heavy morning traffic, her average speed on the return trip Monday morning is only 80% of her average speed going home, and the trip takes 36 min longer. Find her average speed each way.

31. An inlet valve can fill a tank in 20 h, while it takes 25 h to drain the tank. If the tank is initially empty and both the valve and the drain are open, how long will it take to fill the tank? (See Example 6.)

32. A storage tank can be filled in 20 min and drained in 15 min. If the tank is initially full and both drain and inlet valve are open, how long will it take to empty the tank?

33. A tank is equipped with two drains. One drain can empty the tank in 16.0 min and the other in 18.0 min. If both drains are open, how long will it take to drain a full tank?

34. One inlet valve can fill a tank in 10 h, and a second valve can fill the same tank in 16 h. How long will it take to fill the empty tank if both valves are open?

35. A chemical tank can be filled in 12.0 min and drained in 15.0 min. If both the drain and the valve are open and if the tank is initially one-fourth full, how long does it take to fill the rest of the tank?

36. A technician opens the inlet valve of an empty tank, which normally takes 20 min to fill. Eight minutes later he closes the drain, having discovered that it was accidently left open. Given that the tank can be drained in 25 min, determine the total time required to fill the tank.

37. In 3 years Joan will be twice as old as she was 5 years ago. How old is she now?

38. Four years from now a computer will be twice as old as it was two years ago. How old is the computer?

39. A man earns $80 plus board every day that he works. On idle days he is charged $25 for board. In one 30-day period he received $1,560. How many days did he work?

40. Sue has taken three out of the first five tests in her physics class, and her test average is 91%. What will the average grade on her last two tests have to be for her to receive an A (93%)?

41. Two cars traveling toward each other at 36 mi/h and 44 mi/h, respectively, are 200 mi apart. How much of the 200-mi distance has each car covered when they meet?

42. How much water must be evaporated from a 1-gal can containing a 20% salt solution to produce a 25% salt solution?

2.6 Ratio and Proportion

In this section we are going to discuss two important concepts—**ratio** and **proportion**.

Ratio

A **ratio** is a quotient of two quantities.

So if the quantities are denoted by a and b, then the ratio is a/b.

Ratios occur frequently in everyday life. For example, if you peddle your bicycle at the rate of 15 ft/s, then you change your distance by 15 ft every second. So the velocity

$$v = \frac{15 \text{ ft}}{1 \text{ s}} = 15 \text{ ft/s}$$

is a ratio. Similarly, 30 mi/gal is a ratio. Since the number π is defined to be the circumference of a circle divided by its diameter, π is also a ratio.

E X A M P L E **1** If Joan travels 100 miles in 2 h, then her average velocity, which is the ratio of distance to time, is given by

$$\frac{100 \text{ mi}}{2 \text{ h}} = 50 \frac{\text{mi}}{\text{h}}$$

◄

As shown in Example 1, if the ratio involves dissimilar quantities, then proper units must be assigned to the ratio. If the ratio involves measurements of the same kind, then it should be expressed as a *dimensionless number*. For example, the definition of π involves two lengths measured in the same units of measure. Consider another example.

E X A M P L E **2**   The *specific gravity* of a substance is the ratio of the density of the substance to the density of water (1.94 slugs/ft^3). Given that the density of copper is 17.25 slugs/ft^3, what is its specific gravity?

Solution. Since the ratio involves units of the same kind, we get the following dimensionless number:

$$\frac{17.25}{1.94} = 8.9$$

(Other examples of standard ratios will be mentioned in the exercises.)   ◄

Proportion

Many problems in science and technology lead to two equal ratios. This equality is called a **proportion.**

> The equality of two ratios is called a **proportion.**

It follows from the definition that a proportion has the form

$$\frac{a}{b} = \frac{c}{d}$$

For example, $\frac{3}{4}$ is equal to $\frac{9}{12}$. This equality can be expressed as a proportion:

$$\frac{3}{4} = \frac{9}{12}$$

Proportions are of interest to us because in many technical problems one ratio is known while only part of another ratio is known. By setting up a proportion, the remaining part can be determined.

E X A M P L E **3**   The ratio of an unknown number to 11 is the same as the ratio of 18 to 33. Find the number.

Solution. Let x be the unknown number. We are given that the ratio of x to 11 is the same as the ratio of 18 to 33. So

$$\frac{x}{11} = \frac{18}{33}$$

Solving for x, we get

$$\frac{x}{11} \cdot 11 = \frac{18}{33} \cdot 11 \qquad \text{\textbf{multiplying by 11}}$$

$$\frac{x}{\cancel{11}} \cdot \cancel{11} = \frac{18}{\underset{3}{\cancel{33}}} \cdot \cancel{11}$$

$$x = 6 \qquad\qquad \blacktriangleleft$$

Proportions can be used for simple conversions. A particularly useful case is the conversion from feet per second to miles per hour and vice versa by using the relationship

$$60\,\frac{\text{mi}}{\text{h}} = 88\,\frac{\text{ft}}{\text{s}}$$

(Other conversion techniques are discussed in Appendix B.)

Remark. When setting up a proportion, use the same units for the two numerators and the same units for the two denominators.

E X A M P L E **4**   Convert 49.7 ft/s to miles per hour.

Solution. Let x (in miles per hour) be the unknown rate. Then

$$\frac{x}{49.7\,\frac{\text{ft}}{\text{s}}} = \frac{60\,\frac{\text{mi}}{\text{h}}}{88\,\frac{\text{ft}}{\text{s}}}$$

Multiplying both sides by 49.7 ft/s, we get

$$x = \frac{60\,\frac{\text{mi}}{\text{h}}}{88\,\frac{\cancel{\text{ft}}}{\cancel{\text{s}}}} \cdot 49.7\,\frac{\cancel{\text{ft}}}{\cancel{\text{s}}}$$

$$x = 33.9\,\frac{\text{mi}}{\text{h}} \qquad\qquad \blacktriangleleft$$

Other technical applications are given in the remaining examples and exercises.

E X A M P L E **5**   The ratio of the number of teeth of two gears is 5 to 9. If the larger gear has 45 teeth, how many teeth does the smaller gear have?

Solution. Let

$$x = \text{number of teeth of smaller gear}$$

Then

$$\frac{x}{45} = \frac{5}{9}$$

$$x = \frac{5}{9} \cdot 45 = 25 \text{ teeth}$$

◀

 E X A M P L E **6**   The resistance in a wire is proportional to its length. If the resistance in a wire 10.2 ft long is 0.560 Ω, determine the resistance in a wire 16.3 ft long if the wire is made of the same material.

Solution. Let R denote the resistance in the wire. Then

$$\frac{R}{16.3 \text{ ft}} = \frac{0.560 \text{ } \Omega}{10.2 \text{ ft}}$$

and

$$R = \frac{(16.3)(0.560)}{10.2} \Omega = 0.895 \text{ } \Omega$$

◀

E X E R C I S E S / S E C T I O N   **2.6**

1. The *Mach number* is the ratio of the velocity of an object to the velocity of sound. Given that the velocity of sound is 330 m/s, what is the Mach number of a rocket whose velocity is 750 m/s?

2. The *density* of a substance is the ratio of its mass to its volume. Find the density of gold, given that 2.60 cm³ of gold has a mass of 50.18 g.

3. *Pressure* is defined as the ratio of force to area. The force against a horizontal plate with an area of 5.00 ft² submerged 10.0 ft below the surface of the water is 3,120 lb. Find the pressure on the plate.

4. The *compression ratio* of a car engine is the ratio of the cylinder volume to the compressed volume. If the cylinder volume is 50.0 in.³ when the piston is at the bottom of

its stroke and 5.75 in.³ when it is at the top, what is the compression ratio?

5. The *intelligence quotient* (IQ) for children is defined as the ratio of mental age to chronological age multiplied by 100. If the scholastic ability of a boy aged 6 years and 9 months corresponds to that of a child aged $8\frac{1}{2}$ years, what is his IQ?

6. Convert 65.4 ft/s to miles per hour. (See Example 4.)

7. Convert 120 ft/s to miles per hour.

8. Convert 10.0 ft/s to miles per hour.

9. Convert 55.00 mi/h to feet per second.

10. Convert 130 mi/h to feet per second.

11. Given that 1 in. = 2.54 cm, convert 10.0 cm to inches.

12. Given that 1 oz = 28.35 g, convert 2.25 oz to grams.

13. The resistance (in ohms) in a wire is proportional to its length. If a wire 12 ft long has a resistance of 0.10 Ω, what is the resistance in a wire 19 ft long if it is made of the same material?

14. If a car uses 1.5 qt of oil in 670 mi, how much oil will it use in 940 mi?

15. It takes 21 lb of a certain base to neutralize 35 lb of sulfuric acid. How many pounds of this base is needed to neutralize 15 lb of this acid?

16. Medieval records of the city of Goslar, a former seat of the Holy Roman Empire situated in northern Germany, show that 1.2 silver marks had the buying power of $1,500 (in 1950 dollars). Determine the buying power of 20 silver marks, a typical yearly income of a medieval count.

17. A plumber charges $24 for 30 min of labor. How much would the plumber charge for a job taking 1 h and 46 min?

18. The owner of a small car gets 29.4 mi/gal of gasoline. How far can he travel on a full tank of 15.0 gal?

19. Mr. Veldboom figures that a 4-day vacation will cost his family $460. How much would a 9-day vacation cost?

20. The cost of a casting is proportional to its weight. If a 15-lb casting costs $1.46, how much does a 23-lb casting cost?

21. A restaurant manager knows that she needs 3 lb of rice to make rice pudding for 21 people. How much rice does she need to serve 50 people?

22. Suppose you were pedaling your bicycle at the rate of 2.2 revolutions per second to attain a speed of 10 mi/h. How fast would you have to pedal to attain a speed of 16 mi/h?

23. We know from geometry that two triangles are similar if the corresponding parts are proportional. Suppose a man 6 ft tall casts a shadow 4 ft long. At the same time a tree casts a shadow 20 ft long. How tall is the tree?

24. A recipe for cocoa fudge calls for $\frac{2}{3}$ cup of cocoa and $1\frac{1}{2}$ cups of milk, in addition to other ingredients. If you had $3\frac{1}{2}$ cups of cocoa to use up, how much milk would you need?

25. A Wheatstone bridge (Figure 2.5) is a convenient instrument for measuring resistance. It was invented by the

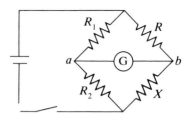

Figure 2.5

English scientist Charles Wheatstone in 1843. In the figure both R_1 and R_2 have a known constant resistance, R is an adjustable resistor, and X is the unknown resistance. R is adjusted so that the current from a to b is zero, as measured by galvanometer G. The relationship between the resistances is given by the proportion

$$\frac{R_1}{R_2} = \frac{R}{X}$$

If $R_1 = 10\,\Omega$, $R_2 = 100\,\Omega$, and R was found to be $7\,\Omega$, determine the resistance X.

26. A transformer consists of two coils. The current in one coil induces a current in the other coil. (See Figure 2.6.) The following proportion gives the relationship between the voltages in each coil and the number of windings:

$$\frac{V_1}{V_2} = \frac{N_1}{N_2}$$

If $V_1 = 80$ V, $N_1 = 500$ (first coil), and $N_2 = 1,200$ (second coil), show that $V_2 = 192$ V. Because of the increase in voltage, this is called a step-up transformer.

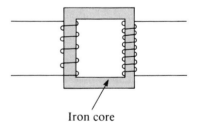

Iron core

Figure 2.6

2.7   Variation

Many situations in science involve variable quantities. In this section we will be concerned with a special group of relationships called **variations**. The first type, called **direct variation**, is really a proportion.

Direct Variation: If two variables x and y are related so that $y = kx$, then y is said to vary directly as x or to be directly proportional to x. The constant k is called the **constant of proportionality**.

(Observe that if $y = kx$, y increases whenever x increases.)

If we are told that y varies directly as x, we write $y = kx$, where k is the constant of proportionality. If we are given a pair of values for x and y, then k can be evaluated, thereby yielding a specific equation. For example, given that

$$y = kx, \qquad x = 2, \qquad \text{and} \qquad y = 4$$

we get

$$4 = k \cdot 2, \qquad \text{or} \qquad k = 2$$

So the relationship is $y = 2x$. Consider another example.

E X A M P L E **1** The force F against a horizontal plate varies directly as the depth d. If the force is 1,560 lb at a depth of 2.5 ft, what is the force at a depth of 7.4 ft?

Solution. By definition, $F = kd$. The first task in a problem on variation is to compute the constant k. From the given information, $F = 1,560$ lb and $d = 2.5$ ft. We now get

$$1,560 = 2.5k$$

so that $k = 624$. The resulting equation is

$$F = 624d$$

Finally, if $d = 7.4$, $F = (624)(7.4) = 4,600$ lb, to two significant digits. ◀

The **constant of proportionality** occurs in all variations. If the variable quantities have units, then so does k. In Example 1, if F is in pounds and d in feet, then k is expressed in pounds per foot.

The variable on the right side can take on various forms. For example, "y is directly proportional to x^2" is written $y = kx^2$, "y varies directly as x^3" is written $y = kx^3$, and "y varies directly as $\sqrt{x}$" is written $y = k\sqrt{x}$. Consider another example.

E X A M P L E **2** A freely falling body is subject to a retarding force due to air resistance. This force depends on the size and shape of the object. For some objects this force is directly proportional to the square of the velocity v, or

$$f = kv^2$$ ◀

If y is directly proportional to $1/x$, we write

$$y = k\left(\frac{1}{x}\right) \quad \text{or} \quad y = \frac{k}{x}$$

This relationship is called an **inverse variation**. In an inverse variation, when x increases then y decreases, and when x decreases then y increases.

Inverse Variation: If two variables x and y are related so that $y = k/x$, then we say that y varies inversely as x or that y is inversely proportional to x.

As in the case of direct variations, the variables can take on various forms. For example, if y varies inversely as the cube root of x, we write

$$y = \frac{k}{\sqrt[3]{x}}$$

Consider another example.

E X A M P L E **3**   The gravitational force between two objects is inversely proportional to the square of the distance between them, or

$$F = \frac{k}{x^2}$$

If the force is 3 lb at a distance of 100 mi, what is the force at a distance of 150 mi?

Solution. The first step is to calculate the constant k from the given information. Thus

$$F = \frac{k}{x^2}$$

$$3 = \frac{k}{(100)^2} \quad \text{or} \quad k = 30{,}000 \qquad \text{F = 3 lb, } d = \text{100 mi}$$

and

$$F = \frac{30{,}000}{x^2}$$

If $x = 150$, then

$$F = \frac{30{,}000}{(150)^2} = 1\frac{1}{3}\text{ lb}$$

◀

The last type of variation we will discuss here is called **joint variation**.

> **Joint Variation:** If $z = kxy$, then we say that z varies jointly as x and y or that z is directly proportional to the product of x and y.

E X A M P L E **4** The rate of heat W developed in a circuit varies jointly as the resistance R and the square of the current i. If an element is drawing 2 A and develops heat at the rate of 8 cal/s when $R = 10\ \Omega$, what is the heat developed when R is decreased to $5\ \Omega$ and the current is doubled?

Solution. By definition

$$W = kRi^2$$

Then

$$8 = k(10)(2)^2 \qquad W = 8\ \text{cal/s},\ R = 10\ \Omega,\ i = 2\ \text{A}$$

or $k = \frac{1}{5}$. Thus

$$W = \frac{1}{5} Ri^2$$

From the given information, $R = 5\ \Omega$ and $i = 2(2\text{ A}) = 4$ A. So

$$W = \frac{1}{5}(5)(4)^2 = 16\ \text{cal/s}$$

◀

The variations considered may occur in combination, as shown in the next example.

E X A M P L E **5** The gravitational force F between two masses m_1 and m_2 varies jointly as the masses and inversely as the square of the distance r between them. Thus

$$F = k\frac{m_1 m_2}{r^2}$$

Here k is called the *gravitational constant*.

◀

E X E R C I S E S / S E C T I O N **2.7**

In Exercises 1–12, express each statement in the form of an equation.

1. y is directly proportional to x.

2. y varies directly as x^2.

3. w varies directly as i^2.

4. y is directly proportional to x^3.

5. y varies inversely as x^2.

6. y is inversely proportional to x^3.

7. y varies jointly as x and w.

8. y is directly proportional to the product of w and z.

9. A varies directly as a and inversely as b.

10. Z varies jointly as s and t and inversely as r.

11. F is directly proportional to the product of m and M and inversely proportional to r^2.

12. E varies jointly as m and v^2.

In Exercises 13–16, find the equation, including the constant of proportionality.

13. y varies directly as x;· $y = 8$ when $x = 2$.

14. s is directly proportional to t^2; $s = 64$ when $t = 2$.

15. P varies inversely as V; $P = 1$ when $V = 2$.

16. F varies directly as s; $F = 15$ when $s = 1.5$.

17. The force exerted by a spring is directly proportional to the spring's extension. If a force of 2.5 lb is required to stretch a spring 1.8 in., what is the force required to stretch it 5.0 in.?

18. The acceleration of a particle varies directly as the force applied (Newton's second law). If for some object an acceleration of 30.0 m/s^2 results from a force of 10.0 N, what is the force required to yield an acceleration of 1.5 m/s^2?

19. The strength S of a rectangular beam varies jointly as the width w and the square of the depth d. Find an expression for S.

20. The voltage V across a resistor varies directly as the current I. If $V = 2.4\ \Omega$ when $I = 1.9$ A, find V when $I = 5.2$ A.

21. Within a small range the demand for a product varies inversely as the price. If a store can sell 90 units of a certain commodity per week at $1.50 apiece, how many units per week can it expect to sell if the price is reduced to $1.35?

22. For two gears in mesh, the speed of each is inversely proportional to the number of teeth. If a gear having 30 teeth and turning at 100 rpm (revolutions per minute) is in mesh with a gear having 50 teeth, what is the speed of the second gear?

23. The repulsive force f between two charged particles having like charges q_1 and q_2, respectively, varies jointly as q_1 and q_2 and inversely as the square of the distance r between them. Find an expression for the repulsive force.

24. The kinetic energy of a moving particle is directly proportional to the square of the velocity. If the kinetic energy of a particle moving at 12.0 m/s is 25.0 J, what is the kinetic energy if the velocity is increased to 15.0 m/s?

25. Boyle's law states that the pressure of an ideal gas varies inversely as the volume, provided that the temperature remains constant. If a pressure of 15.0 lb/ft^2 corresponds to a volume of 10.5 ft^3, what is the pressure when the volume is increased to 13.6 ft^3?

26. The volume V of a sphere is directly proportional to the cube of the radius r. Determine the relationship if $V = 36\pi$ when $r = 3$.

27. The simple interest earned during a certain time period varies jointly as the principal and the interest rate. If $500 earned $120 at 8% interest during a certain time period, find the interest earned on $666.67 at 6%.

28. One of Kepler's three laws states that the square of the time required for a planet to make one revolution about the sun is directly proportional to the cube of its average distance from the sun. Given that Mars is approximately $1\frac{1}{2}$ times as far from the sun as the earth, find the number of days required to make one revolution.

29. The resistance R in a wire varies directly as the length L of the wire and inversely as the square of the diameter D. If a wire 90.5 ft long with diameter 0.0808 in. has a resistance of 8.50 Ω, find the resistance in a wire 75.0 ft long if it has a diameter of 0.1019 in. and is made of the same material. (*Note:* there is no need to change units.)

30. The resistance in a wire varies directly as the length and inversely as the square of the diameter. If a wire 100 ft long with a diameter of 0.028 in. has a resistance of 10 Ω, find the resistance in a wire 150 ft long with a diameter of 0.016 in. if it is made of the same material.

2.8   Functions

Many problems in technology deal with variable quantities. An important relationship between variable quantities is that of a *function*.

Definition of a Function

Suppose a resistor of 30 Ω is in series with a variable resistor R. Then the combined resistance R_T is given by $R_T = 30 + R$ (in ohms). This formula gives the resistance of the combination for any value of R. Furthermore, for any value of R, we get a *unique* value for R_T. For example, if $R = 10 \, \Omega$, then $R_T = (30 + 10)\Omega = 40 \, \Omega$, and if $R = 60 \, \Omega$, then $R_T = (30 + 60) \, \Omega = 90 \, \Omega$. Because of this uniqueness, $R_T = 30 + R$ is called a *function*.

We already touched on the idea of a function in the last section. For example, if $y = 3x$ (y varies directly as x), then we have, for every x, a *unique* value for y. Consider the following definition:

> **Definition of a Function:** If two variables x and y are so related that for every value of the variable x there corresponds one, and only one, value of the variable y, then we call y a **function** of x. The variable x is called the **independent variable** and y is called the **dependent variable**.

The variables x and y are representative and can stand for any physical quantity. However, as we can see from the function $R_T = 30 + R$, it is sometimes convenient to use different letters altogether.

E X A M P L E 1 The equation $y = 2x^2 + 1$ is a function of x: For every value of x we get one, and only one, value of y. The independent variable is x and the dependent variable is y. ◀

E X A M P L E 2 The formula for the area of a circle is $A = \pi r^2$. This formula defines a function, since for every $r > 0$ there exists a unique value for A. (A also exists for $r \leq 0$, but using such values makes no sense in this problem.) Note that the independent variable is r and the dependent variable is A (π is a constant). ◀

E X A M P L E 3 The power P developed in a 10-Ω resistor as a function of the current I is $P = 10I^2$. The formula is a function because every value of I yields a unique value of P. The independent variable is I, and the dependent variable is P. ◀

The main purpose of the function concept is to describe an operation on the independent variable that yields a unique value of the dependent variable, regardless of the letters used. For example, $R_T = 30 + R$ and $y = 30 + x$ represent the same function, even though the letters are different.

Notation for Functions

To make our discussion of functions easier, we need a convenient notation for expressing them.

> **Notation for Functions:** If y is a function of x, we write $y = f(x)$, which is read "y equals f of x."

Thus $y = x^2$ can be written $f(x) = x^2$ and $y = \sqrt{x + 1}$ can be written $f(x) = \sqrt{x + 1}$.

E X A M P L E 4   By the functional notation $y = f(x)$, the function

$$y = 5 - x^2$$

can be written

$$f(x) = 5 - x^2$$

In other words, $y = 5 - x^2$ and $f(x) = 5 - x^2$ are two ways of representing the same function.   ◀

One of the main advantages of the functional notation is that a function value can be given in a natural way. Consider, for example, the function $y = x^2$. If $x = 2$, then $y = 2^2 = 4$. Using the form

$$f(x) = x^2$$

we simply write $f(2) = 4$.

E X A M P L E 5   Given the function $f(x) = \sqrt{1 - x}$, find $f(-3)$, $f(0)$, and $f(1)$.

Solution. In each case, we assign the specified value to the independent variable x. To find $f(-3)$, we let $x = -3$ in the expression $\sqrt{1 - x}$ to obtain

$$f(x) = \sqrt{1 - x}$$
$$f(-3) = \sqrt{1 - (-3)} = \sqrt{4} = 2 \qquad x = -3$$

Similarly,

$$f(0) = \sqrt{1 - 0} = 1 \qquad\qquad x = 0$$

and

$$f(1) = \sqrt{1 - 1} = 0 \qquad\qquad x = 1 \qquad\qquad ◀$$

If more than one function occurs in a particular discussion, we need to use different letters. For example, $y = G(x)$, $w = h(y)$, and $z = f(r)$ all represent functions.

E X A M P L E 6   If $H(z) = 2 - 3z^2$, find $H(2)$ and $H(-1)$.

Solution. $H(2) = 2 - 3(2)^2 = -10$

$$H(-1) = 2 - 3(-1)^2 = -1 \qquad\qquad ◀$$

Sometimes functions are evaluated in terms of letters, as shown in the next example.

E   X   A   M   P   L   E   **7**   If $f(x) = \sqrt{1 - x^2}$, find $f(t)$ and $f(a^2)$.

Solution. If we replace x by t, we get

$$f(t) = \sqrt{1 - t^2}$$

Replacing x by a^2, we have

$$f(a^2) = \sqrt{1 - (a^2)^2} = \sqrt{1 - a^4}$$ ◀

Functions in BASIC (Optional)

We saw in Section 1.12 that SQR(X) is the square root function in BASIC. Later we will see other examples of functions that are stored in the computer (library functions). This section is concerned with functions that you define yourself.

To define a function, use the DEF (define) statement, which has the following form:

line number   DEF FNA(X)

The function name, FNA, must consist of three letters, the first two of which are **FN** and the third any letter from **A** to **Z**. For the variable, any legal variable name can be used in most systems.

E   X   A   M   P   L   E   **8**   Write the function

$$f(x) = 3x^3 - 2x + 3$$

in BASIC.

Solution. Using the name FNA, we get

DEF FNA(X) $= 3 * X \uparrow 3 - 2 * X + 3$

The same function can be defined by the following statement:

DEF FNB(Y) $= 3 * Y \uparrow 3 - 2 * Y + 3$ ◀

The computer will evaluate function values in the usual way, as shown in the next example.

E   X   A   M   P   L   E   **9**   Consider the programming sequence

50 DEF FNA(X) $= 2 + 3 * X \uparrow 2$

60 PRINT 4 * SQR(FNA (2))

Line 50 defines the function

$$f(x) = 2 + 3x^2$$

Line 60 causes the value 2 to be substituted for x in line 50. The number printed is therefore

$$4\sqrt{f(2)} = 4\sqrt{2 + 3(2)^2} = 14.9666295471 \qquad \blacktriangleleft$$

A convenient way to generate a table of values is by means of the FOR-NEXT statement. (See Appendix C.)

E X E R C I S E S / S E C T I O N   2.8

In Exercises 1–8, identify the independent and dependent variables. (See Examples 1–3.)

1. $y = x^2 + 2$

2. $w = 1 - 2z^3$

3. $E = 2R$

4. $s = 9.8t^2$

5. $A = 4\pi r^2$

6. $L = 3t + 1$

7. $C = \dfrac{5}{9}(F - 32)$

8. $n = \sqrt{m + 4}$

In Exercises 9–16, write each function in the form of $f(x)$. (See Example 4.)

9. $y = x^2 - 3$

10. $y = x^2 - \sqrt{x}$

11. $z = \sqrt{6w + 2}$

12. $v = t^2 - 3t + 1$

13. $P = 1 - 3v^2$

14. $Z = \dfrac{5}{X + 4}$

15. $N = \sqrt{4n + 6}$

16. $A = 7r^2 + 4$

17. If $f(x) = x^3 - x + 1$, find $f(1)$, $f(-1)$, $f(0)$, $f(4)$.

18. If $f(x) = \sqrt{x - 3}$, find $f(3)$, $f(4)$, $f(7)$, $f(10)$.

19. If $f(x) = \sqrt{2 - x}$, find $f(2)$, $f(-7)$.

20. If $g(x) = \sqrt{x} - 3$, find $g(0)$, $g(4)$.

21. If $h(x) = \dfrac{1}{x + 2}$, find $h(0)$, $h(-1)$.

22. If $f(x) = \dfrac{1}{\sqrt{x}}$, find $f(1)$, $f(9)$.

23. If $i(t) = 2t + 1$, find $i(0)$, $i(1)$.

24. If $q(t) = \sqrt{1 + t}$, find $q(0)$, $q(3)$.

25. If $H(z) = z^3 - 1$, find $H(1)$, $H(-3)$.

26. If $G(w) = \sqrt{2w + 3}$, find $G(0)$, $G(7)$.

27. If $f(x) = x^2 + 2$, find $f(a)$, $f(a^2)$.

28. If $g(x) = 1 - 2x^2$, find $f(t)$, $f(t^3)$.

29. The resistance in ohms in a certain wire as a function of temperature in degrees Celsius is given by

$$R(T) = 2.00 + 0.132T + 0.00100T^2$$

Find $R(10.0°)$ and $R(100.0°)$.

30. The relation between the tensile strength (in pounds) of a piece of material and the temperature (in degrees Fahrenheit) is given by

$$S(T) = 515.0 - 0.07500\sqrt{T}$$

Find $S(112.1°)$ and $S(248.3°)$.

In Exercises 31–36, write each statement as a function.

31. The area A of a square as a function of the side s.

32. The volume V of a cube as a function of the edge s.

33. The volume V of a sphere as a function of its radius r.

34. The perimeter P of a square as a function of its side s.

35. The volume V of a cylinder of radius 2 as a function of its height h.

36. The area A of a trapezoid with bases 2 and 4 as a function of its altitude h.

37. The cost of finishing a metal plate is $3/cm^2$. Express the cost C as a function of the area A (in square centimeters).

38. The demand D of a certain product is 2.5 times the reciprocal of the price P. Express D as a function of P.

39. A consultant charges a base fee of $50 plus $25 per hour. Express the cost C of consultation as a function of time t (in hours).

40. The energy E radiated by a filament is equal to the fourth power of the temperature T multiplied by a constant k. Express E as a function of T.

2.9 The Graph of a Function

In the last section we introduced the concept of a function. In this section we are going to study graphs of functions.

The graph of a given function is represented on a reference system called the **rectangular** or **Cartesian coordinate system**. The coordinate system is named after the French mathematician and philosopher René Descartes (1601–1665).

Recall from Chapter 1 that the set of real numbers can be made to correspond to the points on a line. To locate points in a plane, we construct a reference system consisting of two number lines that intersect at right angles at their respective origins. (See Figure 2.7.) The horizontal line is called the **x-axis** and the vertical line the **y-axis**; together these lines are known as the **coordinate axes**. The positive numbers are on the right side of the origin on the x-axis and above the origin on the y-axis. Points can be described in the plane by giving their perpendicular distances and directions from the coordinate axes. The distance from the y-axis, with the proper sign, is called the **x-coordinate** or **abscissa**, and the distance from the x-axis the **y-coordinate** or **ordinate**.

To designate a point, it is customary to place the two coordinates in parentheses with the x-coordinate given first. The two coordinates are separated by a comma. For example, the point P in Figure 2.7 is denoted by $(1, 4)$, Q by $(-2, 3)$, and R by $(-4, -2)$.

The coordinate axes divide the plane into four parts, called **quadrants**, numbered I, II, III, and IV (see Figure 2.7). Note that both coordinates are positive in the first quadrant and negative in the third quadrant. In the second quadrant the x-coordinates are negative and the y-coordinates positive; in the fourth quadrant the x-coordinates are positive and the y-coordinates negative.

By using the rectangular coordinate system, functions can be graphed in a simple and natural way. Consider, for example, the function

$$y = 2x + 1$$

x-axis, *y*-axis
Coordinate axes

Abscissa, Ordinate

Quadrant

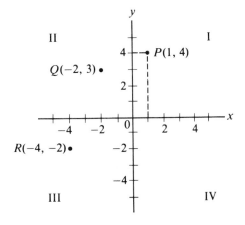

Figure 2.7

Since for any x there exists a unique value for y, we can find any number of such pairs and plot them. For example, if $x = \frac{1}{2}$, then $y = 2(\frac{1}{2}) + 1 = 2$. A few such pairs are given in the following table:

Plotting points

x:	-2	-1	0	1	2
y:	-3	-1	1	3	5

Once these points are plotted, they can be connected by a smooth curve, which turns out to be a straight line (Figure 2.8).

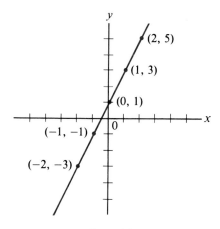

Figure 2.8

E X A M P L E **1** Graph the function $y = x^2 - 1$.

Solution. First we need to construct a table of values:

x:	-3	-2	-1	0	1	2	3
y:	8	3	0	-1	0	3	8

For example, if $x = -3$, then $y = (-3)^2 - 1 = 8$. These points are now plotted and connected by a smooth curve; see Figure 2.9. ◄

E X A M P L E **2** Graph the function $y = \sqrt{1 - x}$.

Solution. Recall that the symbol $\sqrt{}$ stands for the principal, or positive, square root. Hence for every x there is only one y, so that the equation is indeed a function. However, if $x > 1$, then the values of $1 - x$ become negative, so that the resulting y values are imaginary. (See Section 1.8.) We conclude that $x \leq 1$.

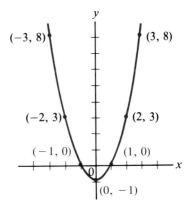

Figure 2.9

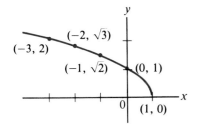

Figure 2.10

x:	1	0	-1	-2	-3
y:	0	1	$\sqrt{2}$	$\sqrt{3}$	2

The graph is shown in Figure 2.10. ◄

Example 2 shows that a function may not be defined for all values of x in the equation. The set of all x values for which the y values exist is called the **domain** of the function. The domain of the function $y = \sqrt{1 - x}$ in Example 2 is $x \leq 1$ (to avoid imaginary values).

Domain

E X A M P L E **3** State the domain of the function

$$y = \frac{1}{x - 2}$$

Solution. To avoid division by 0, x cannot be equal to 2. So the domain is the set of all x such that $x \neq 2$. ◄

Range

If y is a function of x, then the set of admissible y values is called the **range** of the function. For example, the range of the function $y = x^2$ is $y \geq 0$.

Representing a function in graphical form can be very useful in technology, since a graph gives a revealing picture of the behavior of the function. Consider the next example.

 E X A M P L E **4** The output P (in watts) of a certain battery is given by $P = 4I - I^2$, where I is measured in amperes. Sketch the graph. (For physical reasons, the domain is $0 \leq I \leq 4$.)

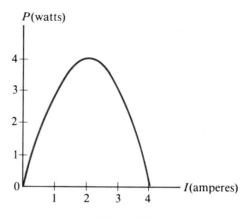

Figure 2.11

Solution. The graph (Figure 2.11) is plotted from the following table of values:

I (amperes):	0	1	2	3	4
P (watts):	0	3	4	3	0

Note that the graph shows the output to be at a maximum when $I = 2$ A. ◀

While functions play an important role in our later work, *not every equa-tion in x and y represents a function of x.* Consider, for example, the equation $y^2 = x$. If $x = 4$, then $y^2 = 4$; thus, $y = 2$ or $y = -2$ (usually written $y = \pm 2$). In fact, for every $x > 0$, we obtain two y values. Some of these values are listed in the following table:

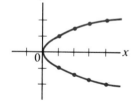

Figure 2.12

x:	0	1	2	3	4
y:	0	± 1	$\pm\sqrt{2}$	$\pm\sqrt{3}$	± 2

The graph, shown in Figure 2.12, confirms that every $x > 0$ yields two y values. So y is not a function of x.

E X E R C I S E S / S E C T I O N **2.9**

In Exercises 1–10, find the domain of each function.

1. $y = 2x$

2. $y = 1 - 3x$

3. $y = \dfrac{1}{x}$

4. $y = \dfrac{1}{x - 2}$

5. $y = \dfrac{1}{x^2 - 1}$

6. $y = \dfrac{1}{x^3 - 1}$

7. $y = \sqrt{x + 2}$

8. $y = \sqrt{2 - x}$

9. $y = \sqrt{4 - 2x}$

10. $y = \sqrt{2x + 4}$

In Exercises 11–32, graph each of the functions.

11. $y = x - 1$ **12.** $y = 3x + 2$ **13.** $y = 3x$

14. $y = 1 - 2x$ **15.** $y = x^2 + 1$ **16.** $y = 4x^2 - 1$

17. $y = 1 - x^2$ **18.** $y = x(x - 1)$ **19.** $y = x^3$

20. $y = x^3 - 1$ **21.** $y = \sqrt{x}$ **22.** $y = \sqrt{x - 1}$

23. $y = \sqrt{4 - x}$ **24.** $y = \sqrt{x + 3}$ **25.** $y = \dfrac{1}{x}$

26. $y = \dfrac{1}{x - 1}$ **27.** $y = \dfrac{1}{x^2}$ **28.** $y = \dfrac{1}{x^2 + 4}$

29. $y = 1 - \sqrt{x}$ **30.** $y = \dfrac{1}{x^2 - 1}$ **31.** $y = \dfrac{x}{x^2 + 1}$

32. $y = \dfrac{x^2}{x^2 + 1}$

33. The power P (in watts) drawn by a certain variable resistor is given by

$$P = \frac{2R}{(R + 1)^2} \qquad (R \geq 0)$$

where R is measured in ohms. Sketch P as a function of R and estimate the setting on R for which the power drawn is a maximum.

34. For a projectile shot directly upward with a velocity of 10 m/s, the distance s (in meters) from the ground as a function of t (in seconds) is $s = 10t - 5t^2$ ($t \geq 0$). Sketch the curve.

REVIEW EXERCISES / CHAPTER 2

In Exercises 1–12, solve each equation for x.

1. $2x - 3 = x + 6$ **2.** $4x - 5 = 2x - 7$

3. $2(x + 2) = 5x + 1$ **4.** $x - 3 = 3(x + 1)$

5. $\dfrac{1}{2}x - \dfrac{1}{4} = \dfrac{1}{4}x + \dfrac{1}{2}$ **6.** $\dfrac{x}{3} + \dfrac{1}{6} = \dfrac{2x}{3} + \dfrac{1}{3}$

7. $\dfrac{1}{x} = \dfrac{2}{x} + \dfrac{1}{2}$ **8.** $\dfrac{1}{2x} = \dfrac{1}{2} + \dfrac{1}{4x}$

9. $5ax + b = 4ax + 2b$ **10.** $2(ax + 1) = 4(ax - 3)$

11. $3(ax + b) = 5ax - 3$ **12.** $4(ax - 2b) = 2(ax - 2b)$

In Exercises 13–16, solve each of the given formulas for the letter indicated.

13. $V = IR,\quad R$ **14.** $A = an + bn,\quad a$

15. $Q = \dfrac{3a}{b} + C,\quad b$ **16.** $A = p + prt,\quad t$

17. The sum of two electric currents is 3.2 A. If one current is 1.8 A more than the other, find the values of the two currents.

18. Two cars 190 mi apart are traveling toward each other at speeds of 45 mi/h and 50 mi/h, respectively. When will they meet?

19. How many pounds of an alloy containing 20% brass must be combined with 20 lb of an alloy containing 14% brass to produce an alloy containing 18% brass?

20. Jane has $1.75 in change. She has twice as many dimes as quarters and two more nickels than dimes. How many of each does she have?

21. Otto requires 50 min to wash the family car, and his brother Otis only 40 min. How long would it take for them to wash the car together?

22. If the side of a square is increased by 4 ft, the area of the square is increased by 48 ft². How long is the side of the square?

23. The *efficiency* of an engine is defined as the ratio of output to input; this ratio is commonly expressed as a percentage. If the output of a certain engine is 9,000 W and the input 15,000 W, find its efficiency.

24. The *relative error* is defined as the ratio of the error of measurement to the correct value. If the side of a square is 10.0 ft with a possible error of 0.2 ft, find the resulting relative error in the area.

25. Given that 1 in. $= 2.54$ cm, convert 11.7 in. to centimeters.

26. A flagpole casts a shadow 8 ft long. At the same time a yardstick casts a shadow 6 in. long. Determine the height of the flagpole.

In Exercises 27–32, express each statement in the form of an equation.

27. y varies inversely as r and jointly as s and t.

28. y is directly proportional to x^2; $y = 9$ when $x = 3$.

29. The weight w of a body above the earth's surface varies inversely as the square of the distance d of the body from the center of the earth.

30. The capacitance C of two parallel plates varies inversely as the distance d between the plates.

31. The resistance R in a wire varies directly as its length L and inversely as the square of its diameter D.

32. The period P of a pendulum varies directly as the square root of its length L.

33. Graph the function $y = 2\sqrt{x - 4}$.

34. Graph the function $y = 1/(x^2 + 1)$.

35. State the domain of the function in Exercise 33.

36. If $g(z) = \sqrt{1 - z}$, find $g(0)$, $g(1)$.

Systems of Linear Equations and Introduction to Determinants

3.1 Simultaneous Linear Equations

In Chapter 2 we studied the solution of first-order equations. However, many applications in technology require the solution of systems containing two or more equations. In this section we will study the geometric basis of systems containing two equations, as well as the solution of such systems by drawing graphs.

It is shown in analytic geometry that the general equation of a straight line is

$$ax + by = c \tag{3.1}$$

(A detailed discussion of the line is given in Chapter 19.) Since equation (3.1) represents a straight line, it is called a **linear equation** or a **linear equation in two variables.** For example, the equation $3x - 2y = 6$ fits form (3.1) and therefore represents a line. Suppose we graph this line from the following table of values:

Linear equation

x:	0	1	2
y:	-3	$-\dfrac{3}{2}$	0

The graph is shown in Figure 3.1. Note that the coordinates of the points $(0, -3)$, $(1, -\frac{3}{2})$, and $(2, 0)$ in the table satisfy the equation. Moreover, the coordinates of *every* point on the line satisfy the equation.

Now consider another line, $x + y = 2$, shown in Figure 3.2. As before, the coordinates of *every* point on this line satisfy the equation $x + y = 2$.

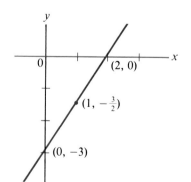

Figure 3.1

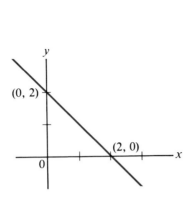

Figure 3.2   **Figure 3.3**

According to Figure 3.3, which shows both lines, the two lines intersect at (2, 0). So the coordinates of the point (2, 0) satisfy *both* equations. Consequently, $x = 2$ and $y = 0$ is called the **common solution** of the system

$$3x - 2y = 6$$
$$x + \ y = 2$$

In general, two linear equations in two unknowns are referred to as a system of two **simultaneous linear equations.**

Simultaneous Linear Equations:

$$a_1 x + b_1 y = c_1$$
$$a_2 x + b_2 y = c_2$$

(3.2)

Common solution

Any pair (x, y) of values that satisfies both equations is called a **common solution** of the system, or simply a **solution**. If a common solution exists, then it follows from the foregoing discussion that the point (x, y) is the intersection of the two lines. If the lines are distinct, then this point, and hence the solution, is necessarily unique. On the other hand, since two lines may be parallel, a system may not have any solution.

Graphical Solution: To find the common solution of a system of two linear equations graphically, draw the two lines and determine the point of intersection from the graph.

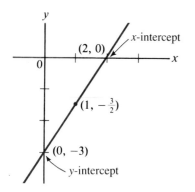

Figure 3.4

To be able to sketch the graphs of linear equations rapidly, we need the notion of **intercept**, which is defined next.

Definition of Intercept: An **intercept** is a point at which the graph crosses a coordinate axis. A point $(a, 0)$ is called an **x-intercept**, and a point $(0, b)$ is called a **y-intercept**.

To find the x-intercept, we let $y = 0$ and solve the resulting equation for x. To find the y-intercept, we let $x = 0$ and solve the resulting equation for y.

Consider, for example, the equation $3x - 2y = 6$. If $y = 0$, then $x = 2$. So the point **(2, 0)** is the x-intercept. (See Figure 3.4.) If $x = 0$, then $y = -3$, so that **(0, -3)** is the y-intercept. Since two distinct points determine a straight line, these are the only points needed to draw the graph. A third point, $(1, -\frac{3}{2})$, is included only as a check. (See Figure 3.4.)

E X A M P L E **1**   Determine the common solution of the system

$$2x + \; y = 5$$
$$x + 3y = 5$$

by drawing the graph of each line and estimating the coordinates of the point of intersection.

Solution. As indicated earlier, the simplest way to draw a line is to find the intercepts. Letting $x = 0$ in the first equation, we find that $y = 5$. Similarly, if we let $y = 0$, then $x = \frac{5}{2}$. (See Figure 3.5.) For the second equation, if $x = 0$, then $y = \frac{5}{3}$; and if $y = 0$, then $x = 5$. (See Figure 3.5.) To get a check point for the first equation, we let $x = 1$, so that $y = 3$. A check point for the second line is $(-1, 2)$; both points are shown in Figure 3.5.

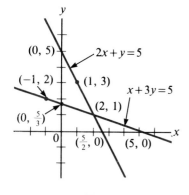

Figure 3.5

$2x + y = 5$		
x	y	
0	5	y-intercept
$\frac{5}{2}$	0	x-intercept
1	3	check point

$x + 3y = 5$		
x	y	
0	$\frac{5}{3}$	
5	0	
-1	2	

We now draw the two lines and observe that they appear to cross at $(2, 1)$, at least as closely as can be determined from the graph. As a check, let us substitute the coordinates of this point into the given equations:

$$2(2) + 1 = 5 \qquad x = 2, y = 1$$
$$2 + 3(1) = 5$$

So the common solution is indeed given by

$$x = 2 \quad \text{and} \quad y = 1$$ ◀

E X A M P L E 2 Determine the solution of the system

$$2x - y = -8$$
$$x - 3y = 3$$

graphically to the nearest tenth of a unit.

Solution. To obtain the intercepts for the first equation, let $x = 0$, so that $y = 8$. If $y = 0$, then $x = -4$. The intercepts are $(-4, 0)$ and $(0, 8)$.

Now assign some arbitrary value to x, such as $x = 1$. Then $2(1) - y = -8$, and $y = 10$. So a check point for the first equation is $(1, 10)$.

Similarly, for the second equation we find that the intercepts are $(3, 0)$ and $(0, -1)$. A check point is $(6, 1)$.

Now we draw the two lines and estimate the coordinates of the point of intersection. To the nearest tenth the coordinates appear to be $(-5.4, -2.8)$. (See Figure 3.6.)

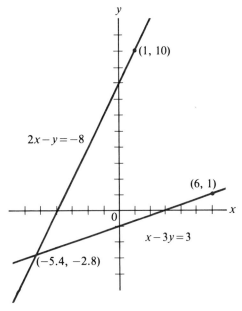

Figure 3.6

$2x - y = -8$			$x - 3y = 3$	
x	y		x	y
0	8	y-intercept	3	0
-4	0	x-intercept	0	-1
1	10	check point	6	1

◀

We know from geometry that two distinct lines are either parallel or intersecting. If they are parallel, then a common solution cannot exist. Whether two lines really *are* parallel cannot be determined graphically, since two lines may look parallel and yet intersect at some distant point. Moreover, Example 2 shows that the graphical method for solving simultaneous equations is awkward at best. Fortunately, there are algebraic methods for solving systems directly, which we will study in the next section. However, the graphical approach has provided us with the necessary geometric background.

E X E R C I S E S / S E C T I O N　3.1

Solve the following systems of equations graphically. Estimate the answers to the nearest tenth of a unit.

1. $2x - y = 4$
$\quad x + y = 2$

2. $x - 3y = 6$
$\quad x + 2y = -4$

3. $x - y = -3$
$\quad x + y = -1$

4. $2x - y = 2$
$\quad x + y = 1$

5. $x + 2y = 12$
$\quad x - 3y = 3$

6. $2x - y = 4$
$\quad x + 3y = 6$

7. $x - 2y = 4$
$\quad 2x - y = 6$

8. $2x - y = 6$
$\quad 4x + y = 4$

9. $x + 3y = -4$
$\quad 2x + y = 6$

10. $5x - 2y = 9$
$\quad 4x - 3y = 4$

11. $2x - 3y = 6$
$\quad x - 4y = 12$

12. $x - 3y = 10$
$\quad 3x - y = 3$

3.2　Algebraic Solutions

In the last section we considered the graphical solution of two simultaneous linear equations. In this section we will turn our attention to solving such systems algebraically.

The real difficulty in solving two equations simultaneously is that two different unknowns are involved. Algebraic solutions resolve this difficulty by eliminating one of the unknowns, thereby reducing the problem to solving a single equation with one unknown.

Addition or Subtraction

The first method to be considered is the **method of addition or subtraction**.

Method of Addition or Subtraction:

1. Multiply both sides of the equations (if necessary) by constants so chosen that the coefficients of one of the unknowns are numerically equal (having the same absolute value).
2. If the coefficients have opposite signs, add the corresponding members of the equations. If the coefficients have like signs, subtract the corresponding members of the equations.
3. Solve the resulting equation in one unknown for the unknown.
4. Substitute the value of the unknown in either of the original equations and solve for the second unknown.
5. Check the solution in the original system.

To see how addition or subtraction can eliminate one of the unknowns, consider the system

$$2x + 3y = 1$$

$$x + 3y = 2$$

Note that the y coefficients are the same. If the second equation is subtracted from the first, y is eliminated:

$$
\begin{array}{ll}
2x + 3y = \quad 1 & \\
\underline{x + 3y = \quad 2} & \text{subtract} \\
x \qquad\quad = -1 & 2x - x = x; \; 3y - 3y = 0; \; 1 - 2 = -1
\end{array}
$$

We conclude that $x = -1$. From the second equation $(x + 3y = 2)$ we get $(-1) + 3y = 2$, so that $y = 1$. The common solution is therefore $x = -1$ and $y = 1$.

Consider another example.

E X A M P L E **1**   Solve the following system:

$$2x - y = \quad 1$$

$$x - 3y = -2$$

Solution. We can eliminate x as follows: Multiply both sides of the second equation by 2, thereby making the coefficients of x the same, and then *subtract* the second equation from the first. Thus

$$
\begin{array}{lll}
\qquad\quad 2x - y = \quad 1 & \\
\textit{Step 1.} \; 2x - 6y = -4 & 2(x - 3y) = 2(-2) \\
\textit{Step 2.} \; \underline{\quad 0 + 5y = \quad 5} & -y - (-6y) = 5y; \; 1 - (-4) = 5
\end{array}
$$

Step 3. Solving the resulting equation, we get $y = 1$

Step 4. To find the corresponding x value, substitute $y = 1$ into either of the given equations and solve for x. Using the second equation, we get

$$x - 3(1) = -2$$

which yields $x = 1$. The solution is therefore given by $(1, 1)$.

Step 5. As a check, let us substitute these values into the given equations:

$$2(1) - 1 = 1 \qquad \text{and} \qquad 1 - 3(1) = -2$$

The solution checks. This system can also be solved by eliminating y: Multiplying the first equation by 3, we get

$$
\begin{array}{lll}
6x - 3y = \quad 3 & \quad 3(2x - y) = 3(1) \\
\underline{x - 3y = -2} & \quad \text{second equation} \\
5x \qquad\quad = \quad 5 & \quad \text{subtracting} \\
x = \quad 1 &
\end{array}
$$

From the second equation, we have $1 - 3y = -2$, and $y = 1$.   ◀

If the coefficients of one of the two variables are numerically equal (same absolute value) but have opposite signs, this variable is eliminated by adding the two equations.

E X A M P L E 2   Solve the system of equations

$$3x - 2y = 5$$
$$5x + 2y = 1$$

Solution.

Step 1. Not necessary for this system since the y coefficients have the same absolute values.

Step 2. $3x - 2y = 5$
 $\underline{5x + 2y = 1}$
 $8x = 6$ **adding**

Step 3. Solving the resulting equation, we get $x = \frac{3}{4}$

Step 4. Substituting in the first equation, we get

$$3\left(\frac{3}{4}\right) - 2y = 5$$

$$-2y = 5 - \frac{9}{4} \qquad \textbf{transposing}$$

$$-2y = \frac{11}{4} \qquad \textbf{simplifying}$$

$$y = -\frac{11}{8} \qquad \textbf{dividing by } -2$$

Step 5. As a check, substitute $(\frac{3}{4}, -\frac{11}{8})$ into the second equation:

$$5\left(\frac{3}{4}\right) + 2\left(-\frac{11}{8}\right) = \frac{15}{4} - \frac{11}{4} = 1$$

in agreement with the right side. The first equation is checked similarly.   ◄

In some cases both equations have to be multiplied by a constant before one of the variables can be eliminated.

E X A M P L E 3   Solve the system

$$3x + 2y = 1$$
$$4x - 3y = 7$$

Solution. In this example direct addition or subtraction will not eliminate either variable. The simplest approach is to eliminate y by making the y coefficients numerically equal: Multiply the first equation by 3 and the second by 2. Since the resulting y terms are $+6y$ and $-6y$, respectively, we add the equations:

$$
\begin{array}{ll}
9x + 6y = 3 & \textbf{3(3x + 2y) = 3(1)} \\
8x - 6y = 14 & \textbf{2(4x - 3y) = 2(7)} \\
\hline
17x = 17 & \textbf{adding}
\end{array}
$$

It follows that $x = 1$ and $y = -1$.   ◀

Sometimes a system of equations does not have any solution, as we can see in the next example.

E X A M P L E **4**   Solve the system

$$
\begin{aligned}
-5x + 2y &= 7 \\
10x - 4y &= 5
\end{aligned}
$$

Solution. In this example y appears to be the easier of the two unknowns to eliminate:

$$
\begin{array}{ll}
-10x + 4y = 14 & \textbf{multiplying the first equation by 2} \\
10x - 4y = 5 & \textbf{second equation} \\
\hline
 0 = 19 & \textbf{adding}
\end{array}
$$

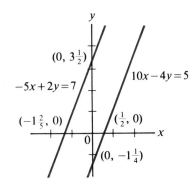

$(0, 3\frac{1}{2})$

$10x - 4y = 5$

$-5x + 2y = 7$

$(-1\frac{2}{5}, 0)$ $(\frac{1}{2}, 0)$

$(0, -1\frac{1}{4})$

Figure 3.7

There is something wrong, since 0 cannot be equal to 19. It follows that the system has no solution. Geometrically, the equations represent two parallel lines. (See Figure 3.7.)   ◀

E X A M P L E **5**   Compare the system

$$
\begin{aligned}
-5x + 2y &= -\frac{5}{2} \\
10x - 4y &= 5
\end{aligned}
$$

to that in Example 4.

Solution.
$$
\begin{array}{ll}
-10x + 4y = -5 & \textbf{multiplying the first equation by 2} \\
10x - 4y = 5 & \textbf{second equation} \\
\hline
 0 = 0 & \textbf{adding}
\end{array}
$$

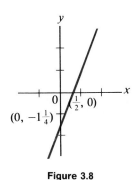

$(\frac{1}{2}, 0)$

$(0, -1\frac{1}{4})$

Figure 3.8

This time no contradiction results. In fact, we have merely shown that the two equations represent exactly the same line. As a result, the coordinates of any point on this line satisfy the given system. (See Figure 3.8.)   ◀

The system in Example 4 is said to be **inconsistent**, which means that the system has no solution. Geometrically, an inconsistent system consists of two parallel lines. The system in Example 5 is said to be **dependent**, that is, the lines coincide, and the system has infinitely many solutions.

We can see from Examples 4 and 5 that a system is inconsistent or dependent if the respective coefficients of the variables are multiples of each other. This means that the ratios of the respective coefficients are equal. If this common ratio is also equal to the ratio of the constants, then the system is dependent. If the common ratio is different from the ratio of the constants, then the system is inconsistent.

Summary.

$$a_1 x + b_1 y = c_1$$ **system of equations**
$$a_2 x + b_2 y = c_2$$

I. The system is **consistent** (having a unique solution) if

$$\frac{a_1}{a_2} \neq \frac{b_1}{b_2}$$

II. The system is **inconsistent** (having no solution) if

$$\frac{a_1}{a_2} = \frac{b_1}{b_2} \neq \frac{c_1}{c_2}$$

III. The system is **dependent** (having infinitely many solutions) if

$$\frac{a_1}{a_2} = \frac{b_1}{b_2} = \frac{c_1}{c_2}$$

Substitution

At this point we seem to have covered all cases. Indeed, the method of addition or subtraction works with any system of linear equations. However, if one equation is easily solved for one of the unknowns, it may be more convenient to solve by the **method of substitution.**

Method of Substitution:

1. Solve one of the equations for one of the unknowns in terms of the other.
2. Substitute the expression obtained into the other equation.
3. Solve the resulting equation in one unknown for the unknown.
4. Substitute the value of the unknown in either of the original equations and solve for the second unknown.
5. Check the solution in the original system.

To see how the method of substitution can be used to eliminate one of the unknowns, consider the system

$$2x - y = 2$$
$$6x + 2y = 1$$

Note that the first equation is readily solved for y to yield

Step 1. $y = 2x - 2$

Substituting this expression for y in the second equation results in an equation containing only one unknown:

$$6x + 2y = 1 \qquad \text{second equation}$$

Step 2. $6x + 2(2x - 2) = 1 \qquad \text{substituting } 2x - 2 \text{ for } y$

Step 3. Solve for x:

$$6x + 4x - 4 = 1$$
$$10x = 5$$
$$x = \frac{1}{2}$$

Step 4. From the first equation, rewritten as $y = 2x - 2$, we get

$$y = 2\left(\frac{1}{2}\right) - 2 = -1 \qquad x = \frac{1}{2}$$

The solution is therefore given by $(\frac{1}{2}, -1)$.

Step 5. *Check:*

$$2\left(\frac{1}{2}\right) - (-1) = 2 \qquad 6\left(\frac{1}{2}\right) + 2(-1) = 1 \quad \checkmark \qquad x = \frac{1}{2}, y = -1$$

This example shows that the method of substitution is most convenient if one of the variables has a coefficient of 1. (Otherwise solving for one of the unknowns may lead to an expression involving fractions.)

Some equations have unknowns occurring in the denominator:

$$\frac{1}{z} + \frac{2}{w} = 1$$

$$\frac{3}{z} - \frac{2}{w} = 7$$

Such systems can be solved by the usual method if we let $x = 1/z$ and $y = 1/w$. Consider the following example:

E X A M P L E   6   Use the method of substitution to solve the system

$$\frac{3}{s_1} - \frac{2}{s_2} = 1$$

$$\frac{16}{s_1} - \frac{12}{s_2} = 5$$

Solution. This system can be written in the usual form by letting $x = 1/s_1$ and $y = 1/s_2$:

$$3x - 2y = 1$$
$$16x - 12y = 5$$

Suppose we solve the first equation for y in terms of x. Then

$$-2y = 1 - 3x \qquad \textbf{first equation}$$

$$y = -\frac{1}{2}(1 - 3x) \qquad\qquad\qquad (3.3)$$

Substituting into the second equation ($16x - 12y = 5$), we get

$$16x - 12\left[-\frac{1}{2}(1 - 3x)\right] = \quad 5 \qquad \textbf{substituting } -\frac{1}{2}(1 - 3x) \textbf{ for } y$$

$$16x + 6(1 - 3x) = \quad 5 \qquad \textbf{-12}(-\tfrac{1}{2}) = \textbf{6}$$

$$16x + 6 - 18x = \quad 5 \qquad \textbf{removing parentheses}$$

$$-2x = -1 \qquad \textbf{simplifying}$$

$$x = \quad \frac{1}{2} \qquad \textbf{dividing both sides by } -\textbf{2}$$

Since $y = -\frac{1}{2}(1 - 3x)$, we now get

$$y = -\frac{1}{2}\left[1 - 3\left(\frac{1}{2}\right)\right] = \frac{1}{4} \qquad x = \frac{1}{2}$$

Finally, since $x = 1/s_1$, it follows that $s_1 = 1/x = 2$. Similarly, $s_2 = 1/y = 4$. ◀

Remark. The method of substitution is particularly important for solving equations of the second degree, where the variables have the form x^2 and y^2.

Since these will be taken up in Chapter 13, you need to become familiar with the method of substitution.

E X E R C I S E S / S E C T I O N   **3.2**

In Exercises 1–10, solve each system of equations by the method of addition or subtraction.

1. $x + y = 4$
$2x - y = 5$

2. $3x - 2y = 6$
$3x - 4y = -2$

3. $3x - 2y = 1$
$4x - 3y = 4$

4. $2x + 7y = 0$
$3x - 2y = 25$

5. $3x - 2y = 21$
$4x - 5y = 42$

6. $3x - 2y = 7$
$4x - 6y = 11$

7. $4x - 3y = -11$
$12x + 25y = 69$

8. $4x + 3y = 1$
$5x + 8y = 10$

9. $2x + 2y = 1$
$5x - 5y = 1$

10. $3x + 5y = 10$
$5x + 3y = 10$

In Exercises 11–20, solve each system of equations by the method of substitution.

11. $x - 3y = 4$
$2x - y = 3$

12. $x + 2y = 12$
$x - 3y = 2$

13. $3x + 4y = 21$
$-x + 2y = 3$

14. $x + 3y = 1$
$2x - y = -5$

15. $2x + y = 1$
$x + 3y = 8$

16. $x + 2y = 13$
$3x - y = -31$

17. $8x - 10y = -13$
$x + 2y = 0$

18. $-x - 2y = 3$
$2x + y = 4$

19. $5x + 2y = 3$
$6x + 3y = 2$

20. $2x - 3y = 3$
$5x - 4y = 1$

In Exercises 21–40, solve each system of equations by either method.

21. $6x - 7y = 49$
$8x - 9y = 63$

22. $3x + 5y = 0$
$x + 4y = 0$

23. $3x - 2y = 1$
$6x - 4y = 5$

24. $2x + 4y = 1$
$6x + 12y = -1$

25. $\dfrac{4}{x} - \dfrac{3}{y} = 1$
$\dfrac{5}{x} - \dfrac{4}{y} = 1$

26. $\dfrac{5}{x} - \dfrac{25}{y} = 51$
$\dfrac{10}{x} - \dfrac{55}{y} = 112$

27. $\dfrac{2}{x} - \dfrac{3}{y} = 1$
$\dfrac{3}{x} - \dfrac{2}{y} = 2$

28. $\dfrac{2}{x} - \dfrac{1}{y} = 2$
$\dfrac{4}{x} + \dfrac{5}{y} = 6$

29. $\dfrac{2}{p} - \dfrac{1}{q} = 1$
$\dfrac{4}{p} - \dfrac{3}{q} = 4$

30. $\dfrac{2}{R} - \dfrac{3}{P} = 12$
$\dfrac{6}{R} - \dfrac{5}{P} = 15$

31. $3F_1 - 2F_2 = -7$
$3F_1 - 12F_2 = -37$

32. $12s - 8t = 19$
$4s - 12t = 25$

33. $3A_1 + 2A_2 = 2$
$5A_1 + 3A_2 = 3$

34. $-2y + 4w = 1$
$-3y + 5w = 2$

35. $2w - 3z = 5$
$4w - 6z = 10$

36. $7m_1 + 2m_2 = 20$
$9m_1 + 3m_2 = -15$

37. $-2v + 5w = 10$
$4v - 10w = 15$

38. $2v_0 - 6v_1 = 3$
$-3v_0 + 3v_1 = 1$

39. $3I + 2I_0 = 7$
$2I + I_0 = 4$

40. $-2r + 4s = -6$
$r - 2s = 3$

3.3 **Introduction to Determinants**

So far we have considered three methods for solving systems of equations. In this section we are going to study yet another, the method of **determinants**. This method is not only elegant but quite easy to use and to apply to larger systems.

Determinants were discovered by Gottfried Leibniz, the codiscoverer of the calculus, and promptly forgotten. They were rediscovered by the Swiss mathematician Gabriel Cramer (1704–1752) in 1750. Some time later the English algebraist Arthur Cayley (1821–1895) undertook a systematic study of determinants and matrices. Cayley's work eventually led to a separate branch of mathematics called *linear algebra*.

To see the value of determinants, consider the general solution of the class of systems

$$
\begin{aligned}
a_1 x + b_1 y &= c_1 \\
a_2 x + b_2 y &= c_2
\end{aligned}
\tag{3.4}
$$

We can eliminate y by multiplying the first equation by b_2 and the second by b_1 to obtain

$$
\begin{aligned}
a_1 b_2 x + b_1 b_2 y &= c_1 b_2 \\
a_2 b_1 x + b_2 b_1 y &= c_2 b_1
\end{aligned}
$$

Subtracting, we get

$$
a_1 b_2 x - a_2 b_1 x = c_1 b_2 - c_2 b_1
$$

It now follows from the distributive law that

$$
(a_1 b_2 - a_2 b_1)x = c_1 b_2 - c_2 b_1
$$

and therefore

$$
x = \frac{c_1 b_2 - c_2 b_1}{a_1 b_2 - a_2 b_1}
\tag{3.5}
$$

By eliminating x, we can show that

$$
y = \frac{a_1 c_2 - a_2 c_1}{a_1 b_2 - a_2 b_1}
\tag{3.6}
$$

We have found the solution of the general system (3.4). Thus anyone who memorizes formulas (3.5) and (3.6) would never again have to solve such a system. Unfortunately, memorizing such formulas is probably more trouble than solving the system unless, of course, some kind of simple pattern can be discovered. Such a pattern is provided by a square array of numbers called a **determinant**. Thus $a_1 b_2 - a_2 b_1$, which appears in each of the denominators, is a quantity denoted by the symbol given next.

Definition of a 2 × 2 Determinant:

$$
\begin{vmatrix} a_1 & b_1 \\ a_2 & b_2 \end{vmatrix} = a_1 b_2 - a_2 b_1
\tag{3.7}
$$

The determinant

$$\begin{vmatrix} a_1 & b_1 \\ a_2 & b_2 \end{vmatrix}$$

Elements
Row and column

is called a 2×2 (two-by-two) determinant. (Larger determinants will be taken up later.) The entries are called **elements**. The elements a_1 and b_1 form the first **row**, and the elements a_1 and a_2 the first **column**. (The second row and column are then defined in a similar way.) Writing the square array as a number, called the **expansion** of the determinant, can best be remembered by means of the following diagram:

Expansion of a
2×2 determinant

$$= a_1 b_2 - a_2 b_1 \tag{3.8}$$

The arrows indicate which elements are to be multiplied, and the sign at the tip of each arrow indicates which sign is to be affixed to the product.

E X A M P L E **1** Expand the determinants

a. $\begin{vmatrix} 0 & 2 \\ -3 & 5 \end{vmatrix}$ **b.** $\begin{vmatrix} 2 & -1 \\ -6 & -4 \end{vmatrix}$

Solution. By expansion (3.8)

a. $\begin{vmatrix} 0 & 2 \\ -3 & 5 \end{vmatrix} = (0)(5) - (-3)(2) = 0 + 6 = 6$

b. $\begin{vmatrix} 2 & -1 \\ -6 & -4 \end{vmatrix} = (2)(-4) - (-6)(-1) = -8 - 6 = -14$ ◀

Let us now return to the system of equations (3.4):

$$a_1 x + b_1 y = c_1$$
$$a_2 x + b_2 y = c_2 \tag{3.9}$$

Observe that the solution, given in statements (3.5) and (3.6), can be written

$$x = \frac{\begin{vmatrix} c_1 & b_1 \\ c_2 & b_2 \end{vmatrix}}{\begin{vmatrix} a_1 & b_1 \\ a_2 & b_2 \end{vmatrix}}, \qquad y = \frac{\begin{vmatrix} a_1 & c_1 \\ a_2 & c_2 \end{vmatrix}}{\begin{vmatrix} a_1 & b_1 \\ a_2 & b_2 \end{vmatrix}}$$

Now notice the pattern: The entries in the two denominators are the coefficients of the unknowns arranged as in the original system (3.9). The entries in the numerator of the x value are obtained by replacing the coefficients of x by the constants on the right side of the system. Similarly, the entries in the numerator of the y value are obtained by replacing the coefficients of y by the constants on the right. This method of solution is known as **Cramer's rule** and can be extended to larger systems of equations.

Before continuing with an example, let us summarize the solution of systems of two equations by Cramer's rule.

Cramer's Rule: The solution of the system

$$a_1 x + b_1 y = c_1$$
$$a_2 x + b_2 y = c_2$$

is given by

$$x = \frac{\begin{vmatrix} c_1 & b_1 \\ c_2 & b_2 \end{vmatrix}}{\begin{vmatrix} a_1 & b_1 \\ a_2 & b_2 \end{vmatrix}}, \qquad y = \frac{\begin{vmatrix} a_1 & c_1 \\ a_2 & c_2 \end{vmatrix}}{\begin{vmatrix} a_1 & b_1 \\ a_2 & b_2 \end{vmatrix}}$$

E X A M P L E **2**   Solve the system

$$5x - 2y = 10$$
$$-3x + 7y = \ 0$$

by Cramer's rule.

Solution. By Cramer's rule, the determinant in both denominators consists of the coefficients of the unknowns arranged as in the given system:

$$\begin{vmatrix} 5 & -2 \\ -3 & 7 \end{vmatrix}$$

The numerator for x can be constructed from this determinant by replacing the first column, the coefficients of x, by the constants on the right:

$$\begin{vmatrix} 10 & -2 \\ 0 & 7 \end{vmatrix}$$

Similarly, to find y we construct the determinant in the numerator by replacing the second column, the coefficients of y, by the constants on the right.

$$\begin{vmatrix} 5 & 10 \\ -3 & 0 \end{vmatrix}$$

The solution is now written as follows:

$$x = \frac{\begin{vmatrix} 10 & -2 \\ 0 & 7 \end{vmatrix}}{\begin{vmatrix} 5 & -2 \\ -3 & 7 \end{vmatrix}} = \frac{(10)(7) - (0)(-2)}{(5)(7) - (-3)(-2)} = \frac{70 - 0}{35 - 6} = \frac{70}{29}$$

$$y = \frac{\begin{vmatrix} 5 & 10 \\ -3 & 0 \end{vmatrix}}{\begin{vmatrix} 5 & -2 \\ -3 & 7 \end{vmatrix}} = \frac{(5)(0) - (-3)(10)}{29} = \frac{0 + 30}{29} = \frac{30}{29}$$

◀

Returning to Cramer's rule, note that the determinant occurring in each denominator has to be different from 0 to avoid division by 0. If the determinant is different from 0, then the solution of system (3.9) is necessarily unique, since the value of a determinant is unique. If

$$\begin{vmatrix} a_1 & b_1 \\ a_2 & b_2 \end{vmatrix} = 0$$

then the system does not have a unique solution. In that case the system is either dependent or inconsistent.

If the arithmetic involved in expanding a determinant is awkward, a calculator can make the computation more convenient. Consider the determinant

$$\begin{vmatrix} -3.59 & 5.26 \\ 7.35 & 3.98 \end{vmatrix} = -3.59 \times 3.98 - 7.35 \times 5.26$$

Since scientific calculators perform multiplication and division before addition and subtraction, the sequence is

$$3.59 \boxed{+/-} \boxed{\times} 3.98 \boxed{-} 7.35 \boxed{\times} 5.26 \boxed{=} \rightarrow -52.9492$$

For calculators that do not automatically perform the operations in the above order, an alternate sequence is

$$3.59 \boxed{+/-} \boxed{\times} 3.98 \boxed{=} \boxed{STO} 7.35 \boxed{\times} 5.26$$
$$\boxed{=} \boxed{+/-} \boxed{+} \boxed{MR} \boxed{=} \rightarrow -52.9492$$

EXERCISES/SECTION 3.3

In Exercises 1–16, expand each determinant.

1. $\begin{vmatrix} 0 & 0 \\ 2 & 1 \end{vmatrix}$

2. $\begin{vmatrix} 1 & -2 \\ 4 & -8 \end{vmatrix}$

3. $\begin{vmatrix} -2 & 4 \\ 4 & -8 \end{vmatrix}$

4. $\begin{vmatrix} 1 & -2 \\ 0 & 2 \end{vmatrix}$

5. $\begin{vmatrix} 2 & 1 \\ 7 & 4 \end{vmatrix}$

6. $\begin{vmatrix} 1 & 5 \\ 2 & 3 \end{vmatrix}$

7. $\begin{vmatrix} 2 & 1 \\ 4 & -3 \end{vmatrix}$

8. $\begin{vmatrix} -2 & 6 \\ 8 & -4 \end{vmatrix}$

9. $\begin{vmatrix} -2 & -1 \\ 12 & 5 \end{vmatrix}$

10. $\begin{vmatrix} -3 & -5 \\ -6 & -4 \end{vmatrix}$

11. $\begin{vmatrix} -6 & -17 \\ 3 & 20 \end{vmatrix}$

12. $\begin{vmatrix} 12 & -3 \\ -15 & 6 \end{vmatrix}$

13. $\begin{vmatrix} 32 & 21 \\ -17 & 16 \end{vmatrix}$

14. $\begin{vmatrix} 21 & 5 \\ 20 & -8 \end{vmatrix}$

15. $\begin{vmatrix} 18 & -6 \\ 75 & 0 \end{vmatrix}$

16. $\begin{vmatrix} -12 & -11 \\ 5 & -18 \end{vmatrix}$

In Exercises 17–41, solve each system of equations by using Cramer's rule.

17. $3x + 4y = 1$
$2x + 3y = 4$

18. $-x + 3y = 5$
$-2x - y = 0$

19. $3x - 2y = 4$
$-7x + 5y = 1$

20. $-2x - 6y = 4$
$5x + 10y = 5$

21. $-5x + 6y = 7$
$4x - 5y = 8$

22. $3x - 4y = 20$
$5x - 6y = 8$

23. $6x + 3y = -1$
$5x + 3y = 2$

24. $-3x + 4y = 11$
$6x - 8y = 8$

25. $\dfrac{2}{x} - \dfrac{3}{y} = 7$
$\dfrac{1}{x} + \dfrac{5}{y} = 3$

26. $\dfrac{4}{x} - \dfrac{4}{y} = 9$
$\dfrac{3}{x} + \dfrac{5}{y} = 8$

27. $\dfrac{8}{x} - \dfrac{7}{y} = 20$
$\dfrac{3}{x} - \dfrac{5}{y} = 16$

28. $\dfrac{2}{x} + \dfrac{5}{y} = 3$
$\dfrac{5}{x} - \dfrac{7}{y} = 6$

29. $4T_1 - 7T_2 = 10$
$5T_1 - 8T_2 = 20$

30. $2w_1 + w_2 = 5$
$w_1 + 3w_2 = 5$

31. $F_1 + 2F_2 = 5$
$2F_1 + F_2 = 6$

32. $2W_1 + 3W_2 = 12$
$2W_1 + 4W_2 = 15$

33. $3R_1 + 4R_2 = 20$
$4R_1 + 2R_2 = 15$

34. $2I_1 + 8I_2 = 25$
$4I_1 + 7I_2 = 30$

35. $3C_1 - 7C_2 = 3$
$-9C_1 + 21C_2 = 5$

36. $2p - 7q = 10$
$5p - q = 15$

(The remaining problems should be solved with a calculator.)

37. $2.73x - 1.52y = 5.02$
$0.130x + 2.49y = 2.98$

38. $0.980x + 0.730y = 1.21$
$-1.32x - 5.21y = -1.11$

39. $-2.10x + 3.64y = 1.32$
$1.00x + 1.78y = -4.05$

40. $6.52x + 3.98y = -1.25$
$1.35x - 1.44y = 2.73$

41. $-7.63x - 5.02y = 1.31$
$2.84x - 1.54y = 3.87$

3.4   Applications of Systems of Linear Equations

Systems of linear equations have many applications in science and technology. In addition, many problems that were solved in Chapter 2 by using a single equation can be solved more readily by using a system of equations.

EXAMPLE **1**   The sum of two numbers is 15.5. One number is 1.7 more than twice the other. Find the two numbers.

Solution. Since we are looking for *two* numbers, let us denote the first number by x and the other (second) number by y.

The sum of the two numbers is 15.5. So the first equation is

$$x + y = 15.5$$

Now, twice the other number is $2y$ and 1.7 more than twice the other number is $2y + 1.7$. So the second equation is

$$x = 2y + 1.7$$

The resulting system of equations is

$$x + y = 15.5$$
$$x = 2y + 1.7$$

Since the second equation is $x = 2y + 1.7$, let us solve the system by substitution:

$x = 2y + 1.7$	second equation
$x + y = 15.5$	first equation
$2y + 1.7 + y = 15.5$	substituting for x
$3y + 1.7 = 15.5$	$2y + y = 3y$
$3y = 15.5 - 1.7$	subtracting 1.7
$3y = 13.8$	
$y = 4.6$	dividing by 3

Substituting $y = $ **4.6** into the second equation ($x = 2y + 1.7$), we get

$$x = 2(\textbf{4.6}) + 1.7 = 10.9$$

We conclude that the numbers are 10.9 and 4.6.

To check the solution, we return to the given problem. The sum of the two numbers is $10.9 + 4.6 = 15.5$. Twice the second number ($y = 4.6$) plus 1.7 is $2(4.6) + 1.7 = 10.9$, which is the first number. ◀

E X A M P L E **2**   Two resistors are connected in series. The resistance of the second is 5.6 Ω less than that of the first. The total resistance is 50.2 Ω. Find the resistance of each resistor.

Solution. Recall that the total resistance of two or more resistors in series is equal to the sum of the individual resistances. Letting R_1 and R_2 be the individual resistances, we get

$$R_1 - R_2 = 5.6$$
$$\underline{R_1 + R_2 = 50.2}$$
$$2R_1 \qquad = 55.8 \qquad \text{adding}$$
$$R_1 = 27.9\ \Omega$$
$$R_2 = 22.3\ \Omega$$

◀

E   X   A   M   P   L   E   **3**   A technician wants to make a rectangular metal plate meeting the following requirements: The length is to be 1.5 cm less than 3 times the width, and the perimeter is to be 22.6 cm. Find the dimensions.

Solution. Let

$$x = \text{width of rectangle}$$

and

$$y = \text{length of rectangle}$$

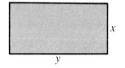

Figure 3.9

By Figure 3.9, the perimeter is $x + x + y + y = 2x + 2y = 22.6$. The condition "length is 1.5 cm less than 3 times the width" is now written

$$y = 3x - 1.5$$

We have the following system:

$$2x + 2y = 22.6$$
$$y = 3x - 1.5$$

Solving this system by substitution, we have

$y = 3x - 1.5$	**second equation**
$2x + 2y = 22.6$	**first equation**
$2x + 2(3x - 1.5) = 22.6$	**substitution**
$2x + 6x - 3.0 = 22.6$	**removing parentheses**
$8x = 22.6 + 3.0$	**adding 3.0**
$8x = 25.6$	
$x = 3.2$ cm	**dividing by 8**

From the second equation,

$$y = 3x - 1.5 = 3(3.2) - 1.5 = 8.1 \text{ cm}$$

We conclude that the length is 8.1 cm and the width is 3.2 cm.   ◀

E   X   A   M   P   L   E   **4**   A portion of $13,580 was invested at 8% interest and the rest at 10%. If the total interest income was $1,253, how much was invested at each rate?

Solution. Let

$$x = \text{amount invested at } 8\%$$
$$y = \text{amount invested at } 10\%$$

Then

$$x + y = \$13{,}580 \qquad \text{total amount invested}$$

$$0.08x + 0.10y = \$1{,}253 \qquad \text{total interest (8\% of x + 10\% of y)}$$

are the equations to be solved.

$$\begin{array}{ll} 8x + 8y = 108{,}640 & \text{multiplying first equation by 8} \\ 8x + 10y = 125{,}300 & \text{multiplying second equation by 100} \\ \hline -2y = -16{,}660 & \text{subtracting} \\ y = \$8{,}330 & \\ x = \$5{,}250 & \end{array}$$

As usual, the solution of a problem can be checked against the given information. Thus 8% of $5,250 equals $(0.08)(\$5{,}250) = \$420$ and 10% of $8,330 is $833, for a total of $1,253.   ◀

Mixture problems can also be solved more easily by using systems of equations.

E X A M P L E 5 One alloy contains 20% copper, and another 25% copper. How many pounds of each must be combined to form 60 lb of an alloy containing 22% copper?

Solution. Let x be the number of pounds of the first alloy and y the number of pounds of the second. Then $x + y = 60$. Recall that it is best to work with the quantities directly: For example, $(0.22)(60) = 13.2$ lb, the weight of copper in the 22% alloy. Thus

$$0.20x + 0.25y = (0.22)(60)$$

$$x + y = 60$$

are the resulting equations, which can be solved by addition or subtraction:

$$\begin{array}{ll} 20x + 25y = (22)(60) & \text{multiplying first equation by 100} \\ 20x + 20y = 1200 & \text{multiplying second equation by 20} \\ \hline 5y = 120 & \text{subtracting} \\ y = 24 \text{ lb} & \\ x = 36 \text{ lb} & \end{array}$$

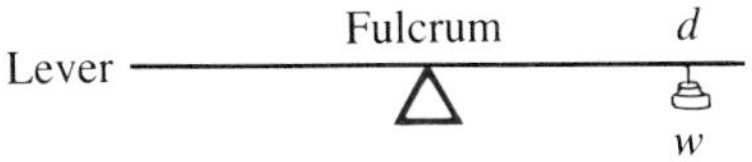

◀

Many problems in technology involve **moments**. Consider the **lever** in Figure 3.10 supported at a point called the **fulcrum**. Neglecting the weight of the lever, a weight w at a distance d from the fulcrum (Figure 3.10) has a **moment** given by weight times distance, or wd. The distance d is called the **moment arm**. If two weights w_1 and w_2 are placed on opposite sides of the fulcrum at respective distances d_1 and d_2, then

Figure 3.10

$$w_1 d_1 = w_2 d_2$$

whenever the weights balance on the lever.

One of the fundamental principles of levers is that moments are additive; that is, $w_1 d_1 + w_2 d_2$ is equal to the total moment. Consider, for example, the weights on the lever in Figure 3.11. For the weights to be balanced on the lever, the moment on the left must be equal to the moment on the right, or

Figure 3.11

$$w_1 d_1 + w_2 d_2 = w_3 d_3$$

This formula can be extended to any number of weights.

E X A M P L E 6   A weight of 1 lb and a lever are to be used to determine two other weights. Referring to Figure 3.11, the following measurements were taken: Given $w_3 = 1$ lb, the lever balances when $d_3 = 31$ in., $d_1 = 5$ in., and $d_2 = 4$ in. Another balance is obtained when $d_3 = 33$ in., $d_1 = 3$ in., and $d_2 = 6$ in.

Solution. From the relationship

$$w_1 d_1 + w_2 d_2 = w_3 d_3$$

and the given measurements, we get the system

$$5w_1 + 4w_2 = 31 \cdot 1$$
$$3w_1 + 6w_2 = 33 \cdot 1$$

By Cramer's rule

$$w_1 = \frac{\begin{vmatrix} 31 & 4 \\ 33 & 6 \end{vmatrix}}{\begin{vmatrix} 5 & 4 \\ 3 & 6 \end{vmatrix}} \quad \text{and} \quad w_2 = \frac{\begin{vmatrix} 5 & 31 \\ 3 & 33 \end{vmatrix}}{\begin{vmatrix} 5 & 4 \\ 3 & 6 \end{vmatrix}}$$

Hence

$$w_1 = \frac{186 - 132}{30 - 12} = \frac{54}{18} = 3 \text{ lb}$$

and

$$w_2 = \frac{165 - 93}{18} = \frac{72}{18} = 4 \text{ lb}$$

◀

E X E R C I S E S / S E C T I O N   3.4

1. The velocities v_1 and v_2 (in feet per second) of two colliding bodies satisfy the following relationships:

$$3v_1 + 2v_2 = 5$$
$$4v_1 + 3v_2 = 6$$

Determine the two velocities.

2. Measurements of the tension (in pounds) of two support cables produced the following equations:

$$0.25T_1 + 0.45T_2 = 37$$
$$0.34T_1 + 0.54T_2 = 46$$

Find T_1 and T_2.

3. The relationship between the tensile strength S (measured in pounds) of a certain metal rod and the temperature T (in degrees Celsius) has the form $S = a - bT$. Experimenters found that if $T = 50.0°C$, then $S = 565.9$ lb; if $T = 100°C$, then $S = 565.8$ lb. Find the relationship.

4. The relationship between the length of a certain bar (measured in centimeters) and its temperature (in degrees Celsius) is known to be $L = aT + b$. Tests show that if $T = 15°C$, then $L = 50.0$ cm; if $T = 60°C$, then $L = 50.8$ cm. Find the relationship.

In Exercises 5 and 6, refer to Figure 3.12 and find w_1 and w_2 in each case.

Figure 3.12

5. $w_3 = 2.0$ N; a balance is obtained if $d_1 = 2.0$ m, $d_2 = 2.0$ m, and $d_3 = 3.5$ m and if $d_1 = 2.0$ m, $d_2 = 1.0$ m, and $d_3 = 2.0$ m. (See Example 6.)

6. $w_3 = 4.000$ lb; a balance is obtained if $d_1 = 3.000$ in., $d_2 = 4.000$ in., and $d_3 = 17.25$ in. and if $d_1 = 5.000$ in., $d_2 = 2.000$ in., and $d_3 = 18.25$ in.

7. The sum of two numbers is 44 and their difference is 8. Find the numbers. (See Example 1.)

8. The sum of two numbers is 76 and their difference is 14. Find the numbers.

9. One number is twice the other and their sum is 12.9. What are the numbers?

10. One number is 1.1 less than three times the other. If their sum is 5.3, determine the numbers.

11. How must a 54-ft cable be cut so that one part is 2 ft shorter than three times the other part?

12. Two separate squares are to be made from a piece of wire 54.0 cm long. If the perimeter of one square is to be 3.0 cm larger than that of the other, how must the wire be cut?

13. Two gears have a total of 51 teeth. If the number of teeth of one is 3 more than twice the number of the other, find the number of teeth of each.

14. Two machines have a total of 62 moving parts. If one machine has 2 more than 3 times as many moving parts as the other, how many moving parts does each machine have?

15. Two resistors connected in series have a combined resistance of 150 Ω. If the resistance of one resistor is 10 Ω less

than that of the other, find the resistance of each. (See Example 2.)

16. The combined resistance of two resistors in a series is 130 Ω. If the resistance of one resistor is 20 Ω less than that of the other, find the resistance of each.

17. The sum of the voltages across two resistors is 55.1 V. It was found that 3 times the first voltage is 9.7 V less than 4 times the second. What are the two voltages?

18. The sum of two currents is 4.4 A. Given that one current is 1.0 A less than twice the other, determine the two currents.

19. A machinist has an order for a rectangular metal plate with the following specifications: The length is 1.0 in. less than 3 times the width, and the perimeter is 24.4 in. Find the dimensions of the plate. (See Example 3.)

20. The length of a metal plate is 1.8 cm more than twice the width. The perimeter is 18.6 cm. Find the dimensions.

21. A portion of $8,500 is invested at 12% interest and the remainder at 11%. The total interest income is $976.10. Determine the amount invested at each rate. (See Example 4.)

22. A woman invests a certain amount of money at 10% interest and the rest at 8%. If the first investment is $2,000 more than the second and her total interest income is $740, find how much was invested at each rate.

23. A technician needs 100 mL of a 16% nitric acid solution (by volume). He has a 20% and a 10% solution (by volume) in stock. How many milliliters of each must he mix to obtain the required solution? (See Example 5.)

24. How many liters of a 5% solution (by volume) must be added to a 10% solution to obtain 20 L of an 8% solution?

25. One alloy contains 6% brass (by weight) and another 12% brass (by weight). How many pounds of each must be combined to form 50 lb of an alloy containing 10% brass?

26. How many pounds of an alloy containing 8% copper (by weight) must be combined with an alloy containing 14% copper to form 36.5 lb of an alloy containing 10% copper?

27. The manager of a shop spent $189 in his budget to buy 80 small castings. Some cost $1.95 apiece, and the rest cost $2.50 apiece. How many of each did he buy?

28. Tickets for an industrial exhibit cost $5.00 for regular admission and $4.00 for senior citizens. On one day 215 tickets were sold, for a total intake of $1,050. How many tickets of each type were sold?

29. John has $6.10 in dimes and quarters. If he has five more dimes than quarters, how many of each does he have?

30. Jane has \$3.30 in nickels and dimes. If she has nine more nickels than dimes, determine the number of each.

31. An office building has 20 offices. The smaller offices rent for \$300 per month, and the larger offices for \$420 per month. If the rental income is \$7,440 per month, how many of each type of office are there?

32. One consultant to a firm charges \$200 per day, and another consultant charges \$250 per day. After 13 days the total charged by the two consultants came to \$2,950. Assuming that only one of the two consultants was called in on any one day, how many days did each one work?

3.5   Systems of Linear Equations with More Than Two Unknowns

Our method for solving systems of two equations can be readily extended to systems of three or more equations. We will concentrate on the **method of addition or subtraction** in this section and return to the method of determinants in Section 3.6.

To solve a system of three equations

$$a_1 x + b_1 y + c_1 z = d_1$$
$$a_2 x + b_2 y + c_2 z = d_2 \qquad\qquad (3.10)$$
$$a_3 x + b_3 y + c_3 z = d_3$$

algebraically, we eliminate one of the unknowns between any two of the equations. Then, taking a different pair of equations, we eliminate the same unknown. The resulting system of two equations can then be solved by one of the earlier methods. Consider the following example.

E X A M P L E **1**   Solve the system

  1. $\quad x - 3y - z = -2$
  2. $4x - y - 2z = 8$
  3. $3x + 2y + 2z = 1$

Solution. A glance at the different coefficients tells us that z is the easiest of the three unknowns to eliminate.

4. $\quad 2x - 6y - 2z = -4$	**multiplying equation (1) by 2**
5. $\quad 4x - y - 2z = 8$	**repeating (2) and (3)**
6. $\quad 3x + 2y + 2z = 1$	
7. $-2x - 5y = -12$	**subtracting (5) from (4)**
8. $\quad 7x + y = 9$	**adding (5) and (6)**
9. $-2x - 5y = -12$	**repeating (7)**
10. $\ 35x + 5y = 45$	**multiplying (8) by 5**
11. $\ 33x = 33$	**adding (9) and (10)**
12. $ x = 1$	
13. $\quad 7(1) + y = 9$	**substituting in (8)**
14. $ y = 2$	
15. $\quad 1 - 3(2) - z = -2$	**substituting $x = 1$ and $y = 2$ in (1)**
16. $ z = -3$	

The solution is therefore given by $x = 1$, $y = 2$, and $z = -3$. As a check, substitute these values into equations (2) and (3). Then

$$4(1) - 2 - 2(-3) = 8$$

and

$$3(1) + 2(2) + 2(-3) = 1 \quad \checkmark$$

◀

E X A M P L E 2   Solve the system

1. $4R_1 - 2R_2 + R_3 = 8$
2. $3R_1 - 3R_2 + 4R_3 = 8$
3. $R_1 + R_2 + 2R_3 = 6$

Solution. We eliminate R_2 as follows:

4. $12R_1 - 6R_2 + 3R_3 = 24$	multiplying (1) by 3	
5. $6R_1 - 6R_2 + 8R_3 = 16$	multiplying (2) by 2	
6. $6R_1 + 6R_2 + 12R_3 = 36$	multiplying (3) by 6	
7. $6R_1 \qquad - 5R_3 = 8$	subtracting (5) from (4)	
8. $12R_1 \qquad + 20R_3 = 52$	adding (5) and (6)	
9. $24R_1 \qquad - 20R_3 = 32$	multiplying (7) by 4	
10. $12R_1 \qquad + 20R_3 = 52$	repeating (8)	
11. $36R_1 \qquad\qquad = 84$	adding (9) and (10)	

12. $$R_1 = \frac{84}{36} = \frac{7}{3}$$

13. $$6\left(\frac{7}{3}\right) \qquad - 5R_3 = 8 \qquad \text{substituting into (7)}$$

14. $$R_3 = \frac{6}{5}$$

15. $$\frac{7}{3} + R_2 + 2\left(\frac{6}{5}\right) = 6 \qquad \text{substituting } R_1 = \frac{7}{3} \text{ and } R_3 = \frac{6}{5} \text{ in (3)}$$

16. $$R_2 = \frac{19}{15}$$

As a check, we substitute the values $R_1 = \frac{7}{3}$, $R_2 = \frac{19}{15}$, and $R_3 = \frac{6}{5}$ into equation (1):

$$4\left(\frac{7}{3}\right) - 2\left(\frac{19}{15}\right) + \frac{6}{5} = \frac{140}{15} - \frac{38}{15} + \frac{18}{15} = \frac{120}{15} = 8 \quad \checkmark$$

Equations (2) and (3) are checked similarly.   ◀

To solve a system with four unknowns, proceed by eliminating one of the unknowns among three different pairs of equations. The resulting system of three equations can then be solved by the methods of this section.

E X E R C I S E S / S E C T I O N   **3.5**

Solve the following systems of equations.

1. $\begin{aligned} x + y + 2z &= 9 \\ x \quad\ - z &= -2 \\ 2x - y \quad\ &= 0 \end{aligned}$

2. $\begin{aligned} x - \ y \quad\ &= 2 \\ x + 2y - z &= -3 \\ -y + z &= 3 \end{aligned}$

3. $\begin{aligned} 3x \quad\ + 2z &= -1 \\ 4x - y - 2z &= 7 \\ x + y \quad\ &= 2 \end{aligned}$

4. $\begin{aligned} 2x - 3y \quad\ &= 8 \\ 3x - \ y + 2z &= 8 \\ 2x + 3y \quad\ &= -4 \end{aligned}$

5. $\begin{aligned} x - 2y - \ z &= 1 \\ 2x - \ y - 2z &= -1 \\ 3x + \ y + 2z &= 6 \end{aligned}$

6. $\begin{aligned} 3x - 3y - 2z &= 1 \\ -x + \ y - 6z &= 3 \\ x + \ y + 2z &= 3 \end{aligned}$

7. $\begin{aligned} 2x - \ y + 3z &= 16 \\ 3x + 4y + 2z &= 7 \\ x - 2y - 2z &= -2 \end{aligned}$

8. $\begin{aligned} x - 2y + \ z &= 1 \\ 2x + \ y + 2z &= 2 \\ 3x + 3y - 3z &= 2 \end{aligned}$

9. $\begin{aligned} 3R_1 - \ 3R_2 + 7R_3 &= 35 \\ 6R_1 + \ 3R_2 - \ R_3 &= -5 \\ 9R_1 - 12R_2 + 3R_3 &= 38 \end{aligned}$

10. $\begin{aligned} 2F_1 - \ 5F_2 + 10F_3 &= 21 \\ 3F_1 + \ 5F_2 + 10F_3 &= 6 \\ 4F_1 - 10F_2 + 40F_3 &= 64 \end{aligned}$

11. $\begin{aligned} \frac{2}{x} - \frac{1}{y} + \frac{2}{z} &= 2 \\[4pt] -\frac{4}{x} + \frac{5}{y} - \frac{3}{z} &= 1 \\[4pt] \frac{3}{x} - \frac{4}{y} + \frac{1}{z} &= 3 \end{aligned}$

12. $\begin{aligned} \frac{2}{x} + \frac{1}{y} - \frac{1}{z} &= 0 \\[4pt] \frac{8}{x} - \frac{2}{y} + \frac{1}{z} &= 1 \\[4pt] \frac{4}{x} - \frac{4}{y} - \frac{1}{z} &= -1 \end{aligned}$

13. $\begin{aligned} 2y + 3z + \ w &= 1 \\ 2x - 2y \quad\ + \ w &= 1 \\ 3x \quad\ - 2z + 2w &= 13 \\ x + 3y - \ z \quad\ &= 9 \end{aligned}$

14. $\begin{aligned} x \quad\ + 2z - 3w &= 12 \\ 3x - 2y + \ z \quad\ &= -3 \\ 3y - 3z - 5w &= 12 \\ 2x - \ y \quad\ - 2w &= 3 \end{aligned}$

15. $\begin{aligned} 2x - \ y - 2z - 2w &= -7 \\ 4x - 2y + 3z \quad\ &= 3 \\ 6x + 2y + 4z - 3w &= 16 \\ 8x \quad\ - 2z - 4w &= -6 \end{aligned}$

3.6   More on Determinants

The method of **determinants** can be extended to systems of more than two equations. In this section we will confine ourselves to systems of three equations. Larger systems will be discussed in Chapter 15.

The easiest way to see the extension of the determinant method is to solve the system

$$a_1x + b_1y + c_1z = d_1$$
$$a_2x + b_2y + c_2z = d_2 \tag{3.11}$$
$$a_3x + b_3y + c_3z = d_3$$

by the method of addition or subtraction. Although straightforward, the calculation is quite lengthy. The expression for x turns out to be

$$x = \frac{d_1b_2c_3 - d_1b_3c_2 - d_2b_1c_3 + d_3b_1c_2 + d_2b_3c_1 - d_3b_2c_1}{a_1b_2c_3 - a_1b_3c_2 - a_2b_1c_3 + a_3b_1c_2 + a_2b_3c_1 - a_3b_2c_1} \tag{3.12}$$

A determinant of the third order is defined with Cramer's rule in mind: If Cramer's rule is to carry over, then the denominator of solution (3.12) should be a determinant whose elements are the coefficients of the unknown. In other words,

$$\begin{vmatrix} a_1 & b_1 & c_1 \\ a_2 & b_2 & c_2 \\ a_3 & b_3 & c_3 \end{vmatrix} = a_1 b_2 c_3 - a_1 b_3 c_2 - a_2 b_1 c_3 + a_3 b_1 c_2 \\ + a_2 b_3 c_1 - a_3 b_2 c_1 \tag{3.13}$$

The last expression may look like a jumble of symbols without any discernible pattern, but if you look more closely, you will see that *every product consists of exactly one element from each row and one from each column.* (The same holds true for higher-order determinants.) This observation enables us to express a third-order determinant in terms of second-order determinants. For example,

$$a_1 \begin{vmatrix} b_2 & c_2 \\ b_3 & c_3 \end{vmatrix} = a_1(b_2 c_3 - b_3 c_2) = a_1 b_2 c_3 - a_1 b_3 c_2$$

which are the first two terms in expansion (3.13). The determinant

$$\begin{vmatrix} b_2 & c_2 \\ b_3 & c_3 \end{vmatrix}$$

is called the **minor** of the element a_1.

Definition of a Minor: The **minor** of a given element is the determinant formed by deleting all the elements in the row and column in which the element lies.

Thus in determinant (3.13) the minor of a_2 is

$$\begin{vmatrix} b_1 & c_1 \\ b_3 & c_3 \end{vmatrix}$$

and the minor of b_2 is

$$\begin{vmatrix} a_1 & c_1 \\ a_3 & c_3 \end{vmatrix}$$

Now observe that, in terms of minors, the determinant (3.13) can be written

$$\begin{vmatrix} a_1 & b_1 & c_1 \\ a_2 & b_2 & c_2 \\ a_3 & b_3 & c_3 \end{vmatrix} = a_1 \begin{vmatrix} b_2 & c_2 \\ b_3 & c_3 \end{vmatrix} - b_1 \begin{vmatrix} a_2 & c_2 \\ a_3 & c_3 \end{vmatrix} + c_1 \begin{vmatrix} a_2 & b_2 \\ a_3 & b_3 \end{vmatrix}$$

In other words, the determinant is expanded by forming the products of the elements in the first row with their corresponding minors and affixing either a plus or minus sign. Moreover, the same expression on the right side of expansion (3.13) can be obtained by using the elements in some other row or even in a column. For example, using the second column, we get the expansion given next.

Typical Expansion by Minors:

$$\begin{vmatrix} a_1 & b_1 & c_1 \\ a_2 & b_2 & c_2 \\ a_3 & b_3 & c_3 \end{vmatrix} = -b_1 \begin{vmatrix} a_2 & c_2 \\ a_3 & c_3 \end{vmatrix} + b_2 \begin{vmatrix} a_1 & c_1 \\ a_3 & c_3 \end{vmatrix} - b_3 \begin{vmatrix} a_1 & c_1 \\ a_2 & c_2 \end{vmatrix}$$

$$= -b_1 a_2 c_3 + b_1 a_3 c_2 + b_2 a_1 c_3 - b_2 a_3 c_1$$
$$- b_3 a_1 c_2 + b_3 a_2 c_1 \qquad (3.14)$$

Rule for signs

We still need a rule for affixing the sign. It turns out that the sign depends only on the position of the element. Consider the row and column in which the element lies. *If the sum of the number of the row and the number of the column is even, affix a plus sign; if the sum is odd, affix a minus sign.*

E X A M P L E **1**   Expand the determinant

$$\begin{vmatrix} 2 & -3 & 1 \\ -4 & 0 & -7 \\ -3 & -1 & 1 \end{vmatrix}$$

Solution. As already noted, we can expand the determinant along any row or column. Suppose we arbitrarily choose the first column. Then we get

$$\begin{vmatrix} 2 & -3 & 1 \\ -4 & 0 & -7 \\ -3 & -1 & 1 \end{vmatrix}$$

$$= +(2)\begin{vmatrix} 0 & -7 \\ -1 & 1 \end{vmatrix} - (-4)\begin{vmatrix} -3 & 1 \\ -1 & 1 \end{vmatrix} + (-3)\begin{vmatrix} -3 & 1 \\ 0 & -7 \end{vmatrix}$$

Take a closer look at how the signs were determined. The first element, **2**, lies in row 1, column 1, and $1 + 1 = 2$, which is even. Hence the element is given a plus sign. The next element, **−4**, lies in row 2, column 1, and $2 + 1 = 3$, which is odd. So the element −4 is given a minus sign to become −(−4). Finally, the third element, -3, lies in row 3, column 1, and $3 + 1 = 4$, which is even. So the element is given a plus sign to become $+(-3)$.

No particular row or column offers any obvious advantage with one notable exception: A row or a column containing one or more zeros reduces the number of calculations required. For example, expanding the last determinant along the second row yields

$$
\begin{vmatrix} 2 & -3 & 1 \\ -4 & 0 & -7 \\ -3 & -1 & 1 \end{vmatrix}
$$

$$
= -(-4)\begin{vmatrix} -3 & 1 \\ -1 & 1 \end{vmatrix} + (0)\begin{vmatrix} 2 & 1 \\ -3 & 1 \end{vmatrix} - (-7)\begin{vmatrix} 2 & -3 \\ -3 & -1 \end{vmatrix}
$$

$$
= 4(-3 + 1) + 0 + 7(-2 - 9) = -85 \qquad \blacktriangleleft
$$

E X A M P L E **2**   Expand the determinant

$$
\begin{vmatrix} -3 & 2 & 1 \\ 4 & -2 & 3 \\ -3 & 1 & 0 \end{vmatrix}
$$

Solution. Because of the 0, we expand along the third row, starting with the element -3. This element is situated in row 3, column 1. Since $3 + 1 = 4$, we affix a plus sign, which yields $+(-3)$:

$$
\begin{vmatrix} -3 & 2 & 1 \\ 4 & -2 & 3 \\ -3 & 1 & 0 \end{vmatrix}
$$

$$
= +(-3)\begin{vmatrix} 2 & 1 \\ -2 & 3 \end{vmatrix} - (1)\begin{vmatrix} -3 & 1 \\ 4 & 3 \end{vmatrix} - (0)\begin{vmatrix} -3 & 2 \\ 4 & -2 \end{vmatrix}
$$

$$
= -3(6 + 2) - (-9 - 4) + 0 = -11
$$

(Note that the signs necessarily alternate, so that only the first one needs to be determined.) $\qquad \blacktriangleleft$

E X A M P L E **3**   Expand the determinant

$$
\begin{vmatrix} 7 & -2 & -3 \\ -11 & 0 & 5 \\ 2 & 0 & 2 \end{vmatrix}
$$

Solution. If we expand along the second column, then only one minor needs to be evaluated. Note also that the element -2 lies in row 1, column 2, and is therefore given a minus sign.

$$\begin{vmatrix} 7 & -2 & -3 \\ -11 & 0 & 5 \\ 2 & 0 & 2 \end{vmatrix} = -(-2) \begin{vmatrix} -11 & 5 \\ 2 & 2 \end{vmatrix} = 2(-22 - 10) = -64 \qquad \blacktriangleleft$$

Having defined a third-order determinant, we can now return to Cramer's rule and the solution of equations.

Cramer's Rule: The solution of the system

$$a_1x + b_1y + c_1z = d_1$$
$$a_2x + b_2y + c_2z = d_2 \qquad\qquad (3.15)$$
$$a_3x + b_3y + c_3z = d_3$$

is given by

$$x = \frac{\begin{vmatrix} d_1 & b_1 & c_1 \\ d_2 & b_2 & c_2 \\ d_3 & b_3 & c_3 \\ a_1 & b_1 & c_1 \\ a_2 & b_2 & c_2 \\ a_3 & b_3 & c_3 \end{vmatrix}}{}, \qquad y = \frac{\begin{vmatrix} a_1 & d_1 & c_1 \\ a_2 & d_2 & c_2 \\ a_3 & d_3 & c_3 \\ a_1 & b_1 & c_1 \\ a_2 & b_2 & c_2 \\ a_3 & b_3 & c_3 \end{vmatrix}}{},$$

$$z = \frac{\begin{vmatrix} a_1 & b_1 & d_1 \\ a_2 & b_2 & d_2 \\ a_3 & b_3 & d_3 \\ a_1 & b_1 & c_1 \\ a_2 & b_2 & c_2 \\ a_3 & b_3 & c_3 \end{vmatrix}}{} \qquad\qquad (3.16)$$

As in the case of two equations, the denominators are the same for all the unknowns. The determinants in the numerators are found by replacing the column of coefficients of the unknown to be found by the column of numbers on the right side of the equation. We will see in Chapter 15 that the rule can be applied to systems of equations of any size.

If the determinant

$$\begin{vmatrix} a_1 & b_1 & c_1 \\ a_2 & b_2 & c_2 \\ a_3 & b_3 & c_3 \end{vmatrix}$$

Independent system

in (3.16) is different from zero, then the system has a unique solution, since the values of the determinants are unique. Such a system is called *independent*. If the determinant is zero, then the system is not independent and no unique solution exists. (The system is either inconsistent, having no solution, or dependent, having infinitely many solutions.)

E X A M P L E **4**   Solve the given system by determinants.

$$2x - y + 3z = 2$$

$$x + 2y - z = 1$$

$$3x - 2y = 4$$

Solution. First we evaluate the determinant whose elements consist of the coefficients of the unknowns; this determinant will occur in all the denominators. Expanding along the third row:

$$\begin{vmatrix} 2 & -1 & 3 \\ 1 & 2 & -1 \\ 3 & -2 & 0 \end{vmatrix} = +3 \begin{vmatrix} -1 & 3 \\ 2 & -1 \end{vmatrix} - (-2) \begin{vmatrix} 2 & 3 \\ 1 & -1 \end{vmatrix} + 0 \begin{vmatrix} 2 & -1 \\ 1 & 2 \end{vmatrix}$$

$$= 3(1 - 6) + 2(-2 - 3) + 0 = -25$$

So by Cramer's rule,

$$x = \frac{\begin{vmatrix} 2 & -1 & 3 \\ 1 & 2 & -1 \\ 4 & -2 & 0 \end{vmatrix}}{-25} \qquad \begin{array}{l} \textbf{replacing the coefficients of } x \\ \textbf{by the numbers on the right side} \end{array}$$

$$= \frac{\begin{vmatrix} 2 & -1 & 3 \\ 1 & 2 & -1 \\ 4 & -2 & 0 \end{vmatrix}}{-25}$$

Expanding along the third column, we get

$$x = -\frac{1}{25} \left[3 \begin{vmatrix} 1 & 2 \\ 4 & -2 \end{vmatrix} - (-1) \begin{vmatrix} 2 & -1 \\ 4 & -2 \end{vmatrix} + 0 \right]$$

$$= -\frac{1}{25} [3(-2 - 8) + 1(-4 + 4)]$$

$$= -\frac{1}{25} (-30)$$

$$= \frac{6}{5}$$

Next,

$$y = -\frac{1}{25} \begin{vmatrix} 2 & 2 & 3 \\ 1 & 1 & -1 \\ 3 & 4 & 0 \end{vmatrix} \qquad \text{replacing the coefficients of } y \\ \text{by the numbers on the right side}$$

$$= -\frac{1}{25} \left[3 \begin{vmatrix} 1 & 1 \\ 3 & 4 \end{vmatrix} - (-1) \begin{vmatrix} 2 & 2 \\ 3 & 4 \end{vmatrix} + 0 \right] \qquad \text{expanding along third} \\ \text{column}$$

$$= -\frac{1}{25} \left[3(4-3) + 1(8-6) \right]$$

$$= -\frac{1}{25} (5)$$

$$= -\frac{1}{5}$$

Finally,

$$z = -\frac{1}{25} \begin{vmatrix} 2 & -1 & 2 \\ 1 & 2 & 1 \\ 3 & -2 & 4 \end{vmatrix} \qquad \text{replacing the coefficients of } z \\ \text{by the numbers on the right side}$$

$$= -\frac{1}{25} \left[2 \begin{vmatrix} 2 & 1 \\ -2 & 4 \end{vmatrix} - (-1) \begin{vmatrix} 1 & 1 \\ 3 & 4 \end{vmatrix} + 2 \begin{vmatrix} 1 & 2 \\ 3 & -2 \end{vmatrix} \right] \qquad \text{expanding} \\ \text{along} \\ \text{first row}$$

$$= -\frac{1}{25} \left[2(8+2) + 1(4-3) + 2(-2-6) \right]$$

$$= -\frac{1}{25} (5)$$

$$= -\frac{1}{5}$$

As a check, suppose we substitute $x = \frac{6}{5}$, $y = -\frac{1}{5}$, and $z = -\frac{1}{5}$ in the second equation. Then

$$\frac{6}{5} + 2\left(-\frac{1}{5}\right) - \left(-\frac{1}{5}\right) = \frac{5}{5} = 1 \quad \checkmark$$

The other equations are checked similary.   ◀

The expansion of a 3 × 3 determinant can also be accomplished by rewriting the first two columns to the right of the determinant and then forming products of elements along the resulting full diagonals, as shown in the following scheme:

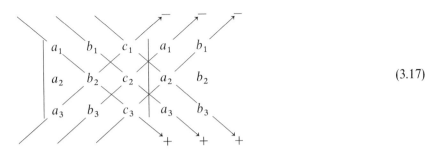

$$(3.17)$$

The leftmost arrow pointing upward yields the product $-a_3b_2c_1$, the leftmost arrow pointing downward yields $+a_1b_2c_3$, and so forth. Comparing the resulting six products to the expansion by minors shows that the terms are identical. Unfortunately, *this scheme does not work with higher-order determinants.*

E X A M P L E **5**   Expand the determinant in Example 2 by using the expansion scheme (3.17).

Solution. Rewriting the first two columns, we get

$$
\begin{vmatrix} -3 & 2 & 1 \\ 4 & -2 & 3 \\ -3 & 1 & 0 \end{vmatrix} \begin{matrix} -3 & 2 \\ 4 & -2 \\ -3 & 1 \end{matrix}
\begin{aligned}
&= +(-3)(-2)(0) + (2)(3)(-3) + (1)(4)(1) \\
&\quad -(-3)(-2)(1) - (1)(3)(-3) - (0)(4)(2) \\
&= 0 - 18 + 4 - 6 + 9 - 0 \\
&= -11
\end{aligned}
$$

You will have to judge whether this method is more convenient than expansion by minors. ◀

 An important application of systems of equations is the analysis of basic electrical circuits by means of Kirchhoff's laws. Since a discussion of setting up systems of equations using Kirchhoff's laws is too lengthy to consider here, we will work with the given system of equations corresponding to a particular circuit in the next example and the exercises.

E X A M P L E **6**   Consider the circuit in Figure 3.13. If the directions of the currents are as indicated in the diagram, then from Kirchhoff's laws the system of equations is given by

$$
\begin{aligned}
I_1 + I_2 - I_3 &= 0 \\
10I_1 - 3I_2 &= 4 \\
3I_2 + 5I_3 &= 2
\end{aligned}
$$

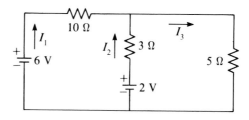

Figure 3.13

By Cramer's rule the solution is found to be $I_1 = \frac{38}{95}$ A, $I_2 = 0$ A, and $I_3 = \frac{38}{95}$ A. (See Exercise 34 in the following exercise set.)

Since the calculated values of I_1 and I_3 are positive, the directions of these currents agree with the directions originally assigned. (If a calculated current is negative, its direction is actually opposite to the direction originally assigned.)

◀

EXERCISES / SECTION 3.6

In Exercises 1–12, evaluate each determinant.

1.
$$\begin{vmatrix} 1 & 0 & 0 \\ 1 & -2 & 3 \\ 1 & -4 & 4 \end{vmatrix}$$

2.
$$\begin{vmatrix} 0 & -1 & 0 \\ 3 & 0 & -6 \\ -7 & 3 & -4 \end{vmatrix}$$

3.
$$\begin{vmatrix} 2 & 1 & 3 \\ 0 & 5 & 0 \\ 3 & 2 & -1 \end{vmatrix}$$

4.
$$\begin{vmatrix} -2 & 3 & -5 \\ 3 & -1 & 4 \\ 0 & -2 & 0 \end{vmatrix}$$

5.
$$\begin{vmatrix} 2 & -1 & 3 \\ 3 & 0 & -5 \\ 10 & 5 & -10 \end{vmatrix}$$

6.
$$\begin{vmatrix} -5 & 12 & 3 \\ 7 & -2 & 1 \\ -3 & 0 & 2 \end{vmatrix}$$

7.
$$\begin{vmatrix} 2 & 3 & 8 \\ -1 & 3 & -2 \\ 5 & -6 & -12 \end{vmatrix}$$

8.
$$\begin{vmatrix} -15 & 20 & 10 \\ 3 & 1 & 7 \\ 9 & -3 & 15 \end{vmatrix}$$

9.
$$\begin{vmatrix} -3 & 2 & 4 \\ 10 & 18 & -20 \\ -11 & -15 & 8 \end{vmatrix}$$

10.
$$\begin{vmatrix} -5 & 0 & 20 \\ 30 & 3 & 40 \\ 5 & -40 & 10 \end{vmatrix}$$

11.
$$\begin{vmatrix} -3 & -4 & -7 \\ 3 & 0 & -6 \\ 10 & 15 & 18 \end{vmatrix}$$

12.
$$\begin{vmatrix} -6 & 13 & -3 \\ -2 & 17 & 6 \\ 0 & 19 & 2 \end{vmatrix}$$

In Exercises 13–22, solve each system by Cramer's rule. (Exercises 13–20 are the same as Exercises 1–8 in Section 3.5, respectively.)

13.
$$\begin{aligned} x + y + 2z &= 9 \\ x \quad\;\; - z &= -2 \\ 2x - y \quad &= 0 \end{aligned}$$

14.
$$\begin{aligned} x - y \quad &= 2 \\ x + 2y - z &= -3 \\ -y + z &= 3 \end{aligned}$$

15.
$$\begin{aligned} 3x \quad\; + 2z &= -1 \\ 4x - y - 2z &= 7 \\ x + y \quad &= 2 \end{aligned}$$

16.
$$\begin{aligned} 2x - 3y \quad &= 8 \\ 3x - y + 2z &= 8 \\ 2x + 3y \quad &= -4 \end{aligned}$$

17.
$$\begin{aligned} x - 2y - z &= 1 \\ 2x - y - 2z &= -1 \\ 3x + y + 2z &= 6 \end{aligned}$$

18.
$$\begin{aligned} 3x - 3y - 2z &= 1 \\ -x + y - 6z &= 3 \\ x + y + 2z &= 3 \end{aligned}$$

19.
$$\begin{aligned} 2x - y + 3z &= 16 \\ 3x + 4y + 2z &= 7 \\ x - 2y - 2z &= -2 \end{aligned}$$

20.
$$\begin{aligned} x - 2y + z &= 1 \\ 2x + y + 2z &= 2 \\ 3x + 3y - 3z &= 2 \end{aligned}$$

21.
$$\begin{aligned} 2x - 3y + z &= 1 \\ x - 2y - 3z &= 1 \\ x - 4y + 2z &= 2 \end{aligned}$$

22.
$$\begin{aligned} 3x - y + 4z &= 4 \\ -x + 2y - 3z &= 6 \\ -2x - 3y + z &= 10 \end{aligned}$$

Solve Exercises 23 and 24; refer to Figure 3.14.

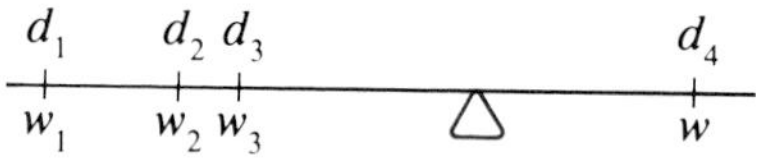

$$d_1 \qquad d_2\; d_3 \qquad\qquad\qquad d_4$$

$$w_1 \qquad w_2\; w_3 \qquad\qquad\qquad w$$

Figure 3.14

23. Find the weights w_1, w_2, and w_3 given the following sets of measurements: $w = 2$ lb and
   a. $d_1 = 4$ ft, $d_2 = 3$ ft, $d_3 = 2$ ft, $d_4 = 5.5$ ft
   b. $d_1 = 3$ ft, $d_2 = 2$ ft, $d_3 = 1$ ft, $d_4 = 3.5$ ft
   c. $d_1 = 5$ ft, $d_2 = 4$ ft, $d_3 = 1$ ft, $d_4 = 5.5$ ft

24. Find the weights w_1, w_2, and w_3 given the following sets of measurements: $w = 11$ N
   a. $d_1 = 2$ m, $d_2 = 1$ m, $d_3 = 3$ m, $d_4 = 2$ m
   b. $d_1 = 4$ m, $d_2 = 3$ m, $d_3 = 3$ m, $d_4 = 3\frac{6}{11}$ m
   c. $d_1 = 6$ m, $d_2 = 2$ m, $d_3 = 6$ m, $d_4 = 4\frac{7}{11}$ m

25. A portion of $5,950 was invested at 8%, another portion at 10%, and the rest at 12%. The total interest income was $635. If the sum of the second investment and twice the first investment was $750 more than the third investment, find the amount invested at each rate.

26. A woman has $16,750 to invest. She decides to invest the largest portion at 8.25% in a safe investment. She invests the smallest portion at 12% in a high-risk investment and the rest at 10%. In fact, the safe investment is only $750 less than the other two combined. Determine the amount invested at each rate, given that the total interest income is $1,605.

27. Three machine parts cost a total of $40. The first part costs as much as the other two together, while the cost of 6 times the second is $2 more than the total cost of the other two. Find the cost of each part.

28. The combined ages of three brothers, Richard, Paul, and Craig, total 24 years. Three years ago Richard's age was one-sixth of his brothers' combined present ages, while Paul's age was one-fourth of what his brothers' combined ages will be two years from now. Find their respective ages.

In Exercises 29–33, find the currents in each of the circuits by solving the system of equations given in each case.

29.
$$\begin{aligned} I_1 + I_2 - I_3 &= 0 \\ I_1 - 2I_2 \quad &= 4 \\ 2I_2 + 3I_3 &= 2 \end{aligned}$$

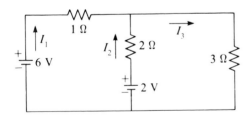

Figure 3.15

30.
$$\begin{aligned} I_1 - I_2 + I_3 &= 0 \\ 2I_1 + 3I_2 \quad &= 2 \\ -3I_2 - I_3 &= -4 \end{aligned}$$

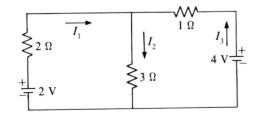

Figure 3.16

31.
$$\begin{aligned} I_1 - I_2 + I_3 &= 0 \\ I_1 + 2I_2 \quad &= 10 \\ -2I_2 - I_3 &= -5 \end{aligned}$$

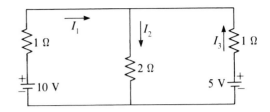

Figure 3.17

32.
$$\begin{aligned} -I_1 + I_2 + I_3 &= 0 \\ -2I_1 - 5I_2 \quad &= -20 \\ 5I_2 - 10I_3 &= 10 \end{aligned}$$

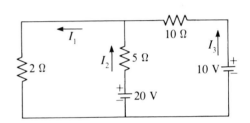

Figure 3.18

33.
$$\begin{aligned} I_1 + I_2 + I_3 - I_4 &= 0 \\ 2I_1 - I_2 \quad &= 1 \\ I_2 - 3I_3 \quad &= 3 \\ 3I_3 + I_4 &= -2 \end{aligned}$$

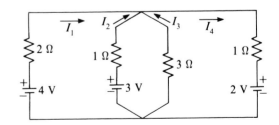

Figure 3.19

34. Solve the system of equations in Example 6.

REVIEW EXERCISES/CHAPTER 3

In Exercises 1–8, evaluate each determinant.

1. $\begin{vmatrix} -2 & 1 \\ 5 & 1 \end{vmatrix}$ **2.** $\begin{vmatrix} 1 & 10 \\ -5 & -3 \end{vmatrix}$

3. $\begin{vmatrix} -8 & 0 \\ 30 & -20 \end{vmatrix}$ **4.** $\begin{vmatrix} 1 & -7 \\ -10 & 1 \end{vmatrix}$

5. $\begin{vmatrix} 1 & 3 & -2 \\ 0 & 0 & -3 \\ 1 & 4 & 5 \end{vmatrix}$ **6.** $\begin{vmatrix} 2 & 7 & 1 \\ 3 & -2 & 0 \\ 7 & 8 & -10 \end{vmatrix}$

7. $\begin{vmatrix} 5 & 6 & -1 \\ -4 & 7 & 2 \\ -3 & 0 & 5 \end{vmatrix}$ **8.** $\begin{vmatrix} 7 & -2 & -11 \\ 10 & 4 & 13 \\ 12 & 6 & 15 \end{vmatrix}$

In Exercises 9–12, solve each system of equations graphically, giving the values to the nearest tenth of a unit.

9. $2x - y = -5$
$2x + 3y = 3$

10. $3x - 6y = 1$
$6x - 3y = 5$

11. $2x - y = 3$
$2x - 5y = -5$

12. $2x - y = 7$
$4x - y = 11$

In Exercises 13–16, solve each system by the method of substitution.

13. $x - 3y = 1$
$3x + 2y = 4$

14. $2x - y = 9$
$x + 2y = 7$

15. $-4x + 2y = 7$
$x - 3y = 2$

16. $5x - 4y = 5$
$2x - y = 1$

In Exercises 17–20, solve each system by the method of addition or subtraction.

17. $3x + 2y = 5$
$x - y = 2$

18. $2x - 4y = 2$
$3x - 5y = 4$

19. $4x + 3y = 9$
$5x + 6y = 12$

20. $2x + 6y = 17$
$-4x + 7y = 23$

In Exercises 21–24, solve each system by the method of determinants.

21. $5x - 7y = 8$
$4x + 2y = 3$

22. $-x + 3y = 2$
$2x + 4y = 3$

23. $2x - 2y = -7$
$2x + 3y = 6$

24. $x - 3y = 5$
$3x + 2y = 2$

In Exercises 25–36, solve the given systems by any method.

25. $x - 3y = 5$
$2x - y = 5$

26. $\dfrac{1}{x} - \dfrac{3}{y} = 4$

$\dfrac{2}{x} - \dfrac{1}{y} = 5$

27. $-x + 2y = 3$
$2x + 4y = 5$

28. $3x - 4y = 1$
$-2x + 6y = 1$

29. $\dfrac{5}{x} - \dfrac{4}{y} = 9$

$\dfrac{2}{x} - \dfrac{1}{y} = 11$

30. $\dfrac{3}{x} + \dfrac{2}{y} = 3$

$\dfrac{4}{x} + \dfrac{7}{y} = 5$

31. $s_1 + 3s_2 = 14$
$2s_1 - s_2 = 0$

32. $2T_1 - T_2 = 0$
$3T_1 + 2T_2 = 7$

33. $2d_1 + 3d_2 = 13$
$d_1 + 2d_2 = 8$

34. $2C_1 + 2C_2 = 9$
$4C_1 + 3C_2 = 14$

35. $2m + 3n = 8$
$4m + 3n = 7$

36. $r + 2s = 0$
$12r - 6s = -5$

In Exercises 37–40, solve the given systems algebraically.

37. $x - y = 2$
$2x + 3z = 11$
$3x + 2y + 4z = 13$

38. $2w_1 + 2w_2 = 3$
$-w_2 + w_3 = 1$
$4w_1 + 2w_3 = 4$

39. $\dfrac{1}{y} + \dfrac{2}{z} = 7$

$\dfrac{2}{x} + \dfrac{3}{z} = 5$

$\dfrac{3}{x} + \dfrac{4}{y} - \dfrac{1}{z} = -5$

40. $2T_1 - 3T_2 + T_3 = 8$
$6T_1 + T_2 - 3T_3 = -2$
$4T_1 - 4T_2 + 3T_3 = 13$

In Exercises 41–44, solve the given systems by using Cramer's rule.

41. $V_1 - V_2 + 4V_3 = -12$
$3V_1 + 2V_2 = 3$
$-2V_1 + 3V_2 - 3V_3 = 17$

42. $a + b + c = 1$
$2a - b + 2c = 2$
$a - b - 4c = 3$

43. $\dfrac{1}{x} - \dfrac{1}{y} - \dfrac{2}{z} = 3$

$\dfrac{2}{x} - \dfrac{4}{z} = 5$

$\dfrac{1}{x} - \dfrac{3}{y} + \dfrac{2}{z} = 2$

44. $T_1 - T_2 = 6$
$2T_1 - 2T_2 + T_3 = 5$
$3T_1 - T_2 + 2T_3 = 1$

45. Determine the value of a that makes the system

$2x + ay = 3$
$4x - 2y = 5$

inconsistent.

46. Show that the system

$$x + 2y + 3z = 3$$

$$4x + 5y + 6z = 1$$

$$7x + 8y + 9z = 2$$

cannot have a unique solution.

47. A portion of $1,014.80 is invested at 10%, another at 8%, and the rest at $12\frac{1}{2}\%$. The $12\frac{1}{2}\%$ investment yields as much interest as the other two investments combined. Find the amount invested at each rate if the total interest income is $103.20.

48. Experimenters found that the velocities (in feet per second) of two moving parts in a machine satisfy the relations

$$\frac{1}{v_1} + \frac{3}{v_2} = 5$$

$$\frac{2}{v_1} + \frac{1}{v_2} = 6$$

Determine v_1 and v_2.

49. Two kinds of milk containing 1% butterfat and 4% butterfat by volume, respectively, are to be mixed to obtain 90 gal of milk containing 2% butterfat. Determine the number of gallons of each required.

50. The perimeter of a triangle is 14 in. The longest side is twice as long as the shortest side and 2 in. less than the sum of the two shorter sides. Find the length of each side.

In Exercises 51 and 52, find the currents given in Figures 3.20 and 3.21, respectively.

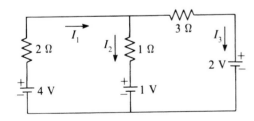

Figure 3.20

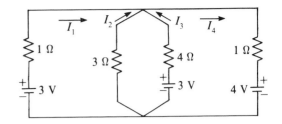

Figure 3.21

51. $I_1 - I_2 - I_3 = 0$

$2I_1 + I_2 \qquad = 3$

$\qquad -I_2 + 3I_3 = -1$

52. $I_1 + I_2 + I_3 - I_4 = 0$

$I_1 - 3I_2 \qquad\qquad = 3$

$\qquad 3I_2 - 4I_3 \qquad = -3$

$\qquad\qquad 4I_3 + I_4 = -1$

53. An experimenter determined that the lever in the figure balances if the weights are positioned as follows:
  a. $d_1 = 3$ in., $d_2 = 2$ in., $d_3 = 1$ in., $d = 2\frac{1}{4}$ in.
  b. $d_1 = 2$ in., $d_2 = 1$ in., $d_3 = 3$ in., $d = 2\frac{1}{2}$ in.
  c. $d_1 = 3$ in., $d_2 = 2$ in., $d_3 = 2$ in., $d = 2\frac{3}{4}$ in.
  Find the weights.

Figure 3.22

54. Two resistors connected in series have a combined resistance of 57.2 Ω. The resistance of one resistor is 15.6 Ω more than that of the other. Find the resistance of each.

1. Subtract $5T_a + 2T_b - 3T_c$ from the sum of $-9T_a - 2T_b + 6T_c$ and $7T_a - 6T_b - 7T_c$.

2. Simplify: $-\{L - [(-L - C) - C] - 4L\}$

3. Write $\sqrt{150}$ in its simplest radical form.

4. Simplify by rationalizing the denominator: $\dfrac{L}{2\sqrt{\pi}}$

5. Simplify: $\dfrac{3p^{-2}q^{-1}}{p^{-6}q^{-4}}$

6. Simplify: $\dfrac{28A^2(-AB)^3}{-7A^3B^3}$

7. Perform the following multiplication:

$$(a - 2b) \cdot (a^2 - 3ab - 2b^2)$$

8. Perform the following division:

$$(2x^3 - 2x - 12) \div (2x - 4)$$

9. Convert the numbers to scientific notation before multiplying: $(0.000000721) \cdot (0.000000089)$

10. Solve for x: $2(2 - 3x) = 5(x - 5)$

11. Solve for x: $\dfrac{1}{4}x - \dfrac{1}{2} = \dfrac{1}{3}x - 2$

12. Solve for t_1: $L = L_1[1 + \beta(t_2 - t_1)]$

13. State the domain of the function $f(x) = \sqrt{4 - x}$.

14. If $f(x) = \sqrt{x^2 + 2}$, find $f(0)$ and $f(\sqrt{2})$.

15. Solve the following system graphically:

$$2x - 3y = 4$$
$$x + 2y = 2$$

16. Solve the following system algebraically:

$$2r_1 - 3r_2 = 7$$
$$3r_1 - 5r_2 = 3$$

17. Solve the following system by determinants:

$$3x + 2y - z = 16$$
$$2x + 3y + 4z = 7$$
$$8x + 5y - 6z = 47$$

18. The respective currents in milliamperes (mA) in the two loops of a certain circuit are $1.00I_1 - 2.00I_2$ and $5.00I_1 - 3.00I_2$. Find the sum of the currents.

19. Simplify the following expression from a problem in statics:

$$-[F_1 + 2F_2 - 4(F_1 + F_2)]$$

20. The distance to the moon is approximately 239,000 mi. Write the distance in scientific notation, using two significant digits.

21. A rectangular computer chip is 3 times as long as it is wide. If the perimeter is 9.6 mm, find the dimensions.

22. The force exerted by a spring is directly proportional to the extension. If a force of 2.72 lb stretches a spring 1.70 in., determine the force required to stretch it 4.10 in.

Introduction to Trigonometry*

4.1 What Is Trigonometry?

The literal meaning of **trigonometry** is "triangle measurement." Although the study of triangles is certainly part of the subject, another important aspect is the study of trigonometric functions and their relationships. This part of the subject is called *analytic trigonometry* and will be of great concern to us later. In this chapter we will concentrate on the basic functions and their applications to right triangles.

Trigonometry originally developed in Egypt and Babylonia, but a systematic study was not undertaken until the second half of the second century B.C. The true originator was Hipparchus of Nicaea (a Greek who lived from 180? to 125? B.C.). Hipparchus constructed the first trigonometric table, which was apparently intended for use in astronomy, and he was also the first mathematician to have used degree measure systematically. Degree measure is discussed in the next section.

4.2 Angles

The purpose of this section is to introduce some of the basic concepts used in trigonometry, particularly angle measurement and angles in standard position. (For a brief review of geometry, see Appendix D.)

Angles in Trigonometry

Angle

An **angle** is defined as a geometric figure consisting of two rays with a common endpoint. (A ray is that portion of a line on one side of a fixed point on the

---

* *Note:* This chapter may be postponed until the end of Chapter 6.

line.) Angle measurement is of primary importance in trigonometry, where an angle is better defined in terms of rotation: Start with a single ray, whose original position is called the **initial side**, and rotate this ray about its endpoint to a new position. This new position is called the **terminal side**. The common endpoint is called the **vertex** of the angle (Figure 4.1). The **measure**, or **size**, of the angle is the amount of rotation.

Initial side
Terminal side
Vertex
Measure

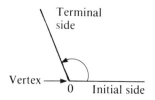

Figure 4.1

Angle Measurement

The most common units of measurement of angles are the **degree** and the **radian**; in this chapter we will consider only degree measure.

> A **degree** is defined as $\frac{1}{360}$ of a complete rotation.

It follows from this definition that a complete rotation, called a *whole angle*, is 360 degrees. Why the number 360 was chosen is not known. However, Hipparchus, a student of Babylonian astronomy, may have chosen the number since it is close to the number of days in a year.

Minute
Second

The Babylonians used a number system with base 60, a system that had a remarkably long life. So it was natural at one time to subdivide an angle of 1 degree into 60 parts, each part called a **minute**. One minute was further subdivided into 60 **seconds**. The symbols °, ′, and ″ are used to denote degrees, minutes, and seconds, respectively. Angles are often denoted by Greek letters, such as θ (theta), α (alpha), β (beta), or γ (gamma).

Angles in Standard Position

Another fruitful idea in trigonometry results from combining the concept of an angle with that of a coordinate system.

> **Standard Position of an Angle:** An angle is said to be in **standard position** if its vertex is at the origin and its initial side is the positive x-axis. (See Figure 4.2.)

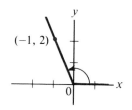

Figure 4.2

Once in standard position, an angle can be described by specifying one point on the terminal side. For example, in Figure 4.2, the point $(-1, 2)$ uniquely determines an angle (between $0°$ and $360°$).

Quadrantal angle

If the terminal side coincides with one of the axes, the angle is called a **quadrantal** angle. For example, the angle θ whose terminal side passes through $(0, -2)$ is quadrantal (Figure 4.3). Note that $\theta = 270°$. Some other quadrantal angles are $0°$, $90°$, $180°$, and $360°$.

Positive and negative angles

Since an angle is defined in terms of rotation, there is no limit to its size. For example, the angle $\theta = 380°$ indicates $20°$ more than one complete rotation. (See Figure 4.4.) If the rotation is **counterclockwise**, the angle is considered **positive**. If the rotation is **clockwise**, the angle is considered **negative**. For example, the angle α in Figure 4.5 is negative.

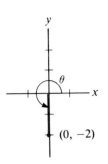

Figure 4.3

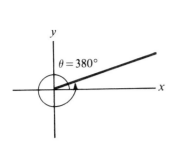

Figure 4.4

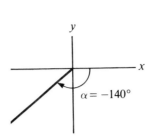

Figure 4.5

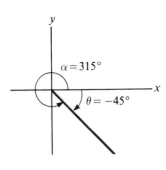

Figure 4.6

Two different angles can have a common terminal side. For example, the angle $\theta = -45°$ has the same terminal side as $\alpha = 315°$ (Figure 4.6). Such angles are called **coterminal**.

Coterminal angles

E X A M P L E **1** Find the smallest positive angle α so that α and $\theta = -45°25'$ are coterminal.

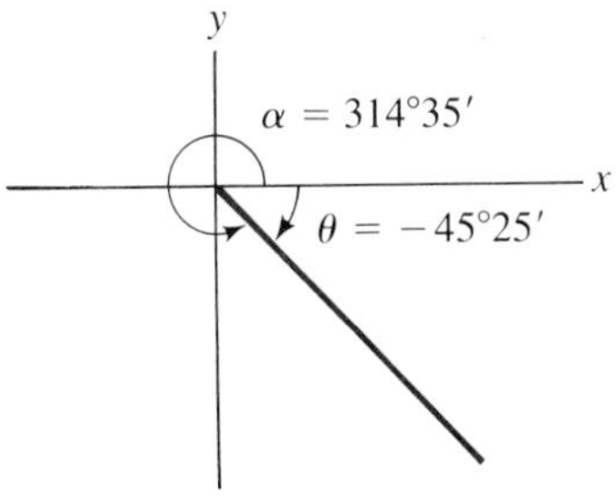

Solution. Write $360°$ as $359°60'$. Now add $359°60'$ to $-45°25'$:

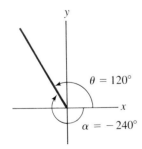

$$\begin{array}{r} 359°60' \\ +(-\ 45°25') \\ \hline \alpha = 314°35' \end{array}$$

(See Figure 4.7.) In general, to find the smallest positive angle coterminal with a given negative angle, add $360°$ or a multiple of $360°$ ($2 \cdot 360°$, $3 \cdot 360°$, and so on). ◀

Figure 4.7

E X A M P L E **2** If $\theta = 120°$, find all negative angles α so that θ and α are coterminal.

Solution. If we subtract $360°$ from $120°$, we get $120° - 360° = -240°$. So $\theta = 120°$ and $\alpha = -240°$ are coterminal. (See Figure 4.8.) Similarly, θ and $\alpha = 120° - 2 \cdot 360° = -600°$ are coterminal. In general, $\alpha = 120° - n \cdot 360°$, $n = 1, 2, 3, \ldots$. ◀

We can see from Examples 1 and 2 that all angles of the form

$$\theta + n \cdot 360°, n = 0, \pm 1, \pm 2, \pm 3, \ldots$$

are coterminal.

Figure 4.8

E X E R C I S E S / S E C T I O N **4.2**

1. Draw the following angles in standard position:
   a. $45°$ **b.** $-45°$ **c.** $150°$ **d.** $-180°$
   e. $-110°$ **f.** $250°$ **g.** $450°$ **h.** $-400°$

2. Draw the following angles in standard position so that the terminal side passes through the given point.
   a. $(-1, 3)$ **b.** $(2, 1)$ **c.** $(3, -4)$ **d.** $(-5, -6)$

In Exercises 3–16, find the smallest positive angle coterminal with the angle specified. (See Example 1.)

3. $-155°$

4. $-30°30'$

5. $-72°40'$

6. $-314°23'$

7. $-179°57'$

8. $-90°12'$

9. $-146°5'$

10. $-89°15'$

11. $-181°54'$

12. $-269°16'$

13. $-325°41'$

14. $-359°2'$

15. $-365°$

16. $-400°$

17. Find a negative angle coterminal with:
   a. $27°$ **b.** $151°$ **c.** $216°40'$ **d.** $312°25'$
   (See Example 2.)

18. Find a negative angle coterminal with:
   a. $65°$ **b.** $118°$ **c.** $226°10'$ **d.** $343°50'$

4.3   Trigonometric Functions

The study of trigonometry, as well as the applications of trigonometry, is based on the ratios of the lengths of two sides of a right triangle. These ratios are used in defining the trigonometric functions of an angle, introduced in this section.

To understand the idea behind the trigonometric ratios, we need to recall the following property of a 30°–60° right triangle.

> In a 30°–60° right triangle, the side opposite the 30° angle is one-half the hypotenuse.

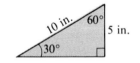

Figure 4.9

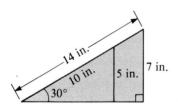

Figure 4.10

Now consider the triangle in Figure 4.9. Note that the side opposite the 30° angle is 5 in. long. Consequently, the hypotenuse must be 10 in. long. Similarly, if the side opposite the 30° angle is 7 in. long, then the hypotenuse must be 14 in. long. (See Figure 4.10.) Now consider the ratio of the side opposite the 30° angle to the hypotenuse. In the first case the ratio is

$$\frac{5 \text{ in.}}{10 \text{ in.}} = \frac{1}{2}$$

and in the second case

$$\frac{7 \text{ in.}}{14 \text{ in.}} = \frac{1}{2}$$

The size of the triangle doesn't matter; the ratio of the side opposite the 30° angle to the hypotenuse is always $\frac{1}{2}$.

A similar statement can be made for any angle. If θ is an angle of a right triangle, then the ratio of the side opposite this angle to the hypotenuse is *always fixed*. This ratio has been given a name: It is called the **sine of angle θ**, abbreviated **sin θ**. Referring to the descriptive labels in Figure 4.11, we now write

$$\sin \theta = \frac{\text{opposite side}}{\text{hypotenuse}}$$

For every angle θ, sin θ is a unique number. (Since θ is assumed to be an angle of a right triangle, this definition makes sense only for acute angles—angles between $0°$ and $90°$. Chapter 7 discusses the trigonometric functions of arbitrary angles.)

The sine of θ is only one of several possible ratios. For example, the ratio of the opposite side to the adjacent side (see Figure 4.11) is called the **tangent of θ**, abbreviated **tan θ**. The ratio of the adjacent side to the hypotenuse is called the **cosine of θ**, abbreviated **cos θ**.

These ratios turn out to be entirely adequate for the study of right triangles. However, for convenience, the reciprocals of the above ratios have also been named. They are called, respectively, the **cosecant of θ (csc θ)**, the **cotangent of θ (cot θ)**, and the **secant of θ (sec θ)**. These ratios should be carefully memorized.

Since the ratios corresponding to each angle are unique, each ratio is a function of the angle. For that reason, we refer to the ratios as the **trigonometric functions**.

Trigonometric Functions:

Name	Abbreviation	Value	
sine of θ	$\sin \theta$	$\sin \theta = \dfrac{\text{opposite}}{\text{hypotenuse}}$	(4.1)
cosine of θ	$\cos \theta$	$\cos \theta = \dfrac{\text{adjacent}}{\text{hypotenuse}}$	(4.2)
tangent of θ	$\tan \theta$	$\tan \theta = \dfrac{\text{opposite}}{\text{adjacent}}$	(4.3)
cosecant of θ	$\csc \theta$	$\csc \theta = \dfrac{\text{hypotenuse}}{\text{opposite}}$	(4.4)
secant of θ	$\sec \theta$	$\sec \theta = \dfrac{\text{hypotenuse}}{\text{adjacent}}$	(4.5)
cotangent of θ	$\cot \theta$	$\cot \theta = \dfrac{\text{adjacent}}{\text{opposite}}$	(4.6)

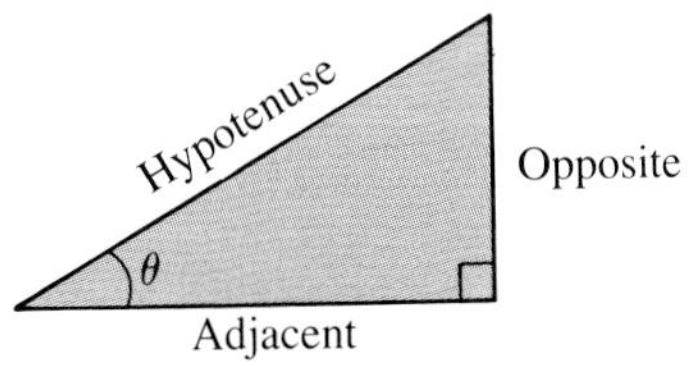

Figure 4.11

In the next section, we will find the values of the functions of an angle. But first we must become acquainted with the trigonometric functions by doing some appropriate exercises.

Let θ be an angle whose terminal side passes through the point $P(x, y)$, which we will place in the first quadrant (Figure 4.12). Dropping a perpendicular line from P to the x-axis, we determine a right triangle. To find the six ratios, we need to compute the length r of the hypotenuse by using the **Pythagorean theorem**:

Pythagorean theorem

$$r^2 = x^2 + y^2 \qquad \text{or} \qquad r = \sqrt{x^2 + y^2}$$

Radius vector

The distance r from P to the origin is called the **radius vector** and is always taken as positive.

Using the coordinates of P and the radius vector r in Figure 4.12, the definitions of the six trigonometric functions can also be stated in the following form:

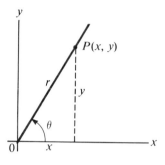

Figure 4.12

Trigonometric Functions of an Angle in Standard Position:

$$\sin \theta = \frac{y}{r} \qquad \csc \theta = \frac{r}{y} \qquad \text{(See Figure 4.12.)}$$

$$\cos \theta = \frac{x}{r} \qquad \sec \theta = \frac{r}{x}$$

$$\tan \theta = \frac{y}{x} \qquad \cot \theta = \frac{x}{y}$$

These definitions will now be illustrated by several examples.

E X A M P L E 1   Find the values of the six trigonometric functions of the angle in standard position whose terminal side passes through $(3, 4)$.

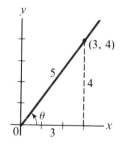

Figure 4.13

Solution. Drop a perpendicular from the point $(3, 4)$ to the x-axis (Figure 4.13). Placing the numbers 3 and 4 along the sides, we see that $r = \sqrt{3^2 + 4^2} = \sqrt{25} = 5$. It now follows directly from the definitions that

$$\sin \theta = \frac{\text{opposite}}{\text{hypotenuse}} = \frac{4}{5} \qquad \text{or} \qquad \sin \theta = \frac{y}{r} = \frac{4}{5}$$

and

$$\cos \theta = \frac{\text{adjacent}}{\text{hypotenuse}} = \frac{3}{5} \qquad \text{or} \qquad \cos \theta = \frac{x}{r} = \frac{3}{5}$$

We also have

$$\tan \theta = \frac{y}{x} = \frac{4}{3} \qquad \csc \theta = \frac{r}{y} = \frac{5}{4}$$

$$\sec \theta = \frac{r}{x} = \frac{5}{3} \qquad \cot \theta = \frac{x}{y} = \frac{3}{4}$$

◄

E X A M P L E **2**   Find the values of the six trigonometric functions of the angle in standard position whose terminal side passes through $(5, \sqrt{3})$.

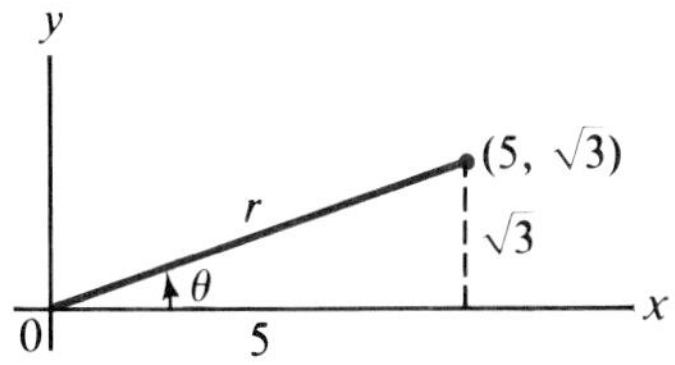

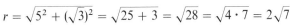

Figure 4.14

Solution. Draw Figure 4.14. From the Pythagorean theorem

$$r = \sqrt{5^2 + (\sqrt{3})^2} = \sqrt{25 + 3} = \sqrt{28} = \sqrt{4 \cdot 7} = 2\sqrt{7}$$

Hence

$$\sin \theta = \frac{y}{r} = \frac{\sqrt{3}}{2\sqrt{7}} \qquad \cos \theta = \frac{x}{r} = \frac{5}{2\sqrt{7}} \qquad \tan \theta = \frac{y}{x} = \frac{\sqrt{3}}{5}$$

$$\csc \theta = \frac{r}{y} = \frac{2\sqrt{7}}{\sqrt{3}} \qquad \sec \theta = \frac{r}{x} = \frac{2\sqrt{7}}{5} \qquad \cot \theta = \frac{x}{y} = \frac{5}{\sqrt{3}} \qquad \blacktriangleleft$$

Although the values in Example 2 are perfectly satisfactory, recall that it is customary to rationalize denominators. For example,

$$\cot \theta = \frac{5}{\sqrt{3}} = \frac{5\sqrt{3}}{\sqrt{3}\sqrt{3}} = \frac{5\sqrt{3}}{(\sqrt{3})^2} = \frac{5\sqrt{3}}{3}$$

and

$$\sin \theta = \frac{\sqrt{3}}{2\sqrt{7}} = \frac{\sqrt{3} \cdot \sqrt{7}}{2\sqrt{7} \cdot \sqrt{7}} = \frac{\sqrt{3}\sqrt{7}}{2(\sqrt{7})^2} = \frac{\sqrt{3 \cdot 7}}{2 \cdot 7} = \frac{\sqrt{21}}{14}$$

and so forth.

In Example 2 we assumed that the given numbers are exact, so the answers are left in radical form. In the next example the numbers are approximate, so the answers must be rounded off to the proper number of significant digits.

E X A M P L E **3**   Use a calculator to find $\sin \theta$ and $\sec \theta$ for θ in Figure 4.15.

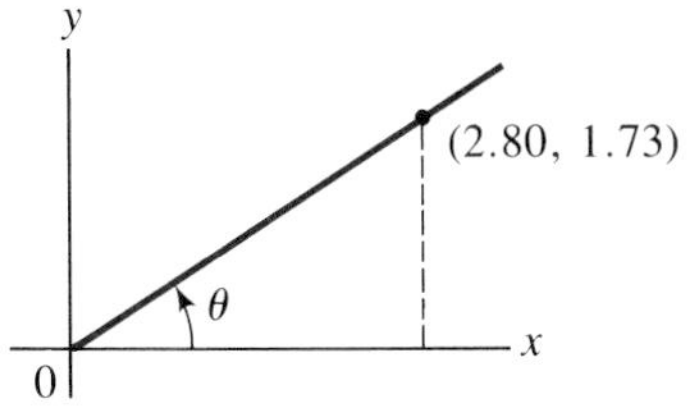

Solution. From the Pythagorean theorem,

$$r = \sqrt{(2.80)^2 + (1.73)^2} = 3.291$$

So

$$\sin \theta = \frac{y}{r} = \frac{1.73}{3.291} = 0.526 \qquad \sin \theta = \frac{\textbf{opposite}}{\textbf{hypotenuse}}$$

and

$$\sec \theta = \frac{r}{x} = \frac{3.291}{2.80} = 1.18 \qquad \sec \theta = \frac{\textbf{hypotenuse}}{\textbf{adjacent}}$$

Figure 4.15

(Both answers are rounded off to three significant digits.)

To avoid round-off errors in the intermediate calculations, it is best to store the value of r in the memory and press $\boxed{\text{MR}}$ whenever the value is needed.

r: 2.80 $\boxed{x^2}$ $\boxed{+}$ 1.73 $\boxed{x^2}$ $\boxed{=}$ $\boxed{\sqrt{}}$ $\boxed{\text{STO}}$

$\sin \theta$:   1.73 $\boxed{\div}$ $\boxed{\text{MR}}$ $\boxed{=}$ $\rightarrow 0.5256222$

$\sec \theta$: $\boxed{\text{MR}}$ $\boxed{\div}$ 2.80 $\boxed{=}$ $\rightarrow 1.1754775$ ◀

If the value of one trigonometric function of an angle is known, the values of the remaining functions can be found, as shown in the next example.

E X A M P L E **4**   Given that $\cos \theta = \frac{3}{7}$, θ in quadrant I, find $\cot \theta$ and $\csc \theta$.

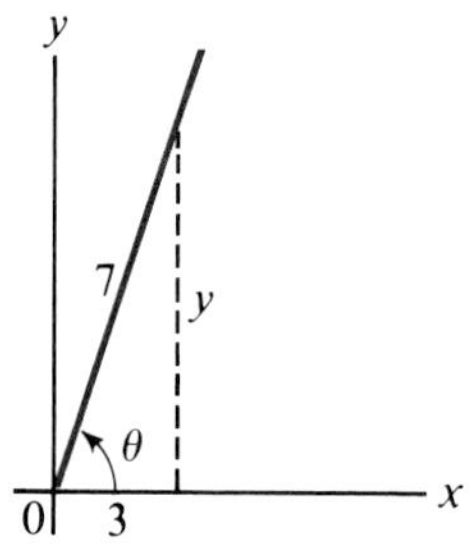

Solution. Since $\cos \theta = $ adjacent/hypotenuse, we need to draw the figure with 3 along the adjacent side and 7 along the hypotenuse (Figure 4.16). Now denote the opposite side by y and use the Pythagorean theorem again. Thus

Figure 4.16

$$y^2 + 3^2 = 7^2$$

$$y^2 = 49 - 9 = 40$$

$$y = \sqrt{40} = \sqrt{4 \cdot 10} = 2\sqrt{10} \qquad \textbf{positive root}$$

Hence

$$\cot \theta = \frac{3}{2\sqrt{10}} = \frac{3\sqrt{10}}{2\sqrt{10} \cdot \sqrt{10}} = \frac{3\sqrt{10}}{2 \cdot 10} = \frac{3\sqrt{10}}{20} \qquad \cot \theta = \frac{\textbf{adjacent}}{\textbf{opposite}}$$

and

$$\csc \theta = \frac{7}{2\sqrt{10}} = \frac{7\sqrt{10}}{2\sqrt{10} \cdot \sqrt{10}} = \frac{7\sqrt{10}}{2 \cdot 10} = \frac{7\sqrt{10}}{20} \qquad \csc \theta = \frac{\textbf{hypotenuse}}{\textbf{opposite}}$$ ◀

Cofunctions

Recall from your study of geometry that two angles A and B are called **complementary angles** if $A + B = 90°$. For example, 30° and 60° are complementary. Suppose α and β are the two acute angles of a right triangle. Since the sum of the angles of a right triangle is 180°, it follows that $\alpha + \beta = 90°$. So the acute angles of a right triangle are complementary.

As we will see in a moment, if α and β are complementary, then $\sin \alpha = \cos \beta$. For that reason the functions sine and cosine are called **cofunctions** (*cosine* means "complementary sine"). Also, the tangent and cotangent are cofunctions, as are the secant and cosecant. These facts yield the following result.

Any trigonometric function of an acute angle is equal to the corresponding cofunction of its complement.

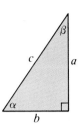

Figure 4.17

To see why this is so, consider angles α and β in Figure 4.17. Note that

$$\sin \alpha = \frac{a}{c}$$

Since a is adjacent to angle β and b is opposite angle β, it follows that

$$\cos \beta = \frac{a}{c}$$

Thus $\sin \alpha = \cos \beta$. Using Figure 4.17, we can also show that $\tan \alpha = \cot \beta$ and $\sec \alpha = \csc \beta$.

E X A M P L E **5** Referring to Figure 4.18, show that $\tan A = \cot B$.

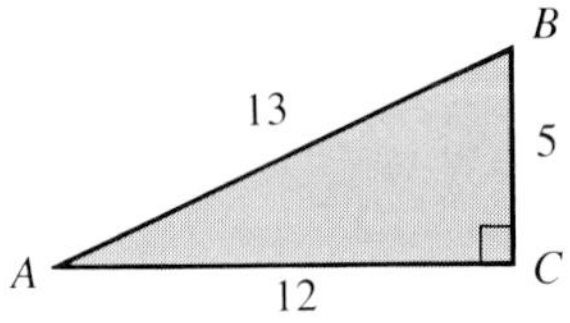

Solution. Since 5 is the side opposite angle A and 12 the side adjacent to angle A, we get

$$\tan A = \frac{5}{12} \qquad \tan A = \frac{\text{opposite}}{\text{adjacent}}$$

Figure 4.18

Since 5 is adjacent to angle B and 12 opposite angle B, we get

$$\cot B = \frac{5}{12} \qquad \cot A = \frac{\text{adjacent}}{\text{opposite}}$$

Thus $\tan A = \cot B$. ◄

Remark. The word *sine* originated from a mistranslation. A half-chord of a circle was called *jiva* by the ancient Hindus. The Arabs took over the word as *jiba*, which resembles *jaib*, meaning "bay." Confusing the two terms, Robert of Chester, a twelfth-century English mathematician, called the half-chord *sinus*, the Latin word for "bay," in his Latin translation. To see the significance of the half-chord, consider the circle of unit radius in Figure 4.19. Since the radius is 1 unit long, the length of the half-chord is numerically equal to $\sin \theta$.

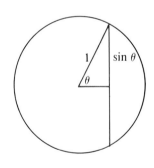

Figure 4.19

E X E R C I S E S / S E C T I O N **4.3**

In Exercises 1–28, find the values of the six trigonometric functions for angles in standard position and having the given points on their terminal sides. (Use a calculator in Exercises 21–28.)

1. $(1, 2)$
2. $(3, 4)$
3. $(5, 2)$
4. $(3, 7)$
5. $(4, 2)$
6. $(8, 2)$
7. $(\sqrt{2}, 4)$
8. $(5, \sqrt{3})$
9. $(6, \sqrt{6})$
10. $(\sqrt{3}, 3)$
11. $(7, \sqrt{10})$
12. $(6, \sqrt{11})$
13. $(\sqrt{11}, 1)$
14. $(1, \sqrt{19})$
15. $(\sqrt{3}, \sqrt{5})$
16. $(\sqrt{5}, \sqrt{7})$
17. $(3\sqrt{2}, \sqrt{7})$
18. $(3\sqrt{5}, 2)$
19. $(3\sqrt{7}, 2\sqrt{2})$
20. $(4\sqrt{2}, 2\sqrt{3})$
21. $(1.50, 2.60)$
22. $(3.60, 7.20)$
23. $(6.14, 2.85)$
24. $(5.07, 9.10)$
25. $(12.4, 16.1)$
26. $(14.2, 8.69)$
27. $(0.52, 0.13)$
28. $(0.72, 0.97)$

In Exercises 29–51, θ is in the first quadrant.

29. If $\sin \theta = \frac{3}{5}$, find $\tan \theta$.   **30.** If $\tan \theta = \frac{3}{2}$, find $\cos \theta$.

31. If $\sec \theta = 2$, find $\sin \theta$.   **32.** If $\csc \theta = 3$, find $\tan \theta$.
Note: $\sec \theta = 2 = \frac{2}{1}$.

33. If $\tan \theta = \frac{5}{3}$, find $\csc \theta$.   **34.** If $\cos \theta = \frac{1}{2}$, find $\cot \theta$.

35. If $\sec \theta = \frac{9}{4}$, find $\cos \theta$.   **36.** If $\cot \theta = \frac{5}{3}$, find $\sec \theta$.

37. If $\csc \theta = \frac{7}{6}$, find $\sin \theta$.   **38.** If $\tan \theta = \frac{4}{7}$, find $\cot \theta$.

39. If $\tan \theta = \sqrt{3}/4$, find $\cos \theta$.

40. If $\sin \theta = \sqrt{3}/2$, find $\cot \theta$.

41. If $\csc \theta = \sqrt{3}$, find $\cos \theta$.

42. If $\cot \theta = 2\sqrt{2}/5$, find $\csc \theta$.

43. If $\cos \theta = \sqrt{3}/6$, find $\sin \theta$.

44. If $\sec \theta = 2\sqrt{7}/3$, find $\tan \theta$.

45. If $\sin \theta = 3\sqrt{10}/10$, find $\sec \theta$.

46. If $\tan \theta = 6\sqrt{7}$, find $\cot \theta$.

47. If $\cos \theta = \sqrt{11}/11$, find $\tan \theta$.

48. If $\sec \theta = \sqrt{17}$, find $\tan \theta$.

49. Given that $\tan \theta = 2$, show that $\sin^2 \theta + \cos^2 \theta = 1$.
Note: $\sin^2 \theta = (\sin \theta)^2$.

50. Given that $\sin \theta = \frac{3}{5}$, show that $1 + \tan^2 \theta = \sec^2 \theta$.

51. Given that $\cos \theta = \frac{1}{3}$, show that $\cot^2 \theta = \csc^2 \theta - 1$.

52. Referring to Figure 4.20, find
a. $\sin A$ and $\cos B$
b. $\tan A$ and $\cot B$
c. $\sec B$ and $\csc A$ (See Example 5.)

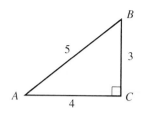

Figure 4.20

53. Referring to Figure 4.21, find
a. $\sin B$ and $\cos A$
b. $\tan B$ and $\cot A$
c. $\sec A$ and $\csc B$

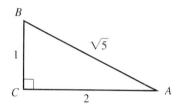

Figure 4.21

4.4   Values of Trigonometric Functions

So far we have considered only the definitions of the trigonometric functions. We now turn to the problem of finding the values of trigonometric functions of specific angles. We will first consider certain special angles and then discuss arbitrary angles.

Special Angles

The values of trigonometric functions of certain special angles—$0°$, $30°$, $45°$, $60°$, and $90°$—can be found by means of diagrams. Consider again the $30°$–$60°$ triangle (Figure 4.22). Since the ratios of the sides are the same regardless of the size of the triangle, we can assign a value of 1 unit to the length of the side opposite the $30°$ angle, thereby making the hypotenuse 2 units long. The length of the remaining side is $x = \sqrt{2^2 - 1} = \sqrt{3}$. For the $60°$ angle, the adjacent side is 1 unit long and the opposite side $\sqrt{3}$.

Special angles

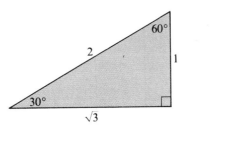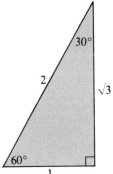

Figure 4.22

Given this information, we can find the values of the trigonometric functions of 30° and 60°.

E X A M P L E **1**   From Figure 4.22

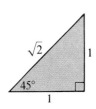

Figure 4.23

$$\sin 30° = \frac{1}{2}, \qquad \cos 30° = \frac{\sqrt{3}}{2}, \qquad \tan 30° = \frac{1}{\sqrt{3}} = \frac{\sqrt{3}}{3}, \qquad \text{and so forth}$$

$$\sin 60° = \frac{\sqrt{3}}{2}, \qquad \cos 60° = \frac{1}{2}, \qquad \tan 60° = \sqrt{3}, \qquad \text{and so forth} \qquad \blacktriangleleft$$

For $\theta = 45°$, we construct the right triangle in Figure 4.23. Note that the hypotenuse is $h = \sqrt{1^2 + 1^2} = \sqrt{2}$.

E X A M P L E **2**   From Figure 4.23

$$\tan 45° = 1, \qquad \sin 45° = \frac{1}{\sqrt{2}} = \frac{\sqrt{2}}{2}, \qquad \text{and so forth} \qquad \blacktriangleleft$$

For the quadrantal angles 0° and 90°, we pick a point on the terminal side and apply the definitions for angles in standard position.

E X A M P L E **3**   Find sec 0°, sin 0°, and cot 0°.

Solution. Since the angle is 0°, the terminal side coincides with the initial side, so no triangle is formed. (See Figure 4.24.)

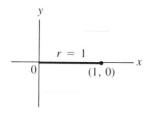

Figure 4.24

We can obtain the desired values anyway by picking a point, say $(1, 0)$, on the terminal side and applying the definitions:

$$\sec 0° = \frac{r}{x} = \frac{1}{1} = 1 \qquad r = 1, x = 1$$

$$\sin 0° = \frac{y}{r} = \frac{0}{1} = 0 \qquad y = 0, r = 1$$

$$\cot 0° = \frac{x}{y} = \frac{1}{0} \qquad \text{(undefined)}$$

(Recall that division by 0 is not allowed, so that $\frac{1}{0}$ is undefined.) ◄

E X A M P L E **4**   Find $\cos 90°$, $\csc 90°$, and $\tan 90°$.

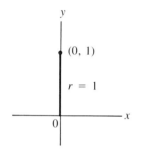

Figure 4.25

Solution. Since 90° is a quadrantal angle, we do not get a triangle. (See Figure 4.25.) As in the case of 0°, we pick a point on the terminal side, say $(0, 1)$, and apply the definitions:

$$\cos 90° = \frac{x}{r} = \frac{0}{1} = 0 \qquad x = 0, r = 1$$

$$\csc 90° = \frac{r}{y} = \frac{1}{1} = 1 \qquad r = 1, y = 1$$

$$\tan 90° = \frac{y}{x} = \frac{1}{0} \qquad \text{(undefined)}$$ ◄

The following table lists all of the special angles just discussed. (Special angles beyond 90° will be studied in Chapter 7.)

	Special Angles				
θ	$0°$	$30°$	$45°$	$60°$	$90°$
$\sin \theta$	0	$\dfrac{1}{2}$	$\dfrac{\sqrt{2}}{2}$	$\dfrac{\sqrt{3}}{2}$	1
$\cos \theta$	1	$\dfrac{\sqrt{3}}{2}$	$\dfrac{\sqrt{2}}{2}$	$\dfrac{1}{2}$	0
$\tan \theta$	0	$\dfrac{\sqrt{3}}{3}$	1	$\sqrt{3}$	undefined
$\csc \theta$	undefined	2	$\sqrt{2}$	$\dfrac{2\sqrt{3}}{3}$	1
$\sec \theta$	1	$\dfrac{2\sqrt{3}}{3}$	$\sqrt{2}$	2	undefined
$\cot \theta$	undefined	$\sqrt{3}$	1	$\dfrac{\sqrt{3}}{3}$	0

Although the table lists the values of all the special angles, it would be ludicrous to memorize the whole set. Even if you could remember them, it is much better in the long run to obtain the values directly from a diagram that you have constructed. In addition, other special angles exist having terminal sides in the different quadrants, which makes memorization of all special angles infeasible anyway.

Common error

Mislabeling the diagram for the $30°$–$60°$ right triangle, such as placing the 1 or $\sqrt{3}$ along the hypotenuse. With a little care, such errors can be avoided since the hypotenuse is necessarily the longest side and the side opposite the $30°$ angle the shortest.

Arbitrary Acute Angles and Using Tables

Our next task is to determine the values of trigonometric functions of arbitrary acute angles. Because these values cannot be obtained from a diagram except for the special angles, tables of values have been devised: One such table is given in Appendix E, a portion of which is shown below. Note that angles from $0°$ to $45°$ are listed in the left column, while the names of the functions are given on top in the first row. Angles from $45°$ to $90°$ are listed in the right column with the names of the functions along the bottom row. The angles themselves are listed in intervals of $10'$.

Tables

Table 4.1

Degrees	Sin θ	Cos θ	Tan θ	Cot θ	Sec θ	Csc θ	
39°00′	0.6293	0.7771	0.8098	1.235	1.287	1.589	51°00′
10	0.6316	0.7753	0.8146	1.228	1.290	1.583	50
20	0.6338	0.7735	0.8195	1.220	1.293	1.578	40
30	0.6361	0.7716	0.8243	1.213	1.296	1.572	30
40	0.6383	0.7698	0.8292	1.206	1.299	1.567	20
50	0.6406	0.7679	0.8342	1.199	1.302	1.561	10
40°00′	0.6428	0.7660	0.8391	1.192	1.305	1.556	50°00′
	Cos θ	Sin θ	Cot θ	Tan θ	Csc θ	Sec θ	Degrees

E X A M P L E **5** Find sec $39°50'$.

Solution. Since the angle is listed in the left column, we need to refer to the secant column listed at the top of the table to get

$$\sec 39°50' = 1.302 \qquad \blacktriangleleft$$

E X A M P L E **6** Find cos $50°20'$.

Solution. Since the angle is listed in the right column, we refer to the names at the bottom of the table. In locating the angle, note that $50°20'$ is *above* $50°$.

$$\cos 50°20' = 0.6383 \qquad \blacktriangleleft$$

It is often necessary to reverse the process: Find the angle, given a function value of the angle.

E X A M P L E **7** Find θ, given that sec $\theta = 1.578$.

Solution. We locate the value 1.578 in the secant column (Table 4.1). Since the name of the function is given at the bottom of the table, we read off the angle in the right column to obtain

$$\theta = 50°40'$$

◄

Interpolation

If the angle is not listed precisely, we obtain the value by "reading between the lines," a process called **interpolation**. Some cases can be seen directly. For example, to find cos 19°15', we would expect the value to be about midway between

$$\cos 19°10' = 0.9446 \qquad \text{and} \qquad \cos 19°20' = 0.9436$$

or

$$\cos 19°15' = 0.9441$$

Strictly speaking, the desired value does not lie exactly between the other two, but the approximation is totally adequate for most purposes.

If the desired value does not lie midway between two given entries, then we must set up a proportion, as illustrated in the next example.

E X A M P L E **8** Find tan 26° 26' from Table 1 in Appendix E.

Solution. The desired value is between tan 26°20' = 0.4950 and tan 26°30' = 0.4986. Now, since 26' is $\frac{6}{10}$ of the way from 20' to 30', we would expect tan 26°26' to be $\frac{6}{10}$ of the way from 0.4950 to 0.4986. Note that 0.4986 − 0.4950 = 0.0036. The calculation can be done systematically as follows:

$$10\left[6\left[\begin{array}{l}\tan 26°20' = 0.4950\\\tan 26°26' = \ldots\\\tan 26°30' = 0.4986\end{array}\right]x\right]0.0036$$

From the diagram we set up the proportion

$$\frac{x}{0.0036} = \frac{6}{10} \qquad \text{or} \qquad x = \frac{6}{10}(0.0036)$$

So $x = 0.00216$. Since we cannot obtain accuracy to more than four decimal places, x is rounded off to 0.0022. So 0.0022 is the difference between 0.4950 and the desired value. It follows that

$$\tan 26°26' = 0.0022 + 0.4950$$
$$= 0.4972 \qquad \blacktriangleleft$$

If the function values decrease as θ increases, special care must be taken when interpolating, as shown in the next example.

E X A M P L E **9**   Find cot 55°3′.

Solution. $10 \left[3 \begin{bmatrix} \cot 55° \ 0' = 0.7002 \\ \cot 55° \ 3' = \ \dots \\ \cot 55°10' = 0.6959 \end{bmatrix} x \right] 0.0043$

Note that $0.7002 - 0.6959 = 0.0043$. The resulting proportion is

$$\frac{x}{0.0043} = \frac{3}{10} \qquad \text{or} \qquad x = \frac{3}{10}(0.0043)$$

So $x = 0.00129 \approx 0.0013$. Since the values of $\cot \theta$ are decreasing, the number 0.0013 has to be subtracted from 0.7002 to ensure that the desired value falls between 0.7002 and 0.6959. Since $0.7002 - 0.0013 = 0.6989$,

$$\cot 55°3' = 0.6989 \qquad \blacktriangleleft$$

Interpolation may also be necessary when determining the angle from the value of a trigonometric function.

E X A M P L E **10**   Find θ such that $\cos \theta = 0.6869$, θ in the first quadrant.

Solution. We look along the cosine column to find the entry or nearest entry. The given value lies between 0.6884 and 0.6862. This observation suggests the following diagram:

$$10 \left[x \begin{bmatrix} \cos 46°30' = 0.6884 \\ \cos \theta \quad\ \ = 0.6869 \\ \cos 46°40' = 0.6862 \end{bmatrix} 0.0015 \right] 0.0022$$

For the proportion we get

$$\frac{x}{10} = \frac{0.0015}{0.0022}$$

yielding $x = 6.8 \approx 7$. It follows that

$$\theta = 46°37' \qquad \blacktriangleleft$$

Finding Values of Trigonometric Functions with a Calculator

Scientific calculators are programmed to evaluate the basic trigonometric functions sine, cosine, and tangent: Put the calculator in **degree mode**, enter the degree measure of the angle, and press

$$\boxed{\text{SIN}}, \quad \boxed{\text{COS}}, \quad \text{or} \quad \boxed{\text{TAN}}$$

To find csc θ, sec θ, or cot θ, we use the following reciprocal relations:

> **Reciprocal Relations:**
>
> $$\csc \theta = \frac{1}{\sin \theta}, \qquad \sec \theta = \frac{1}{\cos \theta}, \qquad \cot \theta = \frac{1}{\tan \theta}$$

Reciprocals are found by pressing $\boxed{1/x}$.

E X A M P L E **11**   Use a calculator to find

a. sin 28°   **b.** cos 72.4°

Solution.

a. To find sin 28°, enter 28 and press $\boxed{\text{SIN}}$. The sequence is

$$28 \ \boxed{\text{SIN}} \rightarrow 0.4694716$$

So sin 28° = 0.4695 to four significant digits.
b. For cos 72.4°, the sequence is

$$72.4 \ \boxed{\text{COS}} \rightarrow 0.3023699$$

Thus cos 72.4° = 0.3024 to four significant digits. ◄

E X A M P L E **12**   Find

a. csc 36.7°   **b.** cot 68.3°

Solution.

a. To find csc 36.7°, we use the reciprocal relation csc θ = 1/sin θ:

$$36.7 \ \boxed{\text{SIN}} \boxed{1/x} \rightarrow 1.6732897$$

So csc 36.7° = 1.673 to four significant digits.
b. Since cot θ = 1/tan θ, the sequence is

$$68.3 \ \boxed{\text{TAN}} \boxed{1/x} \rightarrow 0.3979483$$

◄

Many calculators will not accept angles expressed in degrees and minutes. In that case, the angle must be changed to decimal form. For example, to change $32°15'$ to decimal form, recall that $1° = 60'$. Thus

$$32°15' = 32° + \left(\frac{15}{60}\right)° = \left(32 + \frac{15}{60}\right)° = 32.25°$$

E X A M P L E **13**   Find

a. tan 25°17'   **b.** sec 52°49'

Solution.

a. Since $25°17' = (25 + \frac{17}{60})°$, the sequence is

$$25 \boxed{+} 17 \boxed{÷} 60 \boxed{=} \boxed{\text{TAN}} \rightarrow 0.472342$$

Thus tan 25°17' = 0.4723.

b. Since $52°49' = (52 + \frac{49}{60})°$, the sequence is

$$52 \boxed{+} 49 \boxed{÷} 60 \boxed{=} \boxed{\text{COS}} \boxed{1/x} \rightarrow 1.6546227$$

So sec 52°49' = 1.655 to four significant digits.   ◀

To obtain the angle θ, given the value of a trigonometric function, use

$$\boxed{\text{INV}} \boxed{\text{SIN}}, \qquad \boxed{\text{ARCSIN}}, \qquad \text{or} \qquad \boxed{\text{SIN}^{-1}}$$
$$\boxed{\text{INV}} \boxed{\text{COS}}, \qquad \boxed{\text{ARCCOS}}, \qquad \text{or} \qquad \boxed{\text{COS}^{-1}}$$
$$\boxed{\text{INV}} \boxed{\text{TAN}}, \qquad \boxed{\text{ARCTAN}}, \qquad \text{or} \qquad \boxed{\text{TAN}^{-1}}$$

or similar key. For the remaining functions, we enter the function value and press $\boxed{1/x}$, followed by one of the above operations.

E X A M P L E **14**   Find θ, given that tan θ = 1.289.

Solution. The sequence is

$$1.289 \boxed{\text{INV}} \boxed{\text{TAN}} \rightarrow 52.195862$$

So $\theta = 52.20°$ to the nearest hundredth of a degree.   ◀

E X A M P L E **15**   Find θ, given that sec θ = 2.762.

Solution. Since sec $\theta = 1/\cos \theta$, it follows that $\cos \theta = 1/2.762$. So the sequence is

$$2.762 \boxed{1/x} \boxed{\text{INV}} \boxed{\text{COS}} \rightarrow 68.773455$$

So $\theta = 68.77°$ to the nearest hundredth of a degree.   ◀

If the angle is to be expressed in degrees and minutes, we need to change the decimal form, as shown next.

E X A M P L E **16**   Find θ in degrees and minutes, given that csc $\theta = 2.238$.

Solution. Since csc $\theta = 1/\sin \theta$, we have sin $\theta = 1/2.238$, and the sequence is

$$2.238 \boxed{1/x} \boxed{\text{INV}} \boxed{\text{SIN}} \rightarrow 26.540323$$

Thus $\theta = 26.54°$ in decimal form.

To change the fractional part (0.54) to minutes, we set up the following proportion:

$$\frac{54}{100} = \frac{x}{60} \qquad 0.54 = \frac{54}{100}$$

or

$$x = \frac{54}{100} \cdot 60 = (0.54)(60) = 32.4 \approx 32$$

So $\theta = 26°32'$.   ◀

The availability of scientific calculators seems to have made interpolation obsolete. However, since there are many tables besides trigonometric tables, interpolation is still a valuable technique to master. (See, for example, Exercise 89.)

E X E R C I S E S / S E C T I O N **4.4**

In Exercises 1–20, draw a diagram and determine the exact value of each trigonometric function. (See Examples 1–4.)

1. tan 30°	**2.** sec 60°	**3.** cos 30°	**4.** cos 45°
5. sin 90°	**6.** tan 0°	**7.** csc 30°	**8.** sec 30°
9. sec 0°	**10.** cot 90°	**11.** csc 60°	**12.** csc 90°
13. tan 45°	**14.** csc 45°	**15.** sec 45°	**16.** tan 60°
17. cos 60°	**18.** sin 30°	**19.** sin 45°	**20.** cot 30°

In Exercises 21–28, use Table 1 in Appendix E to find the value of each trigonometric function.

21. sin 21°40'	**22.** cos 39°10'
23. sec 53°50'	**24.** tan 70°30'
25. sin 19°37'	**26.** cos 36°43'
27. cot 63°52'	**28.** sec 75°14'

In Exercises 29–32, use Table 1 in Appendix E to find θ.

29. csc $\theta = 3.356$	**30.** tan $\theta = 2.850$
31. cos $\theta = 0.8956$	**32.** sin $\theta = 0.6764$

In Exercises 33–48, use a calculator to find the value of each trigonometric function. Round off the answers to four significant digits.

33. sin 12.4°	**34.** sin 38.7°
35. cos 47.2°	**36.** cos 19.3°
37. tan 75.26°	**38.** tan 63.91°
39. csc 82.04°	**40.** csc 28.16°
41. sec 1.59°	**42.** sec 5.42°
43. cot 65.42°	**44.** cot 21.16°
45. csc 11.4°	**46.** sec 33.4°
47. cot 52.4°	**48.** cos 41.4°

In Exercises 49–68, use a calculator to find the value of each trigonometric function. Round off the answers to four significant digits.

49. sin 10°40′   **50.** tan 79°30′

51. cot 37°10′   **52.** cos 11°20′

53. sec 84°50′   **54.** csc 15°40′

55. csc 55°15′   **56.** cot 45°5′

57. sec 17°12′   **58.** cos 22°31′

59. cot 18°16′   **60.** sin 85°22′

61. tan 17°35′   **62.** sec 51°43′

63. cos 5°6′   **64.** sin 2°28′

65. tan 38°29′   **66.** csc 46°11′

67. csc 11°55′   **68.** tan 39°14′

In Exercises 69–78, use a calculator to find θ to the nearest hundredth of a degree.

69. cot θ = 0.9377   **70.** tan θ = 0.9217

71. sec θ = 1.595   **72.** csc θ = 1.295

73. sin θ = 0.9971   **74.** sin θ = 0.1984

75. cos θ = 0.1492   **76.** cot θ = 0.1776

77. tan θ = 0.1066   **78.** cos θ = 0.1812

In Exercises 79–88, use a calculator to find θ in degrees and minutes.

79. csc θ = 1.618   **80.** sec θ = 1.933

81. cos θ = 0.9476   **82.** sin θ = 0.8265

83. cot θ = 0.9574   **84.** sec θ = 2.421

85. csc θ = 1.368   **86.** tan θ = 1.100

87. sin θ = 0.3120   **88.** csc θ = 1.091

89. In the table below, W (in grams) represents the experimentally determined weight of potassium chloride that will dissolve in 50 g of water at temperature T (in degrees Celsius).

T	0	10	20	30	40	50	60	70	80
W	13.6	15.4	16.8	18.3	19.7	21.1	22.4	24.0	25.2

Use interpolation to estimate W if
a. $T = 15°C$ **b.** $T = 22°C$ **c.** $T = 37°C$
d. $T = 66°C$ **e.** $T = 79°C$

90. The *index of refraction* μ of a medium is given by

$$\mu = \frac{\sin \theta_i}{\sin \theta_r}$$

where θ_i is the angle of incidence and θ_r is the angle of refraction. If $\theta_i = 47.0°$ and $\theta_r = 24.1°$, find μ.

91. The instantaneous voltage in a coil rotating in a magnetic field is given by

$$V = V_m \cos \phi$$

where V_m is the maximum voltage and ϕ is the angle that the coil makes with the magnetic field. Given that $V_m = 120$ V, find V at the instant when $\phi = 67.4°$.

92. The largest weight W that can be pulled up an inclined plane making an angle θ with the horizontal by a force F is

$$W = \frac{F}{\mu}(\cos \theta + \mu \sin \theta)$$

where μ is the coefficient of friction. Find the largest weight if $F = 49.5$ lb, $\theta = 15.6°$, and $\mu = 0.276$.

4.5   Applications of Right Triangles

In this section we will study some applications of trigonometric functions to problems involving right triangles.

For our first example, suppose a technician needs to find the height of a condemned building. Because of the condition of the building, it is too dangerous to measure the vertical wall directly. At this point trigonometry comes to the rescue: Our technician measures the distance d (Figure 4.26) and the angle θ determined by the ground and the top of the building. Once d and θ have been measured, the height x of the building can be calculated by observing that

$$\tan \theta = \frac{x}{d}$$

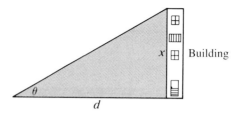

Figure 4.26

so that

$$x = d \tan \theta$$

From this example, we can write the rule for finding the length of an un-known side of a right triangle.

> **To Find the Unknown Side of a Right Triangle:**
>
> **1.** Form a ratio involving the unknown side and a known side of the triangle.
> **2.** Set this ratio equal to the corresponding function of the known angle of the triangle.
> **3.** Solve the resulting equation for the unknown.

We need to illustrate these ideas with some examples. The next three ex-amples use the labels in Figure 4.27. Note that capital letters are used to denote the angles and lowercase letters their respective opposite sides. (Angle C is the right angle.)

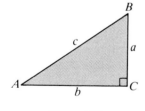

Figure 4.27

> *Reminder:* The terms *opposite* and *adjacent* are relative. For angle A, a is the opposite side and b the adjacent side. For angle B, b is the opposite side and a the adjacent side. Thus $\tan A = a/b$ and $\tan B = b/a$.

E X A M P L E **1**   Given that $B = 36.7°$ and $c = 1.30$, find a. (See Figure 4.28.)

Solution. We first observe that for angle B, **a is the adjacent side** (and b the opposite side). Since **c is the hypotenuse**, we have from the definition of cosine that

$$\frac{a}{c} = \frac{a}{1.30} = \cos 36.7° \qquad \frac{\text{adjacent side}}{\text{hypotenuse}} = \cos \theta$$

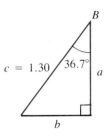

Figure 4.28

It follows that

$$a = 1.30 \cos 36.7° = (1.30)(0.8018) = 1.04$$

rounded off to three significant digits. The sequence is

$$1.30 \boxed{\times} 36.7 \boxed{\text{COS}}\boxed{=} \rightarrow 1.0423083 \qquad \blacktriangleleft$$

Significant digits. Two significant digits in the side measurements corresponds to the nearest degree in the angle measurements; three significant digits corresponds to the nearest multiple of 10′ or tenth of a degree; four significant digits corresponds to the nearest minute or hundredth of a degree. The rules are summarized in the following table:

Table 4.2

Significant digits

Accuracy of degree measurement		Significant digits in side measurement
1°	1°	2
10′	0.1°	3
1′	0.01°	4

E X A M P L E **2**   If $A = 59°34'$ and $a = 5.320$, find b (Figure 4.29).

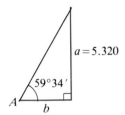

Figure 4.29

Solution. The two sides involved are a and b, so the appropriate ratio is either $\tan A$ or $\cot A$. Thus

$$\frac{b}{5.320} = \cot 59°34' \qquad \frac{\textbf{adjacent side}}{\textbf{opposite side}} = \cot \theta$$

or

$$b = 5.320 \cot 59°34'$$

A sequence is

$$59 \boxed{+} 34 \boxed{\div} 60 \boxed{=} \boxed{\text{TAN}}\boxed{1/x}\boxed{\times} 5.320 \boxed{=} \rightarrow 3.1253873$$

So $b = 3.125$ to four significant digits. $\blacktriangleleft$

E X A M P L E **3**   If $a = 3.78$ and $b = 5.289$, find B (Figure 4.30).

Solution. In this problem two sides are known. Angle B can be determined by noting that

$$\tan B = \frac{b}{a} = \frac{5.289}{3.78} \qquad \tan \theta = \frac{\textbf{opposite side}}{\textbf{adjacent side}}$$

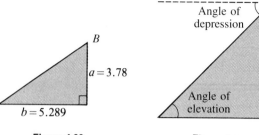

Figure 4.30　　　　　　　　　**Figure 4.31**

The sequence is

$$5.289 \boxed{\div} 3.78 \boxed{=} \boxed{\text{INV}} \boxed{\text{TAN}} \rightarrow 54.446954$$

Since 3.78 has three significant digits, $B = 54.4°$ to the nearest tenth of a degree.

◀

Angle of elevation

Angle of depression

　　To make some of the applied problems easier to state, we need to introduce two basic terms. If an object is located above a horizontal plane, then the angle between the horizontal and the line of sight is called the **angle of elevation**. (See Figure 4.31.) If the observer is looking down at an object, then the angle between the horizontal and the line of sight is called the **angle of depression**. (See Figure 4.31.) Note that the angle of elevation in Figure 4.31 is equal to the angle of depression.

E　X　A　M　P　L　E　**4**　An observer standing on top of a building is looking down at an intersection. He wants to know the distance from the foot of the building to the middle of the intersection. If the building is 22.0 ft high and the angle of depression is 25.3°, what is the desired distance?

Solution. In Figure 4.32 let us denote the unknown side by x. From the given angle of depression, note that the angle on the left is also 25.3°. It follows that

$$\frac{x}{22.0} = \cot 25.3° \qquad \frac{\textbf{adjacent}}{\textbf{opposite}} = \cot \theta$$

$$x = 22.0 \cot 25.3°$$
$$= (22.0)(2.1155) = 46.541$$

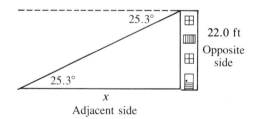

Figure 4.32

Since 22.0 has three significant digits, the desired distance is rounded off to 46.5 ft. ◄

E X A M P L E 5 A 40-ft ladder is leaning against a wall, making an angle of 71° with the ground. How high does the ladder reach up the wall?

Solution. Draw Figure 4.33 and denote the unknown side by x. It follows that

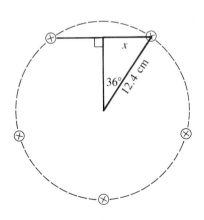

$$\frac{x}{40} = \sin 71° \qquad \frac{\text{opposite}}{\text{hypotenuse}} = \sin \theta$$

$$x = 40 \sin 71°$$

$$= (40)(0.9455)$$

$$= 37.82 \text{ ft}$$

Assuming the ladder is of standard length (accurate to the nearest foot), we get $x = 38$ ft to two significant digits.

Note that in this problem we could also use the cosecant to find x:

$$\frac{40}{x} = \csc 71°$$

or

$$x = \frac{40}{\csc 71°} = 38 \text{ ft}$$ ◄

40 ft Hypotenuse x Opposite side 71°

Figure 4.33

E X A M P L E 6 Five screws are equally spaced on a circle of radius 12.4 cm. Find the center-to-center distance between two adjacent screws.

Solution. The arrangement of screws is shown in Figure 4.34.

Let x denote half the distance between the two adjacent screws on top in Figure 4.34. The central angle (vertex at the center of the circle) determined by the two adjacent screws is $360° \div 5 = 72°$. So for the triangle in the figure, the angle is $\frac{1}{2}(72°) = 36°$. We now get

$$\frac{x}{12.4} = \sin 36° \qquad \frac{\text{opposite}}{\text{hypotenuse}} = \sin \theta$$

Solving for x:

$$x = 12.4 \sin 36° = 7.29 \text{ cm}$$

It follows that the center-to-center distance is $2(7.29 \text{ cm}) = 14.6$ cm. ◄

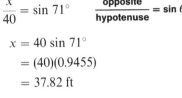

x 36° 12.4 cm

Figure 4.34

Some problems in trigonometry involve two triangles, each with an unknown side. The resulting relationships can lead to a system of two equations, as shown in the next example.

E X A M P L E **7**   The angle of depression from the top of a building to a park bench is 19°25′, and that from a window 20.00 ft below the top to the bench is 10°43′. How far is the bench from the foot of the building?

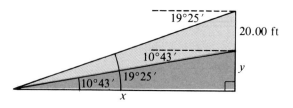

19°25′

20.00 ft

10°43′

y

10°43′ 19°25′

10°43′

x

Figure 4.35

Solution. Draw Figure 4.35 and label the angles on the left. (Recall that the angle of depression is equal to the corresponding angle of elevation.) There are two unknown sides, labeled x and y, and two triangles. Consequently, we need to find two relationships that lead to a system of two equations. From the respective triangles, we get

$$\frac{y + 20.00}{x} = \tan 19°25′$$

$$\frac{y}{x} = \tan 10°43′ \qquad \tan \theta = \frac{\text{opposite}}{\text{adjacent}}$$

$$\frac{y + 20.00}{x} = 0.35248$$

$$\frac{y}{x} = 0.18925$$

$$y + 20.00 = 0.35248x \qquad \text{multiplying both sides by } x$$
$$y = 0.18925x$$

$$20.00 = 0.16323x \qquad \text{subtracting}$$
$$x = 122.5 \text{ ft}$$

rounded off to four significant digits. (The other unknown is not needed.) ◄

E X E R C I S E S / S E C T I O N   **4.5**

In Exercises 1–12, refer to Figure 4.36.

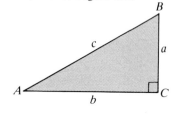

B

c

a

A

b

C

Figure 4.36

1. $A \doteq 37.3°$, $c = 2.76$; find a

2. $B = 65.7°$, $a = 3.96$; find c

3. $B = 9.5°$, $b = 12.3$; find a

4. $A = 15.8°$, $b = 19.2$; find c

In Exercises 5–12, find the remaining parts of each triangle.

5. $A = 30.52°$, $b = 3.520$ **6.** $B = 50.43°$, $c = 5631$

7. $a = 524$, $c = 942$ **8.** $b = 22.6$, $c = 38.3$

9. $A = 22°40'$, $c = 2.00$

10. $B = 65°9'$, $b = 0.09410$

11. $B = 7°55'$, $a = 0.09000$

12. $A = 5°13'$, $a = 0.008200$

13. From a point 60.4 ft from the base of a building, the angle of elevation of the top is 26.1°. Find the height of the building.

14. A pole casts a shadow 11.6 ft long. The angle of elevation of the sun is 63.7°. Find the height of the pole.

15. The angle of depression from the top of a building 19.4 m high to a rock on the ground is 15.7°. Find the straight-line distance from the top of the building to the rock.

16. The angle of depression from a window 15 m above the ground to a small bush is 28°. Determine the straight-line distance from the window to the bush.

17. A tree casts a shadow 58.2 ft long. If the angle of elevation of the sun is 48°10', how tall is the tree?

18. From a vertical cliff 119 m high, the angle of depression of a buoy is 15°50'. Find the distance from the buoy to the base of the cliff.

19. An 8.0 ft ladder is leaning against a building. If the ladder makes an angle of 61° with the ground, how high does the ladder reach on the wall?

20. A guy wire attached to the top of a tower makes an angle of 63°20' with the ground. If the tower is known to be 50.2 m high, find the length of the wire.

21. Suppose a 19.0-ft telephone pole leans 11.3° from the vertical. What is the length of the shadow if the sun is directly overhead?

22. The Great Pyramid of Cheops has a base of 755 ft and originally had a height of 481 ft. (The top 31 ft have been destroyed.) What angle do the sides make with the ground?

23. In an alternating current circuit containing a resistance R and an inductive reactance X_L, the *phase angle* ϕ is the angle between the impedance Z and the resistance. Find the phase angle if $R = 10.20\ \Omega$ and $X_L = 6.21\ \Omega$. (See Figure 4.37.)

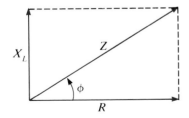

Figure 4.37

24. A rectangular chip in a computer circuit measures 1.90 mm by 1.08 mm. Find the angle that the diagonal makes with the longer side.

25. Determine angle α in Figure 4.38.

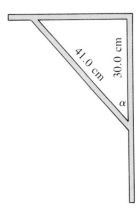

Figure 4.38

26. Find the volume of the water in the conical tank in Figure 4.39.

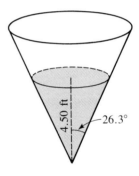

Figure 4.39

27. If a pendulum 17.9 in. long swings through an arc of 15°20', how high will the bob rise above its lowest position?

28. The metal plate in Figure 4.40 has the shape of a regular pentagon. Find the perimeter of the plate.

Figure 4.40

29. From a point 245 ft from the base of the Empire State Building and in the same horizontal plane, the angle of elevation of the top of the building is 78.9°. Find the height of the building.

30. Eight screws are equally spaced on the rim of a circular plate. The distance from center to center between two adjacent screws is 2.58 cm. Find the radius of the plate.

31. A platform 15.0 ft high is 211 ft from a building. From the top of the platform, the angle of elevation of the top of the building is 29°30′. Find the height of the building.

32. A woman 5 ft 8 in. tall observes that the angle of elevation of the top of a flagpole is 71°. If she is standing 20 ft from the base, how high is the flagpole?

33. A building is located 151.0 ft from a tall lookout tower. From the top of the building, the angle of depression of the tower base is 25.53°, and the angle of elevation of the top is 61.08°. Find the height of the tower.

34. A war memorial features the figure of a soldier on a pedestal. At a point 19.5 ft from the base of the pedestal, the angles of elevation of the foot and head of the figure are 46.2° and 57.0°, respectively. Find the height of the figure.

35. From the top of a lighthouse 29.30 m above the surface of the water, the angles of depression of two ships due east are 12.42° and 15.53°, respectively. Find the distance between the ships.

36. The angle of elevation of a mountaintop is observed to be 4°50′. From a second point 9.20 mi closer to the mountain, the angle of elevation is 6°50′. How high above the plane is the mountaintop? (See Example 7.)

37. In finding the height of the Egyptian pyramid at Chephren, it is observed that the sides make an angle of 53.0° with the ground. From a point 99.9 ft from the base of the pyramid and on a line perpendicular to the midpoint of the base, the angle of elevation of the top is measured to be 46.0°. Find the height of the pyramid. (See Example 7.)

38. An observer in a balloon measures the respective angles of depression to two distant intersections on the same side of the balloon as 18°21′ and 25°32′. If the intersections are

known to be 1.10 mi apart and in the same vertical plane with the balloon, how high is the balloon?

39. A swimming pool measures 20.1 ft by 45.0 ft. (See Figure 4.41.) If the bottom is flat and inclined 14.0° with the horizontal, what is its area?

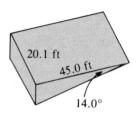

20.1 ft

45.0 ft

14.0°

Figure 4.41

40. A dealer in sand and gravel wants to estimate the volume of a conical pile of sand. The circumference of the base measures 20 ft. From one point on the ground, the angle of elevation of the top is 26°; from another point 10 ft further back and in line with the first point and the center of the base, the angle of elevation is 20°. Find the volume of the pile.

41. A castle wall is surrounded by a moat of unknown width. An architect wants to measure the height of the wall by marking off two points 10.0 m apart along a line perpendicular to the wall. The respective angles of elevation of the top were found to be 24.5° and 16.3°. Find the height of the wall.

42. Archimedes of Syracuse (a Greek who lived between 287? and 212 B.C.) made the best estimate of π in antiquity. He obtained his estimate by using a polygon of 96 sides inscribed in a circle. Show that this method yields the value to two decimal places, that is, 3.14. Assuming the radius of the circle to be 1 unit long, show that for an inscribed polygon of n sides, $\pi \approx n \sin(180°/n)$.

REVIEW EXERCISES/CHAPTER 4

In Exercises 1–4, find the values of the six trigonometric functions of the angle whose terminal side passes through the given point.

1. (3.1, 1.6)

2. $(1, \sqrt{7})$

3. $(\sqrt{5}, 2)$

4. $(1, 2\sqrt{2})$

5. If $\sin \theta = \frac{1}{5}$, find $\cos \theta$.

6. If $\sec \theta = \sqrt{7}/2$, find $\csc \theta$.

7. If $\cot \theta = \sqrt{3}/5$, find $\sin \theta$.

8. If $\csc \theta = \sqrt{11}$, find $\tan \theta$.

9. If $\cos \theta = \sqrt{3}/2$, find $\cot \theta$.

10. If $\sin \theta = \sqrt{5}/3$, find $\tan \theta$.

11. If $\tan \theta = \sqrt{2}/6$, find $\sec \theta$.

12. If $\csc \theta = \sqrt{7}/2$, find $\sin \theta$.

13. If $\sin \theta = a/2$, find $\cos \theta$.

14. If $\sec \theta = b$, find $\tan \theta$.

15. Given that $\sin \theta = \frac{1}{3}$, show that $\sin^2 \theta + \cos^2 \theta = 1$.

16. Given that $\sec \theta = 3$, show that $\tan^2 \theta = \sec^2 \theta - 1$.

In Exercises 17–24, draw a diagram and determine the exact value of each trigonometric function.

17. sin 60°   **18.** sin 90°   **19.** cot 45°

20. sec 30°   **21.** tan 0°   **22.** csc 45°

23. cos 90°   **24.** cot 60°

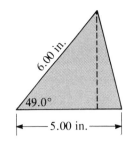

Figure 4.42

In Exercises 25–30, use Table 1 in Appendix E to determine the value of each trigonometric function.

25. sin 32°15′   **26.** tan 62°28′   **27.** sec 55°19′

28. cos 9°58′   **29.** csc 8°8′   **30.** cot 75°31′

In Exercises 31–36, use Table 1 in Appendix E to find θ.

31. sin θ = 0.1865   **32.** cos θ = 0.1914

33. tan θ = 0.1941   **34.** sec θ = 3.157

35. csc θ = 1.975   **36.** cot θ = 0.2525

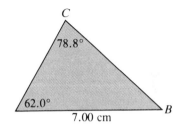

Figure 4.43

In Exercises 37–44, use a calculator to find the value of each trigonometric function. Round off the answers to four significant digits.

37. sin 53.7°   **38.** cos 75.21°   **39.** sec 21.48°

40. csc 43.2°   **41.** tan 81.36°   **42.** cot 9.43°

43. cos 16°37′   **44.** csc 24°42′

In Exercises 45–50, use a calculator to find θ to the nearest hundredth of a degree (θ in quadrant I).

45. cos θ = 0.6819   **46.** csc θ = 3.614

47. tan θ = 1.843   **48.** sec θ = 2.074

49. cot θ = 0.5039   **50.** sin θ = 0.6706

In Exercises 51 and 52, use a calculator to find θ in degrees and minutes (θ in quadrant I).

51. cot θ = 1.374   **52.** sec θ = 2.708

53. A television tower 251 ft high is supported by guy wires extending from the top to points on the ground 94.0 ft away from the base. What angle do the wires make with the ground?

54. A regular pentagon is inscribed in a circle of radius 2.13 cm. Find the length of one side.

55. Find the area of the triangle in Figure 4.42.

56. Find the length of side *BC* in Figure 4.43.

57. Find the radius of the circular portion of the machine part pictured in Figure 4.44. (The dotted line is tangent to the circle.)

58. The angle of depression from the top of a building to a fire hydrant 56.30 ft from the base of the building is 66°39′. How tall is the building?

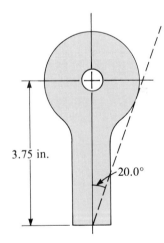

Figure 4.44

59. John wants to determine the height of his apartment building, which is located near an observation tower 175 ft high. From the top of the building he measures the angle of elevation of the top of the tower to be 38.0° and the angle of depression of the tower base to be 25.0°. What is the height of his apartment building?

60. A tree is standing on top of a cliff 162.0 ft high. The angle of elevation of the bottom of the tree is 41°38′, and the angle of elevation of the top of the tree is 52°17′. How tall is the tree?

61. Two observers located 2,100 ft apart measure the angle of elevation of a weather balloon located between them and directly over the line joining them. If the angles are 42°10′ and 51°20′, find the height of the balloon above the ground.

62. Find the area of the triangular machine part shown in Figure 4.45.

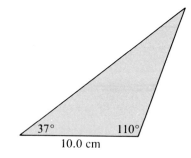

Figure 4.45

Factoring and Fractions

At this point we need to discuss some methods that make certain computations much faster and more accurate. In particular, we need to improve our ability to perform certain multiplications, usually called *special products*. These products occur so often in technology and mathematics that we must be thoroughly familiar with them. Special products lead to the important inverse operation called *factoring*, which is also discussed in this chapter.

5.1 Special Products

Two special products frequently encountered in mathematics and its applications are the *distributive law*

$$a(b + c) = ab + ac$$

and the *product of the sum and difference of two terms*

$$(a - b)(a + b) = a^2 - b^2$$

The Distributive Law

$$a(b + c) = ab + ac$$

E X A M P L E **1**   **a.** $2a(r + s) = \mathbf{2a}(r) + \mathbf{2a}(s) = 2ar + 2as$
                              b. $xy^2(x^2 - xy) = \mathbf{xy^2}(x^2) + \mathbf{xy^2}(-xy) = x^3y^2 - x^2y^3$ ◀

E X A M P L E **2** $ab(3a - 2b - 3ab^2) = \mathbf{ab}(3a - 2b - 3ab^2)$
$$= \mathbf{ab}(3a) + \mathbf{ab}(-2b) + \mathbf{ab}(-3ab^2)$$
$$= 3a^2b - 2ab^2 - 3a^2b^3$$ ◀

Product of the Sum and Difference of Two Terms

$$(a - b)(a + b) = a^2 - b^2$$

Note that upon multiplying $(a - b)(a + b)$, the middle terms add to zero:

$$(a - b)(a + b) = a^2 + ab - ab - b^2 = a^2 - b^2$$

Thus, *the product of the sum and difference of two terms is equal to the difference of their squares.*

E X A M P L E **3** $(3V - 2R)(3V + 2R) = (3V)^2 - (2R)^2$ **a = 3V, b = 2R**

$$= 9V^2 - 4R^2$$ ◀

E X A M P L E **4** $(2x^2 - 5y^3)(2x^2 + 5y^3) = (2x^2)^2 - (5y^3)^2$ **a = 2x², b = 5y³**

$$= 4x^4 - 25y^6$$ ◀

E X E R C I S E S / S E C T I O N   **5.1**

Find the following products in one step.

1. $2(x - 3y)$

2. $4x(x^2 - 2y)$

3. $2a(a^3 - ab^2 - 3ac)$

4. $5x^2y^2(2x^2y^2 - 1)$

5. $V(R_1^2 - R_2^2)$

6. $V^2(3S^3 - 2V^3)$

7. $2P(Q^2 - P^2 + 1)$

8. $3T_1(2T_1 + aT_2 - bT_3)$

9. $2b^2c(a - 2b - 1)$

10. $rs(rs^2 + 3rs + 1)$

11. $(x - 2y)(x + 2y)$

12. $(a + 3b)(a - 3b)$

13. $(6A - B)(6A + B)$

14. $(5P - 3Q)(5P + 3Q)$

15. $(a^2 + b)(a^2 - b)$

16. $(m - n^2)(m + n^2)$

17. $(2x^2 - y^2)(2x^2 + y^2)$

18. $(3M + 2N)(3M - 2N)$

19. $(S - 2T^2)(S + 2T^2)$

20. $(5L^2 + T^3)(5L^2 - T^3)$

21. $(xy + 1)(xy - 1)$

22. $(1 - pq)(1 + pq)$

23. $(3RS^2 - 1)(3RS^2 + 1)$

24. $(5NP - x)(5NP + x)$

5.2   Common Factors and Factors of a Binomial

Factoring plays an important role in algebra. We must be able to factor in order to simplify certain algebraic expressions, to reduce a fraction to lowest terms, and to carry out the four fundamental operations with fractions. In this section we will see how factoring can be performed on some types of polynomials. All cases considered involve certain special products in reverse: common factors, the difference of two squares, and the sum and difference of two cubes.

Factoring is essentially multiplication in reverse: We start with the product and go backward to recover the factors. For example, we know from the last section that

$$3(a - 2b) = 3a - 6b$$

Therefore *factoring* the binomial $3a - 6b$ means to write $3a - 6b$ as $3(a - 2b)$.

Prime factors

We say that a polynomial has been *factored* if it is written as a product of prime factors. A factor is **prime** or **irreducible** if it contains no other factors than itself and 1.

Common Factors

If every term of a polynomial contains the same factor, this **common factor** may be factored out of the polynomial. Consider, for example, the expression $2x + 4y$. Since each term has a factor 2, we may write

$$2x + 4y = 2(x) + 2(2y) = 2(x + 2y)$$

by the distributive law in reverse. Similarly, the expression $6x^2 - 9x^3$ has two terms with the common factor $3x^2$. So

$$6x^2 - 9x^3 = 3x^2(2) - 3x^2(3x) = 3x^2(2 - 3x)$$

The general case is just the distributive law stated in reverse.

Common Factor:

$$ab + ac = a(b + c)$$

E X A M P L E **1**

a. $2x - 8y = 2x + 2(-4y)$
$\qquad\qquad = 2(x - 4y) = 2(x - 4y)$ **common factor 2**

b. $2x^2 - 8xy = 2x(x) + 2x(-4y)$
$\qquad\qquad = 2x(x - 4y) = 2x(x - 4y)$ **common factor 2x**   ◀

Removing a common factor may leave the factor 1, as shown in the next example.

E X A M P L E **2**   Factor

$\qquad$ **a.** $3xy - 3$ $\qquad$ **b.** $2y^2 + y$

Solution.

a. $3xy - 3 \cdot 1 = 3(xy - 1) = 3(xy - 1)$

b. To factor $2y^2 + y$, observe that the coefficient of y is 1, even though the 1 is not explicitly written:

$$2y^2 + y = 2y^2 + 1y = y(2y) + y(1) = y(2y + 1) = y(2y + 1)$$ ◀

Common factors can be removed even if the polynomial contains more than two terms.

E   X   A   M   P   L   E   **3**   **a.** $2x^2y - 4xy^2 + 6x^2y^2 = 2xy(x) + 2xy(-2y) + 2xy(3xy)$
$$= 2xy(x - 2y + 3xy)$$

b. $9a^3b^3 - 15a^2bc + 3a^2b = 3a^2b(3ab^2) + 3a^2b(-5c) + 3a^2b(1)$
$$= 3a^2b(3ab^2 - 5c + 1)\qquad◄$$

Common error

Failing to factor out all the common factors. For example,

$$5a^2bc - 15ab^2c - 5a^2b^2c^2 = abc(5a - 15b - 5abc)$$

While this result is not really incorrect, the second factor is not prime. To obtain the necessary *prime* factors, the 5 must also be factored out to yield

$$5abc(a - 3b - abc)$$

Difference of Two Squares

Another factorable expression comes from the special product of the sum and difference of two terms

$$(a + b)(a - b) = a^2 - b^2$$

We see that the product is the difference of the squares of the terms. Consequently, factoring the difference of two squares yields the product of the sum and difference of the terms. For example, $x^2 - 9y^2 = x^2 - (3y)^2$. So x^2 is the square of x and $9y^2$ is the square of $3y$. We therefore get

$$x^2 - (3y)^2 = (x - 3y)(x + 3y)$$

Difference of Two Squares:

$$a^2 - b^2 = (a + b)(a - b) \qquad\qquad (5.1)$$

E   X   A   M   P   L   E   **4**   Factor $4x^2 - b^2$.

Solution. Since $4x^2 - b^2 = (2x)^2 - b^2$, we see that $4x^2$ is the square of $2x$ and that b^2 is the square of b. It follows that

$$4x^2 - b^2 = (2x + b)(2x - b)$$

or

$$4x^2 - b^2 = (2x - b)(2x + b) \qquad\qquad ◄$$

If the expressions are more complicated, a preliminary step can be helpful. In particular, a binomial that is a difference of two squares may need to be rewritten to make this fact clear. Some examples follow.

E   X   A   M   P   L   E   **5**   **a.** $9a^2 - 16b^4 = (3a)^2 - (4b^2)^2 = (3a - 4b^2)(3a + 4b^2)$
b. $25a^2b^4 - 1 = (5ab^2)^2 - 1^2 = (5ab^2 - 1)(5ab^2 + 1)$ ◄

E   X   A   M   P   L   E   **6**   Factor $1 - (A - B)^2$.

Solution. Since $1 = 1^2$, the form is a difference of two squares with $a = 1$ and $b = A - B$ in formula (5.1). *In the first step, keep the parentheses around the second term* to avoid possible errors with signs:

$$1^2 - (A - B)^2 = [1 + (A - B)][1 - (A - B)]$$
$$= (1 + A - B)(1 - A + B)$$ ◄

The sum of two squares cannot be factored. Thus $x^2 + 4y^2$ is nonfactorable. A nonfactorable polynomial is called **prime** or **irreducible**.

Sum or Difference of Two Cubes

The factoring cases given next are called, respectively, the **sum** and **difference of two cubes**. They are sometimes listed as special products.

Sum of Two Cubes:

$$a^3 + b^3 = (a + b)(a^2 - ab + b^2) \qquad (5.2)$$

Difference of Two Cubes:

$$a^3 - b^3 = (a - b)(a^2 + ab + b^2) \qquad (5.3)$$

To check the sum and difference of two cubes, we multiply the factors on the right in each case:

$$
\begin{array}{ll}
\quad a^2 - ab + b^2 & \quad a^2 + ab + b^2 \\
\quad\quad\quad a + b & \quad\quad\quad a - b \\
\hline
a^3 - a^2b + ab^2 & a^3 + a^2b + ab^2 \\
\quad\quad a^2b - ab^2 + b^3 & \quad\quad - a^2b - ab^2 - b^3 \\
\hline
a^3 \quad\quad\quad\quad + b^3 & a^3 \quad\quad\quad\quad - b^3
\end{array}
$$

To remember these forms, note that the binomials have identical signs in each case: The factors of $a^3 + b^3$ are $a + b$ and a trinomial, and those of $a^3 - b^3$ are $a - b$ and a trinomial. Note especially that the trinomials are irreducible.

E X A M P L E **7**   Factor the following expressions:

a. $8x^3 + 1$ **b.** $x^3y^3 - 64t^6$

Solution. All the terms should be written as cubes for two reasons: (1) to check if the binomial really *is* a sum or difference of two cubes and (2) to facilitate the factoring procedure.

a. First note that $8x^3 = (2x)^3$ and $1 = 1^3$, so that $8x^3 + 1 = (2x)^3 + 1^3$. Then in form (5.2), $a = 2x$ and $b = 1$; so $a^2 = (2x)^2$, $ab = (2x)(1)$, and $b^2 = 1^2$. It follows that

$$8x^3 + 1 = (\mathbf{2x})^3 + 1^3 \qquad \text{a = 2x, b = 1 by (5.2)}$$
$$= (2x + 1)[(2x)^2 - (2x)(1) + 1^2]$$
$$= (2x + 1)(4x^2 - 2x + 1)$$

b. $x^3y^3 - 64t^6 = (\mathbf{xy})^3 - (\mathbf{4t^2})^3 \qquad \text{a = xy, b = 4t}^2 \text{ by (5.3)}$
$$= (xy - 4t^2)[(xy)^2 + (xy)(4t^2) + (4t^2)^2]$$
$$= (xy - 4t^2)(x^2y^2 + 4xyt^2 + 16t^4) \qquad \blacktriangleleft$$

As in the case of the difference of two squares, the terms may themselves be polynomials. Such cases are more complicated, at least in appearance, and extra care must be taken when writing the factors.

E X A M P L E **8**   Factor $z^3 - (x + y)^3$.

Solution. This expression fits form (5.3) with $a = z$ and $b = x + y$; so $a^2 = z^2$, $ab = z(x + y)$, and $b^2 = (x + y)^2$. In the first step, we keep the parentheses around $x + y$. Thus

$$z^3 - (\mathbf{x + y})^3 = [z - (x + y)][z^2 + z(x + y) + (x + y)^2] \qquad \text{by (5.3)}$$
$$= (z - x - y)(z^2 + zx + zy + x^2 + 2xy + y^2) \qquad \blacktriangleleft$$

Sometimes the factored expressions themselves are factorable, as can be seen in the remaining examples. (Note especially that an even power is always a perfect square.)

E X A M P L E **9**   Factor $x^8 - 16y^{12}$.

Solution. $x^8 - 16y^{12}$
$$= (\mathbf{x^4})^2 - (\mathbf{4y^6})^2$$
$$= (x^4 - 4y^6)(x^4 + 4y^6) \qquad \text{difference of two squares}$$
$$= [(\mathbf{x^2})^2 - (\mathbf{2y^3})^2](x^4 + 4y^6)$$
$$= (x^2 - 2y^3)(x^2 + 2y^3)(x^4 + 4y^6) \qquad \text{difference of two squares}$$

$\blacktriangleleft$

E X A M P L E **10**   Factor $2x^2 - 8y^4$.

Solution. At first glance this expression does not appear to fit any of the forms considered. However, if the common factor **2** is factored out, we get

$$2x^2 - 8y^4 = \mathbf{2}(x^2) + \mathbf{2}(-4y^4)$$
$$= \mathbf{2}(x^2 - 4y^4)$$
$$= 2[x^2 - (2y^2)^2]$$
$$= 2(x - 2y^2)(x + 2y^2)$$

◀

Rule: In all factoring problems, first check for common factors.

Common error

Writing

$$a^3 + b^3 \qquad \text{as} \qquad (a + b)^3$$

Note that

$$a^3 + b^3 = (a + b)(a^2 - ab + b^2)$$

while

$$(a + b)^3 = (a + b)(a + b)(a + b) = (a + b)(a + b)^2$$
$$= (a + b)(a^2 + 2ab + b^2)$$

Thus

$$a^3 + b^3 \neq (a + b)^3$$

 E X A M P L E **11**   A cylindrical lead pipe of length L has inner radius r_1 and outer radius r_2. Factor the expression for the volume of the wall of the pipe, which is given by $V = \pi r_2^2 L - \pi r_1^2 L$.

Solution. $V = \pi r_2^2 L - \pi r_1^2 L = \pi L(r_2^2 - r_1^2)$
$$= \pi L(r_2 - r_1)(r_2 + r_1)$$

◀

E X E R C I S E S / S E C T I O N   **5.2**

In Exercises 1–61, factor the expressions completely.

1. $2a + 2b$

2. $3x - 3y$

3. $5x - 5$

4. $ax + ay$

5. $3x^2 - 6x$

6. $3x^3 - 6xy + 9x$

7. $4x^2y^2 - 6x^2y + 12xy^2$

8. $9a^2b^3 - 21a^3b^4 - 12a^2b^2$

9. $14c^3d^2 + 28c^3d^3 + 14c^2d^3$

10. $6R^2C - 15R^3C^2 - 3RC$

11. $11R_1^2R_2 - 22R_1R_2^2 + 11R_1R_2$

12. $mc^2 - m_0c^2$

13. $x^2 - y^2$

14. $4x^2 - y^2$

15. $4x^2 - 4y^2$

16. $x^2 - 16y^2$

17. $64m^2 - 16n^2$

18. $25V^2 - 4C^2$

19. $8f_1^2 - 2f_2^2$

20. $25f^2 - 16g^2$

21. $50x^2 - 32z^2$

22. $D^4 - 1$

23. $\dfrac{1}{4}x^2 - y^2$

24. $a^2 - \dfrac{1}{16}b^2$

25. $\dfrac{1}{9}x^2 - \dfrac{1}{16}y^2$

26. $\dfrac{1}{25}x^2 - \dfrac{1}{49}y^2$

27. $\dfrac{x^2}{a^2} - B^2$

28. $\dfrac{s^2}{t^2} - c^2$

29. $1 - \dfrac{b^2}{a^2}$

30. $4 - \dfrac{x^2}{v^2}$

31. $t_1^4 - t_2^4$

32. $3s^4 - 3$

33. $3x^2 - 27a^2$

34. $a^4 - 16$

35. $64a^4 - 1$

36. $125x^2 - 5$

37. $x^3 - y^3$

38. $z^3 + b^3$

39. $2x^3 + 2a^3$

40. $8x^3 + y^3$

41. $27a^3 - y^3$

42. $64s^3 - 8t^3$

43. $125m^3 - 27a^3$

44. $64D^3 - 1$

45. $a^3b^3 - 1$

46. $1 - x^3y^3$

47. $1 + x^3y^6$

48. $a^3 + x^3y^3$

49. $(a + b)^2 - 1$

50. $(a + b)^3 - 1$

51. $1 - (x + y)^2$

52. $1 - (x + y)^3$

53. $(x - y)^3 + z^3$

54. $(x - y)^2 - z^2$

55. $(v_1 - v_2)^3 - v_3^3$

56. $(l_1 + l_2)^3 + 1$

57. $(F_1 + F_2)^3 - 8$

58. $27 - (m + 2n)^3$

59. $8 + (3r + 2s)^3$

60. $x^4 - 81y^4$

61. $16x^3 - 128a^3$

62. The increase in the momentum of an object when the velocity increases from v_1 to v_2 is

$$M = mv_2 - mv_1$$

Factor the right side.

63. The energy required to lift an object from level y_1 to level y_2 is

$$E = mgy_2 - mgy_1$$

Factor the right side.

64. When the velocity of a rocket increases from v_1 to v_2, then the force due to air resistance increases by

$$kv_2^2 - kv_1^2$$

where k is a constant. Factor this expression.

65. The cross-sectional area A of a pipe with inner radius r_1 and outer radius r_2 is

$$A = \pi r_2^2 - \pi r_1^2$$

Factor the expression for A.

66. The energy E radiated by a blackbody is $E = kT^4 - kT_0^4$, where T is the temperature of the body, T_0 the temperature of the surrounding medium, and k a constant. Factor the expression for E.

67. The difference E_d in the energy radiated by a filament at temperatures T_1 and T_2, respectively, is given by $E_d = aT_1^4 - aT_2^4$, where a is a constant. Factor this expression.

68. The power delivered to a resistor is given by $P = I^2R$ (in watts). Factor the expression that represents the difference in the power delivered to a resistor when the current increases from I_1 to I_2.

69. The kinetic energy of a particle of mass m traveling at velocity v is given by $\frac{1}{2}mv^2$. Factor the expression for the change in the kinetic energy of a particle of mass m whose velocity increases from v_1 to v_2.

70. The voltage in a certain circuit was found to be $4IR_1 - 6IR_2 + 12IR_3$. Factor this expression.

71. The volume of a spherical shell is $V = \frac{4}{3}\pi r_2^3 - \frac{4}{3}\pi r_1^3$, where r_1 is the inner radius and r_2 the outer radius. Factor the expression for V.

72. The area of a rectangle is $x^3 + 27$ square units. One side is $x + 3$ units in length. Find an expression for the length of the other side.

5.3   More Special Products

In Section 5.1 we started our discussion of special products. Now we will look at two more special products, the square of a binomial and the product of two distinct binomials.

Square of a Binomial

$$(a + b)^2 = a^2 + 2ab + b^2$$
$$(a - b)^2 = a^2 - 2ab + b^2$$

The square of the sum (or difference) of two terms is the square of the first term, plus (or minus) twice the product of the two terms, plus the square of the second term.

E X A M P L E **1**   To expand $(x + 3y)^2$, we let $a = x$ and $b = 3y$, so that $2ab = 6xy$. Thus

$$(x + 3y)^2 = x^2 + 2(x)(3y) + (3y)^2 \qquad a^2 + 2ab + b^2$$
$$= x^2 + 6xy + 9y^2$$
◀

E X A M P L E **2** $$(3x^2 - 4y)^2 = (3x^2)^2 - 2(3x^2)(4y) + (4y)^2 \qquad a^2 - 2ab + b^2$$
$$= 9x^4 - 24x^2y + 16y^2$$

by the second form $(a - b)^2$. The first form $(a + b)^2$ could also be used if the binomial is treated as the square of $3x^2 + (-4y)$:

$$[3x^2 + (-4y)]^2 = (3x^2)^2 + 2(3x^2)(-4y) + (-4y)^2$$
$$= 9x^4 - 24x^2y + 16y^2$$
◀

Product of Two Binomials

The product of two distinct binomials has the following form:

$$(ax + by)(cx + dy) = acx^2 + (ad + bc)xy + bdy^2$$

The best way to carry out this multiplication directly is with the help of a diagram. First observe that the right side consists of the product of the first terms, plus the product of the last terms, plus a middle term consisting of a sum of two products. These two products are the products of the outer terms ax and dy and the inner terms by and cx. Schematically,

FOIL method

First

$$(ax + by)(cx + dy) = acx^2 + adxy + bcxy + bdy^2$$

The scheme can be remembered by the acronym **FOIL**, where F stands for "first," O for "outer," I for "inner," and L for "last."

E X A M P L E **3**   Multiply $(2x - y)(x + 3y)$.

Solution. Using the acronym FOIL, we get the following diagram:

$$(2x - y)(x + 3y) = (2x)(x) + [(2x)(3y) + (-y)(x)] + (-y)(3y)$$

Hence

$$(2x - y)(x + 3y) = 2x^2 + (6xy - xy) - 3y^2$$
$$= 2x^2 + 5xy - 3y^2 \quad \blacktriangleleft$$

E X A M P L E **4** Multiply $(3x - 2a)(2x - 6a)$.

Solution.

$$\overset{F}{\overbrace{}} \quad \overset{L}{} \quad \overset{F}{} \quad \overset{O}{} \quad \overset{I}{} \quad \overset{L}{}$$

$$(3x - 2a)(2x - 6a) = (3x)(2x) + [(3x)(-6a) + (-2a)(2x)] + (-2a)(-6a)$$

Thus

$$(3x - 2a)(2x - 6a) = 6x^2 + (-18ax - 4ax) + 12a^2$$
$$= 6x^2 - 22ax + 12a^2$$

In practice, *the middle terms should be added mentally.* $\quad \blacktriangleleft$

Larger polynomials can also be multiplied directly, but this may require an additional step. Of course, the larger the expression, the greater the difficulty in obtaining the product directly. If the factors get too complicated, it is undoubtedly best to return to the method studied in Chapter 1.

E X A M P L E **5** Multiply $7(x - 5y)(x - 6y)$.

Solution. We multiply the two binomials first and then use the distributive law to multiply by 7:

$$7(x - 5y)(x - 6y) = 7(x^2 - 6xy - 5xy + 30y^2)$$
$$= 7(x^2 - 11xy + 30y^2)$$
$$= 7x^2 - 77xy + 210y^2 \quad \blacktriangleleft$$

E X A M P L E **6** Multiply $(s + t - w)^2$.

Solution. This product can be treated as the square of a binomial if the terms are first grouped as follows:

$$[(s + t) - w]^2 = (s + t)^2 - 2(s + t)(w) + w^2$$
$$= s^2 + 2st + t^2 - 2sw - 2tw + w^2 \quad \blacktriangleleft$$

E X A M P L E **7**   Multiply $(a + 2b - c)(a + 2b + 5c)$.

Solution. First group the terms as follows:

$$[(a + 2b) - c][(a + 2b) + 5c]$$

Now treat the quantity $a + 2b$ as a single term and multiply by the FOIL scheme:

$$[(a + 2b) - c][(a + 2b) + 5c] = (a + 2b)^2 + [(a + 2b)(5c) + (-c)(a + 2b)] - 5c^2$$
$$= a^2 + 4ab + 4b^2 + (5ac + 10bc - ac - 2bc) - 5c^2$$
$$= a^2 + 4ab + 4b^2 + 4ac + 8bc - 5c^2 \quad \blacktriangleleft$$

Common error   Forgetting to take twice the product for the middle term or forgetting the middle term altogether. Remember that

$$(a + b)^2 \neq a^2 + ab + b^2$$

and

$$(a + b)^2 \neq a^2 + b^2$$

Instead,

$$(a + b)^2 = a^2 + \mathbf{2ab} + b^2$$

For easy reference, the following box summarizes the different special products.

Special Products:

$$a(b + c) = ab + ac \tag{5.4}$$
$$(a + b)^2 = a^2 + 2ab + b^2 \tag{5.5}$$
$$(a - b)^2 = a^2 - 2ab + b^2 \tag{5.6}$$
$$(a + b)(a - b) = a^2 - b^2 \tag{5.7}$$
$$(ax + b)(cx + d) = acx^2 + (ad + bc)x + bd \tag{5.8}$$
$$(ax + by)(cx + dy) = acx^2 + (ad + bc)xy + bdy^2 \tag{5.9}$$

EXERCISES / SECTION **5.3**

In Exercises 1–26, find the special products in *one step*.

1. $(x + 2)^2$ 　　**2.** $(y - 4)^2$ 　　**3.** $(C_1 + C_2)^2$

4. $(v - w)^2$ 　　**5.** $(x + 1)(x - 2)$ 　　**6.** $(x - 4)(x - 1)$

7. $(3x - 2)(5x + 1)$ 　　　　**8.** $(8x - 3)(4x - 2)$

9. $(4x - y)(3x - 2y)$ 　　　　**10.** $(2a + 5b)(a - 4b)$

11. $(s - 2t)(s - 3t)$ 　　　　**12.** $(f_1 - 2f_2)(3f_1 + 5f_2)$

13. $(7i_1 + 2i_2)^2$ 　　　　**14.** $(6A - 5B)^2$

15. $(3x - 2y)^2$ 　　　　**16.** $(f + 3g)^2$

17. $(2i + 5k)(i + 3k)$ 　　　　**18.** $(-6x + 7y)(x - 2y)$

19. $(2x - 10y)(x + y)$ 　　　　**20.** $(3x - y)(7x - 2y)$

21. $(2x - 5y)(3x - 8y)$ 　　　　**22.** $(v_1 + 2v_2)^2$

23. $(v_1 - 2v_2)(3v_1 + 6v_2)$ 　　　**24.** $(3x + 2y)(3x - 2y)$

25. $(9x - 6y)(9x + 6y)$ 　　**26.** $(10x - 3y)(10x + 3y)$

In Exercises 27–44, multiply the given polynomials. (More than one step may be required.)

27. $7(x - 2y)(x + 2y)$ 　　**28.** $3(2x - 5y)(2x + 5y)$

29. $x(3x - y)(3x + y)$ 　　**30.** $2x(4x - 2y)(4x + 2y)$

31. $2(x - 3y)(x - 2y)$ 　　**32.** $x(2x + y)(3x + 2y)$

33. $a(4x - y)(3x + 5y)$ 　　**34.** $a(2C_1 - 3C_2)(C_1 + 4C_2)$

35. $(x + y + 2z)^2$ 　　**36.** $(2x - 3y + w)^2$

37. $(a + 2b + c)^2$ 　　**38.** $(t - 3w - 1)^2$

39. $(x + y + 3z)(x + y - 2z)$ 　　**40.** $(x - 3b + c)(x - 3b + 2c)$

41. $(p + q - 2r)(p + q - 5r)$ 　　**42.** $(s + 2t + 4w)(s + 2t - w)$

43. $(x + a + b)(2x + a + b)$ 　　**44.** $(y + 2a - b)(3y + 2a - b)$

5.4　Factoring Trinomials

If two binomials of first degree are multiplied, the result is a trinomial. For example, $(x - 2)(x + 4) = x^2 + 2x - 8$. In this section we will develop techniques for the inverse operation—factoring a trinomial into two factors of first degree.

Perfect-Square Trinomials

A trinomial is a perfect square if it can be written as the square of a binomial. By reversing the special products (5.5) and (5.6), we can obtain the forms of perfect-square trinomials.

> **Perfect-Square Trinomials:**
>
> $$a^2 + 2ab + b^2 = (a + b)^2 \tag{5.10}$$
>
> $$a^2 - 2ab + b^2 = (a - b)^2 \tag{5.11}$$

Perfect squares

　　These forms suggest how to recognize a perfect square: *The first and last terms of the trinomial must be perfect squares, and the middle term must be equal to twice the product of a and b.*

$$\overset{\displaystyle \text{2 times } ab}{\underset{\displaystyle \text{perfect squares}}{a^2 + 2ab + b^2}}$$

E X A M P L E **1**   Factor:

    a. $x^2 + 4xy + 4y^2$ **b.** $9x^2 - 30xy + 25y^2$

Solution.

a. The first term x^2 and the last term $4y^2 = (2y)^2$ are both perfect squares. To check the middle term, note that $2(x)(2y) = 4xy$. It follows that the given trinomial is indeed a perfect square. Hence

$$x^2 + 4xy + 4y^2 = x^2 + 2(x)(2y) + (2y)^2$$
$$= (x + 2y)^2 \quad \text{by form (5.10)}$$

b. The first term $9x^2 = (3x)^2$ and the last term $25y^2 = (5y)^2$ are both perfect squares; also, $-2(3x)(5y) = -30xy$, which agrees with the middle term. The form fits (5.11):

$$9x^2 - 30xy + 25y^2 = (3x)^2 - 30xy + (5y)^2$$
$$= (3x)^2 - 2(3x)(5y) + (5y)^2$$
$$= (3x - 5y)^2 \quad \text{by form (5.11)} \quad \blacktriangleleft$$

Some expressions have a common factor, which must be factored first.

E X A M P L E **2**   Factor $128a^2 - 352ab^2 + 242b^4$.

Solution. It was emphasized earlier that you should always be on the alert for common factors. Only after removing the common factor **2** do the first and last terms of the trinomial become perfect squares. We now have

$$2(64a^2 - 176ab^2 + 121b^4) = 2[(8a)^2 - 2(88ab^2) + (11b^2)^2]$$
$$= 2[(8a)^2 - 2(8a)(11b^2) + (11b^2)^2]$$
$$= 2(8a - 11b^2)^2 \quad \text{by form (5.11)} \quad \blacktriangleleft$$

General Trinomials

Factoring trinomials other than perfect squares is somewhat more difficult. The forms for the general trinomial arise from the special products (5.8) and (5.9).

General Trinomials:

$$acx^2 + (ad + bc)x + bd = (ax + b)(cx + d) \tag{5.12}$$

$$acx^2 + (ad + bc)xy + bdy^2 = (ax + by)(cx + dy) \tag{5.13}$$

The problem here is that the middle term is really a combination of four different terms. Consequently, it may be necessary to juggle several numbers until the right combination is found. Let's consider the trinomial

$$x^2 + x - 6$$

At this point we know only that the two factors are first-degree binomials (if the expression is factorable). Keeping in mind the acronym FOIL, consider the following diagram:

FOIL method

$$\text{F} \quad \text{L} \quad \overbrace{\qquad}^{\text{F}} \text{L}$$
$$x^2 + x - 6 = (x \qquad)(x \qquad)$$

Since $x \cdot x = x^2$, we are satisfied with the F terms. The L terms have several possibilities. Suppose we try -6 and 1. Then the diagram becomes

$$x^2 + x - 6 = (x - 6)(x + 1)$$

Now the L terms check, but the sum of O and I is $-5x$, which does not agree with the middle term x. The combination 6 and -1 also fails.

For the next combination let us try -3 and 2. Then we get

$$x^2 + x - 6 = (x - 3)(x + 2)$$

Although the L terms check, the sum of the inner and outer products is $-x$, which has the wrong sign. This suggests switching signs on the last combination. The result is

$$x^2 + x - 6 = (x + 3)(x - 2)$$

Since we checked all possible combinations in this demonstration, the procedure looks more difficult than it really is. The FOIL scheme tells us that to factor $x^2 + x - 6$, we need two numbers whose *product* is -6 and whose *sum* is 1. These numbers are 3 and -2.

Let's consider some more examples.

E X A M P L E **3**   Factor $x^2 - 6xy + 8y^2$.

Solution. Since the last term is preceded by a plus sign, the two L terms must agree in sign. This common sign must be a minus because the middle term is preceded by a minus sign. So we need two numbers whose *product* is 8 and whose *sum* is -6. The numbers are -2 and -4.

$$x^2 - 6xy + 8y^2 = (x - 2y)(x - 4y)$$

The middle term checks.   ◄

E X A M P L E **3** Factor $x^2 + 5x - 24$.

Solution. This time we need two numbers whose *product* is -24 and whose *sum* is 5. These numbers are 8 and -3, so that

$$x^2 + 5x - 24 = (x - 3)(x + 8)$$ ◄

Examples 3 and 4 suggest that there is a simple method for factoring trinomials. Unfortunately, the coefficient of x^2 may be a number other than 1, thereby increasing the number of possible combinations.

E X A M P L E **5** Factor $4x^2 + 21xy - 18y^2$.

Solution. Since $4x^2$ can be split in more than one way, let us try

$$(2x \qquad)(2x \qquad)$$

Since $18 = 9 \cdot 2$, $18 = 6 \cdot 3$, and $18 = 18 \cdot 1$, we have the following possibilities:

$$(2x + 9y)(2x - 2y) \qquad (2x + 6y)(2x - 3y) \qquad (2x + 18y)(2x - y)$$

In the first case the resulting middle term is $14xy$, in the second case $6xy$, and in the third case $34xy$. None of these is correct. Reversing the signs produces negative middle terms, which makes matters worse. For example,

$$(2x - 9y)(2x + 2y) = 4x^2 - 14xy - 18y^2$$

For our next trial let us start with

$$(4x \qquad)(x \qquad)$$

Using the same factors of 18, we might try

$$4x^2 + 21xy - 18y^2 = (4x - 3y)(x + 6y)$$

This result is correct.   ◄

E X A M P L E **6**   Factor $9x^2 - 219x + 72$.

Solution. First note the common factor **3**:

$$9x^2 - 219x + 72 = 3(3x^2 - 73x + 24)$$

Now we need a combination that produces a very large middle term. As a consequence, even though 24 has many different divisors, only the factors 24 and 1 could produce a sufficiently large middle term. It follows that

$$3(3x^2 - 73x + 24) = 3(3x - 1)(x - 24)$$ ◄

Prime factor

As noted earlier, not every polynomial is factorable. A nonfactorable polynomial is called **prime** or **irreducible**. For example, $x^2 + x + 7$ is prime.

 E X A M P L E **7**   A stone is hurled upward from a height of 32 ft at 56 ft/s. Its distance s (in feet) above the ground is given by $s = -16t^2 + 56t + 32$, where t is measured in seconds. Factor the expression for s and determine when the stone hits the ground. (Assume that the motion starts at $t = 0$ s.)

Solution.
$$s = -16t^2 + 56t + 32$$
$$= -8(2t^2 - 7t - 4)$$
$$= -8(2t + 1)(t - 4)$$

Since the stone is at ground level when $s = 0$, we need to find the value of t for which

$$-8(2t + 1)(t - 4) = 0$$

This product is 0 when $t - 4 = 0$ or when $2t + 1 = 0$. Note that $t - 4 = 0$ when $t = 4$. Now, $2t + 1 = 0$ only when $t = -\frac{1}{2}$. Since the motion begins at $t = 0$, the only admissible value is $t = 4$. We conclude that the stone takes 4 s to reach the ground.   ◄

E X E R C I S E S / S E C T I O N **5.4**

In Exercises 1–50, factor each expression.

1. $x^2 + 4xy + 4y^2$

2. $4x^2 - 4xz + z^2$

3. $9x^2 - 24xy + 16y^2$

4. $16x^2 - 8x + 1$

5. $x^2 - 4x + 3$

6. $x^2 - 5x + 6$

7. $x^2 - x - 12$

8. $x^2 - 2x - 15$

9. $2a^2 - 8ab + 8b^2$

10. $3m^2 + 18mn + 27n^2$

11. $2x^2 + 14x + 12$

12. $ax^2 - 5ax - 14a$

13. $x^2 - 5xy - 6y^2$

14. $a^2 + 3ab - 10b^2$

15. $D^2 + 5D - 14$

16. $T_1{}^2 + 7T_1T_2 + 12T_2{}^2$

17. $2x^2 - 3xy + y^2$

18. $3x^2 + xy - 2y^2$

19. $5x^2 - 11xy + 2y^2$

20. $4x^2 - 7xy + 3y^2$

21. $4x^2 + 13xy + 3y^2$

22. $4x^2 - 8xy + 3y^2$

23. $6x^2 - xy - 12y^2$

24. $6p^2 - 13pq + 5q^2$

25. $5w_1{}^2 - 22w_1w_2 + 8w_2{}^2$

26. $18x^2 - 21xy - 15y^2$

27. $8L^2 - 8LC - 48C^2$

28. $12R^2 - 18RC - 12C^2$

29. $18f^2 - 48fg + 32g^2$

30. $4\alpha A^2 + 12\alpha AB + 9\alpha B^2$

31. $x^4 - 4x^3 + 4x^2$

32. $\alpha^2 x^2 - 2\alpha\beta x^2 + \beta^2 x^2$

33. $D^2 - D - 8$

34. $y^2 - 3yz + 9z^2$

35. $(a + b)^2 - (a + b) - 6$

36. $(x - y)^2 - (x - y) - 12$

37. $(n + m)^2 - 3(n + m) + 2$

38. $(\beta + \gamma)^2 + 7(\beta + \gamma) + 12$

39. $2(a + b)^2 - 9(a + b) + 4$

40. $3(t_1 - t_2)^2 - 14(t_1 - t_2) + 8$

41. $(f_1 + 2f_2)^3 - (f_1 + 2f_2)^2$

42. $(c - 3d)^2 - (c - 3d)^3$

43. $1 - (x - y)^3$

44. $8 - (x + 2y)^3$

45. $28a^2 - ab - 2b^2$

46. $54x^2 - 3xy - 35y^2$

47. $40x^2 + 7xy - 3y^2$

48. $24x^2 - 38xy + 15y^2$

49. $12\alpha^2 - 23\alpha\beta + 10\beta^2$

50. $42D^2 - 11D - 20$

51. If a stone is hurled upward at a velocity of 24 ft/s from a height of 72 ft (starting at $t = 0$), then the distance s (in feet) above the ground is given by $s = -16t^2 + 24t + 72$, where t is in seconds. Factor the expression to determine when the stone hits the ground. (See Example 7.)

52. The deflection of a certain beam is $3a^2x^3 - 5ax^2 - 2x$. Factor this expression.

53. The resistance of a variable resistor is given by $R = 18.0t^2 - 48.0t + 32.0$, where t is measured in seconds. Factor the expression for R and determine when the resistance is 0.

5.5 Factoring by Grouping

In this section we are going to discuss **factoring by grouping**. In some cases the terms in a polynomial can be grouped so that the expression can be factored by one of the earlier methods. (This technique is ordinarily used when the polynomial has more than three terms.)

Consider, for example, the expression

$$ax + ay + 3x + 3y$$

Note that the first two terms have a common factor, suggesting that the terms be grouped as follows:

$$(ax + ay) + (3x + 3y)$$

We now see that the last two terms also have a common factor; thus

$$(ax + ay) + (3x + 3y) = a(x + y) + 3(x + y)$$

The resulting expression consists of two terms with the common factor $x + y$. Factoring out $x + y$, we obtain

$$ax + ay + 3x + 3y = a(x + y) + 3(x + y) = (x + y)(a + 3)$$

E X A M P L E **1**   Factor $RV_1 - RV_2 + V_1 - V_2$.

Solution. The common factor R in the first two terms suggests the following grouping:

$$(RV_1 - RV_2) + (V_1 - V_2) = R(V_1 - V_2) + (V_1 - V_2)$$

When removing the common factor $V_1 - V_2$, note that the coefficient of the second term is 1, even though the 1 is not explicitly written. We now get

$$RV_1 - RV_2 + V_1 - V_2 = R(V_1 - V_2) + 1(V_1 - V_2) = (V_1 - V_2)(R + 1)$$
$$= (V_1 - V_2)(R + 1) \qquad \blacktriangleleft$$

The idea in the next example is similar.

E X A M P L E **2**   Factor $a^2 + ab - 2b^2 - 2a + 2b$.

Solution. The first three terms look like a typical trinomial, which is indeed factorable. At the same time, the last two terms have the common factor 2 or -2. Let us group the terms as follows:

$$(a^2 + ab - 2b^2) + (-2a + 2b) \qquad \text{inserting parentheses}$$
$$= (a - b)(a + 2b) + (-2a + 2b) \qquad \text{factoring the trinomial}$$

Now we need to remove -2 from the second expression to obtain a common factor $a - b$:

$$(a - b)(a + 2b) - 2(a - b) \qquad \text{common factor } -2$$
$$= (a - b)(a + 2b) - 2(a - b)$$
$$= (a - b)(a + 2b - 2) \qquad \text{common factor } a - b \qquad \blacktriangleleft$$

Some expressions containing several terms can be grouped in different ways, some of which may not work. Consider the next example.

E X A M P L E **3**   Factor $xz + xy - 2yz - x^2 + 2y^2$.

Solution. If we group the first three terms as in Example 2, we get a nonfactorable combination. A more productive approach is to group the first two terms which contain a common factor x. Then we get

$$(xz + xy) + (-2yz - x^2 + 2y^2)$$

Unfortunately, the trinomial contains $-2yz$, which is completely different from the other two terms. This observation provides the clue: All the x and y terms should be grouped together. Thus

$$xz - 2yz - x^2 + xy + 2y^2 \qquad \text{regrouping}$$
$$= (xz - 2yz) - (x^2 - xy - 2y^2) \qquad \text{inserting parentheses}$$
$$= z(x - 2y) - (x - 2y)(x + y) \qquad \text{factoring}$$
$$= z(x - 2y) - (x - 2y)(x + y)$$
$$= (x - 2y)[z - (x + y)] \qquad \text{common factor } x - 2y$$
$$= (x - 2y)(z - x - y) \qquad \blacktriangleleft$$

In some cases even the methods of Section 5.2 apply if the terms are suitably grouped.

E X A M P L E **4** Factor $a^2 - 4x^2 + 4xy - y^2$.

Solution. Since the first two terms are a difference of two squares, we try

$$(a^2 - 4x^2) + (4xy - y^2) = (a - 2x)(a + 2x) + y(4x - y)$$

Although a perfectly logical procedure, this attempt leads to a dead end. An alternative is to keep a^2 apart and to group the remaining terms as follows:

$$a^2 - (4x^2 - 4xy + y^2) \qquad \text{inserting parentheses}$$
$$= a^2 - [(2x)^2 - 2(2xy) + y^2] \qquad \text{perfect-square trinomial}$$
$$= a^2 - (2x - y)^2 \qquad \text{factoring the trinomial}$$
$$= [a - (2x - y)][a + (2x - y)] \qquad \text{difference of two squares}$$
$$= (a - 2x + y)(a + 2x - y) \qquad \blacktriangleleft$$

 E X A M P L E **5** Suppose the distance s from the origin as a function of time of a particle traveling along a line is $s = 2t(t - 1)^3$. It can be shown by the methods of calculus that the velocity v is given by

$$v = 6t^3 - 12t^2 + 6t + 2(t - 1)^3$$

Write this expression as a product.

Solution.
$$v = 6t^3 - 12t^2 + 6t + 2(t - 1)^3$$
$$= 6t(t^2 - 2t + 1) + 2(t - 1)^3 \qquad \text{common factor } 6t$$
$$= 6t(t - 1)^2 + 2(t - 1)^3 \qquad \text{factoring the trinomial}$$
$$= 2(t - 1)^2(3t + t - 1) \qquad \text{common factor } 2(t - 1)^2$$
$$= 2(t - 1)^2(4t - 1) \qquad \blacktriangleleft$$

E X E R C I S E S / S E C T I O N   **5.5**

In Exercises 1–34, factor the given expressions by grouping.

1. $ax - ay + bx - by$

2. $xy + x + 3y + 3$

3. $2x^2 + 6xy + x + 3y$

4. $3x - 3y + cx - cy$

5. $4ac - 4bc + a - b$

6. $5bx - 5by + x - y$

7. $5bx - 5by - x + y$

8. $ax + ay - x - y$

9. $2ax + 2ay - 2cx - 2cy$

10. $4ax - 4bx - 4ay + 4by$

11. $3aR - 3ar - 6bR + 6br$

12. $4a_1v_1 + 4a_2v_1 - 4a_1v_2 - 4a_2v_2$

13. $x^2 - y^2 - xz - yz$

14. $x^2 - y^2 - xz + yz$

15. $ax + ay - x^2 + y^2$

16. $zx - zy - x^2 + y^2$

17. $x^2 - 2xz - y^2 + 2yz$

18. $x^2 - 3xy + 2y^2 + 2x - 2y$

19. $x^2 + 3xy - 4y^2 + x - y$

20. $x^2 + 3xy - 4y^2 - x + y$

21. $4x^2 - 4xy + y^2 - z^2$

22. $x^2 - 6xy + 9y^2 - w^2$

23. $x^2 - z^2 + 4y^2 + 4xy$

24. $a^2 - c^2 + 16b^2 - 8ab$

25. $9a^2 - 4b^2 - c^2 - 4bc$

26. $2a - 2b + a^2 + ab - 2b^2$

27. $3x - 3y - x^2 - 3xy + 4y^2$

28. $2x^2 - 3xy - 2y^2 + 6x + 3y$

29. $ax + xy + 2ay - x^2 + 6y^2$

30. $xz + 2xy + 3yz - x^2 + 15y^2$

31. $2xy + x^2 - yz - 3y^2 + xz$

32. $2R^2 + 2VC - 2C^2 + RV + 3RC$

33. $V_1{}^2 - V_3{}^2 + V_2{}^2 - 2V_1V_2$

34. $R_1R_2 + R_1R_3 + V_1R_2 + V_1R_3$

35. In a problem on cost analysis, the following expression has to be simplified:

$$a^2(A - 1) - [2 + (A - 1)a]^2 + 4 + 2a^2(A - 1)^2$$

Carry out the simplification by factoring.

36. Factor the following expression, which arose in a problem on the strength of materials:

$$2a(x - 2) + 3a(x - 2)^2$$

5.6   Equivalent Fractions

In the earlier chapters we considered the four fundamental operations of addition, subtraction, multiplication, and division on signed numbers and polynomials. We will now learn how to perform these operations on fractions. Just as arithmetic fractions play an important role in everyday life, algebraic fractions play an important role in mathematics and its technical applications.

Before we get into the four fundamental operations, we must learn to *reduce* a given fraction. For example, we know that

$$\frac{6}{9} = \frac{2}{3}$$

So we would expect that

$$\frac{6x}{9x} = \frac{2}{3} \qquad (x \neq 0)$$

In other words, algebraic fractions can be reduced by dividing the numerator and denominator by the same factor.

> A fraction is said to be **reduced to lowest terms** if its numerator and denominator have no common factors except 1.

Equivalent fractions

A given fraction and the corresponding reduced fraction are said to be **equivalent fractions**.

E X A M P L E **1**   Reduce the fraction

$$\frac{14a^3b^2c^4}{21ab^2c^2}$$

to lowest terms.

Solution. After removing the common factor $7ab^2c^2$ from numerator and denominator, we get

$$\frac{14a^3b^2c^4}{21ab^2c^2} = \frac{(7ab^2c^2)(2a^2c^2)}{(7ab^2c^2)(3)}$$

$$= \frac{2a^2c^2}{3}$$

obtained by dividing the numerator and denominator by $7ab^2c^2$.   ◄

Reducing fractions

The procedure in Example 1 can be applied to fractions containing polynomials. Since fractions are always reduced by dividing the numerator and denominator by the same factor, *both numerator and denominator must first be factored*. This operation is one of the most important applications of the factoring procedure.

(From now on, the numerical values of the variables in a fraction are understood to be restricted so that the denominator is different from zero. This restriction eliminates division by zero.)

E X A M P L E **2**   Reduce the fraction

$$\frac{x^2 + xy - 2y^2}{x^2 - 4xy + 3y^2}$$

to lowest terms.

Solution. We first factor both the numerator and denominator to obtain

$$\frac{x^2 + xy - 2y^2}{x^2 - 4xy + 3y^2} = \frac{(x - y)(x + 2y)}{(x - y)(x - 3y)}$$

After dividing both numerator and denominator by the common factor $x - y$, the fraction is reduced to

$$\frac{x + 2y}{x - 3y}$$

The simplest and safest way to reduce this fraction is to cross out the factor $x - y$. This procedure is called **cancellation**. In our problem

Cancellation

$$\frac{x^2 + xy - 2y^2}{x^2 - 4xy + 3y^2} = \frac{\overset{1}{\cancel{(x - y)}}(x + 2y)}{\underset{1}{\cancel{(x - y)}}(x - 3y)} = \frac{x + 2y}{x - 3y}$$

The 1 next to each canceled factor is often omitted:

$$\frac{\cancel{(x - y)}(x + 2y)}{\cancel{(x - y)}(x - 3y)} = \frac{x + 2y}{x - 3y}$$

In this case the line through the canceled factor has to be thought of as a 1.

◀

E X A M P L E **3**   Reduce the fraction

$$\frac{a^2 + ab - 6b^2}{2a^2 + 7ab + 3b^2}$$

by cancellation.

Solution. $\dfrac{a^2 + ab - 6b^2}{2a^2 + 7ab + 3b^2} = \dfrac{(a + 3b)(a - 2b)}{(2a + b)(a + 3b)}$ **factoring**

$$= \frac{\cancel{(a + 3b)}(a - 2b)}{(2a + b)\cancel{(a + 3b)}}$$ **cancellation**

$$= \frac{a - 2b}{2a + b}$$ ◀

To avoid possible difficulties with signs, observe that the quantities $a - b$ and $b - a$ are numerically equal but opposite in sign. Indeed, $-(b - a) = -b + a = a - b$, or

$a - b = -(b - a)$	(5.14)

Also, note that

$$\frac{-a}{b} = \frac{a}{-b} = -\frac{a}{b} \qquad\qquad (5.15)$$

E X A M P L E **4**   Simplify the fraction

$$\frac{x^3 - y^3}{xy - x^2}$$

Solution. The numerator is a difference of two cubes, and the terms in the denominator have a common factor x. Thus

$$\frac{x^3 - y^3}{xy - x^2} = \frac{(x - y)(x^2 + xy + y^2)}{(y - x)x}$$

$$= \frac{(x - y)(x^2 + xy + y^2)}{-(x - y)x} \qquad \text{since } a - b = -(b - a)$$

$$= -\frac{(x - y)(x^2 + xy + y^2)}{(x - y)x} \qquad \text{since } \frac{a}{-b} = -\frac{a}{b}$$

$$= -\frac{\cancel{(x - y)}(x^2 + xy + y^2)}{\cancel{(x - y)}x} \qquad \text{cancellation}$$

$$= -\frac{x^2 + xy + y^2}{x} \qquad\qquad\qquad\qquad \blacktriangleleft$$

Common error   Forgetting that

$$\frac{a + b}{a} \neq 1 + b$$

Factors can be canceled only if both numerator and denominator are written as products.

   Although the fraction is already in simplest form, an acceptable alternative is

$$\frac{a + b}{a} = \frac{a}{a} + \frac{b}{a} = 1 + \frac{b}{a}$$

E X A M P L E **5**   Simplify the fraction

$$\frac{x + 2}{(x + 3)x + 2}$$

Solution. In this example it is particularly tempting to cancel $x + 2$. However, the denominator is not a product, since $(x + 3)x + 2$ is not the same as $(x + 3)(x + 2)$. The denominator needs to be factored first.

$$\frac{x + 2}{(x + 3)x + 2} = \frac{x + 2}{x^2 + 3x + 2} \qquad \text{distributive law}$$

$$= \frac{x + 2}{(x + 2)(x + 1)} \qquad \text{factoring}$$

$$= \frac{\overset{1}{\cancel{x + 2}}}{\underset{1}{\cancel{(x + 2)}}(x + 1)} \qquad \text{canceling}$$

$$= \frac{1}{x + 1}$$

Note that since the only factor appearing in the numerator is canceled, it is wise to place a 1 next to the canceled factors. (However, as mentioned earlier, the line through a canceled factor may be thought of as a 1.) ◀

 E X A M P L E **6**　If a mass m_1 strikes a stationary mass m_2, then the ratio of the kinetic energy of the system with respect to the floor to the kinetic energy with respect to the center of mass of the system can be expressed as

$$\frac{\frac{1}{2}m_1 v^2(m_1 + m_2) - \frac{1}{2}m_1{}^2 v^2}{\frac{1}{2}m_1 v^2(m_1 + m_2)}$$

Reduce this expression.

Solution.　$\dfrac{\frac{1}{2}m_1 v^2(m_1 + m_2) - \frac{1}{2}m_1{}^2 v^2}{\frac{1}{2}m_1 v^2(m_1 + m_2)} = \dfrac{\frac{1}{2}m_1{}^2 v^2 + \frac{1}{2}m_1 m_2 v^2 - \frac{1}{2}m_1{}^2 v^2}{\frac{1}{2}m_1 v^2(m_1 + m_2)}$

$$= \frac{\cancel{\frac{1}{2}m_1 m_2 v^2}}{\cancel{\frac{1}{2}m_1 v^2}(m_1 + m_2)}$$

$$= \frac{m_2}{m_1 + m_2}$$ ◀

E X E R C I S E S / S E C T I O N　**5.6**

In Exercises 1–42, reduce the fractions to lowest terms.

1. $\dfrac{3x^2 y^3}{6xy^3}$

2. $\dfrac{-6x^3 y^2 z}{9x^2 y^2 z}$

3. $\dfrac{-14a^2 b^3 x^2}{-28ab^3 x^3}$

4. $\dfrac{9a^4 bc^3}{-15a^2 b^3 c}$

5. $\dfrac{x^2 + x}{x + 1}$

6. $\dfrac{x - 3}{2x - 6}$

7. $\dfrac{2x^2 + 8x}{x^3 + 4x^2}$

8. $\dfrac{ax^2 - 6ax}{ax + 6a}$

9. $\dfrac{x^2 - 16}{x - 4}$

10. $\dfrac{2x^2 - 8}{x + 2}$

11. $\dfrac{x^3 + y^3}{x + y}$

12. $\dfrac{x^3 - y^3}{x^2 + xy + y^2}$

13. $\dfrac{x^2 - y^2}{3x + 3y}$

14. $\dfrac{s^2 - t^2}{vs + vt}$

15. $\dfrac{R^3 - 8}{R - 2}$

16. $\dfrac{v^3 - 27}{v^2 + 3v + 9}$

17. $\dfrac{x^2 - 6x + 8}{x - 2}$

18. $\dfrac{4x^2 + 11x - 3}{x + 3}$

19. $\dfrac{i^2 + 7i + 10}{i + 2}$

20. $\dfrac{3R^2 + 13R + 4}{R + 4}$

21. $\dfrac{6x^2 - 11x + 4}{2x^2 + 5x - 3}$

22. $\dfrac{4x^2 - 15x - 4}{4x^2 + 13x + 3}$

23. $\dfrac{7x^2 - 40x - 12}{2x^2 - 11x - 6}$

24. $\dfrac{3v^2 - 2v - 16}{3v^2 - 11v + 8}$

25. $\dfrac{m_1{}^2 - m_2{}^2}{m_1{}^2 - 3m_1 m_2 + 2m_2{}^2}$

26. $\dfrac{R^3 + r^3}{R^2 + 4Rr + 3r^2}$

27. $\dfrac{(x - 3)(x + 4)}{(x + 4)(3 - x)}$

28. $\dfrac{(\beta - 2)(2\beta - 3)}{(3 - 2\beta)(2 - \beta)}$

29. $\dfrac{2L^2 - 3LC - 2C^2}{3L^2 - 10LC + 8C^2}$

30. $\dfrac{15y^2 - 7xy - 2x^2}{4x^2 - 4xy - 3y^2}$

31. $\dfrac{(x + 2)(x^2 - 7x + 12)}{(x - 3)(x^2 - 4x - 12)}$

32. $\dfrac{(c - d)(2c^2 - 3cd - 2d^2)}{(c - 2d)(3c^2 - cd - 2d^2)}$

33. $\dfrac{ac + ad - 4bc - 4bd}{2ac + 2ad - bc - bd}$

34. $\dfrac{ac + bc - 3ad - 3bd}{ac - 3ad - bc + 3bd}$

35. $\dfrac{x + 2}{(2x + 5)x + 2}$

36. $\dfrac{f + 4}{(4f + 17)f + 4}$

37. $\dfrac{l - 3}{(2l - 5)l - 3}$

38. $\dfrac{s - 2}{(3s - 5)s - 2}$

39. $\dfrac{t + 2}{(3t + 7)t + 2}$

40. $\dfrac{v_0 + 5}{(3v_0 + 16)v_0 + 5}$

41. $\dfrac{2E + 1}{3E(2E + 1) - 2(2E + 1)}$

42. $\dfrac{2\pi - 3}{2\pi(2\pi - 3) + 2(2\pi - 3)}$

43. The area of a rectangle is $PL - PM - QL + QM$, and the length is $L - M$. Find the width.

44. A common phenomenon in mechanical and electrical systems is simple harmonic motion (which will be discussed in Chapter 8). If a particle is subject to two such motions at right angles, then the resulting motion can be described by

$$y^2 = \frac{4x^2}{A^2}(A^2 - x^2)$$

under certain conditions. Simplify the expression for the ratio of y^2 to $(A - x)/A^2$, which is given by

$$\frac{4x^2(A^2 - x^2)A^2}{A^2(A - x)}$$

45. Suppose a neutron of mass m traveling at velocity v_0 collides with a stationary nucleus of mass M. If the two particles are scattered at right angles to the original motion, then the velocity v of the neutron after the collision satisfies the relation

$$v^2 = \frac{v_0{}^2 M^2 + v_0{}^2 m^2}{(M + m)^2}$$

Find the ratio of the kinetic energy of the neutron before the collision to its kinetic energy after the collision; that is, simplify the expression for

$$\frac{E}{E_0} = \frac{mv^2/2}{mv_0{}^2/2} = \frac{\frac{1}{2}m(v_0{}^2 M^2 + v_0{}^2 m^2)}{(M + m)^2(\frac{1}{2}mv_0{}^2)}$$

46. The voltage drop (in volts) across a certain resistor as a function of time is $V = 16t^2 - 4$ (t in seconds). The variable resistance is given by $R = 4t + 2$ (in ohms). Find a simplified expression for the current $I = V/R$ (in amperes).

47. Repeat Exercise 46 for $V = 2t^2 + 4t - 30$ and $R = t + 5$.

5.7 Multiplication and Division of Fractions

The rules for multiplication and division of fractions carry over from arithmetic:

Multiplication:

$$\frac{a}{b} \cdot \frac{c}{d} = \frac{ac}{bd} \qquad\qquad (5.16)$$

Division:

$$\frac{a}{b} \div \frac{c}{d} = \frac{a}{b} \cdot \frac{d}{c} = \frac{ad}{bc} \qquad\qquad (5.17)$$

Multiplication and division of algebraic fractions should be performed by a procedure similar to the reduction of fractions in the previous section. Consider the product

$$\frac{x}{x+y} \cdot \frac{x^2 - y^2}{x^2}$$

If the multiplication rule were taken literally, the numerators and denominators would be multiplied together and the resulting fraction simplified. However, multiplying the polynomials is inadvisable, since in the very next step the resulting products must be factored again. Instead, the expressions can be combined as follows:

$$\frac{x(x^2 - y^2)}{(x+y)x^2} = \frac{x(x-y)(x+y)}{(x+y)x^2} = \frac{x-y}{x}$$

It should now be apparent that even the first step is unnecessary. The expressions in the given fractions should be factored and the appropriate factors canceled. Thus

$$\frac{x}{x+y} \cdot \frac{x^2 - y^2}{x^2} = \frac{\cancel{x}}{\cancel{x+y}} \cdot \frac{(x-y)\cancel{(x+y)}}{x^{\cancel{2}}} = \frac{x-y}{x}$$

Multiplication

To multiply two or more fractions, factor the expression in the numerator and denominator of each fraction and cancel. Then multiply the resulting fractions by rule (5.16).

E X A M P L E 1   Carry out the following multiplication:

$$\frac{x^2 - 3xy}{y - x} \cdot \frac{x^3 - y^3}{x^3 + 2x^2y - 15xy^2}$$

Solution. Following the plan just discussed, let us factor all the polynomials before multiplying. Then we get

$$\frac{x^2 - 3xy}{y - x} \cdot \frac{x^3 - y^3}{x^3 + 2x^2y - 15xy^2}$$

$$= \frac{x(x - 3y)}{y - x} \cdot \frac{(x - y)(x^2 + xy + y^2)}{x(x^2 + 2xy - 15y^2)} \qquad \text{common factors;}\\ \text{difference of two cubes}$$

$$= \frac{x(x - 3y)}{-(x - y)} \cdot \frac{(x - y)(x^2 + xy + y^2)}{x(x + 5y)(x - 3y)} \qquad \text{factoring the trinomial;}\\ a - b = -(b - a)$$

$$= -\frac{\cancel{x}\cancel{(x-3y)}}{\cancel{x-y}} \cdot \frac{\cancel{(x-y)}(x^2 + xy + y^2)}{\cancel{x}(x + 5y)\cancel{(x - 3y)}} \qquad \text{cancellation; } \frac{a}{-b} = -\frac{a}{b}$$

$$= -\frac{x^2 + xy + y^2}{x + 5y}$$

◀

Division

Division, according to rule (5.17), requires one more step than multiplication: *We invert the divisor and then proceed as in multiplication.* In other words, we multiply by the reciprocal of the divisor.

E X A M P L E   2   Perform the following division:

$$\frac{x^2 + 2xy + y^2}{x - 2y} \div \frac{x + y}{x^2 - 5xy + 6y^2}$$

Solution. $\dfrac{x^2 + 2xy + y^2}{x - 2y} \div \dfrac{x + y}{x^2 - 5xy + 6y^2}$

$$= \frac{x^2 + 2xy + y^2}{x - 2y} \cdot \frac{x^2 - 5xy + 6y^2}{x + y} \qquad \text{inverting the divisor}$$

$$= \frac{(x + y)^{\cancel{2}}}{x - 2y} \cdot \frac{(x - 3y)(x - 2y)}{\cancel{x + y}} \qquad \text{factoring and canceling}$$

$$= (x + y)(x - 3y) \qquad\qquad\qquad \blacktriangleleft$$

E X A M P L E   3   Carry out the following division:

$$\frac{a - 4b}{2a - 4b} \div \frac{a^2 - 16b^2}{aL - aC - 2bL + 2bC}$$

Solution.   After inverting the divisor, we get

$$\frac{a - 4b}{2a - 4b} \cdot \frac{aL - aC - 2bL + 2bC}{a^2 - 16b^2}$$

$$= \frac{a - 4b}{2a - 4b} \cdot \frac{(aL - aC) + (-2bL + 2bC)}{a^2 - 16b^2} \qquad \text{grouping}$$

$$= \frac{a - 4b}{2(a - 2b)} \cdot \frac{a(L - C) - 2b(L - C)}{(a - 4b)(a + 4b)} \qquad \begin{array}{l}\text{common factors; difference}\\\text{of two squares}\end{array}$$

$$= \frac{\cancel{a - 4b}}{2(a - 2b)} \cdot \frac{(L - C)(a - 2b)}{\cancel{(a - 4b)}(a + 4b)} \qquad \begin{array}{l}\text{common factor } L - C;\\\text{cancellation}\end{array}$$

$$= \frac{L - C}{2(a + 4b)} \qquad\qquad\qquad\qquad \blacktriangleleft$$

 E X A M P L E   4   The discovery of deuterium depended on the fact that the motion of the nucleus reduces the mass of an electron in an atom. If m is the mass of an electron, then the reduced mass m' is given by

$$m' = \frac{mM}{M + m}$$

where M is the mass of the nucleus. Determine the ratio m'/m.

Solution. $\dfrac{m'}{m} = \dfrac{\dfrac{mM}{M+m}}{m} = \dfrac{mM}{M+m} \cdot \dfrac{1}{m} = \dfrac{M}{M+m}$ ◀

E X E R C I S E S / S E C T I O N **5.7**

In Exercises 1–34, carry out the indicated operations and simplify.

1. $\dfrac{2x^2y^2}{3z^3w^3} \cdot \dfrac{9z^2w}{4xy}$

2. $\dfrac{2a^3b^2}{5c^2d^3} \cdot \dfrac{5c^3d^3}{a^3b^3}$

3. $\dfrac{6xy}{ab} \cdot \dfrac{8x^2y}{9a^2b} \cdot \dfrac{ab}{2xy}$

4. $\dfrac{5a^4b^6}{x^3y^3} \div \dfrac{15a^2b^5}{x^3y}$

5. $\dfrac{9x^4y^2z^2}{2a^2b^3c^5} \div \dfrac{15x^3yz^2}{4a^2b^4c^6}$

6. $\dfrac{x^2+xy}{a} \cdot \dfrac{b^2}{x^3+x^2y}$

7. $\dfrac{cd^2}{ax-2ay} \cdot \dfrac{x^2-2xy}{c^3d}$

8. $\dfrac{ay^2z^3}{x^2y-x^2z} \cdot \dfrac{xy-xz}{a^2yz^4}$

9. $\dfrac{x^2-y^2}{2x+y} \cdot \dfrac{4x^2+2xy}{x+y}$

10. $\dfrac{x-2y}{x^3-3x^2y} \div \dfrac{x^2-4y^2}{x-3y}$

11. $\dfrac{x^3+y^3}{x^2-y^2} \cdot \dfrac{x+y}{x^2-xy+y^2}$

12. $\dfrac{x^2-16y^2}{x^3-y^3} \cdot \dfrac{x-y}{x-4y}$

13. $\dfrac{a^2-4b^2}{3a-9b} \cdot \dfrac{a-3b}{2b-a}$

14. $\dfrac{a^2-16b^2}{1-2a} \cdot \dfrac{4a^2-1}{a+4b}$

15. $\dfrac{1-9a^2}{x+3y} \div \dfrac{3a-1}{x^3+27y^3}$

16. $\dfrac{x^2-16y^2}{a^2-1} \div \dfrac{x+4y}{a+1}$

17. $\dfrac{2a^2-ab-b^2}{6x^2+x-1} \div \dfrac{a^2-b^2}{8x+4}$

18. $\dfrac{e_1^2+3e_1e_2-10e_2^2}{e_1^2-e_2^2} \cdot \dfrac{e_1-e_2}{2e_1^2+11e_1e_2+5e_2^2}$

19. $\dfrac{3v_0^2-7v_0+4}{4v_1^2-8v_1-5} \div \dfrac{2v_0^2-5v_0+3}{2v_1^2+7v_1-30}$

20. $\dfrac{12s_0^2-13s_0-4}{5s_0^2+14s_0-3} \cdot \dfrac{10s_0^2-27s_0+5}{3s_0^2+2s_0-8}$

21. $\dfrac{6T^2-JT-J^2}{2K^2-9K-35} \div \dfrac{8T^2-2JT-J^2}{10K^2+13K-30}$

22. $\dfrac{2P^2-15PQ+18Q^2}{12R^2-RS-63S^2} \cdot \dfrac{3R^2-13RS+14S^2}{10P^2-33PQ+27Q^2}$

23. $\dfrac{x^2-5xy+6y^2}{x^2-y^2} \div \dfrac{x^3-8y^3}{x-y}$

24. $\dfrac{x^3-27}{x^3+3xy^2} \cdot \dfrac{x^2y+3y^3}{x^2+3x+9}$

25. $\dfrac{x^2-y^2}{x^3+8y^3} \div \dfrac{x+y}{x^2-2xy+4y^2}$

26. $\dfrac{s^2-t^2+2mt-m^2}{s^2+st} \cdot \dfrac{st+t^2}{s-t+m}$

27. $\dfrac{R+r+1}{R^2-Rr-6r^2} \div \dfrac{R^2+2Rr+r^2-1}{R-3r}$

28. $\dfrac{xy-4y+2x-8}{2x+6} \cdot \dfrac{x+3}{2x^2-3x-20}$

29. $\dfrac{2L^2-L-21}{L^2-16} \div \dfrac{LC-5L+3C-15}{2L-8}$

30. $\dfrac{\theta+2\rho}{\theta^2+4\theta\rho+3\rho^2} \cdot \dfrac{2\theta+2\rho}{\theta+3\rho} \cdot \dfrac{\theta^2-9\rho^2}{2\theta+4\rho}$

31. $\dfrac{c^2-4d^2}{c^2-d^2} \cdot \dfrac{c-d}{2c-4d} \cdot \dfrac{4c+4d}{c^2+cd-2d^2}$

32. $\dfrac{2x^2-6xy}{x^2-10xy+24y^2} \cdot \dfrac{x^2-3xy-4y^2}{4x-12y} \cdot \dfrac{2x-12y}{x^2+3xy+2y^2}$

33. $\left(\dfrac{x^2-16}{x^2-25} \cdot \dfrac{x+5}{x-4}\right) \div \dfrac{x+4}{2x-10}$

34. $\left(\dfrac{mv_1^2-mv_2^2}{v_1^2+2v_1v_2} \cdot \dfrac{2v_1v_2^2}{m^2}\right) \div \dfrac{v_1^2+2v_1v_2-3v_2^2}{3mv_1+6mv_2}$

35. The resistance R (in ohms) of a certain variable resistor is

$$R = \dfrac{2t^2+7t+3}{t+4}$$

and the voltage drop V (in volts) across the resistor is $V = t+3$ (t measured in seconds). Find a simplified expression for the current $i = V/R$.

36. The mass of an object is $(4a^2-4b^2)/(ab)$ and its volume is $(2a+2b)/a^2$. Find a simplified expression for the density (density = mass/volume).

37. Two unequal weights w_1 and w_2 are hanging on a cord passing over a pulley. The tension T on the cord due to the weights, which is the same for both weights, is given by

$$T = \dfrac{2w_1w_2}{w_1+w_2}$$

The acceleration of the system is

$$a = \dfrac{(w_1-w_2)g}{w_1+w_2}$$

Write an expression for $gT(1/a)$ and simplify.

38. In the study of the dispersion of X-rays, the expression

$$A = \frac{Ne^2}{\pi m(f_0{}^2 - f^2)}$$

arises. Multiply A by $m(f_0 + f)$ and simplify.

39. A perfectly flexible cable, suspended from two points at the same height, hangs under its own weight. The tension T_0 at its lowest point is

$$T_0 = \frac{w(s^2 - 4H^2)}{8H}$$

where s is the length of the cable, H the sag, and w the weight per unit length. Find an expression for $T_0[H/(s + 2H)]$.

5.8   Addition and Subtraction of Fractions

In this section we discuss addition and subtraction of algebraic fractions.

In considering the addition and subtraction of algebraic fractions, recall from arithmetic that the fractions $\frac{1}{3}$ and $\frac{1}{6}$ can be added if $\frac{1}{3}$ is changed to $\frac{2}{6}$, so that

$$\frac{2}{6} + \frac{1}{6} = \frac{3}{6} = \frac{1}{2}$$

The number 6 is called the **lowest common denominator (LCD)**.

Since algebraic fractions are added by the same rules, let us first state the definition of the lowest common denominator.

> **Lowest Common Denominator:** The lowest common denominator (LCD) of two or more fractions is an expression that is divisible by every denominator and does not have any more factors than needed to satisfy this condition.

The LCD can be found by the precedure described next.

> **To construct the lowest common denominator** for a set of algebraic fractions, factor each of the denominators. Then the LCD is the product of the factors of the denominators, each with an exponent equal to the largest of the exponents of that factor.

To add (or subtract) two or more fractions, we find the LCD for the fractions and change each fraction to an equivalent fraction having the LCD for its denominator. Next, we add (or subtract) the numerators of the fractions, placing the result over the LCD. Finally, we simplify the resulting fraction.

For example, the LCD of $\frac{3}{4}$ and $\frac{1}{8}$ is 8. To add these fractions, we change $\frac{3}{4}$ to an equivalent fraction with denominator 8:

$$\frac{3}{4} = \frac{3 \cdot 2}{4 \cdot 2} = \frac{6}{8}$$

We now get

$$\frac{3}{4} + \frac{1}{8} = \frac{6}{8} + \frac{1}{8} = \frac{7}{8}$$

In general, we can build up a fraction to an equivalent fraction by *multiplying numerator and denominator by the same nonzero quantity.* Consider the next example.

E X A M P L E **1**   Combine:

$$\frac{x}{x - y} - \frac{x^2}{x^2 - y^2}$$

Solution. Factoring the denominator of the second fraction, we get

$$\frac{x}{x - y} - \frac{x^2}{(x - y)(x + y)}$$

The LCD is the product of these factors, each with exponent 1:

$$\text{LCD} = (x - y)(x + y)$$

Since the denominator of the second fraction is the LCD, we need to adjust only the first fraction: Multiplying numerator and denominator by $x + y$, we get

$$\frac{x}{x - y} = \frac{x}{x - y} \cdot \frac{x + y}{x + y}$$

The resulting fractions have identical denominators. We now get

$$\frac{x}{x - y} - \frac{x^2}{(x - y)(x + y)}$$

$$= \frac{x(x + y)}{(x - y)(x + y)} - \frac{x^2}{(x - y)(x + y)}$$

$$= \frac{x(x + y) - x^2}{(x - y)(x + y)} \qquad \text{combining numerators}$$

$$= \frac{x^2 + xy - x^2}{(x - y)(x + y)} \qquad \text{simplifying the numerator}$$

$$= \frac{xy}{(x - y)(x + y)} \qquad \text{leave denominator in factored form} \qquad \blacktriangleleft$$

E X A M P L E **2**   Combine the following fractions and simplify:

$$\frac{x}{x + 3} - \frac{6x + 6}{x^2 - 9} + \frac{x + 1}{x - 3}$$

Solution. The denominator of the middle fraction is

$$x^2 - 9 = (x - 3)(x + 3)$$

which contains the other denominators. So the factors are $(x - 3)$ and $(x + 3)$, each with exponent 1. Thus

$$\text{LCD} = (x + 3)(x - 3)$$

We now get

$$\frac{x}{x + 3} - \frac{6x + 6}{x^2 - 9} + \frac{x + 1}{x - 3}$$

$$= \frac{x}{x + 3} \frac{x - 3}{x - 3} - \frac{6x + 6}{(x + 3)(x - 3)} + \frac{x + 1}{x - 3} \frac{x + 3}{x + 3}$$

$$= \frac{x(x - 3)}{(x + 3)(x - 3)} - \frac{6x + 6}{(x + 3)(x - 3)} + \frac{(x + 1)(x + 3)}{(x - 3)(x + 3)}$$

$$= \frac{x(x - 3) - (6x + 6) + (x + 1)(x + 3)}{(x + 3)(x - 3)}$$

$$= \frac{x^2 - 3x - 6x - 6 + x^2 + 4x + 3}{(x + 3)(x - 3)} \tag{5.18}$$

$$= \frac{2x^2 - 5x - 3}{(x + 3)(x - 3)} \qquad \textbf{combining like terms}$$

$$= \frac{(2x + 1)(x - 3)}{(x + 3)(x - 3)} \qquad \textbf{factoring the numerator}$$

$$= \frac{2x + 1}{x + 3}$$

The numerator should always be factored in order to reduce the fraction, if possible. ◀

Another method of combining fractions is to supply each fraction with any factors missing in the denominator until all the denominators are the same.

E X A M P L E **2**
(Alternate)

The fractions in Example 2 are

$$\frac{x}{x + 3}, \qquad \frac{6x + 6}{(x + 3)(x - 3)}, \qquad \text{and} \qquad \frac{x + 1}{x - 3}$$

The denominator of the first fraction contains $x + 3$ but not $x - 3$. Supplying this missing factor, the first fraction becomes

$$\frac{x}{x + 3} = \frac{x(x - 3)}{(x + 3)(x - 3)}$$

Similarly, the denominator of the last fraction contains $x - 3$ but not $x + 3$. Supplying this factor, we get

$$\frac{x + 1}{x - 3} = \frac{(x + 1)(x + 3)}{(x - 3)(x + 3)}$$

Note that the denominator of the middle fraction already contains both factors. We now have

$$\frac{x(x - 3)}{(x + 3)(x - 3)} - \frac{6x + 6}{(x + 3)(x - 3)} + \frac{(x + 1)(x + 3)}{(x - 3)(x + 3)}$$

The simplification now proceeds as in Example 2.   ◀

Common errors

1. Failing to change all the signs when subtracting the numerator. For example, in step (5.18) of Example 2, both signs in the numerator of the middle fraction are changed, to become $-6x - 6$.
2. Forgetting that

$$\frac{1}{x} + \frac{1}{y} \neq \frac{1}{x + y}$$

3. Forgetting that

$$\frac{1}{x + y} \neq \frac{1}{x} + \frac{1}{y}$$

In case **2** the correct procedure is

$$\frac{1}{x} + \frac{1}{y} = \frac{y}{xy} + \frac{x}{xy} = \frac{x + y}{xy}$$

In case **3** the fraction

$$\frac{1}{x + y}$$

is already in simplest form and cannot be split up.

E X A M P L E **3**   Combine the following fractions and simplify:

$$\frac{2}{x - 3y} - \frac{1}{x + 5y} - \frac{8y}{x^2 + 2y - 15y^2}$$

Solution. After factoring the denominator of the third fraction, we get

$$\frac{2}{x - 3y} - \frac{1}{x + 5y} - \frac{8y}{(x - 3y)(x + 5y)}$$

Next, we supply the missing factors in the first two fractions:

$$\frac{2}{x - 3y}\frac{x + 5y}{x + 5y} - \frac{1}{x + 5y}\frac{x - 3y}{x - 3y} - \frac{8y}{(x - 3y)(x + 5y)}$$

$$= \frac{2(x + 5y)}{(x - 3y)(x + 5y)} - \frac{x - 3y}{(x + 5y)(x - 3y)} - \frac{8y}{(x - 3y)(x + 5y)}$$

$$= \frac{2(x + 5y) - (x - 3y) - 8y}{(x - 3y)(x + 5y)}$$

$$= \frac{2x + 10y - x + 3y - 8y}{(x - 3y)(x + 5y)}$$

$$= \frac{\overset{1}{\cancel{x + 5y}}}{(x - 3y)\cancel{(x + 5y)}} = \frac{1}{x - 3y}$$

◀

  E X A M P L E **4**   Two perfectly elastic balls collide with a common velocity v. If their respective masses are m and M, then the velocity of m after the collision is

$$v\left(\frac{M}{M + m} - \frac{m}{M + m}\right) - \frac{2vm}{M + m}$$

Simplify this expression.

Solution. $v\left(\dfrac{M}{M + m} - \dfrac{m}{M + m}\right) - \dfrac{2vm}{M + m}$

$$= \frac{vM}{M + m} - \frac{vm}{M + m} - \frac{2vm}{M + m}$$

$$= \frac{vM - vm - 2vm}{M + m}$$

$$= \frac{vM - 3vm}{M + m}$$

◀

E X E R C I S E S / S E C T I O N   **5.8**

In Exercises 1–35, combine the given fractions and simplify.

1. $\dfrac{1}{2} - \dfrac{1}{18} + \dfrac{5}{9}$

2. $\dfrac{5}{36} - \dfrac{3}{108} + \dfrac{1}{9}$

3. $\dfrac{2x + 1}{9} + \dfrac{x}{2} - \dfrac{x}{6}$

4. $\dfrac{1}{x} - \dfrac{3}{2x} - \dfrac{1}{3x}$

5. $\dfrac{3x + 3}{4x} - \dfrac{1}{2x} - \dfrac{1}{x}$

6. $\dfrac{x - 1}{5y} + \dfrac{x}{10y} - \dfrac{7x}{30y}$

7. $\dfrac{3a}{4b} - \dfrac{4a}{9b} + \dfrac{a - 3}{36b}$

8. $\dfrac{x}{x + y} - \dfrac{x - y}{x}$

9. $\dfrac{x + 3y}{x - y} - \dfrac{x - 3y}{x + y}$

10. $\dfrac{x}{2x + y} - \dfrac{x}{x + y}$

11. $\dfrac{2}{xy} - \dfrac{1}{x} - \dfrac{y^2 + 2x - 2y}{xy(x - y)}$

12. $\dfrac{x}{x - 3y} - \dfrac{y}{2x + y}$

13. $\dfrac{x + y}{x + 2y} + \dfrac{x}{x - y}$

14. $\dfrac{3}{x - 2} + \dfrac{4}{x - 3}$

15. $\dfrac{2}{x-3}+\dfrac{1}{x+2}-\dfrac{2x-1}{(x-3)(x+2)}$

16. $\dfrac{x}{(x+y)(x+2y)}+\dfrac{1}{x+y}$

17. $\dfrac{1}{x+3y}+\dfrac{4y}{(x+3y)(x-y)}$

18. $\dfrac{x}{2x-y}-\dfrac{y}{y-3x}$

19. $\dfrac{2x}{3x-y}-\dfrac{2y}{3y-2x}$

20. $\dfrac{1}{x}+\dfrac{2}{y}+\dfrac{1}{x+y}$

21. $\dfrac{1}{x}-\dfrac{1}{y}-\dfrac{y}{x(x+y)}$

22. $\dfrac{y}{x+y}-\dfrac{2y^2}{x(x+y)}+\dfrac{x+2y}{x}$

23. $\dfrac{a+b}{b}-\dfrac{a^2}{b(a+2b)}+\dfrac{a}{a+2b}$

24. $\dfrac{3ab}{a^2-b^2}+\dfrac{a}{b-a}+\dfrac{a-b}{a+b}$

25. $\dfrac{A}{A+B}-\dfrac{B^2}{A(A+B)}+\dfrac{2B}{A}$

26. $\dfrac{3s-t}{s(s+t)}-\dfrac{1}{s+t}+\dfrac{1}{s}$

27. $\dfrac{n}{n-m}-\dfrac{m}{n+m}-\dfrac{2m^2}{n^2-m^2}$

28. $\dfrac{4}{(R+3r)(R+r)}-\dfrac{1}{(R+3r)(R+2r)}-\dfrac{1}{(R+2r)(R+r)}$

29. $\dfrac{3}{(x-y)(x+2y)}-\dfrac{1}{(x+y)(x+2y)}+\dfrac{1}{(y-x)(x+y)}$

30. $\dfrac{2}{x+1}-\dfrac{2x}{x^2-1}+\dfrac{1}{x-1}$

31. $\dfrac{x}{x-2}-\dfrac{2}{x+2}-\dfrac{x^2}{x^2-4}$

32. $\dfrac{T_0}{T_0-2}-\dfrac{2}{T_0+2}+\dfrac{2T_0^2}{4-T_0^2}$

33. $\dfrac{c}{c+2d}-\dfrac{c}{2c+d}$

34. $\dfrac{2}{x^2-y^2}-\dfrac{3}{x^2+xy-2y^2}+\dfrac{1}{x^2+3xy+2y^2}$

35. $\dfrac{1}{2x^2+3xy+y^2}-\dfrac{1}{x^2+4xy+3y^2}+\dfrac{1}{2x^2+7xy+3y^2}$

36. A light rod of length 1 is clamped at both ends and carries a load W at the center. If x is the distance from one end of the rod, then the deflection is

$$y=\frac{W}{k}\left(\frac{x^3}{12}-\frac{x^2}{16}\right),\qquad 0\le x\le\frac{1}{2}$$

where k is a constant. Write y as a single fraction.

37. A hydrogen atom dropping from the nth energy level to the $(n-p)$th energy level emits a photon whose frequency is

$$k\left[\frac{1}{(n-p)^2}-\frac{1}{n^2}\right]$$

where k is a constant. Simplify this expression.

38. The distance between two heat sources with respective intensities a and b is L. The total intensity I of heat at a point between the sources is

$$I=\frac{a}{x^2}+\frac{b}{(L-x)^2}$$

where x is the distance from one of the sources. Express I as a single fraction.

39. In determining the force required to rotate a body about a fixed axis, the expression

$$1-\frac{L^2}{k^2+L^2}$$

arises. Simplify.

40. A neutron of mass m moving with velocity v_0 collides with a stationary nucleus of mass M. The resulting velocity V of the system is known to be

$$V=v_0-\frac{Mv_0}{M+m}$$

Combine the two terms on the right side.

41. The combined resistance R of the resistors in Figure 5.1 is

$$R=\frac{R_1R_2}{R_1+R_2}+R_3$$

Write R as a single fraction.

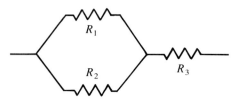

Figure 5.1

5.9   Complex Fractions

If the terms in a fraction are themselves fractions, then the whole expression is called a **complex fraction**. In this section we will see how to simplify such fractions.

The procedure is best seen by means of an arithmetic example. Consider the complex fraction

$$\frac{\dfrac{1}{2} - \dfrac{1}{3}}{\dfrac{1}{4} - \dfrac{1}{2}}$$

We combine the fractions in the numerator and denominator separately and divide the resulting simple fractions. Thus

$$\frac{\dfrac{1}{2} - \dfrac{1}{3}}{\dfrac{1}{4} - \dfrac{1}{2}} = \frac{\dfrac{3}{6} - \dfrac{2}{6}}{\dfrac{1}{4} - \dfrac{2}{4}} = \frac{\dfrac{1}{6}}{-\dfrac{1}{4}} = \left(\frac{1}{6}\right)\left(-\frac{4}{1}\right) = -\frac{2}{3}$$

The procedure for algebraic fractions is similar.

E X A M P L E **1**   Simplify:

$$\frac{\dfrac{1}{x} - \dfrac{1}{y}}{\dfrac{1}{xy}}$$

Solution. The terms in the numerator may be combined in the usual way:

$$\frac{1}{x} - \frac{1}{y} = \frac{y}{xy} - \frac{x}{xy} = \frac{y - x}{xy} \qquad \textbf{LCD = xy}$$

The division can now be performed:

$$\frac{\dfrac{1}{x} - \dfrac{1}{y}}{\dfrac{1}{xy}} = \frac{\dfrac{y - x}{xy}}{\dfrac{1}{xy}} \qquad \textbf{combine fractions in the numerator}$$

$$= \frac{y - x}{xy} \cdot \frac{xy}{1} = y - x \qquad \textbf{invert and multiply}$$

◀

E X A M P L E **2**   Simplify the fraction:

$$\frac{\dfrac{1}{x+y}+\dfrac{1}{x-y}}{\dfrac{x}{(x-y)(x+y)}}$$

Solution. For the terms in the numerator we have

$$\frac{1}{x+y}+\frac{1}{x-y}=\frac{1(x-y)}{(x+y)(x-y)}+\frac{1(x+y)}{(x+y)(x-y)}$$

$$=\frac{x-y+x+y}{(x+y)(x-y)}$$

$$=\frac{2x}{(x+y)(x-y)}$$

We now divide numerator by denominator to get

$$\frac{2x}{\cancel{(x+y)(x-y)}}\cdot\frac{\cancel{(x-y)(x+y)}}{x}=2 \qquad \blacktriangleleft$$

E X A M P L E **2**
(Alternate)

Since the LCD of all the fractions in Example 2 is $(x-y)(x+y)$, a simple alternative to the above method is to multiply the numerator and denominator of the complex fraction by $(x-y)(x+y)$, thereby reducing the complex fraction to a simple fraction. *Remember to multiply all the terms.*

$$\frac{\dfrac{1}{x+y}\cdot\dfrac{(x-y)(x+y)}{1}+\dfrac{1}{x-y}\cdot\dfrac{(x-y)(x+y)}{1}}{\dfrac{x}{(x-y)(x+y)}\cdot\dfrac{(x-y)(x+y)}{1}}$$

$$=\frac{\dfrac{1}{\cancel{x+y}}\cdot\dfrac{(x-y)\cancel{(x+y)}}{1}+\dfrac{1}{\cancel{x-y}}\cdot\dfrac{\cancel{(x-y)}(x+y)}{1}}{\dfrac{x}{\cancel{(x-y)(x+y)}}\cdot\dfrac{\cancel{(x-y)(x+y)}}{1}}$$

$$=\frac{(x-y)+(x+y)}{x}=\frac{2x}{x}=2 \qquad \blacktriangleleft$$

The remaining examples further illustrate the alternate technique discussed in the previous example.

E X A M P L E **3** Simplify the fraction:

$$\frac{\dfrac{R}{R+3}+\dfrac{R}{R^2-9}}{\dfrac{1}{R-3}+1}$$

Solution. Since $R^2 - 9 = (R - 3)(R + 3)$, it follows that

$$\text{LCD} = (R - 3)(R + 3)$$

As before, multiplying the numerator and denominator of the complex fraction by the LCD will reduce the fraction directly.

$$\frac{\dfrac{R}{\cancel{R+3}}\cdot\dfrac{(R-3)\cancel{(R+3)}}{1}+\dfrac{R}{\cancel{R^2-9}}\cdot\dfrac{\cancel{(R-3)(R+3)}}{1}}{\dfrac{1}{\cancel{R-3}}\cdot\dfrac{\cancel{(R-3)}(R+3)}{1}+1\cdot(R-3)(R+3)}$$

$$= \frac{R(R-3)+R}{(R+3)+(R-3)(R+3)}$$

$$= \frac{R^2 - 3R + R}{R + 3 + R^2 - 9} \qquad \text{removing parentheses}$$

$$= \frac{R^2 - 2R}{R^2 + R - 6} \qquad \text{collecting like terms}$$

$$= \frac{R\cancel{(R-2)}}{(R+3)\cancel{(R-2)}} \qquad \text{factoring and canceling}$$

$$= \frac{R}{R+3} \qquad \blacktriangleleft$$

 E X A M P L E **4** Just as electrical components can be connected in parallel, blood vessels that branch out and come together again are said to be connected in parallel. Each such vessel offers a certain resistance to the flow of blood. If r_1, r_2, and r_3 are the respective resistance forces of the blood vessels in parallel, then the combined resistance is given by

$$\frac{1}{\dfrac{1}{r_1}+\dfrac{1}{r_2}+\dfrac{1}{r_3}}$$

Simplify this expression.

Solution.

$$\frac{1}{\dfrac{1}{r_1}+\dfrac{1}{r_2}+\dfrac{1}{r_3}} = \frac{1}{\dfrac{1}{r_1}+\dfrac{1}{r_2}+\dfrac{1}{r_3}} \cdot \frac{r_1 r_2 r_3}{r_1 r_2 r_3}$$

$$= \frac{r_1 r_2 r_3}{r_2 r_3 + r_1 r_3 + r_1 r_2} \qquad \blacktriangleleft$$

E X E R C I S E S / S E C T I O N   **5.9**

In Exercises 1–26, simplify the complex fractions.

1. $\dfrac{1+\dfrac{1}{3}}{2+\dfrac{2}{3}}$

2. $\dfrac{\dfrac{1}{7}+\dfrac{2}{7}}{1+\dfrac{2}{7}}$

3. $\dfrac{1+\dfrac{1}{3}}{2-\dfrac{1}{6}}$

4. $\dfrac{\dfrac{1}{3}+\dfrac{1}{2}}{1-\dfrac{5}{6}}$

5. $\dfrac{3-\dfrac{1}{x}}{9-\dfrac{1}{x^2}}$

6. $\dfrac{\dfrac{1}{x}-2}{4-\dfrac{1}{x^2}}$

7. $\dfrac{1-\dfrac{16}{x^2}}{1+\dfrac{4}{x}}$

8. $\dfrac{1-\dfrac{9}{x^2}}{1-\dfrac{3}{x}}$

9. $\dfrac{\dfrac{1}{C_1}+\dfrac{1}{C_2}}{\dfrac{1}{C_1 C_2}}$

10. $\dfrac{v_1-\dfrac{v_2^{\,2}}{v_1}}{1-\dfrac{v_2}{v_1}}$

11. $\dfrac{1+\dfrac{3}{x}-\dfrac{10}{x^2}}{1-\dfrac{4}{x}+\dfrac{4}{x^2}}$

12. $\dfrac{2-\dfrac{11}{y}-\dfrac{6}{y^2}}{2-\dfrac{5}{y}-\dfrac{3}{y^2}}$

13. $\dfrac{3-\dfrac{17}{h}+\dfrac{10}{h^2}}{3+\dfrac{10}{h}-\dfrac{8}{h^2}}$

14. $\dfrac{s-\dfrac{3s}{s+6}}{s+\dfrac{9}{s+6}}$

15. $\dfrac{w-\dfrac{w}{w-5}}{w-\dfrac{6}{w-5}}$

16. $\dfrac{z+\dfrac{4z}{z-3}}{z-\dfrac{4}{z-3}}$

17. $\dfrac{\dfrac{1}{\beta+1}+\dfrac{1}{\beta^2-1}}{\dfrac{\beta}{\beta-1}}$

18. $\dfrac{a-\dfrac{3a+2}{a+2}}{a-\dfrac{2a+1}{a+2}}$

19. $\dfrac{\dfrac{1}{E-1}+\dfrac{1}{E-2}}{1+\dfrac{1}{E-2}}$

20. $\dfrac{\dfrac{1}{t+2}-\dfrac{t}{t-3}}{\dfrac{1}{t+2}-2}$

21. $\dfrac{\dfrac{k}{k+1}-\dfrac{6}{(k+1)^2}}{1-\dfrac{9}{(k+1)^2}}$

22. $\dfrac{\dfrac{y}{y-3}+\dfrac{1}{y+1}}{\dfrac{y}{y+1}-\dfrac{1}{y+1}}$

23. $\dfrac{\dfrac{x}{x-2}-\dfrac{2}{(x-1)(x-2)}}{\dfrac{x-4}{x-1}}$

24. $\dfrac{\dfrac{b}{a-2b}+1}{\dfrac{b^2}{a-2b}+a}$

25. $\dfrac{\dfrac{1}{t-3}-1}{\dfrac{1}{t+1}-2}$

26. $\dfrac{1-\dfrac{1}{x-1}}{1-\dfrac{1}{1-\dfrac{1}{x}}}$

27. If two resistors R_1 and R_2 (in ohms) are connected in parallel, then the combined resistance (in ohms) is given by

$$\frac{1}{\dfrac{1}{R_1}+\dfrac{1}{R_2}}$$

Simplify this expression.

28. When the rate of decrease of the function $f(x)=1/x$ is studied in calculus, the following expression has to be simplified:

$$\frac{\dfrac{1}{x+h}-\dfrac{1}{x}}{h}$$

Carry out this simplification.

29. If p is the distance of an object from a lens whose focal length is f, then the distance of the image from the lens is given by

$$\frac{1}{\dfrac{1}{f}-\dfrac{1}{p}}$$

Simplify this expression.

30. Simplify the following expression from a problem on vibrating membranes:

$$\frac{1}{1 + \left(\dfrac{y}{x}\right)^2}\left(-\frac{y}{x^2}\right)$$

31. The total resistance R of the circuit in Figure 5.2 is given by

$$R = \frac{1}{\dfrac{1}{R_1} + \dfrac{1}{R_2 + R_3}}$$

Simplify the expression for R.

32. A light cord wrapped around the rim of a wheel of radius r has a mass m hanging on the end. The acceleration of the mass is given by

$$a = \frac{gm}{m + \dfrac{I}{r^2}}$$

where g is the acceleration due to gravity and I the "moment of inertia" of the wheel. Simplify this expression.

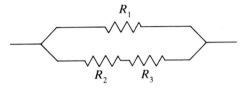

Figure 5.2

5.10 Fractional Equations

In this section we will continue the study of fractional equations begun in Section 2.2. First let us return to the equation in Example 5, Section 2.2:

$$\frac{1}{5}x + 1 = \frac{1}{15}x - \frac{1}{5}$$

Recall that the simplest way to solve such an equation is to **clear fractions** by multiplying both sides of the equation by the LCD, in this case **15**. Then we get

$$15\left(\frac{1}{5}x + 1\right) = 15\left(\frac{1}{15}x - \frac{1}{5}\right)$$
$$3x + 15 = x - 3$$
$$3x - x = -3 - 15$$
$$2x = -18$$
$$x = -9$$

> **Clearing Fractions:** To solve an equation containing fractions, *clear the fractions* by multiplying both sides of the equation by the LCD of all the fractions in the equation.

E X A M P L E **1**   Solve the equation:

$$\frac{1}{2(x-1)} + \frac{1}{x-1} = \frac{3}{2}$$

Solution. We multiply both sides of the equation by the LCD, which is $2(x - 1)$
Then

$$\frac{2(x - 1)}{1}\left(\frac{1}{2(x - 1)} + \frac{1}{x - 1}\right) = \frac{2(x - 1)}{1} \cdot \frac{3}{2}$$

$$\frac{2(x - 1)}{1}\frac{1}{2(x - 1)} + \frac{2(x - 1)}{1}\frac{1}{x - 1} = \frac{2(x - 1)}{1} \cdot \frac{3}{2}$$

$$\frac{\cancel{2(x - 1)}}{1}\frac{1}{\cancel{2(x - 1)}} + \frac{\cancel{2(x - 1)}}{1}\frac{1}{\cancel{x - 1}} = \frac{\cancel{2}(x - 1)}{1} \cdot \frac{3}{\cancel{2}}$$

$$1 + 2 = 3(x - 1)$$

$$3 = 3x - 3$$

$$x = 2$$

Check:

$$\frac{1}{2(2 - 1)} + \frac{1}{2 - 1} = \frac{1}{2} + 1 = \frac{3}{2}$$

which is equal to the right side.   ◄

E X A M P L E **2**   Solve the equation:

$$\frac{1}{x - 3} - \frac{4}{x + 4} = \frac{x}{x^2 + x - 12}$$

Solution. The trinomial can be written

$$x^2 + x - 12 = (x + 4)(x - 3)$$

and turns out to be the LCD. Multiplying both sides of the equation, we get

$$\frac{(x + 4)(x - 3)}{1}\left(\frac{1}{x - 3} - \frac{4}{x + 4}\right) = \frac{(x + 4)(x - 3)}{1}\frac{x}{(x + 4)(x - 3)}$$

$$\frac{(x + 4)\cancel{(x - 3)}}{1}\frac{1}{\cancel{x - 3}} + \frac{\cancel{(x + 4)}(x - 3)}{1}\left(-\frac{4}{\cancel{x + 4}}\right) = \frac{\cancel{(x + 4)}\cancel{(x - 3)}}{1}\frac{x}{\cancel{(x + 4)}\cancel{(x - 3)}}$$

$$(x + 4) - 4(x - 3) = x$$

$$x + 4 - 4x + 12 = x$$

$$-3x + 16 = x$$

$$-4x = -16$$

$$x = 4$$

Check:

Left Side	Right Side

$$\frac{1}{4-3} - \frac{4}{4+4} = 1 - \frac{1}{2} = \frac{1}{2} \qquad \frac{4}{4^2 + 4 - 12} = \frac{1}{2} \quad \checkmark \qquad \blacktriangleleft$$

Some fractional equations do not have any solution, as shown in the next two examples.

E X A M P L E **3**   Solve the equation:

$$\frac{x-5}{x-6} - 1 = \frac{2}{x-6}$$

Solution. Clearing fractions, we get

$$(x-5) - (x-6) = 2$$
$$x - 5 - x + 6 = 2$$
$$1 = 2$$

which is impossible. We conclude that the equation has no solution. $\blacktriangleleft$

E X A M P L E **4**   Solve the equation:

$$\frac{x+4}{x+3} + 2 = \frac{1}{x+3}$$

Solution. Clearing fractions, we have

$$(x+4) + 2(x+3) = 1$$
$$x + 4 + 2x + 6 = 1$$
$$3x = -9$$
$$x = -3$$

Extraneous roots

Check: Since substituting into the given equation leads to division by zero, the equation has no solution. The (apparent) root $x = -3$ is called **extraneous.** $\blacktriangleleft$

Remark on extraneous roots. Multiplying both sides of an equation by an expression containing the unknown may give rise to extraneous roots. For example, the equation

$$x - 2 = 0$$

has only one root, $x = 2$. If both sides are multiplied by $x + 1$, we get

$$(x + 1)(x - 2) = 0$$

which has two roots, $x = 2$ and $x = -1$. The root $x = -1$ does not, of course, satisfy the original equation $x - 2 = 0$. While we normally avoid multiplying both sides of an equation by an expression containing an unknown, such multiplications are necessary with fractional equations. Consequently, extraneous roots can result.

If two resistors R_1 and R_2 (in ohms) are connected in parallel, then the combined resistance R_T (in ohms) is given by

$$\frac{1}{R_T} = \frac{1}{R_1} + \frac{1}{R_2} \tag{5.19}$$

(See Figure 5.3)

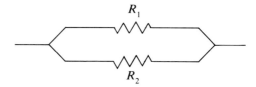

Figure 5.3

 E X A M P L E **5** The combined resistance of two resistors connected in parallel is measured to be 4.0 Ω. If one of the resistors has a resistance of 6.0 Ω, what is the resistance of the other?

Solution. By formula (5.19)

$$\frac{1}{R_1} + \frac{1}{6.0} = \frac{1}{4.0}$$

To clear fractions, we multiply both sides by $12R_1$:

$$12 + 2.0R_1 = 3.0R_1$$
$$12 = 3.0R_1 - 2.0R_1$$
$$R_1 = 12 \ \Omega$$

Check: Leaving out final zeros, we get

$$\frac{1}{12} + \frac{1}{6} = \frac{1}{12} + \frac{2}{12} = \frac{3}{12} = \frac{1}{4}$$

which equals the right side.   ◄

E X E R C I S E S / S E C T I O N   5.10

In Exercises 1–32 solve each equation for x.

1. $\dfrac{4}{x} + \dfrac{2}{3} = 2$

2. $1 - \dfrac{2}{x} = \dfrac{1}{2}$

3. $\dfrac{1}{x-2} = \dfrac{1}{2x}$

4. $\dfrac{x}{x-7} = 2$

5. $\dfrac{x-3}{x+4} = \dfrac{x-4}{x+2}$

6. $\dfrac{x-6}{x+4} = \dfrac{x-4}{x+2}$

7. $\dfrac{x-2}{x} = \dfrac{x-3}{x+1}$

8. $\dfrac{x+6}{x-5} = \dfrac{x}{x-2}$

9. $\dfrac{x-7}{x-4} = \dfrac{x-2}{x}$

10. $\dfrac{1}{x} = \dfrac{1}{x-2}$

11. $\dfrac{x-3}{x+2} = \dfrac{x-4}{x+1}$

12. $\dfrac{2x-3}{x+2} = \dfrac{2x-4}{x-3}$

13. $\dfrac{2x}{x-2} + \dfrac{1}{x+3} = 2$

14. $\dfrac{1}{x+1} + \dfrac{2x}{2-x} + 2 = 0$

15. $\dfrac{x+3}{x-5} - 2 = \dfrac{1}{x-5}$

16. $\dfrac{x-5}{x-6} - 2 = \dfrac{1}{x-6}$

17. $\dfrac{x+1}{x-8} = 4 + \dfrac{9}{x-8}$

18. $\dfrac{x}{x-4} - \dfrac{8}{x^2-4x} = 1$

19. $\dfrac{x}{x-1} - \dfrac{3}{2x^2-2x} = 1$

20. $\dfrac{x}{x-2} - \dfrac{x+1}{x^2-4} = 1$

21. $\dfrac{1}{x-1} - \dfrac{x+2}{x^2-1} = \dfrac{1}{x+1}$

22. $\dfrac{x+3}{x^2-4} - \dfrac{1}{x+2} = \dfrac{1}{x-2}$

23. $\dfrac{2}{x-3} - \dfrac{2}{x+3} = \dfrac{x+2}{x^2-9}$

24. $\dfrac{4}{x-2} - \dfrac{2}{x+2} = \dfrac{x+3}{x^2-4}$

25. $\dfrac{2}{x+3} - \dfrac{5}{x-2} = \dfrac{x+1}{x^2+x-6}$

26. $\dfrac{5}{x-1} - \dfrac{6}{x+3} = \dfrac{x+7}{x^2+2x-3}$

27. $\dfrac{4}{x-3} - \dfrac{5}{x+4} = \dfrac{x+13}{x^2+x-12}$

28. $\dfrac{3}{x+4} + \dfrac{x-8}{x^2+6x+8} = \dfrac{1}{x+}$

29. $\dfrac{5}{2x+6} - \dfrac{3}{x+3} = \dfrac{1}{x}$

30. $\dfrac{2}{x+2} - \dfrac{1}{x} = \dfrac{1}{2x+4}$

31. $\dfrac{2}{3x-6} + \dfrac{1}{6x} = \dfrac{1}{2x-4}$

32. $\dfrac{4}{3x-18} + \dfrac{2}{x} = \dfrac{4}{3x}$

In Exercises 33–39, solve each formula for the indicated letter.

33. $\dfrac{1}{c} + \dfrac{1}{d} = 3; \quad d$

34. $\dfrac{c_1}{c_1+c_2} = \dfrac{1}{q}; \quad c_1$

35. $\dfrac{1}{R} = \dfrac{R_1+3}{3R_1}; \quad R_1$

36. $\dfrac{1}{V_1} - \dfrac{2}{V_2} = \dfrac{a}{b}; \quad V_2$

37. $\dfrac{p}{q} = \dfrac{f}{q-f}; \quad f$

38. $\dfrac{1}{p} + \dfrac{1}{q} = \dfrac{2}{R}; \quad p$

39. $\dfrac{1}{R_1} + \dfrac{1}{R_2} = \dfrac{1}{R}; \quad R$

40. The combined capacitance C of two capacitors C_1 and C_2 connected in series (Figure 5.4) is given by

$$\frac{1}{C} = \frac{1}{C_1} + \frac{1}{C_2}$$

Solve for C.

Figure 5.4

41. Two resistors are connected in parallel and have a combined resistance of 5.0 Ω. If one is a 10-Ω resistor, find the resistance of the other.

42. Two blood vessels are connected in parallel (Figure 5.5). If r_1 and r_2 denote the respective resistances of the two vessels to the flow of blood, then the combined resistance r is given by

Figure 5.5

$$\frac{1}{r} = \frac{1}{r_1} + \frac{1}{r_2}$$

If r and r_1 are measured to be 15.0 dynes and 25.0 dynes, respectively, find r_2.

43. According to Hooke's law, the force F required to stretch a spring x units is $F = kx$; k is called the *spring constant*. If two springs with respective spring constants k_1 and k_2 are connected as shown in Figure 5.6, then the spring constant k of the combination is related to k_1 and k_2 by the formula

$$\frac{1}{k} = \frac{1}{k_1} + \frac{1}{k_2}$$

Solve this formula for k.

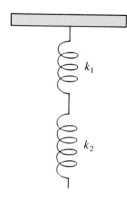

Figure 5.6

REVIEW EXERCISES/CHAPTER 5

In Exercises 1–12, find each product directly (an intermediate step may be needed in some cases).

1. $4s^2t^3(2s - 3st^2 + 5s^2t)$

2. $(F_1 + 4F_2)^2$

3. $(2i_1 - 3i_2)^2$

4. $(x - 3z)(x + 3z)$

5. $(2s - 5t)(2s + 5t)$

6. $(5w_1 - 11w_2)^2$

7. $(2c - 5d)(c + 3d)$

8. $(3a - 7y)(3a + y)$

9. $xy(x - 3y)(x + 3y)$

10. $6(a - 4b)(a - 2b)$

11. $(v + 2w + 1)^2$

12. $(a - 3b + 1)^2$

In Exercises 13–42, factor each expression.

13. $4ax - 4bx$

14. $8R^3C^2 + 12R^2C^3 - 4R^2C^2$

15. $3p^3q^4 + pq$

16. $A^2 - W^2$

17. $4\beta^2 - \gamma^2$

18. $2a^2 - 18b^2$

19. $16L^2 - 64C^2$

20. $b^4 - 1$

21. $27h^3 + g^3$

22. $8F^3 - 1$

23. $(a + 2b)^3 - 1$

24. $1 - (x - y)^3$

25. $a^6 - 1$

26. $x^2 + 2x - 8$

27. $v_0{}^2 - v_0 - 12$

28. $2a^2 + 7ab - 4b^2$

29. $3v_1{}^2 + 2v_1v_2 - v_2{}^2$

30. $P^3S^6 + 1$

31. $C_1{}^2 - 2C_1C_2 - 15C_2{}^2$

32. $4x^2 - 9xw + 2w^2$

33. $9y^2 - 3yz + z^2$

34. $2a^2 - 2ac - 24c^2$

35. $x^2 - 2xy + y^2 - 1$

36. $a^2 + 2ad + d^2 - 4$

37. $s^2 - st - 2t^2 + s + t$

38. $(x + b)^2 - (x + b) - 6$

39. $ax + bx - ay - by$

40. $sv - 2sw + 2tv - 4tw$

41. $4x^2 - a^2 + 4xy + y^2$

42. $1 - a^2 - 4b^2 + 4ab$

In Exercises 43–61, carry out the operations and simplify.

43. $\dfrac{x^2 - 9y^2}{x - 3y}$

44. $\dfrac{a^3 - 27c^3}{a^2 + 3ac + 9c^2}$

45. $\dfrac{q^2 + 6qr - 7r^2}{q^2 + 8qr - 9r^2}$

46. $\dfrac{a - 2t}{a - t} \cdot \dfrac{3a - 3t}{a^2 - 4t^2}$

47. $\dfrac{a^2 + 2ad + d^2}{a^3d^3} \div \dfrac{a + d}{2ad^2}$

48. $\dfrac{x^2 - 16w^2}{x^3 - 8w^3} \cdot \dfrac{x - 2w}{x - 4w}$

49. $\left(\dfrac{ax + ay - bx - by}{x - y} \cdot \dfrac{x^2 - y^2}{a - b} \right) \div (x + y)$

50. $\dfrac{4f^2 - 4fg + g^2 - 1}{3f - 6g} \div \dfrac{2f - g - 1}{2f - 4g}$

51. $\dfrac{y}{x + 2y} - \dfrac{2y^2}{x(x + 2y)} + \dfrac{x + y}{x}$

52. $\dfrac{a - 3c}{c} - \dfrac{a^2}{c(a - 3c)} + \dfrac{a}{a - 3c}$

53. $\dfrac{y + w}{y - w} - \dfrac{3yw}{y^2 - w^2} - \dfrac{y}{y + w}$

54. $\dfrac{2y}{x + y} - \dfrac{2xy}{x^2 - y^2} + \dfrac{y}{x - y}$

55. $\dfrac{L}{L - 1} - \dfrac{1}{L + 1} - \dfrac{2}{L^2 - 1}$

56. $\dfrac{2}{n^2 - m^2} - \dfrac{3}{n^2 + nm - 2m^2} + \dfrac{2}{n^2 + 3nm + 2m^2}$

57. $\dfrac{1}{a^2 - b^2} + \dfrac{1}{a^2 + 3ab + 2b^2} - \dfrac{1}{a^2 + ab - 2b^2}$

58. $\dfrac{x}{x - 3y} + \dfrac{y}{2x - y}$

59. $\dfrac{v - w}{v - 2w} - \dfrac{v}{v + w}$

60. $\dfrac{2}{\theta + 3} + \dfrac{1}{\theta - 3}$

61. $\dfrac{2}{\omega + 4} + \dfrac{3}{\omega - 2} - \dfrac{4\omega + 4}{(\omega + 4)(\omega - 2)}$

In Exercises 62–68, simplify each fraction.

62. $\dfrac{1 + \dfrac{2}{y} - \dfrac{3}{y^2}}{2 + \dfrac{3}{y} - \dfrac{5}{y^2}}$

63. $\dfrac{i + \dfrac{2i}{i - 4}}{i + \dfrac{4}{i - 4}}$

64. $\dfrac{v - \dfrac{8}{v - 2}}{v + \dfrac{3v - 20}{v - 2}}$

65. $\dfrac{\dfrac{r_2}{r_1 - 2r_2} + 1}{\dfrac{r_1 r_2}{r_1 - 2r_2} + r_1}$

66. $\dfrac{\dfrac{G}{G - 1} - \dfrac{6}{(G - 1)^2}}{1 - \dfrac{4}{(G - 1)^2}}$

67. $\dfrac{\dfrac{V}{T - V} + \dfrac{T}{T + V}}{\dfrac{T^2 + V^2}{T^2 - V^2}}$

68. $\dfrac{\dfrac{2}{A - 2C} - \dfrac{2A}{(A + C)(A - 2C)}}{\dfrac{4}{A + C}}$

In Exercises 69–77, solve each equation for x.

69. $\dfrac{1}{x + 3} = \dfrac{1}{2x}$

70. $\dfrac{2x - 2}{x - 3} = \dfrac{2x + 3}{x - 1}$

71. $\dfrac{x}{x - 4} + \dfrac{1}{x + 4} = 1$

72. $\dfrac{x - 2}{x + 5} - 2 = \dfrac{2}{x + 5}$

73. $\dfrac{3}{2(x + 3)} - \dfrac{x}{x + 3} + 1 = 0$

74. $\dfrac{x}{x - 3} - \dfrac{6}{x^2 - 3x} = 1$

75. $\dfrac{2}{x - 2} - \dfrac{x}{x^2 + x - 6} = \dfrac{3}{x + 3}$

76. $\dfrac{2}{x - 2} + \dfrac{x}{(2 - x)(x + 2)} = \dfrac{3}{x + 2}$

77. $\dfrac{x - 1}{(x - 3)(x + 3)} + \dfrac{1}{3 - x} = \dfrac{2}{x + 3}$

78. Solve for R:
$$\dfrac{1}{R} = \dfrac{1}{R_1} + \dfrac{1}{R_2} + \dfrac{1}{R_3}$$

79. Solve for a:
$$S = \dfrac{a - ar^n}{1 - r}$$

80. Solve for m_1:
$$w = \dfrac{m_1 m_2}{m_1 + m_2}$$

81. The current (in amperes) in a circuit at any time t (in seconds) is $2t^2 - 19t + 45$. Factor this expression and determine when the current is zero.

82. The combined resistance r of two blood vessels in parallel is given by
$$r = \dfrac{1}{\dfrac{1}{r_1} + \dfrac{1}{r_2}}$$
where r_1 and r_2 are the respective resistances of the individual vessels. Simplify the expression for r.

83. The focal length f of a lens is related to the object distance q and the image distance p by the formula
$$\dfrac{1}{f} = \dfrac{1}{p} + \dfrac{1}{q}$$
If $f = 3.0$ cm and $p = 4.0$ cm, find q.

84. If a body travels along the x-axis so that its position as a function of time is $x(t) = t(t^2 - 4)^4$, the velocity v is given by $v = (t^2 - 4)^4 + 8t^2(t^2 - 4)^3$. Simplify the expression for v.

85. If a mass m_1 strikes a stationary mass m_2, the ratio of the kinetic energy of the system with respect to the floor to the kinetic energy with respect to the center of mass of the system is
$$\dfrac{T - \dfrac{1}{2}(m_1 + m_2)V^2}{T}$$
where $T = \frac{1}{2}m_1 v^2$ and $V = m_1 v/(m_1 + m_2)$. Simplify this expression.

86. If two thin lenses having respective focal lengths f_1 and f_2 are placed in contact, the focal length of the combination is given by
$$\dfrac{1}{f} = \dfrac{1}{f_1} + \dfrac{1}{f_2}$$
Solve this equation for f.

Quadratic Equations

6.1 Solution by Factoring and Pure Quadratic Equations

All the equations introduced in the earlier chapters were of first degree (involving only the first power of the unknown). Many technical problems, however, lead to equations of second degree, involving an unknown raised to the second power. For example, if an object is tossed upward from the ground with initial velocity v_0, then the distance above the ground as a function of time is given by $s = v_0 t - 16t^2$, where s is measured in feet and t in seconds. Equations of second degree are called **quadratic equations**.

In this chapter we study only quadratic equations. Equations of degree higher than two will be taken up in Chapter 14.

A **quadratic** equation has the form

$$ax^2 + bx + c = 0, \qquad a \neq 0 \qquad\qquad (6.1)$$

The equation $ax^2 + bx + c = 0$ is called the **standard form** of the quadratic equation.

In this section we will confine ourselves to those cases in which the left side of equation (6.1) is factorable. The method of solution depends on the following property of real numbers:

$ab = 0$ if and only if $a = 0$ or $b = 0$.

In words, the product of two real numbers is zero if and only if one of the factors is zero.

Consider, for example, the equation

$$x^2 = 8 - 2x$$

For the equation to fit the standard form (6.1), all the terms have to be collected on the left side and written in descending powers of x. Adding $-8 + 2x$ to both sides, we get

$$\text{Step 1. } x^2 + 2x - 8 = 0 \qquad\qquad\qquad (6.2)$$

If we now factor the left side, the equation becomes

$$\text{Step 2. } (x + 4)(x - 2) = 0 \qquad\qquad\qquad (6.3)$$

Next, we set each factor equal to 0:

$$\text{Step 3. } \mathbf{x + 4 = 0} \qquad \mathbf{x - 2 = 0}$$

Finally, we solve each of the resulting linear equations:

$$\text{Step 4. } x = -4 \qquad x = 2$$

The solution is therefore given by two values, $x = -4$ and $x = 2$.

As a check, let us substitute both values into the original equation $x^2 = 8 - 2x$:

$$(\mathbf{-4})^2 = 8 - 2(\mathbf{-4}) \qquad \mathbf{2}^2 = 8 - 2(\mathbf{2})$$
$$16 = 8 + 8 \qquad\qquad 4 = 8 - 4$$
$$16 = 16 \qquad\qquad\quad 4 = 4$$

Since both values check, we see that the equation has two distinct roots.

Finally, note that the roots are unique:

$$(x + 4)(x - 2) = 0$$

if and only if $x + 4 = 0$ or $x - 2 = 0$. But $x + 4 = 0$ if and only if $x = -4$; and $x - 2 = 0$ if and only if $x = 2$.

To illustrate the solution of the equation $x^2 + 2x - 8 = 0$ in (6.2) graphically, consider the function

$$y = x^2 + 2x - 8$$

whose graph appears in Figure 6.1. The solution of $x^2 + 2x - 8 = 0$ consists of the x-intercepts (where $y = 0$).

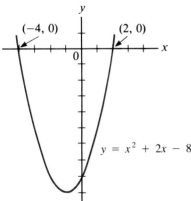

Figure 6.1

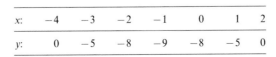

x:	-4	-3	-2	-1	0	1	2
y:	0	-5	-8	-9	-8	-5	0

> **Solution by Factoring:**
>
> 1. Write the equation in standard form by collecting all the terms on the left side.
> 2. Factor the expression on the left side.
> 3. Set each of the factors equal to zero.
> 4. Solve the resulting two linear equations.

E X A M P L E **1** Solve the equation $2x^2 + 15 = 13x$.

Solution. The first step is to write the equation in standard form by transposing $13x$.

Step 1. $2x^2 - 13x + 15 = 0$	**collecting terms on the left side**
Step 2. $(2x - 3)(x - 5) = 0$	**factoring the left side**
Step 3. $2x - 3 = 0 \qquad x - 5 = 0$	**setting each factor equal to 0**
Step 4. $\qquad 2x = 3 \qquad\qquad x = 5$	**solving the resulting linear equations**
$\qquad\qquad x = \dfrac{3}{2} \qquad\qquad x = 5$	

Check:

Left Side	Right Side
$x = \dfrac{3}{2}: \quad 2\left(\dfrac{3}{2}\right)^2 + 15 = \dfrac{39}{2}$	$13\left(\dfrac{3}{2}\right) = \dfrac{39}{2}$
$x = 5: \quad 2(5)^2 + 15 = 65$	$13(5) = 65 \qquad \checkmark$

◄

E X A M P L E **2** Solve the equation $x^2 + 7x = 0$.

Solution. In this equation $c = 0$. As a result, the terms on the left have a common factor x.

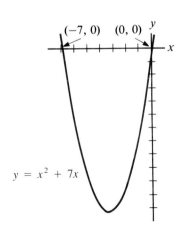

Step 1. $x^2 + 7x = 0$	**given equation (already in standard form)**
Step 2. $x(x + 7) = 0$	**common factor x**
Step 3. $x = 0 \qquad x + 7 = 0$	**setting each factor equal to zero**
Step 4. $x = 0 \qquad\qquad x = -7$	**solution**

The solution can be checked as in Example 1.

The graph of the function $y = x^2 + 7x$ is shown in Figure 6.2. Note that the solution consists of the x-intercepts (where $y = 0$).

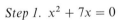

x:	-7	-6	-5	-4	-3	-2	-1	0
y:	0	-6	-10	-12	-12	-10	-6	0

◄

Figure 6.2

Pure quadratic equations

An equation for which $b = 0$ is called a **pure quadratic equation** and may be solved by a different method. For example, to solve the equation

$$x^2 - 4 = 0$$

we first solve for x^2 to obtain

$$x^2 = 4$$

Taking the square root of both sides, we find that $x = \pm\sqrt{4} = \pm 2$, where $\pm$ means *plus or minus*. So the equation has two roots, $x = 2$ and $x = -2$.
Alternatively,

$$x^2 - 4 = (x - 2)(x + 2) = 0$$

leads to the same solution. However, factoring does not work for the equation in the next example.

E X A M P L E **3**   Solve the equation $x^2 - 5 = 0$.

Solution. Adding 5 to both sides of the equation, we get

$$x^2 - 5 + 5 = 0 + 5$$
$$x^2 = 5$$
$$x = \pm\sqrt{5} \qquad \text{taking the square root of each side} \qquad \blacktriangleleft$$

Remark. The left side of the equation $x^2 + 4x + 4 = 0$ is a perfect-square trinomial leading to identical factors:

$$(x + 2)(x + 2) = 0$$

Double root

The solution $x = -2$ and $x = -2$ is called a **repeating root** or a **double root**.

Common errors

1. Forgetting the negative root when solving $x^2 - a^2 = 0$. The roots are $x = \pm a$.
2. Attempting to solve by factoring when the right side is not zero. For example, if

$$(x - 3)(x + 2) = 6$$

we may *not* conclude that $x - 3 = 6$ and $x + 2 = 6$. Instead, we need to write the equation in the form $ax^2 + bx + c = 0$:

$$x^2 - x - 6 = 6$$
$$x^2 - x - 12 = 0$$
$$(x - 4)(x + 3) = 0$$
$$x = 4, -3$$

Historical note. As mentioned in Chapter 1, the first systematic study of second-degree equations was undertaken by al-Khowarizmi in Baghdad. He gave an exhaustive exposition of various cases using ingenious geometric arguments, in the manner of the ancient Greeks. Consequently, al-Khowarizmi's algebra was rhetorical, using words and drawings instead of symbols. Further progress in algebra was slow until algebraic notation was introduced. This far-reaching innovation was due to the French lawyer Franciscus Vieta (1540–1603), who recognized the advantage of using letters to denote both known and unknown quantities.

Other now-familiar symbols were introduced only gradually. In medieval times the letters *p* and *m* were widely used to denote addition and subtraction, and the Latin word *cosa* for the unknown. The signs $+$ and $-$ first appeared in print in a book published in 1489 by Johann Widman, a lecturer in Leipzig. The English mathematician William Oughtred (1574–1660) popularized the symbol $\times$ for multiplication, and Johan Rahn of Switzerland first used the sign $\div$ for division in 1659. The French philosopher René Descartes (1596–1650) introduced the exponential notation, and the Englishman Robert Recorde (1510–1558) used the symbol $=$ (two parallel lines) for equality because, as he put it, "noe 2. thynges can be moare equalle."

E X E R C I S E S / S E C T I O N   **6.1**

In Exercises 1–12, solve the given pure quadratic equations.

1. $x^2 - 1 = 0$

2. $x^2 - 25 = 0$

3. $x^2 - 36 = 0$

4. $x^2 - 121 = 0$

5. $x^2 - 9 = 0$

6. $x^2 - 16 = 0$

7. $x^2 - 10 = 0$

8. $x^2 - 12 = 0$

9. $2x^2 - 32 = 0$

10. $4x^2 - 1 = 0$

11. $36x^2 - 25 = 0$

12. $49x^2 - 16 = 0$

In Exercises 13–40, solve the given quadratic equations by the method of factoring.

13. $x^2 + x - 2 = 0$

14. $x^2 + 7x + 10 = 0$

15. $x^2 + 2x - 24 = 0$

16. $x^2 + 4x - 21 = 0$

17. $2x^2 - 5x - 3 = 0$

18. $2x^2 - 7x - 15 = 0$

19. $3x^2 + 7x + 2 = 0$

20. $6x^2 + 11x + 4 = 0$

21. $4x^2 + 5x - 6 = 0$

22. $5x^2 + 16x + 12 = 0$

23. $5x^2 + 8x - 21 = 0$

24. $4x^2 + 29x - 24 = 0$

25. $4x^2 + 4x = 15$

26. $6x^2 + 7x = 5$

27. $30x^2 = 7x + 15$

28. $12x^2 = 7x + 10$

29. $8x^2 = 2x + 45$

30. $40x^2 = 67x - 28$

31. $7x^2 + 4x = 11$

32. $14x^2 + 53x + 45 = 0$

33. $18x^2 + 17x = 15$

34. $3x^2 = 10x + 13$

35. $11x^2 = 76x + 7$

36. $4x^2 = 49x - 90$

37. $9x^2 + 24x + 16 = 0$

38. $25x^2 - 10x + 1 = 0$

39. $16x^2 - 8x + 1 = 0$

40. $4x^2 - 20x + 25 = 0$

41. The path of an object tossed at an angle of 45° to the ground is $y = x - 32x^2/v_0{}^2$, where v_0 is the initial velocity in feet per second and y is the distance in feet above the ground. How far from the starting point ($x = 0$) will the object land?

42. The weekly profit P of a company is $P = x^4 - 30x^3, x \geq 1$, where x is the week in the year. During what week is the profit equal to zero? (*Hint:* Factor out x^3.)

43. The load on a beam of length L is such that the deflection is given by $d = 3x^4 - 4Lx^3 + L^2x^2$, where x is the distance from one end. Determine where the deflection is zero. (*Hint:* Factor out x^2.)

44. The formula for the output of a battery is $P = VI - RI^2$. For what values of I is the output equal to zero?

6.2   Solution by Completing the Square

In the last section we restricted our attention to factorable quadratic equations. In this section we will take up a general method for solving any given quadratic equation. The procedure is referred to as **completing the square**.

Completing the square depends on the fact that any quadratic equation can be written in the form

$$(x + b)^2 = a \qquad\qquad (6.4)$$

To understand this, recall that the square of a binomial is given by

$$(x + b)^2 = x^2 + 2bx + b^2$$

Perfect-square trinomial

Looking at the right side, observe that *a trinomial in x with a coefficient of x^2 equal to 1 is a perfect square if the square of one-half the coefficient of x is equal to the third term.* For example,

$$x^2 + 6x + 9$$

is a perfect square, since $(\frac{1}{2} \cdot 6)^2 = 9$; that is, the square of one-half of 6 is equal to the third term, 9. Thus

$$x^2 + 6x + 9 = (x + 3)^2 \qquad \text{(x + 3)}^2 = x^2 + 2(x \cdot 3) + 3^2$$

Similarly,

$$x^2 - \frac{3}{2}x + \frac{9}{16}$$

is a perfect square since

$$\frac{9}{16} = \left[\frac{1}{2} \cdot \left(-\frac{3}{2} \right) \right]^2$$

so that

$$x^2 - \frac{3}{2}x + \frac{9}{16} = \left(x - \frac{3}{4} \right)^2 \qquad \left(x - \frac{3}{4} \right)^2 = x^2 + 2\left(-\frac{3}{4}x \right) + \left(-\frac{3}{4} \right)^2$$

> The method of completing the square consists of rewriting one side of the equation so that it forms a perfect square.

Consider the next example.

E X A M P L E 1   Solve the equation $x^2 - 6x + 8 = 0$ by completing the square.

Solution. The first step is to transpose the 8 (or add -8 to both sides) in order to retain only the x^2 and x terms on the left side. Thus

$$x^2 - 6x = -8$$

The critical step is to complete the square on the left side by adding to both sides the square of one-half the coefficient of x, or $[\frac{1}{2} \cdot (-6)]^2 = 9$. We then get

$$x^2 - 6x + 9 = -8 + 9 \qquad \textbf{adding 9 to both sides}$$

The left side is now a perfect square, so that the equation can be written

$$(x - 3)^2 = 1 \qquad \textbf{(x − 3)}^2 = \textbf{x}^2 + \textbf{2(−3x)} + \textbf{(−3)}^2$$

The resulting pure quadratic form can be solved by taking the square root of both sides, yielding the two linear equations

$$\sqrt{(x - 3)^2} = \pm \sqrt{1} \qquad \textbf{taking the square root of each side}$$
$$x - 3 = \pm 1 \qquad \sqrt{\textbf{a}^2} = \textbf{a}$$

Solving, we get

$$x = 3 \pm 1 \qquad \textbf{adding 3 to both sides}$$

which gives $x = 4$ and $x = 2$.

Check:

$$x = 4: \quad (4)^2 - 6(4) + 8 = 0$$
$$x = 2: \quad (2)^2 - 6(2) + 8 = 0 \quad \checkmark$$

◄

Let us now summarize the procedure for completing the square.

Solution by Completing the Square:

1. Write the equation in the form $ax^2 + bx = -c$.
2. Multiply each side by $1/a$ (or divide by a).
3. Complete the square on the left side by adding the square of one-half the coefficient of x to both sides.
4. Write the left side as a square; simplify the right side.
5. Take the square root of both sides.
6. Solve the resulting two linear equations.

E X A M P L E **2**   Solve the equation $2x^2 + 6x - 3 = 0$ by completing the square.

Solution. Following the procedure for completing the square, we get

$$2x^2 + 6x - 3 = 0 \qquad \textbf{given equation}$$
$$\textit{Step 1.} \ \ 2x^2 + 6x \quad = 3 \qquad \textbf{transposing −3}$$
$$\textit{Step 2.} \ \ x^2 + 3x \qquad = \frac{3}{2} \qquad \textbf{multiplying by } \frac{1}{2} \textbf{ (or dividing by 2)}$$

Note that the square of one-half the coefficient of x is

$$\left(\frac{1}{2} \cdot 3\right)^2 = \frac{9}{4}$$

This number has to be added to both sides to complete the square. It follows that

$$Step\ 3.\quad x^2 + 3x + \frac{9}{4} = \frac{3}{2} + \frac{9}{4} \qquad \text{adding } \frac{9}{4} \text{ to both sides}$$

and

$$Step\ 4.\quad \left(x + \frac{3}{2}\right)^2 = \frac{6}{4} + \frac{9}{4} = \frac{15}{4} \qquad \begin{array}{l}\text{factoring the left side and}\\ \text{simplifying the right side}\end{array}$$

Taking the square root of both sides, we get

$$Step\ 5.\quad x + \frac{3}{2} = \pm\sqrt{\frac{15}{4}} = \pm\frac{\sqrt{15}}{2} \qquad \sqrt{\frac{x}{y}} = \frac{\sqrt{x}}{\sqrt{y}}$$

Solving for x,

$$Step\ 6.\quad x = -\frac{3}{2} \pm \frac{\sqrt{15}}{2} \qquad \text{transposing } +\frac{3}{2}$$

or

$$x = \frac{-3 \pm \sqrt{15}}{2} \qquad \text{combining the fractions}$$

The roots can also be written separately as

$$x = \frac{-3 + \sqrt{15}}{2} \qquad \text{and} \qquad x = \frac{-3 - \sqrt{15}}{2}$$

(Although highly desirable, checking the solution is difficult at this point, since we do not discuss the multiplication of radical expressions of this complexity until Chapter 10.) ◀

Complex Roots

It was noted in Section 1.8 that numbers fall into two categories, real and complex. Complex numbers will be studied in detail in Chapter 11. In this chapter we need only to understand the basic concepts and notations.

Consider the pure quadratic equation

$$x^2 + 4 = 0$$

Solving for x, we get $x = \pm\sqrt{-4}$. Since we cannot find a (real) number whose square is -4, $\sqrt{-4}$ is called a **pure imaginary number**.

For the past two hundred years the letter i (for "imaginary") has been used to denote the imaginary number $\sqrt{-1}$. When imaginary numbers were first introduced in electrical circuit theory, the convention of using i for instantaneous current had already become well established. Consequently, using the letter j to denote $\sqrt{-1}$ became standard in physics and technology, and we will observe this convention here.

Returning now to $\sqrt{-4}$, observe that

$$\sqrt{-4} = \sqrt{(4)(-1)} = \sqrt{4}\sqrt{-1} = 2\sqrt{-1} = 2j$$

Thus $\sqrt{-4} = 2j$, where $j = \sqrt{-1}$. Imaginary numbers are always written in this form.

Basic Imaginary Unit:

$$j = \sqrt{-1} \quad \text{or} \quad j^2 = -1$$

$\sqrt{-a}, a > 0$, is written $\sqrt{a}j$

If a and b are real numbers, then $a + bj$ is called a *complex number*. So a pure imaginary number is a complex number for which $a = 0$; a real number is a complex number for which $b = 0$.

Complex Number: A **complex number** has the form $a + bj$, where a and b are real numbers and $j = \sqrt{-1}$.

Some quadratic equations lead to complex roots, as shown in the remaining examples.

E X A M P L E   **3**   Solve the equation $x^2 + 4x + 16 = 0$ by completing the square.

Solution.

$$x^2 + 4x + 16 = 0 \qquad \textbf{given equation}$$

Step 1. $x^2 + 4x \qquad = -16 \qquad \textbf{transposing} + \textbf{16}$

Step 2. Not necessary, since the coefficient of x is 1.

Step 3. $x^2 + 4x + 4 = -16 + 4 \qquad \textbf{adding } \left(\dfrac{1}{2} \cdot 4\right)^2 \textbf{ to both sides}$

Step 4. $\qquad (x + 2)^2 = -12 \qquad \textbf{factoring the left side}$

Step 5. $\qquad x + 2 = \pm\sqrt{-12} \qquad \textbf{taking the square root of each side}$

Step 6. Note that $\sqrt{-12} = \sqrt{(-1) \cdot 4 \cdot 3} = \sqrt{4 \cdot 3}\sqrt{-1} = 2\sqrt{3}\sqrt{-1} = 2\sqrt{3}j$. It follows that $x = -2 \pm 2\sqrt{3}j$. ◀

E X A M P L E   **4**   Solve the equation $2x^2 - 3x + 4 = 0$ by completing the square.

Solution.

$$2x^2 - 3x + 4 = 0 \qquad\qquad \text{given equation}$$

$$x^2 - \frac{3}{2}x + 2 = 0 \qquad\qquad \text{dividing by 2}$$

$$x^2 - \frac{3}{2}x \quad\quad = -2 \qquad\qquad \text{transposing the 2}$$

$$x^2 - \frac{3}{2}x + \frac{9}{16} = -2 + \frac{9}{16} \qquad\qquad \text{adding } \left[\frac{1}{2}\cdot\left(-\frac{3}{2}\right)\right]^2 \text{ to both sides}$$

$$\left(x - \frac{3}{4}\right)^2 = -\frac{32}{16} + \frac{9}{16} = -\frac{23}{16} \qquad\qquad \text{factoring the left side}$$

$$x - \frac{3}{4} = \pm\sqrt{-\frac{23}{16}} = \pm\frac{\sqrt{-23}}{4} \qquad\qquad \sqrt{\frac{x}{y}} = \frac{\sqrt{x}}{\sqrt{y}}$$

$$= \pm\frac{\sqrt{23(-1)}}{4} = \pm\frac{\sqrt{23}\sqrt{-1}}{4}$$

$$= \pm\frac{\sqrt{23}j}{4} \qquad\qquad j = \sqrt{-1}$$

$$x = \frac{3}{4} \pm \frac{\sqrt{23}}{4}j \qquad\qquad \text{solving for } x \qquad\qquad \blacktriangleleft$$

Remark. We will see in the next section that a nonfactorable quadratic equation can be solved directly by a formula. Once you learn this formula, you may feel that completing the square is a waste of time. However, *completing the square is an algebraic technique that arises in contexts other than solving equations.* In fact, solving quadratic equations is merely a convenient way to introduce this technique. It is therefore very important for you to practice solving equations by completing the square in the next exercise set.

E X E R C I S E S / S E C T I O N   **6.2**

Solve each equation by completing the square.

1. $x^2 - 6x + 8 = 0$

2. $x^2 - 6x + 5 = 0$

3. $x^2 + 4x - 12 = 0$

4. $x^2 + 2x - 15 = 0$

5. $x^2 + x - 12 = 0$

6. $x^2 + 3x - 28 = 0$

7. $x^2 + 7x + 10 = 0$

8. $x^2 - 9x + 20 = 0$

9. $x^2 + 5x + 2 = 0$

10. $x^2 - 4x - 6 = 0$

11. $x^2 + 6x + 6 = 0$

12. $x^2 + 3x + 1 = 0$

13. $2x^2 - 6x + 1 = 0$

14. $2x^2 + 5x + 2 = 0$

15. $2x^2 + 3x - 3 = 0$

16. $2x^2 - 3x - 5 = 0$

17. $3x^2 + 2x - 1 = 0$

18. $3x^2 - 2x - 3 = 0$

19. $3x^2 - 4x - 5 = 0$

20. $2x^2 + 5x - 2 = 0$

21. $4x^2 - x - 3 = 0$

22. $5x^2 - 2x - 1 = 0$

23. $6x^2 + x + 2 = 0$

24. $5x^2 + 9x + 1 = 0$

25. $x^2 - 4x + 5 = 0$ **26.** $3x^2 - 2x + 1 = 0$ **33.** $6x^2 + 5x - 50 = 0$ **34.** $8x^2 - 7x + 2 = 0$

27. $4x^2 - 5x + 3 = 0$ **28.** $2x^2 + 3x + 2 = 0$ **35.** $5x^2 - x + 1 = 0$ **36.** $5x^2 + 2x - 3 = 0$

29. $7x^2 + 2x - 1 = 0$ **30.** $8x^2 + 3x + 1 = 0$ **37.** $x^2 - bx + 2 = 0$ **38.** $x^2 - x + c = 0$

31. $6x^2 - 5x - 2 = 0$ **32.** $7x^2 - 19x - 6 = 0$ **39.** $ax^2 + 5x - 1 = 0$ **40.** $x^2 - 3bx + 5 = 0$

6.3   The Quadratic Formula

Completing the square can be used to obtain a general formula for solving any quadratic equation. We start with the standard form:

$$ax^2 + bx + c = 0$$

$$ax^2 + bx = -c \qquad \text{transposing } +c$$

$$x^2 + \frac{b}{a}x = -\frac{c}{a} \qquad \text{dividing by } a$$

$$x^2 + \frac{b}{a}x + \left(\frac{b}{2a}\right)^2 = -\frac{c}{a} + \left(\frac{b}{2a}\right)^2 \qquad \text{adding } \left(\frac{1}{2}\cdot\frac{b}{a}\right)^2 \text{ to each side}$$

$$\left(x + \frac{b}{2a}\right)^2 = -\frac{c}{a} + \frac{b^2}{4a^2} \qquad \text{factoring the left side and simplifying the right side}$$

$$= -\frac{4ac}{4a^2} + \frac{b^2}{4a^2} \qquad \text{LCD} = 4a^2$$

$$= \frac{b^2 - 4ac}{4a^2}$$

$$x + \frac{b}{2a} = \pm\sqrt{\frac{b^2 - 4ac}{4a^2}} \qquad \text{taking the square root of each side}$$

$$= \pm\frac{\sqrt{b^2 - 4ac}}{2a} \qquad \sqrt{\frac{x}{y}} = \frac{\sqrt{x}}{\sqrt{y}}$$

$$x = -\frac{b}{2a} \pm \frac{\sqrt{b^2 - 4ac}}{2a} \qquad \text{transposing } +\frac{b}{2a}$$

$$x = \frac{-b \pm \sqrt{b^2 - 4ac}}{2a}$$

This formula, known as the **quadratic formula**, should be carefully memorized.

Quadratic Formula: The roots of the quadratic equation

$$ax^2 + bx + c = 0, \qquad a \neq 0 \tag{6.5}$$

are given by

$$x = \frac{-b \pm \sqrt{b^2 - 4ac}}{2a} \tag{6.6}$$

By using the quadratic formula, the solutions of a quadratic equation can be written directly, but they usually have to be simplified. The first three examples illustrate the technique.

E X A M P L E 1 Solve the equation $6x^2 = 2x + 1$ by means of the quadratic formula.

Solution. The equation is first written in the standard form

$$6x^2 - 2x - 1 = 0$$

By equation (6.5), $a = 6$, $b = -2$, and $c = -1$. So by the quadratic formula (6.6), the solution is

$$x = \frac{-(-2) \pm \sqrt{(-2)^2 - 4(6)(-1)}}{2 \cdot 6}$$

$$= \frac{2 \pm \sqrt{28}}{12}$$

To simplify the radical, note that $\sqrt{28} = \sqrt{4 \cdot 7} = 2\sqrt{7}$. Thus

$$x = \frac{2 \pm 2\sqrt{7}}{12}$$

$$= \frac{2(1 \pm \sqrt{7})}{12} \qquad \text{factoring 2}$$

$$x = \frac{1 \pm \sqrt{7}}{6} \qquad \text{reducing} \qquad \blacktriangleleft$$

E X A M P L E 2 Solve the equation $5x^2 + 2x + 4 = 0$ by the quadratic formula.

Solution. From the standard form (6.5), we see that $a = 5$, $b = 2$, and $c = 4$ So by the quadratic formula (6.6), we get

$$x = \frac{-2 \pm \sqrt{2^2 - 4(5)(4)}}{2 \cdot 5}$$

$$= \frac{-2 \pm \sqrt{4 - 80}}{10}$$

$$= \frac{-2 \pm \sqrt{-76}}{10}$$

Since $\sqrt{-76} = \sqrt{(4)(19)(-1)} = \sqrt{4 \cdot 19}\sqrt{-1} = 2\sqrt{19}j$, we get $\qquad \sqrt{-1} = j$

$$x = \frac{-2 \pm 2\sqrt{19}j}{10} = \frac{2(-1 \pm \sqrt{19}j)}{10} \qquad \text{common factor 2}$$

$$= \frac{-1 \pm \sqrt{19}j}{5} \qquad \text{reducing} \qquad \blacktriangleleft$$

E X A M P L E **3**   Solve the equation $4x^2 - 12x + 9 = 0$.

Solution. Since $a = 4$, $b = -12$, and $c = 9$, we get

$$x = \frac{-(-12) \pm \sqrt{(-12)^2 - 4(4)(9)}}{2 \cdot 4}$$

$$= \frac{12 \pm \sqrt{144 - 144}}{8}$$

$$= \frac{12 \pm 0}{8}$$

$$= \frac{3 \pm 0}{2}$$

Hence $x = \frac{3}{2}, \frac{3}{2}$. (Whenever the radical is 0, we get a double root.)   ◄

These examples show that the radical in the quadratic formula

$$x = \frac{-b \pm \sqrt{b^2 - 4ac}}{2a}$$

Discriminant

determines whether the roots are real or complex. The expression $b^2 - 4ac$ under the radical sign is called the **discriminant**. We have seen that a given equation has two distinct real roots if $b^2 - 4ac > 0$ (Example 1) and complex roots if $b^2 - 4ac < 0$ (Example 2). If $b^2 - 4ac = 0$, then the equation has a double root (Example 3).

E X A M P L E **4**   Use a calculator to solve the equation:

CALCULATOR COMMENT
$$3.17x^2 - 1.98x - 6.83 = 0$$

Solution. Since $a = 3.17$, $b = -1.98$, and $c = -6.83$, we get

$$x = \frac{-(-1.98) \pm \sqrt{(-1.98)^2 - 4(3.17)(-6.83)}}{2(3.17)}$$

The simplest way to carry out this calculation is to find the value of the radical and store it in the memory. Now add 1.98 to the positive value of the radical and divide the sum by $(2 \times 3.17) = 6.34$ to get

$$x = 1.81$$

to two decimal places. Next, transfer the content of the memory to the register, change the sign to minus, and proceed as before. The second root is

$$x = -1.19$$

again to two decimal places.

The sequences are

$$1.98 \boxed{+/-} \boxed{x^2} \boxed{-} 4 \boxed{\times} 3.17 \boxed{\times} 6.83 \boxed{+/-} \boxed{=}$$
$$\boxed{\sqrt{}} \boxed{STO} \boxed{+} 1.98 \boxed{=} \boxed{\div} 2 \boxed{\div} 3.17 \boxed{=} \rightarrow 1.8130051$$
$$\boxed{MR} \boxed{+/-} \boxed{+} 1.98 \boxed{=} \boxed{\div} 2 \boxed{\div} 3.17 \boxed{=} \rightarrow -1.1883994 \qquad \blacktriangleleft$$

Fractional equations may also lead to quadratic equations, as illustrated in the next example.

E X A M P L E **5** Solve the equation

$$\frac{1}{x} - \frac{1}{x+1} = \frac{1}{20}$$

Solution. Recall that the simplest way to solve a fractional equation is to clear the fractions by multiplying both sides by the LCD—in this case **$20x(x + 1)$** Then

$$20x(x + 1)\left(\frac{1}{x} - \frac{1}{x+1}\right) = 20x(x + 1)\frac{1}{20}$$

$$20x(x + 1)\frac{1}{x} + 20x(x + 1)\left(-\frac{1}{x+1}\right) = 20x(x + 1)\frac{1}{20}$$

$$20\cancel{x}(x + 1)\frac{1}{\cancel{x}} + 20x\cancel{(x+1)}\left(-\frac{1}{\cancel{x+1}}\right) = 2\cancel{0}x(x + 1)\frac{1}{\cancel{20}}$$

$$20(x + 1) - 20x = x(x + 1)$$

$$20x + 20 - 20x = x^2 + x$$

$$20 = x^2 + x$$

$$0 = x^2 + x - 20$$

$$(x + 5)(x - 4) = 0$$

$$x = 4, -5 \qquad \blacktriangleleft$$

E X E R C I S E S / S E C T I O N **6.3**

In Exercises 1–34, solve each equation by the quadratic formula.

1. $x^2 + x - 6 = 0$

2. $x^2 - 3x - 4 = 0$

3. $x^2 - 9x + 20 = 0$

4. $x^2 + 8x + 15 = 0$

5. $2x^2 = 5x - 2$

6. $3x^2 = 13x + 10$

7. $6x^2 - x = 2$

8. $5x^2 - 7x = 6$

9. $2x^2 = 3x + 1$

10. $2x^2 + 2x = 1$

11. $3x^2 + 2x = 2$

12. $x^2 = 3x + 2$

13. $x^2 - 2x + 2 = 0$

14. $x^2 + 3x + 3 = 0$

15. $x^2 + 2x + 4 = 0$

16. $x^2 + 3x + 5 = 0$

17. $3x^2 + 3x + 1 = 0$

18. $3x^2 + 5x + 2 = 0$

19. $4x^2 + 2x + 1 = 0$

20. $4x^2 + 3x - 2 = 0$

21. $4x^2 + 5x = 3$

22. $5x^2 + 3x = 3$

23. $5x^2 + 1 = 0$

24. $3x^2 + 4 = 0$

25. $2x^2 + 3x = 0$

26. $3x^2 - 5x = 0$

27. $2x^2 - 3cx + 1 = 0$

28. $3x^2 + 3x - 2a = 0$

29. $bx^2 + 3x + 1 = 0$

30. $cx^2 + bx - 4 = 0$

31. $4x^2 - 12x + 9 = 0$

32. $9x^2 + 12x + 4 = 0$

33. $4x^2 - 20x + 25 = 0$

34. $9x^2 + 42x + 49 = 0$

In Exercises 35–40, solve the given equations using a calculator. (Find the roots to two decimal places.)

35. $2.00x^2 + 3.12x - 3.19 = 0$

36. $4.12x^2 - 1.30x - 12.1 = 0$

37. $1.79x^2 - 10.0x - 1.91 = 0$

38. $7.179x^2 + 2.862x - 1.998 = 0$

39. $10.103x^2 - 1.701x - 3.28 = 0$

40. $1.738x^2 - 10.162x - 11.773 = 0$

In Exercises 41–48, solve each equation for x.

41. $\dfrac{1}{x} - \dfrac{1}{x+1} = \dfrac{1}{20}$

42. $\dfrac{1}{x} - \dfrac{1}{x+2} = \dfrac{1}{4}$

43. $\dfrac{1}{x+2} + \dfrac{1}{x} = \dfrac{5}{12}$

44. $\dfrac{1}{x} + \dfrac{1}{x+8} = \dfrac{1}{3}$

45. $\dfrac{1}{x} + \dfrac{1}{x-4} = \dfrac{3}{8}$

46. $\dfrac{1}{x} - \dfrac{1}{x+3} = 1$

47. $\dfrac{1}{x} + \dfrac{2}{x-4} = 1$

48. $\dfrac{2}{x} - \dfrac{3}{x+1} = 2$

49. By treating y as a constant, solve the equation
$$x^2 - 2x + 1 - y^2 = 0$$
for x.

6.4   Applications of Quadratic Equations

Many physical problems lead quite naturally to quadratic equations. One such case was already mentioned at the beginning of the chapter. Now we will consider a similar example.

E X A M P L E 1   A rock is hurled upward at the rate of 24 m/s from a height of 5 m. The distance s (in meters) above the ground as a function of time t (in seconds) is given by $s = -5t^2 + 24t + 5$. (The instant at which the rock is hurled upward corresponds to $t = 0$ s.) When will the rock strike the ground?

Solution. Since s is the distance above the ground, the problem is to find the value of t for which $s = 0$. Thus

$$-5t^2 + 24t + 5 = 0$$

$$5t^2 - 24t - 5 = 0 \qquad \textbf{multiplying by } -1$$

$$(5t + 1)(t - 5) = 0 \qquad \textbf{factoring the left side}$$

$$t = -\frac{1}{5}, 5$$

Since $t = 0$ corresponds to the instant when the motion begins, the root $t = -\frac{1}{5}$ has no meaning here. We conclude that the rock hits the ground in 5 s. ◄

E X A M P L E 2   Find, to three significant digits, two numbers whose sum is 15.4 and whose product is 57.6.

Solution. Let

$$x = \text{first number}$$

Then

$$15.4 - x = \text{second number}$$

Since the product is 57.6, we obtain the equation

$$x(15.4 - x) = 57.6$$
$$15.4x - x^2 = 57.6$$
$$x^2 - 15.4x + 57.6 = 0$$
$$x = \frac{-(-15.4) \pm \sqrt{(-15.4)^2 - 4(57.6)}}{2}$$

Using a calculator, we get for the first number

$$x = 9.00 \quad \text{or} \quad x = 6.40$$

If $x = 9.00$, then $15.4 - x = 6.40$; if $x = 6.40$, then $15.4 - x = 9.00$. Either way, we conclude that the numbers are 9.00 and 6.40. ◄

E X A M P L E **3**   If the length of a square is increased by 6.0 in., the area becomes 4 times as large. Find the original length of the side.

Solution. Let x be the original length of the side. Then $x + 6.0$ in. is the length of the side when increased. So x^2 is the original area and $(x + 6.0)^2$ is the new area. Since the new area is equal to 4 times the old area, we get (in square inches, omitting final zeros)

$$(x + 6)^2 = 4x^2 \qquad \text{new area = 4 × original area}$$
$$x^2 + 12x + 36 = 4x^2 \qquad \text{expanding the left side}$$
$$-3x^2 + 12x + 36 = 0 \qquad \text{subtracting } 4x^2 \text{ from both sides}$$
$$x^2 - 4x - 12 = 0 \qquad \text{dividing by } -3$$
$$(x - 6)(x + 2) = 0 \qquad \text{factoring the left side}$$
$$x = 6, -2$$

Since the root $x = -2$ has no meaning here, the original side is 6.0 in. long (using two significant digits). ◄

E X A M P L E **4**   Two resistors connected in parallel have a combined resistance of 4 Ω, and the resistance of one resistor is 6 Ω more than that of the other. Find the resistance of each.

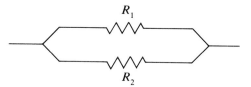

If R_T is the combined resistance, then

$$\frac{1}{R_T} = \frac{1}{R_1} + \frac{1}{R_2}$$

Figure 6.3

Solution. Let

R = resistance of first resistor

Then

$R + 6$ = resistance of second resistor

Since the resistors are connected in parallel (Figure 6.3), we have

$$\frac{1}{R} + \frac{1}{R + 6} = \frac{1}{4}$$

This is an example of a fractional equation that reduces to a quadratic equation after clearing fractions. Multiplying both sides of the equation by the LCD $= 4R(R + 6)$, we get

$$4R(R + 6)\left(\frac{1}{R} + \frac{1}{R + 6}\right) = 4R(R + 6)\frac{1}{4}$$

$$4R(R + 6)\frac{1}{R} + 4R(R + 6)\frac{1}{R + 6} = 4R(R + 6)\frac{1}{4}$$

$$4\cancel{R}(R + 6)\frac{1}{\cancel{R}} + 4R\cancel{(R + 6)}\frac{1}{\cancel{R + 6}} = \cancel{4}R(R + 6)\frac{1}{\cancel{4}}$$

$$4(R + 6) + 4R = R(R + 6)$$

$$4R + 24 + 4R = R^2 + 6R$$

$$8R + 24 = R^2 + 6R$$

$$R^2 - 2R - 24 = 0$$

$$(R - 6)(R + 4) = 0$$

$$R = -4, 6$$

Since the negative root has no meaning here, we conclude that $R = 6 \, \Omega$. The resistance of the other resistor is therefore $12 \, \Omega$. ◀

E X A M P L E **5**   A tank can be filled by two inlet valves in 2 h. One inlet valve requires $7\frac{1}{2}$ h longer to fill the tank than the other. How long does it take for each valve alone to fill the tank?

Solution. Let x equal the time taken for the faster valve to fill the tank. Then $x + 7\frac{1}{2} = x + \frac{15}{2}$ is the time required for the slower valve to fill the tank.
   Now recall from Section 2.5 that an equation can be readily obtained from this information by finding expressions for the fractional part of the tank that can be filled in one time unit, in this case 1 h. So.

$$\frac{1}{x} + \frac{1}{x + \dfrac{15}{2}} = \frac{1}{2} \qquad \text{fractional part filled in 1 h}$$

To clear fractions, note that the LCD equals $2x(x + \frac{15}{2})$. We now get

$$2x\left(x + \frac{15}{2}\right)\left[\frac{1}{x} + \frac{1}{x + \dfrac{15}{2}}\right] = 2x\left(x + \frac{15}{2}\right)\cdot\frac{1}{2}$$

$$2\left(x + \frac{15}{2}\right) + 2x = x\left(x + \frac{15}{2}\right)$$

$$2x + 15 + 2x = x^2 + \frac{15}{2}x$$

Multiplying both sides by 2, we then have

$$8x + 30 = 2x^2 + 15x$$

$$2x^2 + 7x - 30 = 0$$

$$(2x - 5)(x + 6) = 0$$

$$x = -6, \frac{5}{2}$$

Again, the negative root has no meaning, so we conclude that the times are $2\frac{1}{2}$ h and 10 h, respectively.   ◀

E X A M P L E **6**   An executive drives to a conference early in the day. Due to heavy morning traffic, her average speed for the first 120 mi is 10 mi/h less than for the second 120 mi and requires 1 h more time. Find the two average speeds.

Solution. Recall from Section 2.5 that the basic relationship is distance = rate × time. Therefore, time = distance/rate.
   If we let x equal the slower rate, then $x + 10$ equals the faster rate. Since the distance is the same in both cases, it follows that $120/x$ is the time required

to cover the first 120 mi, and $120/(x + 10)$ the time required to cover the second 120 mi. The difference in the two times is 1 h. Hence

$$\frac{120}{x} - \frac{120}{x + 10} = 1$$

Clearing fractions,

$$120(x + 10) - 120x = x(x + 10) \qquad \textbf{LCD = x(x + 10)}$$

$$120x + 1{,}200 - 120x = x^2 + 10x$$

$$x^2 + 10x - 1{,}200 = 0$$

$$(x - 30)(x + 40) = 0$$

$$x = 30, \ -40$$

Taking the positive root again, we conclude that 30 mi/h is the slower rate and $x + 10 = 40$ mi/h the faster rate. ◀

E X E R C I S E S / S E C T I O N   6.4

1. A rock is hurled upward from the ground at the rate of 64 ft/s. Its distance d (in feet) above the ground is given by $d = -16t^2 + 64t$, where t is measured in seconds. When will the rock strike the ground? (See Example 1.)

2. The current i (in amperes) in a certain circuit at any time t (in seconds) is given by $i = 9.5t^2 - 4.7t$. At what time is the current equal to zero?

3. The voltage (in volts) across a certain resistor is given by $v = 2.0t^2 - 8.3t + 5.1$, where t is measured in seconds. Determine when the voltage is 0.

4. If an object is hurled vertically downward with velocity v_0, then the distance s that the object falls at any time is $s = v_0 t + \frac{1}{2}gt^2$. Find an expression for t.

5. The power P (in watts) delivered to the resistor R_L in Figure 6.4 is given by

$$P = VI - I^2 R$$

If $V = 84.3$ V and $R = 60.0$ Ω, find the current I (in amperes) for which $P = 26.8$ W.

6. Repeat Exercise 5 with $V = 75.6$ V, $R = 50.0$ Ω, and $P = 23.7$ W.

7. Find two positive numbers whose difference is 8 and whose product is 105. (See Example 2.)

8. Find two positive numbers whose product is 144 and whose difference is 7.

9. Find, to three significant digits, two numbers whose sum is 16.2 and whose product is 63.4.

10. Find, to three significant digits, two positive numbers whose difference is 2.64 and whose product is 15.7.

11. One number is 2.00 more than another number and their product is 66.4. Find the numbers.

12. The difference between a positive integer and its reciprocal is $2\frac{2}{3}$. Find the number.

13. A rectangular metal plate is twice as long as it is wide. When heated, each side increases by 2.0 mm. As a result, the increased area is 1.1 times as large as the original area. Find the original dimensions. (See Example 3.)

14. To cover the floor of a new storage area, 100 square tiles of a certain size are needed. If square tiles 2 in. longer on each side are used, only 64 tiles are needed. What is the size of the smaller tile?

15. The length of a rectangle is 2.00 cm more than the width. The area is 66.4 cm². Find the dimensions. (See Exercise 11.)

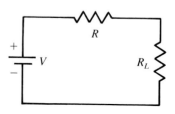

Figure 6.4

16. A metal shop has an order for a rectangular metal plate of area 84 in.^{2}. Find its dimensions if the length exceeds the width by 5.0 in.

17. A parallelogram has an area of 149.0 in.^{2}, and the base exceeds the height by 10.00 in. Find the base and height.

18. A technician has to order a frame meeting the following specifications: It has the shape of a right triangle with a hypotenuse 26.0 cm long, while the sum of the lengths of the other two sides is 34.0 cm. Find the dimensions.

19. Two resistors connected in parallel have a combined resistance of 6 Ω. If the resistance of one resistor is 5 Ω more than that of the other, find the resistance of each. (See Example 4.)

20. Two resistors connected in parallel have a combined resistance of 4.2 Ω. Determine the resistance of each, given that the resistance of one resistor is 8.0 Ω more than that of the other.

21. If two capacitors are connected in series, then the combined capacitance C_T is related to C_1 and C_2 by the formula

$$\frac{1}{C_T} = \frac{1}{C_1} + \frac{1}{C_2}$$

(See Figure 6.5.) If the combined capacitance is 7.2 μF and the capacitance of one is 6.0 μF more than that of the other, find the capacitance of each.

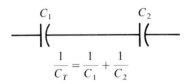

$$\frac{1}{C_T} = \frac{1}{C_1} + \frac{1}{C_2}$$

Figure 6.5

22. Recall that the relationship of the focal length f of a lens to the object distance q and the image distance p is

$$\frac{1}{f} = \frac{1}{p} + \frac{1}{q}$$

If $f = 2.0$ cm and p is 3.0 cm longer than q, find p.

23. If both inlets of a tank are open, then the tank can be filled in 2 h. One of the inlets alone requires $5\frac{1}{3}$ h more than the other to fill the tank. How long does each one take? (See Example 5.)

24. Two machines are used to print labels for a large mailing; the job normally takes 2 h. One day the faster machine breaks down and the slower machine, which takes 3 h longer than the faster machine to do the job alone, has to be used. How long will it take to complete the job?

25. Two card sorters used simultaneously can sort a set of cards in 24 min. If only one machine is used, then the slower one requires 20 min more than the faster one. How long does each one take?

26. Working together, two painters can paint a room in 3.6 h. Working alone, one man requires 3.0 h more than the other. How long does each one take to do the job alone?

27. In city traffic, a car travels 15 mi/h faster than a bicycle. The car can travel 50 mi in 3 h less time than the bicycle. Find the rate of each. (See Example 6.)

28. A heavy machine is delivered by truck to a factory 200 mi away. The empty truck makes the return trip 10 mi/h faster and gets back in 1 h less time. Find the rate each way.

29. A rectangular casting 0.500 in. thick is to be made from 44.0 in. of forming (Figure 6.6). If 42.5 in.^{3} are poured into the form, find the dimensions of the casting.

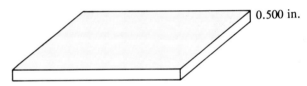

0.500 in.

Figure 6.6

30. Suppose the casting in Exercise 29 is 3.00 in. thick and its length is 2.00 in. more than its width. If the volume is 72.0 in.^{3}, find its dimensions.

31. A rectangular enclosure is to be fenced along four sides and divided into two parts by a fence parallel to one of the sides. (See Figure 6.7.) If 170 ft of fence are available and the total area is 1,200 ft^2, what are the dimensions? (There are two possible solutions.)

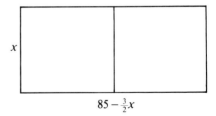

x

$85 - \frac{3}{2}x$

Figure 6.7

32. A tool shed is adjacent to a building, which serves as the back wall of the tool shed. The total length of the other three walls is 39 ft, and the area of the floor is 180 ft^2. Find the dimensions.

33. The cost of carpeting an office at \$10/ft^2 was \$1,500. If the length exceeds the width by 5 ft, what are the dimensions of the office?

34. Maggie's car gets 5 mi/gal less in the city than on the highway. Driving 300 mi in the city requires 2 gal more gas than driving the same distance on the highway. Determine the gas mileage in the city.

35. An engineer wants to buy $2,400 worth of stock. One stock costs $10 more per share than another. If she decides to buy the cheaper stock, she can afford 40 more shares.

How many shares of the more expensive stock can she buy?

36. An investor purchases a number of shares of stock for $600. If the investor had paid $2 less per share, the number of shares would have been increased by 10. How many shares of the cheaper stock can he buy?

REVIEW EXERCISES/CHAPTER   6

In Exercises 1–8, solve the equations by factoring.

1. $x^2 - 3x - 10 = 0$ **2.** $2x^2 - 7x - 4 = 0$

3. $5x^2 - 7x - 6 = 0$ **4.** $2x^2 + 15x - 27 = 0$

5. $6x^2 + 7x = -2$ **6.** $4x^2 = 17x - 15$

7. $6x^2 = 11x - 5$ **8.** $10x^2 + 3x = 18$

In Exercises 9–16, solve the equations by completing the square.

9. $x^2 - 2x - 8 = 0$ **10.** $2x^2 + 3x - 5 = 0$

11. $x^2 + 5x + 3 = 0$ **12.** $2x^2 - 4x + 1 = 0$

13. $2x^2 - 5x + 4 = 0$ **14.** $3x^2 - 2x + 3 = 0$

15. $4x^2 - x + 3 = 0$ **16.** $2x^2 + 5x + 2 = 0$

In Exercises 17–24, solve the equations by applying the quadratic formula.

17. $x^2 - 6x + 8 = 0$ **18.** $6x^2 + 7x = 3$

19. $3x^2 - 3x + 1 = 0$ **20.** $x^2 - 8x + 6 = 0$

21. $2x^2 = 4x + 3$ **22.** $2x^2 - 4x + 3 = 0$

23. $6x^2 - 8x - 3 = 0$ **24.** $5x^2 + x = -1$

In Exercises 25–30, solve the equations using any method.

25. $1.72x^2 + 1.89x - 2.64 = 0$

26. $3.98x^2 + 0.46x - 0.42 = 0$

27. $x^2 - 4x + 4 - y^2 = 0$

28. $x^2 - 4x = a^2$

29. $\dfrac{1}{x} + \dfrac{1}{x+1} = \dfrac{9}{20}$

30. $\dfrac{1}{x-1} + \dfrac{1}{x+2} = 1$

31. Two resistors in parallel have a combined resistance of $2\ \Omega$, while the resistance of one resistor is $3\ \Omega$ more than that of the other. Find the resistance of each.

32. A rectangular enclosure is to be fenced along four sides and divided into three parts by two fences that are parallel to one side. If 80 ft of fence are available and the total area is $200\ \text{ft}^2$, what are the dimensions of the enclosure?

33. Working together, two men can unload a boxcar in 4 h. Working alone, one man requires 6 h more than the other. How long does it take for each man to do the job alone?

34. The cost of tiling a kitchen is $5/ft². If the length of the kitchen exceeds the width by 4 ft and the total cost of tiling comes to $960, what are the dimensions of the kitchen?

35. Early one morning a shop assistant delivers a motor to a garage 70 mi from downtown. Because of lighter traffic, his average speed on the return trip is 15 mi/h more than on the delivery run, and he returns in 90 min less time. Find his average speed each way.

1. Find the smallest positive angle coterminal with $-237°41'$.

2. Find the values of the six trigonometric functions of the angle in standard position whose terminal side passes through $(\sqrt{13}, \sqrt{3})$.

3. If $\cos\theta = \dfrac{\sqrt{2}}{3}$, find $\csc\theta$.

4. Use a diagram to find $\cot 60°$.

5. Use a calculator to find $\csc 36°19'$.

6. If $\sec\theta = 1.831$, use a calculator to find θ to the nearest hundredth of a degree.

7. If $\cot\theta = 2.672$, use a calculator to find θ in degrees and minutes.

8. Factor the expression $sV_a - sV_b - V_a + V_b$.

9. Reduce the fraction $\dfrac{L^3 - C^3}{2L - 2C}$

10. Perform the following multiplication:
$$\frac{x^2 - y^2}{x^2 + xy - 6y^2} \cdot \frac{x - 2y}{2x - 2y}$$

11. Perform the following division:
$$\frac{x - 2y}{ax + 2ay} \div \frac{2x^2 - 3xy - 2y^2}{2ax + 4ay - x - 2y}$$

12. Combine the following fractions:
$$\frac{s}{s - t} + \frac{1}{s^2 - t^2} - \frac{s - t}{s + t}$$

13. Reduce the following complex fraction:
$$\frac{\dfrac{L}{L + 4} + \dfrac{L}{L^2 - 16}}{1 + \dfrac{1}{L - 4}}$$

14. Solve for x:
$$\frac{1}{x - 2} - \frac{1}{x} = \frac{1}{4}$$

15. Solve for x by the method of factoring:
$$4x^2 + 11x - 3 = 0$$

16. Solve for x by completing the square:
$$x^2 - 4x + 3 = 0$$

17. Solve for x using the quadratic formula:
$$x^2 - 2x + 2 = 0$$

18. Use a calculator to solve the following equation for x:
$$3.12x^2 + 1.71x - 1.30 = 0$$

19. The current in a rotating coil is $i = 0.462 \cos\phi$. Determine the current at the instant when $\phi = 20.4°$.

20. Nine rivets are equally spaced on the rim of a circular metal plate. The distance from center to center between adjacent rivets is 1.06 in. Find the radius of the plate.

21. Three resistors R_1, R_2, and R_3 are connected in parallel. The combined resistance is
$$\frac{1}{\dfrac{1}{R_1} + \dfrac{1}{R_2} + \dfrac{1}{R_3}}$$

Simplify this expression.

22. The cost of finishing a metal plate is $4.00 per square inch. If the length of the plate exceeds the width by 2.0 in. and the total cost comes to $73.44, what are the dimensions of the plate?

More on Trigonometric Functions

7.1 Algebraic Signs of Trigonometric Functions

Trigonometric Functions of Any Angle

In Chapter 4 we defined the six trigonometric functions for angles with terminal sides in the first quadrant. In this section we will extend these definitions to apply to angles with terminal sides in any quadrant.

Recall from Chapter 4 that if an angle is in standard position, then the functions can be expressed in terms of the coordinates and radius vector of a point on the terminal side. Consider the angle θ whose terminal side passes through (x, y) (Figure 7.1). The **radius vector** r is the distance from (x, y) to the origin. Then $\sin \theta = y/r$, $\cos \theta = x/r$, and so on.

With these definitions, we can find the function values of angles whose terminal sides do not lie in the first quadrant. (Refer to Figure 7.2.)

Radius vector

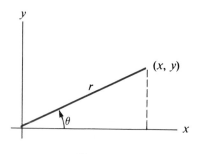

Figure 7.1

The Trigonometric Functions:

$$\sin \theta = \frac{y}{r} \qquad \csc \theta = \frac{r}{y}$$

$$\cos \theta = \frac{x}{r} \qquad \sec \theta = \frac{r}{x}$$

$$\tan \theta = \frac{y}{x} \qquad \cot \theta = \frac{x}{y}$$

(7.1)

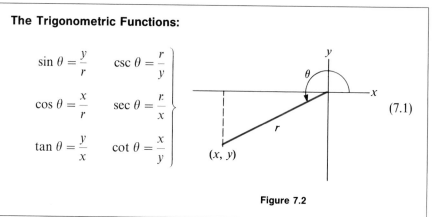

Figure 7.2

Note that all the definitions from Chapter 4 carry over automatically. The only difference is that some of the function values are negative, since the coordinates of (x, y) are signed numbers. However, *the radius vector is always positive.* Moreover, by the Pythagorean theorem,

$$r = \sqrt{x^2 + y^2} \qquad (7.2)$$

regardless of the position of the point (x, y). Consider the example below.

E X A M P L E **1**   Find the values of the six trigonometric functions for the angle θ in Figure 7.3.

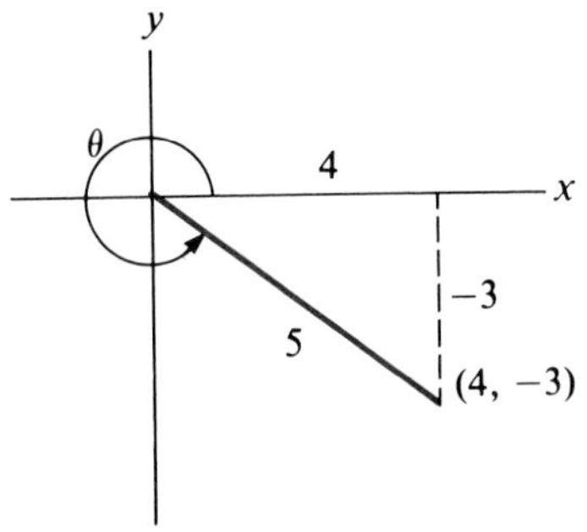

Solution. By formula (7.2), $r = \sqrt{x^2 + y^2}$, we have

$$r = \sqrt{4^2 + (-3)^2} = 5 \qquad \text{*r is always positive*}$$

Since $x = 4$ and $y = -3$, we obtain directly from the definitions

$$\sin \theta = \frac{y}{r} = \frac{-3}{5} = -\frac{3}{5} \qquad \csc \theta = \frac{r}{y} = \frac{5}{-3} = -\frac{5}{3}$$

$$\cos \theta = \frac{x}{r} = \frac{4}{5} \qquad \sec \theta = \frac{r}{x} = \frac{5}{4}$$

$$\tan \theta = \frac{y}{x} = \frac{-3}{4} = -\frac{3}{4} \qquad \cot \theta = \frac{x}{y} = \frac{4}{-3} = -\frac{4}{3} \qquad \blacktriangleleft$$

Figure 7.3

The values of trigonometric functions can also be found by means of *reference angles.* For any angle θ, the **reference angle** is the positive acute angle formed by the terminal side of θ and the *x*-axis.

For example, the reference angle of $140°$ is $180° - 140° = 40°$ (Figure 7.4). The reference angle of $215°$ is $215° - 180° = 35°$ (Figure 7.5).

Reference angle

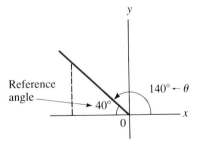

Figure 7.4

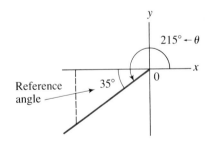

Figure 7.5

E X A M P L E **1**
(Alternate)

The descriptive definitions of the trigonometric ratios—using the opposite side, the adjacent side, and the hypotenuse—have proved useful in applied problems and have already become familiar. Can these definitions be preserved? They

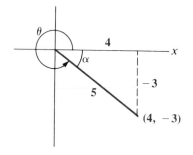

Figure 7.6

can indeed by means of the following scheme: Drop a perpendicular from the point $(4, -3)$ to the x-axis and place the numbers 4, -3, and 5 right on the resulting triangle, as shown in Figure 7.6. It now follows that

$$\sin \theta = \frac{\text{opposite}}{\text{hypotenuse}} = \frac{-3}{5} = -\frac{3}{5}$$

$$\cos \theta = \frac{\text{adjacent}}{\text{hypotenuse}} = \frac{4}{5}$$

and so forth. The triangle in Figure 7.6 is called the **reference triangle**, and, as already noted, the acute angle α is called the **reference angle**. To use the descriptive definitions, note that *the opposite side is the side opposite the reference angle and the adjacent side is the side adjacent to the reference angle.*

As a result, a trigonometric function of θ has the same absolute (numerical) value as the trigonometric function of α. The algebraic sign of the function of θ depends on the quadrant in which the terminal side of θ lies. (See Figure 7.6.)

◀

E X A M P L E **2**   Find the values of the six trigonometric functions of the angle θ in standard position whose terminal side passes through $(-\sqrt{5}, -2)$.

Solution. From $x = -\sqrt{5}$ and $y = -2$, we obtain

$$r = \sqrt{(-\sqrt{5})^2 + (-2)^2} = 3$$

Now drop a perpendicular from $(-\sqrt{5}, -2)$ to the x-axis and label the sides of the resulting triangle as shown in Figure 7.7. The reference angle, shown in the figure, is the positive acute angle formed by the terminal side of θ and the x-axis. Recalling that the terms *opposite* and *adjacent* refer to the reference angle, we get

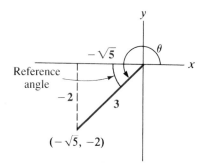

Figure 7.7

$$\sin \theta = \frac{\text{opposite}}{\text{hypotenuse}} = \frac{-2}{3} = -\frac{2}{3} \qquad \csc \theta = \frac{3}{-2} = -\frac{3}{2}$$

$$\cos \theta = \frac{\text{adjacent}}{\text{hypotenuse}} = \frac{-\sqrt{5}}{3} = -\frac{\sqrt{5}}{3} \qquad \sec \theta = \frac{3}{-\sqrt{5}} = -\frac{3\sqrt{5}}{5}$$

$$\tan \theta = \frac{\text{opposite}}{\text{adjacent}} = \frac{-2}{-\sqrt{5}} = \frac{2\sqrt{5}}{5} \qquad \cot \theta = \frac{-\sqrt{5}}{-2} = \frac{\sqrt{5}}{2}$$

Alternatively, using the definitions in statement (7.1), we get

$$\sin \theta = \frac{y}{r} = \frac{-2}{3} = -\frac{2}{3} \qquad \csc \theta = \frac{r}{y} = \frac{3}{-2} = -\frac{3}{2}$$

$$\cos \theta = \frac{x}{r} = \frac{-\sqrt{5}}{3} = -\frac{\sqrt{5}}{3} \qquad \sec \theta = \frac{r}{x} = \frac{3}{-\sqrt{5}} = -\frac{3\sqrt{5}}{5}$$

$$\tan \theta = \frac{y}{x} = \frac{-2}{-\sqrt{5}} = \frac{2\sqrt{5}}{5} \qquad \cot \theta = \frac{x}{y} = \frac{-\sqrt{5}}{-2} = \frac{\sqrt{5}}{2}$$

◀

Determining the Signs of Trigonometric Functions

We can see from Examples 1 and 2 that the various trigonometric functions are either positive or negative, depending on which quadrant contains the terminal side of the angle. This sign can be easily determined from the reference angle. (From now on, we will say that an angle is in the first quadrant if its terminal side lies in the first quadrant, in the second quadrant if its terminal side lies in the second quadrant, and so on.)

Since $\sin \theta$ and $\csc \theta$ are defined in terms of y and r, both functions are positive when y is positive (Figure 7.8). (Recall that r is always positive.)

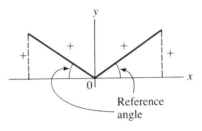

$\sin \theta > 0$ and $\csc \theta > 0$
in quadrants I and II

Figure 7.8

Since $\cos \theta$ and $\sec \theta$ are defined in terms of x and r, both functions are positive when x is positive (Figure 7.9).

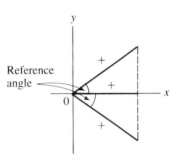

$\cos \theta > 0$ and $\sec \theta > 0$
in quadrants I and IV

Figure 7.9

Finally, $\tan \theta$ and $\cot \theta$ are defined in terms of x and y. As a result, these functions are positive when x and y have the same sign (Figure 7.10).

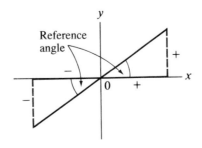

$\tan \theta > 0$ and $\cot \theta > 0$
in quadrants I and III

Figure 7.10

All other functions are negative. Thus $\sin \theta$ and $\csc \theta$ are negative in quadrants III and IV; $\cos \theta$ and $\sec \theta$ are negative in quadrants II and III; and $\tan \theta$ and $\cot \theta$ are negative in quadrants II and IV.

For example,

$$\csc 160° > 0 \qquad \text{(quadrant II)}$$

$$\cos 345° > 0 \qquad \text{(quadrant IV)}$$

$$\sec 220° < 0 \qquad \text{(quadrant III)}$$

$$\cot 140° < 0 \qquad \text{(quadrant II)}$$

If a function value and the quadrant of an angle are known, the values of the remaining functions can be determined, as shown in the next example.

E X A M P L E **3** Determine $\sin \theta$, given that $\cos \theta = \frac{1}{3}$ and $\tan \theta < 0$.

Solution. The first step is to determine the quadrant in which θ lies. We are given that $\cos \theta = x/r$ is positive. Since r is always positive, x must also be positive. So θ is in quadrant I or quadrant IV. [See Figure 7.11(a).]

Since $\tan \theta = y/x$ is negative, x and y have opposite signs. So θ is in quadrant II or quadrant IV. [See Figure 7.11(b).]

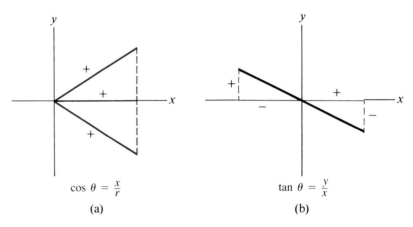

$$\cos \theta = \frac{x}{r}$$

(a)

$$\tan \theta = \frac{y}{x}$$

(b)

Figure 7.11

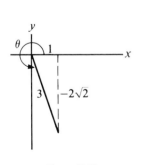

Figure 7.12

To satisfy both conditions (quadrants I or IV *and* quadrants II or IV), θ must be in quadrant IV. (See Figure 7.12.)

From $\cos \theta = \frac{1}{3} = x/r$, we have $x = 1$ and $r = 3$. The numerical value of y is found by using the Pythagorean theorem:

$$y^2 + 1^2 = 3^2 \qquad \text{or} \qquad y = \pm\sqrt{8} = \pm 2\sqrt{2}$$

and since θ is in the fourth quadrant, $y = -2\sqrt{2}$. Now draw the reference triangle (Figure 7.12). It follows that

$$\sin\theta = \frac{y}{r} = \frac{-2\sqrt{2}}{3} = -\frac{2\sqrt{2}}{3}$$

◀

E X A M P L E **4**   Given that $\csc\theta = -\sqrt{13}/2$ and $\cos\theta < 0$, find $\sec\theta$.

Solution. Since $\csc\theta = r/y$ is negative and r positive, y must be negative. So θ is in quadrant III or quadrant IV [Figure 7.13(a)]. Since $\cos\theta = x/r$ is negative, x must be negative. So θ is in quadrant II or quadrant III [Figure 7.13(b)]. To satisfy both conditions (quadrants III or IV *and* quadrants II or III), θ must be in quadrant III (Figure 7.14).

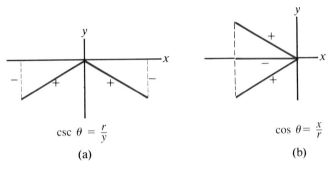

$$\csc\theta = \frac{r}{y}$$

(a)

$$\cos\theta = \frac{x}{r}$$

(b)

Figure 7.13

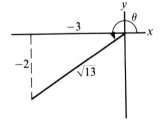

Figure 7.14

From $\csc\theta = \dfrac{r}{y} = -\dfrac{\sqrt{13}}{2} = \dfrac{\sqrt{13}}{-2}$, we get $y = -2$ and $r = \sqrt{13}$. (Recall that r is always positive.) By the Pythagorean theorem

$$x^2 + (-2)^2 = (\sqrt{13})^2 \qquad \text{or} \qquad x = \pm 3$$

Since θ is in the third quadrant, $x = -3$. Now draw the reference triangle in Figure 7.14. Thus

$$\sec\theta = \frac{r}{x} = \frac{\sqrt{13}}{-3} = -\frac{\sqrt{13}}{3}$$

◀

E X A M P L E **5**   Use a calculator to find $\cos\theta$, given that $\cot\theta = -1.728$, θ in quadrant II. (The numbers are approximate.)

Solution. Since θ is in quadrant II, $x < 0$ and $y > 0$:

$$-1.728 = \cot\theta = \frac{x}{y} = \frac{-1.728}{1}$$

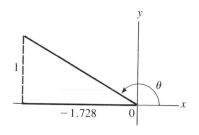

Figure 7.15

So $x = -1.728$ and $y = 1$. (See Figure 7.15.) Now store the value of $r = \sqrt{(-1.728)^2 + 1^2}$:

$$r: \quad 1.728 \;\boxed{+/-}\;\boxed{x^2}\;\boxed{+}\;1\;\boxed{=}\;\boxed{\sqrt{}}\;\boxed{\text{STO}} \to 1.9964929$$

Finally, we use the stored value of r to find $\cos \theta$:

$$\cos \theta = \frac{x}{r}: \quad 1.728 \;\boxed{+/-}\;\boxed{\div}\;\boxed{\text{MR}}\;\boxed{=} \to -0.8655177$$

Thus $\cos \theta = -0.8655$, rounded off to four significant digits. ◀

E X E R C I S E S / S E C T I O N   7.1

In Exercises 1–20, find the values of the six trigonometric functions for the angle in standard position whose terminal side passes through the given point.

1. $(-4, 3)$ **2.** $(-3, -4)$ **3.** $(5, -12)$

4. $(-5, -12)$ **5.** $(1, 2\sqrt{2})$ **6.** $(-3, 2)$

7. $(-2, -6)$ **8.** $(3, -5)$ **9.** $(-1, \sqrt{15})$

10. $(2, \sqrt{5})$ **11.** $(-3, 6)$ **12.** $(-4, -7)$

Use a calculator in Exercises 13–20. (The numbers are approximate.)

13. $(0.72, -1.8)$ **14.** $(-3.6, -2.4)$

15. $(-17, 21)$ **16.** $(230, 420)$

17. $(-0.476, -0.248)$ **18.** $(3.84, -5.05)$

19. $(669, -807)$ **20.** $(-823, 617)$

21. If $\tan \theta = \frac{5}{2}$ and $\cos \theta > 0$, find $\sin \theta$.

22. If $\sin \theta = \frac{1}{3}$ and $\tan \theta < 0$, find $\cot \theta$.

23. If $\sec \theta = -\sqrt{6}$ and $\cot \theta > 0$, find $\csc \theta$.

24. If $\cot \theta = -\frac{3}{2}$ and $\sin \theta < 0$, find $\cos \theta$.

25. If $\sin \theta = -\sqrt{7}/4$ and $\sec \theta > 0$, find $\tan \theta$.

26. If $\cos \theta = -\sqrt{7}/3$ and $\sin \theta > 0$, find $\sin \theta$.

27. If $\csc \theta = -\sqrt{3}$ and $\cot \theta < 0$, find $\sec \theta$.

28. If $\cos \theta = -3/\sqrt{21}$ and $\sin \theta > 0$, find $\tan \theta$.

29. If $\cos \theta = -\sqrt{3}/4$ and $\csc \theta < 0$, find $\sin \theta$.

30. If $\tan \theta = -\frac{7}{2}$ and $\cos \theta > 0$, find $\sec \theta$.

31. If $\cot \theta = \frac{1}{3}$ and $\cos \theta < 0$, find $\cos \theta$.

32. If $\sec \theta = \frac{7}{3}$ and $\cot \theta > 0$, find $\sin \theta$.

33. If $\sec \theta = -\sqrt{6}/2$ and $\tan \theta < 0$, find $\csc \theta$.

34. If $\sin \theta = \sqrt{2}/2$ and $\cot \theta < 0$, find $\tan \theta$.

35. If $\cos \theta = -\sqrt{15}/5$ and $\sin \theta < 0$, find $\cot \theta$.

36. If $\sin \theta = \sqrt{6}/3$ and $\cos \theta > 0$, find $\sec \theta$.

Use a calculator in Exercises 37–46. (The numbers are approximate.)

37. If $\sin \theta = 0.8914$, θ in quadrant II, find $\sec \theta$.

38. If $\cos \theta = 0.5041$, θ in quadrant IV, find $\sin \theta$.

39. If $\tan \theta = 1.607$ θ in quadrant III, find $\csc \theta$.

40. If $\sin \theta = 0.4309$, θ in quadrant II, find $\cot \theta$.

41. If $\cot \theta = -3.163$, θ in quadrant II, find $\sin \theta$.

42. If $\tan \theta = -1.846$, θ in quadrant II, find $\cos \theta$.

43. If $\sec \theta = 4.892$, θ in quadrant IV, find $\tan \theta$.

44. If $\csc \theta = -3.067$, θ in quadrant III, find $\tan \theta$.

45. If $\csc \theta = 1.926$, θ in quadrant II, find $\sec \theta$.

46. If $\sec \theta = -2.047$, θ in quadrant III, find $\sin \theta$.

47. Verify the following relationships:

$$\sin(-\theta) = -\sin \theta$$
$$\tan(-\theta) = -\tan \theta$$
$$\cos(-\theta) = \cos \theta$$

7.2   Special Angles

We will now learn to find the trigonometric functions of special angles. Because these angles are used so frequently in the examples and exercises of more advanced courses, as well as trigonometry, you must be thoroughly familiar with them.

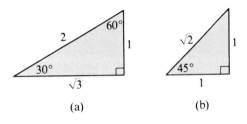

Special angles

Figure 7.16

Recall that the triangles involving special angles are those illustrated in Figure 7.16. As we saw in Chapter 4, these triangles are adequate for discussing special angles in the first quadrant. Since we want to consider special angles in any quadrant, the triangles in Figure 7.16 will serve as reference triangles.

The first two examples illustrate the technique for finding the function values of special angles by means of diagrams.

E X A M P L E **1**   Find $\tan 135°$ and $\sec 135°$.

Solution. The angle is shown in Figure 7.17. To find the reference angle α, note that $\alpha + 135° = 180°$; so $\alpha = 180° - 135° = 45°$. If we now drop a perpendicular from a point on the terminal side to the x-axis and use the numbers in the triangle in Figure 7.16(b), we obtain the reference triangle in Figure 7.18. It follows that

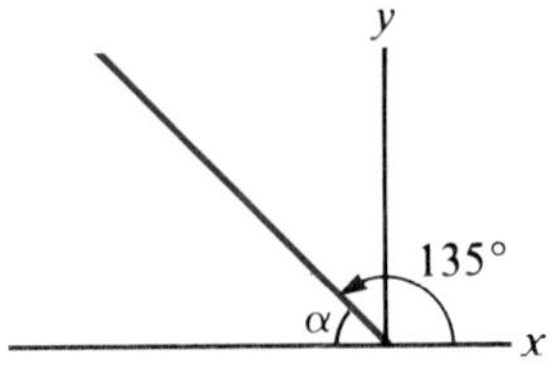

$$\tan 135° = \frac{\text{opposite}}{\text{adjacent}} = \frac{1}{-1} = -1$$

and

$$\sec 135° = \frac{\text{hypotenuse}}{\text{adjacent}} = \frac{\sqrt{2}}{-1} = -\sqrt{2}$$

◀

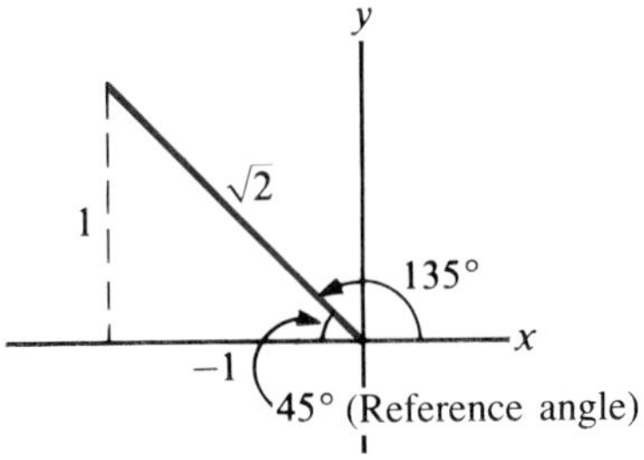

Figure 7.17

Figure 7.18

$\sqrt{2}$

$135°$

-1

$45°$ (Reference angle)

E X A M P L E **2**   Find $\sin 240°$ and $\cos 240°$.

Solution. Since $240°$ is between $180°$ and $270°$, the angle is in the third quadrant. To find the reference angle α, note that $\alpha + 180° = 240°$. So $\alpha = 240° - 180° = 60°$. The resulting reference triangle, shown in Figure 7.19, is a $30°$–$60°$ right triangle. (Refer to Figure 7.16(a).) It follows that

$$\sin 240° = \frac{\text{opposite}}{\text{hypotenuse}} = -\frac{\sqrt{3}}{2} \quad \text{and} \quad \cos 240° = \frac{\text{adjacent}}{\text{hypotenuse}} = -\frac{1}{2}$$

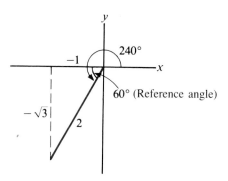

Figure 7.19   ◄

We can see from Examples 1 and 2 that if $f(\theta)$ is a trigonometric function of θ and if α is the reference angle, then $f(\theta) = \pm f(\alpha)$. The sign can be determined from the quadrant in which θ lies. Example 1, for instance, shows that

$$\tan 135° = -\tan 45° \quad \text{and} \quad \sec 135° = -\sec 45°$$

From Example 2 we have

$$\sin 240° = -\sin 60° \quad \text{and} \quad \cos 240° = -\cos 60°$$

Quadrantal angle

As noted in Chapter 4, if the terminal side coincides with an axis, the angle is called a **quadrantal angle**. (For example, $0°$, $270°$, and $360° + 90° = 450°$ are quadrantal angles.) The values of trigonometric functions of quadrantal angles can also be found by using diagrams, as shown in Examples 3 and 4.

E   X   A   M   P   L   E   **3**   Find $\cos 180°$ and $\csc 180°$.

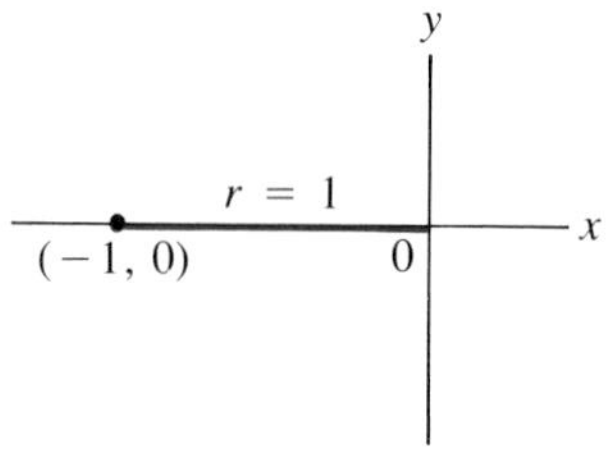

Solution. Since the terminal side coincides with the negative x-axis, we do not get a triangle. So we pick a point, say $(-1, 0)$, on the terminal side (Figure 7.20) and apply the definitions in (7.1). Note that $r = 1$:

$$\cos 180° = \frac{x}{r} = \frac{-1}{1} = -1$$

$$\csc 180° = \frac{r}{y} = \frac{1}{0} \quad \text{(undefined)}$$

◄

Figure 7.20

E   X   A   M   P   L   E   **4**   Find $\cot 270°$, $\tan 270°$, and $\sin 270°$.

Solution. Let $(0, -1)$ be a point on the terminal side (Figure 7.21). Noting that $r = 1$, we get, by the definitions of the trigonometric functions (7.1),

$$\cot 270° = \frac{x}{y} = \frac{0}{-1} = 0$$

$$\tan 270° = \frac{y}{x} = \frac{-1}{0} \quad \text{(undefined)}$$

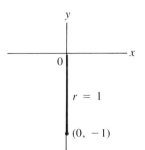

Figure 7.21

and

$$\sin 270° = \frac{y}{r} = \frac{-1}{1} = -1$$

◀

In the remaining examples we consider the problem of finding special angles, given the value of a trigonometric function.

E   X   A   M   P   L   E   **5**   Find all angles θ such that $0° \le \theta < 360°$ if $\cos \theta = \frac{1}{2}$.

Solution. Since $\cos \theta > 0$ in quadrants I and IV, we draw the reference triangles shown in Figure 7.22 with $x = 1$ and $r = 2$ (since $\cos \theta = x/r$). From the resulting 30°–60° right triangle, we see that the reference angle is 60° (Figure 7.22). It follows that $\theta = 60°$ and 300°.

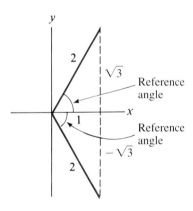

$$x^2 + y^2 = r^2$$
$$1^2 + y^2 = 2^2$$
$$y^2 = 3$$
$$y = \pm\sqrt{3}$$

Figure 7.22   ◀

E   X   A   M   P   L   E   **6**   Find all angles θ such that $0° \le \theta < 360°$ if $\sec \theta = -2\sqrt{3}/3$.

Solution. First note that $\sec \theta < 0$ in quadrants II and III. Since

$$-\frac{2\sqrt{3}}{3} = \sec \theta = \frac{r}{x} = \frac{2\sqrt{3}}{-3}\qquad \textbf{since } r \textbf{ is positive}$$

we have in either instance $x = -3$ and $r = 2\sqrt{3}$. The numerical value of y is obtained from the equation $x^2 + y^2 = r^2$:

$$(-3)^2 + y^2 = (2\sqrt{3})^2 \qquad \text{which yields} \qquad y = \pm\sqrt{3}$$

Now draw Figure 7.23. Observe that in either reference triangle, the hypotenuse is twice as long as the opposite side. So the reference angle is 30°. It follows that $\theta = 150°$ and $\theta = 210°$.   ◀

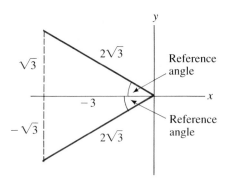

Figure 7.23

E X A M P L E **7**   Find θ, given that sec $\theta = -1$.

Solution. $-1 = \sec \theta = \dfrac{r}{x} = \dfrac{1}{-1}$ **since r is positive**

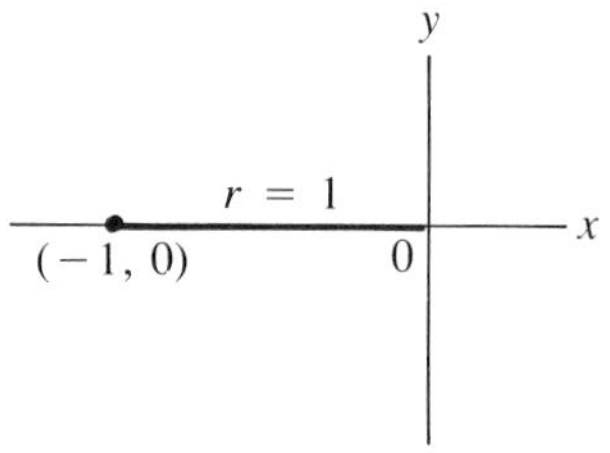

Thus $r = 1$ and $x = -1$. From $x^2 + y^2 = r^2$, we have

$$(-1)^2 + y^2 = 1^2 \qquad x = -1, r = 1$$

$$1 + y^2 = 1 \qquad \text{and} \qquad y = 0$$

So the point $(-1, 0)$ lies on the terminal side of θ (Figure 7.24). It follows that $\theta = 180°$.

Figure 7.24

E X E R C I S E S / S E C T I O N   **7.2**

In Exercise 1–40, find the value of each trigonometric function by means of a diagram. (Do not use a table or a calculator.)

1. sin 30°

2. cos 120°

3. tan 330°

4. sec 240°

5. cos 135°

6. sec 225°

7. csc 30°

8. cot 315°

9. sin($-30°$)

10. cos 0°

11. csc($-270°$)

12. sin 180°

13. sec($-60°$)

14. sec 180°

15. cot 90°

16. cos 240°

17. csc 270°

18. cos($-90°$)

19. cot 120°

20. cot($-120°$)

21. csc 180°

22. tan 240°

23. sin 300°

24. cos 150°

25. sec($-150°$)

26. tan($-180°$)

27. cot 300°

28. sin 0°

29. sec 210°

30. cos($-45°$)

31. cos($-270°$)

32. cot 210°

33. sin 315°

34. tan 300°

35. tan 150°

36. csc 210°

37. sec 390°

38. sin 405°

39. cos 420°

40. tan 450°

41. Show that $\sin^2 150° + \cos^2 150° = 1$.

42. Show that $1 + \tan^2 210° = \sec^2 210°$.

43. Show that $(\cos 300°)/(\sin 300°) = \cot 300°$.

44. Show that $\sec^2(-30°) = \tan^2(-30°) + 1$.

In Exercises 45–75, find all angles θ such that $0° \leq \theta < 360°$ for the given value of the trigonometric function. (Do not use a table or a calculator.)

45. $\sin \theta = \dfrac{1}{2}$

46. $\cos \theta = -\dfrac{1}{2}$

47. $\csc \theta = 2$

48. $\sec \theta = -2$

49. $\sin \theta = -\dfrac{\sqrt{3}}{2}$

50. $\sec \theta = \sqrt{2}$

51. $\cos \theta = -1$

52. $\sin \theta = 0$

53. $\tan \theta = \sqrt{3}$

54. $\cot \theta = -\sqrt{3}$ **55.** $\sec \theta = 2$ **56.** $\sin \theta = -\dfrac{1}{2}$ **68.** $\csc \theta$ undefined   **69.** $\tan \theta$ undefined

70. $\cos \theta = \dfrac{\sqrt{2}}{2}$ **71.** $\cot \theta = -\dfrac{\sqrt{3}}{3}$

57. $\sin \theta = -\dfrac{\sqrt{2}}{2}$ **58.** $\cos \theta = -\dfrac{\sqrt{3}}{2}$ **59.** $\cot \theta = 0$

60. $\csc \theta = 1$ **61.** $\tan \theta = -1$ **62.** $\cot \theta = 1$ **72.** $\cos \theta = 0$ **73.** $\sin \theta = \dfrac{\sqrt{2}}{2}$

63. $\cos \theta = -\dfrac{\sqrt{2}}{2}$ **64.** $\tan \theta = \dfrac{\sqrt{3}}{3}$ **65.** $\csc \theta = \dfrac{2\sqrt{3}}{3}$ **74.** $\tan \theta = -\dfrac{\sqrt{3}}{3}$ **75.** $\sec \theta = \dfrac{2\sqrt{3}}{3}$

66. $\csc \theta = -\dfrac{2\sqrt{3}}{3}$ **67.** $\tan \theta = -\sqrt{3}$

7.3   Trigonometric Functions of Arbitrary Angles

In this section we will (1) determine the values of trigonometric functions by means of tables and calculators and (2) determine an angle when given the value of a trigonometric function.

The first example illustrates how to find the value of a trigonometric function of a given angle.

E X A M P L E **1**   Use Table 1 of Appendix E to find $\cos 220°10'$.

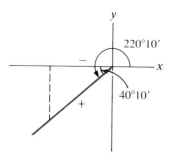

Figure 7.25

Solution. Since Table 1 lists function values only for angles between $0°$ and $90°$, we need to work with the reference angle. Since the angle is in quadrant III, the reference angle is $220°10' - 180° = 40°10'$. (See Figure 7.25.) So $\cos 220°10'$ has the same absolute value as $\cos 40°10'$. Since $\cos \theta$ is negative in quadrant III, it follows that

$$\cos 220°10' = -\cos 40°10' = -0.7642 \qquad \blacktriangleleft$$

CALCULATOR COMMENT

If the trigonometric functions are obtained with a calculator, the procedure of Chapter 4 carries over directly: *Enter the angle measurement and press the appropriate function key.* For example, to find $\csc 316.4°$, use the sequence

$$316.4 \;\boxed{\text{SIN}}\;\boxed{1/x} \to -1.4500749$$

However, if the value of the function is given and the angle is to be found, then special problems arise both with the table and the calculator.

E X A M P L E **2**   Find θ, given that $\tan \theta = -1.311$ and θ is in quadrant II by using **(a)** Table 1 of Appendix E; **(b)** a calculator.

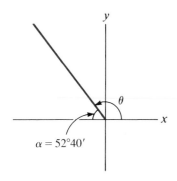

Figure 7.26

CALCULATOR COMMENT

Solution.

a. Since Table 1 of Appendix E lists function values only for angles between $0°$ and $90°$, the most we can get is the reference angle. (Recall that reference angles are always acute.) From

$$\tan \alpha = 1.311$$

we get $\alpha = 52°40'$, the reference angle (Figure 7.26). Since θ is in quadrant II, we obtain θ by subtracting α from 180:

$$
\begin{array}{r}
179°60' \\
- 52°40' \\
\hline
\theta = 127°20'
\end{array}
$$

b. When finding θ with a calculator, it is important to note that θ is always given in a definite range. For the tangent function, this range is $-90° < \theta < 90°$. So for $\tan \theta = -1.311$, the sequence

$$1.311 \boxed{+/-} \boxed{\text{INV}} \boxed{\text{TAN}}$$

yields $\theta = -52.664463 \approx -52°40'$. So the desired reference angle is $\theta = 52°40'$ and $\theta = 127°20'$ by part **a**.

A good alternative is to ignore the negative sign and obtain the reference angle directly from the sequence

$$1.311 \boxed{\text{INV}} \boxed{\text{TAN}} \rightarrow 52.664463 \qquad \blacktriangleleft$$

Scientific calculators are programmed to yield θ in the following range:

Function	Angle
$\sin \theta$	$-90° \leq \theta \leq 90°$
$\cos \theta$	$0° \leq \theta \leq 180°$
$\tan \theta$	$-90° < \theta < 90°$

E X A M P L E **3**   Use a calculator to find θ, given that $\sec \theta = -2.987$, θ in quadrant III.

Solution. Let α be the reference angle. Then $\sec \alpha = 2.987$. Since $\sec \alpha = 1/\cos \alpha$, $\cos \alpha = 1/2.987$; so α is obtained from the sequence

$$2.987 \boxed{1/x} \boxed{\text{INV}} \boxed{\text{COS}} \rightarrow 70.440592$$

Thus $\alpha = 70.44°$. Since θ is in quadrant III, θ is $70.44°$ more than $180°$ (Figure 7.27):

$$\theta = 180° + 70.44° = 250.44° \qquad \blacktriangleleft$$

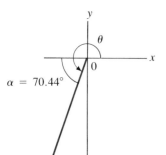

Figure 7.27

E X A M P L E   **4**   Use a calculator to find all angles θ such that $0° \le \theta < 360°$ if $\sin \theta = -0.2745$.

Solution. We obtain the reference angle from the sequence

$$0.2745 \; \boxed{\text{INV}} \; \boxed{\text{SIN}} \; \to \; 15.93222$$

So the reference angle is 15.93°. Since $\sin \theta < 0$ in quadrants III and IV, we get

$$\theta = 180° + 15.93° = 195.93°$$

and

$$\theta = 360° - 15.93° = 344.07°$$

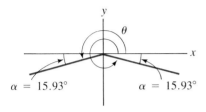

$\alpha = 15.93°$ $\alpha = 15.93°$

Figure 7.28

(See Figure 7.28.)   ◀

   E X A M P L E   **5**   If a projectile is hurled with velocity v (in feet per second) along an inclined plane making a constant angle α with the horizontal, then the projectile's range up the plane is given by

$$R = \frac{2v^2 \cos \theta \sin(\theta - \alpha)}{32 \cos^2 \alpha}$$

where θ is the angle with the horizontal at which the projectile is aimed. If $v = 100$ ft/s, $\alpha = 10.4°$, and $\theta = 16.7°$, find the range.

Solution. From the given formula we get

$$R = \frac{2(100)^2 \cos 16.7° \sin(16.7° - 10.4°)}{32 \cos^2 10.4°}$$

The sequence is

$$2 \; \boxed{\times} \; 100 \; \boxed{x^2} \; \boxed{\times} \; 16.7 \; \boxed{\text{COS}} \; \boxed{\times} \; \boxed{(} \; 16.7$$
$$\boxed{-} \; 10.4 \; \boxed{)} \; \boxed{\text{SIN}} \; \boxed{\div} \; 32 \; \boxed{\div} \; 10.4 \; \boxed{\text{COS}}$$
$$\boxed{x^2} \boxed{=} \; \to \; 67.904045$$

So the projectile lands 67.9 ft away along the plane.   ◀

E X E R C I S E S / S E C T I O N   **7.3**

In Exercises 1–16, use a calculator to find the value of each trigonometric function. Round off the answers to four significant digits.

1. $\sin 60.2°$

2. $\cos 129.5°$

3. $\sec 259.7°$

4. $\tan(-30.3°)$

5. $\csc 318.25°$

6. $\csc 280.3°$

7. $\cot 215.9°$

8. $\cot 155.73°$

9. $\sin 140.53°$

10. $\cos 320.15°$

11. $\sec 320.08°$

12. $\tan 18.27°$

13. $\tan 19.64°$

14. $\sin 24.93°$

15. $\cos(-25.9°)$

16. $\sec 333.55°$

In Exercises 17–30 use Table 1 or a calculator to obtain the value of each trigonometric function.

17. $\sin 60°10'$

18. $\cos 129°30'$

19. $\sec 259°40'$

20. $\tan(-30°20')$

21. $\csc 318°15'$

22. $\sin 280°18'$

23. $\cot 215°54'$

24. $\cos 155°44'$

25. $\tan 140°32'$

26. $\cot 320°9'$

27. $\sec(-32°5')$

28. $\tan 18°16'$

29. $\cot 19°39'$

30. $\csc 24°56'$

In Exercises 31–42, use a calculator to find all angles θ such that $0° \le \theta < 360°$ to the nearest hundredth of a degree.

31. $\sin \theta = 0.2924$

32. $\tan \theta = 0.1492$

33. $\cos \theta = 0.1776$

34. $\cot \theta = 0.1812$

35. $\sin \theta = -0.1933$

36. $\cos \theta = -0.1945$

37. $\csc \theta = -1.130$

38. $\sec \theta = 2.316$

39. $\cot \theta = -3.859$

40. $\tan \theta = -2.893$

41. $\csc \theta = 2.754$

42. $\sin \theta = -0.5163$

In Exercises 43–50, use Table 1 or a calculator to find all angles θ such that $0° \le \theta < 360°$ expressed in degrees and minutes.

43. $\cos \theta = -0.7318$

44. $\cot \theta = -3.896$

45. $\sec \theta = 2.374$

46. $\csc \theta = -1.478$

47. $\tan \theta = 1.179$

48. $\sin \theta = 0.3146$

49. $\cos \theta = -0.6438$

50. $\cot \theta = 0.3842$

In Exercises 51–58, use a calculator to find θ to the nearest hundredth of a degree.

51. Find θ, given that $\cos \theta = -0.4712$, θ in III.

52. Find θ, given that $\sec \theta = -2.093$, θ in III.

53. Find θ, given that $\sec \theta = -1.816$, θ in III.

54. Find θ, given that $\cos \theta = -0.6493$, θ in III.

55. Find θ, given that $\tan \theta = -0.8301$, θ in II.

56. Find θ, given that $\sin \theta = -0.2562$, θ in III.

57. Find θ, given that $\sin \theta = -0.6594$, θ in III.

58. Find θ, given that $\tan \theta = -2.804$ θ in II.

59. The range R (in meters) along the ground of a projectile fired at velocity v (in meters per second) at an angle θ with the horizontal is given by

$$R = \frac{v^2}{9.8} \sin 2\theta$$

Find the range if $\theta = 32.1°$ and $v = 24.2$ m/s.

60. Repeat Exercise 59 for $\theta = 53.0°$ and $v = 40.3$ m/s.

61. The largest weight that can be pulled up a plane inclined at an angle θ with the horizontal by a force F is

$$W = \frac{F}{\mu} (\cos \theta + \mu \sin \theta)$$

where μ is the coefficient of friction. If $\theta = 22°20'$, $F = 30.3$ lb, and $\mu = 0.350$, find W.

62. A weight W is to be dragged along a horizontal plane by a force whose line of action makes an angle θ with the plane. The force required to move the weight is given by

$$F = \frac{\mu W}{\mu \sin \theta + \cos \theta}$$

where μ is the coefficient of friction. If $W = 55.2$ lb, $\theta = 29.3°$, and $\mu = 0.200$, find F.

63. The formula for the magnetic intensity is

$$B = \frac{F}{qv \sin \theta}$$

where q is the magnitude of the charge, v its velocity, θ the angle between the direction of motion and the direction of the magnetic field, and F the force acting on the moving charge. If $B = 10$ webers/m^2, $F = 3.2 \times 10^{-11}$ N, $v = 2.0 \times 10^7$ m/s, and $q = 1.6 \times 10^{-19}$ C, find θ.

64. Repeat Exercise 63 with $q = 1.6 \times 10^{-19}$ C, $F = 4.2 \times 10^{-11}$ N, $v = 3.2 \times 10^7$ m/s, and $B = 10$ webers/m^2.

7.4   Radians

In this section we will study an alternate unit of angle measurement called the **radian**.

Definition of Radian Measure

As noted in Chapter 4, degree measure is so old that its origins are obscure. Why 360° was chosen for a whole angle is not clear. Apart from the fact that the number 360 has many divisors, it is a quite arbitrary choice. For someone

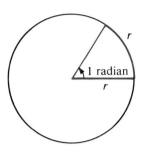

Figure 7.29

accustomed to the decimal system, even 100 units for a whole angle seems more natural. However, an even more natural unit of measure was eventually found. This unit, called the *radian*, is based on certain properties of the circle and is ideally suited for theoretical work. In fact, when trigonometric functions are discussed in calculus, radian measure is used almost exclusively.

Consider the circle in Figure 7.29. As the length of the radius r is measured off along the circumference, a central angle is created. The measure of this angle is called 1 *radian* (abbreviated "rad"), and this measure is independent of the size of the circle.

> **Definition of Radian Measure:** One **radian** is the size of an angle whose vertex is at the center of a circle and whose intercepted arc on the circumference is equal in length to the radius of the circle.

Relationship Between Radian and Degree Measure

To see the relationship between radian and degree measure, consider the angles of measure 2 rad, 3 rad, and 6 rad, respectively, shown in Figure 7.30. To obtain the radian measure of one complete rotation (360°), we need to determine how many times r can be measured off along the circumference. According to Figure 7.30(c), this appears to be just over six times. We can obtain the exact number from the formula for the circumference of a circle: $C = 2\pi r$. (Note that $C \approx 6.28r$.) So r can be measured off exactly 2π times along the circumference. It follows that

$$2\pi \text{ rad} = 360° \qquad \text{or} \qquad \pi \text{ rad} = 180° \tag{7.3}$$

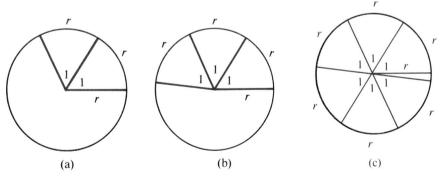

(a)   (b)   (c)

Figure 7.30

This relationship, in turn, gives us some of the special angles. For example, $\pi/2$ rad = 90°, $\pi/3$ rad = 60°, $\pi/4$ rad = 45°, and so on. Some of these angles are listed next.

Degrees:	0°	30°	45°	60°	90°	180°	270°	360°
Radians:	0	$\dfrac{\pi}{6}$	$\dfrac{\pi}{4}$	$\dfrac{\pi}{3}$	$\dfrac{\pi}{2}$	π	$\dfrac{3\pi}{2}$	2π

These measurements may appear strange at first, but most people do not find radian measure any more peculiar than degree measure once they get used to it. For example, just as 90° brings to mind a right angle, so does $\pi/2$ if it is seen often enough.

To convert degree measure to radian measure and vice versa, we need two other relationships, which follow directly from the relationship π rad = 180°: Dividing both sides by 180, we get $1° = (\pi/180)$ rad, or

$$1° = 1°\left(\frac{\pi}{180°}\right)\text{rad} = \frac{\pi}{180}\text{rad} \approx 0.01745\,\text{rad} \tag{7.4}$$

Similarly, dividing both sides of π rad = 180° by π, we get

$$1\,\text{rad} = \frac{180°}{\pi} = 57°17'45'' \approx 57.296° \tag{7.5}$$

From these relationships we obtain a simple method for converting from one system to the other. For example,

$$15° = 15°\,\frac{\pi}{\mathbf{180°}} = \frac{\pi}{12}\,\text{rad} \qquad \mathbf{15° = 15(1°)}$$

Conversely,

$$\frac{2\pi}{9}\,\text{rad} = \frac{2\pi}{9} \times \mathbf{1\,rad} = \frac{2\pi}{9}\left(\frac{\mathbf{180°}}{\boldsymbol{\pi}}\right) = 40°$$

> To convert degree measure to radian measure, multiply by $\pi/180°$.
> To convert radian measure to degree measure, multiply by $180°/\pi$.

In a pure conversion problem, it is customary to express radian measure in terms of π. However, if the problem contains physical data, the measure may have to be expressed in decimal form so that it can be rounded off to the proper number of significant digits.

Radian Measure of Central Angles

Before proceeding with any further examples, let us make another observation about the radian as a unit of measure. Consider the circle in Figure 7.31, which has a radius of 2 cm. As already noted, the central angle in radians is equal to

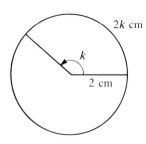

Figure 7.31

the number of times that the radius can be measured off along the circumference. In Figure 7.31, the arc length is $2k$ cm for $k \geq 0$. So k is equal to the number of times that $r = 2$ cm can be measured off along the circumference. It follows that k is the measure of the central angle in radians. Expressed in another way,

$$\frac{2k \text{ cm}}{2 \text{ cm}} = \frac{\cancel{2}k \cancel{\text{ cm}}}{\cancel{2} \cancel{\text{ cm}}} = k$$

Since the units cancel, radian measure is free of units.

> Whenever an angle is expressed in radians, no units are indicated.

For example, $30°$ is expressed simply as $\pi/6$.

In general, if s is the arc length and θ the central angle (Figure 7.32), then $\theta = s/r$, where θ is measured in radians.

> **Radian Measure of Central Angles:**
>
>
> $$\theta = \frac{s}{r} \qquad (7.6)$$ 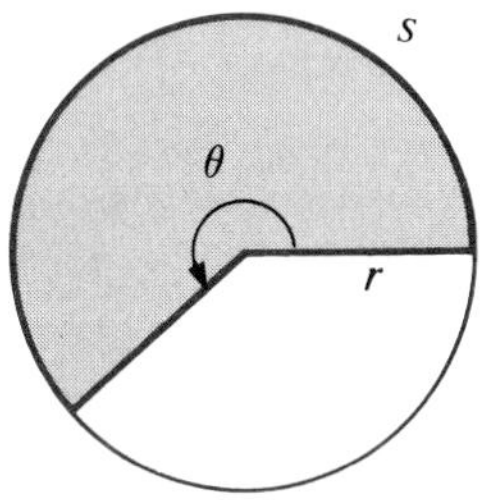
>
> Figure 7.32

Examples of Conversions

E X A M P L E **1** Convert $12°$ and $150°$ to radian measure.

Solution. Multiplying each angle by $\pi/180°$, we obtain

$$12° = 12° \frac{\pi}{180°} = \frac{12\pi}{180} = \frac{\pi}{15} \qquad \textbf{degrees cancel}$$

$$150° = 150° \frac{\pi}{180°} = \frac{150\pi}{180} = \frac{5\pi}{6} \qquad \textbf{degrees cancel}$$

Note that both measures are expressed in terms of π. ◀

E X A M P L E **2**   Convert $\pi/9$ and $7\pi/6$ to degree measure.

Solution. We multiply by $180°/\pi$ in each case:

$$\frac{\pi}{9} = \frac{\pi}{9} \times \frac{180°}{\pi} = 20°$$

$$\frac{7\pi}{6} = \frac{7\pi}{6} \times \frac{180°}{\pi} = 210°$$

◀

CALCULATOR COMMENT   Conversions can also be done with a calculator.

E X A M P L E **3**   Convert to radian measure:

    a. $26.3°$ **b.** $38°47'$

Solution. We multiply by $\pi/180°$ in each case.

  a. The sequence is

$$26.3 \boxed{\times} \boxed{\pi} \boxed{\div} 180 \boxed{=} \rightarrow 0.4590216$$

So $26.3° = 0.4590$ to four decimal places.

  b. The sequence is

$$38 \boxed{+} 47 \boxed{\div} 60 \boxed{=} \boxed{\times} \boxed{\pi} \boxed{\div} 180 \boxed{=} \rightarrow 0.6768969$$

So $38°47' = 0.6769$.

◀

E X A M P L E **4**   Convert 2.46 to degrees.

Solution. We multiply by $180°/\pi$.

$$2.46 \boxed{\times} 180 \boxed{\div} \boxed{\pi} \boxed{=} \rightarrow 140.94762$$

So $2.46 = 140.9°$ to the nearest tenth of a degree.

◀

Functions of Angles in Radians

When finding the value of a trigonometric function of an angle in radians, take

Radian mode   care to set your calculator in the **radian mode**.

E X A M P L E **5**   Evaluate:

    a. $\sin 1.921$ **b.** $\sec(4\pi/7)$

Solution. Set your calculator in the radian mode and use the sequences

  a. $1.921 \boxed{\text{SIN}} \rightarrow 0.9393028$

  b. $4 \boxed{\times} \boxed{\pi} \boxed{\div} 7 \boxed{=} \boxed{\text{COS}} \boxed{1/x} \rightarrow -4.4939592$

◀

Trigonometric Functions in BASIC (Optional)

To find the values of the trigonometric functions in BASIC, use the library functions

$$SIN(X), COS(X), \text{ and } TAN(X)$$

In most systems, the angles must be expressed in radians.

E X A M P L E **6**   Evaluate:

      a. sin 1.5   **b.** tan 2   **c.** cos(−2)

      Solution.

Function	Result
a. SIN(1.5)	0.9974949866
b. TAN(2)	−2.1850398633
c. COS(−2)	−0.4161468365

◀

E X E R C I S E S / S E C T I O N   **7.4**

In Exercises 1–24, convert each degree measure to radian measure expressed in terms of π.

1. 30°   **2.** 45°   **3.** 60°   **4.** 15°

5. 20°   **6.** 32°   **7.** 72°   **8.** −45°

9. −60°   **10.** 150°   **11.** 210°   **12.** 225°

13. 99°   **14.** 135°   **15.** 220°   **16.** 315°

17. 108°   **18.** 144°   **19.** 38°   **20.** −112°

21. −336°   **22.** 276°   **23.** 117°   **24.** 189°

In Exercises 25–40, convert each radian measure to degree measure.

25. $\dfrac{\pi}{4}$ **26.** $\dfrac{4\pi}{3}$ **27.** $-\dfrac{7\pi}{6}$ **28.** $\dfrac{5\pi}{3}$

29. $\dfrac{5\pi}{36}$ **30.** $\dfrac{7\pi}{12}$ **31.** $\dfrac{11\pi}{10}$ **32.** $\dfrac{25\pi}{18}$

33. $\dfrac{16\pi}{9}$ **34.** $-\dfrac{17\pi}{12}$ **35.** $\dfrac{21\pi}{10}$ **36.** $\dfrac{17\pi}{18}$

37. $\dfrac{\pi}{60}$ **38.** $-\dfrac{17\pi}{90}$ **39.** $\dfrac{7\pi}{9}$ **40.** $\dfrac{5\pi}{12}$

In Exercises 41–48, use a calculator to convert the degree measures to radian measures accurate to four decimal places.

41. 17.58°   **42.** 74.17°   **43.** 132.80°   **44.** 218.88°

45. 320°24′   **46.** −25°5′   **47.** −100°10′   **48.** 170°51′

In Exercises 49–56, use a calculator to convert the radian measures to degree measures to the nearest tenth of a degree.

49. 0.4160   **50.** 1.7320   **51.** 2.6810   **52.** 2.9460

53. 3.2572   **54.** −4.8417   **55.** −0.9026   **56.** 2.1136

In Exercises 57–68, use a calculator to find the values of the given trigonometric functions. (Remember to set your calculator in the radian mode.)

57. sin(0.8642)   **58.** cos(1.3246)

59. sec(−0.9174)   **60.** csc(2.1385)

61. tan(−1.7392)   **62.** cot(1.9127)

63. cos(2.0176)   **64.** tan(2.4014)

65. $\sin\left(\dfrac{2\pi}{3}\right)$ **66.** $\tan\left(\dfrac{5\pi}{12}\right)$

67. $\sec\left(\dfrac{7\pi}{15}\right)$ **68.** $\cot\left(\dfrac{3\pi}{13}\right)$

69. The current i in an alternating-series circuit as a function of time t (in seconds) is $i = 1.50 \cos 265t$. Find i when $t = 0.00650$ s.

70. The voltage across an inductor as a function of time t (in seconds) is $v = 120 \sin 350t$. Find v when $t = 0.0025$ s.

7.5   Applications of Radian Measure

The basic applications of radian measure we will consider are the determination of the length of a circular arc, the area of a circular sector, and angular and linear velocity.

Arc Length

To obtain the formula for the length of a circular arc, we return to the definition of radian measure, formula (7.6): $\theta = s/r$. From this formula it follows that

$$s = r\theta$$

where θ is the central angle shown in Figure 7.33.

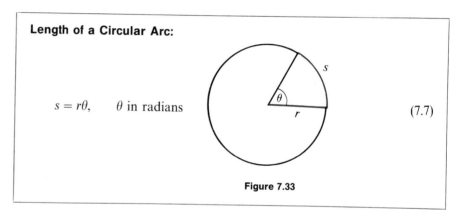

Length of a Circular Arc:

$$s = r\theta, \qquad \theta \text{ in radians}$$

(7.7)

Figure 7.33

E X A M P L E **1**   Find the arc length s in Figure 7.34.

Solution. First we need to change $20°$ to radians:

$$20° = \frac{20\pi}{180} = \frac{\pi}{9}$$

Then by formula (7.7)

$$s = (3.0 \text{ in.})\left(\frac{\pi}{9}\right) = 1.0 \text{ in.}$$

(Note that $\pi/9$ is treated as a dimensionless number.)
    A good alternative is to write s in the form

$$s = (3.0 \text{ in.})\left(20° \frac{\pi}{180°}\right)$$

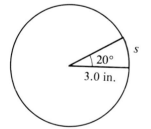

Figure 7.34

CALCULATOR COMMENT

and use the sequence

$$3.0 \boxed{\times} 20 \boxed{\times} \boxed{\pi} \boxed{\div} 180 \boxed{=} \rightarrow 1.0471976$$

So $s = 1.0$ in. to two significant digits.   ◀

The arc length formula can be used to find the approximate height of a distant object, as shown in the next example.

E X A M P L E **2**   A tree 193 ft from an observer on the ground intercepts an angle of 6.24°. (See Figure 7.35.) Find the approximate height of the tree.

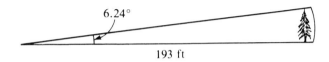

6.24°

193 ft

Figure 7.35

Solution. The height of the tree is approximately equal to the length of the intercepted arc. (See Figure 7.35.) To find the length of the arc, we use the formula $s = r\theta$. Since

$$\theta = 6.24° = 6.24\left(\frac{\pi}{180}\right) \approx 0.1089$$

it follows that the height is approximately equal to

$$(193 \text{ ft})(0.1089) = 21.0 \text{ ft}$$ ◀

Area of a Circular Sector

Radian measure also enables us to calculate the area of a sector of a circle. Since the area of a circular sector is proportional to the central angle, we have for the sectors in Figure 7.36,

$$\frac{A}{\theta} = \frac{A'}{\theta'} \tag{7.8}$$

Now suppose sector A' consists of the entire circle. Then $A' = \pi r^2$ and $\theta' = 2\pi$, and equation (7.8) becomes

$$\frac{A}{\theta} = \frac{\pi r^2}{2\pi}$$

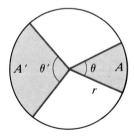

Figure 7.36

or

$$A = \frac{1}{2}r^2\theta$$

The sector is shown in Figure 7.37.

Area of a Circular Sector:

$$A = \frac{1}{2}r^2\theta, \qquad \theta \text{ in radians}$$

Figure 7.37

(7.9)

E X A M P L E **3**   Find the area of the circular sector in Figure 7.38.

Solution. Converting θ to radian measure,

$$\theta = 120.43° \left(\frac{\pi}{180°}\right) = 2.1019$$

Figure 7.38

So by formula (7.9), we get

$$A = \frac{1}{2}(15.61 \text{ cm})^2(2.1019) = 256.1 \text{ cm}^2$$

Alternatively, we can write A in the form

$$A = \frac{1}{2}(15.61 \text{ cm})^2 \left(120.43° \frac{\pi}{180°}\right)$$

A possible sequence is

$$0.5 \;\boxed{\times}\; 15.61 \;\boxed{x^2}\;\boxed{\times}\; 120.43 \;\boxed{\times}\;\boxed{\pi}\;\boxed{\div}\; 180 \;\boxed{=}\; \rightarrow 256.0872 \qquad \blacktriangleleft$$

Linear and Angular Velocity

A particularly interesting application of radian measure involves rotational motion. Recall that for motion along a path, distance = rate × time, and rate = distance/time, where the rate is assumed to be constant. Suppose that a particle is moving around a circle at a constant rate. If we denote the distance along the circle by s and the rate by v (for velocity), then $v = s/t$. Since $s = r\theta$, we have $v = (r\theta)/t = r(\theta/t)$. The ratio θ/t (in radians per unit time) is called the **angular velocity** and is usually denoted by ω (omega). So $v = r\omega$; v is called the **linear velocity**.

Angular velocity
Linear velocity

> **Relationship between Linear and Angular Velocity:**
>
> $$v = \omega r \qquad\qquad (7.10)$$
>
> Units for ω: radians per unit time

E X A M P L E **4** A particle is moving about a circle of radius 3.0 in. with an angular velocity of 2.0 rad/min. Find the linear velocity v.

Solution. By formula (7.10), $v = \omega r$:

$$v = 2.0 \text{ rad/min} \times 3.0 \text{ in.} = 6.0 \text{ in./min}$$

Note that since a radian is a dimensionless number, it is not included in the final result. Consequently, v can be expressed simply as 6.0 in./min, which is the velocity of the particle along the rim. ◄

In many problems ω is expressed in terms of revolutions per unit time. In such a case ω has to be converted to radians per unit time.

E X A M P L E **5** A computer disk rotates at the rate of 335 rev/min (revolutions per minute). If the radius of the disk is 6.15 cm, find the linear velocity of a point on the rim in meters per second.

Solution. To convert the angular velocity ω from revolutions per minute to radians per minute, we use the relationship

$$1 \text{ rev} = 2\pi \text{ rad}$$

As a result,

$$1 \frac{\text{rev}}{\text{min}} = 2\pi \frac{\text{rad}}{\text{min}}$$

So

$$335 \frac{\text{rev}}{\text{min}} = 335 \cdot 1 \frac{\text{rev}}{\text{min}} = 335 \cdot 2\pi \frac{\text{rad}}{\text{min}}$$

and

$$v = \omega r = (335 \cdot 2\pi) \frac{\text{rad}}{\text{min}} \cdot 6.15 \text{ cm}$$

Converting to meters per second, we get

$$v = (335)(2\pi)\frac{\text{rad}}{\text{min}} \cdot \frac{6.15\ \text{cm}}{1} \cdot \frac{1\ \text{m}}{100\ \text{cm}} \cdot \frac{1\ \text{min}}{60\ \text{s}} = 2.16\ \frac{\text{m}}{\text{s}}$$

(A discussion of conversion of units can be found in Appendix B.)   ◀

E X A M P L E **6**   A bicycle is moving at 18.0 mi/h. Determine the angular velocity of the wheels in revolutions per minute, given that each wheel has a diameter of 30.0 in.

Solution. Recall from Section 2.6 that 60 mi/h = 88 ft/s. Let x be the speed of the bicycle in feet per second. Then

$$\frac{x}{18.0\ \text{mi/h}} = \frac{88\ \text{ft/s}}{60\ \text{mi/h}}$$

or

$$x = \frac{88\ \text{ft/s}}{60\ \text{mi/h}} \cdot 18.0\ \text{mi/h} = 26.4\ \frac{\text{ft}}{\text{s}}$$

So the speed of the bicycle is 26.4 ft/s.

From the relationship $v = \omega r$, we get $\omega = v/r$. Since the radius is 15.0 in., this formula yields

$$\omega = \frac{26.4\ \text{ft/s}}{15.0\ \text{in.}} = 26.4\ \frac{\text{ft}}{\text{s}} \cdot \frac{1}{15.0\ \text{in.}}$$

To convert ω to revolutions per minute, we use the relationship 1 rev = 2π rad:

$$v = 26.4\ \frac{\text{ft}}{\text{s}} \cdot \frac{1}{15.0\ \text{in.}} \cdot \frac{1\ \text{rev}}{2\pi} \cdot \frac{12\ \text{in.}}{1\ \text{ft}} \cdot \frac{60\ \text{s}}{1\ \text{min}} = 202\ \frac{\text{rev}}{\text{min}}$$ ◀

E X E R C I S E S / S E C T I O N **7.5**

In each of Exercises 1–10, find the arc length determined by the given central angle. (See Example 1.)

1. $\theta = 1.98, r = 10.0$ cm   **2.** $\theta = 2.50, r = 2.78$ in.

3. $\theta = 29.5°, r = 5.40$ m   **4.** $\theta = 37.93°, r = 4.178$ cm

5. $\theta = 62°40', r = 1.39$ ft   **6.** $\theta = 51°22', r = 28.6$ m

7. $\theta = 26.74°, r = 2.070$ m   **8.** $\theta = 93°, r = 12$ ft

9. $\theta = 120°, r = 0.25$ ft   **10.** $\theta = 80.0°, r = 63.4$ in.

In Exercises 11–20, find the area of each sector. (See Example 3.)

11. $\theta = 0.176, r = 13.6$ cm   **12.** $\theta = 0.408, r = 5.80$ in.

13. $\theta = 43.7°, r = 12.6$ m   **14.** $\theta = 27.08°, r = 18.61$ cm

15. $\theta = 26°50', r = 9.31$ ft   **16.** $\theta = 67°17', r = 2.865$ cm

17. $\theta = 19.29°, r = 4.087$ m   **18.** $\theta = 140.0°, r = 0.725$ ft

19. $\theta = 170.0°, r = 0.419$ ft   **20.** $\theta = 150.0°, r = 12.0$ in.

21. A tree 520 ft away intercepts an angle of 1.5°. Find the approximate height of the tree. (See Example 2.)

22. A building 1,010 ft away intercepts an angle of 3.0°. Find the approximate height.

23. Find the degree measure of the central angle that intercepts an arc of length 6.5 ft on a circle of radius 6.0 ft.

24. Find the degree measure of the central angle that intercepts an arc of length 20.0 m on a circle of radius 8.0 m.

25. The full moon intercepts an angle of 0.518°. Find the diameter of the moon, given that the distance from the earth to the moon is about 239,000 mi.

26. The mean distance from the earth to the sun is 93 million miles. The diameter of the sun is 866,000 mi. Find its angular size, that is, the intercepted angle. Why do the sun and moon appear to be the same size to the naked eyes (referring to Exercise 25)?

27. Find the area swept out by a pendulum of length 24.70 cm that swings through an angle of 12.75°.

28. A pendulum of length 15.0 in. swings through an arc of 5°10'. Find the distance covered by the end of the pendulum as it swings from one end of the arc to the other.

29. Two circles are concentric if they have the same center. Suppose two concentric circles have radii 8.00 cm and 10.00 cm, respectively. Find the area of the portion of the sector inside the larger circle and outside the smaller circle if the central angle is 125°50'.

30. The first reasonably accurate estimate of the radius of the earth was obtained by Eratosthenes (276–198 B.C.) in Egypt. He observed that whenever the sun was directly overhead in the town of Aswan, it made an angle of 7.2° with a line perpendicular to the ground in Alexandria 495 mi away (see Figure 7.39). Use Eratosthenes' method to estimate the radius.

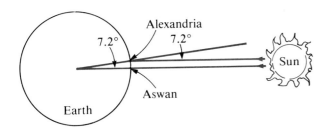

Figure 7.39

31. An object is moving about a circle of radius 10.0 cm with an angular velocity of 0.50 rad/s. Find its linear velocity in centimeters per second.

32. A circular disk 8.0 in. in diameter rotates at the rate of 5.2 rad/s. Find the linear velocity of a point on the rim in inches per second.

33. A wheel having a diameter of 10.7 ft rotates at the rate of 4.80 rev/s. Find the velocity of a point on the rim in feet per second.

34. A flywheel with radius 7.0 in. rotates at the rate of 15 rev/min. Find the velocity of a point on the rim in inches per minute.

35. A wheel with a radius of 3.50 in. rotates at the rate of 5.70 rev/s. Find the linear velocity of a point on the rim in feet per second.

36. A flywheel spins at the rate of 312 rev/min. If the radius of the flywheel is 8.15 cm, find the velocity of a point on the rim in meters per minute.

37. A wheel of radius 12.0 cm rotates at the rate of 5.50 rev/s. Find the velocity of a point on the rim in meters per minute.

38. A wheel with radius 6.00 in. rotates at the rate of 20.0 rev/s. Find the velocity of a point on the rim in feet per minute.

39. Find the velocity in miles per hour of a point on the equator due to the earth's rotation. (The radius of the earth is approximately 4,000 mi.)

40. A floppy disk has a radius of 2.50 in. and rotates at the rate of 452 rev/min. Find the velocity of a point on the rim in feet per second.

41. The radius of each wheel on a truck is 17.0 in. If the angular velocity of the wheels is 544 rev/min, find the velocity of the truck in miles per hour. (Recall that 88 ft/s = 60 mi/h.)

42. The angular velocity of the wheels of a car is 9.42 rev/s. If the radius of each wheel is 14.0 in., what is the speed of the car in miles per hour?

43. The wheels on a bicycle have a diameter of 32.0 in. If the wheels rotate at 3.53 rev/s, determine the speed of the bicycle in miles per hour.

44. A bicycle is traveling at the rate of 1,600 ft/min. Find the angular velocity of the wheels in radians per second, given that each wheel has a diameter of 32 in.

45. Determine the angular velocity of the wheels of a car in revolutions per second if each wheel has a radius of 14 in. and the car is traveling at 30 mi/h.

46. A pulley belt is 10 ft long and takes 45 s to make a complete circuit. If the radius of the pulley is 12 in., determine its angular velocity in radians per second.

47. A pulley belt 38.0 cm long takes 3.45 s to make a complete circuit. Determine the angular velocity of the pulley in revolutions per minute, given that the radius of the pulley is 5.90 cm.

48. Determine the velocity of the moon in miles per hour relative to the earth. (Assume that the moon takes 28 days to make one revolution about the earth and that its distance from the earth is 240,000 mi.)

49. Determine the velocity in miles per second of the earth relative to the sun. (Assume that the distance from the earth to the sun is 93 million miles.)

REVIEW EXERCISES/CHAPTER 7

In Exercises 1–4, find the values of the six trigonometric functions for the angle whose terminal side passes through the point indicated.

1. $(-2, \sqrt{5})$ **2.** $(-3, -1)$

3. $(2, -4)$ **4.** $(-5.36, 12.0)$

5. If $\sin \theta = -\frac{1}{3}$ and $\cot \theta > 0$, find $\cos \theta$.

6. If $\csc \theta = -\sqrt{6}/2$ and $\cos \theta < 0$, find $\cot \theta$.

7. If $\cos \theta = \sqrt{3}/4$ and $\tan \theta < 0$, find $\sin \theta$.

8. It $\tan \theta = 0.257$ and $\sec \theta < 0$, find $\csc \theta$.

In Exercises 9–20, find the values of the trigonometric functions by means of diagrams. (Do not use a table or a calculator.)

9. $\sin 135°$ **10.** $\tan 180°$ **11.** $\cos 240°$

12. $\sec 315°$ **13.** $\tan 225°$ **14.** $\csc 150°$

15. $\sec 210°$ **16.** $\cot 330°$ **17.** $\csc 90°$

18. $\tan 270°$ **19.** $\cos 150°$ **20.** $\csc 300°$

In Exercises 21–28, use a diagram to find all angles θ such that $0° \le \theta < 360°$. (Do not use a table or a calculator.)

21. $\cos \theta = \dfrac{1}{2}$ **22.** $\csc \theta = -2$

23. $\cot \theta = -\dfrac{\sqrt{3}}{3}$ **24.** $\tan \theta = \dfrac{\sqrt{3}}{3}$

25. $\tan \theta$ undefined **26.** $\cos \theta = -1$

27. $\sec \theta = -\dfrac{2\sqrt{3}}{3}$ **28.** $\sin \theta = -\dfrac{\sqrt{2}}{2}$

In Exercises 29–32, use a calculator or Table 1 of Appendix E to find each value.

29. $\tan 123.25°$ **30.** $\csc 250.30°$

31. $\cos 318°50'$ **32.** $\sec(-107°5')$

In Exercises 33–38, use a calculator or Table 1 of Appendix E to find θ such that $0° \le \theta < 360°$.

33. $\cos \theta = -0.8107$ **34.** $\csc \theta = 1.397$

35. $\tan \theta = -0.4170$ **36.** $\sin \theta = 0.2713$

37. $\cot \theta = 1.786$ **38.** $\sec \theta = -3.162$

In Exercises 39–44, convert the given degree measures to radian measures expressed in terms of π.

39. $40°$ **40.** $160°$ **41.** $-100°$

42. $-36°$ **43.** $54°$ **44.** $236°$

In Exercises 45–50, convert the given radian measures to degree measures.

45. $\dfrac{5\pi}{6}$ **46.** $\dfrac{7\pi}{36}$ **47.** $\dfrac{13\pi}{10}$

48. $\dfrac{5\pi}{12}$ **49.** $\dfrac{17\pi}{9}$ **50.** $\dfrac{23\pi}{18}$

In Exercises 51–58, use a calculator to convert the given degree measures to radian measures. (Give answers to four decimal places.)

51. $25.30°$ **52.** $79.67°$ **53.** $144.10°$

54. $-23.85°$ **55.** $-257°10'$ **56.** $340°26'$

57. $390°$ **58.** $400°$

In Exercises 59–62, use a calculator to convert the given radian measures to degree measures to the nearest tenth of a degree.

59. 0.7162 **60.** 1.438

61. 2.607 **62.** 3.861

In Exercises 63–68, find the value of each trigonometric function. (Set your calculator in the radian mode.)

63. $\tan 2.161$ **64.** $\sin(-0.3680)$ **65.** $\cos(1.783)$

66. $\tan(2.349)$ **67.** $\csc 1.507$ **68.** $\sec\left(\dfrac{7\pi}{10}\right)$

69. A circle has a radius of 6.00 in. Find the length of the arc intercepted by a central angle of $32.0°$.

70. Find the area of the circular sector in Exercise 69.

71. A tower 860 ft away intercepts an angle of $3.2°$. Find the approximate height of the tower.

72. Find the degree measure of the central angle that intercepts an arc length of 13.3 ft on a circle of radius 7.80 ft.

73. The wheels of a bicycle have a radius of 16.0 in. and rotate at the rate of 2.20 rev/s. Find the speed of the bicycle in miles per hour.

74. An artificial satellite travels around the earth in a circular orbit at an altitude of 300 mi. Find its velocity (in miles per hour) if it makes one revolution every 90 min. (The radius of the earth is approximately 4,000 mi.)

75. Find the velocity in miles per hour of a communications satellite that remains 22,300 mi above a point on the equator at all times. (See Exercise 74.)

Graphs of Trigonometric Functions

8.1 Graphs of Sine and Cosine Functions

The brief discussion in Chapter 2 showed that a graph illustrates the behavior of a given function. In this chapter we will study the graphs of trigonometric functions, especially the sine and cosine functions, which are particularly useful in physical applications. Our first task is to draw the graphs of $y = \sin x$ and $y = \cos x$ by constructing a table of values and plotting enough points to obtain a smooth curve.

Since it is customary to use radian measure when graphing trigonometric functions, let us first recall the radian measure of certain special angles. From

$$\pi = 180°$$

we get $\pi/6 = 30°$. This relationship yields all special angles that are multiples of $30°$. For example,

$$150° = 5 \cdot 30° = 5\left(\frac{\pi}{6}\right) = \frac{5\pi}{6} \qquad \text{and} \qquad 330° = 11 \cdot 30° = 11\left(\frac{\pi}{6}\right) = \frac{11\pi}{6}$$

Similarly, since $60° = \pi/3$,

$$120° = 2 \cdot 60° = 2\left(\frac{\pi}{3}\right) = \frac{2\pi}{3} \qquad \text{and} \qquad 300° = 5 \cdot 60° = 5\left(\frac{\pi}{3}\right) = \frac{5\pi}{3}$$

From $45° = \pi/4$, we get

$$225° = 5 \cdot 45° = 5\left(\frac{\pi}{4}\right) = \frac{5\pi}{4}$$

and so forth. Using these special angles, we construct a table of values for the function $y = \sin x$. Plotting these points on the rectangular coordinate system, we obtain the graph shown in Figure 8.1. (When labeling the horizontal axis, remember that π is approximately 3 units.)

$y = \sin x$

x (deg):	$0°$	$30°$	$45°$	$60°$	$90°$	$120°$	$150°$	$180°$	$210°$	$240°$	$270°$	$330°$	$360°$	$390°$
x (rad):	0	$\dfrac{\pi}{6}$	$\dfrac{\pi}{4}$	$\dfrac{\pi}{3}$	$\dfrac{\pi}{2}$	$\dfrac{2\pi}{3}$	$\dfrac{5\pi}{6}$	π	$\dfrac{7\pi}{6}$	$\dfrac{4\pi}{3}$	$\dfrac{3\pi}{2}$	$\dfrac{11\pi}{6}$	2π	$2\pi + \dfrac{\pi}{6}$
y (exact):	0	$\dfrac{1}{2}$	$\dfrac{1}{\sqrt{2}}$	$\dfrac{\sqrt{3}}{2}$	1	$\dfrac{\sqrt{3}}{2}$	$\dfrac{1}{2}$	0	$-\dfrac{1}{2}$	$-\dfrac{\sqrt{3}}{2}$	-1	$-\dfrac{1}{2}$	0	$\dfrac{1}{2}$
y (decimal):	0	0.5	0.7	0.87	1	0.87	0.5	0	-0.5	-0.87	-1	-0.5	0	0.5

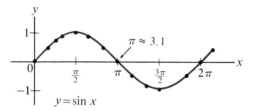

Figure 8.1

Observe that as x increases from 0 to $\pi/2$, the values of $\sin x$ increase from 0 to 1. As x continues to increase from $\pi/2$ to π, the values of $\sin x$ repeat in reverse order from 1 to 0. As x increases from π to $3\pi/2$, the values of $\sin x$ become negative, decreasing from 0 to -1. Finally, as x increases from $3\pi/2$ to 2π, the values of $\sin x$ increase once again, from -1 back to 0. *Starting at 2π, the values of $\sin x$ repeat.*

Note especially the zero values of $\sin x$ as well as the largest and smallest values:

x:	0	$\dfrac{1}{4}(2\pi) = \dfrac{\pi}{2}$	$\dfrac{1}{2}(2\pi) = \pi$	$\dfrac{3}{4}(2\pi) = \dfrac{3\pi}{2}$	$1 \cdot (2\pi) = 2\pi$
$\sin x$:	0	1	0	-1	0

To obtain the graph of $y = \cos x$, we construct the following table:

$y = \cos x$

x (deg):	$0°$	$30°$	$45°$	$60°$	$90°$	$120°$	$150°$	$180°$	$210°$	$240°$	$270°$	$300°$	$330°$	$360°$
x (rad):	0	$\dfrac{\pi}{6}$	$\dfrac{\pi}{4}$	$\dfrac{\pi}{3}$	$\dfrac{\pi}{2}$	$\dfrac{2\pi}{3}$	$\dfrac{5\pi}{6}$	π	$\dfrac{7\pi}{6}$	$\dfrac{4\pi}{3}$	$\dfrac{3\pi}{2}$	$\dfrac{5\pi}{3}$	$\dfrac{11\pi}{6}$	2π
y (exact):	1	$\dfrac{\sqrt{3}}{2}$	$\dfrac{1}{\sqrt{2}}$	$\dfrac{1}{2}$	0	$-\dfrac{1}{2}$	$-\dfrac{\sqrt{3}}{2}$	-1	$-\dfrac{\sqrt{3}}{2}$	$-\dfrac{1}{2}$	0	$\dfrac{1}{2}$	$\dfrac{\sqrt{3}}{2}$	1
y (decimal):	1	0.87	0.7	0.5	0	-0.5	-0.87	-1	-0.87	-0.5	0	0.5	0.87	1

Plotting these points, we get the graph shown in Figure 8.2.

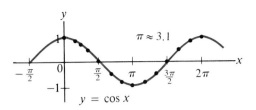

Figure 8.2

Note especially the zero values of cos x as well as the largest and smallest values:

x:	0	$\frac{1}{4}(2\pi) = \frac{\pi}{2}$	$\frac{1}{2}(2\pi) = \pi$	$\frac{3}{4}(2\pi) = \frac{3\pi}{2}$	$1 \cdot (2\pi) = 2\pi$
cos x:	1	0	-1	0	1

A closer inspection of the tables for $y = \sin x$ and $y = \cos x$ suggests that the values of the sine and cosine functions and hence the shapes of their graphs are identical except for their positions. Indeed, if the graph of the cosine function is moved $\pi/2$ units to the right, it becomes the graph of the sine function.

Given these basic shapes, our real goal in this section is to sketch the graphs of $y = a \sin bx$ and $y = a \cos bx$ without plotting points. To this end, observe that the graph of $y = a \cos x$ can be obtained from the graph of $y = \cos x$ by multiplying each value of cos x by a. The number $|a|$ is called the *amplitude*. Consider the following examples.

E X A M P L E 1 Sketch the graph of $y = 2 \cos x$.

Solution. First note that the coefficient 2 doubles the values of $y = \cos x$; otherwise, the shape is essentially the same as that of the basic cosine function. The graph, shown in Figure 8.3, looks like a tall version of the graph of $y = \cos x$.

x:	0	$\frac{\pi}{6}$	$\frac{\pi}{4}$	$\frac{\pi}{3}$	$\frac{\pi}{2}$	$\frac{2\pi}{3}$	$\frac{5\pi}{6}$	π	$\frac{7\pi}{6}$	$\frac{4\pi}{3}$	$\frac{3\pi}{2}$	$\frac{5\pi}{3}$	2π
y:	2	$\sqrt{3}$	$\sqrt{2}$	1	0	-1	$-\sqrt{3}$	-2	$-\sqrt{3}$	-1	0	1	2

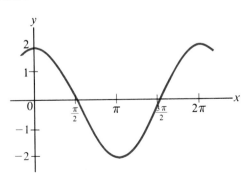

Figure 8.3

E X A M P L E **2** Sketch the graph of $y = -3 \sin x$.

Solution. The effect of the coefficient -3 is twofold: It multiplies the values of the basic sine function by 3 and changes the sign of each value. The given

x:	0	$\dfrac{\pi}{2}$	π	$\dfrac{3\pi}{2}$	2π
y:	0	-3	0	3	0

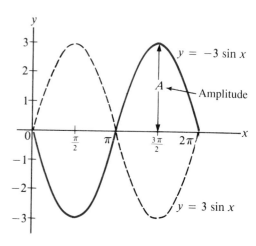

Figure 8.4

Intercepts

curve is therefore the "mirror image" of $y = 3 \sin x$ with respect to the x-axis. Since the shape of the basic sine curve is already known, we can sketch the curve by plotting only the highest and lowest points on the curve and the points where the curve crosses the x-axis (called the x-intercepts). The graph of $y = 3 \sin x$ is the dashed curve in Figure 8.4. The reflection of the dashed curve is the graph of $y = -3 \sin x$, shown by the solid curve in Figure 8.4.

◀

The Graphs of $y = a \sin bx$ and $y = a \cos bx$

Examples 1 and 2 show that in the graphs of $y = a \sin x$ and $y = a \cos x$ the numerical value of the coefficient a is the maximum distance from the curve to the x-axis. This distance is called the **amplitude**, which we will denote by A. Thus $A = |a|$. (See Figure 8.4.)

Another important property of trigonometric functions is the **periodicity**. This term refers to the fact that the function values eventually repeat. In the case of the sine and cosine functions, the y values repeat every 2π radians; this interval is called the **period**. Thus $y = \sin x$ and $y = \cos x$ are said to have a period of 2π. The general case is given next.

Amplitude and period of $y = a \sin bx$ and $y = a \cos bx$:

$$\textbf{Amplitude:} \quad A = |a| \qquad \textbf{Period:} \quad P = \frac{2\pi}{b} \tag{8.1}$$

To see why the period is $2\pi/b$, note first that the graph of $y = \sin x$ passes through the origin and crosses the x-axis at

$$x = \pi, 2\pi, 3\pi, \ldots \tag{8.2}$$

The *second* intercept, 2π, is equal to the period, denoted by P. To obtain the period of $y = a \sin bx$, we observe that

$$a \sin bx = 0$$

at the origin and whenever

$$bx = \pi, 2\pi, 3\pi, \ldots$$

or

$$x = \frac{\pi}{b}, \frac{2\pi}{b}, \frac{3\pi}{b}, \ldots \tag{8.3}$$

Comparing the intercepts given in statements (8.3) and (8.2) shows that $P = 2\pi/b$ (second intercept). By a similar argument, $2\pi/b$ is also the period of the cosine function. (Note that $b = 1$ for $y = \sin x$ and $y = \cos x$.)

Sinusoidal function

For convenience, the functions $y = a \sin bx$ and $y = a \cos bx$ are both called **sinusoidal functions** and their graphs **sinusoidal curves**. (Slightly more general forms of the sinusoidal curves will be discussed in the next section.)

To graph a sinusoidal curve, we determine the period and amplitude and sketch the curve from the basic shape. The procedure is summarized next.

To Sketch a Sinusoidal Curve: $y = a \sin bx$ or $y = a \cos bx$

1. Determine the period $P = 2\pi/b$. Start at $x = 0$ and mark off the distance $2\pi/b$.
2. Divide the interval from 0 to $2\pi/b$ into *four* equal parts.
3. Locate the x-intercepts.
   a. For $y = a \sin bx$, the intercepts are the beginning, midpoint, and end of the interval.
   b. For $y = a \cos bx$, the intercepts are one-fourth and three-fourths of the way through the interval.
4. Determine the amplitude and mark off the highest and lowest points on the curve.
5. Sketch the curve over one period.
6. Extend the graph over additional periods, if desired.

The remaining examples illustrate the technique for graphing sinusoidal curves.

E X A M P L E **3** Sketch the graph of $y = 0.8 \sin 2x$.

Solution. $y = \mathbf{0.8} \sin 2x$

Step 1. $P = \dfrac{2\pi}{2} = \pi$. Mark off $x = \pi$ (Figure 8.5).

Step 2. Since $P = \pi$, one-fourth of the period is $\frac{1}{4}\pi = \pi/4$. So the four equally spaced points (and $x = 0$) are

$$x = 0, \frac{\pi}{4}, \frac{2\pi}{4}, \frac{3\pi}{4}, \frac{4\pi}{4}$$

or

$$x = 0, \frac{\pi}{4}, \frac{\pi}{2}, \frac{3\pi}{4}, \pi$$

Step 3. The intercepts are

$x = 0$ (beginning of interval)

$x = \dfrac{\pi}{2}$ (midpoint)

$x = \pi$ (end of interval)

Step 4. The highest and lowest points occur at $x = \pi/4$ and $x = 3\pi/4$, the points midway between the intercepts. (These are the points in Step 2 that are not in Step 3.) Note that

$$A = |0.8| = \mathbf{0.8} \qquad \text{(Figure 8.5)}$$

Step 5. The graph is shown in Figure 8.5.
Step 6. The graph is extended to the right over one additional period.

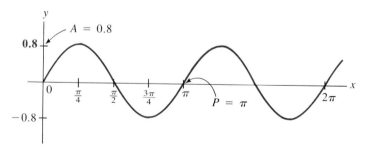

Figure 8.5

E X A M P L E **4**   Sketch the graph of $y = \frac{1}{2}\cos 2x$.

Solution. $y = \dfrac{1}{2}\cos 2x$

Step 1. $P = \dfrac{2\pi}{2} = \pi$. Mark off $x = \pi$ (Figure 8.6).

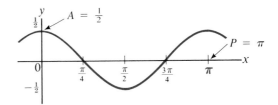

Figure 8.6

Step 2. The four equally spaced points (and $x = 0$) are

$$x = 0, \frac{\pi}{4}, \frac{2\pi}{4}, \frac{3\pi}{4}, \frac{4\pi}{4}$$

or

$$x = 0, \frac{\pi}{4}, \frac{\pi}{2}, \frac{3\pi}{4}, \pi$$

Step 3. The intercepts are

$$x = \frac{\pi}{4} \qquad \text{(one-fourth of the way)}$$

$$x = \frac{3\pi}{4} \qquad \text{(three-fourths of the way)}$$

Step 4. The highest and lowest points on the curve occur at $x = 0$, $\pi/2$, and π, the points in Step 2 that are not in Step 3. Also note that

$$A = \left|\frac{1}{2}\right| = \frac{1}{2} \qquad \text{(Figure 8.6)}$$

Step 5. The graph is shown in Figure 8.6.   ◄

E X A M P L E **5**   Sketch the graph of $y = 2 \sin 3x$.

Solution. $y = \textbf{2} \sin \textbf{3}x$

Step 1. $P = \dfrac{2\pi}{3}$. Mark off $x = \dfrac{2\pi}{3}$ (Figure 8.7).

Step 2. Since $P = 2\pi/3$, the four equally spaced points (and $x = 0$) are

$$x = 0, \frac{1}{4}\left(\frac{2\pi}{3}\right) = \frac{\pi}{6}, \frac{2\pi}{6}, \frac{3\pi}{6}, \frac{4\pi}{6}$$

or

$$x = 0, \frac{\pi}{6}, \frac{\pi}{3}, \frac{\pi}{2}, \frac{2\pi}{3}$$

Step 3. The intercepts are

$x = 0$ (beginning of interval)

$x = \dfrac{\pi}{3}$ (midpoint)

$x = \dfrac{2\pi}{3}$ (end of interval)

Step 4. The highest and lowest points occur at $x = \pi/6$ and $x = \pi/2$, the points midway between the intercepts. Note that

$A = |2| = \textbf{2}$ (Figure 8.7)

Step 5. The graph is shown in Figure 8.7.
Step 6. The graph is extended to the right over one additional period.

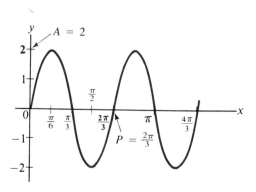

Figure 8.7

E X A M P L E **6**   Sketch the graph of $y = 5 \sin \frac{1}{2}x$.

Solution. $A = |5| = 5$, $P = 2\pi/\frac{1}{2} = 2\pi \cdot 2 = 4\pi$. See Figure 8.8.

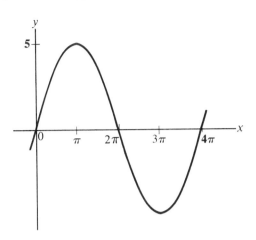

Figure 8.8

E X A M P L E **7**   Sketch the graph of $y = -6 \cos \frac{1}{3}x$.

Solution. $A = |-6| = 6$, $P = 2\pi/\frac{1}{3} = 2\pi \cdot 3 = 6\pi$. Because of the negative co-efficient, the graph is the reflection of the graph of $y = 6 \cos \frac{1}{3}x$ (Figure 8.9).

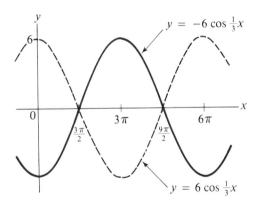

Figure 8.9

E X E R C I S E S / S E C T I O N   **8.1**

State the amplitude and period of each function; sketch the curve.

1. $y = 2 \sin x$

2. $y = 3 \cos x$

3. $y = -\sin x$

4. $y = -\cos x$

5. $y = \dfrac{1}{2} \sin x$

6. $y = \dfrac{1}{3} \cos 2x$

7. $y = 2 \cos 3x$

8. $y = \dfrac{1}{2} \sin \dfrac{1}{2} x$

9. $y = -5 \cos 2x$

10. $y = 6 \sin \frac{1}{2} x$

11. $y = \frac{1}{2} \cos 3x$

12. $y = 2 \sin 3x$

13. $y = 4 \sin \frac{2}{3} x$

14. $y = 6 \cos \frac{3}{4} x$

15. $y = -10 \sin \frac{1}{4} x$

16. $y = -12 \cos \frac{1}{8} x$

17. $y = 5 \sin \frac{1}{5} x$

18. $y = -\frac{3}{4} \sin \frac{1}{2} x$

19. $y = -4 \cos \frac{7}{3} x$

20. $y = 6 \sin \frac{3}{5} x$

21. $y = \sin \pi x$

22. $y = \cos \pi x$

23. $y = \frac{1}{3} \cos 3\pi x$

24. $y = 2 \sin 2\pi x$

8.2 Phase Shifts

The sinusoidal curves discussed in the last section are not the most general possible curves of this type. In addition to having a definite period and amplitude, the curves may be shifted to the left or right.

To see this behavior, consider the function

$$y = 2 \sin\left(x - \frac{\pi}{4} \right)$$

Suppose we compare the given curve to that of the same function but without the shift: $y = 2 \sin x$, which has amplitude 2 and period 2π. The y values of $y = 2 \sin(x - \pi/4)$ are identical to the values of $y = 2 \sin x$, but they correspond to different x values. Note especially that to get $\sin 0$, we must let $x = \pi/4$. So the point $(\pi/4, 0)$ corresponds to the origin. The other points are shifted similarly, so that the y values start repeating at $x = 2\pi + \pi/4$ (instead of $x = 2\pi$ for the basic sine function). The graph is shown in Figure 8.10.

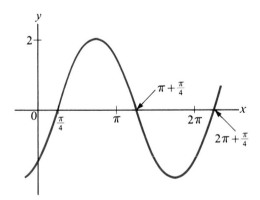

Figure 8.10

Let's consider another case.

E X A M P L E **1**   Sketch the graph of $y = 2 \sin(x + \pi/4)$.

Solution. This function is similar to the function above, but this time we need to let $x = -\pi/4$ to obtain sin 0. Consequently, the point $(-\pi/4, 0)$ corresponds to the origin, and the entire curve is shifted to the left (Figure 8.11).

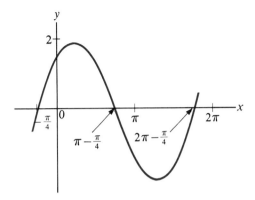

Figure 8.11 ◀

More generally, it is shown in analytic geometry that the graph of $y = f(x - h)$, $h > 0$, is the graph of $y = f(x)$ shifted h units to the right, while the graph of $y = f(x + h)$ is the graph of $y = f(x)$ shifted h units to the left.

To see how the graph of $y = a \sin(bx + c)$ is shifted, we first factor b and write $y = a \sin b(x + c/b)$, which shows that the shift is given by c/b. Similarly, from $y = a \cos b(x + c/b)$ we see that the shift is also given by c/b. The number c/b is commonly referred to as the **phase shift**.

We will now summarize the basic features of sinusoidal functions.

Amplitude, period, and phase shift of $y = a \sin(bx - c)$ and $y = a \cos (bx - c)$, $(b, c > 0)$:

$$\text{Amplitude:}\quad A = |a| \qquad \text{Period:}\quad P = \frac{2\pi}{b}$$

$$\text{Phase shift:}\quad \frac{c}{b} \text{ units to } \textit{right}$$

Amplitude, period, and phase shift of $y = a \sin (bx + c)$ and $y = a \cos (bx + c)$, $(b, c > 0)$:

$$\text{Amplitude:}\quad A = |a| \qquad \text{Period:}\quad P = \frac{2\pi}{b}$$

$$\text{Phase shift:}\quad \frac{c}{b} \text{ units to } \textit{left}$$

Sketching Sinusoidal Curves

To sketch the graph of $y = a \sin(bx \pm c)$ or $y = a \cos(bx \pm c)$, we first sketch the curve of $y = a \sin bx$ or $y = a \cos bx$ by the method discussed in Section 8.1. We then shift the curve c/b units to the left or right.

A simple way to determine the phase shift, including the direction, is to set $bx \pm c$ equal to 0 and solve for x. The x value determines the point where the "first" period begins. This procedure is referred to as the *alternate method* in Examples 2 and 3.

E X A M P L E **2**   Sketch the graph of $y = 3 \sin(2x - \pi)$.

Solution. The equation has the form $y = a \sin(bx - c)$:

$$\overset{a}{} \quad \overset{b}{} \quad \overset{c}{}$$
$$y = 3 \sin(2x - \pi)$$

$$A = 3, \qquad P = \frac{2\pi}{b} = \frac{2\pi}{2} = \pi, \qquad \textbf{phase shift:} \quad \frac{c}{b} = \frac{\pi}{2} \qquad \text{(right)}$$

Suppose we first sketch the curve $y = 3 \sin 2x$. Since $A = 3$ and $P = \pi$, we obtain the dashed curve in Figure 8.12. This graph is then shifted $\pi/2$ units to the right, shown as the solid curve in Figure 8.12.

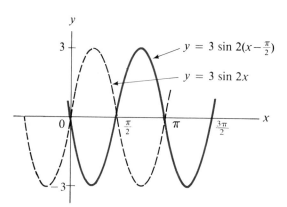

Figure 8.12

Alternate method: A good alternative to this procedure is to determine where the curve $y = 3 \sin(2x - \pi)$ "begins" by setting $2x - \pi$ equal to 0 and solving for x:

$$2x - \pi = 0$$
$$2x = \pi$$
$$x = \frac{\pi}{2}$$

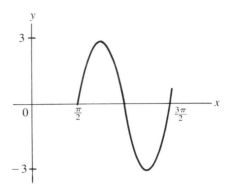

Figure 8.13

So the "first" period begins at $x = \pi/2$. Since $P = \pi$, the "first" period ends at $x = \pi/2 + \pi = 3\pi/2$. Since $A = 3$, we obtain the curve in Figure 8.13. (The curve can now be extended back to the origin, as shown in Figure 8.12.) ◀

E X A M P L E **3** Sketch of the graph of

$$y = 4 \cos\left(\frac{1}{2}x + \frac{\pi}{16}\right)$$

Solution. The equation has the form $y = a \cos(bx + c)$:

$$\begin{matrix} a & b & c \end{matrix}$$

$$y = \mathbf{4} \cos\left(\frac{\mathbf{1}}{\mathbf{2}}x + \frac{\boldsymbol{\pi}}{\mathbf{16}}\right)$$

$$A = \mathbf{4}, \qquad P = \frac{2\pi}{b} = \frac{2\pi}{1/2} = 4\pi, \qquad \textbf{phase shift:} \quad \frac{c}{b} = \frac{\pi/16}{1/2} = \frac{\pi}{8} \qquad \text{(left)}$$

Now sketch the curve $y = 4 \cos \frac{1}{2}x$. Since $A = 4$ and $P = 4\pi$, we obtain the dashed curve shown in Figure 8.14. This graph is now shifted $\pi/8$ units to the left, shown as the solid curve in Figure 8.14.

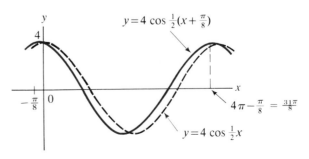

Figure 8.14

Alternate method: Determine where the "first" period begins:

$$\frac{1}{2}x + \frac{\pi}{16} = 0$$

$$\frac{1}{2}x = -\frac{\pi}{16}$$

$$x = -\frac{\pi}{8} \qquad \textbf{multiplying by 2}$$

So the "first" period begins at $x = -\pi/8$ and ends (since $P = 4\pi$) at $x = -\pi/8 + 4\pi = 31\pi/8$. Since $A = 4$, we obtain the curve in Figure 8.15.

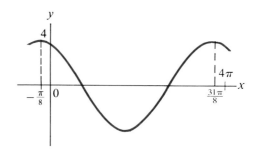

Figure 8.15

E X E R C I S E S / S E C T I O N   **8.2**

State the amplitude, period, and phase shift of each function. Sketch each curve over one period.

1. $y = 2\sin\left(x - \frac{\pi}{4}\right)$

2. $y = 2\cos\left(x + \frac{\pi}{4}\right)$

3. $y = \frac{1}{2}\sin\left(x + \frac{\pi}{8}\right)$

4. $y = \frac{1}{3}\cos\left(x - \frac{\pi}{4}\right)$

5. $y = \frac{1}{2}\sin\left(\frac{1}{2}x - \frac{\pi}{8}\right)$

6. $y = \frac{1}{2}\cos\left(\frac{1}{2}x + \frac{\pi}{8}\right)$

7. $y = 3\sin\left(x + \frac{3\pi}{2}\right)$

8. $y = \frac{1}{2}\cos\left(2x - \frac{\pi}{4}\right)$

9. $y = 2\cos\left(x - \frac{\pi}{3}\right)$

10. $y = 4\sin(2x - 1)$

11. $y = -3\cos(x - 2)$

12. $y = -4\sin(3x - 3)$

13. $y = 3\sin\left(2x + \frac{\pi}{2}\right)$

14. $y = 3\cos\left(3x + \frac{3\pi}{4}\right)$

15. $y = 10\cos\left(\frac{1}{3}x - \frac{4\pi}{5}\right)$

16. $y = 2\sin\left(3x + \frac{\pi}{4}\right)$

17. $y = 2\cos(\pi x + \pi)$

18. $y = \sin(\pi x - \pi)$

19. $y = 3\sin\left(\frac{1}{2}\pi x - 1\right)$

20. $y = 2\sin(\pi x + 1)$

8.3   **Applications of Sinusoidal Functions**

Many natural phenomena are sinusoidal, even such diverse phenomena as water waves, sound waves, and alternating current. To understand why this is so, we must study a phenomenon called **simple harmonic motion**.

Simple Harmonic Motion

Consider a wheel of radius r with a handle on the rim rotating at a constant rate. If a light source is placed some distance away, we can study the movement of the shadow of the handle on the wall (Figure 8.16). The problem is equivalent to finding the projection of the point P onto the y-axis in Figure 8.17. In fact, observe that the distance d from P to the horizontal axis can be found from the relation $\sin \theta = d/r$, so that

$$d = r \sin \theta$$

Now assume that the motion starts at a and proceeds in a counterclockwise direction with angular velocity ω. By definition, $\omega = \theta/t$, so that $\theta = \omega t$. (See Section 7.5.) Hence

$$d = r \sin \omega t \tag{8.4}$$

This equation expresses the distance d from the horizontal axis as a function of time. Moreover, the projection onto the y-axis in Figure 8.17 is the y-coordinate of the graph of $y = r \sin \omega t$. Consequently, the graph of $y = r \sin \omega t$ gives a pictorial representation of the variable distance from P to the t-axis. Observe also that $A = r$ and $P = 2\pi/\omega$; that is, the amplitude is the radius of the circle, and the period corresponds to one complete revolution.

The motion just described is called **simple harmonic motion**. If the motion begins at b, then it is described by $y = r \cos \omega t$.

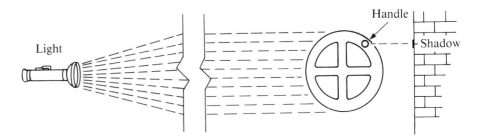

Figure 8.16

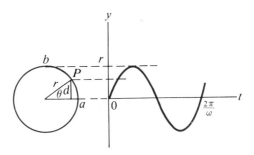

Figure 8.17

Simple harmonic motion is described by

$$y = r \sin \omega t \qquad \text{or} \qquad y = r \cos \omega t \qquad (8.5)$$

This type of motion turns out to be the common trait among the phenomena mentioned at the beginning of this section. For example, the motion of a weight hanging on a spring and oscillating vertically is simple harmonic motion; that is, the motion of the weight is identical to the motion of the shadow in Figure 8.16. Other examples are the vertical motion of a floating object caused by water waves and the sound-producing vibration of the end of a tuning fork.

E X A M P L E **1**   An object is moving around a circle of radius **5.0** cm with a constant angular velocity of **2** rad/s. If the motion starts at point a in Figure 8.17, sketch the graph of the function describing the motion of the vertical projection.

Solution. By formula (8.5), $y = \mathbf{5.0} \sin 2t$. Note that $A = 5.0$ cm and $P = \pi$ s. The graph is shown in Figure 8.18 over two periods.

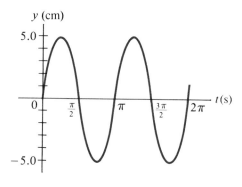

Figure 8.18   ◀

Alternating Current

Equations similar to those for simple harmonic motion arise in the study of alternating current. If a wire is moved rapidly through a magnetic field so that it "cuts" the lines of force, then a current is induced in the wire. More precisely, if B is the flux density of the magnetic field (in webers per square meter), L the length of the wire (in meters), and v the velocity of the wire (in meters per second), and if the wire cuts the lines of force at right angles, then the induced emf (electromotive force in volts) is given by

$$e = BLv \qquad (8.6)$$

The velocity comes into play because it is proportional to the number of lines of force that the wire is able to cut in a given time interval.

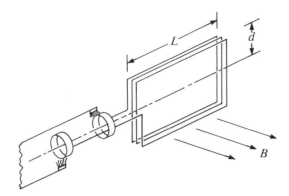

Figure 8.19

Formula (8.6) is precisely the principle used in the operation of a generator. Consider the schematic diagram in Figure 8.19. The rectangular coil is rotated so that the wires cut the lines of force of the magnetic field, and the terminals of the coil are connected to the rings on the left. Brushes bearing against the rings connect the coil to an external circuit.

Suppose that the angular velocity of the coil is ω. Then by formula (7.10) in Section 7.5,

$$v = \omega d \tag{8.7}$$

At the instant that the plane of the coil is parallel to the lines of force, we have

$$e = BL\omega d$$

by formula (8.6), since the lines of force are cut at right angles. If the angle that the plane of the coil makes with the lines of force is $\pi/2 - \omega t$ (Figure 8.20), then e is reduced, since the wires now cut fewer lines of force during a given small time interval. Suppose we think of B as the number of lines of force cut in the parallel position during the same small time interval; now examine Figure 8.20. A little reflection shows that the number of lines of force cut is reduced to $B \sin \omega t$ during this time interval. Thus e varies according to

$$e = BL\omega d \sin \omega t$$

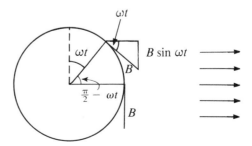

Figure 8.20

As already noted, if $\omega t = \pi/2$, then $\sin \omega t = 1$ and e attains its maximum value $BL\omega d$, denoted by e_{max}. It follows that the emf is given by

$$e = e_{max} \sin \omega t \tag{8.8}$$

If the motion starts at $t = -\alpha/\omega$ s, then equation (8.8) becomes

$$e = e_{max} \sin(\omega t + \alpha) \tag{8.9}$$

From the relation $i = V/R$, it follows that the current in the external circuit is given by

$$i = i_{max} \sin(\omega t + \alpha) \tag{8.10}$$

E X A M P L E 2 If $i_{max} = $ **5.00** A, $\omega = $ **100 π** rad/s, and $\alpha = \pi/3$, sketch the graph of the current.

Solution. By equation (8.10),

$$i = \textbf{5.00} \sin\left(\textbf{100}\pi t + \frac{\pi}{3}\right) \qquad \frac{c}{b} = \frac{\pi/3}{\textbf{100}\pi} = \frac{1}{\textbf{300}} \quad \text{(left)}$$

So the amplitude is $i_{max} = 5.00$ A, $P = 2\pi/100\pi = (1/50)$ s, and the phase shift is $c/b = \frac{1}{300}$ s to the left. The graph is shown in Figure 8.21. At

$$t = \frac{1}{100} - \frac{1}{300} = \frac{1}{150} \text{ s} \qquad \text{phase shift} = \frac{1}{300}\text{ s} \quad \text{(left)}$$

i becomes negative, indicating that the current has reversed direction. This corresponds to the instant when the coil in Figure 8.19 passes the horizontal position. The current is said to be **alternating**.
The graph crosses the t-axis again at

$$\frac{1}{50} - \frac{1}{300} = \frac{1}{60} \text{ s} \qquad P = \frac{1}{50}\text{ s}$$

At $t = \frac{1}{60}$ s, the current becomes positive (Figure 8.21).   ◀

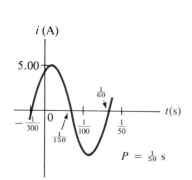

i (A)

5.00

$\frac{1}{60}$

$-\frac{1}{300}$ 0 $\frac{1}{150}$ $\frac{1}{100}$ $\frac{1}{50}$ $t(s)$

$P = \frac{1}{50}$ s

Figure 8.21

Finally, let us note that the reciprocal of the period is called the **frequency**. This is a natural idea, for if a wave has a period of $\frac{1}{20}$ s, then one cycle is completed in $\frac{1}{20}$ s. Thus 20 cycles are completed in 1 s. We now say that the frequency is 20 cycles/s, or 20 **hertz** (abbreviated "Hz"). This unit was named after the German physicist Heinrich Hertz (1857–1894), a pioneer in the study of electromagnetic phenomena.

E X A M P L E 3

Sound

The sound of a tuning fork may be expressed in the form $y = A \sin 2\pi ft$, where f is the frequency. If a tuning fork vibrates at 190 Hz and has an amplitude of 0.001 in., then $y = 0.001 \sin 2\pi(\textbf{190})t$. Note that the period is $1/f = \frac{1}{190}$ s. This waveform can actually be observed on an oscilloscope.   ◀

E X E R C I S E S / S E C T I O N 8.3

1. An object moves around a circle of radius 3.0 in. starting at point (3.0, 0) with angular velocity 2 rad/s. Sketch the graph of the resulting motion of the vertical projection. (See Example 1.)

2. Repeat Exercise 1 with the motion starting at (0, 3.0).

3. An object moves around a circle of radius 5 cm at the rate of 4 rev/min. Sketch the motion of the vertical projection, which is assumed to start at (0, 5). (*Note:* 1 rev = 2π radians)

4. Repeat Exercise 3 with the motion starting at (5, 0).

5. If $e_{max} = 6.00$ V, $\omega = 80\pi$ rad/s, and $\alpha = \pi/4$, sketch the graph of e as a function of time.

6. If $e_{max} = 10.0$ V, $\omega = 120\pi$ rad/s, and $\alpha = \pi/6$, sketch the graph of e as a function of time.

7. If $i_{max} = 6.0$ A, $\omega = 120\pi$ rad/s, and $\alpha = \pi/4$, sketch the current i as a function of time.

8. If $i_{max} = 9.0$ A, $\omega = 150\pi$ rad/s, and $\alpha = \pi/6$, sketch the current i as a function of time.

9. A weight is hanging on a spring and oscillating vertically. The displacement of the weight from the equilibrium position as a function of time is given by $y = 5\cos 2t$. Sketch the curve of this motion.

10. Under certain conditions the vertical displacement of a horizontal string as a function of position and time is given by $y = A\sin(t - x/v)$. If $A = 2$ cm and $v = -10$ cm/s, sketch the graph of the string at the instant when $t = 1$ s.

11. A tuning fork vibrates at 224 Hz and has an amplitude of 0.002 in. Sketch the curve produced on an oscilloscope. (See Example 3.)

12. The form of a water wave is given by $y = A\sin 2\pi(ft - r/\lambda)$, where f is the frequency, r the distance from the source, and λ the wavelength, which is approximately $5.12T^2$ ft, where T is the period. Find the equation for the vertical motion of a floating body as a function of time if the body is 10.0 ft from the source and the period and amplitude are 8.0 s and 2.8 ft, respectively. Sketch the curve.

13. Repeat Exercise 12 if the body is 20.0 ft from the source and if the wave has a period of 6.0 s and an amplitude of 2.1 ft.

8.4 Graphs of Other Trigonometric Functions

The graphs of the remaining trigonometric functions occur less often in technical applications than the sinusoidal curves but play an important role in more advanced mathematics. Thus we should at least become familiar with them.

To obtain the graph of $y = \tan x$, we make up a table of values.

$y = \tan x$

x:	0	$\dfrac{\pi}{6}$	$\dfrac{\pi}{4}$	$\dfrac{\pi}{3}$	$\dfrac{\pi}{2}$	$-\dfrac{\pi}{6}$	$-\dfrac{\pi}{4}$	$-\dfrac{\pi}{3}$	$-\dfrac{\pi}{2}$
y:	0	0.6	1	1.7	undefined	-0.6	-1	-1.7	undefined

In the neighborhood of $x = \pi/2$ and $x = -\pi/2$, the values of $\tan x$ become numerically large, as a calculator will readily confirm. (See Figure 8.22.) The dashed vertical lines at

$$x = \pm\frac{\pi}{2}, \pm\frac{3\pi}{2}, \dots$$

are approached by the graph and are called **vertical asymptotes**. Although the table of values does not extend beyond $x = \pm\pi/2$, the values can be seen to repeat themselves, indicating that the function $y = \tan x$ has a period of π. The

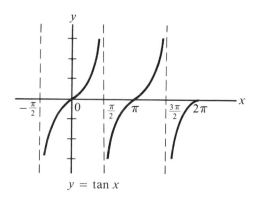

$y = \tan x$

Figure 8.22

concept of amplitude is not defined since the curve extends indefinitely in the vertical direction.

The graphs of $y = \cot x$, $y = \sec x$, and $y = \csc x$ are shown in Figures 8.23 through 8.25, respectively. The graphs of $y = \sec x$ and $y = \csc x$ can be obtained from the graphs of $y = \cos x$ and $y = \sin x$ by using the reciprocal relations

$$\sec x = \frac{1}{\cos x} \quad \text{and} \quad \csc x = \frac{1}{\sin x}$$

(See Example 1.) In other words, the graph of $y = \sec x$ is obtained by starting with $y = \cos x$ and plotting the reciprocals of the values of $\cos x$. When $\cos x = 1$, we have $\sec x = 1$; when $0 < \cos x < 1$, then $\sec x > 1$; and when $\cos x = 0$, then $\sec x = \frac{1}{0}$, which is undefined. To show how the curves are related, the graphs of $y = \cos x$ and $y = \sin x$ are shown as dashed curves in

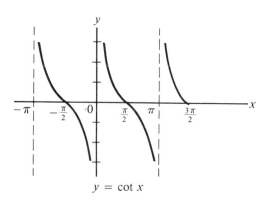

$y = \cot x$

Figure 8.23

Figures 8.24 and 8.25, respectively. Note that $y = \sec x$ and $y = \csc x$ have period 2π, the same as for $\cos x$ and $\sin x$.

x	$\cos x$	$\sec x = \dfrac{1}{\cos x}$
$-\dfrac{\pi}{2}$	0	$\dfrac{1}{0}$ (undefined)
$-\dfrac{\pi}{3}$	$\dfrac{1}{2}$	$\dfrac{1}{1/2} = 2$
$-\dfrac{\pi}{4}$	$\dfrac{1}{\sqrt{2}}$	$\dfrac{1}{1/\sqrt{2}} = \sqrt{2}$
0	1	$\dfrac{1}{1} = 1$
$\dfrac{\pi}{4}$	$\dfrac{1}{\sqrt{2}}$	$\dfrac{1}{1/\sqrt{2}} = \sqrt{2}$
$\dfrac{\pi}{3}$	$\dfrac{1}{2}$	$\dfrac{1}{1/2} = 2$
$\dfrac{\pi}{2}$	0	$\dfrac{1}{0}$ (undefined)

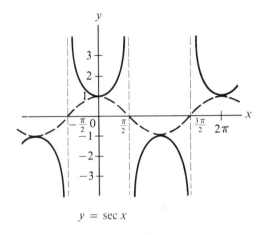

$y = \sec x$

Figure 8.24

x	$\sin x$	$\csc x = \dfrac{1}{\sin x}$
0	0	$\dfrac{1}{0}$ (undefined)
$\dfrac{\pi}{6}$	$\dfrac{1}{2}$	$\dfrac{1}{1/2} = 2$
$\dfrac{\pi}{4}$	$\dfrac{1}{\sqrt{2}}$	$\dfrac{1}{1/\sqrt{2}} = \sqrt{2}$
$\dfrac{\pi}{2}$	1	$\dfrac{1}{1} = 1$
$\dfrac{3\pi}{4}$	$\dfrac{1}{\sqrt{2}}$	$\dfrac{1}{1/\sqrt{2}} = \sqrt{2}$
$\dfrac{5\pi}{6}$	$\dfrac{1}{2}$	$\dfrac{1}{1/2} = 2$
π	0	$\dfrac{1}{0}$ (undefined)

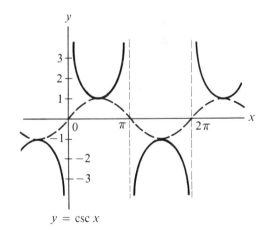

$y = \csc x$

Figure 8.25

E X A M P L E **1**   Sketch the graph of $y = 2 \csc x$.

Solution. The most direct way to sketch this graph is to start with Figure 8.25 and then multiply the values by 2.

Alternatively, we can start with the graph of $y = 2 \sin x$ (the dashed curve in Figure 8.26) and obtain some of the reciprocal values. For example, when $x = \pi/2$, then

$$y = 2 \csc \frac{\pi}{2} = \frac{2}{\sin \dfrac{\pi}{2}} = \frac{2}{1} = 2$$

When $\sin x = 0$, then $y = \csc x = 1/\sin x$ is undefined, confirming the existence of asymptotes at $x = 0,\ \pm \pi,\ \pm 2\pi,\ \ldots$. (See Figure 8.26.)

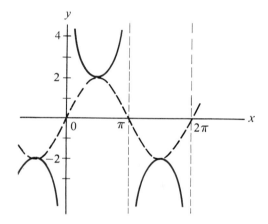

Figure 8.26

x:	0	$\dfrac{\pi}{6}$	$\dfrac{\pi}{4}$	$\dfrac{\pi}{3}$	$\dfrac{\pi}{2}$	$\dfrac{5\pi}{6}$	π
$\dfrac{2}{\sin x}$:	undefined	4	$2\sqrt{2}$	$\dfrac{4}{\sqrt{3}}$	2	4	undefined

◀

General Forms

To sketch the graph of $y = a \tan bx$, recall that $y = \tan x$ has period π and that the first asymptote to the right of the origin is at $x = \pi/2$. The distance from the origin to this asymptote is therefore half a period. For $y = \tan bx$, the first asymptote is at $x = \pi/(2b)$; so the period must be

$$P = \frac{2\pi}{2b} = \frac{\pi}{b}$$

The **period** of $y = a \tan bx$ and $y = a \cot bx$ is

$$P = \frac{\pi}{b} \tag{8.11}$$

The **period** of $y = a \sec bx$ and $y = a \csc bx$ is

$$P = \frac{2\pi}{b} \tag{8.12}$$

E X A M P L E **2**   Sketch the graph of $y = 3 \cot 2x$.

Solution. The period is $\pi/2$. Since the basic shape of the cotangent function is known from Figure 8.23, we readily obtain the desired sketch (Figure 8.27).

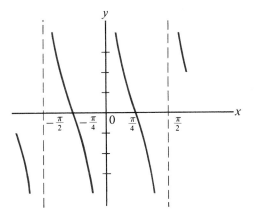

Figure 8.27

x:	0	$\dfrac{\pi}{8}$	$\dfrac{\pi}{4}$	$\dfrac{3\pi}{8}$	$\dfrac{\pi}{2}$
y:	undefined	3	0	-3	undefined

E X E R C I S E S / S E C T I O N **8.4**

In Exercises 1–20, state the period of each function and sketch the curve.

1. $y = 2 \tan x$

3. $y = 4 \cot x$

2. $y = 3 \sec x$

4. $y = 2 \csc x$

5. $y = \sec 2x$

7. $y = \csc 2x$

9. $y = \dfrac{1}{2} \tan 3x$

6. $y = \tan 3x$

8. $y = 2 \cot 2x$

10. $y = \dfrac{1}{3} \sec x$

11. $y = -\csc 2x$

12. $y = -\tan 2x$

13. $y = 3 \sec \dfrac{1}{2} x$

14. $y = 4 \tan \dfrac{1}{3} x$

15. $y = 5 \csc \dfrac{2}{3} x$

16. $y = 6 \cot \dfrac{4}{3} x$

17. $y = \tan\left(x + \dfrac{\pi}{6}\right)$

18. $y = \sec\left(2x - \dfrac{\pi}{6}\right)$

19. $y = \dfrac{1}{2} \cot 3\pi x$

20. $y = 2 \csc \pi x$

21. A dam has a V-shaped notch with angle θ. If the water reaches a level 1 m above the bottom of the notch, then the rate of flow in meters per second across the notch is given by $Q = 2.506 \tan(\theta/2)$. Sketch the graph for $0 \le \theta < \pi$.

22. If friction is taken into account, then the heaviest weight that can be pulled up an inclined plane depends only on the angle θ that the plane makes with the horizontal. The methods of differential calculus show that the maximum is attained when $\mu = \tan \theta$, where μ is the coefficient of friction. Sketch the graph of $\mu = \tan \theta$ for $0 \le \theta < \pi/2$.

8.5   Addition of Coordinates

Many applications of trigonometric functions involve combinations of two or more functions. Therefore, in this section we will study the **addition of coordinates**. To graph a function that is the sum of two other functions, we graph the individual functions and add their y values graphically. Because of their importance, we will confine ourselves to sinusoidal and simple algebraic functions.

E X A M P L E **1**   Sketch the graph of $y = x + \sin x$.

Solution. We first graph the functions $y = x$ and $y = \sin x$, shown as dashed curves in Figure 8.28. When adding the coordinates, care must be taken to treat the y-coordinates of the points below the x-axis as negative. Certain key points are particularly useful. For example, the graph of $y = \sin x$ crosses the x-axis at $x = 0, \pi, 2\pi, \ldots$. Since the corresponding y values are 0, the points on the graph of $y = x + \sin x$ lie on $y = x$. (See Figure 8.28.) Point P_1 is a typical

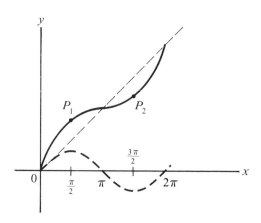

Figure 8.28

point between the zeros of $y = \sin x$; its coordinates are $(\pi/2, \pi/2 + 1)$ since, at $x = \pi/2$,

$$y = x + \sin x = \frac{\pi}{2} + \sin \frac{\pi}{2} = \frac{\pi}{2} + 1$$

Similarly, the coordinates of P_2 are $(3\pi/2, 3\pi/2 - 1)$. We locate as many points as necessary to obtain a smooth graph, shown as the solid curve in Figure 8.28.   ◀

As pointed out in Section 8.3, the sound of a tuning fork can be seen as a wave on an oscilloscope; this wave takes the form $y = A \sin 2\pi ft$. It is interesting to note that, theoretically, complex sounds can be reproduced by a proper combination of tuning forks. Therefore, the functions describing such sounds are necessarily combinations of sines and cosines. One such combination is a *Fourier series*, useful in the study of vibration, heat flow, electrical circuits, and so on. An example of a Fourier series is

$$y = \sin t + \frac{1}{2} \sin 2t + \frac{1}{3} \sin 3t + \frac{1}{4} \sin 4t + \cdots \tag{8.13}$$

The three dots indicate that the sum continues indefinitely.

E X A M P L E **2**   Graph the first two terms of equation (8.13) by addition of coordinates.

Solution. The graphs of $y = \sin t$ and $y = \frac{1}{2} \sin 2t$ are shown as dashed curves in Figure 8.29. Note that the graph of $y = \frac{1}{2} \sin 2t$ crosses the t-axis at

$$t = \frac{\pi}{2}, \pi, \frac{3\pi}{2}, 2\pi, \ldots$$

At these t values the points on the curve to be sketched lie on the curve $y = \sin t$. Using this information, we obtain the solid curve in Figure 8.29.

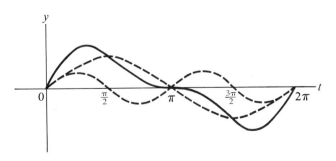

Figure 8.29

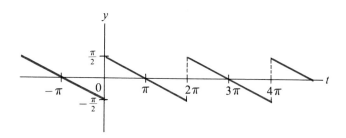

Figure 8.30

Graphing more and more terms of equation (8.13) results in a graph that gets ever closer to the "sawtooth function" shown in Figure 8.30.   ◄

E X E R C I S E S / S E C T I O N   **8.5**

In Exercises 1–10, sketch the graph of each function by addition of coordinates.

1. $y = x + \cos x$

2. $y = 2x + \sin x$

3. $y = 2 \sin x + \cos x$

4. $y = 2 \sin x + 3 \cos x$

5. $y = 2 \cos x + \sin 2x$

6. $y = 3 \cos x + \cos 2x$

7. $y = \dfrac{1}{2} x - \sin x$

8. $y = \sin 2x - \cos x$

9. $y = \cos x + 2 \sin 2x$

10. $y = \cos \dfrac{1}{2} x + 2 \sin x$

11. Graph the first two terms of the Fourier series

$$f(t) = \sin \pi t + \frac{1}{3} \sin 3\pi t + \frac{1}{5} \sin 5\pi t + \cdots$$

12. Graph the first two terms of the Fourier series

$$f(t) = \cos t + \frac{1}{9} \cos 3t + \frac{1}{25} \cos 5t + \cdots$$

13. The current in a certain circuit is given by $i = 4.0 \cos 60\pi t + 2.0 \sin 60\pi t$. Sketch the graph of the current (in amperes) as a function of time (in seconds).

14. A weight of mass 1 slug is oscillating on a spring with spring constant 64 lb/ft. If the weight has an initial velocity of 4 ft/s and an initial displacement of 2 ft, its displacement y (in feet) as a function of t (in seconds) is given by $y = 2 \cos 8t + \frac{1}{2} \sin 8t$. Sketch the curve.

R E V I E W   E X E R C I S E S / C H A P T E R   **8**

In Exercises 1–8, state the amplitude and period and sketch the curves.

1. $y = 2 \cos x$

2. $y = \dfrac{1}{2} \sin 2x$

3. $y = 4 \cos 4x$

4. $y = -2 \cos x$

5. $y = -\dfrac{1}{2} \sin 2x$

6. $y = 3 \sin \dfrac{1}{2} x$

7. $y = 4 \sin \dfrac{1}{2} x$

8. $y = 8 \cos \dfrac{1}{4} x$

In Exercises 9–16, state the amplitude, period, and phase shift. Sketch the curves.

9. $y = 2 \sin \left(x + \dfrac{\pi}{4} \right)$

10. $y = 2 \cos \left(x - \dfrac{\pi}{4} \right)$

11. $y = \dfrac{1}{2} \cos \left(x + \dfrac{\pi}{8} \right)$

12. $y = \dfrac{4}{3} \sin \left(x - \dfrac{\pi}{4} \right)$

13. $y = 4 \cos \left(\dfrac{1}{2} x - \dfrac{\pi}{8} \right)$

14. $y = 2 \sin \left(\dfrac{1}{2} x + \dfrac{\pi}{8} \right)$

15. $y = \dfrac{1}{2} \sin \left(2x - \dfrac{\pi}{4} \right)$

16. $y = 3 \sin(2x - 2)$

In Exercises 17–20, sketch the given curves.

17. $y = \sec 2x$

18. $y = 3 \tan 2x$

19. $y = 4 \cot \dfrac{1}{2} x$

20. $y = \dfrac{1}{4} \csc 3x$

21. An object moving around a circle of radius 4 cm starts at $(4, 0)$ with an angular velocity of 2 rad/s. Sketch the graph of the motion of the vertical projection.

22. Repeat Exercise 21 if the motion starts at (0, 4).

23. An object moves around a circle of radius 6 cm at the rate of 5 rev/min. If the motion starts at (0, 6), sketch the graph of the motion of the vertical projection.

24. Repeat Exercise 23 if the motion starts at (6, 0).

25. If $e_{max} = 5.00$ V, $\omega = 60\pi$ rad/s, and $\alpha = \pi/3$, sketch the graph of e as a function of time.

26. If $i_{max} = 4.00$ A, $\omega = 90\pi$ rad/s, and $\alpha = \pi/6$, sketch the current i as a function of time.

27. A tuning fork vibrates at 236 Hz and has an amplitude of 0.003 in. Sketch the corresponding curve appearing on an oscilloscope.

28. The current (in amperes) is given by $i = 2.0 \sin 120\pi t + 3.0 \cos 120\pi t$. (Note that the frequency is $1/P = 60$ Hz.) Sketch the curve.

29. A weight of mass 1 slug is oscillating on a spring with spring constant 16 lb/ft. If the weight has an initial velocity of 8 ft/s and an initial displacement of 1 ft, its displacement y (in feet) as a function of time t (in seconds) is given by $y = \cos 4t + 2 \sin 4t$. Sketch the curve.

30. Graph the first two terms of the Fourier series

$$f(t) = \sin t - \frac{1}{2} \sin 2t + \frac{1}{3} \sin 3t - \frac{1}{4} \sin 4t + \cdots$$

by adding the coordinates between $-\pi$ and π.

Vectors and Oblique Triangles

9.1 Vectors

Scalar

Vector

Initial and terminal point

Most quantities discussed so far can be fully described by a number. Examples are areas, volumes, and temperature. Quantities that have only magnitude are called **scalar** quantities. Other physical phenomena have both **magnitude and direction**. For example, the description of the velocity of a moving body must account for both the magnitude of the velocity and its direction. To describe a force, we need to know both the magnitude of the force and the direction in which the force is acting. Such entities are called **vectors**.

An arrow is a convenient way to represent a vector graphically. The arrow points in the *direction* of the vector, and the *length* of the arrow represents the magnitude. A vector is denoted by a boldfaced letter, while the same letter in italic represents the magnitude. Thus **A** represents a vector and A or $|\mathbf{A}|$ its magnitude. (In handwriting, a common notation for **A** is $\vec{A}$.) The base of the arrow is called the **initial point**, and the tip of the arrow the **terminal point**. (See Figure 9.1.)

So if vector **A** in Figure 9.1 has magnitude $|\mathbf{A}| = 10$ lb, then **A** represents a force of 10 lb acting in the direction of the arrow.

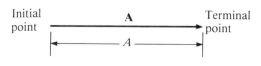

Figure 9.1

Addition of Vectors

If two vectors are to be added, both their magnitude and direction must be taken into account. Let **A** and **B** represent two vectors with the same initial

Resultant

point O (Figure 9.2). In the parallelogram determined by **A** and **B**, the vector sum **A** + **B**, called the **resultant**, is the vector with initial point O coinciding with the diagonal of the parallelogram. The initial point was introduced here only for convenience, for *the initial point of a vector can be placed anywhere in the plane*. For this reason the sum can also be obtained by placing the initial point of **B** at the terminal point of **A**. The resultant **R** is the vector from the initial point of **A** to the terminal point of **B**, as shown in Figure 9.3.

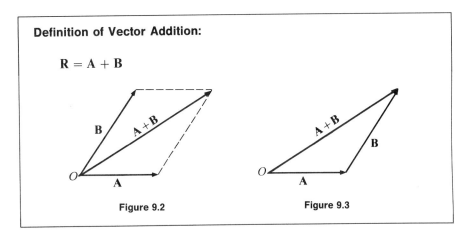

Definition of Vector Addition:

$$R = A + B$$

Figure 9.2   Figure 9.3

If more than two vectors have to be added, the procedure shown in Figure 9.3 is by far the more convenient. Consider the example below.

E X A M P L E **1**   Find the sum of the vectors **A**, **B**, **C**, and **D** shown in Figure 9.4.

Solution. Draw vector **A**. Place the initial point of **B** at the terminal point of **A**. Now place the initial point of **C** at the terminal point of **B**, and the initial point of **D** at the terminal point of **C**. The final resultant **R** is the vector from the initial point of **A** to the terminal point of **D**, as shown in Figure 9.5.

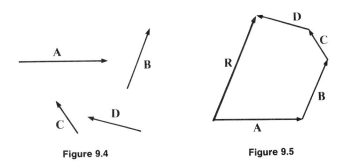

Figure 9.4   Figure 9.5

Commutativity

Since vector addition is commutative, the vectors can be added in any order. For example, the same resultant **R** is obtained if the vectors are added as shown in Figure 9.6. **R is always drawn from the initial point of the first vector in the sum to the terminal point of the last vector.**   ◄

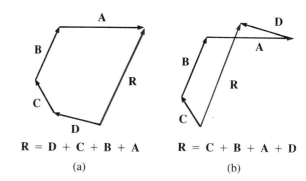

$$\mathbf{R} = \mathbf{D} + \mathbf{C} + \mathbf{B} + \mathbf{A} \qquad \mathbf{R} = \mathbf{C} + \mathbf{B} + \mathbf{A} + \mathbf{D}$$

(a)   (b)

Figure 9.6

Scalar Multiplication

Scalar product

If a vector **A** is multiplied by a number c, the resulting vector $c\mathbf{A}$, called the **scalar product**, has the same direction as **A** and magnitude cA. For example, the vector 3**A** is three times as long as **A** and has the same direction as **A**.

E X A M P L E **2**   Given the vectors **A** and **B** in Figure 9.7, draw $2\mathbf{A} + 3\mathbf{B}$.

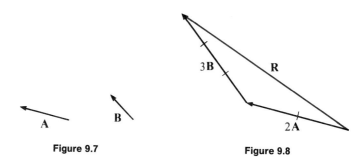

Figure 9.7   **Figure 9.8**

Solution. From vector **A** construct the vector **2A**, which has the same direction as **A** but twice the magnitude. (See Figure 9.8.) In a similar manner we construct the vector **3B** from **B**. The resultant $\mathbf{R} = 2\mathbf{A} + 3\mathbf{B}$, shown in Figure 9.8, is the vector from the initial point of **2A** to the terminal point of **3B**.   ◀

Subtraction of Vectors

Figure 9.9

To subtract vectors, define $-\mathbf{B}$ as equal to $(-1)\mathbf{B}$, which has the same magnitude as **B** but is oppositely directed. (See Figure 9.9.)

As a result, $\mathbf{A} - \mathbf{B} = \mathbf{A} + (-\mathbf{B})$, which can be added in the usual way (Figure 9.10). It also follows that $\mathbf{A} - \mathbf{B}$ is the second diagonal in the parallelogram determined by **A** and **B**, as shown in Figure 9.11.

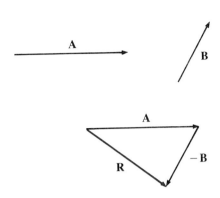

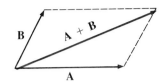

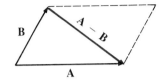

Figure 9.10   Figure 9.11

E X E R C I S E S / S E C T I O N   9.1

Carry out each of the indicated operations with the vectors
given in Figure 9.12 by means of a diagram.

7. $\mathbf{A} - \mathbf{C}$ 8. $2\mathbf{B} - \mathbf{C}$ 9. $2\mathbf{A} - 3\mathbf{B}$

10. $\mathbf{A} + 2\mathbf{B} - 2\mathbf{E}$ 11. $2\mathbf{D} - 4\mathbf{C}$ 12. $2\mathbf{C} - 3\mathbf{E}$

1. $\mathbf{A} + \mathbf{C}$ 2. $\mathbf{C} + \mathbf{E}$ 3. $\mathbf{A} + \mathbf{B} + \mathbf{C}$

13. $2\mathbf{C} + \mathbf{E}$ 14. $\mathbf{A} + 2\mathbf{E}$ 15. $\mathbf{A} + \mathbf{E} + 2\mathbf{C}$

4. $\mathbf{B} + \mathbf{C} + \mathbf{D}$ 5. $2\mathbf{D} + \mathbf{E}$ 6. $2\mathbf{B} + 2\mathbf{C} + \mathbf{D}$

16. $2\mathbf{D} - \mathbf{E}$ 17. $\mathbf{A} + 2\mathbf{B} + \mathbf{C}$ 18. $2\mathbf{B} + \mathbf{C} - \mathbf{A}$

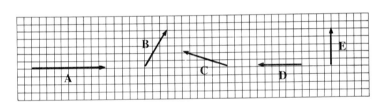

Figure 9.12

9.2   Vector Components

While vector addition by means of diagrams is conceptually important, it is
impractical for computations. In this section we will see how the resultant can
be determined exactly. To this end we need to combine the vector concept with
that of the coordinate system: We place the initial point of the vector at the
origin so that the terminal point uniquely describes the vector. For example,
the vector in Figure 9.13 can be specified by the point $(-2, 2)$. The magnitude,
which can be computed directly by the Pythagorean theorem, turns out to be
$\sqrt{(-2)^2 + 2^2} = \sqrt{8} = 2\sqrt{2}$. The direction is given by the angle $\theta = 135°$ (since
the reference angle is $45°$).

In general, if the terminal point of vector $\mathbf{A}$ is (A_x, A_y), we write

Figure 9.13

$$\mathbf{A} = (A_x, A_y) \tag{9.1}$$

The magnitude of **A**, denoted by $|\mathbf{A}|$ or by A, is

$$|\mathbf{A}| = \sqrt{A_x^{\,2} + A_y^{\,2}} = A$$

The direction θ is the angle determined by **A** and the positive x-axis, so that

$$\tan \theta = \frac{A_y}{A_x}$$

Magnitude and Direction of A $= (A_x, A_y)$:

  Magnitude: $|\mathbf{A}| = A = \sqrt{A_x^{\,2} + A_y^{\,2}}$ (9.2)

Direction: The direction is obtained from the relation

  $$\tan \theta = \frac{A_y}{A_x}$$ (9.3)

These ideas are illustrated in the examples below.

E X A M P L E **1** Find the magnitude and direction of the vector **A** whose terminal point is $(-3, -5)$

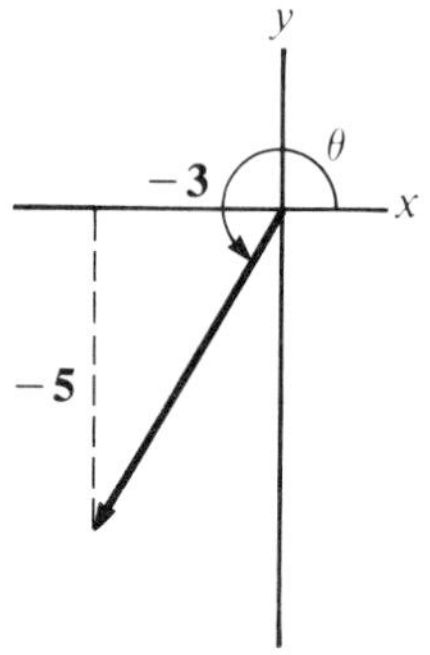

Figure 9.14

Solution. The vector is shown in Figure 9.14. The magnitude is

$$|\mathbf{A}| = \sqrt{(-3)^2 + (-5)^2} = \sqrt{9 + 25} = \sqrt{34}$$

To find the direction, note that

$$\tan \theta = \frac{-5}{-3} = \frac{5}{3}$$

From the sequence

$$5 \boxed{\div} 3 \boxed{=} \boxed{\text{INV}} \boxed{\text{TAN}} \rightarrow 59.036243$$

we see that the reference angle is $59.0°$. Since θ is in quadrant III,

$$\theta = 180° + 59.0° = 239.0°$$

◀

E X A M P L E **2** Find the magnitude and direction of the vector $\mathbf{A} = (\sqrt{7}, -3)$.

Solution. $|\mathbf{A}| = \sqrt{(\sqrt{7})^2 + (-3)^2} = \sqrt{7 + 9} = 4$

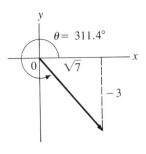

Figure 9.15

and

$$\tan \theta = \frac{-3}{\sqrt{7}} = -\frac{3}{\sqrt{7}}, \qquad \theta \text{ in quadrant IV (Figure 9.15)}$$

Using a calculator, the reference angle is found to be 48.6°, so that $\theta = 360° - 48.6° = 311.4°$. ◀

It is sometimes desirable to work in reverse: Starting with the magnitude and direction of a vector, we express the vector in the form $\mathbf{A} = (A_x, A_y)$. This process is called **resolving the vector into its components**.

E X A M P L E **3**   A vector $\mathbf{A}$ has magnitude **3** and direction **320°**. Resolve the vector into its components.

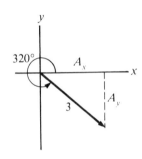

Figure 9.16

Solution. The vector is shown in Figure 9.16. To find A_x, we note that $\cos \theta = A_x/3$. Hence $A_x = \mathbf{3} \cos \mathbf{320°} = 2.30$. Similarly, $A_y = \mathbf{3} \sin \mathbf{320°} = -1.93$. It follows that

$$A_x = 2.30 \qquad \text{and} \qquad A_y = -1.93$$

or

$$\mathbf{A} = (2.30, -1.93)$$

◀

If the direction of $\mathbf{A}$ is θ, then

$$A_x = |\mathbf{A}| \cos \theta \qquad \text{and} \qquad A_y = |\mathbf{A}| \sin \theta$$

or

$$\mathbf{A} = (A \cos \theta, A \sin \theta) \qquad\qquad (9.4)$$

Vector Addition by Components

To add two or more vectors by components, we proceed as follows:

1. Resolve each vector into its components.
2. Add all x-components and all y-components to obtain the components of the resultant.
3. Find the magnitude and direction of the resultant from the components obtained in Step 2.

E X A M P L E **4**   Add the vectors in Figure 9.17. Find $|\mathbf{R}|$ to two significant digits and the direction of $\mathbf{R}$ to the nearest tenth of a degree. ($\mathbf{A}$ has magnitude 3 and direction 73°; $\mathbf{B}$ has magnitude 2 and direction 170°; the numbers are assumed to be exact.)

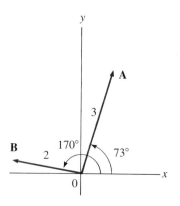

Figure 9.17   Figure 9.18

Solution. We resolve the vectors **A** and **B** into their components. We then add the corresponding components to obtain the components of **R**:

$$A_x = 3 \cos 73° = 0.8771 \qquad A_y = 3 \sin 73° = 2.8689$$
$$\underline{B_x = 2 \cos 170° = -1.9696} \qquad \underline{B_y = 2 \sin 170° = 0.3473}$$
$$\text{adding:} \quad R_x = -1.0925 \qquad \text{adding:} \quad R_y = 3.2162$$

It follows that

$$\mathbf{R} = (R_x, R_y) = (-1.0925, 3.2162)$$

and

$$|\mathbf{R}| = \sqrt{(-1.0925)^2 + (3.2162)^2} = 3.4$$

Since R_x is negative and R_y positive, θ is in quadrant II. From

$$\tan \theta = \frac{3.2162}{-1.0925}, \qquad \theta \text{ in quadrant II}$$

we get $71.2°$ for the reference angle. So $\theta = 180° - 71.2° = 108.8°$. (See Figure 9.18.)   ◄

The calculator operations are shown in the next example.

E X A M P L E   **5**   Add the vectors in Figure 9.19.

Solution.

$$A_x = 3 \cos 141° = -2.3314 \qquad A_y = 3 \sin 141° = 1.8880$$
$$\underline{B_x = 7 \cos 252° = -2.1631} \qquad \underline{B_y = 7 \sin 252° = -6.6574}$$
$$\text{adding:} \quad R_x = -4.4945 \qquad \text{adding:} \quad R_y = -4.7694$$

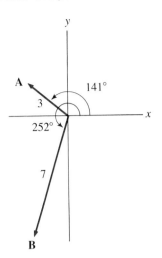

Figure 9.19

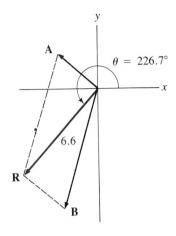

Figure 9.20

$$\mathbf{R} = (R_x, R_y) = (-4.4945, -4.7694)$$
$$|\mathbf{R}| = \sqrt{(-4.4945)^2 + (-4.7694)^2} = 6.6$$
$$\tan \theta = \frac{-4.7694}{-4.4945}, \qquad \theta \text{ in quadrant III}$$

(θ is in quadrant III because R_x and R_y are both negative.) Since the reference angle is 46.7°, $\theta = 226.7°$. (See Figure 9.20.)

CALCULATOR COMMENT

It is much simpler to find $\mathbf{R}$ in one continuous operation with a calculator:

R_x: 3 $\boxed{\times}$ 141 $\boxed{\text{COS}}$ $\boxed{+}$ 7 $\boxed{\times}$ 252 $\boxed{\text{COS}}$ $\boxed{=}$ $\boxed{\text{STO}}$ → −4.4945568

R_y: 3 $\boxed{\times}$ 141 $\boxed{\text{SIN}}$ $\boxed{+}$ 7 $\boxed{\times}$ 252 $\boxed{\text{SIN}}$ $\boxed{=}$ → −4.7694344

        (record this value)

R: $\boxed{x^2}$ $\boxed{+}$ $\boxed{\text{MR}}$ $\boxed{x^2}$ $\boxed{=}$ $\boxed{\sqrt{}}$ → 6.553514

Reference angle:

4.7694344 $\boxed{+/-}$ $\boxed{\div}$ $\boxed{\text{MR}}$ $\boxed{=}$ $\boxed{\text{INV}}$ $\boxed{\text{TAN}}$ → 46.69956

To the nearest tenth, $\theta = 180° + 46.7° = 226.7°$, as before.   ◄

The next example illustrates addition of three vectors by components.

E   X   A   M   P   L   E   **6**   Add the vectors **A**, **B** and **C** in Figure 9.21.

Vector **A**:   magnitude 1, direction 15°

Vector **B**:   magnitude 2, direction 65°

Vector **C**:   magnitude 4, direction 260°

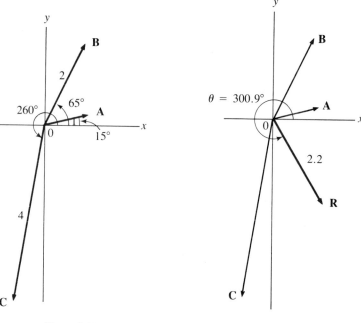

Figure 9.21   Figure 9.22

Solution.

$$
\begin{aligned}
A_x &= 1 \cos 15° = 0.9659 & A_y &= 1 \sin 15° = 0.2588 \\
B_x &= 2 \cos 65° = 0.8452 & B_y &= 2 \sin 65° = 1.8126 \\
C_x &= 4 \cos 260° = -0.6946 & C_y &= 4 \sin 260° = -3.9392 \\
\text{adding:}\quad R_x &= 1.1165 & \text{adding:}\quad R_y &= -1.8678
\end{aligned}
$$

$$
\mathbf{R} = (R_x, R_y) = (1.1165, -1.8678)
$$

$$
R = \sqrt{(1.1165)^2 + (-1.8678)^2} = 2.2
$$

$$
\tan \theta = \frac{-1.8678}{1.1165}, \qquad \theta \text{ in quadrant IV}
$$

$$
\theta = 300.9° \qquad \text{(See Figure 9.22.)}
$$

◀

E X E R C I S E S / S E C T I O N 9.2

In Exercises 1–16, find the magnitude of each vector and the direction to the nearest tenth of a degree.

1. $(1, 1)$

2. $(-1, \sqrt{3})$

3. $(-3, -3)$

4. $(\sqrt{3}, -1)$

5. $(2, 3)$

6. $(-4, 5)$

7. $(-\sqrt{3}, -\sqrt{6})$

8. $(-1, \sqrt{15})$

9. $(-\sqrt{7}, -3)$

10. $(1, -5)$

11. $(-\sqrt{13}, \sqrt{3})$

12. $(2, \sqrt{2})$

13. $(1, -2)$

14. $(-2, 5)$

15. $(-1, -4)$

16. $(-2, -7)$

In Exercises 17–26, resolve each vector into its components. (Give answers to two decimal places.)

17. $|\mathbf{A}| = 3, \theta = 80°$

18. $|\mathbf{A}| = 2, \theta = 110°$

19. $|\mathbf{A}| = \sqrt{6}, \theta = 145°$

20. $|\mathbf{A}| = \sqrt{3}, \theta = 185°$

21. $|\mathbf{A}| = 4$, $\theta = 261°$

22. $|\mathbf{A}| = \sqrt{5}$, $\theta = 312°$

23. $|\mathbf{A}| = 3$, $\theta = 350°$

24. $|\mathbf{A}| = 7$, $\theta = 73°$

25. $|\mathbf{A}| = \sqrt{2}$, $\theta = 112°$

26. $|\mathbf{A}| = \sqrt{3}$, $\theta = 121°$

In Exercises 27–36, add the given vectors by components. Find $|\mathbf{R}|$ to two significant digits and θ to the nearest tenth of a degree. (The given numbers are assumed to be exact.)

27. A: $A = 3$, direction $15°$
B: $B = 2$, direction $81°$

28. A: $A = 3$, direction $20°$
B: $B = 5$, direction $100°$

29. A: $A = 2$, direction $112°$
B: $B = 4$, direction $156°$

30. A: $A = 8$, direction $35°$
B: $B = 6$, direction $250°$

31. A: $A = 5$, direction $245°$
B: $B = 3$, direction $290°$

32. A: $A = 10$, direction $25°$
B: $B = 15$, direction $300°$

33. A: $A = 1$, direction $10°$
B: $B = 2$, direction $70°$
C: $C = 4$, direction $340°$

34. A: $A = 2$, direction $80°$
B: $B = 6$, direction $136°$
C: $C = 7$, direction $215°$

35. A: $A = 6$, direction $165°$
B: $B = 4$, direction $221°$
C: $C = 5$, direction $340°$

36. A: $A = 4$, direction $10°$
B: $B = 3$, direction $212°$
C: $C = 2$, direction $350°$

9.3 Basic Applications of Vectors

This section is devoted to various physical applications of the vector concepts introduced so far.

First let us recall the rules for significant digits for triangles:

Degree measurement to nearest	Significant digits for measurement of a side
1°	2
10' or 0.1°	3
1' or 0.01°	4

Vectors at Right Angles

The first three examples illustrate the problem of finding the resultant of two given vectors.

E X A M P L E 1 Two forces act on the same object at right angles to each other. One force is 25.0 N (newtons) and the other is 16.1 N. Find the resultant.

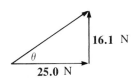

16.1 N

θ

25.0 N

Figure 9.23

Solution. This problem can be solved without referring to a coordinate system. From Figure 9.23, the magnitude of the resultant is $\sqrt{(25.0)^2 + (16.1)^2} = 29.7$ N. To obtain the direction, note that $\tan \theta = \frac{16.1}{25.0}$, so that $\theta = 32°50'$, rounded off to the nearest 10'. (θ may be described as the angle made with the larger force.) ◀

E X A M P L E 2 A small plane is headed east at 150.00 mi/h. The wind is from the north at 20.10 mi/h. What is the resulting velocity and direction of the plane?

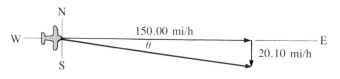

Figure 9.24

Solution. From the diagram (Figure 9.24), the resulting velocity is $\sqrt{(150.00)^2 + (20.10)^2} = 151.3$ mi/h (four significant digits). Also, $\tan \theta = \frac{20.10}{150.00}$, so that $\theta = 7°38'$. The direction is said to be $7°38'$ *south of east.* ◀

The next example features a force acting on an object hanging on a rope. In this and similar problems, we assume the system to be in *equilibrium*. A system is in equilibrium if there are no net forces acting on it. In particular, in the next example the tension on the rope is the resultant of the two forces acting on it.

E X A M P L E **3**

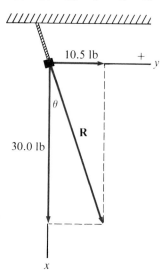

Figure 9.25

A 30.0-lb weight hanging on a rope is pulled sideways by a force of 10.5 lb. Determine the resulting tension on the rope and the angle that the rope makes with the vertical.

Solution. In the diagram (Figure 9.25) the tension must be equal to the resultant **R** of the two forces, the 30.0-lb downward force and the 10.5-lb pull to the side. Relative to the coordinate system shown in Figure 9.25,

$$\mathbf{R} = (30.0, 10.5)$$

So the tension on the rope is

$$|\mathbf{R}| = \sqrt{(30.0)^2 + (10.5)^2} = 31.8 \text{ lb}$$

The angle θ with the vertical is found from $\tan \theta = \frac{10.5}{30.0}$, or $\theta = 19.3°$ to the nearest tenth of a degree. ◀

The next example shows how resolving a vector into its components can be of interest for physical reasons.

E X A M P L E **4**

A block of ice weighing 95.0 lb is resting on a plane inclined 15.0° to the horizontal. What is the force required to keep the block from sliding down the plane? (See Figure 9.26; the weight of an object always acts vertically downward.)

Solution. In this problem the downward force **F**, where $|\mathbf{F}| = 95.0$ lb, has to be split into two components, one directed down the plane and the other perpendicular to the plane. (The components must obey the parallelogram law.)

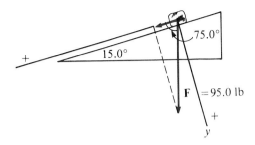

Figure 9.26

The component along the plane represents the tendency of the ice to slide downward. In other words

$$F_x = 95.0 \cos 75.0° = 24.6 \text{ lb}$$

So a force of 24.6 lb is required to keep the block from slipping.   ◀

Vectors Not at Right Angles

The next example involves vectors not positioned at right angles to each other. As a result, the given vectors have to be resolved into their components before the addition can be performed.

E X A M P L E **5**   A ship sailing at the rate of 21 knots in still water is heading 16° *west of south.* It runs into a strong current of 5.3 knots from the south. Find the resultant velocity vector. (See Figure 9.27.)

Solution. First we place the vectors with initial points at the origin, as shown in Figure 9.28.

Let **S** represent the velocity of the ship and **C** the velocity of the current. The direction of the ship (16° west of south) is now represented by $270° - 16° = 254°$. (See Figure 9.28.)

To find the resulting magnitude and direction, we resolve the vectors **S** and **C** into their components:

$$
\begin{array}{ll}
S_x = 21 \cos 254° = -5.788 & S_y = 21 \sin 254° = -20.19 \\
C_x = 5.3 \cos 90° = 0 & C_y = 5.3 \sin 90° = 5.3 \\
\hline
\text{adding: } \quad -5.788 & \text{adding: } \quad -14.89
\end{array}
$$

$$\mathbf{R} = (-5.788, -14.89)$$
$$|\mathbf{R}| = \sqrt{(-5.788)^2 + (-14.89)^2} = 16 \text{ knots}$$

To find θ, observe that

$$\tan \theta = \frac{-14.89}{-5.788}, \qquad \theta \text{ in quadrant III}$$

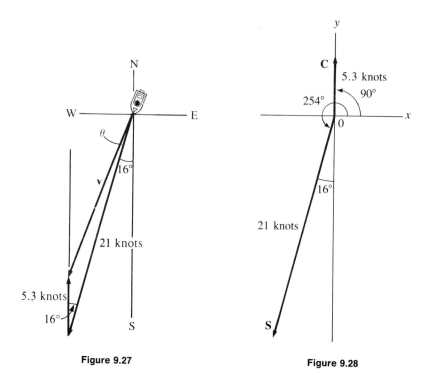

Figure 9.27

Figure 9.28

The reference angle is 69°. Since the angle made with the *x*-axis is 69°, the angle made with the *y*-axis is 90° − 69° = 21°. We conclude that the direction is 21° west of south. The velocity is 16 knots.   ◀

E X E R C I S E S / S E C T I O N   9.3

In Exercises 1–4, two forces with respective magnitudes F_1 and F_2 are acting on the same object at right angles. Determine the resultant.

1. $F_1 = 10.2$ lb, $F_2 = 5.70$ lb

2. $F_1 = 24.3$ lb, $F_2 = 15.6$ lb

3. $F_1 = 5.800$ N, $F_2 = 19.300$ N

4. $F_1 = 30.00$ N, $F_2 = 21.20$ N

5. A boat is heading directly across a river at 15.7 mi/h. If the river flows at 4.10 mi/h, find the actual speed (relative to the bank) and direction of the boat.

6. A boat is heading directly across a river at 14.3 km/h. The river is flowing at 6.20 km/h. What is the direction and the speed of the boat (relative to the bank)?

7. A boy and a girl are pulling a cart in mutually perpendicular directions with forces of 32 lb and 45 lb, respectively. What single force will have the same effect?

8. Two tractors are pulling on a tree stump with forces of 12,000 lb and 14,000 lb, respectively, in mutually perpendicular directions. What single force will have the same effect?

9. With the sun directly overhead, a jet is taking off at 112 mi/h at an angle of 42.0° with the ground. How fast is its shadow moving along the ground?

10. Two forces of 75.0 lb and 32.0 lb, respectively, are acting on an object in mutually perpendicular directions. What is the magnitude of the resultant?

11. A boat crosses a river flowing at 4.30 mi/h and lands at a point directly across. If the velocity of the boat is 10.00 mi/h in still water, find its velocity with respect to the bank and the direction in which the boat must head.

12. A boat crossing a river landed at a point directly across from where it started. Due to the flow of the river, it headed in a direction 25° upstream from the line directly across. Its velocity in still water is 17 mi/h. Find the rate at which the river was flowing.

13. A jet is heading due north at 400 km/h with the wind from the west at 29 km/h. Find the resultant velocity and direction. (See Example 2.)

14. A small plane is heading due east at 121.0 mi/h with the wind from the north at 25.3 mi/h. Find the resultant velocity and direction.

15. A pilot wishes to fly directly south. His air speed is 225.0 km/h, and the wind blows from the east at 30.1 km/h. In what direction should he orient his plane, and what is the resultant speed relative to the ground?

16. A 107-lb weight is supported as illustrated in Figure 9.29. Find the tension on the rope.

17. Find the force required to keep a 3,125-lb car parked on a hill that makes an angle of 11°31′ with the horizontal.

18. What is the force required to keep a 295-lb cart on a ramp inclined at an angle of 15.0° to the horizontal?

19. A force of 473 lb is required to pull a boat up a ramp inclined at 19.0° to the horizontal. Find the weight of the boat.

20. A woman is dragging a crate across the floor by means of a rope. She is able to pull with a force of 79 lb at an angle of 41° with the ground. What force parallel to the floor would have the same effect?

21. A 98-lb weight hanging on a rope is pulled sideways by a force of 25 lb. Assuming the system to be in equilibrium, what is the tension on the rope and the angle of the rope with the vertical?

22. A 31-lb weight hanging on a rope is to be pulled sideways so that the rope makes an angle of 17° with the vertical. What is the force required to accomplish this?

23. Determine the force against the horizontal support in Figure 9.30.

24. A 51-lb weight hanging on a rope is pulled sideways by a force of 15 lb. If the system is in equilibrium, determine the tension on the rope and the angle that the rope makes with the vertical.

The remaining exercises involve vectors not at right angles.

25. A small plane is flying 12.4° north of east at 156 mi/h. The wind is from the north at 27.6 mi/h. Find the resulting velocity and direction of the plane.

26. A plane heading 13.2° east of south at 465 km/h encounters a wind from the west at 20.9 km/h. Find the resulting velocity and direction.

27. A plane is heading 10.3° south of west at 452.0 km/h. The direction of the wind is 7.8° south of east and the velocity is 15.6 km/h. Find the resulting velocity and direction.

28. A small plane is heading 20.8° south of east at 138.0 mi/h. The wind is out of the northeast at 35.1 mi/h. Determine the resulting velocity and direction of the plane.

29. Two girls are pulling a cart with forces of 36.5 lb and 23.7 lb, respectively. If the forces are 47.0° apart, what single force will have the same effect?

30. Two men are pulling a crate with forces of 104 lb and 131 lb, respectively. Determine what single force will have the same effect, given that the forces are 53.0° apart.

31. A box is resting on the floor. Two forces are pulling on the box from the right. A force of 56.0 lb acts at an angle of 30.0° with the horizontal and a force of 40.0 lb acts at an angle of 35.0° with the horizontal. What single force will have the same effect?

32. An object resting on the floor is pulled by two forces on opposite sides, as shown in Figure 9.31. Find the resultant.

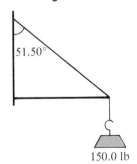

107 lb

Figure 9.29

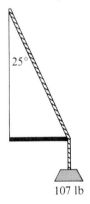

51.50°

150.0 lb

Figure 9.30

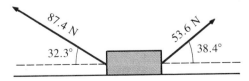

87.4 N   53.6 N
32.3°   38.4°

Figure 9.31

33. A small ship is pulled by three tug boats, as shown in Figure 9.32. Determine the resultant force.

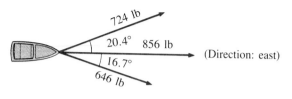

Figure 9.32

34. A crate resting on the floor is pulled by three forces, as shown in Figure 9.33. Find the resultant.

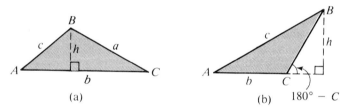

Figure 9.33

9.4   The Law of Sines

So far all of our applications have involved right triangles. We will now consider oblique triangles (triangles not containing a right angle) and thereby greatly increase the usefulness of our previous methods. We will begin by studying the *law of sines* in this section and then the *law of cosines* in the next.

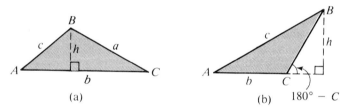

Figure 9.34

To derive the law of sines, consider the triangles in Figure 9.34. Now pick an arbitrary vertex such as B and drop a perpendicular to the side opposite. (The triangles in the figure illustrate the different cases.) Making use of the resulting right triangles, we get from Figure 9.34(a)

$$\sin A = \frac{h}{c} \quad \text{and} \quad \sin C = \frac{h}{a}$$

or $h = c \sin A$ and $h = a \sin C$. From Figure 9.34(b),

$$\boldsymbol{h} = c \sin A \quad \text{and} \quad \boldsymbol{h} = a \sin(180° - C) = a \sin C$$

[Since $\sin \theta > 0$ in the second quadrant, $\sin \theta = \sin(180° - \theta)$.] So we have in both cases

$$c \sin A = a \sin C$$

and

$$\frac{c \, \cancel{\sin A}}{\cancel{\sin A} \, \sin C} = \frac{a \, \cancel{\sin C}}{\sin A \, \cancel{\sin C}} \qquad \text{dividing by sin } A \text{ sin } C$$

$$\frac{a}{\sin A} = \frac{c}{\sin C} \tag{9.5}$$

If we drop a perpendicular from C to c, we get by the same argument

$$\frac{a}{\sin A} = \frac{b}{\sin B} \tag{9.6}$$

Combining equations (9.5) and (9.6), we get

$$\frac{a}{\sin A} = \frac{b}{\sin B} = \frac{c}{\sin C}$$

called the **law of sines** or simply the **sine law**.

Law of Sines:

$$\frac{\sin A}{a} = \frac{\sin B}{b} = \frac{\sin C}{c} \tag{9.7}$$

or

$$\frac{a}{\sin A} = \frac{b}{\sin B} = \frac{c}{\sin C} \tag{9.8}$$

Solving a triangle

The law of sines enables us to solve many oblique triangles. **Solving a triangle** means finding the measures of all sides and angles of the triangle.
Note. Examples 1 and 2 refer to the labels in Figure 9.34.

E X A M P L E   **1**   If $A = 45.68°$, $C = 110.43°$, and $b = 5.000$, find B, a, and c. (See Figure 9.35.)

Solution. Since the sum of the angles of a triangle is 180°, we get for B,

$$B = 180° - (45.68° + 110.43°) = 23.89°$$

By the law of sines we now have

$$\frac{a}{\sin 45.68°} = \frac{5.000}{\sin 23.89°} \qquad \frac{a}{\sin A} = \frac{b}{\sin B}$$

or

$$a = \frac{5.000 \sin 45.68°}{\sin 23.89°} \qquad \textbf{multiplying both sides by sin 45.68°}$$

CALCULATOR COMMENT

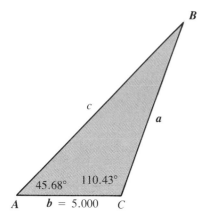

Figure 9.35

The sequence is

$$5.000 \boxed{\times} 45.68 \boxed{\text{SIN}} \boxed{\div} 23.89 \boxed{\text{SIN}} \boxed{=} \rightarrow 8.8330947$$

So $a = 8.833$ (four significant digits). The side c is found similarly:

$$\frac{c}{\sin 110.43°} = \frac{5.000}{\sin 23.89°} \qquad \frac{c}{\sin C} = \frac{b}{\sin B}$$

or

$$c = \frac{5.000 \sin 110.43°}{\sin 23.89°} = 11.57$$

So $B = 23.89°$, $a = 8.833$, and $c = 11.57$.   ◄

The next example shows how the sine law can be used to find an unknown angle.

E X A M P L E 2   Given that $B = 54°10'$, $b = 15.1$, and $c = 10.0$, find the remaining parts (Figure 9.36).

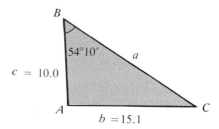

Figure 9.36

Solution. Since only one angle is known, one of the ratios must be

$$\frac{\sin 54°10'}{15.1} \qquad \text{or} \qquad \frac{15.1}{\sin 54°10'}$$

We now find angle C as follows:

$$\frac{\sin C}{10.0} = \frac{\sin 54°10'}{15.1} \qquad \frac{\sin C}{c} = \frac{\sin B}{b}$$

Thus

$$\sin C = \frac{10.0 \sin 54°10'}{15.1} = 0.5369$$

and $C = 32°30'$ (to the nearest $10'$). It follows that $A = 180° - (32°30' + 54°10') = 179°60' - (86°40') = 93°20'$.

Using angle A, we now find a:

$$\frac{a}{\sin 93°20'} = \frac{15.1}{\sin 54°10'} \qquad \frac{a}{\sin A} = \frac{b}{\sin B}$$

or

$$a = \frac{15.1 \sin 93°20'}{\sin 54°10'} = 18.6$$

Thus $A = 93°20'$, $C = 32°30'$, and $a = 18.6$. ◄

If three sides of a triangle are known, then the sine law is of no help: All the ratios in the sine law involve an angle, at least one of which has to be known. We run into a similar problem if $a = 5$, $b = 6$, and $C = 20°$, even though one angle is known:

$$\frac{c}{\sin 20°} = \frac{6}{\sin B} \qquad \text{and} \qquad \frac{c}{\sin 20°} = \frac{5}{\sin A}$$

Both equations involve two unknowns. These cases will be studied in the next section. The cases to which the sine law does apply are summarized next.

To solve an oblique triangle by the **law of sines**, we need either:

1. Two angles and one side, or
2. Two sides and the angle opposite one of them.

The sine law can be applied to certain problems involving vectors, as shown in the next example.

 E X A M P L E 3 A plane has an air speed of 525.0 mi/h and wants to fly on a course 20.0° east of north. If the wind is out of the north at 40.1 mi/h, determine the direction in which the plane must head to stay on the proper course. What is the resultant velocity relative to the ground?

Solution. Denote the desired velocity vector by **v**. We can see from the diagram in Figure 9.37 that the angle x opposite **v** has to be known in order to determine $|\mathbf{v}| = v$. To this end we first find angle y:

$$\frac{\sin y}{40.1} = \frac{\sin 160°}{525.0} \qquad \text{or} \qquad \sin y = \frac{40.1 \sin 160°}{525.0} = 0.0261$$

Thus $y = 1.5°$. It follows that $x = 180° - 161.5° = 18.5°$. Finally,

$$\frac{v}{\sin 18.5°} = \frac{525.0}{\sin 160°} \qquad \text{or} \qquad v = \frac{525.0 \sin 18.5°}{\sin 160°}$$

The calculator sequence is

CALCULATOR COMMENT

$$525.0 \; \boxed{\times} \; 18.5 \; \boxed{\text{SIN}} \; \boxed{\div} \; 160 \; \boxed{\text{SIN}} \; \boxed{=} \; \rightarrow 487.06179$$

If this is rounded off to three significant digits, we get 487.

Thus the plane must head 18.5° east of north, and it will fly at the rate of 487 mi/h relative to the ground. ◀

Caution. Suppose we want to find the obtuse angle C (greater than 90°) in Figure 9.38. By the sine law

$$\frac{\sin C}{4} = \frac{\sin 35°}{3}$$

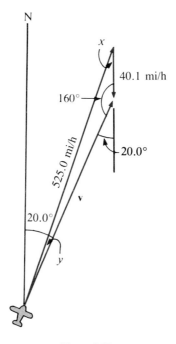

Figure 9.37

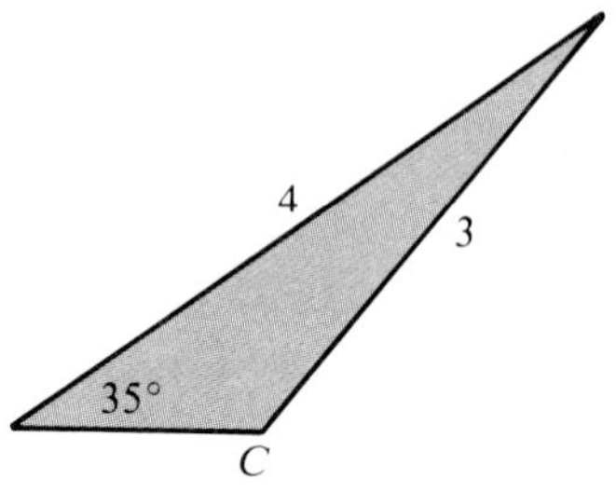

Figure 9.38

or sin $C = 0.7648$. If we now use a calculator to find C, we might conclude that $C = 49.9°$. However, since angle C is obtuse, the correct value is $180° - 49.9° = 130.1°$. Consequently, *we have to know in advance whether an angle to be found by the sine law is obtuse or acute.* (Recall that an angle is *obtuse* if its measure is more than $90°$ and *acute* if its measure is less than $90°$.)

Ambiguous case

The case just described is called the *ambiguous case.* If for some reason the relevant information is withheld, then there may exist two solutions or no solution at all. The different cases are summarized in the following chart:

Number of possible triangles	Sketch	Condition
0	Figure 9.39	$a < h$ or $a < b \sin A$
1	Figure 9.40	$a = h$ or $a = b \sin A$
2	Figure 9.41	$b > a > h$ or $b > a > b \sin A$

EXERCISES / SECTION 9.4

In Exercises 1–20, solve the triangles from the given information. (Refer to Figure 9.34 on page 290.)

1. $A = 20.2°$, $C = 50.7°$, $b = 4.00$
2. $A = 40.3°$, $B = 80.5°$, $a = 5.32$
3. $B = 25.8°$, $C = 130.3°$, $c = 15.1$
4. $A = 112.1°$, $C = 10.5°$, $c = 36.0$
5. $A = 19°50'$, $a = 102$, $b = 46.5$
6. $C = 47°38'$, $a = 0.7980$, $c = 1.3200$

7. $C = 31.5°$, $a = 13.30$, $c = 6.82$
8. $A = 56°$, $a = 8.1$, $c = 10.0$
9. $A = 29.3°$, $a = 71.6$, $c = 136$; angle C acute
10. Same as Exercise 9 with angle C obtuse
11. $B = 46.0°$, $C = 54.0°$, $a = 236$
12. $A = 10.1°$, $C = 22.7°$, $c = 0.450$
13. $C = 63.6°$, $a = 12.4$, $c = 11.6$; angle A acute
14. Same as Exercise 13 with angle A obtuse

15. $B = 33°45'$, $a = 1.146$, $b = 2.805$

16. $B = 33°45'$, $a = 1.1460$, $b = 0.6200$

17. $B = 33°45'$, $a = 1.146$, $b = 1.000$; angle C obtuse

18. $A = 72.3°$, $a = 86.0$, $c = 73.0$

19. $A = 126.50°$, $C = 10.40°$, $a = 136.4$

20. $B = 28.0°$, $C = 16.7°$, $c = 1.03$

21. A plane maintaining an air speed of 560 mi/h is heading 10° west of north. A north wind causes the actual course to be 11° west of north. Find the velocity of the plane with respect to the ground.

22. Two forces act on the same object in directions that are 40.0° apart. If one force is 48.0 lb, what must the other force be so that the combined effect is equivalent to a force of 65.0 lb?

23. A surveyor wants to find the width of a river from a certain point on the bank. Since no other points on the bank nearby are accessible, he takes the measurements shown in Figure 9.42. Find the width of the river.

24. A building 81 ft tall stands on top of a hill. From a point at the foot of the hill, the angles of elevation of the top and bottom of the building are 39°10' and 37°40', respectively. What is the distance from the point to the bottom of the building?

25. From a point on the ground, the angle of elevation of a balloon is 49.0°. From a second point 1,250 ft away on the opposite side of the balloon and in the same vertical plane as the balloon and the first point, the angle of ele-

vation is 33.0°. Find the distance from the second point to the balloon.

26. Find the area of the triangle in Figure 9.43.

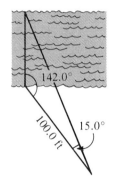

Figure 9.42

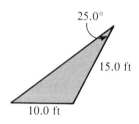

Figure 9.43

9.5   The Law of Cosines

We saw in the previous section that not all triangles can be solved by the law of sines. In this section we will consider the solution of these triangles by means of the *law of cosines*.

Use the law of cosines to solve an oblique triangle given:

1. Two sides and the included angle, or

2. Three sides.

To derive the law of cosines, consider the triangle in Figure 9.44. Observe that the height h divides AC into lengths x and $b - x$.

From the Pythagorean theorem we get for the respective right triangles

$$c^2 = h^2 + x^2 \qquad \text{and} \qquad a^2 = h^2 + (b - x)^2$$

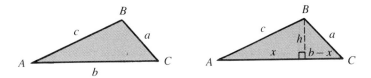

Figure 9.44

Solving each equation for h^2, we get

$$h^2 = c^2 - x^2 \quad \text{and} \quad h^2 = a^2 - (b - x)^2$$

Substituting the expression for h^2 on the left into the equation on the right, we get

$$c^2 - x^2 = a^2 - (b - x)^2$$
$$c^2 - x^2 = a^2 - b^2 + 2bx - x^2 \qquad \text{expanding } (b - x)^2$$
$$a^2 = b^2 + c^2 - 2bx \qquad \text{solving for } a^2$$

Finally, since $\cos A = x/c$, we have $x = c \cos A$. Substituting in the last equation, we get

$$a^2 = b^2 + c^2 - 2bc \cos A$$

This formula is known as the **law of cosines** or simply the **cosine law**.

By dropping the perpendicular from the other vertices, we get two other forms of the cosine law, as summarized next.

Law of Cosines:

$$a^2 = b^2 + c^2 - 2bc \cos A \qquad\qquad (9.9)$$
$$b^2 = a^2 + c^2 - 2ac \cos B \qquad\qquad (9.10)$$
$$c^2 = a^2 + b^2 - 2ab \cos C \qquad\qquad (9.11)$$

The law of cosines can also be stated verbally.

Law of Cosines (Verbal Form): The square of any side of a triangle equals the sum of the squares of the other two sides minus twice the product of those two sides and the cosine of the angle between them.

Note especially that if $A = 90°$, then formula (9.9) reduces to

$$a^2 = b^2 + c^2 - 2bc \cdot 0 = b^2 + c^2$$

It follows that the cosine law is a generalization of the Pythagorean theorem.

The next example illustrates how the law of cosines is used to solve oblique triangles.

E   X   A   M   P   L   E   **1**   Given that $A = 117.5°$, $b = 7.50$, and $c = 3.90$, find the remaining parts. (See Figure 9.45.)

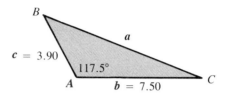

Figure 9.45

Solution. Note that this triangle cannot be solved by the sine law. However, using the cosine law, we get

$$a^2 = b^2 + c^2 - 2bc \cos A$$

$$a^2 = (7.50)^2 + (3.90)^2 - 2(7.50)(3.90) \cos 117.5°$$

The sequence is

$$7.50 \boxed{x^2} \boxed{+} 3.90 \boxed{x^2} \boxed{-} 2 \boxed{\times} 7.50 \boxed{\times} 3.90$$

$$\boxed{\times} 117.5 \boxed{\text{COS}} \boxed{=} \boxed{\sqrt{}} \to 9.9233207$$

So $a = 9.92$ (three significant digits).

Now that the side opposite angle A is known, we can find angle B (or angle C) by the sine law:

$$\frac{\sin B}{7.50} = \frac{\sin 117.5°}{9.92} \qquad \text{or} \qquad B = 42.1°$$

Finally, $C = 180° - (42.1° + 117.5°) = 20.4°$. So

$$a = 9.92, \qquad B = 42.1°, \qquad \text{and} \qquad C = 20.4°$$

Note. Angle B can also be found by the cosine law:

$$b^2 = a^2 + c^2 - 2ac \cos B$$

$$(7.50)^2 = (9.92)^2 + (3.90)^2 - 2(9.92)(3.90) \cos B$$

or

$$\cos B = \frac{(7.50)^2 - (9.92)^2 - (3.90)^2}{-2(9.92)(3.90)}$$

The sequence is

$$7.50 \boxed{x^2} \boxed{-} 9.92 \boxed{x^2} \boxed{-} 3.90 \boxed{x^2} \boxed{=} \boxed{\div} 2 \boxed{+/-} \boxed{\div} 9.92$$
$$\boxed{\div} 3.90 \boxed{=} \boxed{INV} \boxed{COS} \rightarrow 42.149373$$

However, the calculation using the sine law is easier.   ◄

Example 1 shows that even though the cosine law may be needed to find the first of the unknown parts, it is easier to find the remaining parts by the sine law. As we saw in the last section, however, to find an angle by the sine law, we have to know in advance whether the angle is acute or obtuse. We can handle this difficulty in one of two ways: We can find all angles by the cosine law (if $\cos \theta$ is negative, a calculator yields an obtuse angle automatically). A simpler alternative is the following:

> First use the cosine law to find the angle opposite the longest side. Then use the sine law to find the remaining angles.

This procedure is illustrated in the next example.

E X A M P L E   **2**   Given that $a = 4.3$, $b = 5.2$, and $c = 8.2$, find the three angles (Figure 9.46).

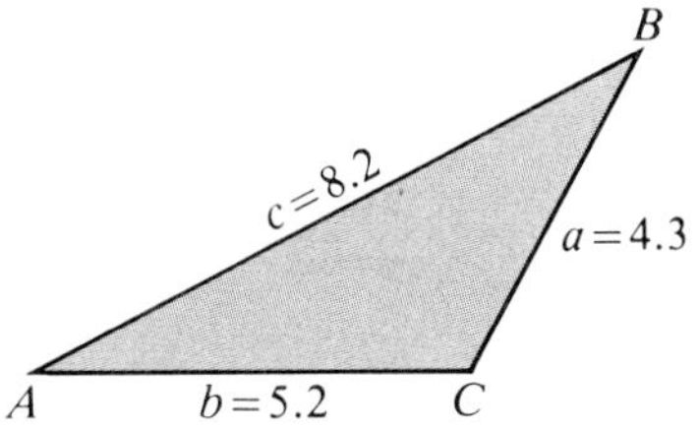

Figure 9.46

Solution. Since angle C is the angle opposite the longest side, we find angle C by the cosine law:

$$c^2 = a^2 + b^2 - 2ab \cos C$$
$$(8.2)^2 = (4.3)^2 + (5.2)^2 - 2(4.3)(5.2) \cos C$$
$$\cos C = \frac{(8.2)^2 - (4.3)^2 - (5.2)^2}{-2(4.3)(5.2)}$$

CALCULATOR COMMENT   The sequence is

$$8.2 \boxed{x^2} \boxed{-} 4.3 \boxed{x^2} \boxed{-} 5.2 \boxed{x^2} \boxed{=} \boxed{\div} 2 \boxed{+/-} \boxed{\div} 4.3 \boxed{\div} 5.2$$
$$\boxed{=} \boxed{INV} \boxed{COS} \rightarrow 119.04295$$

So $C = 119°$. We can now safely apply the sine law to obtain angle A:

$$\frac{\sin A}{4.3} = \frac{\sin 119°}{8.2} \qquad \text{or} \qquad A = 27°$$

Finally, $B = 180° - (27° + 119°) = 34°$.   ◄

E X A M P L E **3**

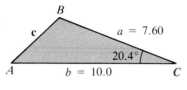

Figure 9.47

Given that $C = 20.4°$, $a = 7.60$, and $b = 10.0$, find the remaining parts (Figure 9.47).

Solution. First we need to find side c by the cosine law.

$$c^2 = a^2 + b^2 - 2ab \cos C$$

$$c^2 = (7.60)^2 + (10.0)^2 - 2(7.60)(10.0) \cos 20.4$$

which yields $c = 3.91$.

   Since B is the angle opposite the longest side, it may be an obtuse angle. (Judging from the figure, it is.) So the safest procedure is to use the cosine law again:

$$b^2 = a^2 + c^2 - 2ac \cos B$$

$$(10.0)^2 = (7.60)^2 + (3.91)^2 - 2(7.60)(3.91) \cos B$$

$$\cos B = \frac{(10.0)^2 - (7.60)^2 - (3.91)^2}{-2(7.60)(3.91)}$$

or

$$B = 117.0°$$

Finally, $A = 180° - (117.0° + 20.4°) = 42.6°$ ◀

   Caution. It happens occasionally that using the sine law (instead of the cosine law) to find an angle leads to a slightly different result due to round-off errors. For example, if angle B in Example 3 is found by the sine law, we get

$$\frac{\sin B}{10.0} = \frac{\sin 20.4°}{3.91} \qquad \text{or} \qquad \sin B = \frac{10.0 \sin 20.4°}{3.91}$$

Assuming B to be obtuse, this leads to $116.9°$, instead of $117.0°$ from the cosine law.

   We saw in Sections 9.2 and 9.3 that the sum of two vectors that are not at right angles can be found by adding components. Using the cosine law, some of these problems can be solved in a simpler way. Let us rework Example 5, Section 9.3.

    E X A M P L E **4**

(Same as Example 5, Section 9.3). A ship sailing at the rate of 21 knots in still water is heading 16° west of south. It runs into a strong current of 5.3 knots from the south. Find the resultant velocity vector.

Solution. In Figure 9.48, denote the desired velocity vector by **v**. Then

$$|\mathbf{v}|^2 = (5.3)^2 + (21)^2 - 2(5.3)(21) \cos 16°$$

N

W ──────┼────── E

θ

16°

v

21 knots

5.3 knots

16°

S

Figure 9.48

Using a calculator, we find that $|\mathbf{v}| = 16$ knots. To obtain the direction, we need to find θ in Figure 9.48. By the cosine law:

$$(5.3)^2 = (21)^2 + (16)^2 - 2(21)(16)\cos\theta$$

from which $\theta = 5°$. (The angle θ can also be found by the sine law.) So the direction is $16° + 5° = 21°$ west of south. ◄

E X A M P L E **5**   Two forces of 55.0 lb and 37.0 lb, respectively, are acting on the same object. If the angle between their directions is 23.4°, what single force would produce the same effect?

Solution. Denote the resultant force by $\mathbf{F}$ and consider the diagram in Figure 9.49. By the law of cosines

$$|\mathbf{F}|^2 = (55.0)^2 + (37.0)^2 - 2(55.0)(37.0)\cos 156.6°$$

Using a calculator we get $|\mathbf{F}| = 90.2$ lb. Since the angle opposite the longest side is already known, we can safely apply the sine law to find angle θ, the direction of $\mathbf{F}$ relative to the 55.0-lb force. By the sine law, $\theta = 9.4°$. (If the cosine law is used, the calculated value turns out to be 9.3°.)

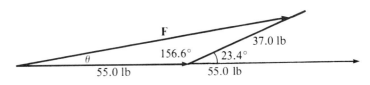

F

37.0 lb

θ   156.6°   23.4°

55.0 lb   55.0 lb

Figure 9.49

So the magnitude of $\mathbf{F}$ is 90.2 lb and its direction is 9.4° relative to the larger force. ◄

E X E R C I S E S / S E C T I O N   **9.5**

In Exercises 1–12, solve the given triangles. (See Examples 1–3.)

1. $a = 20.1$, $b = 30.4$, $c = 25.7$

2. $a = 2.46$, $b = 1.97$, $c = 4.10$

3. $a = 21.46$, $b = 12.85$, $c = 9.179$

4. $a = 20$, $b = 23$, $c = 18$

5. $A = 46.3°$, $b = 1.00$, $c = 2.30$

6. $B = 62.7°$, $a = 7.00$, $c = 10.00$

7. $C = 125°10'$, $a = 178$, $b = 137$

8. $A = 100.0°$, $b = 2.36$, $c = 1.97$

9. $C = 39.4°$, $a = 126.0$, $b = 80.1$

10. $A = 14°40'$, $b = 11.70$, $c = 7.80$

11. $A = 63.0°$, $b = 35.1$, $c = 86.1$

12. $a = 0.471$, $b = 0.634$, $c = 0.239$

13. A civil engineer wants to find the length of a proposed tunnel. From a distant point he observes that the respective distances to the ends of the tunnel are 585 ft and 624 ft. The angle between the lines of sight is 33.4°. Find the length of the proposed tunnel.

14. Two forces of 150.0 lb and 80.0 lb produce a resultant force of 209 lb. Find the angle between the forces.

15. A ship sails 16.0 km due east, turns 20.0° north of east, and then continues for another 11.5 km. Find its distance from the starting point.

16. A car travels 85 km due west of point A and then northwest for another 50 km. Find its distance from point A.

17. A small plane is heading 5.0° north of east with an air speed of 151.0 mi/h. The wind is from the south at 35.3 mi/h. Find the actual course and the velocity with respect to the ground.

18. In Figure 9.50 the point P is moving around the circle. Find d as a function of θ.

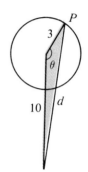

Figure 9.50

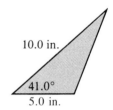

Figure 9.51

19. Find the perimeter of the triangle in Figure 9.51.

20. A river is flowing east at 4.10 mi/h. A boat crosses the river in the direction 29.0° east of north. If the speedometer reads 10.0 mi/h, what is the boat's actual direction and the speed relative to the bank?

21. A force of 157.3 N and a force of 120.6 N are acting on the same object in directions that are 28.71° apart. Determine the resultant.

R E V I E W   E X E R C I S E S / C H A P T E R   **9**

In Exercises 1–8, find the magnitude of each vector and the direction to the nearest tenth of a degree.

1. $(-1, 1)$ **2.** $(-1, -\sqrt{3})$ **3.** $(1, 2)$

4. $(1, -3)$ **5.** $(\sqrt{15}, 1)$ **6.** $(-3, \sqrt{7})$

7. $(-\sqrt{2}, -2)$ **8.** $(\sqrt{2}, 3)$

In Exercises 9–12, resolve each vector into its components. (Give answers to two decimal places.)

9. $|\mathbf{A}| = 5$, $\theta = 75°$ **10.** $|\mathbf{A}| = 3$, $\theta = 115°$

11. $|\mathbf{A}| = \sqrt{7}$, $\theta = 220°$ **12.** $|\mathbf{A}| = \sqrt{3}$, $\theta = 328°$

In Exercises 13 and 14, add the given vectors by components.

13. A: $A = 2$, direction 18°
    B: $B = 3$, direction 158°

14. A: $A = 1$, direction 12°
    B: $B = 2$, direction 205°
    C: $C = 3$, direction 345°

In Exercises 15–24, solve each triangle from the given information.

15. $A = 46.3°$, $C = 53.7°$, $b = 5.26$

16. $A = 29.4°$, $B = 115.2°$, $c = 63.5$

17. $B = 10°40'$, $C = 130°20'$, $c = 236$

18. $A = 41°$, $B = 52°$, $b = 22$

19. $A = 26°$, $a = 25$, $c = 37$ (angle C is obtuse)

20. $C = 31.6°$, $a = 38.4$, $b = 62.0$

21. $A = 19°23'$, $b = 11.23$, $c = 30.04$

22. $B = 51°$, $a = 1.9$, $c = 1.4$

23. $a = 3.74$, $b = 5.86$, $c = 5.50$

24. $a = 321.7$, $b = 276.4$, $c = 248.2$

25. A railroad car weighing 9.8 tons is resting on a track inclined 7° with the horizontal. What is the force needed to keep it from rolling downhill?

26. A 320-lb cart is resting on a ramp inclined 12.0° to the horizontal. How much of the weight does the ramp support?

27. A surveyor wants to determine the width of a river. He can reach one point on the bank, but no other point nearby is accessible. Instead, he takes the measurements shown in Figure 9.52. Find the width of the river.

28. Find the perimeter of the triangle in Figure 9.53.

29. A 12.0-lb weight hanging from a rope is pushed sideways so that the rope makes a 15.0° angle with the vertical. If the system is in equilibrium, find the tension on the rope.

30. A 463-lb weight is supported as shown in Figure 9.54. Find the compression force on the angled bracket PQ.

31. A force of 25.1 lb and a force of 39.4 lb produce a resultant force of 59.9 lb. Find the angle between the two forces.

32. From a point on the ground, the angle of elevation of the top of a tower is 26.3°. From a second point 48.3 ft closer to the tower, the angle of elevation of the top is 37.4°. Find the height of the tower.

33. A racing car is traveling at 90.0 mi/h along a straight road inclined at 7.0° to the horizontal. Determine the rate of ascent.

34. A plane is heading 20°0′ south of west at 311 mi/h. If the wind is from the east at 20.0 mi/h, determine its velocity with respect to the ground and the resulting direction by **(a)** addition of components and **(b)** the law of cosines.

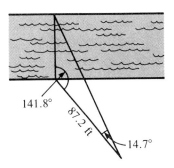

141.8°   87.2 ft   14.7°

Figure 9.52

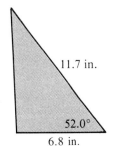

11.7 in.

52.0°

6.8 in.

Figure 9.53

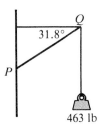

Q

31.8°

P

463 lb

Figure 9.54

1. If $\sec \theta = -\sqrt{5}$ and $\cot \theta < 0$, find $\csc \theta$.

2. Use a diagram to find **(a)** $\sec 210°$; **(b)** $\cot 300°$.

3. Find all angles between $0°$ and $360°$ such that **(a)** $\cos \theta = \dfrac{\sqrt{2}}{2}$; **(b)** $\tan \theta = -\dfrac{\sqrt{3}}{3}$

4. Use a calculator to find $\cot 215.78°$.

5. Use a calculator to find all angles between $0°$ and $360°$, given that $\csc \theta = -1.341$.

6. Change $112°$ to radian measure.

7. Change $\dfrac{25\pi}{9}$ to degree measure.

8. A circle has radius 6.00 cm. Find the length of the arc intercepted by a central angle of $10.6°$.

9. Find the area of the circular sector whose central angle is $78.0°$ if the radius of the circle is 12.0 in.

10. The range R (in meters) along the ground of a projectile fired at velocity v (m/s) at angle θ with the horizontal is

$$R = \frac{v^2}{9.8} \sin 2\theta$$

Find the range if $\theta = 25.6°$ and $v = 32.7$ m/s.

11. The cylindrical shaft on a motor rotates at the rate of 35 rev/s. If the radius of the shaft is 8.5 cm, find the velocity of a point on the rim in meters per second.

12. Sketch the graph of $y = -10 \sin \dfrac{1}{4} x$.

13. Sketch the graph of $y = 2 \cos\left(x - \dfrac{\pi}{3}\right)$.

14. Find the magnitude and direction of the vector $(-2\sqrt{3}, 2)$.

15. A block of ice weighing 50.0 lb is resting on an inclined plane which makes an angle of $26.8°$ with the horizontal. What force is required to keep the block from slipping?

16. A small plane is heading west at 145 mi/h. The wind is from the south at 20.3 mi/h. Find the resulting direction and the velocity with respect to the ground.

17. Two tractors are pulling a heavy machine in directions that are $20.0°$ apart. If one force is 619 lb, what must the other force be so that the combined effect is a force of $1,150$ lb?

18. A ship sailing at the rate of 28.0 knots in still water is heading in the direction $15.0°$ east of south. Find the resultant velocity and direction if it runs into a strong current of 4.60 knots from the south by **(a)** addition of components and **(b)** the law of cosines.

C H A P T E R 10

Exponents and Radicals

10.1 Review of Exponents

The purpose of this section is to review the basic operations involving exponents, which are listed next.

Laws of Exponents:

$$a^m a^n = a^{m+n} \tag{10.1}$$

$$\frac{a^m}{a^n} = a^{m-n}, \qquad a \neq 0, \qquad m > n \tag{10.2}$$

$$(a^m)^n = a^{mn} \tag{10.3}$$

$$(ab)^n = a^n b^n \tag{10.4}$$

$$\left(\frac{a}{b}\right)^n = \frac{a^n}{b^n}, \qquad b \neq 0 \tag{10.5}$$

The following examples illustrate the basic laws:

E X A M P L E 1 $x^3 x^6 x^{10} = x^{3+6+10} = x^{19}$ **adding exponents**   ◀

E X A M P L E 2 $\dfrac{a^{12}}{a^8} = a^{12-8} = a^4$ **subtracting exponents**   ◀

E X A M P L E **3** $(a^3)^4 = a^{3 \cdot 4} = a^{12}$ **multiplying exponents**   ◀

E X A M P L E **4** $(2x)^3 = 2^3 x^3 = 8x^3$ **by rule (10.4)**

$$\left(\frac{1}{x}\right)^2 = \frac{1^2}{x^2} = \frac{1}{x^2}$$ **by rule (10.5)**   ◀

The basic laws are frequently used in combination, as illustrated in the remaining examples.

E X A M P L E **5** Simplify:

$$\frac{(2x^3 a^4)^2}{3x^2 a^3}$$

Solution. $\dfrac{(2x^3 a^4)^2}{3x^2 a^3} = \dfrac{2^2 (x^3)^2 (a^4)^2}{3x^2 a^3}$ $(ab)^n = a^n b^n$ [rule (10.4)]

$$= \frac{4x^6 a^8}{3x^2 a^3}$$ $2^2 = 4$; multiplying exponents

$$= \frac{4x^4 a^5}{3}$$ subtracting exponents   ◀

E X A M P L E **6** Perform the following multiplication and simplify:

$$\left(\frac{6v^7 L^6 c^4}{15v^4 L^4 c}\right)^4 \cdot \left(\frac{5a}{v^8 L^4 c^2}\right)^2$$

Solution.

$$\left(\frac{6v^7 L^6 c^4}{15v^4 L^4 c}\right)^4 \cdot \left(\frac{5a}{v^8 L^4 c^2}\right)^2 = \left(\frac{2v^3 L^2 c^3}{5}\right)^4 \cdot \left(\frac{5a}{v^8 L^4 c^2}\right)^2$$

$$= \frac{(2v^3 L^2 c^3)^4}{5^4} \cdot \frac{(5a)^2}{(v^8 L^4 c^2)^2}$$ $\left(\dfrac{a}{b}\right)^n = \dfrac{a^n}{b^n}$

$$= \frac{2^4 (v^3)^4 (L^2)^4 (c^3)^4}{5^4} \cdot \frac{5^2 a^2}{(v^8)^2 (L^4)^2 (c^2)^2}$$ $(ab)^n = a^n b^n$

$$= \frac{16 v^{12} L^8 c^{12}}{5^4} \cdot \frac{5^2 a^2}{v^{16} L^8 c^4} = \frac{16 a^2 c^8}{5^2 v^4} = \frac{16 a^2 c^8}{25 v^4}$$ ◀

E X A M P L E **7** Simplify:

$$\frac{x^{3a+1} y^{a+3}}{x^{3a} y^3}$$

Solution. Even though the exponents are letters, the basic laws can be applied: Subtracting exponents, we get

$$x^{(3a+1)-3a}y^{(a+3)-3} = xy^a \qquad \blacktriangleleft$$

E X E R C I S E S / S E C T I O N 10.1

Perform the indicated operations and simplify.

1. $3^2 \cdot 2^3$

2. $\dfrac{5^6}{5^4}$

3. $\left(\dfrac{3}{4}\right)^3$

4. $(2^2 3^2)^2$

5. $(-3x^4y^6)(-2xy^3)$

6. $(-4a^3b^2c^5)(3abc^4)$

7. $\dfrac{-24a^7b^4}{-12a^3b}$

8. $\dfrac{25c^8d^4}{-15c^6d^3}$

9. $(-RS^2)^4$

10. $(-2v_1{}^2v_2{}^3)^3$

11. $(-4C_1{}^2C_2)^3(2C_1C_2{}^3)^3$

12. $\dfrac{3u^3v^2}{5w^4} \cdot \dfrac{10v^2w^5}{9u^4}$

13. $\dfrac{9s^2t^3}{3v} \cdot \dfrac{3v^2}{21st^4}$

14. $\dfrac{7a^3b^4}{8c^5d^2} \div \dfrac{21a^7b}{16c^3d^7}$

15. $\dfrac{9mn^3}{16s^5} \div \dfrac{27mn^2}{16s^6}$

16. $(2xy)^3 \div xy$

17. $\dfrac{(3RV^2)^2}{3R^2V}$

18. $\dfrac{9w^4x^6y^8}{(3wx^2y^2)^3}$

19. $\dfrac{8x^{10}y^5z^6}{(-2x^2y^2z^3)^4}$

20. $\dfrac{18a^2bc^2}{(-3a^2bc^2)^3}$

21. $\dfrac{(3abc)^3}{(3abc)^2}$

22. $\dfrac{(5p^2q^3r)^2}{(3pqr^5)^3}$

23. $\dfrac{(7v_0s_1{}^2s_2)^3}{(7v_0s_1{}^3s_2{}^4)^2}$

24. $\dfrac{(3x^2y^3z)^4}{(3x^2y^3z^2)^4}$

25. $\left(\dfrac{6x^3y^6}{9x^2y^7}\right)^3$

26. $\left(\dfrac{8v_1{}^4v_2{}^2v_3}{12v_1{}^3v_2{}^4v_3{}^2}\right)^2$

27. $\left(\dfrac{14p^3q^4r^5}{21p^5qr^6}\right)^4$

28. $\left(\dfrac{27a^{10}c^8x^5}{45a^{15}c^3x^4}\right)^3$

29. $(-3x^2y^2z^3)^2(5xy^3z)^4$

30. $(-4\alpha^2\beta^3)^2(-\alpha^2\beta\omega)^5$

31. $\left(\dfrac{2ab^2}{3xy}\right)^2\left(\dfrac{3xy^2}{2a^2b}\right)^3$

32. $\left(\dfrac{5a^2b}{m^2n^2}\right)^3\left(\dfrac{m^3n^2}{10ab}\right)^2$

33. $\left(\dfrac{2R^2V^2}{C^3}\right)^3\left(\dfrac{C^2}{4RV}\right)^4$

34. $\left(\dfrac{4xy}{3st}\right)^2\left(\dfrac{3st^2}{8x^2y}\right)^4$

35. $\dfrac{x^{a+1}}{x^a}$

36. $\dfrac{y^{b+4}}{y^{b+2}}$

37. $\dfrac{y^{a+2}}{y^{a+1}}$

38. $\dfrac{y^{2a}}{y^a}$

39. $\dfrac{x^{a+1}y^{3a}}{x^ay^{2a}}$

40. $\dfrac{a^{x+2}b^{5x}}{a^{x+1}b^{2x}}$

41. $\dfrac{a^{x-2}}{a^{x-3}}$

42. $\left(\dfrac{y^{b-1}}{y^{b-3}}\right)^2$

43. $\left(\dfrac{x^{a+4}}{x^{a-1}}\right)^b$

44. $\left(\dfrac{y^{2b+1}}{y^{b+1}}\right)^2$

45. $\left(\dfrac{x^{3b+2}}{x^{b+2}}\right)^3$

10.2 Zero and Negative Integral Exponents

So far we have assumed exponents to be positive integers. In this section we will introduce zero exponents and exponents that are negative integers.

Let us first consider the zero exponent a^0, $a \neq 0$. There are several ways to make sense out of this expression. For example, note that

$$\frac{a^n}{a^n} = 1, \qquad a \neq 0$$

If the law of division is to hold, then

$$\frac{a^n}{a^n} = a^{n-n} = a^0$$

We conclude that $a^0 = 1$. As a check, observe that

$$a^n a^0 = a^{n+0} = a^n$$

by the rule for multiplication. This product makes sense only if we define $a^0 = 1$.

   Similar checks for consistency can be made for negative exponents. Consider, for example,

$$\frac{a^3}{a^5} = \frac{1}{a^2}$$

If we decide to carry out the division by subtracting the exponents, we get

$$\frac{a^3}{a^5} = a^{3-5} = a^{-2}$$

We conclude that $a^{-2} = 1/a^2$. (In general, we define $a^{-n} = 1/a^n$). As another check, note that

$$a^{-2}a^6 = a^{-2+6} = a^4$$

by the rule for multiplication. This result agrees with

$$a^{-2}a^6 = \frac{1}{a^2} \cdot a^6 = \frac{a^6}{a^2} = a^4$$

obtained from the definition. In fact, rules (10.1) through (10.5) in Section 10.1 hold for zero and negative exponents as well. Let us now summarize the new definitions.

Zero and Negative Exponents:

$$a^0 = 1, \qquad a \neq 0 \tag{10.6}$$

$$a^{-n} = \frac{1}{a^n}, \qquad a \neq 0 \tag{10.7}$$

   The following examples illustrate how the basic operations can be performed with zero and negative exponents.

E X A M P L E **1**

$$\frac{x^{-4}}{x^{-2}} = x^{-4-(-2)} = x^{-2} \qquad \frac{a^m}{a^n} = a^{m-n} \qquad \textbf{(subtracting exponents)} \qquad \blacktriangleleft$$

E X A M P L E **2**

$$(2x^{-3})^4 = 2^4(x^{-3})^4 = 2^4 x^{(-3)(4)} = 16x^{-12} \qquad \textbf{(ab)}^n = \textbf{a}^n\textbf{b}^n; \ (\textbf{a}^m)^n = \textbf{a}^{mn} \qquad \blacktriangleleft$$

E X A M P L E 3 $(x^{-2})^{-4} = x^{(-2)(-4)} = x^8$ $(a^m)^n = a^{mn}$ **(multiplying exponents)**   ◀

E X A M P L E 4 $x^{-3}x^4x^{-6} = x^{-3+4-6} = x^{-5}$ $a^m a^n = a^{m+n}$ **(adding exponents)**   ◀

E X A M P L E 5 $\dfrac{a^0 x^4}{x^9} = 1 \cdot x^{4-9} = x^{-5}$ $a^0 = 1$; **subtracting exponents**   ◀

Before attempting more involved simplifications, observe that

$$\frac{1}{a^{-n}} = \frac{1}{\dfrac{1}{a^n}} = a^n$$

or $a^n = 1/a^{-n}$.

When a **factor** is moved from the numerator of a fraction to the denominator or from the denominator to the numerator, the sign on the exponent is changed:

$$a^n = \frac{1}{a^{-n}} \quad \text{and} \quad a^{-n} = \frac{1}{a^n} \tag{10.8}$$

These rules are illustrated in the examples that follow.

E X A M P L E 6 $\dfrac{a^{-3}b}{c^{-4}} = \dfrac{bc^4}{a^3}$ since $a^{-3} = \dfrac{1}{a^3}$ and $\dfrac{1}{c^{-4}} = c^4$ ◀

E X A M P L E 7 $\dfrac{2^{-1}x^{-2}}{y^4} = \dfrac{1}{2x^2 y^4}$ **by rule (10.8):** $2^{-1} = \dfrac{1}{2}$ and $x^{-2} = \dfrac{1}{x^2}$ ◀

E X A M P L E 8   Evaluate:

   a. $(2^{-1})^{-3}$ **b.** $\dfrac{3^{-2}5^0}{2^{-3}}$ **c.** $\dfrac{9^{-2}}{3^{-6}}$

Solution.

   a. Multiplying exponents, we get

$$(2^{-1})^{-3} = 2^3 = 8$$

b. $\dfrac{3^{-2}5^0}{2^{-3}} = \dfrac{3^{-2} \cdot 1}{2^{-3}} = \dfrac{2^3}{3^2}$ $3^{-2} = \dfrac{1}{3^2}; \dfrac{1}{2^{-3}} = 2^3$

$\qquad\qquad = \dfrac{8}{9}$

c. Since $9 = 3^2$, the expression can be written

$$\dfrac{9^{-2}}{3^{-6}} = \dfrac{(3^2)^{-2}}{3^{-6}} = \dfrac{3^{-4}}{3^{-6}} \qquad\qquad \textbf{multiplying exponents}$$

$$\qquad\qquad = \dfrac{3^6}{3^4} = 3^2 = 9 \qquad 3^{-4} = \dfrac{1}{3^4}; \dfrac{1}{3^{-6}} = 3^6$$ ◀

Although negative exponents are useful in making calculations, final results are ordinarily written without them.

E X A M P L E **9**   Simplify and write without negative exponents:

$$\dfrac{2^{-3}T_a^{-3}T_b^{-4}}{4^{-4}T_a^2 T_b^{-6}}$$

Solution. Since $4 = 2^2$, $4^{-4} = (2^2)^{-4} = 2^{-8}$. Thus

$$\dfrac{2^{-3}T_a^{-3}T_b^{-4}}{4^{-4}T_a^2 T_b^{-6}} = \dfrac{2^{-3}T_a^{-3}T_b^{-4}}{2^{-8}T_a^2 T_b^{-6}}$$

$$= \dfrac{2^8 T_a^{-3}T_b^{-4}}{2^3 T_a^2 T_b^{-6}} \qquad 2^{-3} = \dfrac{1}{2^3}, \dfrac{1}{2^{-8}} = 2^8$$

$$= \dfrac{2^8 T_a^{-3}T_b^{-4}T_b^{6}}{2^3 T_a^2} \qquad \dfrac{1}{T_b^{-6}} = T_b^6$$

$$= \dfrac{2^8 T_b^{6}}{2^3 T_a^2 T_a^3 T_b^4} \qquad T_a^{-3} = \dfrac{1}{T_a^3}, T_b^{-4} = \dfrac{1}{T_b^4}$$

$$= \dfrac{2^8 T_b^{6}}{2^3 T_a^5 T_b^4} \qquad T_a^2 T_a^3 = T_a^5$$

$$= \dfrac{2^5 T_b^{2}}{T_a^5} \qquad\qquad \textbf{subtracting exponents}$$

$$= \dfrac{32 T_b^{2}}{T_a^5} \qquad\qquad 2^5 = 32$$

The same result could be obtained by subtracting exponents in the original expression. For example

$$\dfrac{T_a^{-3}}{T_a^2} = T_a^{-3-2} = T_a^{-5} = \dfrac{1}{T_a^5}$$

and

$$\frac{T_b^{-4}}{T_b^{-6}} = T_b^{-4-(-6)} = T_b^2 \qquad \blacktriangleleft$$

E X A M P L E 10   Simplify:

$$(2^{-1}R^{-3}r^{-4}V^2)^{-2}$$

Solution. The expression can be written

$$\left(\frac{V^2}{2R^3r^4}\right)^{-2}$$

However, because so many of the exponents are negative, it is better to multiply the exponents directly. Thus

$$(2^{-1}R^{-3}r^{-4}V^2)^{-2} = (2^{-1})^{-2}(R^{-3})^{-2}(r^{-4})^{-2}(V^2)^{-2} \qquad \textbf{(ab)}^n = \textbf{a}^n\textbf{b}^n$$

$$= 2^2R^6r^8V^{-4} \qquad \textbf{multiplying exponents}$$

$$= \frac{4R^6r^8}{V^4} \qquad 2^2 = 4;\ V^{-4} = \frac{1}{V^4} \qquad \blacktriangleleft$$

Fractions containing sums of terms with negative exponents have to be changed to complex fractions and then simplified, as shown in the next example.

E X A M P L E 11   Simplify:

$$\frac{x^{-1}}{x^{-1} + y^{-1}}$$

Solution. Since the denominator is a sum of two terms (not a product!), the terms in the denominator cannot be "moved up." In other words,

$$\frac{x^{-1}}{x^{-1} + y^{-1}} \neq \frac{x^{-1}(x + y)}{1}$$

Instead, we write the given fraction as a complex fraction:

$$\frac{x^{-1}}{x^{-1} + y^{-1}} = \frac{\dfrac{1}{x}}{\dfrac{1}{x} + \dfrac{1}{y}} \qquad x^{-1} = \frac{1}{x};\ y^{-1} = \frac{1}{y}$$

$$= \frac{\dfrac{1}{x}}{\dfrac{y}{xy} + \dfrac{x}{xy}} = \frac{\dfrac{1}{x}}{\dfrac{x + y}{xy}} \qquad \textbf{LCD} = \textbf{xy}$$

$$= \frac{1}{x} \cdot \frac{xy}{x + y} = \frac{y}{x + y} \qquad \textbf{invert and multiply} \qquad \blacktriangleleft$$

Common errors

1. Forgetting that

$$a^{-1} \neq -\frac{1}{a}$$

Instead,

$$a^{-1} = \frac{1}{a}$$

2. Forgetting that it is not correct to write

$$\frac{1}{a^{-1} + b^{-1}} \quad \text{as } a + b$$

since a^{-1} and b^{-1} are *not factors*. Instead,

$$\frac{1}{a^{-1} + b^{-1}} = \frac{1}{\dfrac{1}{a} + \dfrac{1}{b}} = \frac{1}{\dfrac{a + b}{ab}} = \frac{ab}{a + b}$$

E X A M P L E **12**   Simplify:

$$(a^{-2} + 1)^{-1}$$

Solution. Since the expression inside the parentheses is a sum, the rule $(ab)^n = a^n b^n$ does *not* apply. Instead, we write a^{-2} as $1/a^2$ and add the resulting fractions:

$$(a^{-2} + 1)^{-1} = \left(\frac{1}{a^2} + 1 \right)^{-1} \qquad a^{-2} = \frac{1}{a^2}$$

$$= \left(\frac{1}{a^2} + \frac{a^2}{a^2} \right)^{-1} \qquad \text{LCD} = a^2$$

$$= \left(\frac{1 + a^2}{a^2} \right)^{-1} \qquad \text{adding fractions}$$

$$= \frac{a^2}{1 + a^2} \qquad \left(\frac{x}{y} \right)^{-1} = \frac{y}{x} \qquad \blacktriangleleft$$

 E X A M P L E **13**   In the mathematics of finance, the *present value* is the amount of money that must be invested in order to accumulate to a prescribed amount in a given time interval. An *annuity* is a sequence of equal payments R made at equal time intervals. The present value A of an annuity, which is the total value of all payments before the first payment is made, can be written

$$A = R \frac{(1 + i)^{-n-1} - (1 + i)^{-1}}{(1 + i)^{-1} - 1}$$

where i is the interest rate. Simplify the expression for A.

Solution. $A = R \dfrac{(1 + i)^{-n-1} - (1 + i)^{-1}}{(1 + i)^{-1} - 1}$

$= R \dfrac{(1 + i)^{-n-1} - (1 + i)^{-1}}{(1 + i)^{-1} - 1} \cdot \dfrac{(1 + i)^{n+1}}{(1 + i)^{n+1}}$

$= R \dfrac{1 - (1 + i)^n}{(1 + i)^n - (1 + i)^{n+1}}$ **adding exponents**

$= R \dfrac{1 - (1 + i)^n}{(1 + i)^n[1 - (1 + i)]}$ **common factor $(1 + i)^n$**

$= R \dfrac{1 - (1 + i)^n}{(1 + i)^n(-i)}$ **simplifying the denominator**

$= R \dfrac{(1 + i)^n - 1}{i(1 + i)^n}$ **multiplying numerator and denominator by -1**   ◀

E X E R C I S E S / S E C T I O N   10.2

In Exercises 1–8, evaluate the given expressions.

1. 4^{-2} **2.** $3^{-1} \cdot 3^0$ **3.** $(2^{-1})^{-4}$ **4.** $(2^{-2})^3$

5. $\dfrac{6^{-2}}{6^{-4}}$ **6.** $\dfrac{9^{-2}}{3^{-2}}$ **7.** $\dfrac{4^{-4}}{16^{-2}}$ **8.** $\dfrac{2^0 \cdot 2^2}{8^{-2}}$

In Exercises 9–66, combine the expressions and write the results without zero or negative exponents.

9. $\dfrac{x^2}{x^{-3}}$

10. $\dfrac{a^{-3}}{a^{-4}}$

11. $\dfrac{a^{-3}b^2}{b^{-4}}$

12. $\dfrac{2^0 z^2 w^{-3}}{2^{-1}z^{-2}w^{-4}}$

13. $2^{-2}x^0 z^{-1}$

14. $\dfrac{5^{-1}V^2 W^{-2}}{2V^0 W^3}$

15. $\dfrac{2^{-4}f_1^{-2}f_2^4 f_3^{-1}}{4^{-3}f_1^{-3}f_2^{-1}f_3^4}$

16. $\dfrac{3a^{-2}b^4 t^{-7}}{9^{-1}a^3 b^{-5} t^{-5}}$

17. $\dfrac{3^{-4}s^{-3}t^4 u^{-7}}{9^{-2}s^{-2}t^2}$

18. $\dfrac{5^0 e^0 n^{-3} m^{-1}}{5^{-2}e^{-2}n^{-7}m}$

19. $\dfrac{3^{-1}q^3 r^{-2}s^{-5}}{4^{-2}q^{-4}s^2}$

20. $\dfrac{7^{-2}a^0 x^{-4}w^3}{3^0 a^{-6}x^0 w^{-1}}$

21. $\left(\dfrac{a}{b}\right)^{-1}$

22. $\left(\dfrac{x}{y}\right)^{-2}$

23. $(2x^{-2}y^2)^{-3}$

24. $(3^{-1}a^{-2}b^{-6})^{-3}$

25. $(2^{-2}x^{-1}b^2)^4$

26. $(2^{-3}x^{-12}y^{16})^{-1}$

27. $(2x^{-7}y^0)^0$

28. $(2V^{-2}R^{-3})^{-2}$

29. $(4^{-1}\pi^2\omega^{-2})^{-3}$

30. $(5e^{-3}\pi^{-4})^{-1}$

31. $\left(\dfrac{a^{-2}}{2b^2}\right)^4$

32. $\left(\dfrac{2^{-1}x^{-2}y^4}{x^{-6}y^{-1}}\right)^{-2}$

33. $\left(\dfrac{2^{-3}L^{-2}R^3 C^{-2}}{4^{-1}L^2 R^{-1}C^3}\right)^{-3}$

34. $\left(\dfrac{a^2 p^{-4}e^{-1}}{a^{-2}p^{-3}e^0}\right)^{-2}$

35. $\left(\dfrac{2a^3 c_1^{-1}c_2^{-2}}{3^{-1}a^{-1}c_1^{-2}c_2^{-1}}\right)^3$

36. $\left(\dfrac{4R^{-3}r^2}{2^{-1}R^{-7}r^{-6}}\right)^{-1}$

37. $\left(\dfrac{s^{-5}r^{-6}t^{-2}}{s^{-3}r^5 t}\right)^{-2}$

38. $\left(\dfrac{2^{-1}f^{-3}g^3 h^{-4}}{f^3 g^{-1}h^{-2}}\right)^4$

39. $\left(\dfrac{a^2 b^{-2}y^{-3}}{3^{-2}a^{-2}b^{-3}y^{-4}}\right)^{-2}$

40. $\left(\dfrac{7^{-1}x^{-2}y^3 w^{-1}}{2^0 x^0 y^{-3}w^2}\right)^{-2}$

41. $(\pi^{-e})^{-2}$

42. $(2x^{-a})^{-1}$

43. $(3a^x b^{-y})^{-1}$

44. $(2^{-1}q^{-2r})^{-3}$

45. $(a^b x^c)^b$

46. $(c^{-x}d^y)^{-t}$

47. $x^2 + \dfrac{1}{x^{-2}}$

48. $2a^{-1} + \dfrac{1}{a}$

49. $x^{-1} + y^{-1}$

50. $2^{-1} - 3^{-1}$

51. $\dfrac{a^{-1} + b}{a^{-1}}$

52. $\dfrac{1 + x^{-2}}{x}$

53. $(1 + b^{-1})^{-1}$

54. $(1 - a^{-2})^{-1}$

55. $(a^{-1} + b^{-1})^{-1}$

56. $(y^{-1} - x^{-1})^{-2}$

57. $(1 - z^{-2})^{-2}$

58. $\dfrac{a^{-1}}{1 + a^{-1}}$

59. $\dfrac{a^{-1}b^{-1}}{a^{-1} - b^{-1}}$

60. $\left(\dfrac{1}{1 - a^{-1}}\right)^{-2}$

61. $\left(\dfrac{1 + b^{-2}}{b}\right)^{-1}$

62. $\left(\dfrac{x^{-2} + 2}{x^{-1}}\right)^{-2}$

63. $\left(\dfrac{a^{-1}b^{-1}}{a^{-1} - b^{-1}}\right)^2$

64. $\left(\dfrac{x^{-1}y^{-1}}{x^{-1} + y^{-1}}\right)^{-1}$

65. $\left(\dfrac{a^{-2}b^{-2}}{a^{-2} - b^{-2}}\right)^{-1}$

66. $\left(\dfrac{x^{-2}y^{-2}}{x^{-2} + y^{-2}}\right)^{-2}$

67. The diameter of a certain molecule is 3.8×10^{-8} cm. Write this number as a decimal.

68. If h is the height in meters above sea level, then the atmospheric pressure P in millimeters of mercury is

   $$P = 760(2.71)^{-0.00013h}$$

   Write this formula without the negative exponent.

69. Show that the combined resistance of two resistors R and r in parallel can be written $(R^{-1} + r^{-1})^{-1}$.

70. The relationship between the object distance q, the image distance p, and the focal length f of a lens can be written

   $$f^{-1} = p^{-1} + q^{-1}$$

   Find f if $q = \frac{3}{2}$ cm and $p = 3$ cm.

71. In determining the period of a pendulum, the expression $(1 - n^{-1})^{-1}$ has to be simplified. Carry out this simplification.

72. The total resistance of the circuit in Figure 10.1 is given by

   $$R = [R_1^{-1} + (R_2 + R_3)^{-1}]^{-1}$$

   Simplify this expression.

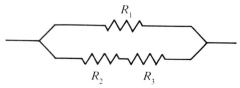

Figure 10.1

73. The *present value* of an investment is the amount of money P that needs to be invested in order to yield a certain amount S at the end of a prescribed time period. The equation for the relationship is

   $$P = S(1 + i)^{-n}$$

   where i is the interest rate compounded annually and n the number of years. How much has to be invested in order to obtain $10,000 at the end of eight years if the interest rate is 12% compounded annually?

74. Determine the size of an investment that will yield $15,000 at the end of 10 years if the interest rate is 10% compounded annually. (Refer to Exercise 73.)

10.3   Fractional Exponents

In this section we will study **fractional exponents**. As in the case of zero and negative integral exponents, the definitions will be formulated to preserve the basic laws of exponents. In particular, to define $a^{1/2}$, we recall that $(a^m)^n = a^{mn}$ and require that

   $$(a^{1/2})^2 = a$$

Now let $b = a^{1/2}$. Then $b^2 = (a^{1/2})^2 = a$, so that $b = \sqrt{a}$. It follows that $a^{1/2} = \sqrt{a}$. In general, since $(a^{1/n})^n = a$, $a^{1/n} = \sqrt[n]{a}$. To define $a^{m/n}$, observe that

   $$a^{m/n} = (a^{1/n})^m = (a^m)^{1/n}$$

These results are summarized next.

Fractional Exponents:

$$a^{1/n} = \sqrt[n]{a} \tag{10.9}$$

$$a^{m/n} = \sqrt[n]{a^m} = (\sqrt[n]{a})^m \tag{10.10}$$

Since fractional exponents are really radicals in disguise, numerical expressions with fractional exponents are evaluated by changing the expression to the radical form. It is usually easier to find the root first and then the power. For example, $8^{2/3} = (\sqrt[3]{8})^2 = 2^2 = 4$. Consider another example.

E X A M P L E **1**   Evaluate:

a. $16^{3/2}$ **b.** $32^{4/5}$ **c.** $8^{-1/3}$ **d.** $27^{-2/3}$

Solution.

a. $16^{3/2} = (\sqrt{16})^3 = 4^3 = 64$ $16^{3/2} = (16^{1/2})^3$

b. $32^{4/5} = (\sqrt[5]{32})^4 = 2^4 = 16$

c. $8^{-1/3} = \dfrac{1}{8^{1/3}} = \dfrac{1}{2}$

d. $27^{-2/3} = \dfrac{1}{27^{2/3}} = \dfrac{1}{(\sqrt[3]{27})^2} = \dfrac{1}{3^2} = \dfrac{1}{9}$ ◀

Algebraic expressions with fractional exponents are simplified by using the laws of exponents, as shown in the remaining examples.

E X A M P L E **2**   Simplify:

a. $x^{3/2}x^{1/3}$ **b.** $\dfrac{a^{3/4}}{a^{5/2}}$ **c.** $(16x^6y^2)^{1/2}$

Solution.

a. $x^{3/2}x^{1/3} = x^{3/2 + 1/3} = x^{11/6}$ **adding exponents**

b. $\dfrac{a^{3/4}}{a^{5/2}} = a^{3/4 - 5/2} = a^{-7/4} = \dfrac{1}{a^{7/4}}$ **subtracting exponents**

c. $(16x^6y^2)^{1/2} = (16)^{1/2}(x^6)^{1/2}(y^2)^{1/2}$ **since $(ab)^n = a^n b^n$**

$\qquad\qquad\quad = 4x^3y$ **$(16)^{1/2} = \sqrt{16} = 4$; multiplying exponents** ◀

E X A M P L E **3**   Simplify and write without negative exponents:

$$\dfrac{5^{-1}R^{-1/4}C^{1/2}}{5^{-3/2}R^{-3/4}C^4}$$

Solution. Subtracting exponents, we get

$$5^{-1-(-3/2)}R^{-1/4-(-3/4)}C^{1/2-4}$$

Note that

$$-1-\left(-\frac{3}{2}\right) = -\frac{2}{2}+\frac{3}{2} = \frac{1}{2} \qquad \textbf{exponent of 5}$$

$$-\frac{1}{4}-\left(-\frac{3}{4}\right) = -\frac{1}{4}+\frac{3}{4} = \frac{2}{4} = \frac{1}{2} \qquad \textbf{exponent of } R$$

and

$$\frac{1}{2}-4 = \frac{1}{2}-\frac{8}{2} = -\frac{7}{2} \qquad \textbf{exponent of } C$$

The resulting expression is

$$5^{1/2}R^{1/2}C^{-7/2} = \frac{5^{1/2}R^{1/2}}{C^{7/2}} \qquad \blacktriangleleft$$

E X A M P L E **4**   Simplify and write without zero or negative exponents:

$$\frac{27^{-2/3}a^{-1/7}b^{-1/5}c^{-3/8}d^{0}}{8^{1/3}a^{3/7}b^{-2/3}c^{-1/4}}$$

Solution. As in the case of integral exponents, the various factors can be combined by subtracting exponents. For example,

$$\frac{a^{-1/7}}{a^{3/7}} = a^{-1/7-3/7} = a^{-4/7} = \frac{1}{a^{4/7}}$$

A good alternative is to write the expression without negative exponents first. Since $d^0 = 1$, we get

$$\frac{b^{2/3}c^{1/4}\cdot 1}{8^{1/3}27^{2/3}a^{3/7}a^{1/7}b^{1/5}c^{3/8}} \qquad \frac{1}{a^n} = a^{-n}, \frac{1}{a^{-n}} = a^n$$

$$= \frac{b^{2/3}c^{1/4}}{8^{1/3}\mathbf{27}^{2/3}a^{3/7}a^{1/7}b^{1/5}c^{3/8}}$$

$$= \frac{b^{2/3-1/5}c^{1/4-3/8}}{\mathbf{2\cdot 9}a^{4/7}} = \frac{b^{7/15}c^{-1/8}}{18a^{4/7}} = \frac{b^{7/15}}{18a^{4/7}c^{1/8}} \qquad \blacktriangleleft$$

E X A M P L E **5**   Simplify and write without negative exponents:

$$\left(\frac{8^{-1}c_1^{-3/4}c_2^{-6}}{27^{-3}c_1^{-3/2}c_2^{9/4}}\right)^{-2/3}$$

Solution. The factors inside the parentheses could be combined first. However, because of the large number of negative exponents, it may be easier to multiply exponents first. Then we get

$$\frac{(8^{-1})^{-2/3}(c_1{}^{-3/4})^{-2/3}(c_2{}^{-6})^{-2/3}}{(27^{-3})^{-2/3}(c_1{}^{-3/2})^{-2/3}(c_2{}^{9/4})^{-2/3}} = \frac{8^{2/3}c_1{}^{1/2}c_2{}^{4}}{27^2 c_1 c_2{}^{-3/2}} = \frac{4c_2{}^{4}c_2{}^{3/2}}{729 c_1 c_1{}^{-1/2}} = \frac{4c_2{}^{11/2}}{729 c_1{}^{1/2}} \quad \blacktriangleleft$$

E X A M P L E **6**   Simplify: $2(x+2)^{-1/2} + (x+2)^{1/2}$.

Solution. Note that the exponent $-\frac{1}{2}$ is on the factor $(x+2)$, not on the coefficient 2. We therefore have

$$= \frac{2}{(x+2)^{1/2}} + (x+2)^{1/2}$$

$$= \frac{2}{(x+2)^{1/2}} + \frac{(x+2)^{1/2}}{1}\frac{(x+2)^{1/2}}{(x+2)^{1/2}} \qquad \text{LCD} = (x+2)^{1/2}$$

$$= \frac{2 + (x+2)^{1/2}(x+2)^{1/2}}{(x+2)^{1/2}} \qquad \text{adding fractions}$$

$$= \frac{2 + (x+2)}{(x+2)^{1/2}} \qquad a^{1/2}a^{1/2} = a^{1/2+1/2} = a$$

$$= \frac{x+4}{(x+2)^{1/2}} \qquad\qquad \blacktriangleleft$$

CALCULATOR COMMENT

To evaluate a given power with a calculator, use $\boxed{y^x}$ or $\boxed{x^y}$ For example, to evaluate $(\pi + 4)^{-5/6}$, use the sequence

$$\boxed{\pi}\;\boxed{+}\;\boxed{4}\;\boxed{=}\;\boxed{y^x}\;\boxed{(}\;\boxed{5}\;\boxed{+/-}\;\boxed{\div}\;\boxed{6}\;\boxed{)}\;\boxed{=} \rightarrow 0.194314$$

(Most calculators will not evaluate y^x if y is negative.)

 E X A M P L E **7**   The amount of water A flowing across a dam with a V-shaped notch is given by

$$A = 2.506\left(\tan\frac{\theta}{2}\right)h^{2.47}$$

where A is the volume of the flow in cubic meters per second, h the height in meters above the bottom of the notch, and θ the angle of the notch. If $h = 2.30$ m and $\theta = 20.3°$, find A.

Solution. From the given formula,

$$A = 2.506\left(\tan\frac{20.3°}{2}\right)(2.30)^{2.47}$$

The sequence is

$$2.506 \boxed{\times} \boxed{(} 20.3 \boxed{\div} \boxed{2} \boxed{)} \boxed{\text{TAN}} \boxed{\times} 2.30 \boxed{y^x} 2.47 \boxed{=} \rightarrow 3.5104991$$

So the water flows across at the rate of 3.51 m³/s. ◀

E X E R C I S E S / S E C T I O N **10.3**

In Exercises 1–20, evaluate the given expressions.

1. $81^{1/2}$ **2.** $64^{1/3}$ **3.** $27^{2/3}$ **4.** $16^{3/4}$

5. $49^{-1/2}$ **6.** $27^{-1/3}$ **7.** $(0.09)^{1/2}$ **8.** $\left(\dfrac{1}{4}\right)^{1/2}$

9. $\left(\dfrac{1}{9}\right)^{-1/2}$ **10.** $32^{2/5}$ **11.** $(-64)^{1/3}$ **12.** $(-8)^{1/3}$

13. $(-32)^{3/5}$ **14.** $(-32)^{4/5}$ **15.** $49^{-3/2}$ **16.** $27^{-4/3}$

17. $125^{-1/3}$ **18.** $(-125)^{-1/3}$

19. $(-125)^{-2/3}$ **20.** $(0.064)^{1/3}$

In Exercises 21–68, simplify the given expressions and write the results without zero or negative exponents.

21. $a^{1/2}a^{-1/2}$ **22.** $x^{3/4}x^{1/4}$

23. $m^{1/7}m^{2/7}$ **24.** $p^{-1/3}p^{4/3}$

25. $\dfrac{x^{7/3}}{x^{1/4}}$ **26.** $\dfrac{Q^{1/5}}{Q^{-2/5}}$

27. $\dfrac{\omega^{-1/4}}{\omega^{-3/4}}$ **28.** $\dfrac{y^{-1/3}}{y^{1/6}}$

29. $(-5x^{1/2})(-2x^{-3/4})$ **30.** $(3a^{-5/9})(10a^{1/3})$

31. $\dfrac{2a^{1/3}b^{-1/4}}{4a^{-1/6}b^{-1/5}}$ **32.** $\dfrac{6V^{1/5}I^{1/12}}{9V^{1/10}I^{1/6}}$

33. $\dfrac{15R_1^{-1/3}R_2^{2/3}}{25R_1^{1/4}R_2^{3/4}}$ **34.** $\dfrac{21v_1^{1/2}v_2^{-3/4}}{18v_1^{-2/9}v_2^{2/5}}$

35. $\dfrac{36^{1/2}a^{-3/5}w^{-1/8}}{27^{1/3}a^{-5/6}w^{3/4}}$ **36.** $\dfrac{6^{1/3}s^{-3/7}t^{2/3}}{3^{1/3}s^{-1/7}t^{-2/5}}$

37. $(2^{1/2}x^{3/2}y^8)^2$ **38.** $(2^{2/3}x^{-1/3}y^{1/3})^3$

39. $(9a^4y^{-6})^{-1/2}$ **40.** $(27L^{-6}R^9)^{-1/3}$

41. $(32^{-1}x^5y^{-10}z^{-15})^{1/5}$ **42.** $(2^{-3}v^{-6}q^{-12})^{-1/3}$

43. $\left(\dfrac{2^0a^2c^{-3}}{2^{-1}a^{1/3}c^{-1/4}}\right)^{-1}$ **44.** $\left(\dfrac{27R^{-9}C}{8R^3C^{-3}}\right)^{-2/3}$

45. $\left(\dfrac{64^{-1}m^9d^0}{m^{-6/5}d^{-12}}\right)^{-1/6}$ **46.** $\left(\dfrac{16a^{-2}c^{2/5}}{25a^{4/3}c^{-4}}\right)^{-1/2}$

47. $\left(\dfrac{81m^{-8}v^{-4}s^0}{16^{-1}m^4v^{4/5}}\right)^{-1/4}$ **48.** $\left(\dfrac{4^{-1}x^{-2/3}y^{4/5}t^0}{16^{-1}x^{-2}y^{-4}}\right)^{3/2}$

49. $\left(\dfrac{L^{-4/3}T^{-8/3}}{16^{-1}L^{-16/3}T^4}\right)^{3/4}$ **50.** $\left(\dfrac{32r^{-10}s^{-2/5}}{r^{5/8}s^{1/5}}\right)^{-4/5}$

51. $x(x+2)^{-1/2}+(x+2)^{1/2}$ **52.** $2x(x+1)^{1/2}+(x+1)^{-1/2}$

53. $(x+2)(x-3)^{-1/2}+(x-3)^{1/2}$

54. $(x-4)(x+3)^{-1/2}+(x+3)^{1/2}$

55. $x(x+1)^{-2/3}+(x+1)^{1/3}$

56. $(x+2)(x-3)^{-2/3}+(x-3)^{1/3}$

57. $(x+6)(x-1)^{-2/3}-(x-1)^{1/3}$

58. $x(x+2)^{-2/3}-(x+2)^{1/3}$

59. $(x-3)(x-1)^{-1/3}-2(x-1)^{2/3}$

60. $x(x-2)^{-3/4}+2(x-2)^{1/4}$

61. $x(x+2)^{-4/5}-3(x+2)^{1/5}$

62. $(x+3)(x-1)^{-5/6}-(x-1)^{1/6}$

63. $(x-1)(x+2)^{-6/7}-(x+2)^{1/7}$

64. $\left[x^{1/(a-1)}\right]^{a-1}$ **65.** $\left[x^{1/(a-1)}\right]^{a^2-1}$

66. $\left(\dfrac{x^{3a-b}}{x^a}\right)^{1/(2a-b)}$ **67.** $\left(\dfrac{x^{1/a}}{x^{-1}}\right)^{-a/(1+a)}$

68. $\left(\dfrac{x^{4a-1}}{x^{2a}}\right)^{1/(4a^2-1)}$

In Exercises 69–76, evaluate the given expressions with a calculator.

69. $(9.84-1.62)^{1/4}$ **70.** $(2.87)^{2/7}+6.46$

71. $10.37-(4.432)^{-1.732}$ **72.** $(5.043)^{-0.8013}+9.347$

73. $[(17.4)^{1/5}+1.84]^{0.902}$ **74.** $[10.8-(19.8)^{1/4}]^{2.36}$

75. $[(0.891)^{-1.87}+(8.39)^{0.592}]^{2/3}$

76. $[12.7-(15.6)^{0.760}]^{-1.53}$

77. When a gas is compressed adiabatically (with no gain or loss of heat), the pressure and volume of the gas are related by the formula $p = kv^{-7/5}$, where k is a constant. Express the formula in the form of a radical.

78. Two hallways of width a and b intersect at right angles (Figure 10.2). The longest rod that can be carried horizontally around the corner is of length $L = (a^{2/3} + b^{2/3})^{3/2}$.

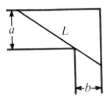

Figure 10.2

If $a = 27$ ft and $b = 8$ ft, show that the length of the longest rod is $\sqrt{2{,}197}$ ft.

79. The distance between the centers of two spheres of radii a and b, respectively, is d. From a point P on the line of centers AB, the greatest amount of surface area is visible if the distance D from P to A is

$$D = da^{3/2}(a^{3/2} + b^{3/2})^{-1}$$

Find D if $a = 4$ cm, $b = 9$ cm, and $d = 20$ cm.

80. The equation $E = T/(x^2 + a^2)^{3/2}$ gives the electric-field intensity on the axis of a uniformly charged ring, where T is the total charge on the ring and a the radius of the ring. Write E in the form of a radical.

81. The rate of runoff of rainfall from an area of land to an inlet reaches its maximum after t minutes, where

$$t = C\sqrt[3]{L}\, i^{-2/3}(\tan\theta)^{-1/3}$$

Here L is the distance from the most remote point in the area, θ the angle the area makes with the horizontal, i the rain intensity, and C a constant. Show that t can be written

$$t = C\left(\frac{L}{i^2 \tan\theta}\right)^{1/3}$$

82. If V_1 and V_2 are the respective maximum and minimum volumes of air in the cylinder of an internal combustion engine, then V_1/V_2 is called the *compression ratio*. The *efficiency* Eff (expressed as a percentage) is given by

$$\text{Eff} = 100\left(1 - \frac{1}{\left(\dfrac{V_1}{V_2}\right)^{k-1}}\right)$$

where k is a constant. Show that the formula is equivalent to

$$\text{Eff} = 100\left(\frac{V_1^{k-1} - V_2^{k-1}}{V_1^{k-1}}\right)$$

83. If the distance $x(t)$ from the origin of a particle moving along the x-axis as a function of time is $x(t) = 2t^2(1 + t)^{1/3}$, we get from calculus that the velocity is

$$v = \frac{2}{3}t^2(1 + t)^{-2/3} + 4t(1 + t)^{1/3}$$

Write v as a single fraction.

10.4   The Simplest Radical Form and Addition/Subtraction of Radicals

In this section we will learn how to simplify radicals. Also, the addition and subtraction of radicals, which are natural by-products of simplification, are considered briefly.

First recall the laws of radicals from Chapter 1.

Laws of Radicals:

$$\sqrt[n]{a^n} = (\sqrt[n]{a})^n = a \tag{10.11}$$

$$\sqrt[n]{a}\,\sqrt[n]{b} = \sqrt[n]{ab} \qquad (a > 0 \quad \text{or} \quad b > 0) \tag{10.12}$$

$$\frac{\sqrt[n]{a}}{\sqrt[n]{b}} = \sqrt[n]{\frac{a}{b}} \tag{10.13}$$

$$\sqrt[m]{\sqrt[n]{a}} = \sqrt[mn]{a} \tag{10.14}$$

Radicand; order of a radical

Note. In the radical expression $\sqrt[n]{a}$, the number under the radical sign is called the **radicand** and the index n is called the **order** of the radical.

The laws of radicals follow directly from the laws of exponents, as shown next.

(10.11): $\sqrt[n]{a^n} = (a^n)^{1/n} = a$ and $(\sqrt[n]{a})^n = (a^{1/n})^n = a$

(10.12): $\sqrt[n]{a}\sqrt[n]{b} = a^{1/n}b^{1/n} = (ab)^{1/n} = \sqrt[n]{ab}$

(10.13): $\dfrac{\sqrt[n]{a}}{\sqrt[n]{b}} = \dfrac{a^{1/n}}{b^{1/n}} = \left(\dfrac{a}{b}\right)^{1/n} = \sqrt[n]{\dfrac{a}{b}}$

(10.14): $\sqrt[m]{\sqrt[n]{a}} = (a^{1/n})^{1/m} = a^{1/mn} = \sqrt[mn]{a}$

Removing Perfect nth-Root Factors

In Chapter 1 we used law (10.12) to simplify certain radicals. Let's recall the procedure.

E X A M P L E **1**

To simplify $\sqrt{50}$, first note that $50 = 25 \cdot 2$ and $\sqrt{25} = 5$. So $\sqrt{50} = \sqrt{25 \cdot 2} = \sqrt{25}\sqrt{2} = 5\sqrt{2}$.
   Similarly,

$$\sqrt[5]{64} = \sqrt[5]{32 \cdot 2} = \sqrt[5]{32}\sqrt[5]{2} = 2\sqrt[5]{2}$$

◄

> To simplify a radical of the nth order, remove all perfect nth-root factors from the radical.

A factor of the form $\sqrt[n]{a^m}$ is a perfect root if m is divisible by n. For example,

$$\sqrt[7]{a^{21}} = (a^{21})^{1/7} = a^{21/7} = a^3$$

So to simplify $\sqrt[7]{a^{25}}$, we remove the factor a^{21} as follows:

$$\sqrt[7]{a^{25}} = \sqrt[7]{a^{21}a^4} = \sqrt[7]{a^{21}}\sqrt[7]{a^4} = a^3\sqrt[7]{a^4}$$

E X A M P L E **2**

Simplify:

a. $\sqrt{81x^5y^3}$ b. $\sqrt[4]{32a^5b^{10}}$ c. $\sqrt[6]{128a^{11}b^{20}}$

Solution.

a. $\sqrt{81x^5y^3} = \sqrt{9^2x^4xy^2y} = \sqrt{(9^2x^4y^2)(xy)} = \sqrt{9^2x^4y^2}\sqrt{xy} = 9x^2y\sqrt{xy}$

b. First observe that $\sqrt[4]{a^4} = a$ and $\sqrt[4]{b^8} = (b^8)^{1/4} = b^2$. Hence

$$\sqrt[4]{32a^5b^{10}} = \sqrt[4]{16 \cdot 2a^4ab^8b^2} = \sqrt[4]{16a^4b^8}\sqrt[4]{2ab^2} = 2ab^2\sqrt[4]{2ab^2}$$

c. Here we use the fact that $\sqrt[6]{a^6} = a$ and $\sqrt[6]{b^{18}} = (b^{18})^{1/6} = b^3$. Thus

$$\sqrt[6]{128a^{11}b^{20}} = \sqrt[6]{64 \cdot 2a^6a^5b^{18}b^2} = \sqrt[6]{64a^6b^{18}}\sqrt[6]{2a^5b^2} = 2ab^3\sqrt[6]{2a^5b^2}$$

◀

Rationalizing the Denominator

Now recall from Chapter 1 that $1/\sqrt{a}$ can be written

$$\frac{1}{\sqrt{a}} = \frac{1}{\sqrt{a}} \cdot \frac{\sqrt{a}}{\sqrt{a}} = \frac{\sqrt{a}}{a}$$

Rationalizing the denominator   This operation is called **rationalizing the denominator**. To generalize this operation to radicals of any order, it is best to change the radicand to an equivalent fraction whose denominator is a perfect square, as shown in the following example.

E X A M P L E **3**   Rationalize the denominators:

    a. $\sqrt{\dfrac{3}{2}}$ **b.** $\sqrt{\dfrac{2a}{b}}$

Solution.

  a. To simplify $\sqrt{3/2}$, we multiply the numerator and the denominator inside the radical sign by **2**:

$$\sqrt{\frac{3}{2}} = \sqrt{\frac{3 \cdot 2}{2 \cdot 2}} = \sqrt{\frac{6}{2^2}} = \frac{\sqrt{6}}{\sqrt{2^2}} = \frac{\sqrt{6}}{2}$$

  b. Here we multiply the numerator and denominator inside the radical by **b**:

$$\sqrt{\frac{2a}{b}} = \sqrt{\frac{2ab}{bb}} = \sqrt{\frac{2ab}{b^2}} = \frac{\sqrt{2ab}}{\sqrt{b^2}} = \frac{\sqrt{2ab}}{b}$$

◀

    The method of Example 3 works with radicals of any order: We change the radicand to an equivalent fraction whose denominator is a perfect nth root.

E X A M P L E **4**   Rationalize the denominators:

    a. $\sqrt[3]{\dfrac{3}{2}}$ **b.** $\sqrt[3]{\dfrac{2a}{b}}$ **c.** $\sqrt[4]{\dfrac{x}{y^6}}$

Solution.

a. We multiply the numerator and denominator inside the radical by 2^2:

$$\sqrt[3]{\frac{3}{2}} = \sqrt[3]{\frac{3 \cdot 2^2}{2 \cdot 2^2}} = \frac{\sqrt[3]{3 \cdot 2^2}}{\sqrt[3]{2^3}} \qquad \text{by rule (10.13)}$$

$$= \frac{\sqrt[3]{12}}{2} \qquad \text{by rule (10.11)}$$

b.
$$\sqrt[3]{\frac{2a}{b}} = \sqrt[3]{\frac{2ab^2}{bb^2}} = \frac{\sqrt[3]{2ab^2}}{\sqrt[3]{b^3}} \qquad \text{by rule (10.13)}$$

$$= \frac{\sqrt[3]{2ab^2}}{b} \qquad \text{by rule (10.11)}$$

c.
$$\sqrt[4]{\frac{x}{y^6}} = \sqrt[4]{\frac{xy^2}{y^6y^2}} = \frac{\sqrt[4]{xy^2}}{\sqrt[4]{y^8}} = \frac{\sqrt[4]{xy^2}}{y^2} \qquad \blacktriangleleft$$

Radicals containing negative exponents

If the radical contains **negative exponents**, we use rule (10.8) to rewrite the expression so it contains only positive exponents. We then rationalize the denominator and remove all perfect nth roots from the radical in the numerator.

E X A M P L E **5**   Simplify:

$$\sqrt{\frac{3^{-2}u^{-1}v^{-4}w^3}{27u^0v^{-1}}}$$

Solution.

$$\sqrt{\frac{3^{-2}u^{-1}v^{-4}w^3}{27u^0v^{-1}}} = \sqrt{\frac{3^{-2}u^{-1}v^{-4}w^3}{3^3v^{-1}}} \qquad 27 = 3^3;\ u^0 = 1$$

$$= \sqrt{\frac{w^3}{3^5uv^3}} \qquad \text{eliminating negative exponents}$$

$$= \sqrt{\frac{w^3(3uv)}{3^5uv^3(3uv)}} \qquad \text{rationalizing the denominator}$$

$$= \sqrt{\frac{3uvw^3}{3^6u^2v^4}} \qquad \text{adding exponents}$$

$$= \frac{\sqrt{3uvw^3}}{\sqrt{3^6u^2v^4}} \qquad \sqrt{\frac{x}{y}} = \frac{\sqrt{x}}{\sqrt{y}}$$

$$= \frac{w\sqrt{3uvw}}{27uv^2} \qquad \text{removing perfect squares} \qquad \blacktriangleleft$$

E X A M P L E **6**   Simplify:

$$\sqrt[4]{\frac{3^{-1}R^{-1}C}{8C^3}}$$

Solution.

$$\sqrt[4]{\frac{3^{-1}R^{-1}C}{8C^3}} = \sqrt[4]{\frac{3^{-1}R^{-1}C}{2^3C^3}} \qquad 8 = 2^3$$

$$= \sqrt[4]{\frac{C}{2^3 \cdot 3RC^3}} \qquad \text{eliminating negative exponents}$$

$$= \sqrt[4]{\frac{C(2 \cdot 3^3R^3C)}{2^3 \cdot 3RC^3(2 \cdot 3^3R^3C)}} \qquad \text{rationalizing the denominator}$$

$$= \frac{\sqrt[4]{54R^3C^2}}{\sqrt[4]{2^4 \cdot 3^4R^4C^4}} \qquad \sqrt[4]{\frac{x}{y}} = \frac{\sqrt[4]{x}}{\sqrt[4]{y}}$$

$$= \frac{\sqrt[4]{54R^3C^2}}{6RC} \qquad \sqrt[4]{a^4} = a \qquad \blacktriangleleft$$

Reducing the Order of the Radical

It is sometimes possible to **reduce the order of a radical** by means of fractional exponents: Write the radicand as a single positive power (if possible), write the entire radical as a fractional power, multiply the exponents, and reduce the resulting fractional exponent (if possible). The result is a radical with a smaller index. Consider the next example.

E X A M P L E **7**   **a.** $\sqrt[4]{9} = \sqrt[4]{3^2} = (3^2)^{1/4} = 3^{2/4} = 3^{1/2} = \sqrt{3}$
b. $\sqrt[6]{16} = \sqrt[6]{2^4} = (2^4)^{1/6} = 2^{4/6} = 2^{2/3} = \sqrt[3]{2^2} = \sqrt[3]{4}$
c. $\sqrt[10]{x^6} = (x^6)^{1/10} = x^{6/10} = x^{3/5} = \sqrt[5]{x^3}$
d. $\sqrt[15]{x^3y^9} = \sqrt[15]{(xy^3)^3} = [(xy^3)^3]^{1/15} = (xy^3)^{1/5} = \sqrt[5]{xy^3}$ $\blacktriangleleft$

E X A M P L E **8**   Reduce the order of the radical:

$$\sqrt[6]{x^4y^8z^{16}}$$

Solution. The reduction is best accomplished by means of fractional exponents:

$$\sqrt[6]{x^4y^8z^{16}} = (x^4y^8z^{16})^{1/6} = [(xy^2z^4)^4]^{1/6} = (xy^2z^4)^{4/6}$$
$$= (xy^2z^4)^{2/3} = \sqrt[3]{(xy^2z^4)^2} = \sqrt[3]{x^2y^4z^8}$$

or

$$\sqrt[6]{x^4y^8z^{16}} = (x^4y^8z^{16})^{1/6} = x^{4/6}y^{8/6}z^{16/6} = x^{2/3}y^{4/3}z^{8/3}$$
$$= (x^2y^4z^8)^{1/3} = \sqrt[3]{x^2y^4z^8}$$ ◄

Summary

We will now summarize the procedures for simplifying radicals that have been considered so far.

A radical is in simplest form if the following conditions are met:

1. No negative exponents appear in the radical.
2. No power in the radical exceeds the order of the radical.
3. No radical appears in the denominator.
4. The power of the radicand and the index of the radical have no common factors.

If any of the conditions are not met, proceed as follows:

To simplify a radical:

1. Write the radicand without negative exponents.
2. Remove all perfect nth-root factors from the radical of order n.
3. Rationalize the denominator.
4. Reduce the order of the radical.

Addition and Subtraction of Radicals

In adding and subtracting radicals, recall that *only similar terms can be combined.*

To add or subtract radicals, we express each radical in its simplest form and combine similar terms.

For example,

$$9\sqrt{2} + 5\sqrt{2} = (9 + 5)\sqrt{2} = 14\sqrt{2}$$

On the other hand, $9\sqrt{2} + 5\sqrt[3]{2}$ cannot be combined since the radicals have different orders.

E X A M P L E **9**   Combine the following radicals:

$$\sqrt[3]{54} + \sqrt{25y} - \sqrt[3]{16} - \sqrt{36y}$$

Solution. The radicals can be written as follows:

$$\sqrt[3]{27 \cdot 2} + \sqrt{25y} - \sqrt[3]{8 \cdot 2} - \sqrt{36y}$$

Removing perfect nth roots from each radical, we get

$$3\sqrt[3]{2} + 5\sqrt{y} - 2\sqrt[3]{2} - 6\sqrt{y}$$

Combining similar terms, we have

$$(3\sqrt[3]{2} - 2\sqrt[3]{2}) + (5\sqrt{y} - 6\sqrt{y}) = \sqrt[3]{2} - \sqrt{y}$$ ◀

E X A M P L E **10**   Combine the following radicals:

$$\sqrt{8s^3t} - \sqrt{2s^5t^7} + \sqrt{32s^3t^5}$$

Solution.

$$\begin{aligned}\sqrt{8s^3t} - \sqrt{2s^5t^7} + \sqrt{32s^3t^5} &= \sqrt{2 \cdot 4s^2st} - \sqrt{2s^4st^6t} + \sqrt{2 \cdot 16s^2st^4t} \\ &= \sqrt{4s^2 \cdot 2st} - \sqrt{s^4t^6 \cdot 2st} + \sqrt{16s^2t^4 \cdot 2st} \\ &= 2s\sqrt{2st} - s^2t^3\sqrt{2st} + 4st^2\sqrt{2st} \\ &= (2s - s^2t^3 + 4st^2)\sqrt{2st}\end{aligned}$$ ◀

Common error

Forgetting that

$$\sqrt{a} + \sqrt{b} \neq \sqrt{a + b}$$

$\sqrt{a} + \sqrt{b}$ is already in simplest form. For example,

$$\sqrt{9} + \sqrt{16} \neq \sqrt{9 + 16}$$

Also,

$$\sqrt{x^2 + y^2} \neq x + y$$

For example, $\sqrt{3^2 + 4^2} \neq 3 + 4$. Remember that $\sqrt{x^2 + y^2}$ is already in simplest form and should not be written as $x + y$.

CALCULATOR COMMENT

Radicals can be evaluated with a calculator by writing the radical as a fractional exponent.

E X A M P L E **11**   The tensile strength (in pounds) of a certain wire varies with the temperature according to the formula

$$S = 875.0 - 0.6700\sqrt[5]{T}$$

Find S when $T = 325.0°F$.

Solution. Writing the radical as a fractional exponent and substituting $T = 325.0$, we get

$$S = 875.0 - 0.6700T^{1/5} = 875.0 - 0.6700(325.0)^{1/5}$$

A correct sequence is

$$875.0 \boxed{-} 0.6700 \boxed{\times} 325.0 \boxed{y^x} 0.2 \boxed{=} \rightarrow 872.86965$$

So $S = 872.9$ lb   ◀

E X E R C I S E S / S E C T I O N   **10.4**

In Exercises 1–4, write each expression as a single radical. [See formula (10.14).]

1. $\sqrt{\sqrt{x}}$ **2.** $\sqrt{\sqrt[3]{a}}$ **3.** $\sqrt[3]{\sqrt[4]{b}}$ **4.** $\sqrt[5]{\sqrt{c}}$

In Exercises 5–12, reduce the order of the given radicals. (See Examples 7 and 8.)

5. $\sqrt[4]{a^2b^2c^6}$ **6.** $\sqrt[6]{s^6t^9u^{12}}$ **7.** $\sqrt[8]{u^4v^6w^8}$

8. $\sqrt[12]{f^3g^6h^9}$ **9.** $\sqrt[9]{v^6m^6g^9}$ **10.** $\sqrt[6]{8x^3y^6z^9}$

11. $\sqrt[8]{25p^4q^8r^{10}}$ **12.** $\sqrt[15]{27x^9y^{12}v^{15}}$

In Exercises 13–20, simplify the given radicals. (See Examples 1 and 2.)

13. $\sqrt{4x^3}$ **14.** $\sqrt{8x^3y^5}$

15. $\sqrt[3]{54a^4c^8}$ **16.** $\sqrt[4]{48V^5W^{10}}$

17. $\sqrt[4]{32u^{13}v^{18}}$ **18.** $\sqrt[5]{64L^7R^{13}C^{18}}$

19. $\sqrt{18m^2n^3s^7t^{11}}$ **20.** $\sqrt{24v^8x^3y^4z^7}$

In Exercises 21–40, rationalize the denominator in each expression. (See Examples 3 and 4.)

21. $\sqrt{\dfrac{1}{5}}$ **22.** $\sqrt{\dfrac{2}{7}}$ **23.** $\sqrt{\dfrac{3}{x}}$ **24.** $\sqrt{\dfrac{4}{b}}$

25. $\sqrt{\dfrac{a}{bc}}$ **26.** $\sqrt{\dfrac{k}{mn}}$ **27.** $\sqrt{\dfrac{v}{3pq}}$ **28.** $\sqrt{\dfrac{s}{5uv}}$

29. $\sqrt[3]{\dfrac{2}{3}}$ **30.** $\sqrt[4]{\dfrac{1}{5}}$ **31.** $\sqrt[3]{\dfrac{v}{4\pi}}$ **32.** $\sqrt[3]{\dfrac{B}{9A}}$

33. $\sqrt[4]{\dfrac{3}{8L^5V^2}}$ **34.** $\sqrt[3]{\dfrac{3}{2ab}}$ **35.** $\sqrt{\dfrac{7}{3x^3y^5}}$ **36.** $\sqrt[5]{\dfrac{1}{6}}$

37. $\sqrt[3]{\dfrac{aR}{9C}}$ **38.** $\sqrt[4]{\dfrac{a^2c}{8b^2}}$ **39.** $\dfrac{1}{\sqrt{2pqr^2}}$ **40.** $\sqrt[3]{\dfrac{2q}{9c^2d}}$

In Exercises 41–56, simplify the given expressions. Write without zero or negative exponents and rationalize denominators. (See Examples 5 and 6.)

41. $\sqrt{\dfrac{4a^4b^{-2}}{3a^0b}}$ **42.** $\sqrt{\dfrac{2^{-1}c^{-4}d^2}{2c^{-2}d^{-1}}}$ **43.** $\sqrt{\dfrac{w^{-4}z^{-1}}{9^{-1}w}}$

44. $\sqrt{\dfrac{3^2T^{-1}W^{-2}}{9^{-2}T^2W}}$ **45.** $\sqrt{\dfrac{2^{-3}t^0xz}{4^{-3}x^{-1}z^4}}$ **46.** $\sqrt{\dfrac{3V^0R^{-2}C}{9^2R^3C^5}}$

47. $\sqrt{\dfrac{3^{-1}u^{-2}v^2w^{-2}}{uv^{-1}w^5}}$ **48.** $\sqrt{\dfrac{5^{-3}aL^{-6}}{2a^{-4}c^0L}}$

49. $\sqrt[3]{\dfrac{2^{-1}a^{-1}}{a^2c}}$ **50.** $\sqrt[3]{\dfrac{16^{-1}cx^{-1}}{c^3x^4}}$

51. $\sqrt[3]{\dfrac{3u^{-4}v^{-1}}{4u^{-2}}}$ **52.** $\sqrt[3]{\dfrac{st^{-1}z^{-2}}{9s^6t}}$

53. $\sqrt[4]{\dfrac{x^{-1}y^{-2}}{8x^2y^5z^{-1}}}$ **54.** $\sqrt[5]{\dfrac{2^{-6}A^{-2}B^{-6}}{4AB^{-1}}}$

55. $\sqrt{\dfrac{7^{-2}s^2t^{-3}u^{-7}}{7s^5t^2u^{-2}}}$ **56.** $\sqrt[4]{\dfrac{2^{-1}x^{-3}y^2}{y^{-1}}}$

In Exercises 57–68, combine the given radicals. (See Examples 9 and 10.)

57. $\sqrt{24} + \sqrt{54}$ **58.** $\sqrt{12} - \sqrt{27} + \sqrt{48}$

59. $2\sqrt{5} + \sqrt{80} - \sqrt{45}$ **60.** $\sqrt{28} + \sqrt{63} - \sqrt{112}$

61. $\sqrt[3]{24} - \sqrt[3]{81}$

62. $\sqrt{20} + \sqrt{4x} - \sqrt{80} + \sqrt{9x}$

63. $\sqrt{20} - \sqrt[3]{16} + \sqrt{125} - \sqrt[3]{54}$

64. $\sqrt[3]{250} + \sqrt[3]{8x} - \sqrt[3]{54} - \sqrt[3]{27x}$

65. $3b\sqrt{a^3b} - 4a\sqrt{ab^3}$

66. $\sqrt{8uv^4} - \sqrt{2u^3} - \sqrt{2u^3v^2}$

67. $\sqrt{25m^3n^3} - \sqrt{m^5n^3} + \sqrt{4m^5n^5}$

68. $\sqrt{16p^4q^3} + \sqrt{4p^2q^3} + \sqrt{p^8q^5}$

69. The average speed of a molecule of an ideal gas is $\sqrt{8kT/\pi m}$, where m is the mass, T the absolute temperature, and k the Boltzmann constant. Simplify this expression by rationalizing the denominator.

70. According to the special theory of relativity, the mass m of a body moving at velocity v relative to a stationary object is

$$m = \frac{m_0}{\sqrt{1 - \dfrac{v^2}{c^2}}}$$

where m_0 is the mass of the body at rest and c the velocity of light. Simplify this expression.

71. The velocity in meters per second of sound in air at temperature T (in degrees Celsius) is given by

$$v = 331.7\sqrt{1 + \frac{T}{273}}$$

Simplify this formula by combining terms and rationalizing the denominator.

72. Simplify the expression

$$\frac{1}{2L}\sqrt{R^2 - \frac{4L}{C}}$$

which arises in the study of electrical circuits.

73. The flow of water around an obstacle often results in *stagnation points* (where the velocity is zero). For a cylindrical obstacle placed at the origin, the x-coordinates of the stagnation points are

$$x = \pm\sqrt{1 - \frac{k^2}{16\pi^2}}$$

where k depends on the nature of the unimpeded flow. Simplify the expression for x.

74. A manufacturing process produces $100p\%$ defectives on the average. The process is *out of control* if a sample of size n contains more than

$$3\sqrt{\frac{p(1 - p)}{n}}$$

defectives above the average. Simplify this expression by rationalizing the denominator.

75. By rationalizing the denominator, simplify the formula

$$t = C\sqrt[3]{\frac{L}{mi^2}}$$

which arises in the study of soil mechanics.

76. Evaluate $\sqrt[8]{(2.71)^4 + 3.12}$. (See Example 11.)

77. Evaluate $\sqrt[9]{(3.84)^6 + (2.56)^3}$.

78. Evaluate $\sqrt[5]{4.96} + \sqrt[4]{0.831}$.

79. Evaluate $\sqrt[4]{10.00} - \sqrt[3]{\pi}$.

10.5 Multiplication of Radicals

In the last section we briefly considered the addition and subtraction of radicals. We still need to discuss multiplication and division. Since these operations are somewhat more involved, we will discuss them in separate sections.

Formula (10.12), $\sqrt[n]{a}\,\sqrt[n]{b} = \sqrt[n]{ab}$, was used in the last section to simplify certain radicals. The same formula can be used to multiply single radicals, as shown in the next example.

E X A M P L E 1

a. $\sqrt{2x}\sqrt{3x} = \sqrt{(2x)(3x)} = \sqrt{6x^2} = x\sqrt{6}$

b. $\sqrt[3]{6a^2}\sqrt[3]{4a^2} = \sqrt[3]{(6a^2)(4a^2)} = \sqrt[3]{24a^4} = \sqrt[3]{3\cdot 8a^3a} = \sqrt[3]{(8a^3)(3a)} = 2a\sqrt[3]{3a}$

c. $\sqrt[4]{4u^3v^2w}\sqrt[4]{8uv^3w^2} = \sqrt[4]{32u^4v^5w^3} = \sqrt[4]{2\cdot 16u^4v^4vw^3}$
$= \sqrt[4]{(16u^4v^4)(2vw^3)} = 2uv\sqrt[4]{2vw^3}$ ◀

The next two examples illustrate the multiplication of multinomials containing radical expressions.

E X A M P L E **2**   Multiply: $(\sqrt{3} - 2)(\sqrt{2} + \sqrt{6})$.

Solution. By the FOIL scheme

$$
\begin{aligned}
(\sqrt{3} - 2)(\sqrt{2} + \sqrt{6}) &= \sqrt{3}\sqrt{2} + \sqrt{3}\sqrt{6} - 2\sqrt{2} - 2\sqrt{6} \\
&= \sqrt{6} + \sqrt{18} - 2\sqrt{2} - 2\sqrt{6} \\
&= \sqrt{6} + 3\sqrt{2} - 2\sqrt{2} - 2\sqrt{6} \\
&= (3\sqrt{2} - 2\sqrt{2}) + (\sqrt{6} - 2\sqrt{6}) \\
&= \sqrt{2} - \sqrt{6} \qquad 3x - 2x = x;\ x - 2x = -x
\end{aligned}
$$

◄

E X A M P L E **3**   Show that $a - b = (\sqrt{a} - \sqrt{b})(\sqrt{a} + \sqrt{b})$.

Solution. $(\sqrt{a} - \sqrt{b})(\sqrt{a} + \sqrt{b}) = (\sqrt{a})^2 - (\sqrt{b})^2 = a - b$

Note that this product shows that $a - b$ can be factored as a difference of two squares.   ◄

If radicals of different orders are to be multiplied and combined into a single radical, then the order of one or both radicals must be changed. This change can be accomplished by using fractional exponents, as shown in the next example.

E X A M P L E **4**   Multiply:

a. $\sqrt[3]{3}\sqrt[5]{3}$ **b.** $\sqrt[3]{a}\sqrt[4]{b}$

Solution.

a. $\sqrt[3]{3}\sqrt[5]{3} = 3^{1/3}3^{1/5} = 3^{5/15}3^{3/15} = 3^{(5/15 + 3/15)} = 3^{8/15} = (3^8)^{1/15} = \sqrt[15]{3^8}$
$= \sqrt[15]{6{,}561}$

b. $\sqrt[3]{a}\sqrt[4]{b} = a^{1/3}b^{1/4} = a^{4/12}b^{3/12} = (a^4b^3)^{1/12} = \sqrt[12]{a^4b^3}$ ◄

 E X A M P L E **5**   A topic in calculus is finding the center of mass of a plate. Suppose a plate of uniform density k (weight per unit area) is bounded by the graphs of $y = \sqrt{2x}$ and $y = \sqrt{x}$ and by a vertical line. In determining the y-coordinate of its center of mass, the following expression arises:

$$
k\frac{\sqrt{2x} + \sqrt{x}}{2}(\sqrt{2x} - \sqrt{x})
$$

Simplify this expression.

Solution. The expression can be written

$$\frac{k}{2}(\sqrt{2x} + \sqrt{x})(\sqrt{2x} - \sqrt{x}) = \frac{k}{2}[(\sqrt{2x})^2 - (\sqrt{x})^2]$$

$$= \frac{k}{2}(2x - x) = \frac{1}{2}kx \qquad \blacktriangleleft$$

EXERCISES / SECTION 10.5

In Exercises 1–12, perform the multiplications and simplify. (See Example 1.)

1. $\sqrt{2y}\sqrt{6y}$ **2.** $\sqrt{16x}\sqrt{2y}$

3. $\sqrt{3ab}\sqrt{2ab}$ **4.** $\sqrt{5uv}\sqrt{6u}$

5. $\sqrt[3]{2x^2y}\sqrt[3]{4xy}$ **6.** $\sqrt[4]{4c^3d}\sqrt[4]{4c^2d^5}$

7. $\sqrt[5]{8m^3c^2}\sqrt[5]{8m^2c^4}$ **8.** $\sqrt[3]{4R^4C^3}\sqrt[3]{8RC^4}$

9. $\sqrt{2st}\sqrt{2s}\sqrt{5t}$ **10.** $\sqrt[3]{18a^2b}\sqrt[3]{3a^2b}\sqrt[3]{ab^2}$

11. $\sqrt[4]{9\pi^3\omega^2}\sqrt[4]{3\pi\omega^2}\sqrt[4]{3\pi^3\omega}$ **12.** $\sqrt[3]{4e^2f^2}\sqrt[3]{4ef^2}\sqrt[3]{ef}$

In Exercises 13–20, multiply the given radicals. (See Example 4.)

13. $\sqrt{3}\sqrt[3]{3}$ **14.** $\sqrt[3]{x}\sqrt[4]{x}$ **15.** $\sqrt{x}\sqrt[4]{y}$ **16.** $\sqrt[3]{a}\sqrt[5]{b}$

17. $\sqrt{c}\sqrt[5]{d}$ **18.** $\sqrt{m}\sqrt[6]{n}$ **19.** $\sqrt[4]{\pi}\sqrt[6]{e}$ **20.** $\sqrt[6]{f}\sqrt[8]{g}$

In Exercises 21–46, multiply the given multinomials and simplify. (See Examples 2 and 3.)

21. $(\sqrt{3} + \sqrt{7})(\sqrt{3} - \sqrt{7})$ **22.** $(\sqrt{5} - \sqrt{3})(\sqrt{5} + \sqrt{3})$

23. $(\sqrt{7} - \sqrt{2})(\sqrt{7} + \sqrt{2})$ **24.** $(\sqrt{11} - \sqrt{5})(\sqrt{11} + \sqrt{5})$

25. $(\sqrt{2} + \sqrt{3})^2$ **26.** $(1 - \sqrt{7})^2$

27. $(2 + \sqrt{13})^2$ **28.** $(\sqrt{5} - \sqrt{2})^2$

29. $(2\sqrt{3} - 4\sqrt{5})(2\sqrt{3} + 4\sqrt{5})$

30. $(3\sqrt{5} - \sqrt{11})(3\sqrt{5} + \sqrt{11})$

31. $(2\sqrt{13} - 2\sqrt{2})(2\sqrt{13} + 2\sqrt{2})$

32. $(\sqrt{17} - 2\sqrt{3})(\sqrt{17} + 2\sqrt{3})$

33. $(x - \sqrt{y})(x + \sqrt{y})$

34. $(\sqrt{a} - a)(\sqrt{a} + a)$

35. $(1 - \sqrt{k})(1 + \sqrt{k})$

36. $(2 - 2\sqrt{b})(2 + 2\sqrt{b})$

37. $(2 + \sqrt{q})^2$

38. $(3 - \sqrt{L})^2$

39. $(\sqrt{a} + 2\sqrt{b})(2\sqrt{a} - 3\sqrt{b})$

40. $(\sqrt{s} - 3\sqrt{t})(3\sqrt{s} - 4\sqrt{t})$

41. $(2\sqrt{5} - 3\sqrt{7})(\sqrt{5} + 4\sqrt{7})$

42. $(3\sqrt{6} + 2\sqrt{5})(2\sqrt{6} - 4\sqrt{5})$

43. $(\sqrt{a} + \sqrt{b})(a - \sqrt{ab} + b)$

44. $(\sqrt{x} - \sqrt{y})(x - \sqrt{xy} - y)$

45. $(\sqrt{a} - \sqrt{b})(a + \sqrt{ab} - b)$

46. $(\sqrt{a} + \sqrt{b} - \sqrt{ab})(\sqrt{a} - \sqrt{b} + \sqrt{ab})$

47. One root of the quadratic equation $x^2 + 2x - 2 = 0$ is $x = -1 + \sqrt{3}$, obtained by using the quadratic formula. Check this root by substituting the value of x into the equation.

48. Repeat Exercise 47 for the other root, $x = -1 - \sqrt{3}$.

49. The roots of the equation $x^2 - 4x + 2 = 0$ are $x = 2 \pm \sqrt{2}$, obtained by using the quadratic formula. Check the roots by substituting the values of x into the equation.

10.6 Division of Radicals

In this section we take up division involving radical expressions. We will consider two basic types, division by a single radical and division by a multinomial. Recall that

$$\frac{\sqrt[n]{a}}{\sqrt[n]{b}} = \sqrt[n]{\frac{a}{b}} \qquad (10.15)$$

Thus, to divide two radicals of the same index, combine the two radicals and rationalize the denominator.

E X A M P L E **1**

a. $\dfrac{\sqrt{3ab}}{\sqrt{2b}} = \sqrt{\dfrac{3ab}{2b}} = \sqrt{\dfrac{3a}{2}} = \sqrt{\dfrac{3a(2)}{2(2)}} = \dfrac{\sqrt{6a}}{2}$

b. $\dfrac{\sqrt[4]{5\pi^3}}{\sqrt[4]{8r}} = \sqrt[4]{\dfrac{5\pi^3}{8r}} = \sqrt[4]{\dfrac{5\pi^3(2r^3)}{2^3r(2r^3)}} = \dfrac{\sqrt[4]{10\pi^3 r^3}}{2r}$ ◀

Dividing by a multinomial, such as $2 \div (\sqrt{3} - 4)$, is essentially a problem in rationalizing the denominator. Suppose we multiply the numerator and denominator of the fraction $2/(\sqrt{3} - 4)$ by $\sqrt{3} + 4$:

$$\frac{2}{\sqrt{3} - 4} = \frac{2}{\sqrt{3} - 4} \cdot \frac{\sqrt{3} + 4}{\sqrt{3} + 4} = \frac{2(\sqrt{3} + 4)}{3 - 16} = -\frac{2}{13}(\sqrt{3} + 4)$$

The expression $\sqrt{3} + 4$ is called the **conjugate** of $\sqrt{3} - 4$.

This procedure suggests the following definition.

Expressions of the form

$$a + b \qquad \text{and} \qquad a - b \qquad\qquad (10.16)$$

are called **conjugates**.

This definition can be used in stating the rule for rationalizing binomial denominators containing square roots.

To **rationalize** a fraction in which the denominator contains a sum or difference of two terms, at least one of which is a square root, multiply the numerator and denominator by the conjugate of the denominator.

This technique is illustrated in the remaining examples.

E X A M P L E **2**   Simplify the fraction:

$$\frac{1}{\sqrt{2} + 1}$$

Solution. We multiply the numerator and denominator by $\sqrt{2} - 1$, the conjugate of $\sqrt{2} + 1$:

$$\frac{1}{\sqrt{2} + 1} = \frac{1}{\sqrt{2} + 1} \cdot \frac{\sqrt{2} - 1}{\sqrt{2} - 1} = \frac{\sqrt{2} - 1}{(\sqrt{2})^2 - 1^2} = \frac{\sqrt{2} - 1}{2 - 1} = \sqrt{2} - 1$$ ◀

E   X   A   M   P   L   E   **3**   Simplify the fraction:

$$\frac{1}{\sqrt{x} - 2\sqrt{y}}$$

Solution. The conjugate of $\sqrt{x} - 2\sqrt{y}$ is $\sqrt{x} + 2\sqrt{y}$

$$\frac{1}{\sqrt{x} - 2\sqrt{y}} = \frac{1}{\sqrt{x} - 2\sqrt{y}} \cdot \frac{\sqrt{x} + 2\sqrt{y}}{\sqrt{x} + 2\sqrt{y}}$$

$$= \frac{\sqrt{x} + 2\sqrt{y}}{(\sqrt{x})^2 - (2\sqrt{y})^2} = \frac{\sqrt{x} + 2\sqrt{y}}{x - 4y}$$ ◄

E   X   A   M   P   L   E   **4**   Perform the following division:

$$(\sqrt{15} - 2) \div (\sqrt{5} + \sqrt{3})$$

Solution. From

$$\frac{\sqrt{15} - 2}{\sqrt{5} + \sqrt{3}}$$

we get

$$\frac{\sqrt{15} - 2}{\sqrt{5} + \sqrt{3}} \cdot \frac{\sqrt{5} - \sqrt{3}}{\sqrt{5} - \sqrt{3}} = \frac{\sqrt{75} - \sqrt{45} - 2\sqrt{5} + 2\sqrt{3}}{(\sqrt{5})^2 - (\sqrt{3})^2}$$

$$= \frac{\sqrt{25 \cdot 3} - \sqrt{9 \cdot 5} - 2\sqrt{5} + 2\sqrt{3}}{5 - 3}$$

$$= \frac{5\sqrt{3} - 3\sqrt{5} - 2\sqrt{5} + 2\sqrt{3}}{2}$$

$$= \frac{7\sqrt{3} - 5\sqrt{5}}{2}$$ ◄

   E   X   A   M   P   L   E   **5**   The time required to drain the contents of a vessel through an opening in the bottom from level L_1 to level L_2 is given by

$$t = \frac{k(L_1 - L_2)}{\sqrt{2g}(\sqrt{L_1} + \sqrt{L_2})}$$

where g is the acceleration due to gravity and k is a constant. Simplify the expression for t.

Solution. Multiplying the numerator and denominator of the fraction by the conjugate of the denominator, we get

$$t = \frac{k(L_1 - L_2)}{\sqrt{2g}(\sqrt{L_1} + \sqrt{L_2})} \cdot \frac{\sqrt{L_1} - \sqrt{L_2}}{\sqrt{L_1} - \sqrt{L_2}}$$

$$= \frac{k(L_1 - L_2)(\sqrt{L_1} - \sqrt{L_2})}{\sqrt{2g}[(\sqrt{L_1})^2 - (\sqrt{L_2})^2]}$$

$$= \frac{k(L_1 - L_2)(\sqrt{L_1} - \sqrt{L_2})}{\sqrt{2g}(L_1 - L_2)} = \frac{k(\sqrt{L_1} - \sqrt{L_2})}{\sqrt{2g}}$$

$$= \frac{k(\sqrt{L_1} - \sqrt{L_2})}{\sqrt{2g}} \cdot \frac{\sqrt{2g}}{\sqrt{2g}} = \frac{k\sqrt{2g}(\sqrt{L_1} - \sqrt{L_2})}{2g}$$ ◀

EXERCISES / SECTION 10.6

In Exercises 1–36, carry out the indicated divisions and simplify.

1. $\dfrac{\sqrt{2a}}{\sqrt{b}}$

2. $\dfrac{\sqrt{x}}{\sqrt{3y}}$

3. $\dfrac{\sqrt{4c}}{\sqrt{d}}$

4. $\dfrac{\sqrt{9ac}}{\sqrt{c}}$

5. $\dfrac{\sqrt{2xy}}{\sqrt{3x}}$

6. $\dfrac{\sqrt[3]{7x^2z^2}}{\sqrt[3]{4xy^2}}$

7. $\dfrac{\sqrt[3]{2v_1^2 v_2}}{\sqrt[3]{9v_1^2 v_3^2}}$

8. $\dfrac{\sqrt[3]{C_1^2 C_2^2}}{\sqrt[3]{6C_1 C_2}}$

9. $\dfrac{\sqrt[4]{3\pi^3 r}}{\sqrt[4]{8\pi^2}}$

10. $\dfrac{\sqrt[5]{m^3 v^2}}{\sqrt[5]{8vw^3}}$

11. $\dfrac{\sqrt{a}}{\sqrt[3]{a}}$

12. $\dfrac{\sqrt[4]{a}}{\sqrt{a}}$

13. $\dfrac{1}{1 - \sqrt{2}}$

14. $\dfrac{2}{\sqrt{3} - 1}$

15. $\dfrac{4}{\sqrt{5} + 2}$

16. $\dfrac{2}{2 + \sqrt{6}}$

17. $\dfrac{\sqrt{2} + 2}{\sqrt{2} + 1}$

18. $\dfrac{3 - \sqrt{5}}{3 + \sqrt{5}}$

19. $\dfrac{1 + \sqrt{2}}{3 - \sqrt{2}}$

20. $\dfrac{1 - \sqrt{5}}{4 + \sqrt{5}}$

21. $\dfrac{\sqrt{3} - \sqrt{2}}{\sqrt{3} + \sqrt{2}}$

22. $\dfrac{\sqrt{5} + 2\sqrt{3}}{\sqrt{5} + \sqrt{3}}$

23. $\dfrac{\sqrt{10} - 2}{\sqrt{5} - \sqrt{2}}$

24. $\dfrac{1 - \sqrt{21}}{\sqrt{3} - \sqrt{7}}$

25. $\dfrac{\sqrt{18} + 3}{\sqrt{6} - \sqrt{3}}$

26. $\dfrac{2\sqrt{3} + 2}{\sqrt{2} - \sqrt{6}}$

27. $\dfrac{2}{1 - \sqrt{a}}$

28. $\dfrac{5}{\sqrt{x} + 2}$

29. $\dfrac{3}{\sqrt{b} - 2}$

30. $\dfrac{1}{\sqrt{a} - \sqrt{b}}$

31. $\dfrac{\sqrt{a} + \sqrt{b}}{\sqrt{a} - \sqrt{b}}$

32. $\dfrac{\sqrt{c} - \sqrt{d}}{\sqrt{c} + \sqrt{d}}$

33. $\dfrac{1}{\sqrt{a} - \sqrt{a - b}}$

34. $\dfrac{1}{\sqrt{R} + \sqrt{R - C}}$

35. $\dfrac{1}{\sqrt[3]{a} - \sqrt[3]{b}}$ [*Hint:* $a - b = (a^{1/3})^3 - (b^{1/3})^3$
$= (a^{1/3} - b^{1/3})(a^{2/3} + a^{1/3}b^{1/3} + b^{2/3})$]

36. $\dfrac{1}{\sqrt[3]{a} + \sqrt[3]{b}}$

37. By rationalizing the numerator, show that the function $f(x) = x - \sqrt{x^2 - 2}$ approaches zero as x gets large.

38. Suppose the position of a particle moving along the x-axis as a function of time is $x(t) = \sqrt{t}$. From calculus, the velocity is

$$v = \lim_{h \to 0} \frac{\sqrt{t + h} - \sqrt{t}}{h}$$

where the symbol $\lim_{h \to 0}$ means "h approaches 0." Show that $v = 1/(2\sqrt{t})$ by multiplying the fraction by

$$\frac{\sqrt{t + h} + \sqrt{t}}{\sqrt{t + h} + \sqrt{t}}$$

and letting h be equal to 0.

39. The *coefficient of reflection R* is defined as the ratio of the amplitude of the reflected wave to the amplitude of the incident wave. If two ropes whose masses per unit length are μ_1 and μ_2, respectively, lie along the *x*-axis, and are joined at the origin, then

$$R = \frac{\sqrt{\mu_1} - \sqrt{\mu_2}}{\sqrt{\mu_1} + \sqrt{\mu_2}}$$

Simplify the expression for *R*.

40. Simplify the following expression from a problem in hydrostatics:

$$\frac{I}{1 - \sqrt{V_\rho}}$$

41. Two heat sources *A* and *B* having intensities *a* and *b*, respectively, are 1 unit apart. From a point *P* between *A* and *B*, the temperature is lowest if the distance from *P* to *A* is

$$\frac{\sqrt[3]{a}}{\sqrt[3]{a} + \sqrt[3]{b}}$$

Simplify this expression. (See Exercise 36.)

42. Simplify the following expression from a problem in mechanics:

$$\frac{\sqrt{v}}{\sqrt{v} - \sqrt{m}}$$

REVIEW EXERCISES/CHAPTER 10

In Exercises 1–6, evaluate the expressions.

1. $5^0 5^{-2}$

2. $(3^{-2})^{-1}$

3. $\dfrac{4^{-3}}{16^{-3}}$

4. $64^{-1/3}$

5. $(0.008)^{1/3}$

6. $(-27)^{-1/3}$

In Exercises 7–30, simplify the given expressions and write the results without zero or negative exponents.

7. $\dfrac{3^{-2}V^3W^{-6}}{2^{-1}V^0W^4}$

8. $\dfrac{4^{-1}\pi^{-6}r^2}{4\pi^{-4}r^{-1}}$

9. $\left(\dfrac{2C_1}{C_2}\right)^{-1}$

10. $(3^{-1}p^{-6}q^{-3}c^{-4})^{-3}$

11. $(2^{-2}v^{-3}w^2z^{-1})^4$

12. $\left(\dfrac{2^{-2}f^{-3}g^4}{3^{-1}f^{-4}g^{-2}}\right)^{-2}$

13. $\left(\dfrac{3^{-2}R^{-4}r^3v^0}{3^{-1}R^2r^{-6}}\right)^2$

14. $\left(\dfrac{2v_1^{-2}v_2^{3}v_3^{4}}{7^{-2}v_1^{-1}v_2^{-1}v_3^{6}}\right)^{-1}$

15. $\left(\dfrac{x^{a+1}}{x^{2a+1}}\right)^{-1}$

16. $\dfrac{V^{-1}+I}{V^{-1}}$

17. $(1 - C^{-1})^{-1}$

18. $\left(\dfrac{a^{-1}b^{-1}}{a^{-1}+b^{-1}}\right)^{-2}$

19. $\dfrac{4^{1/2}v^{3/2}w^{-1/2}}{v^{1/2}}$

20. $(4w^4v^{-1})^{3/2}$

21. $(-5x^{3/2})(-2x^{-3/4})(3x)^0$

22. $\dfrac{2^{-1}A^{1/4}B^{-2/3}}{A^{-1/6}B^{-1/5}}$

23. $\dfrac{3^{-1/3}F_1^{-1/7}F_2^{1/3}}{9^{-1/3}F_1^{4/7}F_2^{1/4}}$

24. $(16a^0b^4c^{-8})^{-1/2}$

25. $\left(\dfrac{m^{12}n^{-9}}{64m^{-3/4}n^{-8}}\right)^{-1/6}$

26. $x(x-3)^{-1/2} + (x-3)^{1/2}$

27. $(x+1)(x-1)^{-2/3} + (x-1)^{1/3}$

28. $x(x+2)^{-3/4} + 2(x+2)^{1/4}$

29. $(x+1)(x-2)^{-4/5} - 4(x-2)^{1/5}$

30. $\left[x^{1/(a+1)}\right]^{(a^2-1)}$

31. Simplify $\sqrt[3]{\sqrt{x}}$.

32. Simplify $\sqrt[3]{\sqrt[3]{b}}$.

In Exercises 33–34, reduce the order of the given radicals.

33. $\sqrt[6]{64a^6b^9c^{15}}$

34. $\sqrt[8]{16x^4y^6z^{12}}$

In Exercises 35–38, simplify the given radicals.

35. $\sqrt{12R^3r^7}$

36. $\sqrt[3]{24\pi^{10}q^8}$

37. $\sqrt[4]{16a^5b^{11}}$

38. $\sqrt[4]{32x^6y^8z^{13}}$

In Exercises 39–42, combine the radicals.

39. $\sqrt{8} + \sqrt{18} - \sqrt{50}$

40. $\sqrt{20} - \sqrt{80} + \sqrt{180}$

41. $\sqrt{45} - \sqrt[3]{24} + \sqrt{125} - \sqrt[3]{81}$

42. $\sqrt{36x^3y^3} - \sqrt{x^5y^5} + \sqrt{16xy^3}$

In Exercises 43–52, simplify each expression. (Write without zero or negative exponents and rationalize the denominator.)

43. $\sqrt{\dfrac{4a^{-4}b^{-3}}{5a^{-2}b}}$

44. $\sqrt{\dfrac{3^{-1}R^{-6}r^3}{3R^{-3}r^{-1}}}$

45. $\sqrt{\dfrac{3^{-1}s^{-1}tc^0}{2^{-1}t^2}}$

46. $\sqrt{\dfrac{3ab^2c^0}{5a^{-2}b}}$

47. $\sqrt[3]{\dfrac{3^{-1}R^{-1}}{R^2s}}$

48. $\sqrt[3]{\dfrac{2x^{-2}y^{-1}}{x^{-4}}}$

49. $\sqrt[3]{\dfrac{ab^{-1}}{9a^5b}}$

50. $\sqrt[4]{\dfrac{3u^{-1}v^{-1}}{8}}$

51. $\sqrt[5]{\dfrac{3^{-4}\pi\omega^{-2}}{\pi^4}}$ **52.** $\sqrt[4]{\dfrac{3^{-1}a^{-3}b^{-2}}{b}}$

In Exercises 53–60, carry out the multiplications and simplify.

53. $\sqrt[3]{3xy}\,\sqrt[3]{9x^2y}$ **54.** $\sqrt[5]{4R^2V^2}\,\sqrt[5]{8R^3V}$

55. $\sqrt[3]{a}\,\sqrt[4]{a}$ **56.** $\sqrt[4]{a}\,\sqrt[3]{b}$

57. $(\sqrt{6}-\sqrt{2})(\sqrt{6}+\sqrt{2})$ **58.** $(\sqrt{5}-\sqrt{3})^2$

59. $(1-3\sqrt{x})(1+3\sqrt{x})$ **60.** $(\sqrt{a}-2\sqrt{b})(3\sqrt{a}+5\sqrt{b})$

In Exercises 61–66, perform each indicated division by rationalizing the denominator.

61. $\dfrac{4}{3-\sqrt{6}}$ **62.** $\dfrac{4}{3+\sqrt{5}}$

63. $\dfrac{1-\sqrt{10}}{\sqrt{2}-\sqrt{5}}$ **64.** $\dfrac{1}{\sqrt{R}+\sqrt{C}}$

65. $\dfrac{1}{\sqrt{\pi-1}+\sqrt{\pi}}$ **66.** $\dfrac{1}{\sqrt{x-2}-\sqrt{x}}$

67. In the mathematics of finance, the *present value* P is the amount that must be invested at interest rate i to yield S dollars at the end of n years. If the interest is compounded annually, then

$$P = S(1+i)^{-n}$$

How much money must be invested at present to yield $10,000 after 5 years at 8% interest compounded annually?

68. The volume of a sphere is given by $V = \frac{4}{3}\pi r^3$. Solve for r and express the result in the simplest radical form.

69. A Borda's pendulum consists of a spherical bob of radius r and a long, thin wire of negligible weight (Figure 10.3). If the length of the wire from the point of suspension to the center of the bob is L, then the period P (in seconds) for one cycle of the pendulum for small oscillations is

$$P = 2\pi\sqrt{\dfrac{L}{g}\left(1+\dfrac{2r^2}{5L^2}\right)}$$

where g is the acceleration due to gravity. Simplify this expression by combining the terms and rationalizing the denominator.

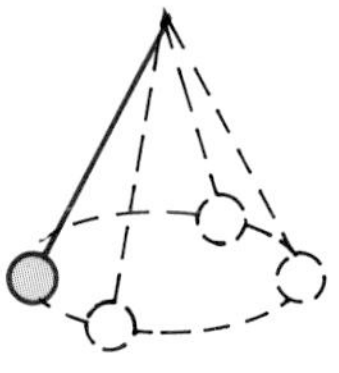

Figure 10.3

70. In an adiabatic expansion of a gas (no gain or loss of heat) the temperature and volume are related by the formula

$$\dfrac{T_1}{T_2} = \left(\dfrac{V_1}{V_2}\right)^{1-a}$$

Show that this formula can be written

$$\dfrac{T_1}{T_2} = \left(\dfrac{V_2}{V_1}\right)^{a-1}$$

C H A P T E R **11**

Complex Numbers

11.1 Basic Concepts and Definitions

The Basic Imaginary Unit

When complex numbers arose in our discussion of quadratic equations in Chapter 6, we concentrated mostly on notation. In this chapter we will study the meaning and properties of complex numbers in greater detail. A knowledge of these properties is essential in more advanced technical applications. (One such application is discussed in Section 11.6.)

Let us first recall that the square root of a negative number is called an *imaginary number*. This term, introduced by Descartes, suggests that such numbers cannot exist. Indeed, we cannot find a *real* number a such that $a = \sqrt{-4}$. (If $a = \sqrt{-4}$, then $a^2 = -4$, but the square of a real number cannot be negative.) However, square roots of negative numbers can be extremely useful in certain applications. To realize this advantage, we have to *extend* the real number system in such a way that square roots of negative numbers are included. This can be done by introducing a number that is *not a real number*. This number, denoted by j, has the property

$$j^2 = -1$$

and is called the **basic imaginary unit**.

Basic Imaginary Unit:

$$j = \sqrt{-1} \qquad \text{or} \qquad j^2 = -1 \tag{11.1}$$

With this number, the square root of a negative number can be written as a product of j and a real number. For example,

$$\sqrt{-4} = \sqrt{4(-1)} = \sqrt{4}\sqrt{-1} = 2j$$

In general,

$$\sqrt{-a},\ a > 0,\ \text{is written } \sqrt{a}j \qquad\qquad (11.2)$$

To repeat: j *is not a real number* and must therefore be viewed as a new kind of number. By introducing j, we can extend the real number system to include square roots of negative numbers.

As has been emphasized repeatedly, the symbol $\sqrt[n]{\ }$ denotes the principal nth root. In particular, $\sqrt{16} = 4$, not -4. Because of this convention, the rule

$$\sqrt{a}\sqrt{b} = \sqrt{ab}$$

holds only if $a > 0$ or $b > 0$. For example, while

$$\sqrt{-4}\sqrt{-9} = (2j)(3j) = 6j^2 = -6$$

the law of radicals, if applied here, would yield $\sqrt{(-4)(-9)} = \sqrt{36} = 6$, not -6. This observation leads to the following rule.

$\sqrt{-a},\ a > 0,$ should always be written $\sqrt{a}j$ before performing any algebraic operation.

By using $\sqrt{a}j$ instead of $\sqrt{-a}$, problems with signs are avoided.

E X A M P L E **1**   Write $\sqrt{-44}$ in the form bj.

Solution. $\sqrt{-44} = \sqrt{44(-1)} = \sqrt{4 \cdot 11(-1)} = \sqrt{4 \cdot 11}\sqrt{-1} = 2\sqrt{11}j$ ◀

Complex Numbers

Even more general than imaginary numbers are **complex numbers**, which have the form

$$a + bj, \qquad a \text{ and } b \text{ real numbers}$$

Complex numbers include real numbers as special cases.

Definition of a Complex Number:

If a and b are real numbers and $j = \sqrt{-1}$, then $a + bj$ is called a **complex number**; a is called the **real part** and b the **imaginary part**.

Remark. You may prefer to call a the *real part* and bj the *imaginary part*. Because of the interpretation of a complex number as a vector from the origin to (a, b) in Figure 11.1, this book adheres to the usual convention of calling b the imaginary part.

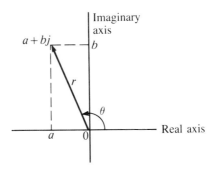

Figure 11.1

If $a = 0$, then the complex number has the form bj and is called a **pure imaginary number**. For example, $2j = 0 + 2j$ is a pure imaginary number. If $b = 0$, then the complex number is real. For example, $3 = 3 + 0j$ is real. From this point of view *a real number is a complex number whose imaginary part is zero*.

Complex Numbers as Vectors

A complex number can be given a geometric interpretation as a vector, as shown in Figure 11.1. The initial point of the arrow representing $a + bj$ is placed at the origin and the tip at the point (a, b). The horizontal axis is then called the **real axis** and the vertical axis the **imaginary axis**, although the terms *x-axis* and *y-axis* are also commonly used. This coordinate system used for plotting complex numbers is called the **complex plane**.

Real axis
Imaginary axis
Complex plane
Absolute value
Argument
Conjugate

The length of the arrow is $r = \sqrt{a^2 + b^2}$, called the **absolute value** or **modulus**. The angle θ made with the positive real axis is called the **argument**. (See Figure 11.1.) The **conjugate** of the complex number $a + bj$, denoted by $\overline{a + bj}$, is the reflection of $a + bj$ in the real axis. (See Figure 11.2.) Thus $\overline{a + bj} = a - bj$.

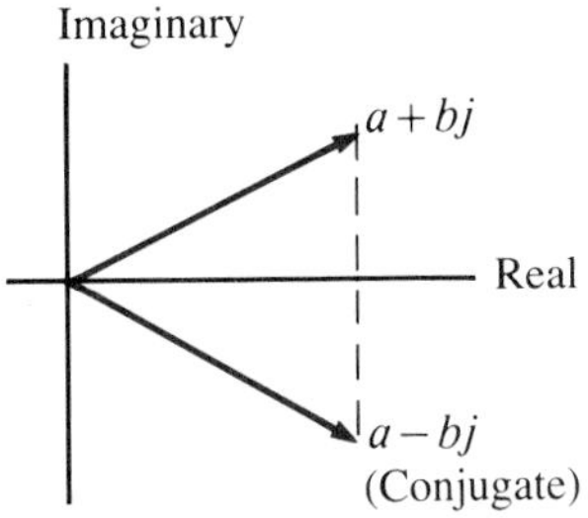

Imaginary

$a + bj$

Real

$a - bj$
(Conjugate)

Figure 11.2

A Complex Number as a Vector: The complex number $a + bj$ is represented as a **vector** extending from the origin to (a, b) in the complex plane (Figure 11.1).

Absolute value $r = \sqrt{a^2 + b^2}$

Argument   angle θ (Figure 11.1)

Conjugate $\overline{a + bj} = a - bj$ (Figure 11.2)

E   X   A   M   P   L   E   **2**   Find the absolute value and conjugate of the complex number $-3 - 4j$.

Solution. The absolute value r is given by

$$r = \sqrt{(-3)^2 + (-4)^2} = \sqrt{9 + 16} = 5$$

The conjugate of $-3 - 4j$ is $-3 + 4j$, shown in Figure 11.3. (As already noted, the axes are often labeled x-axis and y-axis, respectively.)   ◀

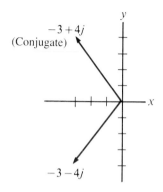

Figure 11.3

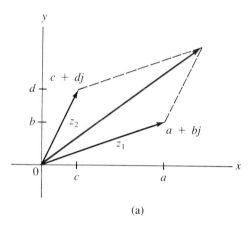

(a)

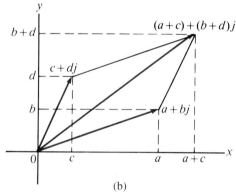

(b)

Figure 11.4

Addition and Subtraction

Addition

Addition of complex numbers is defined so that it is identical to vector addition. In other words, the sum of $z_1 = a + bj$ and $z_2 = c + dj$ is the resultant obtained from the parallelogram determined by z_1 and z_2 [Figure 11.4(a)]. The resultant can also be obtained algebraically by adding the "components" of z_1 and z_2; that is,

$$z_1 + z_2 = (a + bj) + (c + dj) = (a + c) + (b + d)j \tag{11.3}$$

The sum is shown in Figure 11.4(b).

Subtraction

As would be the case with ordinary vectors, the difference $z_1 - z_2$ is the second diagonal of the parallelogram in Figure 11.5(a). In terms of components, we have

$$z_1 - z_2 = z_1 + (-z_2) = (a + bj) + (-c - dj)$$

or

$$z_1 - z_2 = (a - c) + (b - d)j \tag{11.4}$$

Note, however, that the vector representing $z_1 - z_2$ has to be drawn with the initial point at the origin to conform to the usual convention. [See Figure 11.5(b).]

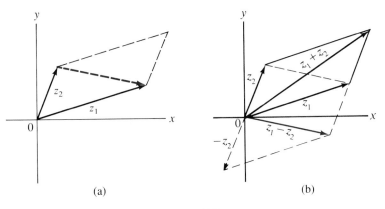

(a)   (b)

Figure 11.5

E X A M P L E **3**   Add the complex numbers $2 - 3j$ and $-3 - 4j$ graphically.

Solution. The sum is shown in Figure 11.6. Note that the sum is $(2 - 3j) + (-3 - 4j) = (2 - 3) + (-3 - 4)j = -1 - 7j$. ◄

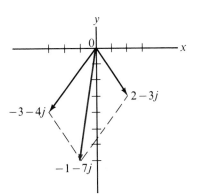

Figure 11.6

E X A M P L E **4**   Subtract $1 - 3j$ from $4 + 2j$ graphically.

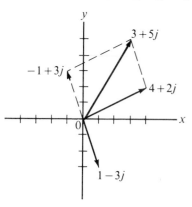

Solution. Subtracting $1 - 3j$ is equivalent to adding $-1 + 3j$, shown in Figure 11.7. Note that the result is $(4 + 2j) + (-1 + 3j) = (4 - 1) + (2 + 3)j = 3 + 5j$. ◄

Equality

To complete this section, we will define the equality of complex numbers.

> **Equality of Two Complex Numbers:** Two complex numbers are equal if and only if their real parts are equal and their imaginary parts are equal:
>
> $$a + bj = c + dj \qquad \text{if and only} \qquad \text{if } a = c \qquad \text{and} \qquad b = d \quad (11.5)$$

Figure 11.7

E X A M P L E **5**   Determine the values of x and y if

$$(-x + 3) + (x + 4)j = 2y - yj$$

Solution. $(-x + 3) + (x + 4)j = 2y - yj$

By definition (11.5) we get the following system of equations:

$$
\begin{array}{ll}
-x + 3 = 2y & \text{equating real parts} \\
\underline{x + 4 = -y} & \text{equating imaginary parts} \\
7 = y & \text{adding}
\end{array}
$$

Substituting $y = 7$ into the second equation, we get $x + 4 = -7$, or $x = -11$. So the two complex numbers are equal if $x = -11$ and $y = 7$. ◄

Remark. As mentioned in Chapter 6, i is another common designation for $\sqrt{-1}$. However, when complex numbers were first applied to the study of electricity, i had already become the standard notation for instantaneous current. As a result, j came to be used for the basic imaginary unit in physics and technology.

E X E R C I S E S / S E C T I O N   **11.1**

In Exercises 1–8, write the given numbers in the form bj. (See Example 1.)

1. $\sqrt{-9}$ **2.** $\sqrt{-16}$ **3.** $\sqrt{-8}$ **4.** $\sqrt{-18}$

5. $\sqrt{-20}$ **6.** $\sqrt{-32}$ **7.** $\sqrt{-75}$ **8.** $\sqrt{-125}$

In Exercises 9–12, draw each complex number and its conjugate in the complex plane.

9. $-4 - 6j$ **10.** $5 - 10j$

11. $-2 + 7j$ **12.** $-3 - j$

In Exercises 13–28, perform the indicated operations graphically and check algebraically.

In Exercises 29–38, determine the values of x and y for which equality holds. (See Example 5.)

13. $(1 - 3j) + (4 + 2j)$

14. $(2 + j) + (3 - 4j)$

15. $(-4 - 2j) + (6 + 5j)$

16. $(-3 + 7j) + (-6 + 2j)$

17. $(7 + 5j) - (2 - j)$

18. $(3 + 2j) - (1 - 2j)$

19. $(-2 + 3j) - (2 - 3j)$

20. $(1 - 4j) - (-2 - 4j)$

21. $(5 + 4j) - (5 - 3j)$

22. $(-4 - j) + (-3 + 2j)$

23. $(6 + 7j) + (4 - 7j)$

24. $(10 - 6j) + (-5 - 3j)$

25. $(-12 + 4j) - (-5 + 5j)$

26. $(7 + 12j) + (1 - 11j)$

27. $(-5 - 9j) - (-3 + 5j)$

28. $(-1 - 4j) - (10 + 2j)$

29. $2 + 3j = x + yj$

30. $x + 4j = 2 + yj$

31. $x + 3j = 2 - yj$

32. $-x + 2j = -3 - yj$

33. $(x + 1) + yj = 4 - j$

34. $(x + y) + (2x + y)j = 2 + 3j$

35. $(x - 2y) + (x - y)j = 4 + 5j$

36. $(x + 1) + (x + 3)j = y - yj$

37. $(x - 3) + (y + 2)j = -y + xj$

38. $(2x + y) + (-x + y)j = x - yj$

11.2 The Fundamental Operations

To study the four fundamental operations involving complex numbers, we will examine several examples and summarize the results at the end.

Addition and subtraction

Let us return to the operations of **addition** and **subtraction**, already considered in Section 11.1.

E X A M P L E **1**   Perform the indicated operations:

    a. $(-5 + 2j) + (-4 + 7j)$

    b. $(-5 + 2j) - (-4 + 7j)$

    c. $(2 - 3\sqrt{2}j) - (5 - 2\sqrt{2}j) - (-6 + 10\sqrt{2}j)$

Solution.

    a. $(-5 + 2j) + (-4 + 7j) = (-5 - 4) + (2j + 7j) = -9 + 9j$

    b. $(-5 + 2j) - (-4 + 7j) = -5 + 2j + 4 - 7j = -1 - 5j$

    c. $(2 - 3\sqrt{2}j) - (5 - 2\sqrt{2}j) - (-6 + 10\sqrt{2}j)$
      $= 2 - 3\sqrt{2}j - 5 + 2\sqrt{2}j + 6 - 10\sqrt{2}j = 3 - 11\sqrt{2}j$ ◀

Multiplication

The usual procedure for multiplying binomials can be applied to the **multiplication** of complex numbers, but j^2 has to be changed to -1.

E X A M P L E **2**   Multiply:

    a. $(2 - 3j)(4 + 2j)$ **b.** $(-5 + 7j)(2 + 6j)$

Solution.

    a. $(2 - 3j)(4 + 2j) = 8 - 8j - 6j^2$ **by FOIL**
                 $= 8 - 8j - 6(-1)$ $j^2 = -1$
                  $= 8 - 8j + 6 = 14 - 8j$

b. $(-5 + 7j)(2 + 6j) = -10 - 16j + 42j^2$
$$= -10 - 16j - 42 = -52 - 16j \qquad \blacktriangleleft$$

A special case of multiplying complex numbers involves powers of j. Since $j^2 = -1$, we have $j^3 = j^2 \cdot j = (-1)j = -j$. Similarly, $j^4 = j^3 \cdot j = -j \cdot j = -j^2 = -(-1) = 1$. Now the cycle repeats: $j^5 = j^4 \cdot j = 1 \cdot j = j$, $j^6 = j^5 \cdot j = j \cdot j = -1$, and so forth. Note especially that $j^4 = 1$.

E X A M P L E **3** Simplify:

    a. j^{20} **b.** j^{41} **c.** j^{10} **d.** j^{31}

Solution.

    a. $j^{20} = (j^4)^5 = 1^5 = 1$ $j^4 = 1$

    b. $j^{41} = j^{40}j = (j^4)^{10}j = 1^{10}j = j$ $j^4 = 1$

    c. $j^{10} = j^8 j^2 = (j^4)^2 j^2 = 1^2 j^2 = j^2 = -1$

    d. $j^{31} = j^{28} j^3 = (j^4)^7 j^3 = 1^7 j^3 = j^3 = -j$ $\blacktriangleleft$

Division

    Since a complex number is a radical expression in disguise, **division** can be carried out by the same method used in Section 10.6 to divide radicals. In other words, *we rationalize the denominator by multiplying numerator and denominator by the conjugate of the denominator.*

E X A M P L E **4** Perform the following divisions:

    a. $\dfrac{1}{1 - 2j}$ **b.** $\dfrac{2 - j}{3 + 4j}$

Solution.

    a. $\dfrac{1}{1 - 2j} = \dfrac{1}{1 - 2j} \cdot \dfrac{1 + 2j}{1 + 2j}$ $\overline{1 - 2j} = 1 + 2j$

         $= \dfrac{1 + 2j}{1 - 4j^2}$ **multiplication**

         $= \dfrac{1 + 2j}{1 - (-4)}$ $j^2 = -1$

         $= \dfrac{1 + 2j}{5}$

         $= \dfrac{1}{5} + \dfrac{2}{5}j$ **form:** $a + bj$

b. $\dfrac{2-j}{3+4j} = \dfrac{2-j}{3+4j} \cdot \dfrac{3-4j}{3-4j}$

$$= \frac{6-11j+4j^2}{9-16j^2} = \frac{6-11j-4}{9+16}$$

$$= \frac{2-11j}{25}$$

$$= \frac{2}{25} - \frac{11}{25}j \qquad \text{form:} \quad \textbf{a + bj}$$ ◀

Let us now summarize the four fundamental operations.

Fundamental Operations for Complex Numbers:

Addition $(a + bj) + (c + dj) = (a + c) + (b + d)j$

Subtraction $(a + bj) - (c + dj) = (a - c) + (b - d)j$

Multiplication $(a + bj)(c + dj) = (ac - bd) + (ad + bc)j$

Division (procedure) $\dfrac{a + bj}{c + dj} = \dfrac{(a + bj)(\overline{c + dj})}{(c + dj)(\overline{c + dj})} = \dfrac{(a + bj)(c - dj)}{(c + dj)(c - dj)}$

E X E R C I S E S / S E C T I O N   **11.2**

In Exercises 1–47, perform the indicated operations and express the results in the form $a + bj$.

1. $(5 + 3j) + (4 + 7j)$

2. $(-2 + j) - (10 + 2j)$

3. $(\sqrt{5} - j) + (2\sqrt{5} - 3j)$

4. $(\sqrt{2} + j) + 2j$

5. $2 + (4 - 6j)$

6. $(2 - 5j) - (7 + 2j) - (8 + 6j)$

7. $(-1 - 7j) + (2 - 8j) - (-3 - 9j)$

8. $(\sqrt{2} + 2\sqrt{3}j) - (2\sqrt{2} - 3\sqrt{3}j)$

9. $(-3 - j) + (-2 + 3j) - (4 - 11j)$

10. $(\sqrt{3} + j) + (\sqrt{3} - 3j) - (2\sqrt{3} + 5j)$

11. $(-1 - 2j) - (6 - 4j) - (5 + j)$

12. $-3j + (1 - 9j) - (11 + 15j)$

13. $(2 + j)(3 - j)$ **14.** $(1 - 3j)(2 + 2j)$

15. $(2 - j)(3 - 2j)$ **16.** $(3 - 4j)(1 + j)$

17. $(3 - 4j)(3 + 4j)$

18. $(4 - 6j)(4 + 6j)$

19. $(4 - 5j)(1 - 3j)$

20. $(6 + j)(-5 + 2j)$

21. $(1 + j)^2$

22. $(3 - 4j)^2$

23. $(\sqrt{3} - 2j)^2$

24. $(2 + \sqrt{3}j)^2$

25. $(\sqrt{2} - j)^2$

26. $(2 - 3j)(1 + 7j)$

27. $(5 - 7j)(2 + 4j)$

28. $(4 + 6j)(5 + 10j)$

29. $\dfrac{1}{1 - 3j}$

30. $\dfrac{1}{2 - j}$

31. $\dfrac{2}{2 + j}$

32. $\dfrac{3}{2 + 3j}$

33. $\dfrac{1 - 2j}{1 + 2j}$

34. $\dfrac{2 + 3j}{2 - 3j}$

35. $\dfrac{3 - 4j}{3 + 4j}$

36. $\dfrac{1 - 6j}{1 + 6j}$

37. $\dfrac{2 + 3j}{1 - 4j}$

38. $\dfrac{1 + j}{4 + 5j}$

39. $\dfrac{3 - 2j}{6 + 10j}$

40. $\dfrac{2 - 3j}{4 + 2j}$

47. $\dfrac{(2 - 3j)(2 - j)}{(2 + j)(1 - 4j)}$

41. $\dfrac{(1 - j)(1 + 2j)}{2 + j}$

42. $\dfrac{(2 - 3j)(2 - j)}{1 + 2j}$

In Exercises 48–60, simplify the given powers of j.

43. $\dfrac{(3 - j)(2 - 3j)}{2 + 2j}$

44. $\dfrac{(2 - j)(3 + 2j)}{2 - 3j}$

48. j^7 **49.** j^{11} **50.** j^{17} **51.** j^{21}

52. j^{22} **53.** j^{26} **54.** j^{69} **55.** j^{71}

45. $\dfrac{(1 - 2j)(1 + j)}{(1 - j)(3 + j)}$

46. $\dfrac{(1 - 3j)(3 - 2j)}{(1 - j)(2 + j)}$

56. j^{100} **57.** j^{102} **58.** j^{500} **59.** j^{48}

60. j^{55}

11.3   Polar and Exponential Forms

Complex numbers have been successfully applied to problems in science and technology, in part because of their geometric interpretation. To exploit this interpretation fully, we need to be able to represent a complex number in terms of its absolute value and argument. This representation leads to the **polar** and **exponential forms** of complex numbers.

Polar Form

In Section 11.1, we defined the absolute value r and the argument θ of a complex number. (See Figure 11.1.) Now consider the complex number $x + yj$ in Figure 11.8. Again denoting the absolute value by r and the argument by θ, $\cos \theta = x/r$ and $\sin \theta = y/r$. It follows that $x = r \cos \theta$ and $y = r \sin \theta$. Also, $x^2 + y^2 = r^2$ and $\tan \theta = y/x$. These ideas are summarized below.

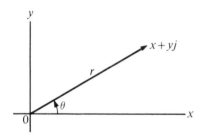

Figure 11.8

$$x = r \cos \theta \quad \text{and} \quad y = r \sin \theta \tag{11.6}$$

$$r = \sqrt{x^2 + y^2} \quad \text{and} \quad \tan \theta = \frac{y}{x} \tag{11.7}$$

where the signs of x and y determine the quadrant of the argument θ.

If we substitute the expressions for x and y into the form $x + yj$, we get by (11.6)

$$x + yj = r \cos \theta + jr \sin \theta$$

or, by factoring r,

$$x + yj = r(\cos \theta + j \sin \theta) \tag{11.8}$$

The right side of equation (11.8) is called the **polar form** of the complex number $x + yj$.

Since it is a standard notation in electronics, we will frequently use the abbreviation $r/\underline{\theta}$.

The form $x + yj$, which we used earlier, is called the **rectangular form**.

Rectangular Form	$x + yj$	(11.9)
Polar Form	$r(\cos\theta + j\sin\theta) = r\underline{/\theta}$	(11.10)

Converting from the rectangular to the polar form requires only basic trigonometry, as illustrated in the first two examples. (In the examples and exercises, we will make frequent use of special angles.)

E X A M P L E   **1**   Change the complex number $-1 + \sqrt{3}j$ to polar form.

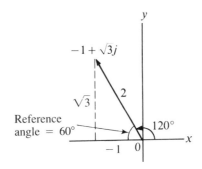

$$r^2 = x^2 + y^2$$
$$r^2 = (-1)^2 + (\sqrt{3})^2$$
$$r^2 = 1 + 3 = 4$$
$$r = 2$$

Figure 11.9

Solution. The number is shown in Figure 11.9.

After dropping a perpendicular from $(-1, \sqrt{3})$ to the x-axis, we get from the resulting $30°$–$60°$ reference triangle $r = 2$ and $\theta = 120°$. Hence

$$-1 + \sqrt{3}j = 2(\cos 120° + j\sin 120°) = 2\underline{/120°}$$

(To review the special angles, look at Section 7.2.)   ◀

E X A M P L E   **2**   Change $2 - 5j$ to polar form.

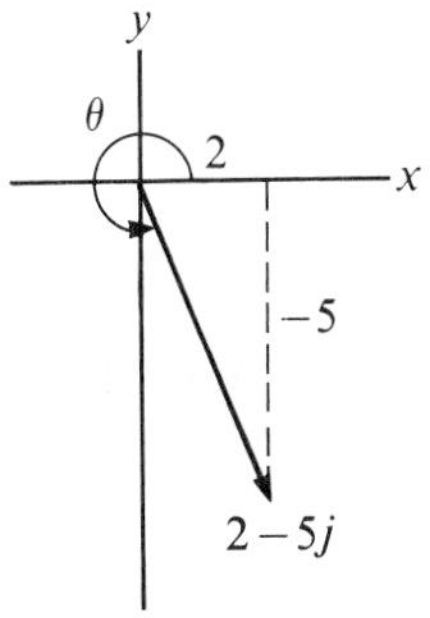

Solution. The number is shown in Figure 11.10.
The absolute value is

$$r = \sqrt{2^2 + (-5)^2} = \sqrt{29}$$

Note that $\tan\theta = -\frac{5}{2}$. From

$$5\boxed{+/-}\boxed{\div}2\boxed{=}\boxed{INV}\boxed{TAN} \rightarrow -68.198591$$

we obtain the polar form

$$\sqrt{29}\underline{/-68.2°} = \sqrt{29}\underline{/291.8°} \qquad 360° - 68.2° = 291.8°$$

Figure 11.10

In fact,

$$2 - 5j = \sqrt{29}\,\underline{/291.8° + k \cdot 360°}, \qquad k = 0, \pm 1, \pm 2, \ldots \qquad (11.11)$$

(See remark below.)   ◀

Remark. The polar form (11.11) lists all possible arguments of the complex number $2 - 5j$. In this and the next section, there is no compelling reason to specify all arguments. For uniformity, we choose θ such that $0° \leq \theta < 360°$. However, we will see in Section 11.5 that in some problems the general polar form

$$r\,\underline{/\theta + k \cdot 360°}, \qquad k = 0, \pm 1, \pm 2, \ldots$$

is needed.

The next example discusses certain special cases.

E X A M P L E   **3**

The arguments of real and pure imaginary numbers are quadrantal angles. For example,

$$3 = 3 + 0j = 3\,\underline{/0°} \qquad \text{tip of arrow at (3, 0)}$$

$$4j = 0 + 4j = 4\,\underline{/90°} \qquad \text{tip of arrow at (0, 4)}$$

$$-6 = -6 + 0j = 6\,\underline{/180°} \qquad \text{r is always positive}$$

$$-7j = 0 - 7j = 7\,\underline{/270°} \qquad \text{r is always positive} \qquad ◀$$

The next example illustrates the conversion of a complex number in polar form to rectangular form.

E   X   A   M   P   L   E   **4**   Change $2\,\underline{/150°}$ to rectangular form.

Solution. $2\,\underline{/150°} = 2(\cos 150° + j \sin 150°)$

$$= 2\left(-\frac{\sqrt{3}}{2} + \frac{1}{2}j\right)$$

$$= -\sqrt{3} + j \qquad\qquad ◀$$

We can see from Example 4 that the polar form of a complex number can be changed to rectangular form by replacing $\cos \theta$ and $\sin \theta$ by their values. If θ is a special angle, then the rectangular form can be expressed in exact form by using radicals. If θ is not a special angle, then we will express the function values of θ to four decimal places. For uniformity, other values will be rounded off to three significant digits.

E   X   A   M   P   L   E   **5**   Change $3\,\underline{/230°}$ to rectangular form.

Solution. $3\underline{/230^\circ} = 3(\cos 230^\circ + j \sin 230^\circ)$

$$= 3(-0.6428 - 0.7660j)$$

$$= -1.93 - 2.30j$$ ◄

Exponential Form

A third form of a complex number, called the **exponential form**, is particularly important in applications to electronics. This form is based on a relationship called *Euler's identity*:

Euler's identity

$$e^{j\theta} = \cos \theta + j \sin \theta \tag{11.12}$$

The letter e represents an irrational number whose approximate value is 2.71828. The origin and significance of this number are explored in the study of calculus.

Comparing the polar and exponential forms, it follows that the exponential form can be obtained by replacing $\underline{/\theta}$ by $e^{j\theta}$. Thus

$$r\underline{/\theta} = re^{j\theta} \tag{11.13}$$

The forms are summarized next.

Three Forms of a Complex Number:

Rectangular form	$x + yj$
Polar form	$r\underline{/\theta}$
Exponential form	$re^{j\theta}$

In the exponential form, θ is usually expressed in radians. For θ in radians, $j\theta$ is a complex power for which the laws of exponents hold. It is precisely this property that makes the exponential form so useful.

E X A M P L E   **6**   Write $-1 + j$ in exponential form.

Solution. $r = \sqrt{(-1)^2 + 1^2} = \sqrt{2}$ and $\theta = 135^\circ$

(Reference angle $= 45^\circ$, θ in quadrant II.) In radians,

$$\theta = 135^\circ = \frac{135\pi}{180} = \frac{3\pi}{4}$$

Thus

$$-1 + j = \sqrt{2}e^{(3\pi/4)j} \qquad r = \sqrt{2};\, \theta = \frac{3\pi}{4}$$ ◄

E X E R C I S E S / S E C T I O N   **11.3**

In Exercises 1–16, express the given complex numbers in polar form ($0° \leq \theta < 360°$).

1. 1 (See Example 3.)   **2.** -2

3. -4 **4.** j

5. $2j$ **6.** $-3j$

7. $-5j$ **8.** 5

9. $1 + j$ **10.** $-\sqrt{3} + j$

11. $1 - \sqrt{3}j$ **12.** $\sqrt{3} + j$

13. $-3 - 3j$ **14.** $-2 + 2\sqrt{3}j$

15. $-3\sqrt{3} + 3j$ **16.** $4 - 4j$

In Exercises 17–24, change the given complex numbers to exponential form ($0 \leq \theta < 2\pi$). (See Example 6.)

17. $2j$ **18.** $-4j$

19. $1 - j$ **20.** $-3 + 3\sqrt{3}j$

21. $\sqrt{3} - j$ **22.** $-1 - \sqrt{3}j$

23. $-2 + 2j$ **24.** $6\sqrt{3} + 6j$

In Exercises 25–36, express the given complex numbers in rectangular form. (Use a diagram and express each result in exact form.)

25. $2(\cos 45° + j \sin 45°)$

26. $4(\cos 210° + j \sin 210°)$

27. $2(\cos 120° + j \sin 120°)$

28. $\sqrt{2}(\cos 135° + j \sin 135°)$

29. $3\underline{/240°}$ **30.** $6\underline{/300°}$

31. $5\underline{/150°}$ **32.** $\sqrt{2}\underline{/225°}$

33. $3\underline{/0°}$ **34.** $4\underline{/180°}$

35. $5\underline{/90°}$ **36.** $6\underline{/270°}$

In Exercises 37–42, use a calculator to convert each complex number to polar form. Express the argument to the nearest tenth of a degree ($0° \leq \theta < 360°$). (See Example 2.)

37. $2 + 5j$ **38.** $-1 + 4j$

39. $-2 - 4j$ **40.** $-1 - \sqrt{11}j$

41. $\sqrt{14} - 2j$ **42.** $\sqrt{5} + \sqrt{3}j$

In Exercises 43–50, use a calculator to convert the given complex numbers to rectangular form. (Use three significant digits.)

43. $2.00(\cos 37.1° + j \sin 37.1°)$

44. $3.12(\cos 136.0° + j \sin 136.0°)$

45. $0.361\underline{/221.3°}$ **46.** $4.96\underline{/310.8°}$

47. $6.34\underline{/274.6°}$ **48.** $10.3\underline{/162.9°}$

49. $3.03\underline{/25.6°}$ **50.** $4.96\underline{/173.4°}$

51. We will see in Section 11.6 that the *impedance Z* in an alternating-current circuit is commonly expressed as a complex number. For a certain circuit, $Z = 7.9 - 5.6j$. Express Z in a polar form.

52. For a certain circuit, $Z = 20.6 - 15.8j$. Express Z in polar form. (Refer to Exercise 51.)

11.4   Products and Quotients of Complex Numbers

As we saw in Section 11.1, vector interpretation of complex numbers yields simple and natural geometric interpretations of addition and subtraction. Unfortunately, geometric interpretations of multiplication and division are not given as easily. However, the polar form is particularly convenient for these operations: It yields the desired geometric interpretation and even paves the way for the calculation of powers and roots.

Consider the following arbitrary complex numbers in polar and rectangular forms:

$$r_1(\cos \theta_1 + j \sin \theta_1) = r_1 e^{j\theta_1}$$

and

$$r_2(\cos \theta_2 + j \sin \theta_2) = r_2 e^{j\theta_2}$$

Their product is given by

$$r_1(\cos\theta_1 + j\sin\theta_1) \cdot r_2(\cos\theta_2 + j\sin\theta_2)$$
$$= r_1 e^{j\theta_1} \cdot r_2 e^{j\theta_2} = r_1 r_2 e^{j(\theta_1 + \theta_2)} \qquad \textbf{adding exponents}$$
$$= r_1 r_2 [\cos(\theta_1 + \theta_2) + j\sin(\theta_1 + \theta_2)]$$

Product of Two Complex Numbers: The absolute value of the product of two complex numbers is the product of their absolute values. The argument of the product is the sum of their arguments:

$$(r_1 \underline{/\theta_1})(r_2 \underline{/\theta_2}) = r_1 r_2 \underline{/\theta_1 + \theta_2} \qquad (11.14)$$

This delightfully terse and simple derivation illustrates the great power of the exponential form.

For the quotient of two complex numbers we get

$$r_1 e^{j\theta_1} \div r_2 e^{j\theta_2} = \frac{r_1 e^{j\theta_1}}{r_2 e^{j\theta_2}} = \frac{r_1}{r_2} e^{j(\theta_1 - \theta_2)} \qquad \textbf{subtracting exponents}$$

Quotient of Two Complex Numbers: The absolute value of the quotient of two complex numbers is the quotient of their absolute values. The argument of the quotient is the difference of their arguments in the proper order:

$$\frac{r_1 \underline{/\theta_1}}{r_2 \underline{/\theta_2}} = \frac{r_1}{r_2} \underline{/\theta_1 - \theta_2} \qquad (11.15)$$

E X A M P L E **1** Multiply:

$$[2(\cos 20° + j\sin 20°)][2(\cos 30° + j\sin 30°)]$$

Solution. By formula (11.14), we multiply the absolute values and add the arguments:

$$[2(\cos 20° + j\sin 20°)][2(\cos 30° + j\sin 30°)]$$
$$= 2 \cdot 2[\cos(20° + 30°) + j\sin(20° + 30°)]$$
$$= 4(\cos 50° + j\sin 50°)$$

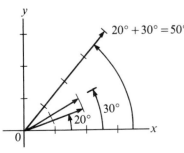

Figure 11.11

(See Figure 11.11.) ◀

E X A M P L E **2**   Carry out the following division:

$$\frac{12/150°}{4/220°}$$

Solution. By formula (11.15), we divide the absolute values and subtract the arguments:

$$\frac{12/150°}{4/220°} = \frac{12}{4}/150° - 220° = 3/-70° = 3/360° - 70° = 3/290°$$

(See Figure 11.12.) ◀

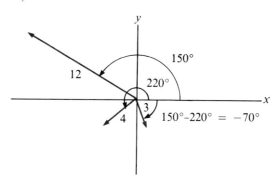

Figure 11.12

E X E R C I S E S / S E C T I O N   **11.4**

Carry out the indicated operations. (Leave answers in polar form.)

1. $[2(\cos 31° + j \sin 31°)][3(\cos 62° + j \sin 62°)]$

2. $[4(\cos 81° + j \sin 81°)][8(\cos 22° + j \sin 22°)]$

3. $(6/130°)(3/220°)$

4. $(10/265°)(4/10°)$

5. $(15/230°)(2/300°)$

6. $(20/328°)(30/275°)$

7. $[18(\cos 139° + j \sin 139°)][3(\cos 46° + j \sin 46°)]$

8. $[24(\cos 318° + j \sin 318°)][4(\cos 342° + j \sin 342°)]$

9. $\dfrac{36(\cos 100° + j \sin 100°)}{9(\cos 40° + j \sin 40°)}$

10. $\dfrac{50(\cos 280° + j \sin 280°)}{10(\cos 150° + j \sin 150°)}$

11. $\dfrac{39(\cos 160° + j \sin 160°)}{13(\cos 260° + j \sin 260°)}$

12. $\dfrac{8(\cos 70° + j \sin 70°)}{24(\cos 90° + j \sin 90°)}$

13. $\dfrac{3/172°}{2/321°}$

14. $\dfrac{5/216°}{8/284°}$

15. $\dfrac{(25/102°)(1/87°)}{15/36°}$

16. $\dfrac{(4/25°)(6/230°)}{8/125°}$

17. $\dfrac{(6/60°)(4/148°)}{12/342°}$

18. $\dfrac{10/119°}{(4/73°)(5/156°)}$

19. $\dfrac{2(\cos 165° + j \sin 165°)}{[7(\cos 43° + j \sin 43°)][4(\cos 301° + j \sin 301°)]}$

20. $\dfrac{3(\cos 20° + j \sin 20°)}{[8(\cos 70° + j \sin 70°)][10(\cos 117° + j \sin 117°)]}$

11.5   Powers and Roots of Complex Numbers

We now have enough information to find powers and roots of complex numbers. In addition, we will study a method for finding all the roots of a number by means of *De Moivre's theorem.*

Raising a complex number to an integral power involves only repeated multiplication. Thus

$$[r(\cos\theta + j\sin\theta)]^3$$
$$= [r(\cos\theta + j\sin\theta)][r(\cos\theta + j\sin\theta)][r(\cos\theta + j\sin\theta)]$$
$$= r \cdot r \cdot r[\cos(\theta + \theta + \theta) + j\sin(\theta + \theta + \theta)]$$
$$= r^3(\cos 3\theta + j\sin 3\theta)$$

So in general,

$$[r(\cos\theta + j\sin\theta)]^n = r^n(\cos n\theta + j\sin n\theta)$$

or

$$(r\underline{/\theta})^n = r^n\underline{/n\theta}$$

This formula is known as **De Moivre's theorem**, after Abraham De Moivre (1667–1754). De Moivre, a contemporary of Isaac Newton, was a major figure in the development of probability theory. His theorem is actually valid for all real numbers. If n is rational, it can be used to find roots, as we will see later in this section.

De Moivre's Theorem:

$$[r(\cos\theta + j\sin\theta)]^n = r^n(\cos n\theta + j\sin n\theta) \qquad (11.16)$$

or

$$(r\underline{/\theta})^n = r^n\underline{/n\theta}, \qquad \text{for all real } n \qquad (11.17)$$

The first two examples illustrate the use of De Moivre's theorem for finding the nth power of a complex number.

E X A M P L E **1** Use De Moivre's theorem to find $(1 + \sqrt{3}j)^{10}$.

Solution. $(1 + \sqrt{3}j)^{10} = (2\underline{/60°})^{10} = 2^{10}\underline{/10 \cdot 60°}$

by De Moivre's theorem. It follows that

$$(1 + \sqrt{3}j)^{10} = 2^{10}\underline{/600°} = 2^{10}\underline{/600° - 360°}$$
$$= 2^{10}\underline{/240°} = 2^{10}(\cos 240° + j\sin 240°)$$
$$= 2^{10}\left(-\frac{1}{2} - \frac{\sqrt{3}}{2}j\right) = 2^{10}\left(\frac{-1 - \sqrt{3}j}{2}\right)$$
$$= 2^9(-1 - \sqrt{3}j) = -512 - 512\sqrt{3}j \qquad \blacktriangleleft$$

E X A M P L E **2**   Use De Moivre's theorem to find $(-2 + 3j)^5$.

Solution. First we need to convert $-2 + 3j$ to polar form: $r = \sqrt{(-2)^2 + 3^2} = \sqrt{13}$; since $\tan \theta = -\frac{3}{2}$ and θ is in the second quadrant, we get $\theta = 123.69°$ to two decimal places. Then

$$
\begin{aligned}
(-2 + 3j)^5 &= (13^{1/2}\underline{/123.69°})^5 \\
&= (13^{1/2})^5\underline{/5 \cdot 123.69°} \\
&= 13^{5/2}(\cos 618.45° + j \sin 618.45°) \\
&= 13^{5/2}(\cos 258.45° + j \sin 258.45°) \\
&= -122 - 597j
\end{aligned}
$$

$$
\begin{aligned}
&618.45° \\
&-360.00° \\
\hline
&258.45°
\end{aligned}
$$

◀

Roots

To use De Moivre's theorem for finding roots of complex numbers, we need to list all the arguments in order to obtain all the roots. In other words, we write $r\underline{/\theta}$ in the form

$$r\underline{/\theta + k \cdot 360°}$$

Roots of a Complex Number: The n nth roots of a complex number $r(\cos \theta + j \sin \theta)$ are given by

$$
\{r[\cos(\theta + k \cdot 360°) + j \sin(\theta + k \cdot 360°)]\}^{1/n}
$$
$$
= r^{1/n}\left[\cos \frac{1}{n}(\theta + k \cdot 360°) + j \sin \frac{1}{n}(\theta + k \cdot 360°)\right],
$$
$$
k = 0, 1, 2, \ldots, n - 1
$$

or

$$
(r\underline{/\theta + k \cdot 360°})^{1/n} = r^{1/n}\underline{\left/\frac{1}{n}(\theta + k \cdot 360°)\right.}, \qquad k = 0, 1, 2, \ldots, n - 1
$$

In the remaining examples, De Moivre's theorem is used to find all the roots of a complex number.

E X A M P L E **3**   Find all the roots of the equation $x^6 - 1 = 0$.

Solution. From $x^6 - 1 = 0$, we get

$$x^6 = 1$$

So we need to find the 6 sixth roots of 1. (Every number has n nth roots.)

To express 1 in polar form, note that the absolute value is 1 and the argument $0°$. Now recall that all the arguments differ by a multiple of $360°$. So the complete polar form is

$$1 = 1\underline{/0^\circ + k \cdot 360^\circ}, \qquad k = 0, 1, 2, \ldots$$

(Listing only positive multiples turns out to be sufficient, as we will see.) Next, by De Moivre's theorem

$$(1\underline{/0^\circ + k \cdot 360^\circ})^{1/6}, \qquad k = 0, 1, 2, \ldots$$

$$= 1^{1/6}\underline{\left|\frac{1}{6}(0^\circ + k \cdot 360^\circ)\right.}, \qquad k = 0, 1, 2, \ldots$$

$$= 1^{1/6}\underline{/k \cdot 60^\circ}, \qquad k = 0, 1, 2, \ldots$$

Since the absolute value is a positive real number, it is understood that $1^{1/6}$ is the principal sixth root; that is, $1^{1/6} = 1$. So the sixth roots are

$$1\underline{/k \cdot 60^\circ}, \qquad k = 0, 1, 2, \ldots$$

Converting back to the rectangular form, we get:

$k = 0$: $1\underline{/0 \cdot 60^\circ} = \cos 0^\circ + j \sin 0^\circ = 1$

$k = 1$: $1\underline{/1 \cdot 60^\circ} = \cos 60^\circ + j \sin 60^\circ = \dfrac{1}{2} + \dfrac{\sqrt{3}}{2}j$

$k = 2$: $1\underline{/2 \cdot 60^\circ} = \cos 120^\circ + j \sin 120^\circ = -\dfrac{1}{2} + \dfrac{\sqrt{3}}{2}j$

$k = 3$: $1\underline{/3 \cdot 60^\circ} = \cos 180^\circ + j \sin 180^\circ = -1$

$k = 4$: $1\underline{/4 \cdot 60^\circ} = \cos 240^\circ + j \sin 240^\circ = -\dfrac{1}{2} - \dfrac{\sqrt{3}}{2}j$

$k = 5$: $1\underline{/5 \cdot 60^\circ} = \cos 300^\circ + j \sin 300^\circ = \dfrac{1}{2} - \dfrac{\sqrt{3}}{2}j$

Starting with $k = 6$, the cycle begins again: $\cos 360^\circ + j \sin 360^\circ = 1$, which corresponds to $k = 0$. If $k = -1$, we get $\cos(-60^\circ) + j \sin(-60^\circ) = \cos 300^\circ + j \sin 300^\circ$, corresponding to $k = 5$. In other words, the values **$k = 0, 1, \ldots, 5$ generate all six roots**.

Note that only 1 and -1 could have been obtained directly by inspection. That the remaining numbers are also roots of unity can be checked by direct multiplication. For example,

$$\left(\frac{1}{2} + \frac{\sqrt{3}}{2}j\right)^6 = 1$$

◀

E X A M P L E **4**   Find the 5 fifth roots of $1 - j$.

Solution. Since $r = \sqrt{2}$ and $\theta = 315^\circ$, we obtain

$$1 - j = \sqrt{2}\underline{/315^\circ + k \cdot 360^\circ}, \qquad k = 0, 1, 2, 3, 4$$

By De Moivre's theorem the roots are given by

$$(1 - j)^{1/5} = (2^{1/2} \underline{/315° + k \cdot 360°})^{1/5}$$

$$= (2^{1/2})^{1/5} \left\lfloor \frac{1}{5}(315° + k \cdot 360°) \right.$$

$$= 2^{1/10} \underline{/63° + k \cdot 72°}, \qquad k = 0, 1, 2, 3, 4$$

Since $2^{1/10} = \sqrt[10]{2}$, the principal tenth root, the roots of $1 - j$ can now be listed:

k = 0: $\sqrt[10]{2}(\cos 63° + j \sin 63°) = 0.487 + 0.955j$

k = 1: $\sqrt[10]{2}(\cos 135° + j \sin 135°) = -0.758 + 0.758j$

k = 2: $\sqrt[10]{2}(\cos 207° + j \sin 207°) = -0.955 - 0.487j$

k = 3: $\sqrt[10]{2}(\cos 279° + j \sin 279°) = 0.168 - 1.06j$

k = 4: $\sqrt[10]{2}(\cos 351° + j \sin 351°) = 1.06 - 0.168j$

If $k = 5$, we get $\sqrt[10]{2}\underline{/423°} = \sqrt[10]{2}\underline{/63°}$, which corresponds to $k = 0$.   ◀

These examples show that the *n* nth roots of $a + bj$ are equally spaced along the circumference of a circle of radius $\sqrt[n]{r}$ centered at the origin and $360°/n$ apart. Thus the tips of the vectors form the vertices of a regular polygon of n sides. For example, the 5 fifth roots of $1 - j$ (Example 4) are shown in Figure 11.13.

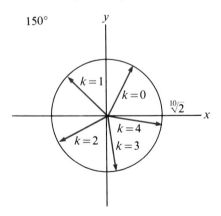

Figure 11.13

E X A M P L E **5**   Find the 7 seventh roots of $2j$ in polar form.

Solution. $2j = 2\underline{/90° + k \cdot 360°}$

The seven roots are therefore given by

$$(2\underline{/90° + k \cdot 360°})^{1/7} = 2^{1/7}\left\lfloor \frac{1}{7}(90° + k \cdot 360°) \right., \qquad k = 0, 1, 2, 3, 4, 5, 6 \quad ◀$$

In Exercises 1–12, use De Moivre's theorem to find the indicated powers. Express the answers in rectangular form.

1. $(1 + j)^5$ **2.** $(-1 + j)^6$ **3.** $(2 - 2j)^8$
4. $(-3 - 3j)^4$ **5.** $(-1 + \sqrt{3}j)^5$ **6.** $(\sqrt{3} + j)^6$
7. $(-\sqrt{3} - j)^6$ **8.** $(1 - \sqrt{3}j)^7$ **9.** $(-2 + 2\sqrt{3}j)^4$
10. $(-3\sqrt{3} - 3j)^5$ **11.** $\left(\frac{1}{2} - \frac{\sqrt{3}}{2}j\right)^6$ **12.** $\left(\frac{\sqrt{3}}{2} + \frac{1}{2}j\right)^8$

In Exercises 13–16, use a calculator to find the indicated powers. (See Example 2.)

13. $(1 + 2j)^4$ **14.** $(2 - 3j)^4$
15. $(-1 + 4j)^5$ **16.** $(-2 - 4j)^5$

In Exercises 17–20, find the powers in polar form.

17. $(3\underline{/75.3°})^{10}$ **18.** $(4\underline{/118.4°})^{12}$
19. $(3\underline{/137.4°})^{15}$ **20.** $(3\underline{/236.1°})^{16}$

In Exercises 21–28, find the indicated roots. Express the results in exact rectangular form. (See Example 3.)

21. Cube roots of 1 **22.** Fourth roots of -16
23. Square roots of $-j$ **24.** Cube roots of j
25. Fourth roots of -81 **26.** Fourth roots of 1
27. Square roots of $16j$ **28.** Sixth roots of -64

In Exercises 29–32, find the indicated roots in rectangular form. (See Example 4.)

29. Fifth roots of $32j$ **30.** Fourth roots of $1 + \sqrt{3}j$
31. Sixth roots of $1 + \sqrt{3}j$ **32.** Fifth roots of $-1 + j$

In Exercises 33–44, find the indicated roots in polar form. (See Example 5.)

33. Fifth roots of $-32j$ **34.** Sixth roots of $-2 + 2\sqrt{3}j$
35. Eighth roots of $-1 - j$ **36.** Seventh roots of $1 - j$
37. Sixth roots of $-2\sqrt{3} - 2j$ **38.** Ninth roots of $2j$
39. Eighth roots of $3 - 3\sqrt{3}j$ **40.** Fifth roots of $1 + j$
41. Ninth roots of $-3j$ **42.** Fifth roots of $-\sqrt{3} - j$
43. Tenth roots of $-1 - \sqrt{3}j$ **44.** Tenth roots of $\sqrt{3} - j$

11.6 Phasors (Optional)

A particularly interesting application of complex numbers is the study of phase relations between voltage and current in a simple alternating-current circuit. The problem arises because, as we saw in Chapter 8, the current produced by a generator is sinusoidal. While the current and voltage reach their peak periodically, they do not ordinarily peak at the same time. Since the phase relations are known for the individual elements, they can be determined for the circuit as a whole.

Consider the circuit in Figure 11.14 containing a resistor R, an inductor L, and a capacitor C in series with a generator E. Each of the elements offers a type of resistance to the current flow, called the **reactance**. The reactance is denoted by X and is measured in ohms. We already know that $X = R$ for the

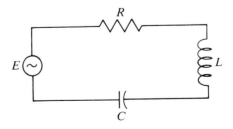

Figure 11.14

resistor. The voltage across the resistor is given by $V = IR$. An analogous form for the voltage exists for all three elements: The voltages across the resistor, capacitor, and inductor are given by

$$V_R = IX_R = IR, \qquad V_C = IX_C, \qquad \text{and} \qquad V_L = IX_L \qquad (11.18)$$

respectively.

Now consider the following facts: The voltage across the resistor reaches its peak at the same time as the current. The voltage and current are said to be **in phase**. The voltage across the inductor reaches its peak before the current and is said to **lead** the current (by 90°). Similarly, the voltage across the capacitor **lags** the current (by 90°). To obtain the phase relation for the combination, we use complex numbers in the following way: V_R is represented by a positive real number, V_L by the pure imaginary number $V_L j$, and V_C by the pure imaginary number $-V_C j$, as shown in Figure 11.15. The total voltage across the combination of elements is represented by

$$V = IR + IX_L j - IX_C j = I[R + j(X_L - X_C)] \qquad (11.19)$$

If we let

$$Z = R + (X_L - X_C)j \qquad (11.20)$$

(Figure 11.16), we get

$$V = IZ \qquad (11.21)$$

which has the same form as $V = IX$. Z is called the **impedance**. The impedance is the total effective resistance that takes into account the phase relation between the voltage and the current. More precisely, the magnitude of Z,

$$|Z| = \sqrt{R^2 + (X_L - X_C)^2} \qquad (11.22)$$

is the *total effective resistance measured in ohms*. The argument of Z is the phase angle between the voltage and the current (Figure 11.16). If θ is positive, then the voltage leads the current; if θ is negative, then the voltage lags the current.

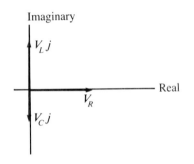

Figure 11.15

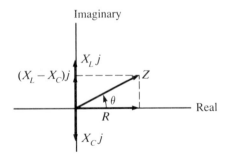

Figure 11.16

E X A M P L E **1** If $R = 10.0\ \Omega$, $X_L = 5.00\ \Omega$, and $X_C = 20.0\ \Omega$, find **(a)** the impedance and **(b)** the magnitude of the voltage across the combination at the instant when $I = 4.21$ A.

Solution.

a. Draw the diagram in Figure 11.17. By formula (11.20)

$$Z = R + (X_L - X_C)j$$

$$Z = 10.0 + (5.00 - 20.0)j = 10.0 - 15.0j$$

So the magnitude of Z is

$$|Z| = \sqrt{(10.0)^2 + (-15.0)^2} = 18.0\ \Omega$$

From $\tan\theta = -\frac{15.0}{10.0}$, we get $\theta = -56.3°$. So the voltage lags the current by $56.3°$.

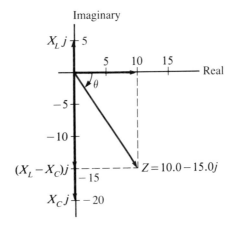

Figure 11.17

b. It follows from formula (11.21) that

$$|V| = I|Z| = (4.21)(18.0) = 75.8\ \text{V}$$ ◀

E X E R C I S E S / S E C T I O N **11.6**

Find Z, $|Z|$, and θ for the given values of R, X_L, and X_C. Refer to the circuit in Figure 11.14.

1. $R = 3.0\ \Omega$, $X_L = 8.0\ \Omega$, $X_C = 4.0\ \Omega$
2. $R = 4.0\ \Omega$, $X_L = 7.0\ \Omega$, $X_C = 4.0\ \Omega$
3. $R = 12.0\ \Omega$, $X_L = 17.0\ \Omega$, $X_C = 8.00\ \Omega$
4. $R = 13.1\ \Omega$, $X_L = 15.3\ \Omega$, $X_C = 10.2\ \Omega$
5. $R = 20.0\ \Omega$, $X_L = 12.3\ \Omega$, $X_C = 29.7\ \Omega$

6. $R = 55.3\ \Omega$, $X_L = 25.7\ \Omega$, $X_C = 42.6\ \Omega$

In Exercises 7–10, find the voltage across the combination, given that $I = 1.50$ A.

7. $R = 15.6\ \Omega$, $X_L = 12.0\ \Omega$
8. $R = 25.0\ \Omega$, $X_C = 14.3\ \Omega$
9. $R = 36.0\ \Omega$, $X_L = 25.7\ \Omega$, $X_C = 17.8\ \Omega$
10. $R = 11.6\ \Omega$, $X_L = 21.6\ \Omega$, $X_C = 29.4\ \Omega$

REVIEW EXERCISES/CHAPTER 11

In Exercises 1–4, perform the indicated operations graphically. Check your answers algebraically.

1. $(2 + 3j) + (1 - 2j)$ **2.** $(-2 + j) - (1 - 4j)$

3. $(2 - j) - (1 + 2j)$ **4.** $(1 - 3j) + (2 + 4j)$

In Exercises 5–8, determine the values of x and y for which equality holds.

5. $(x - 1) + yj = 3 + 2j$

6. $(x + 2) + (2x + 1)j = -y + yj$

7. $(x + y) + (x - y)j = 2 + 4j$

8. $(x - 4) + (2y + 2)j = y + xj$

In Exercises 9–18, perform the indicated operations and express the results in rectangular form.

9. $(\sqrt{2} - j) + (2\sqrt{2} + j)$ **10.** $(4\sqrt{5} + 2j) - (2\sqrt{5} - j)$

11. $(3 - 4j)(3 + 4j)$ **12.** $(1 - 7j)(1 + 7j)$

13. $(1 - 3j)^2$ **14.** $(2 + 3j)(7 - 4j)$

15. $\dfrac{3}{4 - 3j}$ **16.** $\dfrac{2}{4 - j}$ **17.** $\dfrac{1 - 2j}{2 + j}$ **18.** $\dfrac{2 - 3j}{1 + 2j}$

In Exercises 19–26, express the given complex numbers in polar form ($0° \leq \theta < 360°$).

19. 2 **20.** $2j$

21. -3 **22.** $-4j$

23. $2 + 2j$ **24.** $2 - 2j$

25. $-1 + \sqrt{3}j$ **26.** $-2 - 2\sqrt{3}j$

In Exercises 27–32, express the given complex numbers in exponential form ($0 \leq \theta < 2\pi$).

27. $3j$ **28.** -2

29. $-3 + 3j$ **30.** $-2\sqrt{3} + 2j$

31. $3 - 3\sqrt{3}j$ **32.** $-\sqrt{3} - j$

In Exercises 33–38, write the given complex numbers in rectangular form. (Use a diagram and express the results in terms of radicals.)

33. $2(\cos 135° + j \sin 135°)$ **34.** $2(\cos 240° + j \sin 240°)$

35. $4(\cos 225° + j \sin 225°)$ **36.** $6 / 330°$

37. $3 / 210°$ **38.** $4 / 300°$

In Exercises 39–42, use a calculator to convert the given complex numbers to rectangular form. (Use three significant digits.)

39. $3.20(\cos 136.0° + j \sin 136.0°)$

40. $1.32(\cos 250.1° + j \sin 250.1°)$

41. $6.03 / 25.1°$

42. $4.96 / 317.0°$

In Exercises 43–46, use a calculator to convert each complex number to polar form. Express the argument to the nearest tenth of a degree.

43. $3 - 5j$ **44.** $-2 - 7j$ **45.** $-3 + j$ **46.** $4 - 3j$

In Exercises 47–54, carry out each indicated operation. Leave the answer in polar form.

47. $[2(\cos 110° + j \sin 110°)][3(\cos 32° + j \sin 32°)]$

48. $[6(\cos 151° + j \sin 151°)][2(\cos 32° + j \sin 32°)]$

49. $[2(\cos 300° + j \sin 300°)][\sqrt{5}(\cos 72° + j \sin 72°)]$

50. $\dfrac{7(\cos 280° + j \sin 280°)}{2(\cos 142° + j \sin 142°)}$

51. $\dfrac{21 / 312°}{3 / 76°}$ **52.** $\dfrac{(6 / 281°)(12 / 39°)}{24 / 124°}$

53. $\dfrac{(25 / 53°)(2 / 132°)}{100 / 334°}$ **54.** $\dfrac{(3\sqrt{2} / 17°)(2 / 44°)}{\sqrt{2} / 216°}$

In Exercises 55–58, use De Moivre's theorem to find the indicated powers. Express the results in exact rectangular form.

55. $(1 + j)^7$ **56.** $(-2 + 2j)^8$

57. $(-\sqrt{3} - j)^6$ **58.** $(-1 + \sqrt{3}j)^7$

In Exercises 59–62, use a calculator to find the indicated powers.

59. $(2 + j)^5$ **60.** $(-3 - 2j)^5$

61. $(-1 + 3j)^4$ **62.** $(3 - j)^4$

63. Find the cube roots of $8j$ in rectangular form.

64. Find the fourth roots of -1 in rectangular form.

In Exercises 65–67, find the roots in polar form.

65. Fifth roots of 1

66. Square roots of $1 + j$

67. Sixth roots of $\sqrt{3} + j$

C H A P T E R **12**

Logarithmic and Exponential Functions

12.1 The Definition of Logarithm

Logarithms were first introduced by John Napier (1550–1617), a Scottish nobleman, as a device for carrying out lengthy arithmetic computations in a relatively fast way. At a time when arithmetic had to be done by hand, this enormous breakthrough was a blessing for the whole scientific community, particularly astronomy. Even the slide rule was an offshoot of logarithms. In the days of the electronic calculator, the importance of logarithms as a computational device has all but vanished, but their importance in theoretical work and in advanced mathematics is greater than ever.

To see how logarithms are defined, let us consider a typical expression containing an exponent:

$$2^3 = 8$$

Recall that 2 is called the *base* and 3 the *exponent*, and we say "2 cubed equals 8." If we start with the number 8, then we could also say "8 is written as a power of 2" or "the exponent corresponding to 8 is 3, provided that 2 is the base." Now, a **logarithm** is merely an exponent. Using this term, we could say "the logarithm of 8 is 3, provided that the base is 2." More simply, "the logarithm of 8 to the base 2 is 3." This statement is abbreviated

Logarithm

$$\log_2 8 = 3$$

where *log* stands for *logarithm*. In summary,

$$2^3 = 8 \quad \text{and} \quad \log_2 8 = 3$$

mean exactly the same thing. In general,

$$x = b^y \qquad \text{means the same as} \qquad y = \log_b x$$

Definition of Logarithm:

$$y = \log_b x \text{ means the same as } b^y = x, \qquad b > 0, \qquad b \neq 1 \qquad (12.1)$$

The equation $y = \log_b x$ is read *y is equal to the logarithm of x to the base b*, or *y equals log x to the base b*.

E X A M P L E **1**   **a.** $\log_3 9 = 2$ means $3^2 = 9$.
b. $\log_4 2 = \frac{1}{2}$ means $4^{1/2} = 2$.
c. $\log_8 \frac{1}{2} = -\frac{1}{3}$ means $8^{-1/3} = \frac{1}{2}$.   ◀

As we will see throughout this chapter, logarithms have many applications. In this section, however, we are concerned only with the basic definitions. In the examples and exercises, we will convert exponential forms ($x = b^y$) to logarithmic forms ($y = \log_b x$) and vice versa. Again for practice, we will solve simple equations involving logarithms. More complicated equations will be taken up in Section 12.7.

E X A M P L E **2**   Write $2^{-3} = \frac{1}{8}$ in logarithmic form.

Solution. Since the base is 2 and the exponent -3, we write

$$\log_2 \frac{1}{8} = -3$$

◀

E X A M P L E **3**   Write $\log_9 3 = \frac{1}{2}$ in exponential form.

Solution. Since 9 is the base and $\frac{1}{2}$ the exponent, we get

$$9^{1/2} = 3$$

◀

E X A M P L E **4**   Find the value of the unknown:

$$\textbf{a. } \log_{27} x = -\frac{1}{3} \qquad \textbf{b. } \log_b \frac{1}{16} = -2$$

Solution.

a. If we change $\log_{27} x = -\frac{1}{3}$ to the exponential form, we get

$$27^{-1/3} = x$$

Thus

$$x = 27^{-1/3} = \frac{1}{27^{1/3}} = \frac{1}{\sqrt[3]{27}} = \frac{1}{3}$$

b. By writing

$$\log_b \frac{1}{16} = -2 \quad \text{as} \quad b^{-2} = \frac{1}{16}$$

we get

$$\frac{1}{b^2} = \frac{1}{16}$$

$$\frac{1}{b^2} = \frac{1}{4^2}$$

$$b = 4$$

(From $b^2 = 16$ we actually get $b = \pm 4$, but since $b > 0$, we choose the positive square root.)   ◀

E X A M P L E **5**   Find x:

a. $\log_3 x = 1$ **b.** $\log_3 x = 0$

Solution.

a. $\log_3 x = 1$ can be written $3^1 = x$; thus $x = 3$.
b. $\log_3 x = 0$ is equivalent to $3^0 = x$; thus $x = 1$.   ◀

The results of Example 5 can be generalized.

$$\boxed{\log_b b = 1, \qquad \log_b 1 = 0} \qquad\qquad (12.2)$$

E X E R C I S E S / S E C T I O N   **12.1**

In Exercises 1–16, change each exponential form to the logarithmic form.

1. $3^4 = 81$ **2.** $2^6 = 64$ **3.** $10^3 = 1{,}000$

4. $7^2 = 49$ **5.** $4^4 = 256$ **6.** $5^{-2} = \dfrac{1}{25}$

7. $2^{-4} = \dfrac{1}{16}$ **8.** $49^{1/2} = 7$ **9.** $3^0 = 1$

10. $2^0 = 1$ **11.** $\left(\dfrac{3}{4}\right)^2 = \dfrac{9}{16}$ **12.** $\left(\dfrac{1}{2}\right)^{-2} = 4$

13. $\left(\dfrac{3}{2}\right)^{-3} = \dfrac{8}{27}$ **14.** $10^{-4} = 0.0001$

15. $3^{-4} = \dfrac{1}{81}$ **16.** $\left(\dfrac{1}{10}\right)^{-2} = 100$

In Exercises 17–32, change each logarithmic form to the exponential form.

17. $\log_5 125 = 3$

18. $\log_{10} 100 = 2$

19. $\log_5 1 = 0$

20. $\log_6 6 = 1$

21. $\log_2 2 = 1$

22. $\log_2 \dfrac{1}{64} = -6$

23. $\log_3 \dfrac{1}{27} = -3$

24. $\log_6 216 = 3$

25. $\log_{1/2} \dfrac{1}{4} = 2$

26. $\log_{1/3} 27 = -3$

27. $\log_{10} \dfrac{1}{1,000} = -3$

28. $\log_4 \dfrac{1}{256} = -4$

29. $\log_6 \dfrac{1}{36} = -2$

30. $\log_{15} 225 = 2$

31. $\log_{25} \dfrac{1}{25} = -1$

32. $\log_7 \dfrac{1}{7} = -1$

In Exercises 33–50, find the value of the unknown.

33. $\log_3 x = 4$

34. $\log_4 x = 2$

35. $\log_3 27 = a$

36. $\log_5 125 = a$

37. $\log_{10} x = 4$

38. $\log_b 49 = 2$

39. $\log_b 32 = 5$

40. $\log_{3/2} x = 2$

41. $\log_{25} x = \dfrac{1}{2}$

42. $\log_5 x = -2$

43. $\log_4 x = -3$

44. $\log_6 x = -1$

45. $\log_{3/4} x = -2$

46. $\log_b 16 = -2$

47. $\log_b \dfrac{27}{8} = -3$

48. $\log_b \dfrac{1}{64} = -6$

49. $\log_{49} \dfrac{1}{7} = a$

50. $\log_{27} \dfrac{1}{3} = a$

12.2   Graphs of Exponential and Logarithmic Functions

We have seen repeatedly that the behavior of a function can be observed from its graph. In this section we will discuss two new functions, the exponential and logarithmic functions, with the aid of their graphs.

> **Exponential Function:** An exponential function has the form
>
> $$y = b^x, \qquad b > 0, \qquad b \neq 1 \tag{12.3}$$

The exponential function appears to have a simple, even familiar, form and yet is quite different from the exponents discussed in Chapter 10. The reason is that the variable x in equation (12.3) need not be an integer or even a rational number—it can take on any real value. So we need to assume without proof that $y = b^x$ is defined for all real x.

As usual, we will construct our graphs by plotting points from a table of values. Consider the next example.

E X A M P L E **1**   Sketch the graph of $y = 2^x$.

Solution. To graph the function, we construct a table of values. For example, if $x = -3$, then $y = 2^x = 2^{-3} = 1/2^3 = 1/8$. Some other pairs of values are given next.

x:	-3	-2	-1	0	1	2	3
y:	$\dfrac{1}{8}$	$\dfrac{1}{4}$	$\dfrac{1}{2}$	1	2	4	8

By plotting the points from the values in the table and connecting them by a smooth curve, we obtain the graph in Figure 12.1.   ◀

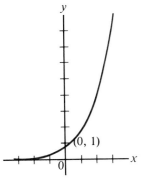

Figure 12.1

Since $b^0 = 1$, all graphs of the form $y = b^x$ have a single y-intercept at $(0, 1)$. If $b > 1$, then the graph is always similar to Figure 12.1, *approaching the negative x-axis* and rising rapidly through positive x values. For example, the y values of $y = 3^x$ increase rapidly as x increases.

We now turn to the logarithmic function, defined next.

Logarithmic Function: A logarithmic function has the form

$$y = \log_b x, \qquad b > 0, \qquad b \neq 1 \qquad\qquad (12.4)$$

A logarithmic function can be graphed by converting it to the form

$$x = b^y$$

Using this form, a table of values can readily be constructed, as shown in the next example.

E X A M P L E **2**   Sketch the graph of $y = \log_2 x$.

Solution. The function $y = \log_2 x$ can be written

$$x = 2^y$$

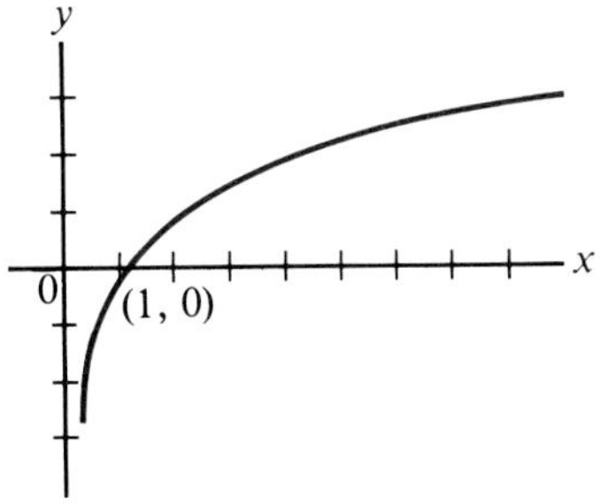

To obtain the table of values, we assign various values to y and compute x. Note that the resulting table is identical to the table in Example 1 with x and y interchanged.

y:	-3	-2	-1	0	1	2	3
x:	$\dfrac{1}{8}$	$\dfrac{1}{4}$	$\dfrac{1}{2}$	1	2	4	8

Figure 12.2

The graph is shown in Figure 12.2.   ◀

All graphs of the form $y = \log_b x$ have a single x-intercept at $(1, 0)$. If $b > 1$, the graph will be similar to Figure 12.2, *approaching the negative y-axis* and rising as x increases. Also, since $b^y > 0$, x is always positive. Since the graph crosses the x-axis at $(1, 0)$, $\log_b x < 0$ for $0 < x < 1$, and $\log_b x > 0$ for $x > 1$. For example, $\log_2 \frac{1}{2}$ is negative, while $\log_2 3$ is positive. (Although far from obvious, $\log_b x$ is complex for $x < 0$. Consequently, we exclude the case $x < 0$ from our discussion.)

> **Properties of the Logarithmic Function $y = \log_b x$:**
>
> **1.** If $0 < x < 1$, $\log_b x < 0$. $(b > 1)$
> **2.** If $x > 1$, $\log_b x > 0$. $(b > 1)$
> **3.** $\log_b 1 = 0$, $\log_b b = 1$.

For the exponential function, the case $0 < b < 1$ arises occasionally. For this case $y = b^x$ differs from the shape in Figure 12.1, as shown in the next example.

E X A M P L E   **3**

Half-life

The *half-life* of a radioactive element is the time required for half of a given amount of the element to decay. If N_0 is the given amount and H the half-life, then the amount left as a function of time t is given by

$$N = N_0\left(\frac{1}{2}\right)^{t/H}, \qquad t \geq 0$$

For example, the half-life of radium is 1,590 years, so that

$$N = N_0\left(\frac{1}{2}\right)^{t/1,590}$$

As another example, suppose a radioactive isotope has a half-life of 1.0 day. If the initial amount is 1.0 kg, then the amount after t days is

$$N = 1.0\left(\frac{1}{2}\right)^{t/1.0} \qquad \text{(in kilograms)}$$

Sketch the curve.

Solution. The curve (Figure 12.3) is plotted from the given table of values. Thus for $t = 0$,

$$N = 1.0\left(\frac{1}{2}\right)^{0/1.0} = 1.0\left(\frac{1}{2}\right)^{0} = 1.0(1) = 1.0$$

t (days):	0	1	2	3
N (kg):	1.0	0.5	0.25	0.125

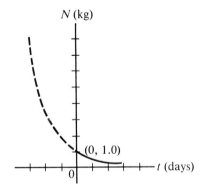

Figure 12.3

Note that the curve approaches the horizontal axis on the positive side. To show the behavior of this exponential function, the curve in Figure 12.3 is drawn for negative as well as positive values. However, since the original amount of 1.0 kg corresponds to $t = 0$, the N values are meaningful only for $t \geq 0$. Consequently, the portion of the curve for negative values is drawn as a dashed curve. ◀

E X E R C I S E S / S E C T I O N   **12.2**

In Exercises 1–20, sketch the graph of each function.

1. $y = 3^x$ **2.** $y = 4^x$ **3.** $y = 5^x$

4. $y = 6^x$ **5.** $y = \left(\frac{1}{3}\right)^x$ **6.** $y = \left(\frac{1}{4}\right)^x$

7. $y = 2^{-x}$ **8.** $y = 3^{-x}$ **9.** $y = 2^{2x}$

10. $y = \left(\frac{1}{2}\right)^{3x}$ **11.** $y = \log_3 x$ **12.** $y = \log_4 x$

13. $y = \log_5 x$ **14.** $y = \log_6 x$ **15.** $y = \log_{1/2} x$

16. $y = \log_{1/3} x$ **17.** $y = \log_{1/4} x$ **18.** $y = 2\log_2 x$

19. $y = 3\log_4 x$ **20.** $y = 3\log_5 x$

21. A certain radioactive isotope has a half-life of 2.0 years. If 10.0 g are present initially, then the amount left after t years is

$$N = 10.0\left(\frac{1}{2}\right)^{t/2.0}, \qquad t \geq 0$$

Sketch N as a function of t.

22. The number of bacteria in a culture doubles every hour. If there are 1,000 bacteria initially, then the number of bacteria after t hours is given by $N = 1,000(2^t)$, $t \geq 0$. Sketch the curve.

23. In a certain chemical reaction, a substance is converted into another substance. Starting with 1 kg of unconverted substance, the time required for the amount of unconverted substance to shrink to N $(0 < N \leq 1)$ is given by

$$t = -10 \log_{10} N, \qquad 0 < N \leq 1$$

where t is in seconds. Sketch t as a function of N.

24. An object at room temperature (20°C) is placed in an oven kept at constant temperature 100°C. If after 1 min the temperature has risen to 40°C, it can be shown that the temperature at any time is given by

$$T = 100 - 80\left(\frac{3}{4}\right)^t, \qquad t \geq 0$$

Sketch T as a function of t.

25. The tension T (in pounds) on a certain rope around a cylindrical post at a point P is related to the central angle θ determined by P by the formula $\log_{10} T = 0.5\theta$, where $\theta \geq 0$ is measured in radians. Write this formula in exponential form and sketch the graph.

26. In a certain chemical reaction, the amount N (in kilograms) of unconverted substance varies with time t (in seconds) according to the relation

$$\log_{10} N = -0.5t, \qquad t \geq 0$$

Write this formula in exponential form and sketch the graph.

12.3 Properties of Logarithms

In this section we will study the basic properties of logarithms. A knowledge of these properties is essential in using logarithms effectively.

Let us first recall the laws of exponents, expressed below in the forms that we will need:

$$b^M b^N = b^{M+N} \tag{12.5}$$

$$\frac{b^M}{b^N} = b^{M-N} \tag{12.6}$$

$$(b^M)^r = b^{Mr} \tag{12.7}$$

Since logarithms are exponents, the laws of logarithms are just the laws of exponents stated in a different form. For example, formula (12.5) corresponds to

$$\log_b xy = \log_b x + \log_b y \tag{12.8}$$

To see why, let $M = \log_b x$ and $N = \log_b y$. Then

$$b^M = x \quad \text{and} \quad b^N = y$$

and

$$xy = b^M b^N = b^{M+N}$$

Converting $b^{M+N} = xy$ to a logarithmic form, we get

$$\log_b xy = M + N$$

Substituting back, we get (since $M = \log_b x$ and and $N = \log_b y$)

$$\log_b xy = \log_b x + \log_b y$$

To obtain the form corresponding to rule (12.6), we take the quotient of x and y to get

$$\frac{x}{y} = \frac{b^M}{b^N} = b^{M-N}$$

Converting $b^{M-N} = x/y$ to logarithmic form, we get

$$\log_b \frac{x}{y} = M - N = \log_b x - \log_b y \tag{12.9}$$

The form corresponding to rule (12.7) can be obtained similarly. Let $M = \log_b x$, so that $b^M = x$. The rth power of x is

$$x^r = (b^M)^r = b^{Mr}$$

Converting $b^{Mr} = x^r$ to logarithmic form, we have (since $M = \log_b x$)

$$\log_b x^r = Mr = r \log_b x \tag{12.10}$$

Thus we obtain the three formulas listed in the following box.

Properties of Logarithms: For any positive real number b, $b \neq 1$, for any *positive* real numbers x and y, and for any real number r:

1. $\log_b x + \log_b y = \log_b xy$ $\qquad\qquad$ (12.11)

2. $\log_b x - \log_b y = \log_b \dfrac{x}{y}$ $\qquad\qquad$ (12.12)

3. $\log_b x^r = r \log_b x$ $\qquad\qquad$ (12.13)

These properties are illustrated in the next example.

E X A M P L E **1**   **a.** $\log_3 35 = \log_3 5 \cdot 7 = \log_3 5 + \log_3 7$ **by property (12.11)**

b. $\log_5 \dfrac{7}{11} = \log_5 7 - \log_5 11$ **by property (12.12)**

c. $\log_{10} 16 = \log_{10} 2^4 = \mathbf{4} \log_{10} 2$ **by property (12.13)**   ◀

Certain special cases involve the following properties of logarithms: $\log_b b = 1$ and $\log_b 1 = 0$. [See statement (12.2) in Section 12.1.]

E X A M P L E **2**   Evaluate:

a. $\log_5 25$ **b.** $\log_4 \frac{1}{4}$

Solution.

a. Since $25 = 5^2$, we may write

$$\log_5 25 = \log_5 5^2$$

Since $\log_b x^r = r \log_b x$, we have

$$\log_5 5^2 = \mathbf{2} \log_5 5$$

Finally, since $\log_b b = 1$, it follows that

$$2 \log_5 5 = 2 \cdot \mathbf{1} = 2$$

b. $\log_4 \dfrac{1}{4} = \log_4 1 - \log_4 \mathbf{4}$ **by property (12.12)**

$$= 0 - 1 = -1$$ **$\log_b 1 = 0$, $\log_b b = 1$**   ◀

Now that we have seen how the properties of logarithms can be used to simplify certain elementary expressions, we can break up more complicated logarithmic expressions into sums, differences, and multiples of logarithms. Consider the next example.

E X A M P L E **3**   Break up the given logarithms into sums, differences, or multiples of logarithms. If possible, use the fact that $\log_b b = 1$ and $\log_b 1 = 0$.

a. $\log_3 3x^2$ **b.** $\log_7 \dfrac{1}{x^3}$ **c.** $\log_3 \sqrt[5]{3}$

Solution.

a. $\log_3 3x^2 = \log_3 3 + \log_3 x^2$ **log of a product**

$$= 1 + \log_3 x^2$$ **$\log_b b = 1$**

$$= 1 + \mathbf{2} \log_3 x$$ **property (12.13)**

b. $\log_7 \dfrac{1}{x^3} = \log_7 1 - \log_7 x^3$ **log of a quotient**

$$= 0 - \log_7 x^3 \qquad \textbf{log}_b \textbf{ 1} = \textbf{0}$$

$$= -3 \log_7 x \qquad \textbf{property (12.13)}$$

c. $\log_3 \sqrt[5]{3} = \log_3 3^{1/5}$

$$= \frac{1}{5} \log_3 3 \qquad \textbf{log}_b \textbf{ } x^r = r \textbf{ log}_b \textbf{ } x$$

$$= \frac{1}{5}(1) \qquad \textbf{log}_b \textbf{ } b = \textbf{1}$$

$$= \frac{1}{5} \qquad\qquad\qquad\qquad \blacktriangleleft$$

The properties of logarithms can also be used to combine expressions involving logarithms. Consider the next example.

E X A M P L E   **4**   Combine $\frac{1}{2} \log_5 x - 2 \log_5 x$ into a single logarithm.

Solution. $\dfrac{1}{2} \log_5 x - \mathbf{2} \log_5 x = \log_5 x^{1/2} - \log_5 x^2$ $r \textbf{ log}_b \textbf{ } x = \textbf{log}_b \textbf{ } x^r$

$$= \log_5 \frac{x^{1/2}}{x^2} \qquad \textbf{log of a quotient}$$

$$= \log_5 \frac{1}{x^{3/2}} \qquad \textbf{reducing} \qquad \blacktriangleleft$$

Common error   Writing

$$\log_b(x + y) \qquad \text{as} \qquad \log_b x + \log_b y$$

By the first property of logarithms,

$$\log_b x + \log_b y = \log_b xy$$

The expression $\log_b(x + y)$ cannot be written as a sum of two logarithms.

 E X A M P L E   **5**   The loudness β of a sound (in decibels) is given by

$$\beta = 10(\log_{10} I - \log_{10} I_0)$$

where I is the intensity of the sound in watts per square meter and I_0 the intensity of the faintest audible sound. Write β as a single logarithm.

Solution. $\beta = 10(\log_{10} I - \log_{10} I_0)$

$$= 10 \log_{10} \frac{I}{I_0}$$

$$= \log_{10} \left(\frac{I}{I_0}\right)^{10}$$

◄

E X E R C I S E S / S E C T I O N **12.3**

In Exercises 1–40, write each given expression as the sum, difference, or multiple of logarithms. Whenever possible, use $\log_b b = 1$ and $\log_b 1 = 0$ to simplify the result. (See Example 3.)

1. $\log_{10} 81$ **2.** $\log_2 27$ **3.** $\log_3 27$

4. $\log_5 21$ **5.** $\log_7 36$ **6.** $\log_7 49$

7. $\log_{10} 100$ **8.** $\log_5 125$ **9.** $\log_2 16$

10. $\log_3 81x^3$ **11.** $\log_3 \sqrt{27}$ **12.** $\log_{10} 1,000$

13. $\log_5 25x^2$ **14.** $\log_7 \sqrt{7x}$ **15.** $\log_3 15x^2$

16. $\log_4 4x^3$ **17.** $\log_4 20\sqrt{a}$ **18.** $\log_2 \frac{1}{2x}$

19. $\log_3 \frac{1}{9a^2}$ **20.** $\log_4 \frac{1}{16V^2}$ **21.** $\log_3 \frac{1}{\sqrt{3C}}$

22. $\log_2 \frac{1}{\sqrt[3]{4v}}$ **23.** $\log_5 \frac{1}{30\sqrt{u}}$ **24.** $\log_5 \frac{1}{50r^2}$

25. $\log_a \frac{1}{a^2}$ **26.** $\log_c \frac{1}{\sqrt{c}}$ **27.** $\log_e \frac{1}{\sqrt[3]{e}}$

28. $\log_e 36e^2$ **29.** $\log_e \frac{25}{e^3}$ **30.** $\log_e \frac{49}{\sqrt{e}}$

31. $\log_e \frac{1}{\sqrt{\pi e}}$ **32.** $\log_e 9\sqrt{e}$ **33.** $\log_e 2\sqrt[3]{e}$

34. $\log_{10} 900$ **35.** $\log_{10} \frac{1}{400}$ **36.** $\log_{10} \frac{1}{16,000}$

37. $\log_{10} 0.009$ **38.** $\log_{10} 0.00025$ **39.** $\log_{10} \frac{x}{300}$

40. $\log_{10} \frac{x^2}{75}$

In Exercises 41–56, write each expression as a single logarithm. (See Example 4.)

41. $\log_2 3 + \log_2 5$ **42.** $\log_5 10 - \log_5 5$

43. $\log_4 30 - \log_4 3$ **44.** $\log_6 14 - \log_6 7$

45. $\log_2 7 + \log_2 5 - \log_2 3$ **46.** $\log_e 2 + \log_e 5 - \log_e 7$

47. $2 \log_2 5 - \log_2 3$ **48.** $3 \log_{10} 4 - \log_{10} 32$

49. $\frac{1}{2} \log_{10} 16$ **50.** $\frac{1}{3} \log_6 27$

51. $\frac{1}{2} \log_2 x - \log_2 3$ **52.** $\frac{1}{2} \log_5 V + \log_5 L$

53. $\frac{1}{2} \log_e m + \frac{1}{2} \log_e v$ **54.** $\frac{1}{3} \log_e s + \frac{1}{4} \log_e 16$

55. $\frac{1}{2} \log_7 s + \frac{1}{2} \log_7 t - \log_7 2$

56. $\frac{1}{3} \log_5 a^6 - \frac{1}{3} \log_5 64$

57. If p_0 is the pressure at sea level and p the pressure at the top of a column of air h meters in height and having a uniform temperature T (in Kelvin), then

$$\log_{10} p - \log_{10} p_0 = -0.0149 \frac{k}{T}$$

where k is a constant. Combine the terms on the left side and write the equation in exponential form.

58. The fallout from a nuclear explosion can be written $\log_{10} R - \log_{10} R_0 = kt$, where R is the amount of radiation after the explosion, R_0 the amount before the explosion, and k a constant. Combine the terms on the left side and write R as a function of time t.

59. If p (in millimeters of mercury) is the vapor pressure of carbon tetrachloride, then

$$\log_{10} p + \log_{10}(5.97 \times 10^7) = -\frac{1706.4}{T}$$

where T is the temperature in Kelvin. Combine the terms on the left side and write the equation in exponential form.

12.4　Common Logarithms

Although logarithms were defined for an arbitrary base $b > 0$, $b \neq 1$, in practice only two bases are generally used. Logarithms to base 10, called **common logarithms**, are suitable for numerical work, since our number system is in base 10. They also arise in certain applications. Logarithms to base e are called **natural logarithms**, since they arose quite naturally in the development of the calculus. In this section we discuss common logarithms.

Before calculators became widely available, common logarithms of numbers were obtained from a table. (Table 2 in Appendix E is a table of common logarithms.) The table lists the logarithms only for numbers between 1 and 10. For example, to find $\log_{10} 5.36$ we look under n and find 5.3 (the first two significant digits). In this row, under the 6 (the third significant digit), we find .7292. So

$$\log_{10} 5.36 = 0.7292$$

To see the practical importance of base 10, let us find $\log_{10} 536$. In scientific notation,

$$536 = 5.36 \times 10^2$$

As a result,

$$
\begin{aligned}
\log_{10} 536 &= \log_{10}(5.36 \times 10^2) \\
&= \log_{10} 5.36 + \log_{10} 10^2 \\
&= \log_{10} 5.36 + 2 \log_{10} 10 \\
&= \log_{10} 5.36 + 2 \cdot 1 \\
&= 0.7292 + 2 = 2.7292
\end{aligned}
$$

In the statement

$$\log_{10} 536 = 0.7292 + 2$$

the number 0.7292 is the *mantissa* and the number 2 is the *characteristic*.

In general, for the number

$$M = N \times 10^k \qquad (1 \leq N < 10)$$

$\log_{10} N$ is the **mantissa** and k the **characteristic**. In summary, to find the common logarithm of a number using Table 2, write the number in scientific notation, obtain the mantissa from the table, and add the characteristic to the mantissa.

It is customary to omit the number 10 indicating the base. Thus $\log N$ is understood to mean $\log_{10} N$.

$$\boxed{\quad \log N \quad \text{means} \quad \log_{10} N \quad}$$

E X A M P L E **1**   Find log 3,610.

Solution. $3,610 = \mathbf{3.61} \times 10^3$. So the mantissa is **log 3.61** and the characteristic **3** From Table 2 of Appendix E, log 3.61 = **0.5575**. Hence

$$\log 3,610 = 0.5575 + 3 = 3.5575 \qquad \blacktriangleleft$$

E X A M P L E **2**   Find log 74,600.

Solution. $74,600 = \mathbf{7.46} \times 10^4$

From the table, **log 7.46 = 0.8727**. Since the characteristic is **4**,

$$\log 74,600 = 0.8727 + 4 = 4.8727 \qquad \blacktriangleleft$$

E X A M P L E **3**   Find log 0.00813.

Solution. $0.00813 = 8.13 \times 10^{-3}$

From the table, log 8.13 = 0.9101; since the characteristic is -3, we get

$$\log 0.00813 = -3 + 0.9101 = -2.0899$$

To preserve the mantissa obtained from the table, the logarithm is also written $-3 + 0.9101 = 7 - 10 + 0.9101 = (7 + 0.9101) - 10$, or

$$\log 0.00813 = 7.9101 - 10 \qquad \blacktriangleleft$$

CALCULATOR COMMENT   Logarithms can be found using a calculator. For example, to find log 0.000581, enter 0.000581 and press the $\boxed{\text{LOG}}$ key.

E X A M P L E **4**   Use a calculator to find:

a. log 4,936   **b.** log 0.0003716

Solution.

a. The sequence is

$$4936 \; \boxed{\text{LOG}} \; \rightarrow \; 3.6933752$$

or log 4,936 = 3.6934 to four decimal places.

b. The sequence is

$$0.0003716 \; \boxed{\text{LOG}} \; \rightarrow \; -3.4299243$$

So log 0.0003716 = -3.4299 to four decimal places. $\qquad \blacktriangleleft$

Antilogarithms

Antilogarithm

If log N is known, we can use the table to find N. The number N is called the **antilogarithm**.

E X A M P L E **5** If log $N = 3.8943$, find N, the antilogarithm.

Solution. We look up the number 0.8943 in the table and find that it corresponds to 7.84. Since the characteristic is 3,

$$N = 7.84 \times 10^3 = 7,840$$ ◄

E X A M P L E **6** If log $N = 0.9560$, find N.

Solution. Since the number 0.9560 is not listed, we need to interpolate by the same method used in Chapter 4:

$$0.01 \left[x \begin{bmatrix} \log 9.03 = 0.9557 \\ \\ \log N \quad = 0.9560 \\ \log 9.04 = 0.9562 \end{bmatrix} 0.0003 \right] 0.0005$$

Thus

$$\frac{x}{0.01} = \frac{0.0003}{0.0005} \quad \text{and} \quad x = 0.006$$

We conclude that $N = 9.03 + 0.006 = 9.036$

CALCULATOR COMMENT

As with most calculations, it is simpler to obtain the antilogarithm with a calculator: enter 0.9560, press the $\boxed{\text{INV}}$ key, and then press the $\boxed{\text{LOG}}$ key. The sequence is

$$0.9560 \boxed{\text{INV}} \boxed{\text{LOG}} \to 9.0364947$$

On some calculators the proper sequence is

$$0.9560 \boxed{10^x}$$ ◄

E X A M P L E **7** If log $N = -3.6747$, find N.

Solution. The sequence is

$$3.6747 \boxed{+/-} \boxed{\text{INV}} \boxed{\text{LOG}} \to 2.1149495 \quad -04$$

or

$$3.6747 \boxed{+/-} \boxed{10^x}$$

The result is $N = 2.1149 \times 10^{-4}$. ◄

E X A M P L E **8**   The atmospheric pressure P (in pounds per square inch) can be obtained from the equation

$$\log_{10} P = 0.434(2.69 - 0.21h)$$

where h is the altitude in miles above sea level. Find the atmospheric pressure at an altitude of **45** mi.

Solution. From the given formula,

$$\log_{10} P = 0.434(2.69 - 0.21 \cdot \mathbf{45})$$

The sequence is

$$2.69\;\boxed{-}\;0.21\;\boxed{\times}\;45\;\boxed{=}\;\boxed{\times}\;0.434\;\boxed{=}\;\boxed{\text{INV}}\;\boxed{\text{LOG}} \rightarrow 0.001164555$$

So the atmospheric pressure is 0.0012 lb/in.^{2} to 2 significant digits.   ◀

E X E R C I S E S / S E C T I O N   **12.4**

In Exercises 1–10, find the common logarithm of each number by using Table 2 of Appendix E.

1. log 5.59

2. log 559

3. log 83,400

4. log 613,000

5. log 0.00524

6. log 1,290,000

7. log 2,478 (interpolate)

8. log 72,870

9. log 853,200

10. log 0.002463

In Exercises 11–20, use a calculator to find each common logarithm to four decimal places.

11. log 18.2   **12.** log 0.1763   **13.** log 0.006417

14. log 280.4   **15.** log 29,654   **16.** log 0.01834

17. log 2.846   **18.** log 3.074   **19.** log 63.712

20. log 21.53

In Exercises 21–30, use Table 2 of Appendix E to find the antilogarithm in each case.

21. 1.8639   **22.** 2.5599   **23.** 3.7300

24. 8.7745 − 10   **25.** 9.9325 − 10   **26.** 2.8497

27. 3.6472   **28.** 8.7221 − 10   **29.** 7.9308 − 10

30. 6.9615 − 10

In Exercises 31–40, use a calculator to find the antilogarithm in each case. Round off the answers to four decimal places.

31. 0.8962   **32.** 0.4018   **33.** 1.6326

34. 2.9304   **35.** 2.0439   **36.** 0.09432

37. −1.743   **38.** −1.1917   **39.** −0.8436

40. −0.09432

41. The acidity of a chemical solution is determined by the concentration of the hydrogen ion H^+, written $[H^+]$, and is measured in moles per liter (mol/L). The hydrogen potential pH is defined by

$$pH = -\log_{10}[H^+]$$

Given that the acid concentration of water is 10^{-7} mol/L, determine the pH of water.

42. The acid concentration of blood is 3.98×10^{-8} mol/L. Determine the pH value.

43. A certain satellite has a power supply whose output P (in watts) is given by $\log_{10} P = -t/901$, where t is the number of days that the battery has operated. What will the output be after 379 days?

44. A record company estimates that its monthly profit (in thousands of dollars) from a new hit record is $P = 12 - 15 \log_{10}(1 + t)$, where t is the number of months after release. Calculate the profit (or loss) for the first six months and determine when the record should be withdrawn.

12.5   Computations with Logarithms (Optional)

In this section we will briefly see how logarithms are used in numerical calculations. As indicated earlier, this application of logarithms has lost much of its importance since the advent of the scientific calculator.

E   X   A   M   P   L   E   **1**   Multiply $396,100 \times 0.0005686$ by means of logarithms.

Solution. Let

$$N = (396,100)(0.0005686)$$

Then, by property (12.11),

$$\log N = \log 396,100 + \log 0.0005686$$
$$= \log(3.961 \times 10^5) + \log(5.686 \times 10^{-4})$$

From Table 2 of Appendix E,

$$
\begin{array}{lll}
\log 396,100 & = 5.5978 & \textbf{characteristic} = \textbf{5} \\
\log 0.0005686 = & 6.7548 - 10 & \textbf{characteristic} = -\textbf{4} = \textbf{6} - \textbf{10} \\
\hline
\log N & = 12.3526 - 10 & \text{(adding)}
\end{array}
$$

which gives $\log N = 2.3526$. The desired product is the antilogarithm $N = 225.2$.   ◀

E   X   A   M   P   L   E   **2**   Calculate $\sqrt[6]{2,746}$.

Solution. The extraction of a root requires the third law of logarithms (12.13),

$$\log_b x^r = r \log_b x$$

Let $N = \sqrt[6]{2,746} = (2,746)^{1/6}$. Then $\log N = \frac{1}{6} \log 2,746 = \frac{1}{6}(3.4387) = 0.5731$. The desired root is the antilogarithm $N = 3.742$.   ◀

E   X   A   M   P   L   E   **3**   Calculate $\sqrt[4]{0.007140}$.

Solution. Let $N = (0.007140)^{1/4}$. From Table 2,

$$
\begin{aligned}
\log 0.007140 &= \log(7.140 \times 10^{-3}) \\
&= \log 7.140 + \log 10^{-3} \\
&= 0.8537 - 3 \\
&= 7.8537 - 10
\end{aligned}
$$

This is the usual form for negative characteristics. To obtain this form *after* the division by 4, we need to write

$$7.8537 - 10 = 37.8537 - 40 \qquad 7 - 10 = 37 - 40 = -3$$

Now dividing by 4 leaves -10. Thus

$$\log N = \frac{1}{4}(37.8537 - 40) = 9.4634 - 10 \qquad \textbf{characteristic} = -1$$

and $N = 2.907 \times 10^{-1} = 0.2907$. ◀

Logarithms are particularly useful when several operations are combined, as in the next example.

E X A M P L E **4**   Calculate:

$$\frac{3.87 \sqrt[3]{7,462}}{\sqrt{0.0321}}$$

Solution. Let

$$N = \frac{3.87 \sqrt[3]{7,462}}{\sqrt{0.0321}}$$

Then $\log N = \log 3.87 + \frac{1}{3}\log 7{,}462 - \frac{1}{2}\log 0.0321$

$$
\begin{array}{ll}
\log 3.87 \quad= 0.5877 & \\
\frac{1}{3}\log 7{,}462 \ = 1.2909 & \textbf{log 7,462 = 3.8728 by interpolation} \\
\hline
\qquad\qquad 1.8786 & \textbf{adding} \\
\frac{1}{2}\log 0.0321 = 9.2533 - 10 & \textbf{log 0.0321 = 8.5065 − 10 = 18.5065 − 20} \\
\hline
\qquad\qquad 2.6253 & \textbf{subtracting: 1.8786 − (9.2533 − 10)}
\end{array}
$$

Thus $\log N = 2.6253$ and $N = 422$ to three significant digits. ◀

E X E R C I S E S / S E C T I O N   **12.5**

Use logarithms to perform the indicated calculations.

1. $(0.00962)(8{,}464)$

2. $(0.6416)(72.11)$

3. $\dfrac{79.60}{24.94}$

4. $\dfrac{6.477}{0.03420}$

5. $\sqrt[3]{8{,}642}$

6. $\sqrt[3]{0.0005128}$

7. $\sqrt[10]{0.000000176}$

8. $\sqrt[3]{6{,}416}$

9. $\dfrac{\sqrt[3]{46.71}}{\sqrt[4]{3.068}}$

10. $\dfrac{\sqrt[3]{1.36}}{\sqrt[5]{0.0846}}$

11. $47.62\sqrt{5.620}\sqrt[3]{7.301}$

12. $\sqrt[4]{(0.001230)(0.009647)}$

13. $\dfrac{2{,}765\sqrt[3]{0.0003620}}{5.247}$

14. $(0.007648\sqrt[3]{2313})^{1/2}$

15. $\dfrac{9.623\sqrt[3]{5.128}}{\sqrt{0.07225}}$

16. $\dfrac{0.0000123}{\sqrt[6]{0.000926}}$

12.6   Natural Logarithms, Change of Base, and Powers of e

The last two sections were devoted to common logarithms. In this section we will briefly discuss logarithms to base e, which are called **natural logarithms** and play an important role in theoretical work.

We already saw that $e \approx 2.71828$ is used in the exponential form of a complex number. Although the origin of this number is clarified in the study of calculus, the examples and exercises that follow will indicate how such logarithms arise.

To avoid having to write the base each time, $\log_e x$ is denoted by "ln x."

Notation for Natural Logarithm:

$\log_e x \qquad$ is denoted by $\qquad \ln x$

where $e \approx 2.71828$.

The natural logarithm of a number can be obtained from the common logarithm. This method is a special case of the more general problem of changing from one base to another, which is stated next.

Formula for Changing the Base of a Logarithm:

$$\log_b x = \frac{\log_a x}{\log_a b} \qquad\qquad (12.14)$$

where a is any new base.

To derive this formula, let $M = \log_b x$, so that $b^M = x$. Take the logarithm to base a of both sides to obtain

$$\log_a b^M = \log_a x$$

$$M \log_a b = \log_a x \qquad \mathbf{\log_b x^r = r \log_b x}$$

$$M = \frac{\log_a x}{\log_a b} \qquad \textbf{dividing by } \mathbf{\log_a b}$$

Since $M = \log_b x$, we get

$$\log_b x = \frac{\log_a x}{\log_a b}$$

E X A M P L E **1**   Find ln 3.21 by means of common logarithms.

Solution. By formula (12.14)

$$\ln 3.21 = \log_e \mathbf{3.21} = \frac{\log_{10} \mathbf{3.21}}{\log_{10} e}$$

$$= \frac{\log_{10} 3.21}{\log_{10} 2.718}$$

$$= \frac{0.5065}{0.4343} = 1.166 \qquad \text{(using Table 2 of Appendix E)} \qquad \blacktriangleleft$$

It is true in general that

$$\ln x = \frac{\log_{10} x}{\log_{10} e} = \frac{\log x}{0.4343}$$

Since $1/0.4343 = 2.3026$, we now have the following relationships:

$$\ln x = 2.3026 \log x \qquad\qquad\qquad (12.15)$$

$$\log x = 0.4343 \ln x \qquad\qquad\qquad (12.16)$$

CALCULATOR COMMENT

Natural logarithms can also be obtained by using a scientific calculator, as shown in the next example.

E X A M P L E **2**   Use a calculator to find:

a. ln 4.7125   **b.** $\log_6$ 3.8926

Solution.

a. Enter 4.7125 and press $\boxed{\text{LN}}$ or $\boxed{\ln x}$ to get ln 4.7125 = 1.5502. The sequence is

$$4.7125 \boxed{\text{LN}} \rightarrow 1.5502186$$

b. By (12.14),

$$\log_6 \mathbf{3.8926} = \frac{\ln \mathbf{3.8926}}{\ln \mathbf{6}} = 0.7585$$

The sequence is

$$3.8926 \boxed{\text{LN}} \boxed{\div} 6 \boxed{\text{LN}} \boxed{=} \rightarrow 0.7585155 \qquad \blacktriangleleft$$

 E X A M P L E **3**

Half-life

The *half-life* of a radioactive substance is the time required for half of a given amount to decay. If a radioactive substance has a half-life H, then the time taken for the initial amount N_0 to shrink to N is given by

$$t = \frac{H}{\ln 2} \ln \frac{N_0}{N} \qquad (12.17)$$

How long will **10** g of polonium, which has a half-life of **140** days, take to shrink to **7.0** g?

Solution. By formula (12.17),

$$t = \frac{140}{\ln 2} \ln \frac{10}{7.0}$$

A possible sequence is

$$140 \;\boxed{\div}\; 2 \;\boxed{\text{LN}}\;\boxed{\times}\;\boxed{(}\; 10 \;\boxed{\div}\; 7.0 \;\boxed{)}\;\boxed{\text{LN}}\;\boxed{=} \rightarrow 72.040244$$

The final result is 72 days. ◀

The Exponential Function

Since e is the base for natural logarithms, $\ln N = x$ is equivalent to $N = e^x$. Because of this close relationship between logarithms and exponents, the exponential form $y = e^{bx}$ also occurs frequently in applications. The function values can be obtained with a scientific calculator by pressing $\boxed{e^x}$ or $\boxed{\text{INV}}\;\boxed{\ln x}$.

CALCULATOR COMMENT

To find natural logarithms in BASIC, use the library function LOG(X) [that is, LOG(X) is the function $\ln x$]. The library function EXP(X) is the exponential function e^x.

 E X A M P L E **4**

Just as interest can be compounded quarterly or daily, it can also be compounded continuously. The formula is

$$P = P_0 e^{rt} \qquad (12.18)$$

where P_0 is the initial amount invested, t the time in years, and r the rate of interest. To what amount will $1,000 accumulate after 9.5 years at 10.25% interest compounded continuously?

Solution. Since $10.25\% = 0.1025$, we get, by formula (12.18),

$$P = 1,000e^{(0.1025)(9.5)} = \$2,647.86$$

The sequence is

$$0.1025 \;\boxed{\times}\; 9.5 \;\boxed{=}\;\boxed{e^x}\;\boxed{\times}\; 1000 \;\boxed{=} \rightarrow 2647.8553$$

or

$$\boxed{(}\; 0.1025 \;\boxed{\times}\; 9.5 \;\boxed{)}\;\boxed{e^x}\;\boxed{\times}\; 1000 \;\boxed{=}$$

◀

E X E R C I S E S / S E C T I O N   **12.6**

In Exercises 1–8, use a calculator to find the natural logarithms to four decimal places.

1. ln 7.32   **2.** ln 27.601   **3.** ln 0.5173

4. ln 0.00123   **5.** ln 0.1736   **6.** ln 2.3194

7. ln 3.8314   **8.** ln 0.089636

In Exercises 9–14, use a calculator to find the indicated logarithms to four decimal places. (See Example 2.)

9. $\log_5 3.864$ **10.** $\log_3 27.164$ **11.** $\log_4 0.00713$

12. $\log_7 0.03926$ **13.** $\log_6 126.77$ **14.** $\log_2 0.000127$

In Exercises 15–24, use a calculator to find the exponential values to four decimal places.

15. e^3 **16.** $e^{1.5}$ **17.** e^{-2} **18.** $e^{0.162}$

19. $e^{-0.92}$ **20.** $e^{2.68}$ **21.** $e^{0.013}$ **22.** $e^{-1.79}$

23. $e^{-0.03}$ **24.** $e^{-0.056}$

25. A radioactive isotope has a half-life of 50 years. Use formula (12.17) to find the time required for 50 g of the substance to shrink to 40 g.

26. A woman deposits $2,150 into a long-term savings account that pays 11% compounded continuously if left for 8 years. How much will she have at the end of this period? [Use formula (12.18).]

27. The present value is the amount of money that has to be invested in order to receive a specified amount at the end of a certain time period. The equation is given by $P_0 = Pe^{-rt}$. If you wish to have $10,000 at the end of 6 years, how much will you have to invest at 10.2% compounded continuously? (Compare this problem with Example 4.)

28. The quantity $n! = n \cdot (n - 1) \cdot (n - 2) \cdots \cdot 2 \cdot 1$, read "$n$ factorial," can be approximated by Stirling's formula

$$n! \approx \sqrt{2\pi n}\left(\frac{n}{e}\right)^n$$

Approximate 30!.

29. For a certain gas, an enclosed volume of 10 cm³ is gradually increased to v. The average pressure (in atmospheres) is given by

$$P_{av} = \frac{4}{v - 10}\ln\frac{v}{10}$$

Find the average pressure if the volume is increased to 50 cm³.

30. An electric circuit consists of a resistor of $R = 10.0\ \Omega$ and an inductor of $L = 0.0123$ H. When the current is $I_0 = 3.50$ A, the current source is removed and the current dies out quickly according to the equation

$$I = I_0 e^{-Rt/L}$$

Determine the current after 1.00 ms.

31. The distance traveled by a motorboat after the engine is shut off is

$$x = \frac{1}{k}\ln\left(v_0 kt + 1\right)$$

where v_0 is the velocity when the motor is running and k a constant. If $k = 0.00300$, a typical value, determine how far a motorboat traveling at 20.0 ft/s will continue in the first 15.0 s after the motor is shut off.

32. An object of weight w moving around a circle of radius a is subject to a retarding force. If the initial velocity is v_0, then the angular displacement as a function of time t is given by

$$\theta = \frac{w}{kga}\ln\left(1 + \frac{kgv_0 t}{w}\right)$$

where k is a constant. Suppose $k = 0.0040$. Determine the angular displacement of an object after 8.0 s if $v_0 = 25$ ft/s, $w = 10$ lb, and $a = 50$ ft ($g = 32$ ft/s²).

33. A certain satellite has a power supply whose output in watts is given by $P = 30.0e^{-t/700.0}$, where t is the number of days that the battery has operated. (Note that $P = 30.0$ W when $t = 0$.) What is the output after 1 year?

34. The radioactive element strontium-90 decays according to the formula

$$N = N_0 e^{-0.025t}$$

where N_0 is the initial amount and t the time in years. If $N_0 = 100$ g, determine the amount left after **(a)** 1 year, **(b)** 10 years, **(c)** 100 years.

12.7   Special Exponential and Logarithmic Equations

One of the most interesting applications of logarithms is the solution of equations in which the unknown is an exponent. To solve such equations, we write the equation in the form $a^x = b$ and take the logarithm of both sides:

$$\log a^x = \log b$$

By the third law of logarithms,

$$x \log a = \log b$$

and

$$x = \frac{\log b}{\log a}$$

E X A M P L E **1**   Solve the equation $3^x = 12$.

Solution. First we take common logarithms of both sides:

$$\log 3^x = \log 12$$

By (12.13), the third property of logarithms, we get

$$x \log 3 = \log 12$$

Then, dividing $x \log 3 = \log 12$ by log 3, we get

$$x = \frac{\log 12}{\log 3} = 2.262$$

CALCULATOR COMMENT

The sequence is

$$12 \boxed{\text{LOG}} \boxed{\div} 3 \boxed{\text{LOG}} \boxed{=} \rightarrow 2.2618595 \qquad \blacktriangleleft$$

Natural logarithms can be used equally well, as shown in the next example.

E X A M P L E **2**   Solve the equation

$$3(4^{x+1}) = 20$$

Solution. The given equation is equivalent to

$$4^{x+1} = \frac{20}{3} \qquad \textbf{dividing both sides by 3}$$

Taking natural logarithms, we get

$$\ln 4^{x+1} = \ln \frac{20}{3}$$

$$(x + 1) \ln 4 = \ln \frac{20}{3} \qquad \log_b x^r = r \log_b x$$

$$(x + 1) \mathbf{\ln 4} = \ln \frac{20}{3}$$

$$x + 1 = \frac{\ln \dfrac{20}{3}}{\mathbf{\ln 4}} \qquad \text{dividing by ln 4}$$

$$x = -1 + \frac{\ln \dfrac{20}{3}}{\ln 4} = 0.3685$$

A possible sequence is

$$20 \boxed{\div} 3 \boxed{=} \boxed{\text{LN}} \boxed{\div} 4 \boxed{\text{LN}} \boxed{=} \boxed{-} 1 \boxed{=} \rightarrow 0.3684828 \qquad \blacktriangleleft$$

If two or more logarithms are contained in an equation, they must first be combined by using the properties of logarithms, as shown in the next example.

E X A M P L E **3**   Solve the equation

$$2 \ln x - \ln(x + 1) = 0$$

Solution. The left side must be written as a single logarithm:

$$2 \ln x - \ln(x + 1) = 0$$

$$\ln x^2 - \ln(x + 1) = 0 \qquad r \log_b x = \log_b x^r$$

$$\ln \frac{x^2}{x + 1} = \log_e \frac{x^2}{x + 1} = 0 \qquad \text{difference of two logs}$$

By using the definition of a logarithm, this equation can be written in exponential form. Since e is the base and $\mathbf{0}$ the exponent, we get

$$\frac{x^2}{x + 1} = e^0$$

$$\frac{x^2}{x + 1} = 1$$

$$x^2 = x + 1$$

$$x^2 - x - 1 = 0$$

$$x = \frac{1 \pm \sqrt{1 + 4}}{2} = \frac{1 \pm \sqrt{5}}{2} \qquad \textbf{by the quadratic formula}$$

Since the given equation contains the term $2 \ln x$, x must be positive. Taking the positive root, we conclude that $x = \frac{1}{2}(1 + \sqrt{5})$ is the solution of the equation.   ◀

Exponential and logarithmic equations arise in various technical fields. Some of these applications are illustrated in the examples and exercises that follow.

 E X A M P L E **4**   In certain chemical reactions, the amount of unconverted substance is given by $x = x_0 e^{kt}$, where x_0 is the initial amount, t the time, and k a negative constant that depends on the substance. Solve this equation for t.

Solution. From the given equation, we have

$$\frac{x}{x_0} = e^{kt}$$

Taking natural logarithms of both sides, we get

$$\ln \frac{x}{x_0} = \ln e^{kt}$$

Recalling that $\log_b b = 1$, it follows that

$$\ln \frac{x}{x_0} = kt \ln e \qquad \log_b x^r = r \log_b x$$

$$\ln \frac{x}{x_0} = kt \cdot 1 \qquad \log_b b = 1$$

$$t = \frac{1}{k} \ln \frac{x}{x_0} \qquad \text{dividing both sides by } k$$

◀

 E X A M P L E **5**   Stars are classified according to their visual brightness, which is called *magnitude*. Stars of the first magnitude are the brightest. Let b_n and b_m denote the measured (actual) brightness of two stars having magnitudes n and m, respectively. They are related by the formula

$$m - n = 2.5 \log \frac{b_n}{b_m} \tag{12.19}$$

a. If one star is 10 times brighter than another, what is the difference in their magnitudes?
b. Sirius has magnitude 1.4. How much brighter is Sirius than a star of magnitude 12?

Solution.

a. Let $b_n/b_m = 10$. Then by formula (12.19), the difference in magnitudes is

$$m - n = 2.5 \log 10 = 2.5$$

b. Let $m = 12$ and $n = 1.4$. Now use formula (12.19) to find the ratio b_n/b_m:

$$12 - 1.4 = 2.5 \log \frac{b_n}{b_m}$$

$$10.6 = 2.5 \log \frac{b_n}{b_m}$$

$$\log_{10} \frac{b_n}{b_m} = \frac{10.6}{2.5} = 4.24$$

Since 10 is the base and 4.24 the exponent, we now get from the definition of logarithm

$$\frac{b_n}{b_m} = 10^{4.24} = 17,000$$

We conclude that Sirius is 17,000 times brighter than the other star. ◀

E X E R C I S E S / S E C T I O N   12.7

In Exercises 1–24, solve each equation for x.

1. $2^x = 5$
2. $3^x = 10$
3. $4^x = 20$
4. $6^x = 5$
5. $3^x = \frac{1}{2}$
6. $3^{x+2} = \frac{1}{4}$
7. $5^{x-1} = 2$
8. $2(3^x) = 7$
9. $2(6^x) = 21$
10. $3(7^{x-2}) = 59$
11. $9^{2x+3} = 5$
12. $15^{3x-1} = 6$
13. $2 \log_3 x = 4$
14. $\log_5(x + 2) = 2$
15. $\log_2 x = 0.64590$
16. $\log_{3/2} x = 5$
17. $\log_2 x + \log_2(x - 3) = 2$
18. $2 \log x - \log(x + 2) = 0$
19. $\frac{1}{2} \ln x + \ln 2 = 1$
20. $2 \log x = 1 - \log 5$
21. $\log x + \log x^2 = \log 8$
22. $2 \ln x - \ln(x + 3) = 0$
23. $2 \ln x - 2 \ln(x - 1) = 0$
24. $2 \ln x - 2 \ln(x + 1) = 0$

In Exercises 25–28, solve for the indicated letter.

25. $i = \frac{E}{R} e^{-t/RC}$, for t

26. $\log \frac{I}{I_0} = -\beta x$, for I

27. $Q = P_1 V_1 \log \frac{V_2}{V_1}$, for V_2

28. $T = T_2 + (T_1 - T_2)e^{-at}$, for a

29. What is the difference in magnitudes if one star is brighter than another by (a) 20 times? (b) 100 times? (Refer to Example 5.)

30. How much brighter is a star of magnitude 1 than a star of magnitude 10? (Refer to Example 5.)

31. The subjective impression of sound, measured in decibels, is related to the intensity of sound (measured in watts per square meter) by the formula

$$\beta = 10 \log \frac{I}{I_0}$$

where I_0 is an arbitrary reference level corresponding roughly to the faintest sound that can be heard (10^{-12} W/m^2). Rustling leaves have an intensity of 10^{-11} W/m^2. What is the loudness in decibels?

32. An elevated train has a loudness of 90 decibels. How many times more intense is the sound of an elevated train than the sound of rustling leaves (10 decibels)? (Refer to Exercise 31.)

33. Determine the intensity level of riveting, which has a loudness of 95 decibels. (Refer to Exercise 31.)

34. A certain body cools according to the equation

$$T = 20 + 60e^{-t/100}$$

where T is the temperature in degrees Celsius and t the time in minutes. Find an expression for t as a function of T.

35. Another logarithmic scale, used for measuring the magnitude of an earthquake, is the Richter scale

$$R = \log \frac{I}{I_0}$$

where I_0 is some arbitrary reference level. How much more intense was the famous San Francisco earthquake of 1906, which measured 8.25 on the Richter scale, than a minor tremor measuring 3.0 on the Richter scale?

36. How much more intense was the 1964 Alaska earthquake, which registered 7.5 on the Richter scale, than an earthquake measuring only 5.0 on the Richter scale? (Refer to Exercise 35.)

37. A certain bacteria population doubles every hour. If there are 1,000 bacteria initially, then the number of bacteria after t hours is given by $N = 1,000(2^t)$.
   a. Determine the bacteria population after 3.5 h.
   b. Determine when the population will reach 20,000.

38. In a body of water, the light intensity I diminishes with the depth. If I_0 is the intensity level at the surface, then at depth d (in feet)

$$I = I_0e^{-kd}$$

where k is a constant. For a chlorinated swimming pool, a typical value for k is 0.0080. At what depth is the intensity 95% of the intensity at the surface? (*Hint:* Find d such that $I = 0.95I_0$.)

39. Logarithms may be used to determine the age of a fossil in a procedure called *carbon dating*. This method is based on the fact that carbon-14 is found in all organisms in a fixed percentage. When an organism dies, the carbon-14 decays according to the formula

$$P = (P_0)(2^{-t/5,580})$$

where P_0 is the initial amount. Suppose a fossil contains only 25% of the original amount. How long has the organism been dead? (Find t such that $P = 0.25P_0$.)

40. If a fossil contains 42% of the original amount of carbon-14, how long ago did the organism die? (Refer to Exercise 39.)

41. The decay of uranium-238, which has a half-life of 4.5 billion years, is

$$N = N_0e^{-(1.54 \times 10^{-10})t}$$

Determine how long 0.1% of the initial amount N_0 takes to decay. (*Hint:* If 0.1% has decayed, then $0.999N_0$ remains.)

42. From the formula pH $= -\log_{10} [H^+]$, find the concentration of hydrogen ions (in moles per liter) of milk whose hydrogen potential pH is 6.39. (Refer to Exercise 41 of Section 12.4.)

43. The atmospheric pressure (in millimeters of mercury) is given by

$$P = 760e^{-0.00013h}$$

where h is the height (in meters) above sea level. At what height above sea level is the atmospheric pressure equal to 605 mm of mercury?

12.8   Graphs on Logarithmic Paper (Optional)

So far all the graphs we have studied were drawn using the **linear scale**, on which the integers are all equally spaced. For some equations it is more convenient to use graph paper on which one axis (usually the y-axis) or both axes use a **logarithmic scale**. To see what is meant by a logarithmic scale, consider the following table:

Logarithmic scale

x:	1	2	3	4	5	6	7	8	9	10	20	30	100
$\log x$:	0	0.301	0.477	0.602	0.699	0.778	0.845	0.903	0.954	1	1.301	1.477	2

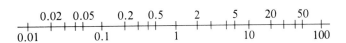

Figure 12.4

The resulting scale, using log x instead of x, is shown in Figure 12.4. Since the logarithmic scale has cycles, the first number in each cycle must be labeled as a power of 10: 0.01, 0.1, 1, 10, 100, 1,000, and so on. (See Figure 12.5.)

Figure 12.5

Log paper

If one axis is linear and one logarithmic (usually the y-axis), it is called *semilog paper*. If both axes have a logarithmic scale, it is called *log-log paper*. (Graph paper using logarithmic scales is available in many bookstores.) When the y-variable has a range that is much larger than that of the x-variable, it is often advantageous to use semilog paper. Log-log paper is frequently used if both variables have large scales.

We will concentrate mostly on the important special case where the graph to be drawn appears as a straight line on semilog or log-log paper. To this end, let us recall from Chapter 3 that the graph of

$$ax + by = c \qquad (12.20)$$

is a straight line. This equation can also be written in the form

$$y = mx + b \qquad (12.21)$$

To graph an equation on logarithmic paper, we take the logarithm of both sides and graph the resulting equation. Consider the next example.

E X A M P L E **1**   Graph $y = 5^x$ on semilog paper.

Solution. To graph an equation on either type of log paper, we first take the logarithm of both sides of the equation. Thus, in this case,

$$\log y = x \log 5$$

If we let $u = \log y$, the equation becomes

$$u = (\log 5)x \qquad m = \log 5,\ b = 0$$

This equation has the same form as equation (12.21). The equation will therefore appear as a line when graphed on semilog paper. (Semilog paper is used because u is a logarithm but x is not.) Next, note that when $x = 0$, then $y = 1$. To obtain the graph, we need only one other point, but it is best to include a check point. So if $x = 3$, then $y = 125$, and if $x = 4$, $y = 625$. The graph is shown in Figure 12.6.

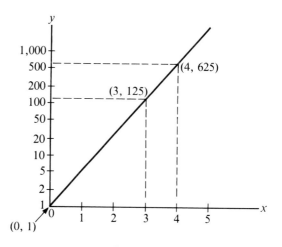

Figure 12.6 ◀

E X A M P L E **2**   Graph $y = 2(3^{-x})$ on semilog paper.

Solution. Taking logarithms again, we get

$$\log y = \log 2 - (\log 3)x$$

If we let $u = \log y$, then the resulting equation is

$$u = -(\log 3)x + \log 2 \qquad m = -\log 3,\ b = \log 2$$

This equation has the same form as equation (12.21). The graph, which is a line, is constructed from the following short table of values:

x:	-1	0	1	2	3	4
y:	6	2	0.67	0.22	0.074	0.025

See Figure 12.7.

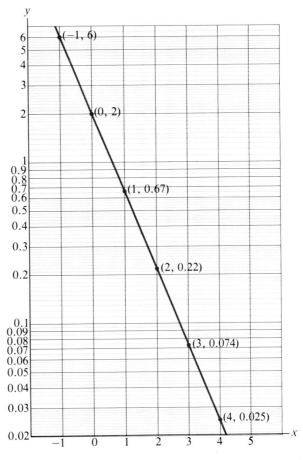

Figure 12.7

E   X   A   M   P   L   E   **3**   Graph $y^{1/2} = 3x^{1/3}$ on log-log paper.

Solution. From the given equation,

$$\frac{1}{2} \log y = \log 3 + \frac{1}{3} \log x$$

We now let $u = \log y$ and $v = \log x$. The equation becomes

$$\frac{1}{2} u = \frac{1}{3} v + \log 3$$

or

$$\frac{1}{2} u - \frac{1}{3} v = \log 3 \qquad a = \frac{1}{2}, b = -\frac{1}{3}, c = \log 3$$

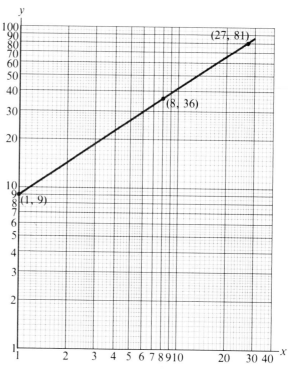

Figure 12.8

This equation has the form of the equation of a line, equation (12.20). Since both variables are logarithms, log-log paper is appropriate. The graph, shown in Figure 12.8, is constructed from the following short table:

x:	1	8	27
y:	9	36	81

◀

E X A M P L E **4**   Determine the type of graph paper for which the following equations are straight lines:

a. $x^4y^3 = 1$ **b.** $3^xy = 2$

Solution.

a. Taking logarithms, we get

$$4 \log x + 3 \log y = 0$$

Letting $u = \log y$ and $v = \log x$, we get $4v + 3u = 0$, which has the form of a line. Since both u and v are logarithms, log-log paper is required.

b. From $3^x y = 2$, we get

$$x \log 3 + \log y = \log 2$$

Letting $u = \log y$, we get $u + (\log 3)x = \log 2$, which fits form (12.20). Since u is a logarithm and x is not, semilog paper is required. ◀

Occasionally experimental data are plotted on logarithmic paper to give a clearer picture of relationships. (See Exercise 13.)

E X E R C I S E S / S E C T I O N **12.8**

In Exercises 1–4, graph each equation on semilog paper.

1. $y = 3^x$ **2.** $y = 3(2^x)$

3. $y = 3(4^{-x})$ **4.** $y = 2^{-x}$

In Exercises 5–8, graph each equation on log-log paper.

5. $y = 3x^4$ **6.** $y = 2x^{2/3}$

7. $y^3 = 3x^{2/3}$ **8.** $x^2 y^3 = 1$

In Exercises 9–12, determine the type of graph paper for which the graph of each given equation will be a straight line; sketch the graph.

9. $y(3^x) = 1$ **10.** $y = 2^{x/2}$

11. $\sqrt{x}\,y^3 = 1$ **12.** $x\sqrt{y} = 2$

13. Experimenters determined the relationship between the pressure P (in atmospheres) and the volume V (in cubic inches) of a certain gas. The following results were ob-

tained:

P:	1	3	5	7
V:	2.69	1.23	0.85	0.67

Plot the resulting curve on log-log paper.

14. The period P of a pendulum is directly proportional to the square root of its length L, a fact first discovered by Galileo. An experiment to test this relationship yielded the following results:

L (*feet*):	1	2	3	4	5
P (*seconds*):	1.13	1.59	1.95	2.24	2.50

Plot the resulting curve on log-log paper. Do the data confirm the relationship $P = k\sqrt{L}$?

R E V I E W E X E R C I S E S / C H A P T E R **12**

In Exercises 1–4, change each exponential form to the logarithmic form.

1. $3^{-3} = \dfrac{1}{27}$ **2.** $\left(\dfrac{1}{2}\right)^2 = \dfrac{1}{4}$

3. $7^{-2} = \dfrac{1}{49}$ **4.** $a^3 = b$

In Exercises 5–8, change each logarithmic form to the exponential form.

5. $\log_4 4 = 1$ **6.** $\log_3 \dfrac{1}{9} = -2$

7. $\log_{1/4} \dfrac{1}{16} = 2$ **8.** $\log_{1/2} 2 = -1$

In Exercises 9–16, find the value of the unknown.

9. $\log_5 x = -2$ **10.** $\log_{1/2} x = -3$

11. $\log_2 \dfrac{1}{16} = a$ **12.** $\log_b 36 = 2$

13. $\log_{3/2} x = 3$ **14.** $\log_b 9 = -2$

15. $\log_b \dfrac{1}{32} = -5$ **16.** $\log_8 \dfrac{1}{2} = a$

In Exercises 17–22, sketch each function.

17. $y = 3^x$

18. $y = e^x$

19. $y = e^{-x}$

20. $y = \log_5 x$

21. $y = \log_{1/2} x$

22. $y = \left(\dfrac{1}{3}\right)^x$

In Exercises 23–36, write the given expressions as sums, differences, or multiples of logarithms. Whenever possible, simplify the result by using $\log_b 1 = 0$ and $\log_b b = 1$.

23. $\log_2 32$

24. $\log_5 35$

25. $\log_{10} 1,000$

26. $\log_3 81$

27. $\log_5 125x^3$

28. $\log_b b^3$

29. $\log_6 \sqrt{6x}$

30. $\log_3 \dfrac{1}{3b}$

31. $\log_2 \dfrac{1}{16a^2}$

32. $\log_{10} \dfrac{1}{\sqrt{10a}}$

33. $\ln \dfrac{16}{e^2}$

34. $\ln \sqrt{\pi e}$

35. $\ln \dfrac{1}{\sqrt[3]{e}}$

36. $\log_{10} 0.004$

In Exercises 37–40, write each expression as a single logarithm.

37. $\log_5 26 - \log_5 13$

38. $\ln a + \dfrac{1}{2} \ln b$

39. $\dfrac{1}{2} \log_6 x - 2 \log_6 b$

40. $\dfrac{1}{3} \ln 27$

In Exercises 41–46, use Table 2 of Appendix E to find the common logarithm. Check with a calculator.

41. $\log 2.58$

42. $\log 2,580$

43. $\log 9,142$

44. $\log 73.28$

45. $\log 0.005765$

46. $\log 0.0004517$

In Exercises 47–50, use Table 2 of Appendix E to find the antilogarithm. Check with a calculator.

47. 2.6517

48. 3.8324

49. $9.5327 - 10$

50. $7.4385 - 10$

In Exercises 51–66, use a calculator to find the values to four decimal places.

51. $\log 2.64$

52. $\ln 2.64$

53. $\ln 37.42$

54. $\log 37.42$

55. $\ln 10.74$

56. $\ln 8.03$

57. $e^{-1.2}$

58. $e^{-2.3}$

59. $e^{2.7}$

60. $e^{0.365}$

61. $e^{-1.6}$

62. $e^{-0.342}$

63. $\log_5 3.92$

64. $\log_2 35.6$

65. $\log_7 0.728$

66. $\log_6 0.00312$

In Exercises 67–71, solve each equation for x.

67. $2^x = 6$

68. $3^{x-1} = \dfrac{1}{2}$

69. $2(3^{-x}) = 5$

70. $\log x + 2 \log x = \log 27$

71. $\log_3 x + \log_3(x + 1) = 1$

72. If $L = R \ln(V_1/V_2)$, show that $V_1 = V_2 e^{L/R}$.

73. Recall that the formula relating the loudness of a sound in decibels to its intensity is

$$\beta = 10 \log \dfrac{I}{I_0}$$

A quiet automobile has a loudness of 50 decibels, and busy street traffic 70 decibels. How many times more intense is the latter?

74. If one star is 30 times brighter than another, what is their difference in magnitude? [Refer to formula (12.19).]

75. Newton's law of cooling states that the rate of change of the temperature of a body is directly proportional to the difference of the temperature of the body and the temperature of the surrounding medium. If T_0 is the initial temperature and M_T the temperature of the medium, then the temperature of the body as a function of time is

$$T = M_T + (T_0 - M_T)e^{-kt}$$

where k is a constant. For a certain body, $M_T = 10°C$, $T_0 = 50°C$, and $k = 0.13$. Find the temperature of the body after 5.0 min.

76. The current i as a function of time is given by $i = 3.0e^{-2t}$. Graph i (in amperes) as a function of time (in seconds).

77. Solve the equation in Exercise 76 for t.

78. Determine the present value of $5,000 if the money is to be invested for 7.5 years at 9.75% compounded continuously.

79. If P dollars is compounded n times a year, then the value of the investment at the end of t years is $A = P(1 + r/n)^{nt}$, where r is the interest rate. Determine how long an initial investment will take to double if invested at 10% compounded quarterly.

80. A body cools according to the formula $T = 10(1 + e^{-0.11t})$, where T is in degrees Celsius and t in minutes. Since

$T = 20$ when $t = 0$, the initial temperature is 20°C. Determine how long the temperature will take to drop to 18°C.

81. A bacteria culture grows according to $N = 10,000(1.75)^t$, where t is measured in hours. Find how long the bacteria population will take to reach 50,000.

82. Carbon-14 decays according to the formula

$$P = P_0(2^{-t/5,580})$$

If a fossil contains 30% of the original carbon, how long has the organism been dead?

83. From the formula $pH = -\log_{10}[H^+]$, determine the concentration of hydrogen ions (in moles per liter) of wine grapes ($pH = 3.15$).

84. The atmospheric pressure (in millimeters of mercury) is given by

$$P = 760e^{-0.00013h}$$

where h is the height (in meters) above sea level. At what height is the atmospheric pressure equal to 550 mm of mercury?

In Exercises 1–6, simplify the given expressions and write the results without zero or negative exponents.

1. $\dfrac{2^0 x^2 y^{-1/2}}{x^{-2} y^{3/2}}$

2. $\dfrac{3^{-1} g_1^{-2} g_2^{2} g_3^{-5}}{6 g_1^{4} g_2^{-3} g_3^{-3}}$

3. $\dfrac{a^{-1} - b^{-1}}{a^{-1}}$

4. $(64 L^{12} R^{-9})^{-1/3}$

5. $\dfrac{27 x^0 x^{-1/2} y^{1/5}}{9 x^{1/3} y^{-1/15}}$

6. $(a - 2)(a + 3)^{-1/2} - (a + 3)^{1/2}$

7. Rationalize the denominator:

$$\sqrt[4]{\dfrac{1}{ab^3}}$$

8. Simplify and rationalize the denominator:

$$\sqrt[3]{\dfrac{4^{-1} G^{-1} H^{-2}}{G^3 H^3}}$$

9. Rationalize the denominator:

$$\dfrac{\sqrt{2} - 1}{\sqrt{5} - \sqrt{2}}$$

10. Combine the following complex numbers:

$$(-2 + 3j) + (6 - 2j) - (2 - j)$$

11. Perform the indicated multiplication:

$$(2 - 3j)(-1 + j)$$

12. Divide $1 - j$ by $2 + 3j$.

13. Change the complex number $-2\sqrt{3} - 2j$ to **(a)** polar form, **(b)** exponential form.

14. Divide $3\underline{/10^\circ}$ by $6\underline{/150^\circ}$.

15. Find the 4 fourth roots of $-1 + \sqrt{3}j$. (Leave answer in polar form.)

16. Combine into a single logarithm:

$$\dfrac{1}{2} \log_6 P + \log_6 Q - 2 \log_6 R$$

17. Solve for x:

$$3^x = 7$$

18. Solve the formula

$$S = S_1 - S_2 e^{-kt}$$

for t.

19. Recall that the equation relating the object distance q and the image distance p to the focal length f of a thin lens is

$$f^{-1} = p^{-1} + q^{-1}.$$

Solve this equation for p.

20. Use the formula $P = S(1 + i)^{-n}$ (for obtaining the present value) to find the amount of money that has to be invested to yield \$3,000 in 6 years at 8.5% interest compounded annually.

21. The amount N of a certain radioactive isotope varies with time according to the relation

$$\log_{10} N = -0.4t, \qquad t \geq 0.$$

Write this formula in exponential form.

22. The power supply P (in watts) of a certain satellite after t days is $P = 20.0(e^{-0.00341t})$. Note that $P = 20.0$ W when $t = 0$. How many days will it take for the power supply to diminish to 10.0 W?

More on Equations and Systems of Equations

13.1 Solving Systems of Equations Graphically

The main purpose of this section is to learn how to solve certain systems of equations graphically. Recall that we have already studied the solution of systems of two linear equations in Chapter 3. In this section we will move on to systems of two quadratic equations, as well as a few other types. However, first we need to look at the graphs of second-order equations in two variables x and y, which are defined next.

General Quadratic Equation in x and y:

$$Ax^2 + Bxy + Cy^2 + Dx + Ey + F = 0 \qquad (13.1)$$

where A, B, and C are not all zero.

Conic sections

The graphs of equations of form (13.1) are collectively known as **conic sections**, which include the **circle**, **ellipse**, **parabola**, and **hyperbola**. Our purpose here is to classify the different conic sections according to their equations; we will take up the details in Chapter 19. As a starting point, consider the following example of a quadratic equation.

E X A M P L E **1** Graph the equation $x^2 + y^2 = 25$.

Solution. As usual, we obtain the graph by plotting enough points to get a smooth curve. So let us solve the given equation for y in terms of x:

$$x^2 + y^2 = 25$$
$$y^2 = 25 - x^2$$
$$y = \pm\sqrt{25 - x^2}$$

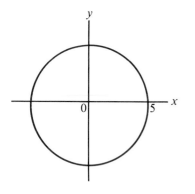

Figure 13.1

Now observe that whenever $x > 5$ or $x < -5$, the radicand is negative and the corresponding y values imaginary. Since complex numbers cannot be plotted on the Cartesian coordinate system, we must exclude them here. The x values must therefore lie between -5 and $+5$. Observe also that for both $x = 4$ and $x = -4$, we get $y = \pm 3$, written compactly as $x = \pm 4$, $y = \pm 3$. Some of the other values are listed in the following table:

x:	0	± 1	± 2	± 3	± 4	± 5
y:	± 5	± 4.9	± 4.6	± 4	± 3	0

The graph, shown in Figure 13.1, is a *circle*. ◄

Example 1 has shown that $x^2 + y^2 = 25$ is a circle. The general case is given next.

A **circle** of radius r centered at the origin is represented by the equation

$$x^2 + y^2 = r^2 \qquad\qquad (13.2)$$

If the coefficients of x^2 and y^2 (in the equation of the circle) are positive but have different values, we get a curve called an *ellipse*.

E X A M P L E **2**   Graph the equation $x^2 + 2y^2 = 16$.

Solution. As in the case of the circle in Example 1, we solve the equation for y in terms of x and construct a table of values:

$$x^2 + 2y^2 = 16$$

$$2y^2 = 16 - x^2 \qquad \text{subtracting } x^2 \text{ from both sides}$$

$$y^2 = \frac{16 - x^2}{2} \qquad \text{dividing by 2}$$

$$y = \pm\sqrt{\frac{16 - x^2}{2}} \qquad \text{taking square roots}$$

So if $x = 0$, $y = \pm\sqrt{\frac{16}{2}} = \pm\sqrt{8} = \pm 2.8$.

x:	0	± 1	± 2	± 3	± 4
y:	± 2.8	± 2.7	± 2.4	± 1.9	0

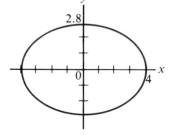

Figure 13.2

Note that the values of x numerically greater than 4 (that is, $|x| > 4$) lead to imaginary values for y and must be excluded. The graph, shown in Figure 13.2, is an ellipse. ◄

> **Ellipse:** An equation of the form
>
> $$ax^2 + by^2 = c, \qquad a \neq b \qquad \text{and} \qquad a, b, c > 0 \qquad (13.3)$$
>
> represents an ellipse centered at the origin.

The next example illustrates the graph of another conic section, called a *parabola*.

E X A M P L E 3   Graph the equation $y = 4x - 2x^2$.

Solution. For this equation, y is defined for all values of x. The graph, shown in Figure 13.3, is a parabola.

x:	-2	-1	0	1	2	3	4
y:	-16	-6	0	2	0	-6	-16

◀

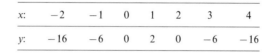

Figure 13.3

> **Parabola:** An equation of the form
>
> $$y = ax^2 + bx + c \qquad \text{or} \qquad x = ay^2 + by + c, \qquad a \neq 0 \qquad (13.4)$$
>
> represents a parabola.

Our final example of a conic section is a *hyperbola*.

E X A M P L E 4   Graph the equation $2x^2 - y^2 = 8$.

Solution. Once again we solve the equation for y in terms of x:

$$-y^2 = -2x^2 + 8$$
$$y^2 = 2x^2 - 8$$
$$y = \pm\sqrt{2x^2 - 8}$$

To avoid imaginary values, we must have $x \geq 2$ or $x \leq -2$.

x:	± 2	± 2.1	± 2.5	± 3	± 4	± 5
y:	0	± 0.9	± 2.1	± 3.2	± 4.9	± 6.5

The curve, shown in Figure 13.4, is a hyperbola.

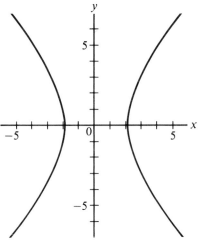

Figure 13.4

Hyperbola: An equation of the form

$$ax^2 - by^2 = c, \qquad ay^2 - bx^2 = c, \qquad \text{or} \qquad xy = k \qquad (a, b > 0)$$
$$(13.5)$$

represents a hyperbola.

Solving Systems of Equations Graphically

Our brief discussion of conic sections has provided enough information so that we can solve certain systems of equations graphically. To solve such systems, we graph the given equations and estimate the points of intersection.

E X A M P L E **5** Graphically solve the system:

$$xy = 4$$
$$y = 4x^2$$

Solution. By form (13.5) the first equation is a hyperbola, plotted from the following table of values:

x:	-4	-3	-2	-1	0	1	2	3	4
y:	-1	$-\dfrac{4}{3}$	-2	-4	undefined	4	2	$\dfrac{4}{3}$	1

(See Figure 13.5.) The second equation is a parabola by form (13.4).

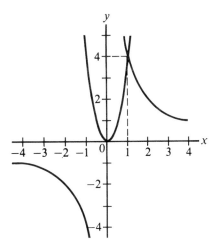

Figure 13.5

x:	-2	-1	$-\dfrac{1}{2}$	0	$\dfrac{1}{2}$	1	2
y:	16	4	1	0	1	4	16

According to Figure 13.5, the graphs intersect at (1.0, 4.0). So the solution of the system is $x = 1.0$, $y = 4.0$.   ◀

E X A M P L E **6** Graphically solve the system:

$$x^2 - y^2 = 4$$
$$x^2 + 2y^2 = 64$$

Solution. The hyperbola $x^2 - y^2 = 4$ (Figure 13.6) is plotted from the following tables of values:

x:	2	2.5	3	4	5
y:	0	± 1.5	± 2.2	± 3.5	± 4.6

The second equation represents an ellipse by form (13.3).

x:	0	± 2	± 4	± 6	± 8
y:	± 5.7	± 5.5	± 4.9	± 3.7	0

According to Figure 13.6, there are four points of intersection and hence four solutions to the system. The approximate solutions are (4.9, 4.5), (4.9, -4.5), (-4.9, 4.5), and (-4.9, -4.5).

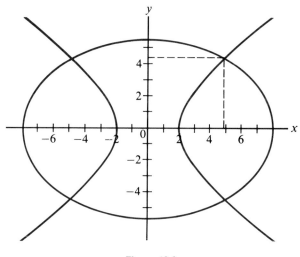

Figure 13.6

After discussing graphical methods of solution in Chapter 3, we found that algebraic methods were far more convenient. For linear equations and, to a large extent, quadratic equations, this is true, as we will see in the next section. However, many systems have no algebraic methods of solution. An example of such a system is given next.

E X A M P L E **7**   Solve the following system graphically:

$$y - 4x = 4$$

$$y = e^{-x}$$

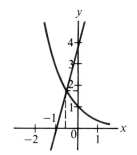

Figure 13.7

Solution. The equation $y - 4x = 4$ is a line passing through $(-1, 0)$ and $(0, 4)$, as shown in Figure 13.7. The exponential function is plotted from the following table of values:

x:	-2	-1	-0.8	-0.6	-0.4	0	1
y:	7.4	2.7	2.2	1.8	1.5	1	0.4

From the graphs, the approximate solution is $(-0.6, 1.8)$.   ◀

E X E R C I S E S / S E C T I O N   **13.1**

Solve each system of equations graphically. Estimate the solutions to the nearest tenth of a unit.

1. $x^2 + y^2 = 8$
$x - y = 0$

2. $2x + y = 3$
$y = x^2$

3. $y^2 - 3x + 1 = 0$
$3x - 2y = 0$

4. $x - 2y = 1$
$y^2 - x^2 = 4$

5. $2y - x = 2$
$x^2 - y^2 = 2$

6. $2x^2 + y^2 = 8$
$x - y = 1$

7. $xy = 6$
$y^2 = x$

8. $x^2 - 2y^2 = 4$
$xy = 1$

13. $3y^2 - 5x^2 = 12$
$2y + x^2 = 0$

14. $2x^2 - 4y^2 = 9$
$4y^2 - x^2 = 12$

9. $3x + 2y = 6$
$xy = 1$

10. $2x^2 + y^2 = 32$
$y = x^2$

15. $y^2 = x$
$y = e^{-x}$

16. $y = \sin x$
$y = x^2$

11. $x^2 + 4y^2 = 16$
$x^2 + y^2 = 9$

12. $x^2 - 2y^2 = 4$
$2x^2 + y^2 = 18$

17. $y = \log_3 x$
$y = 2^{-x}$

18. $y = \cos x$
$y = \sqrt{x}$

13.2 Solving Systems of Equations Algebraically

In the previous section we saw that graphical methods can be used to obtain approximate solutions to systems of equations. However, if both equations are algebraic, the exact solution can often be obtained algebraically. In fact, the methods for solving systems of quadratic equations are identical to the methods used in Chapter 3 for solving systems of linear equations, namely, the method of **substitution** and the method of **addition or subtraction**.

To solve the system

$$y = 4 - x^2$$

$$y - x = 1$$

Substitution

by the *method of substitution*, we solve the second equation for y in terms of x to get $y = x + 1$ and then substitute this expression for y in the first equation. Thus

$$y = 4 - x^2 \qquad \text{first equation}$$

$$x + 1 = 4 - x^2 \qquad \text{substituting } x + 1 \text{ for } y$$

$$x^2 + x - 3 = 0$$

By the quadratic formula,

$$x = \frac{-1 \pm \sqrt{13}}{2} \qquad \text{see formula (6.6)}$$

From the second equation, $y = x + 1$, we now find the corresponding values for y:

$$x = \frac{-1 + \sqrt{13}}{2}: \qquad x = \frac{-1 - \sqrt{13}}{2}:$$

$$y = \frac{-1 + \sqrt{13}}{2} + 1 \qquad y = \frac{-1 - \sqrt{13}}{2} + 1$$

$$= \frac{-1 + \sqrt{13}}{2} + \frac{2}{2} \qquad = \frac{-1 - \sqrt{13}}{2} + \frac{2}{2}$$

$$= \frac{-1 + \sqrt{13} + 2}{2} \qquad = \frac{-1 - \sqrt{13} + 2}{2}$$

$$= \frac{1 + \sqrt{13}}{2} \qquad = \frac{1 - \sqrt{13}}{2}$$

So the solution is given by

$$\left(\frac{-1+\sqrt{13}}{2}, \frac{1+\sqrt{13}}{2}\right), \qquad \left(\frac{-1-\sqrt{13}}{2}, \frac{1-\sqrt{13}}{2}\right)$$

As a check, we substitute $x = \frac{1}{2}(-1+\sqrt{13})$ in the equation $y = 4 - x^2$:

$$y = 4 - \left(\frac{-1+\sqrt{13}}{2}\right)^2$$

$$= 4 - \frac{1 - 2\sqrt{13} + 13}{4}$$

$$= \frac{16}{4} - \frac{14 - 2\sqrt{13}}{4}$$

$$= \frac{16 - (14 - 2\sqrt{13})}{4}$$

$$= \frac{2 + 2\sqrt{13}}{4}$$

$$= \frac{1 + \sqrt{13}}{2} \qquad \text{(solution checks)}$$

The other pair is checked similarly. (Note that the solution gives us the coordinates of the intersection of a parabola and a straight line.)

E X A M P L E **1**   Solve the system:

$$2x - y = 2$$
$$xy = 4$$

Solution. As before, we will use the method of substitution. From the second equation, $y = 4/x$. Substituting into the first equation, we get

$2x - y = 2$	**first equation**
$2x - \dfrac{4}{x} = 2$	**substituting $\dfrac{4}{x}$ for y**
$2x^2 - 4 = 2x$	**clearing fractions**
$2x^2 - 2x - 4 = 0$	**collecting terms on left side**
$x^2 - x - 2 = 0$	**dividing by 2**
$(x - 2)(x + 1) = 0$	**factoring**
$x = 2, \ -1$	

From $xy = 4$, we get the following: If $x = 2$, then $y = 2$; if $x = -1$, then $y = -4$. So the solution is given by $(2, 2)$ and $(-1, -4)$.

As a check, let us substitute the values obtained into the first equation, $2x - y = 2$:

$$2(2) - 2 = 2; \qquad 2(-1) - (-4) = 2 \quad \checkmark$$

(The solution yields the coordinates of the points of intersection of a hyperbola and a straight line.) ◀

Addition or subtraction

Sometimes one of the unknowns can be eliminated by *addition or subtraction*, as shown in the next example.

E X A M P L E **2** Solve the system:

$$2x^2 - 3y^2 = 19$$
$$3x^2 + y^2 = 1$$

Solution. We multiply both sides of the second equation by 3 and add:

$$
\begin{aligned}
2x^2 - 3y^2 &= 19 \\
9x^2 + 3y^2 &= 3 \qquad &&3(3x^2 + y^2) = 3(1) \\
\hline
11x^2 &= 22 \qquad &&\text{adding} \\
x^2 &= 2 \qquad &&\text{dividing by 11} \\
x &= \pm\sqrt{2} \qquad &&\text{taking square roots}
\end{aligned}
$$

To find the corresponding y values, let us substitute the x values into the second equation:

$$
\begin{array}{ll}
3(+\sqrt{2})^2 + y^2 = 1 \qquad & 3(-\sqrt{2})^2 + y^2 = 1 \\
6 + y^2 = 1 & 6 + y^2 = 1 \\
y^2 = -5 & y^2 = -5 \\
y = \pm\sqrt{5}j & y = \pm\sqrt{5}j
\end{array}
$$

Since each x value yields two y values, we obtain the following four solutions:

$$(\sqrt{2}, \sqrt{5}j), \qquad (\sqrt{2}, -\sqrt{5}j), \qquad (-\sqrt{2}, \sqrt{5}j), \qquad (-\sqrt{2}, -\sqrt{5}j),$$

Since all members of this solution set involve pure imaginary numbers, the graphs of these equations, a hyperbola and an ellipse, do not intersect at any point. However, algebraically speaking, all four pairs satisfy the system. ◀

E X A M P L E **3**   Recall that two resistances R_1 and R_2 in *series* have a combined resistance of $R_1 + R_2$ (in ohms). If the same resistances are connected in *parallel*, then the combined resistance R_T (in ohms) is such that $1/R_T = 1/R_1 + 1/R_2$. Suppose the resistors have a combined resistance of $15\ \Omega$ when connected in series and $\frac{10}{3}\ \Omega$ when connected in parallel. Find the two resistances.

Solution. From the given information

$$R_1 + R_2 = 15$$

$$\frac{1}{R_1} + \frac{1}{R_2} = \frac{3}{10} \qquad \frac{1}{R_1} + \frac{1}{R_2} = \frac{1}{10/3}$$

The second equation can be written

$$\frac{R_1 + R_2}{R_1 R_2} = \frac{3}{10} \qquad\qquad \text{adding fractions on left side}$$

$$10R_1 R_2 \left(\frac{R_1 + R_2}{R_1 R_2} \right) = 10(R_1 R_2) \frac{3}{10} \qquad \text{clearing fractions}$$

$$10(R_1 + R_2) = 3R_1 R_2 \qquad\qquad \text{second equation}$$

From the first equation, $R_2 = 15 - R_1$. Substituting into the second equation, $10R_1 + 10R_2 = 3R_1 R_2$, we get

$$10R_1 + 10(15 - R_1) = 3R_1(15 - R_1) \qquad \text{substitution}$$

$$10R_1 + 150 - 10R_1 = 45R_1 - 3R_1{}^2 \qquad \text{removing parentheses}$$

$$3R_1{}^2 - 45R_1 + 150 = 0 \qquad\qquad \text{collecting terms on the left side}$$

$$R_1{}^2 - 15R_1 + 50 = 0 \qquad\qquad \text{dividing by 3}$$

$$(R_1 - 10)(R_1 - 5) = 0 \qquad\qquad \text{factoring the left side}$$

$$R_1 = 10,\ 5$$

From $R_2 = 15 - R_1$ we have the following: If $R_1 = 10$, then $R_2 = 5$; if $R_1 = 5$, then $R_2 = 10$. We conclude that the two resistances are $5\ \Omega$ and $10\ \Omega$. ◀

E X E R C I S E S / S E C T I O N   **13.2**

In Exercises 1–28, solve the systems of equations.

1. $y = x + 2$
$y = x^2$

2. $y = x - 3$
$y^2 = x - 1$

3. $y - x^2 = 1$
$y - x = 3$

4. $y = 4 - x$
$y = x^2 - 4x + 4$

5. $y + x^2 - 4 = 0$
$y - x - 2 = 0$

6. $x^2 + y = 3$
$x + y = 1$

7. $x^2 - x + y = 0$
$x + y = 0$

8. $x + y^2 - 4y + 2 = 0$
$x - y^2 + 4y - 4 = 0$

9. $y^2 + x = 3$
$y - x + 1 = 0$

10. $4x - 3y - 4 = 0$
$4x - y^2 = 0$

11. $y = x^2 - 4$
$2y = x + 2$

12. $y = x - 2$
$y^2 - x^2 = 3$

13. $y = x + 2$
$x^2 + y^2 = 4$

14. $y = x + 1$
$y^2 = 3x + 2$

15. $x + 2y = 3$
$x^2 + 2y^2 = 2$

16. $2x + y = 1$
$3x^2 - 2y^2 = 2$

17. $x + y + 2 = 0$
$xy = 1$

18. $x + y = 5$
$xy = 6$

19. $2y^2 + xy + x^2 = 7$
$x - 2y = 5$

20. $2x - 3y = 2$
$xy = 4$

21. $x^2 + y^2 = 5$
$3x^2 - 2y^2 = -5$

22. $2x^2 + y^2 = 4$
$x^2 + y^2 = 3$

23. $3x^2 + 3y^2 = 2$
$7x^2 + 3y^2 = 10$

24. $x^2 - 2y^2 = 1$
$4x^2 + y^2 = 5$

25. $4y^2 - 3x^2 = 29$
$y^2 + 7x^2 = 46$

26. $5x^2 + 2y^2 = 37$
$x^2 + y^2 = 5$

27. $x^2 + y^2 = 16$
$y^2 - 2x^2 = 10$

28. $4x^2 + 3y^2 = 24$
$3x^2 - 2y^2 = 35$

29. The area of a metal plate is 25 cm², while the length is 4 times the width. Find its dimensions.

30. Recall that the focal length f of a lens is related to the object distance q and the image distance p by

$$\frac{1}{f} = \frac{1}{p} + \frac{1}{q}$$

If the focal length is 2.0 cm and $p + q = 9.0$ cm, find p and q.

31. Starting at $t = 0$, two particles travel along a line. The distances s from the origin are given by $s = 80 - 16t^2$ and $s = 16t^2 - 100t$, respectively, where s is measured in feet and t in seconds. At what time do the particles meet?

32. Two resistors have a combined resistance of 18 Ω when connected in series and of 4 Ω when connected in parallel. Determine the two resistances. (See Example 3.)

33. Repeat Exercise 32 if the combined resistance is 9.00 Ω when connected in series and 1.82 Ω when connected in parallel.

34. A rectangular pasture is to be divided into two equal parts by a fence parallel to the smaller sides. The area of each subdivision is 2,500 ft² and the total amount of fence used is 350 ft. Find the dimensions of the pasture.

35. The kinetic energy of a body of mass m and velocity v is given by $\frac{1}{2}mv^2$ and its momentum by mv. Both are conserved in perfectly elastic collisions. Two elastic balls of mass 0.5 kg each collide with respective velocities v_1 and v_2 (in meters per second). The following relationships were found experimentally:

$$\frac{1}{2}(0.5)v_1{}^2 + \frac{1}{2}(0.5)v_2{}^2 = 1.25 \quad \text{(joules)}$$

$$0.5v_1 + 0.5v_2 = 1.5 \quad \text{(kg·m/s)}$$

Find the two velocities.

36. The laws of conservation of energy and momentum give rise to the following system of equations for two colliding bodies:

$$m_1v_1 + m_2v_2 = m_1V_1 + m_2V_2$$

$$\frac{1}{2}m_1v_1{}^2 + \frac{1}{2}m_2v_2{}^2 = \frac{1}{2}m_1V_1{}^2 + \frac{1}{2}m_2V_2{}^2$$

Show that $V_1 - V_2 = v_2 - v_1$.

13.3 Equations in Quadratic Form

Some equations that don't appear to be quadratic really are quadratic equations in disguise. The reason is that the unknown in a quadratic equation may be any quantity. Denoting the unknown quantity by $f(x)$, the equation

$$a[f(x)]^2 + b[f(x)] + c = 0$$

Quadratic form

is called an **equation in quadratic form**.

Quadratic Form

$$a[f(x)]^2 + b[f(x)] + c = 0 \qquad (13.6)$$

Such equations can be solved by letting $z = f(x)$ and then solving the resulting quadratic equation

$$az^2 + bz + c = 0$$

for z. The roots of the given equation are then found by solving the equation $z = f(x)$ for x.

To be a quadratic form, the equation must contain some quantity [called $f(x)$ in equation (13.6)], the square of this quantity, and perhaps a constant. To solve a quadratic form, we let z be the literal part of the second term; z^2 will then be the literal part of the first term.

For example, the equation

$$2x^{-4} + 5x^{-2} - 3 = 0$$

is a quadratic form because

$$f(x) = x^{-2} \quad \text{and} \quad [f(x)]^2 = (x^{-2})^2 = x^{-4}$$

So if $z = x^{-2}$, the literal part of the second term, the equation becomes

$$2z^2 + 5z - 3 = 0$$

which is a regular quadratic equation.

The examples given next illustrate how to solve a variety of equations in quadratic form, starting with $z = x^2$.

E X A M P L E **1**   Solve the equation $x^4 - 4x^2 + 4 = 0$.

Solution. This equation is not quadratic, but suppose we let $z = x^2$. Then $z^2 = (x^2)^2 = x^4$, and we get from $(x^2)^2 - 4x^2 + 4 = 0$

$$z^2 - 4z + 4 = 0 \qquad z = x^2$$

which is quadratic after all! Solving for z, we now have

$$(z - 2)^2 = 0$$
$$z = 2, 2$$

Since $z = x^2$, it follows that

$$x^2 = 2, 2$$

and

$$x = \pm\sqrt{2}, \pm\sqrt{2}$$

which are repeating roots.

Check: $(\pm\sqrt{2})^4 - 4(\pm\sqrt{2})^2 + 4 = 4 - 8 + 4 = 0$ ✓ ◀

E X A M P L E 2 Solve the equation $2x + 7\sqrt{x} - 15 = 0$.

Solution. As in Example 1, we let z be the literal part of the second term. Thus $z = \sqrt{x}$ and $z^2 = x$. After substituting, we get the following quadratic equation:

$$2z^2 + 7z - 15 = 0 \qquad\qquad 2(\sqrt{x})^2 + 7\sqrt{x} - 15 = 0$$

$$(2z - 3)(z + 5) = 0$$

$$z = \frac{3}{2}, -5$$

From $z = \sqrt{x}$, it follows that $\sqrt{x} = \frac{3}{2}$ and $\sqrt{x} = -5$. However, since $\sqrt{x}$ is the principal square root, which is always positive, the value $\sqrt{x} = -5$ must be excluded. This leaves $\sqrt{x} = \frac{3}{2}$, or $(\sqrt{x})^2 = (\frac{3}{2})^2$ and $x = \frac{9}{4}$

Check: If $x = \frac{9}{4}$,

$$2x + 7\sqrt{x} - 15 = 2\left(\frac{9}{4}\right) + 7\sqrt{\frac{9}{4}} - 15$$

$$= \frac{9}{2} + 7\left(\frac{3}{2}\right) - 15$$

$$= \frac{9}{2} + \frac{21}{2} - \frac{30}{2} = 0 \; ✓ \qquad ◀$$

E X A M P L E 3 Solve the equation $2x^{2/3} + x^{1/3} - 6 = 0$.

Solution. Letting $z = x^{1/3}$, $z^2 = (x^{1/3})^2 = x^{2/3}$, and we have

$$2z^2 + z - 6 = 0 \qquad\qquad 2(x^{1/3})^2 + x^{1/3} - 6 = 0$$

$$(2z - 3)(z + 2) = 0$$

$$z = \frac{3}{2}, -2$$

Since $x^{1/3} = z$, it follows that $(x^{1/3})^3 = z^3$, or $x = z^3$. So

$$x = \left(\frac{3}{2}\right)^3 = \frac{27}{8} \qquad z = \frac{3}{2}$$

and

$$x = (-2)^3 = -8 \qquad z = -2 \qquad\qquad\qquad \blacktriangleleft$$

E X A M P L E **4**   Solve the equation:

$$(x^2 - x)^2 - 4(x^2 - x) - 12 = 0$$

Solution. Let $z = x^2 - x$, so that $z^2 = (x^2 - x)^2$:

$$z^2 - 4z - 12 = 0$$

$$(z - 6)(z + 2) = 0$$

$$z = 6, \, -2$$

Since $z = x^2 - x$, we obtain the following equations:

$$x^2 - x = 6 \qquad\qquad\qquad x^2 - x = -2$$

$$x^2 - x - 6 = 0 \qquad\qquad x^2 - x + 2 = 0$$

$$(x - 3)(x + 2) = 0 \qquad x = \frac{1 \pm \sqrt{1 - 8}}{2}$$

$$x = 3, \, -2 \qquad\qquad\qquad x = \frac{1 \pm \sqrt{7}j}{2}$$

$$x = \frac{1}{2} \pm \frac{\sqrt{7}}{2}j$$

$$\qquad\qquad\qquad\qquad\qquad\qquad\qquad\qquad\qquad \blacktriangleleft$$

      Systems of quadratic equations sometimes lead to quadratic forms if the solution is carried out by the method of substitution. (See Exercise 26.)
      Quadratic forms may also occur in equations containing trigonometric, exponential, or logarithmic functions, as shown in the next example.

E X A M P L E **5**   Find the real roots of the equation $e^{-2x} - e^{-x} - 2 = 0$.

Solution. Let $z = e^{-x}$; then $z^2 = (e^{-x})^2 = e^{-2x}$, and we get

$$z^2 - z - 2 = 0 \qquad (e^{-x})^2 - e^{-x} - 2 = 0$$

$$(z - 2)(z + 1) = 0$$

$$z = 2, \, -1$$

From $z = e^{-x}$, we get

$$e^{-x} = 2 \qquad z = 2$$

Taking natural logarithms of both sides, we have

$$\ln e^{-x} = \ln 2$$

$$-x \ln e = \ln 2 \qquad \text{ln } x^r = r \text{ ln } x$$

$$-x = \ln 2 \qquad \text{ln } e = 1$$

$$x = -\ln 2 \approx -0.693$$

The other root, $z = -1$, leads to $e^{-x} = -1$, which is impossible since e^{-x} is positive for all real x.   ◀

E X E R C I S E S / S E C T I O N   **13.3**

In Exercises 1–25, solve the given equations.

1. $x^4 - 5x^2 + 4 = 0$

2. $x^4 - 7x^2 + 12 = 0$

3. $x^4 - 3x^2 - 4 = 0$

4. $x^4 - 4x^2 - 12 = 0$

5. $x^6 - 7x^3 - 8 = 0$

6. $x^6 - 28x^3 + 27 = 0$

7. $x^{-2} - x^{-1} - 6 = 0$

8. $x^{-2} + x^{-1} - 12 = 0$

9. $x^{-4} + 2x^{-2} - 24 = 0$

10. $2x^{-4} + x^{-2} - 6 = 0$

11. $4x^{-4} + 7x^{-2} - 2 = 0$

12. $6x^{-4} - 7x^{-2} + 2 = 0$

13. $x - 6\sqrt{x} + 5 = 0$

14. $x + 2\sqrt{x} - 48 = 0$

15. $6x + 7\sqrt{x} - 20 = 0$

16. $3x - 13\sqrt{x} + 14 = 0$

17. $(x - 1) - 7\sqrt{x - 1} + 10 = 0$

18. $2(x + 1) - 9\sqrt{x + 1} + 10 = 0$

19. $x^{2/3} + x^{1/3} - 6 = 0$

20. $6x^{2/3} + 13x^{1/3} - 5 = 0$

21. $(x^2 - 4x)^2 + 2(x^2 - 4x) - 8 = 0$

22. $(x^2 + 2x)^2 + 4(x^2 + 2x) + 3 = 0$

23. $24x^4 + 10x^2 - 21 = 0$

24. $24x^4 - 41x^2 + 12 = 0$

25. $\sqrt{x} - 3\sqrt[4]{x} + 2 = 0$

26. In designing a machine, a technician needs a rectangular metal plate whose area is 2 ft². If the diagonal is to be 2 ft in length, what must the dimensions be?

27. A rectangular metal plate has an area of 120 in.² and a diagonal of 20 in. Find the dimensions.

In Exercises 28–31, solve each equation.

28. $e^{2y} - e^y = 0$

29. $\sin^2 \theta - \sin \theta = 0, 0° \le \theta < 360°$

30. $2 \sin^2 \theta + \sin \theta - 1 = 0, 0° \le \theta < 360°$

31. $2 \cos^2 \theta - \cos \theta - 1 = 0, 0° \le \theta < 360°$

13.4   Radical Equations

Our final topic in this chapter is the solution of equations containing radicals, which we will call **radical equations**.

Except for certain quadratic forms discussed in the previous section, radical equations are generally solved by eliminating the radical. If all the radicals are square roots, we square both sides of the equation. If the radicals have index n, we raise both sides to the nth power.

Unfortunately, when both sides of an equation are raised to a power, the solutions obtained may not all satisfy the given equation and must therefore be checked. Solutions that do not check are called **extraneous solutions** or **roots**. Extraneous roots may also occur when solving fractional equations, as we already saw in Section 5.10. [Recall that extraneous roots may occur whenever both sides of an equation are multiplied by an expression containing the unknown. Raising both sides to a power and clearing fractions (as in Section 5.10) are instances of such multiplications.]

Extraneous roots

The first two examples illustrate the solution of equations containing a single square root.

E X A M P L E **1**   Solve the equation $\sqrt{x+4} = x - 2$.

Solution. We can eliminate the radical by squaring both sides of the equation:

$$\sqrt{x+4} = x - 2 \qquad \text{original equation}$$

$$(\sqrt{x+4})^2 = (x-2)^2 \qquad \text{squaring both sides}$$

$$x + 4 = x^2 - 4x + 4 \qquad (a-b)^2 = a^2 - 2ab + b^2$$

$$0 = x^2 - 5x \qquad \text{combining like terms}$$

$$x(x-5) = 0 \qquad \text{factoring}$$

$$x = 0, 5$$

Check: Let $x = \mathbf{5}$; then

$$\sqrt{x+4} = x - 2$$

$$\sqrt{\mathbf{5}+4} = \mathbf{5} - 2$$

$$3 = 3 \quad \checkmark$$

Let $x = \mathbf{0}$:

Left side: $\sqrt{x+4} = \sqrt{\mathbf{0}+4} = 2$

Right side: $x - 2 = \mathbf{0} - 2 = -2$

The two values do not agree, so $x = 0$ is an *extraneous root*. Therefore, the only solution is $x = 5$.   ◀

E X A M P L E **2**   Solve the equation $\sqrt{3x+4} = -1$.

Solution. Since the principal square root is positive, this equation cannot have a solution. If this observation is overlooked, we get

$$(\sqrt{3x+4})^2 = (-1)^2$$

$$3x + 4 = 1$$

$$x = -1$$

Check: If $x = -1$, $\sqrt{3x+4} = \sqrt{1} = 1$, which does not agree with the right side. The only (apparent) root is therefore *extraneous*.   ◀

Isolating the radical

If an equation contains a radical as well as other terms on the same side, we first **isolate** the radical; that is, we write the radical on one side of the equation and collect the remaining terms on the opposite side. Squaring both sides of the equation then eliminates the radical.

E X A M P L E **3**   Solve the equation $\sqrt{x+3} - x - 3 = 0$.

Solution. Squaring both sides of this equation directly would lead to a complicated expression containing $\sqrt{x+3}$. Thus, we must first *isolate* the radical by keeping the radical on the left side and collecting the remaining terms on the right side:

$$\sqrt{x+3} = x + 3 \qquad \text{isolating the radical}$$

Then we get

$$(\sqrt{x+3})^2 = (x+3)^2 \qquad \text{squaring both sides}$$

$$x + 3 = x^2 + 6x + 9 \qquad (a+b)^2 = a^2 + 2ab + b^2$$

$$x^2 + 5x + 6 = 0 \qquad \text{combining like terms}$$

$$(x+2)(x+3) = 0 \qquad \text{factoring}$$

$$x = -2, -3$$

Check: Let $x = -2$ Then

$$\sqrt{x+3} - x - 3 = \sqrt{-2+3} - (-2) - 3 = \sqrt{1} + 2 - 3 = 0$$

Let $x = -3$. Then

$$\sqrt{x+3} - x - 3 = \sqrt{-3+3} - (-3) - 3 = 0$$

Since both values check, this equation has no extraneous roots. The roots are therefore -2 and -3.   ◄

The equation in the next example contains a sum of two radicals, of which only one can initially be isolated.

E X A M P L E **4**   Solve the equation $\sqrt{x+2} + \sqrt{3x+4} = 2$.

Solution. In this equation, eliminating both radicals at once is impossible. The best procedure is placing the radicals on opposite sides and then squaring. Thus

$$\sqrt{x+2} = 2 - \sqrt{3x+4} \qquad \text{isolating one radical}$$

$$(\sqrt{x+2})^2 = (2 - \sqrt{3x+4})^2 \qquad \text{squaring both sides}$$

$$x + 2 = 2^2 - 2 \cdot 2\sqrt{3x+4} + (\sqrt{3x+4})^2 \qquad (a-b)^2 = a^2 - 2ab + b^2$$

$$x + 2 = 4 - 4\sqrt{3x+4} + (3x+4)$$

$$x + 2 = 4 - 4\sqrt{3x+4} + 3x + 4$$

Although one radical is now gone, the other remains. We now isolate this radical by collecting all other terms on the opposite side:

$$x + 2 - 4 - 3x - 4 = -4\sqrt{3x + 4} \qquad \text{isolating the remaining radical}$$

$$-2x - 6 = -4\sqrt{3x + 4} \qquad \text{simplifying the left side}$$

$$x + 3 = 2\sqrt{3x + 4} \qquad \text{dividing by } -2$$

$$(x + 3)^2 = (2\sqrt{3x + 4})^2 \qquad \text{squaring both sides}$$

$$x^2 + 6x + 9 = 4(3x + 4) \qquad (a + b)^2 = a^2 + 2ab + b^2$$

$$x^2 - 6x - 7 = 0 \qquad \text{collecting terms on left side}$$

$$(x - 7)(x + 1) = 0 \qquad \text{factoring}$$

$$x = 7, -1$$

The root $x = -1$ checks, but $x = 7$ is extraneous.   ◄

Common error

Forgetting the middle term when squaring. In Example 4

$$(2 - \sqrt{3x + 4})^2 \qquad \text{is not} \qquad 4 + (3x + 4)$$

The correct expression is obtained by recalling that

$$(a - b)^2 = a^2 - 2ab + b^2$$

so that

$$(2 - \sqrt{3x + 4})^2 = 4 - 2 \cdot 2\sqrt{3x + 4} + (3x + 4)$$

 E X A M P L E **5**   The base of a right triangle is 3 ft shorter than the height, and the perimeter is 36 ft. Find the dimensions.

Solution. In Figure 13.8, x is the length of the base and $x + 3$ the height. By the Pythagorean theorem, the length of the hypotenuse is $\sqrt{x^2 + (x + 3)^2}$. Since the perimeter is 36 ft, we obtain the equation

$$x + (x + 3) + \sqrt{x^2 + (x + 3)^2} = 36$$

$$2x + 3 + \sqrt{2x^2 + 6x + 9} = 36$$

Isolating the radical, we get

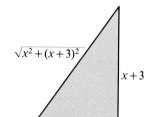

Figure 13.8

$$\sqrt{2x^2 + 6x + 9} = 33 - 2x$$

$$2x^2 + 6x + 9 = 1{,}089 - 132x + 4x^2 \qquad \text{squaring both sides}$$

$$2x^2 - 138x + 1{,}080 = 0 \qquad \text{collecting terms on one side}$$

$$x^2 - 69x + 540 = 0 \qquad \text{dividing by 2}$$

$$(x - 9)(x - 60) = 0 \qquad \text{factoring}$$

$$x = 9, 60$$

The root $x = 60$ is extraneous. The base is therefore 9 ft and the height $(9 + 3)$ ft $= 12$ ft.   ◀

E X E R C I S E S / S E C T I O N   **13.4**

In Exercises 1–28, solve the given equations. Check for possible extraneous roots.

1. $\sqrt{x - 4} = 2$

2. $\sqrt{2x - 1} = 3$

3. $\sqrt{2x + 6} = -2$

4. $\sqrt{2x + 6} = 2$

5. $\sqrt{x + 9} = x + 3$

6. $\sqrt{2x + 4} = x - 2$

7. $\sqrt{x + 6} = 2\sqrt{x}$

8. $\sqrt{x + 4} = \sqrt{2x + 3}$

9. $\sqrt{3x + 4} = \sqrt{4x - 5}$

10. $\sqrt[3]{x + 3} = 2$

11. $\sqrt{15 - 2x} = x$

12. $\sqrt{48 - 2x} = x$

13. $\sqrt{2x - 3} = x - 3$

14. $\sqrt{x} + \sqrt{x + 12} = 6$

15. $\sqrt{x} + \sqrt{x - 1} = 1$

16. $\sqrt{x - 1} + \sqrt{x + 3} = 2$

17. $\sqrt{x + 2} + \sqrt{x + 4} = 1$

18. $\sqrt{4 - x} + \sqrt{7 - x} = 3$

19. $\sqrt{x - 2} - \sqrt{x - 5} = 1$

20. $x + 6 - 5\sqrt{x} = 0$

21. $\sqrt{x - 1} - \sqrt{6 - x} = 1$

22. $\sqrt{4 + x} + \sqrt{6 - x} = 4$

23. $\sqrt{2x - 1} - \sqrt{x - 4} = 2$

24. $\sqrt{3x + 4} - \sqrt{2x + 1} = 1$

25. $\sqrt{2x - 3} - \sqrt{x - 2} = 1$

26. $\sqrt{5x - 1} - \sqrt{x + 2} = 1$

27. $\sqrt[3]{2x - 5} = \sqrt[3]{6x + 7}$

28. $\sqrt[4]{x + 4} = \sqrt{x - 2}$

29. Solve the equation $L = \sqrt{v} - \sqrt{v - 1}$ for v.

30. Recall that the magnitude of the impedance in an alternating-current circuit is $|Z| = \sqrt{R^2 + (X_L - X_C)^2}$. Solve this equation for R.

31. The base of a right triangle is 14 in. longer than the height, while the perimeter is 60 in. Find the lengths of the base and the height.

32. According to the special theory of relativity, the kinetic energy K of a body moving at velocity v relative to an observer is

$$K = \frac{m_0 c^2}{\sqrt{1 - \dfrac{v^2}{c^2}}} - m_0 c^2$$

where c is the velocity of light and m_0 the mass of the body at rest (relative to the observer). Show that

$$v = c\sqrt{1 - \left(\frac{m_0 c^2}{K + m_0 c^2}\right)^2}$$

33. In an experiment on the refraction of light (Figure 13.9), a light ray from a point A in medium M_1 to a point B in medium M_2 was observed to follow the path shown, a distance of 15 cm. Determine where the light enters M_2. (Find x in Figure 13.9; there are two possible solutions, depending on which medium has the larger index of refraction.)

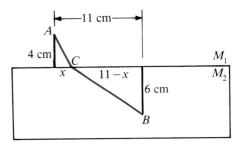

Figure 13.9

34. A lighthouse is situated 3 mi from a straight shore. The lighthouse keeper wants to travel to a point A on the shore 10 mi from P, the point on the shore nearest the lighthouse (Figure 13.10). Since A is not accessible by boat, the keeper lands at a point between A and P and walks the rest of the way, traveling a total of 11 mi. Where does he land?

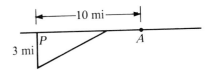

Figure 13.10

35. Under certain conditions, the impedance in a circuit is given by

$$Z = \frac{1}{\sqrt{R + \left(\dfrac{1}{2\pi f C}\right)^2}}$$

Solve this formula for C.

36. One form of the equation of the parabola is $\sqrt{x} + \sqrt{y} = \sqrt{a}$. Solve this equation for y.

REVIEW EXERCISES/CHAPTER   **13**

In Exercises 1–2, solve each system graphically to the nearest tenth of a unit.

1. $x^2 - y^2 = 2$
$y + 1 = x$

2. $y = \cos x$
$y = \sqrt{2x}$

In Exercises 3–10, solve the systems of equations algebraically.

3. $y = x + 2$
$y = 2x^2 + 1$

4. $y = x^2 - 2$
$y - x + 1 = 0$

5. $y = x^2 - 4$
$y - 3x + 6 = 0$

6. $x + y + 3 = 0$
$xy = 2$

7. $x^2 + 3y^2 = 4$
$2x^2 + 4y^2 = 5$

8. $x^2 - 2y^2 = 4$
$3x^2 + 4y^2 = 6$

9. $x^2 + y^2 = 4$
$xy = 2$

10. $x^2 - y^2 = 3$
$xy = 2$

In Exercises 11–30, solve each equation.

11. $x^4 - 20x^2 + 64 = 0$

12. $4x^4 - 5x^2 - 9 = 0$

13. $4x^4 + 13x^2 - 12 = 0$

14. $10x^{-2} - 13x^{-1} - 3 = 0$

15. $2x^{-2} + x^{-1} - 15 = 0$

16. $x^{-4} + 12x^{-2} - 64 = 0$

17. $4x^{-4} - 17x^{-2} + 4 = 0$

18. $2x + \sqrt{x} - 6 = 0$

19. $6x + 13\sqrt{x} - 28 = 0$

20. $15x - 7\sqrt{x} - 2 = 0$

21. $(x + 1) - 6\sqrt{x + 1} + 8 = 0$

22. $x^{2/3} + 2x^{1/3} - 8 = 0$

23. $(x^2 + 2x)^2 + 3(x^2 + 2x) + 2 = 0$

24. $\sqrt{x + 10} = -3$

25. $\sqrt{2 - x} = x + 4$

26. $\sqrt{x} + \sqrt{x - 2} = 2$

27. $\sqrt[3]{x - 3} = -2$

28. $\sqrt{x + 11} - \sqrt{x} = 1$

29. $\sqrt{x - 3} - \sqrt{8 - x} = 1$

30. $\sqrt{4x - 3} - \sqrt{x + 1} = 1$

31. Two resistors connected in series have a combined resistance of 10.0 Ω. The same resistors connected in parallel have a combined resistance of 2.172 Ω. Find the two resistances.

32. Two bodies weighing 100 g and 80 g collide with velocities v_1 and v_2 (in centimeters per second), respectively. From the laws of conservation of energy and momentum, the following relationships are obtained:

$$100v_1{}^2 + 80v_2{}^2 = 11{,}600$$
$$50v_1 + 40v_2 = 700$$

Find v_1 and v_2.

33. A technician wants to design a rectangular metal plate having an area of 540 in.² and a diagonal 39 in. long. What must the dimensions be?

34. The base of a right triangle is 4 in. longer than the height, and the perimeter is 48 in. Find the dimensions.

35. According to the theory of relativity, the mass of a body moving at velocity v relative to an observer is given by

$$m = \frac{m_0}{\sqrt{1 - \dfrac{v^2}{c^2}}}$$

where c is the velocity of light and m_0 the mass of the body at rest (relative to the observer). Solve for $v > 0$.

Higher-Order Equations in One Variable

14.1 Introduction

So far we have confined ourselves to equations of first and second degree except for some special quadratic forms. In this chapter we will turn our attention to equations of higher degree, which are defined next.

A polynomial equation of nth degree has the form

$$a_0 x^n + a_1 x^{n-1} + \cdots + a_{n-1} x + a_n = 0, \qquad a_0 \neq 0 \qquad (14.1)$$

To see how far we have come, let us first note that equations of third degree were studied by the Persian poet Omar Khayyám in the twelfth century and equations of fourth degree by the Italian mathematician and physician Geronimo Cardano (1501–1576) in 1545. Although there is some doubt about the true originator, a publication attributed to Cardano had such an impact on the development of algebra that the year 1545 is frequently considered the beginning of the modern period in mathematics.

The next great step was taken by the German mathematician Karl F. Gauss (1777–1855), who proved in his doctoral dissertation of 1799 that every polynomial equation has a root. This statement, known as the *fundamental theorem of algebra*, will be discussed in greater detail in Section 14.4.

Many attempts were made to apply Cardano's method to equations of degree greater than four, but up to the time of Gauss's famous theorem, and afterwards, none of these efforts was successful. Imagine the shock when in 1824 a young Norwegian named Niels Abel (1802–1829) proved that no such method exists. In other words, there is no formula that will yield the roots of an equation of degree larger than four in terms of radicals (unlike the situation with quadratic equations, for which such a formula does exist).

In spite of this general problem, many equations can be solved by special methods. Techniques also exist for finding approximate solutions in decimal form. These are the topics to be discussed in this chapter.

14.2　The Remainder and Factor Theorems

As noted in the introduction, the purpose of this chapter is to develop special techniques for solving certain higher-order equations. These techniques depend on two theorems, the **remainder theorem** and the **factor theorem**, which will be discussed in this section.

Given the form (14.1) of a polynomial equation, we can define a **polynomial function** $f(x)$.

A polynomial function has the form

$$f(x) = a_0 x^n + a_1 x^{n-1} + \cdots + a_{n-1} x + a_n \tag{14.2}$$

where $a_0, a_1, \ldots, a_n$ are real or complex constants.

If $f(x)$ is divided by $x - r$, we obtain a polynomial $Q(x)$ and remainder R. These are related by the form

$$f(x) = Q(x)(x - r) + R \tag{14.3}$$

as illustrated in the next example.

E X A M P L E　1　Divide $f(x) = 2x^2 - 4x + 4$ by $x - 3$ and write $f(x)$ in the form $f(x) = Q(x)(x - r) + R$.

Solution.

$$
\begin{array}{r}
2x + 2 \qquad\quad Q(x) \\
x - 3 \overline{)\, 2x^2 - 4x + 4} \quad\ f(x) \\
\underline{2x^2 - 6x \qquad\qquad} \\
2x + 4 \qquad\quad \\
\underline{2x - 6 \qquad\quad} \\
10 \qquad R
\end{array}
$$

Thus,

$$f(x) = (2x + 2)(x - 3) + 10 \qquad f(x) = Q(x)(x - r) + R$$

which can be checked by direct multiplication.　　　　◀

An immediate consequence of formula (14.3) is the **remainder theorem**, which will be used in Section 14.6 to find approximate roots of a given polynomial equation $f(x) = 0$. Let $x = r$ in the formula $f(x) = Q(x)(x - r) + R$. Then

$$f(r) = Q(r)(r - r) + R = Q(r) \cdot 0 + R$$

or

$$f(r) = R \hspace{6cm} (14.4)$$

In other words, R is equal to the function value of $f(x)$ at $x = r$.

> **Remainder Theorem:** If a polynomial function is divided by $x - r$, then the remainder is equal to $f(r)$.

E X A M P L E **2**   If $f(x) = x^3 - 4x^2 + 5x - 3$, evaluate $f(2)$ by using the remainder theorem.

Solution. To get the proper remainder, we must divide $f(x)$ by $x - 2$. The division yields

$$f(x) = (x^2 - 2x + 1)(x - 2) - 1 \qquad R = -1$$

Thus $f(2) = -1$.
$$\text{Check: } f(2) = 2^3 - 4 \cdot 2^2 + 5 \cdot 2 - 3 = -1 \qquad\blacktriangleleft$$

E X A M P L E **3**   Use the remainder theorem to find R when $f(x) = 2x^3 - x^2 + 2x - 1$ is divided by $x + 3$.

Solution. Since the divisor is $x + 3$, we have $r = -3$. Hence

$$R = f(-3) = 2(-3)^3 - (-3)^2 + 2(-3) - 1 = -70$$

So $R = -70$. $\qquad\blacktriangleleft$

Another immediate consequence of formula (14.3) is the **factor theorem:** If $R = 0$ in the formula $f(x) = Q(x)(x - r) + R$, then

$$f(x) = Q(x)(x - r)$$

and $f(r) = 0$ by (14.4); that is, r is a root of $f(x) = 0$.

> **Factor Theorem:** If r is a root of $f(x) = 0$, then $x - r$ is a factor of $f(x)$. Conversely, if $x - r$ is a factor of $f(x)$, then r is a root of $f(x) = 0$. In symbols,
>
> $$f(x) = Q(x)(x - r)$$

(The factor theorem will be used in Sections 14.4 and 14.5 to find one or more factors of a given polynomial equation $f(x) = 0$. Finding such factors greatly simplifies the computations involved in solving the equation.)

E X A M P L E **4**   Determine whether $x + 5$ is a factor of $f(x) = x^4 + 5x^3 + 2x^2 + 11x + 5$ by means of the factor theorem.

Solution. Since $x + 5$ is the divisor, $r = -5$. Since

$$f(-5) = (-5)^4 + 5(-5)^3 + 2(-5)^2 + 11(-5) + 5 = 0$$

it follows that $x + 5$ is a factor of $f(x)$. ◀

E X A M P L E **5**   Determine whether $x - 4$ is a factor of $f(x) = 3x^4 - 2x^3 + 3x - 5$.

Solution. Since

$$f(4) = 3(4)^4 - 2(4)^3 + 3(4) - 5 = 647 \neq 0$$

$x - 4$ is not a factor. ◀

Summary. The remainder theorem says that whenever a polynomial function $f(x)$ is divided by $x - r$, the remainder R is equal to $f(r)$, the value of $f(x)$ at $x = r$. If the remainder R is zero, then $f(r) = 0$ and r is therefore a root of $f(x) = 0$. By the factor theorem, r is a root of $f(x) = 0$ if and only if $x - r$ is a factor of $f(x)$. These observations will be used in later sections to solve a given polynomial equation $f(x) = 0$.

E X E R C I S E S / S E C T I O N   **14.2**

In Exercises 1–20, find the remainder in each division problem by the remainder theorem. Check by long division. (See Example 3.)

1. $(x^3 + 2x^2 - x + 1) \div (x - 1)$
2. $(x^3 + 3x^2 + x - 3) \div (x + 2)$
3. $(x^3 - 2x^2 - 2x + 3) \div (x + 3)$
4. $(x^3 - x^2 + x - 2) \div (x - 1)$
5. $(x^3 - 4x^2 - x + 2) \div (x - 3)$
6. $(2x^3 - x^2 + x + 3) \div (x - 2)$
7. $(2x^3 + 3x - 100) \div (x - 4)$
8. $(2x^3 - x^2 + 10) \div (x + 3)$
9. $(x^4 + x^3 - x^2 + x + 1) \div (x - 1)$
10. $(x^4 + x^3 - 3x - 1) \div (x + 2)$
11. $(x^4 - 2x^3 + x^2 + x - 3) \div (x - 3)$

12. $(x^3 - x^2 - x - 50) \div (x - 5)$
13. $(2x^4 + 3x^3 + x - 38) \div (x + 3)$
14. $(4x^3 - 10x^2 + 5x + 3) \div (x - 2)$
15. $(4x^3 + 20x^2 - 18x - 30) \div (x + 5)$
16. $(5x^3 - 32x^2 + 20x - 40) \div (x - 6)$
17. $(5x^4 - 28x^3 - 20x - 300) \div (x - 6)$
18. $(6x^4 - 25x^3 + 4x^2 + 7x - 25) \div (x - 4)$
19. $(3x^5 + 5x^4 - 12x^3 + 4x^2 + 10x - 6) \div (x + 3)$
20. $(4x^5 + 20x^4 + 8x^2 + 35x - 25) \div (x + 5)$

In Exercises 21–30, use the factor theorem to determine whether the second expression is a factor of the first. (See Examples 4 and 5.)

21. $x^3 - 5x^2 + 8x - 4, x - 2$
22. $x^3 - 2x^2 - 11x - 5, x + 2$

23. $x^4 - 2x^3 - 9x^2 + x + 12, x + 2$
24. $x^4 - 5x^2 + 15x + 9, x + 3$
25. $2x^4 - 7x^3 - 19x^2 + 23x - 15, x - 5$
26. $2x^4 + 4x^3 - 13x^2 + 10x - 10, x + 4$

27. $3x^5 - 7x^4 - 6x^3 + x^2 - 2x - 3, x - 3$
28. $6x^5 - 36x^4 - x^3 + 6x^2 + x - 6, x - 6$
29. $3x^5 + 17x^4 + 10x^3 + x^2 + 7x + 5, x + 5$
30. $5x^5 + 16x^4 + 2x^3 - 2x^2 + 4x + 3, x + 3$

14.3 Synthetic Division

The remainder and factor theorems point out the close relationship between the value of a polynomial function $f(x)$ at r and the remainder due to division by $x - r$. To use these theorems effectively later on, a fast and efficient means of dividing $f(x)$ by $x - r$ is extremely helpful. **Synthetic division** provides us with this means. After we study synthetic division, we will return to the factor and remainder theorems.

Synthetic Division

To derive the method of synthetic division, let us consider a regular division problem and then see what steps can be eliminated:

$$
\begin{array}{r}
x^3 - x^2 - 4x - 7 \\
x - 2 \overline{)\, x^4 - 3x^3 - 2x^2 + x + 5} \\
\underline{x^4 - 2x^3} \\
-x^3 - 2x^2 \\
\underline{-x^3 + 2x^2} \\
-4x^2 + x \\
\underline{-4x^2 + 8x} \\
-7x + 5 \\
\underline{-7x + 14} \\
-9
\end{array}
$$

First, observe that the entire division procedure depends only on the coefficients. The powers of x serve as placeholders, provided that the given polynomial is written in descending powers of x. Let's see what the division problem looks like without the powers of x:

$$
\begin{array}{r}
1 - 1 - 4 - 7 \\
1 - 2 \overline{)\, 1 - 3 - 2 + 1 + 5} \\
\underline{1 - 2} \\
-1 - 2 \\
\underline{-1 + 2} \\
-4 + 1 \\
\underline{-4 + 8} \\
-7 + 5 \\
\underline{-7 + 14} \\
-9
\end{array}
$$

Since the division procedure we are developing applies only to the division by $x - r$, the number 1 in the divisor $1 - 2$ can be omitted. Even more significantly, the division process contains a number of repeated terms (shown in color) that can also be eliminated. The division now has the following form:

$$
\begin{array}{r}
1 - 1 - 4 - 7 \\
-2\,\overline{)\,1 - 3 - 2 + 1 +\ \ 5} \\
-2 \\
\overline{-1} \\
+2 \\
\overline{-4} \\
+8 \\
\overline{-7} \\
+14 \\
\overline{-9}
\end{array}
$$

This format can now be condensed by eliminating the empty spaces:

$$
\begin{array}{r}
1 - 1 - 4 - 7 \\
-2\,\overline{)\,1 - 3 - 2 + 1 +\ \ 5} \\
-2 + 2 + 8 + 14 \\
\overline{-1 - 4 - 7 -\ \ 9}
\end{array}
$$

Observe next that the numbers in the first row are contained in the fourth row, except for the first term. By repeating this term in the fourth row, the first row can be eliminated:

$$
\begin{array}{r}
-2\,\overline{)\,1 - 3 - 2 + 1 +\ \ 5} \\
-2 + 2 + 8 + 14 \\
\overline{1 - 1 - 4 - 7 -\ \ 9}
\end{array}
$$

To simplify even further, note first that the division can actually be carried out by the last scheme: Bring down the 1 in the first row and multiply by -2. Then place the product -2 in the second row under the -3. Subtract -2 from -3 and write the resulting -1 in the third row, and so on. Now, in subtraction we change the sign of the subtrahend and add. This change in sign becomes automatic if we replace the -2 in the divisor by 2.

We now have a greatly simplified method of division. All we need to do in addition is observe the custom of placing the 2 on the right:

Synthetic division

$$
\begin{array}{r}
1 - 3 - 2 + 1 +\ \ 5\,\underline{)\,2} \\
2 - 2 - 8 - 14 \\
\overline{1 - 1 - 4 - 7 -\ \ 9}
\end{array}
$$

The steps can be described as follows: Bring down the first 1, multiply by 2, place the product 2 under the -3, add 2 and -3, and place the sum -1 below the 2. Then repeat the process.

In general, to divide a polynomial function $f(x)$ by $x - r$ using synthetic division, follow the procedure outlined next.

Synthetic Division: To divide a polynomial $f(x)$ by $x - r$:

1. Write $f(x)$ in descending powers of x and supply a zero for each missing power.
2. Omit the powers of x and set up the division scheme in the following manner:

$$a_0 + a_1 + a_2 + \cdots + a_n \,) \, r$$

3. Bring down a_0, multiply a_0 by r, place the product beneath a_1, and add the product to a_1. Multiply the sum by r, place the product beneath a_2, and add the product to a_2. Continue the process to the end of the row.
4. The numbers in the bottom row are, reading from left to right, the coefficients of the quotient, which is of degree one less than $f(x)$. The last number in the bottom row is $R = f(r)$.

E X A M P L E 1 Use synthetic division to divide $2x^3 - 3x + 2$ by $x + 3$.

Solution. Since $x - r = x + 3$, we have $r = -3$. Now supply a zero for the missing power x^2 and proceed as follows:

$$
\begin{array}{r}
2 + 0 - 3 + 2 \,) \, {-3} \\
\underline{-6 + 18 - 45} \\
2 - 6 + 15 - 43
\end{array}
$$

The quotient, obtained from the bottom row, is therefore $2x^2 - 6x + 15$, while the remainder is $f(-3) = -43$. ◀

E X A M P L E 2 Determine whether $x - 4$ is a factor of $13x - 14x^2 + 3x^3 - 20$.

Solution. Writing the polynomial in descending powers of x, we get $3x^3 - 14x^2 + 13x - 20$. By synthetic division:

$$
\begin{array}{r}
3 - 14 + 13 - 20 \,) \, 4 \\
\underline{12 - 8 + 20} \\
3 - 2 + 5 + 0
\end{array}
$$

Since the remainder is 0, $x - 4$ is a factor. ◀

Function Values, Zeros, and Factors

We learned in Section 14.2 that $x - r$ is a factor of $f(x)$ if r is a root of $f(x) = 0$, that is, if $f(r) = 0$. Consequently, computing $f(r)$ appeared to be the quickest way to determine whether $x - r$ was a factor. However, as we have seen, synthetic division is usually simpler than carrying out the necessary arithmetic and is actually a convenient method for computing $f(r)$. Moreover, if $f(r) = 0$, then r is called a **zero** of $f(x)$. So synthetic division can be used to determine whether a number is a zero of $f(x)$. [Observe that a zero of $f(x)$ is really a root of the polynomial equation $f(x) = 0$.]

Function Values, Zeros, and Factors:

1. To compute $f(r)$, divide $f(x)$ by $x - r$ by means of synthetic division. The remainder R is equal to $f(r)$.
2. If $f(r) = 0$, then the root r is called a **zero of $f(x)$**.
3. If r is a zero of $f(x)$, then $x - r$ is a factor of $f(x)$.

Example 2 illustrates part 3: Since $x = 4$ is a zero of $f(x) = 13x - 14x^2 + 3x^3 - 20$, $x - 4$ is a factor of $f(x)$. The other parts are illustrated in following examples.

E X A M P L E **3**   Given that $f(x) = 9x^4 - 7x^2 + x + 4$, find $f(-\frac{1}{3})$ by synthetic division.

Solution. Since $f(-\frac{1}{3})$ is the remainder when $f(x)$ is divided by $x + \frac{1}{3}$, we proceed by synthetic division as follows:

$$
\begin{array}{r}
9 + 0 - 7 + 1 + 4 \,\big)\!-\dfrac{1}{3} \\
\underline{-3 + 1 + 2 - 1} \\
9 - 3 - 6 + 3 + \mathbf{3}
\end{array}
$$

Thus $f(-\frac{1}{3}) = \mathbf{3}$.   ◀

E X A M P L E **4**   Determine whether the number -2 is a zero of $f(x) = 2x^5 - x^4 - 20x + 40$.

Solution. Supplying zeros for the missing powers x^3 and x^2, we get the following scheme:

$$
\begin{array}{r}
2 - 1 + 0 + 0 - 20 + 40 \,\big)\!-\mathbf{2} \\
\underline{-4 + 10 - 20 + 40 - 40} \\
2 - 5 + 10 - 20 + 20 + \ \ \mathbf{0}
\end{array}
$$

Since the remainder is **0**, we have shown that $f(-2) = \mathbf{0}$. [We have also shown that $x + 2$ is a factor of $f(x)$ and that $x = -2$ is a root of the equation $f(x) = 0$.]   ◀

EXERCISES / SECTION 14.3

In Exercises 1–20, find the quotient and remainder by means of synthetic division.

1. $(x^3 + 2x^2 - x + 1) \div (x - 1)$
2. $(x^3 + 3x^2 + x - 3) \div (x + 2)$
3. $(x^3 - 4x^2 - x + 2) \div (x - 3)$
4. $(2x^3 - x^2 + x + 3) \div (x - 2)$
5. $(x^4 + x^3 - 3x - 1) \div (x + 2)$
6. $(2x^4 + 3x^3 + x - 38) \div (x + 3)$
7. $(4x^3 + 20x^2 - 18x - 30) \div (x + 5)$
8. $(5x^3 - 32x^2 + 20x - 40) \div (x - 6)$
9. $(4x^5 + 20x^4 + 8x^2 + 35x - 25) \div (x + 5)$
10. $(3x^5 + 5x^4 - 12x^3 + 4x^2 + 10x - 6) \div (x + 3)$
11. $(8x^3 + 2x^4 + 15x + 3x^2 + 8) \div (x + 4)$
12. $(3x^4 + 2 - 5x - 25x^2) \div (x - 3)$
13. $(2x^3 + x - 4x^2 + 4x^4 + 2) \div \left(x - \dfrac{1}{2}\right)$
14. $(4x^5 + 6x^2 - x^3 + 5x - 3) \div \left(x + \dfrac{1}{2}\right)$
15. $(6 - 4x + x^5 - 8x^2) \div (x - 2)$
16. $(1 + x^5 - 35x^2 - 30x^3 - 6x) \div (x - 6)$
17. $(2x^3 + 9x^5 - 5x - 4x^2 + 2) \div \left(x - \dfrac{1}{3}\right)$
18. $(14x - 3x^4 + x^5 - 15x^3 - 20x^2 - 10) \div (x - 6)$
19. $(20x^2 + 10x^3 + x^4 + 35x + 20) \div (x + 8)$
20. $(8x^5 + 4x - 3 - x^2) \div \left(x - \dfrac{1}{2}\right)$

In Exercises 21–32, use synthetic division to determine whether the second expression is a factor of the first. (See Example 2.)

21. $2x^3 - 8x^2 + 5x + 3, x - 3$
22. $5x^3 + 12x^2 - x - 10, x + 2$
23. $x^4 + 4x^3 + 2x^2 - x - 20, x + 4$
24. $5x^3 + 20x^2 + 30x + 24, x + 5$

25. $2x^4 + 4x^3 - 2x - 4, x + 2$
26. $3x^3 - 10x^2 + 2x - 40, x - 4$
27. $5x + 4x^2 + 2x^5 - 20x^3 + 3, x - 3$
28. $x^2 + 3x^4 - 6x - 15x^3 + 4, x - 5$
29. $6 - x - 12x^3 + 2x^4, x - 6$
30. $x^2 - 16 + 3x^5 + 11x^4 - 4x^3, x + 4$
31. $x^5 + 32, x + 2$
32. $x^6 - 1, x - 1$

In Exercises 33–38, determine whether the indicated value is a zero of the polynomial. (See Example 4.)

33. $4x^3 - 10x^2 - 6x, 3$
34. $3x^4 + 9x^3 - 10x^2 + 10x + 8, -4$
35. $4x^4 + 4x^3 + x^2 - 6x + 1, \dfrac{1}{2}$
36. $6x^3 + 15x^2 + 9x + 7, -2$
37. $4x^4 + 5x^3 - 6x^2 + 4x - 3, \dfrac{3}{4}$
38. $8x^5 - 6x^4 + 10x^3 - x^2 + 3x + 1, -\dfrac{1}{4}$

In Exercises 39–45, find $f(r)$ for the indicated value of r. (See Example 3.)

39. $f(x) = 2x^3 - 4x^2 + x - 2; r = 3$
40. $f(x) = 3x^3 + 5x^2 - 11x + 2; r = \dfrac{1}{3}$
41. $f(x) = 4x^3 - 3x^2 - 7x + 1; r = -\dfrac{1}{4}$
42. $f(x) = x^3 - x^2 - 30x + 10; r = -6$
43. $f(x) = x^4 + x^3 - 10x^2 - 20x + 30; r = 5$
44. $f(x) = 4x^4 - 7x^3 + 2x^2 - x + 1; r = -\dfrac{1}{4}$
45. $f(x) = 6x^4 - 8x^3 - x^2 - 2x - 6; r = \dfrac{1}{3}$

14.4 The Fundamental Theorem of Algebra

To solve higher-order equations, we need to employ special tools such as the factor theorem and synthetic division. Another useful tool, discussed in this section, is the criterion for determining the number of roots of a polynomial equation. This criterion follows from the **fundamental theorem of algebra**, first mentioned in Section 14.1, and enables us to solve polynomial equations under certain special conditions.

Fundamental Theorem of Algebra: Every polynomial equation

$$f(x) = a_0 x^n + a_1 x^{n-1} + \cdots + a_n = 0$$

has at least one real or complex root.

The fundamental theorem of algebra leads to a much more general statement by repeated application of the factor theorem. Let r_1 be the root of $f(x) = 0$ guaranteed by the fundamental theorem. Then by the factor theorem

$$f(x) = Q_1(x)(x - r_1)$$

where $Q_1(x)$ is of degree $n - 1$. Since $Q_1(x) = 0$ is also a polynomial equation, it has a root r_2 by the fundamental theorem. Thus

$$Q_1(x) = Q_2(x)(x - r_2)$$

where $Q_2(x)$ is of degree $n - 2$. It follows that

$$f(x) = Q_2(x)(x - r_1)(x - r_2)$$

Continuing this process, we obtain n linear (first-degree) factors of $f(x)$ and hence n roots of the equation $f(x) = 0$. Thus

$$f(x) = (x - r_1)(x - r_2) \cdots (x - r_n) \tag{14.5}$$

where $r_1, r_2, \ldots, r_n$ are the roots of $f(x) = 0$. Unfortunately, the numbers $r_1, r_2, \ldots, r_n$ may not all be distinct; thus, the number of distinct roots may be less than n. For this reason, we need to introduce the following definition:

Multiplicity of a Root: If a root r_k occurs k times ($1 \leq k \leq n$) then r_k is said to be a **root of multiplicity k.**

With this definition, we can make the following general statements:

---

1. A polynomial of degree n can be factored into n linear factors.
2. A polynomial equation of degree n has n roots if a root of multiplicity k is counted as k roots.

E X A M P L E **1**   Consider the following equation of degree 11:

$$(x - 1)(x + 2)^4(x - 5)^6 = 0$$

The roots of the equation are: $x = 1$ (single root), $x = -2$ (multiplicity 4), and $x = 5$ (multiplicity 6). The total number of roots, counting their multiplicities, is $1 + 4 + 6 = 11$, which is equal to the degree of the equation.   ◀

Recall from Section 14.1 that no general method exists for solving by algebraic means an equation of degree 5 or higher. However, if one or more of the roots is already known, the remaining roots may possibly be found by the factor theorem, as shown in the remaining examples.

E X A M P L E **2**   One zero of the polynomial function $f(x) = 2x^3 - 7x^2 + x + 1$ is $\frac{1}{2}$. Determine the remaining zeros.

Solution. By synthetic division

$$
\begin{array}{r}
2 - 7 + 1 + 1 \left) \dfrac{1}{2} \right. \\
\underline{1 - 3 - 1 } \\
2 - 6 - 2 + 0
\end{array}
$$

Hence

$$f(x) = (2x^2 - 6x - 2)\left(x - \frac{1}{2}\right)$$

$$= 2(x^2 - 3x - 1)\left(x - \frac{1}{2}\right)$$

It follows that $f(x) = 0$ if $x - \frac{1}{2} = 0$ or $x^2 - 3x - 1 = 0$. The roots of $x^2 - 3x - 1 = 0$ are (by the quadratic formula),

$$x = \frac{3 \pm \sqrt{9 + 4}}{2} = \frac{3 \pm \sqrt{13}}{2}$$

So the zeros of $f(x)$ are $\frac{1}{2}$, $\frac{1}{2}(3 + \sqrt{13})$, and $\frac{1}{2}(3 - \sqrt{13})$.   ◀

E X A M P L E **3**   Solve the equation:

$$x^4 - x^3 - 26x^2 + 84x - 72 = 0$$

Double root

given that 2 is a root of multiplicity 2, also called a **double root**.

Solution. By synthetic division

$$
\begin{array}{r}
1 - 1 - 26 + 84 - 72 \,\underline{)\,2} \\
2 + \ \ 2 - 48 + 72 \\
\hline
1 + 1 - 24 + 36 + \ \ 0
\end{array}
$$

This division shows that

$$x^4 - x^3 - 26x^2 + 84x - 72 = (x^3 + x^2 - 24x + 36)(x - 2)$$

Since 2 is a double root, it must also be a root of the equation $x^3 + x^2 - 24x + 36 = 0$. Thus we repeat the synthetic division process:

$$
\begin{array}{r}
\mathbf{1 + 1 - 24 + 36}\,\underline{)\,2} \\
2 + \ \ 6 - 36 \\
\hline
1 + 3 - 18 + \ \ 0
\end{array}
$$

Dividing by a known factor always leaves a remainder of 0. If we avoid writing down the zero remainder, then synthetic divisions can be carried out successively:

$$
\begin{array}{r}
1 - 1 - 26 + 84 - 72 \,\underline{)\,2} \\
2 + \ \ 2 - 48 + 72 \\
\hline
1 + 1 - 24 + 36 \qquad \underline{)\,2} \\
2 + \ \ 6 - 36 \\
\hline
1 + 3 - 18
\end{array}
$$

The resulting quotient $x^2 + 3x - 18$ factors into $(x + 6)(x - 3)$. It follows that

$$
\begin{aligned}
x^4 - x^3 - 26x^2 + 84x - 72 &= (x - 2)^2(x + 6)(x - 3) \\
&= 0
\end{aligned}
$$

The roots are therefore given by $x = 2, 2, -6, 3$.   ◄

Complex roots

This procedure is by no means restricted to real roots. Moreover, if the coefficients of a polynomial equation are real and if $a + bj$ is a root, then so is $a - bj$. This observation follows from the quadratic formula:

$$x = \frac{-b \pm \sqrt{b^2 - 4ac}}{2a} = -\frac{b}{2a} \pm \frac{\sqrt{b^2 - 4ac}}{2a}$$

In other words, complex roots will always occur as pairs of complex conjugates.

E X A M P L E **4**   Solve the equation:

$$x^4 - 6x^3 + 15x^2 - 18x + 10 = 0$$

given that $2 - j$ is a root.

Solution. Since $2 - j$ is a root, so is $2 + j$. By synthetic division we get

$$
\begin{array}{rrrrr}
1 & -6 & +15 & -18 & +10 \,)\,\mathbf{2 - j} \\
  & 2 - j & -9 + 2j & 14 - 2j & -10 \\
\hline
1 & -4 - j & 6 + 2j & -4 - 2j & \,)\,\mathbf{2 + j} \\
  & 2 + j & -4 - 2j & +4 + 2j & \\
\hline
1 & -2 & +2 & &
\end{array}
$$

The zeros of the resulting quadratic factor $x^2 - 2x + 2$ can be found by the quadratic formula:

$$
x = \frac{2 \pm \sqrt{4 - 8}}{2} = 1 \pm j
$$

The roots are therefore $x = 2 \pm j,\ 1 \pm j$. ◀

E X E R C I S E S / S E C T I O N **14.4**

In Exercises 1–8, state the degree of the equation, each root, and its multiplicity.

1. $(x + 2)^2(x - 3)^2(x + 1) = 0$

2. $(x - 4)^3(x + 1)^2(x + 2) = 0$

3. $x^2(x + 4)^3(x - 5)^2 = 0$

4. $x^2(x - 7)^3(x + 4)^4 = 0$

5. $(2x + 1)^4(x - 10)^5 = 0$

6. $(3x + 2)^5(x - 3)^2 = 0$

7. $x^2(4x - 3)^3(x + 7)^4 = 0$

8. $(x - 10)^3(x + 9)^2(x - 3)^4 = 0$

In Exercises 9–32, find the roots not given.

9. $x^3 - 7x + 6 = 0$; one root is 2

10. $x^3 + 4x^2 + x - 6 = 0$; one root is -3

11. $x^3 + x^2 - 16x - 16 = 0$; one root is -4

12. $x^3 - x^2 - 17x - 15 = 0$; one root is 5

13. $x^3 - 4x + 3 = 0$; one root is 1

14. $2x^3 + 6x^2 + 3x - 2 = 0$; one root is -2

15. $2x^3 - 2x^2 - 9x - 9 = 0$; one root is 3

16. $x^3 - 14x - 8 = 0$; one root is 4

17. $21x^2 + x^4 - 8x^3 - 22x + 8 = 0$; 1 is a double root

18. $6x^3 + x^4 + 9x^2 - 4x - 12 = 0$; -2 is a double root

19. $x^4 + 4x^2 + 3x^3 + 3x + 1 = 0$; -1 is a root of multiplicity 2

20. $3x^4 - 13x^3 + 39x - 2x^2 + 9 = 0$; 3 is a root of multiplicity 2

21. $x^4 - x^3 - x^2 - x - 2 = 0$; one root is j

22. $x^4 - 2x^3 - 7x^2 - 2x - 8 = 0$; one root is $-j$

23. $x^4 + x^3 - 2x^2 + 4x - 24 = 0$; one root is $-2j$

24. $x^4 + 2x^3 - 15x^2 + 18x - 216 = 0$; one root is $3j$

25. $x^4 + 2x^3 + 5x^2 + 2x + 4 = 0$; one root is j

26. $x^4 - x^3 + 13x^2 - 16x - 48 = 0$; one root is $4j$

27. $x^4 - 5x^3 + 10x^2 - 10x + 4 = 0$; one root is $1 + j$

28. $x^4 - 23x^2 + 68x - 60 = 0$; one root is $2 - j$

29. $4x^4 - 16x^3 + 21x^2 - 11x + 2 = 0$; $\dfrac{1}{2}$ is a root of multiplicity 2

30. $9x^4 - 21x^3 - 107x^2 - 63x - 10 = 0$; $-\dfrac{1}{3}$ is a root of multiplicity 2

31. $x^4 - 7x^3 + 13x^2 - 2x - 8 = 0$; two roots are 2 and 4

32. $x^4 + 10x^3 - 31x^2 - 280x + 300 = 0$; two roots are 5 and -6

14.5 Rational Roots

The technique for solving equations discussed in the last section has one serious limitation: At least one root has to be known in advance. In this section we will learn how to determine one or more rational roots from a list of numbers that includes the rational roots. The remaining roots can then be found by the factor theorem. If no rational roots exist, the method of this section cannot be used. This case will be discussed in Section 14.6.

Let us first consider a special type of polynomial equation (with integral coefficients) for which the coefficient of the nth power is unity:

$$f(x) = x^n + a_1 x^{n-1} + \cdots + a_n = 0 \tag{14.6}$$

By the discussion in Section 14.4

$$f(x) = (x - r_1)(x - r_2) \cdots (x - r_n) = 0 \tag{14.7}$$

Multiplying out the middle expression shows that the constant a_n is numerically equal to a product of integers:

$$a_n = r_1 r_2 \cdots r_n \tag{14.8}$$

So a_n is divisible by each r_i ($i = 1, 2, 3, \ldots, n$). Furthermore, since each r_i is a root of $f(x) = 0$, a_n is divisible by each root. It follows, then, that writing down all the factors of a_n gives a list of all possible integral roots. However, these factors do not indicate the sign of a root or give any irrational root.

E X A M P L E 1   Solve the equation $x^3 - 7x + 6 = 0$.

Solution. The integral divisors of 6 are ± 1, ± 2, ± 3, and ± 6. This is a list of all possible integral roots. Suppose we try $x = 1$ by synthetic division:

$$\begin{array}{r}
1 + 0 - 7 + 6 \underline{)\, 1} \\
1 + 1 - 6 \quad\quad \\
\hline
1 + 1 - 6 + 0 \quad\quad
\end{array}$$

Since the remainder is 0, $x = 1$ is a root. The factor $x^2 + x - 6$ can be written $(x + 3)(x - 2)$. So the roots are -3, 2, and 1.   ◀

If the coefficient of the highest power is different from 1, then the list of all possible rational roots may be considerably larger. Note that any polynomial equation

$$f(x) = a_0 x^n + a_1 x^{n-1} + \cdots + a_n = 0, \qquad a_0 \neq 0$$

can be written

$$f(x) = a_0 \left(x^n + \frac{a_1}{a_0} x^{n-1} + \cdots + \frac{a_n}{a_0} \right) = 0 \tag{14.9}$$

By a similar argument we can now obtain a much more general result.

> If the coefficients of
>
> $$a_0 x^n + a_1 x^{n-1} + \cdots + a_{n-1} x + a_n = 0$$
>
> are integers, then each **rational** root, after being reduced to lowest terms, has a factor of a_n for its numerator and a factor of a_0 for its denominator.

This criterion can be used to obtain a list of all possible rational roots. Consider the example below.

E X A M P L E **2**   Solve the equation $4x^3 - 24x^2 + 35x - 12 = 0$.

Solution. To construct a list of all possible rational roots, we first list the factors of the numerator and denominator separately:

Factors of numerator: $\pm 1, \pm 2, \pm 3, \pm 4, \pm 6, \pm 12$ **factors of 12**

Factors of denominator: $\pm 1, \pm 2, \pm 4$ **factors of 4**

Now we form all possible ratios. Omitting repetitions, we get

$$\pm 1, \pm 2, \pm 3, \pm 4, \pm 6, \pm 12, \pm\frac{1}{4}, \pm\frac{1}{2}, \pm\frac{3}{2}, \pm\frac{3}{4} \tag{14.10}$$

This long list would take quite a while to check, even with synthetic division. Fortunately, the number of possibilities can be reduced considerably. Let us postpone solving this equation and see how we can further limit the possible roots. ◄

One way to reduce the number of possible roots is by **Descartes' rule of signs**.

> **Descartes' Rule of Signs:** The number of positive roots of a polynomial equation $f(x) = 0$ with real coefficients is equal to the number of sign changes of $f(x)$ or less than this number by an even integer. The number of negative roots is equal to the number of sign changes of $f(-x)$ or less than this number by an even integer.

To see what is meant by sign changes, consider the polynomial equation $3x^4 - 2x^3 + 3x^2 - x - 1 = 0$. Since the sign changes from positive to negative, then to positive, and then back to negative, there are three sign changes. So the equation has either three positive roots or only $3 - 2 = 1$ positive root. The equation $4x^3 + 10x^2 - 5x + 3 = 0$ has two sign changes and hence either two positive roots or none. The equation $7x^3 + 3x + 2 = 0$ has no variations in signs and hence no positive roots.

Even more useful than Descartes' rule of signs is the criterion for restricting the range in which the real roots can lie. Any number that is greater than or equal to the greatest root of an equation is called an **upper bound** of the roots. Any number that is less than or equal to the least root of an equation is called a **lower bound** of the roots.

> **Upper and Lower Bounds:** If the coefficient of x^n in a polynomial equation $f(x) = 0$ is positive, and if no negative terms are in the third row of the synthetic division of $f(x)$ by $x - k$, $k > 0$, then no roots are larger than k.
>
> If the signs of the third row of the synthetic division of $f(x)$ by $x - (-k) = x + k$ are alternating plus and minus (zero can be called either plus or minus), then no roots are less than $-k$.

To check the first part of this criterion, recall that

$$f(x) = Q(x)(x - k) + R$$

Now, the coefficients of $Q(x)$ and the number R are precisely the entries in the third row of the synthetic division. If these numbers are positive and if $x > k > 0$, then $Q(x)(x - k) + R > 0$ (since $x - k > 0$). So $f(x) > 0$ for $x > k$. The argument for the second part is similar.

Let us now see how these ideas can be used to finish solving the equation in Example 2.

E X A M P L E 2
(Continuation)

Recall that the list of all positive rational roots is given in statement (14.10). Note next that the equation $4x^3 - 24x^2 + 35x - 12 = 0$ has three sign changes. By Descartes' rule of signs, the total number of positive roots must be either three or one.

$$\begin{aligned} f(-x) &= 4(-x)^3 - 24(-x)^2 + 35(-x) - 12 \\ &= -4x^3 - 24x^2 - 35x - 12 \end{aligned}$$

We see that $f(-x)$ has no variations in sign. Consequently, there cannot be any negative roots at all, and we need to check only positive roots.

Of the positive roots, $x = 1$ is the easiest to check:

$$\begin{array}{r} \mathbf{4 - 24 + 35 - 12}\,\underline{)\,1} \\ 4 - 20 + 15 \\ \hline 4 - 20 + 15 + 3 \end{array}$$

Since $R = 3$, $x = 1$ is not a root, but the signs in the third row are not all positive, which means that we may have a root larger than 1. Let us try $x = 4$:

$$\begin{array}{r} \mathbf{4 - 24 + 35 - 12}\,\underline{)\,4} \\ 16 - 32 + 12 \\ \hline 4 - 8 + 3 + \mathbf{0} \end{array}$$

Since $R = 0$, **$x = 4$ is a root**. The resulting factor $4x^2 - 8x + 3$ can be written $(2x - 1)(2x - 3)$, so that the remaining roots are $x = \frac{3}{2}$ and $x = \frac{1}{2}$.   ◀

Before summarizing the procedure in this section, let us recall that if r is a root of $f(x) = 0$, then by the factor theorem we have

$$f(x) = Q(x)(x - r)$$

where $Q(x)$ is of degree one less than $f(x)$. The equation

$$Q(x) = 0$$

Depressed equation   is called the **depressed equation**.

To Find the Rational Roots of $f(x) = 0$:

1. Use Descartes' rule of signs to determine the possible number of positive and negative roots.
2. List all possible rational roots.
3. Test the positive integers.
   a. If one of the values is an upper bound, disregard all larger values.
   b. If one of the values is a root, continue by working with the depressed equation. Remember to check for multiple roots.
4. Test the fractions that remain after the upper bound is obtained. If a value is a root, continue by working with the depressed equation. Check for multiple roots.
5. Repeat Steps 3 and 4 for negative roots using lower bounds.
6. Continue the process until a **quadratic equation is obtained**; solve the quadratic equation.

Remark. While the procedure under discussion is designed to find rational roots, the quadratic equation obtained in the last step may have irrational or complex roots.

The remaining examples illustrate the procedure for finding the rational roots of $f(x) = 0$.

E X A M P L E **3**   Solve the equation $4x^4 - 19x^2 + 3x + 18 = 0$.

Solution. Since $f(x)$ has two sign changes, there exist either two positive roots or none. Also,

$$\begin{aligned} f(-x) &= 4(-x)^4 - 19(-x)^2 + 3(-x) + 18 \\ &= 4x^4 - 19x^2 - 3x + 18 \end{aligned}$$

which has two sign changes. So there are either two negative roots or none.

Factors of numerator: $\pm1, \pm2, \pm3, \pm6, \pm9, \pm18$ **factors of 18**

Factors of denominator: $\pm1, \pm2, \pm4$ **factors of 4**

All possible distinct rational roots:

$$\pm 1, \ \pm 2, \ \pm 3, \ \pm 6, \ \pm 9, \ \pm 18, \ \pm\frac{1}{2}, \ \pm\frac{1}{4}, \ \pm\frac{3}{2}, \ \pm\frac{3}{4}, \ \pm\frac{9}{2}, \ \pm\frac{9}{4}$$

Let us avoid fractions for now and try $x = 2$ to see if any roots are larger than 2:

$$
\begin{array}{r}
4 + 0 - 19 + 3 + 18 \,)\,2 \\
8 + 16 - 6 - 6 \\
\hline
4 + 8 - 3 - 3 + 12
\end{array}
$$

The value $x = 2$ is not a root. Since the signs in the last row are not all positive, we may have a root larger than 2. Let's try $x = 3$:

$$
\begin{array}{r}
4 + 0 - 19 + 3 + 18 \,)\,3 \\
12 + 36 + 51 + 162 \\
\hline
4 + 12 + 17 + 54 + 180
\end{array}
\qquad \textbf{upper bound}
$$

Even though 3 is not a root, at least we know that no root can be larger than 3 (since the numbers in the third row are all positive). Since 2 also failed, let's try $x = \frac{3}{2}$:

$$
\begin{array}{r}
4 + 0 - 19 + 3 + 18 \,\big)\,\tfrac{3}{2} \\
6 + 9 - 15 - 18 \\
\hline
4 + 6 - 10 - 12 \quad\big)\,\tfrac{3}{2} \\
6 + 18 + 12 \\
\hline
4 + 12 + 8
\end{array}
\qquad
\begin{array}{l}
\textbf{checking for double root using} \\
\textbf{the depressed equation}
\end{array}
$$

We see that $x = \frac{3}{2}$ is a double root. (*A root should always be checked for such repetitions.*) From

$$4x^2 + 12x + 8 = 4(x^2 + 3x + 2) = 0 \qquad \textbf{depressed equation}$$

we get

$$4(x + 2)(x + 1) = 0$$
$$x = -1, \ -2$$

The roots are therefore $x = \frac{3}{2}, \ \frac{3}{2}, \ -1, \ -2$. ◀

E X A M P L E 4   Solve the equation $3x^3 + 4x^2 + 14x - 12 = 0$.

Solution. Since $f(x)$ has *one* sign change, there is exactly one positive root. Also, $f(-x) = -3x^3 + 4x^2 - 14x - 12 = 0$. Since $f(-x)$ has two sign changes, there are either two negative roots or none. (If none, then two of the roots are complex.)

Factors of the numerator: $\pm1, \pm2, \pm3, \pm4, \pm6, \pm12$ **factors of 12**

Factors of the denominator: $\pm1, \pm3$ **factors of 3**

All possible rational roots:

$$\pm1, \pm2, \pm3, \pm4, \pm6, \pm12, \pm\frac{1}{3}, \pm\frac{2}{3}, \pm\frac{4}{3}$$

Avoiding fractions for now, let's try $x = 1$:

$$
\begin{array}{r}
3 + 4 + 14 - 12 \,\overline{)\, 1} \\
+3 + 7 + 21 \\
\hline
3 + 7 + 21 + 9
\end{array}
\quad \text{**upper bound**}
$$

Since 1 is an upper bound, the positive root must be less than 1. So let's try $x = \frac{2}{3}$:

$$
\begin{array}{r}
3 + 4 + 14 - 12 \,\overline{)\, \dfrac{2}{3}} \\
+2 + 4 + 12 \\
\hline
3 + 6 + 18 + 0
\end{array}
$$

It follows that $x = \frac{2}{3}$ is a root. From the resulting factor we get the depressed equation

$$3x^2 + 6x + 18 = 0 \qquad \text{**depressed equation**}$$

$$x^2 + 2x + 6 = 0 \qquad \text{**dividing by 3**}$$

$$x = \frac{-2 \pm \sqrt{4 - 24}}{2} = \frac{-2 \pm \sqrt{-20}}{2} \qquad \text{**by the quadratic formula**}$$

$$= \frac{-2 \pm 2\sqrt{5}j}{2} = -1 \pm \sqrt{5}j$$

So the roots are $x = \frac{2}{3}$ and $x = -1 \pm \sqrt{5}j$.   ◀

E X A M P L E **5**   Solve the equation $4x^4 + 4x^3 + 9x^2 + 8x + 2 = 0$.

Solution. Since there are no sign changes, the equation has no positive roots. Since

$$f(-x) = 4x^4 - 4x^3 + 9x^2 - 8x + 2$$

$f(x)$ has either four negative roots, two negative roots, or none. (If none, then all roots are complex.)

Factors of numerator: $\pm1, \pm2$ **factors of 2**

Factors of denominator: $\pm1, \pm2, \pm4$ **factors of 4**

All possible rational roots:

$$\pm 1, \pm 2, \pm \frac{1}{2}, \pm \frac{1}{4}$$

Let us try $x = -1$:

$$\begin{array}{r} 4 + 4 + 9 + 8 + 2 \overline{)\,-1} \\ \underline{-4 + 0 - 9 + 1} \\ 4 - 0 + 9 - 1 + 3 \end{array}$$

Reminder. Zero can be regarded as either positive or negative. As written, the signs in the last row alternate, indicating that no roots can be less than -1. So we try $x = -\frac{1}{2}$:

$$\begin{array}{r} 4 + 4 + 9 + 8 + 2 \overline{)\,-\dfrac{1}{2}} \\ \underline{-2 - 1 - 4 - 2} \\ \end{array}$$

$$\begin{array}{r} 4 + 2 + 8 + 4 \overline{)\,-\dfrac{1}{2}} \qquad \textbf{checking for double root using} \\ \underline{-2 + 0 - 4} \qquad\quad \textbf{the depressed equation} \\ 4 + 0 + 8 \end{array}$$

It follows that $x = -\frac{1}{2}$ is a double root. From the resulting depressed equation, we get

$$4x^2 + 8 = 0$$

$$4(x^2 + 2) = 0$$

$$x = \pm \sqrt{2}j$$

So the roots are $x = -\frac{1}{2}, -\frac{1}{2}, \pm \sqrt{2}j$.   ◄

E X E R C I S E S / S E C T I O N   14.5

In Exercises 1–30, solve each equation.

1. $x^3 - 2x^2 - x + 2 = 0$

2. $x^3 - 7x + 6 = 0$

3. $x^3 + 3x^2 - 4x - 12 = 0$

4. $x^3 + x^2 - 14x - 24 = 0$

5. $x^4 + x^3 - 7x^2 - x + 6 = 0$

6. $x^4 - 2x^3 - 7x^2 + 20x - 12 = 0$

7. $x^4 + 5x^3 + x^2 - 21x - 18 = 0$

8. $x^4 - 8x^3 + 17x^2 + 2x - 24 = 0$

9. $x^4 - 3x^3 - 8x^2 + 21x + 9 = 0$

10. $x^4 + 5x^3 - 27x - 27 = 0$

11. $x^4 - 10x^3 + 35x^2 - 56x + 48 = 0$

12. $x^4 - 7x^3 + 16x^2 - 15x + 9 = 0$

13. $3x^3 + 2x^2 + 17x - 6 = 0$

14. $6x^3 - x^2 + 10x - 8 = 0$

15. $2x^3 + 3x^2 + 22x - 12 = 0$

16. $4x^3 + 7x^2 + 22x - 6 = 0$

17. $4x^3 + 4x^2 + 5x + 2 = 0$

18. $x^3 + \dfrac{10}{3}x^2 + \dfrac{20}{3}x + \dfrac{16}{3} = 0$ (Clear fractions first.)

19. $2x^4 + 4x^3 - \dfrac{3}{2}x^2 - \dfrac{5}{2}x + 1 = 0$

20. $2x^4 - 7x^3 + 5x^2 + 3x - 2 = 0$

21. $4x^4 + 8x^3 - 3x^2 - 7x - 2 = 0$

22. $4x^4 - 8x^3 - 19x^2 + 23x - 6 = 0$

23. $6x^4 - 11x^3 - 8x^2 + 25x - 12 = 0$

24. $9x^4 - 15x^3 - 17x^2 + 40x - 16 = 0$

25. $6x^4 + 5x^3 - 22x^2 - 11x + 10 = 0$

26. $3x^4 + 13x^3 + 14x^2 - 4x - 8 = 0$

27. $4x^4 - 8x^3 - 11x^2 + 33x - 18 = 0$

28. $4x^4 - 8x^3 - 15x^2 + 29x - 10 = 0$

29. $6x^4 + 7x^3 - 57x^2 + 42x - 8 = 0$

30. $6x^4 - 19x^3 - 37x^2 + 62x + 24 = 0$

31. A sphere of radius r and specific gravity c will sink in water to a depth h according to the formula $h^3 - 3rh^2 + 4r^3c = 0$. Find the depth to which a wooden ball having a radius of 3 in. and specific gravity 0.5 will sink.

32. If $c = \frac{1}{2}$ in the formula $h^3 - 3rh^2 + 4r^3c = 0$ in Exercise 31, show that r is a root. Why is $h = r$ the only root that is physically realistic?

33. A box with a capacity of 32 ft^3 has a square base and a height that is 2 ft less than the length of the base. Find the dimensions.

34. The length of one side of a rectangle is equal to the cube of the length of the side adjacent. If the perimeter is $\frac{20}{27}$ cm, find the dimensions.

35. If three capacitors C_1, C_2, and C_3 are connected in series, then the capacitance C_T of the combination satisfies the relationship

$$\frac{1}{C_T} = \frac{1}{C_1} + \frac{1}{C_2} + \frac{1}{C_3}$$

The capacitance of the second is 3 μF (microfarads) more than that of the first. The capacitance of the third is 6 μF more than that of the second. If $C_T = \frac{12}{7}\,\mu$F, find the capacitance of each.

14.6   Irrational Roots

If all the roots of an equation are irrational, we have no way of making a list of possible roots. Fortunately, there are numerical procedures by which an irrational root can be approximated to any desired degree of accuracy.

The method we will study in this section is called the **method of successive linear approximations**. The idea behind the approximation scheme can be seen geometrically. Consider the graph of the function $y = f(x)$. Any value of x for which $y = 0$ is an *x-intercept*, that is, a point where the graph crosses the x-axis. For example, the graph in Figure 14.1 crosses the x-axis at $(p, 0)$, $(q, 0)$, and

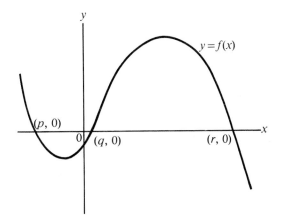

Figure 14.1

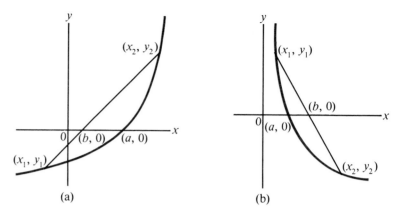

Figure 14.2

$(r, 0)$. So $x = p$, $x = q$, and $x = r$ are the real roots of the equation $f(x) = 0$. Now let (x_1, y_1) and (x_2, y_2) be two points on the graph close to an intercept but on opposite sides of the x-axis. (See Figure 14.2.) Then the line segment joining (x_1, y_1) and (x_2, y_2) will cross the x-axis at a point that is very close to the intercept. In other words, in Figure 14.2, $x = b$ is approximately equal to the root $x = a$. As we will see, this approximation can be found by repeated use of synthetic division. This process can be facilitated by a formula expressing b in terms of the coordinates of (x_1, y_1) and (x_2, y_2), which is given below.

$$b = x_1 - \frac{y_1(x_2 - x_1)}{y_2 - y_1} \tag{14.11}$$

Formula (14.11) is merely a convenience since b can also be found from a scale drawing.

E X A M P L E **1**   Find the irrational root of $x^3 - 6x + 2 = 0$ that lies between 2 and 3.

Solution. First we compute the function values for $x = 2$ and $x = 3$ by synthetic division:

$$
\begin{array}{r}
1 + 0 - 6 + 2\,\underline{)\,2} \\
2 + 4 - 4 \\
\hline
1 + 2 - 2 - 2
\end{array}
\qquad
\begin{array}{r}
1 + 0 - 6 + 2\,\underline{)\,3} \\
3 + 9 + 9 \\
\hline
1 + 3 + 3 + \mathbf{11}
\end{array}
$$

Since $f(2) = -2$ and $f(3) = 11$, the graph does indeed cross the x-axis between $(2, 0)$ and $(3, 0)$. As indicated earlier, we can estimate the intersection by a scale drawing, as shown in Figure 14.3. According to the figure, the root is approximately equal to 2.2. The same estimate can be obtained from formula (14.11), letting $(x_1, y_1) = (2, -2)$ and $(x_2, y_2) = (3, 11)$:

$$b = 2 - \frac{-2(3 - 2)}{11 - (-2)} = 2 + \frac{2}{13} = 2.2$$

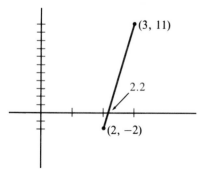

Figure 14.3

The next step is to place the root between two values close to 2.2. To this end, we compute the function values for 2.1, 2.2, and 2.3. Carrying two significant digits for now, we obtain the following:

$$\begin{array}{r} 1 + 0 \ -6 \ +2 \ \overline{)\, 2.1} \\ 2.1 + 4.4 - 3.4 \\ \hline 1 + 2.1 - 1.6 - \mathbf{1.4} \end{array} \qquad \begin{array}{r} 1 + 0 \ -6 \ +2 \ \overline{)\, 2.2} \\ 2.2 + 4.8 - 2.6 \\ \hline 1 + 2.2 - 1.2 - \mathbf{0.6} \end{array}$$

$$\begin{array}{r} 1 + 0 \ -6 \ +2 \ \overline{)\, 2.3} \\ 2.3 + 5.3 - 1.6 \\ \hline 1 + 2.3 - 0.7 + \mathbf{0.4} \end{array}$$

So $f(2.1) = -\mathbf{1.4}$, $f(2.2) = -\mathbf{0.6}$, and $f(2.3) = \mathbf{0.4}$. It follows that the graph crosses the x-axis between 2.2 and 2.3. To estimate the location, we can use another scale drawing (Figure 14.4) or formula (14.11) with $(x_1, y_1) = (2.2, -0.6)$ and $(x_2, y_2) = (2.3, 0.4)$:

$$b = 2.2 - \frac{-0.6(2.3 - 2.2)}{0.4 - (-0.6)} = 2.2 + 0.06 = 2.26$$

As always, this value is only an estimate. To refine the estimate, let us locate the root between two values close to 2.26. Suppose we try 2.26 and 2.27, carrying the calculations to three significant digits this time:

$$\begin{array}{r} 1 + 0 \ -6 \ +2 \ \overline{)\, 2.26} \\ 2.26 + 5.11 - 2.01 \\ \hline 1 + 2.26 - 0.89 - \mathbf{0.01} \end{array} \qquad \begin{array}{r} 1 + 0 \ -6 \ +2 \ \overline{)\, 2.27} \\ 2.27 + 5.15 - 1.93 \\ \hline 1 + 2.27 - 0.85 + \mathbf{0.07} \end{array}$$

Since $f(2.26) = -\mathbf{0.01}$ and $f(2.27) = \mathbf{0.07}$, the root lies between 2.26 and 2.27. Since the function value (remainder) is *numerically smaller* for $x = 2.26$, the root must be closer to 2.26 than to 2.27. We conclude that $x = \mathbf{2.26}$ is the root accurate to two decimal places.

The process can be continued to obtain even greater accuracy. ◀

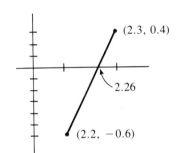

(2.3, 0.4)

2.26

(2.2, −0.6)

Figure 14.4

E X A M P L E **2**   Find the irrational root of $2x^3 - 5x^2 + 4 = 0$ located between -1 and 0.

Solution. Since $f(-1) = -3$ and $f(0) = 4$, the graph appears to be rising and does indeed cross the x-axis between the values indicated. By formula (11.14), for $(x_1, y_1) = (-1, -3)$ and $(x_2, y_2) = (0, 4)$,

$$b = -1 - \frac{-3[0 - (-1)]}{4 - (-3)} = -1 + 0.4 = -0.6$$

As before, we try to locate the root between two values close to -0.6. For this value we get

$$\begin{array}{r} 2 - 5 \ +0 \ +4 \ \overline{)\, -0.6} \\ -1.2 + 3.72 - 2.2 \\ \hline 2 - 6.2 + 3.72 + \mathbf{1.8} \end{array}$$

This calculation can also be done with a calculator. To do so, we store the value being tested (-0.6) in the memory:

$$0.6 \boxed{+/-} \boxed{\text{STO}} \boxed{\times} 2 \boxed{-} 5 \boxed{=} \boxed{\times} \boxed{\text{MR}} \boxed{\times} \boxed{\text{MR}} \boxed{+} 4 \boxed{=} \rightarrow 1.768$$

Thus $f(-0.6) = \mathbf{1.8}$. Since the graph is rising as we move to the right, the function values should decrease as we move to the left. So let's try -0.7:

$$
\begin{array}{rrrr|r}
2 - 5 & +0 & +4 &)-0.7 \\
& -1.4 & +4.48 & -3.14 \\
\hline
2 - 6.4 & +4.48 & +\mathbf{0.86}
\end{array}
$$

The sequence (after 0.6) is the same:

$$0.7 \boxed{+/-} \boxed{\text{STO}} \boxed{\times} 2 \boxed{-} 5 \boxed{=} \boxed{\times} \boxed{\text{MR}} \boxed{\times} \boxed{\text{MR}} \boxed{+} 4 \boxed{=} \rightarrow 0.864$$

Since the remainder is still positive, let's move further to the left to -0.8:

$$
\begin{array}{rrrr}
2 - 5 & +0 & +4 &)-0.8 \\
& -1.6 & +5.28 & -4.22 \\
\hline
2 - 6.6 & +5.28 & -\mathbf{0.22}
\end{array}
$$

Now the remainder is negative, so the root must lie between -0.8 and -0.7. By formula (14.11)

$$b = -0.8 - \frac{-0.22[-0.7 - (-0.8)]}{0.86 - (-0.22)} = -0.8 + 0.02 = -0.78$$

The remaining trials are

$$
\begin{array}{rrrr} \qquad \qquad
2 - 5 & +0 & +4 &)-0.78 \\
& -1.56 & +5.12 & -3.99 \\
\hline
2 - 6.56 & +5.12 & +\mathbf{0.01}
\end{array}
\qquad
\begin{array}{rrrr}
2 - 5 & +0 & +4 &)-0.79 \\
& -1.58 & +5.20 & -4.11 \\
\hline
2 - 6.58 & +5.20 & -\mathbf{0.11}
\end{array}
$$

Since the function value corresponding to -0.78 is the smaller numerically, we conclude that $x = \mathbf{-0.78}$ is the desired root to two decimal places.   ◄

If the location of the root is not indicated or if more than one root must be found, the initial estimate must be obtained from a graph. In most cases plotting points for integral values of x is sufficient.

E X A M P L E 3   Solve the equation $x^3 - 2x - 5 = 0$.

Solution. First we construct the following short table for $y = x^3 - 2x - 5$:

x:	-2	-1	0	1	2	3
y:	-9	-4	-5	-6	-1	16

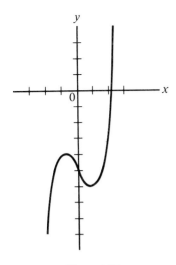

Figure 14.5

The graph is shown in Figure 14.5. The root evidently lies between 2 and 3. By formula (14.11)

$$b = 2 - \frac{-1(1)}{16 + 1} = 2 + \frac{1}{17} = 2.1$$

We now proceed as before:

$$1 + 0 - 2 - 5 \,)\,2.0$$
$$\underline{2 + 4 + 4}$$
$$1 + 2 + 2 - 1$$

$$1 + 0 \quad -2 \quad -5 \quad)\,2.1$$
$$\underline{2.1 + 4.4 + 5.04}$$
$$1 + 2.1 + 2.4 + 0.04$$

$$b = 2.0 - \frac{-1(0.1)}{0.04 + 1} = 2.10$$

$$1 + 0 \quad -2 \quad -5 \quad)\,2.09$$
$$\underline{2.09 + 4.37 + 4.95}$$
$$1 + 2.09 + 2.37 - 0.05$$

$$1 + 0 \quad -2 \quad -5 \quad)\,2.10$$
$$\underline{2.10 + 4.41 + 5.06}$$
$$1 + 2.10 + 2.41 + 0.06$$

Thus $x = 2.09$ to two decimal places. ◀

E X E R C I S E S / S E C T I O N   **14.6**

In Exercises 1–12, find the irrational root between the indicated values to two decimal places.

1. $x^3 - 4x - 2 = 0$; $[2, 3]$

2. $x^3 - 8x + 1 = 0$; $[2, 3]$

3. $2x^3 + 3x - 6 = 0$; $[1, 2]$

4. $3x^3 + x - 1 = 0$; $[0, 1]$

5. $x^3 - 3x^2 - x + 2 = 0$; $[0, 1]$

6. $x^3 - 5x^2 + 2x + 3 = 0$; $[1, 2]$

7. $2x^3 - 3x^2 - 10x - 5 = 0$; $[3, 4]$

8. $2x^3 - 9x^2 + x + 10 = 0$; $[1, 2]$

9. $x^3 + 4x + 14 = 0$; $[-2, -1]$

10. $4x^3 + 2x^2 + 2x + 2 = 0$; $[-1, 0]$

11. $x^4 + 2x^3 + x - 3 = 0$; $[0, 1]$

12. $x^4 - x^2 - 3x - 2 = 0$; $[-1, 0]$

13. Determine the root of the equation $x^3 - 3x^2 + 2x - 3 = 0$ located between 2 and 3 to three decimal places.

14. Determine the root of the equation $x^3 - 4x^2 + 4 = 0$ located between 1 and 2 to three decimal places.

15. Find the negative root of the equation

$$x^3 - 4x^2 + 3x + 2 = 0$$

to three decimal places.

16. Solve the equation $x^3 + 2x + 6 = 0$ to two decimal places.

17. Find the largest root of the equation

$$x^4 - 4x^3 - 2x^2 + 12x + 8 = 0$$

to two decimal places.

18. A sphere of radius r and specific gravity c will sink in water to a depth h according to the equation $h^3 - 3rh^2 + 4r^3c = 0$. Find to two decimal places the depth to which a ball having a radius of 1.00 ft and specific gravity 0.75 will sink.

19. A box is to be constructed from a square piece of cardboard 10 in. on each side by cutting equal squares from each corner and bending up the sides. (See Figure 14.6.) Determine to two decimal places the size of the square that must be cut out if the volume is to be 73 in.³.

20. Use the method for finding irrational roots to determine the approximate value of $\sqrt[3]{6}$ to two decimal places.

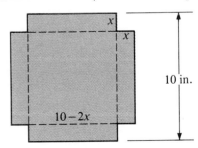

Figure 14.6

REVIEW EXERCISES/CHAPTER 14

In Exercises 1–4, determine whether the second expression is a factor of the first by means of the factor theorem.

1. $x^3 - x^2 - 4x + 4,\ x - 2$

2. $x^4 + 2x^3 - x^2 + 7x + 3,\ x + 3$

3. $x^5 + 3x^4 + 15x^2 - 4x + 1,\ x + 4$

4. $x^5 - 5x^4 + x^3 - 5x^2 + 4x - 20,\ x - 5$

In Exercises 5–10, perform the indicated divisions by synthetic division.

5. $(x^3 + 3x - 2x^2 + 1) \div (x - 1)$

6. $(2x - 3x^2 - 10 + x^4) \div (x - 3)$

7. $(2x^5 + 10x^4 - 2x^3 - 20x^2 + 60x - 60) \div (x + 4)$

8. $(3x^4 - 15x^3 + x^2 - 7x + 12) \div (x - 5)$

9. $(4x^5 - 2x^3 - 2x^2 + 3x - 4) \div (x - 1)$

10. $(3x^5 + 8x^4 + 10x^2 + 5x + 4) \div (x + 3)$

In Exercises 11–16, determine whether the given value is a zero of the function.

11. $6 - 5x + 4x^3 + x^4,\ -2$

12. $x^4 + 3x^3 - 6x + 4,\ -2$

13. $4x^3 + 10x^2 - 8x + 6,\ -3$

14. $3x^4 - 15x^3 + 10x^2 + 7x + 4,\ 4$

15. $2x^4 + 10x^3 + x^2 + 6x - 5,\ -5$

16. $x^5 + 5x^4 - 12x^3 - 30x^2 + 40x + 24,\ -6$

In Exercises 17–20, use synthetic division to find $f(r)$ for the given value of r.

17. $f(x) = 2x^3 - 3x^2 + 7x - 5,\ r = \dfrac{1}{2}$

18. $f(x) = 6x^4 + 5x^3 + 6x^2 + 10x - 10,\ r = -\dfrac{3}{2}$

19. $f(x) = 6x^4 - 2x^3 + 3x^2 - 4x + 5,\ r = \dfrac{1}{3}$

20. $f(x) = 3x^4 + 5x^3 - x^2 + 4x - 2,\ r = -\dfrac{2}{3}$

In Exercises 21–26, find the roots not given.

21. $x^4 - 11x^2 - 12x + 2x^3 + 36 = 0;\ -3$ is a double root

22. $2x^3 + 3x^2 + 5x + 2 = 0;\ -\dfrac{1}{2}$ is a root

23. $4x^2 + 4x - 1 + 9x^4 - 24x^3 = 0;\ \dfrac{1}{3}$ is a double root

24. $x^3 + 4x^2 + x + 4 = 0;\ -j$ is a root

25. $x^4 + 2x^3 - 14x^2 + 2x - 15 = 0;\ j$ is a root

26. $x^3 + x^2 - 4x + 6 = 0;\ 1 + j$ is a root

In Exercises 27–34, solve the given equations by the method described in Section 14.5.

27. $x^3 + 7x^2 + 4x - 12 = 0$

28. $x^3 - 5x^2 + 6x - 8 = 0$

29. $x^4 - 9x^2 - 4x + 12 = 0$

30. $x^4 - 4x^3 + x^2 - 6x + 36 = 0$

31. $4x^3 + 8x^2 - 11x + 3 = 0$

32. $9x^3 - 6x^2 - 20x - 8 = 0$

33. $9x^4 - 15x^3 - 20x^2 + 17x - 3 = 0$

34. $4x^4 - 24x^3 + 21x^2 + 41x + 12 = 0$

In Exercises 35–38, find the irrational roots between the indicated values to two decimal places.

35. $2x^3 + x - 1 = 0;\ [0, 1]$

36. $2x^3 - 9x^2 + 2x + 12 = 0;\ [1, 2]$

37. $x^4 - x^2 - 2x - 1 = 0;\ [-1, 0]$

38. $2x^4 + 3x^3 + 2x - 6 = 0;\ [0, 1]$

39. Find all the real roots of the equation $x^3 - 3x - 4 = 0$ accurate to three decimal places.

40. Find all the real roots of the equation

$$x^3 - 6x^2 - 4x + 6 = 0$$

accurate to two decimal places.

C H A P T E R 15

Determinants and Matrices

15.1 Introduction

In the first part of this chapter we will continue our study of determinants, which we began in Chapter 3. Our goal is to develop various techniques for evaluating higher-order determinants and to use determinants for solving larger systems of equations. The second part of the chapter is devoted to a study of matrix algebra and the application of matrix methods to solving equations.

15.2 Review of Determinants

Recall from Chapter 3 that a 2×2 determinant is defined by

$$\begin{vmatrix} a_1 & b_1 \\ a_2 & b_2 \end{vmatrix} = a_1 b_2 - a_2 b_1 \tag{15.1}$$

EXAMPLE 1

$$\begin{vmatrix} -4 & -3 \\ 4 & 1 \end{vmatrix} = (-4)(1) - (4)(-3) = -4 + 12 = 8 \qquad \blacktriangleleft$$

A third-order determinant

$$\begin{vmatrix} a_1 & b_1 & c_1 \\ a_2 & b_2 & c_2 \\ a_3 & b_3 & c_3 \end{vmatrix} \tag{15.2}$$

Minor

is expanded by minors. A **minor** of a given element is the determinant found by deleting all the elements in the row and column in which that element lies.

One possible expansion of determinant (15.2) by minors is

$$
\begin{vmatrix} a_1 & b_1 & c_1 \\ a_2 & b_2 & c_2 \\ a_3 & b_3 & c_3 \end{vmatrix} = a_1 \begin{vmatrix} b_2 & c_2 \\ b_3 & c_3 \end{vmatrix} - b_1 \begin{vmatrix} a_2 & c_2 \\ a_3 & c_3 \end{vmatrix} + c_1 \begin{vmatrix} a_2 & b_2 \\ a_3 & b_3 \end{vmatrix} \quad (15.3)
$$

Rule for signs

The sign depends on the position of the element: If the sum of the number of the row of the element and the number of the column is even, affix a plus sign; if the sum is odd, affix a minus sign. For example, in expansion (15.3), the element b_1 lies in the first row, second column; since $1 + 2 = 3$, b_1 is preceded by a minus sign. A determinant can be expanded along any row or column.

E X A M P L E **2**   Expand the determinant:

$$
\begin{vmatrix} -2 & 4 & 3 \\ 1 & 3 & -2 \\ -1 & 5 & -3 \end{vmatrix}
$$

Solution. Arbitrarily choosing the third column, we get

$$
\begin{vmatrix} -2 & 4 & \mathbf{3} \\ 1 & 3 & \mathbf{-2} \\ -1 & 5 & \mathbf{-3} \end{vmatrix} = \mathbf{3} \begin{vmatrix} 1 & 3 \\ -1 & 5 \end{vmatrix} - (\mathbf{-2}) \begin{vmatrix} -2 & 4 \\ -1 & 5 \end{vmatrix}
$$
$$
+ (\mathbf{-3}) \begin{vmatrix} -2 & 4 \\ 1 & 3 \end{vmatrix}
$$

Note that the element 3 lies in row 1, column 3; since $1 + 3 = 4$, which is even, we affix a plus sign. The element -2 lies in row 2, column 3; since $2 + 3 = 5$, which is odd, we affix a minus sign to get $-(-2)$. Finally, -3 lies in row 3, column 3, and $3 + 3 = 6$, which is again even. So we affix a plus sign to get $+(-3)$. Continuing the expansion, we have

$$
3 \begin{vmatrix} 1 & 3 \\ -1 & 5 \end{vmatrix} + 2 \begin{vmatrix} -2 & 4 \\ -1 & 5 \end{vmatrix} - 3 \begin{vmatrix} -2 & 4 \\ 1 & 3 \end{vmatrix}
$$
$$
= 3(5 + 3) + 2(-10 + 4) - 3(-6 - 4) = 42 \quad \blacktriangleleft
$$

E X A M P L E **3**   Expand the determinant:

$$
\begin{vmatrix} -5 & 3 & 6 \\ -2 & 0 & 0 \\ 7 & 1 & 4 \end{vmatrix}
$$

Solution. Because of the zeros, it is best to expand the determinant along the second row. Thus

$$\begin{vmatrix} -5 & 3 & 6 \\ -2 & 0 & 0 \\ 7 & 1 & 4 \end{vmatrix} = -(-2)\begin{vmatrix} 3 & 6 \\ 1 & 4 \end{vmatrix} + 0 + 0 = 2(12 - 6) = 12 \quad \blacktriangleleft$$

E X E R C I S E S / S E C T I O N **15.2**

Evaluate each determinant.

1. $\begin{vmatrix} 2 & 3 \\ -4 & 0 \end{vmatrix}$

2. $\begin{vmatrix} 0 & 1 \\ -3 & 2 \end{vmatrix}$

3. $\begin{vmatrix} 1 & -4 \\ 7 & -8 \end{vmatrix}$

4. $\begin{vmatrix} -3 & -5 \\ -4 & 6 \end{vmatrix}$

5. $\begin{vmatrix} 1 & 0 & 0 \\ 2 & -3 & 4 \\ 3 & -2 & 1 \end{vmatrix}$

6. $\begin{vmatrix} 2 & 2 & 1 \\ 0 & -3 & 4 \\ 0 & 6 & 0 \end{vmatrix}$

7. $\begin{vmatrix} 0 & 1 & 6 \\ 0 & 0 & -3 \\ -7 & -4 & 3 \end{vmatrix}$

8. $\begin{vmatrix} 3 & 4 & 2 \\ 0 & -1 & 0 \\ 8 & 3 & -5 \end{vmatrix}$

9. $\begin{vmatrix} 2 & -7 & 10 \\ 3 & 0 & 1 \\ -2 & 4 & 3 \end{vmatrix}$

10. $\begin{vmatrix} 2 & 8 & 9 \\ 0 & 3 & 6 \\ -2 & -4 & 0 \end{vmatrix}$

11. $\begin{vmatrix} 4 & 6 & 4 \\ -7 & 0 & 3 \\ 11 & -5 & 1 \end{vmatrix}$

12. $\begin{vmatrix} -3 & -2 & 1 \\ 6 & 10 & 3 \\ -4 & -3 & 11 \end{vmatrix}$

13. $\begin{vmatrix} 3 & 6 & 0 \\ -5 & 7 & 3 \\ -8 & 0 & 10 \end{vmatrix}$

14. $\begin{vmatrix} 2 & -6 & 8 \\ -7 & 12 & 3 \\ 2 & 0 & 4 \end{vmatrix}$

15.3 **Properties of Determinants**

So far we have discussed expansion by minors only for third-order determinants. In this section we will extend the expansion by minors to determinants of any order. To do this, however, we must first know several properties of determinants, which are discussed next.

To get an overview of the problem, consider the following expansion of a fourth-order determinant by minors:

$$\begin{vmatrix} a_1 & b_1 & c_1 & d_1 \\ a_2 & b_2 & c_2 & d_2 \\ a_3 & b_3 & c_3 & d_3 \\ a_4 & b_4 & c_4 & d_4 \end{vmatrix} = a_1 \begin{vmatrix} b_2 & c_2 & d_2 \\ b_3 & c_3 & d_3 \\ b_4 & c_4 & d_4 \end{vmatrix} - b_1 \begin{vmatrix} a_2 & c_2 & d_2 \\ a_3 & c_3 & d_3 \\ a_4 & c_4 & d_4 \end{vmatrix}$$

$$+ c_1 \begin{vmatrix} a_2 & b_2 & d_2 \\ a_3 & b_3 & d_3 \\ a_4 & b_4 & d_4 \end{vmatrix} - d_1 \begin{vmatrix} a_2 & b_2 & c_2 \\ a_3 & b_3 & c_3 \\ a_4 & b_4 & c_4 \end{vmatrix}$$

Given the amount of arithmetic required to expand a single third-order determinant, the expansion of a fourth-order determinant looks like a discouraging task. For even higher order determinants, the amount of work becomes prohibitive. Fortunately, the work can be greatly simplified by making use of several properties of determinants, which are stated next.

Properties of Determinants:

1. If all the entries in a row or column are 0, then the value of the determinant is 0.
2. If two rows (or columns) of a given determinant are interchanged, then the value of the resulting determinant is the negative of the value of the given determinant.
3. If a determinant has two identical rows or columns, then the value of the determinant is 0.
4. Whenever the elements of a row or a column have a common factor, this factor can be removed from the row or column and placed in front of the determinant as a multiplicative constant.
5. If all the elements of a row (or column) are multiplied by k and the resulting numbers added to the corresponding elements of another row (or column), then the value of the determinant is unchanged.

A proof of these properties requires a more sophisticated study of determinants than is needed for our purposes. However, to illustrate the type of argument used, we will check Property 2 for an arbitrary third-order determinant and then use Property 2 to prove Property 3. Thus, we start with the following:

$$\begin{vmatrix} a_1 & b_1 & c_1 \\ a_2 & b_2 & c_2 \\ a_3 & b_3 & c_3 \end{vmatrix} = a_1 \begin{vmatrix} b_2 & c_2 \\ b_3 & c_3 \end{vmatrix} - b_1 \begin{vmatrix} a_2 & c_2 \\ a_3 & c_3 \end{vmatrix} + c_1 \begin{vmatrix} a_2 & b_2 \\ a_3 & b_3 \end{vmatrix}$$

If the first two rows are interchanged, we get by expanding along the second row

$$\begin{vmatrix} a_2 & b_2 & c_2 \\ a_1 & b_1 & c_1 \\ a_3 & b_3 & c_3 \end{vmatrix} = -a_1 \begin{vmatrix} b_2 & c_2 \\ b_3 & c_3 \end{vmatrix} + b_1 \begin{vmatrix} a_2 & c_2 \\ a_3 & c_3 \end{vmatrix} - c_1 \begin{vmatrix} a_2 & b_2 \\ a_3 & b_3 \end{vmatrix}$$

$$= - \begin{vmatrix} a_1 & b_1 & c_1 \\ a_2 & b_2 & c_2 \\ a_3 & b_3 & c_3 \end{vmatrix}$$

This confirms Property 2.

Given Property 2, Property 3 can actually be proved. Let the value of the determinant be D. If two identical rows (or columns) are interchanged and the value of the new determinant denoted by D', then $D' = -D$ by Property 2. However, since the rows (or columns) that were interchanged are identical, $D' = D$. Hence $D = -D$, which is possible only if $D = 0$.

Although all five properties have their uses, Property 5 is by far the most important.

Use **Property 5** to change all the elements but one in a row (or column) to zeros. Expand along this row (or column) to obtain a determinant of order one less.

Let's illustrate Property 5 with an example.

E X A M P L E **1**   Change all the elements except the 1 in the first column to zeros and evaluate the resulting determinant:

$$\begin{vmatrix} 1 & 2 & -1 \\ 2 & 0 & 3 \\ -3 & -4 & -2 \end{vmatrix}$$

Solution. By **Property 5** we may multiply the first row by -2 and add the resulting numbers to the corresponding elements in the second row. Thus

$$\begin{vmatrix} 1 & 2 & -1 \\ 2 & 0 & 3 \\ -3 & -4 & -2 \end{vmatrix} = \begin{vmatrix} 1 & 2 & -1 \\ 2+(-2)(1) & 0+(-2)(2) & 3+(-2)(-1) \\ -3 & -4 & -2 \end{vmatrix}$$

$$= \begin{vmatrix} 1 & 2 & -1 \\ 0 & -4 & 5 \\ -3 & -4 & -2 \end{vmatrix}$$

Similarly, we may multiply the first row of the last determinant by **3** and add the resulting numbers to the third row to obtain

$$\begin{vmatrix} 1 & 2 & -1 \\ 0 & -4 & 5 \\ -3 & -4 & -2 \end{vmatrix} = \begin{vmatrix} 1 & 2 & -1 \\ 0 & -4 & 5 \\ -3+3(1) & -4+3(2) & -2+3(-1) \end{vmatrix}$$

$$= \begin{vmatrix} 1 & 2 & -1 \\ 0 & -4 & 5 \\ 0 & 2 & -5 \end{vmatrix}$$

Expanding along the first column, we get

$$1\begin{vmatrix} -4 & 5 \\ 2 & -5 \end{vmatrix} + 0 + 0 = 20 - 10 = 10 \qquad \blacktriangleleft$$

Notation. To make the operation involving Property 5 easier to describe, we will adopt the notation

$$kR_j + R_i \qquad\qquad\qquad (15.4)$$

to indicate that the *j*th row is multiplied by *k* and the resulting elements added to the *i*th row. The analogous operation with columns will be denoted by

$$kC_j + C_i \qquad\qquad\qquad (15.5)$$

With this notation, the operations in Example 1 can now be described compactly as follows:

$$\begin{vmatrix} 1 & 2 & -1 \\ 2 & 0 & 3 \\ -3 & -4 & -2 \end{vmatrix} \begin{matrix} \\ -2R_1 + R_2 \\ 3R_1 + R_3 \end{matrix} = \begin{vmatrix} 1 & 2 & -1 \\ 0 & -4 & 5 \\ 0 & 2 & -5 \end{vmatrix}$$

This notation, in conjunction with Property 5, is further illustrated in the next example.

E X A M P L E **2**   Evaluate the determinant:

$$\begin{vmatrix} -2 & 5 & -2 & 1 \\ 1 & 0 & -3 & 2 \\ 2 & -1 & 3 & -4 \\ 3 & 6 & -2 & 0 \end{vmatrix}$$

Solution. Remember that *our immediate goal is to change all elements but one in a given row or column to zeros.* This can be done in several ways. For example, retaining the -2 in the upper left-hand corner, we can obtain a zero for the element directly below by the operation $\frac{1}{2} R_1 + R_2$:

$$\begin{vmatrix} -2 & 5 & -2 & 1 \\ 1 & 0 & -3 & 2 \\ 2 & -1 & 3 & -4 \\ 3 & 6 & -2 & 0 \end{vmatrix} \begin{matrix} \\ \frac{1}{2}R_1 + R_2 \\ \\ \\ \end{matrix} = \begin{vmatrix} -2 & 5 & -2 & 1 \\ 0 & \frac{5}{2} & -4 & \frac{5}{2} \\ 2 & -1 & 3 & -4 \\ 3 & 6 & -2 & 0 \end{vmatrix}$$

In a similar manner, we can obtain zeros for the remaining elements in the first column.

Avoiding a fractional k

However, working with $k = \frac{1}{2}$, a fraction, is inconvenient. We can avoid a fractional k by working with 1 or -1 instead of -2. Suppose we retain the 1 in the first column and obtain zeros for the remaining elements in the first column as follows:

$$
\begin{vmatrix} -2 & 5 & -2 & 1 \\ 1 & 0 & -3 & 2 \\ 2 & -1 & 3 & -4 \\ 3 & 6 & -2 & 0 \end{vmatrix}
\begin{array}{l} 2R_2 + R_1 \\ \\ -2R_2 + R_3 \\ -3R_2 + R_4 \end{array}
=
\begin{vmatrix} 0 & 5 & -8 & 5 \\ 1 & 0 & -3 & 2 \\ 0 & -1 & 9 & -8 \\ 0 & 6 & 7 & -6 \end{vmatrix}
$$

In expanding the new determinant along the first column, note that the element 1 is in row 2, column 1. The expansion yields

$$
-(1) \begin{vmatrix} 5 & -8 & 5 \\ -1 & 9 & -8 \\ 6 & 7 & -6 \end{vmatrix}
$$

Now retain the -1 in the first column and obtain zeros for the remaining elements in this column.

$$
(-1) \begin{vmatrix} 5 & -8 & 5 \\ -1 & 9 & -8 \\ 6 & 7 & -6 \end{vmatrix}
\begin{array}{l} 5R_2 + R_1 \\ \\ 6R_2 + R_3 \end{array}
= (-1) \begin{vmatrix} 0 & 37 & -35 \\ -1 & 9 & -8 \\ 0 & 61 & -54 \end{vmatrix}
$$

$$
= (-1) \left\{ -(-1) \begin{vmatrix} 37 & -35 \\ 61 & -54 \end{vmatrix} \right\}
$$

$$
= -[(37)(-54) - (61)(-35)] = -137 \qquad \blacktriangleleft
$$

Obtaining a 1 or −1

Example 2 shows that to obtain zeros with relative ease, we need either a 1 or a -1. (Otherwise, the constant k is a fraction.) If such a number is lacking, it may be possible to obtain a 1 or -1 by one of the operations. Consider the next example.

E X A M P L E **3**   Evaluate:

$$
\begin{vmatrix} 2 & 1 & -3 & 4 \\ 0 & 2 & -2 & 3 \\ -4 & 0 & 5 & -6 \\ 2 & -3 & 4 & 6 \end{vmatrix}
$$

Solution. For the sake of variation, let us obtain zeros in the first row except for the 1. The operations are $-2C_2 + C_1$, $3C_2 + C_3$, and $-4C_2 + C_4$:

$$
\begin{vmatrix}
2 & 1 & -3 & 4 \\
0 & 2 & -2 & 3 \\
-4 & 0 & 5 & -6 \\
2 & -3 & 4 & 6
\end{vmatrix}
=
\begin{vmatrix}
0 & 1 & 0 & 0 \\
-4 & 2 & 4 & -5 \\
-4 & 0 & 5 & -6 \\
8 & -3 & -5 & 18
\end{vmatrix}
$$

where the column operations are $-2C_2 + C_1$, $3C_2 + C_3$, $-4C_2 + C_4$.

$$
= -(1)
\begin{vmatrix}
-4 & 4 & -5 \\
-4 & 5 & -6 \\
8 & -5 & 18
\end{vmatrix}
$$

The simplest way to create a -1 (or 1) is to factor 4 (or -4) from each element in the first column (Property 4). Thus

$$
(-1)
\begin{vmatrix}
-4 & 4 & -5 \\
-4 & 5 & -6 \\
8 & -5 & 18
\end{vmatrix}
= (-1)(4)
\begin{vmatrix}
-1 & 4 & -5 \\
-1 & 5 & -6 \\
2 & -5 & 18
\end{vmatrix}
$$

$$
= (-1)(4)
\begin{vmatrix}
-1 & 4 & -5 \\
-1 & 5 & -6 \\
2 & -5 & 18
\end{vmatrix}
\quad
\begin{matrix}
-1R_1 + R_2 \\
2R_1 + R_3
\end{matrix}
$$

$$
= (-4)
\begin{vmatrix}
-1 & 4 & -5 \\
0 & 1 & -1 \\
0 & 3 & 8
\end{vmatrix}
$$

$$
= (-4)(-1)
\begin{vmatrix}
1 & -1 \\
3 & 8
\end{vmatrix}
= 44 \qquad \blacktriangleleft
$$

E X A M P L E 4   In some cases a 1 or -1 can be obtained by using Property 5. For example,

$$
\begin{vmatrix}
3 & 7 & 7 \\
13 & -5 & 6 \\
14 & 0 & 20
\end{vmatrix}
\quad -4R_1 + R_2 \quad =
\begin{vmatrix}
3 & 7 & 7 \\
1 & -33 & -22 \\
14 & 0 & 20
\end{vmatrix}
$$

or

$$
\begin{vmatrix}
3 & 7 & 7 \\
13 & -5 & 6 \\
14 & 0 & 20
\end{vmatrix}
\quad -3R_1 + R_3 \quad =
\begin{vmatrix}
3 & 7 & 7 \\
13 & -5 & 6 \\
5 & -21 & -1
\end{vmatrix}
\qquad \blacktriangleleft
$$

E X A M P L E **5** Evaluate:

$$\begin{vmatrix} 1 & -1 & 2 & 4 & -2 \\ 1 & 0 & -1 & -3 & 1 \\ 0 & -3 & 0 & 2 & 2 \\ 0 & 4 & 3 & 0 & -3 \\ 0 & -6 & 5 & 6 & 0 \end{vmatrix}$$

Solution. By $-1R_1 + R_2$ we get

$$\begin{vmatrix} 1 & -1 & 2 & 4 & -2 \\ 0 & 1 & -3 & -7 & 3 \\ 0 & -3 & 0 & 2 & 2 \\ 0 & 4 & 3 & 0 & -3 \\ 0 & -6 & 5 & 6 & 0 \end{vmatrix} = (1) \begin{vmatrix} 1 & -3 & -7 & 3 \\ -3 & 0 & 2 & 2 \\ 4 & 3 & 0 & -3 \\ -6 & 5 & 6 & 0 \end{vmatrix} \quad \begin{matrix} 3R_1 + R_2 \\ -4R_1 + R_3 \\ 6R_1 + R_4 \end{matrix}$$

$$= \begin{vmatrix} 1 & -3 & -7 & 3 \\ 0 & -9 & -19 & 11 \\ 0 & 15 & 28 & -15 \\ 0 & -13 & -36 & 18 \end{vmatrix}$$

$$= (1) \begin{vmatrix} -9 & -19 & 11 \\ 15 & 28 & -15 \\ -13 & -36 & 18 \end{vmatrix}$$

To obtain a -1, we use the operation $-2C_1 + C_2$:

$$\begin{matrix} & -2C_1 + C_2 & \\ \end{matrix}$$

$$\begin{vmatrix} -9 & -19 & 11 \\ 15 & 28 & -15 \\ -13 & -36 & 18 \end{vmatrix} = \begin{vmatrix} -9 & -1 & 11 \\ 15 & -2 & -15 \\ -13 & -10 & 18 \end{vmatrix}$$

$$= \begin{vmatrix} -9 & -1 & 11 \\ 15 & -2 & -15 \\ -13 & -10 & 18 \end{vmatrix} \quad \begin{matrix} -2R_1 + R_2 \\ -10R_1 + R_3 \end{matrix}$$

$$= \begin{vmatrix} -9 & -1 & 11 \\ 33 & 0 & -37 \\ 77 & 0 & -92 \end{vmatrix}$$

$$= -(-1) \begin{vmatrix} 33 & -37 \\ 77 & -92 \end{vmatrix} = -187$$

◄

Evaluate each determinant by using the properties of determinants.

1.
$$\begin{vmatrix} 1 & 3 & -5 \\ 2 & -4 & 0 \\ 0 & -2 & 2 \end{vmatrix}$$

2.
$$\begin{vmatrix} 1 & 4 & 10 \\ 0 & -6 & -3 \\ -3 & 5 & -20 \end{vmatrix}$$

3.
$$\begin{vmatrix} 3 & 0 & 1 \\ -4 & 7 & 2 \\ 8 & 6 & 5 \end{vmatrix}$$

4.
$$\begin{vmatrix} 2 & 3 & 2 \\ 4 & 5 & 1 \\ -6 & -4 & -2 \end{vmatrix}$$

5.
$$\begin{vmatrix} 1 & 4 & 8 \\ -2 & -7 & 3 \\ 3 & 6 & -5 \end{vmatrix}$$

6.
$$\begin{vmatrix} 3 & 1 & -2 \\ -2 & 10 & 2 \\ 10 & -8 & 4 \end{vmatrix}$$

7.
$$\begin{vmatrix} 1 & 12 & 7 \\ 6 & -5 & -2 \\ 5 & -12 & 6 \end{vmatrix}$$

8.
$$\begin{vmatrix} -1 & 2 & -5 \\ 2 & 0 & 3 \\ -3 & 6 & 2 \end{vmatrix}$$

9.
$$\begin{vmatrix} -8 & 3 & 5 \\ 7 & 2 & -1 \\ 0 & 6 & 4 \end{vmatrix}$$

10.
$$\begin{vmatrix} 0 & -2 & 6 \\ 7 & 3 & -5 \\ 2 & -1 & -3 \end{vmatrix}$$

11.
$$\begin{vmatrix} 9 & -13 & -6 \\ -4 & 5 & 3 \\ 7 & 2 & 5 \end{vmatrix}$$
(See Example 5.)

12.
$$\begin{vmatrix} -5 & 7 & 11 \\ -7 & 11 & 9 \\ 4 & 6 & 11 \end{vmatrix}$$

13.
$$\begin{vmatrix} 2 & 1 & -4 & 0 \\ 2 & -1 & 3 & -4 \\ 0 & -3 & 5 & 2 \\ 5 & 2 & 2 & 2 \end{vmatrix}$$

14.
$$\begin{vmatrix} -3 & 0 & -1 & 3 \\ 1 & 2 & 0 & -1 \\ 3 & 4 & -2 & 5 \\ 2 & -2 & -4 & 6 \end{vmatrix}$$

15.
$$\begin{vmatrix} 2 & -3 & 0 & 5 \\ 1 & 2 & -1 & 3 \\ -2 & 3 & -4 & 0 \\ 3 & -2 & 3 & 5 \end{vmatrix}$$

16.
$$\begin{vmatrix} -1 & 2 & -3 & 0 \\ -5 & 0 & 2 & -2 \\ 0 & 4 & 0 & -1 \\ 2 & -1 & 6 & -3 \end{vmatrix}$$

17.
$$\begin{vmatrix} 4 & 1 & 3 & 1 \\ 3 & -2 & -2 & 4 \\ 2 & 3 & 1 & -2 \\ -2 & 5 & 1 & 3 \end{vmatrix}$$

18.
$$\begin{vmatrix} 3 & 1 & -2 & 1 \\ -2 & -1 & 1 & 2 \\ 1 & 2 & -3 & 1 \\ -1 & -3 & 3 & 2 \end{vmatrix}$$

19.
$$\begin{vmatrix} -2 & -1 & 0 & 3 \\ 0 & 2 & 6 & 4 \\ 1 & 2 & 0 & 4 \\ 6 & 0 & 1 & 2 \end{vmatrix}$$

20.
$$\begin{vmatrix} 4 & -1 & 0 & 3 \\ 3 & 0 & 2 & 0 \\ 3 & -1 & 2 & 1 \\ 1 & -3 & 0 & 2 \end{vmatrix}$$

21.
$$\begin{vmatrix} 1 & 3 & -4 & 0 & -5 \\ 0 & -1 & 5 & 0 & -3 \\ 0 & -3 & 2 & 3 & 1 \\ 0 & 2 & -2 & 1 & 0 \\ 0 & 0 & 4 & -6 & 0 \end{vmatrix}$$

22.
$$\begin{vmatrix} -2 & 1 & 0 & -3 & 1 \\ 1 & 1 & -2 & 2 & 0 \\ 0 & 1 & 1 & 1 & 0 \\ 3 & 2 & 0 & 4 & 0 \\ 0 & -4 & -1 & -2 & 0 \end{vmatrix}$$

23.
$$\begin{vmatrix} 2 & 1 & 2 & 3 & 1 \\ 3 & 2 & 1 & 1 & 1 \\ 1 & 1 & 2 & 2 & 3 \\ 1 & 1 & 3 & 1 & 1 \\ 2 & 1 & 1 & 1 & 3 \end{vmatrix}$$

24.
$$\begin{vmatrix} 1 & -1 & -2 & 0 & 2 \\ 1 & 1 & 0 & 2 & 0 \\ 0 & 0 & 1 & 2 & 1 \\ -1 & -1 & 1 & 0 & 1 \\ -2 & 2 & 1 & 0 & -1 \end{vmatrix}$$

15.4   Cramer's Rule

In Section 3.6 we discussed solving systems of equations by **Cramer's rule**. Since we can now evaluate determinants of higher order, we can also use Cramer's rule to solve systems with more than three equations and three unknowns.

Even though Cramer's rule holds for systems of equations of any size, let us state the rule for systems of three equations and three unknowns.

Cramer's Rule: The solution of the system

$$a_1 x + b_1 y + c_1 z = d_1$$
$$a_2 x + b_2 y + c_2 z = d_2$$
$$a_3 x + b_3 y + c_3 z = d_3$$

is given by

$$x = \frac{\begin{vmatrix} d_1 & b_1 & c_1 \\ d_2 & b_2 & c_2 \\ d_3 & b_3 & c_3 \end{vmatrix}}{\begin{vmatrix} a_1 & b_1 & c_1 \\ a_2 & b_2 & c_2 \\ a_3 & b_3 & c_3 \end{vmatrix}}, \qquad y = \frac{\begin{vmatrix} a_1 & d_1 & c_1 \\ a_2 & d_2 & c_2 \\ a_3 & d_3 & c_3 \end{vmatrix}}{\begin{vmatrix} a_1 & b_1 & c_1 \\ a_2 & b_2 & c_2 \\ a_3 & b_3 & c_3 \end{vmatrix}},$$

$$z = \frac{\begin{vmatrix} a_1 & b_1 & d_1 \\ a_2 & b_2 & d_2 \\ a_3 & b_3 & d_3 \end{vmatrix}}{\begin{vmatrix} a_1 & b_1 & c_1 \\ a_2 & b_2 & c_2 \\ a_3 & b_3 & c_3 \end{vmatrix}}$$

Note that the denominator, which is the same for each of the unknowns, consists of the determinant whose elements are the coefficients of the unknowns. The respective numerators are formed by replacing the column of coefficients of the unknown to be found by the column of numbers on the right side.

Cramer's rule can be stated more compactly by means of the following notation:

Notation for Cramer's rule

$$D = \begin{vmatrix} a_1 & b_1 & c_1 \\ a_2 & b_2 & c_2 \\ a_3 & b_3 & c_3 \end{vmatrix}$$

$$D_x = \begin{vmatrix} d_1 & b_1 & c_1 \\ d_2 & b_2 & c_2 \\ d_3 & b_3 & c_3 \end{vmatrix} \qquad D_y = \begin{vmatrix} a_1 & d_1 & c_1 \\ a_2 & d_2 & c_2 \\ a_3 & d_3 & c_3 \end{vmatrix}$$

$$D_z = \begin{vmatrix} a_1 & b_1 & d_1 \\ a_2 & b_2 & d_2 \\ a_3 & b_3 & d_3 \end{vmatrix}$$

The solution can now be written in the form given next.

Notation for Cramer's Rule:

$$x = \frac{D_x}{D}, \qquad y = \frac{D_y}{D}, \qquad z = \frac{D_z}{D} \qquad\qquad (15.6)$$

Analogous notations are used for larger systems.

E X A M P L E **1**   Solve the following system of equations by using Cramer's rule:

$$2x_1 - x_2 + 2x_3 - x_4 = 1$$
$$x_1 \qquad - 2x_3 + x_4 = 0$$
$$x_1 - 2x_2 + x_3 - 3x_4 = 2$$
$$-x_2 + x_3 - x_4 = -1$$

Solution. The denominator D, common to all the unknowns, will be evaluated first:

$$D = \begin{vmatrix} 2 & -1 & 2 & -1 \\ 1 & 0 & -2 & 1 \\ 1 & -2 & 1 & -3 \\ 0 & -1 & 1 & -1 \end{vmatrix} \begin{matrix} -2R_2 + R_1 \\ \\ -1R_2 + R_3 \\ \\ \end{matrix} = \begin{vmatrix} 0 & -1 & 6 & -3 \\ 1 & 0 & -2 & 1 \\ 0 & -2 & 3 & -4 \\ 0 & -1 & 1 & -1 \end{vmatrix}$$

$$= - \begin{vmatrix} -1 & 6 & -3 \\ -2 & 3 & -4 \\ -1 & 1 & -1 \end{vmatrix} \begin{matrix} -2R_1 + R_2 \\ \\ -1R_1 + R_3 \end{matrix}$$

$$= -\begin{vmatrix} -1 & 6 & -3 \\ 0 & -9 & 2 \\ 0 & -5 & 2 \end{vmatrix} = -(-1)\begin{vmatrix} -9 & 2 \\ -5 & 2 \end{vmatrix} = -18 + 10 = -8$$

Now by Cramer's rule

$$x_1 = \frac{D_{x_1}}{D} = -\frac{1}{8}\begin{vmatrix} 1 & -1 & 2 & -1 \\ 0 & 0 & -2 & 1 \\ 2 & -2 & 1 & -3 \\ -1 & -1 & 1 & -1 \end{vmatrix} \begin{matrix} \\ \\ -2R_1 + R_3 \\ 1R_1 + R_4 \end{matrix}$$

$$= -\frac{1}{8}\begin{vmatrix} 1 & -1 & 2 & -1 \\ 0 & 0 & -2 & 1 \\ 0 & 0 & -3 & -1 \\ 0 & -2 & 3 & -2 \end{vmatrix}$$

$$= -\frac{1}{8}\begin{vmatrix} 0 & -2 & 1 \\ 0 & -3 & -1 \\ -2 & 3 & -2 \end{vmatrix} = -\frac{1}{8}(-2)\begin{vmatrix} -2 & 1 \\ -3 & -1 \end{vmatrix}$$

$$= \frac{1}{4}(2 + 3) = \frac{5}{4}$$

The procedure for finding x_2 is similar:

$$x_2 = \frac{D_{x_2}}{D} = -\frac{1}{8}\begin{vmatrix} 2 & 1 & 2 & -1 \\ 1 & 0 & -2 & 1 \\ 1 & 2 & 1 & -3 \\ 0 & -1 & 1 & -1 \end{vmatrix} \begin{matrix} -2R_2 + R_1 \\ \\ -1R_2 + R_3 \\ \\ \end{matrix}$$

$$= -\frac{1}{8}\begin{vmatrix} 0 & 1 & 6 & -3 \\ 1 & 0 & -2 & 1 \\ 0 & 2 & 3 & -4 \\ 0 & -1 & 1 & -1 \end{vmatrix}$$

$$= -\frac{1}{8}(-1)\begin{vmatrix} 1 & 6 & -3 \\ 2 & 3 & -4 \\ -1 & 1 & -1 \end{vmatrix} \begin{matrix} -2R_1 + R_2 \\ 1R_1 + R_3 \end{matrix}$$

$$= \frac{1}{8}\begin{vmatrix} 1 & 6 & -3 \\ 0 & -9 & 2 \\ 0 & 7 & -4 \end{vmatrix}$$

$$= \frac{1}{8}\begin{vmatrix} -9 & 2 \\ 7 & -4 \end{vmatrix} = \frac{1}{8}(36 - 14) = \frac{1}{8}(22) = \frac{11}{4}$$

For x_3 we have

$$x_3 = \frac{D_{x_3}}{D} = -\frac{1}{8} \begin{vmatrix} 2 & -1 & 1 & -1 \\ 1 & 0 & 0 & 1 \\ 1 & -2 & 2 & -3 \\ 0 & -1 & -1 & -1 \end{vmatrix} = -\frac{1}{2}$$

The simplest way to find x_4 is to substitute the values already found in one of the equations and to solve for x_4. From the fourth equation,

$$-x_2 + x_3 - x_4 = -1$$

we get

$$-\frac{11}{4} - \frac{1}{2} - x_4 = -1$$

which gives $x_4 = -\frac{9}{4}$. As a check, let us substitute the values into the first equation $2x_1 - x_2 + 2x_3 - x_4 = 1$:

$$2\left(\frac{5}{4}\right) - \frac{11}{4} + 2\left(-\frac{1}{2}\right) - \left(-\frac{9}{4}\right) = \frac{10}{4} - \frac{11}{4} - \frac{4}{4} + \frac{9}{4} = \frac{4}{4} = 1 \quad \checkmark$$

(The other equations are checked similarly.)   ◀

Since every determinant has a unique value, Cramer's rule guarantees a unique solution for every linear system, provided that $D \neq 0$ (to avoid division by zero). Such a system is called **consistent**.

The system of equations

$$a_1 x + b_1 y + c_1 z = d_1$$
$$a_2 x + b_2 y + c_2 z = d_2$$
$$a_3 x + b_3 y + c_3 z = d_3$$

has a unique solution if and only if $D \neq 0$. An analogous statement holds for systems of n equations with n unknowns.

E X E R C I S E S / S E C T I O N   15.4

In Exercises 1–20, solve each system of equations by using Cramer's rule.

1.
$$x_1 + x_2 + 2x_3 = 9$$
$$x_1 \quad\quad - x_3 = -2$$
$$2x_1 - x_2 \quad\quad = 0$$

2.
$$x_1 - x_2 \quad\quad = 2$$
$$x_1 + 2x_2 - x_3 = -3$$
$$-x_2 + x_3 = 3$$

3.
$$3x_1 \quad\quad + 2x_3 = -1$$
$$4x_1 - x_2 - 2x_3 = 7$$
$$x_1 + x_2 \quad\quad = 2$$

4.
$$2x_1 - 3x_2 \quad\quad = 8$$
$$3x_1 - x_2 + 2x_3 = 8$$
$$2x_1 + 3x_2 \quad\quad = -4$$

5.
$$x_1 - 2x_2 - x_3 = 1$$
$$2x_1 - x_2 - 2x_3 = -1$$
$$3x_1 + x_2 + 2x_3 = 6$$

6.
$$3x_1 - 3x_2 - 2x_3 = 1$$
$$-x_1 + x_2 - 6x_3 = 3$$
$$x_1 + x_2 + 2x_3 = 3$$

7. $2x_1 - x_2 + 3x_3 = 16$
   $3x_1 + 4x_2 + 2x_3 = 7$
   $5x_1 - 6x_2 + 8x_3 = 47$

8. $x_1 - 2x_2 + x_3 = 1$
   $2x_1 + x_2 + 2x_3 = 2$
   $3x_1 + 3x_2 - 3x_3 = 2$

9. $3x_1 - 3x_2 + 7x_3 = 35$
   $6x_1 + 3x_2 - x_3 = -5$
   $9x_1 - 12x_2 + 3x_3 = 38$

10. $2x_1 - 5x_2 + 10x_3 = 21$
    $3x_1 + 5x_2 + 10x_3 = 6$
    $4x_1 - 10x_2 + 40x_3 = 64$

11. $2x_1 - x_2 + 2x_3 = 2$
    $-4x_1 + 5x_2 - 3x_3 = 1$
    $3x_1 - 4x_2 + x_3 = 3$

12. $2x_1 + x_2 - x_3 = 0$
    $8x_1 - 2x_2 + x_3 = 1$
    $4x_1 - 4x_2 + x_3 = 0$

13. $2x_2 + 3x_3 + x_4 = 1$
    $2x_1 - 2x_2 + x_4 = 1$
    $3x_1 - 2x_3 + 2x_4 = 13$
    $x_1 + 3x_2 - x_3 = 9$

14. $x_1 + 2x_3 - 3x_4 = 12$
    $3x_1 - 2x_2 + x_3 = -3$
    $3x_2 - 3x_3 - 5x_4 = 12$
    $2x_1 - x_2 - 2x_4 = 3$

15. $2x_1 - x_2 - 2x_3 - 2x_4 = -7$
    $4x_1 - 2x_2 + 3x_3 = 3$
    $6x_1 + 2x_2 + 4x_3 - 3x_4 = 16$
    $8x_1 - 2x_3 - 4x_4 = -6$

16. $2x_1 + x_3 - x_4 = 0$
    $x_1 - x_2 + x_4 = 1$
    $x_1 + x_2 - x_4 = 0$
    $2x_2 + x_3 + x_4 = -1$

17. $3x_1 - 2x_2 + x_3 - x_4 = 2$
    $-x_1 - x_3 + 2x_4 = 1$
    $x_1 - x_2 + x_3 = 0$
    $x_1 + 2x_2 + 3x_3 - x_4 = -2$

18. $x_1 - x_2 - 3x_3 + x_4 = 2$
    $2x_1 + x_3 - x_4 = 0$
    $3x_1 - 2x_2 + 3x_3 + x_4 = -1$
    $x_1 + x_2 - 2x_4 = 0$

19. $x_1 - x_2 + x_4 - x_5 = 1$
    $x_2 - x_3 + x_4 + 2x_5 = 0$
    $x_1 - x_2 + x_4 - 2x_5 = -1$
    $x_2 + x_3 - x_4 = 0$
    $2x_2 - x_3 + x_5 = 2$

20. $x_1 - 2x_2 + x_3 - x_5 = 0$
    $2x_1 - x_3 + 2x_4 - x_5 = 1$
    $x_2 - 2x_3 - x_4 + 2x_5 = 1$
    $x_1 + x_3 + x_4 = 0$
    $x_1 + x_3 + x_5 = 1$

21. The system of equations for finding the currents in the circuit in Figure 15.1 is

$$I_1 + I_2 + I_3 - I_4 = 0$$
$$5I_1 - 4I_2 = -1$$
$$4I_2 - 3I_3 = -1$$
$$3I_3 + 6I_4 = 7$$

Solve this system by using Cramer's rule.

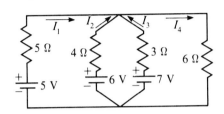

Figure 15.1

22. Use Cramer's rule to solve the system

$$I_1 + I_2 + I_3 - I_4 = 0$$
$$2I_1 - I_2 = 1$$
$$I_2 - 3I_3 = 3$$
$$3I_3 + I_4 = -2$$

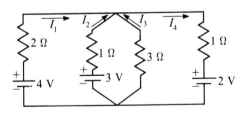

Figure 15.2

which is the system for finding the currents in the circuit shown in Figure 15.2. (See Exercise 33, Section 3.6.)

23. Four machine parts cost a total of $55. The first part costs $5 less than the other three combined, while the second part costs as much as the last two combined. Finally, the first part costs 5 times as much as the fourth part. Find the cost of each.

15.5 Matrix Algebra

We will now turn to the other important concept in this chapter, that of a *matrix*. Matrices were first studied systematically by the English mathematician Arthur Cayley (1821–1895) and have grown in importance ever since. They are particularly well suited for solving systems of equations by means of computers, but matrix methods have been making inroads into many other fields as well, such as statistics, mechanics, and economics.

In this section we will introduce the basic definitions and then study the algebra of matrices. The solution of equations by matrix methods will be left to a later section.

Basic Definitions

A **matrix** is a rectangular array of numbers. The array is enclosed in brackets or parentheses to distinguish it from a determinant. For example, the array

$$\begin{bmatrix} 1 & 0 & -4 & 5 \\ 6 & -3 & 2 & 0 \end{bmatrix}$$

is a matrix. The distinction between a matrix and a determinant is an important one, for *a matrix does not have a numerical value attached to it*. A matrix with *m* **rows** and *n* **columns** is said to have **dimensions** $m \times n$. The numbers in the array are called the **elements** of the matrix. Capital letters such as *A*, *B*, and *C* will be used to denote a matrix.

m × *n* matrix

Double subscripts are customarily used to denote the individual elements of a matrix. For example, a general 3×3 and a general 2×4 matrix are written, respectively,

$$\begin{bmatrix} a_{11} & a_{12} & a_{13} \\ a_{21} & a_{22} & a_{23} \\ a_{31} & a_{32} & a_{33} \end{bmatrix} \quad \text{and} \quad \begin{bmatrix} a_{11} & a_{12} & a_{13} & a_{14} \\ a_{21} & a_{22} & a_{23} & a_{24} \end{bmatrix} \qquad (15.7)$$

Meaning of a_{ij}

The subscript a_{ij} refers to the *i*th **row** and *i*th **column**. (This convention can be remembered by thinking of "RC," as in "Royal Crown.") For example, the element a_{25} is located in row 2, column 5. Similarly, a 3×5 matrix has 3 rows and 5 columns, again thinking of "RC."

A matrix can have a single row or column. Thus

$$\begin{bmatrix} a_{11} \\ a_{21} \\ a_{31} \end{bmatrix}$$

is a 3×1 matrix, while

$$\begin{bmatrix} a_{11} & a_{12} & a_{13} & a_{14} \end{bmatrix}$$

Column matrix
Row matrix

is a 1×4 matrix. The first is called a **column matrix** and the second a **row matrix**.

Equality

To define equality in a natural way, note that a vector $\mathbf{A} = (a, b)$ can be viewed as a row or column matrix. Now recall that

$$(a, b) = (a', b')$$

if $a = a'$ and $b = b'$. So it seems reasonable to define the equality of matrices as follows:

Equality: Two matrices are equal if and only if the corresponding elements are equal.

E X A M P L E **1**

$$\begin{bmatrix} a & b \\ c & 1 \end{bmatrix} = \begin{bmatrix} 3 & -4 \\ 2 & d \end{bmatrix}$$

if and only if $a = 3$, $b = -4$, $c = 2$, and $d = 1$. ◀

Two matrices can be equal only if they have the same dimensions.

E X A M P L E **2**

$$[1 \quad -3] \neq \begin{bmatrix} 1 & -3 \\ 2 & 0 \end{bmatrix}$$

since they have different dimensions. ◀

Addition and Subtraction

The definitions of addition and subtraction of matrices can also be motivated by the corresponding operations for vectors:

$$(a, b) \pm (a', b') = (a \pm a', b \pm b')$$

Addition and Subtraction:

The sum (difference) of two $m \times n$ matrices is obtained by adding (subtracting) the corresponding elements.

E X A M P L E **3**

$$\begin{bmatrix} 2 & 3 \\ -1 & 0 \\ 2 & 6 \end{bmatrix} + \begin{bmatrix} -2 & 5 \\ 6 & -4 \\ 0 & -3 \end{bmatrix} = \begin{bmatrix} 2-2 & 3+5 \\ -1+6 & 0-4 \\ 2+0 & 6-3 \end{bmatrix} = \begin{bmatrix} 0 & 8 \\ 5 & -4 \\ 2 & 3 \end{bmatrix}$$

Addition and subtraction are not defined for matrices of different dimensions. ◀

Zero Matrix

In ordinary algebra

$$a + 0 = a \qquad \text{for all } a$$

In matrix algebra, a matrix whose elements are all zero is called the **zero matrix** and is denoted by O. It has the analogous property

$$A + O = A$$

provided that A and O have the same dimensions.

Zero Matrix:

$$O = \begin{bmatrix} 0 & 0 & \cdots & 0 \\ 0 & 0 & \cdots & 0 \\ \vdots & & & \\ 0 & 0 & \cdots & 0 \end{bmatrix} \qquad (15.8)$$

E X A M P L E **4** If

$$A = \begin{bmatrix} -2 & 3 & 0 \\ 1 & 4 & -3 \end{bmatrix}$$

then

$$A + \boldsymbol{O} = \begin{bmatrix} -2 & 3 & 0 \\ 1 & 4 & -3 \end{bmatrix} + \begin{bmatrix} \mathbf{0} & \mathbf{0} & \mathbf{0} \\ \mathbf{0} & \mathbf{0} & \mathbf{0} \end{bmatrix} = \begin{bmatrix} -2 & 3 & 0 \\ 1 & 4 & -3 \end{bmatrix} \qquad \blacktriangleleft$$

Scalar Product

A type of multiplication in matrix algebra involves a number k and a matrix A, which is defined next.

Scalar Product: The product kA of a number k and a matrix A is called the **scalar product** and is obtained by multiplying each element of A by k.

E X A M P L E **5** If $k = -2$ and

$$A = \begin{bmatrix} 1 & 2 \\ -3 & 4 \\ -7 & 8 \end{bmatrix}$$

then

$$kA = -2\begin{bmatrix} 1 & 2 \\ -3 & 4 \\ -7 & 8 \end{bmatrix} = \begin{bmatrix} -2(1) & -2(2) \\ -2(-3) & -2(4) \\ -2(-7) & -2(8) \end{bmatrix} = \begin{bmatrix} -2 & -4 \\ 6 & -8 \\ 14 & -16 \end{bmatrix}$$ ◄

E X A M P L E **6**   If $k = 3$,

$$A = \begin{bmatrix} -3 & 4 & 6 \\ 0 & 1 & -2 \end{bmatrix}, \quad \text{and} \quad B = \begin{bmatrix} 0 & 2 & 4 \\ -1 & -2 & 5 \end{bmatrix}$$

find $kA + B$.

Solution. $kA + B = 3\begin{bmatrix} -3 & 4 & 6 \\ 0 & 1 & -2 \end{bmatrix} + \begin{bmatrix} 0 & 2 & 4 \\ -1 & -2 & 5 \end{bmatrix}$

$$= \begin{bmatrix} -9 & 12 & 18 \\ 0 & 3 & -6 \end{bmatrix} + \begin{bmatrix} 0 & 2 & 4 \\ -1 & -2 & 5 \end{bmatrix}$$

$$= \begin{bmatrix} -9 & 14 & 22 \\ -1 & 1 & -1 \end{bmatrix}$$ ◄

Multiplication

Multiplying two matrices is harder to describe than the other operations. To motivate this operation, let us write the system of equations

$$a_{11}x_1 + a_{12}x_2 = b_1$$
$$a_{21}x_1 + a_{22}x_2 = b_2$$ (15.9)

in matrix form. Let

$$A = \begin{bmatrix} a_{11} & a_{12} \\ a_{21} & a_{22} \end{bmatrix}, \quad X = \begin{bmatrix} x_1 \\ x_2 \end{bmatrix}, \quad \text{and} \quad B = \begin{bmatrix} b_1 \\ b_2 \end{bmatrix}$$

Then the left side of

$$AX = B$$

must be

$$AX = \begin{bmatrix} a_{11} & a_{12} \\ a_{21} & a_{22} \end{bmatrix}\begin{bmatrix} x_1 \\ x_2 \end{bmatrix} = \begin{bmatrix} a_{11}x_1 + a_{12}x_2 \\ a_{21}x_1 + a_{22}x_2 \end{bmatrix}$$

by equations (15.9). This means that the product of A and X is formed by multiplying the first (and then second) row of A by the corresponding elements in the column matrix X and adding the products.

This definition can be generalized. Let $AB = C$, where A has dimensions $m \times n$ and B has dimensions $n \times k$. The element c_{ij} in C is found by the scheme given next.

Matrix Multiplication:

$$
\text{ith row} \rightarrow
\begin{bmatrix}
a_{11} & a_{12} & \cdots & a_{1n} \\
\vdots & \vdots & & \vdots \\
a_{i1} & a_{i2} & \cdots & a_{in} \\
\vdots & \vdots & & \vdots \\
a_{m1} & a_{m2} & \cdots & a_{mn}
\end{bmatrix}
\times
\begin{bmatrix}
b_{11} & \cdots & b_{1j} & \cdots & b_{1k} \\
b_{21} & \cdots & b_{2j} & \cdots & b_{2k} \\
\vdots & & \vdots & & \vdots \\
b_{n1} & \cdots & b_{nj} & \cdots & b_{nk}
\end{bmatrix}
= i \rightarrow
\begin{bmatrix}
& & \downarrow j & & \\
& & \vdots & & \\
\cdots & & c_{ij} & & \cdots \\
& & \vdots & &
\end{bmatrix}
$$

jth column ↓

where $c_{ij} = a_{i1}b_{1j} + a_{i2}b_{2j} + \cdots + a_{in}b_{nj}$.

To be able to form the product AB, the number of columns in A must be equal to the number of rows in B; otherwise multiplication is undefined. Observe also that if A is an $m \times n$ matrix and B an $n \times k$ matrix, then $C = AB$ is an $m \times k$ matrix.

E X A M P L E **7** If

$$
A = \begin{bmatrix} 2 & -1 \\ -3 & 4 \\ 2 & -6 \end{bmatrix}
\quad \text{and} \quad
B = \begin{bmatrix} -1 & 2 & -3 & 1 \\ 5 & -6 & 2 & 2 \end{bmatrix}
$$

find $C = AB$.

Solution. To find c_{11}, we multiply the elements of the first row of A by the corresponding elements in the first column of B and add the products:

$$
\begin{bmatrix} 2 & -1 \\ -3 & 4 \\ 2 & -6 \end{bmatrix}
\begin{bmatrix} -1 & 2 & -3 & 1 \\ 5 & -6 & 2 & 2 \end{bmatrix}
= \begin{bmatrix} c_{11} & \cdots \\ \vdots & \end{bmatrix}
$$

Thus $c_{11} = (2)(-1) + (-1)(5) = -7$.

The element c_{14} is found as follows:

$$
\begin{bmatrix} 2 & -1 \\ -3 & 4 \\ 2 & -6 \end{bmatrix}
\begin{bmatrix} -1 & 2 & -3 \\ 5 & -6 & 2 \end{bmatrix}
= \begin{bmatrix} \cdots & c_{14} \\ & \vdots \end{bmatrix}
$$

Thus $c_{14} = (2)(1) + (-1)(2) = 0$.

For c_{32} we have

$$\begin{bmatrix} 2 & -1 \\ -3 & 4 \\ 2 & -6 \end{bmatrix} \begin{bmatrix} -1 & 2 & -3 & 1 \\ 5 & -6 & 2 & 2 \end{bmatrix} = \begin{bmatrix} \vdots & \cdots \\ c_{32} \end{bmatrix}$$

Thus $c_{32} = (2)(2) + (-6)(-6) = 40$.

For c_{23}:

$$\begin{bmatrix} 2 & -1 \\ -3 & 4 \\ 2 & -6 \end{bmatrix} \begin{bmatrix} -1 & 2 & -3 & 1 \\ 5 & -6 & 2 & 2 \end{bmatrix} = \begin{bmatrix} & \vdots \\ \cdots & c_{23} \\ & \vdots \end{bmatrix}$$

Thus $c_{23} = (-3)(-3) + (4)(2) = 17$.

Continuing in this manner, we obtain the remaining elements of C:

$$\begin{bmatrix} 2 & -1 \\ -3 & 4 \\ 2 & -6 \end{bmatrix} \begin{bmatrix} -1 & 2 & -3 & 1 \\ 5 & -6 & 2 & 2 \end{bmatrix} = \begin{bmatrix} -7 & 10 & -8 & 0 \\ 23 & -30 & 17 & 5 \\ -32 & 40 & -18 & -10 \end{bmatrix} \quad \blacktriangleleft$$

One significant difference between matrix algebra and ordinary algebra can be seen in Example 7. We found the product $C = AB$. However,

$$BA = \begin{bmatrix} -1 & 2 & -3 & 1 \\ 5 & -6 & 2 & 2 \end{bmatrix} \begin{bmatrix} 2 & -1 \\ -3 & 4 \\ 2 & -6 \end{bmatrix}$$

Noncommutativity

is undefined. Hence $AB \neq BA$, and we conclude that matrix multiplication is not commutative.

The Identity Matrix

In ordinary algebra, the number 1 has the property

$$1 \cdot a = a \cdot 1 = a \quad \text{for all } a$$

In matrix algebra the matrix I with the analogous property

$$IA = AI = A \quad \text{for all } A \tag{15.10}$$

is called the **multiplicative identity** or simply the **identity matrix**. The identity matrix is a **square matrix** $(n \times n)$ with 1's along the diagonal starting in the upper left-hand corner and 0's elsewhere. For example, the 4×4 identity matrix is

$$I = \begin{bmatrix} 1 & 0 & 0 & 0 \\ 0 & 1 & 0 & 0 \\ 0 & 0 & 1 & 0 \\ 0 & 0 & 0 & 1 \end{bmatrix}$$

The diagonal referred to is called the **main diagonal**.

Identity Matrix:

$$I = \begin{bmatrix} 1 & 0 & 0 & \cdots & 0 \\ 0 & 1 & 0 & \cdots & 0 \\ \vdots & & & & \\ 0 & 0 & 0 & \cdots & 1 \end{bmatrix}$$

(15.11)

E X A M P L E **8**   Show that $AI = IA = A$ if

$$A = \begin{bmatrix} -2 & 1 \\ -3 & 4 \end{bmatrix}$$

Solution.

$$\begin{bmatrix} -2 & 1 \\ -3 & 4 \end{bmatrix} \begin{bmatrix} 1 & 0 \\ 0 & 1 \end{bmatrix}$$

$$= \begin{bmatrix} (-2)(1) + (1)(0) & (-2)(0) + (1)(1) \\ (-3)(1) + (4)(0) & (-3)(0) + (4)(1) \end{bmatrix} = \begin{bmatrix} -2 & 1 \\ -3 & 4 \end{bmatrix}$$

$$\begin{bmatrix} 1 & 0 \\ 0 & 1 \end{bmatrix} \begin{bmatrix} -2 & 1 \\ -3 & 4 \end{bmatrix}$$

$$= \begin{bmatrix} (1)(-2) + (0)(-3) & (1)(1) + (0)(4) \\ (0)(-2) + (1)(-3) & (0)(1) + (1)(4) \end{bmatrix}$$

$$= \begin{bmatrix} -2 & 1 \\ -3 & 4 \end{bmatrix}$$

◀

According to equation (15.10), $AI = IA$, even though matrix multiplication is not commutative in general. The exercises will illustrate other differences between ordinary and matrix algebra.

 E X A M P L E **9**   One form of the Lorentz transformation in special relativity is

$$x_1' = x_1$$
$$x_2' = x_2$$
$$x_3' = \gamma x_3 + j\beta\gamma x_4$$
$$x_4' = -j\beta\gamma x_3 + \gamma x_4$$

The transformation is frequently written in matrix form using the definition of multiplication:

$$\begin{bmatrix} x_1' \\ x_2' \\ x_3' \\ x_4' \end{bmatrix} = \begin{bmatrix} 1 & 0 & 0 & 0 \\ 0 & 1 & 0 & 0 \\ 0 & 0 & \gamma & j\beta\gamma \\ 0 & 0 & -j\beta\gamma & \gamma \end{bmatrix} \begin{bmatrix} x_1 \\ x_2 \\ x_3 \\ x_4 \end{bmatrix}$$

Here $\gamma = 1/\sqrt{1 - \beta^2}$ and $\beta = v/c$, where v is the velocity of the body and c the velocity of light.   ◀

E X E R C I S E S / S E C T I O N   **15.5**

In Exercises 1–4, find the values of the unknowns in each case by using the definition of matrix equality.

1. $\begin{bmatrix} x & y \\ 1 & x \end{bmatrix} = \begin{bmatrix} -3 & 2 \\ 1 & -3 \end{bmatrix}$

2. $\begin{bmatrix} a & x \\ b & y \end{bmatrix} = \begin{bmatrix} 2 & 6 \\ -7 & 3 \end{bmatrix}$

3. $\begin{bmatrix} x & 0 & a+b \\ -4 & a & y \end{bmatrix} = \begin{bmatrix} 6 & 0 & -5 \\ -4 & 2 & 3 \end{bmatrix}$

4. $\begin{bmatrix} x+y & 1 & x-y \\ -2 & 2 & y \end{bmatrix} = \begin{bmatrix} 4 & 1 & 0 \\ -2 & 2 & 2 \end{bmatrix}$

In Exercises 5–12, refer to the following matrices:

$$A = \begin{bmatrix} 1 & -3 & -2 \\ 2 & 0 & 6 \end{bmatrix}, \quad B = \begin{bmatrix} 2 & -6 & 5 \\ 7 & 3 & 0 \end{bmatrix},$$

$$C = \begin{bmatrix} -6 & -7 & 0 \\ 5 & 2 & 1 \end{bmatrix}$$

Perform the indicated operations.

5. $A + B$ **6.** $A - C$ **7.** $C - B$

8. $B + C$ **9.** $-3A$ **10.** $-2A - C$

11. $2B + C$

12. Verify:
   a. $A + B = B + A$
   b. $A + (B - C) = (A + B) - C$

In Exercises 13–24, perform the indicated multiplications.

13. $\begin{bmatrix} -1 & 2 \\ 0 & 1 \end{bmatrix} \begin{bmatrix} 3 \\ 4 \end{bmatrix}$

14. $\begin{bmatrix} 2 & 3 & 0 \\ 1 & -4 & 6 \end{bmatrix} \begin{bmatrix} -3 \\ 5 \\ 2 \end{bmatrix}$

15. $\begin{bmatrix} -1 & 2 \end{bmatrix} \begin{bmatrix} 3 & -4 \\ -6 & 0 \end{bmatrix}$

16. $\begin{bmatrix} 2 & 0 & -3 & 4 \end{bmatrix} \begin{bmatrix} 1 & 2 \\ -3 & 0 \\ 4 & -6 \\ 0 & 1 \end{bmatrix}$

17. $\begin{bmatrix} 10 & -2 \\ 0 & 5 \end{bmatrix} \begin{bmatrix} 6 & 8 \\ -1 & 0 \end{bmatrix}$

18. $\begin{bmatrix} 5 & -3 \\ 5 & 2 \\ 10 & 3 \end{bmatrix} \begin{bmatrix} 2 & 5 \\ -3 & -4 \end{bmatrix}$

19. $\begin{bmatrix} 0 & -1 \\ 3 & -8 \\ -6 & 7 \end{bmatrix} \begin{bmatrix} 1 & -5 & 0 \\ 2 & 4 & 3 \end{bmatrix}$

20. $\begin{bmatrix} -1 & 0 \\ 2 & 8 \end{bmatrix} \begin{bmatrix} 1 & 5 & -12 \\ -2 & 10 & 0 \end{bmatrix}$

21. $\begin{bmatrix} -5 & 10 & 3 \\ 2 & 0 & -6 \end{bmatrix} \begin{bmatrix} 6 & 7 \\ -10 & 3 \\ 12 & 6 \end{bmatrix}$

22. $\begin{bmatrix} 2 & 5 & 0 \\ 8 & 0 & 1 \\ -2 & 3 & 4 \end{bmatrix} \begin{bmatrix} 1 & 7 \\ -6 & 2 \\ 0 & 3 \end{bmatrix}$

23. $\begin{bmatrix} 2 & 0 & -5 & 6 \\ 7 & 3 & 0 & 2 \end{bmatrix} \begin{bmatrix} 1 & 2 \\ 0 & 3 \\ -5 & 6 \\ 10 & 12 \end{bmatrix}$

24. $\begin{bmatrix} -2 & 0 & 3 \\ 0 & -5 & 6 \\ 3 & -8 & 10 \end{bmatrix} \begin{bmatrix} 7 & 6 & 0 \\ -3 & 4 & 2 \\ -7 & 0 & 1 \end{bmatrix}$

In Exercises 25–28, find AB and, if possible, BA.

25. $A = \begin{bmatrix} 1 & -2 \\ 2 & 1 \end{bmatrix}, \quad B = \begin{bmatrix} -2 & 3 \\ 4 & 0 \end{bmatrix}$

26. $A = \begin{bmatrix} 3 & -4 \\ 2 & 6 \\ 0 & 1 \end{bmatrix}, \quad B = \begin{bmatrix} 6 & 0 & 2 \\ 2 & -1 & 3 \end{bmatrix}$

27. $A = \begin{bmatrix} 5 & -10 \\ 8 & 6 \end{bmatrix}, \quad B = \begin{bmatrix} -4 \\ 2 \end{bmatrix}$

28. $A = \begin{bmatrix} -2 & 0 & 6 \\ 7 & 8 & 12 \end{bmatrix}, \quad B = \begin{bmatrix} 7 \\ 2 \\ 0 \end{bmatrix}$

In Exercises 29–32, show that $AI = IA = A$.

29. $\begin{bmatrix} -2 & 0 \\ 1 & 8 \end{bmatrix}$

30. $\begin{bmatrix} 1 & 0 \\ 6 & 1 \end{bmatrix}$

31. $\begin{bmatrix} 0 & -3 & 4 \\ 1 & 3 & 7 \\ -4 & 0 & 1 \end{bmatrix}$

32. $\begin{bmatrix} 0 & 0 & -1 & 1 \\ 2 & 1 & 0 & 3 \\ 0 & 1 & 0 & 2 \\ 0 & 0 & 1 & 6 \end{bmatrix}$

33. If $ab = 0$, then $a = 0$ or $b = 0$ for any real or complex numbers a and b. This fundamental property of the number system does not carry over to matrix algebra. If

$$A = \begin{bmatrix} 1 & 1 & 1 \\ 1 & 3 & 2 \\ 2 & 0 & 1 \end{bmatrix} \quad \text{and}$$

$$B = \begin{bmatrix} -1 & 2 & -4 \\ -1 & 2 & -4 \\ 2 & -4 & 8 \end{bmatrix}$$

show that $AB = 0$ (even though $A \neq 0$ and $B \neq 0$).

34. If $ab = ac$, $a \neq 0$, then $b = c$ for real or complex numbers (law of cancellation). If

$$A = \begin{bmatrix} 2 & -1 & 3 \\ 3 & 2 & 1 \\ 1 & -4 & 5 \end{bmatrix}, \quad B = \begin{bmatrix} 1 & -1 & -3 & 2 \\ -1 & 1 & 4 & -1 \\ -1 & 2 & 3 & -1 \end{bmatrix},$$

$$C = \begin{bmatrix} -2 & -3 & 2 & -2 \\ 2 & 3 & -1 & 3 \\ 2 & 4 & -2 & 3 \end{bmatrix}$$

show that $AB = AC$. This example shows that the law of cancellation does not necessarily hold.

35. If

$$A = \begin{bmatrix} 1 & -2 \\ 3 & -1 \end{bmatrix} \quad \text{and} \quad B = \begin{bmatrix} -1 & 3 \\ -2 & 1 \end{bmatrix}$$

show that $A^2 - B^2 \neq (A - B)(A + B)$. [*Note:* $A^n = A \cdot A \cdots A$ (n times).]

36. If

$$A = \begin{bmatrix} j & 0 \\ 0 & j \end{bmatrix}$$

show that $A^4 = I$.

37. In quantum mechanics, electron spin can be studied in terms of the *Pauli matrices*

$$s_x = \frac{1}{2}\begin{bmatrix} 0 & 1 \\ 1 & 0 \end{bmatrix},$$

$$s_y = \frac{1}{2}\begin{bmatrix} 0 & -j \\ j & 0 \end{bmatrix},$$

$$s_z = \frac{1}{2}\begin{bmatrix} 1 & 0 \\ 0 & -1 \end{bmatrix}$$

Verify that

$$2s_y s_z = j s_x, \quad 2s_z s_x = j s_y, \quad \text{and} \quad 2s_x s_y = j s_z$$

15.6   Inverse of a Matrix and Row Operations

The purpose of this section is twofold: to introduce the *inverse* of a matrix and to find the inverse by a matrix technique called *row operations*. This technique can also be used to solve systems of equations, as we will see in the next section.

The inverse matrix plays a role similar to division in algebra: Although division as such is not defined in matrix algebra, an operation does exist analogous to

$$a^{-1}a = 1, \qquad a \neq 0$$

This operation involves the **inverse matrix** of A, denoted by A^{-1}, which has the property

$$A^{-1}A = AA^{-1} = I$$

As we will see shortly, not every matrix has an inverse.

Inverse of a Matrix: The **inverse matrix** of A, denoted by A^{-1}, is the matrix such that

$$A^{-1}A = AA^{-1} = I \qquad (15.12)$$

where I is the identity matrix.

The definition of an inverse matrix is illustrated in the next example.

E X A M P L E 1   Given

$$A = \begin{bmatrix} 1 & 1 \\ 1 & 0 \end{bmatrix}, \qquad \text{show that} \qquad B = \begin{bmatrix} 0 & 1 \\ 1 & -1 \end{bmatrix}$$

is the inverse of A.

Solution. $AB = \begin{bmatrix} 1 & 1 \\ 1 & 0 \end{bmatrix}\begin{bmatrix} 0 & 1 \\ 1 & -1 \end{bmatrix} = \begin{bmatrix} 1 & 0 \\ 0 & 1 \end{bmatrix}$

and

$$BA = \begin{bmatrix} 0 & 1 \\ 1 & -1 \end{bmatrix}\begin{bmatrix} 1 & 1 \\ 1 & 0 \end{bmatrix} = \begin{bmatrix} 1 & 0 \\ 0 & 1 \end{bmatrix}$$

Since $AB = I$ and $BA = I$, B is the inverse of A, or $B = A^{-1}$.
    To show that $B = A^{-1}$ for a given matrix A, it is sufficient to check just one of the products.   ◀

    Two methods are commonly employed for finding the inverse of a matrix, the proofs of which are too lengthy for us here. The first of these involves determinants. Let A be a 3×3 matrix, and denote by $|A|$ the determinant of A. Thus

$$A = \begin{bmatrix} a_{11} & a_{12} & a_{13} \\ a_{21} & a_{22} & a_{23} \\ a_{31} & a_{32} & a_{33} \end{bmatrix} \quad \text{and} \quad |A| = \begin{vmatrix} a_{11} & a_{12} & a_{13} \\ a_{21} & a_{22} & a_{23} \\ a_{31} & a_{32} & a_{33} \end{vmatrix}$$

Then the inverse matrix is given by

Inverse matrix

$$
A^{-1} = \frac{1}{|A|}
\begin{bmatrix}
\begin{vmatrix} a_{22} & a_{23} \\ a_{32} & a_{33} \end{vmatrix} &
-\begin{vmatrix} a_{12} & a_{13} \\ a_{32} & a_{33} \end{vmatrix} &
\begin{vmatrix} a_{12} & a_{13} \\ a_{22} & a_{23} \end{vmatrix} \\[10pt]
-\begin{vmatrix} a_{21} & a_{23} \\ a_{31} & a_{33} \end{vmatrix} &
\begin{vmatrix} a_{11} & a_{13} \\ a_{31} & a_{33} \end{vmatrix} &
-\begin{vmatrix} a_{11} & a_{13} \\ a_{21} & a_{23} \end{vmatrix} \\[10pt]
\begin{vmatrix} a_{21} & a_{22} \\ a_{31} & a_{32} \end{vmatrix} &
-\begin{vmatrix} a_{11} & a_{12} \\ a_{31} & a_{32} \end{vmatrix} &
\begin{vmatrix} a_{11} & a_{12} \\ a_{21} & a_{22} \end{vmatrix}
\end{bmatrix}
\qquad (15.13)
$$

Since $1/|A|$ is a constant, A^{-1} has the form of a scalar product. Formula (15.13) can be extended to matrices of any size.

Since the elements of the matrix (15.13) are determinants, which always exist and have unique values, A^{-1} always exists, provided that $|A| \neq 0$ (to avoid division by 0). We now have the criterion below.

Existence of the Inverse of a Matrix: If A is a square matrix, then A^{-1} exists and is unique if and only if $|A| \neq 0$. If A is not a square matrix, then $|A|$ is undefined and A^{-1} does not exist.

Matrix Inversion by Row Operations

Since matrix (15.13) takes too long to evaluate for larger matrices, we will concentrate mostly on another method for determining the inverse of a matrix. This method is based on **row operations**, which are defined next.

Row Operations on a Matrix:

1. Interchange any two rows.
2. Multiply any row by a nonzero constant.
3. Add a nonzero multiple of any row to any other row.

Equivalence

If matrix A is changed to matrix B by one of the row operations, then matrix A is said to be **equivalent** to matrix B, designated by $A \sim B$. The term *equivalent* is borrowed from the theory of equations, for two systems of equations are also called equivalent if they have the same solution set. In fact, the row operations are precisely those operations used to solve a given system by addition or subtraction.

Note especially that Row Operation 3 is identical to Property 5 for determinants applied to rows. For that reason, we are going to employ the same notation we used earlier:

$$kR_j + R_i$$

[See statement (15.4) in Section 15.3.]

Row operations provide a convenient method for matrix inversion. The technique consists of transforming the given matrix by successive row operations to the identity matrix and simultaneously performing the same row operations on the identity matrix. The converted identity matrix is the inverse of the given matrix. In practice, these steps are best carried out by **augmenting** the given matrix with the identity matrix. For example, if the given matrix is

$$
\begin{bmatrix} a_{11} & a_{12} & a_{13} \\ a_{21} & a_{22} & a_{23} \\ a_{31} & a_{32} & a_{33} \end{bmatrix}, \quad \text{then} \quad \begin{bmatrix} a_{11} & a_{12} & a_{13} & 1 & 0 & 0 \\ a_{21} & a_{22} & a_{23} & 0 & 1 & 0 \\ a_{31} & a_{32} & a_{33} & 0 & 0 & 1 \end{bmatrix}
$$

Augmented matrix

is called the **augmented matrix**. Loosely speaking, the two matrices are "put together."

Matrix Inversion by Row Operations: Transform the augmented matrix

given matrix

$$
A
$$

$$
\begin{bmatrix} a_{11} & a_{12} & \cdots & a_{1n} & 1 & 0 & \cdots & 0 \\ a_{21} & a_{22} & \cdots & a_{2n} & 0 & 1 & \cdots & 0 \\ \vdots & & & & & & & \\ a_{n1} & a_{n2} & \cdots & a_{nn} & 0 & 0 & \cdots & 1 \end{bmatrix}
$$

by successive row operations to

inverse matrix

$$
A^{-1}
$$

$$
\begin{bmatrix} 1 & 0 & \cdots & 0 & b_{11} & b_{12} & \cdots & b_{1n} \\ 0 & 1 & \cdots & 0 & b_{21} & b_{22} & \cdots & b_{2n} \\ \vdots & & & & & & & \\ 0 & 0 & \cdots & 1 & b_{n1} & b_{n2} & \cdots & b_{nn} \end{bmatrix}
$$

E X A M P L E **2** Find A^{-1} for

$$
\begin{bmatrix} 1 & 1 & 0 \\ 2 & 3 & 3 \\ -1 & 1 & 1 \end{bmatrix}
$$

Solution. We form the augmented matrix by putting the given matrix together with the identity matrix:

$$
\begin{bmatrix} 1 & 1 & 0 & 1 & 0 & 0 \\ 2 & 3 & 3 & 0 & 1 & 0 \\ -1 & 1 & 1 & 0 & 0 & 1 \end{bmatrix}
$$

We need to transform this matrix to the form

$$
\begin{bmatrix}
1 & 0 & 0 & \cdots \\
0 & 1 & 0 & \cdots \\
0 & 0 & 1 & \cdots
\end{bmatrix}
\tag{15.14}
$$

by successive row operations. By Row Operations 3, adding a nonzero multiple of one row to another row, we get

$$
\begin{bmatrix}
\mathbf{1} & 1 & 0 & 1 & 0 & 0 \\
\mathbf{2} & 3 & 3 & 0 & 1 & 0 \\
\mathbf{-1} & 1 & 1 & 0 & 0 & 1
\end{bmatrix}
\begin{matrix}
\\
-2R_1 + R_2 \\
1R_1 + R_3
\end{matrix}
$$

$$
\sim
\begin{bmatrix}
\mathbf{1} & 1 & 0 & 1 & 0 & 0 \\
\mathbf{0} & 1 & 3 & -2 & 1 & 0 \\
\mathbf{0} & 2 & 1 & 1 & 0 & 1
\end{bmatrix}
$$

The first column now has the desired form. (Always complete a column before moving on to the next column.)

To change the second column, we first obtain the 1 on the main diagonal and then the 0's. Since we already have a 1, we can proceed:

$$
\begin{bmatrix}
1 & \mathbf{1} & 0 & 1 & 0 & 0 \\
0 & \mathbf{1} & 3 & -2 & 1 & 0 \\
0 & \mathbf{2} & 1 & 1 & 0 & 1
\end{bmatrix}
\begin{matrix}
-1R_2 + R_1 \\
\\
-2R_2 + R_3
\end{matrix}
$$

$$
\sim
\begin{bmatrix}
1 & \mathbf{0} & -3 & 3 & -1 & 0 \\
0 & \mathbf{1} & 3 & -2 & 1 & 0 \\
0 & \mathbf{0} & -5 & 5 & -2 & 1
\end{bmatrix}
$$

These operations complete the second column. Moving on to the third column, we first obtain the 1 on the main diagonal and then the 0's. Multiplying the third row by $-\frac{1}{5}$ (Row Operation 2), we have

$$
\begin{bmatrix}
1 & 0 & -3 & 3 & -1 & 0 \\
0 & 1 & 3 & -2 & 1 & 0 \\
0 & 0 & -5 & 5 & -2 & 1
\end{bmatrix}
$$

$$
\sim
\begin{bmatrix}
1 & 0 & -3 & 3 & -1 & 0 \\
0 & 1 & 3 & -2 & 1 & 0 \\
0 & 0 & 1 & -1 & \dfrac{2}{5} & -\dfrac{1}{5}
\end{bmatrix}
\qquad \left(\text{multiply } R_3 \text{ by } -\frac{1}{5}\right)
$$

$$
=
\begin{bmatrix}
1 & 0 & -3 & 3 & -1 & 0 \\
0 & 1 & 3 & -2 & 1 & 0 \\
0 & 0 & 1 & -1 & \dfrac{2}{5} & -\dfrac{1}{5}
\end{bmatrix}
\begin{matrix}
3R_3 + R_1 \\
-3R_3 + R_2 \\
\\
\end{matrix}
$$

$$\sim \begin{bmatrix} 1 & 0 & \mathbf{0} & 0 & \dfrac{1}{5} & -\dfrac{3}{5} \\ 0 & 1 & \mathbf{0} & 1 & -\dfrac{1}{5} & \dfrac{3}{5} \\ 0 & 0 & 1 & -1 & \dfrac{2}{5} & -\dfrac{1}{5} \end{bmatrix}$$

Since the augmented matrix has now been transformed to form (15.14), we conclude that

$$A^{-1} = \begin{bmatrix} 0 & \dfrac{1}{5} & -\dfrac{3}{5} \\ 1 & -\dfrac{1}{5} & \dfrac{3}{5} \\ -1 & \dfrac{2}{5} & -\dfrac{1}{5} \end{bmatrix}$$

By the definition of a scalar product, we can remove the factor $\frac{1}{5}$ from all the elements and write

$$A^{-1} = \frac{1}{5} \begin{bmatrix} 0 & 1 & -3 \\ 5 & -1 & 3 \\ -5 & 2 & -1 \end{bmatrix}$$

Check:

$$A^{-1}A = \frac{1}{5} \begin{bmatrix} 0 & 1 & -3 \\ 5 & -1 & 3 \\ -5 & 2 & -1 \end{bmatrix} \begin{bmatrix} 1 & 1 & 0 \\ 2 & 3 & 3 \\ -1 & 1 & 1 \end{bmatrix}$$

$$= \frac{1}{5} \begin{bmatrix} 5 & 0 & 0 \\ 0 & 5 & 0 \\ 0 & 0 & 5 \end{bmatrix} = \begin{bmatrix} 1 & 0 & 0 \\ 0 & 1 & 0 \\ 0 & 0 & 1 \end{bmatrix}$$

Remember. When working on a given column, obtain the 1 first and then the 0's. Always complete the column before moving on to the next column. ◀

In our next example we will find the inverse of a 4 × 4 matrix.

E   X   A   M   P   L   E   **3**   Find A^{-1} for

$$A = \begin{bmatrix} 0 & 1 & 0 & 2 \\ 1 & 2 & 0 & 0 \\ 0 & 0 & 3 & 0 \\ 0 & 1 & 2 & 5 \end{bmatrix}$$

Solution. As before, we start with the augmented matrix:

$$\begin{bmatrix} 0 & 1 & 0 & 2 & \mathbf{1} & \mathbf{0} & \mathbf{0} & \mathbf{0} \\ 1 & 2 & 0 & 0 & \mathbf{0} & \mathbf{1} & \mathbf{0} & \mathbf{0} \\ 0 & 0 & 3 & 0 & \mathbf{0} & \mathbf{0} & \mathbf{1} & \mathbf{0} \\ 0 & 1 & 2 & 5 & \mathbf{0} & \mathbf{0} & \mathbf{0} & \mathbf{1} \end{bmatrix} \quad \text{(interchange } R_1 \text{ and } R_2)$$

$$\sim \begin{bmatrix} 1 & 2 & 0 & 0 & 0 & 1 & 0 & 0 \\ 0 & 1 & 0 & 2 & 1 & 0 & 0 & 0 \\ 0 & 0 & 3 & 0 & 0 & 0 & 1 & 0 \\ 0 & 1 & 2 & 5 & 0 & 0 & 0 & 1 \end{bmatrix} \quad \begin{matrix} -2R_2 + R_1 \\ \\ \\ -1R_2 + R_4 \end{matrix}$$

$$\sim \begin{bmatrix} 1 & 0 & 0 & -4 & -2 & 1 & 0 & 0 \\ 0 & 1 & 0 & 2 & 1 & 0 & 0 & 0 \\ 0 & 0 & 3 & 0 & 0 & 0 & 1 & 0 \\ 0 & 0 & 2 & 3 & -1 & 0 & 0 & 1 \end{bmatrix} \quad \left(\text{multiply } R_3 \text{ by } \frac{1}{3}\right)$$

$$\sim \begin{bmatrix} 1 & 0 & 0 & -4 & -2 & 1 & 0 & 0 \\ 0 & 1 & 0 & 2 & 1 & 0 & 0 & 0 \\ 0 & 0 & 1 & 0 & 0 & 0 & \frac{1}{3} & 0 \\ 0 & 0 & 2 & 3 & -1 & 0 & 0 & 1 \end{bmatrix} \quad -2R_3 + R_4$$

$$\sim \begin{bmatrix} 1 & 0 & 0 & -4 & -2 & 1 & 0 & 0 \\ 0 & 1 & 0 & 2 & 1 & 0 & 0 & 0 \\ 0 & 0 & 1 & 0 & 0 & 0 & \frac{1}{3} & 0 \\ 0 & 0 & 0 & 3 & -1 & 0 & -\frac{2}{3} & 1 \end{bmatrix} \quad \left(\text{multiply } R_4 \text{ by } \frac{1}{3}\right)$$

$$\sim \begin{bmatrix} 1 & 0 & 0 & -4 & -2 & 1 & 0 & 0 \\ 0 & 1 & 0 & 2 & 1 & 0 & 0 & 0 \\ 0 & 0 & 1 & 0 & 0 & 0 & \frac{1}{3} & 0 \\ 0 & 0 & 0 & 1 & -\frac{1}{3} & 0 & -\frac{2}{9} & \frac{1}{3} \end{bmatrix} \quad \begin{matrix} 4R_4 + R_1 \\ -2R_4 + R_2 \end{matrix}$$

$$\sim \begin{bmatrix} 1 & 0 & 0 & 0 & -\frac{10}{3} & 1 & -\frac{8}{9} & \frac{4}{3} \\ 0 & 1 & 0 & 0 & \frac{5}{3} & 0 & \frac{4}{9} & -\frac{2}{3} \\ 0 & 0 & 1 & 0 & 0 & 0 & \frac{1}{3} & 0 \\ 0 & 0 & 0 & 1 & -\frac{1}{3} & 0 & -\frac{2}{9} & \frac{1}{3} \end{bmatrix}$$

Thus

$$
A^{-1} = \begin{bmatrix} -\dfrac{10}{3} & 1 & -\dfrac{8}{9} & \dfrac{4}{3} \\[2mm] \dfrac{5}{3} & 0 & \dfrac{4}{9} & -\dfrac{2}{3} \\[2mm] 0 & 0 & \dfrac{1}{3} & 0 \\[2mm] -\dfrac{1}{3} & 0 & -\dfrac{2}{9} & \dfrac{1}{3} \end{bmatrix} = \dfrac{1}{9}\begin{bmatrix} -30 & 9 & -8 & 12 \\ 15 & 0 & 4 & -6 \\ 0 & 0 & 3 & 0 \\ -3 & 0 & -2 & 3 \end{bmatrix}
$$

Check:

$$
A^{-1}A = \dfrac{1}{9}\begin{bmatrix} -30 & 9 & -8 & 12 \\ 15 & 0 & 4 & -6 \\ 0 & 0 & 3 & 0 \\ -3 & 0 & -2 & 3 \end{bmatrix}\begin{bmatrix} 0 & 1 & 0 & 2 \\ 1 & 2 & 0 & 0 \\ 0 & 0 & 3 & 0 \\ 0 & 1 & 2 & 5 \end{bmatrix}
$$

$$
= \dfrac{1}{9}\begin{bmatrix} 9 & 0 & 0 & 0 \\ 0 & 9 & 0 & 0 \\ 0 & 0 & 9 & 0 \\ 0 & 0 & 0 & 9 \end{bmatrix} = \begin{bmatrix} 1 & 0 & 0 & 0 \\ 0 & 1 & 0 & 0 \\ 0 & 0 & 1 & 0 \\ 0 & 0 & 0 & 1 \end{bmatrix} \quad \blacktriangleleft
$$

E X E R C I S E S / S E C T I O N   **15.6**

In Exercises 1–24, find the inverse matrix of each given matrix by means of row operations.

1. $\begin{bmatrix} 1 & 3 \\ 2 & 4 \end{bmatrix}$

2. $\begin{bmatrix} 3 & -2 \\ 1 & 1 \end{bmatrix}$

3. $\begin{bmatrix} 2 & 4 \\ 3 & 2 \end{bmatrix}$

4. $\begin{bmatrix} 3 & -2 \\ 4 & 0 \end{bmatrix}$

5. $\begin{bmatrix} 1 & 0 & 0 \\ 0 & 1 & 0 \\ 1 & 0 & 1 \end{bmatrix}$

6. $\begin{bmatrix} 1 & 2 & 5 \\ 2 & 3 & 8 \\ -1 & 1 & 2 \end{bmatrix}$

7. $\begin{bmatrix} 1 & 0 & 2 \\ 0 & -1 & 1 \\ 1 & -2 & 1 \end{bmatrix}$

8. $\begin{bmatrix} 1 & 2 & 3 \\ -1 & 1 & 0 \\ 0 & 2 & 1 \end{bmatrix}$

9. $\begin{bmatrix} 0 & 1 & 2 \\ 2 & 0 & 4 \\ 0 & -1 & 1 \end{bmatrix}$

10. $\begin{bmatrix} 2 & -1 & 1 \\ 1 & 0 & -1 \\ 0 & -1 & 2 \end{bmatrix}$

11. $\begin{bmatrix} 3 & 3 & 2 \\ 2 & 2 & 1 \\ 0 & 1 & -1 \end{bmatrix}$

12. $\begin{bmatrix} 2 & 0 & 0 \\ 2 & 4 & 4 \\ 2 & 0 & -1 \end{bmatrix}$

13. $\begin{bmatrix} 0 & 2 & 1 \\ 3 & 1 & 3 \\ 4 & -2 & -1 \end{bmatrix}$

14. $\begin{bmatrix} 1 & 4 & 0 \\ 3 & 3 & 1 \\ 2 & -2 & 0 \end{bmatrix}$

15. $\begin{bmatrix} 1 & 2 & 3 \\ 0 & 1 & -2 \\ 3 & 2 & 1 \end{bmatrix}$

16. $\begin{bmatrix} 2 & 1 & 0 \\ 0 & -1 & 0 \\ 4 & 3 & 1 \end{bmatrix}$

17. $\begin{bmatrix} 1 & 2 & 0 & 0 \\ -2 & -3 & 1 & 2 \\ 0 & 1 & 0 & -1 \\ 0 & 0 & 1 & 2 \end{bmatrix}$

18. $\begin{bmatrix} 3 & 2 & 0 & 0 \\ 0 & 1 & -2 & 0 \\ -2 & 0 & -3 & 0 \\ 0 & 1 & 0 & 1 \end{bmatrix}$

19. $\begin{bmatrix} 1 & 0 & 1 & 2 \\ 0 & 1 & 2 & -1 \\ 0 & 0 & 1 & 3 \\ 2 & 3 & 0 & 1 \end{bmatrix}$

20. $\begin{bmatrix} 3 & -2 & 0 & 1 \\ 0 & 1 & 2 & 0 \\ 2 & 1 & -2 & 0 \\ 1 & 0 & 0 & 1 \end{bmatrix}$

21. $\begin{bmatrix} 2 & 0 & 1 & 2 \\ 2 & 1 & 3 & 0 \\ 4 & 4 & 1 & 2 \\ 0 & 1 & 0 & -1 \end{bmatrix}$

22. $\begin{bmatrix} 1 & 1 & 2 & 1 \\ 2 & -1 & -1 & 2 \\ 0 & 0 & -1 & 2 \\ 2 & 3 & 1 & 0 \end{bmatrix}$

23. $\begin{bmatrix} 2 & -1 & 0 & 1 \\ 3 & -2 & 2 & 1 \\ 0 & 3 & 4 & 1 \\ 2 & 0 & 0 & 2 \end{bmatrix}$ **24.** $\begin{bmatrix} 1 & -1 & 0 & 0 \\ 3 & -2 & -1 & 0 \\ 2 & 2 & 0 & 0 \\ 0 & 3 & 1 & 1 \end{bmatrix}$ **25.** Find A^{-1} for the matrix in Exercise 5 by using formula (15.13).

26. Find A^{-1} for the matrix in Exercise 6 by using formula (15.13).

15.7   Solving Systems of Equations by Matrix Methods

In this section we will learn two methods for solving systems of equations by means of matrices. The first method uses the inverse of the coefficient matrix; the second yields the solution directly by row operations.

Using the Inverse of the Coefficient Matrix

To describe the first method, consider a typical system of three equations and three unknowns:

$$a_{11}x_1 + a_{12}x_2 + a_{13}x_3 = b_1$$
$$a_{21}x_1 + a_{22}x_2 + a_{23}x_3 = b_2 \qquad (15.15)$$
$$a_{31}x_1 + a_{32}x_2 + a_{33}x_3 = b_3$$

Let

$$A = \begin{bmatrix} a_{11} & a_{12} & a_{13} \\ a_{21} & a_{22} & a_{23} \\ a_{31} & a_{32} & a_{33} \end{bmatrix}, \qquad X = \begin{bmatrix} x_1 \\ x_2 \\ x_3 \end{bmatrix}, \qquad B = \begin{bmatrix} b_1 \\ b_2 \\ b_3 \end{bmatrix}$$

Coefficient matrix

where A is called the **coefficient matrix**. If we write the system in the form $AX = B$, the method of solution can be compared with that of the single linear equation $ax = b$.

$ax = b$	$AX = B$
$a^{-1}ax = a^{-1}b$	$A^{-1}AX = A^{-1}B$
$1x = a^{-1}b$	$IX = A^{-1}B$
$x = a^{-1}b$	$X = A^{-1}B$ $\qquad (15.16)$

Once A^{-1} is found, the solution can be obtained by multiplying A^{-1} by the column matrix B.

The solution of the system	
$AX = B$ $\quad$ is given by $\quad$ $X = A^{-1}B$	$\qquad (15.17)$

E X A M P L E **1** Solve the system

$$
\begin{aligned}
x_1 + x_2 \qquad\quad &= 0 \\
2x_1 + 3x_2 + 3x_3 &= 1 \\
-x_1 + x_2 + x_3 &= 1
\end{aligned}
$$

by using A^{-1}.

Solution. For this system

$$
A = \begin{bmatrix} 1 & 1 & 0 \\ 2 & 3 & 3 \\ -1 & 1 & 1 \end{bmatrix}, \qquad X = \begin{bmatrix} x_1 \\ x_2 \\ x_3 \end{bmatrix}, \qquad B = \begin{bmatrix} 0 \\ 1 \\ 1 \end{bmatrix}
$$

By the method of Section 15.6

$$
A^{-1} = \frac{1}{5} \begin{bmatrix} 0 & 1 & -3 \\ 5 & -1 & 3 \\ -5 & 2 & -1 \end{bmatrix}
$$

So by formula (15.17)

$$
X = A^{-1}B = \frac{1}{5} \begin{bmatrix} 0 & 1 & -3 \\ 5 & -1 & 3 \\ -5 & 2 & -1 \end{bmatrix} \begin{bmatrix} 0 \\ 1 \\ 1 \end{bmatrix} = \frac{1}{5} \begin{bmatrix} -2 \\ 2 \\ 1 \end{bmatrix}
$$

and

$$
X = \begin{bmatrix} -\dfrac{2}{5} \\[2mm] \dfrac{2}{5} \\[2mm] \dfrac{1}{5} \end{bmatrix}
$$

◀

Considering the number of steps necessary to find A^{-1}, the method used in solving Example 1 does not seem as practical as employing Cramer's rule. However, as pointed out earlier, matrix algebra has found its way into ever more areas of application beyond the solution of equations. Systems of equations merely provide a simple and natural way to introduce the concept.

Solution by Row Operations

To describe the method of solution by row operations, let us return to the system (15.15). By **augmenting the coefficient matrix** A with the column matrix

B, we get the matrix

Augmented matrix

$$\begin{bmatrix} a_{11} & a_{12} & a_{13} & b_1 \\ a_{21} & a_{22} & a_{23} & b_2 \\ a_{31} & a_{32} & a_{33} & b_3 \end{bmatrix}$$

Recall that any row operation transforms this matrix into an equivalent matrix. So if the matrix is transformed to

$$\begin{bmatrix} 1 & 0 & 0 & s_1 \\ 0 & 1 & 0 & s_2 \\ 0 & 0 & 1 & s_3 \end{bmatrix}$$

we get the equivalent system of equations

$$x_1 + 0 \ + 0 \ = s_1$$
$$0 \ + x_2 + 0 \ = s_2$$
$$0 \ + 0 \ + x_3 = s_3$$

Hence $x_1 = s_1$, $x_2 = s_2$, and $x_3 = s_3$.

Solution by Row Operations:

Transform the augmented matrix

$$\begin{bmatrix} a_{11} & a_{12} & \cdots & a_{1n} & b_1 \\ a_{21} & a_{22} & \cdots & a_{2n} & b_2 \\ \vdots & & & & \\ a_{n1} & a_{n2} & \cdots & a_{nn} & b_n \end{bmatrix}$$

by successive row operations to

solution column

$$\begin{bmatrix} 1 & 0 & \cdots & 0 & s_1 \\ 0 & 1 & \cdots & 0 & s_2 \\ \vdots & & & & \\ 0 & 0 & \cdots & 1 & s_n \end{bmatrix} \tag{15.18}$$

This method of solving systems of equations is known as the *Gauss-Jordan method*. The next example illustrates the technique.

E X A M P L E **2**   Solve the system in Example 1,

$$x_1 + x_2 \qquad\quad = 0$$
$$2x_1 + 3x_2 + 3x_3 = 1$$
$$-x_1 + x_2 + x_3 = 1$$

by means of row operations.

Solution. From

$$A = \begin{bmatrix} 1 & 1 & 0 \\ 2 & 3 & 3 \\ -1 & 1 & 1 \end{bmatrix} \quad \text{and} \quad B = \begin{bmatrix} \mathbf{0} \\ \mathbf{1} \\ \mathbf{1} \end{bmatrix}$$

we form the augmented matrix

$$\begin{bmatrix} 1 & 1 & 0 & \mathbf{0} \\ 2 & 3 & 3 & \mathbf{1} \\ -1 & 1 & 1 & \mathbf{1} \end{bmatrix}$$

Then we transform this matrix to form (15.18):

$$\begin{bmatrix} 1 & 1 & 0 & 0 \\ 2 & 3 & 3 & 1 \\ -1 & 1 & 1 & 1 \end{bmatrix} \begin{matrix} \\ -2R_1 + R_2 \\ 1R_1 + R_3 \end{matrix}$$

$$\sim \begin{bmatrix} 1 & 1 & 0 & 0 \\ 0 & 1 & 3 & 1 \\ 0 & 2 & 1 & 1 \end{bmatrix} \begin{matrix} -1R_2 + R_1 \\ \\ -2R_2 + R_3 \end{matrix}$$

$$\sim \begin{bmatrix} 1 & 0 & -3 & -1 \\ 0 & 1 & 3 & 1 \\ 0 & 0 & -5 & -1 \end{bmatrix} \left(\text{multiply } R_3 \text{ by } -\frac{1}{5}\right)$$

$$\sim \begin{bmatrix} 1 & 0 & -3 & -1 \\ 0 & 1 & 3 & 1 \\ 0 & 0 & 1 & \dfrac{1}{5} \end{bmatrix} \begin{matrix} 3R_3 + R_1 \\ -3R_3 + R_2 \\ \\ \end{matrix}$$

$$\sim \begin{bmatrix} 1 & 0 & 0 & -\dfrac{2}{5} \\ 0 & 1 & 0 & \dfrac{2}{5} \\ 0 & 0 & 1 & \dfrac{1}{5} \end{bmatrix}$$

The solution is therefore given by $x_1 = -\frac{2}{5}$, $x_2 = \frac{2}{5}$, and $x_3 = \frac{1}{5}$.   ◀

E X A M P L E **3**   Solve the following system of equations by means of row operations:

$$2x_1 - x_2 \qquad + 2x_4 = \quad 0$$
$$x_1 + 2x_2 - x_3 - x_4 = \quad 1$$
$$x_1 \qquad + 2x_3 + x_4 = -1$$
$$x_1 - x_2 - x_3 + x_4 = \quad 2$$

Solution.

$$\begin{bmatrix} 2 & -1 & 0 & 2 & 0 \\ 1 & 2 & -1 & -1 & 1 \\ 1 & 0 & 2 & 1 & -1 \\ 1 & -1 & -1 & 1 & 2 \end{bmatrix} \quad \text{(interchange } R_1 \text{ and } R_2)$$

$$\sim \begin{bmatrix} 1 & 2 & -1 & -1 & 1 \\ 2 & -1 & 0 & 2 & 0 \\ 1 & 0 & 2 & 1 & -1 \\ 1 & -1 & -1 & 1 & 2 \end{bmatrix} \quad \begin{matrix} -2R_1 + R_2 \\ -1R_1 + R_3 \\ -1R_1 + R_4 \end{matrix}$$

$$\sim \begin{bmatrix} 1 & 2 & -1 & -1 & 1 \\ 0 & -5 & 2 & 4 & -2 \\ 0 & -2 & 3 & 2 & -2 \\ 0 & -3 & 0 & 2 & 1 \end{bmatrix} \quad -2R_4 + R_2 \qquad \text{(to obtain a 1 for } a_{22})$$

$$\sim \begin{bmatrix} 1 & 2 & -1 & -1 & 1 \\ 0 & 1 & 2 & 0 & -4 \\ 0 & -2 & 3 & 2 & -2 \\ 0 & -3 & 0 & 2 & 1 \end{bmatrix} \quad \begin{matrix} -2R_2 + R_1 \\ \\ 2R_2 + R_3 \\ 3R_2 + R_4 \end{matrix}$$

$$\sim \begin{bmatrix} 1 & 0 & -5 & -1 & 9 \\ 0 & 1 & 2 & 0 & -4 \\ 0 & 0 & 7 & 2 & -10 \\ 0 & 0 & 6 & 2 & -11 \end{bmatrix} \quad -1R_4 + R_3 \qquad \text{(to obtain a 1 for } a_{33})$$

$$\sim \begin{bmatrix} 1 & 0 & -5 & -1 & 9 \\ 0 & 1 & 2 & 0 & -4 \\ 0 & 0 & 1 & 0 & 1 \\ 0 & 0 & 6 & 2 & -11 \end{bmatrix} \quad \begin{matrix} 5R_3 + R_1 \\ -2R_3 + R_2 \\ \\ -6R_3 + R_4 \end{matrix}$$

$$\sim \begin{bmatrix} 1 & 0 & 0 & -1 & 14 \\ 0 & 1 & 0 & 0 & -6 \\ 0 & 0 & 1 & 0 & 1 \\ 0 & 0 & 0 & 2 & -17 \end{bmatrix} \quad \text{(divide } R_4 \text{ by 2)}$$

$$\sim \begin{bmatrix} 1 & 0 & 0 & -1 & 14 \\ 0 & 1 & 0 & 0 & -6 \\ 0 & 0 & 1 & 0 & 1 \\ 0 & 0 & 0 & 1 & -\dfrac{17}{2} \end{bmatrix} \quad 1R_4 + R_1$$

$$\sim \begin{bmatrix} 1 & 0 & 0 & 0 & \dfrac{11}{2} \\ 0 & 1 & 0 & 0 & -6 \\ 0 & 0 & 1 & 0 & 1 \\ 0 & 0 & 0 & 1 & -\dfrac{17}{2} \end{bmatrix}$$

The solution is therefore given by $x_1 = \frac{11}{2}$, $x_2 = -6$, $x_3 = 1$, and $x_4 = -\frac{17}{2}$. ◀

Since the method of row operations is systematic and has certain operations repeating, it is relatively easy to program. Consequently, it is a good method to use when solving systems of equations on a computer.

EXERCISES / SECTION 15.7

In Exercises 1–32, **(a)** solve each of the given systems of equations by mean of row operations; **(b)** solve each system of equations by using A^{-1}. (In Exercises 1–24, A is identical to the matrix in the corresponding exercises in Section 15.6.)

1. $\quad x_1 + 3x_2 = 2$
$\quad 2x_1 + 4x_2 = 1$

2. $\quad 3x_1 - 2x_2 = 7$
$\quad x_1 + x_2 = 4$

3. $\quad 2x_1 + 4x_2 = 6$
$\quad 3x_1 + 2x_2 = -1$

4. $\quad 3x_1 - 2x_2 = 2$
$\quad 4x_1 = 5$

5. $\quad x_1 = 2$
$\quad x_2 = 3$
$\quad x_1 + x_3 = 4$

6. $\quad x_1 + 2x_2 + 5x_3 = 1$
$\quad 2x_1 + 3x_2 + 8x_3 = 2$
$\quad -x_1 + x_2 + 2x_3 = 3$

7. $\quad x_1 + 2x_3 = 3$
$\quad -x_2 + x_3 = -1$
$\quad x_1 - 2x_2 + x_3 = 7$

8. $\quad x_1 + 2x_2 + 3x_3 = -2$
$\quad -x_1 + x_2 = -1$
$\quad 2x_2 + x_3 = 1$

9. $\quad x_2 + 2x_3 = -1$
$\quad 2x_1 + 4x_3 = 6$
$\quad -x_2 + x_3 = 1$

10. $\quad 2x_1 - x_2 + x_3 = 3$
$\quad x_1 - x_3 = 2$
$\quad -x_2 + 2x_3 = 2$

11. $\quad 3x_1 + 3x_2 + 2x_3 = 1$
$\quad 2x_1 + 2x_2 + x_3 = 2$
$\quad x_2 - x_3 = 3$

12. $\quad 2x_1 = 1$
$\quad 2x_1 + 4x_2 + 4x_3 = 1$
$\quad 2x_1 - x_3 = -1$

13. $\quad 2x_2 + x_3 = 2$
$\quad 3x_1 + x_2 + 3x_3 = -1$
$\quad 4x_1 - 2x_2 - x_3 = 2$

14. $\quad x_1 + 4x_2 = 4$
$\quad 3x_1 + 3x_2 + x_3 = 7$
$\quad 2x_1 - 2x_2 = 3$

15. $\quad x_1 + 2x_2 + 3x_3 = 1$
$\quad x_2 - 2x_3 = 1$
$\quad 3x_1 + 2x_2 + x_3 = -1$

16. $\quad 2x_1 + x_2 = 2$
$\quad -x_2 = 2$
$\quad 4x_1 + 3x_2 + x_3 = 4$

17. $\quad x_1 + 2x_2 = 1$
$\quad -2x_1 - 3x_2 + x_3 + 2x_4 = 0$
$\quad x_2 - x_4 = 2$
$\quad x_3 + 2x_4 = -3$

18. $\quad 3x_1 + 2x_2 = 1$
$\quad x_2 - 2x_3 = 1$
$\quad -2x_1 - 3x_3 = 3$
$\quad x_2 + x_4 = 2$

19. $\quad x_1 + x_3 + 2x_4 = 2$
$\quad x_2 + 2x_3 - x_4 = 1$
$\quad x_3 + 3x_4 = 1$
$\quad 2x_1 + 3x_2 + x_4 = -1$

20. $\quad 3x_1 - 2x_2 + x_4 = 1$
$\quad x_2 + 2x_3 = 1$
$\quad 2x_1 + x_2 - 2x_3 = 1$
$\quad x_1 + x_4 = 1$

21. $\quad 2x_1 + x_3 + 2x_4 = 1$
$\quad 2x_1 + x_2 + 3x_3 = 0$
$\quad 4x_1 + 4x_2 + x_3 + 2x_4 = 1$
$\quad x_2 - x_4 = 0$

22.
$$\begin{aligned} x_1 + x_2 + 2x_3 + x_4 &= -1 \\ 2x_1 - x_2 - x_3 + 2x_4 &= 1 \\ -x_3 + 2x_4 &= -1 \\ 2x_1 + 3x_2 + x_3 &= 1 \end{aligned}$$

23.
$$\begin{aligned} 2x_1 - x_2 + x_4 &= 2 \\ 3x_1 - 2x_2 + 2x_3 + x_4 &= 0 \\ 3x_2 + 4x_3 + x_4 &= 0 \\ 2x_1 + 2x_4 &= 3 \end{aligned}$$

24.
$$\begin{aligned} x_1 - x_2 &= 1 \\ 3x_1 - 2x_2 - x_3 &= 0 \\ 2x_1 + 2x_2 &= 3 \\ 3x_2 + x_3 + x_4 &= 2 \end{aligned}$$

25.
$$\begin{aligned} x_1 + x_2 &= 1 \\ x_1 + 2x_2 + x_3 &= 1 \\ x_1 - 2x_3 &= 3 \end{aligned}$$

26.
$$\begin{aligned} 2x_1 - x_2 + x_3 &= 1 \\ -2x_2 + 2x_3 &= -2 \\ x_1 + x_3 &= 3 \end{aligned}$$

27.
$$\begin{aligned} x_1 + x_3 &= 0 \\ 2x_1 + 2x_2 + 2x_3 &= 0 \\ 3x_1 - x_2 + x_3 &= 1 \end{aligned}$$

28.
$$\begin{aligned} x_1 + x_2 &= 2 \\ x_2 - x_3 &= 3 \\ x_1 + x_2 - x_3 &= 4 \end{aligned}$$

29.
$$\begin{aligned} x_1 - x_3 + x_4 &= -1 \\ x_1 - 2x_2 + 2x_4 &= 0 \\ x_1 + x_2 - x_3 &= 0 \\ x_1 - x_2 + 4x_4 &= 1 \end{aligned}$$

30.
$$\begin{aligned} x_1 - x_3 + 2x_4 &= 1 \\ 2x_1 + x_2 - 2x_3 + 4x_4 &= 1 \\ -3x_2 + x_3 + x_4 &= 2 \\ x_1 - x_3 + 3x_4 &= 0 \end{aligned}$$

31.
$$\begin{aligned} x_1 - 2x_3 + 3x_4 &= -2 \\ 2x_1 + x_2 - 4x_3 + 5x_4 &= -1 \\ 2x_2 + 3x_4 &= 1 \\ 2x_1 - 5x_3 + 4x_4 &= 1 \end{aligned}$$

32.
$$\begin{aligned} 3x_1 - x_2 + 2x_3 + 2x_4 &= 2 \\ x_1 + x_3 + x_4 &= 1 \\ -x_2 - x_4 &= 1 \\ -x_1 + x_2 + x_3 - x_4 &= 1 \end{aligned}$$

33. Solve the given system by means of row operations. (See Exercise 22, Section 15.4.)

$$\begin{aligned} I_1 + I_2 + I_3 - I_4 &= 0 \\ 2I_1 - I_2 &= 1 \\ I_2 - 3I_3 &= 3 \\ 3I_3 + I_4 &= -2 \end{aligned}$$

34. Three machine parts cost a total of $40. The first part costs as much as the other two together, and the cost of 6 times the second is $2 more than the total cost of the other two. Find the cost of each part.

35. An experimenter determined that three resistors satisfy the following relationships:

$$\begin{aligned} R_1 + 2R_2 + 2R_3 &= 14 \\ 2R_1 + R_2 + R_3 &= 10 \\ R_1 + 2R_2 + 3R_3 &= 18 \end{aligned}$$

Find the three resistances (in ohms).

REVIEW EXERCISES/CHAPTER 15

In Exercises 1–6, evaluate the given determinants.

1.
$$\begin{vmatrix} -10 & 2 \\ -3 & 1 \end{vmatrix}$$

2.
$$\begin{vmatrix} 2 & -2 & 0 \\ 3 & -4 & 2 \\ 2 & 2 & 3 \end{vmatrix}$$

3.
$$\begin{vmatrix} 4 & 2 & 4 & 3 \\ 2 & 3 & 1 & 1 \\ 4 & 4 & 3 & 2 \\ 4 & 4 & 3 & 5 \end{vmatrix}$$

4.
$$\begin{vmatrix} 1 & 1 & 3 & 3 \\ 1 & -3 & 2 & -1 \\ 3 & 1 & 6 & -3 \\ 3 & 2 & -3 & 10 \end{vmatrix}$$

5.
$$\begin{vmatrix} 2 & 0 & 0 & 0 & 5 \\ 1 & 0 & 1 & 0 & 2 \\ 1 & 0 & 0 & 0 & 3 \\ 3 & 3 & 2 & 0 & 1 \\ 2 & 1 & 3 & 2 & 0 \end{vmatrix}$$

6.
$$\begin{vmatrix} 1 & 0 & 3 & 2 & 1 \\ 3 & 1 & 3 & 2 & 1 \\ 2 & 2 & 2 & 4 & 2 \\ 2 & 2 & -2 & -2 & -1 \\ 4 & 1 & 6 & 3 & 3 \end{vmatrix}$$

In Exercises 7–10, solve each system of equations by using Cramer's rule.

7.
$$\begin{aligned} 2x_1 + x_2 - x_3 &= 0 \\ x_1 - x_2 + x_3 &= 1 \\ x_1 - 2x_2 - 3x_3 &= 2 \end{aligned}$$

8.
$$\begin{aligned} x_1 - x_2 + 2x_3 &= 1 \\ x_1 - 3x_3 &= 2 \\ 2x_1 + 2x_2 &= 3 \end{aligned}$$

9.
$$\begin{aligned} 5x_1 + 6x_2 + 4x_3 + 3x_4 &= 0 \\ 3x_1 - 4x_2 + 2x_3 + 8x_4 &= -9 \\ 2x_1 + 3x_2 + 5x_3 + 4x_4 &= 4 \\ 3x_1 + 5x_2 - x_3 + x_4 &= 2 \end{aligned}$$

10. $2x_1 - x_2 - x_3 + 2x_4 = 1$
$4x_1 - 2x_2 + x_3 + 3x_4 = 6$
$x_1 + x_2 + 2x_3 + 2x_4 = 3$
$2x_1 + 2x_2 - 3x_3 + x_4 = 2$

In Exercises 11–12, use the definition of matrix equality to determine the value of each of the unknowns.

11. $\begin{bmatrix} x+y & 1 & y \\ z & 2 & 1 \end{bmatrix} = \begin{bmatrix} 3 & 1 & 1 \\ -3 & 2 & 1 \end{bmatrix}$

12. $\begin{bmatrix} a-b & a+b & c \\ c+d & 4 & 5 \end{bmatrix} = \begin{bmatrix} 2 & 6 & -1 \\ 2 & 4 & 5 \end{bmatrix}$

In Exercises 13–16, refer to the following matrices:

$$A = \begin{bmatrix} 1 & -1 \\ 0 & 2 \\ 4 & 5 \end{bmatrix}, \quad B = \begin{bmatrix} -2 & 6 \\ 0 & -3 \\ 2 & 7 \end{bmatrix}, \quad C = \begin{bmatrix} 0 & 1 \\ 2 & -4 \\ 3 & 5 \end{bmatrix}$$

Find

13. $A + B - C$

14. $C - B$

15. $-2A + C$

16. $A - 3C$

In Exercises 17–20, perform the indicated multiplications.

17. $\begin{bmatrix} 1 & -2 & 3 \\ 2 & -5 & 6 \end{bmatrix} \begin{bmatrix} -2 \\ 2 \\ -1 \end{bmatrix}$

18. $\begin{bmatrix} -1 & 3 & -2 & 0 \end{bmatrix} \begin{bmatrix} 2 & -2 \\ -1 & 5 \\ 0 & -2 \\ 1 & 3 \end{bmatrix}$

19. $\begin{bmatrix} -1 & 2 \\ 0 & 6 \\ 4 & -1 \end{bmatrix} \begin{bmatrix} -1 & 4 & -3 \\ 0 & -8 & 5 \end{bmatrix}$

20. $\begin{bmatrix} 2 & -4 \\ 0 & -2 \end{bmatrix} \begin{bmatrix} 4 & 3 \\ -1 & 2 \end{bmatrix}$

In Exercises 21–27, find the inverse of each matrix by means of row operations.

21. $\begin{bmatrix} 2 & 6 \\ 3 & 11 \end{bmatrix}$

22. $\begin{bmatrix} 5 & 6 \\ 3 & 9 \end{bmatrix}$

23. $\begin{bmatrix} -3 & -5 & 0 \\ 2 & 6 & 1 \\ 0 & 7 & 3 \end{bmatrix}$

24. $\begin{bmatrix} 3 & -1 & 2 \\ 2 & 0 & 1 \\ 1 & 2 & 0 \end{bmatrix}$

25. $\begin{bmatrix} 2 & -3 & 4 \\ 1 & 0 & -5 \\ 0 & -2 & 1 \end{bmatrix}$

26. $\begin{bmatrix} 1 & 0 & 0 & 0 \\ 0 & 2 & 0 & -1 \\ 2 & -1 & 0 & 0 \\ 0 & 0 & -1 & 1 \end{bmatrix}$

27. $\begin{bmatrix} 3 & -1 & 0 & 2 \\ 1 & 0 & -1 & 3 \\ 0 & 1 & 0 & 1 \\ 2 & 2 & 2 & 1 \end{bmatrix}$

28. Find the inverse of the matrix in Exercise 25 by using formula (15.13).

In Exercises 29–33, **(a)** solve each of the given systems of equations by using row operations; **(b)** solve each of the systems by using A^{-1} (the coefficient matrices are identical to the matrices in Exercises 23–27, respectively).

29. $-3x_1 - 5x_2 \qquad = 2$
$2x_1 + 6x_2 + x_3 = -1$
$7x_2 + 3x_3 = 1$

30. $3x_1 - x_2 + 2x_3 = 2$
$2x_1 + x_3 = 1$
$x_1 + 2x_2 = 4$

31. $2x_1 - 3x_2 + 4x_3 = 1$
$x_1 - 5x_3 = 2$
$-2x_2 + x_3 = -1$

32. $x_1 \qquad = 1$
$2x_2 - x_4 = 3$
$2x_1 - x_2 \qquad = -5$
$-x_3 + x_4 = -2$

33. $3x_1 - x_2 + 2x_4 = 2$
$x_1 - x_3 + 3x_4 = 2$
$x_2 + x_4 = 0$
$2x_1 + 2x_2 + 2x_3 + x_4 = -1$

In Exercises 34–35, solve the systems by using row operations.

34. In a problem involving static forces, four forces (in pounds) are related by the following equations:

$$2f_1 + f_2 + 3f_3 \qquad = 15$$
$$2f_2 + f_3 + 4f_4 = 23$$
$$f_1 \qquad + 2f_3 + 3f_4 = 20$$
$$2f_1 + 3f_2 + 3f_3 \qquad = 19$$

Find the forces.

35. Find the currents (in amperes) satisfying the following relationships:

$$I_1 + I_2 - I_3 - I_4 = 0$$
$$2I_1 + 3I_2 \qquad = 2$$
$$-2I_2 + 2I_3 \qquad = -2$$
$$-3I_3 - I_4 = 3$$

1. Solve the given system of equations graphically. Estimate the solution to the nearest tenth of a unit.

$$x + y = 1$$
$$y = x^2$$

In Exercises 2 and 3, solve the given systems algebraically.

2. $x + y = 6$
   $xy = 8$

3. $x^2 - 2y^2 = 10$
   $x^2 + y^2 = 16$

4. Solve the following equation in quadratic form:

$$x^{-4} - 3x^{-2} - 4 = 0$$

5. Solve for x:

$$\sqrt{x - 5} - \sqrt{10 - x} = 1$$

In Exercises 6–11, use synthetic division.

6. Determine whether $x - \frac{1}{2}$ is a factor of the polynomial

$$4x^4 + 4x^3 + x^2 - 6x + 1$$

7. Determine whether $x = -2$ is a zero of $f(x) = 2x^4 - 10x^2 - 2x + 4$.

8. Given $f(x) = 2x^3 - x + 6$, find $f(-1)$.

9. Solve the equation

$$x^4 - 8x^3 + 21x^2 - 22x + 8 = 0$$

given that 1 is a double root.

10. Find the rational roots of the equation

$$6x^4 - 11x^3 - 8x^2 + 25x - 12 = 0$$

11. Estimate the irrational root of

$$x^3 - 3x^2 - x + 2 = 0$$

located between 0 and 1 to two decimal places.

12. Evaluate

$$\begin{vmatrix} 2 & 2 & 2 & 5 \\ -4 & -1 & 3 & 2 \\ 2 & -3 & 5 & 0 \\ 0 & 1 & -4 & 2 \end{vmatrix}$$

13. Solve the following system by Cramer's rule:

$$I_1 + 3I_2 - I_3 \qquad = 9$$
$$3I_1 \qquad - 2I_3 + 2I_4 = 13$$
$$2I_1 - 2I_2 \qquad + I_4 = 1$$
$$2I_2 + 3I_3 + I_4 = 1$$

14. Perform the following addition of matrices:

$$\begin{bmatrix} 0 & -3 & 4 \\ 7 & 0 & 5 \\ -9 & 6 & 5 \end{bmatrix} + \begin{bmatrix} 2 & 0 & -6 \\ -8 & 7 & 0 \\ 3 & 0 & 6 \end{bmatrix}$$

15. Perform the following multiplication of matrices:

$$\begin{bmatrix} 6 & 0 & -2 \\ 1 & 3 & 5 \end{bmatrix} \begin{bmatrix} 1 & 0 \\ -2 & 1 \\ 3 & -4 \end{bmatrix}$$

16. Find the inverse of the following matrix:

$$\begin{bmatrix} 3 & 2 & 0 \\ 3 & 2 & 1 \\ 2 & 1 & -1 \end{bmatrix}$$

17. An experimenter determined that three forces F_1, F_2, and F_3 on a structure satisfy the following relationships:

$$F_1 + 2F_2 + 3F_3 = 1$$
$$F_2 - 2F_3 = 1$$
$$3F_1 + 2F_2 + F_3 = -1$$

Use the method of row operations to solve this system.

18. Two resistors have a combined resistance of $10.0\ \Omega$ if connected in series and $2.14\ \Omega$ if connected in parallel. Determine the two resistances.

CHAPTER **16**

Additional Topics in Trigonometry

16.1 Fundamental Identities

We saw in earlier chapters that solving triangles is an integral part of trigonometry. Another branch, called **analytic trigonometry**, deals mainly with **identities**. This aspect of the subject plays a major role in more advanced areas of mathematics, especially calculus.

Most of this chapter is devoted to the study of trigonometric identities. Identities are then used in Section 16.6 to help solve trigonometric equations. The chapter ends with a brief study of inverse trigonometric functions.

Some Basic Identities

Identity

Recall that an **identity** is an equation that is satisfied for every value of the variable. For example, $x^2 - 1 = (x - 1)(x + 1)$ is an identity. In trigonometry, identities arise almost as soon as the basic definitions are given. For example, $\sin \theta = 1/\csc \theta$ is valid for every $\theta \neq 0 \pm n\pi$. To obtain other identities, let us recall the basic definitions of the trigonometric functions.

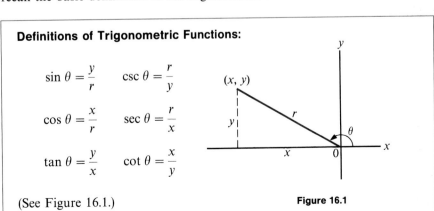

Definitions of Trigonometric Functions:

$$\sin \theta = \frac{y}{r} \qquad \csc \theta = \frac{r}{y}$$

$$\cos \theta = \frac{x}{r} \qquad \sec \theta = \frac{r}{x}$$

$$\tan \theta = \frac{y}{x} \qquad \cot \theta = \frac{x}{y}$$

(See Figure 16.1.)

Figure 16.1

From these definitions we get the following reciprocal relations:

$$\sin \theta = \frac{1}{\csc \theta}, \qquad \cos \theta = \frac{1}{\sec \theta}, \qquad \tan \theta = \frac{1}{\cot \theta} \tag{16.1}$$

For example,

$$\sin \theta = \frac{y}{r} = \frac{1}{\dfrac{r}{y}} = \frac{1}{\csc \theta}$$

$$\cos \theta = \frac{x}{r} = \frac{1}{\dfrac{r}{x}} = \frac{1}{\sec \theta}$$

Since $\tan \theta = y/x = (y/r)/(x/r)$, we get from the definitions of sine and cosine the identity

$$\tan \theta = \frac{y}{x} = \frac{\dfrac{y}{r}}{\dfrac{x}{r}} = \frac{\sin \theta}{\cos \theta} \tag{16.2}$$

Finally, since $\cot \theta = 1/\tan \theta$, we have

$$\cot \theta = \frac{1}{\dfrac{\sin \theta}{\cos \theta}} = \frac{\cos \theta}{\sin \theta} \tag{16.3}$$

Note especially that the secant, cosecant, tangent, and cotangent functions can be expressed in terms of sines and cosines:

$$\sec \theta = \frac{1}{\cos \theta}, \qquad \csc \theta = \frac{1}{\sin \theta}, \qquad \tan \theta = \frac{\sin \theta}{\cos \theta}, \qquad \cot \theta = \frac{\cos \theta}{\sin \theta}$$

It follows that any combination of these functions can be expressed in terms of sines and cosines. For example,

$$2 \tan \theta + \frac{1}{2} \cot \theta = 2 \left(\frac{\sin \theta}{\cos \theta} \right) + \frac{1}{2} \left(\frac{\cos \theta}{\sin \theta} \right) = \frac{2 \sin \theta}{\cos \theta} + \frac{\cos \theta}{2 \sin \theta}$$

Consider another example.

E X A M P L E **1**   Change the following expressions to equivalent expressions involving sines and cosines:

a. $\sec \theta + \tan \theta$ **b.** $\dfrac{1}{\csc \theta} + \cot \theta \sin \theta$

Solution.

a. $\sec \theta + \tan \theta = \dfrac{1}{\cos \theta} + \tan \theta$ **by statement (16.1)**

$\qquad\qquad\qquad = \dfrac{1}{\cos \theta} + \dfrac{\sin \theta}{\cos \theta}$ **by identity (16.2)**

$\qquad\qquad\qquad = \dfrac{1 + \sin \theta}{\cos \theta}$

b. Since $1/\csc \theta = \sin \theta$,

$$\frac{1}{\csc \theta} + \cot \theta \sin \theta = \sin \theta + \cot \theta \sin \theta$$

$$= \sin \theta + \frac{\cos \theta}{\sin \theta} \sin \theta \qquad \textbf{by (16.3):}\quad \cot \theta = \frac{\cos \theta}{\sin \theta}$$

$$= \sin \theta + \cos \theta \qquad\qquad\qquad\qquad\qquad \blacktriangleleft$$

Other Fundamental Identities

The definitions of the trigonometric functions yield other basic relationships, but the derivations can be carried out in a more interesting way. Consider the unit circle ($r = 1$) in Figure 16.2 and any point (a, b) on the circle. Note that

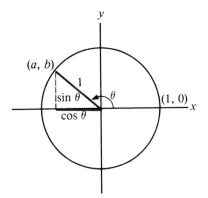

Figure 16.2

$$\cos \theta = \frac{a}{1} \qquad \text{and} \qquad \sin \theta = \frac{b}{1}$$

or

$$a = \cos \theta \qquad \text{and} \qquad b = \sin \theta$$

Since $a^2 + b^2 = 1$, it follows that

$$(\cos \theta)^2 + (\sin \theta)^2 = 1$$

Since we usually write $(\cos \theta)^2 = \cos^2 \theta$ and $(\sin \theta)^2 = \sin^2 \theta$, the identity becomes

$$\sin^2 \theta + \cos^2 \theta = 1 \qquad\qquad\qquad\qquad (16.4)$$

The last identity can also be obtained by observing that

$$(\sin \theta)^2 + (\cos \theta)^2 = \left(\frac{y}{r}\right)^2 + \left(\frac{x}{r}\right)^2$$

$$= \frac{y^2}{r^2} + \frac{x^2}{r^2} = \frac{y^2 + x^2}{r^2} = \frac{r^2}{r^2} = 1$$

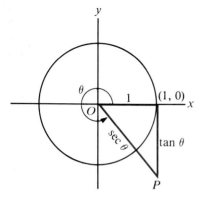

Figure 16.3

Similar identities hold for the remaining trigonometric functions. In Figure 16.3, the y-coordinate of P is equal to $\tan \theta$ and the length of PO is numerically equal to $\sec \theta$. By the Pythagorean theorem

$$1 + \tan^2 \theta = \sec^2 \theta \tag{16.5}$$

Of particular interest in Figure 16.3 is that the names *tangent* and *secant* suggest themselves quite naturally. (Recall that a secant line intersects a circle at two points and a tangent line at one point.)

To obtain the remaining identity,

$$1 + \cot^2 \theta = \csc^2 \theta \tag{16.6}$$

observe that

$$1 + (\cot \theta)^2 = 1 + \left(\frac{x}{y}\right)^2$$

$$= 1 + \frac{x^2}{y^2} = \frac{y^2 + x^2}{y^2} = \frac{r^2}{y^2} = \left(\frac{r}{y}\right)^2 = (\csc \theta)^2$$

E X A M P L E 2   Use the fundamental identities to simplify the given expressions. Write the expressions in terms of sines and cosines, if necessary.

a. $1 - \sin^2 \alpha$ **b.** $\dfrac{\cot \beta}{\csc \beta}$ **c.** $\csc x(1 - \cos^2 x)$ **d.** $\dfrac{\sec^2 \theta - \tan^2 \theta}{\cot \theta}$

Solution.

a. $1 - \sin^2 \alpha = \cos^2 \alpha$

by identity (16.4).

b. $\dfrac{\cot \beta}{\csc \beta} = \dfrac{\cot \beta}{1} \cdot \dfrac{1}{\csc \beta}$

$\qquad = \dfrac{\cos \beta}{\sin \beta} \cdot \dfrac{1}{\csc \beta}$ **by identity (16.3)**

$\qquad = \dfrac{\cos \beta}{\sin \beta} \cdot \dfrac{\sin \beta}{1} = \cos \beta$ **by identity (16.1)**

c. $\csc x(1 - \cos^2 x) = \csc x \sin^2 x$ **by identity (16.4)**

$\qquad\qquad = \dfrac{1}{\sin x} \cdot \dfrac{\sin^2 x}{1}$ **csc x = 1/sin x**

$\qquad\qquad = \sin x$

d. $\dfrac{\sec^2 \theta - \tan^2 \theta}{\cot \theta} = \dfrac{(1 + \tan^2 \theta) - \tan^2 \theta}{\cot \theta}$ $\sec^2 \theta = 1 + \tan^2 \theta$

$$= \dfrac{1}{\cot \theta}$$

$$= \tan \theta \qquad\qquad 1/\cot \theta = \tan \theta \qquad \blacktriangleleft$$

For easy reference, the basic identities are given in the following box.

Fundamental Trigonometric Identities:

$$\sin \theta = \dfrac{1}{\csc \theta} \qquad \cos \theta = \dfrac{1}{\sec \theta} \qquad \tan \theta = \dfrac{1}{\cot \theta}$$

$$\tan \theta = \dfrac{\sin \theta}{\cos \theta} \qquad \cot \theta = \dfrac{\cos \theta}{\sin \theta}$$

$$\sin^2 \theta + \cos^2 \theta = 1$$

$$1 + \tan^2 \theta = \sec^2 \theta$$

$$1 + \cot^2 \theta = \csc^2 \theta$$

Alternate forms of the identities can be obtained by rearranging the terms. The following sets of identities are equivalent:

$$\sin^2 \theta + \cos^2 \theta = 1, \qquad 1 - \cos^2 \theta = \sin^2 \theta, \qquad 1 - \sin^2 \theta = \cos^2 \theta$$

$$1 + \tan^2 \theta = \sec^2 \theta, \qquad \sec^2 \theta - \tan^2 \theta = 1, \qquad \sec^2 \theta - 1 = \tan^2 \theta$$

$$1 + \cot^2 \theta = \csc^2 \theta, \qquad \csc^2 \theta - \cot^2 \theta = 1, \qquad \csc^2 \theta - 1 = \cot^2 \theta$$

E X E R C I S E S / S E C T I O N   **16.1**

In Exercises 1–14, change each expression to an equivalent expression involving sines and cosines. Simplify if possible.

1. $\cot \beta$

2. $\sin \alpha \cot \alpha$

3. $\cos \theta \tan \theta$

4. $\sec \theta$

5. $\cos \gamma \sec \gamma$

6. $1 - \tan \beta \cot \beta$

7. $\cot x + \dfrac{1}{\sin x}$

8. $\dfrac{1}{\sec \omega} + \cos \omega$

9. $\tan \theta \cos \theta \cot \theta$

10. $\cot^2 t \sin^2 t$

11. $\cot^2 s(1 + \tan^2 s)$

12. $\tan^2 x - \sec^2 x$

13. $1 - \sec^2 \theta$

14. $\dfrac{1 + \tan^2 x}{\sec^2 x}$

In Exercises 15–31, use the fundamental identities to simplify each given expression. Convert to an expression involving sines and cosines if necessary.

15. $\dfrac{\cos^2 x + \sin^2 x}{\sin x}$

16. $\dfrac{\tan \beta}{\sec \beta}$

17. $\dfrac{1}{1 - \cos^2 \theta}$

18. $\sin^2 \gamma (1 + \tan^2 \gamma)$

19. $\sin \theta (1 + \cot^2 \theta)$

20. $\csc \beta (1 - \cos^2 \beta)$

21. $\dfrac{\tan \theta \csc \theta}{\sec \theta}$

22. $\dfrac{\sec y}{\csc y}$

23. $1 + \dfrac{\csc^2 t}{\sec^2 t}$

24. $\sin^2 x \sec^2 x$

25. $\dfrac{\sin \theta}{\tan \theta}$

26. $\dfrac{\tan^2 \omega - \sec^2 \omega}{\sec \omega}$

27. $\csc^2 \alpha - \cot^2 \alpha$

28. $\tan^2 y - \dfrac{\sec^2 y}{\csc^2 y}$

29. $\dfrac{1 + \tan^2 x}{\cos x}$

30. $\tan \theta \sin^2 \theta + \tan \theta \cos^2 \theta$

31. $\cot \theta \cos^2 \theta + \cot \theta \sin^2 \theta$

16.2 Proving Identities

In this section we will use the fundamental identities to verify more complicated identities. Writing trigonometric expressions in alternate form is a skill required in more advanced work in mathematics.

In one respect, proving identities is similar to solving word problems: Each identity has its own features and must be verified in its own way. A facility for verifying identities can be developed only through practice. Although no general method can be given, the guidelines that follow will help you decide what approach to take.

Guidelines for Proving Identities:

1. Memorize the fundamental identities and use them whenever possible.
2. Start with the more complicated side of the identity and try to reduce it to the simpler side.
3. Perform any algebraic operation indicated. For example, it may help to multiply out the terms in an expression, to factor an expression, to add fractions, and so on.
4. If everything else fails, try expressing all functions in terms of sines and cosines.
5. When working on one side of the identity, always keep the other side in mind for possible clues on how to proceed.

Caution. When proving an identity, the given relationship may not be treated as an equation—establishing equality is the very purpose of the verification. For this reason, it is not permissible to transpose terms, to multiply both sides by an expression, and so on. Instead, work on one side of the identity until the other side is obtained.

To see how to use the guidelines, consider the identity

$$\cot \theta \sin \theta = \cos \theta$$

In accordance with Guideline 2, start with the left side of the equation, since it is more complicated. While no algebraic operations come to mind (Guideline 3), the identity $\cot \theta = \cos \theta / \sin \theta$ may be useful (Guideline 1). This identity converts the left side to sines and cosines (Guideline 4), so that

$$\cot \theta \sin \theta = \frac{\cos \theta}{\sin \theta} \sin \theta = \cos \theta$$

The identity is thereby verified.

The examples below further illustrate these guidelines.

E X A M P L E **1** Prove the identity:

$$\cos^4 \theta - \sin^4 \theta = \cos^2 \theta - \sin^2 \theta$$

Solution. The left side, which is the more complicated side (Guideline 2), is factorable as a difference of two squares (Guideline 3). Thus

$$\cos^4 \theta - \sin^4 \theta = (\cos^2 \theta)^2 - (\sin^2 \theta)^2$$
$$= (\cos^2 \theta - \sin^2 \theta)(\cos^2 \theta + \sin^2 \theta) \qquad \text{difference of two squares}$$
$$= (\cos^2 \theta - \sin^2 \theta)(1) \qquad \text{replacing } \cos^2 \theta + \sin^2 \theta \text{ by 1}$$
$$= \cos^2 \theta - \sin^2 \theta$$

which is the right side. Note that Guideline 1 was also used. ◀

E X A M P L E **2** Prove the identity:

$$\sin^3 \beta \cos \beta + \cos^3 \beta \sin \beta = \sin \beta \cos \beta$$

Solution. The left side, which is more complicated, contains a common factor $\sin \beta \cos \beta$ (Guideline 3). Thus

$$\sin^3 \beta \cos \beta + \cos^3 \beta \sin \beta$$
$$= \sin \beta \cos \beta(\sin^2 \beta + \cos^2 \beta) \qquad \text{common factor}$$
$$= \sin \beta \cos \beta(\sin^2 \beta + \cos^2 \beta)$$
$$= \sin \beta \cos \beta(1) \qquad \text{from Guideline 1}$$
$$= \sin \beta \cos \beta \qquad \qquad ◀$$

E X A M P L E **3** Show that

$$\left(\csc \gamma + \frac{\cos^2 \gamma}{\sin^3 \gamma} \right) \sin \gamma = \csc^2 \gamma$$

Solution. We multiply the expression on the left side (Guideline 3) to obtain

$$\left(\csc \gamma + \frac{\cos^2 \gamma}{\sin^3 \gamma} \right) \sin \gamma = \csc \gamma \sin \gamma + \frac{\cos^2 \gamma}{\sin^2 \gamma}$$

$$= \frac{1}{\sin \gamma} \sin \gamma + \left(\frac{\cos \gamma}{\sin \gamma} \right)^2 \qquad \text{csc } \gamma = 1/\text{sin } \gamma$$

$$= 1 + \cot^2 \gamma \qquad \qquad \text{cos } \gamma/\text{sin } \gamma = \text{cot } \gamma$$

$$= \csc^2 \gamma \qquad \qquad 1 + \text{cot}^2 \, \gamma = \text{csc}^2 \, \gamma \quad \blacktriangleleft$$

E X A M P L E **4**   Show that

$$\frac{1}{1 - \sin x} + \frac{1}{1 + \sin x} = 2 \sec^2 x$$

Solution. The left side is more complicated and contains two fractions, which should be combined:

$$\frac{1}{1 - \sin x} + \frac{1}{1 + \sin x} \qquad \qquad \text{LCD} = (1 - \text{sin } x)(1 + \text{sin } x)$$

$$= \frac{1(1 + \sin x)}{(1 - \sin x)(1 + \sin x)} + \frac{1(1 - \sin x)}{(1 + \sin x)(1 - \sin x)}$$

$$= \frac{(1 + \sin x) + (1 - \sin x)}{(1 - \sin x)(1 + \sin x)} \qquad \qquad \text{adding numerators}$$

$$= \frac{2}{1 - \sin^2 x}$$

$$= \frac{2}{\cos^2 x} \qquad \qquad \text{since sin}^2 \, x + \text{cos}^2 \, x = 1$$

$$= 2 \left(\frac{1}{\cos^2 x} \right)$$

$$= 2 \left(\frac{1}{\cos x} \right)^2$$

$$= 2(\sec x)^2 \qquad \qquad \text{replacing 1/cos } x \text{ by sec } x$$

$$= 2 \sec^2 x \qquad \qquad \blacktriangleleft$$

E X A M P L E **5**   Show that

$$\tan \theta + \cot \theta = \sec \theta \csc \theta$$

Solution. Since the left side contains two terms, it must be considered the more complicated side. Since no algebraic operations and no fundamental identities come to mind, we write both terms as expressions involving sines and cosines (Guideline 4). Thus

$$\tan \theta + \cot \theta = \frac{\sin \theta}{\cos \theta} + \frac{\cos \theta}{\sin \theta}$$

$$= \frac{\sin \theta \, \sin \theta}{\cos \theta \, \sin \theta} + \frac{\cos \theta \, \cos \theta}{\sin \theta \, \cos \theta} \qquad \text{LCD} = \cos \theta \, \sin \theta$$

$$= \frac{\sin^2 \theta + \cos^2 \theta}{\cos \theta \, \sin \theta} = \frac{1}{\cos \theta \, \sin \theta} \qquad \sin^2 \theta + \cos^2 \theta = 1$$

$$= \frac{1}{\cos \theta} \cdot \frac{1}{\sin \theta} = \sec \theta \, \csc \theta \qquad \blacktriangleleft$$

E X A M P L E   **6**   Show that

$$\frac{\sin^2 \theta}{1 + \cos \theta} = 1 - \cos \theta$$

Solution. Apart from the fact that the left side is more complicated than the right side, none of the guidelines we have used so far seem to apply. Thinking of Guideline 5, see if the right side offers a clue. It does suggest one possibility: Multiply the numerator and denominator of the left side by $1 - \cos \theta$, not only to introduce this expression but also to reduce the denominator. Thus

$$\frac{\sin^2 \theta}{1 + \cos \theta} \cdot \frac{1 - \cos \theta}{1 - \cos \theta} = \frac{\sin^2 \theta (1 - \cos \theta)}{1 - \cos^2 \theta}$$

$$= \frac{\cancel{\sin^2 \theta} (1 - \cos \theta)}{\cancel{\sin^2 \theta}} \qquad 1 - \cos^2 \theta = \sin^2 \theta$$

$$= 1 - \cos \theta$$

Guidelines 1 and 3 could actually be used to prove this identity: Since $\sin^2 \theta + \cos^2 \theta = 1$, the numerator of the left side can be written $\sin^2 \theta = 1 - \cos^2 \theta = (1 - \cos \theta)(1 + \cos \theta)$, so that the factor $1 + \cos \theta$ cancels. However, the above technique using Guideline 5 is needed in Exercises 32 and 33. $\blacktriangleleft$

E X A M P L E   **7**   Show that

$$(\sec \beta - \tan \beta)^2 = \frac{1 - \sin \beta}{1 + \sin \beta}$$

Solution. The most promising approach is to multiply out the expression on the left side (Guideline 3):

$$(\sec \beta - \tan \beta)^2$$

$$= \sec^2 \beta - 2 \sec \beta \tan \beta + \tan^2 \beta \qquad (a - b)^2 = a^2 - 2ab + b^2$$

$$= \frac{1}{\cos^2 \beta} - 2 \left(\frac{1}{\cos \beta} \right) \left(\frac{\sin \beta}{\cos \beta} \right) + \left(\frac{\sin \beta}{\cos \beta} \right)^2 \qquad \begin{matrix} \text{converting to} \\ \text{sines and cosines} \end{matrix}$$

$$= \frac{1}{\cos^2 \beta} - \frac{2 \sin \beta}{\cos^2 \beta} + \frac{\sin^2 \beta}{\cos^2 \beta}$$

$$= \frac{1 - 2 \sin \beta + \sin^2 \beta}{\cos^2 \beta} \qquad \text{combining fractions}$$

$$= \frac{(1 - \sin \beta)^2}{1 - \sin^2 \beta} \qquad \begin{matrix} \text{factoring the numerator} \\ \sin^2 \beta + \cos^2 \beta = 1 \end{matrix}$$

$$= \frac{(1 - \sin \beta)^{\cancel{2}}}{\cancel{(1 - \sin \beta)}(1 + \sin \beta)} \qquad \text{difference of two squares}$$

$$= \frac{1 - \sin \beta}{1 + \sin \beta} \qquad \blacktriangleleft$$

 E X A M P L E   **8**   The current in a certain circuit as a function of time is given by

$$i = \sqrt{0.04 \cos^2 \omega t - 0.04 + 2.0 \sin^2 \omega t}$$

Simplify this expression.

Solution. $i = \sqrt{0.04 \cos^2 \omega t - 0.04 + 2.0 \sin^2 \omega t}$

$$= \sqrt{-0.04(1 - \cos^2 \omega t) + 2.0 \sin^2 \omega t} \qquad \text{common factor } -0.04$$

$$= \sqrt{-0.04 \sin^2 \omega t + 2.0 \sin^2 \omega t} \qquad \begin{matrix} \text{replacing } 1 - \cos^2 \omega t \text{ by} \\ \sin^2 \omega t \end{matrix}$$

$$= \sqrt{1.96 \sin^2 \omega t}$$

$$= 1.4 \sin \omega t \qquad \blacktriangleleft$$

E X E R C I S E S / S E C T I O N   **16.2**

In Exercises 1–40, prove the given identities.

1. $\cot \theta \sin \theta = \cos \theta$

2. $\sin \theta \cot \theta \sec \theta = 1$

3. $\tan \theta \csc \theta = \sec \theta$

4. $\cos \theta + \sin \theta \tan \theta = \sec \theta$

5. $\dfrac{\cos^2 \beta}{\sin \beta} + \sin \beta = \csc \beta$

6. $\sin^3 x + \sin x \cos^2 x = \sin x$

7. $\dfrac{1 - \cos^2 \gamma}{\cos^2 \gamma} = \tan^2 \gamma$

8. $\sin^2 y + \tan^2 y + \cos^2 y = \sec^2 y$

9. $\dfrac{1 + \tan^2 \omega}{1 + \cot^2 \omega} = \tan^2 \omega$

10. $\tan \theta + \cot \theta = \dfrac{\tan \theta}{\sin^2 \theta}$

11. $(1 + \tan^2 x) \cos^2 x = 1$

12. $\dfrac{\csc \theta}{\tan \theta + \cot \theta} = \cos \theta$

13. $\dfrac{1}{\cot \gamma + \tan \gamma} = \sin \gamma \cos \gamma$

14. $(\tan \theta + \cot \theta)^2 = \sec^2 \theta + \csc^2 \theta$

15. $\dfrac{\sin \beta + \tan \beta}{1 + \cos \beta} = \tan \beta$

16. $\dfrac{\sin \theta}{1 + \cos \theta} + \dfrac{\sin \theta}{1 - \cos \theta} = 2 \csc \theta$

17. $\dfrac{\sin \theta}{1 - \cos \theta} - \dfrac{1 - \cos \theta}{\sin \theta} = 2 \cot \theta$

18. $\sec^2 x + \csc^2 x = \sec^2 x \csc^2 x$

19. $\cot^2 \theta - \cos^2 \theta = \cot^2 \theta \cos^2 \theta$

20. $\tan^4 \alpha + \tan^2 \alpha = \sin^2 \alpha \sec^4 \alpha$

21. $\dfrac{1 + \sin \theta}{\cos \theta} + \dfrac{\cos \theta}{1 + \sin \theta} = 2 \sec \theta$

22. $\dfrac{\cos \theta}{\tan \theta + \sec \theta} + \dfrac{\cos \theta}{\tan \theta - \sec \theta} = -2 \sin \theta$

23. $2 \csc x - \cot x \cos x = \sin x + \csc x$

24. $(1 - \cos \beta)(1 + \cos \beta) = \dfrac{1}{1 + \cot^2 \beta}$

25. $\dfrac{1 - \tan \gamma}{1 + \tan \gamma} = \dfrac{\cot \gamma - 1}{\cot \gamma + 1}$

26. $\dfrac{\cos \omega}{\cos \omega - \sin \omega} = \dfrac{1}{1 - \tan \omega}$

27. $\dfrac{\tan \theta}{\csc \theta - \cot \theta} - \dfrac{\sin \theta}{\csc \theta + \cot \theta} = \sec \theta + \cos \theta$

28. $\dfrac{1 - \tan^2 \alpha}{1 + \tan^2 \alpha} = 1 - 2 \sin^2 \alpha$

29. $\dfrac{1 + \tan^2 \theta}{\csc^2 \theta} = \tan^2 \theta$

30. $\dfrac{\cot \theta + \tan \theta}{\sec \theta} = \csc \theta$

31. $\dfrac{1 + \cos \gamma - \sin^2 \gamma}{\sin \gamma (1 + \cos \gamma)} = \cot \gamma$

32. $\dfrac{\cos \theta}{1 - \sin \theta} = \dfrac{1 + \sin \theta}{\cos \theta}$ (See Example 6.)

33. $\sec \theta + \tan \theta = \dfrac{\cos \theta}{1 - \sin \theta}$ (See Example 6.)

34. $\dfrac{1}{\sec \theta - \tan \theta} = \sec \theta + \tan \theta$

35. $\cos^4 x - \sin^4 x = 2 \cos^2 x - 1$

36. $\dfrac{\sin^4 \alpha - \cos^4 \alpha}{\sin^2 \alpha - \cos^2 \alpha} = 1$

37. $\dfrac{\tan \theta + \sec^2 \theta - 1}{\tan \theta - \sec^2 \theta + 1} = \dfrac{\cos \theta + \sin \theta}{\cos \theta - \sin \theta}$

38. $\left(\dfrac{\sec \beta + \csc \beta}{1 + \tan \beta} \right)^2 = \csc^2 \beta$

39. $\dfrac{\tan x - \sin x}{\sin^3 x} = \dfrac{\sec x}{1 + \cos x}$

40. $\dfrac{\tan \alpha + \cot \alpha}{\cos^2 \alpha} - \sin \alpha \sec^3 \alpha = \sec \alpha \csc \alpha$

41. An object traveling along a circle of radius a (in feet) at an angular velocity of $\omega/(2\pi)$ rev/s has linear velocity

$$\sqrt{(a\omega \sin \omega t)^2 + (a\omega \cos \omega t)^2}$$

Simplify this expression. (See Example 8.)

42. Suppose a particle moves along a line with velocity $v = 2 \cos t + 2 \sin t$ (in meters per second). The methods of calculus show that the acceleration is given by $a = -2 \sin t + 2 \cos t$ (in meters per second per second). Show that $a = 0$ whenever $\tan t = 1$.

43. Neglecting air resistance, the equation of the path of a missile projected at velocity v_0 at an angle α with the horizontal is

$$y = x \tan \alpha - \dfrac{gx^2}{2{v_0}^2 \cos^2 \alpha}$$

Write y as a single fraction.

44. In the study of the motion of a pendulum, the expression $\sqrt{1 - k^2 \sin^2 \theta}$ arises. Show that

$$1 - k^2 \sin^2 \theta = k^2 \cos^2 \theta + 1 - k^2$$

45. A beam of length L (in inches) weighing w (pounds per inch) and clamped at the left end is subjected to a compressive force P at the free end. The minimum deflection is given by

$$y_{\min} = \dfrac{wEI}{P^2} \left(1 - \dfrac{1}{2} \theta^2 - \sec \theta + \theta \tan \theta \right)$$

where $\theta = L\sqrt{P/EI}$. Show that

$$y_{\min} = \dfrac{wEI}{P^2} \dfrac{2 \cos \theta - \theta^2 \cos \theta - 2 + 2\theta \sin \theta}{2 \cos \theta}$$

46. If a vertical plate is partly submerged in a liquid, then the capillarity will cause the liquid to rise on the plate to a height of

$$h = c\sqrt{\frac{1 - \sin \theta}{2}}$$

where θ is the contact angle between the liquid and the plate and c a constant that depends on the surface tension

and specific gravity of the liquid. Show that

$$h = \frac{c \cos \theta}{\sqrt{2(1 + \sin \theta)}}$$

47. In some problems on the motion of a pendulum, the expression $1/\sqrt{1 - \cos x}$ arises. Show that this expression is equivalent to $\sqrt{1 + \cos x}/\sin x$.

16.3   The Sum and Difference Formulas

It is sometimes useful to write a trigonometric function of the sum of two angles in terms of trigonometric functions of each angle. For example, $\sin(A + B)$ can be expressed in terms of $\sin A$, $\cos A$, $\sin B$, and $\cos B$.

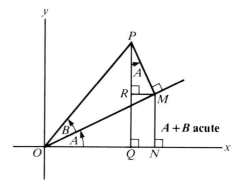

Figure 16.4

To do so, let A and B be two acute angles. Then $A + B$ may be either acute (Figure 16.4) or obtuse (Figure 16.5). In both figures, PQ and MN are perpendicular to the x-axis, PM is perpendicular to OM, and MR is perpendicular to PQ. Note that $\angle MPQ = \angle A$, since the two angles have their sides perpendicular, right side to right side and left side to left side.

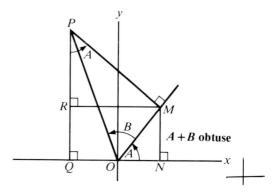

Figure 16.5

In both figures we have

$$\sin(A + B) = \frac{PQ}{OP} = \frac{PR + RQ}{OP} = \frac{PR}{OP} + \frac{RQ}{OP} = \frac{PR}{OP} + \frac{MN}{OP}$$

The last two fractions do not define functions of either A or B. However, if we multiply numerator and denominator of the first fraction by PM and the second by OM, each of the resulting ratios is a function of A or B:

$$\frac{PR}{OP} \cdot \frac{PM}{PM} + \frac{MN}{OP} \cdot \frac{OM}{OM}$$

$$= \frac{PR}{PM} \cdot \frac{PM}{OP} + \frac{MN}{OM} \cdot \frac{OM}{OP} = \cos A \sin B + \sin A \cos B$$

or

$$\sin(A + B) = \sin A \cos B + \cos A \sin B \tag{16.7}$$

For the corresponding identity involving $\cos (A + B)$, we get

$$\cos(A + B) = \frac{OQ}{OP} = \frac{ON - QN}{OP} = \frac{ON}{OP} - \frac{QN}{OP} = \frac{ON}{OP} - \frac{RM}{OP}$$

$$= \frac{ON}{OM} \cdot \frac{OM}{OP} - \frac{RM}{PM} \cdot \frac{PM}{OP}$$

or

$$\cos(A + B) = \cos A \cos B - \sin A \sin B \tag{16.8}$$

Since $\sin(-B) = -\sin B$ and $\cos(-B) = \cos B$ (see Exercise 47, Section 7.1),

$$\sin(A - B) = \sin A \cos B - \cos A \sin B \tag{16.9}$$

and

$$\cos(A - B) = \cos A \cos B + \sin A \sin B \tag{16.10}$$

These four formulas can be written more compactly in the forms given below. (The combination "$\pm$" and "$\mp$" in formula (16.12) indicates that the terms have opposite signs.)

Sum and Difference Formulas:

$$\sin(A \pm B) = \sin A \cos B \pm \cos A \sin B \tag{16.11}$$

$$\cos(A \pm B) = \cos A \cos B \mp \sin A \sin B \tag{16.12}$$

As noted earlier, these identities enable us to express a function of a sum of two angles in terms of functions of the angles themselves. To illustrate these identities, let us find the values of certain trigonometric functions without tables or calculators.

E X A M P L E **1**   Find the exact value of $\cos 75°$ by means of the sum and difference formulas.

Solution. Since $75°$ is not a special angle, $\cos 75°$ cannot be found from a diagram. However, $75° = 30° + 45°$, a sum of two special angles having known function values. So it follows from identity (16.8) that

$$\cos 75° = \cos(30° + 45°)$$
$$= \cos 30° \cos 45° - \sin 30° \sin 45°$$
$$= \frac{\sqrt{3}}{2} \cdot \frac{1}{\sqrt{2}} - \frac{1}{2} \cdot \frac{1}{\sqrt{2}} = \frac{\sqrt{3} - 1}{2\sqrt{2}} \cdot \frac{\sqrt{2}}{\sqrt{2}} = \frac{\sqrt{6} - \sqrt{2}}{4}$$
◀

E X A M P L E **2**   Find the exact value of

$$\sin 25° \cos 20° + \cos 25° \sin 20°$$

Solution. By identity (16.7)

$$\sin 25° \cos 20° + \cos 25° \sin 20° = \sin(25° + 20°)$$
$$= \sin 45° = \frac{\sqrt{2}}{2}$$
◀

The sum and difference identities are sometimes used to combine certain expressions, as shown in the next example.

E X A M P L E **3**   Combine

$$\sin 3x \cos 2x - \cos 3x \sin 2x$$

into a single term.

Solution. By identity (16.9) we get directly

$$\sin(3x - 2x) = \sin x \qquad A = 3x, B = 2x$$
◀

In our study of the graphs of sinusoidal functions, we found that the graph of $y = \sin(x \pm c)$ can be obtained from the graph of $y = \sin x$ by translating the latter graph by c units. If c is a special angle, the relationship between the two functions can be obtained readily by means of the sum and difference identities.

E X A M P L E **4**   Simplify $\sin(x + \pi/2)$.

Solution. By identity (16.7)

$$\sin\left(x + \frac{\pi}{2}\right) = \sin x \cos\frac{\pi}{2} + \cos x \sin\frac{\pi}{2}$$

$$= (\sin x)(0) + (\cos x)(1)$$

$$= \cos x \qquad\qquad\blacktriangleleft$$

E X A M P L E **5**   Simplify $\cos(2x - \pi)$.

Solution. By identity (16.10)

$$\cos(2x - \pi) = \cos 2x \cos \pi + \sin 2x \sin \pi$$

$$= (\cos 2x)(-1) + (\sin 2x)(0)$$

$$= -\cos 2x \qquad\qquad\blacktriangleleft$$

The sum and difference identities for the tangent occur less frequently and are listed mainly for completeness.
By identities (16.7) and (16.8),

$$\tan(A + B) = \frac{\sin (A + B)}{\cos(A + B)} = \frac{\sin A \cos B + \cos A \sin B}{\cos A \cos B - \sin A \sin B}$$

Dividing numerator and denominator by $\cos A \cos B$, we get

$$\tan(A + B) = \frac{\dfrac{\sin A \cos B}{\cos A \cos B} + \dfrac{\cos A \sin B}{\cos A \cos B}}{\dfrac{\cos A \cos B}{\cos A \cos B} - \dfrac{\sin A \sin B}{\cos A \cos B}}$$

or

$$\tan(A + B) = \frac{\tan A + \tan B}{1 - \tan A \tan B} \qquad\qquad (16.13)$$

Similarly,

$$\tan(A - B) = \frac{\tan A - \tan B}{1 + \tan A \tan B} \qquad\qquad (16.14)$$

E X A M P L E **6**   Verify the identity:

$$\tan\left(2x + \frac{\pi}{4}\right) = \frac{1 + \tan 2x}{1 - \tan 2x}$$

Solution. By identity (16.13)

$$\tan\left(2x + \frac{\pi}{4}\right) = \frac{\tan 2x + \tan \dfrac{\pi}{4}}{1 - \tan 2x \tan \dfrac{\pi}{4}} = \frac{\tan 2x + 1}{1 - (\tan 2x) \cdot 1}$$

$$= \frac{1 + \tan 2x}{1 - \tan 2x}$$ ◄

Remark. The identities $\tan \theta = \sin \theta/\cos \theta$ and $\cot \theta = \cos \theta/\sin \theta$ were known to the Arabs. The Hindus knew the fundamental identity $\sin^2 \theta + \cos^2 \theta = 1$, while the formula for $\sin(A + B)$ was discovered by the Belgian mathematician Romanus (1561–1615).

Common error

Writing

$$\sin(A + B) \qquad \text{as} \qquad \sin A + \sin B$$

Instead, $\sin(A + B)$ should be written as

$$\sin A \cos B + \cos A \sin B$$

Similarly,

$$\cos(A + B) \qquad \text{should not be written} \qquad \cos A + \cos B.$$

 E X A M P L E 7   If $i = 3 \sin(\omega t - \pi/2)$ is the current in a circuit and $e = 5 \sin \omega t$ the voltage, find an expression for the power $P = ei$ as a function of time and simplify the result.

Solution. The power is given by

$$P = ei = (5 \sin \omega t)\left[3 \sin\left(\omega t - \frac{\pi}{2}\right)\right]$$

$$= (5 \sin \omega t) \cdot 3\left[\left(\sin \omega t \cos \frac{\pi}{2} - \cos \omega t \sin \frac{\pi}{2}\right)\right]$$

$$= 15 \sin \omega t(\sin \omega t \cdot 0 - \cos \omega t \cdot 1)$$

$$= -15 \sin \omega t \cos \omega t$$ ◄

E X E R C I S E S / S E C T I O N 16.3

In Exercises 1–14, use the sum and difference identities to find each given value without using a table or a calculator. (See Examples 1 and 2.)

1. $\cos 15°$

2. $\sin 105°$

3. $\sin 75°$

4. $\sin 285°$

5. $\cos 165°$

6. $\cos 195°$

7. $\sin(-15°)$

8. $\cos 255°$

9. $\sin 50° \cos 10° + \cos 50° \sin 10°$

10. $\cos 16° \cos 29° - \sin 16° \sin 29°$

11. $\cos 55° \cos 10° + \sin 55° \sin 10°$

12. $\sin 76° \cos 16° - \cos 76° \sin 16°$

13. $\sin 39° \cos 6° + \cos 39° \sin 6°$

14. $\cos 18° \cos 12° - \sin 18° \sin 12°$

In Exercises 15–24, write each expression as a single term. (See Example 3.)

15. $\sin 4x \cos 2x - \cos 4x \sin 2x$

16. $\sin 6x \cos 3x - \cos 6x \sin 3x$

17. $\sin x \cos 2x + \cos x \sin 2x$

18. $\sin 3x \cos x + \cos 3x \sin x$

19. $\cos 3x \cos x + \sin 3x \sin x$

20. $\cos 5x \cos 3x + \sin 5x \sin 3x$

21. $\cos 5x \cos 4x - \sin 5x \sin 4x$

22. $\cos 2x \cos 3x - \sin 2x \sin 3x$

23. $\sin(x + y) \cos y - \cos(x + y) \sin y$

24. $\cos(2x - y) \cos(y - x) - \sin(2x - y) \sin(y - x)$

In Exercises 25–42, write each expression as a function of x or $2x$. (See Examples 4 and 5.)

25. $\cos(x + 30°)$

26. $\sin(x - 30°)$

27. $\cos\left(2x + \dfrac{\pi}{2}\right)$

28. $\sin(x - \pi)$

29. $\cos(x + \pi)$

30. $\cos(x - \pi)$

31. $\sin(2x - 2\pi)$

32. $\cos\left(x - \dfrac{\pi}{2}\right)$

33. $\sin\left(2x + \dfrac{\pi}{2}\right)$

34. $\cos\left(x - \dfrac{\pi}{3}\right)$

35. $\sin\left(x - \dfrac{\pi}{3}\right)$

36. $\sin\left(x + \dfrac{\pi}{6}\right)$

37. $\cos\left(x + \dfrac{\pi}{6}\right)$

38. $\cos\left(2x - \dfrac{\pi}{6}\right)$

39. $\sin\left(x + \dfrac{\pi}{4}\right)$

40. $\cos\left(x - \dfrac{\pi}{4}\right)$

41. $\tan\left(x + \dfrac{\pi}{4}\right)$

42. $\tan\left(2x - \dfrac{\pi}{4}\right)$

In Exercises 43–52, prove the given identities.

43. $\sin\left(\theta - \dfrac{\pi}{4}\right) = -\cos\left(\theta + \dfrac{\pi}{4}\right)$

44. $\sin 2\theta = \sin(\theta + \theta) = 2 \sin \theta \cos \theta$

45. $\cos 2\theta = \cos(\theta + \theta) = \cos^2 \theta - \sin^2 \theta$

46. $\sin(x + y) + \sin(x - y) = 2 \sin x \cos y$

47. $\sin(x + y) - \sin(x - y) = 2 \cos x \sin y$

48. $\sin\left(x + \dfrac{\pi}{6}\right) + \cos\left(x + \dfrac{\pi}{3}\right) = \cos x$

49. $\sin(x + y) \sin(x - y) = \sin^2 x - \sin^2 y$

50. $\cos(x + y) \cos(x - y) = \cos^2 x - \sin^2 y$

51. $\tan(x - y) - \tan(y - x) = \dfrac{2(\tan x - \tan y)}{1 + \tan x \tan y}$

52. $\dfrac{\tan(x + y) - \tan y}{1 + \tan(x + y) \tan y} = \tan x$

In Exercises 53–55, we will obtain a few other standard identities. The first four are known as the *product-to-sum formulas* and the last four as the *sum-to-product formulas.*

53. Derive the following **product-to-sum formulas** by adding and subtracting formulas (16.7) and (16.9):

$$\sin A \cos B = \frac{1}{2}\left[\sin(A + B) + \sin(A - B)\right] \quad (16.15)$$

$$\cos A \sin B = \frac{1}{2}\left[\sin(A + B) - \sin(A - B)\right] \quad (16.16)$$

54. Derive the following **product-to-sum formulas** by adding and subtracting formulas (16.8) and (16.10):

$$\cos A \cos B = \frac{1}{2}\left[\cos(A + B) + \cos(A - B)\right] \quad (16.17)$$

$$\sin A \sin B = \frac{1}{2}\left[\cos(A - B) - \cos(A + B)\right] \quad (16.18)$$

55. The **sum-to-product formulas** can be obtained from identities (16.15) through (16.18) by letting $A + B = x$ and $A - B = y$. Thus

$$A = \frac{x + y}{2} \quad \text{and} \quad B = \frac{x - y}{2}$$

By substituting show that

$$\sin x + \sin y = 2 \sin\left(\frac{x + y}{2}\right) \cos\left(\frac{x - y}{2}\right) \quad (16.19)$$

$$\sin x - \sin y = 2 \cos\left(\frac{x + y}{2}\right) \sin\left(\frac{x - y}{2}\right) \quad (16.20)$$

$$\cos x + \cos y = 2 \cos\left(\frac{x + y}{2}\right) \cos\left(\frac{x - y}{2}\right) \quad (16.21)$$

$$\cos x - \cos y = -2 \sin\left(\frac{x + y}{2}\right) \sin\left(\frac{x - y}{2}\right) \quad (16.22)$$

56. Prove the following identity occurring in the study of alpha particle scattering:

$$\sin \frac{1}{2}(\pi - \theta) = \cos \frac{\theta}{2}$$

57. The equation of a standing wave may be obtained by adding the displacements of two waves of equal amplitude and wavelength but traveling in opposite directions. Given that at some particular instant

$$y_1 = A \sin 2\left(x - \frac{\pi}{4}\right)$$

is the equation of a wave traveling in the positive x-direction and

$$y_2 = A \sin 2\left(x + \frac{\pi}{4}\right)$$

is the equation of the corresponding wave traveling in the negative x-direction, show that $y_1 + y_2 = 0$; that is, that the waves cancel each other at the instant in question.

58. The current in a certain electric circuit is given by

$$i = A \sin\left(\omega t - \frac{\pi}{4}\right) + B \cos\left(\omega t + \frac{\pi}{4}\right)$$

Simplify this expression.

59. If a force $F_0 \cos \omega t$ is applied to a weight oscillating on a spring, then the energy supplied to the system can be written in the form

$$E = A\omega F_0 \cos(\omega t - \gamma) \cos \omega t$$

Show that

$$E = A\omega F_0(\cos^2 \omega t \cos \gamma + \cos \omega t \sin \omega t \sin \gamma)$$

60. A light ray strikes a glass plate of thickness a at an angle of incidence ϕ. If ϕ' is the angle of refraction within the glass, then the lateral displacement D of the emerging beam is given by

$$D = \frac{a \sin(\phi - \phi')}{\cos \phi'}$$

Show that $D = a(\sin \phi - \cos \phi \tan \phi')$.

61. Given that

$$y_1 = A \cos 2\pi\left(\frac{t}{T} - \frac{x}{\lambda}\right)$$

is the equation of a wave traveling in the positive x-direction and

$$y_2 = A \cos 2\pi\left(\frac{t}{T} + \frac{x}{\lambda}\right)$$

is the equation of the corresponding wave traveling in the negative x-direction, find $y = y_1 + y_2$, the equation of a standing wave. (Refer to Exercise 57.)

62. In the development of the theory of Fourier series (see Section 8.5) the product

$$\cos \frac{m\pi t}{p} \cos \frac{n\pi t}{p}$$

has to be written as a sum. Carry out this operation.

63. Show that the product of two complex numbers $r_1/\underline{\theta_1}$ and $r_2/\underline{\theta_2}$ is

$$r_1 r_2[(\cos \theta_1 \cos \theta_2 - \sin \theta_1 \sin \theta_2)$$
$$+ j(\sin \theta_1 \cos \theta_2 + \cos \theta_1 \sin \theta_2)]$$

Simplify this expression to obtain the standard form $r_1 r_2/\underline{\theta_1 + \theta_2}$.

16.4   Double-Angle Formulas

Some special cases of the sum and difference formulas occur often enough to warrant separate classification. One such classification includes the **double-angle formulas**.

Let $A = B$ in the identity

$$\sin(A + B) = \sin A \cos B + \cos A \sin B$$

Then

$$\sin(A + A) = \sin A \cos A + \cos A \sin A$$

or

$$\sin 2A = 2 \sin A \cos A$$

If $A = B$ in the identity

$$\cos(A + B) = \cos A \cos B - \sin A \sin B$$

then

$$\cos(A + A) = \cos A \cos A - \sin A \sin A$$

or

$$\cos 2A = \cos^2 A - \sin^2 A$$

If we let $\cos^2 A = 1 - \sin^2 A$, then $\cos 2A = 1 - \sin^2 A - \sin^2 A = 1 - 2 \sin^2 A$. Similarly, $\cos 2A = \cos^2 A - (1 - \cos^2 A) = 2 \cos^2 A - 1$.

Double-Angle Formulas:

$$\sin 2A = 2 \sin A \cos A \qquad\qquad (16.23)$$

$$\cos 2A = \cos^2 A - \sin^2 A \qquad\qquad (16.24)$$

$$= 2 \cos^2 A - 1 \qquad\qquad (16.25)$$

$$= 1 - 2 \sin^2 A \qquad\qquad (16.26)$$

The double-angle formulas can be used to express the sine or cosine of twice an angle in terms of functions of a single angle. In particular, if $\sin \theta$ or $\cos \theta$ are known, we can use the identities to find $\sin 2\theta$ and $\cos 2\theta$. Consider the following examples.

E X A M P L E **1**   Use the double-angle formulas to find $\sin 2\theta$ and $\cos 2\theta$, given that $\sin \theta = \frac{5}{13}$, θ in quadrant II.

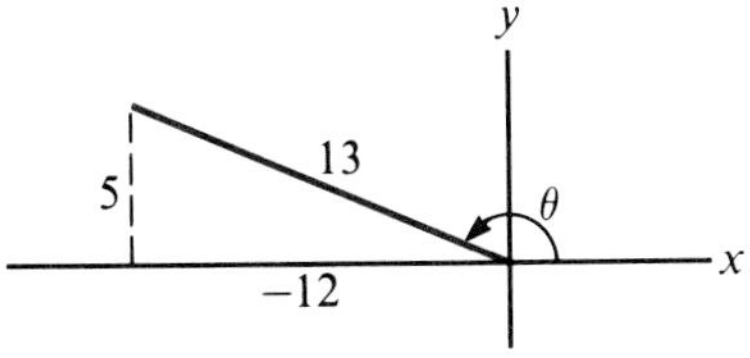

Solution. Since $\sin \theta = \frac{5}{13}$, θ in quadrant II, we construct Figure 16.6 with 5 on the opposite side and 13 on the hypotenuse. From $x^2 + y^2 = r^2$, we have

$$x^2 + 5^2 = 13^2 \qquad \text{or} \qquad x = \pm 12$$

Since θ is in quadrant II, $x = -12$, and

$$\cos \theta = \frac{x}{r} = -\frac{12}{13}$$

Since $\sin \theta = \frac{5}{13}$, we get

$$\sin 2\theta = 2 \sin \theta \cos \theta = 2\left(\frac{5}{13}\right)\left(-\frac{12}{13}\right) = -\frac{120}{169}$$

Figure 16.6

and

$$\cos 2\theta = \cos^2 \theta - \sin^2 \theta = \left(-\frac{12}{13}\right)^2 - \left(\frac{5}{13}\right)^2$$

$$= \frac{144}{169} - \frac{25}{169} = \frac{119}{169} \qquad \blacktriangleleft$$

E X A M P L E **2** Find $\sin 2\theta$ and $\cos 2\theta$, given that $\cos \theta = -\frac{2}{5}$, θ in quadrant III.

Solution. From $\cos \theta = -\frac{2}{5}$, we have $x = -2$ and $r = 5$ (Figure 16.7). Since θ is in quadrant III, $y = -\sqrt{21}$. It follows that $\sin \theta = -\sqrt{21}/5$. Hence

$$\sin 2\theta = 2 \sin \theta \cos \theta = 2\left(-\frac{\sqrt{21}}{5}\right)\left(-\frac{2}{5}\right) = \frac{4\sqrt{21}}{25}$$

and

$$\cos 2\theta = \cos^2 \theta - \sin^2 \theta = \left(\frac{-2}{5}\right)^2 - \left(\frac{-\sqrt{21}}{5}\right)^2 = \frac{4}{25} - \frac{21}{25} = -\frac{17}{25}$$

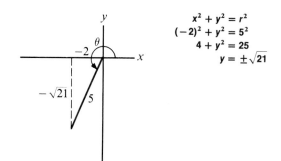

Figure 16.7 $\qquad\qquad\blacktriangleleft$

The double-angle formulas are also applicable to trigonometric functions of multiple angles. For example, from the identity $\sin 2A = 2 \sin A \cos A$, it follows that

$$\sin 16\theta = \sin 2(8\theta) = 2 \sin 8\theta \cos 8\theta$$

Similarly, since $\cos 2A = 2 \cos^2 A - 1$, we have

$$2 \cos^2 6x - 1 = \cos 2(6x) = \cos 12x$$

E X A M P L E **3** Change $\cos^2 4x - \sin^2 4x$ to a single term.

Solution. By formula (16.24), $\cos 2A = \cos^2 A - \sin^2 A$, we get

$$\cos^2 4x - \sin^2 4x = \cos 8x \qquad \textbf{\textit{A} = 4x and 2\textit{A} = 8x}$$

$\qquad\qquad\blacktriangleleft$

E X A M P L E **4**   Prove the identity:

$$\cos^4 2\theta - \sin^4 2\theta = \cos 4\theta$$

Solution.   Factoring the left side, we get

$$
\begin{aligned}
\cos^4 2\theta &- \sin^4 2\theta \\
&= (\cos^2 2\theta)^2 - (\sin^2 2\theta)^2 \qquad &&\textbf{difference of two squares} \\
&= (\cos^2 2\theta - \sin^2 2\theta)(\cos^2 2\theta + \sin^2 2\theta) \\
&= \cos^2 2\theta - \sin^2 2\theta \qquad &&\textbf{cos}^2\ \textbf{2}\theta + \textbf{sin}^2\ \textbf{2}\theta = \textbf{1} \\
&= \cos 2(2\theta) = \cos 4\theta
\end{aligned}
$$

by formula (16.24) with $A = 2\theta$.   ◀

Common error

Equating $\sin 2A$ with $2 \sin A$ and $\cos 2A$ with $2 \cos A$. As we have seen,

$$\sin 2A = 2 \sin A \cos A$$

and

$$\cos 2A = \cos^2 A - \sin^2 A$$

 E X A M P L E **5**   The range R of a projectile fired with velocity v at an angle θ with the ground is given by

$$R = \frac{2v^2}{g} \sin \theta \cos \theta$$

Write R as a single trigonometric function of θ.

Solution. $R = \dfrac{2v^2}{g} \sin \theta \cos \theta$

$$= \frac{v^2}{g} (2 \sin \theta \cos \theta) = \frac{v^2}{g} \sin 2\theta$$

by the double-angle formula (16.23).   ◀

E X E R C I S E S / S E C T I O N   **16.4**

1. Find $\sin 2\theta$, given that $\sin \theta = \frac{3}{5}$, θ in quadrant I.

2. Find $\sin 2\theta$, given that $\sin \theta = \frac{4}{5}$, θ in quadrant II.

3. Find $\cos 2\theta$, given that $\sin \theta = -\frac{3}{5}$, θ in quadrant III.

4. Find $\cos 2\theta$, given that $\cos \theta = \frac{5}{13}$, θ in quadrant I.

5. Find $\sin 2\theta$, given that $\cos \theta = -\frac{12}{13}$, θ in quadrant II.

6. Find $\cos 2\theta$, given that $\sin \theta = -\frac{5}{13}$, θ in quadrant III.

7. Find $\sin 2\theta$, given that $\cos \theta = \frac{1}{2}$, θ in quadrant IV.

8. Find $\cos 2\theta$, given that $\sin \theta = -\frac{1}{2}$, θ in quadrant IV.

9. Find $\cos 2\theta$, given that $\sin \theta = \frac{2}{3}$, θ in quadrant II.

10. Find $\sin 2\theta$, given that $\sin \theta = \frac{1}{3}$, θ in quadrant I.

11. Find $\cos 2\theta$, given that $\cos \theta = -\frac{3}{7}$, θ in quadrant III.

12. Find $\sin 2\theta$, given that $\cos \theta = -\frac{2}{3}$, θ in quadrant II.

In Exercises 13–24, write each expression as a single trigonometric function. (See Examples 3 and 5.)

13. $\cos^2 3y - \sin^2 3y$

14. $\sin^2 x - \cos^2 x$

15. $2 \sin 3\theta \cos 3\theta$

16. $1 - 2 \sin^2 5x$

17. $2 \cos^2 2\beta - 1$

18. $\sin 2x \cos 2x$

19. $1 - 2 \cos^2 4y$

20. $2 \sin^2 A - 1$

21. $\sin 4\omega \cos 4\omega$

22. $\sin 3\theta \cos 3\theta$

23. $4 \sin 2x \cos 2x$

24. $6 \sin 5x \cos 5x$

In Exercises 25–35, prove the given identities.

25. $\cos^4 x - \sin^4 x = \cos 2x$

26. $\sin 2\theta = (\tan \theta)(1 + \cos 2\theta)$

27. $1 - \cos 2\beta = \tan \beta \sin 2\beta$

28. $\sin 2\beta = \dfrac{2 \tan \beta}{1 + \tan^2 \beta}$

29. $\dfrac{\cos 2\theta + \cos \theta + 1}{\sin 2\theta + \sin \theta} = \cot \theta$

30. $\sin 4x = 4 \sin x \cos x \cos 2x$

31. $\dfrac{\cos^2 \gamma + 1}{2 \cos^4 \gamma + \cos^2 \gamma - 1} = \sec 2\gamma$

32. $(\cos x + \sin x)^2 = 1 + \sin 2x$

33. $\dfrac{1 + \cos 2\omega}{\sin 2\omega} = \cot \omega$

34. $\dfrac{\cot^2 y - 1}{2 \cot y} = \cot 2y$

35. $\dfrac{\csc^2 \theta - 2}{\csc^2 \theta} = \cos 2\theta$

36. By letting $A = B$ in identity (16.13), show that

$$\tan 2A = \frac{2 \tan A}{1 - \tan^2 A}$$

which is the *double-angle formula for the tangent.*

37. Suppose a particle is traveling along a line according to the equation $s = 4 \sin^2 t$, where s is measured in meters and t in seconds. Calculus shows that the velocity is given by $v = 8 \sin t \cos t$. Write v as a single trigonometric function of t.

38. Prove the following identity from the derivation of Rutherford's scattering formula:

$$2\pi r^2 \sin \theta = 4\pi r^2 \sin \frac{\theta}{2} \cos \frac{\theta}{2}$$

39. An axle is placed through the center of a circular disk at an angle α. The magnitude T of the torque on the bearings holding the axle has the form $T = k\omega^2 \sin \alpha \cos \alpha$, where ω is the angular velocity. Show that

$$T = \frac{1}{2} k\omega^2 \sin 2\alpha$$

40. The equation of the path of a missile projected at velocity v at an angle θ with the ground is

$$y = x \tan \theta - \frac{gx^2}{2v^2 \cos^2 \theta}$$

Show that

$$y = \frac{x(v^2 \sin 2\theta - gx)}{2v^2 \cos^2 \theta}$$

16.5   Half-Angle Formulas

The identities in the previous section allow us to write a function of $2A$ in terms of functions of A. In this section we will study the **half-angle formulas**, which enable us to express a function of $\frac{1}{2}A$ in terms of functions of A.

The half-angle formulas can be obtained from the double-angle formulas by properly rearranging the terms. If we start with

$$\cos 2x = 1 - 2 \sin^2 x$$

we get

$$2 \sin^2 x = 1 - \cos 2x$$

$$\sin^2 x = \frac{1 - \cos 2x}{2}$$

$$\sin x = \pm \sqrt{\frac{1 - \cos 2x}{2}}$$

Letting $x = A/2$, we have

$$\sin \frac{A}{2} = \pm \sqrt{\frac{1 - \cos A}{2}}$$

Similarly, from $\cos 2x = 2 \cos^2 x - 1$, we obtain

$$\cos \frac{A}{2} = \pm \sqrt{\frac{1 + \cos A}{2}}$$

The algebraic sign depends on the quadrant in which the terminal side of $A/2$ lies.

Half-Angle Formulas:

$$\sin \frac{A}{2} = \pm \sqrt{\frac{1 - \cos A}{2}} \tag{16.27}$$

$$\cos \frac{A}{2} = \pm \sqrt{\frac{1 + \cos A}{2}} \tag{16.28}$$

The half-angle identities can be used to express the sine and cosine of a given half-angle in terms of the cosine of the angle, as shown in the first two examples.

E   X   A   M   P   L   E   **1**   Use the appropriate half-angle formula to find the exact value of $\cos 165°$.

Solution. By identity (16.28)

$$\cos 165° = \pm \sqrt{\frac{1 + \cos[(2)(165°)]}{2}} = \pm \sqrt{\frac{1 + \cos 330°}{2}}$$

$$= \pm \sqrt{\frac{1 + \frac{\sqrt{3}}{2}}{2}} = \pm \sqrt{\frac{1 + \frac{\sqrt{3}}{2}}{2} \cdot \frac{2}{2}}$$

$$= \pm \sqrt{\frac{2 + \sqrt{3}}{4}} = \pm \frac{\sqrt{2 + \sqrt{3}}}{2}$$

Since $165°$ is in the second quadrant, $\cos 165°$ is negative. So

$$\cos 165° = - \frac{\sqrt{2 + \sqrt{3}}}{2}$$

◀

E X A M P L E **2**   Find $\cos \theta/2$, given that $\sin \theta = -\frac{12}{13}$, θ in quadrant IV.

Solution. If $\sin \theta = -\frac{12}{13}$, then $\cos \theta = \frac{5}{13}$ for θ in quadrant IV. (See Figure 16.8). Moreover, since $270° < \theta < 360°$, it follows that $135° < \theta/2 < 180°$. Thus $\cos \theta/2 < 0$ and

Figure 16.8

$$\cos \frac{\theta}{2} = -\sqrt{\frac{1 + \cos \theta}{2}} = -\sqrt{\frac{1 + \frac{5}{13}}{2}} = -\sqrt{\frac{1 + \frac{5}{13}}{2} \cdot \frac{13}{13}}$$

$$= -\sqrt{\frac{13 + 5}{26}} = -\sqrt{\frac{9}{13}}$$

$$= -\frac{3}{\sqrt{13}} = -\frac{3\sqrt{13}}{13}$$

◀

Sometimes the half-angle formulas are used to simplify certain radical expressions.

E X A M P L E **3**   Simplify $\sqrt{6 - 6 \cos 4\theta}$.

Solution. The expression resembles the right side of identity (16.27). To get the proper form, we need to remove 6 from the radical and obtain a 2 in the denominator. Thus

$$\sqrt{6 - 6 \cos 4\theta} = \sqrt{6(1 - \cos 4\theta)} = \sqrt{6}\sqrt{1 - \cos 4\theta}$$

$$= \sqrt{6} \sqrt{\frac{2(1 - \cos 4\theta)}{2}}$$

$$= \sqrt{6}\sqrt{2} \sqrt{\frac{1 - \cos 4\theta}{2}} = \sqrt{12} \sin 2\theta$$

$$= 2\sqrt{3} \sin 2\theta$$

◀

Alternate Forms

In the study of calculus, the forms of the half-angle formulas stated below are sometimes more useful.

$$\sin^2 A = \frac{1 - \cos 2A}{2} \tag{16.29}$$

$$\cos^2 A = \frac{1 + \cos 2A}{2} \tag{16.30}$$

Note that these formulas are really the double-angle identities (16.25) and (16.26) slightly rewritten.

E X A M P L E 4   Write $4 \sin^2 3x$ without the square.

Solution. By identity (16.29) with $A = 3x$,

$$4 \sin^2 3x = 4\left(\frac{1 - \cos 6x}{2}\right) = 2(1 - \cos 6x)$$

◀

E X A M P L E 5   Write $6 \cos^2 2x$ without the square.

Solution. By identity (16.30) with $A = 2x$,

$$6 \cos^2 2x = 6\left(\frac{1 + \cos 4x}{2}\right) = 3(1 + \cos 4x)$$

◀

E X A M P L E 6   Prove the identity:

$$2 \cos^2 \frac{\theta}{2} = \frac{\sin^2 \theta}{1 - \cos \theta}$$

Solution. Since $\sin^2 \theta = 1 - \cos^2 \theta$, we have

$$\frac{\sin^2 \theta}{1 - \cos \theta} = \frac{1 - \cos^2 \theta}{1 - \cos \theta} = \frac{(1 - \cos \theta)(1 + \cos \theta)}{1 - \cos \theta} \qquad \text{difference of two squares}$$

$$= 1 + \cos \theta = \frac{2(1 + \cos \theta)}{2} = 2\left(\frac{1 + \cos \theta}{2}\right)$$

$$= 2 \cos^2 \frac{\theta}{2} \qquad \text{by (16.30) with } A = \frac{\theta}{2}$$

◀

E X A M P L E 7   The motion of a planet or comet about the sun can be described by an equation of the form

$$r = \frac{c}{a - b \cos \theta}$$

where a, b, and c are constants. (See Figure 16.9.) This equation represents a conic section. For example, the equation of the elliptic path of Mercury is

$$r = \frac{3.442 \times 10^7}{1 - 0.206 \cos \theta}$$

where r is measured in miles.

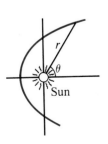

Figure 16.9

Some comets follow a path that is nearly parabolic:

$$r = \frac{A}{1 - \cos \theta}$$

Show that

$$r = \frac{A}{2} \csc^2 \frac{\theta}{2}$$

Solution. $r = \dfrac{A}{1 - \cos \theta} = \dfrac{A}{2} \cdot \dfrac{2}{1 - \cos \theta} = \dfrac{A}{2} \cdot \dfrac{1}{\dfrac{1 - \cos \theta}{2}}$

$$= \frac{A}{2} \cdot \frac{1}{\sin^2 \dfrac{\theta}{2}} = \frac{A}{2} \csc^2 \frac{\theta}{2}$$

◀

E X E R C I S E S / S E C T I O N   16.5

In Exercises 1–5, find the exact value of each trigonometric function by means of the half-angle formulas.

1. $\sin 15°$ **2.** $\cos 75°$ **3.** $\cos 22.5°$

4. $\cos 105°$ **5.** $\sin 112.5°$

6. Find $\sin(\theta/2)$, given that $\cos \theta = \frac{4}{5}$, θ in quadrant I.

7. Find $\sin(\theta/2)$, given that $\cos \theta = -\frac{24}{25}$, θ in quadrant III.

8. Find $\cos(\theta/2)$, given that $\sin \theta = \frac{3}{5}$, θ in quadrant II.

9. Find $\sin(\theta/2)$, given that $\cos \theta = \frac{5}{13}$, θ in quadrant IV.

10. Find $\cos(\theta/2)$, given that $\sin \theta = -\frac{7}{25}$, θ in quadrant IV.

11. Find $\cos(\theta/2)$, given that $\cos \theta = \frac{12}{13}$, θ in quadrant IV.

12. Find $\sin(\theta/2)$, given that $\cos \theta = -\frac{3}{5}$, θ in quadrant III.

13. Find $\cos(\theta/2)$, given that $\sin \theta = -\frac{5}{13}$, θ in quadrant III.

14. Find $\cos(\theta/2)$, given that $\sin \theta = \frac{24}{25}$, θ in quadrant II.

In Exercises 15–24, simplify each given expression. (See Example 3.)

15. $\sqrt{\dfrac{1 - \cos 4\theta}{2}}$

16. $\sqrt{\dfrac{1 + \cos 6\theta}{2}}$

17. $\sqrt{1 + \cos 6\theta}$

18. $\sqrt{4 - 4 \cos 8\theta}$

19. $\sqrt{5 - 5 \cos 4\theta}$

20. $\sqrt{6 + 6 \cos 8\theta}$

21. $\sqrt{2 + 2 \cos 6\theta}$

22. $\sqrt{3 - 3 \cos 4\theta}$

23. $\sqrt{7 - 7 \cos \theta}$

24. $\sqrt{9 + 9 \cos \theta}$

In Exercises 25–32, eliminate the exponent. (See Examples 4 and 5.)

25. $\sin^2 4x$ **26.** $\sin^2 3x$ **27.** $\cos^2 2x$

28. $\cos^2 3x$ **29.** $2 \sin^2 3x$ **30.** $4 \cos^2 4x$

31. $12 \sin^2 x$ **32.** $2 \cos^2 x$

In Exercises 33–36, prove the given identities.

33. $\dfrac{\sin 2\theta}{2 \sin \theta} = \cos^2 \dfrac{\theta}{2} - \sin^2 \dfrac{\theta}{2}$

34. $\cos^2 \dfrac{x}{4} = \dfrac{1 + \cos \dfrac{x}{2}}{2}$

35. $\csc^2 \theta = \dfrac{2}{1 - \cos 2\theta}$

36. $2 \cos \dfrac{\beta}{2} = (1 + \cos \beta) \sec \dfrac{\beta}{2}$

37. Simplify the following expression from a problem in the study of the pendulum:

$$\frac{1}{\sqrt{1 - \cos x}}$$

38. In the study of the motion of a pendulum, the expression

$$A = \sqrt{\frac{1 - \cos \theta}{1 - \cos \alpha}}$$

needs to be simplified. Show that

$$A = \frac{\sin \dfrac{\theta}{2}}{\sin \dfrac{\alpha}{2}}$$

39. In determining the length of the path along which a particle will slide from a higher to a lower point in minimum time, the expression $\sqrt{2 - 2\cos \theta}$ needs to be simplified. Carry out this simplification.

40. A common exercise in calculus is determining the area under a curve. To find the area under one arch of the

curve $y = \sin^2 x$, the equation must be written without the exponent. Rewrite this equation.

41. The index of refraction n of a prism with apex angle A whose minimum angle of refraction is δ is given by

$$n = \frac{\sin \dfrac{A + \delta}{2}}{\sin \dfrac{A}{2}}, \qquad n \geq 0$$

Show that the expression is equivalent to

$$n = \sqrt{\frac{1 + \sin A \sin \delta - \cos A \cos \delta}{1 - \cos A}}$$

16.6 Trigonometric Equations

So far we have concentrated only on identities, equations that are valid for all values of the variable. Now we will turn to **conditional equations**, which are valid only for certain values of the angle.

For example, the equation

$$\sin \theta = 0$$

is not an identity, since equality holds only if

$$\theta = 0°, \pm 180°, \pm 360°, \text{ and so on}$$

To solve an equation containing a single trigonometric function, we solve the equation for this function and then determine the values of the angle for which equality holds. Consider the next example.

E X A M P L E **1** Solve the equation $2 \cos x - 1 = 0, 0 \leq x < 2\pi$.

Solution. The first step is to solve the given equation for $\cos x$. Thus

$$2 \cos x - 1 = 0 \qquad \textbf{given equation}$$
$$2 \cos x = 1 \qquad \textbf{transposing } -1$$
$$\cos x = \frac{1}{2} \qquad \textbf{dividing by 2}$$

The angles between 0 and 2π whose cosine is $\frac{1}{2}$ are 60° and 300°. In radians,

$$x = \frac{\pi}{3} \qquad \text{and} \qquad x = \frac{5\pi}{3}$$

Substituting into the given equation shows that the solutions check. ◄

If an equation involves different functions, we can often use the identities to convert it to an equation involving only one function, as shown in the next example.

E X A M P L E **2**   Solve the equation $\sec^2 x - 4 \tan^2 x = 0$, $0 \le x < 2\pi$.

Solution. Since the equation involves two different functions, no direct solution is possible. However, if we recall that $1 + \tan^2 x = \sec^2 x$, we can convert one of the functions. Thus

$$\sec^2 x - 4 \tan^2 x = 0$$

$$1 + \tan^2 x - 4 \tan^2 x = 0$$

$$1 - 3 \tan^2 x = 0$$

$$\tan^2 x = \frac{1}{3} \qquad \text{solving for } \tan^2 x$$

$$\sqrt{\tan^2 x} = \pm \sqrt{\frac{1}{3}} \qquad \text{taking square roots}$$

$$\tan x = \pm \frac{1}{\sqrt{3}}$$

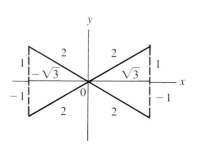

Figure 16.10

It follows that $x = 30°$, $150°$, $210°$, and $330°$. (See Figure 16.10.) In radian measure

$$x = \frac{\pi}{6}, \frac{5\pi}{6}, \frac{7\pi}{6}, \frac{11\pi}{6}$$

◀

Quadratic forms

Some trigonometric equations are actually in quadratic form, as shown in the next example.

E X A M P L E **3**   Solve the equation $2 \csc^2 x + 3 \csc x - 2 = 0$, $0 \le x < 2\pi$.

Solution. Let $y = \csc x$. Then the equation becomes

$$2y^2 + 3y - 2 = 0 \qquad \text{quadratic form}$$

$$(2y - 1)(y + 2) = 0 \qquad \text{factoring}$$

$$y = -2, \frac{1}{2} \qquad \text{setting each factor equal to zero}$$

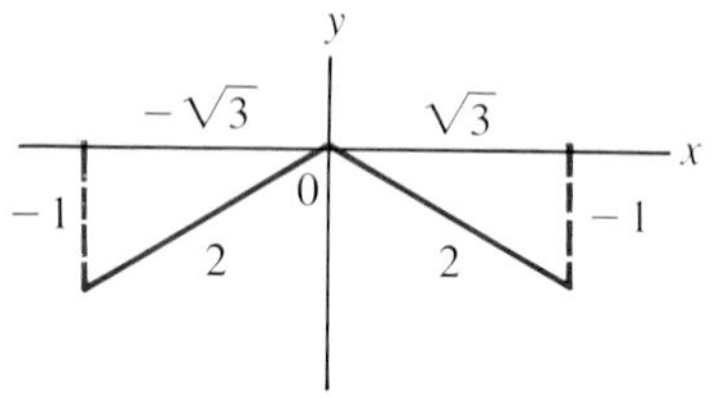

It follows from $y = \csc x$ that

$$\csc x = -2 \qquad \text{and} \qquad \csc x = \frac{1}{2}$$

Since a value of $\csc x$ cannot be less than unity, the equation $\csc x = \frac{1}{2}$ has no solution. From $\csc x = -2$, we obtain $x = 210°$ and $330°$ (Figure 16.11). In

Figure 16.11

radian measure

$$x = \frac{7\pi}{6}, \frac{11\pi}{6}$$ ◀

In most cases, equations involving functions of multiple angles should be solved by using an appropriate identity, as shown in the next example.

E X A M P L E **4**   Solve the equation $\cos 2x - \cos x = 0, 0 \le x < 2\pi$.

Solution. Because of the double angle, $\cos 2x$ must first be changed to a function of x by one of the double-angle formulas. (*Reminder:* $\cos 2x \ne 2 \cos x$.) Since the other term on the left is $-\cos x$, we choose the formula $\cos 2A = 2 \cos^2 A - 1$, so that the equation contains only cosines. Thus

$$\cos 2x - \cos x = 0$$

$$2 \cos^2 x - 1 - \cos x = 0$$

$$2 \cos^2 x - \cos x - 1 = 0$$

$$(2 \cos x + 1)(\cos x - 1) = 0 \qquad 2z^2 - z - 1 = (2z + 1)(z - 1)$$

$$\cos x = -\frac{1}{2}, 1 \qquad \text{setting each factor equal to zero}$$

From $\cos x = -\frac{1}{2}$, we obtain $x = 120°, 240°$. From $\cos x = 1$, we get $x = 0°$.
In radians

$$x = 0, \frac{2\pi}{3}, \frac{4\pi}{3}$$ ◀

E X A M P L E **5**   Use a calculator to solve the equation

$$4 \sin 2x - 5 \sin x = 0$$

to the nearest tenth of a degree ($0° \le x < 360°$).

Solution. By the double-angle formula for the sine function, $\sin 2x = 2 \sin x \cos x$, we get

$$4 \sin 2x - 5 \sin x = 0 \qquad \text{given equation}$$

$$4(2 \sin x \cos x) - 5 \sin x = 0$$

$$8 \sin x \cos x - 5 \sin x = 0$$

$$\sin x(8 \cos x - 5) = 0 \qquad \text{common factor } \sin x$$

$$\sin x = 0 \qquad 8 \cos x - 5 = 0$$

$$\cos x = \frac{5}{8}$$

From $\sin x = 0$, we get $x = 0°$, $180°$. Using a calculator, $\cos x = \frac{5}{8}$ yields $x = 51.3°$, $308.7°$ (since $\cos x > 0$ in quadrants I and IV). ◄

 E X A M P L E **6** The range R (in feet) of a projectile hurled at an angle θ with the horizontal at velocity v (in feet per second) is given by

$$R = \frac{2v^2 \cos \theta \sin \theta}{g}$$

where $g = 32$ ft/s². (See Figure 16.12.) If $v = 40$ ft/s, determine the angle θ at which the projectile has to be aimed to hit an object 45 ft away.

Figure 16.12

Solution. Substituting into the given equation, we get

$$\frac{2(40)^2 \cos \theta \sin \theta}{32} = 45$$

$$\frac{(40)^2}{32}(2 \sin \theta \cos \theta) = 45$$

$$2 \sin \theta \cos \theta = \frac{(45)(32)}{(40)^2}$$

$$\sin 2\theta = 0.9$$

$$2\theta = 64°, 116°$$

$$\theta = 32°, 58°$$

So the projectile can be aimed at either 32° or 58° to land 45 ft away. ◄

E X E R C I S E S / S E C T I O N   **16.6**

In Exercises 1–31, solve the given equations for x, $0 \le x < 2\pi$.

1. $2 \sin x - 1 = 0$

2. $3 \sin x + 3 = 0$

3. $4 \tan x + 4 = 8$

4. $2 \sec x + 4 = 0$

5. $\cos^2 x - 1 = 0$

6. $2 \sin^2 x - 1 = 0$

7. $\sin^2 x - \sin x = 0$

8. $\tan x(\csc x + 1) = 0$

9. $4 \cos^2 x - 3 = 0$

10. $(\sec x - 1)(\tan x + 1) = 0$

11. $(\cot x - 1)(\cos x + 1) = 0$

12. $(2 \cos x - 1)(\csc x - 2) = 0$

13. $2 \sin^2 x - \sin x - 1 = 0$

14. $2 \sin^2 x - \sin x - 3 = 0$

15. $3 \cos^2 x - 7 \cos x + 4 = 0$

16. $2 \sec^2 x + 3 \sec x - 2 = 0$

17. $\sin x - \cos x = 0$

18. $\cot^2 x - \tan^2 x = 0$

19. $2 \sin^2 x + \cos x + 1 = 0$

20. $6 \cos^2 x - 5 \sin x - 2 = 0$

21. $2 \tan^2 x - \sec^2 x = 0$ **22.** $\cos 2x = 1$

23. $\cos 2x = 0$ **24.** $\sin x + \sin 2x = 0$

25. $\sin^2 x + \cos 2x = 0$ **26.** $\sin 2x + \cos x = 0$

27. $\cos 2x + \cos x = 0$ **28.** $\cos 2x - \sin x = 0$

29. $\sin \dfrac{x}{2} = \cos \dfrac{x}{2}$ **30.** $\sin 2x = 0$

31. $\sin x \cos x - \sin 2x = 0$

In Exercises 32–40, use a calculator to solve the given equations to the nearest tenth of a degree ($0 \le x < 360°$).

32. $3 \sin x \cos x - \cos x = 0$ **33.** $\tan^2 x - 2 = 0$

34. $2 \cos^2 x = 1 + \sin^2 x$ **35.** $2 \sin 2x + \cos x = 0$

36. $\sec^2 x - 2 \tan x - 4 = 0$ **37.** $2 \sin 2x = 3 \sin x$

38. $\cos^2 x - 2 \sin^2 x = 0$ **39.** $\csc^2 x - 3 \cot^2 x = 0$

40. $5 \sin^2 x + 8 \sin x - 4 = 0$

41. Certain problems in mechanics are simplified by rotating the coordinate axes. In the process, the following equa-

tion has to be solved:

$$2(C - A) \sin \theta \cos \theta + B(\cos^2 \theta - \sin^2 \theta) = 0$$

Solve this equation for θ ($0 \le \theta < 180°$), given that $A = B = 1$ and $C = 0$.

42. Suppose a projectile hurled at a velocity of 80 ft/s is to hit a target 100 ft away. At what angle with respect to the ground does the projectile have to be hurled? (See Example 6.)

43. The current in a certain circuit is given by

$$i = e^{-5t}(\cos 4.0t - \sqrt{3} \sin 4.0t)$$

Find the smallest positive value of t (in seconds) for which the current is zero.

44. For a certain mass oscillating on a spring, the vertical displacement is given by

$$x = 2.0 \cos 2t - 1.0 \sin 2t, \qquad t \ge 0$$

where x is measured in centimeters and t in seconds. Find the smallest value of t for which the displacement is zero. (Set your calculator in the radian mode.)

45. Starting at $t = 0$, the current in a circuit is

$$i = 2 \sin^2 \omega t + 3 \sin \omega t$$

Find the smallest value of t (in seconds) for which $i = 2$ A.

16.7   Inverse Trigonometric Relations

We know from our study of equations that it is often desirable to solve a given equation for one of the variables in terms of the other variables. To solve a trigonometric equation $y = f(x)$ for the variable x, we need the concept of an *inverse trigonometric relation*.

Consider, for example, the function $y = \sin x$. To solve this equation for x in terms of y, we introduce the following notation:

$$x = \arcsin y$$

Expressed verbally, "x is a number (or angle measure) whose sine is y." Following the usual convention of placing y on the left side of the equal sign, we write

$$y = \arcsin x \tag{16.31}$$

The equation says that y is a number (or angle measure) whose sine is x; $y = \arcsin x$ is called an **inverse trigonometric relation**.

To illustrate the meaning of this kind of notation, let us evaluate $y = \arcsin x$ for certain values of x.

E X A M P L E **1**   Find y if $y = \arcsin \frac{1}{2}$, $0 \le y < 2\pi$.

Solution. By the definition of arcsin $\frac{1}{2}$ we need to find a number y whose sine is $\frac{1}{2}$. Two such numbers exist between 0 and 2π, namely,

$$y = \frac{\pi}{6} \quad \text{and} \quad y = \frac{5\pi}{6}$$

As a check, observe that $\sin(\pi/6) = \frac{1}{2}$ and $\sin(5\pi/6) = \frac{1}{2}$.   ◄

E X A M P L E **2**   Find y if $y = \arcsin 0$.

Solution. We are looking for all angles y such that $\sin y = 0$. We know that

$$\sin 0 = 0, \quad \sin(\pm \pi) = 0, \quad \sin(\pm 2\pi) = 0, \quad \text{and so on}$$

So

$$y = n\pi, \quad n = 0, \pm 1, \pm 2, \ldots$$ ◄

Remark. Recall that a relation between two variables x and y is called a **function** if for every value of x there exists a unique value of y, denoted by $y = f(x)$. By this definition, $y = \arcsin x$ is *not* a function, as we can see from Example 2: The value $x = 0$ does not yield a unique value for y. To obtain a function, the values of y must be suitably restricted. That is the topic of the next section.

The other trigonometric functions have similar inverse relations, as shown in the next two examples.

E X A M P L E **3**   Find y if $y = \arccos(\sqrt{3}/2)$, $0 \le y < 2\pi$.

Solution. The notation $\arccos(\sqrt{3}/2)$ has an analogous meaning as an angle y whose cosine is $\sqrt{3}/2$; that is, $\cos y = \sqrt{3}/2$. For $0 \le y < 2\pi$ we have

$$y = \frac{\pi}{6} \quad \text{and} \quad y = \frac{11\pi}{6} \qquad \textbf{cos y > 0 in quadrants I and IV}$$ ◄

E X A M P L E **4**   Find y if $y = \arctan(-1)$, $0 \le y < 2\pi$.

Solution. Between 0 and 2π the only angles y whose tangents are equal to -1 are

$$y = \frac{3\pi}{4} \quad \text{and} \quad y = \frac{7\pi}{4} \qquad \textbf{tan y < 0 in quadrants II and IV}$$ ◄

As noted at the beginning of this section, the notation for the inverse relationship enables us to solve a trigonometric equation for x in terms of y, as shown in the next example.

E X A M P L E **5**   Solve the equation $y = 1 + \sin 2x$ for x in terms of y.

Solution. The equation $y = 1 + \sin 2x$ can also be written

$$\sin 2x = y - 1$$

Using the inverse relationship,

$$2x = \arcsin(y - 1)$$

we get

$$x = \frac{1}{2}\arcsin(y - 1) \qquad\qquad\blacktriangleleft$$

E X E R C I S E S / S E C T I O N **16.7**

In Exercises 1–16, find $y(0 \le y < 2\pi)$ without using a table or calculator.

1. $y = \arcsin \dfrac{\sqrt{3}}{2}$

2. $y = \arcsin(-1)$

3. $y = \arctan 1$

4. $y = \text{arccot}(-1)$

5. $y = \text{arccsc}\, 1$

6. $y = \arccos(-1)$

7. $y = \arcsin 0$

8. $y = \arcsin\left(-\dfrac{1}{2}\right)$

9. $y = \arccos \dfrac{1}{2}$

10. $y = \arcsin\left(-\dfrac{\sqrt{2}}{2}\right)$

11. $y = \arccos 0$

12. $y = \text{arcsec}\, 1$

13. $y = \text{arccsc}(-2)$

14. $y = \arctan\left(-\dfrac{1}{\sqrt{3}}\right)$

15. $y = \text{arccot}\, \dfrac{1}{\sqrt{3}}$

16. $y = \arctan 0$

In Exercises 17–30, solve each equation for x in terms of y.

17. $y = \sin 3x$

18. $y = \cos 4x$

19. $y = 1 - \tan 2x$

20. $y = \sec 2x + 1$

21. $y = 2 + \dfrac{1}{3}\csc x$

22. $y = 1 - 2\cot x$

23. $y = 4\cos(x - 2)$

24. $y = 5\sin(x + 4)$

25. $y = \arctan x$

26. $y = \arccos 3x$

27. $y = 1 - \arcsin x$

28. $y = 2 + \text{arcsec}\, x$

29. $y = \text{arccsc}(x + 1)$

30. $y = 3\,\text{arccot}\, 3x$

16.8   Inverse Trigonometric Functions

We learned in our study of logarithms that the equation $y = b^x$ can be written $x = \log_b y$. While the two equations mean the same thing, the first expresses y as a function of x and the second expresses x as a function of y; $y = b^x$ and $y = \log_b x$ are called **inverse functions**. An analogous situation exists in trigonometry in the sense that every trigonometric function has an inverse function.

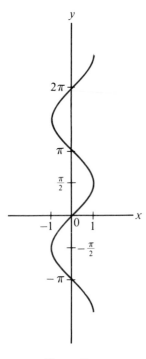

Figure 16.13

In the last section we introduced the customary notation for inverse trigonometric relations. We also noted that a relation such as $y = \arcsin x$ does not represent a function. Given the importance of the function concept, this state of affairs is unsatisfactory. The variable y must be suitably restricted so that every value of x yields a unique value of y. This restriction leads to the definition of an **inverse trigonometric function.**

First we need to consider the graph of the relation $y = \arcsin x$. Writing this equation in the form $x = \sin y$, we get the graph of the sine function with x and y interchanged, as shown in Figure 16.13. This graph shows why the relation $y = \arcsin x$ is not a function: For every x such that $-1 \le x \le 1$, we get infinitely many values for y.

We can also see from the graph that y becomes unique if all but a small section of the graph is eliminated. This elimination can be done in several ways. The restriction that has become standard is $-\pi/2 \le y \le \pi/2$, which corresponds to the portion of the graph through the origin, drawn as the solid curve in Figure 16.14. To distinguish between the solid portion and the entire curve, the equation of the solid curve is written

$$y = \text{Arcsin } x$$

using the capital letter A. Note especially that

$y = \text{Arcsin } x$ is a function

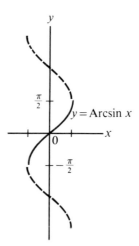

Figure 16.14

E X A M P L E **1**   Find the exact value of y in each case:

a. $y = \arcsin \dfrac{1}{2} (0 \le y < 2\pi)$ **b.** $y = \text{Arcsin } \dfrac{1}{2}$

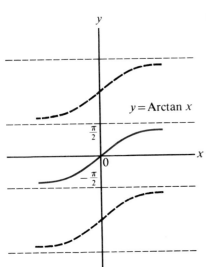

$y = \text{Arctan } x$

Figure 16.15

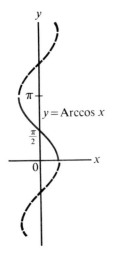

$y = \text{Arccos } x$

Figure 16.16

Solution.

a. As we saw in the previous section,

$$y = \frac{\pi}{6} \quad \text{and} \quad y = \frac{5\pi}{6}$$

b. From the condition $-\pi/2 \le y \le \pi/2$, y has to be an angle between $-\pi/2$ and $\pi/2$. So the only permissible value is

$$y = \frac{\pi}{6}$$

Thus $\text{Arcsin } \frac{1}{2} = \pi/6$, a unique value.   ◀

The inverse functions corresponding to $y = \arctan x$ and $y = \arccos x$ are obtained from the graphs of $x = \tan y$ and $x = \cos y$, shown in Figure 16.15 and Figure 16.16, respectively.

Following the usual conventions, $y = \text{Arctan } x$ is the solid curve in Figure 16.15 and $y = \text{Arccos } x$ the solid curve in Figure 16.16. Note that in all cases the angle y is in quadrant I whenever x is positive. The different cases are summarized next.

Inverse Trigonometric Functions:

$y = \text{Arcsin } x,$	$-\pi/2 \le y \le \pi/2$	(16.32)
$y = \text{Arctan } x,$	$-\pi/2 < y < \pi/2$	(16.33)
$y = \text{Arccos } x,$	$0 \le y \le \pi$	(16.34)

Although inverse trigonometric functions exist for the remaining functions, we will confine ourselves to the cases already presented. For completeness, however, let us list the most commonly accepted ranges:

$$y = \text{Arccsc } x, \qquad -\frac{\pi}{2} \le y \le \frac{\pi}{2}, y \ne 0$$

$$y = \text{Arccot } x, \qquad 0 < y < \pi$$

$$y = \text{Arcsec } x, \qquad 0 \le y \le \pi, y \ne \frac{\pi}{2}$$

E X A M P L E **2**   Find the exact values of

a. $\text{Arccos } \dfrac{1}{2}$ b. $\text{Arccos}\left(-\dfrac{1}{2}\right)$

Solution.

a. Since $x = \frac{1}{2}$ is positive, Arccos $\frac{1}{2}$ is in quadrant I. Thus

$$\text{Arccos}\,\frac{1}{2} = \frac{\pi}{3} \qquad \cos 60° = \frac{1}{2}$$

(Don't forget that Arccos $\frac{1}{2}$ is an angle!)

b. Since $x = -\frac{1}{2}$ is negative, the angle cannot be in quadrant I. To find the proper quadrant, we must refer to the definition of Arccos x. By agreement, *the angle must lie between 0 and π.* Thus

$$\text{Arccos}\left(-\frac{1}{2}\right) = \frac{2\pi}{3} \qquad \cos 120° = -\frac{1}{2} \qquad \blacktriangleleft$$

E X A M P L E **3**   Find the exact value of Arctan(-1).

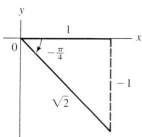

Figure 16.17

Solution. This is a problem that many students find troublesome. If we were looking merely for θ such that $\tan\theta = -1$, we would choose 135° and 315° ($3\pi/4$ and $7\pi/4$). Knowing that the value has to be unique, some students proceed to drop one of the values and keep only $7\pi/4$. Now, while this angle does lie in quadrant IV, this choice still violates the convention in statement (16.33). Since $-\pi/2 < y < \pi/2$, the angle chosen must be negative (between $-\pi/2$ and 0). Thus

$$\text{Arctan}(-1) = -\frac{\pi}{4}$$

(See Figure 16.17.) $\blacktriangleleft$

E X A M P L E **4**   Find the exact value of $\text{Arcsin}\left(-\dfrac{\sqrt{3}}{2}\right)$.

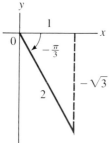

Figure 16.18

Solution. Since x is negative, the angle cannot lie in the first quadrant. By the definition of Arcsin x, we have $-\pi/2 \le y \le \pi/2$ so that

$$\text{Arcsin}\left(-\frac{\sqrt{3}}{2}\right) = -\frac{\pi}{3}$$

(See Figure 16.18.) $\blacktriangleleft$

CALCULATOR COMMENT

To show how strictly these conventions must be observed, let us find some of the values of the inverse trigonometric functions by using a calculator. (If the angles are not special angles, a calculator should be used anyway.)

E X A M P L E **5**   Use a calculator to find Arcsin(0.4278).

Radian mode

Solution. We are looking for an angle whose sine is 0.4278. To obtain this angle in radians, we first set the calculator in the radian mode. Now enter 0.4278 and press the $\boxed{\text{INV}}$ key, followed by the $\boxed{\text{SIN}}$ key, to obtain 0.4421. As expected, the angle is in quadrant I.

The sequence is

$$0.4278 \;\boxed{\text{INV}}\;\boxed{\text{SIN}} \to 0.4420574 \qquad \text{(radians)}$$

Degree mode

By setting the calculator in the degree mode, the same sequence yields Arcsin(0.4278) = 25.3°. ◀

E X A M P L E **6**   Evaluate Arcsin(−0.6845).

Solution. Set the calculator in the radian mode and proceed as in Example 5. We obtain

$$\text{Arcsin}(-0.6845) = -0.7539$$

The result agrees with the convention in statement (16.32). In degree mode, the same sequence yields −43.2°:

$$0.6845 \;\boxed{+/-}\;\boxed{\text{INV}}\;\boxed{\text{SIN}} \to -43.196298 \qquad \text{(degrees)}$$ ◀

The next two examples involve trigonometric functions in a way that is particularly useful in calculus.

E X A M P L E **7**   Find the exact value of $\sin\left[\text{Arctan}\left(-\frac{1}{4}\right)\right]$.

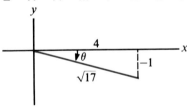

Figure 16.19

Solution. Recall that Arctan$\left(-\frac{1}{4}\right)$ is an angle whose tangent is $-\frac{1}{4}$. Since $-\frac{1}{4}$ is negative, $\theta = \text{Arctan}\left(-\frac{1}{4}\right)$ is an angle between 0° and −90°. Now construct the triangle in Figure 16.19, with −1 on the opposite side and 4 on the adjacent side. The hypotenuse is found to be $\sqrt{17}$. So

$$\sin\left[\text{Arctan}\left(-\frac{1}{4}\right)\right] = \sin\theta = \frac{-1}{\sqrt{17}} = -\frac{\sqrt{17}}{17}$$ ◀

E X A M P L E **8**   Find an algebraic expression equivalent to tan(Arccos 2x).

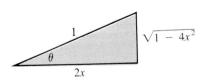

Figure 16.20

Solution. Let $\theta = \text{Arccos }2x$. Thus θ is an angle whose cosine is $2x = 2x/1$. Draw a right triangle with $2x$ on the adjacent side and 1 on the hypotenuse. (See Figure 16.20.) By the Pythagorean theorem, the length of the opposite side is $\sqrt{1-4x^2}$. It follows that

$$\tan(\text{Arccos }2x) = \tan\theta = \frac{\sqrt{1-4x^2}}{2x}$$ ◀

E X A M P L E **9**   The width w of a laser beam at a distance d from the source is given by

$$w = 2d \tan \frac{\alpha}{2} \qquad (0 \le \alpha < \pi)$$

where α is the angle of the beam. Solve this equation for α.

Solution. $w = 2d \tan \dfrac{\alpha}{2}$

$$\frac{w}{2d} = \tan \frac{\alpha}{2}$$

$$\frac{\alpha}{2} = \text{Arctan} \frac{w}{2d}$$

$$\alpha = 2 \,\text{Arctan} \frac{w}{2d}$$ ◀

E X E R C I S E S / S E C T I O N   **16.8**

In Exercises 1–17, find the exact value (in radian measure) of each expression without using a table or a calculator.

1. $\text{Arcsin} \dfrac{\sqrt{3}}{2}$

2. $\text{Arcsin}(-1)$

3. $\text{Arctan } 1$

4. $\text{Arccos}(-1)$

5. $\text{Arcsin } 0$

6. $\text{Arcsin}\left(-\dfrac{1}{2}\right)$

7. $\text{Arccos } 0$

8. $\text{Arctan}\left(-\dfrac{1}{\sqrt{3}}\right)$

9. $\text{Arctan } 0$

10. $\text{Arcsin } 1$

11. $\text{Arcsin}\left(-\dfrac{1}{\sqrt{2}}\right)$

12. $\text{Arcsin}\left(\dfrac{1}{\sqrt{2}}\right)$

13. $\text{Arctan}(-\sqrt{3})$

14. $\text{Arccos}\left(-\dfrac{1}{\sqrt{2}}\right)$

15. $\text{Arctan } \sqrt{3}$

16. $\text{Arccos } \dfrac{1}{\sqrt{2}}$

17. $\text{Arccos}\left(-\dfrac{\sqrt{3}}{2}\right)$

In Exercises 18–40, evaluate the given expressions without a table or a calculator. (See Example 7.)

18. $\sin(\text{Arctan } 2)$

19. $\tan\left[\text{Arccos}\left(-\dfrac{1}{3}\right)\right]$

20. $\tan\left[\text{Arcsin}\left(-\dfrac{1}{3}\right)\right]$

21. $\csc\left[\text{Arcsin}\left(-\dfrac{3}{4}\right)\right]$

22. $\csc\left[\text{Arccos}\left(-\dfrac{3}{4}\right)\right]$

23. $\cos[\text{Arctan}(-2)]$

24. $\sec(\text{Arctan } 3)$

25. $\cos\left(\text{Arcsin } \dfrac{2}{3}\right)$

26. $\sec\left(\text{Arcsin } \dfrac{4}{5}\right)$

27. $\csc\left[\text{Arctan}\left(-\dfrac{3}{4}\right)\right]$

28. $\tan\left[\text{Arcsin}\left(-\dfrac{12}{13}\right)\right]$

29. $\cot\left[\text{Arccos}\left(-\dfrac{5}{13}\right)\right]$

30. $\cot\left[\text{Arctan}\left(-\dfrac{5}{6}\right)\right]$

31. $\sec\left(\text{Arccos } \dfrac{1}{4}\right)$

32. $\csc\left(\text{Arcsin } \dfrac{2}{5}\right)$

33. $\cot\left[\text{Arcsin}\left(-\dfrac{1}{4}\right)\right]$

34. $\sec\left[\text{Arcsin}\left(-\dfrac{3}{7}\right)\right]$

35. $\sin\left[\text{Arccos}\left(-\dfrac{2}{5}\right)\right]$

36. $\csc(\text{Arctan } \sqrt{5})$

37. $\sin\left(\text{Arcsin } \dfrac{1}{5}\right)$

38. $\tan(\text{Arctan } 4)$

39. $\cot(\text{Arctan } 4)$

40. $\cos\left(\text{Arccos } \dfrac{2}{5}\right)$

In Exercises 41–50, for each expression find an equivalent algebraic expression. Use the positive square root in each case. (See Example 8.)

41. tan(Arcsin x)   **42.** cos(Arctan x)   **43.** sec(Arctan x)

44. sin(Arccos x)   **45.** cot(Arcsin $2x$)   **46.** sin(Arccos $2x$)

47. csc(Arctan $3x$)   **48.** tan(Arccos $3x$)   **49.** sin(Arccos $2x$)

50. tan(Arcsin $3x$)

In Exercises 51–60, use a calculator to evaluate each inverse function in **(a)** radians, **(b)** degrees.

51. Arctan 2

52. Arctan(-2)

53. Arcsin$\left(-\dfrac{1}{3}\right)$

54. Arccos$\left(-\dfrac{2}{3}\right)$

55. Arctan(-1.3142)

56. Arcsin(-0.7418)

57. Arccos(-0.4915)

58. Arctan(2.672)

59. Arcsin(0.4970)

60. Arccos(-0.8736)

61. A woman is walking toward a building 100 ft high. Show that when she is x feet from the base of the building, the angle of elevation of the top is given by $\theta = $ Arctan($100/x$).

62. Show that angle $A = $ Arctan$[(b/a)\tan B]$ in Figure 16.21.

63. The formula $\phi = $ Arctan (X/R) arises in the study of alternating current. Solve this formula for R.

64. A small body is revolving in a horizontal circle at the end of a cord of length L, making an angle θ with the vertical (Figure 16.22). The time for one complete revolution is

$$T = 2\pi \sqrt{\dfrac{L \cos \theta}{g}} \quad \left(0 < \theta < \dfrac{\pi}{2}\right)$$

Solve this equation for θ.

65. Recall that the equation of simple harmonic motion is

$$x = A \cos \sqrt{\dfrac{k}{m}}\, t$$

Find the formula for the time t required for the particle to move from its starting position $x = A$ (when $t = 0$) to a new position $(0 \le t \le \pi \sqrt{m/k})$.

66. The formula for *magnetic intensity* is

$$B = \dfrac{F}{qv \sin \phi} \quad \left(0 < \phi \le \dfrac{\pi}{2}\right)$$

where q is the magnitude of the charge, v its velocity, ϕ the angle between the direction of motion and the direction of the magnetic field, and F the force acting on the moving charge. Solve this formula for ϕ.

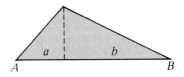

Figure 16.21

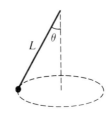

Figure 16.22

REVIEW EXERCISES/CHAPTER 16

In Exercises 1–4, use the appropriate identity to evaluate each function without using a table or a calculator.

1. sin 22.5°

2. cos 112.5°

3. cos 12° cos 18° − sin 12° sin 18°

4. sin 110° cos 20° − cos 110° sin 20°

In Exercises 5–8, combine each expression into a single term.

5. sin $2x$ cos $4x$ + cos $2x$ sin $4x$

6. cos $6x$ cos x + sin $6x$ sin x

7. sin $4x$ cos x − cos $4x$ sin x

8. cos($x − y$) cos y − sin($x − y$) sin y

In Exercises 9–14, write each expression as a function of x or $2x$.

9. cos($2x - \pi$)

10. sin$\left(x - \dfrac{\pi}{2}\right)$

11. cos$\left(2x + \dfrac{\pi}{2}\right)$

12. cos($x − 2\pi$)

13. sin$\left(x - \dfrac{\pi}{6}\right)$

14. cos$\left(x - \dfrac{\pi}{4}\right)$

15. Find sin 2θ, given that cos $\theta = -\frac{4}{5}$, θ in quadrant II.

16. Find sin 2θ, given that sin $\theta = -\frac{1}{3}$, θ in quadrant III.

17. Find cos 2θ, given that sin $\theta = -\frac{5}{13}$, θ in quadrant IV.

18. Find sin($\theta/2$), given that cos $\theta = -\frac{4}{5}$, θ in quadrant II.

19. Find $\cos(\theta/2)$, given that $\sin \theta = -\frac{24}{25}$, θ in quadrant III.

20. Find $\cos(\theta/2)$, given that $\cos \theta = \frac{12}{13}$, θ in quadrant IV.

In Exercises 21–26, write each expression as a single trigonometric function.

21. $\cos^2 3x - \sin^2 3x$

22. $\sin^2 2x - \cos^2 2x$

23. $1 - 2 \sin^2 4x$

24. $2 \cos^2 3\beta - 1$

25. $2 \sin 3x \cos 3x$

26. $\sin 4x \cos 4x$

In Exercises 27–30, simplify the given expressions.

27. $\sqrt{\dfrac{1 - \cos 4\theta}{2}}$

28. $\sqrt{\dfrac{1 + \cos 4\theta}{2}}$

29. $\sqrt{1 + \cos 4\theta}$

30. $\sqrt{2 - 2 \cos 8\theta}$

In Exercises 31–34, eliminate the exponent.

31. $\sin^2 3x$

32. $\cos^2 4x$

33. $2 \cos^2 3x$

34. $4 \sin^2 4x$

In Exercises 35–56, prove the given identities.

35. $\dfrac{\cos \beta \tan \beta + \sin \beta}{\tan \beta} = 2 \cos \beta$

36. $(\sec \alpha - \tan \alpha)(\csc \alpha + 1) = \cot \alpha$

37. $\dfrac{1}{1 + \sin x} - \dfrac{1}{1 - \sin x} = -2 \tan x \sec x$

38. $\dfrac{\sec \theta}{\cot \theta + \tan \theta} = \sin \theta$

39. $\dfrac{1 + \sin^2 \theta \sec^2 \theta}{1 + \cos^2 \theta \csc^2 \theta} = \tan^2 \theta$

40. $\tan^2 \theta - \sin^2 \theta = \tan^2 \theta \sin^2 \theta$

41. $\dfrac{\cos^4 x - \sin^4 x}{1 - \tan^4 x} = \cos^4 x$

42. $\dfrac{1}{\csc x + \cot x} = \dfrac{1 - \cos x}{\sin x}$

43. $\cos y \sin(x - y) + \sin y \cos(x - y) = \sin x$

44. $\cos\left(x - \dfrac{\pi}{6}\right) - \cos\left(x + \dfrac{\pi}{6}\right) = \sin x$

45. $\cos(x + y) + \cos(x - y) = 2 \cos x \cos y$

46. $\cos 2x + 2 \sin^2 x = 1$

47. $\dfrac{2 - \sec^2 y}{\sec^2 y} = \cos 2y$

48. $\dfrac{2 \tan \theta}{\sin 2\theta} = \sec^2 \theta$

49. $\dfrac{1 - \tan^2 \theta}{1 + \tan^2 \theta} = \cos 2\theta$

50. $\cot 2\theta = \dfrac{\cot^2 \theta - 1}{2 \cot \theta}$

51. $\dfrac{1 - \cos 2\gamma + \sin 2\gamma}{1 + \cos 2\gamma + \sin 2\gamma} = \tan \gamma$

52. $\sin \theta = 2 \sin \dfrac{\theta}{2} \cos \dfrac{\theta}{2}$

53. $\csc^2 \theta = \dfrac{2}{1 - \cos 2\theta}$

54. $\dfrac{1}{2} \sec^2 \theta = \dfrac{1}{1 + \cos 2\theta}$

55. $\left(\sin \dfrac{\theta}{2} - \cos \dfrac{\theta}{2}\right)^2 = 1 - \sin \theta$

56. $\sin^2 \alpha = \dfrac{\sin^2 2\alpha}{2(1 + \cos 2\alpha)}$

In Exercises 57–62, solve the given equations ($0 \le x < 2\pi$).

57. $\cos^2 x + 2 \sin x = 1$

58. $\sin 2x - 3 \sin x = 0$

59. $2 \cos^2 x + \cos x = 1$

60. $\tan x = \tan^2 x$

61. $\cos \dfrac{x}{2} + \cos x + 1 = 0$

62. $3 \cos^2 x - 14 \cos x + 8 = 0$

63. Suppose a projectile is fired along an inclined plane making an angle α to the horizontal from a gun making an angle θ to the horizontal. Calculus shows that the range of the projectile along the inclined plane is a maximum when θ satisfies the relation

$$\cos \theta \cos(\theta - \alpha) - \sin \theta \sin(\theta - \alpha) = 0$$

Find angle θ.

64. A body of weight W is dragged along a horizontal plane by a force whose line of action makes an angle θ with the plane. Calculus shows that the pull is least when θ satisfies the equation

$$\mu \cos \theta - \sin \theta = 0$$

where μ is the coefficient of friction. Show that the pull is least when $\theta = \text{Arctan } \mu$.

In Exercises 65–68, find y ($0 \le y < 2\pi$) in each case without using a table or a calculator.

65. $y = \arccos \dfrac{1}{2}$

66. $y = \text{arccsc } 1$

67. $y = \text{arccot}(-\sqrt{3})$

68. $y = \arctan \dfrac{1}{\sqrt{3}}$

In Exercises 69–72, evaluate each expression without using a table or a calculator.

69. $\text{Arccos}\left(-\dfrac{1}{2}\right)$

70. $\text{Arctan}\left(-\dfrac{1}{\sqrt{3}}\right)$

71. $\cot\left[\text{Arcsin}\left(-\dfrac{1}{6}\right)\right]$

72. $\sin[\text{Arctan}(-2)]$

73. Find an algebraic expression for $\sin(\text{Arccos } 2x)$

74. Solve for x: $y = 2 - 3\sin 4x$

75. Solve for x: $y = 2\arctan(x + 2)$

76. The expression $a\sin\theta + b\cos\theta$ can be written in simpler form by noting that

$$\sqrt{a^2 + b^2}\left(\frac{a}{\sqrt{a^2 + b^2}}\sin\theta + \frac{b}{\sqrt{a^2 + b^2}}\cos\theta\right)$$

$$= a\sin\theta + b\cos\theta$$

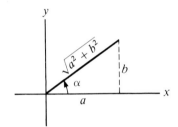

Figure 16.23

and that

$$\sin\alpha = \frac{b}{\sqrt{a^2 + b^2}} \quad \text{and} \quad \cos\alpha = \frac{a}{\sqrt{a^2 + b^2}}$$

where α is an angle determined by a and b. (See Figure 16.23.) Use the identity for $\sin(A + B)$ to show that

$$a\sin\theta + b\cos\theta = k\sin(\theta + \alpha)$$

where $k = \sqrt{a^2 + b^2}$ and α is any angle for which

$$\sin\alpha = \frac{b}{\sqrt{a^2 + b^2}} \quad \text{and}$$

$$\cos\alpha = \frac{a}{\sqrt{a^2 + b^2}}$$

77. Use the formulas in Exercise 76 to write the following expressions in the form $k\sin(\theta + \alpha)$:

  a. $3\sin\theta + 4\cos\theta$

  b. $4\sin\theta + 6\cos\theta$

Inequalities

17.1 Basic Properties of Inequalities

A lot of time and effort in algebra and trigonometry is devoted to solving equations and verifying identities. Yet **inequalities** also play an important role, as we have seen from time to time. For example, we saw in the last chapter that if $y = \text{Arctan } x$, then $-\pi/2 < y < \pi/2$. In this chapter we will study inequalities in detail. This first section describes the basic properties of inequalities.

First recall that $a < b$ means "a is less than b," and $a > b$ means "a is greater than b." The direction of the inequality is called the **sense** of the inequality. For example, the inequalities $a < b$ and $x < y$ have the same sense, while $a < b$ and $x > y$ have the opposite sense. To combine the notions of equality and inequality, we use the notation $a \leq b$, meaning "a is less than or equal to b," and $a \geq b$, meaning "a is greater than or equal to b."

If an inequality contains a variable, then a **solution** of the inequality may exist. For example, the inequality $x + 1 < 2$ is satisfied when $x < 1$, while the inequality $x + 1 \leq 2$ is satisfied when $x \leq 1$.

Inequalities containing an unknown can be solved by a method similar to solving equations. Analogous geometric interpretations can also be given. To do so, however, we need to know the basic properties of inequalities.

Sense of an inequality

Solution of an inequality

Basic Properties of Inequalities:

1. For any real numbers a, b, and c, if $a < b$, then $a + c < b + c$ and $a - c < b - c$.
2. If $c > 0$ and $a < b$, then $ac < bc$ and $a/c < b/c$.
3. If $c < 0$ and $a < b$, then $ac > bc$ and $a/c > b/c$.
4. If a and b are positive and $a < b$, then

$$a^n < b^n \qquad \text{and} \qquad \sqrt[n]{a} < \sqrt[n]{b}$$

5. If $a > 0$ and $b > 0$ (or if $a < 0$ and $b < 0$) and $a < b$, then

$$\frac{1}{a} > \frac{1}{b}$$

Similar statements hold for the opposite sense, $a > b$.

It may be worthwhile now to summarize the basic properties of inequalities:

1. The sense of the inequality is preserved whenever the same number is added to or subtracted from both sides.

If both sides are multiplied or divided by c, then

2. The sense is preserved whenever $c > 0$; and

3. Reversed whenever $c < 0$.

With suitable restrictions,

4. Taking powers or roots preserves the sense of the inequality,

while

5. Taking reciprocals reverses the sense.

Consider the following example.

E   X   A   M   P   L   E   **1**   Since $4 < 7$,

$$2 \cdot 4 < 2 \cdot 7 \qquad \text{or} \qquad 8 < 14 \qquad \qquad \textbf{by Property 2}$$

and

$$-3 \cdot 4 > -3 \cdot 7 \qquad \text{or} \qquad -12 > -21 \qquad \textbf{by Property 3}$$

Also,

$$\sqrt{4} < \sqrt{7} \qquad \text{or} \qquad 2 < \sqrt{7} \approx 2.6 \qquad \textbf{by Property 4}$$

and

$$\frac{1}{4} > \frac{1}{7}$$

by Property 5   ◀

Solving Linear Inequalities

The next example illustrates how to solve a given inequality by using the basic properties. (Since the inequality involves a polynomial of first degree, the inequality is said to be **linear**.)

E X A M P L E 2   Solve the inequality $2x - 4 < 5x + 6$.

Solution.
$$2x - 4 < 5x + 6$$
$$2x - 4 + 4 < 5x + 6 + 4 \qquad \text{by Property 1}$$
$$2x < 5x + 10$$
$$2x - 5x < 5x + 10 - 5x \qquad \text{by Property 1}$$
$$-3x < 10$$
$$x > \frac{10}{-3} = -\frac{10}{3} \qquad \text{by Property 3}$$

We conclude that the given inequality is satisfied whenever $x > -\frac{10}{3}$. In other words, any value of x greater than $-\frac{10}{3}$ satisfies the given inequality.

Note that the sense of the inequality is *reversed* on the last step, since both sides are divided by -3. Otherwise, the procedure for solving (linear) inequalities is essentially the same as the procedure for solving equations.   ◀

E X A M P L E 3   Solve the inequality $2 + 4x \geq 12 - x$.

Solution.
$$2 + 4x \geq 12 - x \qquad \text{given inequality}$$
$$2 + 4x - 2 \geq 12 - x - 2 \qquad \text{subtracting 2}$$
$$4x \geq 10 - x$$
$$4x + x \geq 10 - x + x \qquad \text{adding } x$$
$$5x \geq 10$$
$$\frac{5x}{5} \geq \frac{10}{5} \qquad \text{dividing by 5; sense preserved}$$
$$x \geq 2 \qquad\qquad\qquad ◀$$

Graphing Linear Inequalities

To see how the solution of an inequality can also be shown graphically, consider the inequality $4x + 2y \geq 5$. Let us first graph the equation of the line

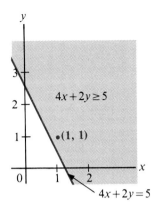

$4x + 2y \geq 5$

•(1, 1)

$4x + 2y = 5$

Figure 17.1

$4x + 2y = 5$ by finding the intercepts. If $x = 0$, then $y = \frac{5}{2}$ and if $y = 0$, then $x = \frac{5}{4}$. So the intercepts are $(\frac{5}{4}, 0)$ and $(0, \frac{5}{2})$. (See Figure 17.1.)

The coordinates of every point on the line satisfy the equation $4x + 2y = 5$. At all other points inequality holds. For example, if $x = 1$ and $y = 1$, then $4x + 2y = 6 > 5$; thus, the coordinates of **(1, 1)** satisfy the inequality. Checking one point is sufficient. The solution consists of the set of all points on the line and on the side of the line containing the point (1, 1), shown as the shaded region in Figure 17.1.

If we choose a point on the other side of the line, the inequality will not be satisfied. For example, choosing $(1, 0)$ leads to $4(1) + 2(0) = 4 < 5$, which violates the condition $4x + 2y \geq 5$.

Let's consider another example.

E X A M P L E **4**   Graph the inequality $2x - 3y < 9$.

Solution. This time the coordinates of the points on the line $2x - 3y = 9$ do not satisfy the inequality, but the line is still needed to determine the boundary of the region. (See Figure 17.2.)

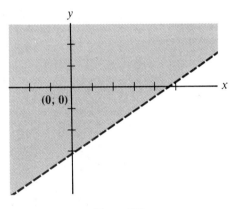

Figure 17.2

If $x = 0$ and $y = 0$, we have

$$0 - 0 < 9$$

So the point **(0, 0)** lies in the region. We conclude that the solution consists of the set of all points above the line (in the region containing the origin). The line itself is *dashed* to indicate that the points on the boundary do not belong to the solution set.   ◀

    E X A M P L E **5**   The current through two variable resistors in series is 2 A. Determine graphically the values of R_1 and R_2 for which the voltage across the two resistors is greater than 7 V.

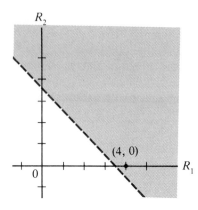

Figure 17.3

Solution. By Ohm's law and the fact that the resistors are connected in series, we get

$$2R_1 + 2R_2 > 7$$

The graph of $2R_1 + 2R_2 = 7$ is the dashed line in Figure 17.3. If $R_1 = 4$ and $R_2 = 0$, we get $8 + 0 > 7$. So the inequality is satisfied for any point in the shaded region. ◀

E X E R C I S E S / S E C T I O N **17.1**

In Exercises 1–16, solve the given inequalities.

1. $x + 2 < 5$ **2.** $x - 3 > 4$

3. $2x + 4 \leq 8$ **4.** $3x - 1 \geq 5$

5. $2x + 1 < x - 5$ **6.** $3x + 2 > 5x - 6$

7. $5x - 2 \leq 6x + 3$ **8.** $2x - 2 < 7x - 3$

9. $-2x + 2 \geq x - 2$ **10.** $3 - 6x < -7x - 3$

11. $1 - 2x < 2x + 7$ **12.** $-1 - 3x > 5 + 2x$

13. $-2x - 1 \leq x - 8$ **14.** $2x + 1 \geq 6x - 3$

15. $5x - 3 > 3x + 2$ **16.** $2x + 2 < -x + 7$

In Exercises 17–26, graph the given inequalities.

17. $2x + 3y \leq 6$ **18.** $x + 4y \geq 4$

19. $x + 3y > 3$ **20.** $x - 2y < 3$

21. $2x - y > 0$ **22.** $-2x + y \leq 6$

23. $3x - 6y \geq 9$ **24.** $-2x - 5y \geq 5$

25. $-3x - 6y < -6$ **26.** $-x + 4y < 3$

27. The cost C (in dollars) of operating a certain machine is $C = 80 + 5t$ (t in hours). If the machine is started at $t = 0$, how long can the machine be run before the cost exceeds $118?

28. The relationship between degrees Fahrenheit and degrees Celsius is given by

$$C = \frac{5}{9}(F - 32)$$

Determine the values of C for which $F \leq 212°$, the boiling point of water.

29. The force F (in pounds) required to pull a certain spring x inches is given by $F = 10.0x$ within the limits $0.0 \leq x \leq 8.0$ (in inches). What values of F does this correspond to?

17.2 More Inequalities

The inequalities discussed in the last section involved polynomials of first degree. In this section we will study inequalities containing polynomials of higher degree. The key to solving such inequalities is to systematically extend the method for solving linear inequalities: We change the given inequality to an equivalent inequality for which the right side is equal to zero. We then determine the values of x for which the left side is either positive or negative. To do this, we first find those values of x for which the left side is either equal to zero or undefined. Such values of x are called **critical values**.

> **Critical Values of a Function:** The **critical values** of a function $y = f(x)$ are the values of x for which $y = f(x)$ is either zero or undefined.

To see how critical values may be used to solve inequalities, consider the inequality

$$x^2 < 4 - 3x$$

By adding $3x - 4$ to both sides, we see that the given inequality is equivalent to

$$f(x) = x^2 + 3x - 4 < 0$$

To find the critical values, we set the left side equal to zero:

$$x^2 + 3x - 4 = 0$$
$$(x - 1)(x + 4) = 0$$
$$\boxed{x = 1, \, -4}$$

Thus $x = 1$ and $x = -4$ are the critical values. Note especially that the critical values are precisely those values at which the function $f(x) = (x - 1)(x + 4)$ changes signs.

To the left of -4, both factors are negative, so that the product is positive. Between -4 and 1 we have $x - 1 < 0$ and $x + 4 > 0$; the product is therefore negative. Finally, to the right of $x = 1$ both factors are positive, so that the product is positive. Consequently, the inequality is satisfied if $-4 < x < 1$. So the solution of the inequality is given by

$$-4 < x < 1$$

This procedure can be made more systematic by using a chart.

	Sign of		
x values	$x - 1$	$x + 4$	$f(x)$
$x < -4$	$-$	$-$	$+$
$-4 < x < 1$	$-$	$+$	$-$
$x > 1$	$+$	$+$	$+$

Test value

While this procedure for solving inequalities is adequate, another shortcut is possible. Since $(x - 1)(x + 4) > 0$ whenever $x < -4$, it follows that $(x - 1)(x + 4) > 0$ for *any* value of x less than -4. For example, if $x = -5$, then $(x - 1)(x + 4) = (-6)(-1) = 6 > 0$. We will call such a value a **test value**.

If we plot the critical values first, we can easily pick out some arbitrary test values between or beyond the critical values. (See Figure 17.4.)

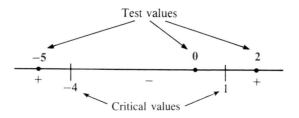

Figure 17.4

By using test values, the chart can be modified as follows:

x values	Test values	Sign of x − 1	x + 4	f(x)
x < −4	−5	−	−	+
−4 < x < 1	0	−	+	−
x > 1	2	+	+	+

We can see from the chart that $f(x)$ is negative whenever $-4 < x < 1$.
Before summarizing the procedure, let's consider another example.

E X A M P L E **1**   Solve the inequality:

$$(x - 2)(x + 3)(x + 5) < 0$$

Solution. From

$$f(x) = (x - 2)(x + 3)(x + 5) = 0$$

we see that the critical values are

$$\boxed{x = 2, -3, -5}$$

For the test values, we use *any* values between or beyond the critical values. (See Figure 17.5.)
We obtain the following chart:

x values	Test values	Sign of x + 5	x + 3	x − 2	f(x)
x < −5	−6	−	−	−	−
−5 < x < −3	−4	+	−	−	+
−3 < x < 2	0	+	+	−	−
x > 2	3	+	+	+	+

Noting where $f(x)$ is negative, we conclude that the solution is given by

$$x < -5, \qquad -3 < x < 2$$ ◀

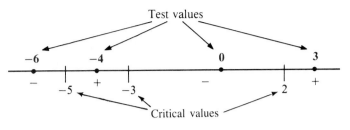

Figure 17.5

The procedure for solving inequalities is summarized next.

Procedure for Solving Inequalities:

1. Change the given inequality to an equivalent inequality for which the right side is zero.
2. Find the critical values.
3. Choose arbitrary values between and beyond the critical values (test values).
4. Use the test values to determine the sign of the left side.

E X A M P L E **2**   Solve the inequality:

$$(x + 1)^2(x - 3)(x - 6) \geq 0$$

Solution. The critical values are $x = -1, 3, 6$ (Figure 17.6).

x values	Test values	Sign of			
		$(x + 1)^2$	$x - 3$	$x - 6$	$f(x)$
$x < -1$	-2	$+$	$-$	$-$	$+$
$-1 < x < 3$	0	$+$	$-$	$-$	$+$
$3 < x < 6$	4	$+$	$+$	$-$	$-$
$x > 6$	7	$+$	$+$	$+$	$+$

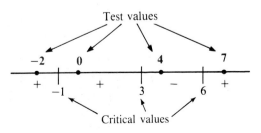

Figure 17.6

Note that since $(x + 1)^2 \geq 0$, $f(x)$ does not undergo a sign change at $x = -1$. So the inequality is satisfied for $x < -1$, as well as for $-1 < x < 3$. Also, because of the equality part, the given inequality is satisfied at the critical values, $x = -1$, 3, and 6. The solution is therefore given by

$$x < -1, \qquad x = -1, \qquad -1 < x < 3, \qquad x = 3, \qquad x = 6, \qquad x > 6$$

This solution can be written more simply as

$$x \leq 3, \qquad x \geq 6 \qquad\qquad\qquad\qquad \blacktriangleleft$$

In the next example, one of the critical values is a value of x for which $f(x)$ is undefined.

E X A M P L E **3**   Solve the inequality:

$$\frac{x^2 + 2x - 8}{x + 1} \leq 0$$

Solution. Factoring the numerator, the inequality becomes

$$f(x) = \frac{(x + 4)(x - 2)}{x + 1} \leq 0$$

The critical values are $x = -4, 2, -1$. (The value $x = -1$ is a critical value because it leads to division by zero.) Note that even though $f(x)$ is undefined at $x = -1$, the sign of $f(x)$ changes at this value (Figure 17.7).

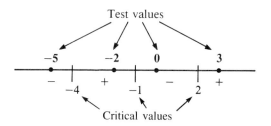

Figure 17.7

We obtain the following chart:

		Sign of			
x values	Test values	$x + 4$	$x + 1$	$x - 2$	$f(x)$
$x < -4$	-5	$-$	$-$	$-$	$-$
$-4 < x < -1$	-2	$+$	$-$	$-$	$+$
$-1 < x < 2$	0	$+$	$+$	$-$	$-$
$x > 2$	3	$+$	$+$	$+$	$+$

The inequality is therefore satisfied whenever

$$x < -4 \quad \text{or} \quad -1 < x < 2$$

In addition, $f(x) = 0$ when $x = -4$ and $x = 2$. So the solution is

$$x \le -4, \quad -1 < x \le 2 \quad \blacktriangleleft$$

E X E R C I S E S / S E C T I O N 17.2

In Exercises 1–32, solve the given inequalities.

1. $2x - 3 < 4x + 3$

2. $5x - 4 \ge 4x - 6$

3. $x^2 - 4 < 0$

4. $x^2 - x - 6 \le 0$

5. $x^2 + 5x \ge -6$

6. $x^2 > 6x - 8$

7. $x^3 < x^2 + 6x$

8. $x^3 + 3x^2 > 4x$

9. $(x + 2)(x - 3)(x + 5) \le 0$

10. $(x + 2)(x + 4)(x - 1) \ge 0$

11. $(x - 6)(x - 7)(x + 8) > 0$

12. $(x + 1)(x + 2)(x + 3) < 0$

13. $\dfrac{x - 1}{x - 2} < 0$

14. $\dfrac{x - 3}{x + 4} \le 0$

15. $\dfrac{x - 6}{x - 7} \ge 0$

16. $\dfrac{x + 3}{x + 5} \ge 0$

17. $\dfrac{x - 3}{4 - x} > 0$

18. $\dfrac{x + 6}{2 - x} < 0$

19. $\dfrac{(x - 3)(x + 4)}{x - 1} \le 0$

20. $\dfrac{(x + 2)(x - 1)}{x + 3} > 0$

21. $\dfrac{(x - 6)(x + 3)}{3 - x} < 0$

22. $\dfrac{(x + 5)(x + 2)}{(x - 3)(x + 4)} \le 0$

23. $\dfrac{(x - 1)(x + 6)}{(x - 2)(x - 3)} \ge 0$

24. $\dfrac{(x + 3)(x - 2)}{(x - 4)(x + 2)} < 0$

25. $(x - 2)^2(x - 4) > 0$

26. $(x + 3)(x - 1)^2 \le 0$

27. $(x - 3)^2(x - 7)^2 \ge 0$

28. $(x + 2)(x - 4)^2 < 0$

29. $(x - 4)^2(x + 1) < 0$

30. $(x - 3)^2(x + 2)^4 < 0$

31. $\dfrac{(x - 2)^2(x - 4)}{(x - 1)(x + 4)} \le 0$

32. $\dfrac{(x + 3)(x - 4)}{(x + 1)^2(x - 6)} \ge 0$

33. The current in a certain circuit is given by $i = 4t - 3t^2$, where i is measured in amperes and t in seconds. Determine the time interval for which $i \ge 0$.

34. The resistance R (in ohms) of a certain wire as a function of temperature T (in degrees Celsius) is given by $R = 10.0 + 0.10T + 0.0060T^2$. Find $T > 0$ such that $R > 20 \ \Omega$.

35. The deflection d (in feet) of a beam 20 ft long is given by $d = \frac{1}{100}(20 - x)x$, where x is the distance from one end of the beam. For what values of x is $d \le \frac{3}{4}$ ft?

36. The current in a circuit is given by $i = 2.0t^2 - 19.0t + 45.0$. In what time interval is $i \le 0$?

37. The weekly profit P of a company is $P = x^4 - 20x^3, x \ge 1$, where x is the week in the year. During what period is the company operating at a loss?

17.3 Absolute Inequalities

Many situations, particularly in calculus, lead to inequalities involving absolute values. In this section we will consider some of the basic cases.

First consider the inequality $|x| < 1$. Any number between 0 and 1 would certainly satisfy this relation, but if we choose $x = -\frac{2}{3}$, then $|x| = \left|-\frac{2}{3}\right| = \frac{2}{3}$, which also satisfies the inequality. In fact, the inequality is satisfied for $-1 < x < 1$.

The general rule for solving absolute inequalities will now be stated.

Absolute Inequalities:

1. If $|f(x)| < r$, then $-r < f(x) < r$. $\hspace{2cm}$ (17.1)

2. If $|f(x)| > r$, then $f(x) < -r$ or $f(x) > r$. $\hspace{1cm}$ (17.2)

Rule (17.1) is illustrated in the following example.

E X A M P L E **1**   Solve the inequality $|x - 2| < 3$.

Solution. By statement (17.1), $|x - 2| < 3$ is equivalent to

$$-3 < x - 2 < 3$$

Adding **2** to each member, we obtain

$$-3 + \mathbf{2} < x - 2 + \mathbf{2} < 3 + \mathbf{2}$$

or

$$-1 < x < 5$$

Figure 17.8

which is the solution (Figure 17.8). ◄

The inequality in the next example is solved with the help of rule (17.2).

E X A M P L E **2**   Solve the inequality $|3x + 2| \geq 6$.

Solution. By statement (17.2), the inequality $|3x + 2| \geq 6$ is equivalent to

$$3x + 2 \leq -6 \quad \text{or} \quad 3x + 2 \geq 6$$
$$3x \leq -8 \quad \text{or} \quad 3x \geq 4 \qquad \text{adding } -2$$
$$x \leq -\frac{8}{3} \quad \text{or} \quad x \geq \frac{4}{3} \qquad \text{dividing by 3}$$

So the given equality is satisfied whenever $x \leq -\frac{8}{3}$ or whenever $x \geq \frac{4}{3}$. The solution is therefore given by

$$x \leq -\frac{8}{3}, \qquad x \geq \frac{4}{3}$$

◄

In the remaining examples, the absolute inequalities lead to two quadratic inequalities.

E X A M P L E **3**   Solve the inequality $|x^2 - x + 1| > 2$.

Solution. By statement (17.2) the given inequality is equivalent to

$$x^2 - x + 1 > 2 \qquad \text{**or**} \qquad x^2 - x + 1 < -2$$

For the inequality on the left side, we have

$$x^2 - x + 1 > 2$$
$$x^2 - x - 1 > 0 \qquad \text{adding } -2 \text{ to both sides}$$

The critical values are obtained by using the quadratic formula:

$$x^2 - x - 1 = 0$$

$$x = \frac{1 \pm \sqrt{1 + 4}}{2}$$

that is,

$$x = \frac{1 - \sqrt{5}}{2} \approx -0.6$$

or

$$x = \frac{1 + \sqrt{5}}{2} \approx 1.6$$

Using the test values -1, 0, and 2, we find that $x^2 - x - 1 > 0$ whenever $x < -0.6$ or $x > 1.6$. Using the exact values,

$$x < \frac{1 - \sqrt{5}}{2} \quad \text{or} \quad x > \frac{1 + \sqrt{5}}{2} \qquad (17.3)$$

For the inequality on the right we have, by adding 2 to each side,

$$x^2 - x + 1 < -2$$
$$x^2 - x + 3 < \quad 0 \qquad (17.4)$$

By the quadratic formula, the critical values are

$$x = \frac{1 \pm \sqrt{1 - 12}}{2} = \frac{1 \pm \sqrt{11}j}{2}$$

which are complex numbers. Thus $x^2 - x + 3 \neq 0$. In other words, either $x^2 - x + 3 > 0$ for all x or $x^2 - x + 3 < 0$ for all x. Since $x^2 - x + 3 > 0$ when $x = 1$, the left side of inequality (17.4) is positive for all x. Thus, inequality (17.4) has no solution.

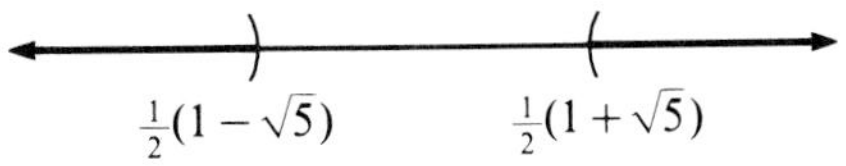

Figure 17.9

The solution of the given inequality, shown in Figure 17.9, is therefore given by statement (17.3):

$$x < \frac{1 - \sqrt{5}}{2}, \quad x > \frac{1 + \sqrt{5}}{2}$$

E X A M P L E **4** Solve the inequality $|x^2 - 2x - 2| < 1$.

Solution. By statement (17.1), we have

$$-1 < x^2 - 2x - 2 < 1$$

Both inequalities must be satisfied.

Suppose we solve the right inequality first by subtracting 1 from each member. Thus

$x^2 - 2x - 2 < 1$	**right inequality**
$x^2 - 2x - 3 < 0$	**adding −1 to each side**
$(x - 3)(x + 1) < 0$	**factoring**

By the method of the previous section, the solution is

$$-1 < x < 3 \qquad \text{**test values: −2, 0, 4**}$$

Solving the left inequality, we get

$x^2 - 2x - 2 > -1$	**left inequality**
$x^2 - 2x - 1 > 0$	

The critical values are

$$x = \frac{2 \pm \sqrt{4 + 4}}{2} = \frac{2 \pm 2\sqrt{2}}{2} = 1 \pm \sqrt{2}$$

So $x \approx -0.4$ and 2.4. Using the test values -1, 0, and 3, we find that the left inequality is satisfied whenever

$$x < 1 - \sqrt{2} \quad (x < -0.4) \qquad \text{or} \qquad x > 1 + \sqrt{2} \quad (x > 2.4)$$

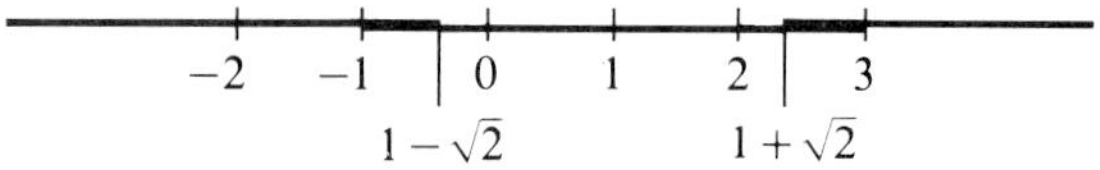

Figure 17.10

The two sets of solutions are shown in Figure 17.10. Since x must satisfy *both* sets of inequalities, we conclude from the diagram that the solution of the given inequality is given by the "overlap"

$$-1 < x < 1 - \sqrt{2}, \qquad 1 + \sqrt{2} < x < 3 \qquad\qquad \blacktriangleleft$$

E X E R C I S E S / S E C T I O N   **17.3**

In Exercises 1–20, solve the given inequalities.

1. $|x - 1| < 2$ **2.** $|x + 1| < 2$ **3.** $|x + 3| \le 1$

4. $|x - 2| < 4$ **5.** $|2x + 1| < 1$ **6.** $|3x - 3| > 2$

7. $|2x - 4| \ge 3$ **8.** $|4x - 8| \le 4$ **9.** $|2x + 7| < 6$

10. $|3x + 2| > 4$ 11. $|2 - x| > 3$ 12. $|1 - 2x| < 5$

13. $|x^2 - x - 9| > 3$ 14. $|x^2 - x - 9| < 3$

15. $|x^2 - 4x + 2| > 2$ 16. $|x^2 - x + 1| > 3$

17. $|x^2 - 2x - 7| < 1$ 18. $|x^2 + 5x + 2| < 2$

19. $|x^2 + 6x + 2| < 3$ 20. $|x^2 + 5x + 2| > 2$

21. The diameter d of a shaft is 2.5550 in., and the tolerance of the shaft is 0.0001 in. Express this statement as an inequality containing absolute values.

22. An athlete can run a mile in approximately 4 min 20 s, expressed more precisely as $|T - (4 \text{ min } 20 \text{ s})| \le 5$ s. What are his fastest and slowest times?

17.4 Graphing Inequalities

In this section we will continue our study of the graphs of inequalities begun in Section 17.1. First we consider single inequalities, and then systems of inequalities.

To recall the basic ideas, consider the example of a linear inequality below.

E X A M P L E 1 Graph the inequality $x - 2y < 4$.

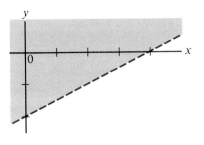

Figure 17.11

Solution. First we need to graph the line $x - 2y = 4$. If $y = 0$, then $x = 4$; if $x = 0$, $y = -2$. So the intercepts are $(4, 0)$ and $(0, -2)$. The resulting graph is the dashed line in Figure 17.11. (Recall that the line is shown as a dashed line to indicate that the points on the line do not belong to the solution set.) Next we check a point not on the line. For example, at $(0, 0)$ we get, from the given inequality, $0 - 0 < 4$, which satisfies the relation. So the coordinates of all the points on the side of the line containing the origin satisfy the inequality. (See Figure 17.11.) ◄

The next example illustrates the graphical solution of a quadratic inequality.

E X A M P L E 2 Graph the inequality $y \ge x^2 + 1$.

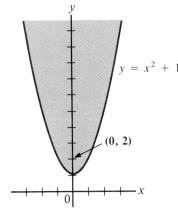

Figure 17.12

Solution. We first graph the curve $y = x^2 + 1$ from the following table:

x:	-3	-2	-1	0	1	2	3
y:	10	5	2	1	2	5	10

The graph is shown as a solid curve in Figure 17.12.

Now we check a point not on the curve. For the point $(0, 2)$, we get

$$y \ge x^2 + 1$$

$$2 > 0 + 1$$

which satisfies the inequality. The inequality is therefore satisfied by the coordinates of all the points in the shaded region shown in Figure 17.12. ◄

Systems of Inequalities

To graph a **system of inequalities**, we first graph the individual inequalities. In each case the solution is a region in which the coordinates of every point satisfy the particular inequality. The common solution of the system is the resulting *intersection*, where all the regions overlap. These ideas are illustrated in the remaining examples.

E X A M P L E **3**   Solve the following system of inequalities graphically:

$$y > x$$

$$y < 2x$$

Solution. The points whose coordinates satisfy the inequality $y > x$ lie in the region above the line $y = x$, shown in Figure 17.13. Similarly, the points whose coordinates satisfy the inequality $y < 2x$ lie below the line $y = 2x$ (Figure 17.13). The solution must therefore be the intersection, where both regions overlap. The intersection is shown in Figure 17.14.

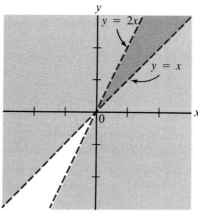

Figure 17.13   Figure 17.14

E X A M P L E **4**   Sketch the region in which the coordinates of the points satisfy the given inequalities:

$$x + y \leq 2$$

$$x \geq 0$$

$$y \geq 0$$

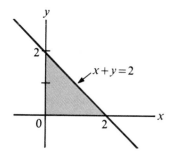

Figure 17.15

Solution. The region defined by $x \geq 0$ and $y \geq 0$ is quadrant I (including the nonnegative coordinate axes). The points whose coordinates satisfy $x + y \leq 2$ lie below the line $x + y = 2$. The points whose coordinates satisfy all three inequalities lie in the shaded region shown in Figure 17.15.   ◄

E   X   A   M   P   L   E   **5**   Graph the following system of inequalities:

$$y > -3x$$

$$x + 4y < 11$$

$$y > \frac{2}{3}x$$

Solution. To find the proper region, we need to solve the corresponding equations in pairs. The first two equations

$$y = -3x$$

$$x + 4y = 11$$

may be solved by the method of substitution:

$$x + 4(-3x) = 11$$

$$x - 12x = 11$$

$$x = -1, \quad y = 3$$

So the lines intersect at $(-1, 3)$. (See Figure 17.16.)
   Similarly, the lines

$$x + 4y = 11$$

$$y = \frac{2}{3}x$$

intersect at $(3, 2)$. Finally, the lines

$$y = -3x$$

$$y = \frac{2}{3}x$$

intersect at the origin.

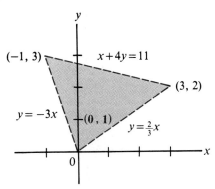

Figure 17.16

The dashed lines and their intersections are shown in Figure 17.16. As can be readily checked, the shaded region in Figure 17.16 is the intersection of the three solution sets. For example, the coordinates of the point **(0, 1)** satisfy all three inequalities.

◀

E X A M P L E **6**   Graph the following system of inequalities:

$$y \le x$$

$$y \ge x^2$$

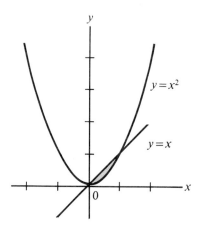

Figure 17.17

Solution. Solving the corresponding system of equations, $y = x$ and $y = x^2$, by substitution, we get

$$x^2 = x$$

$$x^2 - x = 0$$

$$x(x - 1) = 0$$

$$x = 0, 1$$

So the curves intersect at $(0, 0)$ and $(1, 1)$. The curve $y = x^2$ is sketched from the following table of values:

x:	-2	-1	0	1	2
y:	4	1	0	1	4

The shaded region between the graphs, shown in Figure 17.17, is the solution set.

◀

E X E R C I S E S / S E C T I O N   **17.4**

In each of the following exercises, sketch the region in which the coordinates of the points satisfy the given inequalities.

1. $y \le 3x + 6$

2. $y > 4 - 2x$

3. $y > x$
$y < 3x$

4. $y > x^2$

5. $y > x^2 - 1$

6. $y < x^3$

7. $y \le x$
$y \le 2 - x$
$y \ge 0$

8. $1 < x < 2$

9. $0 \le y \le 1$
$y \le 2 - x$
$x \ge 0$

10. $2y + x \le 2$
$x \ge 0$
$y \ge 0$

11. $x + y \le 4$
$x + 2y \le 6$
$x \ge 0$
$y \ge 0$

12. $x + 2y \le 6$
$2x + y \le 6$
$x \ge 0$
$y \ge 0$

13. $2x + y \ge 6$
$x + y \ge 5$
$x \ge 0$
$y \ge 0$

14. $3x + y \ge 4$
$x + 2y \ge 3$
$x \ge 0$
$x \ge 0$

15. $y \le x^2$
$y \ge x^3$

16. $y > x^2 - 4$
$y < 4 - 2x$

17. $y > x^2 - 2$
$y < x$

18. $y \ge x^2$
$y < x + 6$

17.5   Linear Programming (Optional)

Graphs of inequalities have important applications in a field called **linear programming**. Linear programming is a technique for determining which of several

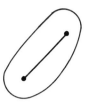

Figure 17.18

Figure 17.19

available options will yield the greatest benefit. For example, the method may be used to determine the maximum profit or the minimum cost, given certain constraints. (The term *linear programming* has nothing to do with computer programming.)

As we saw in the last section, a given set of inequalities determines a region. These inequalities are called the **constraints**, and the region is called the **set of feasible solutions**, or simply the **feasible region**. Linear programming is confined to regions that are **convex**: A line segment connecting two points in the region also lies in the region. For example, the region in Figure 17.18 is convex, but the region in Figure 17.19 is not.

If the boundary of the region consists of straight lines, then the region is called **polygonal**. For example, the two regions shown in Figure 17.20 are convex polygonal regions. Each corner is called a **vertex**.

In linear programming, the quantity whose maximum or minimum value is to be determined is called the **objective function**. The objective function contains two variables and must have the form $P = ax + by + c$. Because of the given constraints, the problem is *to find the maximum or minimum value of P over the feasible region*. The following theorem is fundamental to linear programming.

> The maximum or minimum value of an objective function must occur at a vertex of a convex polygonal region.

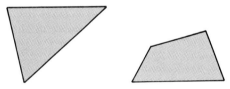

Figure 17.20

The example below illustrates the technique.

E X A M P L E **1**   Given the following constraints:

$$x + 2y \geq 7$$
$$x + y \geq 5$$
$$x \geq 0$$
$$y \geq 0$$

find the minimum value of $P = 4x + 6y$.

Solution. To obtain the feasible region, we need to solve the system of equations

$$x + 2y = 7$$
$$x + y = 5$$

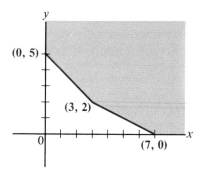

Figure 17.21

simultaneously. Subtracting the second equation from the first, we get $y = 2$. If $y = 2$, then $x = 3$, so that the lines intersect at $(3, 2)$. By the method of the last section, we obtain the feasible region shown in Figure 17.21. (Recall that the feasible region is the solution set of the given system of inequalities.)

By the fundamental theorem, the minimum value of P occurs at one of the vertices. The value at each of the three vertices is given in the following chart:

Vertex	$P = 4x + 6y$
$(0, 5)$	$P = 4 \cdot 0 + 6 \cdot 5 = 30$
$(3, 2)$	$P = 4 \cdot 3 + 6 \cdot 2 = 24$
$(7, 0)$	$P = 4 \cdot 7 + 6 \cdot 0 = 28$

The minimum value of $P = 24$ occurs at the vertex $(3, 2)$.

To see why the fundamental theorem guarantees a minimum or maximum value at a vertex, let P assume some arbitrary values. For example, if $P = 48$, we get

$$4x + 6y = 48$$

If $P = 36$, we get

$$4x + 6y = 36$$

The resulting parallel lines are shown in Figure 17.22.

We can see from the figure that the values of P get smaller as the lines move downward. The minimum value is obtained by moving the line downward as far as possible without leaving the feasible region, which occurs at the vertex $(3, 2)$. (Below this point, the values of x and y do not satisfy the constraints.)

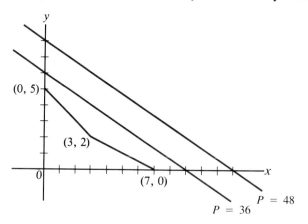

Figure 17.22

E X A M P L E **2**   An electronics manufacturer produces two types of calculators, the basic model and the scientific model. Each model is produced by two sets of operations, electronic assembly (the first operation) and case assembly (the second operation).

For the basic model, the time required for each operation is 5 min. For the scientific model, the first operation requires 10 min, but the second operation only 2 min, since some of the assembly is done in the first operation. The time available for the first operation is at most $133\frac{1}{3}$ h (8,000 min) per month, and for the second at most $73\frac{1}{3}$ h (4,400 min) per month. (During the remaining time, the equipment is used for manufacturing other products.) If the profit is $5 for each basic and $8 for each scientific calculator, how many of each should be produced per month to maximize the profit?

Solution. Let

$$x = \text{number of basic calculators}$$

and

$$y = \text{number of scientific calculators}$$

Then $5x + 10y$ is the total time in minutes required for the first operation and $5x + 2y$ the time required for the second operation. The constraints are therefore given by

$$5x + 10y \le 8,000$$
$$5x + 2y \le 4,400$$
$$x \ge 0$$
$$y \ge 0$$

(The last two conditions say that the number of calculators produced cannot be negative.)

The objective function representing the profit is

$$P = 5x + 8y$$

The feasible region is shown in Figure 17.23.

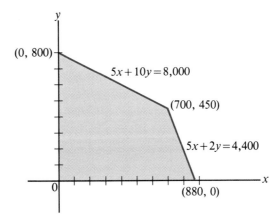

Figure 17.23

The profit evaluated at the different vertices is shown in the following chart:

Vertex	$P = 5x + 8y$
(0, 800)	$P = 5 \cdot 0 + 8 \cdot 800 = \$6{,}400$
(700, 450)	$P = 5 \cdot 700 + 8 \cdot 450 = \$7{,}100$
(880, 0)	$P = 5 \cdot 880 + 8 \cdot 0 = \$4{,}400$

Consequently, the manufacturer should produce 700 basic and 450 scientific calculators per month for a maximum profit of $7,100.   ◄

E X A M P L E **3**   Suppose the management of the firm discussed in Example 2 decides that it cannot sell more than 1,000 calculators per month. How many calculators should then be produced to maximize the profit?

Solution. The extra constraint $x + y \leq 1{,}000$ leads to the following system of inequalities:

$$5x + 10y \leq 8{,}000$$
$$x + y \leq 1{,}000$$
$$5x + 2y \leq 4{,}400$$
$$x \geq 0$$
$$y \geq 0$$

To obtain the feasible region, the corresponding equations have to be solved in pairs. The lines

$$5x + 10y = 8{,}000$$
$$x + y = 1{,}000$$

intersect at (400, 600) and the lines

$$x + y = 1{,}000$$
$$5x + 2y = 4{,}400$$

intersect at (800, 200). The feasible region is shown in Figure 17.24. The values of P at the different vertices are shown in the following chart:

Vertex	$P = 5x + 8y$
(0, 800)	$P = 5 \cdot 0 + 8 \cdot 800 = \$6{,}400$
(400, 600)	$P = 5 \cdot 400 + 8 \cdot 600 = \$6{,}800$
(800, 200)	$P = 5 \cdot 800 + 8 \cdot 200 = \$5{,}600$
(880, 0)	$P = 5 \cdot 880 + 8 \cdot 0 = \$4{,}400$

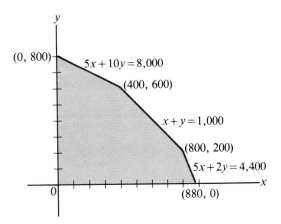

Figure 17.24

We conclude that the manufacturer should produce 400 basic and 600 scientific calculators per month for a maximum profit of \$6,800. ◀

If a linear programming problem contains three variables, then a three-dimensional feasible region is required. More than three variables requires an abstract algebraic approach, which is commonly carried out with the aid of a computer.

E X E R C I S E S / S E C T I O N   **17.5**

In Exercises 1–9, find the minimum or maximum value for P, subject to the given constraints.

1. Maximum P
Constraints: Objective function:
$x + y \le 2$ $P = 2x + y$
$x \ge 0$
$y \ge 0$

2. Maximum P
Constraints: Objective function:
$x + y \le 2$ $P = x + 3y$
$0 \le y \le 1$
$x \ge 0$

3. Minimum P
Constraints: Objective function:
$2x + y \ge 2$ $P = 3x + 2y$
$x \ge 0$
$y \ge 0$

4. Maximum P
Constraints: Objective function:
$x + y \le 3$ $P = 2x + y$
$0 \le x \le 2$
$y \ge 0$

5. Maximum P
Constraints: Objective function:
$x + y \le 4$ $P = 3x + 4y$
$x + 2y \le 5$
$x \ge 0$
$y \ge 0$

6. Minimum P
Constraints: Objective function:
$2x + y \ge 6$ $P = 5x + 4y$
$x + y \ge 5$
$x \ge 0$
$y \ge 0$

7. Minimum P
Constraints: Objective function:
$2x + y \ge 7$ $P = 3x + 4y$
$x + y \ge 5$
$2x + 3y \ge 12$
$x \ge 0$
$y \ge 0$

8. Maximum P

Constraints: Objective function:
$x + 3y \leq 15$ $P = 5x + 12y$
$x + 2y \leq 11$
$x + \ y \leq 8$
$x \geq 0$
$y \geq 0$

9. Maximum P

Constraints: Objective function:
$x + 2y \leq 20$ $P = 3x + 2y$
$x + \ y \leq 11$
$2x + \ y \leq 14$
$3x + \ y \leq 18$
$x \geq 0$
$y \geq 0$

10. Farmer Schmidt and his son wish to raise corn and soybeans on part of their farm. Each acre of corn requires 4 h of labor for the father and 5 h for the son during the season and yields a profit of $140. Each acre of soybeans requires 2 h for the father and 3 h for the son and yields a profit of $75. Because of other farm chores, the father does not wish to spend more than 200 h during the season raising corn and soybeans, and the son not more than 270 h. How many acres of each crop should they plant to maximize their profit? What is the maximum profit?

11. If in Exercise 10 the profit on each acre of corn is $120 and on each acre of soybeans $90, how many acres of each should Farmer Schmidt and his son plant, and what is the maximum profit?

12. A manufacturer produces widgets and gadgets. Each widget requires 4 h of assembly time and 1 h of spraying time. Each gadget requires 3 h assembly time and 2 h spraying time. Assembling is done during the 8-h day shift for a maximum of 40 h per week. Spraying is done during the night shift for a maximum of 20 h per week. The profit on each widget is $15,000 and on each gadget $12,000. How many of each should be produced to maximize the profit? What is the maximum profit?

13. Farmer Armstrong has set aside a section of his farm for growing fruit trees. Each smaller fruit tree requires 150 ft^2 of space and 2 lb of insecticide. Each larger fruit tree requires 200 ft^2 of space but only 1 lb of insecticide. He can spare at most 5,000 ft^2 for the trees, and he cannot afford

more than 50 lb of insecticide. His profit is $290 for each smaller tree and $380 for each larger tree. Determine how many of each kind of tree he should plant for the largest possible profit. What is the profit?

14. A corporation has at most 2,400 ft^2 available to store two types of motors. Each small motor requires 15 ft^2 of storage space, and each large motor 20 ft^2. Storage and maintenance costs are $4 for each small motor and $5 for each large motor. Suppose no more than $620 is to be spent on storage and maintenance. If the profit on the sale of each small motor is $150 and on each large motor $190, how many of each should be stored to obtain a maximum profit? What is the maximum profit?

15. As a sideline, the members of a family produce two types of high-quality lawn mowers, making some of the parts themselves and buying the rest. Type I requires 3 h for making the parts, and Type II, 4 h. Type I requires 3 h for assembly, but Type II only 2 h. Since they hold regular jobs, the members of the family cannot spend more than a total of 56 h per month making parts and 39 h on assembly. The profit is $100 for a Type I mower and $120 for a Type II, and the total number that can be sold each month is no more than 15. How many lawn mowers of each type should be made to maximize the profit? What is the maximum profit?

16. If in Exercise 15 the profit on the Type I lawn mower increases to $135, while the profit on the Type II remains the same, how many of each should be produced and what will the maximum profit be?

17. A certain kind of dog food sold in a bag is to be made from two types of products. Each pound of Product I contains 100 g of protein and 150 g of carbohydrates. Each pound of Product II contains 50 g of protein and 200 g of carbohydrates. The cost of Product I is 45¢ per pound, and of Product II, 30¢ per pound. If each bag must contain at least 400 g of protein and at least 1,100 g of carbohydrates, determine how many pounds of each product must be used per bag in order to minimize the cost. What is the minimum cost per bag?

18. In Exercise 17, how much of each product must be used to minimize the cost if the price of Product I is increased to 50¢ per pound and the price of Product II is decreased to 20¢ per pound? What is the minimum cost?

REVIEW EXERCISES/CHAPTER   17

In Exercises 1–22, solve each inequality.

1. $2x + 2 < x - 3$
2. $5x - 1 \geq x + 4$
3. $x + 1 > 3x - 5$
4. $1 - 3x \geq 2x + 3$
5. $x^2 - 9 \leq 0$
6. $x^2 - x - 6 \leq 0$

7. $(x - 3)(x + 1)(x + 4) > 0$ **8.** $x^2 - x < 0$

9. $\dfrac{x - 1}{x + 2} \le 0$ **10.** $\dfrac{x - 2}{x + 4} \le 0$

11. $\dfrac{x - 6}{x - 1} > 0$ **12.** $\dfrac{x - 5}{2 - x} < 0$

13. $\dfrac{(x - 2)(x - 1)}{x + 4} \ge 0$ **14.** $\dfrac{(x + 3)(x + 5)}{x - 1} \le 0$

15. $(x + 2)(x - 1)^2 \le 0$ **16.** $(x - 1)^2(x + 3) \ge 0$

17. $|x - 4| \le 2$ **18.** $|x - 1| > 3$

19. $|2x + 2| < 3$ **20.** $|1 - x| < 2$

21. $|x^2 - 4x - 4| < 1$ **22.** $|x^2 - 2x - 2| > 1$

In Exercises 23–28, sketch the region in which the coordinates of the points satisfy the given inequalities.

23. $y > 4x + 2$ **24.** $y \le 3 - 6x$ **25.** $y \le \sqrt{x}$
$0 \le x \le 4$
$y \ge 0$

26. $y < x^2$ **27.** $2y > x$ **28.** $2x + y \le 8$
$y < x$ $x + y \le 5$
$x \ge 0$
$y \ge 0$

29. The cost (in dollars) of operating a certain machine is $C = 100 + 4t$, where t is measured in hours. How long can the machine be run before the cost exceeds $126?

30. A technician receives $250 per week plus $18.75/h for overtime, so that his pay can be expressed as $P = 250 + 18.75t$. For what value of t is $P \ge 350$?

31. If an object is hurled upward with a velocity of 80 ft/s from a point 146 ft above the ground, its distance above the ground as a function of time (in seconds) is given by $s = 146 + 80t - 16t^2$. Find the time for which $s \le 50$ ft. (The object is hurled upward when $t = 0$ s.)

32. From past experience, the produce manager of a supermarket estimates that the number S of watermelons sold on July 3 will be $|S - 150| \le 25$. Determine the anticipated sale.

In Exercises 33–34, find the minimum or maximum value for P, subject to the given constraints.

33. Maximum P

Constraints:	Objective function:
$x + 2y \le 20$	$P = 3x + 4y$
$x + y \le 14$	
$x \ge 0$	
$y \ge 0$	

34. Minimum P

Constraints:	Objective function:
$2x + y \ge 11$	$P = x + y$
$x + 2y \ge 10$	
$x \ge 0$	
$y \ge 0$	

35. Anticipating a price increase, the owner of a warehouse stocks up on two types of appliances. Each small appliance requires 10 ft^2 of space and is expected to sell for $100. Each large appliance requires 15 ft^2 of space and is expected to sell for $130. The storage cost is $10 for each small appliance and $12 for each large appliance. If at most 900 ft^2 are available for storage and at most $840 for the cost of storage, how many of each type should be stored to maximize the amount of money received? What is the maximum amount?

Progressions and the Binomial Theorem

18.1 Arithmetic Progressions

Arithmetic progression
Common difference

In science and technology we occasionally encounter a problem involving a sequence of numbers following a definite pattern. One such sequence, called an **arithmetic progression**, will be studied in this section. Another type, called a **geometric progression**, will be taken up in the next section.

In an **arithmetic progression** every term differs from the previous term by a fixed amount. This fixed amount is called the **common difference**. For example, the sequence

$$1, 5, 9, 13, \ldots$$

is an arithmetic progression with common difference 4. (The three dots indicate that the sequence continues indefinitely.)

If the first term is a and the common difference d, then the general form of the arithmetic progression is

$$a, a + d, a + 2d, a + 3d, \ldots$$

The nth Term of an Arithmetic Progression is

$$l = a + (n - 1)d \tag{18.1}$$

E X A M P L E **1** The first three terms of an arithmetic progression are 1, 4, and 7. Find the fourteenth term.

Solution. Note that $a = 1$, $d = 4 - 1 = 3$, and $n = 14$. It follows that

$$l = 1 + (14 - 1)(3) = 40 \qquad \blacktriangleleft$$

E X A M P L E **2** Find the fifteenth term of the following arithmetic progression:

$$-5, -1, 3, 7, 11, \ldots$$

Solution. Since $a = -5$, $d = -1 - (-5) = 4$, and $n = 15$, we get

$$l = -5 + (15 - 1)(4) = 51 \qquad \blacktriangleleft$$

The sum of the first n terms of an arithmetic progression can be found by observing that the sum is equal to n times the average of the first and last terms.

Sum of the First n Terms of an Arithmetic Progression:

$$s = \frac{n}{2}(a + l) \qquad (18.2)$$

E X A M P L E **3** Find the sum of the first 100 positive integers.

Solution. Since $a = 1$ and $l = 100$, we get

$$s = \frac{100}{2}(1 + 100) = 5{,}050 \qquad \blacktriangleleft$$

E X A M P L E **4** Find the sum of the first 20 terms of the following arithmetic progression:

$$-2, 5, 12, 19, \ldots$$

Solution. To find the sum s by the above formula, we first find the 20th term. Since $a = -2$, $d = 5 - (-2) = 7$, and $n = 20$, we get

$$l = -2 + (20 - 1)(7) = 131$$

So

$$s = \frac{n}{2}(a + l) = \frac{20}{2}(-2 + 131) = 1{,}290 \qquad \blacktriangleleft$$

 E X A M P L E **5** The velocity of a certain falling object is 42 ft/s after the first second, 74 ft/s after the next second, 106 ft/s after the third second, and so on. Determine the velocity after 8 s.

Solution. The arithmetic progression is

$$42, 74, 106, \ldots$$

Since $a = 42$, $d = 74 - 42 = 32$, and $n = 8$, we get

$$l = 42 + (8 - 1)(32) = 266 \frac{\text{ft}}{\text{s}}$$
◀

The common difference d may be a negative number, as shown in the next example.

   E X A M P L E   **6**   The interest paid on a loan was $200 for the first month and decreased by \$8 per month as the loan was being paid off. Determine the total interest paid at the end of one year.

Solution. Note that the progression is

$$200, 192, 184, \ldots$$

To find the sum, we must first determine l: $a = 200$, $d = -8$, and $n = 12$. So

$$l = 200 + (12 - 1)(-8) = 112$$

and the sum is

$$s = \frac{12}{2}(200 + 112) = \$1,872$$
◀

E X E R C I S E S / S E C T I O N   18.1

In Exercises 1–10, determine the nth term of each arithmetic progression.

1. $4, 7, 10, 13, \ldots$ $(n = 9)$

2. $5, 9, 13, 17, \ldots$ $(n = 11)$

3. $-6, -4, -2, 0, 2, \ldots$ $(n = 16)$

4. $-6, -3, 0, 3, \ldots$ $(n = 13)$

5. $0, 6, 12, \ldots$ $(n = 12)$

6. $-1, 2, 5, 8, \ldots$ $(n = 14)$

7. $\dfrac{1}{2}, 1, \dfrac{3}{2}, 2, \ldots$ $(n = 10)$

8. $-\dfrac{1}{3}, \dfrac{1}{3}, 1, \dfrac{5}{3}, \ldots$ $(n = 7)$

9. $5, \dfrac{33}{7}, \dfrac{31}{7}, \ldots$ $(n = 8)$

10. $11, 10.7, 10.4, \ldots$ $(n = 14)$

In Exercises 11–16, find the sum of each arithmetic progression.

11. $2, 4, 6, \ldots$ $(n = 16)$

12. $-3, 0, 3, 6, \ldots$ $(n = 12)$

13. $5, 8, 11, 14, \ldots$ $(n = 14)$

14. $97, 93, 89, \ldots$ $(n = 10)$

15. $107, 102, 97, \ldots$ $(n = 10)$

16. $\dfrac{1}{4}, \dfrac{1}{2}, \dfrac{3}{4}, 1, \ldots$ $(n = 12)$

17. Determine the formula for the sum s of the first n positive integers. (See Example 3.)

18. Determine the formula for the sum s of the first n even positive integers.

19. A machine costing $12,000 depreciates 10% during the first year, 9.5% during the second year, 9.0% during the third year, and so on. Determine the value of the machine after 10 years, assuming that all of the percentages apply to the original cost.

20. A consultant charges $120 for the first day, $126 for the second day, $132 for the third day, and so on. Determine the cost of a 12-day consultation.

18.2   Geometric Progressions

Another sequence following a definite pattern is a **geometric progression**.

In a geometric progression, each number after the first in the sequence can be determined by multiplying the preceding term by a number called the **common ratio**. For example, the sequence

Common ratio

$$1, \frac{1}{3}, \frac{1}{9}, \frac{1}{27}, \ldots$$

has a common ratio $\frac{1}{3}$, while the sequence

$$3, \frac{3}{2}, \frac{3}{4}, \frac{3}{8}, \ldots$$

has a common ratio $\frac{1}{2}$. The general definition is stated next.

> A **geometric progression** is a sequence of numbers of the form
>
> $$a, ar, ar^2, ar^3, \ldots, ar^{n-1} \qquad (18.3)$$
>
> where r is called the **common ratio**. The nth term is
>
> $$a_n = ar^{n-1} \qquad (18.4)$$

These definitions are illustrated in the first three examples.

E X A M P L E 1   Write the first six terms of the geometric progression whose first term is 2 and whose common ratio is 3.

Solution. By the sequence (18.3), we have

$$2, 2(3), 2(3^2), 2(3^3), 2(3^4), 2(3^5)$$

Note that the largest exponent is $n - 1$, where $n = 6$. ◀

E X A M P L E 2   Write the tenth term of the geometric progression whose first term is 3 and whose common ratio is $\frac{1}{2}$.

Solution. By formula (18.4), the nth term is ar^{n-1}. Since $a = 3$ and $r = \frac{1}{2}$, we get for the tenth term

$$a_{10} = 3\left(\frac{1}{2}\right)^{10-1} = 3\left(\frac{1}{2}\right)^{9} = \frac{3}{2^9} = \frac{3}{512}$$

◄

 E X A M P L E **3**   A certain radioactive substance has a half-life of 3 years, which means that the amount present is cut in half every 3 years. If there are 20 g of the substance initially, how many grams are left at the end of 24 years?

Solution. Since the amount is cut in half every 3 years, the common ratio is $r = \frac{1}{2}$; the first term is $a = 20$. After 3 years, the given amount of 20 g is cut in half for the first time, leaving $20\left(\frac{1}{2}\right)$ g; this is the second term a_2. After 6 years, the amount is cut in half for the second time, leaving $20\left(\frac{1}{2}\right)^2$ g; this is the third term a_3. After 24 years, the amount is cut in half for the 8th time; the corresponding term is

$$a_9 = 20\left(\frac{1}{2}\right)^{9-1} = 20\left(\frac{1}{2}\right)^{8} = \frac{20}{256} = \frac{5}{64}\ \text{g}$$

◄

It is possible to find the sum of the terms of a geometric progression. Let

$$s = a + ar + ar^2 + \cdots + ar^{n-1}$$

If we multiply s by r and subtract the resulting expression from s, we get

$$
\begin{aligned}
s &= a + ar + ar^2 + \cdots + ar^{n-1} \\
rs &= \qquad\ \ ar + ar^2 + \cdots + ar^{n-1} + ar^n \\
\hline
s - rs &= a \qquad\qquad\qquad\qquad\qquad\ - ar^n \\
s(1 - r) &= a(1 - r^n) \qquad\qquad\qquad \text{factoring} \\
s &= \frac{a(1 - r^n)}{1 - r} \qquad\qquad\qquad \text{dividing by } 1 - r
\end{aligned}
$$

Sum of the First n Terms of a Geometric Progression:

$$s = a + ar + ar^2 + \cdots + ar^{n-1} = \frac{a(1 - r^n)}{1 - r} \qquad (r \neq 1) \qquad (18.5)$$

E X A M P L E **4**   Find the following sum, given that the terms form a geometric progression:

$$s = 3 + 6 + 12 + 24 + 48 + 96$$

Solution. The first term is $a = 3$ and the second term is $ar = 3r = 6$, so that $r = 2$. Since $96 = 3(32) = 3(2^5)$, we have $n - 1 = 5$ and $n = 6$. Since $ar^{n-1} = 3(2)^{6-1}$, the sum is

$$s = \frac{3(1 - 2^6)}{1 - 2} = \frac{3(-63)}{-1} = 189$$ ◀

An *annuity* is a sequence of equal payments at equal time intervals. Formula (18.5) can be used to compute the value of an annuity, as illustrated in the next example.

E X A M P L E **5** One hundred dollars is invested each year at 12% compounded annually. Determine the value of the annuity at the end of 8 years (before the ninth payment is made).

Solution. Suppose we examine the growth of the first payment over the years. At the end of the first year, the first $100 has grown to $100 + 100(0.12) = 100(1 + 0.12) = 100(1.12)$. At the end of the second year, the amount is $[100(1.12)](1.2) = 100(1.12)^2$; and at the end of the third year, $100(1.12)^3$. Continuing in this manner, at the end of the eighth year, the amount has grown to $100(1.12)^8$.

The payment made at the beginning of the second year has only 7 years in which to grow, leading to a final amount of $100(1.12)^7$. The third payment grows to $100(1.12)^6$, and so on.

The last payment draws interest for only 1 year, at which time the amount is $100(1.12)$.

The total amount A of the annuity is given by the sum of the terms of the geometric progression

$$\begin{aligned} A &= 100(1.12) + 100(1.12)^2 + 100(1.12)^3 + \cdots + 100(1.12)^8 \\ &= 100[1.12 + (1.12)^2 + (1.12)^3 + \cdots + (1.12)^8] \\ &= 100[1.12 + 1.12(1.12) + 1.12(1.12)^2 + \cdots + 1.12(1.12)^7] \end{aligned}$$

Note that for the sum inside the brackets, we have $a = 1.12$, $r = 1.12$, and

$$ar^{n-1} = 1.12(1.12)^7 = 1.12(1.12)^{8-1}$$

By formula (18.5),

$$A = 100 \frac{1.12[1 - (1.12)^8]}{1 - 1.12} = \$1,377.57$$

The sequence is

$$100 \boxed{\times} 1.12 \boxed{\times} \boxed{(} \boxed{1} \boxed{-} 1.12 \boxed{y^x} 8 \boxed{)} \boxed{\div} \boxed{(} \boxed{1} \boxed{-}$$
$$1.12 \boxed{)} \boxed{=} \rightarrow 1377.5656$$ ◀

E X A M P L E **6**   Determine the value of the annuity in Example 5 if the fund pays 12% interest compounded quarterly.

Solution. When interest is compounded quarterly, 3% interest is paid every three months. So at the end of 1 year, $100 accumulates to $100(1.03)^4$; at the end of 2 years, to $100\,[(1.03)^4]^2$. At the end of 8 years, the first payment grows to $100\,[(1.03)^4]^8$.

The second payment has only 7 years to grow, leading to a final amount of $100\,[(1.03)^4]^7$.

So the value of the annuity is

$$100\{(1.03)^4 + [(1.03)^4]^2 + [(1.03)^4]^3 + \cdots + [(1.03)^4]^8\}$$
$$= 100\{(1.03)^4 + (1.03)^4(1.03)^4 + (1.03)^4[(1.03)^4]^2 + \cdots$$
$$+ (1.03)^4[(1.03)^4]^7\}$$

For the sum inside the braces, we have

$$a = (1.03)^4, \qquad r = (1.03)^4, \qquad \text{and} \qquad ar^{n-1} = (1.03)^4[(1.03)^4]^{8-1}$$

It follows that

$$A = 100(1.03)^4 \, \frac{1 - [(1.03)^4]^8}{1 - (1.03)^4} = \$1{,}412.47$$

◀

E X E R C I S E S / S E C T I O N   **18.2**

---

1. Find the sixth term of the geometric progression whose first term is 4 and whose common ratio is $\frac{2}{3}$.

2. Find the fourth term of the geometric progression whose first term is $\frac{1}{2}$ and whose common ratio is $\frac{3}{4}$.

In Exercises 3–6, find the nth term of the geometric progression.

3. $a = 5, r = -\dfrac{1}{2}, n = 10$ **4.** $a = 4, r = -\dfrac{2}{3}, n = 6$

5. $a = 2, r = -3, n = 5$ **6.** $a = 3, r = -4, n = 5$

In Exercises 7–18, the terms in each sum form a geometric progression. Find each sum by the method of Example 4.

7. $1 + 2 + 4 + 8 + 16 + 32$

8. $1 + 3 + 9 + 27 + 81$

9. $2 + 6 + 18 + 54 + 162$

10. $3 + 6 + 12 + 24 + 48 + 96 + 192$

11. $1 + \dfrac{1}{2} + \dfrac{1}{4} + \dfrac{1}{8} + \dfrac{1}{16} + \dfrac{1}{32} + \dfrac{1}{64}$

12. $1 + \dfrac{1}{3} + \dfrac{1}{9} + \dfrac{1}{27} + \dfrac{1}{81}$

13. $1 - \dfrac{1}{3} + \dfrac{1}{9} - \dfrac{1}{27} \quad \left(\text{Note: } r = -\dfrac{1}{3}.\right)$

14. $1 - \dfrac{1}{2} + \dfrac{1}{4} - \dfrac{1}{8} + \dfrac{1}{16} - \dfrac{1}{32}$

15. $2 + \dfrac{4}{3} + \dfrac{8}{9} + \dfrac{16}{27} + \dfrac{32}{81}$

16. $3 + \dfrac{9}{4} + \dfrac{27}{16} + \dfrac{81}{64} + \dfrac{243}{256}$

17. $3 - \dfrac{3}{2} + \dfrac{3}{4} - \dfrac{3}{8} + \dfrac{3}{16} - \dfrac{3}{32} + \dfrac{3}{64}$

18. $4 - \dfrac{4}{3} + \dfrac{4}{9} - \dfrac{4}{27} + \dfrac{4}{81}$

19. Five hundred dollars is invested at 8%. Find the value of the investment after one year if interest is compounded **(a)** annually; **(b)** semiannually; **(c)** quarterly; **(d)** monthly.

20. One thousand dollars is invested at 12%. Find the value of the investment after 1 year if interest is compounded **(a)** annually; **(b)** semiannually; **(c)** quarterly; **(d)** daily.

21. What is the value of an investment of $600 after 5 years if it earns 10% interest compounded semiannually?

22. What is the value of an $800 investment after 6 years if it earns 12% interest compounded quarterly?

23. Determine the value of the following annuity: $100 per year for 5 years (before the sixth payment is made). The interest is 10% compounded annually.

24. Determine the value of the following annuity: $200 per year for 6 years (before the seventh payment is made). Interest is 8% compounded annually.

25. What is the value of the annuity in Exercise 23 if the interest is compounded **(a)** semiannually; **(b)** quarterly?

26. Determine the value of the annuity in Exercise 24 if the interest is compounded semiannually.

27. How much will you accumulate at the end of 10 years (before the eleventh payment is made) by depositing $150 per year at 9% interest compounded annually?

28. Show that after n years (before payment $n + 1$ is made), the value of an annuity A_n consisting of P dollars deposited yearly at $100i$% interest compounded annually is given by

$$A_n = \frac{P(1 + i)[1 - (1 + i)^n]}{1 - (1 + i)} = \frac{P[(1 + i)^{n+1} - (1 + i)]}{i}$$

29. The number of bacteria in a culture doubles every 2 h. If there are N bacteria initially, how many will there be 20 h later?

30. The number of bacteria in a culture trebles every hour. If there are N bacteria initially, how many will there be after 8 h?

31. A certain radioactive substance has a half-life of 1 h. If there are N grams initially, how much will be left after 12 h? (See Example 3.)

32. If a radioactive substance has a half-life of 2 h, what fractional part of N grams will be left after 24 h?

33. A ball is tossed upward from the ground to a height of 10 ft. If it rebounds one-half as far as it falls, find the distance traveled when it hits the ground for the fifth time.

34. A ball is tossed upward from the ground to a height of 15 ft. On each rebound it reaches $\frac{1}{3}$ of the height from which it fell. Determine the distance the ball has traveled when it hits the ground for the sixth time.

35. A ball dropped from a height of 12 ft rebounds $\frac{3}{4}$ as far as it falls. Find the distance the ball has traveled when it hits the ground for the seventh time.

18.3 Geometric Series

If a geometric progression is continued indefinitely, it is called an **infinite geometric progression**. If the common ratio is properly restricted, the sum of the terms can be defined, even though the number of terms is infinitely large. Such a sum is called a **geometric series**.

Recall formula (18.5):

$$s = a + ar + ar^2 + \cdots + ar^{n-1} = \frac{a(1 - r^n)}{1 - r}$$

Since the formula is valid for all n, the size of n need not be restricted. Suppose n gets ever larger, beyond any bound; does it still make sense to speak of the sum? It does if we note that r^n gets ever smaller if $|r| < 1$. For example, if $r = \frac{1}{2}$, then $(\frac{1}{2})^n$ approaches 0. Consequently, if $|r| < 1$, then

$$\frac{a(1 - r^n)}{1 - r} \qquad \text{approaches} \qquad \frac{a}{1 - r}$$

This observation suggests the following definition.

Sum of a Geometric Series: The sum of the geometric series

$$S = a + ar + ar^2 + \cdots + ar^{n-1} + \cdots \qquad (18.6)$$

is defined as

$$S = \frac{a}{1 - r} \qquad \text{if} \qquad |r| < 1 \qquad (18.7)$$

To show that formula (18.7) is indeed a reasonable definition, let us check the formula by changing a decimal to a common fraction.

E X A M P L E **1**   Change $0.121212\ldots$ to a common fraction.

Solution. The decimal fraction $N = 0.121212\ldots$ can be written

$$N = 0.12 + 0.0012 + 0.000012 + \cdots$$

$$= \frac{12}{10^2} + \frac{12}{10^4} + \frac{12}{10^6} + \cdots$$

Since $a = 12/10^2$ and $r = 1/10^2$, we get

$$N = \frac{a}{1-r} = \frac{\dfrac{12}{10^2}}{1 - \dfrac{1}{10^2}} = \frac{\left(\dfrac{12}{10^2}\right)(10^2)}{\left(1 - \dfrac{1}{10^2}\right)(10^2)} = \frac{12}{10^2 - 1} = \frac{12}{99} = \frac{4}{33}$$

As a check, long division gives $4 \div 33 = 0.121212\ldots$. ◀

The remaining examples further illustrate the technique for finding the sum of a geometric series.

E X A M P L E **2**   Find the sum of the geometric series

$$S = \frac{2}{5} - \left(\frac{2}{5}\right)^2 + \left(\frac{2}{5}\right)^3 - \left(\frac{2}{5}\right)^4 + \cdots$$

Solution. Note that the series can be written

$$S = \frac{2}{5} + \frac{2}{5}\left(-\frac{2}{5}\right) + \frac{2}{5}\left(-\frac{2}{5}\right)^2 + \frac{2}{5}\left(-\frac{2}{5}\right)^3 + \cdots$$

So by formula (18.6), $a = \frac{2}{5}$ and $r = -\frac{2}{5}$. Thus

$$S = \frac{a}{1-r} = \frac{\frac{2}{5}}{1 - \left(-\frac{2}{5}\right)} = \frac{\frac{2}{5}}{1 + \frac{2}{5}} = \frac{\frac{2}{5}}{\frac{7}{5}} = \frac{2}{5} \cdot \frac{5}{7} = \frac{2}{7}$$

◀

  E X A M P L E **3** A ball is tossed upward from the ground to a height of 6 ft. On each rebound it reaches $\frac{1}{2}$ its former height. How far will it travel before coming to rest?

Solution. The total distance d is given by

$$d = 2(6) + 2\left(\frac{6}{2}\right) + 2\left(\frac{6}{4}\right) + 2\left(\frac{6}{8}\right) + 2\left(\frac{6}{16}\right) + \cdots$$

Thus $a = 2(6) = 12$ and $r = \frac{1}{2}$. Hence

$$d = \frac{12}{1 - \frac{1}{2}} = 24 \text{ ft}$$

◀

E X E R C I S E S / S E C T I O N **18.3**

In Exercises 1–8, convert the decimal fractions to common fractions.

1. 0.4444 . . .

2. 0.6666 . . .

3. 0.131313 . . .

4. 0.212121 . . .

5. 0.312312 . . .

6. 0.613613 . . .

7. 0.1343434 . . .

8. 0.5727272 . . .

In Exercises 9–20, find the sum of the given geometric series.

9. $\dfrac{1}{2} + \dfrac{1}{4} + \dfrac{1}{8} + \dfrac{1}{16} + \cdots$

10. $\dfrac{1}{3} + \dfrac{1}{9} + \dfrac{1}{27} + \dfrac{1}{81} + \cdots$

11. $1 - \dfrac{1}{3} + \dfrac{1}{9} - \dfrac{1}{27} + \cdots$

12. $1 - \dfrac{1}{2} + \dfrac{1}{4} - \dfrac{1}{8} + \dfrac{1}{16} - \cdots$

13. $1 - \dfrac{1}{4} + \dfrac{1}{4^2} - \dfrac{1}{4^3} + \cdots$

14. $2 - 2\left(\dfrac{3}{5}\right) + 2\left(\dfrac{3}{5}\right)^2 - 2\left(\dfrac{3}{5}\right)^3 + \cdots$

15. $3 + 3\left(\dfrac{5}{9}\right) + 3\left(\dfrac{5}{9}\right)^2 + 3\left(\dfrac{5}{9}\right)^3 + \cdots$

16. $\dfrac{3}{4} + \dfrac{9}{16} + \dfrac{27}{64} + \cdots$

17. $\dfrac{3}{5} - \left(\dfrac{3}{5}\right)^2 + \left(\dfrac{3}{5}\right)^3 - \cdots$

18. $\dfrac{2}{3} - \dfrac{2}{9} + \dfrac{2}{27} - \dfrac{2}{81} + \cdots$

19. $4 - 3 + \dfrac{9}{4} - \dfrac{27}{16} + \dfrac{81}{64} - \cdots$

20. $5 - 2 + \dfrac{4}{5} - \dfrac{8}{25} + \dfrac{16}{125} - \cdots$

21. Each swing of a pendulum is 90% of its former swing. If the first swing is 10 in. long, find the distance that the tip of the pendulum travels before coming to rest.

22. A ball is dropped from a height of 9 m. If it reaches $\frac{2}{5}$ of its height on each rebound, how far does it travel before coming to rest?

23. A ball rebounds $\frac{3}{5}$ as far as it falls. If dropped from a height of 12 ft, how far does it travel before coming to rest?

24. A mass hanging on a spring oscillates up and down. Each oscillation is $\frac{3}{5}$ the preceding one. If the first oscillation is 4 in., how far does it travel before coming to rest?

25. A particle oscillates in a magnetic field. Each oscillation is $\frac{5}{8}$ the preceding one. If the first oscillation is 3 mm, how far does the particle travel before coming to rest?

18.4   Binomial Theorem and Series

In this section we will consider another sequence of numbers, one that arises in a formula called the **binomial theorem**. The purpose of this formula is to expand a binomial of the form $(a + b)^n$. This binomial form occurs often enough in certain technical applications and in advanced mathematics to warrant a separate section.

For small values of n, the binomial $(a + b)^n$ can be multiplied directly:

$$(a + b)^0 = 1$$

$$(a + b)^1 = a + b$$

$$(a + b)^2 = a^2 + 2ab + b^2$$

$$(a + b)^3 = a^3 + 3a^2b + 3ab^2 + b^3$$

$$(a + b)^4 = a^4 + 4a^3b + 6a^2b^2 + 4ab^3 + b^4$$

These forms have certain properties in common:

1. The first term is a^n, and the last term is b^n.
2. The exponent of a decreases by 1 from term to term, starting with a^n and ending with a^0.
3. The exponent of b increases by 1 from term to term, starting with b^0 and ending with b^n.
4. The exponents of a and b add up to n in each term.
5. The number of terms is $n + 1$.

In these expansions, only the coefficients do not seem to follow an obvious pattern. Yet a pattern does exist: Each coefficient is the sum of the coefficients immediately above it (when the expansion is laid out in pyramid fashion as shown above). Noted earlier by Chinese mathematicians and rediscovered by the French mathematician Blaise Pascal (1623–1662), the pattern can be described by means of **Pascal's triangle:**

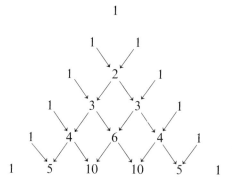

E X A M P L E **1** Expand $(x + y)^6$ by means of Pascal's triangle.

Solution. The next row of the triangle, obtained by adding adjacent numbers is,

$$1 \quad 6 \quad 15 \quad 20 \quad 15 \quad 6 \quad 1$$

Thus

$$(x + y)^6 = x^6 + 6x^5y + 15x^4y^2 + 20x^3y^3 + 15x^2y^4 + 6xy^5 + y^6 \qquad \blacktriangleleft$$

The Binomial Theorem

The pattern emerging from Pascal's triangle can also be described mathematically. To do so, we need the definition of *factorial*:

Factorial:

The product

$$n(n - 1)(n - 2) \cdots (2)(1) \qquad\qquad (18.8)$$

is called **n factorial** and is denoted by **n!**

For example,

$$5! = 5 \cdot 4 \cdot 3 \cdot 2 \cdot 1 = 120$$

$$8! = 8 \cdot 7 \cdot 6 \cdot 5 \cdot 4 \cdot 3 \cdot 2 \cdot 1 = 40{,}320$$

With the factorial notation, the binomial theorem can be stated in the form given next.

Binomial Theorem:

$$(a + b)^n = a^n + na^{n-1}b + \frac{n(n-1)}{2!}a^{n-2}b^2$$

$$+ \frac{n(n-1)(n-2)}{3!}a^{n-3}b^3 + \cdots + b^n \qquad (18.9)$$

The binomial theorem is illustrated in the next two examples.

E X A M P L E **2** Expand $(x + y)^4$ by the binomial theorem.

Solution. By the binomial theorem with $n = 4$,

$$(x + y)^4 = x^4 + 4x^3y + \frac{(4)(3)}{2!}x^2y^2 + \frac{(4)(3)(2)}{3!}xy^3 + \frac{(4)(3)(2)(1)}{4!}y^4$$

$$= x^4 + 4x^3y + \frac{4 \cdot 3}{2 \cdot 1}x^2y^2 + \frac{4 \cdot 3 \cdot 2}{3 \cdot 2 \cdot 1}xy^3 + \frac{4 \cdot 3 \cdot 2 \cdot 1}{4 \cdot 3 \cdot 2 \cdot 1}y^4$$

$$= x^4 + 4x^3y + 6x^2y^2 + 4xy^3 + y^4 \qquad \blacktriangleleft$$

E X A M P L E **3** Expand $(2x + 3y)^5$.

Solution. In this binomial we have $a = 2x$ and $b = 3y$. So by the binomial theorem

$$[(2x) + (3y)]^5 = (2x)^5 + 5(2x)^4(3y) + \frac{(5)(4)}{2!}(2x)^3(3y)^2$$

$$+ \frac{(5)(4)(3)}{3!}(2x)^2(3y)^3 + \frac{(5)(4)(3)(2)}{4!}(2x)(3y)^4$$

$$+ \frac{(5)(4)(3)(2)(1)}{5!}(3y)^5$$

$$= 32x^5 + 240x^4y + 720x^3y^2 + 1080x^2y^3 + 810xy^4 + 243y^5$$

◀

In some applications of the binomial theorem, only the first few terms may be of interest.

E X A M P L E **4** Find the first five terms of the binomial $(v - 2t)^{10}$.

Solution. Writing the binomial in the form $[v + (-2t)]^{10}$, we get

$$[v + (-2t)]^{10} = v^{10} + 10v^9(-2t) + \frac{(10)(9)}{2!}v^8(-2t)^2$$

$$+ \frac{(10)(9)(8)}{3!}v^7(-2t)^3 + \frac{(10)(9)(8)(7)}{4!}v^6(-2t)^4 + \cdots$$

$$= v^{10} - 20v^9t + 180v^8t^2 - 960v^7t^3 + 3{,}360v^6t^4 - \cdots$$

◀

The Binomial Series

It can be shown that the binomial theorem is valid for any rational n under certain conditions. Of particular interest is the special case $a = 1$ in formula (18.9), which is usually called the **binomial series**.

Binomial Series:

$$(1 + b)^n = 1 + nb + \frac{n(n - 1)}{2!}b^2 + \frac{n(n - 1)(n - 2)}{3!}b^3 + \cdots \qquad (18.10)$$

where $|b| < 1$.

Note especially that whenever n is not a positive integer, the expansion does not have a last term. Such an expansion is called an **infinite series**, of which the geometric series is another example. The binomial series is valid only if $|b| < 1$.

E X A M P L E **5** Use the binomial series to approximate $1/\sqrt[3]{0.894}$ to three decimal places.

Solution. To obtain the proper form, we write 0.894 as $1 - 0.106$. So by the binomial series

$$(1 - 0.106)^{-1/3} = 1 + \left(-\frac{1}{3}\right)(-0.106) + \frac{\left(-\frac{1}{3}\right)\left(-\frac{4}{3}\right)}{2!}(-0.106)^2$$

$$+ \frac{\left(-\frac{1}{3}\right)\left(-\frac{4}{3}\right)\left(-\frac{7}{3}\right)}{3!}(-0.106)^3 + \cdots$$

Using the first four terms, we get

$$(0.894)^{-1/3} = 1 + 0.035333 + 0.0024969 + 0.00020585$$
$$= 1.0380$$

Since the next term is

$$\frac{\left(-\frac{1}{3}\right)\left(-\frac{4}{3}\right)\left(-\frac{7}{3}\right)\left(-\frac{10}{3}\right)}{4!}(-0.106)^4 = 0.000018$$

the use of additional terms does not affect the first three decimal places. Thus, $1/\sqrt[3]{0.894} = 1.038$.

If the evaluation of a radical requires more decimal places than a calculator can provide, the binomial series is a convenient method to use. ◀

E X E R C I S E S / S E C T I O N **18.4**

In Exercises 1–4, use Pascal's triangle to expand the given binomials.

1. $(x - y)^4$ **2.** $(a - T)^3$ **3.** $(s + 2v)^4$ **4.** $(a + 3x)^3$

In Exercises 5–16, use the binomial theorem to expand the given binomials.

5. $(x - y)^3$ **6.** $(x - y)^4$ **7.** $(s + 2t)^4$
8. $(M + 3N)^3$ **9.** $(a - b^2)^5$ **10.** $(V^2 - T)^5$
11. $(S - 2W)^6$ **12.** $(2x + y)^6$ **13.** $(R + 2)^6$
14. $(x - y)^7$ **15.** $(R - 3)^6$ **16.** $(3a + b)^6$

In Exercises 17–22, write the first four terms of the binomial expansion.

17. $(x + 2)^{10}$ **18.** $(y + 3)^8$ **19.** $(k - m^2)^{12}$
20. $(r^2 - e)^9$ **21.** $(r + 2\pi)^8$ **22.** $(1 - \gamma)^{10}$

In Exercises 23–26, write the first four terms of the binomial series (18.10).

23. $(1 + x)^{-2}$ **24.** $(1 - R)^{-2}$
25. $(1 + s)^{1/2}$ **26.** $(1 + p)^{1/3}$

27. Write $(1 - x)^{-1}$ as a binomial series and compare it to the geometric series.

28. Write $(1 + x)^{-1}$ as a binomial series and compare it to the geometric series.

29. Approximate $\sqrt[3]{0.912}$ to four decimal places. (See Example 5.)

30. Approximate $\sqrt[3]{0.892}$ to three decimal places.

31. Approximate $1/\sqrt{0.832}$ to three decimal places.

32. Approximate $1/\sqrt{0.917}$ to four decimal places.

33. The period of a pendulum having a large swing is determined by means of so-called elliptic integrals. In the calculation, the expression

$$\frac{1}{\sqrt{1 - \sin^2 \dfrac{\theta}{2} \sin^2 \phi}}$$

has to be expanded. By letting $b = \sin^2(\theta/2) \sin^2 \phi$, show that

$$\frac{1}{\sqrt{1 - \sin^2 \dfrac{\theta}{2} \sin^2 \phi}}$$

$$= 1 + \frac{1}{2} \sin^2 \frac{\theta}{2} \sin^2 \phi + \frac{3}{8} \sin^4 \frac{\theta}{2} \sin^4 \phi + \cdots$$

34. When finding the derivative of the function $f(x) = x^4$ in calculus, the following expression has to be simplified:

$$\frac{(x + h)^4 - x^4}{h}$$

Carry out this simplification.

35. In determining the length of the curve $y = \frac{1}{2}x^2$, $(-1 \le x \le 1)$, the expression $\sqrt{1 + x^2}$ arises. Find an approximation for this expression by writing the first four terms of the binomial series.

36. The deflection of a certain beam is given by $y = x(x - a/2)^4$. Expand this expression.

REVIEW EXERCISES/CHAPTER   **18**

In Exercises 1–2, determine the nth term of each arithmetic progression.

1. 2, 7, 12, ... $(n = 10)$

2. -2, 2, 6, 10, ... $(n = 12)$

In Exercises 3–4, find the sum of each arithmetic progression.

3. 47, 43, 39, ... $(n = 9)$

4. 1.3, 1.9, 2.5, 3.1, ... $(n = 14)$

5. Find the sixth term of the geometric progression whose first term is 2 and common ratio is $\frac{1}{3}$.

6. Find the fifth term of the geometric progression whose first term is 3 and common ratio is $\frac{3}{4}$.

In Exercises 7–12, find each sum, given that the terms form a geometric progression.

7. $1 - \dfrac{1}{2} + \dfrac{1}{4} - \dfrac{1}{8} + \dfrac{1}{16} - \dfrac{1}{32} + \dfrac{1}{64}$

8. $1 - \dfrac{1}{3} + \dfrac{1}{9} - \dfrac{1}{27} + \dfrac{1}{81}$

9. $5 + \dfrac{10}{3} + \dfrac{20}{9} + \dfrac{40}{27} + \dfrac{80}{81}$

10. $2 + \dfrac{4}{5} + \dfrac{8}{25} + \dfrac{16}{125}$

11. $2 - \dfrac{4}{3} + \dfrac{8}{9} - \dfrac{16}{27} + \dfrac{32}{81} - \dfrac{64}{243}$

12. $3 - \dfrac{3}{5} + \dfrac{3}{25} - \dfrac{3}{125}$

In Exercises 13–18, find the sum of the given geometric series.

13. $1 - \dfrac{1}{3} + \dfrac{1}{3^2} - \dfrac{1}{3^3} + \cdots$

14. $1 - \dfrac{2}{3} + \left(\dfrac{2}{3}\right)^2 - \left(\dfrac{2}{3}\right)^3 + \cdots$

15. $2 + \dfrac{4}{5} + \dfrac{8}{25} + \dfrac{16}{125} + \cdots$

16. $3 + \dfrac{3}{4} + \dfrac{3}{16} + \dfrac{3}{64} + \cdots$

17. $3 + \dfrac{9}{4} + \dfrac{27}{16} + \dfrac{81}{64} + \cdots$

18. $3 + \dfrac{6}{5} + \dfrac{12}{25} + \dfrac{24}{125} + \cdots$

In Exercises 19–22, convert the decimal fractions to common fractions.

19. $0.1555\ldots$

20. $0.2777\ldots$

21. $0.383838\ldots$

22. $0.5383838\ldots$

In Exercises 23–30, expand the given expressions by means of the binomial theorem.

23. $(a - b)^3$

24. $(a - b)^4$

25. $(M - 2T)^5$

26. $(\pi + 2)^4$

27. $(e - 2)^4$

28. $(s + t^2)^4$

29. $(m - v^2)^5$

30. $(a + 2)^6$

In Exercises 31–35, write the first four terms of the binomial expansion.

31. $(x - 2)^{10}$

32. $(R - 1)^9$

33. $(1 - t)^{-2}$

34. $(1 + v)^{1/2}$

35. $(1 - c)^{-1/3}$

36. Show that the binomial expansion of $(1 - a)^{-1}$ agrees with the geometric series.

37. Approximate $\sqrt[3]{0.907}$ to four decimal places by means of the binomial series.

38. Approximate $1/\sqrt[3]{0.936}$ to four decimal places by means of the binomial series.

39. Determine the value of a $600 investment after 10 years if it earns 10% interest compounded semiannually.

40. Determine the value of a $1,000 investment after 8 years if it earns 12% interest compounded quarterly.

41. Two hundred dollars is invested yearly at 10% compounded quarterly. Determine the value of the annuity after 4 years (before the fifth payment is made).

42. Determine the value of the following annuity: $100 invested yearly for 5 years at 8% interest compounded quarterly.

43. The number of bacteria in a culture trebles every 2 h. If there are N bacteria initially, how many will there be at the end of 16 h?

44. A certain radioactive substance has a half-life of 2 years. If there are N grams of the substance initially, what fractional part will be left after 20 years?

45. A ball is dropped from a height of 10 ft. On each rebound, it reaches $\frac{1}{3}$ its former height. What is the total distance traveled when it hits the ground for the sixth time?

46. Each swing of a pendulum is 90% of the preceding swing. If the initial swing is 10 in. long, find the total distance that the tip of the pendulum has traveled at the end of 8 swings.

47. Each swing of a pendulum is 95% of the preceding swing. If the first swing is 5 in., how far will the tip travel before coming to rest?

48. A weight on a spring oscillates up and down, each oscillation being $\frac{4}{5}$ the preceding one. If the first oscillation is 6 cm, how far will the weight travel before coming to rest?

49. The *rigidity modulus* of a material is

$$M = \frac{E}{2(s + 1)}$$

Write M as a binomial series, assuming that $|s| < 1$.

In Exercises 1–7, verify the identities.

1. $\cos^3 \theta + \sin^2 \theta \cos \theta = \cos \theta$

2. $\dfrac{1 + \cot^2 x}{1 + \tan^2 x} = \cot^2 x$

3. $\dfrac{1 + \cos \theta}{\sin \theta} + \dfrac{\sin \theta}{1 + \cos \theta} = 2 \csc \theta$

4. $\dfrac{\tan \alpha + \cot \alpha}{\csc \alpha} = \sec \alpha$

5. $\dfrac{1}{\sec x + \tan x} = \sec x - \tan x$

6. $\dfrac{2 - \sec^2 \theta}{\sec^2 \theta} = \cos 2\theta$

7. $\sin^2 \dfrac{\theta}{4} = \dfrac{1 - \cos(\theta/2)}{2}$

In Exercises 8 and 9, find the exact values without tables or calculators.

8. $\cos 60° \cos 15° + \sin 60° \sin 15°$

9. $\cos 15°$

10. Use a calculator to solve the following equation for x $(0° \le \theta < 360°)$:

$$\sec^2 x - 2 \tan x - 1 = 0$$

11. Find the exact values of **(a)** $\text{Arctan}(-\sqrt{3})$; **(b)** $\text{Arccos}(-\tfrac{1}{2})$.

In Exercises 12 and 13, solve the given inequalities for x.

12. $\dfrac{(x - 2)(x + 3)}{(x - 1)(x + 4)} \le 0$ **13.** $|x^2 - 4x + 2| > 2$

14. Find the sum of the geometric series

$$\frac{1}{2} - \frac{1}{4} + \frac{1}{8} - \frac{1}{16} + \cdots$$

15. Expand $(x - 2y)^5$ by the binomial theorem.

16. The range R of a projectile fired at angle θ with the horizontal at velocity v is given by

$$R = \frac{2v^2 \cos \theta \sin \theta}{g}$$

Write R as a single trigonometric function of θ.

17. Starting at $t = 0$ the current in a circuit is given by $i = 2.0 \sin 3.0t$. Find the smallest value of t (in seconds) for which $i = 0.50$ A.

18. The current in a coil rotating in a magnetic field is given by $I = I_{\max} \cos \theta$, where θ is the angle that the coil makes with the field. Solve the equation for θ, $0 \le \theta \le \pi$.

19. A ball is tossed upward from the ground to a height of 10 ft. On each rebound it reaches $\frac{1}{2}$ the height from which it fell. Determine the distance the ball will have traveled when it strikes the ground for the sixth time.

20. Each swing of a pendulum is 95% of its former swing. If the first swing is 5 in. long, find the distance that the tip of the pendulum will travel before coming to rest.

21. An object starts from rest and undergoes a constant acceleration of 2.1 m/s². Find its velocity after 40 s.

Introduction to Analytic Geometry

19.1 Introduction

Analytic geometry may be defined as the study of classical geometry by means of algebra. Credit for the discovery of this method is usually given to René Descartes (1596–1650), the famous French mathematician and philosopher, but the same method was discovered independently and somewhat earlier by Pierre de Fermat (1601–1665), a French lawyer.

Prior to the development of analytic geometry, algebra and geometry were largely independent of each other—lines belonged to geometry and equations belonged to algebra. The fusion of these two branches of mathematics led to entirely new scientific applications, as we will see later in this chapter. Analytic geometry also paved the way for the discovery of the calculus a few decades later.

19.2 The Distance Formula

For our first topic in analytic geometry, we will learn how to find the distance between two points in a plane. First recall from Chapter 2 that points are located in a plane by means of the **rectangular** or **Cartesian coordinate system**. (Please refer to Section 2.9 for the basic definitions.)

In the Cartesian coordinate system, fixed points are usually distinguished by subscripts such as P_1, P_2, P_3 or by (x_1, y_1) or (x_2, y_2). Using this notation, let us find a useful formula for the distance between two arbitrary points $P_1(x_1, y_1)$ and $P_2(x_2, y_2)$, shown in Figure 19.1.

First join P_1 and P_2 by a straight line. Then construct a right triangle by drawing a line through P_1 parallel to the x-axis and a line through P_2 parallel to the y-axis. Denote the point of intersection by P_3. The coordinates of P_3 are (x_2, y_1).

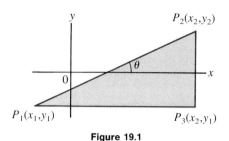

Figure 19.1

The length of the segment P_1P_3 is $x_2 - x_1$ and the length of P_2P_3 is $y_2 - y_1$. (See Exercise 12 at the end of the section.) Since P_1P_2 is the hypotenuse, from the Pythagorean theorem we get

$$(P_1P_2)^2 = (P_1P_3)^2 + (P_2P_3)^2 \qquad \text{or} \qquad P_1P_2 = \sqrt{(x_2 - x_1)^2 + (y_2 - y_1)^2}$$

Distance Formula

The distance d from (x_1, y_1) to (x_2, y_2) is given by

$$d = \sqrt{(x_2 - x_1)^2 + (y_2 - y_1)^2} \tag{19.1}$$

E X A M P L E **1** Find the distance between $(-2, -4)$ and $(6, -2)$.

Solution. Let $(x_2, y_2) = (-2, -4)$ and $(x_1, y_1) = (6, -2)$. Then

$$d = \sqrt{(x_2 - x_1)^2 + (y_2 - y_1)^2} = \sqrt{(-2 - 6)^2 + [-4 - (-2)]^2}$$
$$= \sqrt{64 + 4} = \sqrt{68} = \sqrt{4 \cdot 17} = 2\sqrt{17} \approx 8.25$$

The same result is obtained by letting $(x_2, y_2) = (6, -2)$ and $(x_1, y_1) = (-2, -4)$. ◀

E X E R C I S E S / S E C T I O N **19.2**

In Exercises 1–9, plot each pair of points and find the distance between them.

1. $(2, 4)$ and $(5, 2)$ **2.** $(-3, 2)$ and $(5, -4)$

3. $(-3, -6)$ and $(5, -2)$ **4.** $(0, 0)$ and $(-\sqrt{5}, 2)$

5. $(0, 2)$ and $(\sqrt{3}, 4)$ **6.** $(\sqrt{2}, \sqrt{5})$ and $(\sqrt{2}, 0)$

7. $(1, -\sqrt{2})$ and $(-1, 0)$ **8.** $(1, 6)$ and $(-3, 2)$

9. $(-10, -2)$ and $(-12, 0)$

10. Draw the triangle whose vertices are $(0, 0), (4, 3)$, and $(6, 0)$.

11. If (x, y) is a point, determine in which quadrant x/y is
(a) positive; (b) negative.

12. In the derivation of the distance formula show that the length of P_1P_3 is $x_2 - x_1$ and the length of P_2P_3 is $y_2 - y_1$, no matter in which quadrants the points lie. The distance formula is therefore valid for all points. (See Figure 19.1.)

13. Find the set of all points whose (a) abscissas are zero; (b) ordinates are zero.

14. Find the perimeter of the triangle in Exercise 10.

15. Show that $(-2, 5)$, $(3, 5)$, and $(-2, 2)$ are vertices of a right triangle.

16. Repeat Exercise 15 for $(1, 7)$, $(-2, 1)$, and $(10, -5)$.

17. Show that $(12, 0)$, $(-4, 8)$, and $(-1, -13)$ lie on a circle whose center is $(1, -2)$.

18. Show that $(-2, 10)$, $(3, -2)$, and $(15, 3)$ are vertices of an isosceles triangle.

19. Show that $(-1, -1)$, $(2, 8)$, and $(5, 17)$ are collinear, that is, lie on the same line.

20. Find the relation between x and y so that (x, y) is 3 units from $(-1, 2)$.

21. Find the relation between x and y so that (x, y) is equidistant from the y-axis and the point $(2, 0)$.

19.3 The Slope

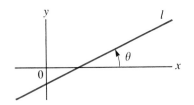

Figure 19.2

We saw in Section 19.2 that every point has a definite location with respect to the coordinate axes. A line also has a definite position. For example, the angle θ in Figure 19.2 is a measure of steepness, just as the steepness of a road is measured with respect to the horizontal. This angle θ $(0° \leq \theta < 180°)$ is called the *angle of inclination*. It turns out to be more convenient, however, to define steepness by the formula $m = \tan \theta$, called the **slope** of the line.

Definition of Slope

$$m = \tan \theta \qquad \text{(where } \theta = \text{angle of inclination)} \tag{19.2}$$

By the definition of the tangent function, m can be viewed as the vertical distance divided by the horizontal distance or by the change in the y-direction with respect to the change in the x-direction. (The importance of this idea will be seen in Chapter 20.)

We can obtain a formula for the slope of a line determined by points $P_1(x_1, y_1)$ and $P_2(x_2, y_2)$, as shown in Figures 19.3 and 19.4. From the previous section, $P_1P_3 = x_2 - x_1$ and $P_2P_3 = y_2 - y_1$. Since $\alpha = \theta$ (so that $\tan \alpha = \tan \theta$),

$$m = \tan \theta = \tan \alpha = \frac{y_2 - y_1}{x_2 - x_1}$$

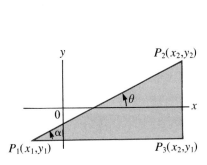

Figure 19.3

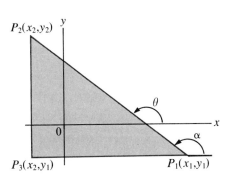

Figure 19.4

Slope of a Line

The slope m of the line passing through the points (x_1, y_1) and (x_2, y_2) is

$$m = \frac{y_2 - y_1}{x_2 - x_1} = \frac{y_1 - y_2}{x_1 - x_2} \qquad (19.3)$$

Note that the angle θ in Figure 19.3 is acute $(0° < \theta < 90°)$, so that the line rises to the right. As a result, $x_2 - x_1 > 0$ and $y_2 - y_1 > 0$, so that m is positive. This also follows from the fact that $m = \tan\theta$ is positive for any angle between $0°$ and $90°$.

In Figure 19.4, θ is obtuse $(90° < \theta < 180°)$, so that the line rises to the left. It follows that m is negative since $x_2 - x_1 < 0$ and $y_2 - y_1 > 0$. This agrees with the fact that $m = \tan\theta$ is negative for any angle between $90°$ and $180°$.

When $\theta = 0°$—that is, when the line is horizontal—$m = 0$. If the line is vertical, then $\theta = 90°$. Since $\tan 90°$ is undefined, m is undefined.

Summary: $m > 0$, or positive, when θ is acute (line rising to the right), $m < 0$, or negative, when θ is obtuse (line rising to the left), $m = 0$ when $\theta = 0°$ (horizontal line), and m is undefined when $\theta = 90°$ (vertical line).

E X A M P L E 1   Find the slope of the line passing through $(-6, -2)$ and $(2, 4)$.

Solution. Let $(x_1, y_1) = (-6, -2)$ and $(x_2, y_2) = (2, 4)$. Then

$$m = \frac{y_2 - y_1}{x_2 - x_1} = \frac{4 - (-2)}{2 - (-6)} = \frac{6}{8} = \frac{3}{4}$$

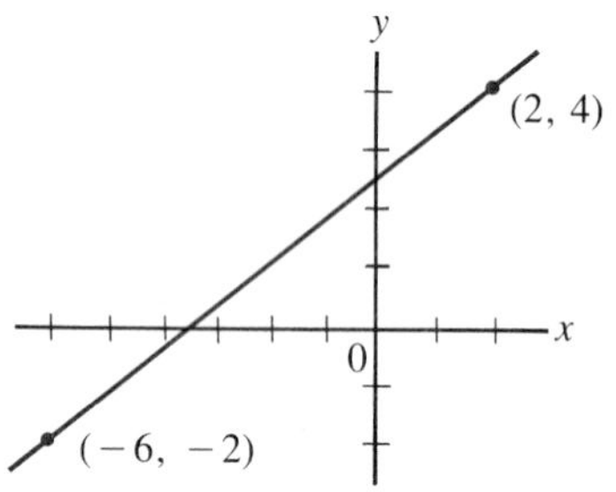

(See Figure 19.5.) Note that m is positive since the line rises to the right (θ acute). We get the same result by letting $(x_1, y_1) = (2, 4)$ and $(x_2, y_2) = (-6, -2)$:

$$m = \frac{y_2 - y_1}{x_2 - x_1} = \frac{-2 - 4}{-6 - 2} = \frac{-6}{-8} = \frac{3}{4}$$

◀

Figure 19.5

E X A M P L E 2   Find the slope of the line determined by $(-2, 5)$ and $(3, -1)$.

Solution. Let $(x_1, y_1) = (3, -1)$ and $(x_2, y_2) = (-2, 5)$. Then

$$m = \frac{y_2 - y_1}{x_2 - x_1} = \frac{5 - (-1)}{-2 - 3} = \frac{6}{-5} = -\frac{6}{5}$$

(See Figure 19.6.) Here m is negative since the line rises to the left (θ obtuse).

◀

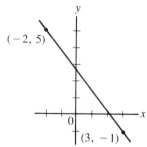

Figure 19.6

E X A M P L E **3**   Determine the slope of the line passing through $(-4, -2)$ and $(5, -2)$.

Solution.

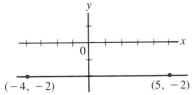

$$m = \frac{-2 - (-2)}{5 - (-4)} = \frac{0}{9} = 0$$

Figure 19.7

Note that $m = 0$ because the line is horizontal. (See Figure 19.7.)   ◀

If the angle of inclination is close to $90°$, then m is numerically large. For example, for the line in Figure 19.8,

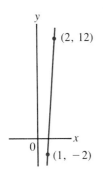

$$m = \frac{12 - (-2)}{2 - 1} = \frac{14}{1} = 14$$

Figure 19.8

As noted earlier, if the line is vertical the slope is undefined (infinitely large). If we calculate the slope of the line in Figure 19.9, for example, we get

$$m = \frac{4 - (-3)}{-1 - (-1)} = \frac{7}{0} \quad \text{(undefined)}$$

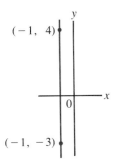

In other words, when a line is vertical, $x_1 = x_2$, so that $x_2 - x_1 = 0$. Since division by zero is not allowed, the slope is undefined.

The slope can be used to determine whether two lines are parallel or perpendicular. If two lines are parallel, they have equal angles of inclination and therefore equal slopes. If two lines are perpendicular, their angles of inclination differ by $90°$—that is, $\theta_2 = \theta_1 + 90°$. From Figure 19.10, we have

Figure 19.9

$$\tan \theta_1 = \frac{b}{a}$$

and

$$\tan \theta_2 = \tan(\theta_1 + 90°) = \frac{a}{-b} = -\frac{1}{b/a} = -\frac{1}{\tan \theta_1}$$

Thus $m_2 = -1/m_1$. These ideas are summarized next.

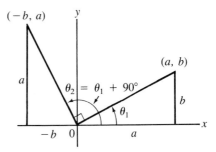

Figure 19.10

> Two lines are **parallel** if
>
> $$m_1 = m_2 \tag{19.4}$$
>
> Two lines are **perpendicular** if
>
> $$m_2 = -\frac{1}{m_1} \quad \text{or} \quad m_1 m_2 = -1 \tag{19.5}$$

E X A M P L E **4**   Show that the line determined by $(-1, 4)$ and $(5, 2)$ is parallel to the line determined by $(-2, -1)$ and $(1, -2)$.

Solution. The lines are shown in Figure 19.11.

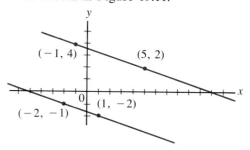

Figure 19.11

The respective slopes are
$$m_1 = \frac{2-4}{5-(-1)} = \frac{-2}{6} = -\frac{1}{3}$$
$$m_2 = \frac{-2-(-1)}{1-(-2)} = \frac{-1}{3} = -\frac{1}{3}$$

$$m_1 = m_2$$

Since the slopes are equal, the lines are parallel.   ◀

E X A M P L E **5**   Show that the line through $(7, -1)$ and $(4, -6)$ is perpendicular to the line through $(2, 0)$ and $(-3, 3)$.

Solution. The respective slopes are

$$m_1 = \frac{-1-(-6)}{7-4} = \frac{5}{3} \quad \text{and} \quad m_2 = \frac{0-3}{2-(-3)} = -\frac{3}{5} \qquad m_1 = -\frac{1}{m_2}$$

Since the slopes are negative reciprocals, the lines are perpendicular. (See Figure 19.12.)   ◀

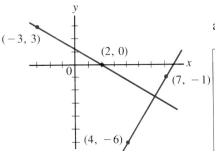

Figure 19.12

We can easily find the midpoint of the line segment with endpoints $P_1(x_1, y_1)$ and $P_2(x_2, y_2)$, as shown in Figure 19.13.

> **Midpoint Formula:** The midpoint of the line segment joining (x_1, y_1) and (x_2, y_2) is
>
> $$\left(\frac{x_1 + x_2}{2}, \frac{y_1 + y_2}{2}\right) \qquad (19.6)$$

To derive this formula, let $P(x, y)$ be the midpoint of the line segment P_1P_2. Draw the triangle shown in Figure 19.13 by constructing lines perpendicular to the x-axis from $P_1(x_1, y_1)$, $P(x, y)$, and $P_2(x_2, y_2)$. It follows that $2(x - x_1) = x_2 - x_1$ and $2(y - y_1) = y_2 - y_1$. Solving, we get $x = \frac{1}{2}(x_1 + x_2)$ and $y = \frac{1}{2}(y_1 + y_2)$, which are the coordinates of the midpoint of the line segment.

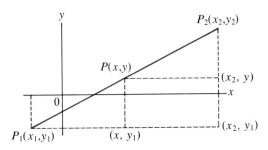

Figure 19.13

EXERCISES / SECTION 19.3

In Exercises 1–13, find the slope of the line through the given points.

1. $(2, 6)$ and $(1, 7)$ **2.** $(-3, -10)$ and $(-5, 2)$

3. $(0, 2)$ and $(-4, -4)$ **4.** $(6, -3)$ and $(4, 0)$

5. $(7, 8)$ and $(-3, -4)$ **6.** $(0, 0)$ and $(8, -4)$

7. $(0, 0)$ and $(-4, -8)$ **8.** $(1, 3)$ and $(-2, 0)$

9. $(3, 4)$ and $(3, -5)$ **10.** $(-4, -3)$ and $(0, 6)$

11. $(5, -3)$ and $(9, -3)$ **12.** $(1, 4)$ and $(5, -7)$

13. $(0, -6)$ and $(8, 0)$

14. Find the slope of the line whose angle of inclination is:
     a. $0°$ **b.** $30°$ **c.** $150°$ **d.** $90°$ **e.** $45°$ **f.** $135°$

15. Draw the line passing through $(0, 2)$ and having a slope of:
     a. 2 **b.** $\frac{1}{2}$ **c.** $-\frac{1}{2}$ **d.** 0 **e.** 3 **f.** $-\frac{2}{3}$

16. The vertices of a triangle are $A(-2, 0)$, $B(-1, 2)$, and $C(2, -3)$. Find the slope of each of its sides.

17. Find the slope of a line perpendicular to the line through $(-7, 1)$ and $(6, -5)$.

18. Show that the points $(-2, 8)$, $(1, -1)$, and $(3, -7)$ are collinear.

19. Without using the distance formula show that the points $(-4, 6)$, $(6, 10)$, and $(10, 0)$ are vertices of a right triangle.

20. Show that $(-4, 2)$, $(-1, 8)$, $(9, 4)$, and $(6, -2)$ are vertices of a parallelogram.

21. Show that $(0, -3)$, $(-2, 3)$, $(7, 6)$, and $(9, 0)$ are vertices of a rectangle.

In Exercises 22–28, find the midpoint of the line segment joining the given points.

22. $(-3, -5)$ and $(-1, 7)$ **23.** $(-2, 6)$ and $(2, -4)$

24. $(-3, 5)$ and $(-2, 9)$ **25.** $(5, 0)$ and $(9, 4)$

26. $(-4, 3)$ and $(-1, -7)$ **27.** $(-3, 10)$ and $(-6, 0)$

28. $(-4, -5)$ and $(-6, 5)$

29. Find the slope of the line through $(5, 6)$ and the midpoint of the line segment from $(-3, 0)$ to $(9, -2)$.

30. Show that $(0, 1)$, $(3, 3)$, $(8, 2)$, and $(-1, -4)$ are vertices of an isosceles trapezoid.

31. Find the slopes of the medians in the triangle in Exercise 16.

32. If the center of a circle is $(1, 2)$ and one end of the diameter is $(4, -3)$, find the other end.

33. Find the value of x so that the line through $(x, 2)$ and $(4, -6)$ is parallel to the line through $(-1, -1)$ and $(3, -5)$.

34. Find x so that the line through $(x, -2)$ and $(4, -7)$ is perpendicular to the line through $(2, -1)$ and $(-3, 2)$.

19.4   The Straight Line

Equations and Graphs

As noted in Section 19.1, analytic geometry combines algebra and geometry. Let us consider a simple example. The equation $x + 2y = 2$ has many solutions. If we let $x = 0$, for example, we get $0 + 2y = 2$, or $y = 1$. Thus $x = 0$ and $y = 1$ is a solution. A few other pairs of solutions are given in the following chart:

x:	-1	0	1	2
y:	$\frac{3}{2}$	1	$\frac{1}{2}$	0

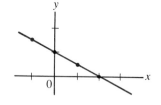

Figure 19.14

If we plot the points $(-1, \frac{3}{2})$, $(0, 1)$, $(1, \frac{1}{2})$, and $(2, 0)$ and connect them by a smooth curve, we get the graph shown in Figure 19.14. The graph looks like a straight line. (It is.) This example shows that the line in Figure 19.14 is the *solution set* of the equation $x + 2y = 2$.

Much of analytic geometry is concerned with geometric figures and their corresponding equations. In this section we will study the following two fundamental problems of analytic geometry for the straight line:

> **1.** Given the geometric figure, find the corresponding equation.
> **2.** Given the equation, draw the figure.

The two problems will be considered separately.

Moving from the Figure to the Equation

Suppose a line has slope m and passes through the point (x_1, y_1). To derive the equation, let (x, y) be any point on the line. Then the slope of the line segment joining (x_1, y_1) and (x, y) is the same regardless of which point is chosen. (See Figure 19.15.) Thus if m is the given slope, then

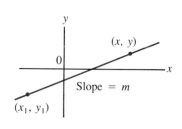

Figure 19.15

$$m = \frac{y - y_1}{x - x_1} \quad \text{or} \quad y - y_1 = m(x - x_1)$$

which is called the **point-slope form** of the line.

> **Point-Slope Form of the Line:** The equation of the line with slope m and passing through the point (x_1, y_1) is given by
>
> $$y - y_1 = m(x - x_1) \qquad\qquad (19.7)$$

The point-slope form is used to write the equation of a given line, as shown in the next example.

E   X   A   M   P   L   E   **1**   Find the equation of the line through $(-3, 5)$ and having a slope of 2

Solution. From Equation (19.7) we get

$$y - 5 = 2[x - (-3)] \qquad \text{or} \qquad 2x - y + 11 = 0 \qquad \blacktriangleleft$$

E   X   A   M   P   L   E   **2**   Find the equation of the line determined by the points $(-5, -1)$ and $(5, 4)$.

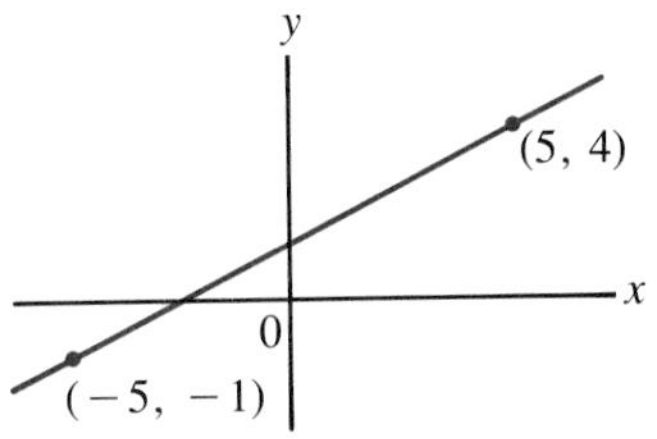

Solution. The slope of the line is found to be $\frac{1}{2}$ The point-slope form can now be obtained by using either point, say **(5, 4)** Thus

$$y - 4 = \tfrac{1}{2}(x - 5) \qquad \text{or} \qquad x - 2y + 3 = 0$$

(See Figure 19.16.) $\qquad \blacktriangleleft$

Figure 19.16

Horizontal line

If the line is horizontal, then $m = 0$ and Equation (19.7) reduces to

$$y = y_1 \qquad\qquad (19.8)$$

If the line is vertical, the slope is undefined. However, all points on a vertical line have a constant abscissa, here called x_1. So the equation is

Vertical line

$$x = x_1 \qquad\qquad (19.9)$$

E   X   A   M   P   L   E   **3**   The horizontal line $y = 3$ is shown in Figure 19.17. The vertical line $x = -2$ is shown in Figure 19.18. $\qquad \blacktriangleleft$

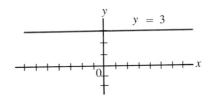

Figure 19.17

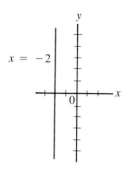

Figure 19.18

Moving from the Equation to the Figure

To see how we can obtain the line from its equation, let us consider the following special case of the point-slope form: suppose that a line has slope m and y-intercept b (that is, the line crosses the y-axis at $(0, b)$). Then, from Equation (19.7),

$$y - b = m(x - 0) \quad \text{or} \quad y = mx + b$$

which is called the **slope-intercept form** of the line.

Slope-Intercept Form of the Line: The equation of the line with slope m and y-intercept $(0, b)$ is given by

$$y = mx + b \tag{19.10}$$

Given the equation of a line, we can use the slope-intercept form to obtain the slope and y-intercept, and hence the graph, of the line.

E X A M P L E   4

Show that the line $x + 2y + 1 = 0$ is perpendicular to the line $2x - y + 3 = 0$.

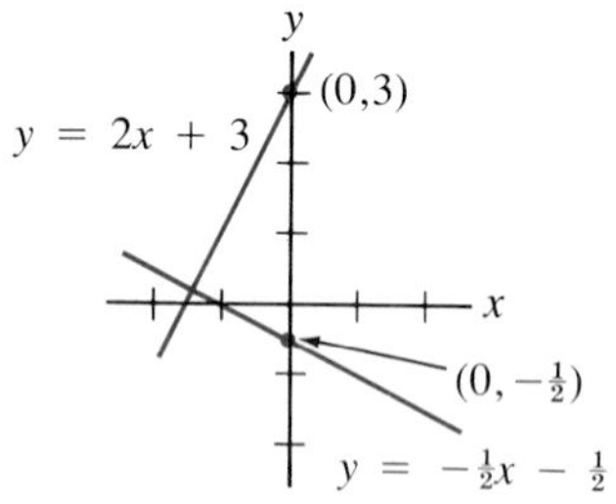

Solution. Solving each equation for y, we get

$$y = -\tfrac{1}{2}x - \tfrac{1}{2} \qquad \text{slope} = -\tfrac{1}{2}; \ \text{y-intercept} = (0, -\tfrac{1}{2})$$

and

$$y = 2x + 3 \qquad \text{slope} = 2; \ \text{y-intercept} = (0, 3)$$

respectively. Since the slopes are negative reciprocals, the lines are perpendicular to each other. The lines are shown in Figure 19.19.   ◀

These examples show that every straight line has an equation that can be written as

---

$$Ax + By + C = 0 \qquad (A \text{ and } B \text{ not both zero})$$

If $B \neq 0$, we can solve the equation for y and obtain the slope-intercept form. If $B = 0$ (and $A \neq 0$), then $x = -C/A$, which is a vertical line. If $A = 0$ (and $B \neq 0$), then $y = -C/B$, which is a horizontal line.

As in algebra, we normally use x and y for the variables. The purpose of this convention is to generalize: since the variables can stand for any physical quantity, no restriction is placed on the problems to which the equations can be applied. However, it is sometimes convenient to use letters that suggest the physical quantities more directly.

Figure 19.19

EXAMPLE 5 If E is the voltage drop across a resistor R, then the current I through the resistor is $I = E/R$ by Ohm's law. If $R = 2\ \Omega$, then $E = 2I$ is the line in Figure 19.20. (For a summary of SI units, see the table on the inside front cover of this book.) ◀

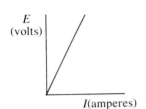

E (volts)

I(amperes)

Figure 19.20

EXERCISES / SECTION 19.4

In Exercises 1–18, write the equation of the line satisfying the given conditions.

1. Passing through $(-7, 2)$ and having a slope of $\frac{1}{2}$
2. Passing through $(0, 3)$ and having a slope of -4
3. Passing through $(3, -4)$ and having a slope of 3
4. Passing through $(1, 2)$ and having a slope of 0
5. Passing through $(0, 0)$ and having a slope of $-\frac{1}{3}$
6. Passing through $(-3, -10)$ and parallel to the y-axis
7. Passing through $(-4, 0)$ and parallel to the line $y = 1$
8. Passing through $(-3, 2)$ and $(7, 6)$
9. Passing through $(-3, 4)$ and $(3, -6)$
10. Passing through $(2, 3)$ and $(-4, 6)$
11. Passing through $(5, 0)$ and $(9, -4)$
12. Passing through $(-3, 5)$ and $(1, 7)$
13. Passing through $(2, 3)$ and $(-6, 4)$
14. Passing through $(1, 2)$ and having an angle of inclination of $135°$
15. Passing through $(0, 10)$ and having an angle of inclination of $45°$
16. Slope of 3 and y-intercept 4
17. Slope of $-\frac{1}{3}$ and y-intercept -2
18. $b = 2$ and $m = 1$

In Exercises 19–24, change each of the straight-line equations to slope-intercept form and draw the lines. (See Example 4.)

19. $6x + 2y = 5$
20. $x - y = 1$
21. $2x = 3y$
22. $\frac{1}{3}x + \frac{1}{12}y = 1$
23. $2y - 7 = 0$
24. $\frac{1}{2}x - 2y = 3$

In Exercises 25–30, state whether the lines are parallel, perpendicular, or neither.

25. $2x - 3y = 1$ and $4x - 6y + 3 = 0$
26. $2x + 4y + 3 = 0$ and $y - 2x = 2$
27. $3x - 4y = 1$ and $3y - 4x = 3$
28. $7x - 10y = 6$ and $y - 4 = 0$
29. $x + 3y = 5$ and $y - 3x - 2 = 0$
30. $2x + 5y = 2$ and $6x + 15y = 1$
31. Find the equation of the line through $(-1, 1)$ and parallel to the line $3x - 4y = 7$.
32. Find the equation of the line through $(-1, 1)$ and perpendicular to the line $3x - 4y = 7$.
33. Find the equation of the line passing through the intersection of the lines $2x - 4y = 1$ and $3x + 4y = 4$ and parallel to $5x + 7y + 3 = 0$.
34. Show that the following lines form the sides of a right triangle: $x - 3y + 3 = 0$, $12x + 4y + 25 = 0$, and $2y + x + 8 = 0$.
35. Sketch the graph of $v = 1 + 10t$, $t \geq 0$, where v is the velocity of an object and t is time, letting the t-axis be horizontal. (Note that the graph is a straight line and gives a pictorial representation of the velocity at any time.)
36. If a spring is stretched x units, it pulls back with a force $F = kx$ by Hooke's law; k is a constant called the *force constant* of the spring. If $k = 3$, draw the graph of the equation.
37. A force of 3 lb is required to stretch a spring $\frac{1}{2}$ ft. Find k and draw the graph. (Refer to Exercise 36.)
38. Deliveries of a certain item are made to a store on the first day of the month. The store manager noticed that the supply y throughout the month was given by

$$y = 60 - 2t \quad \text{(where } t \geq 0 \text{ is measured in days)}$$

Sketch the graph. What is the significance of the y-intercept? The t-intercept?

39. The relationship between temperature measured in degrees Celsius (C) and degrees Fahrenheit (F) is known to be linear—that is, it has the form $F = mC + b$. Water boils at 100°C and 212°F and freezes at 0°C and 32°F. Find the relationship. (For a table of SI units, see the table on the inside front cover.)

40. A consultant charges a flat fee of $50 plus $30/h. Find the equation relating the cost C (in dollars) and the time t (in hours).

41. The resistance R in a certain wire has the form $R = aT + b$, where T is the temperature. A lab technician determines that $R = 51\ \Omega$ when $T = 100°C$ and $R = 54\ \Omega$ when $T = 400°C$. Find the relationship.

19.5   Curve Sketching

We saw in the previous section that a graph gives a revealing geometric picture of the relationship between two quantities. Thus an industrial firm may use graphs to show product and market statistics. A scientist may show graphically how much of 100 g of the radioactive element polonium is left after a certain number of days. (See Figure 19.21.)

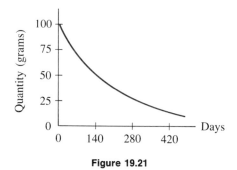

Figure 19.21

In this section we will develop some techniques for sketching the graph of a given equation. The purpose of these techniques is to obtain as much information as possible about the graph from the equation and to use this information to sketch the graph rapidly.

To get an overview of curve sketching, let us first consider a typical example, the equation $y = \frac{1}{9}x^3$. As an algebraic equation it is satisfied by infinitely many pairs of values. A few of these pairs are given in the following table:

x:	-3	-2	-1	0	1	2	3
y:	-3	$-\frac{8}{9}$	$-\frac{1}{9}$	0	$\frac{1}{9}$	$\frac{8}{9}$	3

These pairs can be plotted, as in Figure 19.22, and a smooth curve drawn through them. The more points plotted, the more accurate the graph becomes. In general, a graph of an equation in two dimensions is the set of all possible solutions to that equation. *The graph contains all points, and only those points, whose coordinates satisfy the equation.* Since a graph is a geometric figure, we have established a basic connection between algebra and geometry.

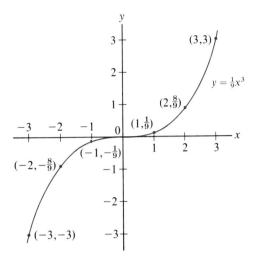

Figure 19.22

The Graph of an Equation: The graph of an equation, which is a set of points called a **locus**, is the solution set of the equation.

Plotting the points of an equation one by one can be a lengthy procedure. What is worse, the curve may have many interesting properties not easily detected by the point-plotting method. As noted earlier, much of this information can be extracted directly from the equation. In fact, the whole idea is to obtain as much information as possible from the equation, thereby reducing point plotting to a minimum. This procedure is referred to as the **discussion of the curve**.

Since the discussion of a curve consists of several parts, let us outline the procedure at this point and then examine each part in detail.

Discussion of a Curve

Step 1. Find the **intercepts**.
Step 2. Determine **symmetry** with respect to the axes.
Step 3. Find the **asymptotes**.
Step 4. Determine the **extent** of the curve.
Step 5. Sketch the curve.

Intercepts

The **intercepts** are the points where the graph crosses the coordinate axes. The points where the graph crosses the y-axis are called the **y-intercepts** and the points where the graph crosses the x-axis are called the **x-intercepts**. To obtain

y-intercepts
x-intercepts

the y-intercepts, we let $x = 0$ and solve for y; similarly, to determine the x-intercepts, we let $y = 0$ and solve for x.

Consider, for example, the curve $y = 4 - x^2$. If $x = 0$, then

$$y = 4 - 0^2 = 4 \qquad y = 4 - x^2$$

Thus $(0, 4)$ is the y-intercept. Letting $y = 0$, we get

$$0 = 4 - x^2 \text{ or } x = \pm 2 \qquad y = 4 - x^2$$

Thus the x-intercepts are $(2, 0)$ and $(-2, 0)$, usually written $(\pm 2, 0)$. The intercepts can tell us a great deal about the graph. In particular, if there are no intercepts, we know that the graph cannot cross the coordinate axes.

Symmetry

We will use the term **symmetry** to denote symmetry with respect to the coordinate axes. Taking the usual meaning of the term, a graph is symmetric with respect to the y-axis if the left half of the graph is the "mirror image" of the right half. More precisely, if the point (a, b) lies on the graph, so does the point $(-a, b)$. To check this condition for all points simultaneously, we replace x by $-x$ in the equation, and if the equation reduces to the given equation, then the graph is symmetric with respect to the y-axis. For example, let $y = x^2 + 1$. Upon replacing x by $-x$, we obtain $y = (-x)^2 + 1$, which reduces to the original equation $y = x^2 + 1$. Hence the graph of this equation is symmetric with respect to the y-axis.

By a similar procedure we can check the equation $y^2 + y^4 = x$ for symmetry with respect to the x-axis. Replacing y by $-y$, we get $(-y)^2 + (-y)^4 = x$, or $y^2 + y^4 = x$. The graph is therefore symmetric with respect to the x-axis.

Finally, a curve is symmetric with respect to the origin if the equation remains unchanged after substituting $-x$ for x and $-y$ for y at the same time. This statement implies that whenever the point (a, b) lies on the graph, so does the point $(-a, -b)$. Symmetry with respect to both axes implies symmetry with respect to the origin, but not conversely. For example, the graph in Figure 19.22 is symmetric with respect to the origin but not symmetric with respect to either axis. Thus

$$y = \tfrac{1}{9}x^3 \qquad \text{given equation}$$

$$-y = \tfrac{1}{9}(-x)^3 \qquad \text{replacing } x \text{ by } -x \text{ and } y \text{ by } -y$$

$$-y = \tfrac{1}{9}(-x^3) \qquad (-x)^3 = -x^3$$

$$y = \tfrac{1}{9}x^3 \qquad \text{reduces to given equation}$$

The curve is therefore symmetric with respect to the origin.

Asymptotes

The notion of asymptote can best be seen from an example. Consider the equation

$$y = \frac{1}{x - 1}$$

We see at once that y is undefined if $x = 1$, since substituting the number 1 for x would lead to division by zero. The real question is: what happens in the vicinity of $x = 1$? Since the denominator is close to zero near this point, the value of y is very large and keeps increasing as x approaches 1. Consequently, the graph will approach the vertical line $x = 1$ as x approaches 1, but without ever reaching it. (See Figure 19.23.) In general, we define an **asymptote** to be a line approached by the graph in the sense that the distance between the line and a point on the graph approaches zero as the point moves to infinity along the graph. For the curves discussed in this section:

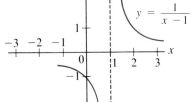

Figure 19.23

> To obtain the vertical asymptotes, solve the equation for y in terms of x. If the solution is a fraction, set the denominator equal to zero and solve for x. At these x values the graph will have vertical asymptotes.

If the asymptote is horizontal, the procedure is a little more complicated. Keeping in mind the definition of asymptote, however, it is clear that we need to study the behavior of the graph as x gets large. It is possible to reverse the foregoing procedure and solve the equation for x in terms of y, but in a case such as

$$y = \frac{3x^2 + 2x - 1}{2x^2 + 2}$$

it would be better to work with the equation directly. Suppose we divide numerator and denominator by the highest power of x found in the equation, in this case x^2. Then

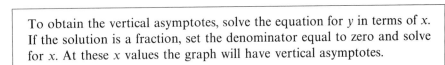

Since $1/x^n$ $(n > 0)$ approaches zero as x gets large, it follows that

$$\frac{2}{x}, \quad -\frac{1}{x^2}, \quad \text{and} \quad \frac{2}{x^2}$$

all approach zero. As a result,

$$y = \frac{3 + \dfrac{2}{x} - \dfrac{1}{x^2}}{2 + \dfrac{2}{x^2}} \quad \text{approaches } \frac{3}{2} \text{ when } x \text{ gets large}$$

Thus $y = \frac{3}{2}$ is a horizontal asymptote.

> In summary, to find a horizontal asymptote, divide numerator and denominator by the highest power of x contained in the equation and let x get large. The value approached by y is the horizontal asymptote. In other words, if y approaches a, then $y = a$ is the horizontal asymptote.

Extent (Along the x-Axis)

The **extent** of a curve along the x-axis consists of those values of x for which y is defined and real. For example, the extent of the curve $y = 1/x$ consists of all x except $x = 0$ (since division by zero is undefined).

The extent of the curve $y = \sqrt{x}$ is the set of all x such that $x \geq 0$. (For $x < 0$, $\sqrt{x}$ is complex; if $x = -4$, for example, then $y = \sqrt{-4} = \sqrt{4(-1)} = \sqrt{4}\sqrt{-1} = 2j$, an imaginary number.)

To determine the extent of a curve, we solve the given equation for y in terms of x. If the resulting expression contains radicals of even index (such as square roots or fourth roots), then the values of x that make the expression under the radical sign negative must be excluded (to avoid complex values). Consider, for example, the equation

$$x^2 + 4y^2 = 1$$

Solving for y, we obtain

$$y^2 = \tfrac{1}{4}(1 - x^2) \quad \text{and} \quad y = \pm\tfrac{1}{2}\sqrt{1 - x^2}$$

So if x exceeds unity in numerical value, then the y-values become imaginary. We would now say that the extent of the curve is from -1 to 1. As noted earlier, we also exclude values of x that lead to division by zero. We will illustrate the ideas in this section by several examples.

E   X   A   M   P   L   E   **1**   Discuss and sketch the curve $y = x^2 - 1$.

Solution.

Step 1. Intercepts: if $x = 0$, then

$$y = 0^2 - 1 = -1$$

If $y = 0$, then

$$0 = x^2 - 1$$
$$x^2 = 1$$
$$x = \pm 1$$

Thus the intercepts are $(\pm 1, 0)$ and $(0, -1)$.

Step 2. Symmetry: replacing x by $-x$, we get

$$y = (-x)^2 - 1$$
$$y = x^2 - 1 \qquad \textbf{note the even power on x}$$

Since the equation reduces to the given equation, the curve is symmetric with respect to the y-axis.

Step 3. Asymptotes: none. (The equation is not in the form of a fraction with a variable in the denominator.)

Step 4. Extent: all x (no radicals).

We now plot the intercepts and four more points and then connect these by a smooth curve. The graph is shown in Figure 19.24. Since the curve is symmetric with respect to the y-axis, it is sufficient to draw the curve through the points on the right—$(0, -1)$, $(1, 0)$, $(2, 3)$, and $(3, 8)$—and then draw the "mirror image" on the left side. ◀

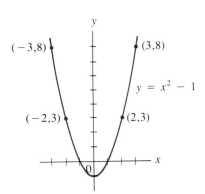

Figure 19.24

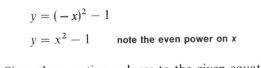

E X A M P L E **2**   Discuss and sketch the curve

$$y = (x + 1)(x - 2)(x - 3)$$

Solution.

Step 1. Intercepts: if $x = 0$, then $y = 6$. If $y = 0$, then

$$(x + 1)(x - 2)(x - 3) = 0$$

and

$$x = -1, 2, 3$$

Step 2. Symmetry: none. (Replacing x by $-x$ or y by $-y$ changes the equation.)

Step 3. Asymptotes: none (no x in the denominator).

Step 4. Extent: all x.

We now plot the intercepts and two more points and connect these by a smooth curve. The graph is shown in Figure 19.25. Two additional points (not shown) are $(4, 10)$ and $(-2, -20)$. ◀

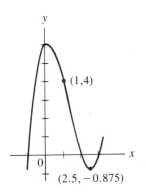

Figure 19.25

The next example illustrates a convenient method for determining the extent of a curve by means of a chart.

E   X   A   M   P   L   E   **3**   Discuss and sketch the curve $y^2 = (x + 1)(x - 2)(x - 3)$.

Solution.

Step 1. Intercepts: if $x = 0$, $y = \pm\sqrt{6}$, and if $y = 0$, $x = -1, 2, 3$.

Step 2. Symmetry: x-axis. (Replacing y by $-y$ leaves the equation unchanged.)

Step 3. Asymptotes: none.

Step 4. Extent: Solving for y, we find that

$$y = \pm\sqrt{(x + 1)(x - 2)(x - 3)}$$

As noted earlier, we must exclude all values of x for which the radicand is negative. We observe that $(x + 1)(x - 2)(x - 3)$ changes signs only at $x = 3, 2$, and -1. If $x > 3$, all the factors are positive, so the radicand is positive. If x is between 2 and 3, then $x - 3 < 0$, so the entire radicand is negative. Hence the interval $2 < x < 3$ must be excluded from the graph (see Figure 19.26). At $x = 2$, the sign of the radicand changes again to become positive. Finally, at $x = -1$ the sign of the radicand changes back to negative, so that all values of x less than -1 must also be excluded. We conclude that the extent of the curve consists of the intervals $-1 \le x \le 2$ and $x \ge 3$.

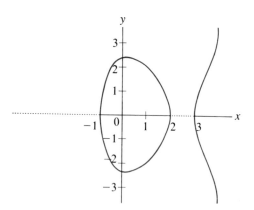

Figure 19.26

Summary. In determining the extent of the curve, note that the radicand is zero only at $x = -1, 2$, and 3 and must therefore be different from zero at all other points. Consequently, it is sufficient to substitute arbitrary "test values" to check signs. When $x = 0$, for example, the radicand is 6 and hence positive in the interval $-1 < x < 2$. These observations are summarized in the following chart:

	Test values	$x + 1$	$x - 2$	$x - 3$	$(x + 1)(x - 2)(x - 3)$
$x > 3$	4	+	+	+	+
$2 < x < 3$	2.5	+	+	−	−
$-1 < x < 2$	0	+	−	−	+
$x < -1$	−2	−	−	−	−

◄

E X A M P L E **4** Discuss and sketch the curve.

$$y = \frac{x^2}{x^2 - 1}$$

Solution.

Step 1. Intercepts: if $x = 0$, $y = 0$, and if $y = 0$, $x = 0$.

Step 2. Symmetry: y-axis. (Note the even power on x.)

Step 3. Vertical asymptotes: setting the denominator equal to zero, we get $x^2 - 1 = 0$ and $x = \pm 1$. Horizontal asymptote: if numerator and denominator are divided by x^2, the equation becomes

$$y = \frac{\dfrac{x^2}{x^2}}{\dfrac{x^2}{x^2} - \dfrac{1}{x^2}} \qquad \text{or} \qquad y = \frac{1}{1 - \dfrac{1}{x^2}}$$

Since $-1/x^2$ approaches zero as x gets large, the right side approaches 1 as x gets large. Thus the line $y = 1$ is a horizontal asymptote.

Step 4. Extent: all x except $x = \pm 1$ (to avoid division by zero).

The graph is shown in Figure 19.27. Note that it follows from the discussion that the left and right branches cannot approach the vertical asymptotes in the downward direction without crossing the x-axis—but the only intercept is the origin. For the middle branch we find that $y \leq 0$ whenever $-1 < x < 1$. ◄

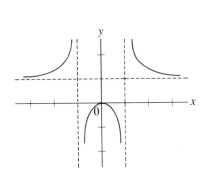

Figure 19.27

E X A M P L E **5** Discuss and sketch the curve

$$y^2 = \frac{x}{x^2 - 4}$$

Solution.

Step 1. Intercepts: if $x = 0$, $y = 0$, and if $y = 0$, $x = 0$.

Step 2. Symmetry: x-axis (Note the even power on y.)

Step 3. Vertical asymptotes: $x = \pm 2$. Horizontal asymptotes: dividing numerator and denominator of the fraction by x^2, we get

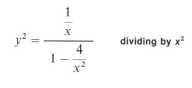

$$y^2 = \frac{\dfrac{1}{x}}{1 - \dfrac{4}{x^2}} \qquad \text{dividing by } x^2$$

As x gets large, y^2 approaches zero. Hence the asymptote is $y = 0$.
Step 4. Extent: solving for y, we find that

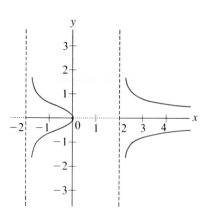

Figure 19.28

$$y = \pm \sqrt{\frac{x}{(x - 2)(x + 2)}} \qquad \text{taking square roots}$$

The radicand $x/[(x - 2)(x + 2)]$ changes signs only at $x = -2$, 0, and 2. If $x > 2$, all the factors are positive. If x is between 0 and 2, then $x - 2 < 0$, so that the radicand is negative. Hence the interval $0 < x < 2$ must be excluded (see Figure 19.28). At $x = 0$ the sign of the radicand changes back to positive. Finally, if $x < -2$ the radicand is negative; so the extent is given by $-2 < x \leq 0$ and $x > 2$.

By using "test values," these results can be summarized as follows:

	Test values	$x + 2$	x	$x - 2$	$\dfrac{x}{(x - 2)(x + 2)}$
$x > 2$	3	+	+	+	+
$0 < x < 2$	1	+	+	−	−
$-2 < x < 0$	−1	+	−	−	+
$x < -2$	−3	−	−	−	−

◀

E X E R C I S E S / S E C T I O N   **19.5**

In Exercises 1–33, use the format in Examples 1–5 to discuss and sketch the graphs of the equations.

1. $y = 2x - 1$

2. $y = x^2 + 1$

3. $y = 1 - x^2$

4. $y = 2x^2 - 1$

5. $y^2 = x$

6. $y^2 = 4x$

7. $y^2 = x + 1$

8. $y = \dfrac{1}{x}$

9. $xy = 3$

10. $y = (x + 1)(x - 2)$

11. $y = (x - 3)(x + 5)$

12. $y = x(x - 1)(x - 2)$

13. $y = x(x - 1)(x - 2)^2$

14. $y = x^2(x + 2)(x - 3)^2$

15. $y = x^2 - x - 6$

16. $y = \sqrt{4 - x^2}$

17. $y = x\sqrt{1 - x^2}$

18. $y = \dfrac{1}{x - 1}$

19. $y = \dfrac{2}{x + 2}$

20. $y = \dfrac{1}{(x + 2)^2}$

21. $y = \dfrac{2}{(x - 1)^2}$

22. $y = \dfrac{x}{x + 2}$

23. $y = \dfrac{x^2}{x - 1}$

24. $y = \dfrac{x}{(x - 3)(x - 2)}$

25. $y = \dfrac{x + 1}{(x - 1)(x + 2)}$

26. $y = \dfrac{x(x - 1)}{(x - 2)(x + 1)}$

27. $y = \dfrac{x^2 - 4}{x^2 - 1}$

28. $y^2 = (x + 1)(x - 2)$

29. $y^2 = (x - 3)(x + 5)$

30. $y^2 = \dfrac{x}{x + 2}$

31. $y^2 = \dfrac{x}{(x - 3)(x - 2)}$

32. $y^2 = \dfrac{x + 1}{(x - 1)(x + 2)}$

33. $y^2 = \dfrac{x^2 - 4}{x^2 - 1}$

34. Recall that by Ohm's law $I = E/R$. Suppose that $E = 3$ volts and that R is a variable resistance. Discuss and sketch the resulting equation.

35. Suppose that two capacitors C_1 and C_2 are connected in series. The equivalent capacitance C is given by the equation

$$C = \frac{C_1 C_2}{C_1 + C_2}$$

If $C_2 = 10^{-2}$ farad, discuss and sketch the resulting graph for $C_1 \geq 0$.

36. The force of gravitational attraction between two point masses m_1 and m_2 is given by

$$F = G \frac{m_1 m_2}{x^2}$$

where x is the distance between the masses and G is a constant. Discuss the curve. What is the significance of the asymptotes?

37. A projectile shot directly upward with a speed of 60 m/s moves according to the law

$$S = 60t - 5t^2 \qquad (t \geq 0)$$

Sketch the curve.

38. The charge on a certain capacitor is $q = 1/(t + 1)^2$. Discuss and sketch the curve for $t \geq 0$.

39. The period P of a pendulum is $P = 1.1\sqrt{L}$, where P is measured in seconds and L in feet. Discuss and sketch the curve.

40. A manufacturer determines that the profit P (in dollars) on a certain item is

$$P = 800{,}000 \frac{x}{(x + 2)^2}$$

where x (in thousands) is the number of units sold. Sketch the graph and estimate the number of units for which the profit is a maximum.

19.6 The Conics

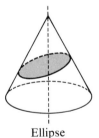

Ellipse

Figure 19.29

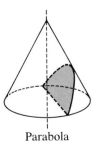

Parabola

Figure 19.30

In the next few sections we are going to return to the two fundamental problems mentioned in Section 19.4:

1. Given the figure, find the corresponding equation.
2. Given the equation, draw the figure.

In the previous section we developed a number of techniques for handling Problem 2, and we are going to study other techniques in Chapter 21. Problem 1 often involves finding a best-fitting curve for a given set of data, a technique usually studied in numerical analysis. On the other hand, when dealing with a special group of figures, such as the conics, it is often possible to go directly from the figure to the equation and vice versa.

The term **conic**, or **conic section**, comes from the fact that the curves for which equations are sought correspond to sections cut from a conic surface by planes. A conic surface consists of two funnel-shaped sheets, called **nappes**, extending indefinitely in both directions from a point (vertex) where the nappes meet. If a plane cuts entirely across one nappe, the intersection of plane and nappe is called an *ellipse* (see Figure 19.29). A special case, the *circle*, is obtained if the plane is perpendicular to the axis of the cone. If the plane is parallel to one line on the surface of the cone, the intersection of plane and nappe is called a *parabola* (see Figure 19.30). If the plane intersects both nappes, the intersection

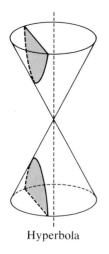

Hyperbola

Figure 19.31

is called a *hyperbola* (see Figure 19.31). Certain special cases occur if the plane passes through the vertex (a point), through the axis (two intersecting lines), or through a single line on the surface.

The conics were well known to the ancient Greeks. Many of the geometric properties of conic sections, including the ones to be used in our later definitions, were studied in antiquity. Although Euclid's own contributions to that study were later lost, they appear to have been incorporated in the work of Apollonius of Perga (*circa* 230 B.C.), to whom the names of the curves have been attributed. Archimedes (287–212 B.C.) was the first to find the area under a parabolic arch. In the remainder of this chapter we will study a number of scientific applications of the conics.

19.7   The Circle

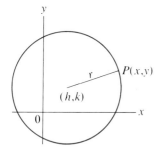

Figure 19.32

We recall from geometry that a **circle** is the locus of all points equidistant from a given point called the *center*. Let (h, k) in Figure 19.32 be the center and let $P(x, y)$ be any point on the circle. If r is the radius, then from the distance formula (19.1) we get

$$\sqrt{(x - h)^2 + (y - k)^2} = r$$

Squaring both sides of this equation, we get the **standard form of the equation of a circle**.

Standard Form of the Equation of a Circle

$$(x - h)^2 + (y - k)^2 = r^2 \tag{19.11}$$

where (h, k) is the center and r is the radius.

E X A M P L E **1**   Find the equation of the circle with center at $(-2, 1)$ and radius 3.

Solution. Since $h = -2$, $k = 1$, and $r = 3$, we have

$$(x - h)^2 + (y - k)^2 = r^2$$
$$[x - (-2)]^2 + (y - 1)^2 = 3^2$$

and

$$(x + 2)^2 + (y - 1)^2 = 9 \qquad \blacktriangleleft$$

If the circle is centered at the origin, then $h = 0$ and $k = 0$, and the equation has the form given next.

Equation of a Circle Centered at the Origin

$$x^2 + y^2 = r^2 \tag{19.12}$$

If we multiply the terms of Equation (19.11), we may conclude that any circle has the following form:

General Form of the Equation of a Circle

$$x^2 + y^2 + ax + by + c = 0 \tag{19.13}$$

E X A M P L E **2** Put $x^2 + y^2 + 6x - 4y + 9 = 0$ into standard form (19.11) and find the center and radius.

Solution. To convert the equation to standard form we complete the square on each quadratic expression. Transposing the 9 and rearranging terms, we get

$$(x^2 + 6x \quad) + (y^2 - 4y \quad) = -9$$

Since the coefficient of x is 6, we must add the square of one-half of 6 to both sides. In other words, since

$$\left[\frac{1}{2} \cdot 6\right]^2 = 9$$

we add 9 to both sides.
Similarly, since the coefficient of y is -4, we add

$$\left[\frac{1}{2}(-4)\right]^2 = 4$$

to both sides. Thus

$$(x^2 + 6x + 9) + (y^2 - 4y + 4) = -9 + 9 + 4$$

Since $x^2 + 6x + 9 = (x + 3)^2$ and $y^2 - 4y + 4 = (y - 2)^2$, we get

$$(x + 3)^2 + (y - 2)^2 = 4$$

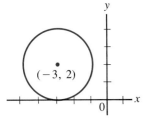

(−3, 2)

Figure 19.33

From the form

$$(x - h)^2 + (y - k)^2 = r^2$$

we conclude that the center is $(-3, 2)$ and the radius $\sqrt{4} = 2$. (See Figure 19.33.) ◀

E X A M P L E **3**   Write $x^2 + y^2 - 2x + 4y + 5 = 0$ in standard form.

Solution. Before completing the square on each quadratic expression, we rearrange the terms thus:

$$(x^2 - 2x \quad) + (y^2 + 4y \quad) = -5$$

One-half of the coefficients of x and y are -1 and 2, respectively. We add the square of each to both sides; hence

$$(x^2 - 2x + \mathbf{1}) + (y^2 + 4y + \mathbf{4}) = -5 + \mathbf{1} + \mathbf{4}$$

or

$$(x - 1)^2 + (y + 2)^2 = 0$$

Point circle

This equation is satisfied by the coordinates of only one real point $(1, -2)$. The locus is sometimes called a **point circle**.
   The equation

$$(x - 1)^2 + (y + 2)^2 = -1$$

is not satisfied by any real point and, for lack of a better term, will be called an *imaginary circle*. ◀

E X E R C I S E S / S E C T I O N   **19.7**

In Exercises 1–10, find the equation of each of the circles.

1. Center at the origin, radius 5

2. Center at the origin, radius 7

3. Center at the origin, passing through $(-6, 8)$

4. Center at the origin, passing through $(1, -4)$

5. Center at $(-2, 5)$, radius 1

6. Center at $(2, -3)$, radius $\sqrt{2}$

7. Center at $(-1, -4)$, passing through origin

8. Center at $(3, 4)$, passing through $(5, 10)$

9. Ends of diameter at $(-2, -6)$ and $(1, 5)$

10. Center on the line $2y = 3x$, tangent to the y-axis, radius 2

In Exercises 11–24, write each equation in standard form and determine the center and radius.

11. $x^2 + y^2 - 2x - 2y - 2 = 0$

12. $x^2 + y^2 - 2x - 4y + 4 = 0$

13. $x^2 + y^2 + 4x - 8y + 4 = 0$

14. $x^2 + y^2 + 2x + 6y + 3 = 0$

15. $x^2 + y^2 + 4x + 2y + 2 = 0$

16. $x^2 + y^2 - 8x + 6y + 20 = 0$

17. $x^2 + y^2 + 4x - 2y - 4 = 0$

18. $x^2 + y^2 + 2x + 8y + 1 = 0$

19. $x^2 + y^2 - x - 2y + \frac{1}{4} = 0$

20. $x^2 + y^2 - 6x - 8y + 19 = 0$

21. $x^2 + y^2 - 4x + y + \frac{9}{4} = 0$

22. $x^2 + y^2 + x - y - \frac{1}{2} = 0$

23. $4x^2 + 4y^2 + 12x + 16y + 5 = 0$

24. $36x^2 + 36y^2 - 144x - 120y + 219 = 0$

In Exercises 25–30, determine which is a point circle and which is an imaginary circle.

25. $4x^2 + 4y^2 - 20x - 4y + 26 = 0$

26. $x^2 + y^2 + 4x - 2y + 7 = 0$

27. $x^2 + y^2 - 6x + 8y + 25 = 0$

28. $x^2 + y^2 + 2x + 4y + 5 = 0$

29. $x^2 + y^2 - 6x - 8y + 30 = 0$

30. $x^2 + y^2 - 6x + 4y + 13 = 0$

31. A washer has inner radius 2.00 cm and outer radius 3.40 cm. Find the equations of the two circles, using the center of the washer as the origin.

32. Find the equation of the balancing hole in Figure 19.34.

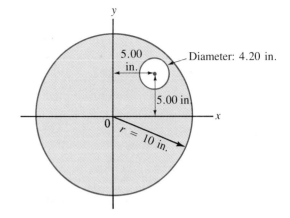

Figure 19.34

19.8 The Parabola

Focus

Directrix

Vertex

Axis

Let F be a fixed point and l be a fixed line. Then the locus of all points equidistant from F and l is called a **parabola** (Figure 19.35). F is called the **focus** of the parabola and l the **directrix**. If we draw a line from F perpendicular to l, intersecting l at Q, then the midpoint V of FQ is called the **vertex**, which is the point on the parabola closest to the line l. The line through F and Q is called the **axis** of the parabola.

To get the equation of the parabola, we select a convenient coordinate system and apply the definition. Let the vertex be at the origin and denote the focus by $(p, 0)$, as in Figure 19.36. If p is positive, then the focus lies to the right of the origin; if p is negative, it lies to the left. The directrix is the line $x = -p$ in each case. Now let $P(x, y)$ be any point on the parabola (Figure 19.36). It

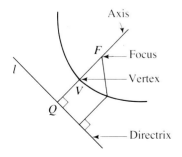

Figure 19.35

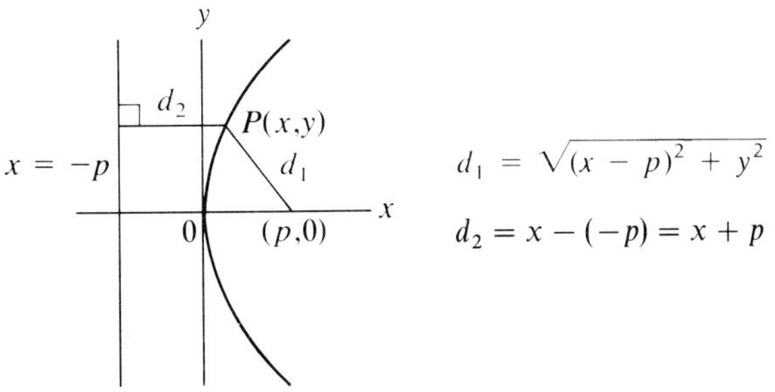

$x = -p$

$P(x, y)$

d_1

$(p, 0)$

d_2

$d_1 = \sqrt{(x - p)^2 + y^2}$

$d_2 = x - (-p) = x + p$

Figure 19.36

follows at once that the distance to the directrix is $x + p$. By the definition of the parabola, the distance d_1 from $P(x, y)$ to the focus is equal to the distance d_2 from $P(x, y)$ to the directrix:

$$d_1 = d_2$$

$$\sqrt{(x - p)^2 + y^2} = x + p \qquad \textbf{by the distance formula}$$

Squaring both sides, we get

$$(x - p)^2 + y^2 = (x + p)^2$$

or

$$x^2 - 2px + p^2 + y^2 = x^2 + 2px + p^2 \qquad \textbf{(a ± b)}^2 = \textbf{a}^2 ± \textbf{2ab} + \textbf{b}^2$$

Combining like terms, the last expression reduces to

$$y^2 = 4px$$

Conversely, any point whose coordinates satisfy this equation lies on the parabola. We owe the simplicity of this equation to the manner in which the coordinate system was chosen. It will be seen later that the equation can be modified to study a parabola whose vertex is not at the origin.

If the focus is at $(0, p)$ and the directrix is the line $y = -p$ (Figure 19.37), then the equation becomes

$$x^2 = 4py$$

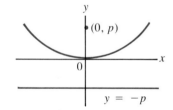

Figure 19.37

These two cases are summarized next.

Equation of a Parabola: The equation of a parabola with vertex at the origin is

$$y^2 = 4px \qquad \text{(axis horizontal)} \tag{19.14}$$

or

$$x^2 = 4py \qquad \text{(axis vertical)} \tag{19.15}$$

To remember the form (and position) of a parabola, note that $y^2 = 4px$ is symmetric with respect to the x-axis, while $x^2 = 4py$ is symmetric with respect to the y-axis.

E X A M P L E 1 Find the focus and directrix of the parabola $y^2 = -9x$ and sketch.

Solution. To find the focus and directrix, we need to write the equation in the form $y^2 = 4px$. Thus

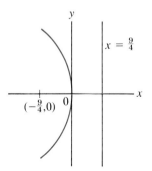

$$y^2 = -9x = 4\left(-\frac{9}{4}\right)x \qquad y^2 = 4px$$

Since $p = -\frac{9}{4}$, the focus is at $(-\frac{9}{4}, 0)$ and the directrix is

$$x = -p = -\left(-\frac{9}{4}\right) = \frac{9}{4}$$

Figure 19.38

(See Figure 19.38.)

E X A M P L E 2 Find the focus and directrix of the parabola $x^2 = 14y$ and sketch.

Solution.

$$x^2 = 14y = 4\left(\frac{14}{4}\right)y$$

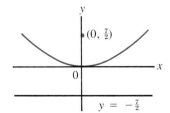

or

$$x^2 = 4\left(\frac{7}{2}\right)y \qquad x^2 = 4py$$

Figure 19.39

Since $p = \frac{7}{2}$, we conclude from Equation (19.15) that the focus is at $(0, \frac{7}{2})$ and the directrix is the line $y = -p = -\frac{7}{2}$ (Figure 19.39).

E X A M P L E 3 Find the equation of the parabola with focus at $(0, 3)$ and the directrix $y = -3$.

Solution. Since the focus is on the y-axis, the form is $x^2 = 4py$. Since $p = 3$, we get

$$x^2 = 4(3)y \qquad x^2 = 4py$$

or

$$x^2 = 12y$$

E X A M P L E **4**   Find the equation of the parabola with focus at $(-5, 0)$ and vertex at the origin.

Solution. Since the focus is on the x-axis, the form of the equation is $y^2 = 4px$ by Equation (19.14). Since $p = -5$, we get

$$y^2 = 4(-5)x \qquad y^2 = 4px$$

or

$$y^2 = -20x \qquad \blacktriangleleft$$

Figure 19.40

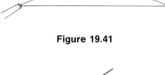

Figure 19.41

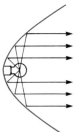

Figure 19.42

Applications of the Parabola

Some common applications of the parabola are listed next:

1. The cable of a suspension bridge whose weight is uniformly distributed over the entire length of the bridge is in the form of a parabola (Figure 19.40).
2. The path of a projectile is a parabola if air resistance is neglected (Figure 19.41). (If the object is heavy and the initial velocity low, the air resistance is negligible.)
3. If a parabola is rotated about its axis, it forms a parabolic surface. If a reflecting surface is parabolic with a light source placed at the focus, the rays will be reflected parallel to the axis. Conversely, light rays coming in parallel to the axis are reflected so they pass through the focus. These principles are employed in the construction of searchlights and reflecting telescopes (Figure 19.42).
4. Another parabolic surface is obtained if a cylindrical vessel is partly filled with water and rotated about the axis of the cylinder: a plane through the axis will cut the surface of the water in a parabola.
5. Steel bridges are frequently built with parabolic arches (Figure 19.43).

Figure 19.43

6. The relationship between the period of a pendulum and its length can be described in the form of a parabolic equation.

E X A M P L E **5**   The entrance to a formal garden is spanned by a parabolic arch 10 ft high and 10 ft across at its base. How wide should the walk through the center of the arch be if a minimum clearance of 7 ft is wanted above the walk?

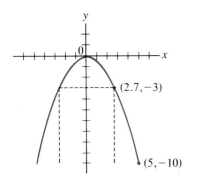

Figure 19.44

Solution. Let the vertex of the arch be at the origin (Figure 19.44). The equation of the parabola has the form $x^2 = 4py$. Since the arch is 10 ft high, we know that the point $(5, -10)$ lies on the curve. Consequently, the coordinates of this point must satisfy the foregoing equation. We begin by substituting:

$$5^2 = 4p(-10) \qquad x^2 = 4py$$

Solving for $4p$, we get

$$4p = \frac{5^2}{-10} = \frac{25}{-10} = -\frac{5}{2}$$

So the equation becomes

$$x^2 = 4py$$

$$x^2 = -\frac{5}{2}\,y$$

Now, a clearance of 7 ft corresponds to a point for which $y = -3$ (see Figure 19.44); its x-coordinate will tell us the width of the walk. Thus

$$x^2 = -\frac{5}{2}(-3) = \frac{15}{2} \qquad y = -3$$

and $x = 2.7$. So the walk must be $2(2.7) = 5.4$ ft wide. ◀

E   X   A   M   P   L   E   **6**   An amateur astronomer constructs a reflecting telescope with a parabolic mirror at one end of a hollow tube. The image is formed at the other end, where it may be viewed through an eyepiece. If the mirror has a radius of 20.0 cm and a depth of 3.33 mm in the center, find the position of the image.

Solution. We may assume the parabolic cross-section to have the form

$$y^2 = 4px$$

We need to find the focus of this parabola. Since 3.33 mm = 0.333 cm, one point in the first quadrant is $(0.333, 20.0)$. The coordinates of this point satisfy the equation. Hence

$$(20.0)^2 = 4p \cdot 0.333$$

and

$$p = \frac{(20.0)^2}{4 \cdot 0.333} = 300 \text{ cm}$$

Therefore, the image is 3 m from the center of the mirror. ◀

E X E R C I S E S / S E C T I O N **19.8**

In Exercises 1–14, write the equations of the parabolas.

1. Vertex at the origin, focus at (3, 0)

2. Vertex at the origin, focus at (−3, 0)

3. Vertex at the origin, focus at (0, −5)

4. Vertex at the origin, focus at (0, 4)

5. Vertex at the origin, focus at (−4, 0)

6. Vertex at the origin, focus at (0, −6)

7. Vertex at the origin, directrix $x = -1$

8. Vertex at the origin, directrix $y = 2$

9. Directrix $x - 2 = 0$, vertex at the origin

10. Directrix $x + 2 = 0$, vertex at the origin

11. Directrix $y - 3 = 0$, focus at (0, −3)

12. Directrix $y = 4$, focus at (0, −4)

13. Vertex at the origin, axis along the x-axis, passing through (−2, −4)

14. Vertex at the origin, axis along the y-axis, passing through (−1, 1)

In Exercises 15–30, find the coordinates of the focus and the equation of the directrix in each exercise and sketch.

15. $x^2 = 8y$ **16.** $x^2 = 16y$ **17.** $x^2 = -12y$

18. $x^2 = -16y$ **19.** $y^2 = 16x$ **20.** $y^2 = 8x$

21. $y^2 = -4x$ **22.** $y^2 = -12x$ **23.** $x^2 - 4y = 0$

24. $x^2 + 8y = 0$ **25.** $y^2 = 9x$ **26.** $y^2 = 10x$

27. $y^2 = -x$ **28.** $2x^2 - 3y = 0$ **29.** $3y^2 + 2x = 0$

30. $y^2 = 2ax$

31. The chord of a parabola that passes through the focus and is parallel to the directrix is called the *focal chord*. Find the equation of the circle having for a diameter the focal chord of the parabola $x^2 = 12y$.

32. Derive Equation (19.15).

33. Use the definition of the parabola to find the equation of the parabola whose focus is at (4, 1) and whose directrix is the y-axis.

34. Repeat Exercise 33 for the parabola having a focus at (4, 7) and directrix $y + 1 = 0$.

35. Find the equations of the two parabolas with vertices at the origin and which pass through (−2, −4).

36. Repeat Exercise 35 for the point (3, −5).

37. The support cables of a suspension bridge hang between towers in the form of a parabolic curve. The towers are 90.0 m high and 200.0 m apart. The roadway is suspended on other cables hung from the support cables. The shortest distance from the road surface to the supporting cable is 20.0 m. Determine the length of the vertical suspension cable which is 30.0 m from the center of the bridge. (Choose for the origin the lowest point on the supporting cable.)

38. The headlight on a car is 20.0 cm in diameter and has a parabolic surface; the headlight is 12.0 cm deep. Where should the light bulb be placed so that the reflected rays are parallel?

39. Determine the distance across the base of a parabolic arch that measures 10.0 m at its highest point if the road through the center of the arch is 16.0 m wide and must have a minimum clearance of 5.00 m.

40. A stream is 60.0 m wide and is spanned by a parabolic arch. Boats pass through the 40.0-m-wide channel in the center of the stream. How high must the arch be so that the minimum clearance above the stream will be 10.0 m?

41. An entrance to a castle is in the form of a parabolic arch 6 m across at the base and 3 m high in the center. What is the length of a beam across the entrance, parallel to the base and 2 m above it?

42. The period T of a pendulum is directly proportional to the square root of its length L, provided that the arc of the swing is small (less than 15°). This can be described by $L = kT^2$. If $T = 6.0$ s when $L = 9.0$ m, find the relationship.

19.9   The Ellipse

Foci

An **ellipse** is the locus of points such that the sum of the distances from two fixed points is constant. Suppose that the two fixed points, called the **foci** of the ellipse, are at $(c, 0)$ and $(−c, 0)$, respectively (Figure 19.45). The distance formula (19.1) can now be used to obtain the equation. A particularly simple form will result

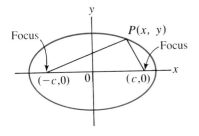

Figure 19.45

if we choose $2a$ for the constant, since the points $(a, 0)$ and $(-a, 0)$ will then lie on the ellipse. From the definition, the sum of the distances from $P(x, y)$ to the foci is

$$\sqrt{(x - c)^2 + y^2} + \sqrt{(x + c)^2 + y^2} = 2a$$

Conversely, any point whose coordinates satisfy this equation will lie on the ellipse. The equation can be simplified by transposing one of the radicals and squaring both sides:

$$[\sqrt{(x - c)^2 + y^2}]^2 = [2a - \sqrt{(x + c)^2 + y^2}]^2$$

or

$$(x - c)^2 + y^2 = 4a^2 - 4a\sqrt{(x + c)^2 + y^2} + (x + c)^2 + y^2$$

Since $(x \pm c)^2 = x^2 \pm 2xc + c^2$, we get

$$x^2 - 2cx + c^2 + y^2 = 4a^2 - 4a\sqrt{(x + c)^2 + y^2} + x^2 + 2cx + c^2 + y^2$$

If we now subtract x^2, c^2, and y^2 from both sides, we obtain

$$-2cx = 4a^2 - 4a\sqrt{(x + c)^2 + y^2} + 2cx$$

and

$$-4cx = 4a^2 - 4a\sqrt{(x + c)^2 + y^2} \qquad \text{subtracting } 2cx$$
$$cx = -a^2 + a\sqrt{(x + c)^2 + y^2} \qquad \text{dividing by } -4$$
$$a^2 + cx = a\sqrt{x^2 + 2cx + c^2 + y^2} \qquad \text{adding } a^2$$

Squaring both sides again, we get

$$a^4 + 2a^2cx + c^2x^2 = a^2(x^2 + 2cx + c^2 + y^2)$$

or

$$a^4 + c^2x^2 = a^2x^2 + a^2c^2 + a^2y^2 \qquad \text{subtracting } 2a^2cx$$

We now collect all x and y terms on one side of the equation and factor the resulting expressions. Hence

$$a^2x^2 - c^2x^2 + a^2y^2 = a^4 - a^2c^2$$

or

$$(a^2 - c^2)x^2 + a^2y^2 = a^2(a^2 - c^2) \qquad \text{common factors } x^2 \text{ and } a^2$$

After dividing both sides by $a^2(a^2 - c^2)$, the equation reduces to

$$\frac{x^2}{a^2} + \frac{y^2}{a^2 - c^2} = 1 \qquad \text{dividing by } a^2(a^2 - c^2)$$

A final simplification of the form can be obtained by letting

$$b^2 = a^2 - c^2$$

so that the equation becomes

$$\frac{x^2}{a^2} + \frac{y^2}{b^2} = 1$$

As a result, the intercepts are simply $x = \pm a$ and $y = \pm b$, while the curve itself is symmetric with respect to both axes. Note that $a > b$.

E X A M P L E **1**   Sketch the ellipse

$$\frac{x^2}{16} + \frac{y^2}{4} = 1$$

and find the foci.

Solution. From the equation

$$\frac{x^2}{a^2} + \frac{y^2}{b^2} = 1$$

we have $a^2 = 16$ and $b^2 = 4$. It follows that $a = 4$ and $b = 2$.
The foci are found by using the equation $b^2 = a^2 - c^2$:

$$b^2 = a^2 - c^2$$
$$4 = 16 - c^2$$
$$c^2 = 12$$
$$c = \pm\sqrt{12} = \pm\sqrt{4 \cdot 3} = \pm\sqrt{4}\sqrt{3} = \pm 2\sqrt{3}$$

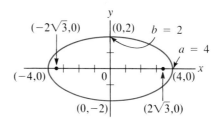

Figure 19.46

or

$$c = \pm 2\sqrt{3}$$

Hence the foci are at $(\pm 2\sqrt{3}, 0)$. (See Figure 19.46.)   ◀

If the foci are at $(0, \pm c)$, we obtain the equation

$$\frac{x^2}{b^2} + \frac{y^2}{a^2} = 1$$

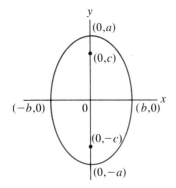

Figure 19.47

(See Figure 19.47 and Exercise 30 at the end of the section.)

The two forms of the ellipse are easily distinguished by examining the intercepts. For the equation

$$\frac{x^2}{a^2} + \frac{y^2}{b^2} = 1$$

the x-intercepts are $x = \pm a$. Since $a > b$, the numerically larger intercepts lie on the x-axis. For the other equation,

$$\frac{x^2}{b^2} + \frac{y^2}{a^2} = 1$$

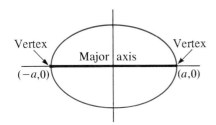

Figure 19.48

the numerically larger intercepts, $y = \pm a$, lie on the y-axis.

The symmetric properties of the ellipse suggest some additional terminology: the line segment through the foci extending between $(a, 0)$ and $(-a, 0)$ or between $(0, a)$ and $(0, -a)$ is called the **major axis**; its length is $2a$. The length a is called the **semimajor axis** (Figure 19.48). The segment of the other line of symmetry between $(b, 0)$ and $(-b, 0)$ or between $(0, b)$ and $(0, -b)$ is called the **minor axis**; its length is $2b$. The length b is called the **semiminor axis** (Figure 19.49). The intersection of the two axes is called the **center** of the ellipse. The ends of the major axis are called the **vertices** (Figure 19.48). These ideas are summarized next.

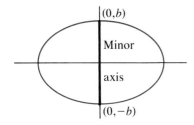

Figure 19.49

Major axis

Minor axis
Center
Vertex

Equation of an Ellipse: The equation of an ellipse with center at the origin is

$$\frac{x^2}{a^2} + \frac{y^2}{b^2} = 1 \qquad \text{(major axis horizontal)} \tag{19.16}$$

or

$$\frac{x^2}{b^2} + \frac{y^2}{a^2} = 1 \qquad \text{(major axis vertical)} \tag{19.17}$$

The foci lie on the major axis, c units from the center, with

$$b^2 = a^2 - c^2 \tag{19.18}$$

E X A M P L E **2**   Find the vertices, foci, and semiminor axis of the ellipse $2x^2 + y^2 = 4$ and sketch.

Solution. To get the equation into standard form, we divide both sides by 4:

$$\frac{x^2}{2} + \frac{y^2}{4} = 1$$

By formula (19.17),

$$\frac{x^2}{b^2} + \frac{y^2}{a^2} = 1 \qquad \text{(major axis vertical)}$$

we have $a^2 = 4$ and $b^2 = 2$, so that $a = 2$ and $b = \sqrt{2}$ Since the major axis is vertical, we conclude that the vertices are at $(0, \pm 2)$. The length of the semiminor axis is $b = \sqrt{2}$. (Note that the x-intercepts are therefore $(\pm\sqrt{2}, 0)$.) From Equation (19.18), we have

$$b^2 = a^2 - c^2$$

$$2 = 4 - c^2 \qquad \text{or} \qquad c = \pm\sqrt{2}$$

Hence the foci are at $(0, \pm\sqrt{2})$ on the major axis. (See Figure 19.50.)   ◄

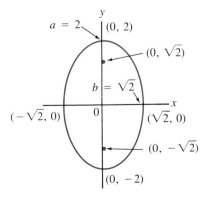

Figure 19.50

E X A M P L E **3**   Find the equation of the ellipse with center at the origin, foci at $(0, \pm 3)$, and major axis 8.

Solution. Since the length of the major axis is $2a = 8$, it follows that $a = 4$ and $a^2 = 16$ Since the foci are at $(0, \pm 3)$, $c = 3$. So by Equation (19.18) we have

$$b^2 = a^2 - c^2 = 4^2 - 3^2 = 7 \qquad a = 4, \; c = 3$$

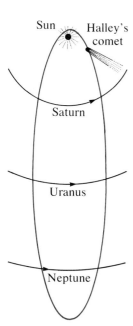

Figure 19.51

Figure 19.52

Since the foci lie on the major axis, we use form (19.17):

$$\frac{x^2}{b^2} + \frac{y^2}{a^2} = 1 \qquad \text{(major axis vertical)}$$

or

$$\frac{x^2}{7} + \frac{y^2}{16} = 1 \qquad b^2 = 7, a^2 = 16 \qquad \blacktriangleleft$$

Applications of the Ellipse

1. The paths of the planets and comets as they move about the sun are ellipses with the sun at one of the foci (Figure 19.51). Similarly, artificial satellites move about the earth in elliptical orbits with the center of the earth at one of the foci.

2. If an ellipse is rotated about its major axis, we obtain a solid called an *ellipsoid*. If a sound wave originates at one of the foci within an ellipsoid enclosure or chamber, the reflected waves converge at the other focus. A room whose ceiling is a semiellipsoid is called a "whispering gallery." A person standing at one focus can hear even a slight noise made at the other focus, while a person standing between does not hear anything. (A famous example of a whispering gallery is the Statuary Hall in the U.S. Capitol in Washington. Guides will position visitors in the proper places to hear faint sounds from another part of the hall. The hall was originally constructed for use as the House of Representatives. Congressmen found the acoustics of the hall distressing since confidential conferences in one place could be overheard easily at the opposite end of the hall.)

3. Arches of stone bridges are frequently semiellipses (Figure 19.52).

4. Elliptical gears, which revolve on a shaft turning through a focus, are sometimes used in power punches since they yield powerful movements with quick returns.

E X A M P L E **4**   A certain whispering gallery has a ceiling with elliptical cross-sections. If the room is 30 ft long and 12 ft high in the center, where should two visitors place themselves to get the whispering effect?

Solution. If the center of the ellipse is placed at the origin, we get $a = 15$ and $b = 12$. Thus

$$c^2 = a^2 - b^2 = 225 - 144 = 81$$

and $c = 9$. So the visitors must place themselves on the two foci, each 9 ft from the center. $\blacktriangleleft$

E X E R C I S E S / S E C T I O N   19.9

In Exercises 1–17, find the vertices, foci, and semiminor axis of each of the ellipses, and sketch each.

1. $\dfrac{x^2}{25} + \dfrac{y^2}{16} = 1$

2. $\dfrac{x^2}{16} + \dfrac{y^2}{9} = 1$

3. $\dfrac{x^2}{9} + \dfrac{y^2}{4} = 1$

4. $\dfrac{x^2}{4} + \dfrac{y^2}{9} = 1$

5. $\dfrac{x^2}{16} + y^2 = 1$

6. $\dfrac{x^2}{2} + \dfrac{y^2}{4} = 1$

7. $16x^2 + 9y^2 = 144$

8. $x^2 + 2y^2 = 4$

9. $5x^2 + 2y^2 = 20$

10. $5x^2 + 9y^2 = 45$

11. $5x^2 + y^2 = 5$

12. $x^2 + 4y^2 = 4$

13. $x^2 + 2y^2 = 6$

14. $9x^2 + 2y^2 = 18$

15. $15x^2 + 7y^2 = 105$

16. $9x^2 + y^2 = 27$

17. $2x^2 + 5y^2 = 50$

In Exercises 18–28, find the equation of each of the ellipses.

18. Foci at $(\pm 1, 0)$, vertices at $(\pm 2, 0)$

19. Foci at $(\pm 3, 0)$, vertices at $(\pm 4, 0)$

20. Foci at $(0, \pm 2)$, major axis 6

21. Foci at $(0, \pm 2)$, major axis 8

22. Foci at $(0, \pm 2)$, minor axis 4

23. Foci at $(0, \pm 3)$, minor axis 6

24. Minor axis 5, vertices at $(\pm 7, 0)$

25. Center at the origin, one vertex at $(0, 8)$, one focus at $(0, -5)$

26. Center at the origin, one focus at $(-3, 0)$, semiminor axis 4

27. Foci at $(\pm 2\sqrt{3}, 0)$, minor axis 4

28. Foci at $(\pm \sqrt{5}, 0)$, vertices at $(\pm \sqrt{7}, 0)$

29. Find the locus of points such that the sum of the distances from $(\pm 6, 0)$ is 16.

30. Derive Equation (19.17).

31. The shape of an ellipse depends on the values of c and a. The fraction $e = c/a$ is called the *eccentricity* of the ellipse. Find the equation of the ellipse with vertices at $(\pm 4, 0)$ and $e = 1/2$.

32. Find the equation of the ellipse centered at the origin, minor axis 6, and passing through the point $(1, 4)$.

33. Find the locus of points for which the distance from $(0, 0)$ is twice the distance from $(3, 0)$. What is the locus?

34. The area of an ellipse is given by $A = \pi ab$. Find the area of the floor of the whispering gallery in Example 4.

35. Find the equation of the ellipse to be inscribed in a rectangle 4 m long and 3 m wide if the center of the rectangle is placed at the origin with the long side horizontal.

36. A gateway is in the form of a semiellipse 10 ft across at the base and 15 ft high in the center. What is the length of the horizontal beam across the gateway 10 ft above the floor?

37. A road 8.0 m wide is spanned by a semielliptical arch 12.0 m across at its base. If the minimum clearance over the road is 4.0 m, what is the height of the arch?

38. Describe a mechanical method for drawing an ellipse inscribed in a rectangle. Refer to Exercise 35.

39. A satellite is in elliptical orbit about the earth. If the maximum altitude of the satellite is 120 mi and the minimum 80 mi, find the equation of the orbit. (The center of the earth is at a focus and the radius of the earth is approximately 4000 mi.)

40. The earth moves about the sun in an elliptical orbit with the center of the sun at one of the foci. The minimum distance to the center of the sun is 9.14×10^7 mi and the maximum distance 9.46×10^7 mi. Show that the eccentricity is only about $\frac{1}{60}$—that is, the path is nearly circular. Refer to Exercise 31.

19.10   The Hyperbola

Starting with two points, $(c, 0)$ and $(-c, 0)$, let us consider the locus of points for which the difference of the distances from the two fixed points is numerically equal to $2a$. The locus is called a **hyperbola** (Figure 19.53).

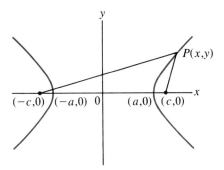

Figure 19.53

The derivation of the equation is similar to that of the ellipse. By the distance formula (19.1):

$$\sqrt{(x - c)^2 + y^2} - \sqrt{(x + c)^2 + y^2} = \pm 2a \qquad (19.19)$$

which reduces to

$$\frac{x^2}{a^2} - \frac{y^2}{c^2 - a^2} = 1 \qquad (19.20)$$

(The simplification will be left as an exercise.) This form suggests the substitution

$$b^2 = c^2 - a^2$$

Hence

$$\frac{x^2}{a^2} - \frac{y^2}{b^2} = 1$$

Conversely, any point whose coordinates satisfy this equation lies on the hyperbola. The x-intercepts of this curve are seen to be $(\pm a, 0)$, but the y-intercepts are lacking. Therefore, there appears to be no obvious interpretation of the constant b—yet the opposite is true. To see why, let us solve the last equation for y in terms of x. Thus

$$\frac{y^2}{b^2} = \frac{x^2}{a^2} - 1$$

and

$$y = \pm b \sqrt{\frac{x^2}{a^2} - 1}$$

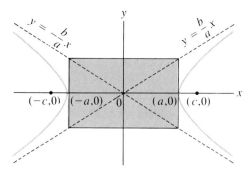

Figure 19.54

Now note that for very large values of x, the 1 in this equation becomes negligible. So y gets closer and closer to $y = \pm b\sqrt{x^2/a^2} = \pm b(x/a)$. In other words, the lines

$$y = \pm \frac{b}{a} x \tag{19.21}$$

Asymptotes
Foci
Vertices
Transverse axis
Conjugate axis
Center

are the **asymptotes** of the hyperbola (Figure 19.54).

As in the case of the ellipse, the points $(\pm c, 0)$ are called the **foci** and $(\pm a, 0)$ the **vertices** of the hyperbola. The line segment of length $2a$ joining the vertices is called the **transverse axis** of the hyperbola and the line segment of length $2b$ joining $(0, b)$ and $(0, -b)$ is called the **conjugate axis**. (See Figure 19.55.) Finally, the point of intersection of the two axes is the **center**.

If the foci lie along the y-axis at $(0, \pm c)$, we obtain the form

$$\frac{y^2}{a^2} - \frac{x^2}{b^2} = 1$$

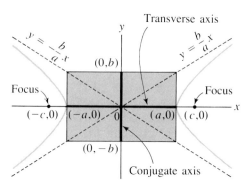

Figure 19.55

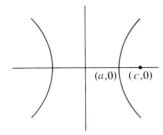

Hyperbola: $c > a$

Figure 19.56

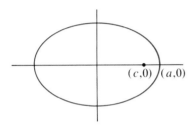

Ellipse: $c < a$

Figure 19.57

Caution. Even though the equations of the hyperbola resemble those of the ellipse, some important differences should be noted. It is clear from the graph of the hyperbola, as well as from the formula $b^2 = c^2 - a^2$, that $c > a$ (Figure 19.56)—just the opposite of the case of the ellipse (Figure 19.57). Moreover, a is not necessarily larger than b; rather, the constant a must be identified from the form of the equation. These ideas are summarized next.

Equation of a Hyperbola: The equation of a hyperbola with center at the origin is

$$\frac{x^2}{a^2} - \frac{y^2}{b^2} = 1 \qquad \text{(transverse axis horizontal)} \qquad (19.22)$$

or

$$\frac{y^2}{a^2} - \frac{x^2}{b^2} = 1 \qquad \text{(transverse axis vertical)} \qquad (19.23)$$

The foci lie on the transverse axis, c units from the center, with

$$b^2 = c^2 - a^2 \qquad (19.24)$$

To distinguish between the two forms of the hyperbola, note that for Equation (19.22) the only intercepts are $x = \pm a$. (If $x = 0$, then y is imaginary.) Similarly, the only intercepts for Equation (19.23) are $y = \pm a$.

E X A M P L E **1** Sketch the hyperbola

$$\frac{x^2}{4} - \frac{y^2}{9} = 1$$

and find the foci, vertices, and asymptotes.

Solution. From Equation (19.22), the transverse axis is horizontal and $a = 2$ and $b = 3$ (Note that $a < b$ in this case.) These numbers can be used to construct the asymptotes, which should always be drawn first to facilitate the sketching. Plot $(\pm 2, 0)$ and $(0, \pm 3)$ and draw the rectangle determined by these points. (See Figure 19.58.) The dotted rectangle is called the **auxiliary rectangle**.

Auxiliary rectangle

Now consider the lines through the vertices of the rectangle. The slope of the line through the upper right-hand corner is $b/a = \frac{3}{2}$, and the slope of the other line is $-\frac{3}{2}$. Thus the lines are indeed the asymptotes. Their respective

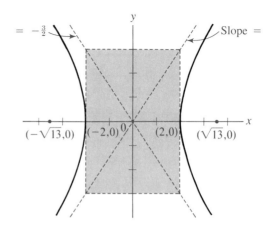

Figure 19.58

equations are

$$y = \frac{3}{2} x \qquad \text{and} \qquad y = -\frac{3}{2} x$$

From Equation (19.24),

$$b^2 = c^2 - a^2$$

we now get

$$9 = c^2 - 4 \qquad \text{and} \qquad c^2 = 13 \qquad b = 3, a = 2$$

so that the foci are at $(\pm\sqrt{13}, 0)$. Since $a = 2$, the vertices are at $(\pm 2, 0)$. (The vertices and foci lie on the transverse axis.)   ◀

E X A M P L E **2**   Sketch the hyperbola $3y^2 - 2x^2 = 12$, including the asymptotes, and find the foci and vertices.

Solution. The equation can be written in the form

$$\frac{y^2}{4} - \frac{x^2}{6} = 1 \qquad \textbf{transverse axis vertical}$$

Therefore, $a = 2$ and $b = \sqrt{6}$ by Equation (19.23) and may be used to draw the auxiliary rectangle in Figure 19.59. The foci are found to be $(0, \pm\sqrt{10})$.

Since $a = 2$, the vertices are at $(0, \pm 2)$ on the transverse axis.   ◀

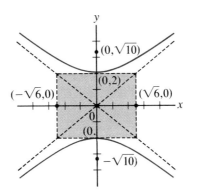

$$b^2 = c^2 - a^2$$
$$(\sqrt{6})^2 = c^2 - 2^2$$
$$c^2 = 6 + 4$$
$$c = \pm\sqrt{10}$$

Figure 19.59

E X A M P L E **3**   Find the equation of the hyperbola satisfying the following conditions: the length of the transverse axis is 4 and the foci are at $(0, \pm 3)$.

Solution. Since the foci are on the y-axis, the form of the equation is

$$\frac{y^2}{a^2} - \frac{x^2}{b^2} = 1$$

The transverse axis, which is the line segment joining the vertices, has length 4, so that $a = 2$ and $a^2 = 4$. Since $c = 3$, from Equation (19.24) we get

$$b^2 = c^2 - a^2 = 3^2 - 2^2 = 5 \qquad \textbf{c = 3, a = 2}$$

The equation is therefore given by

$$\frac{y^2}{4} - \frac{x^2}{5} = 1 \qquad \textbf{a}^2 = 4, \, \textbf{b}^2 = 5 \qquad\qquad \blacktriangleleft$$

E X A M P L E **4**   Find the equation of the hyperbola with asymptotes $3y = \pm 2x$, conjugate axis 8, and foci on the x-axis.

Solution. Since $y = \pm\frac{2}{3}x$, we have $b/a = \frac{2}{3}$ by Equation (19.21). Since the length of the conjugate axis is $2b = 8$, we get $b = 4$. We can solve the last equation for a to obtain

$$\frac{4}{a} = \frac{2}{3} \qquad \text{or} \qquad a = 6$$

So by Equation (19.22),

$$\frac{x^2}{a^2} - \frac{y^2}{b^2} = 1$$

we get

$$\frac{x^2}{36} - \frac{y^2}{16} = 1 \qquad a = 6, b = 4 \qquad \blacktriangleleft$$

Hyperbolas are used in long-range navigation as part of the LORAN system of navigation. A transmitter is located at each focus and radio signals are sent to the navigator simultaneously from each station. The difference in time at which the signals are received enables the navigator to determine his position. To be able to do so readily, however, the navigator uses charts already prepared on which hyperbolas are plotted from these differences in time. This way the navigator can find the hyperbola on which he is located. From a second pair of stations another curve of position is then found, and the navigator then determines his location by noting the point of intersection of the two hyperbolas.

E X E R C I S E S / S E C T I O N 19.10

In Exercises 1–12, find the vertices and foci of each hyperbola; draw each auxiliary rectangle and the asymptotes; and sketch the curves.

1. $\dfrac{x^2}{16} - \dfrac{y^2}{9} = 1$

2. $\dfrac{x^2}{9} - \dfrac{y^2}{4} = 1$

3. $\dfrac{x^2}{9} - \dfrac{y^2}{16} = 1$

4. $\dfrac{x^2}{16} - \dfrac{y^2}{4} = 1$

5. $\dfrac{y^2}{4} - \dfrac{x^2}{4} = 1$

6. $\dfrac{y^2}{4} - \dfrac{x^2}{8} = 1$

7. $x^2 - \dfrac{y^2}{5} = 1$

8. $9y^2 - 2x^2 = 18$

9. $2y^2 - 3x^2 = 24$

10. $x^2 - y^2 = 6$

11. $3y^2 - 2x^2 = 6$

12. $11x^2 - 7y^2 = 77$

In Exercises 13–24, determine the equation of each hyperbola.

13. Transverse axis 6, conjugate axis 4, foci on x-axis, center at origin

14. Transverse axis 6, foci at $(\pm 4, 0)$

15. Conjugate axis 8, foci at $(0, \pm 5)$

16. Vertices at $(0, \pm 5)$, conjugate axis 10

17. Transverse axis 12, foci at $(0, \pm 8)$

18. Transverse axis 8, foci at $(0, \pm 6)$

19. Vertices at $(\pm 4, 0)$, conjugate axis 8

20. Vertices at $(\pm 6, 0)$, conjugate axis 4

21. Vertices at $(\pm 3, 0)$, foci at $(\pm 6, 0)$

22. Vertices at $(\pm 2, 0)$, foci at $(\pm 4, 0)$

23. Asymptotes $y = \pm 2x$, vertices at $(\pm 1, 0)$

24. Asymptotes $y = \pm\frac{1}{2}x$, conjugate axis 3, transverse axis horizontal

25. Find the equation of the locus of points such that the difference of the distances from $(0, \pm 5)$ is ± 6.

26. Reduce Equation (19.19) to (19.20).

27. Find the equation of the locus of points such that the difference of the distances from $(1, 2)$ and $(-3, 2)$ is ± 2.

28. Derive the equation of the hyperbola whose asymptotes are $y = \pm\frac{4}{3}x$ and whose foci are at $(\pm 3, 0)$.

29. Write the equation of the hyperbola whose vertices are at $(0, \pm 12)$ and which passes through the point $(-1, 13)$.

30. Graph the equation $xy = 2$. It can be shown that the graph is a hyperbola.

31. Boyle's law states that for an ideal gas the product of the pressure and volume is always constant at a constant temperature. Mathematically, $pV = k$, where k is a constant. Suppose $V = 3.0 \text{ m}^3$ when $p = 12 \text{ Pa (N/m}^2)$ for some ideal gas. Find k and graph the equation. Refer to Exercise 30.

19.11 Translation of Axes; Standard Equations of the Conics

In this section we are going to obtain more general forms of the equations of the conics by means of *translation of axes*. A translation of axes gives a new set of axes parallel to, and oriented the same way as, the original axes.

Translation of Axes

If the origin, O, is translated to the point $O'(h, k)$, then each point has two sets of coordinates. Suppose (x, y) and (x', y') are the coordinates of a point P with respect to the old and new axes, respectively (Figure 19.60). We see from the diagram that

$$x = OB = OA + AB = h + x'$$

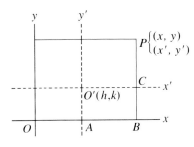

Figure 19.60

and

$$y = BP = BC + CP = k + y'$$

Thus $x' = x - h$ and $y' = y - k$.

Translation Formulas

$$x' = x - h \qquad \text{and} \qquad y' = y - k$$

EXAMPLE **1** Consider the following equation of a circle:

$$(x + 1)^2 + (y - 3)^2 = 4$$

If the origin is translated to $(-1, 3)$, we get

$$x' = x + 1 \qquad \text{and} \qquad y' = y - 3$$

and the equation becomes

$$x'^2 + y'^2 = 4$$

Both forms are familiar from Section 19.7. ◀

EXAMPLE **2** Consider the equation

$$\frac{(x - 2)^2}{9} + \frac{(y + 4)^2}{4} = 1$$

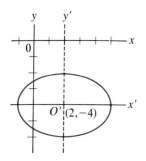

Figure 19.61

Selecting the point $(2, -4)$ for O', we have

$$x' = x - 2 \qquad \text{and} \qquad y' = y + 4$$

and we get

$$\frac{x'^2}{9} + \frac{y'^2}{4} = 1$$

which we recognize to be the equation of an ellipse (Figure 19.61).   ◀

Standard Equations of the Conics

It follows from the preceding discussion that we can obtain a more general form of the conics.

Standard Equations of the Conics

$$\left.\begin{array}{l} (y - k)^2 = 4p(x - h) \\ (x - h)^2 = 4p(y - k) \end{array}\right\} \qquad \text{parabola; vertex at } (h, k)$$

$$\frac{(x - h)^2}{a^2} + \frac{(y - k)^2}{b^2} = 1 \qquad \text{ellipse; center at } (h, k)$$

$$\frac{(x - h)^2}{a^2} - \frac{(y - k)^2}{b^2} = 1 \qquad \text{hyperbola; center at } (h, k)$$

These forms are easily remembered since we are merely replacing x by $x - h$ and y by $y - k$. Geometrically, we have translated the original figure in such a way that the new center or vertex is at the point (h, k).

The translation formulas have nothing to do with conics and can be applied equally well to any graph: if h and k are positive, *replacing x by $x - h$ in an equation will move the corresponding graph h units to the right and replacing y by $y - k$ will move the graph k units up*. Similarly, replacing x by $x + h$ will move the graph h units to the left, and replacing y by $y + k$ will move the graph k units down.

E X A M P L E **3**   Write the following equation in standard form and sketch the curve:

$$y^2 - 4y - 5x - 1 = 0$$

Solution.

$$\begin{array}{ll} y^2 - 4y - 5x - 1 = 0 & \text{given equation} \\ \quad y^2 - 4y = 5x + 1 & \text{adding } 5x + 1 \end{array}$$

We complete the square on the left by adding $[\frac{1}{2}(-4)]^2 = 4$ to both sides:

$$y^2 - 4y + 4 = 5x + 1 + 4 \qquad \text{adding 4 to both sides}$$
$$y^2 - 4y + 4 = 5x + 5$$
$$(y - 2)^2 = 5x + 5 \qquad \text{factoring left side}$$

To obtain the standard form, we need to factor 5 on the right side:

$$(y - 2)^2 = 5(x + 1)$$

which is the equation of a parabola with axis horizontal. From

$$(y - 2)^2 = 4\left(\frac{5}{4}\right)(x + 1)$$

we see that the vertex is at $(-1, 2)$ and that $p = \frac{5}{4}$. It follows that the focus is at

$$\left(-1 + \frac{5}{4}, 2\right) = \left(\frac{1}{4}, 2\right) \qquad \text{axis is horizontal}$$

(See Figure 19.62). The directrix is the line

$$x = -1 - \frac{5}{4} = -\frac{9}{4}$$

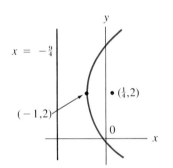

$x = -\frac{9}{4}$

$\bullet (\frac{1}{4}, 2)$

$(-1, 2)$

Figure 19.62

E X A M P L E **4** Write the following equation in standard form and sketch the curve:

$$25x^2 - 9y^2 + 150x + 36y - 36 = 0$$

Solution. We proceed by rearranging the terms and completing the square as we did in Section 19.7; thus

$$25x^2 + 150x - 9y^2 + 36y = 36$$

At this point we need to factor out the coefficients of the squared terms, so that

$$25(x^2 + 6x \quad) - 9(y^2 - 4y \quad) = 36 \qquad \text{factoring 25 and } -9$$

and then complete the square inside the parentheses. Care must be taken when balancing the equation:

$$25(x^2 + 6x + 9) - 9(y^2 - 4y + 4) = 36 + 25 \cdot 9 - 9 \cdot 4$$

or

$$25(x + 3)^2 - 9(y - 2)^2 = 225$$

This equation simplifies to

$$\frac{(x + 3)^2}{9} - \frac{(y - 2)^2}{25} = 1 \qquad \text{dividing by 225}$$

This is the equation of a hyperbola with center at $(-3, 2)$ and transverse axis horizontal. Since $a = 3$ and $b = 5$, the vertices are at $(0, 2)$ and $(-6, 2)$, while the ends of the conjugate axis are at $(-3, 7)$ and $(-3, -3)$. From the formula $b^2 = c^2 - a^2$, we get $c = \sqrt{34}$, so that the foci are at $(-3 \pm \sqrt{34}, 2)$. Using the values of a and b, we now draw the auxiliary rectangle centered at $(-3, 2)$, and the two asymptotes (Figure 19.63). ◀

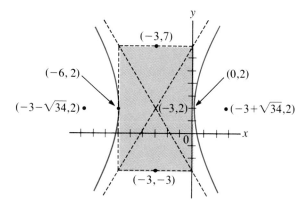

Figure 19.63

E X A M P L E **5**   Find the equation of the following ellipse: center at $(-3, -2)$, one vertex at $(0, -2)$, and one focus at $(-5, -2)$.

Solution. The points are shown in Figure 19.64.
Observe that

$$c = 2 \qquad \text{distance from center to focus}$$

and

$$a = 3 \qquad \text{distance from center to vertex}$$

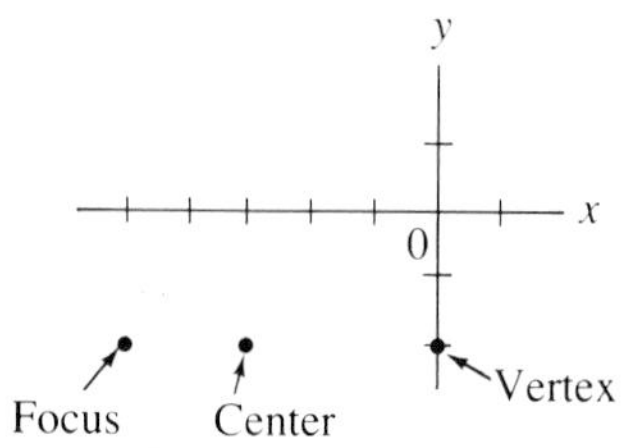

Also,

$$b^2 = a^2 - c^2 = 3^2 - 2^2 = 5$$

Figure 19.64

Since $(h, k) = (-3, -2)$, we get from the form

$$\frac{(x - h)^2}{a^2} + \frac{(y - k)^2}{b^2} = 1 \qquad \text{(major axis horizontal)}$$

the equation

$$\frac{(x + 3)^2}{9} + \frac{(y + 2)^2}{5} = 1$$

◀

It follows from the preceding examples that any equation of the form

$$Ax^2 + By^2 + Cx + Dy + E = 0$$

represents a conic since completing the square will convert the equation to one of the standard forms. The types of loci may be summarized as follows:

1. If $A = B$, the locus is a circle.
2. If $A = 0$ and $B \neq 0$, or if $B = 0$ and $A \neq 0$, then the locus is a parabola.
3. If $A \neq B$ and A and B have like signs, the locus is an ellipse.
4. If A and B have opposite signs, the locus is a hyperbola.

Certain degenerate cases may also occur:

5. The locus may be a point. We encountered this case in Section 19.7, although the possibility exists for type 3 also. Imaginary loci may occur in types 1 and 3.
6. The equation may factor into two linear factors, in which case the locus is two straight lines. As an example of the last case, the equation $x^2 - 1 = 0$ can be written $(x - 1)(x + 1) = 0$, or $x = \pm 1$, which represents two parallel lines.

EXERCISES / SECTION   19.11

In Exercises 1–16, identify the conic, find the center (or vertex), and sketch.

1. $(x - 1)^2 + (y - 2)^2 = 3$

2. $\dfrac{(x - 1)^2}{9} + \dfrac{(y - 2)^2}{5} = 1$

3. $(y + 3)^2 = 8(x - 2)$

4. $\dfrac{(x - 3)^2}{4} - \dfrac{y^2}{9} = 1$

5. $2x^2 - 3y^2 + 8x - 12y + 14 = 0$

6. $4x^2 - 4x - 48y + 193 = 0$

7. $16x^2 + 4y^2 + 64x - 12y + 57 = 0$

8. $y^2 - 12y - 5x + 41 = 0$

9. $x^2 + y^2 + 2x - 2y + 2 = 0$

10. $x^2 + 2y^2 - 6x + 4y + 1 = 0$

11. $2x^2 - 12y^2 + 60y - 63 = 0$

12. $2x^2 + 3y^2 - 8x - 18y + 35 = 0$

13. $64x^2 + 64y^2 - 16x - 96y - 27 = 0$

14. $4x^2 - 4x - 16y + 5 = 0$

15. $3x^2 + y^2 - 18x + 2y + 29 = 0$

16. $100x^2 - 180x - 100y + 81 = 0$

In Exercises 17–33, find the equations of the conics.

17. Parabola: vertex at $(-1, 2)$, focus at $(3, 2)$

18. Parabola: focus at $(2, 5)$, directrix $y + 1 = 0$

19. Ellipse: center at $(-3, 0)$, one vertex at the origin, passing through $(-3, -2)$

20. Ellipse: vertices at $(3, 1)$ and $(3, -5)$, one focus at $(3, 0)$

21. Hyperbola: vertices at $(1, 1)$ and $(-7, 1)$, one focus at $(3, 1)$

22. Hyperbola: center at $(-1, -2)$, one focus at $(2, -2)$, conjugate axis 4

23. Ellipse: center at $(2, 3)$, one vertex at $(-3, 3)$, minor axis 4

24. Parabola: vertex at $(5, -4)$, focus at $(8, -4)$

25. Hyperbola: center at $(1, 0)$, one vertex at $(3, 0)$, one asymptote $x - 2y = 1$

26. Parabola: vertex at $(-1, 3)$, passing through the origin

27. Ellipse: center at $(-3, 1)$, one focus at $(-3, -3)$, one vertex at $(-3, 6)$

28. Ellipse: center at $(4, 0)$, one focus at the origin, passing through $(4, -3)$

29. Parabola: vertex at $(4, -2)$, focus at $(1, -2)$

30. Hyperbola: center at $(3, 1)$, one vertex at $(0, 1)$, one focus at $(-1, 1)$

31. Parabola: focus at $(-2, -8)$, directrix x-axis

32. Ellipse: center at $(1, 1)$, one focus at $(1, 4)$, major axis 8

33. Hyperbola: center at $(-1, 1)$, one focus at $(-1, -2)$, one vertex at $(-1, 3)$

19.12　Polar Coordinates

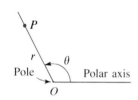

Figure 19.65

So far we have consistently used the rectangular coordinate system. For certain types of curves another coordinate system may be more convenient. Suppose we start with a point O, called the **pole**, and draw a ray with O as the endpoint, called the **polar axis** (Figure 19.65). Using this ray as the initial side, we may generate angles just as we do in trigonometry. Let θ be such an angle and P a point on the terminal side. If r is the distance from P to the pole, then we say that (r, θ) are the **polar coordinates** of P.

> **Polar Coordinates:** Let P be a point in the plane, r the distance from P to a fixed point O (the **pole**), and θ the angle between the ray OP and a fixed ray (the **polar axis**). Then P is said to be represented by the **polar coordinates** (r, θ).

Recall that in trigonometry we always assumed that r is positive. In the polar coordinate system, however, we agree to plot a negative value for r by extending the terminal side of the angle in the opposite direction through the pole and marking off r units on the extended side. Consider the next example.

E X A M P L E **1**　Plot the points whose polar coordinates are $(2, 30°)$, $(3, -\pi/4)$, $(4, 5\pi/6)$, $(-3, 240°)$, $(-1, 120°)$, $(3, 180°)$, and $(4, 0°)$.

Solution. To plot these points we need only the polar axis, but for convenience we keep the vertical axis as a line of reference and also extend the polar axis to the left (Figure 19.66).　◀

We can see from Example 1 that the pair of values (r, θ) determines a unique point. The converse is not true, however. The point $(2, 30°)$, for example,

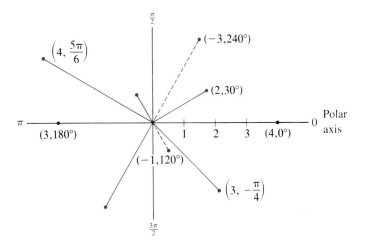

Figure 19.66

can also be represented by (2, 390°), (2, −330°), or (−2, 210°); in fact, there are infinitely many possibilities.

The relationships between the rectangular and polar coordinate systems can be readily seen from the triangle in Figure 19.67.

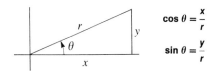

$$\cos \theta = \frac{x}{r}$$

$$\sin \theta = \frac{y}{r}$$

Figure 19.67

Conversion Formulas:

$$x = r \cos \theta \qquad y = r \sin \theta \tag{19.25}$$

$$r^2 = x^2 + y^2 \qquad \tan \theta = \frac{y}{x} \tag{19.26}$$

E X A M P L E **2** Express (−2, 2) in polar coordinates.

Solution. Since $r^2 = x^2 + y^2$, we obtain at once

$$r = \pm\sqrt{(-2)^2 + 2^2} = \pm\sqrt{4 + 4} = \pm2\sqrt{2}$$

and

$$\tan \theta = 2/(-2) = -1$$

Hence $\theta = 135° + n \cdot 360°$ or $315° + n \cdot 360°$. If we choose the positive root for r, then the coordinates are $(2\sqrt{2}, 135°)$ and, more generally, $(2\sqrt{2}, 135° + n \cdot 360°)$. For the negative root we have $(-2\sqrt{2}, 315° + n \cdot 360°)$. ◀

Relationships (19.25) and (19.26) can also be used to convert equations from one system to the other.

E X A M P L E **3**   Write $y^2 = x$ in polar coordinates.

Solution. By direct substitution,

$$(r \sin \theta)^2 = r \cos \theta \qquad y = r \sin \theta, \, x = r \cos \theta$$

$$r^2 \sin^2 \theta - r \cos \theta = 0$$

and

$$r(r \sin^2 \theta - \cos \theta) = 0 \qquad \text{factoring } r$$

from which it follows that $r = 0$ and

$$r \sin^2 \theta - \cos \theta = 0$$

$$r = \frac{\cos \theta}{\sin^2 \theta} = \frac{\cos \theta}{\sin \theta} \frac{1}{\sin \theta} \qquad \text{solving for } r$$

or

$$r = \cot \theta \csc \theta$$

Notice that $r = 0$ may be dropped since it represents only the pole, which is included in the equation $r = \cot \theta \csc \theta$. (For example, $r = 0$ when $\theta = 90°$.) ◀

E X A M P L E **4**   Write the equation $r = 2 \cos 2\theta$ in rectangular form.

Solution. By the double-angle formula,

$$r = 2(\cos^2 \theta - \sin^2 \theta) \qquad \cos 2\theta = \cos^2 \theta - \sin^2 \theta$$

Since $\cos \theta = x/r$ and $\sin \theta = y/r$ by (19.25), we get

$$\pm \sqrt{x^2 + y^2} = 2 \left(\frac{x^2}{r^2} - \frac{y^2}{r^2} \right)$$

$$\pm \sqrt{x^2 + y^2} = 2 \left(\frac{x^2}{x^2 + y^2} - \frac{y^2}{x^2 + y^2} \right) = 2 \frac{x^2 - y^2}{x^2 + y^2} \qquad r^2 = x^2 + y^2$$

and, squaring both sides,

$$x^2 + y^2 = 4\frac{(x^2 - y^2)^2}{(x^2 + y^2)^2}$$

or

$$(x^2 + y^2)^3 = 4(x^2 - y^2)^2 \qquad \text{multiplying by } (x^2 + y^2)^2$$

A simple alternative is to multiply both sides of the equation $r = 2(\cos^2 \theta - \sin^2 \theta)$ by r^2 to get

$$r^2 r = r^2 \cdot 2(\cos^2 \theta - \sin^2 \theta)$$

or

$$r^3 = 2(r^2 \cos^2 \theta - r^2 \sin^2 \theta)$$

In this form the equation converts directly to

$$\pm (x^2 + y^2)^{3/2} = 2(x^2 - y^2) \qquad \begin{array}{l} r\cos\theta = x,\ r\sin\theta = y \\ r = \pm\sqrt{x^2 + y^2} \end{array}$$

or

$$(x^2 + y^2)^3 = 4(x^2 - y^2)^2 \qquad \text{squaring both sides} \qquad \blacktriangleleft$$

E X A M P L E   **5**   Convert the equation $r = 3 \cot \theta$ to rectangular form.

Solution. In this example, inserting r is particularly convenient:

$$r = 3 \cot \theta = 3\,\frac{\cos \theta}{\sin \theta}$$

or

$$r = 3\,\frac{r \cos \theta}{r \sin \theta} \qquad \text{multiplying numerator and denominator by } r$$

This equation converts directly to

$$\pm\sqrt{x^2 + y^2} = 3\,\frac{x}{y} \qquad r\cos\theta = x,\ r\sin\theta = y$$

or

$$x^2 + y^2 = \frac{9x^2}{y^2} \qquad \text{squaring both sides} \qquad \blacktriangleleft$$

E X E R C I S E S / S E C T I O N   19.12

1. Plot the following points: $(1, 60°)$, $(3, 3\pi/4)$, $(2, -50°)$, $(-2, 90°)$, $(-1, -\pi/4)$, $(-2, 270°)$, $(-4, 0°)$.

In Exercises 2–7, express the points in rectangular coordinates.

2. $\left(\sqrt{2}, \dfrac{\pi}{4}\right)$ 3. $(2, 120°)$ 4. $\left(4, -\dfrac{\pi}{6}\right)$

5. $\left(-6, \dfrac{\pi}{3}\right)$ 6. $(1, 50°)$ 7. $(-2, 170°)$

In Exercises 8–12, find a set of polar coordinates for each of the following points in rectangular coordinates.

8. $(\sqrt{3}, 1)$ 9. $(1, -1)$ 10. $(-2, -2\sqrt{3})$

11. $(3, 4)$ 12. $(-2, 5)$

In Exercises 13–20, express the equations in polar coordinates.

13. $x = 2$ 14. $y = 4$

15. $x^2 + y^2 = 2$ 16. $x^2 + y^2 + 3x - 2y = 0$

17. $x^2 - y^2 = 1$ 18. $y = 2x^2$

19. $2x^2 + 4y^2 = 1$ 20. $y = x^3$

In Exercises 21–41, express the equations in rectangular coordinates.

21. $r \sin \theta = 2$ 22. $r = 3 \sec \theta$

23. $\theta = \dfrac{\pi}{4}$ 24. $r = 5$

25. $r = \cos \theta$ 26. $r = \sin \theta$

27. $r = 1 + \cos \theta$ 28. $r = 1 - \sin \theta$

29. $r = 1 - \cos \theta$

30. $r = 3 \sin 2\theta$ (Recall that $\sin 2\theta = 2 \sin \theta \cos \theta$.)

31. $r^2 = \sin 2\theta$ 32. $r = \tan \theta$

33. $r = \dfrac{2}{1 - \cos \theta}$ 34. $r^2 = \cot \theta$

35. $r = \dfrac{4}{1 - \sin \theta}$ 36. $r^2 \cos 2\theta = 4$

37. $r^2 = 3 \cos 2\theta$

38. $r = a \cos 3\theta$ (Use $\cos 3\theta = \cos(2\theta + \theta)$.)

39. $r = a \sin 3\theta$ 40. $r = 4 \sec \theta \tan \theta$

41. $r = 1 - 2 \sin \theta$

19.13   Curves in Polar Coordinates

In this section we are going to study curves in polar coordinates. Without prior experience we are essentially reduced to the point-plotting method. We will soon find out, however, that a number of polar curves can be classified according to type. Once the type is identified, drawing the curve becomes relatively easy.

E X A M P L E 1   Draw the graph of $r = 4(1 + \sin \theta)$.

Solution. First we make a table of values:

θ:	0°	30°	45°	60°	90°	120°	135°	150°	180°	225°	270°	315°
r:	4	6	6.8	7.5	8	7.5	6.8	6	4	1.2	0	1.2

Note that it is not necessary to use radian measure as long as we are just plotting points. As θ ranges from 0° to 90°, r ranges from 4 to 8; from 90° to 180° the values of r repeat in inverse order. Continuing from 180° to 270°, we see that $\sin \theta$ ranges from 0 to -1; hence r decreases from 4 to 0. Finally, as θ ranges from 270° to 360°, r increases from 0 to 4. Once we have a complete

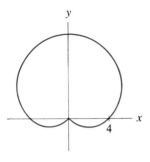

Figure 19.68

circuit, the values of r repeat and it becomes unnecessary to find additional points. The graph is shown in Figure 19.68. ◄

It is easy to check that the curve $r = 4(1 + \cos \theta)$ is the curve in Example 1 rotated by 90° in the clockwise direction.

Figure 19.68 is an example of a curve called a *cardioid* (meaning "heart-shaped").

Cardioid: A cardioid is a curve whose general equation is

$$r = a(1 \pm \sin \theta) \qquad \text{or} \qquad r = a(1 \pm \cos \theta) \qquad (19.27)$$

The coefficient a may be either positive or negative.

E X A M P L E **2** Sketch the graph of $r = 5 + 3 \sin \theta$

Solution. Because of the similarity to Example 1, we locate only the four points analogous to the intercepts: $(5, 0°)$, $(8, 90°)$, $(5, 180°)$, and $(2, 270°)$. The graph is shown in Figure 19.69. Note that at 270°, r is still positive, so that the curve (unlike the cardioid) does not pass through the pole.

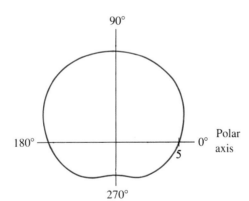

θ:	0°	90°	180°	270°
r:	5	8	5	2

Figure 19.69

◄

The curve in Example 2 is called a *limaçon of Pascal.*

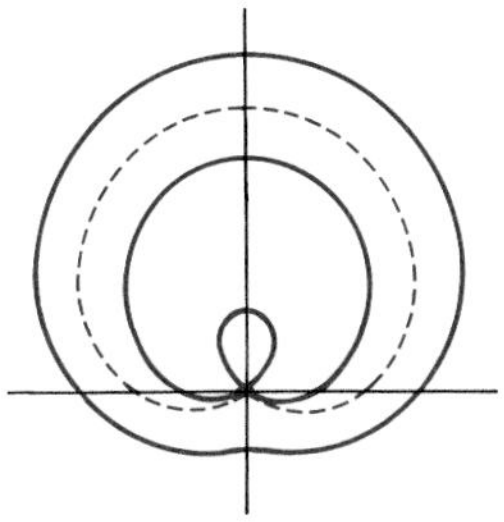

Limaçon: A limaçon is a curve whose general equation is

$$r = a + b \sin \theta \qquad \text{or} \qquad r = a + b \cos \theta \qquad (19.28)$$

Figure 19.70

If $|a| > |b|$ and $a > 0$, r is always positive (as in Example 2). If $|a| = |b|$, then $r = 0$ for one value of θ between 0° and 360°, so that the curve is actually a cardioid. If $|a| < |b|$ and $a > 0$, r is negative for some values of θ, and we obtain a "*limaçon* with a loop." The three cases are shown in Figure 19.70. (If $a < 0$, the conclusion is the same in the sense that the limaçon has a loop if, and only if, $|a| < |b|$.)

E X A M P L E **3**   Sketch the limaçon $r = 1 - 2 \cos \theta$.

Solution. Since $|a| < |b|$, this limaçon has a loop. To find the values of θ tracing out the loop, we must first see where the loop starts and ends; that is, we need to find the values of θ for which the curve passes through the pole. To this end we let $r = 0$ to get

$$1 - 2 \cos \theta = 0 \qquad \text{and} \qquad \cos \theta = \tfrac{1}{2}$$

so that $\theta = \pm 60°$. If $-60° < \theta < 60°$, then $r < 0$. To sketch the rest of the graph, we find the four "intercepts": $(-1, 0°)$, $(1, 90°)$, $(3, 180°)$, and $(1, 270°)$. The graph is shown in Figure 19.71.

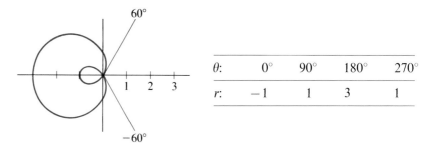

θ:	$0°$	$90°$	$180°$	$270°$
r:	-1	1	3	1

Figure 19.71

E X A M P L E **4**   Sketch the graph of $r = 2 \sin 3\theta$.

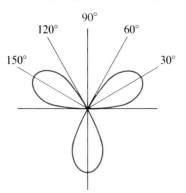

Figure 19.72

Solution. Instead of plotting points, we will discuss the general behavior of this function: r is zero whenever 3θ is coterminal with $0°$ or $180°$. This will occur whenever $\theta = 0°$, $60°$, $120°$, $180°$, $240°$, and $300°$. Numerically, r attains its maximum value of 2 whenever $\theta = 30°$, $90°$, $150°$, $210°$, $270°$, and $330°$. We note especially that when θ ranges from $0°$ to $30°$, 3θ ranges from $0°$ to $90°$, so that $\sin 3\theta$ ranges from 0 to 1; at $\theta = 60°$, r is back to zero. As θ ranges from $60°$ to $120°$, the values of r are negative. The graph is shown in Figure 19.72. ◄

The curve in Figure 19.72 is an example of a *rose*.

Rose: A rose is a curve whose general equation is

$$r = a \sin n\theta \qquad \text{or} \qquad r = a \cos n\theta \qquad\qquad (19.29)$$

Number of leaves

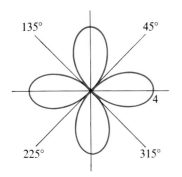

135°   45°

4

225°   315°

Figure 19.73

The coefficient n gives the number of "leaves": the rose has n leaves if n **is odd** and $2n$ leaves if n **is even**. For example, $r = 4 \cos 2\theta$ is a four-leaf rose (Figure 19.73).

Another type of curve, called a *lemniscate of Bernoulli*, has the form given next.

> **Lemniscate:** A lemniscate is a curve whose general equation is
>
> $$r^2 = a \sin 2\theta \quad \text{or} \quad r^2 = a \cos 2\theta \tag{19.30}$$

E X A M P L E **5** Sketch the lemniscate $r^2 = 4 \cos 2\theta$.

Solution. A brief table of values is

θ:	0°	10°	20°	30°	45°
r:	± 2	± 1.9	± 1.75	± 1.4	0

We get the same values for r if θ starts at 0° and goes to $-45°$. In the range $\theta = 45°$ to $\theta = 135°$, $\cos 2\theta < 0$, so that the values of r become imaginary. The graph is shown in Figure 19.74. (A lemniscate always has two "leaves".) ◄

Circle
Line
Spiral

Certain special cases may also occur. The curve $r = a$ is a **circle** centered at the pole, while $\theta = b$ is a **straight line** through the pole. The curve $r = a\theta$ is called a *spiral of Archimedes*. Other spirals will be taken up in the exercises.

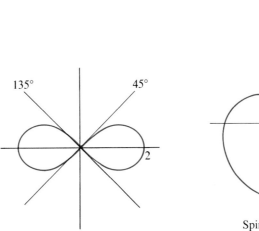

135°   45°

2

Figure 19.74

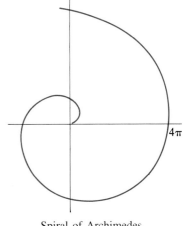

4π

Spiral of Archimedes

Figure 19.75

E X A M P L E **6**   Draw the spiral $r = 2\theta$, $\theta \geq 0$.

Solution. In this case the values of θ must be in radian measure. Since r is directly proportional to θ, the points recede farther and farther from the pole (Figure 19.75).   ◄

E X E R C I S E S / S E C T I O N   **19.13**

Sketch the following curves. Whenever possible, identify the curve before attempting the sketch.

1. $r = 2$

2. $\theta = \dfrac{\pi}{3}$

3. $r = 2 \cos \theta$

4. $r = 3 \sin \theta$

5. $r \cos \theta = 4$

6. $r = 2 \csc \theta$

7. $r = 2(1 + \cos \theta)$

8. $r = -4(1 + \sin \theta)$

9. $r = 9 + 5 \cos \theta$

10. $r = 4 - \sin \theta$

11. $r = 3 - 2 \cos \theta$

12. $r = -1 + \cos \theta$

13. $r = 1 - 2 \sin \theta$

14. $r = 1 + 2 \cos \theta$

15. $r = -1 - 2 \sin \theta$

16. $r = 2 \cos 3\theta$

17. $r = 4 \sin 2\theta$

18. $r = 3 \sin 5\theta$

19. $r = \cos 3\theta$

20. $r^2 = 9 \cos 2\theta$

21. $r^2 = 4 \sin 2\theta$

22. $r = \sqrt{\sin \theta}$

23. $r = 3\sqrt{\sin 2\theta}$, $0° \leq \theta \leq 90°$

24. $r = 4 \cos 2\theta$

25. $r = \cos 3\left(\theta - \dfrac{\pi}{6}\right)$

26. $r^2 = 2 \sin 2\theta$

27. $r = \sin 3\theta$

28. $r = \theta$ (spiral of Archimedes)

29. $r\theta = 2$, $\theta > 0$ (hyperbolic spiral)

30. $r = e^{\theta/3}$, $\theta \geq 0$ (logarithmic spiral)

31. $r = \dfrac{3}{2 - \cos \theta}$ (ellipse)

32. $r = \dfrac{2}{1 - \cos \theta}$ (parabola)

33. $r = \dfrac{6}{2 + 3 \sin \theta}$ (hyperbola)

R E V I E W   E X E R C I S E S / C H A P T E R   **19**

1. Show that $(4, 2)$, $(7, 4)$, and $(3, 10)$ are the vertices of a right triangle.

2. Find the equation of the perpendicular bisector of the line segment joining $(-1, -2)$ and $(5, 4)$.

3. Fahrenheit and Celsius temperatures are related by $C = \frac{5}{9}(F - 32)$. What temperature is the same in both Fahrenheit and Celsius?

4. What is the physical significance of the F-intercept in Exercise 3?

5. Show that $(-1, 5)$, $(3, 1)$, $(3, 9)$, and $(7, 5)$ are the vertices of a square.

6. Find the equation of the line through $(4, 1)$ and perpendicular to the line $x + 2y - 5 = 0$.

7. Find the equation of the line through $(-1, 5)$ and parallel to the line $3x + y = 3$.

8. Show that the following lines form the sides of a right triangle: $4x - y - 7 = 0$, $x + 4y - 1 = 0$, and $y + 2x = 0$.

9. Find the equation of the circle with center at $(1, -2)$ and passing through the origin.

10. Find the equation of the circle centered at $(2, 5)$ and tangent to the x-axis.

11. Find the center and radius of the circle $x^2 + y^2 + 2x + 2y = 0$.

12. Show that the circle $x^2 + y^2 - 10x - 8y + 16 = 0$ is tangent to the y-axis.

In Exercises 13–18, identify the conic, find the vertices and foci, and sketch.

13. $\dfrac{x^2}{9} + \dfrac{y^2}{16} = 1$ **14.** $x^2 + 4y^2 = 1$ **15.** $\dfrac{y^2}{4} - \dfrac{x^2}{7} = 1$

16. $\dfrac{x^2}{9} - \dfrac{y^2}{16} = 1$ **17.** $y^2 = -3x$ **18.** $x^2 = 9y$

In Exercises 19–23, identify the conic, find the center (or vertex), and sketch.

19. $y^2 + 6y + 4x + 1 = 0$

20. $x^2 + y^2 - 8x + 10y - 4 = 0$

21. $16x^2 + 9y^2 - 64x + 18y = 71$

22. $x^2 + 4x + 8y - 20 = 0$

23. $x^2 - y^2 - 4x + 8y - 21 = 0$

In Exercises 24–33, find the equations of the curves indicated.

24. Parabola: vertex at origin, directrix $x = -2$

25. Parabola: vertex at $(1, 3)$, directrix $y = 0$

26. Ellipse: center at $(-2, -4)$, one vertex at $(-2, 0)$, one focus at $(-2, -2)$

27. Ellipse: center at origin, one vertex at $(0, 4)$, one focus at $(0, 3)$

28. Hyperbola: center at origin, one vertex at $(3, 0)$, asymptotes $y = \pm\frac{3}{4}x$

29. Hyperbola: vertices at $(0, 5)$ and $(0, -1)$, one focus at $(0, -2)$

30. Parabola: focus at $(-8, 2)$, directrix y-axis

31. Parabola: focus at $(0, 3)$, directrix $y = 1$

32. Hyperbola: center at $(2, -1)$, one focus at $(2, -4)$, one vertex at $(2, 1)$

33. Ellipse: center at $(4, -1)$, one focus at $(4, 1)$, one vertex at $(4, -5)$

In Exercises 34–40, discuss and sketch the given curves.

34. $y = x^3 + 1$ **35.** $y = (x + 1)^3$ **36.** $y = 2x - x^3$

37. $y^2 = x(x - 4)$ **38.** $y = \dfrac{x^2}{x^2 + 1}$ **39.** $y = \dfrac{x}{x^2 - 4}$

40. $y = x\sqrt{x + 1}$

41. A parabolic reflector has a diameter of 1.20 ft and a depth of 0.90 ft. Where should the light be placed to produce a beam of parallel rays?

42. A semielliptic arch measures 50.0 ft across the base and 15.0 ft high in the center. Find the heights of two vertical pillars each placed 15.0 ft from the center.

43. Equal squares of side x are cut from the corners of a piece of tin 6 inches square, and the edges are folded up to form a box with an open top. Express the volume V of the box in terms of x and draw the graph. Estimate the value of x for which the volume is a maximum.

44. If a stone is thrown upward from the earth's surface with an initial velocity of 96 ft/s, then the distance s above the ground is given by $s = 96t - 16t^2$, where s is measured in feet and t in seconds. Sketch the curve and estimate the greatest height reached by the stone.

45. A comet moves in a parabolic path with the center of the sun as its focus. When the comet is 60 million kilometers from the sun, it makes an angle of $60°$ with the axis of the parabola. How close will the comet come to the sun?

46. Express the equation $y = x$ in polar coordinates.

47. Express the equation $y^2 = 3x$ in polar coordinates.

In Exercises 48—50, express the equations in rectangular coordinates.

48. $r = 4 \csc \theta$ **49.** $r^2 = 4 \tan \theta$

50. $r = 2 - 3 \cos \theta$

In Exercises 51–56, identify the curves and sketch them.

51. $r = \sin \theta$ **52.** $r = 3(1 - \cos \theta)$

53. $r = 2 \cos 2\theta$ **54.** $r = 2 - 5 \sin \theta$

55. $r^2 = 9 \sin 2\theta$ **56.** $r = \sin 3\theta$

Introduction to Calculus:
The Derivative

20.1 Introduction

Calculus was developed during the seventeenth century as a response to problems that were beyond the reach of elementary mathematics. Since mathematics and physics did not yet constitute different branches of knowledge, many of the motivations were based on physical problems. For example, the heliocentric (sun-centered) theory of Nicholas Copernicus (Polish astronomer, 1473–1543) and Johannes Kepler (German astronomer and physicist, 1571–1630) created an interest in problems of motion. However, instead of dealing only with average velocities and accelerations, calculus, once developed sufficiently, was able to handle problems dealing with instantaneous velocities and accelerations. Similarly, the concept of slope of a line was extended to slopes of arbitrary curves, and work done by a variable force could now be determined in addition to work done by a constant force. Equally remarkable were the advances in the study of areas and volumes. Classical geometry is confined to areas bounded by polygons; calculus deals with areas bounded by arbitrary curves. Basic to all these concepts is the notion of *limit*, to be discussed in detail in Section 20.3.

Credit for the discovery of calculus is usually given to Isaac Newton (English physicist and mathematician, 1642–1727) and Gottfried Wilhelm Leibniz (German mathematician and philosopher, 1646–1716). Newton was frail as a child and, unable to participate in the usual games of boys his age, was forced to create his own diversions; he was especially fond of experimentation and showed considerable mechanical ability while still in his early teens. He attended Cambridge University as a young man; his education was interrupted by the great plague of 1664–1665, which he avoided by returning to his home in Lincolnshire for two years. In Lincolnshire, as a result of his meditations, he made his famous discoveries: the nature of colors, the calculus, and the universal law of gravitation; the last of these is probably the most far-reaching physical law ever discovered. He was twenty-three.

Leibniz possessed knowledge so comprehensive that it was said he achieved "universal knowledge" at a time when most scholars could no longer keep track of all new discoveries. He made significant contributions to mathematics, physical science, law, theology, statecraft, history, logic, and philosophy. Only in physics did he lag far behind Newton. Trained in law, Leibniz made his living as a genealogist and family historian to the ancient aristocratic house of Brunswick-Lüneburg, later Hanover (some of whose descendants became kings of England). Since his work required extensive travel, Leibniz came into contact with many of the leading scholars of his day. He did not turn his attention to mathematics until age twenty-six, apparently inspired by discussions with Christian Huygens, the Dutch physicist and mathematician. Leibniz's main contributions to calculus were made around 1675, some time after Newton's. Since Newton had not published his results, an unfortunate and bitter controversy over priority resulted. It is now generally agreed that the discoveries were made independently. The notations dy/dx and $\int f(x)\,dx$ are due to Leibniz.

20.2 Functions and Intervals

In this section we are going to discuss intervals and functions. An **interval** is the set of all points on the number line between a and b. In somes cases we may wish to include either or both endpoints in a designated interval, in other cases not. As you can see from the following chart, we use a square bracket to indicate that the endpoint is included and a parenthesis to indicate that it is not:

Interval	Meaning	Graph
(a, b)	$a < x < b$	
$[a, b]$	$a \leq x \leq b$	
$[a, b)$	$a \leq x < b$	
$(a, b]$	$a < x \leq b$	

These intervals have been given special names. The notation (a, b) denotes an *open* interval; $[a, b]$ is called a *closed* interval; and the remaining two are *half-open* intervals. The endpoints a and b need not be finite; for example, (a, ∞) is the set of all x such that $a < x < \infty$, or simply $x > a$. Since ∞ is not a number, in the present context $(a, \infty]$ has no meaning.

The concept of function is closely related to that of a curve.

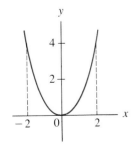

Figure 20.1

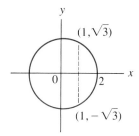

Figure 20.2

Vertical line test

Domain
Range

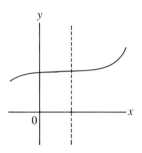

Figure 20.3

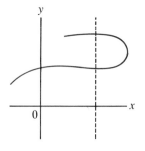

Figure 20.4

> **Definition of a Function:** If for every value of the variable x there corresponds one and only one value of the variable y, then we call y a **function** of x, denoted by $y = f(x)$ (to be read "f of x").

The variable x is called the **independent** variable and y the **dependent** variable. (It is understood in this definition that we use only those values of x for which $f(x)$ is defined.)

To illustrate the definition of a function, consider the equation $y = 2x + 3$. Since for any value of x we obtain a unique value for y, the equation defines a function. The parabola $y = x^2$ also defines a function, since every value of x yields one value for y, and only one; the fact that $x = 2$ and $x = -2$ produce the same y value, namely $y = 4$, is of no consequence (Figure 20.1). Compare this case to that of the circle $x^2 + y^2 = 4$: for every x in the open interval $(-2, 2)$ we obtain two y values. For example, if $x = 1$, then $y = \pm\sqrt{3}$ (Figure 20.2). Hence the equation of the circle does not define a function.

Geometrically, we can identify a function at a glance: a graph represents a function of x if *a vertical line intersects the graph at only one point*. Thus the graph in Figure 20.3 defines a function, but the graph in Figure 20.4 does not.

The set of all values of the independent variable x for which the dependent variable y is defined is called the **domain** of the function. (Note that the domain of the function is the same as the extent of the curve representing the function.) For example, the domain of the function $y = 1/x$ is the set of all x such that $x \neq 0$ (to avoid division by zero). The domain of the function $y = \sqrt{x}$ is the interval $[0, \infty)$. (Whenever $x < 0$, y is not a real number.)

The set of all y values that occur is called the **range** of the function. For example, the range of the function $y = x^2$ is $[0, \infty)$. (See Figure 20.1.)

Many formulas and physical laws are expressed as functions. For instance, if r is the radius of a sphere and V the volume, then $V = \frac{4}{3}\pi r^3$. Since every value of r yields a unique value of V, the relation is a function. (For physical reasons the domain is the interval $[0, \infty)$.) The formula for converting degrees Fahrenheit to degrees Celsius also defines a function:

$$C = \tfrac{5}{9}(F - 32)$$

When writing functions, we can use any letters we choose; the most common letters are x and y for the variables and f for the function. If more than one function is referred to in a discussion, we may have to use a different letter. The most commonly used letters are f, g, F, G, and ϕ. Thus $y = F(x)$, $y = g(x)$, and $y = \phi(x)$ could all represent functions, but so could $y = g(t)$ and $w = \phi(z)$. At times we will simply refer to the functions F, g, or ϕ. Finally, since the letters themselves are arbitrary,

$$F(x) = x^2 + 2x + 1 \qquad \text{and} \qquad g(t) = t^2 + 2t + 1$$

represent the same function.

This functional notation can also be used to specify individual function values. In the example just given, if $x = -3$, then $F(x) = 4$, or simply $F(-3) = 4$. Similarly, $F(0) = 1$ and $F(a) = a^2 + 2a + 1$. In fact, x may be replaced by any number or expression; for example,

$$F(a - 1) = (a - 1)^2 + 2(a - 1) + 1 = a^2$$

The last case may seem like a useless exercise but is actually quite important, as we will see in Section 20.4. A simple way to make a substitution of this type is to leave a blank space for x and fill in the blanks:

$$F(\ \ ) = (\ \ )^2 + 2(\ \ ) + 1$$

We will illustrate the preceding ideas with some examples.

E X A M P L E 1 Write the area of a circle as a function of its radius.

Solution. The formula is $A = \pi r^2$. For every value of r we get a unique value of A. So the relation is indeed a function. ◀

E X A M P L E 2 State the domain and range of the function $y = \sqrt{x^2 - 1}$.

Solution. Since y is real only if $x^2 - 1 \geq 0$, the domain consists of the intervals $(-\infty, -1]$ and $[1, \infty)$. Since y cannot be negative, the range is the interval $[0, \infty)$. ◀

E X A M P L E 3 If $f(x) = x^2 + 3x^3$, find **(a)** $f(1)$; **(b)** $f(a^2)$; **(c)** $f(x - 2)$; **(d)** $f(x + h)$.

Solution. $f(x) = x^2 + 3x^3$
  a. $f(1) = 1^2 + 3 \cdot 1^3 = 4$
  b. $f(a^2) = (a^2)^2 + 3(a^2)^3 = a^4 + 3a^6$
For parts **(c)** and **(d)** we leave a blank space for x:

$$f(\ \ ) = (\ \ )^2 + 3(\ \ )^3$$

c. Placing $x - 2$ in the blank spaces, we get

$$f(x - 2) = (x - 2)^2 + 3(x - 2)^3 = 3x^3 - 17x^2 + 32x - 20$$

d. We fill the blanks with $x + h$ to get

$$f(x + h) = (x + h)^2 + 3(x + h)^3$$ ◀

If $f(x)$ and $g(x)$ are two functions, then $f(g(x))$ is called a **composite function**.

E X A M P L E **4**   If $f(x) = \sqrt{x + 1}$ and $g(x) = x^2 - 1$, find $f(g(x))$.

Solution. From $f(\ ) = \sqrt{(\ )} + 1$, we get

$$f(g(x)) = f(x^2 - 1) = \sqrt{(x^2 - 1) + 1} = x \qquad \blacktriangleleft$$

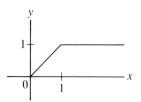

Figure 20.5

A function need not be defined by a single equation or even by an equation at all. The function

$$f(x) = \begin{cases} x & 0 \le x < 1 \\ 1 & x \ge 1 \end{cases}$$

is a perfectly good function whose domain is $[0, \infty)$ (Figure 20.5). Note that $f(0) = 0$, $f(\frac{1}{2}) = \frac{1}{2}$, $f(1) = 1$, and $f(1000) = 1$.

E X A M P L E **5**   Suppose the input voltage in a DC electrical circuit is given by

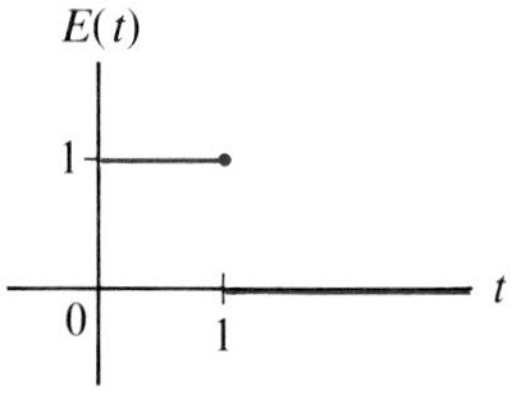

$$E(t) = \begin{cases} 1 & 0 < t \le 1 \\ 0 & t > 1 \end{cases}$$

Draw the graph.

Solution. The graph is shown in Figure 20.6. The domain is the interval $(0, \infty)$.

$\blacktriangleleft$

Figure 20.6

E X E R C I S E S / S E C T I O N   **20.2**

1. Write the area of a square, A, as a function of its side, x.

2. Suppose that $E = 3$ volts in a DC circuit. If a rheostat is used to control the change in R, write I as a function of R.

3. Express the volume V of a cone of height 2 as a function of the radius r.

4. Express the area of the surface of a sphere as a function of the radius.

5. A variable resistor R (in ohms) is in series with a 10-ohm resistor. Write the resistance R_T of the combination as a function of R.

6. A TV repair shop charges a base fee of $15 plus $35/h. Write the cost C of a TV repair as a function of time t (in hours).

7. Express the distance d traveled at 50 mi/h as a function of t (in hours).

8. A bar is 20 ft long. If x is the distance from the left end, write the distance y from the right end as a function of x.

In Exercises 9–20, find the domain and range of the given functions.

9. $y = x + 2$

10. $y = x^2 + 2$

11. $y = \sqrt{x - 2}$

12. $y = \sqrt{1 - x^2}$

13. $y = \sqrt{4 - x^2}$

14. $y = \sqrt{5 - x}$

15. $y = x\sqrt{x - 3}$

16. $y = x^2\sqrt{x - 4}$

17. $y = (x - 1)\sqrt{x - 2}$

18. $y = x^2\sqrt{x^2 - 1}$

19. $y = \dfrac{1}{x - 1}$

20. $y = \dfrac{x}{x^2 - 1}$

21. If $f(x) = 2x$, find $f(0)$, $f(6)$.

22. If $g(x) = 1 - 3x$, find $g(0)$, $g(2)$.

23. If $h(x) = x^2 + 2x$, find $h(1)$, $h(3)$.

24. If $F(x) = -2x^3$, find $F(1)$, $F(-2)$.

25. If $f(x) = x^3 + 1$, find $f(0)$, $f(1)$, $f(-2)$.

26. If $g(x) = \sqrt{x}$, find $g(0)$, $g(4)$, $g(5)$.

27. If $\phi(x) = 1/x$, find $\phi(3)$, $\phi(a)$.

28. If $F(t) = 1/t^2$, find $F(x)$ and $F(a)$.

29. If $G(z) = \sqrt{z^2 - 1}$, find $G(a^2)$, $G(x - 1)$.

30. If $f(t) = 16t^2 - 2t$, find $f(x)$, $f(x + \Delta x)$, where Δx is a constant.

31. If $f(x) = 1 - x^2$, find $f(x + \Delta x)$, $f(x - \Delta x)$.

32. If $g(t) = 1/\sqrt[3]{t}$, find $g(1 - t)$, $g(x - 1)$.

33. If $f(x) = x^2$ and $g(x) = x + 1$, find:
  a. $f(g(x))$ **b.** $g(f(x))$ **c.** $f(f(x))$

34. If $f_1(x) = 1/x$ and $f_2(x) = x^3$, find:
  a. $f_1(f_2(x))$ **b.** $f_2(f_1(x))$ **c.** $f_1(f_1(x))$

35. Graph the function

$$f(x) = \begin{cases} -1 & -\infty < x < 0 \\ 1 & 0 \le x < \infty \end{cases}$$

and find $f(-1)$, $f(0)$, $f(2)$, $f(10)$.

36. Graph the function

$$f(x) = \begin{cases} x & 0 \le x \le 1 \\ x^2 & x > 1 \end{cases}$$

and find $f(0)$, $f(1)$.

20.3 Limits

The notion of **limit** is one of the most important in calculus, since so many concepts are based on it. Fortunately, we have already touched on limits in our discussion of asymptotes. For example, to find the horizontal asymptote of the function $y = 1/x$, we note that y approaches zero as x gets large—that is, the x-axis is an asymptote. This statement can be expressed more simply in symbolic form:

$$\lim_{x \to \infty} \frac{1}{x} = 0$$

Notice that we are not asserting that $1/x = 0$ when $x = \infty$ since ∞ is not a number. All we are claiming is that $1/x$ *approaches* zero as x increases. (See Figure 20.7.)

Now consider the expression

$$\lim_{x \to 4} \frac{x^2}{x + 4}$$

which means: what happens to the values of the function $f(x) = x^2/(x + 4)$ as x approaches 4? Suppose we study the behavior of the function by letting x get closer and closer to 4, as shown in the following table:

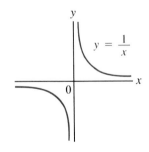

Figure 20.7

x:	4.1	4.01	4.001	4.0001	3.9	3.99	3.999	3.9999
$\dfrac{x^2}{x + 4}$:	2.1	2.01	2.001	2.0001	1.9	1.99	1.999	1.9999

Since the values are getting closer and closer to 2, it is reasonable to conclude that

$$\lim_{x \to 4} \frac{x^2}{x + 4} = 2$$

It seems that the limit could have been obtained more simply by letting $x = 4$ in the function $f(x) = x^2/(x + 4)$, or

$$f(4) = \frac{4^2}{4 + 4} = 2$$

However, the equality $f(4) = 2$ and the statement $\lim_{x \to 4} x^2/(x + 4) = 2$ do not say the same thing. If we say that $f(4) = 2$, we are merely giving the function value at $x = 4$. But if we say that

$$\lim_{x \to 4} \frac{x^2}{x + 4} = 2$$

we are not interested in the value of $f(x)$ at $x = 4$, *but only in the value approached*. (See Figure 20.8.)

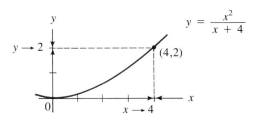

Figure 20.8

An interesting contrast to the limit just discussed is

$$\lim_{x \to 0} \frac{\sin x}{x}$$

Now the question is: what happens to the values of $f(x) = (\sin x)/x$ as x approaches zero? As before, we are not interested in the value of $f(x)$ at $x = 0$, *but only in the value approached*. Moreover, substituting 0 for x is useless anyway since $(\sin 0)/0$ is undefined (because of the division by zero). We will study the behavior of the function by means of a table. (To obtain the values, your calculator must be set in the radian mode.)

x:	0.5	0.25	0.2	0.1	0.05	0.01	0.001
$\dfrac{\sin x}{x}$:	0.9589	0.9896	0.9933	0.9983	0.9996	0.99998	0.9999998

The trend is clear: as x approaches zero, $(\sin x)/x$ approaches 1, or

$$\lim_{x \to 0} \frac{\sin x}{x} = 1$$

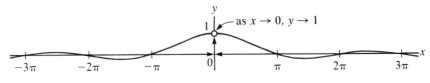

Figure 20.9

(We will see in Chapter 24 that this is indeed the case; merely substituting values does not, of course, constitute a proof.) We see from the table that even though the function $f(x) = (\sin x)/x$ is not defined at $x = 0$, the limit as $x \to 0$ exists just the same. The graph of $y = (\sin x)/x$ (Figure 20.9) tells us why: there is a hole where one would expect to find the point $(0, 1)$!

We are now ready to define the limit of a function in general.

Definition of Limit: Suppose that $f(x)$ becomes arbitrarily close to the number L ($f(x) \to L$) as x approaches a ($x \to a$). Then we say that the limit of $f(x)$ as x approaches a is L and write

$$\lim_{x \to a} f(x) = L$$

The number a may be replaced by ∞.

E X A M P L E **1**   Evaluate

$$\lim_{x \to -1} (x^2 - 3)$$

Solution. This limit is similar to $\lim_{x \to 4} x^2/(x + 4)$ discussed earlier. Since $f(x) = x^2 - 3$ is defined at $x = -1$, we can obtain the limit by inspection: as x gets closer and closer to -1, $f(x) = x^2 - 3$ gets closer and closer to -2. Thus

$$\lim_{x \to -1} (x^2 - 3) = -2 \qquad \blacktriangleleft$$

E X A M P L E **2**   Evaluate

$$\lim_{x \to -2} \frac{x^2 - 4}{x + 2}$$

Solution. This limit is similar to $\lim_{x \to 0} (\sin x)/x$ discussed earlier. Direct substitution of -2 for x in $f(x) = (x^2 - 4)/(x + 2)$ yields $\frac{0}{0}$, which is undefined. We can obtain the limit by simplifying the expression for $f(x)$. Note that

$$\frac{x^2 - 4}{x + 2} = \frac{(x - 2)(x + 2)}{x + 2}$$

This fraction reduces to $x - 2$, provided that x is *not* equal to -2 (to avoid division by zero). But in the statement

$$\lim_{x \to -2} \frac{x^2 - 4}{x + 2}$$

x is never equal to -2; x only *approaches* -2. Since the limit concept avoids division by zero, we may say that

$$\lim_{x \to -2} \frac{x^2 - 4}{x + 2} = \lim_{x \to -2} \frac{(x - 2)(x + 2)}{x + 2}$$

$$= \lim_{x \to -2} (x - 2) = -4$$

Thus

$$\lim_{x \to -2} \frac{x^2 - 4}{x + 2} = -4$$

even though $(x^2 - 4)/(x + 2)$ is undefined at $x = -2$.

It is instructive to compare the graphs of the functions

$$y = \frac{x^2 - 4}{x + 2} \qquad \text{and} \qquad y = x - 2$$

According to Figure 20.10, both functions represent the same straight line, but $y = (x^2 - 4)/(x + 2)$ has a hole at $(-2, -4)$, while $y = x - 2$ does not.   ◀

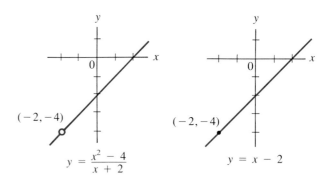

Figure 20.10

E X A M P L E **3**   Evaluate

$$\lim_{x \to 3} \frac{9 - x^2}{x - 3}$$

Solution. Since direct substitution yields $\frac{0}{0}$, we proceed as in Example 2:

$$\lim_{x \to 3} \frac{9 - x^2}{x - 3} = \lim_{x \to 3} \frac{(3 - x)(3 + x)}{x - 3} \qquad \text{factoring}$$

$$= \lim_{x \to 3} \frac{(3 - x)(3 + x)}{-1(3 - x)} \qquad x - 3 = -1(3 - x)$$

$$= \lim_{x \to 3} \frac{3 + x}{-1} \qquad \text{reducing the fraction}$$

$$= \frac{3 + 3}{-1} = -6$$

Note again that

$$\frac{(3 - x)(3 + x)}{-1(3 - x)} = \frac{3 + x}{-1}$$

because x only approaches 3 and is never equal to 3, thereby avoiding division by zero.

The graph of

$$y = \frac{9 - x^2}{x - 3}$$

is shown in Figure 20.11.   ◀

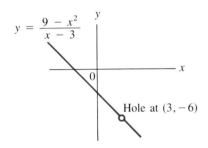

$$y = \frac{9 - x^2}{x - 3}$$

Hole at $(3, -6)$

Figure 20.11

Continuing our general discussion of limits, we use the notation

$$\lim_{x \to a+} f(x) = L$$

to indicate that x approaches a from the right—that is, through values larger than a. Similarly,

$$\lim_{x \to a-} f(x) = L$$

indicates that x approaches a from the left. Together these are referred to as *one-sided limits*.

One-sided limits

E X A M P L E **4** $$\lim_{x \to 2+} \sqrt{x - 2} = 0$$

The restriction $x \to 2+$ is necessary to avoid imaginary values.   ◀

E X A M P L E **5**   Evaluate

$$\lim_{x \to \infty} \frac{3x^2 + x + 1}{2x^2 - x + 2}$$

Solution. Here the technique is identical to that for finding horizontal asymptotes: we divide numerator and denominator by x^2. Hence

$$\lim_{x \to \infty} \frac{3x^2 + x + 1}{2x^2 - x + 2} = \lim_{x \to \infty} \frac{3 + \dfrac{1}{x} + \dfrac{1}{x^2}}{2 - \dfrac{1}{x} + \dfrac{2}{x^2}} = \frac{3}{2}$$
◀

For completeness we will state without proof the following theorems on limits: if $\lim_{x \to a} f(x) = L$ and $\lim_{x \to a} g(x) = M$, then:

A. $\displaystyle\lim_{x \to a} \left[f(x) \pm g(x) \right] = L \pm M$

B. $\displaystyle\lim_{x \to a} f(x)g(x) = LM$

C. $\displaystyle\lim_{x \to a} \frac{f(x)}{g(x)} = \frac{L}{M} \quad (M \neq 0)$

D. $\displaystyle\lim_{x \to a} kf(x) = kL$

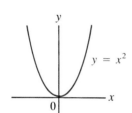

Figure 20.12

(These theorems will be referred to from time to time in later sections.)

Another important concept is continuity. Informally, a function is said to be *continuous* on an interval if its graph has no breaks or gaps. For example, the function $y = x^2$ is certainly continuous since the graph is just a parabola (Figure 20.12). On the other hand,

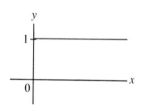

Figure 20.13

$$f(x) = \begin{cases} 0 & x \leq 0 \\ 1 & x > 0 \end{cases}$$

has a break at $x = 0$, since the function values suddenly jump from 0 to 1 (Figure 20.13). The break is called a *discontinuity*.

The function $y = x/(x - 1)$ is clearly discontinuous at $x = 1$, since the line $x = 1$ is a vertical asymptote (Figure 20.14).

Making use of the limit concept, continuity may be defined as follows.

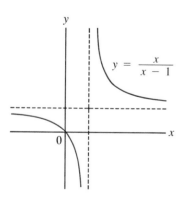

Figure 20.14

Definition of Continuity: A function f is continuous at $x = a$ if

$$\lim_{x \to a} f(x) = f(a)$$

If a function is continuous at all points in an interval, it is said to be continuous in the interval.

E X A M P L E **6** Show that $f(x) = x^2 + x$ is continuous at $x = 1$.

Solution. Since

$$\lim_{x \to 1} (x^2 + x) = 2 = f(1)$$

$f(x)$ is continuous at $x = 1$ by the definition (Figure 20.15).

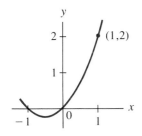

Figure 20.15

E X A M P L E **7** Show that the function

$$f(x) = \begin{cases} 2 & x \geq 1 \\ 1 & x < 1 \end{cases}$$

is discontinuous at $x = 1$.

Solution. The graph of the function is shown in Figure 20.16.
Note that $f(1) = 2$—since $f(x) = 2$ for $x \geq 1$. From the graph, we have

$$\lim_{x \to 1^-} f(x) = 1$$

In other words,

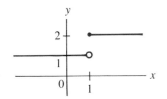

Figure 20.16

$$\lim_{x \to 1^-} f(x) = 1, \text{ while } f(1) = 2$$

This violates the condition

$$\lim_{x \to a} f(x) = f(a)$$

so that the function is not continuous at $x = 1$.

E X E R C I S E S / S E C T I O N **20.3**

In Exercises 1–6, find the limits by using a calculator.

1. $\lim_{x \to 0} \dfrac{\tan x}{x}$

2. $\lim_{x \to 0} \dfrac{1 - \cos x}{x}$

3. $\lim_{x \to 0^+} x^x$

4. $\lim_{x \to 0} (1 + x)^{1/x}$

5. $\lim_{x \to \pi/2} (\sec x - \tan x)$

6. $\lim_{x \to 1^-} 2^{1/(x-1)}$

In Exercises 7–48, find the limits indicated.

7. $\lim_{x \to 0} x^2$

8. $\lim_{x \to 0} (x^3 + 3)$

9. $\lim_{x \to 1} (x^4 - 3x + 2)$

10. $\lim_{x \to 2} \dfrac{x^2 - 4}{x + 2}$

11. $\lim_{x \to 2} \dfrac{x^2 - 4}{x - 2}$

12. $\lim_{x \to 4} \dfrac{x^2 - 16}{x - 4}$

13. $\lim\limits_{x \to 0} \dfrac{x^2 - x}{x}$

14. $\lim\limits_{x \to 0} \dfrac{2x^2 + x}{x}$

15. $\lim\limits_{x \to -4} \dfrac{x^2 - 16}{x + 4}$

16. $\lim\limits_{x \to 8} \dfrac{x^2 - 64}{x - 8}$

17. $\lim\limits_{x \to 2} \dfrac{x^2 - 4x + 4}{x - 2}$

18. $\lim\limits_{x \to 8} \dfrac{x^2 - 64}{x + 8}$

19. $\lim\limits_{x \to 2} \dfrac{x^2 + x - 6}{x - 2}$

20. $\lim\limits_{x \to -2} \dfrac{x^2 + x - 5}{x - 2}$

21. $\lim\limits_{x \to 4} \dfrac{x^2 + x - 8}{x + 4}$

22. $\lim\limits_{x \to -4} \dfrac{x^2 + 2x - 8}{x + 4}$

23. $\lim\limits_{x \to 1} \dfrac{x^3 - 3x^2 + 3x - 1}{x - 1}$

24. $\lim\limits_{x \to 1/2} \dfrac{4x^2 - 4x + 1}{2x - 1}$

25. $\lim\limits_{x \to 3} \dfrac{x^2 - 9}{3 - x}$

26. $\lim\limits_{x \to -5} \dfrac{x^2 - 25}{x + 5}$

27. $\lim\limits_{x \to 5} \dfrac{25 - x^2}{5 - x}$

28. $\lim\limits_{x \to 0} \dfrac{4x^3 - 2x}{x}$

29. $\lim\limits_{x \to 1} \dfrac{4x^3 - 2x}{x}$

30. $\lim\limits_{x \to 3} \dfrac{x^2 - x - 6}{x - 3}$

31. $\lim\limits_{x \to 4} \dfrac{x^2 - 6x + 8}{x - 4}$

32. $\lim\limits_{x \to \infty} \dfrac{2x^2 - 3}{4x^2 + 5x - 7}$

33. $\lim\limits_{x \to \infty} \dfrac{4x^3 - 2x + 1}{5x^3 + 3x^2 - x}$

34. $\lim\limits_{x \to \infty} \dfrac{2x^2 - 3x + 1}{3x^2 - 4}$

35. $\lim\limits_{x \to \infty} \dfrac{3x^2 + 2x}{4x^2 - 3}$

36. $\lim\limits_{x \to \infty} \dfrac{5x - 7}{3x^2 + 5x - 3}$

37. $\lim\limits_{x \to \infty} \dfrac{1 - 4x}{x^2 + 1}$

38. $\lim\limits_{x \to \infty} \dfrac{2x^2 - 3x + 1}{4x^2 - 2}$

39. $\lim\limits_{x \to \infty} \dfrac{3x^2 + 4x}{9x^2 - 7x + 6}$

40. $\lim\limits_{x \to \infty} \dfrac{1 - 3x - 10x^2}{2 - 5x^2}$

41. $\lim\limits_{x \to \infty} \dfrac{1 - 12x^2}{2 + 6x - 6x^2}$

42. $\lim\limits_{x \to \infty} \dfrac{\sqrt{x + 1}}{x + 1}$

43. $\lim\limits_{x \to \infty} \dfrac{\sqrt{x^2 + 6}}{x + 1}$

44. $\lim\limits_{x \to \infty} (\sqrt{x^2 + 4} - x)$ $\left(\text{Hint: multiply by } \dfrac{\sqrt{x^2 + 4} + x}{\sqrt{x^2 + 4} + x}\right)$

45. $\lim\limits_{x \to \infty} (x - \sqrt{x^2 - 1})$

46. $\lim\limits_{x \to 3+} \sqrt{x - 3}$

47. a. $\lim\limits_{x \to 0+} 3^{1/x}$ **b.** $\lim\limits_{x \to 0-} 3^{1/x}$

48. a. $\lim\limits_{x \to 0+} \dfrac{1}{1 + 2^{-1/x}}$ **b.** $\lim\limits_{x \to 0-} \dfrac{1}{1 + 2^{-1/x}}$

20.4   The Derivative

In this section we are going to introduce the concept of the slope of a line tangent to a curve. This concept leads to the definition of a derivative.

The notion of *derivative* originated with Newton's attempt to study the velocity of a falling object and other quantities whose values change continuously. To appreciate the difficulty, suppose a boxlike container with a square base 2 ft by 2 ft is filled with water from a faucet. If water pours in at the rate of 4 ft³/min, then during the course of one minute the volume will increase by 4 ft³. If x stands for the increase in the liquid level during one minute, then we get from $2 \cdot 2 \cdot x = 4$ that $x = 1$ ft. So the liquid level rises at the rate of 1 ft/min. But what if the container has the shape of a cone? Then the water level will rise at an ever-changing rate, a rate that cannot be determined algebraically. (We will consider this case in the next chapter.) The problem of the falling body leads to similar difficulties; both were beyond the scope of the mathematical methods known before Newton's time.

Newton recognized that the problem of the falling body, involving so-called instantaneous rates of change, was equivalent to finding the slope of a line tangent to a curve. For fairly simple functions slopes of tangent lines had already been studied by Fermat, while the acceleration due to gravity had been determined experimentally by Galileo.

To facilitate our discussion we first turn our attention to the geometric problem of the tangent to a curve. We will briefly return to the idea of rate of change at the end of the next section.

First, recall that the slope of a line

$$m = \frac{y_2 - y_1}{x_2 - x_1}$$

through the points (x_1, y_1) and (x_2, y_2) is constant. This is not true of the slope of nonlinear curves. For example, in Figure 20.17 the curve is steeper at point P than at point Q. Yet the two ideas can be connected. Note that the line T, which is drawn tangent to the curve at P, has the same direction as the curve at that point. So the problem of finding the slope of a curve is reduced to finding the slope of the line tangent to the curve.

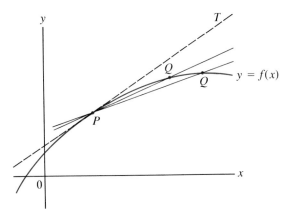

Figure 20.17

We can identify the tangent line to a curve at a given point in the following manner. Consider the function $y = f(x)$ in Figure 20.17. Let P and Q be two distinct points on the graph and let PQ be the chord determined by the two points. Suppose P is held fixed while Q is free to vary. As Q moves toward P, chord PQ moves toward the tangent line T. (See Figure 20.17.) If P and Q coincide, we no longer have two points to determine the chord. In other words: Q only approaches P; it never coincides with P. Thus T is the *limiting position* of chord PQ as Q approaches P. In symbols:

$$\lim_{Q \to P} \text{chord } PQ = T$$

The next step is to convert this limit to a computational formula. To be able to do so, we need a simple notational device: if x is a number to which we add a number Δx (read "delta x"), we obtain $x + \Delta x$ and say that x has been given an **increment** Δx. (One could simply say that Δx is the change in x.) If $y = f(x)$, a change Δx in x induces a corresponding change Δy in y. (See Figure 20.18.)

Increment

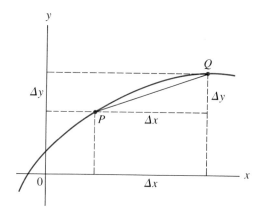

Figure 20.18

Let P be the fixed point (x_0, y_0). Then, making use of the delta notation, Q can be represented by $(x_0 + \Delta x, y_0 + \Delta y)$ (Figure 20.19). As a result,

$$y_0 = f(x_0) \qquad \text{and} \qquad y_0 + \Delta y = f(x_0 + \Delta x)$$

(See Figure 20.19.) Furthermore, the slope of chord PQ can now be written

$$\frac{\Delta y}{\Delta x}$$

From

$$y_0 + \Delta y = f(x_0 + \Delta x)$$

We get

$$\Delta y = f(x_0 + \Delta x) - f(x_0)$$

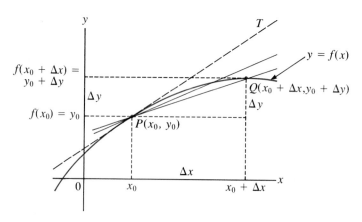

Figure 20.19

(See Figure 20.19.) Hence

$$\text{Slope of chord } PQ = \frac{\Delta y}{\Delta x} = \frac{f(x_0 + \Delta x) - f(x_0)}{\Delta x}$$

The descriptive statement $Q \rightarrow P$ (Q approaches P) can now be replaced by the numerical statement $\Delta x \rightarrow 0$ (Δx approaches zero), since

$$(x_0 + \Delta x, y_0 + \Delta y) \rightarrow (x_0, y_0) \text{ as } \Delta x \rightarrow 0$$

Consequently,

$$\text{Slope of } T = \lim_{\Delta x \to 0} \frac{\Delta y}{\Delta x} = \lim_{\Delta x \to 0} \frac{f(x_0 + \Delta x) - f(x_0)}{\Delta x}$$

Denoting this limit by $f'(x_0)$, we can now write

$$f'(x_0) = \lim_{\Delta x \to 0} \frac{\Delta y}{\Delta x} = \lim_{\Delta x \to 0} \frac{f(x_0 + \Delta x) - f(x_0)}{\Delta x}$$

Derivative

This limit, denoted by $f'(x_0)$ (read "f prime of x_0"), is called the **derivative** of $f(x)$ at x_0. If $y = f(x)$, the derivative is also denoted by y'.

If the limit exists at x_0, then $f(x)$ is said to be *differentiable* at x_0. If $f'(x)$ exists at every point in (a, b), then $f(x)$ is *differentiable* in (a, b). The process of finding $f'(x)$ is called **differentiation**.

Differentiation

20.5   The Derivative by the Four-Step Process

In the previous section we defined the derivative of $y = f(x)$ at $x = x_0$. By omitting the subscript zero, we obtain a general expression for the derivative that is valid for all x.

Definition of the Derivative:

$$f'(x) = \lim_{\Delta x \to 0} \frac{f(x + \Delta x) - f(x)}{\Delta x} \qquad (20.1)$$

To use this definition to find the derivative of a given function $y = f(x)$, we first replace x by $x + \Delta x$. For example, if $f(x) = 2x$, then $f(x + \Delta x) = 2(x + \Delta x)$. Now we subtract $f(x)$ from this expression to get $2(x + \Delta x) - 2x = 2x + 2 \Delta x - 2x = 2 \Delta x$. Dividing by Δx, we get $2 \Delta x / \Delta x = 2$. Finally, $\lim_{\Delta x \to 0} 2 = 2$, so that $f'(x) = 2$.

This procedure, which requires four steps, can be carried out systematically by a process called the *delta process* or **four-step process**.

The Four-Step Process:

Step 1. Replace x by $x + \Delta x$ and y by $y + \Delta y$ in the function $y = f(x)$:

$$y + \Delta y = f(x + \Delta x)$$

Step 2. Subtract $y = f(x)$ from both sides:

$$\Delta y = f(x + \Delta x) - f(x)$$

Step 3. Divide both sides of the resulting expression by Δx:

$$\frac{\Delta y}{\Delta x} = \frac{f(x + \Delta x) - f(x)}{\Delta x}$$

Step 4. Obtain $f'(x)$ by evaluating

$$\lim_{\Delta x \to 0} \frac{\Delta y}{\Delta x}$$

Particular attention should be paid to Step 3. Note that we cannot simply evaluate the limit by substituting zero for Δx since we will always get $\frac{0}{0}$. So the expression has to be simplified (or at least rewritten in some way) to permit the cancelation of Δx in the denominator. This simplification can be done at various points in the process, as the following examples show.

E X A M P L E **1** Differentiate the function $y = 1 - x^2$.

Solution. To obtain an expression for $f(x + \Delta x)$, we leave a blank space for x:

$$f(\ \ ) = 1 - (\ \ )^2$$

Now we substitute $x + \Delta x$:

Step 1. $y + \Delta y = f(x + \Delta x) = 1 - (x + \Delta x)^2$ $f(x + \Delta x)$

$$y + \Delta y = 1 - (x + \Delta x)^2$$

Step 2. $\Delta y = 1 - (x + \Delta x)^2 - (1 - x^2)$ $\Delta y = f(x + \Delta x) - f(x)$

Step 3. $\dfrac{\Delta y}{\Delta x} = \dfrac{1 - (x + \Delta x)^2 - (1 - x^2)}{\Delta x}$ dividing by Δx

Before going to Step 4, we simplify the expression and cancel Δx:

$$\frac{\Delta y}{\Delta x} = \frac{1 - [x^2 + 2x\,\Delta x + (\Delta x)^2] - 1 + x^2}{\Delta x} \qquad (a + b)^2 = a^2 + 2ab + b^2$$

$$= \frac{-2x\,\Delta x - (\Delta x)^2}{\Delta x} = \frac{\Delta x(-2x - \Delta x)}{\Delta x} \qquad \text{factoring } \Delta x$$

$$= -2x - \Delta x \qquad\qquad\qquad \text{reducing}$$

Step 4. $f'(x) = \lim\limits_{\Delta x \to 0} \dfrac{\Delta y}{\Delta x} = \lim\limits_{\Delta x \to 0}\,(-2x - \Delta x) = -2x$

Since $f'(x) = -2x$, we have found a general expression for the derivative, which is itself a function that can be evaluated for any x. For example, $f'(1) = -2$, which is the slope of the tangent line to the curve $y = 1 - x^2$ at the point $(1, 0)$ (Figure 20.20). ◀

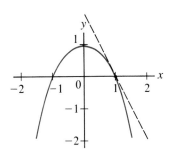

Figure 20.20

E X A M P L E **2**   Find the derivative of $y = 2x^3 - x$.

Solution.

Step 1. $y + \Delta y = f(x + \Delta x) = 2(x + \Delta x)^3 - (x + \Delta x)$
Step 2. $\Delta y = 2(x + \Delta x)^3 - (x + \Delta x) - (2x^3 - x)$

This time it is better to simplify even before going on to Step 3. Expanding the right side by multiplying, we get

$$2[x^3 + 3x^2\,\Delta x + 3x(\Delta x)^2 + (\Delta x)^3] - (x + \Delta x) - (2x^3 - x)$$
$$= 2x^3 + 6x^2\,\Delta x + 6x(\Delta x)^2 + 2(\Delta x)^3 - x - \Delta x - 2x^3 + x$$
$$= 6x^2\,\Delta x + 6x(\Delta x)^2 + 2(\Delta x)^3 - \Delta x$$

Note that all the remaining terms contain at least one Δx, which can be factored. Thus

Step 3. $\dfrac{\Delta y}{\Delta x} = \dfrac{\Delta x[6x^2 + 6x\,\Delta x + 2(\Delta x)^2 - 1]}{\Delta x}$

$$= 6x^2 + 6x\,\Delta x + 2(\Delta x)^2 - 1$$

Step 4. $f'(x) = \lim\limits_{\Delta x \to 0} \dfrac{\Delta y}{\Delta x} = \lim\limits_{\Delta x \to 0}\,[6x^2 + 6x\,\Delta x + 2(\Delta x)^2 - 1] = 6x^2 - 1$

◀

E X A M P L E **3**   Find the derivative of $y = 1/(x - 1)$ at $x = 2$, $x = -1$.

Solution. First we obtain a general expression for y':

Step 1. $y + \Delta y = f(x + \Delta x) = \dfrac{1}{x + \Delta x - 1}$

Step 2. $\Delta y = \dfrac{1}{x + \Delta x - 1} - \dfrac{1}{x - 1}$

Again it is better to simplify before going on to Step 3. In a case like this we need to combine the two fractions:

$$\Delta y = \frac{1}{x + \Delta x - 1}\frac{x - 1}{x - 1} - \frac{1}{x - 1}\frac{x + \Delta x - 1}{x + \Delta x - 1}$$

$$= \frac{(x - 1) - (x + \Delta x - 1)}{(x + \Delta x - 1)(x - 1)} = \frac{-\Delta x}{(x + \Delta x - 1)(x - 1)}$$

Step 3: $\dfrac{\Delta y}{\Delta x} = \dfrac{-\Delta x}{(x + \Delta x - 1)(x - 1)} \cdot \dfrac{1}{\Delta x} = \dfrac{-1}{(x + \Delta x - 1)(x - 1)}$

Step 4: $f'(x) = \lim\limits_{\Delta x \to 0} \dfrac{\Delta y}{\Delta x} = \lim\limits_{\Delta x \to 0} \dfrac{-1}{(x + \Delta x - 1)(x - 1)} = \dfrac{-1}{(x - 1)^2}$

Since

$$f'(x) = \frac{-1}{(x - 1)^2}$$

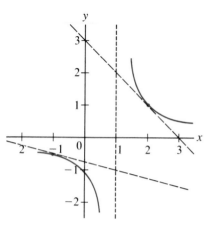

Figure 20.21

the slope of the tangent line at $x = 2$ is $f'(2) = -1/(2 - 1)^2 = -1$. At $x = -1$, $f'(-1) = -\frac{1}{4}$. (See Figure 20.21.)   ◀

E   X   A   M   P   L   E   **4**   Differentiate the function

$$f(x) = \sqrt{x + 2}$$

Solution.

Step 1. $y + \Delta y = f(x + \Delta x) = \sqrt{x + \Delta x + 2}$

Step 2. $\Delta y = \sqrt{x + \Delta x + 2} - \sqrt{x + 2}$

Step 3. $\dfrac{\Delta y}{\Delta x} = \dfrac{\sqrt{x + \Delta x + 2} - \sqrt{x + 2}}{\Delta x}$

This expression cannot really be simplified in the usual sense of algebra. However, to eliminate Δx we can resort to the simple trick of rationalizing the numerator:

$$\frac{\Delta y}{\Delta x} = \frac{\sqrt{x + \Delta x + 2} - \sqrt{x + 2}}{\Delta x} \cdot \frac{\sqrt{x + \Delta x + 2} + \sqrt{x + 2}}{\sqrt{x + \Delta x + 2} + \sqrt{x + 2}}$$

$$= \frac{(\sqrt{x + \Delta x + 2})^2 - (\sqrt{x + 2})^2}{\Delta x(\sqrt{x + \Delta x + 2} + \sqrt{x + 2})}$$

$$= \frac{(x + \Delta x + 2) - (x + 2)}{\Delta x(\sqrt{x + \Delta x + 2} + \sqrt{x + 2})} = \frac{1}{\sqrt{x + \Delta x + 2} + \sqrt{x + 2}}$$

Step 4. $f'(x) = \lim\limits_{\Delta x \to 0} \dfrac{1}{\sqrt{x + \Delta x + 2} + \sqrt{x + 2}} = \dfrac{1}{2\sqrt{x + 2}}$ ◀

Notations for the Derivative: In addition to $f'(x)$ and y', some commonly used notations for the derivative are

$$\frac{dy}{dx} \qquad \frac{df(x)}{dx} \qquad \frac{d}{dx}f(x) \qquad D_x f(x)$$

Newton also employed $\dot{y}$, which is sometimes seen in books on mechanics to denote dy/dt. As noted earlier, dy/dx was introduced by Leibniz. This form suggests that the derivative may be viewed as a quotient; in a limited sense this is indeed the case, as we will see in the section on differentials in Chapter 21.

The Derivative as a Rate of Change

Returning now to Newton's attempt to measure a rate of change, suppose a particle moves according to the equation $s = 2t + 1$, where s is the distance covered as a function of time t. The graph of the function is a straight line (Figure 20.22) with a slope of 2. Consider two points on the graph, say $(0, 1)$ and $(1, 3)$. Calculating the slope again, we get

$$m = \frac{y_2 - y_1}{x_2 - x_1} = \frac{3 - 1}{1 - 0} = \frac{2}{1} = 2$$

by the familiar formula. Now, since the graph has a physical interpretation, what about the slope itself? Measuring s in meters and t in seconds, the preceding calculation tells us that

$$m = \frac{2 \text{ m}}{1 \text{ s}}$$

That is, s changes by 2 meters whenever t changes by 1 second, or *the rate of change of s with respect to t is 2 m/s*. This rate of change is the velocity!

This interpretation, although of some interest, does not tell us what we really need to know about velocity. Suppose, for example, that you are traveling along a road at 30 mi/h and suddenly speed up to 50 mi/h. In that case you will see the speedometer needle move continuously from 30 to 50. In other words, we do not actually have to travel for a whole hour to say that we are doing

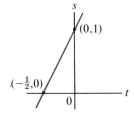

Figure 20.22

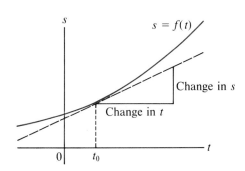

Figure 20.23

40 mi/h; we could be traveling at that speed for only an instant. The graph of a function $s = f(t)$ describing a nonconstant velocity is a smooth curve rather than a straight line. To find the velocity at some instant $t = t_0$, we need the rate of change *only* at that instant. Intuitively, then, this rate of change will be given by the slope of the line tangent to the curve at the point with abscissa $t = t_0$ (Figure 20.23).

Suppose a particle moves along a line according to the function $s = t^2$, where s is the distance (feet) and t the time (seconds). It is easily checked that $ds/dt = 2t$, which is the velocity of the particle as a function of time. At the instant $t = 2$ s, $v = ds/dt = 4$ ft/s; at $t = 5$ s, $v = ds/dt = 10$ ft/s.

As a final comment, the *difference quotient*

$$\frac{\Delta y}{\Delta x} = \frac{f(x + \Delta x) - f(x)}{\Delta x}$$

is called the *average rate of change of y with respect to x*, while $f'(x)$ is called the *instantaneous rate of change of y with respect to x*. As a consequence, the general definition of the derivative is the **instantaneous rate of change of one variable with respect to another**. (Rates of change will be studied in greater detail in Section 20.7.)

E X E R C I S E S / S E C T I O N   **20.5**

In Exercises 1–20, use the four-step process to differentiate each function.

1. $y = 2x + 1$

2. $y = 1 - 5x$

3. $y = 2 - 3x$

4. $y = -x^2$

5. $y = x^2 + 1$

6. $y = 2x^2 + 2$

7. $y = 2x^2 - x$

8. $y = 1 - 4x^3$

9. $y = x^3 - 3x^2$

10. $y = \dfrac{1}{x}$

11. $y = \dfrac{1}{x + 1}$

12. $y = \dfrac{1}{2x - 4}$

13. $y = \dfrac{1}{x^2}$

14. $y = \dfrac{1}{x^2 - 2}$

15. $y = \dfrac{1}{1 - x^2}$

16. $y = \dfrac{1}{x - x^2}$

17. $y = \sqrt{x}$ (see Example 4)

18. $y = \sqrt{x - 2}$

19. $y = \sqrt{1 - x}$

20. $y = \sqrt{2 - 3x}$

In Exercises 21–24, evaluate $f'(x)$ at the given points and interpret each geometrically.

21. $f(x) = x^2 + 2$; $f'(-1)$, $f'(2)$, $f'(3)$

22. $f(x) = \dfrac{1}{x^2 + 1}$; $f'(0)$, $f'(1)$

23. $f(x) = \dfrac{1}{\sqrt{x}}$; $f'(4)$, $f'(9)$

24. $f(x) = \dfrac{1}{\sqrt{x + 1}}$; $f'(0)$, $f'(7)$

20.6   Derivatives of Polynomials

So far we have found derivatives by the four-step process. This process is lengthy even for fairly simple functions. In this section we will develop some rules for differentiating polynomials directly. (However, we will need the four-step process later to obtain additional shortcut methods.)

The first special function to be considered is $y = c$, where c is a constant. Since the graph of $y = c$ is a horizontal line, which is its own tangent, we would expect y' to equal zero. Indeed,

Step 1. $y + \Delta y = f(x + \Delta x) = c$ *Step. 2* $\Delta y = c - c = 0$

Step 3. $\dfrac{\Delta y}{\Delta x} = 0$ *Step.4* $\displaystyle\lim_{\Delta x \to 0} \dfrac{\Delta y}{\Delta x} = 0$

Constant Rule:

$$\frac{dc}{dx} = 0 \qquad \text{where } c \text{ is a constant} \tag{20.2}$$

The function $y = x$ is a straight line having a slope of 1, so that $y' = 1$. Although the function is a special case of Equation (20.4), we will list it separately:

$$\frac{dx}{dx} = 1 \tag{20.3}$$

For the function $y = x^n$, where n is a positive integer, we need the binomial theorem to expand the expression in Step 2.

Step 1. $y + \Delta y = f(x + \Delta x) = (x + \Delta x)^n$

Step 2. $\Delta y = (x + \Delta x)^n - x^n$

$$= x^n + nx^{n-1}\Delta x + \frac{n(n-1)}{2!} x^{n-2}(\Delta x)^2 + \cdots + (\Delta x)^n - x^n$$

$$= nx^{n-1}\Delta x + \frac{n(n-1)}{2!} x^{n-2}(\Delta x)^2 + \cdots + (\Delta x)^n$$

Step 3. $\dfrac{\Delta y}{\Delta x} = nx^{n-1} + \dfrac{n(n-1)}{2!} x^{n-2}\Delta x + \cdots + (\Delta x)^{n-1}$

Step 4. $\displaystyle\lim_{\Delta x \to 0} \dfrac{\Delta y}{\Delta x} = nx^{n-1}$

since all terms except the first contain at least one factor Δx.

Thus we have the following formula, called the *power rule*:

Derivative of x^n, $n > 0$:

$$\frac{dx^n}{dx} = nx^{n-1} \tag{20.4}$$

Since the binomial theorem can be extended to rational powers, formula (20.4) is actually valid for rational n. This case will be checked later by different methods.

The next formula is also needed to differentiate polynomials, but it is really of more general interest.

Suppose $u = f(x)$ and $v = g(x)$ are two differentiable functions. We wish to find the derivative of the sum $f(x) + g(x)$. By definition

$$\frac{d}{dx}[f(x) + g(x)] = \lim_{\Delta x \to 0} \frac{f(x + \Delta x) + g(x + \Delta x) - [f(x) + g(x)]}{\Delta x}$$

$$= \lim_{\Delta x \to 0} \left[\frac{f(x + \Delta x) - f(x)}{\Delta x} + \frac{g(x + \Delta x) - g(x)}{\Delta x} \right]$$

Since the limit of a sum is equal to the sum of the limits (Theorem A on limits), we get

$$\frac{d}{dx}[f(x) + g(x)] = \frac{d}{dx} f(x) + \frac{d}{dx} g(x)$$

In words: *the derivative of a sum is equal to the sum of the derivatives.*

This derivation is quite straightforward, but to make our task easier later on, suppose we return to the delta notation. Since $u = f(x)$ and $v = g(x)$,

$$\Delta u = f(x + \Delta x) - f(x)$$

$$\Delta v = g(x + \Delta x) - g(x)$$

and

$$y = u + v$$

Then the steps above can be written in the following compact form:

Step 1. $y + \Delta y = u + \Delta u + v + \Delta v$

Step 2. $\Delta y = u + \Delta u + v + \Delta v - u - v = \Delta u + \Delta v$

Step 3. $\dfrac{\Delta y}{\Delta x} = \dfrac{\Delta u}{\Delta x} + \dfrac{\Delta v}{\Delta x}$

Step 4. $\dfrac{dy}{dx} = \lim\limits_{\Delta x \to 0} \dfrac{\Delta y}{\Delta x} = \lim\limits_{\Delta x \to 0} \dfrac{\Delta u}{\Delta x} + \lim\limits_{\Delta x \to 0} \dfrac{\Delta v}{\Delta x} = \dfrac{du}{dx} + \dfrac{dv}{dx}$

or $(d/dx)(u + v) = du/dx + dv/dx$, which will be called the *sum rule.*

Sum Rule:
$$\frac{d}{dx}(u + v) = \frac{du}{dx} + \frac{dv}{dx} \tag{20.5}$$

Our final formula in this section is easily obtained (Exercise 23).

Constant Multiplier Rule:

$$\frac{d}{dx}(cu) = c\,\frac{du}{dx} \tag{20.6}$$

We are now ready to differentiate polynomials.

E X A M P L E **1** Differentiate:

    a. $y = x^3$ **b.** $y = 3x^4$

Solution.

    a. Since $y = x^3$,

$$\frac{dy}{dx} = 3x^{3-1} = 3x^2 \qquad \text{by formula (20.4)}$$

    b. By formulas (20.4) and (20.6),

$$\frac{dy}{dx} = 3 \cdot 4x^{4-1} = 12x^3$$

◄

E X A M P L E **2** Differentiate

$$y = 5x^4 - 3x^3 + x^2 - x + 10$$

Solution. Since the derivative of a sum is equal to the sum of the derivatives, we can differentiate each term separately and add the results:

$$y' = 5(4x^3) - 3(3x^2) + 2x - 1 + 0 = 20x^3 - 9x^2 + 2x - 1$$

◄

E X E R C I S E S / S E C T I O N   **20.6**

Differentiate each of the following functions:

1. $y = x + 1$

2. $y = 3x + 2$

3. $y = x^2 - 2$

4. $y = x^3 - 2x$

5. $y = x^2 + x$

6. $y = 4x^2 - x + 2$

7. $y = 3x^2 + 4x$

8. $y = 5x^2 - 3x + 6$

9. $y = 5x^3 - 7x^2 + 2$

10. $y = 10x^5 + 6x^4 - 12x^2$

11. $y = 7x^3 - x^2 - x + 2$

12. $y = 5x^7 - 8x^6 + x^2 - 3$

13. $y = \frac{1}{3}x^3 + \frac{1}{2}x^2 + x$

14. $y = \frac{1}{4}x^4 + \frac{1}{3}x^2 - \sqrt{2}$

15. $y = \frac{x^2}{2} - \frac{x^3}{3}$

16. $y = \frac{x^4}{4} - \frac{x^2}{7}$

17. $y = 20x^{10} - 24x^6 + 2x^3 - \sqrt{3}$

18. $y = \sqrt{5}x^4 - \sqrt{2}x^3 + \pi$

19. $y = \frac{1}{5}x^7 - \frac{1}{\sqrt{2}}x^5 + \frac{1}{3}$

20. $y = \sqrt{5}x^{10} - \sqrt{7}x^7 - 5\sqrt{2}$

21. $y = \frac{1}{6}x^6 + \frac{1}{5}x^4 - \frac{1}{\sqrt{3}}$

22. $y = 3x^{12} + \sqrt{5}x^5 - 2x^3 - \sqrt{3}$

23. Derive formula (20.6).

20.7 Instantaneous Rates of Change

We observed in Section 20.5 that the derivative of a function $y = f(x)$ can be interpreted as the instantaneous rate of change of y with respect to x. For example, if y denotes distance and x time, then y' is the velocity. In this section we will consider several other cases where the instantaneous rate of change has an actual physical meaning.

For our first application we will return to the derivative as a *velocity* and examine *the motion of a particle along a line*.

Velocity:

$$v = \frac{ds}{dt}$$ where s denotes distance and t denotes time

---

E X   A   M   P   L   E   **1**   Suppose a particle moves along the s-axis in Figure 20.24 according to the equation

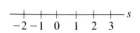

Figure 20.24

$$s = 2t^2 - 4t$$

(Assume s to be in meters and t in seconds.) Describe the motion.

Solution. Since the velocity v is given by

$$v = \frac{ds}{dt}$$

(from Section 20.5), it follows that

$$v = 4t - 4$$

To see how the particle moves, let us construct the following "time table."

t (s)	s (m)	v (m/s)
-1	6	-8
0	0	-4
1	-2	0
2	0	4
3	6	8

To locate the particle, we can either look at the table (as if it were a train schedule) or calculate the position from the function s. The third column gives the velocity. If we interpret $t = 0$ as the present instant, then $t = -1$ means a second ago and $t = 2$ means two seconds from now. We see that the particle moves to the left, turns around, then moves to the right. As we will check below, a negative velocity indicates that it moves to the left. To see precisely where the particle turns around, we need to find the value of t for which $v = 0$; thus

$$v = 4t - 4 = 0$$

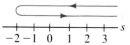

Figure 20.25

or $t = 1$. From the preceding table we see that this value of t corresponds to $s = -2$. The motion is indicated in Figure 20.25. (The path is drawn above the s-axis for easier visualization.)

Another important concept in the study of motion is acceleration. We know from experience that if we step on the accelerator of a car, the car speeds up—that is, the velocity increases. This statement suggests defining the *acceleration* as the rate of change of velocity, or $a = dv/dt$.

Acceleration:

$$a = \frac{dv}{dt} \qquad \text{where } s \text{ denotes distance and } t \text{ denotes time}$$

In our example,

$$a = \frac{dv}{dt} = 4 \text{ m/s}^2$$

so that the velocity increases by 4 m/s every second. If this seems to conflict with the observation that the particle is slowing down near $t = 1$, recall that v is *negative* for $t < 1$, so v actually *increases* from -8 to -4 between $t = -1$ and $t = 0$. Now, the absolute (numerical) value of the velocity is called the **speed**. So while the velocity increases from -8 to -4, the speed decreases from 8 to 4. ◀

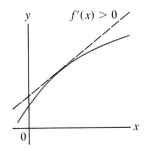

Figure 20.26

Returning now to the sign of the velocity of the particle described in Example 1, observe that whenever the derivative of $y = f(x)$ is positive on an interval, then the slope of the tangent line is positive, and the graph rises as we move to the right. In other words, if x increases then y increases, and f is said to be an **increasing** function (Figure 20.26). Similarly, f is **decreasing** on an interval if f' is negative in the interval. So if v is positive, ds/dt is positive and s increases. Obviously, if s gets larger, then the particle must be moving to the right. Analogous statements apply to negative v.

Rates of change may arise in practically any situation. Consider the following example.

E X A M P L E **2**   Suppose A is the area of a circle and r the radius; then $A = \pi r^2$. If the circle is allowed to expand, then the instantaneous rate of change of the area with respect to the radius is

$$\frac{dA}{dr} = 2\pi r \qquad \text{since } \pi \text{ is a numerical coefficient}$$

(Remember that the derivative is the instantaneous rate of change of one variable with respect to another.)

If $r = 2$ cm, then $dA/dr = 4\pi$ cm^2 per cm; or if $r = 4$ cm, $dA/dr = 8\pi$ cm^2 per cm, and so on. ◀

Rates of change also occur in the study of electrical circuits. The current, for example, is the instantaneous rate of change of the charge. Suppose we let i be the current (amperes), q the charge (coulombs), v the voltage (volts), L the inductance (henrys), and C the capacitance (farads). Then the following list includes some of these relationships:

---

1. **Current** in a circuit: $i = dq/dt$.
2. **Voltage** across an inductor: $v = L\, di/dt$.
3. **Charge** on a capacitor: $q = Cv$, so that $i = C\, dv/dt$, the current to a capacitor.

From this point on we are going to use the following SI (International System of Units) notations:

A for ampere; C for coulomb; V for volt; F for farad; H for henry; and Ω for ohm. Other abbreviations to be used later are J for joule and N for newton.

(For a list of SI units, see the table on the inside front cover.)

E X E R C I S E S / S E C T I O N   20.7

In Exercises 1–12, s is the position of a particle along the s-axis. Find an expression for the velocity and acceleration and determine when $v = 0$.

1. $s = 2t^2$

2. $s = 1 - 3t^2$

3. $s = t^2 - 2t + 1$

4. $s = 2t^2 - 8$

5. $s = 2t - t^2$

6. $s = 4t^2 + 4t - 3$

7. $s = 12t - 2t^2$

8. $s = 3t^2 + 6t + 1$

9. $s = 3t^2 - 6t$

10. $s = t^3 - \dfrac{3}{2}t^2$

11. $s = 3t^3 + 2t^2 + 2$

12. $s = 8t^3 + 2t^2 - 3t$

13. Suppose that $s = 50t^2$ is the distance in meters a rocket has traveled from the lauching pad in t seconds. Find its velocity after 10 s.

14. An object is dropped from a building 100 ft tall. The distance s in feet from the top t seconds later is given by $s = 16t^2$. What is its velocity when it hits the ground?

15. Find the instantaneous rate of change of the area of a square with respect to the length of the side when the side is 2 cm in length. (See Example 2.)

16. Find the instantaneous rate of change of the volume of a sphere with respect to the radius when the radius is 1.50 cm. (Recall that $V = \frac{4}{3}\pi r^3$.)

17. The resistance R (in ohms) in a certain wire as a function of temperature T (in degrees Celsius) is

$$R = 20.0 + 0.520T + 0.00973T^2$$

Find the instantaneous rate of change of R with respect to T when $T = 125°C$.

18. Water is poured into a cylindrical tank of radius 3.0 ft. Find the instantaneous rate of change of the volume with respect to the depth.

19. The power P (in watts) delivered to a 15-Ω resistor is $P = 15i^2$. Find the instantaneous rate of change of P with respect to i when $i = 2.1$ A.

20. Show that the rate of change of the area of a square with respect to the length of a side is one-half the perimeter of the square.

21. Show that the rate of change of the area of a circle with respect to its radius is equal to the circumference of the circle.

22. The rate of change of the cost of producing a commodity with respect to the number of units produced is called the *marginal cost*. If the cost of producing a certain commodity is $f(x) = x^2 - 2x$, find an expression for the marginal cost.

23. For a rotating body, let θ, given in radians, be the amount of rotation from the positive x-axis. Then $\omega = d\theta/dt$ is called the *angular velocity* and $\alpha = d\omega/dt$ the *angular acceleration*. If $\theta = 3t^2$, find ω and α when $t = 1$ s.

24. If the angular displacement of a body is $\theta = 1/(t + 1)$, where $t \geq 0$, find the angular velocity when $t = 2$ s (use the delta process). Refer to Exercise 23.

25. Find the current i in a circuit as a function of time t if the charge $q = 2t^2 + t$.

26. Find the voltage across an inductor of 5.00×10^{-3} H if the current is given by $i = 3.00t$ A.

27. Find the current to a capacitor of 1.0×10^{-2} F after 1 s if the voltage is given by $v = 10t^2$ (for a short time interval).

28. Power is sometimes defined as the time rate of doing work. If W denotes the work done and P the power, then $P = dW/dt$. Suppose the work done in moving an object is $W = 10t^3 + 3t^4$ J (joules). Find the power in joules per second at $t = 2$ s.

29. Refer to Exercise 28. If the work done (in joules) in moving an object is $W = 5t^4 + 2t$, find the power in joules per second when $t = 1$ s.

20.8 Differentiation Formulas

The Product and Quotient Rules

We learned in Section 20.6 that the derivative of a sum of two functions is the sum of their derivatives. In this section we will obtain the formulas for differentiating a product or quotient of two functions.

Suppose u and v are two differentiable functions of x. Then $y = uv$ can be differentiated and the derivative expressed in terms of derivatives of u and v. As in Section 20.5 we will use the delta notation

Step 1. $y + \Delta y = (u + \Delta u)(v + \Delta v)$

(Compare this with the derivation of formula (20.5).)

Step 2. $\Delta y = (u + \Delta u)(v + \Delta v) - uv$
$$= uv + u\,\Delta v + v\,\Delta u + \Delta u\,\Delta v - uv = u\,\Delta v + v\,\Delta u + \Delta u\,\Delta v$$

Step 3. $\dfrac{\Delta y}{\Delta x} = u\dfrac{\Delta v}{\Delta x} + v\dfrac{\Delta u}{\Delta x} + \Delta u\dfrac{\Delta v}{\Delta x}$

Step 4. $\dfrac{dy}{dx} = \lim\limits_{\Delta x \to 0}\dfrac{\Delta y}{\Delta x} = u\lim\limits_{\Delta x \to 0}\dfrac{\Delta v}{\Delta x} + v\lim\limits_{\Delta x \to 0}\dfrac{\Delta u}{\Delta x} + 0$

$$= u\dfrac{dv}{dx} + v\dfrac{du}{dx}$$

by Theorem A on limits.

To see why the last term in Step 3 vanishes, recall that if $u = g(x)$, then

$$\Delta u = g(x + \Delta x) - g(x)$$

Hence

$$\lim_{\Delta x \to 0} \Delta u = \lim_{\Delta x \to 0} \left[g(x + \Delta x) - g(x)\right]$$

$$= \lim_{\Delta x \to 0} \left[\frac{g(x + \Delta x) - g(x)}{\Delta x}\right] \cdot \Delta x$$

$$= \lim_{\Delta x \to 0} \frac{g(x + \Delta x) - g(x)}{\Delta x} \cdot \lim_{\Delta x \to 0} \Delta x = g'(x) \cdot 0 = 0$$

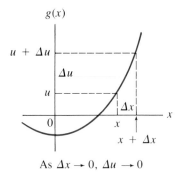

$g(x)$

$u + \Delta u$

Δu

u

Δx

0

x

$x + \Delta x$

x

As $\Delta x \to 0$, $\Delta u \to 0$

Figure 20.27

since the limit of a product is equal to the product of the limits (Theorem B on limits). So $\lim_{\Delta x \to 0} \Delta u = 0$. (See also Figure 20.27.) It follows that

$$\lim_{\Delta x \to 0} \Delta u \frac{\Delta v}{\Delta x} = 0 \cdot \frac{dv}{dx} = 0$$

We have shown that if u and v are differentiable, then $d(uv)/dx = u(dv/dx) + v(du/dx)$, which is called the *product rule*.

Product Rule:

$$\frac{d(uv)}{dx} = u\frac{dv}{dx} + v\frac{du}{dx} \tag{20.7}$$

Using the product rule, we can obtain the *quotient rule*.

Quotient Rule:

$$\frac{d}{dx}\left(\frac{u}{v}\right) = \frac{v\dfrac{du}{dx} - u\dfrac{dv}{dx}}{v^2} \tag{20.8}$$

To derive this rule, we let $w = u/v$. Then $u = vw$ and, by the product rule,

$$\frac{du}{dx} = v\frac{dw}{dx} + w\frac{dv}{dx}$$

$$v\frac{dw}{dx} = \frac{du}{dx} - w\frac{dv}{dx}$$

$$\frac{dw}{dx} = \frac{\dfrac{du}{dx} - w\dfrac{dv}{dx}}{v}$$

$$\frac{d}{dx}\left(\frac{u}{v}\right) = \frac{\dfrac{du}{dx} - \dfrac{u}{v}\dfrac{dv}{dx}}{v} \qquad \text{letting } w = u/v$$

$$\frac{d}{dx}\left(\frac{u}{v}\right) = \frac{v\dfrac{du}{dx} - u\dfrac{dv}{dx}}{v^2} \qquad \text{multiplying numerator and denominator by } v$$

E X A M P L E **1**   Find the derivative of

$$y = (x^3 + 2x^2 - 3x)(x^2 - 4x)$$

Solution. We are certainly not forced to use the product rule here since we could multiply out the terms and obtain a polynomial of fifth degree, but the product rule makes the task easier.

Let $u = x^3 + 2x^2 - 3x$ and $v = x^2 - 4x$; then

$$\frac{dy}{dx} = u\frac{dv}{dx} + v\frac{du}{dx}$$

Since we already know how to differentiate polynomials, we get directly

$$\frac{dy}{dx} = (x^3 + 2x^2 - 3x)(2x - 4) + (x^2 - 4x)(3x^2 + 4x - 3) \qquad \blacktriangleleft$$

EXAMPLE **2**   Differentiate

$$y = \frac{x^3 - 3x}{x + 1}$$

Solution. Since the function has the form of a quotient, we apply the quotient rule:

$$\frac{d}{dx}\left(\frac{u}{v}\right) = \frac{v\dfrac{du}{dx} - u\dfrac{dv}{dx}}{v^2} \qquad u = x^3 - 3x,\ v = x + 1$$

It follows that

$$\frac{dy}{dx} = \frac{(x + 1)\dfrac{d}{dx}(x^3 - 3x) - (x^3 - 3x)\dfrac{d}{dx}(x + 1)}{(x + 1)^2}$$

$$= \frac{(x + 1)(3x^2 - 3) - (x^3 - 3x)\cdot 1}{(x + 1)^2} = \frac{2x^3 + 3x^2 - 3}{(x + 1)^2} \qquad \blacktriangleleft$$

The Generalized Power Rule

Next we will obtain the formula for differentiating the power of a function, thereby extending formula (20.4), the derivative of $y = x^n$.

Let u be a differentiable function of x, and consider the function $y = u^n$, where n is a positive integer.

Step 1. $y + \Delta y = (u + \Delta u)^n$
Step 2. $\Delta y = (u + \Delta u)^n - u^n$

$$= u^n + nu^{n-1}\Delta u + \frac{n(n-1)}{2!}u^{n-2}(\Delta u)^2 + \cdots + (\Delta u)^n - u^n$$

Step 3. $\dfrac{\Delta y}{\Delta x} = nu^{n-1}\dfrac{\Delta u}{\Delta x} + \dfrac{n(n-1)}{2!}u^{n-2}\dfrac{\Delta u}{\Delta x}\Delta u + \cdots + \dfrac{\Delta u}{\Delta x}(\Delta u)^{n-1}$

As noted earlier, $\lim_{\Delta x \to 0} \Delta u = 0$; thus

$$\textit{Step 4. } \frac{dy}{dx} = \lim_{\Delta x \to 0} \frac{\Delta y}{\Delta x} = nu^{n-1} \frac{du}{dx}$$

If $u = x$, we get $dx^n/dx = nx^{n-1}(dx/dx) = nx^{n-1}$, which is formula (20.4).

If n is a negative integer, we let $n = -m$ ($m > 0$) and apply the quotient rule. Then we get

$$y = u^n = u^{-m} = \frac{1}{u^m}$$

and

$$\frac{dy}{dx} = \frac{u^m \dfrac{d}{dx}(1) - \dfrac{du^m}{dx}}{(u^m)^2} = \frac{-mu^{m-1} \dfrac{du}{dx}}{u^{2m}}$$

$$= -mu^{m-1-2m} \frac{du}{dx} = -mu^{-m-1} \frac{du}{dx}$$

Substituting n for $-m$, we have

$$\frac{dy}{dx} = nu^{n-1} \frac{du}{dx}$$

The case where n is a rational number will be left as an exercise in the next section. (Recall that a number is rational if it has the form p/q, where p and q are integers.) The formula is known as the *generalized power rule*.

Generalized Power Rule:

$$\frac{du^n}{dx} = nu^{n-1} \frac{du}{dx} \qquad \text{(for any rational } n\text{)} \qquad\qquad (20.9)$$

E X A M P L E **3**   Differentiate

$$f(x) = \sqrt{x^2 + 1}$$

Solution. Since $f(x) = (x^2 + 1)^{1/2}$, the generalized power rule applies. Thus

$$f'(x) = \tfrac{1}{2}(x^2 + 1)^{-1/2} \frac{d}{dx}(x^2 + 1) \qquad u = x^2 + 1, n = \tfrac{1}{2}$$

$$= \tfrac{1}{2}(x^2 + 1)^{-1/2}(2x) = \frac{x}{\sqrt{x^2 + 1}}$$

Don't forget to multiply by the derivative of $x^2 + 1$. (Even though it is customary to simplify algebraic expressions, the last step is not part of the differentiation procedure.) ◄

The generalized power rule turns out to be a special case of the *chain rule*, which involves composite functions (Section 20.2). If f and g are differentiable functions, and

$$y = f(g(x))$$

then

$$y' = f'(g(x)) \cdot g'(x)$$

Letting $u = g(x)$, a more convenient form of the chain rule can be obtained:

Chain Rule:

$$\frac{dy}{dx} = \frac{dy}{du}\frac{du}{dx} \tag{20.10}$$

We obtain the chain rule as follows:

$$\frac{dy}{dx} = \lim_{\Delta x \to 0} \frac{\Delta y}{\Delta x} = \lim_{\Delta x \to 0} \frac{\Delta y}{\Delta u}\frac{\Delta u}{\Delta x} = \lim_{\Delta x \to 0} \frac{\Delta y}{\Delta u} \cdot \lim_{\Delta x \to 0} \frac{\Delta u}{\Delta x}$$

Since $\Delta u \to 0$ as $\Delta x \to 0$ whenever $u = g(x)$ is continuous,

$$\lim_{\Delta x \to 0} \frac{\Delta y}{\Delta u} = \lim_{\Delta u \to 0} \frac{\Delta y}{\Delta u} = \frac{dy}{du}$$

and formula (20.10) follows. The chain rule will be needed in Chapter 24. (We have assumed that $\Delta u \neq 0$ if $\Delta x \neq 0$. However, it can be shown that the chain rule holds for this case also.)

E X A M P L E **4**   Differentiate $y = u^n$ by the chain rule.

Solution.

$$\frac{dy}{dx} = \frac{du^n}{du}\frac{du}{dx} = nu^{n-1}\frac{du}{dx}$$

which is the generalized power rule. ◄

The remaining examples illustrate the various rules in this section.

E X A M P L E **5**   If $y = x\sqrt{x^2 - 2x}$, find y'.

Solution. We may treat the function as a product and use the product rule with $u = x$ and $v = (x^2 - 2x)^{1/2}$, so that

$$y' = x \frac{d}{dx}(x^2 - 2x)^{1/2} + (x^2 - 2x)^{1/2} \frac{dx}{dx} \qquad \text{product rule}$$

$$= x \cdot \tfrac{1}{2}(x^2 - 2x)^{-1/2}(2x - 2) + (x^2 - 2x)^{1/2} \qquad \text{generalized power rule}$$

$$= x(x^2 - 2x)^{-1/2}(x - 1) + (x^2 - 2x)^{1/2} \qquad \tfrac{1}{2}(2x - 2) = x - 1$$

$$= \frac{x(x - 1)}{(x^2 - 2x)^{1/2}} + \frac{(x^2 - 2x)^{1/2}}{1}$$

Adding these fractions, we get

$$y' = \frac{x(x - 1)}{(x^2 - 2x)^{1/2}} + \frac{(x^2 - 2x)^{1/2}}{1} \frac{(x^2 - 2x)^{1/2}}{(x^2 - 2x)^{1/2}}$$

$$= \frac{x(x - 1)}{(x^2 - 2x)^{1/2}} + \frac{(x^2 - 2x)^1}{(x^2 - 2x)^{1/2}} \qquad a^{1/2} \cdot a^{1/2} = a^1$$

$$= \frac{x^2 - x + x^2 - 2x}{(x^2 - 2x)^{1/2}} = \frac{2x^2 - 3x}{\sqrt{x^2 - 2x}}$$

◀

E X A M P L E **6**   Find the derivative of

$$y = \frac{x}{\sqrt{x^2 + 1}}$$

Solution. We write

$$y = \frac{x}{(x^2 + 1)^{1/2}}$$

and apply the quotient rule:

$$\frac{dy}{dx} = \frac{(x^2 + 1)^{1/2} \dfrac{dx}{dx} - x \dfrac{d}{dx}(x^2 + 1)^{1/2}}{[(x^2 + 1)^{1/2}]^2}$$

Then, by the generalized power rule,

$$\frac{dy}{dx} = \frac{(x^2 + 1)^{1/2} - x \cdot \tfrac{1}{2}(x^2 + 1)^{-1/2}(2x)}{x^2 + 1} = \frac{(x^2 + 1)^{1/2} - \dfrac{x^2}{(x^2 + 1)^{1/2}}}{x^2 + 1}$$

This expression can be simplified by multiplying numerator and denominator by $(x^2 + 1)^{1/2}$; note that $(x^2 + 1)^{1/2}(x^2 + 1)^{1/2} = x^2 + 1$. Thus

$$\frac{dy}{dx} = \frac{(x^2 + 1)^{1/2}(x^2 + 1)^{1/2} - \dfrac{x^2}{(x^2 + 1)^{1/2}}(x^2 + 1)^{1/2}}{(x^2 + 1)(x^2 + 1)^{1/2}}$$

$$= \frac{(x^2 + 1) - x^2}{(x^2 + 1)(x^2 + 1)^{1/2}} = \frac{1}{(x^2 + 1)^{3/2}} \qquad \blacktriangleleft$$

Suggestion: To differentiate a function with a constant numerator

$$y = \frac{k}{g(x)} \qquad (k \text{ a constant})$$

write

$$y = k[g(x)]^{-1}$$

and use the generalized power rule (instead of the quotient rule).

E X A M P L E **7** Differentiate

$$y = \frac{1}{\sqrt[3]{x^3 + x}}$$

Solution. We could use the quotient rule, as in the previous example, but it is much simpler to write the function in the form

$$y = (x^3 + x)^{-1/3}$$

and apply the generalized power rule. We obtain at once

$$\frac{dy}{dx} = -\frac{1}{3}(x^3 + x)^{-4/3}(3x^2 + 1) = -\frac{3x^2 + 1}{3(x^3 + x)^{4/3}} \qquad \begin{array}{l} u = x^3 + x \\ n = -\tfrac{1}{3} \end{array}$$

E X E R C I S E S / S E C T I O N **20.8**

GROUP A   Differentiate the given functions (Exercises 1–32).

1. $y = 4x^4 - 4x^2 + 8$

2. $y = 3 - 8x - 5x^3$

3. $y = \dfrac{1}{x}$ (or $y = x^{-1}$)

4. $y = \dfrac{1}{\sqrt[3]{x}}$ (or $y = x^{-1/3}$)

5. $y = x^5 - 3x^{-3} + 2x^{-2}$

6. $y = x\sqrt{x}$ (or $y = x^{3/2}$)

7. $y = x^2\sqrt{x} + \dfrac{1}{\sqrt{x}}$ (or $y = x^{5/2} + x^{-1/2}$)

8. $y = x^3\sqrt[3]{x}$

9. $y = (2x^2 - 3)^4$

10. $y = (x^2 - 3x + 2)^3$

11. $y = (x^{10} + 1)^{10}$

12. $y = \dfrac{1}{x^2} - \dfrac{1}{x^3}$

13. $y = \dfrac{2}{\sqrt{2x^2 + 3}}$ (See Example 7.)

14. $y = \dfrac{3}{\sqrt{4 - x}}$

15. $y = \dfrac{3}{\sqrt[3]{6 - x}}$

16. $y = \dfrac{6}{\sqrt[3]{x^2 + 7}}$

17. $y = (x^3 - 3x)^{1/2}$

18. $y = \dfrac{1}{\sqrt{1 - x^2}}$

19. $y = \sqrt[3]{x^3 - 3}$

20. $y = \sqrt[4]{x + 2}$

21. $y = x^3(x + 1)^2$

22. $y = x(x - 1)^4$

23. $y = 2x^4(x + 2)^2$

24. $y = 3x^3(x + 3)^4$

25. $y = x^2(x^2 - 5)^2$

26. $y = 3x^3(x - 1)^3$

27. $y = \dfrac{x}{x - 1}$

28. $y = \dfrac{4 + x^2}{4 - x^2}$

29. $y = (x + 1)(x^2 - 2x)^2$

30. $y = \sqrt{1 + \sqrt{x}}$

31. $y = x^2\sqrt{x + 1}$

32. $y = x^2\sqrt{x^2 - 1}$

17. $y = \dfrac{x^2 - 3}{x - 2}$

18. $y = \dfrac{x^2 + x}{x^2 - 4}$

19. $y = \dfrac{x^3 + 2x}{x^2 - 8}$

20. $y = \dfrac{2x^2 - x}{x^3 - 4}$

21. $y = x\sqrt{x^2 - 1}$

22. $y = x\sqrt{x^3 + 2}$

23. $y = \dfrac{\sqrt{x}}{x - 4}$

24. $y = \dfrac{\sqrt{x}}{1 - x}$

25. $y = \dfrac{x^2}{\sqrt{x + 1}}$

26. $y = \dfrac{\sqrt{x + 1}}{x^2}$

27. $y = \dfrac{\sqrt{x^2 - 1}}{x^2}$

28. $y = \dfrac{x\sqrt{x - 1}}{x + 2}$

29. $y = \dfrac{x^2\sqrt{x}}{x^2 + 3}$

30. $y = \dfrac{\sqrt{x}}{1 + \sqrt{x}}$

31. $y = (x - 1)\sqrt{x - 2}$

32. $y = (x - 2)\sqrt{x^2 + 1}$

33. $y = \dfrac{x\sqrt{x - 1}}{2x + 3}$

34. $y = \dfrac{x^2\sqrt{x + 3}}{x - 1}$

35. $y = \dfrac{x^2\sqrt{x - 5}}{\sqrt{x + 3}}$

36. $y = \dfrac{x\sqrt[3]{x^2 + 1}}{x - 8}$

GROUP B   Differentiate the given functions (Exercises 1–36).

1. $y = 3x^{-2} - 2x^{-3}$

2. $y = x^{-1/2} + 4x^{-3/2} + x$

3. $y = 3x\sqrt[3]{x} + 4\sqrt[4]{x} + \sqrt{2}$

4. $y = 10x^{-2} + \dfrac{2}{\sqrt{x}}$

5. $y = (x^3 - 3)(2x^4 - 8x^5)$

6. $y = (2x^2 + 3x - 1)(5x^6 - 2x)$

7. $y = (7x^4 - 3x^2)^6$

8. $y = (x - 5x^4)^8$

9. $y = \sqrt{1 - x}$

10. $y = \sqrt[3]{1 - x^2}$

11. $y = \sqrt{x^2 + 2}$

12. $y = \sqrt[3]{x^3 - 4}$

13. $y = 4\sqrt[4]{x - x^2}$

14. $y = -\dfrac{2}{3\sqrt{x^3 - 6}}$

15. $y = \dfrac{x^2}{x - 1}$

16. $y = \dfrac{x - 1}{x + 1}$

37. Find the slope of the line tangent to the curve $y = (x^3 - 1)(x^2 + 3x + 2)$ at the point $(1, 0)$.

38. Find the point at which the slope of the line tangent to $y = 1/\sqrt{3x^2 + 3}$ is equal to zero.

39. If R ohms resistance is in series with X ohms reactance, then the impedance $Z = \sqrt{R^2 + X^2}$ ohms. If $R = 4\,\Omega$, find the instantaneous rate of change of Z with respect to X.

40. Find the current $i = dq/dt$ in a circuit at $t = 8$ s if the charge $q = 3.3t^{4/3}$ C.

41. Starting at $t = 0$, the charge on a certain capacitor is given by $q(t) = t/(t^2 + 4)$. At what instant is the current to the capacitor equal to zero?

20.9   Implicit Differentiation

So far all our functions have been expressed in the **explicit** form $y = f(x)$. In many relationships y is not expressed explicitly as a function of x, but only in **implicit** form. In this section we will develop a technique for differentiating a function in implicit form, leading to the **implicit derivative**.

To see what all these terms mean, consider the circle $x^2 + y^2 = 25$. On the interval $(-5, 5)$ we get two y values for every x value, so the relation is not a function. If we solve the equation for y in terms of x, we get

$$y = \pm \sqrt{25 - x^2}$$

which can be written as two equations,

$$y = +\sqrt{25 - x^2} \qquad \text{and} \qquad y = -\sqrt{25 - x^2}$$

Each separate equation is now a function; in fact, $y = +\sqrt{25 - x^2}$ is the upper semicircle and $y = -\sqrt{25 - x^2}$ is the lower semicircle in Figure 20.28.

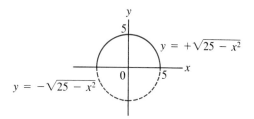

Figure 20.28

If we want to find the slope of the tangent at $(4, -3)$, for example, we take the lower branch $y = -\sqrt{25 - x^2}$ and find dy/dx. The result is

$$\frac{dy}{dx} = -\tfrac{1}{2}(25 - x^2)^{-1/2}(-2x) = \frac{x}{\sqrt{25 - x^2}}$$

and at $x = 4$, we find that $dy/dx = \tfrac{4}{3}$.

The conclusion is that even though the equation $x^2 + y^2 = 25$ is not a function, it actually contains two functions, and we say that y is an **implicit function** of x. With this understanding, it becomes totally unnecessary to solve for y in terms of x—we just use the generalized power rule. For example, while $dx^2/dx = 2x$, as usual, $dy^2/dx = 2y(dy/dx)$, since y *is a function of* x. Many students find this step troublesome, but all we need to do is keep in mind that the u in formula (20.9) has been replaced by y:

Generalized Power Rule:

$$\frac{dy^n}{dx} = ny^{n-1}\frac{dy}{dx}$$ (20.11)

Returning to the equation $x^2 + y^2 = 25$, we now differentiate both sides with respect to x to obtain

$$2x + 2y\frac{dy}{dx} = 0 \qquad \frac{d}{dx}x^2 = 2x \qquad \frac{d}{dx}y^2 = 2y\frac{dy}{dx}$$

whence

$$\frac{dy}{dx} = -\frac{x}{y} \qquad \text{solving for } \frac{dy}{dx}$$

Implicit derivative

which is called the **implicit derivative**. At the point $(4, -3)$,

$$\frac{dy}{dx} = -\frac{4}{-3} = \frac{4}{3}$$

as before. This result should not be surprising, for if we substitute $-\sqrt{25 - x^2}$ in the expression for the implicit derivative, then

$$\frac{dy}{dx} = -\frac{x}{y} = -\frac{x}{-\sqrt{25 - x^2}} = \frac{x}{\sqrt{25 - x^2}}$$

That is, dy/dx collapses to the ordinary derivative.

So far we seem to have been doing the same thing twice, but the real reason for considering implicit derivatives is that some equations are very difficult, or even impossible, to solve for y in terms of x.

E X A M P L E **1**   Find dy/dx for each of the following equations:
  a. $3y^3 = y$ **b.** $x^2y^2 = 1$ **c.** $x^2y = 4$

Solution.

  a. By the power rule we get

$$3 \cdot 3y^2 \frac{dy}{dx} = 1 \qquad \text{since } \frac{d}{dx} y^3 = 3y^2 \frac{dy}{dx}$$

so that

$$\frac{dy}{dx} = \frac{1}{9y^2} \qquad \text{solving for } \frac{dy}{dx}$$

  b. Using the product and power rules, we have

$$x^2 \frac{d(y^2)}{dx} + y^2 \frac{d(x^2)}{dx} = \frac{d}{dx} \quad (1)$$

or

$$x^2 \cdot 2y \frac{dy}{dx} + y^2 \cdot 2x = 0 \qquad \text{and} \qquad 2x^2y \frac{dy}{dx} + 2xy^2 = 0$$

It follows that

$$2x^2y \frac{dy}{dx} = -2xy^2$$

and

$$\frac{dy}{dx} = -\frac{2xy^2}{2x^2y} = -\frac{y}{x}$$

c. Using the product rule to find the implicit derivative of $x^2y = 4$, we get

$$x^2 \frac{d}{dx}(y) + y\frac{d}{dx}(x^2) = \frac{d}{dx}(4)$$

and

$$x^2 \frac{dy}{dx} + 2xy = 0 \qquad \frac{d}{dx}(y) = \frac{dy}{dx}$$

whence

$$\frac{dy}{dx} = -\frac{2xy}{x^2} = -\frac{2y}{x} \qquad\qquad\qquad \blacktriangleleft$$

E X A M P L E **2**   Find dy/dx implicitly:

$$3x^2 + 4x^3y^4 + 2y = 4$$

Solution. Differentiating each term, we get

$$\frac{d}{dx}(3x^2) + \frac{d}{dx}(4x^3y^4) + \frac{d}{dx}(2y) = \frac{d}{dx}(4)$$

Now use the product rule on the second term:

$$6x + 4x^3 \frac{d}{dx}(y^4) + y^4 \frac{d}{dx}(4x^3) + 2\frac{d}{dx}(y) = 0$$

$$6x + 4x^3 \cdot 4y^3 \frac{dy}{dx} + y^4(12x^2) + 2\frac{dy}{dx} = 0$$

Now collect all terms containing dy/dx on the left side:

$$16x^3y^3 \frac{dy}{dx} + 2\frac{dy}{dx} = -6x - 12x^2y^4$$

$$(16x^3y^3 + 2)\frac{dy}{dx} = -6x - 12x^2y^4 \qquad \textbf{factoring } \frac{dy}{dx}$$

$$\frac{dy}{dx} = \frac{-6x - 12x^2y^4}{16x^3y^3 + 2} \qquad \textbf{dividing by } (16x^3y^3 + 2)$$

$$\frac{dy}{dx} = -\frac{3x + 6x^2y^4}{8x^3y^3 + 1} \qquad\qquad\qquad\qquad \blacktriangleleft$$

E X A M P L E **3**   Find the slope of the line tangent to the graph of the equation

$$x^2 - 3xy + y^2 + 4x - 2y = 1$$

at the point $(1, 4)$.

Solution. Differentiating both sides, we obtain

$$2x - 3\left(x\frac{dy}{dx} + y\frac{dx}{dx} \right) + 2y\frac{dy}{dx} + 4 - 2\frac{dy}{dx} = 0$$

or

$$2x - 3x\frac{dy}{dx} - 3y + 2y\frac{dy}{dx} + 4 - 2\frac{dy}{dx} = 0$$

To solve this equation for dy/dx, we keep all the terms containing dy/dx on the left side and transpose the rest:

$$-3x\frac{dy}{dx} + 2y\frac{dy}{dx} - 2\frac{dy}{dx} = -2x + 3y - 4$$

or

$$(-3x + 2y - 2)\frac{dy}{dx} = -2x + 3y - 4$$

and

$$\frac{dy}{dx} = \frac{-2x + 3y - 4}{-3x + 2y - 2} = \frac{2x - 3y + 4}{3x - 2y + 2}$$

Finally, at the point $(1, 4)$,

$$\frac{dy}{dx} = 2$$

◄

E X E R C I S E S / S E C T I O N   **20.9**

In Exercises 1–26, find dy/dx implicitly.

1. $2x + 3y = 3$

2. $4x - 5y = 3$

3. $x^2 - y^2 = 2$

4. $3x^2 + 5y^2 = x$

5. $2x^2 - 3y^2 = 1$

6. $y^2 - x^2 - 2x = 0$

7. $2y^3 + x^2 + 1 = 0$

8. $7y^4 - 3x^2 = 3$

9. $4y^4 - 3x^3 + 2 = 0$

10. $3y^3 - 7x^3 + 2x = 0$

11. $x - 5x^2 - 6y^3 = 0$

12. $x^2 + y^2 = r^2$

13. $\dfrac{x^2}{a^2} + \dfrac{y^2}{b^2} = 1$

14. $y^2 = 4px$

15. $xy = 3$

16. $xy + x = 4$

17. $x^2y = 7$

18. $x^2y = x + 1$

19. $x^2 + x^2y^2 + x = 0$

20. $y^2 = \dfrac{x}{x - 1}$

21. $x^3 - 4x^2y^2 + y^2 = 1$ **22.** $2x^2 + 3x^2y = y^3$

23. $5x^2y^3 - y^4 = 2x^3$ **24.** $2y^4 - x^2y^3 + x^6 = 3$

25. $x^4y^4 - 3y^2 + 5x = 6$

26. $3x^4 + 3xy^3 + y - 3x + 6 = 0$

27. Find the slope of the line tangent to the ellipse $x^2 + 4y^2 = 5$ at the point $(1, -1)$.

28. Find the slope of the line tangent to the parabola $y^2 = -12x$ at the point $(-3, 6)$.

29. Derive the power rule (Equation 20.9) for rational n. (Let $y = u^{p/q}$, where p and q are integers, so that $y^q = u^p$. Now differentiate implicitly and solve for dy/dx.)

20.10   Higher Derivatives

We saw in the earlier sections that the derivative of a function is also a function, as the notation $f'(x)$ suggests. It is sometimes necessary to differentiate a derivative. For example, the acceleration is defined to be the derivative of the velocity, which is itself a derivative, and so the acceleration is called the **second derivative**. The process can be continued indefinitely, as long as the resulting functions are differentiable. The derivative of the second derivative is called the **third derivative**, and so forth. Collectively, these are known as **higher derivatives**.

Notations

If $y = f(x)$, then

$$\frac{d}{dx}\left(\frac{dy}{dx}\right) = \frac{d^2y}{dx^2}$$

and

$$\frac{d}{dx}\left(\frac{d^2y}{dx^2}\right) = \frac{d^3y}{dx^3}$$

nth derivative

In general, d^ny/dx^n is the notation for the **nth derivative**. Alternately, we use the notations

$$f'(x), \ f''(x), \ f'''(x), \ f^{(4)}(x), \ \ldots, \ f^{(n)}(x)$$

or

$$y', \ y'', \ y''', \ y^{(4)}, \ \ldots, \ y^{(n)}$$

For example, if $y = 3x^3 - 2x + 5$, then

$$y' = 9x^2 - 2 \qquad y'' = 18x \qquad y''' = 18 \qquad y^{(4)} = 0$$

Also, if $f(x) = 3/x$, then

$$f'(x) = -\frac{3}{x^2} \qquad f''(x) = \frac{6}{x^3}$$

The main advantage of this notation is that we can write $f'(1) = -3$, $f''(1) = 6$, and so forth.

E X E R C I S E S / S E C T I O N   **20.10**

In Exercises 1–9, find the indicated higher derivatives of the given functions.

1. $y = 5x^4 + 5x^3 - 3x + 1$; find y''.

2. $f(x) = \dfrac{1}{\sqrt{x}}$; find $f''(x)$.

3. $y = \sqrt{x - 1}$; find y''.

4. $f(x) = (x^3 - 2x)^2$; find $f''(x)$.

5. $y = x^6 - 2x^5 - x^4$; find $\dfrac{d^3 y}{dx^3}$.

6. $f(x) = \dfrac{x}{x + 1}$; find $f^{(4)}(x)$.

7. $f(x) = \sqrt{5 + x}$; find $f'''(x)$.

8. $y = \sqrt{x^2 - 1}$; find $\dfrac{d^2 y}{dx^2}$.

9. $y = \dfrac{3 + 2x}{3 - 2x}$; find $\dfrac{d^2 y}{dx^2}$.

10. a. Find $y^{(4)}$ if $y = x^4$.
b. Show that $d^9 x^9/dx^9 = 9!$, where $9! = 9 \cdot 8 \cdot 7 \cdot 6 \cdot 5 \cdot 4 \cdot 3 \cdot 2 \cdot 1$.

R E V I E W   E X E R C I S E S / C H A P T E R   **20**

1. If $f(x) = x^2 - 1$, find $f(0)$, $f(1)$, $f(\sqrt{2})$.

2. If $f(x) = \sqrt[3]{x}$ and $g(x) = x^2 + 2$, find $f(g(x))$ and $g(f(x))$.

3. If $f(x) = \begin{cases} 0 & 0 \le x < 1 \\ 2 & x > 1 \end{cases}$, find $f(0)$, $f\left(\dfrac{1}{2}\right)$, $f\left(\dfrac{5}{2}\right)$, $f(1)$.

4. State the domain and range of the function in Exercise 3.

5. Find the domain and range of the following functions:
   a. $y = \sqrt{x - 1}$ **b.** $y = \sqrt[3]{x - 1}$

In Exercises 6–18, find the limits indicated.

6. $\lim\limits_{x \to 9} \dfrac{x^2 - 81}{x - 9}$

7. $\lim\limits_{x \to 4} \dfrac{16 - x^2}{4 - x}$

8. $\lim\limits_{x \to 3} \dfrac{15 - 2x - x^2}{3 - x}$

9. $\lim\limits_{x \to 0} \dfrac{x^3 - x^2 + 3x}{x}$

10. $\lim\limits_{x \to 0} \dfrac{x^2 + 2x}{x}$

11. a. $\lim\limits_{x \to 2} (1 - x^2)$ **b.** $\lim\limits_{x \to 1} \dfrac{x^2 - 5x + 4}{x - 1}$

12. $\lim\limits_{x \to 3} \dfrac{2x^2 - 5x - 3}{x - 3}$

13. $\lim\limits_{x \to 1} \dfrac{\sqrt{x} - 1}{x - 1}$

14. $\lim\limits_{x \to 0} \dfrac{x^2 - 2}{x - 1}$

15. $\lim\limits_{x \to \infty} \dfrac{2x^2 - 3x + 2}{x^2 - 10x + 1}$

16. $\lim\limits_{x \to \infty} \dfrac{x - 1}{x^2 + x + 2}$

17. $\lim\limits_{x \to 4+} \sqrt{x - 4}$

18. If $f(x) = \begin{cases} 1 & x \le 1 \\ 2 & x > 1, \end{cases}$ find: **a.** $\lim\limits_{x \to 1+} f(x)$ **b.** $\lim\limits_{x \to 1-} f(x)$

In Exercises 19–22, use the four-step process to differentiate the given functions.

19. $y = x - 3x^2$

20. a. $y = x^3$ **b.** $y = \dfrac{2}{x}$

21. a. $y = \dfrac{1}{4 - x}$ **b.** $y = \sqrt{x}$

22. $y = \sqrt{3 - x}$

In Exercises 23–29, differentiate.

23. $y = (x^3 - 2)^4$

24. $y = \dfrac{1}{x^4 + 3}$

25. $y = \dfrac{x - 4}{x + 1}$

26. $y = \dfrac{1}{\sqrt{7 - x^5}}$

27. $y = \dfrac{x^2}{\sqrt{4 - x^2}}$

28. $y = (x^2 + 1)^2(x - 3)$

29. $y = x\sqrt{4 - x^2}$

In Exercises 30–33, find the implicit derivatives.

30. $y^2 + x^2 + 3x = 1$

31. $x^2 y + xy^2 + y^3 = 1$

32. $2x^2 y^2 - 4xy = x$

33. $x^3 y + xy^3 = 5$

34. Find the slope of the line tangent to the hyperbola $3x^2 - y^2 = 23$ at $(3, -2)$.

35. Find the slope of the line tangent to the curve $8x^2 + 4xy + 5y^2 + 28x - 2y + 20 = 0$ at $(-1, \tfrac{6}{5})$.

36. Find $f'''(x)$ if $f(x) = \sqrt{x + 3}$.

37. a. Find the value of x for which the derivative of $f(x) = x/\sqrt{x-1}$ is equal to zero.
  b. If $f(x) = 2x^3 - 6x^2 + 4$, find x such that $f''(x) = 0$.

38. Find the value of x for which the slope of the line tangent to the curve $y = x/\sqrt{4+x^2}$ is $\frac{1}{2}$.

39. The resistance of a certain wire is given by $R = k/r^2$, where k is a constant and r the radius of the wire. Find the expression for the rate of change of R with respect to r.

40. Two curves are perpendicular at their point of intersection if their tangent lines are perpendicular at that point. Show that $y = \frac{1}{3}x^{-3}$ and $y = \frac{1}{3}x^3$ are perpendicular at $(1, \frac{1}{3})$.

41. Recall that the voltage V (volts) across a resistor in a circuit is given by $V = IR$. If $R = 0.010t^2$ Ω and $I = 4.12 + 0.020t$ A, find the rate of change of the voltage with respect to time when $t = 2.5$ s.

42. Suppose that steam flows through a hole with a cross-sectional area of A in.^{2} and approaches the hole under a pressure of P lb/in.^{2}. If w is the weight (in pounds) of steam flowing through the hole each second, then $w = 0.0165AP^{0.95}$, known as "Grashof's formula." If $A = 9.0$ in.^{2}, find the rate of change of w with respect to P when $P = 11$ lb/in.^{2}.

43. It costs the Ace Widget Company

$$C(y) = 50 + \frac{1}{2}y - \frac{1}{500}y^2$$

dollars to produce y widgets. What is the marginal cost of producing 50 widgets?

44. An object is moving vertically according to the equation $s = 100t - t^2$, where t is the time in seconds and s the distance in feet above the ground. Find the value of t for which the velocity is zero.

45. Two unlike charges of 1 electrostatic unit each exert an attractive force of $F = 1/r^2$ dynes on each other at a distance of r cm. What is the instantaneous rate of change of the force with respect to distance at a distance of 5.02 cm?

C H A P T E R **21**

Applications of the Derivative

21.1 Tangents and Normals

In this section we will find the equations of tangent and normal (perpendicular) lines to a given curve. But first we need to introduce another handy notational device. So far, when we have evaluated the derivative $f'(x)$ for some $x = x_0$, we have denoted the resulting value by $f'(x_0)$. A common alternative notation is

$$\left.\frac{dy}{dx}\right|_{x=x_0} \qquad \text{or simply} \qquad \left.\frac{dy}{dx}\right|_{x_0}$$

If both coordinates of the point are specified, then the derivative at (x_0, y_0) is denoted by

$$\left.\frac{dy}{dx}\right|_{(x_0, y_0)}$$

Tangent Lines

Recall from Section 19.4 that the equation of a line having a slope m and passing through a point (x_1, y_1) is given by the point-slope form

$$y - y_1 = m(x - x_1)$$

If $y = f(x)$ is a differentiable function and if (x_1, y_1) is a point on the graph, then the slope of the tangent line at the point is $f'(x_1)$. Consequently, the equation of the tangent line is $y - y_1 = f'(x_1)(x - x_1)$.

Equation of Tangent Line:

$$y - y_1 = f'(x_1)(x - x_1) \tag{21.1}$$

E X A M P L E **1**

Find the equation of the tangent line to the curve $y = 1/\sqrt{x + 1}$ at the point $(3, \frac{1}{2})$.

Solution. Writing the function in the form $y = (x + 1)^{-1/2}$, we find the derivative by the power rule:

$$\frac{dy}{dx} = -\frac{1}{2}(x + 1)^{-3/2} = -\frac{1}{2(x + 1)^{3/2}}$$

Thus

$$\frac{dy}{dx}\bigg|_{(3,1/2)} = -\frac{1}{2(x + 1)^{3/2}}\bigg|_{(3,1/2)} = -\frac{1}{2(4)^{3/2}} = -\frac{1}{16}$$

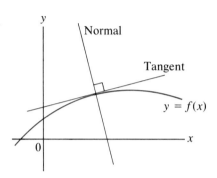

y

Normal

Tangent

$y = f(x)$

x

0

Figure 21.1

Normal line

so that the line has slope equal to $-\frac{1}{16}$ and passes through $(3, \frac{1}{2})$. By the point-slope form, the equation of the line is

$$y - \frac{1}{2} = -\frac{1}{16}(x - 3) \qquad \text{or} \qquad x + 16y - 11 = 0 \qquad \blacktriangleleft$$

Normal Lines

A line is said to be **normal** to the curve $y = f(x)$ at (x_1, y_1) if it is perpendicular to the tangent line through (x_1, y_1). Since two perpendicular lines have slopes that are negative reciprocals, the slope of the normal line is $-1/f'(x_1)$. The equation is given in formula (21.2).

Equation of Normal Line:

$$y - y_1 = -\frac{1}{f'(x_1)}(x - x_1) \qquad f'(x_1) \neq 0 \qquad\qquad (21.2)$$

See Figure 21.1.

E X A M P L E **2**

Find the equations of the tangent and normal lines to the parabola $y^2 = 3x$ at the point $(3, -3)$.

Solution. We differentiate implicitly:

$$2y\frac{dy}{dx} = 3 \qquad \frac{dy}{dx} = \frac{3}{2y}$$

For the slope of the tangent line, we then get

$$\frac{dy}{dx}\bigg|_{(3, -3)} = \frac{3}{2(-3)} = \frac{3}{-6} = -\frac{1}{2}$$

and for the slope of the normal line, we get 2, since $-\frac{1}{2}$ and 2 are negative reciprocals. The equations of the tangent and normal lines are

$$y + 3 = -\tfrac{1}{2}(x - 3) \qquad \text{and} \qquad y + 3 = 2(x - 3)$$

respectively.   ◄

EXERCISES / SECTION   21.1

In Exercises 1–12, find the equations of the tangent and normal lines to the given curves at the points indicated and sketch.

1. $y = 2x$; $(1, 2)$

2. $y = 1 - x$; $(2, -1)$

3. $y = x^2 + 2x$; $(1, 3)$

4. $y = x^2 - 1$; $(-2, 3)$

5. $y = x^2$; $(1, 1)$

6. $y = x - x^2$; $(1, 0)$

7. $y = x^3 + x$; $(-1, -2)$

8. $y = x - x^4$; $(2, -14)$

9. $y = \dfrac{1}{x}$; $\left(4, \dfrac{1}{4}\right)$

10. $y = \dfrac{1}{x^2 + 1}$; $\left(1, \dfrac{1}{2}\right)$

11. $y = \dfrac{1}{\sqrt[3]{x}}$; $\left(8, \dfrac{1}{2}\right)$

12. $y = \dfrac{x + 2}{x - 2}$; $(3, 5)$

13. Find the equation of the tangent to the circle $x^2 + y^2 = 25$ at the point $(-3, -4)$.

14. Find the equation of the normal to the ellipse $(x^2/32) + (y^2/2) = 1$ at $(-4, 1)$.

15. Find the equation of the tangent to the hyperbola $x^2 - 4y^2 = 9$ at $(5, 2)$.

16. Show that the normal to the circle $x^2 + y^2 - 2x - 24 = 0$ at $(4, -4)$ passes through the center.

17. Find the equations of the tangent and normal lines to the curve $3x^2 - xy + y^2 = 3$ at $(1, 1)$.

18. Find the equation of the normal line to the parabola $y^2 = 2 - x$ at $(1, 1)$.

19. For a point mass at the origin, the curves of equal gravitational potential are $x^2 + y^2 = c^2$. The lines of force are normal to these curves. Find the equation of the line of force normal to the curve $x^2 + y^2 = 25$ at the point $(-4, 3)$.

20. In the study of heat flow, *isothermal curves* are curves joining points at the same temperature. For a certain metal plate the isothermal curves are found to be $2x^2 + y^2 = c^2$. Find the line normal to the curve $2x^2 + y^2 = 6$ at $(-1, 2)$. (This line gives the direction in which heat flows at this point.)

21.2   The First-Derivative Test

In this section we will develop a technique for using the derivative as an aid to curve sketching. Mainly, we can use the derivative to determine where a graph reaches its highest and lowest points, called the *maximum* and *minimum points*, respectively. To this end, we need the definition of an increasing and decreasing function.

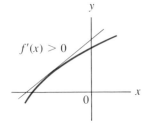

$f(x)$ is increasing

Figure 21.2

> **Definition of Increasing and Decreasing Function:** A function f is **increasing** on an interval if, for any two numbers x_1 and x_2 in the interval,
>
> $$x_1 < x_2 \text{ implies that } f(x_1) < f(x_2)$$
>
> A function f is **decreasing** on the interval if
>
> $$x_1 < x_2 \text{ implies that } f(x_1) > f(x_2)$$

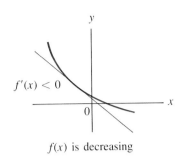

$f'(x) < 0$

$f(x)$ is decreasing

Figure 21.3

Recall from Section 20.7 that a function $f(x)$ is increasing on an interval if $f'(x)$ is positive in the interval (Figure 21.2) and $f(x)$ is decreasing if $f'(x)$ is negative (Figure 21.3).

It is possible for a function to be increasing on one interval and decreasing on another. In that case the graph must rise to a peak and then fall again. Consider the following example.

E X A M P L E **1**   Determine the intervals on which the function $y = 1 - x^2$ is increasing and decreasing and find the highest point on the curve.

Solution. Since $f(x) = 1 - x^2$, $f'(x) = -2x$. To determine where $f(x)$ is increasing and decreasing, we need to determine where $f'(x) = -2x$ is positive and negative. Note that

$$-2x > 0 \qquad \text{if } x < 0$$

and

$$-2x < 0 \qquad \text{if } x > 0$$

Hence f is increasing on $(-\infty, 0)$ and decreasing on $(0, \infty)$. (By our definition, f is actually increasing on $(-\infty, 0]$ and decreasing on $[0, \infty)$.) At $x = 0$, $-2x = 0$, so that the graph has a horizontal tangent line at $x = 0$. Since $f(0) = 1 - 0^2 = 1$, the point is $(0, 1)$. This point has a special property. As we noted above:

1. To the left of the point f is increasing.
2. To the right of the point f is decreasing.

Therefore:

3. f reaches a peak at the point $(0, 1)$.

This peak is called a **maximum point** or simply a **maximum**, and $f(0) = 1$ is called a **maximum value**. This graph is shown in Figure 21.4.   ◄

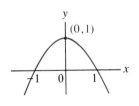

$(0,1)$

Figure 21.4

Just as a graph may have a highest point (Example 1), it may have a lowest point or *minimum*.

Definition of Relative Maximum and Minimum: A point is called a **relative maximum** if it has a larger y-value than any point near it.
   A point is called a **relative minimum** if it has a smaller y-value than any point near it.

E X A M P L E **2** Determine the intervals on which the function $y = x^3 - 3x + 2$ is increasing and decreasing. From this information determine the maximum and minimum points.

Solution. To determine where the function is increasing and decreasing, we find the intervals on which $f'(x) > 0$ and $f'(x) < 0$. To this end, we set $f'(x)$ equal to zero and solve for x. Thus

$$f'(x) = 3x^2 - 3 = 3(x - 1)(x + 1) = 0$$

whence $x = 1$ and -1. Using this information, we find that

$$f'(x) > 0 \text{ on } (-\infty, -1) \qquad (f \text{ is increasing})$$
$$f'(x) < 0 \text{ on } (-1, 1) \qquad (f \text{ is decreasing})$$

and

$$f'(x) > 0 \text{ on } (1, \infty) \qquad (f \text{ is increasing})$$

Moreover, when $x = -1$, $y = (-1)^3 - 3(-1) + 2 = 4$, and when $x = 1$, $y = 0$. Hence the graph has horizontal tangent lines at $(-1, 4)$ and $(1, 0)$. Now observe that:

1. To the left of $(-1, 4)$ f is increasing ($f' > 0$).
2. To the right of $(-1, 4)$ f is decreasing ($f' < 0$).

Therefore:

3. The point $(-1, 4)$ is a relative maximum (Figure 21.5).

For the point $(1, 0)$ we have the following:

1. To the left of $(1, 0)$ f is decreasing ($f' < 0$).
2. To the right of $(1, 0)$ f is increasing ($f' > 0$).

Therefore:

3. The point $(1, 0)$ is a relative minimum.

The graph is shown in Figure 21.5. ◀

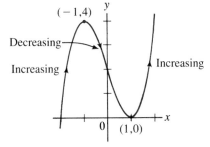

Figure 21.5

We can see from Example 2 that a relative maximum is not necessarily the highest point on the curve—it may be a maximum only in the vicinity of the point. On the other hand, the point $(0, 1)$ in Example 1 is higher than any other point on the curve. Such a point is called an **absolute maximum**. (Similar comments apply to minimum points.)

For convenience we will use the term *maximum* for an absolute or a relative maximum and the term *minimum* for an absolute or a relative minimum. Collectively, maximum and minimum values will be referred to as **extreme values**.

Extreme values

Before summarizing these results we need to consider one more case. Suppose $f(x) = x^{2/3}$; then

$$f'(x) = \frac{2}{3} x^{-1/3} = \frac{2}{3x^{1/3}}$$

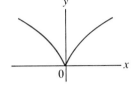

Figure 21.6

Note that for $x = 0$ the derivative is infinitely large (undefined), which means that the graph has a vertical tangent line at $(0, 0)$. To the left of the origin $f'(x) < 0$ and to the right $f'(x) > 0$. Consequently, the origin must be a minimum (Figure 21.6).

It is clear from these examples that to find the extreme values of a function f, we need to examine those values c for which $f'(c) = 0$ or for which $f'(c)$ does not exist. Such values are called **critical values** and the corresponding points $(c, f(c))$ on the graph **critical points**.

> **Definition of Critical Value:** A **critical value** is a number c in the domain of f for which $f'(c) = 0$ or for which $f'(c)$ does not exist. The points $(c, f(c))$ are called **critical points**.

The foregoing procedure for testing the critical points will be referred to as the **first-derivative test**.

> **First-Derivative Test:** If c is a critical value, test the derivative with two values of x, one slightly less and the other slightly more than c. If, as x increases, the sign of the derivative changes from $+$ to $-$, then $f(c)$ is a maximum value and $(c, f(c))$ is a maximum point. If the sign changes from $-$ to $+$, then $f(c)$ is a minimum value. If the sign does not change, then $(c, f(c))$ is neither a minimum nor a maximum.

That a critical point need not lead to an extreme value can be seen from the following example.

E X A M P L E **3**   Test the function $f(x) = x^3$ for extreme values and sketch the graph.

Solution. Since $f'(x) = 3x^2 > 0$ for $x \neq 0$, the function is increasing everywhere. As a consequence, even though $x = 0$ is a critical value, $f(0)$ is neither a minimum nor a maximum value. In sketching the graph we must keep in mind that even though the point $(0, 0)$ is neither a minimum nor a maximum, the line $y = 0$ is a tangent line. (See Figure 21.7.) Note that the tangent line passes through the graph.   ◄

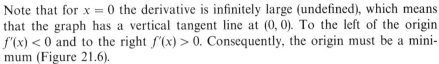

Figure 21.7

These examples show the importance of extreme values in curve sketching: if you know the behavior of the function at the critical points, you have a good idea of the shape of the graph. The next example illustrates a convenient method for testing the critical points by means of a chart.

E X A M P L E **4**   Find the extreme values of the function $y = -\frac{2}{3}x^3 + x^2 + 4x - 5$ and sketch.

Solution. To locate the critical points, we find the derivative of the function and set it equal to zero:

$$\frac{dy}{dx} = -2x^2 + 2x + 4 = 0$$

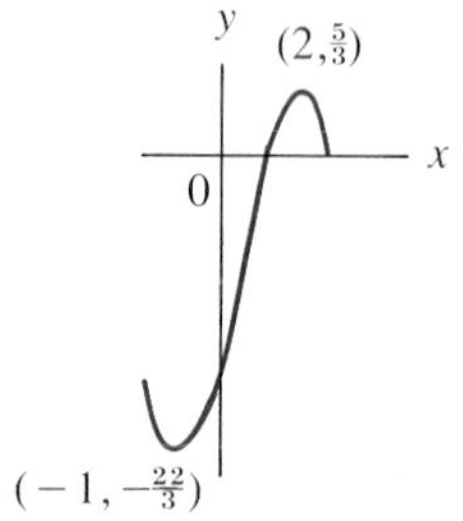

y $(2,\frac{5}{3})$

0 x

$(-1, -\frac{22}{3})$

Figure 21.8

or

$$\frac{dy}{dx} = -2(x^2 - x - 2) = -2(x - 2)(x + 1) = 0$$

Solving for x, we get $x = 2$ and $x = -1$. Substituting in the given equation, we find that $y = \frac{5}{3}$ when $x = 2$ and $y = -\frac{22}{3}$ when $x = -1$. Hence $(2, \frac{5}{3})$ and $(-1, -\frac{22}{3})$ are the critical points. (See Figure 21.8.)

A simple way to determine the signs on the derivative is the following:

$$\frac{dy}{dx} = -2(x - 2)(x + 1) = 0$$

when x is either 2 or -1. Moreover, these values are the *only* values for which $dy/dx = 0$. Consequently, dy/dx is different from zero at all other points. This observation allows us to substitute arbitrary test values for x to determine the signs. For example, if $x = 3$, then $dy/dx = -8$, which is <0, so that $dy/dx < 0$ for *all* $x > 2$. (If this were not so, then there would have to exist another $x > 2$ for which $dy/dx = 0$, but 2 and -1 are the only critical values.) Similarly, if $x = 1$, then $dy/dx = 4$, which is >0. So $dy/dx > 0$ for all x in the interval $(-1, 2)$. Finally, if $x = -3$, then $dy/dx = -20$, which is <0, and $dy/dx < 0$ on $(-\infty, -1)$.

This process can be carried out systematically by means of the following chart:

	Test values	$x - 2$	$x + 1$	$f'(x) = -2(x - 2)(x + 1)$	
$x > 2$	3	+	+	−	*f* decreasing
$-1 < x < 2$	1	−	+	+	*f* increasing
$x < -1$	−3	−	−	−	*f* decreasing

According to the chart:

1. To the left of $x = -1$, f is decreasing ($f' < 0$).
2. To the right of $x = -1$, f is increasing ($f' > 0$).

Therefore:

3. The point $(-1, -\frac{22}{3})$ is a minimum.

Moreover:

1. To the left of $x = 2$, f is increasing ($f' > 0$).
2. To the right of $x = 2$, f is decreasing ($f' < 0$).

Therefore:

3. The point $(2, \frac{5}{3})$ is a maximum.

The graph is shown in Figure 21.8.   ◀

EXERCISES / SECTION 21.2

Find the extreme values of each of the following functions and sketch the curves.

1. $y = x^2 - 2x + 1$
2. $y = x^2 - 4x + 3$
3. $y = 8 - 2x - x^2$
4. $y = -x^2 + 5x - 6$
5. $y = 2x^3 + 3x^2 - 12x + 6$
6. $y = 2x^3 + 7x^2 + 4x + 1$
7. $y = -x^3 + 6x^2 - 9x - 5$
8. $y = x^3 - 3x + 7$
9. $y = x^4 - 2x^2 - 2$
10. $y = (x + 1)^3$
11. $y = x^4 + \frac{4}{3}x^3$
12. $y = x^2 - \frac{1}{2}x^4$
13. $y = 4 - 4x^3 - 3x^4$
14. $y = \sqrt{x}$
15. $y = \sqrt[3]{x}$
16. $y = 4\sqrt{x} - x$

17. A projectile shot directly upward with a velocity of 80 m/s moves according to the equation

$$s = 80t - 5t^2$$

Determine the maximum altitude.

18. If a resistor of 4 Ω is linked parallel with a variable resistor

of R Ω, then the resistance R_T of the combination is given by

$$R_T = \frac{4R}{R + 4}$$

Show that R_T starts at zero and increases steadily as R increases.

19. The power in a circuit with variable resistance R is given by

$$P = \frac{16R}{(R + 2)^2} \text{ watts}$$

Find the setting of the variable resistor that allows it to take maximum power.

20. The output P of a certain battery is given by $P = 20I - 10I^2$, where I is the current in amperes. Find the current for which the output is a maximum.

21.3 The Second-Derivative Test

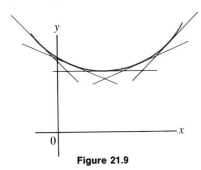

Figure 21.9

We observed in the last section that the critical points give us the kind of information that greatly facilitates curve sketching. In this section we are going to develop a procedure that will give us even more information about the graph by using the second derivative.

Suppose a function f is increasing on some interval. Then $f'(x) > 0$ on that interval. The same statement can be made about f', since f' is itself a function: if f' is increasing on an interval, then $df'(x)/dx = f''(x) > 0$. What does this tell us about the graph? Consider the curve in Figure 21.9. The curve is drawn in such a way that the slope is steadily increasing, which is possible only if the curve remains above the tangent line at each point. In the language of calculus

Concave up
Concave down

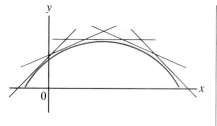

Figure 21.10

we say that the curve in Figure 21.9 is **concave up**. Similarly, if f' is decreasing, then $f''(x) < 0$ and the graph is **concave down** (Figure 21.10).

Concavity and Inflection Points: For all x in an interval $[a, b]$, the graph of a function f is:

1. Concave up if $f''(x) > 0$
2. Concave down if $f''(x) < 0$

   A point on the graph at which the concavity changes is called an **inflection point**.

E   X   A   M   P   L   E   **1**

Suppose we return to the function $f(x) = x^3 - 3x + 2$ whose graph appears in Figure 21.5. From $f''(x) = 6x$ we see that $f''(x) < 0$ if $x < 0$, and $f''(x) > 0$ if $x > 0$. So the graph is concave down to the left of the point $(0, 2)$ and concave up to the right. Since the concavity changes at $(0, 2)$, this point is an inflection point.   ◄

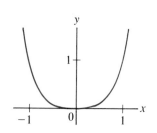

Figure 21.11

   To find the point of inflection, we normally set $f''(x) = 0$, solve for x, and determine the sign of $f''(x)$ to the left and right of the point. (A point $(x, f(x))$ for which $f''(x)$ does not exist may also be an inflection point, as we will see in Exercise 29; we will concentrate mainly on the former case.) Once the values of x for which $f''(x) = 0$ are known, it is a simple matter to determine the sign of $f''(x)$ elsewhere. In Example 1, $f''(x) = 6x = 0$ when $x = 0$. Hence $f''(x) \neq 0$ at all other points. Consequently, since $f''(1) = 6 > 0$, $f''(x)$ must be positive for all $x > 0$ and since $f''(-1) = -6 < 0$, $f''(x)$ must be negative for all $x < 0$.
   Caution. If $(c, f(c))$ is an inflection point, then $f''(c) = 0$ or $f''(c)$ does not exist. The converse is not necessarily true: if $f''(c) = 0$ for some $x = c$, then $(c, f(c))$ is not necessarily an inflection point, as we can see from the function $f(x) = x^4$. (See the graph in Figure 21.11.) Here $f'(x) = 4x^3$ and $f''(x) = 12x^2$; $f''(0) = 0$, but the point $(0, 0)$ is not an inflection point.
   A final key observation from our discussion of concavity concerns the determination of extreme values. Suppose a curve is known to be concave up on some interval. If for some point c inside the interval we have a horizontal tangent line at the corresponding point $(c, f(c))$, then this point must be a minimum. For example, if $f(x) = x^2$, then $f'(x) = 2x$, so that $(0, 0)$ is a critical point. Since $f''(x) = 2 > 0$, the graph is concave up everywhere and $(0, 0)$ must be a minimum. This criterion is known as the **second-derivative test**.

Second-Derivative Test: Suppose $f'(c) = 0$.

1. If $f''(c) > 0$, then $f(c)$ is a minimum value.
2. If $f''(c) < 0$, then $f(c)$ is a maximum value.
3. If $f''(c) = 0$, the test fails.

Caution. Regarding part (3), if the test fails, we mean just that—the second derivative gives us no information about the critical point. For $f(x) = x^4$ we already saw that $f''(0) = 0$; yet f attains a minimum at the critical point (0, 0). In Example 3 of Section 21.2 the function $f(x) = x^3$ has critical value $x = 0$ and again $f''(0) = 0$. This time, however, $f(0)$ is not an extreme value. Occasionally, then, we need to fall back on the first-derivative test. (At times it may also be inconvenient to compute f''.)

Before considering further examples, let us summarize the procedure for sketching curves.

Suggested Procedure for Curve Sketching:

1. Find all critical values and critical points.
2. Test the critical values:
   a. Use the second-derivative test.
   b. If the second-derivative test fails, use the first-derivative test.
3. Use the second derivative to determine the intervals for which the graph is concave up and concave down.
4. Determine the points of inflection from Step 3.
5. Other: Find any easily determined intercepts and asymptotes; test for symmetry.
6. Plot the critical points, inflection points, and (if available) the intercepts. Draw any existing asymptotes. Sketch the curve, using a few additional points.

E X A M P L E **2** Discuss the function $y = x^3 - 3x$ for minima and maxima, concavity, and inflection points, and sketch.

Solution. Derivatives: $f'(x) = 3x^2 - 3$; $f''(x) = 6x$.

Step 1. Critical points: $3x^2 - 3 = 0$ whenever $x = \pm 1$; the points are $(1, -2)$ and $(-1, 2)$, determined from $y = x^3 - 3x$.

Step 2. Test of critical points:

$$f''(1) = 6 > 0; \ (1, -2) \text{ is a minimum point}$$

$$f''(-1) = -6 < 0; \ (-1, 2) \text{ is a maximum point}$$

Step 3. Concavity: we need to determine where $f'' > 0$ and where $f'' < 0$. To this end we first find those values of x for which $f''(x) = 0$:

$$f''(x) = 6x = 0 \text{ when } x = 0$$

Since $x = 0$ is the only value for which $f''(x) = 0$, it follows that $f''(x) \neq 0$ at all other points. Since $f''(1) = 6$, $f''(x) > 0$ whenever $x > 0$. Hence the graph is concave up on $[0, \infty)$. Similarly, since $f''(-1) = -6$, the graph is concave down on $(-\infty, 0]$. (See Figure 21.12)

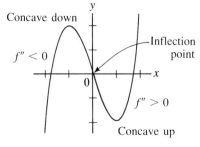

Concave down

$f'' < 0$

Inflection point

$f'' > 0$

Concave up

Figure 21.12

Step 4. Inflection point: because of the change in concavity, there is an inflection point at $x = 0$. If $x = 0$, then $y = 0$; so the inflection point is $(0, 0)$.

Step 5. Other: the intercepts are $(0, 0)$ and $(\pm\sqrt{3}, 0)$; the graph is symmetric with respect to the origin.

Step 6. Plotting all the points found and perhaps a few additional ones, we obtain the graph in Figure 21.12.   ◀

E X A M P L E **3**   Discuss and sketch the graph of $y = x^4 + \frac{4}{3}x^3$.

Solution. Derivatives: $f'(x) = 4x^3 + 4x^2$; $f''(x) = 12x^2 + 8x$.

Step 1. Critical points:

$$4x^3 + 4x^2 = 0 \qquad \text{or} \qquad 4x^2(x + 1) = 0$$

Solving for x, we find that $x = 0$ and $x = -1$. Hence $(-1, -\frac{1}{3})$ and $(0, 0)$ are the critical points.

Step 2. Test of critical points:

$f''(-1) = 4$; $f(-1)$ is a minimum value

$f''(0) = 0$; the test fails

Using the first-derivative test, we conclude that since $f'(x) = 4x^2(x + 1) > 0$ on $(-1, \infty)$, $x = 0$ does not lead to an extreme value.

Step 3. Concavity:

$$f''(x) = 12x^2 + 8x = 0 \qquad \text{or} \qquad 4x(3x + 2) = 0$$

whenever $x = 0$ and $x = -\frac{2}{3}$. These are the only values for which $f''(x) = 0$, so that $f''(x) \neq 0$ at all other points. Since $f''(-1) > 0$, $f''(x) > 0$ on $(-\infty, -\frac{2}{3})$. Similarly, from $f''(-\frac{1}{2}) < 0$, we conclude that $f''(x) < 0$ on $(-\frac{2}{3}, 0)$. Finally, since $f''(1) = 20$, $f''(x) > 0$ on $(0, \infty)$.

These observations are summarized in the following chart:

	Test values	$4x$	$3x + 2$	$f''(x) = 4x(3x + 2)$	
$x > 0$	1	$+$	$+$	$+$	concave up
$-\frac{2}{3} < x < 0$	$-\frac{1}{2}$	$-$	$+$	$-$	concave down
$x < -\frac{2}{3}$	-1	$-$	$-$	$+$	concave up

Step 4. Inflection points: since the concavity changes, the points $(0, 0)$ and $(-\frac{2}{3}, -16/81)$ are inflection points, confirming that $(0, 0)$ is not a minimum or maximum.

Step 5. Other: the intercepts are $(0, 0)$ and $(-\frac{4}{3}, 0)$.

Step 6. The graph is shown in Figure 21.13.   ◀

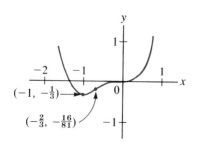

Figure 21.13

E X A M P L E **4**   Discuss and sketch the graph of

$$y = \frac{x}{\sqrt{x - 1}}$$

Solution. Because of the radical, it is wise to start with the general discussion. Since $x - 1$ has to be strictly positive to avoid imaginary numbers and division by zero, the domain of the function is the interval $(1, \infty)$. Setting the denominator equal to zero, we find that $x = 1$ is a vertical asymptote. Finally, $y = 0$ only if $x = 0$, but $x = 0$ is not in the domain. Consequently, the graph has no intercepts.

The derivatives can be found by the quotient rule and will be left as an exercise:

$$f'(x) = \frac{x - 2}{2(x - 1)^{3/2}} \quad \text{and} \quad f''(x) = -\frac{x - 4}{4(x - 1)^{5/2}}$$

Step 1. The only critical value is $x = 2$.
Step 2. Since $f''(2) > 0$, the point $(2, 2)$ is a minimum.
Step 3. Concavity:

$$f''(x) = 0 \text{ when } x = 4$$

It is easy to check that $f''(x) > 0$ on $(1, 4)$ so that the graph is concave up on the interval $(1, 4]$. On the interval $[4, \infty)$ the graph is concave down.

Step 4. Inflection point: since the concavity changes, the point $(4, 4/\sqrt{3})$ is an inflection point.

Steps 5 and 6. The graph is shown in Figure 21.14.

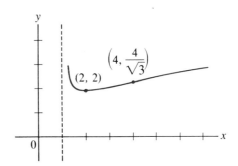

Figure 21.14   ◄

E X A M P L E **5**   Discuss and sketch the graph of $y = (x - 2)^{2/3}$.

Solution. Derivatives:

$$f'(x) = \frac{2}{3(x - 2)^{1/3}} \qquad f''(x) = -\frac{2}{9(x - 2)^{4/3}}$$

Step 1. Critical points. Since $f'(2)$ is undefined, $x = 2$ is a critical value and $(2, 0)$ a critical point; the line $x = 2$ is a vertical tangent.

Step 2. Test of critical points. Since f' is not differentiable at $x = 2$ (that is, $f''(2)$ does not exist), the second-derivative test cannot be employed. It is easy to see, however, that $f'(x) < 0$ for $x < 2$ and $f'(x) > 0$ for $x > 2$. Hence $(2, 0)$ is a minimum by the first-derivative test.

Steps 3 and 4. Concavity and points of inflection. Since $f''(2)$ does not exist, the point $(2, 0)$ is a possible point of inflection. At no point is $f''(x) = 0$. Substituting two values, such as $x = 1$ and $x = 3$, in the expression for $f''(x)$, we find that $f''(x) < 0$ for all $x \neq 2$. Hence $(2, 0)$ is not an inflection point. (The graph is concave down on $(-\infty, 2]$ and $[2, \infty)$.)

Steps 5 and 6. The graph is shown in Figure 21.15.   ◀

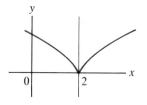

Figure 21.15

E X A M P L E **6**   Discuss and sketch the graph of

$$y = x^2 + \frac{16}{x^2}$$

Solution. Derivatives. From $f(x) = x^2 + 16x^{-2}$ we get

$$f'(x) = 2x - 32x^{-3} \quad \text{and} \quad f''(x) = 2 + 96x^{-4}$$

or

$$f'(x) = 2x - \frac{32}{x^3} \quad \text{and} \quad f''(x) = 2 + \frac{96}{x^4}$$

Step 1. Critical points. Setting $2x - (32/x^3) = 0$, we get

$$2x - \frac{32}{x^3} = 0$$

$$2x^4 - 32 = 0 \quad \text{multiplying by } x^3$$

$$x^4 - 16 = 0 \quad \text{dividing by 2}$$

$$x = \pm 2$$

The critical points are $(\pm 2, 8)$.

Step 2. Test of critical points. See Step 3.

Steps 3 and 4. Concavity. Since $f''(x) = 2 + (96/x^4) > 0$, $x \neq 0$, the graph is concave up everywhere. Consequently, there cannot be any inflection points. Moreover, both critical points are minima.

Step 5. Other: from

$$y = x^2 + \frac{16}{x^2} = \frac{x^4 + 16}{x^2}$$

we see that the graph has no intercepts and that the y-axis is a vertical asymptote. (See Figure 21.16).

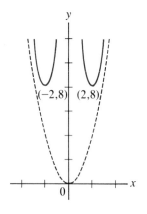

$(-2,8)$ $(2,8)$

Figure 21.16

Step 6. As $x \to \infty$, $16/x^2 \to 0$. So the graph of $y = x^2 + (16/x^2)$ gets closer to the parabola $y = x^2$. The graph of $y = x^2$ may be called an *asymptotic curve*, shown as the dashed curve in Figure 21.16. ◀

E X E R C I S E S / S E C T I O N **21.3**

Find the minima and maxima, the intervals on which the graph is concave up and concave down, and the inflection points of each of the following functions; sketch the curves.

1. $y = 2x^2 - 4x$

2. $y = x^2 - 2x + 1$

3. $y = 6x - 6x^2$

4. $y = 2x - x^2$

5. $y = -4 - 3x - \frac{1}{2}x^2$

6. $y = 3x - 3x^2$

7. $y = 2x^3 - 6x + 1$

8. $y = x^3 - 3x$

9. $y = x^3 - 6x^2 + 9x - 3$

10. $y = 2x^3 + 3x^2 - 12x + 2$

11. $y = x^3 - 3x^2 - 9x + 11$

12. $y = 1 + 9x - 3x^2 - x^3$

13. $y = x^2 - 6x$

14. $y = x^3 - 4x^2 + 4x$

15. $y = 3x^4 - 4x^3 + 2$

16. $y = \frac{1}{x^3}$

17. $y = x^4 + x^3$

18. $y = (x - 2)^3$

19. $y = (x - 3)^4$

20. $y = (x^2 - 9)^2$

21. $y = \dfrac{x}{x - 3}$

22. $y = x + \dfrac{1}{x}$

23. $y = x^2 + \dfrac{8}{x}$

24. $y = x^3 + \dfrac{1}{x^2}$

25. $y = \dfrac{x + 1}{x - 2}$

26. $y = x^{4/3}$

27. $y = x^{3/4}$

28. $y = x^2(4 - x^2)$

29. $y = (x - 3)^{1/3}$

30. $y = x - 2\sqrt{x}$

31. $y = \dfrac{2x}{(x + 1)^2}$

32. $y = \dfrac{x^2}{(x^2 + 1)^2}$

33. $y = \dfrac{6x}{x^2 + 3}$

34. $y = \dfrac{2x}{x^2 + 1}$

35. $y = \dfrac{2}{3}x^3 - \dfrac{2}{5}x^5$

36. $y = x^2 - \dfrac{4}{5}x^5$

37. $y = x^4 + 3x^3 + 6x^2$

21.4 Applications of Minima and Maxima

Have you ever wondered why bubbles are round? The reason is that a bubble encloses a certain volume of air and the surface tension contracts the surface of the bubble to its smallest possible area, which is a spherical surface. Many other situations in nature offer examples of minimum and maximum values. For example, the second law of thermodynamics may be stated in the following form: there is a tendency in nature for all systems to proceed toward a state of maximum molecular disorder. This accounts for the observations that well-formed crystals dissolve in a solvent, organisms decay after death, and rocks weather.

In our study of applied minima and maxima we are naturally confined to problems that can be analyzed with the techniques of Sections 21.2 and 21.3. If the function is known, these techniques can be applied directly, as illustrated in the first two examples.

E X A M P L E **1** The formula for the output P of a battery is given by

$$P = VI - RI^2$$

where V is the voltage, I the current, and R the resistance. Find the current for which the output is a maximum if $V = 12$ V and $R = 5.0 \ \Omega$.

Solution. After substituting the given values, we get

$$P = 12I - 5.0I^2$$

We now maximize P by the method of the last section; that is, we find the derivative with respect to I and set it equal to zero:

$$\frac{dP}{dI} = 12 - 10I = 0$$

so that $I = 1.2$ A is the critical value. Since

$$\frac{d^2P}{dI^2} = -10 < 0$$

it follows that $I = 1.2$ does correspond to the maximum output.   ◀

E   X   A   M   P   L   E   **2**   A turbine for generating power is rotated by means of a high-speed jet of water striking circularly mounted blades. The speed of the jet is normally fixed, but the speed (rate of rotation) of the turbine can be adjusted by changing the blade angle. If J is the speed of the jet and T the speed of the turbine, then the power is given by

$$P = kJT(J - T)$$

where k is a constant. What speed of the turbine will yield maximum power?

Solution. The equation can be written

$$P = kJ^2T - kJT^2$$

Since J is fixed, both J and k are treated as constants. The critical value is found from

$$\frac{dP}{dT} = kJ^2 \frac{d}{dT}(T) - kJ \frac{d}{dT}(T^2) \qquad \text{\textit{k, J} constants}$$

or

$$\frac{dP}{dT} = kJ^2 - 2kJT = 0$$

Solving for T, we obtain

$$2kJT = kJ^2 \qquad \text{and} \qquad T = \frac{kJ^2}{2kJ} = \frac{J}{2}$$

Since $d^2P/dT^2 = -2kJ < 0$, we see that the power is maximal when the speed of the turbine is numerically equal to one-half the speed of the jet.   ◄

In many problems the expression to be minimized or maximized is not known in advance and has to be obtained from the given information. Consider the following example.

E X A M P L E **3**   A rectangle has an area of 100 m². What should the dimensions be so that the perimeter will be as small as possible?

Solution. Let x be the length and y the width of the rectangle. Then the perimeter is given by

$$P = 2x + 2y$$

Before proceeding, we need to eliminate one of the variables: since the area is $xy = 100$, we have $y = 100/x$, so that

$$P = 2x + 2y$$

$$= 2x + 2\left(\frac{100}{x}\right)$$

$$= 2x + \frac{200}{x} = 2x + 200x^{-1}$$

P is now a function of x alone, and we can find the minimum by the usual method. Thus

$$\frac{dP}{dx} = 2 - 200x^{-2} \quad \text{and} \quad \frac{d^2P}{dx^2} = 400x^{-3}$$

Setting $dP/dx = 0$, we obtain

$$2 - 200x^{-2} = 0, \quad 2x^2 - 200 = 0 \qquad \textbf{multiplying by } x^2$$

whence $x = 10$. (The negative root has no meaning in this problem.) Since

$$\frac{d^2P}{dx^2}\bigg|_{x=10} = \frac{2}{5} > 0$$

we see that $x = 10$ leads to a minimum. Since $x = 10$, $y = \frac{100}{10} = 10$. So the desired dimensions are 10 m × 10 m.   ◄

Example 3 is quite typical and gives us a plan of attack for all such problems:

> **To Solve Minimum-Maximum Problems:**
>
> **1.** Write an expression for the quantity F to be minimized or maximized. (To do so you need to use auxiliary variables; drawing a figure may help.)
> **2.** If two variables are involved, eliminate one of them.
> **3.** Minimize or maximize F.

E X A M P L E **4**   The manager of a shop has an order for making a gutter from a long sheet of metal 16 cm wide by bending up equal widths along the edges into vertical positions. The order states that the gutter is to have the largest possible carrying capacity. What should the dimensions be?

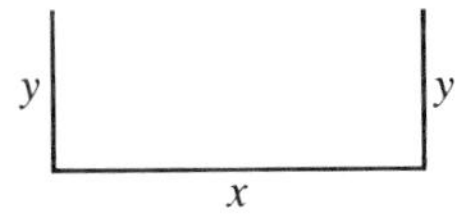

Figure 21.17

Solution. In this problem a figure is indeed helpful (Figure 21.17). Making use of the auxiliary variables, the quantity to be maximized is

$$A = xy$$

That is, the cross-section must be as large as possible. To eliminate one of the variables, we need a relation between them. From the given information,

$$x + 2y = 16 \qquad \text{or} \qquad x = 16 - 2y$$

Substituting in the equation $A = xy$, we obtain

$$A = yx = y(16 - 2y) = 16y - 2y^2$$

which is a function of y alone. Thus

$$\frac{dA}{dy} = 16 - 4y = 0$$

so that $y = 4$ is the critical value. Since $d^2A/dy^2 = -4$, A has a maximum at $y = 4$. Consequently, the desired dimensions are 4 cm by 8 cm. ◀

E X A M P L E **5**   A wholesaler finds that he can make a profit of $5 for each crate of peaches for orders of 100 crates or fewer. Since he gives a discount for large orders, he finds that he makes 2¢ less profit for each crate above 100. (For example, if an order is for 105 crates, he makes $4.90 per crate.) What size order will yield the maximum profit?

Solution. Let x be the number of crates *above* 100. The profit for every crate is $5 - 0.02x$. Since he sells a total of $100 + x$ crates, his profit is

$$P = (100 + x)(5 - 0.02x) = 500 + 3x - 0.02x^2$$

Thus

$$\frac{dP}{dx} = 3 - 0.04x = 0$$

and $x = 75$. Hence an order of $75 + 100 = 175$ crates will yield the maximum profit.   ◀

E X A M P L E **6**

A ray of light from point A reflected to point B from a plane mirror will follow a path requiring the least time (Figure 21.18). Show that the angle of incidence α is equal to the angle of reflection β.

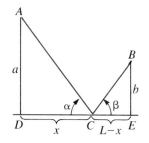

Figure 21.18

Solution. Let D and E be the points on the mirror nearest A and B, respectively, and let C be the point where the ray strikes. (See Figure 21.18.) Denote the distance from D to E by L. So if x is the distance from D to C, then $L - x$ is the distance from C to E. We now let c be the velocity of light and use the fact that distance equals rate times time or time equals distance over rate. Hence

$$T = \text{time of trip along } AC + \text{time of trip along } CB = AC/c + CB/c$$

But $AC = \sqrt{x^2 + a^2}$ and $CB = \sqrt{(L - x)^2 + b^2}$. So

$$T = \frac{\sqrt{x^2 + a^2}}{c} + \frac{\sqrt{(L - x)^2 + b^2}}{c}$$

is the quantity we wish to make a minimum. We now write

$$T = \frac{1}{c}\{(x^2 + a^2)^{1/2} + [(L - x)^2 + b^2]^{1/2}\}$$

Then

$$\frac{dT}{dx} = \frac{1}{c}\left\{\frac{1}{2}(x^2 + a^2)^{-1/2} \cdot 2x + \frac{1}{2}[(L - x)^2 + b^2]^{-1/2} \cdot 2(L - x)(-1)\right\} = 0$$

or

$$\frac{dT}{dx} = \frac{1}{c}\left(\frac{x}{\sqrt{x^2 + a^2}} - \frac{L - x}{\sqrt{(L - x)^2 + b^2}}\right) = 0$$

Normally we would now solve for x, but in this problem that turns out to be entirely unnecessary: referring to Figure 21.18, we see that the last expression implies that

$$\frac{x}{\sqrt{x^2 + a^2}} = \frac{L - x}{\sqrt{(L - x)^2 + b^2}} \qquad \text{or} \qquad \frac{DC}{AC} = \frac{CE}{CB}$$

That is, $\cos \alpha = \cos \beta$, which implies that $\alpha = \beta$.   ◀

E X E R C I S E S / S E C T I O N   21.4

1. According to Kelvin's law, the power lost in a transmission line is inversely proportional to the cross-sectional area (that is, the larger the cross-sectional area, the smaller the heat loss). The cost of the wire, however, is directly proportional to its area. In symbols, the total cost is

$$C = k_1 A + \frac{k_2}{A}$$

where k_1 and k_2 are constants. Show that C is a minimum when $A = \sqrt{k_2/k_1}$. (Since $C = k_1 A + k_2 A^{-1}$, $dC/dA = k_1 + k_2(-1)A^{-2}$.)

2. The drag on an airplane traveling at velocity v is

$$D = av^2 + \frac{b}{v^2}$$

where a and b are positive constants. At what speed does the airplane experience the least drag?

3. The deflection of a beam of length L clamped at both ends and carrying a uniform load F_0 is given by

$$Y(x) = F_0 ax^2(L - x)^2$$

where a is a constant. Find the point at which the deflection is a maximum.

4. The efficiency E of a screw is given by

$$E = \frac{T(1 - Tu)}{T + \mu}$$

where μ is the coefficient of friction and T the tangent of the pitch angle of the screw. For what value of T is the efficiency the greatest?

5. Find the two positive numbers whose sum is 60 and whose product is a maximum.

6. A rectangle has a perimeter of 8 cm. What should the dimensions be so that its area is a maximum?

7. A rectangular area adjacent to a river is to be enclosed by a fence on the other three sides. If the area enclosed is to be 200 m² and if no fencing is needed along the river, what dimensions require the least amount of fencing? (See Figure 21.19.)

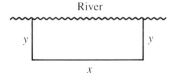

Figure 21.19

8. A rectangular field is to be enclosed by a fence and separated into two parts by a fence parallel to one of the sides. If 600 m of fencing is available, what should the dimensions be so that the area is a maximum? (See Figure 21.20.)

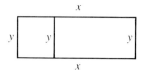

Figure 21.20

9. A box with an open top is to be made from a square piece of cardboard by cutting equal squares from the corners and turning up the sides. If the piece of cardboard measures 12 cm on the side, find the size of the squares that must be cut out to yield the maximum volume for the box. (See Figure 21.21.)

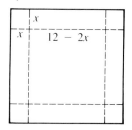

Figure 21.21

10. Find the point on the curve $y = x^2/4$ nearest the point $(1, 2)$. (*Hint:* If d is the distance from $(1, 2)$ to a point on the curve, minimize d^2 to avoid radicals.)

11. Find the point in the first quadrant on the curve $xy = 2$ nearest the origin.

12. Find the largest possible rectangle in the first quadrant such that its sides lie along the axes and it has one vertex on the curve $y = 9 - x^2$. (See Figure 21.22.)

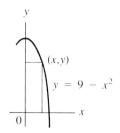

Figure 21.22

13. Find the largest possible rectangle which can be inscribed in a circle of radius 10.

14. By Kirchhoff's voltage law the sum of the voltages across the components of a circuit is equal to the applied voltage in the circuit. Consider the circuit with variable resistor R in Figure 21.23.

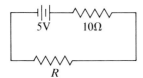

Figure 21.23

a. Find a relationship between I and R in the circuit, using $E = IR$ and Kirchhoff's voltage law.
b. The power to a resistor is given by I^2R watts. Write P as a function of R.
c. Find the setting of the variable resistor R so that it takes maximum power.

15. The total charge in an electrical circuit as a function of time is given by $q = t/(t^2 + 1)$ coulombs. Find the maximum charge q.

16. A closed box is to be a rectangular solid with a square base. If the enclosed volume is 32 in.^{3}, determine the dimensions for which the surface area is a minimum. (See Figure 21.24.)

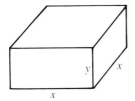

Figure 21.24

17. Repeat Exercise 16 for a box open at the top.

18. A wire of length 50 cm is to be cut into two pieces. One of the pieces is to be bent into the form of a circle and the other into the form of a square. How should the wire be cut so that the sum of the enclosed areas is a minimum?

19. A bus company will take 30 passengers on an excursion trip for $400 per passenger. If more than 30 passengers (up to 50) sign up, the company will reduce the price by $10 for every person above 30. (For example, if 32 passengers sign up, the ticket price is $380 per passenger for all 32 passengers.) What number will maximize the intake?

20. The manager of a store finds that if she charges $20 for an item, she can sell an average of 120 per week. For each $1 increase in price, the average number of sales per week drops by 4 units. What price should she charge for maximum revenue?

21. The *strength* of a beam with rectangular cross-section is directly proportional to the product of the width and the square of the depth (thickness from top to bottom of the beam). Find the shape of the strongest beam that can be cut from a cylindrical log of diameter $d = 3$ ft. (See Figure 21.25.)

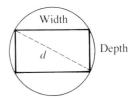

Figure 21.25

22. Show that a tin can having a fixed volume V will require the least amount of material if the height equals the diameter of the base. (*Hint:* Minimize $A = 2\pi rh + 2\pi r^2$ after eliminating h. The critical value is $r^3 = V/(2\pi)$; use this value to calculate $h/r = hr^2/r^3$.)

23. Repeat Exercise 22 for a cylinder with open top.

24. Suppose the area in Exercise 8 is divided by two fences parallel to one of the sides. What dimensions will maximize the area?

25. An arched window is in the shape of a rectangle surmounted by a semicircle. If the perimeter is 5 m, what should the radius of the semicircular part be if the window is to admit as much light as possible?

26. Find the altitude of a cone of maximum volume that can be inscribed in a sphere of radius r.

27. What is the altitude of a cylinder of maximum volume that can be inscribed in a right circular cone?

28. A property owner wants to build a rectangular enclosure around some land that is next to the lot of a neighbor who is willing to pay for half the fence that actually divides the two lots. If the area is A, what should the dimensions of the enclosure be so that the cost to the *owner* is a minimum?

29. In estimating the size n of a sample in statistics, one needs to evaluate $\sqrt{pq/n}$, where $q = 1 - p$ and $p\,(0 \le p \le 1)$ is the proportion of the sample possessing a certain property. (For example, p might be the proportion of students working part time.) Since p is not usually known, it has to be estimated from the sample taken. Alternatively, one can replace pq by its largest possible value to obtain the maximum sample size needed. Find p so that pq is a maximum.

30. For a certain manufacturer, the cost C of producing x machine parts is

$$C(x) = (2.0 \times 10^{-6})x^3 - 0.0015x^2 + 2.5x + 500$$

(in dollars). How many units should be produced to minimize the marginal cost? (The marginal cost is the instantaneous rate of change of the cost.)

31. A person in a boat 6 km from the nearest point on the shore wants to reach a point P on the shore 10 km from that nearest point. He can walk 5 km/h but row only at the

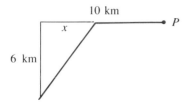

10 km
x
P
6 km

Figure 21.26

rate of 4 km/h. Determine the place where the boat must land if he wants to reach point P in the least possible time. (See Example 6 and Figure 21.26.)

32. A girl on a horse 4 km from a river wants to ride to the stable, which is 2 km from the river. The respective points on the bank nearest the horse and stable are 6 km apart. Where should she stop to water the horse in order to reach the stable in the least time? (See Example 6.)

33. A rectangular area adjacent to a wall is to be enclosed on the other three sides and separated into two parts by a fence perpendicular to the wall. If no fencing is needed along the wall and if the total enclosed area is to be 1200 ft^2, what should the dimensions be so that the total amount of fencing used is a minimum?

34. Suppose the area in Exercise 33 is separated into three parts by two fences perpendicular to the wall. If the total area is 1600 ft^2, what should the dimensions be so that the total amount of fencing used is a minimum?

21.5 Curvilinear Motion

So far all our functions have been of the form $y = F(x)$. Functions, and their corresponding graphs, can also be expressed as *parametric equations.*

Parametric Equations: Parametric equations represent the x and y coordinates of a curve as functions of a third variable t, called the **parameter**:

$$\left. \begin{array}{l} x = f(t) \\ y = g(t) \end{array} \right\} \tag{21.3}$$

Consider the set of parametric equations

$$\left. \begin{array}{l} x = t \\ y = t^2 \end{array} \right\}$$

Both coordinates are expressed as functions of t, so that each value of t determines the coordinates of a point. A few of these coordinates are listed in the following chart:

t:	-2	-1	0	1	2
x:	-2	-1	0	1	2
y:	4	1	0	1	4

Plotting these points and connecting them by a smooth curve, we obtain the graph in Figure 21.27. There is something familiar about the shape of this curve.

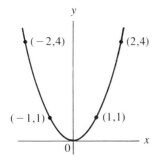

Figure 21.27

Indeed, we may eliminate t by algebraic means: square both sides of the first equation to obtain

$$\left. \begin{array}{l} x^2 = t^2 \\ y = t^2 \end{array} \right\}$$

It follows that $y = x^2$, the equation of a *parabola*.

It is easy to find the slope of a tangent line to a curve represented by parametric equations: the differentials of Equations (21.3) are

$$\left. \begin{array}{l} dx = f'(t)\,dt \\ dy = g'(t)\,dt \end{array} \right\}$$

so that

$$\frac{dy}{dx} = \frac{g'(t)\,dt}{f'(t)\,dt} = \frac{g'(t)}{f'(t)} \qquad (21.4)$$

provided, of course, that f and g are differentiable.

Formula (21.4) is particularly useful in the study of motion by means of vectors. Recall from Chapter 9 that a **vector** has both magnitude and direction. So if the parametric equations are the components of a vector and if the parameter t represents time, then the parametric equations may be interpreted as **curvilinear motion**, that is, the motion of a particle in a plane. Formula (21.4) then yields a method for finding the velocity and acceleration vectors.

E X A M P L E 1 The motion of a particle is described by the parametric equations

$$\left. \begin{array}{l} x = 1 - \dfrac{1}{2}t^3 \\[2mm] y = \dfrac{1}{\sqrt{t}} \end{array} \right\}$$

where x and y are in meters and t is in seconds. Find the velocity vector **v** when $t = 1$.

Solution. Since x and y are the coordinates of the position of the particle at any time, dx/dt and dy/dt are the components of the velocity. We denote these components by v_x and v_y, respectively. Thus

$$\left. \begin{array}{l} v_x = \dfrac{dx}{dt} = -\dfrac{3}{2}t^2 \\[3mm] v_y = \dfrac{dy}{dt} = -\dfrac{1}{2t^{3/2}} \end{array} \right\}$$

so that

$$\mathbf{v} = \left(-\frac{3}{2}t^2, \; -\frac{1}{2t^{3/2}} \right)$$

By formula (21.4)

$$\frac{dy}{dx} = \frac{v_y}{v_x}$$

which implies that the velocity vector is tangent to the curve. (The line determined by the velocity vector has the same slope as the tangent line to the curve.) At $t = 1$, $v_x = -\frac{3}{2}$ and $v_y = -\frac{1}{2}$, or

$$\mathbf{v} = \left(-\frac{3}{2}, \; -\frac{1}{2} \right)$$

When $t = 1$ the particle is located at $(\frac{1}{2}, 1)$. (The curve and the velocity vector **v** are shown in Figure 21.28.) The magnitude is now found to be

$$|\mathbf{v}| = \sqrt{(-\tfrac{3}{2})^2 + (-\tfrac{1}{2})^2} \approx 1.58 \text{ m/s}$$

For the angle α in the figure we have

$$\tan \alpha = \frac{v_y}{v_x} = \frac{-\frac{1}{2}}{-\frac{3}{2}} = \frac{1}{3}$$

or $\alpha = 18.4°$. Hence the direction is given by $\theta = 18.4° + 180° = 198.4°$. ◄

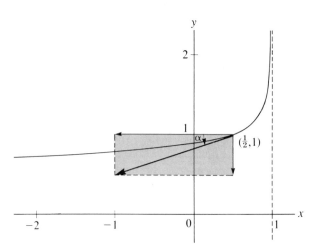

Figure 21.28

E X A M P L E **2**   Find the acceleration vector **a** at $t = 0.5$ s if the motion of the particle is described by the parametric equations

$$\left. \begin{array}{l} x = 1 - t^3 \\ y = 3(t - 1)^2 \end{array} \right\}$$

where x and y are measured in meters and t in seconds.

Solution. We let

$$a_x = \frac{dv_x}{dt} = -6t$$

and

$$a_y = \frac{dv_y}{dt} = 6$$

Thus

$$\mathbf{a} = (a_x, a_y) = (-6t, 6)$$

At $t = 0.5$ s, we have $a_x = -3$ and $a_y = 6$. So the magnitude is

$$|\mathbf{a}| = \sqrt{(-3)^2 + 6^2} = 3\sqrt{5} \approx 6.7 \text{ m/s}^2$$

The direction is given by

$$\tan \theta = \frac{6}{-3} = -2, \quad \theta \text{ in the second quadrant}$$

It follows that $\theta = 116.6°$. At $t = 0.5$ s, the particle is located at $(0.88, 0.75)$. (See Figure 21.29.)

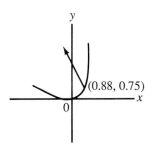

Figure 21.29

E X E R C I S E S / S E C T I O N   **21.5**

In Exercises 1–6, eliminate the parameter.

1. $x = 3t$, $y = t + 1$

2. $x = 4t^2$, $y = t^2 - 2$

3. $x = t^2 + 1$, $y = t$

4. $x = t^2$, $y = t - 1$

5. $x = \ln t$, $y = t + 2$

6. $x = e^{-t}$, $y = e^t$

In Exercises 7–12, find the velocity vector **v** and its magnitude and direction at the indicated value of t. (Assume x and y to be in meters and t in seconds.)

7. $x = (t - 3)^2$, $y = 2t$, $(t = 2)$

8. $x = \dfrac{2}{3}t^{3/2}$, $y = \dfrac{1}{3(t - 2)^3}$, $(t = 3)$

9. $x = \sqrt{t}$, $y = \dfrac{2}{\sqrt{t}}$, $(t = 4)$

10. $x = 3t^{1/3}$, $y = -\dfrac{3}{2}t^{2/3}$, $(t = 8)$

11. $x = 1 - 5t$, $y = 4\sqrt{t - 1}$, $(t = 5)$

12. $x = 1 - 3t^2$, $y = -\sqrt{t}$, $(t = 2)$

In Exercises 13–16, find the magnitude and direction of the velocity and acceleration vectors at the indicated value of t. (Assume x and y to be in meters and t in seconds.)

13. $x = 3t^2 - t$, $y = 3 - t^3$, $(t = 2)$

14. $x = 2t - t^3$, $y = 3t - 5t^2$, $(t = 1)$

15. $x = \dfrac{2}{\sqrt{t}}$, $y = 9\sqrt[3]{t}$, $(t = 1)$

16. $x = \dfrac{8}{t}$, $y = \dfrac{16}{t}$, $(t = 2)$

17. If a projectile is fired from the ground with velocity $\mathbf{v_0}$, where $\mathbf{v_0}$ makes an angle θ with the ground, then the horizontal and vertical motions are completely independent of each other and are given by

$$x = v_0 t \cos \theta, \qquad y = v_0 t \sin \theta - 5t^2$$

($v_0 = |\mathbf{v_0}|$, x and y in meters and t in seconds). If $v_0 = 40$ m/s and $\theta = 30°$, find the magnitude and direction of the velocity and acceleration vectors at the end of the first second.

21.6   Related Rates

In this section we are going to continue our study of rates of change with respect to time, but instead of confining ourselves to motion along a curved path, we will consider rates of change in a general setting. As an example, suppose the radius r of a circle is allowed to expand at some known rate, say 2 cm/min. Since $A = \pi r^2$, it ought to be possible to find the rate of change of A in terms of cm^2/min. To see how this can be done, we must emphasize that in this and all other problems in this section *every variable quantity is a function of time*, so that all derivatives are taken with respect to t. In our example, the equation $A = \pi r^2$ can be written

$$A(t) = \pi [r(t)]^2$$

Differentiating with respect to t, we get

$$\frac{d}{dt} A(t) = \pi \cdot 2[r(t)] \frac{d}{dt} r(t) \qquad \textbf{power rule}$$

If we remember that A and r are *functions of time*, we can get a simpler expression from the original formula $A = \pi r^2$:

$$A = \pi r^2 \qquad \textbf{given formula}$$

$$\frac{dA}{dt} = \pi \cdot 2r \frac{dr}{dt} \qquad \textbf{power rule}$$

$$\frac{dA}{dt} = 2\pi r \frac{dr}{dt}$$

Substituting the known rate of change $dr/dt = 2$, we get

$$\frac{dA}{dt} = 4\pi r$$

So if $r = 1$ cm,

$$\frac{dA}{dt} = 4\pi \ \frac{\text{cm}^2}{\text{min}}$$

Since the rates of change are related, this type of problem is referred to as a problem in **related rates**.

> In a problem in **related rates**, one or more rate is given and another, related rate has to be found.

E X A M P L E **1** An experimenter has determined that the relationship between the tensile strength (in pounds) of a piece of material and the temperature is

$$S = 620 - 0.08\sqrt{T}$$

If the temperature is increasing at the rate of $0.2°\text{F/min}$, how fast is the tensile strength changing when $T = 100°\text{F}$?

Solution. We are given that $dT/dt = 0.2$ and are asked to find dS/dt when $T = 100$. To this end we differentiate both sides of the equation with respect to t. We have

$$S = 620 - 0.08T^{1/2}$$

and

$$\frac{dS}{dt} = (-0.08)\frac{1}{2} T^{-1/2} \frac{dT}{dt} \qquad \frac{d}{dt}[T(t)]^{1/2} = \frac{1}{2}[T(t)]^{-1/2}\frac{dT(t)}{dt}$$

by the power rule. Simplifying, we have

$$\frac{dS}{dt} = \frac{-0.04}{\sqrt{T}} \frac{dT}{dt}$$

Substituting $dT/dt = 0.2$ and $T = 100$, it follows that

$$\frac{dS}{dt} = \frac{-0.04}{\sqrt{100}}(0.2) = \frac{(-0.04)(0.2)}{10} = -0.0008 \text{ lb/min}$$

(The negative sign indicates that the tensile strength is decreasing.)   ◀

Before forming a general plan, consider another example.

E   X   A   M   P   L   E   **2**   The relationship between the resistance in a wire and the temperature is

$$R = a + bT + cT^2 + \cdots$$

For temperatures that are not too great, the terms in T^3 and higher powers may be neglected. Suppose for a certain wire that

$$R = 10.123 + 1.320T + 0.00400T^2$$

If T increases at the rate of $2.00°C/\min$, how fast is R increasing when $T = 20.0°C$?

Solution. As before, we differentiate with respect to t:

$$\frac{dR}{dt} = 1.320\frac{dT}{dt} + (0.00400)(2T)\frac{dT}{dt}$$

Substituting $dT/dt = 2.00$ and $T = 20.0$, we find that

$$\frac{dR}{dt} = 1.320(2.00) + (0.00400)(2)(20.0)(2.00) = 2.96 \text{ } \Omega/\min$$

to three significant digits.   ◀

General Strategy for Solving Problems in Related Rates: Whenever possible:

1. Draw a diagram.
2. Label all quantities that vary with time by letters.
3. Label all numerical quantities—that is, those quantities that remain fixed throughout the problem.
4. Obtain a relationship (equation) between the variables involved.
5. List all given rates of change; state the desired rate of change, as well as the instant at which it is to be found.
6. Differentiate with respect to t.
7. Substitute the known quantities and solve for the unknown rate of change.

E X A M P L E **3** Boyle's law states that for an ideal gas the pressure is inversely proportional to the volume at a constant temperature; that is,

$$P = \frac{k}{V}$$

If a certain gas occupies 3.0 m³ when the pressure is 25 Pa (N/m²) and the volume is increasing at the rate of 0.10 m³/min, how fast is the pressure decreasing at the instant when $V = 4.0$ m³?

Solution. In this problem a diagram is not needed. The constant k in the formula can be calculated from the given information: substituting $V = 3.0$ and $P = 25$, we get

$$25 = \frac{k}{3.0} \quad \text{and} \quad k = 75$$

Thus

$$P = 75V^{-1}$$

We are given that $dV/dt = 0.10$ and must find dP/dt when $V = 4.0$. Differentiating, we get

$$\frac{dP}{dt} = 75(-V^{-2})\frac{dV}{dt} = -\frac{75}{V^2}\frac{dV}{dt}$$

Since $dV/dt = \mathbf{0.10}$ and $V = \mathbf{4.0}$, we get

$$\frac{dP}{dt} = -\frac{75}{(\mathbf{4.0})^2}(\mathbf{0.10}) = -0.47$$

We conclude that the pressure is decreasing at the rate of 0.47 Pa/min at the instant when $V = 4.0$ m³. ◀

E X A M P L E **4** A metal cube contracts when it is cooled. If the edge of the cube decreases at the rate of 1.0 mm/h, how fast is the volume decreasing at the instant when the edge is 50 mm long?

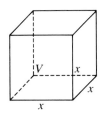

Figure 21.30

Solution. In the diagram (Figure 21.30) we label the edge x and the volume V, since both quantities change continuously. (The length 50 mm does not appear in the diagram since x is not a constant.) We are given that $dx/dt = -1.0$ (since x is decreasing) and are asked to find dV/dt when $x = 50$. From

$$V = x^3$$

we get

$$\frac{dV}{dt} = 3x^2 \frac{dx}{dt}$$

Since $dx/dt = -1.0$, we get

$$\frac{dV}{dt} = 3x^2(-1.0) = -3.0x^2$$

Finally, letting $x = 50$, we get

$$\frac{dV}{dt} = (-3.0)(2500) = -7500 \text{ mm}^3/\text{h}$$

◀

E X A M P L E **5**   A woman is walking at the rate of 4.0 ft/s toward a tower 89.6 ft high. Determine the rate of change of the distance to the top of the tower at the instant when she is 50.0 ft from the base.

Solution. In the diagram (Figure 21.31) we label the distance to the base x and the distance to the top z. These are the variable quantities. The height of the tower does not change and is therefore labeled 89.6 ft. In the language of the diagram, we are given that

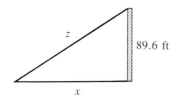

89.6 ft

Figure 21.31

$$\frac{dx}{dt} = -4.0 \qquad \text{(since } x \text{ is decreasing)}$$

and are asked to find dz/dt when $x = 50.0$. The distance 50.0 ft does not appear in the diagram since x is not a constant.

By the Pythagorean theorem,

$$x^2 + (89.6)^2 = z^2$$

Differentiating, we get

$$2x\frac{dx}{dt} + 0 = 2z\frac{dz}{dt}$$

or

$$x\frac{dx}{dt} = z\frac{dz}{dt}$$

Substituting the given rate of change ($dx/dt = -4.0$), we get

$$x(-4.0) = z\frac{dz}{dt}$$

or

$$\frac{dz}{dt} = -4.0\,\frac{x}{z}$$

This is the general expression for dz/dt. To obtain the rate of change at the instant in question, we substitute the values for x and z. From the diagram (Figure 21.32)

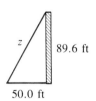

Figure 21.32

$$z^2 = (89.6)^2 + (50.0)^2 \qquad \text{or} \qquad z = 103$$

Hence $z = 103$ ft at the instant when $x = 50.0$ ft. Substituting $x = 50.0$ and $z = 103$, we get

$$\frac{dz}{dt} = -4.0\left(\frac{50.0}{103}\right) = -1.9 \text{ ft/s}$$

(The negative sign shows that z is decreasing.)

Caution. A common error in this kind of problem is premature use of the quantity $x = 50.0$. In most problems on related rates, the desired rate of change is not constant. In our example, we first found a general expression for dz/dt in terms of the variable quantities x and z. From the general expression we obtained the rate of change at the instant in question. As a result, the substitution for x and z was not made until after the differentiation was completed.

◀

E X A M P L E **6**   A chemical is poured into a tank in the shape of a cone with vertex down at the rate of $\frac{1}{5}$ m³/min. If the radius of the cone is 3 m and its height 5 m, how fast is the liquid level rising at the instant when the depth of the liquid is 1 m in the center?

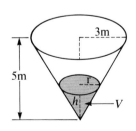

Figure 21.33

Solution   We first draw the diagram in Figure 21.33 and label the variable quantities V, r, and h. The dimensions of the cone do not change and are labeled with constants. The distance $h = 1$ does not appear in the diagram since h is not constant. In the language of the diagram, we are given that $dV/dt = \frac{1}{5}$ and wish to find dh/dt when $h = 1$.

It is desirable to express the volume of the liquid as a function of h. To this end we make use of similar triangles. Note that

$$\frac{r}{h} = \frac{3}{5}$$

so that $r = \frac{3}{5}h$. It follows that

$$V = \tfrac{1}{3}\pi r^2 h = \tfrac{1}{3}\pi(\tfrac{3}{5}h)^2 h \qquad \text{since } r = \tfrac{3}{5}h$$

and

$$V = \frac{3\pi}{25} h^3$$

Differentiating, we get

$$\frac{dV}{dt} = \frac{3\pi}{25} (3h^2) \frac{dh}{dt}$$

We now substitute the given rate of change:

$$\frac{1}{5} = \frac{3\pi}{25} (3h^2) \frac{dh}{dt} \qquad \text{or} \qquad \frac{dh}{dt} = \frac{1}{5} \left(\frac{25}{3\pi}\right) \frac{1}{3h^2} = \frac{5}{9\pi h^2}$$

Finally, when $h = 1$,

$$\frac{dh}{dt} = \frac{5}{9\pi} \frac{m}{min}$$

◀

E X E R C I S E S / S E C T I O N **21.6**

In Exercises 1–4, the respective units are feet and seconds.

1. $y = x^2$. Given: $dx/dt = 1$; find: dy/dt when $x = 2$.

2. $y = x^3 + 2$. Given: $dx/dt = 2$; find: dy/dt when $x = 1$.

3. $x^2 + y^2 = 25$. Given: $dy/dt = 3$; find: dx/dt when $y = 3$ $(x > 0)$.

4. $w^2 = x^2 + y^2$. Given: $dx/dt = dy/dt = 2$; find: dw/dt when $x = 5$ and $y = 12$ $(w > 0)$.

5. Suppose that $E = 100$ V in a DC circuit and that a rheostat is used to increase R at the rate of $2\ \Omega/s$. Find the resulting rate of change of I $(I = E/R)$ when $R = 10\ \Omega$.

6. The impedance Z in a series circuit is given by $Z^2 = R^2 + X^2$, where X is the reactance. If $X = 10.0\ \Omega$ and R increases at the rate of $2.0\ \Omega/s$, find the rate at which Z is changing when $R = 5.0\ \Omega$.

7. If two variable resistances R_1 and R_2 are linked in parallel, then the effective resistance R of the combination is such that $1/R = 1/R_1 + 1/R_2$. If R_1 increases at the rate of $0.33\ \Omega/s$ and R_2 at $0.25\ \Omega/s$, what is the rate of change of R at the instant that $R_1 = 2.00\ \Omega$ and $R_2 = 3.00\ \Omega$?

8. The power, in watts, in a circuit varies according to $P = Ri^2$. If $R = 100\ \Omega$ and i varies at the rate of 0.20 A/s, find the rate of change of the power when $i = 2.0$ A.

9. Suppose a ladder 5 m long is leaning against a wall. If the bottom of the ladder is being pulled away at the rate of 2 m/min, how fast is the top of the ladder slipping

down at the instant that the foot of the ladder is 4 m from the wall? (See Figure 21.34.)

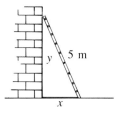

Figure 21.34

10. A balloon is rising from the ground at the rate of 1.5 m/s from a point 30 m from where an observer is standing. How fast is the balloon receding from the observer when it is 40 m above the ground? (See Figure 21.35.)

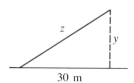

Figure 21.35

11. A balloon is being inflated at the rate of $20.0\ \text{cm}^3/\text{min}$. Find the rate at which the radius is increasing when it is 10.0 cm. (Assume the balloon is a sphere.)

12. A balloon is being inflated in such a way that its radius is increasing at the rate of 1.00 cm/s. How fast is the volume increasing when the radius is 5.00 cm? (Assume the balloon is a sphere.)

13. The adiabatic law (explaining changes in a system that can occur with no change in temperature) for the expansion of a diatomic gas is $PV^{1.4} = k$, where P is the pressure, V the volume of the container, and k a constant. At a certain instant the volume is 2.0 m^3 and decreasing at the rate of 1.0 m^3/min, and the pressure is 76 Pa (N/m^2). Find the rate at which the pressure is changing at this instant.

14. The natural frequency f in an LC circuit is given by

$$ f = \frac{1}{2\pi} \sqrt{\frac{1}{LC}} $$

If $L = 1.00 \times 10^{-2}$ H and C increases at the rate of 1.00×10^{-6} F/s, find the rate of change of f when $C = 1.00 \times 10^{-2}$ F.

15. A student is walking at the rate of 5.0 ft/s toward the Margaret Loock Residence Hall, which is 123 ft high. How fast is the distance to the top of the building changing when he is 60.0 ft from the base of the building?

16. A circular metal plate is being heated and expands so that its radius increases at the rate of 0.25 mm/min. How fast is the area increasing when the radius is 10 cm?

17. An airplane is cruising at 350 km/h at an altitude of 4000 m. If the plane passes directly over an observer on the ground, how fast is the distance from the plane to the observer changing when it is 5000 m away from the observer?

18. A spherical snowball is melting at the rate of 100 cm^3/min. How fast is the radius shrinking when the radius is 10 cm?

19. Two ships are leaving port at the same time. The first ship is sailing due east at 20 km/h and the other due north at 15 km/h. How fast are the ships moving away from each other 2.0 h later?

20. A ladder 4 m long is leaning against a wall. If the top of the ladder is slipping down at the rate of 2 m/s, how fast is the bottom moving away from the wall when it is 3 m from the wall?

21. A point moves along the parabola $y^2 = x$ in such a way that the abscissa is decreasing at the rate of 2 units/min. Find the rate of change of the ordinate at the point (16, 4).

22. A point moves along the hyperbola $x^2 - y^2 = 4$. The ordinate is increasing uniformly at 3 units/s. Find the rate of change of the abscissa at $(-3, \sqrt{5})$.

23. A boat is being pulled toward a wharf by a rope attached to the boat's deck from a point 8.0 m above the deck. If the rope is being pulled in at the rate of 2.0 m/min, how fast is the boat approaching the wharf when it is 12.0 m away?

24. Sand is being poured onto a pile forming a cone whose radius and height are equal at all times. If the sand is poured at the rate of 50 cm^3/s, what is the rate of change of the radius when the pile is 3.0 m high?

25. Wheat is poured on the ground at the rate of 12 ft^3/s. The pile forms a cone whose altitude is always three-fourths of the radius. Find the rate at which the altitude is increasing when the altitude is 6.0 ft.

26. At noon, ship A is 129 km due north of ship B. Ship A is moving due west at 18 km/h and ship B due north at 25 km/h. Find the rate of change of the distance between them at 3 P.M.

27. A baseball player is coming in from third base at 24 ft/s. What is the rate of change of his distance from second base when he is 25 ft from home plate? A baseball diamond is a 90-ft square.

28. The radius r of a cylinder of height 5.0 cm shrinks at the rate of 0.050 cm/min. How fast is the volume shrinking when $r = 3.0$ cm?

29. A man 170 cm tall walks away from a 3.0-m-high street light at the rate of 2.0 m/s. Find the rate at which his shadow is growing.

30. A storage tank is in the shape of a cone with vertex down. Suppose the altitude of the cone is 10.0 m and the radius of the base is 5.0 m. If water is poured into the tank at the rate of 3.0 m^3/min, find the rate at which the water is rising when the water is 4.0 m deep in the center.

31. A conical tank with vertex down has a radius of 6.0 ft and a height of 8.0 ft. Water is poured into the tank at the rate of 9.0 ft^3/min. Find the rate at which the level is increasing at the instant when the water is 2.0 ft deep in the center.

32. A cone has a radius of 10.0 cm. If the altitude of the cone is decreasing at the rate of 2.0 cm/min, find the rate of change of the lateral surface when the altitude is 20.0 cm.

33. A trough 10.0 m long has a triangular cross-section 5.0 m across the top and 4.0 m deep. If water is poured in at the rate of 2.0 m^3/min, how fast is the water level rising when the water is 3.0 m deep?

34. If two ships leave a point at the same time and travel in mutually perpendicular directions at respective velocities a and b, show that they move apart at the constant rate of $\sqrt{a^2 + b^2}$.

21.7   Differentials

In this section we are going to return to the Leibniz notation $dy/dx = f'(x)$ to see in what sense dy/dx may be viewed as a quotient. The question leads quite naturally to the definition of differential. While the concepts studied in calculus could be developed without the differentials, their use has become so common, particularly in physical science, that we need to familiarize ourselves with the basic ideas.

Formally solving $dy/dx = f'(x)$ for dy, we have $dy = f'(x)\,dx$, called the **differential** of f.

By itself the formula $dy = f'(x)\,dx$ has no obvious meaning. But suppose we choose to interpret dx as an increment, usually denoted by Δx. Then it becomes possible to interpret dy geometrically and to establish a connection between Δy and dy. Consider the graph of $y = f(x)$ in Figure 21.36.

The slope of the tangent line at $P(x, y)$ is given by $f'(x)$. Since dx is a finite quantity, we can see from the figure that $f'(x) = dy/dx$, or $dy = f'(x)\,dx$. Thus dy is a vertical distance. Furthermore, $dy \approx \Delta y$ if dx is relatively small. The distinction between dy/dx and dy is now seen: dy/dx is the instantaneous *rate* of change of y with respect to x, while dy is the *amount* of change (approximately equal to Δy).

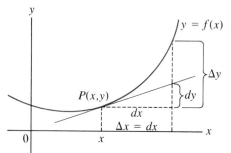

Figure 21.36

Definition of Differential:

$$dy = f'(x)\,dx \qquad \text{(where } dx = \Delta x)\qquad\qquad(21.5)$$

E X A M P L E **1**   If $y = x^2$, compute Δy and dy if $x = 2$ and $dx = \Delta x = 0.01$.

Solution. Recall that

$$\Delta y = f(x + \Delta x) - f(x)$$

Hence

$$\Delta y = (x + \Delta x)^2 - x^2 = (2 + 0.01)^2 - 2^2$$
$$= (2.01)^2 - 2^2 = 4.0401 - 4 = \mathbf{0.0401}$$

In other words, when x increases from 2 to 2.01, the value of y increases from 4 to 4.0401, an increase of 0.0401 unit. Now,

$$dy = f'(x)\,dx = 2x\,dx = 2(2)(0.01) = \mathbf{0.04}$$

Thus dy is indeed a good *approximation* for Δy and is easier to calculate than Δy.   ◄

We now turn to the usual applications of the differential.

E X A M P L E **2**   Resistance is known to vary with temperature. Suppose that for a certain resistor, $R = 3.5 + 0.002T^2$ Ω. If T is measured to be 100°C with a possible error of ± 1°C, what is the approximate maximum possible error in R?

Solution. Here we simply interpret the error in T as an increment dT. The resulting error in R is ΔR, or approximately dR. Since

$$\frac{dR}{dT} = 0.004T$$

we get for the differential

$$dR = 0.004T\, dT$$

We substitute the given values for T and dT to find that

$$\Delta R \approx dR = (0.004)(100)(\pm 1) = \pm 0.4\ \Omega \qquad \blacktriangleleft$$

In many cases the actual error is not as important as the size of the error relative to the quantity being measured. To determine the seriousness of the error, we find the quantity dy/y, called the *approximate relative error*, often expressed as a percentage.

Relative error

E X A M P L E **3**   In Example 2 the actual error was found to be approximately 0.4 Ω. At $T = 100$, $R = 23.5$. So the approximate relative error is

$$\frac{dR}{R} = \frac{0.4}{23.5} = 0.017 = 1.7\% \qquad \blacktriangleleft$$

E X A M P L E **4**   A protective coat of thickness 0.5 mm is applied evenly to the surface of a metal sphere of radius 20.00 cm. Find the approximate number of cubic centimeters of coating used.

Solution. Since $V = \frac{4}{3}\pi r^3$,

$$dV = 4\pi r^2\, dr$$

We let $dr = 0.05$ cm and $r = 20.00$ cm. Then

$$dV = 4\pi(20.00)^2(0.05) = 251\ \text{cm}^3$$

which is the approximate amount of coating used. $\qquad \blacktriangleleft$

E X E R C I S E S / S E C T I O N   21.7

In Exercises 1–4, find the differential of the given functions.

1. $y = x^3 - x$

2. $y = x^2 - \dfrac{1}{x}$

3. $y = \dfrac{x}{x - 1}$

4. $y = \dfrac{1}{\sqrt{x^2 + 1}}$

In Exercises 5 and 6, find Δy and dy for the given values of x and Δx.

5. $y = x^2 - x,\ x = 2,\ dx = 0.1$

6. $y = 1/x^2,\ x = 3,\ dx = 0.2$

7. Find the approximate error and percentage error in the area of a square 6.00 cm on a side if an error of 0.02 cm is made in measuring the edge.

8. The side of a square is measured to be 5.00 inches with an error of 0.04 inch. Find the approximate error and percentage error in the area.

9. The edge of a cube is measured to be 5.00 inches with a possible error of ± 0.01 inch. Find the approximate error and percentage error in the volume.

10. The edge of a cube is measured to be 6.00 inches with a possible error of ± 0.03 inch. Find the approximate error and percentage error in the surface area.

11. Find the approximate percentage error in the volume and the surface of a sphere of radius 10.00 cm, if an error of ± 1 mm is made in measuring the radius.

12. The radius of a sphere is measured to be 3.40 m with an error of ± 0.05 m. Find the approximate error and percentage error in the volume.

13. The period of a simple pendulum is given by

$$T = 2\pi \sqrt{\frac{L}{10}}$$

where L is in meters. If $L = 2.0$ m, with an error in measurement of ± 0.1 m, what is the approximate error and percentage error in T?

14. A DC circuit has a constant voltage source of 20 V. Find the approximate change in the current if the resistance changes from 8.0 to 8.1 Ω.

15. The power P (in watts) delivered to a resistor is $P = 10.0i^2$. If i changes from 2.1 A to 2.2 A, determine the approximate change in P.

16. The resistance in a certain wire is $R = 60.0 + 0.020T^2$. If T changes from $50.4°F$ to $50.7°F$, find the approximate change in R.

17. Find the approximate formula for the area of a circular ring of radius r and width dr.

18. Let A be the area of a square of side s. Draw a figure showing the square, dA, and ΔA.

R E V I E W   E X E R C I S E S / C H A P T E R   21

1. Find the equations of the tangent and normal lines to the curve $y = \sqrt{x - 2}$ at the point $(3, 1)$.

2. Find the equations of the tangent and normal lines to the hyperbola $x^2 - 3y^2 + 23 = 0$ at $(2, -3)$.

In Exercises 3–10, find the minima and maxima, the intervals on which the graph is concave up and concave down, and the inflection points. Sketch the curve in each case.

3. $y = x^2 - 4x + 3$

4. $y = x^3 - 6x^2 + 9x$

5. $y = -x^3 + 12x + 2$

6. $y = 3x^4 - 8x^3 + 9$

7. $y = 3x^4 - 4x^3 + 1$

8. $y = \dfrac{x^2}{x^2 - 1}$

9. $y = x^2 - \dfrac{1}{x}$

10. $y = \dfrac{x^2 - 1}{x^3}$

11. The velocity of air through a bronchial tube under pressure is given by $v = kr^2(a - r)$, where a is the radius of the tube when no pressure is applied and k is a constant. Find the radius of the tube for which the velocity is a maximum. (Experiments by means of x-ray photographs have shown that the velocity is indeed maximized during a cough.)

12. A right circular cylinder is inscribed in a sphere of radius a. Show that its maximum volume is $1/\sqrt{3}$ times the volume of the sphere.

13. The *stiffness* of a beam with rectangular cross-section is directly proportional to the product of the width and the cube of the depth. Show that the ratio of depth to width of the stiffest beam that can be cut from a circular log is $\sqrt{3}$.

14. A farmer has 240 ft of fence to build four adjacent rectangular pig pens. The pens are to be constructed as follows: a large rectangular enclosure is subdivided into four parts by three dividing fences parallel to one of the sides. What should the overall dimensions be in order to maximize the total area?

15. Recall that if $C(x)$ is the cost of producing x units of a certain commodity, then $C'(x)$ is the *marginal cost*. Similarly, if $R(x)$ is the revenue derived from the sale of x units, then $R'(x)$ is the *marginal revenue*. The profit is $P(x) = R(x) - C(x)$. Show that for a company to realize maximum profit, the marginal revenue must equal the marginal cost.

16. Assume that the rate at which a rumor spreads through a company is directly proportional to the product of the number of people who have heard the rumor and the number of those who have not. Show that the rumor spreads most rapidly when half the people have heard it.

17. A nuclear plant is located on the edge of a straight river 20 m wide. An electric cable is to be run to a factory on the opposite edge, 50 m downstream. If it costs \$30 per meter to run the cable on land and \$45 per meter across the river, how should the cable be laid to minimize the cost?

18. A messenger is in a motorboat 24 mi from a straight shore. He wants to reach a point P that is 150 mi along the shore from a point on the shore nearest the boat. His boat can travel at the rate of 11 mi/h and, upon reaching the shore, he will travel by car at 55 mi/h. Where should he land to reach P in minimum time?

19. The electromotive force E in volts that produces a current of I amperes in a wire of diameter d inches is given by $E = 0.1138I/d^2$. If $d = 0.060$ inch and E is increasing at the rate of 3.00 V/s, find the time rate of change of I.

20. The thin-lens equation $1/s_1 + 1/s_2 = 1/f$ relates the distances s_1 and s_2 from an object and its image to the lens. The constant f is called the *focal length*. An object is moving away from a lens with focal length 30.0 cm at the rate of 15.0 cm/min. How fast is the image moving when the object is 90.0 cm from the lens?

21. Water is entering a conical tank 10 m deep and 20 m across the top at 9 m³/min. How fast is the water rising when 3 m deep?

22. The ends of a trough are equilateral triangles with vertex down. The trough is 9 m long. When the water in the trough is 2 m deep, its depth increases at the rate of $\frac{1}{2}$ m/min. At what rate is water being poured in at that instant?

23. The volume V and the pressure P in the cylinder of a diesel engine are related by the equation $PV^{1.4} = k$, where k is a constant. At a certain instant during the compression stroke the cylinder contains 40 in³ of gas vapor under a pressure of 300 lb/in² and the volume is decreasing at the rate of 80 in³/s. At what rate is the pressure increasing at that instant?

24. If $y = x - \sqrt{x}$, compute Δy and dy if $x = 4$ and $dx = \Delta x = 0.02$.

25. The edge of a cube is measured to be 10.00 cm with a possible error of ± 0.2 mm. Find the approximate error and percentage error in the volume.

26. A spherical balloon has a radius of 2.0 ft with a possible error of ± 0.1 ft. What is the approximate percentage error in the volume?

27. The radius of a circle is measured to be 5.00 inches with an error of ± 0.02 inch. Find the approximate error and percentage error in the area.

In Exercises 28 and 29, find the magnitude and direction of the velocity and acceleration vectors at the indicated value of t. (Assume x and y to be in meters and t in seconds.)

28. $x = 4\sqrt{t + 2}$, $y = 1 - \dfrac{1}{6}t^3$, $(t = 2)$

29. $x = -3\sqrt[3]{t}$, $y = 2t^2 + 2$, $(t = 1)$

1. Find the equation of the line passing through $(-2, 1)$ and perpendicular to the line determined by $(3, -2)$ and $(-2, 1)$.

2. Show that the lines $4x - 6y - 7 = 0$ and $3x + 2y + 12 = 0$ are perpendicular.

3. Find the distance from the point $(-3, 4)$ to the center of the circle $x^2 + y^2 - 4x - 6y - 12 = 0$.

4. Find the vertices and foci of the ellipse $3x^2 + y^2 = 9$, and sketch.

5. Find the equation of the parabola whose vertex is at $(1, 0)$ and whose focus is at $(-3, 0)$.

6. Identify the following conic, find its center, and sketch:

$$x^2 - 2y^2 - 2x - 4y - 3 = 0$$

7. Write the equation of the hyperbola $x^2 - y^2 = 4$ in polar coordinates.

In Exercises 8 and 9, find the limits indicated.

8. $\lim\limits_{x \to 4} \dfrac{x^2 - 2x - 8}{x - 4}$

9. $\lim\limits_{x \to \infty} \dfrac{2x^2 - 3}{3x^2 - x + 2}$

10. Find the derivative of $y = 1/(x^2 - 4)$ by the four-step process.

In Exercises 11–13, differentiate each of the functions and simplify.

11. $y = 2\sqrt{1 - 3x^2}$

12. $y = x\sqrt{x + 2}$

13. $y = \dfrac{\sqrt{x}}{x + 3}$

14. **a.** Find dy/dx implicitly: $x^2y - 3xy^2 + 2 = 0$.
    b. Find the equation of the line normal to the curve at $(1, 1)$.

15. Find the maximum and minimum points on the curve $y = 3x^3 - 9x + 2$.

16. Find the maxima and minima, the intervals of concavity, and the inflection points for the following curve and sketch:

$$y = 3x^4 - 4x^3 + 2$$

17. A searchlight with a parabolic surface has a diameter of 2 ft and a depth of 1 ft. Where should the light source be placed to produce a beam of parallel rays?

18. The path of the planet Mercury in polar coordinates is

$$r = \frac{3.442 \times 10^7}{1 - 0.206 \cos \theta}$$

Write this equation in rectangular coordinates.

19. Identify the following curve and sketch: $r = 2(1 + \cos \theta)$.

20. Find the voltage across an inductor of 3.00×10^{-4} H after 1 s if the current is $3.00t^2$ A (for a short time period).

21. A chemical tank is to be constructed with a flat square base, vertical walls, and an open top. If 108 ft^2 of material is to be used, what should the dimensions be so that the volume is a maximum?

22. A spherical metal ball contracts due to cooling. If the radius decreases at the rate of 0.010 in./min, how fast is the volume decreasing when the radius is 8.5 in.?

23. The motion of a particle can be described by the parametric equations

$$x = \frac{2}{t}, \qquad y = \frac{1}{6}t^3$$

Find the magnitude and direction of the velocity vector at $t = 2$. (Assume x and y to be in meters and t in seconds.)

24. The radius of a circle was measured to be 2.50 cm with an error of ± 0.20 cm. Use the differential to determine the approximate percentage error in the circumference.

C H A P T E R **22**

The Integral

22.1 Antiderivatives

Antiderivative

So far we have been concerned primarily with the following problem: given a function, find its derivative. We will now consider the inverse problem: given the derivative of a function, find the function—or, in symbols, given f, find F such that $F' = f$. F is called the **antiderivative** of f. To see how antiderivatives may arise, recall that if v is the velocity of a particle, then the acceleration is $a = dv/dt$. Now g, the acceleration due to gravity, is 32 ft/s^2 or 10 m/s^2. (A more precise value is 9.8 m/s^2.) It follows that the velocity in meters per second is the antiderivative $v = 10t$, since $dv/dt = 10$ m/s^2.

Antiderivatives are not unique. If $a = 10$ m/s^2, then $v = 10t$, as already noted. However, $v = 10t + 1$ and $v = 10t + 3$ are also antiderivatives. In fact, $v = 10t + C$, for an arbitrary constant C, is an antiderivative. (The physical significance of the arbitrary constant will be discussed in Section 22.7.)

Even though antidifferentiation is, in general, much more complicated than differentiation, simple functions can be handled routinely. For example, an antiderivative $F(x)$ of $f(x) = x^2$ has to be a cubic, but since $dx^3/dx = 3x^2$, we must have $F(x) = \frac{1}{3}x^3 + C$. Thus $(d/dx)F(x) = x^2$, as desired. In general, then, if

$$f(x) = x^n \qquad n \text{ rational}$$

then

$$F(x) = \frac{x^{n+1}}{n+1} + C \qquad n \neq -1 \tag{22.1}$$

where C is an arbitrary constant. It is a simple exercise to show that $F'(x) = f(x)$. It follows that for any constant k, if

$$f(x) = kx^n \qquad n \text{ rational}$$

697

then

$$F(x) = \frac{kx^{n+1}}{n+1} + C \qquad n \neq -1 \tag{22.2}$$

Finally, if $f(x) = k$, then

$$F(x) = kx + C \tag{22.3}$$

E X A M P L E **1**   If $f(x) = 2x^2 + x^3$, find an antiderivative $F(x)$.

Solution. By formulas (22.1) and (22.2) we get

$$F(x) = 2 \cdot \frac{x^3}{3} + \frac{x^4}{4} + C = \frac{2x^3}{3} + \frac{x^4}{4} + C$$

As a check, observe that $F'(x) = f(x)$. ◀

E X A M P L E **2**   Find $F(x)$ if $f(x) = 3x^4 + x + 2$.

Solution. In this example, we also need formula (22.3):

$$F(x) = \frac{3x^5}{5} + \frac{x^2}{2} + 2x + C$$ ◀

E X A M P L E **3**   Find $F(x)$ if $f(x) = \sqrt{x} - (2/x^2) - 6$.

Solution. We write

$$f(x) = x^{1/2} - 2x^{-2} - 6$$

and obtain

$$F(x) = \frac{x^{3/2}}{\frac{3}{2}} - \frac{2x^{-1}}{-1} - 6x + C = \tfrac{2}{3}x^{3/2} + \tfrac{2}{x} - 6x + C$$ ◀

E X A M P L E **4**   Find $F(x)$, given that $f(x) = x^2\sqrt{x} - (1/\sqrt[3]{x})$.

Solution. Again the function has to be rewritten:

$$f(x) = x^2 x^{1/2} - \frac{1}{x^{1/3}} = x^{5/2} - x^{-1/3}$$

Hence

$$F(x) = \frac{x^{7/2}}{\frac{7}{2}} - \frac{x^{2/3}}{\frac{2}{3}} + C = \tfrac{2}{7}x^{7/2} - \tfrac{3}{2}x^{2/3} + C$$ ◀

EXERCISES / SECTION **22.1**

Find the antiderivatives of the following functions.

9. $f(x) = x^5 - 6x^4 + 2x^3 + 3$

10. $f(x) = x^{-3} + 2x^{-2} + 4$

1. $f(x) = 3$

2. $f(x) = 2x$

3. $f(x) = 1 - 3x^2$

4. $f(x) = x^2 + x + 4$

11. $f(x) = \dfrac{1}{x^2} - 2$

12. $f(x) = \dfrac{1}{x\sqrt{x}}$

5. $f(x) = 2x^3 - 3x^2 + x$

6. $f(x) = 2x^4 - 6x^2 + x + 5$

13. $f(x) = \dfrac{3}{x^2} + \dfrac{2}{\sqrt[3]{x}}$

14. $f(x) = x - \dfrac{1}{2x^2} + 7$

7. $f(x) = x^3 - 3x^2$

8. $f(x) = x - 7x^4$

15. $f(x) = \dfrac{2}{3x^2} + \dfrac{5}{4x^3} + \sqrt{x}$

16. $f(x) = 2x^{-5/2} + 7x^{-9/2}$

22.2   The Area Problem

In this section we are going to study a limit procedure for determining the area under a curve. This procedure will lead to the definition of the definite integral.

Consider a region bounded by a function, the x-axis, and the vertical lines $x = a$ and $x = b$ (Figure 22.1). Such a region can be approximated by rectangles. The larger the number of rectangles, the better the approximation.

To get an expression for the sum of the areas of the rectangles, we need to introduce the appropriate notation. The number of subdivisions of the interval $[a, b]$ will be denoted by n. Let $x'_0 = a$ and $x'_n = b$, while $x'_1, x'_2, \ldots, x'_{n-1}$ are distinct points between a and b; this way we obtain n subdivisions of the interval. (See Figure 22.2.) Note that the subintervals need not be of equal length. The length of the ith subinterval will be denoted by

$$\Delta x_i = x'_i - x'_{i-1}$$

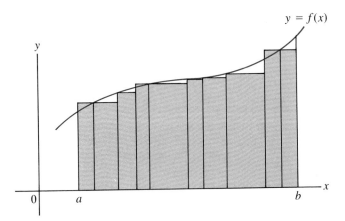

Figure 22.1

Figure 22.2

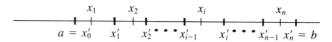

Figure 22.3

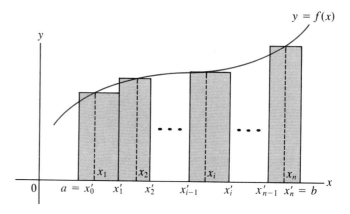

Figure 22.4

To construct the desired rectangles we select an arbitrary point x_i in each subdivision (Figure 22.3). Each x_i can then be used to erect the altitude of the ith rectangle in the region (Figure 22.4). The altitude of each rectangle is the dashed line of length $f(x_i)$. Consequently, the area of the first rectangle is $f(x_1)\Delta x_1$, that of the second $f(x_2)\Delta x_2$, and so on. The sum of all the areas is

$$f(x_1)\Delta x_1 + f(x_2)\Delta x_2 + \cdots + f(x_n)\Delta x_n \tag{22.4}$$

This sum can be written more compactly by means of the *sigma notation*. The sum

$$a_1 + a_2 + \cdots + a_n$$

of n terms can be denoted by

Sigma notation

$$\sum_{i=1}^{n} a_i$$

which means "the summation of all terms a sub i, where i assumes all integral values from 1 to n inclusive." For example:

$$\sum_{i=1}^{n} i^2 = 1^2 + 2^2 + 3^2 + \cdots + n^2$$

$$\sum_{i=0}^{3} 2^i = 2^0 + 2^1 + 2^2 + 2^3$$

$$\sum_{i=2}^{5} \frac{1}{i} = \frac{1}{2} + \frac{1}{3} + \frac{1}{4} + \frac{1}{5}$$

Similarly, the sum (22.4) can be written

$$\sum_{i=1}^{n} f(x_i)\,\Delta x_i \tag{22.5}$$

Our intuition will now lead us to the definition of the area of a region. We have deliberately avoided assigning a specific value to n, so that formula (22.5) remains valid for any number of subdivisions of $[a, b]$. If n gets large, the sum of the areas of the rectangles, $\sum_{i=1}^{n} f(x_i)\,\Delta x_i$, will approximate the area of the region more and more closely. If n increases without limit and if the length of each subinterval approaches zero, then the sum approaches the area of the region in the limit. This can be stated simply as

$$\text{Area under the curve} = \lim_{n \to \infty} \sum_{i=1}^{n} f(x_i)\,\Delta x_i$$

Note: It will be understood from now on that any time we use the notation

$$\lim_{n \to \infty} \sum_{i=1}^{n}$$

all the subintervals will shrink to zero.

Since the region under the curve is bounded by the lines $x = a$ and $x = b$, we will employ the following notation for the area:

$$\text{Area} = \int_{a}^{b} f(x)\,dx = \lim_{n \to \infty} \sum_{i=1}^{n} f(x_i)\,\Delta x_i$$

Definite integral
Integrand
Limits of integration
Integral sign

The middle expression for the area is called the **definite integral of f from a to b** and $f(x)$ is called the **integrand**; the numbers a and b are called the **limits of integration**. The symbol $\int$ is actually an old-fashioned S (for sum) and is called an **integral sign**. The symbolism is due to Leibniz. Although the simplified notation $\int_{a}^{b} f$ would express the definite integral perfectly satisfactorily, it is customary to include letters, such as x, in the expression of the integrand. Since the letters themselves are arbitrary,

$$\int_{a}^{b} f(x)\,dx = \int_{a}^{b} f(y)\,dy = \int_{a}^{b} f(w)\,dw$$

The variables in the integral are sometimes referred to as *dummy variables*.

To clarify the definition, we are going to compute the area of a region by dividing it into subintervals of equal length. The following formulas will be needed:

A. $\displaystyle\sum_{i=1}^{n} i = \frac{n(n + 1)}{2}$

B. $\displaystyle\sum_{i=1}^{n} i^2 = \frac{n(n+1)(2n+1)}{6}$

C. $\displaystyle\sum_{i=1}^{n} i^3 = \left[\frac{n(n+1)}{2}\right]^2$

To illustrate formula A, note that

$$1 + 2 + 3 + \cdots + 15 = \frac{15(15+1)}{2} = 120$$

as can be readily checked.

E X A M P L E 1

Evaluate $\int_0^3 x^2\,dx$ by use of definition (22.6).

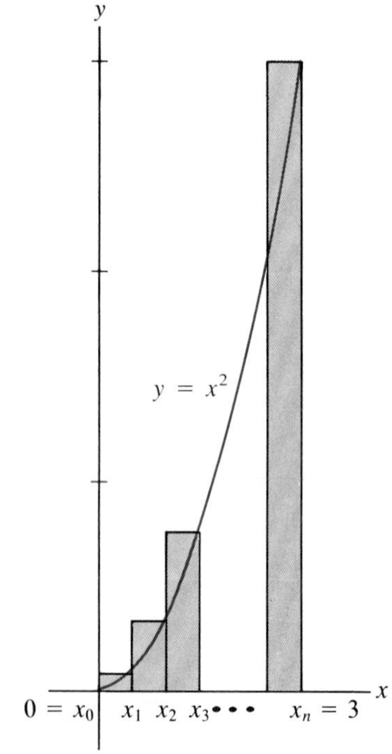

Solution. Subdivide the interval $[0, 3]$ into n subintervals of equal length. Then $\Delta x_i = 3/n$ for all i. We choose the right-hand endpoint for x_i in each subinterval, although the left-hand endpoint could be chosen instead (Figure 22.5). Then the values of x_i are found to be

$$x_1 = 1 \cdot \frac{3}{n}, \, x_2 = 2 \cdot \frac{3}{n}, \ldots, x_i = i \cdot \frac{3}{n}, \ldots, x_n = n \cdot \frac{3}{n} = 3$$

The altitudes of the individual rectangles are computed from the function $f(x) = x^2$. Thus

$$f(x_1) = 1^2 \cdot \frac{9}{n^2} \qquad \text{since } f(x_1) = x_1{}^2 = \left(1 \cdot \frac{3}{n}\right)^2$$

$$f(x_2) = 2^2 \cdot \frac{9}{n^2} \qquad \text{since } f(x_2) = x_2{}^2 = \left(2 \cdot \frac{3}{n}\right)^2$$

$$f(x_i) = i^2 \cdot \frac{9}{n^2}$$

$$f(x_n) = n^2 \cdot \frac{9}{n^2}$$

Since $\Delta x_i = 3/n$, the sum of the areas of the rectangles can now be written

$$\sum_{i=1}^{n} f(x_i)\,\Delta x_i = \sum_{i=1}^{n} i^2 \frac{9}{n^2} \frac{3}{n}$$

This expression reduces to

$$\sum_{i=1}^{n} f(x_i)\,\Delta x_i = \frac{27}{n^3} \sum_{i=1}^{n} i^2 = \frac{27}{n^3} \frac{n(n+1)(2n+1)}{6}$$

$y = x^2$

$0 = x_0 \quad x_1 \, x_2 \, x_3 \bullet \bullet \quad x_n = 3$

Figure 22.5

by formula B. Consequently,

$$\int_0^3 x^2\, dx = \lim_{n \to \infty} \frac{27n(n + 1)(2n + 1)}{6n^3}$$

$$= \lim_{n \to \infty} \frac{27(2n^3 + 3n^2 + n)}{6n^3}$$

$$= \lim_{n \to \infty} \frac{27\left(2 + \dfrac{3}{n} + \dfrac{1}{n^2}\right)}{6} = 9 \qquad \text{dividing by } n^3$$

We conclude that the area of the region is 9 square units. ◀

E X E R C I S E S / S E C T I O N   **22.2 (Optional)**

Use the method of Example 1 to evaluate the following definite integrals.

1. $\int_0^1 x\, dx$ **2.** $\int_0^3 x\, dx$ **3.** $\int_0^1 x^2\, dx$ **4.** $\int_0^1 x^3\, dx$ **5.** $\int_0^2 3x^2\, dx$ **6.** $\int_0^2 (1 + 2x)\, dx$

22.3   The Fundamental Theorem of the Calculus

In Section 22.1 we encountered antiderivatives. In Section 22.2 we considered the definition of an area under a curve. In this section we combine the concepts of antiderivative and area, which leads us to the fundamental theorem of the calculus.

Let B be a fixed number. Consider the area under the curve in Figure 22.6. This area is bounded on the left by the line $x = B$ and on the right by a vertical line through x, where x is allowed to vary. As long as B is fixed, every value of x determines a unique value for the area. Consequently, the area is a function of the variable x, denoted by $A(x)$. Let x take on a small increment Δx, so that $A(x)$ changes by an amount ΔA. These increments are pictured in Figure 22.7.

Now ΔA is the area $PQUT$ under the curve. This area, which is not necessarily rectangular, is wedged between the rectangles $RQUT$ and $PSUT$ in such

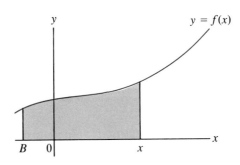

Figure 22.6

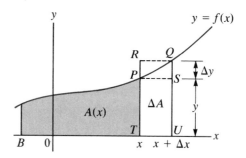

Figure 22.7

a way that

$$\text{area } PSUT \le \Delta A \le \text{area } RQUT$$

If we think about the increment Δx that generated the changed area $PQUT$, we realize that area $PSUT = TP\,\Delta x$ and area $RQUT = UQ\,\Delta x$. It follows that

$$TP\,\Delta x \le \Delta A \le UQ\,\Delta x$$

or

$$TP \le \frac{\Delta A}{\Delta x} \le UQ \qquad \text{dividing by } \Delta x$$

Noting that TP is the value of y at x and that $UQ = y + \Delta y$, the inequalities can be written

$$y \le \frac{\Delta A}{\Delta x} \le y + \Delta y$$

Finally, if $\Delta x \to 0$, then $\Delta y \to 0$, so that $y + \Delta y \to y$. It follows that $\Delta A/\Delta x \to y$ as $\Delta x \to 0$, or

$$y = \lim_{\Delta x \to 0} \frac{\Delta A}{\Delta x} = \frac{dA}{dx}$$

But this is the very definition of derivative of the function $A(x)$. We now reach the surprising conclusion that $dA/dx = f(x)$, which makes $A(x)$ an **antiderivative** of $f(x)$. Moreover, from our definition of area, we have

$$\int_B^x f(x)\,dx = A(x)$$

Using $F(x)$ to denote an antiderivative, we can rewrite the last equation:

$$\int_B^x f(x)\,dx = F(x) + C$$

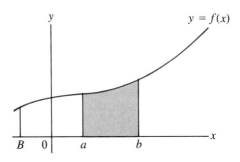

Figure 22.8

(The nonuniqueness of the area function can be explained geometrically by the fact that we could have chosen a different value for B.)

To find the area under the curve between two specific lines $x = a$ and $x = b$ (Figure 22.8), we subtract the smaller from the larger area. The larger area, bounded by B and b, is $F(b) + C$ and the smaller area is $F(a) + C$, so that the area bounded by $x = a$ and $x = b$ is given by

$$[F(b) + C] - [F(a) + C] = F(b) - F(a)$$

Since this area is denoted by the definite integral $\int_a^b f(x)\,dx$, we conclude that

$$\int_a^b f(x)\,dx = F(b) - F(a)$$

This result is known as the **fundamental theorem of the calculus** and was discovered independently by Newton and Leibniz.

Fundamental Theorem of the Calculus: If f is continuous on the interval $[a, b]$, then

$$\int_a^b f(x)\,dx = F(x)\Big|_a^b = F(b) - F(a) \qquad (22.7)$$

where F is a function such that $F' = f$ on $[a, b]$—that is, F is an antiderivative of f.

The fundamental theorem enables us to find areas under a curve by antidifferentiation, thereby establishing a profound connection between areas and antidifferentiation.

According to formula (22.7) we evaluate a definite integral by finding an antiderivative $F(x)$ and evaluating $F(x)$ for $x = b$ and $x = a$. The difference between the two is the value of the integral. Note that the arbitrary constant C does not have to be included since it cancels when the values are subtracted. From now on the process of finding an antiderivative will be called **integration**. (To *integrate* means to find an antiderivative.)

Integration
Integrate

Notation

In formula (22.7) the expression $F(b) - F(a)$ is denoted by

$$F(x)\Big|_a^b$$

The purpose of this notation is to keep track of the limits of integration.

E X A M P L E **1** Let us find the area evaluated in Example 1 of the last section by integration. By formula (22.7)

$$\int_0^3 x^2\, dx = \frac{x^3}{3}\Big|_0^3 = \frac{3^3}{3} - \frac{0^3}{3} = 9$$

The new procedure is somewhat shorter. ◄

E X A M P L E **2** Find the area under the curve $y = 1/x^2$ between the lines $x = 1$ and $x = 3$. (See Figure 22.9.)

Solution. Since $f(x) = 1/x^2$, we have

$$\int_1^3 \frac{1}{x^2}\, dx = \int_1^3 x^{-2}\, dx = \frac{x^{-1}}{-1}\Big|_1^3$$

$$= -\frac{1}{x}\Big|_1^3 = \left(-\frac{1}{3}\right) - (-1) = \frac{2}{3} \text{ square unit}$$ ◄

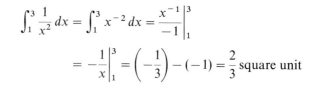

Figure 22.9

E X A M P L E **3** Find the area of the region bounded by $y = 1 - x^2$ and the x-axis.

Solution. Since $1 - x^2 = 0$ when $x = \pm 1$, the x-intercepts are $(-1, 0)$ and $(1, 0)$, so that the region extends from $x = -1$ to $x = 1$ (Figure 22.10). Thus

$$\int_{-1}^1 (1 - x^2)\, dx = x - \frac{x^3}{3}\Big|_{-1}^1$$

$$= \left(1 - \frac{1}{3}\right) - \left(-1 + \frac{1}{3}\right) = \frac{4}{3} \text{ square units}$$ ◄

Figure 22.10

E X E R C I S E S / S E C T I O N   **22.3**

1. Evaluate the definite integrals in the exercises in Section 22.2 by integration.

In Exercises 2–10, find each area bounded by the indicated curves and x-axis.

2. $y = x$, $x = 1$, $x = 4$

3. $y = \frac{1}{2}x$, $x = 0$, $x = 2$

4. $y = x^2$, $x = 1$, $x = 3$

6. $y = x - x^2$

8. $y = \frac{2}{x^2}$, $x = 1$, $x = 2$

10. $y = 1 - x^4$

5. $y = x^3 + 1$, $x = 0$, $x = 1$

7. $y = 4 - x^2$

9. $y = \frac{1}{x^3}$, $x = 1$, $x = 3$

22.4 The Notation for the Integral

Indefinite integral

The purpose of this section is twofold: to discuss the notation used for the integral and to generalize the definition of the definite integral.

Since the fundamental theorem enables us to find areas by antidifferentiation, the antiderivative is denoted by $\int f(x)\,dx$, called the **indefinite integral**. Note especially that the integration symbol uses the differential notation. So if $F(x)$ is the antiderivative of $f(x)$, then

$$dF(x) = d\int f(x)\,dx = f(x)\,dx$$

The differential notation, due to Leibniz, is particularly useful in applications of integration. To see why, recall that we obtained an approximation for the area under a curve by using the sum

$$\sum_{i=1}^{n} f(x_i)\,\Delta x_i$$

Riemann sum

in Section 22.2. A sum of this form is called a **Riemann sum** in honor of the German mathematician G. F. B. Riemann (1826–1866). This concept enables us to generalize the definition of the definite integral.

> **Definition of Definite Integral:** A definite integral is the limit of a Riemann sum:
>
> $$\int_a^b g(x)\,dx = \lim_{n\to\infty} \sum_{i=1}^{n} g(x_i)\,\Delta x_i$$

Typical element

Once this definition has been formulated, *we no longer care how a particular Riemann sum is obtained*—only the form matters. In fact, the Riemann sum enables us to set up different types of integrals in a way that has a definite intuitive appeal. Consider the area under the graph of $y = f(x)$ and draw a **typical element** (Figure 22.11) of height $f(x)$ and thickness dx. The area of the typical element is $f(x)\,dx$, suggesting a shortcut to the Riemann sum procedure.

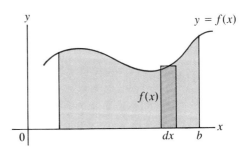

Figure 22.11

> **Shortcut to Riemann Sum:** We can think of $\int_a^b$ as an operation summing up the little areas $f(x)\,dx$ to get
>
> $$\int_a^b f(x)\,dx$$

This shortcut, which is used extensively in Chapter 23, may be called a "sloppy Riemann sum."

   The sloppy Riemann sum is sometimes referred to as the sum of an infinite number of infinitesimally thin rectangles. (See Figure 22.11.) This appears to be the way Leibniz interpreted his own creation. Since the procedure using Riemann sums is always available to us, the reference to infinitesimals can be avoided. On the other hand, the shortcut method is independent of this interpretation of dx. As a heuristic device, thinking of dx as infinitely small cannot really be faulted.

22.5   Basic Integration Formulas

In this section we are going to develop some basic techniques for integrating algebraic functions. Using the notation for the definite integral, formulas (22.1) and (22.3) in Section 22.1 can be written

$$\int x^n\,dx = \frac{x^{n+1}}{n+1} + C \qquad (n \neq -1) \tag{22.8}$$

and

$$\int k\,dx = kx + C \tag{22.9}$$

Formula (22.2) can be expressed in the form

$$\int kx^n\,dx = k\int x^n\,dx = \frac{kx^{n+1}}{n+1} + C \qquad n \neq -1$$

and, more generally,

$$\int kf(x)\,dx = k\int f(x)\,dx \tag{22.10}$$

since $d\int kf(x)\,dx = k[d\int f(x)\,dx] = kf(x)\,dx$.

   Formula (22.10) is a special case of the following property:

$$\int [c_1 f(x) + c_2 g(x)]\,dx = c_1 \int f(x)\,dx + c_2 \int g(x)\,dx$$

Because of this property, the integral is said to be *linear*. (We will see the reason for this name in Chapter 28.)

Most of Chapter 25 is devoted to developing different integration techniques. Being confined to algebraic functions at this point, we will consider only one more special case, *the generalized power rule in reverse*. Recall that for

$$f(x) = (x^2 + 1)^2$$

the generalized power rule yields

$$f'(x) = 2(x^2 + 1)\frac{d}{dx}(x^2 + 1) = 2(x^2 + 1)(2x)$$

or, in differential form,

$$df(x) = 2(x^2 + 1)(2x\,dx)$$

From the last statement it follows that

$$\int 2(x^2 + 1)(2x\,dx) = (x^2 + 1)^2 + C$$

Now consider the integral

$$\int \frac{1}{2}(x^2 + 1)^{-1/2}2x\,dx$$

Keeping the previous example in mind, we first observe that $d(x^2 + 1) = 2x\,dx$. It follows that

$$\int \frac{1}{2}(x^2 + 1)^{-1/2}2x\,dx = (x^2 + 1)^{1/2} + C$$

The general form of this integral, which is readily verified by the generalized power rule, will be called the *general power formula*.

General Power Formula:

$$\int u^n\,du = \frac{u^{n+1}}{n+1} + C \qquad (n \neq -1) \qquad (22.11)$$

Formula (22.11) suggests a systematic procedure for performing an integration of this type.

E X A M P L E 1 Use formula (22.11) to compute the integral $\int \frac{1}{2}(x^2 + 1)^{-1/2}2x\,dx$.

Solution. We let $u = x^2 + 1$ and find that

$$du = 2x\,dx$$

These expressions may be substituted in the integral, which becomes

$$\int \frac{1}{2} u^{-1/2} \, du = \frac{1}{2} \frac{u^{1/2}}{\frac{1}{2}} + C = u^{1/2} + C$$

by formula (22.11). Hence (substituting back)

$$\int \frac{1}{2} (x^2 + 1)^{-1/2} 2x \, dx = (x^2 + 1)^{1/2} + C$$

◄

E X A M P L E **2**   Integrate $\int x^2 \sqrt{x^3 + 1} \, dx$.

Solution. The integral can be written in the form

$$\int (x^3 + 1)^{1/2} x^2 \, dx$$

As before, let $u = x^3 + 1$, whence $du = 3x^2 \, dx$.

Direct substitution in the integral is now impossible since $x^2 \, dx$ does not match du. Yet all that is required here is a simple trick: by formula (22.10)

$$\int (x^3 + 1)^{1/2} x^2 \, dx = \int (x^3 + 1)^{1/2} \frac{1}{3} \cdot 3x^2 \, dx$$

$$= \frac{1}{3} \int (x^3 + 1)^{1/2} 3x^2 \, dx$$

Substitution now gives

$$\frac{1}{3} \int u^{1/2} \, du = \frac{1}{3} \frac{u^{3/2}}{\frac{3}{2}} + C = \frac{2}{9} (x^3 + 1)^{3/2} + C$$

◄

E X A M P L E **3**   Integrate

$$\int \frac{x \, dx}{\sqrt[3]{1 - x^2}}$$

Solution. The integral can be written in the form

$$\int (1 - x^2)^{-1/3} x \, dx$$

Let $u = 1 - x^2$, so that $du = -2x \, dx$. We insert the number -2 and place $-\frac{1}{2}$ in front of the integral:

$$-\frac{1}{2} \int (1 - x^2)^{-1/3} (-2x \, dx) = -\frac{1}{2} \int u^{-1/3} \, du = -\frac{1}{2} \frac{u^{2/3}}{\frac{2}{3}} + C$$

$$= -\frac{3}{4} (1 - x^2)^{2/3} + C$$

◄

Caution. Since we are dealing with a very special type of integral, its form has to match the form in formula (22.11), except for a multiplicative constant. Consider, for example, the integral

$$\int \frac{dx}{\sqrt{1 - x^2}}$$

If we let $u = 1 - x^2$, then $du = -2x\,dx$. Since $-2x$ is not a constant, our previous trick no longer works. In other words, the integral cannot be worked out with our present techniques.

E X A M P L E **4** Integrate $\int (x^2 - 1)^2\,dx$.

Solution. If $u = x^2 - 1$, then $du = 2x\,dx$. As noted above, since $2x$ is not a constant, this integral is not of the form

$$\int u^n\,du$$

Consequently, we must proceed by multiplying out the expression in the integrand and integrating term by term:

$$\int (x^2 - 1)^2\,dx = \int (x^4 - 2x^2 + 1)\,dx = \tfrac{1}{5}x^5 - \tfrac{2}{3}x^3 + x + C \qquad \blacktriangleleft$$

A common error is ignoring the differential du and writing

$$\int (2x^2 + 1)^4 x\,dx \qquad \text{as} \qquad \frac{(2x^2 + 1)^5}{5} + C$$

Since $u = 2x^2 + 1$ and $du = 4x\,dx$, the correct procedure is to insert 4 and place $\tfrac{1}{4}$ in front of the integral. Thus

$$\int (2x^2 + 1)^4 x\,dx = \frac{1}{4} \int (2x^2 + 1)^4 (4x)\,dx = \frac{1}{4} \frac{(2x^2 + 1)^5}{5} + C$$

The next example illustrates the evaluation of a definite integral.

E X A M P L E **5** Evaluate

$$\int_{-\sqrt{6}}^{-1} \frac{x\,dx}{\sqrt{10 - x^2}}$$

Solution. The variable x is changed to the variable u by the substitution

$$u = 10 - x^2$$
$$du = -2x\,dx$$

The equation $u = 10 - x^2$ is also used to change the limits of integration:

Lower limit: if $x = -\sqrt{6}$, then $u = 10 - x^2 = 10 - (-\sqrt{6})^2 = 4$.

Upper limit: if $x = -1$, then $u = 10 - x^2 = 10 - (-1)^2 = 9$.

It follows that

$$\int_{-\sqrt{6}}^{-1} \frac{x\,dx}{\sqrt{10 - x^2}} = -\frac{1}{2}\int_{-\sqrt{6}}^{-1} \frac{-2x\,dx}{\sqrt{10 - x^2}} = -\frac{1}{2}\int_{4}^{9} \frac{du}{\sqrt{u}}$$

$$= -\frac{1}{2}\int_{4}^{9} u^{-1/2}\,du = -u^{1/2}\Big|_{4}^{9}$$

$$= -(9)^{1/2} - [-(4)^{1/2}] = -\sqrt{9} + \sqrt{4}$$

$$= -3 + 2 = -1 \qquad \blacktriangleleft$$

E X E R C I S E S / S E C T I O N   22.5

Perform the following integrations.

1. $\int \sqrt{x}\,dx$

2. $\int (x\sqrt{x} - x)\,dx$ (Recall that $x\sqrt{x} = x^{3/2}$.)

3. $\int \left(\frac{1}{x^3} - \frac{3}{x^2} \right) dx$

4. $\int (x^{-2/3} + 3x^{1/2})\,dx$

5. $\int (2\sqrt{x} - 3x^2 + 1)\,dx$

6. $\int \left(\frac{2}{\sqrt{x}} - 3x\sqrt{x} + 2 \right) dx$

7. $\int \left(\frac{1}{x^4} + \frac{1}{\sqrt{x}} - 4 \right) dx$ **8.** $\int (x^2 + 1)^4(2x)\,dx$

9. $\int (2x^2 - 3)^3(4x)\,dx$ **10.** $\int (2x^2 - 3)^3 x\,dx$

11. $\int (2 - x^2)^4 x\,dx$ **12.** $\int (4 - x^3)^2 x^2\,dx$

13. a. $\int (1 - x)\,dx$ **14.** $\int (x^4 + 1)^3 x^3\,dx$

b. $\int (1 - x)^4\,dx$

15. $\int \frac{x\,dx}{(x^2 - 1)^2}$ **16.** $\int \sqrt{1 + x}\,dx$

17. $\int (2x^2 + x)^3(4x + 1)\,dx$ **18.** $\int (x^3 - 3x)^5(x^2 - 1)\,dx$

19. $\int \frac{dx}{\sqrt{1 - x}}$ **20.** $\int \sqrt{1 - 2x}\,dx$

21. $\int \frac{x\,dx}{\sqrt{1 - x^2}}$ **22.** $\int x\sqrt[3]{x^2 + 1}\,dx$

23. $\int x^2\sqrt[3]{1 - 3x^3}\,dx$ **24.** $\int x^2\sqrt[3]{x}\,dx$

25. $\int \frac{\sqrt[4]{x}}{x}\,dx$ **26.** $\int \sqrt[4]{x^2 - x}\,(2x - 1)\,dx$

27. $\int (2x^3 - 1)\sqrt[5]{x^4 - 2x}\,dx$

28. $\int \frac{2x - 1}{\sqrt{x^2 - x}}\,dx$

29. $\int (x^2 + 1)^2\,dx$ (See Example 4.)

30. $\int 3x\sqrt{x^2 + 1}\,dx$

31. $\int \frac{5x\,dx}{\sqrt{x^2 - 3}}$

32. $\int \frac{1 + \sqrt{x}}{\sqrt{x}}\,dx$ $\left(\text{Recall that } \frac{1 + \sqrt{x}}{\sqrt{x}} = \frac{1}{\sqrt{x}} + 1. \right)$

33. $\int (1 + \sqrt{x})^2\,dx$ **34.** $\int x^2(3x^3 + 1)^{2/3}\,dx$

35. $\int (1 - 5x)^{4/3}\,dx$ **36.** $\int (1 - 3x^2)^2\,dx$

37. $\int \frac{2 - 4x}{\sqrt[4]{x - x^2}}\,dx$ **38.** $\int \frac{1 - 3x}{3\sqrt{x}}\,dx$

39. $\int (x^3 + 1)^2(3x)\,dx$ **40.** $\int 7x\sqrt[3]{x^2 - 10}\,dx$

41. $\int (x^3 + 1)^3(5x^2)\,dx$ **42.** $\int (3 - 2x^3)^2(-6x)\,dx$

43. $\int (4x^3 - 1)^2(12x)\,dx$ **44.** $\int (4x^3 - 1)^2(10x^2)\,dx$

45. $\int (3 - 3x)\sqrt{4x - 2x^2}\,dx$ **46.** $\int \dfrac{x\,dx}{(1 - 5x^2)^5}$

47. $\int \dfrac{\sqrt{1 + \sqrt{x}}}{2\sqrt{x}}\,dx$ **48.** $\int \dfrac{\sqrt{1 - \sqrt{x}}}{\sqrt{x}}\,dx$

49. $\int (2x^3 + 1)^2(6x)\,dx$ **50.** $\int (2x^3 + 1)^2(6x^2)\,dx$

51. $\int (x^4 + 2)^2(4x^3)\,dx$ **52.** $\int (x^4 + 2)^2(4x)\,dx$

53. $\int (x^3 + 1)^2 x\,dx$ **54.** $\int (x^3 + 1)^2 x^2\,dx$

55. $\int (3 - x^4)^2 x^3\,dx$ **56.** $\int (3 - x^4)^2 x^2\,dx$

57. $\int_0^1 (1 - x)\,dx$ **58.** $\int_0^2 (2x - 1)\,dx$

59. $\int_1^8 \sqrt[3]{x}\,dx$ **60.** $\int_1^4 2\sqrt{x}\,dx$

61. $\int_0^1 \sqrt{1 - x}\,dx$ **62.** $\int_1^6 \sqrt{x + 3}\,dx$

63. $\int_0^1 (x^2 - 1)^2\,dx$ **64.** $\int_0^2 \dfrac{x^2 + 2}{2}\,dx$

65. $\int_2^7 \dfrac{dx}{\sqrt{x + 2}}$ **66.** $\int_{-2}^1 \dfrac{dx}{\sqrt{2 - x}}$

67. $\int_{-4}^0 \sqrt{1 - 2x}\,dx$ **68.** $\int_0^1 x\sqrt{1 - x^2}\,dx$

69. $\int_4^9 \dfrac{1 + \sqrt{x}}{\sqrt{x}}\,dx$ **70.** $\int_0^4 (2 + \sqrt{x})^2\,dx$

71. $\int_1^2 x\sqrt{4 - x^2}\,dx$ **72.** $\int_1^3 x\sqrt{x^2 - 1}\,dx$

22.6   Area Between Curves

In this section we are going to continue our study of areas by using the shortcut method discussed in Section 22.4.

E X A M P L E **1**   Find the area of the region bounded by $y = \sqrt{x}$, $x = 4$, and $y = 0$.

Solution. Since $y = 0$ is the x-axis, we obtain the region in Figure 22.12.

Now draw the **typical element** of height y and thickness dx (Figure 22.12). Note that the area of the typical element is

Typical element

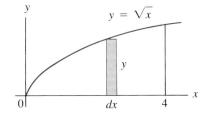

Figure 22.12

$$y\,dx = \sqrt{x}\,dx \qquad \text{area of typical element}$$

The area we wish to find lies between the lines $x = 0$ and $x = 4$. Summing the little areas from $x = 0$ to $x = 4$ by the operation

$$\int_0^4$$

we get

$$A = \int_0^4 \sqrt{x}\,dx = \int_0^4 x^{1/2}\,dx$$

$$= \frac{2}{3} x^{3/2}\Big|_0^4 = \frac{2}{3}(4^{3/2} - 0^{3/2})$$

$$= \frac{2}{3}(8) = \frac{16}{3}$$

◀

Example 1 is a special case of a general problem: finding areas of **regions bounded by two curves**, as shown in Figure 22.13. The typical element of thickness dx now extends between the two curves and has a height given by $f(x) - g(x)$,

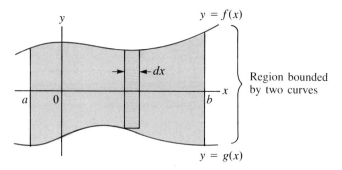

Figure 22.13

regardless of the positions of f and g. Consequently, the area is given by the integral

$$\int_a^b \left[f(x) - g(x) \right] dx$$

Area Between Two Curves: Draw a typical element of thickness dx extending between $f(x)$ and $g(x)$ (Figure 22.13). Sum the small areas $[f(x) - g(x)]\, dx$ by the operation $\int_a^b$ to obtain

$$\int_a^b \left[f(x) - g(x) \right] dx$$

When summing typical elements to get an area, we have to make sure that each such element is positive. This is particularly important when a curve, or part of a curve, lies below the x-axis. Consider the next example.

E X A M P L E **2**   Find the area bounded by the curves $y = -\frac{1}{2}x$, $x = 2$, and the x-axis.

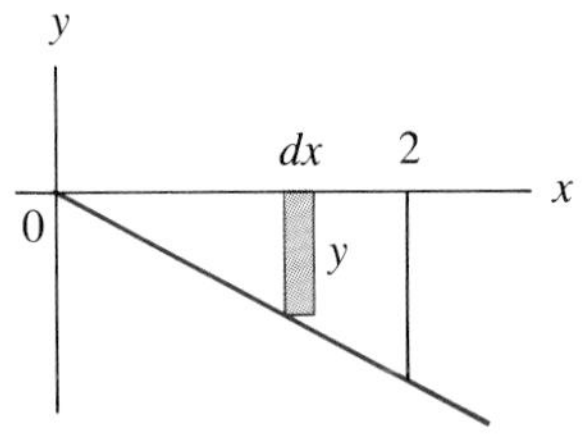

Solution. The region is shown in Figure 22.14.

If we now integrate the function $y = -\frac{1}{2}x$ from $x = 0$ to $x = 2$, we get

$$\int_0^2 \left(-\tfrac{1}{2}x \right) dx = -\frac{1}{2} \frac{x^2}{2} \Big|_0^2 = -1$$

Figure 22.14

The resulting value is negative because each y is negative, yielding a negative sum.

To avoid this problem, we use the principle shown in Figure 22.13. We obtain the height of each element by subtracting $y = -\frac{1}{2}x$ from $y = 0$:

$$0 - \left(-\tfrac{1}{2}x \right) \qquad \text{height of element}$$

(See Figure 22.15.)

The area of the typical element is

$$\left[0 - \left(-\tfrac{1}{2}x \right) \right] dx \qquad \text{height} \cdot \text{base}$$

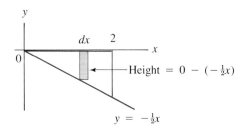

Figure 22.15

Summing from $x = 0$ to $x = 2$, we get

$$A = \int_0^2 \left[0 - (-\tfrac{1}{2}x) \right] dx = \int_0^2 \tfrac{1}{2}x\, dx = 1 \qquad \blacktriangleleft$$

In the next example the curve crosses the x-axis, so that part of the region lies above the x-axis and part of the region below.

E X A M P L E **3**   Find the area of the region bounded by $y = x^3$, $x = -1$, $x = 2$, and the x-axis.

Solution. The region is shown in Figure 22.16.

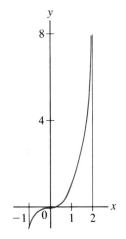

Figure 22.16

Since part of the region lies below the x-axis, we need to find the two areas separately.

Left region: $\displaystyle \int_{-1}^0 (0 - x^3)\, dx = -\frac{x^4}{4}\bigg|_{-1}^0 = \frac{1}{4}$

Right region: $\displaystyle \int_0^2 x^3\, dx = \frac{x^4}{4}\bigg|_0^2 = 4$

Total area: $\displaystyle \frac{1}{4} + 4 = \frac{17}{4}$ $\blacktriangleleft$

Caution. In Example 3 the direct evaluation of the integral yields

$$\int_{-1}^2 x^3\, dx = \frac{x^4}{4}\bigg|_{-1}^2 = 4 - \frac{1}{4} = \frac{15}{4}$$

Although not equal to the area, the value of the definite integral $\int_{-1}^2 x^3\, dx$ is $\frac{15}{4}$, regardless of the geometric interpretation. In other words, *every definite integral is equal to a number*, and this number can have many different interpretations, just as the value of the derivative at a point can have different physical meanings.

E X A M P L E **4**   Find the area of the region bounded by the parabola $y^2 = 4x$ and the line $y = x$.

Solution. To determine the limits of integration we have to solve the two equations simultaneously. From $y = x$ we get $y^2 = x^2$. On substituting in the

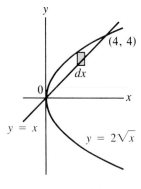

Figure 22.17

other equation we find that

$$x^2 = 4x \qquad x^2 - 4x = 0 \qquad x(x - 4) = 0$$

whence $x = 0$ and $x = 4$. So the points of intersection are $(0, 0)$ and $(4, 4)$. The graphs are now easily sketched (Figure 22.17). The upper half of the parabola $y^2 = 4x$ is the function $y = \sqrt{4x} = 2\sqrt{x}$, so that the typical element has height $2\sqrt{x} - x$. Summing from $x = 0$ to $x = 4$, we get

$$\int_0^4 (2\sqrt{x} - x)\,dx = \frac{4}{3}x^{3/2} - \frac{1}{2}x^2\Big|_0^4 = \frac{8}{3} \qquad \blacktriangleleft$$

So far we have been consistently using the letter x for the independent variable and y for the dependent variable, so that, bowing to the usual convention, y is always a function of x. Yet there are times when it is much more convenient to reverse the roles of x and y. For example, in the equation $x = y^3 - 3y + 1$ we might as well regard x as a function of y instead of solving for y in terms of x. In such a case we write $x = f(y)$, keeping the functional notation.

E X A M P L E **5**   The equation $y = 1/(x - 1)$ is a function of x. Solving for x, we find that $x = (y + 1)/y$, which represents x as a function of y. $\qquad \blacktriangleleft$

For some regions it is convenient to reverse the roles of x and y, as shown in the next example.

E X A M P L E **6**   Find the area of the region bounded by $x = 8y - 2y^2$ and the y-axis.

Solution. To sketch the parabola $x = 8y - 2y^2$, we need to find the y-intercepts. Letting $x = 0$, we get

$$8y - 2y^2 = 0$$
$$2y(4 - y) = 0$$
$$y = 0, 4$$

The region is shown in Figure 22.18.

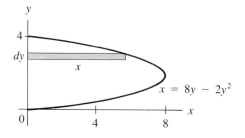

area of typical element:
$x\,dy = (8y - 2y^2)\,dy$

Figure 22.18

The given equation $x = 8y - 2y^2$ represents x as a function of y. So rather than solving the equation for y in terms of x, we interchange the roles of x and y and draw the **typical element in the horizontal position**, as shown in Figure 22.18. Note that the height of the typical element is $x = 8y - 2y^2$ and that the thickness is dy. The area of the typical element is therefore

$$x\, dy = (8y - 2y^2)\, dy \qquad \text{area of typical element}$$

Summing from $y = 0$ to $y = 4$, we get

$$A = \int_0^4 (8y - 2y^2)\, dy$$

$$= 4y^2 - 2\frac{y^3}{3}\bigg|_0^4 = 4\cdot4^2 - 2\cdot\frac{4^3}{3} - 0$$

$$= 4^3 - 2\cdot\frac{4^3}{3} = 4^3\left(1 - \frac{2}{3}\right) = 64\cdot\frac{1}{3} = \frac{64}{3} \qquad \blacktriangleleft$$

E X A M P L E **7**   Find the area of the region bounded by $x = y^2 - y - 2$ and $y = x + 2$.

$x = y^2 - y - 2$

$x = y - 2$

$\overline{0 = y^2 - 2y}$ (subtracting)

$y(y - 2) = 0$

$y = 0, 2$

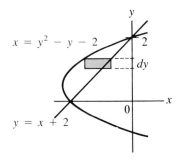

$x = y^2 - y - 2$

$y = x + 2$

Figure 22.19

Solution. The equation $x = y^2 - y - 2 = (y - 2)(y + 1)$ represents a parabola with intercepts $(0, 2)$, $(0, -1)$, and $(-2, 0)$. Solving the equations simultaneously we find that $(-2, 0)$ and $(0, 2)$ are the points of intersection. The graphs are shown in Figure 22.19.

Writing $y = x + 2$ as $x = y - 2$, both equations are functions of y. With this point of view we **must draw the typical element in a horizontal position with thickness dy and height** $(y - 2) - (y^2 - y - 2)$. For the limits of integration we refer to the y-axis: the region we are concerned with lies between the lines $y = 0$ and $y = 2$. So the integral becomes

$$\int_0^2 [(y - 2) - (y^2 - y - 2)]\, dy = \int_0^2 (-y^2 + 2y)\, dy$$

$$= -\frac{y^3}{3} + y^2\bigg|_0^2 = \frac{4}{3} \qquad \blacktriangleleft$$

Improper Integrals

Consider the unbounded region determined by $y = 1/x^2$, $x = 1$, and the x-axis to the right of 1 in Figure 22.20. To find the area of the region we could use our intuition and write the integral formally as

$$\int_1^\infty \frac{1}{x^2}\, dx$$

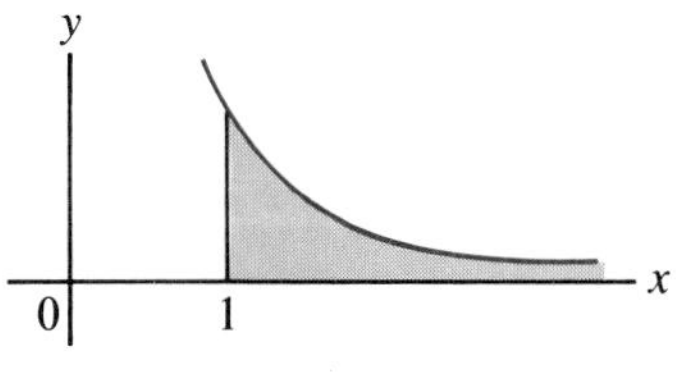

The problem is that given the definition of the integral as a limit of a sum, the interval $[a, b]$ has to be finite. To get around this problem we define

$$\int_1^\infty \frac{dx}{x^2} = \lim_{b \to \infty} \int_1^b \frac{1}{x^2}\, dx$$

Figure 22.20

Improper integral

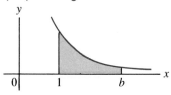

Figure 22.21

which is called an **improper integral**. Geometrically, we are finding the area under the graph bounded by the lines $x = 1$ and $x = b$ (Figure 22.21) and then letting $b \to \infty$. If the limit exists, this limit is defined to be the area. The integration yields

$$\int_1^\infty \frac{dx}{x^2} = \lim_{b \to \infty} \int_1^b x^{-2}\,dx = \lim_{b \to \infty} \frac{x^{-1}}{-1}\Big|_1^b = \lim_{b \to \infty} \left(-\frac{1}{b} + 1\right) = 1$$

E X A M P L E **8**   Evaluate the improper integral

$$\int_0^\infty \frac{dx}{(x+2)^{3/2}}$$

Solution.

$$\int_0^\infty \frac{dx}{(x+2)^{3/2}} = \lim_{b \to \infty} \int_0^b \frac{dx}{(x+2)^{3/2}}$$

$$= \lim_{b \to \infty} \frac{(x+2)^{-1/2}}{-\frac{1}{2}}\Big|_0^b = \lim_{b \to \infty} \frac{-2}{\sqrt{x+2}}\Big|_0^b$$

$$= \lim_{b \to \infty} \left[\frac{-2}{\sqrt{b+2}} + \frac{2}{\sqrt{2}}\right] = 0 + \frac{2}{\sqrt{2}} = \frac{2}{\sqrt{2}}\frac{\sqrt{2}}{\sqrt{2}} = \sqrt{2} \quad \blacktriangleleft$$

An integral is also improper if the integrand has an infinite discontinuity (vertical asymptote) somewhere in the interval $[a, b]$. Consider the next example.

E X A M P L E **9**   Find the area "bounded" by $y = 1/\sqrt{x}$, $x = 1$, and the coordinate axes. (See Figure 22.22.)

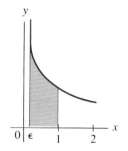

Figure 22.22

Solution. We can avoid the vertical asymptote $x = 0$ by integrating from $x = \epsilon$ to $x = 1$ (as shown in Figure 22.22) and then letting $\epsilon \to 0$:

$$\int_0^1 \frac{1}{\sqrt{x}}\,dx = \lim_{\epsilon \to 0} \int_\epsilon^1 \frac{1}{\sqrt{x}}\,dx = \lim_{\epsilon \to 0} \int_\epsilon^1 x^{-1/2}\,dx$$

$$= \lim_{\epsilon \to 0} 2x^{1/2}\Big|_\epsilon^1 = \lim_{\epsilon \to 0} (2 - 2\epsilon^{1/2}) = 2$$

Since the limit exists, the area is defined to be 2. $\quad \blacktriangleleft$

E X E R C I S E S / S E C T I O N   **22.6**

Find the area of the region bounded by the given curves. Sketch the curves and show a typical horizontal or vertical element of area.

1. $y = 2x$, $x = 1$, x-axis   **2.** $y = \frac{1}{2}x$, $x = 2$, x-axis

3. $y = 2x$, $y = 2$, y-axis   **4.** $y = \frac{1}{2}x$, $y = 2$, y-axis

5. $y = x$, $x = -1$, $x = 1$, $y = 0$

6. $y = 2x$, $x = -1$, $x = 1$, $y = 0$

7. $y = 3x$, $x = 1$, $x = 2$, $y = 0$

8. $y = x^2$, $x = 1$, $y = 0$ **9.** $y = -x$, $x = 1$, $y = 0$

10. $y = x^2$, $x = -1$, $y = 0$ **11.** $y = x^3$, $x = -1$, $y = 0$

12. $y = -x$, $y = 1$, $x = 0$

13. $y = x^2 + 1$, $x = 1$, $x = 3$, x-axis

14. $y = x^3$, $x = -1$, $x = 1$, x-axis

15. $y = \sqrt[3]{x}$, $x = 0$, $x = 8$, x-axis

16. $y = \dfrac{1}{x^2}$, $x = 1$, $x = 2$, x-axis

17. $x = y^2$, $y = 1$, y-axis

18. $y = x$, $y = x^2$ **19.** $y = x^2$, $y = x^3$

20. $y^2 = x$, $y^3 = x$ **21.** $y = x^2 - 4$, x-axis

22. $y^2 = x + y$, y-axis   **23.** $y = x^2 - 1$, $y = 3$

24. $y = \sqrt[3]{x}$, $y = -x$, $x = 0$, $x = 8$

25. $y^2 = x - 1$, $y = x - 3$

26. $x = \sqrt{y}$, $x = -y$, $y = 1$

27. $x^2 + 4y = 0$, $x^2 - 4y - 8 = 0$

28. $x = y^2 - 4y + 2$, $2x - y + 5 = 0$

29. $y = \dfrac{2}{x^2}$, $x = 3$, the x-axis to the right of $x = 3$

30. $y = \dfrac{1}{x^{3/2}}$, $x = 4$, the x-axis to the right of $x = 4$

31. $y = \dfrac{1}{(x-1)^2}$, $x = 2$, the x-axis to the right of $x = 2$

32. $y = \dfrac{1}{(x+2)^2}$, the coordinate axes, area in first quadrant

33. $x^3y = 2$, $x = 1$, the x-axis to the right of $x = 1$

34. $y = x^{-4/3}$, $x = 8$, the x-axis to the right of $x = 8$

35. $y = 1/(x + 1)^{3/2}$, the coordinate axes, area in first quadrant

36. $y = x^{-3/4}$, $x = 0$, $x = 16$, x-axis. *Hint:* Avoid the vertical asymptote by evaluating

$$\lim_{\epsilon \to 0} \int_{\epsilon}^{16} x^{-3/4}dx \qquad \epsilon > 0 \qquad \text{(See Example 9.)}$$

37. $y = \dfrac{1}{\sqrt{x-1}}$, $x = 1$, $x = 5$, x-axis

38. $y = \dfrac{1}{\sqrt[3]{x-2}}$, $x = 2$, $x = 3$, x-axis

39. $y = \dfrac{2}{\sqrt{4-x}}$, $x = 4$, coordinate axes

40. Show that the integral $\int_{1}^{\infty} \dfrac{1}{\sqrt{x}} dx$ does not exist.

41. $y = \dfrac{3}{(3-x)^2}$, $x = 3$, coordinate axes

22.7   The Constant of Integration

The purpose of this section is to discuss some common applications of the indefinite integral.

E X A M P L E **1**   Find the family of curves such that each member has a derivative equal to $2x$.

Solution. For each member of the desired family,

$$\frac{dy}{dx} = 2x$$

By integration,

$$y = x^2 + C$$

Since $y - C = x^2$ has the form $(y - k) = (x - 0)^2$, we need to recall from Section 19.11 that the family is generated by translating the parabola $y = x^2$ by a distance C units in the vertical direction. (Each value of C corresponds to a member of the family.) The graphs are shown in Figure 22.23.

Now if x has a specific value, say $x = 1$, then

$$\frac{dy}{dx} = 2x\Big|_{x=1} = 2$$

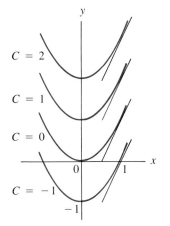

$C = 2$

$C = 1$

$C = 0$

$C = -1$

Figure 22.23

for each member of the family. Consequently, for $x = 1$ all the tangents have the same slope and must be parallel. (See Figure 22.23.) ◄

E X A M P L E **2**   Find the curve in Example 1 passing through the point (2, 5).

Solution. After substituting $x = 2$ and $y = 5$ in the equation $y = x^2 + C$, we get $5 = 4 + C$. It follows that $C = 1$, so that $y = x^2 + 1$ is the desired curve. ◄

Another important application of indefinite integrals is the study of velocity and acceleration. Recall from Section 20.7 that

$$v = \frac{ds}{dt} \quad \text{and} \quad a = \frac{dv}{dt}$$

Since integration is the inverse of differentiation, we may also write

$$s = \int v \, dt \quad \text{and} \quad v = \int a \, dt$$

For objects acting under the influence of gravity, we are going to adopt the following conventions. The acceleration due to gravity is $g = 10$ m/s² (rather than the more accurate value of 9.8 m/s²). All distances are measured from the ground with upward direction positive. Finally, $t = 0$ corresponds to the instant when the motion begins. As a result of these assumptions:

Basic assumptions

1. $g = -10$ m/s².
2. $s = 0$ on the ground.
3. $t = 0$ at the instant when the motion begins.
4. A distance from the ground to a point above the ground is positive.
5. An object moving in the upward direction has a positive velocity.
6. An object moving in the downward direction has a negative velocity.

Remark. We assume in this section that the air resistance is negligible. Motion that takes air resistance into account will be studied in Section 27.5.

E X A M P L E **3**   An object is hurled upward from the ground with a velocity of 25 m/s.
   a. How long does the object stay in the air?
   b. How high does it rise?

Solution. As already noted, the upward direction is positive, so that

$$g = -10 \text{ m/s}^2$$

According to the given conditions, when $t = 0$ the velocity is $+25$ m/s. (The velocity is positive because the object is moving in the upward direction.) The velocity at $t = 0$ is called the **initial velocity** and is denoted by v_0. (See Figure 22.24.)

Initial velocity

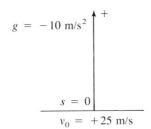

Since upward direction is positive,
$g = -10$ m/s^2 and $v_0 = +25$ m/s.

Figure 22.24

From the formula

$$v = \int a\, dt$$

we get

$$v = \int (-10)\, dt \qquad \text{or} \qquad v = -10t + C$$

where C is an arbitrary constant. To evaluate C, we let $t = 0$ and $v = +25$:

$$v = -10t + C$$
$$+25 = -10 \cdot 0 + C \qquad \text{or} \qquad C = +25$$

It follows that

$$v = -10t + 25$$

From the formula

$$s = \int v\, dt$$

we now get

$$s = \int (-10t + 25)\, dt \qquad \text{or} \qquad s = -5t^2 + 25t + k$$

where k is an arbitrary constant. Since all distances are measured from the ground, we have $s = 0$ when $t = 0$. (See Figure 22.24.) It follows that

$$s = -5t^2 + 25t + k$$
$$0 = -5(0)^2 + 25(0) + k \qquad \text{or} \qquad k = 0$$

so that

$$s = -5t^2 + 25t$$

a. To determine how long the object stays in the air, we let $s = 0$ in the last equation and solve for t:

$$0 = -5t^2 + 25t$$

$$-5t(t - 5) = 0$$

$$t = 0, 5$$

So $s = 0$ when $t = 0$ and $t = 5$. Since $t = 0$ when the motion begins, it takes 5 s for the object to return to the ground.

b. To determine how high the object rises, we need to find the value of t for which $v = 0$ (because at the highest point the object momentarily stops):

$$v = -10t + 25$$

$$0 = -10t + 25$$

$$t = 2.5$$

Thus the object takes **2.5** s to reach the highest point. From

$$s = -5t^2 + 25t$$

we get

$$s = -5(\mathbf{2.5})^2 + 25(\mathbf{2.5}) = 31.25$$

The object therefore reaches a height of 31.25 m.   ◀

E X A M P L E **4**   An object is hurled downward from the top of a building 20 m high with an initial velocity of 5 m/s. Find the time it takes to reach the ground.

Solution. Consider the diagram in Figure 22.25.

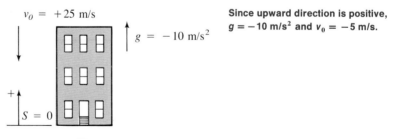

Figure 22.25

Since the upward direction is positive, $g = -10$ m/s². When $t = 0$, $s = 20$ m (since the object starts at a point 20 m above the ground). Since the object is hurled downward—in the negative direction—$v_0 = -5$ m/s. (See Figure 22.25).

We now have

$$v = \int (-10)\, dt = -10t + C$$

Substituting $t = 0$ and $v = -5$:

$$v = -10t + C$$
$$-5 = -10 \cdot 0 + C \quad \text{or} \quad C = -5$$

Hence

$$v = -10t - 5$$

Also,

$$s = \int (-10t - 5)\, dt \quad \text{and} \quad s = -5t^2 - 5t + k$$

Since $s = +20$ when $t = 0$,

$$+20 = -5(0)^2 - 5(0) + k \quad \text{or} \quad k = +20$$

Thus

$$s = -5t^2 - 5t + 20$$

To determine the time required to reach the ground, we let $s = 0$ in the last equation and solve for t:

$$0 = -5t^2 - 5t + 20$$
$$t^2 + t - 4 = 0 \qquad \text{dividing by } -5$$

and by the quadratic formula,

$$t = \frac{-1 \pm \sqrt{17}}{2} \qquad \textbf{See Section 6.3.}$$

Hence $t \approx 1.6$ s. (We ignore the negative root since the motion begins at $t = 0$.)

◀

E X A M P L E 5 If the object in Example 4 is hurled upward at 5 m/s, with what velocity will it strike the ground?

Solution. As before, $g = -10$ m/s². This time, however, $v_0 = +5$ m/s, since the object is moving in the positive direction. We get

$$v = -10t + 5 \quad \text{and} \quad s = -5t^2 + 5t + 20$$

To determine the velocity with which the object strikes the ground, we must first determine the time required to reach the ground. Letting $s = 0$, we get

$$0 = -5t^2 + 5t + 20$$

$$t^2 - t - 4 = 0$$

$$t = \frac{1 + \sqrt{17}}{2} \approx 2.6 \text{ s}$$

It follows that the velocity is

$$v = -10t + 5$$

$$= -10(2.6) + 5 = -21$$

Thus the object attains a velocity of 21 m/s. (The negative sign indicates that the object is moving in the downward direction.) ◄

We recall from Chapter 20 that the current in a circuit is given by $i = dq/dt$. In differential form $dq = i\,dt$, so that $q = \int i\,dt$. Also, since the voltage v across a capacitor is q/C,

$$v = \frac{1}{C} \int i\,dt$$

E X A M P L E **6**   The voltage across a 2.0×10^{-3}-F capacitor is 20 V initially. If the current to the capacitor is $i = 0.3\sqrt{t}$ A, what is the voltage after 1 s?

Solution. From

$$v = \frac{1}{C} \int i\,dt$$

we have

$$v = \frac{1}{2.0 \times 10^{-3}} \int 0.3t^{1/2}\,dt = \frac{0.3}{2.0 \times 10^{-3}} \left(\frac{2}{3} t^{3/2} \right) + k$$

Letting $v = 20$ and $t = 0$, we find that $k = 20$. At $t = 1$,

$$v = \frac{0.3(\frac{2}{3})}{2.0 \times 10^{-3}} + 20 \approx 120 \text{ V}$$ ◄

E X E R C I S E S / S E C T I O N   **22.7**

In Exercises 1–6 find the function satisfying the given conditions.

1. $\frac{dy}{dx} = 3x$, passing through $(0, 1)$

2. $\frac{dy}{dx} = 3x^2$, passing through $(1, 2)$

3. $\frac{dy}{dx} = 6x^2 + 1$, passing through $(-1, 1)$

4. $\dfrac{dy}{dx} = 3x^3 - 1$, passing through $(-2, 3)$

5. $\dfrac{dy}{dx} = 3x^2 + 2$, passing through $(1, 0)$

6. $\dfrac{dy}{dx} = x^3 - 4$, passing through $(2, 3)$

7. A stone is tossed up in the air at 15 m/s. How long does it stay in the air?

8. An object is tossed up in the air at 12.5 m/s. How long does it stay in the air?

9. A ball is thrown upward with a velocity of 30 m/s. How high does it rise?

10. A ball is thrown upward with a velocity of 20 m/s. How high does it rise?

11. An object is dropped from a height of 125 m. How long will it take to hit the ground?

12. An object is dropped from a height of 245 m. How long will it take to hit the ground?

13. An object is hurled downward with an initial velocity of 10 m/s from a height of 50 m. How long does the object take to hit the ground and with what speed does it strike?

14. A stone is hurled downward with an initial velocity of 12 m/s from a height of 80 m. How long does the stone take to hit the ground and with what speed does it strike?

15. A woman standing on a cliff 50 m high tosses a rock upward with an initial velocity of 15 m/s. How long does the rock remain in the air and with what speed does it hit the ground?

16. A man standing on a cliff 60 m high hurls a stone upward at the rate of 20 m/s. How long does the stone remain in the air and with what speed does it strike the ground?

17. A car is moving along a road at 28 m/s. Suddenly the driver sees a child on the road about 60 m ahead of him.

He slams on the brakes and decelerates at the rate of 7 m/s². Will he hit the child?

18. A window washer accidentally falls off his scaffold 110 m above the ground. Two seconds later Superman arrives on the scaffold and dives after the man with an initial velocity of 45 m/s. Will he catch the man before he hits the ground?

19. A student in front of the Roy W. Johnson Residence Hall wants to toss his keys to his roommate, who is leaning out the dorm window 15 m above the student's hand. With what initial velocity must the keys be tossed?

20. Determine the initial velocity of an object that reaches a height of 35 m.

21. An object has an acceleration of $t/(t^2 + 1)^2$. Find an expression for v if $v_0 = 10$.

22. Derive the equations of motion for a particle moving in a straight line with constant acceleration. Use a for acceleration and v_0 and s_0 for the initial velocity and position, respectively.

23. For a short time interval the current in a circuit is given by $i = \sqrt{t + 1.0}$. How many coulombs of charge pass a point in the first 3.0 s if $q = 0$ when $t = 0$?

24. The angular velocity of an object is $\omega = (0.10t^{1/3} + 1.0)$ rad/s. Find the angular displacement θ at $t = 8.0$ s if $\theta = 0$ when $t = 0$.

25. The current to a capacitor initially charged to 0.030 C is given by $i = 0.010t + 0.10$. Find the charge after 3.0 s.

26. The voltage across a 2.0-H inductor is $2.0 - 0.40t$ V. Find the current in the circuit after 5.0 s if $i = 0$ when $t = 0$. (Recall that the voltage is given by $L\, di/dt$.)

27. The voltage across a 0.00300-F capacitor is 30.0 V initially. If the current to the capacitor is $i = 0.0100t$ A, find the voltage after 4.00 s.

28. Find the capacitance of a capacitor with 150 V across it initially and 300 V one second later. The current as a function of time is given by $i = 0.030t^{1/3}$ A.

22.8 Numerical Integration

In this section we are going to study two numerical methods for evaluating definite integrals that cannot be integrated by present techniques.

Trapezoidal Rule

The first technique, called the **trapezoidal rule**, approximates the area under a curve by means of trapezoids. Consider the area under the curve in Figure

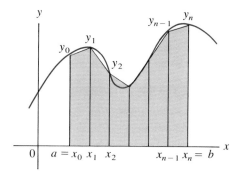

Figure 22.26

22.26. Subdivide the interval $[a, b]$ into n equal parts of length

$$h = \frac{b - a}{n}$$

such that

$$a = x_0 < x_1 < \cdots < x_n = b$$

Now consider the corresponding points $y_0, y_1, \ldots, y_n$ on the curve in Figure 22.26. Connect each pair of adjacent points by a line to form n trapezoids. The area under the curve is approximately equal to the total area of the trapezoids. (Recall that the area of a trapezoid is equal to one-half the product of the altitude and the sum of the bases.) From Figure 22.26,

$$\int_a^b f(x)\,dx \approx \frac{1}{2}(y_0 + y_1)h + \frac{1}{2}(y_1 + y_2)h + \frac{1}{2}(y_2 + y_3)h$$

$$+ \cdots + \frac{1}{2}(y_{n-1} + y_n)h$$

$$= h\left(\frac{1}{2}y_0 + y_1 + y_2 + \cdots + y_{n-1} + \frac{1}{2}y_n\right)$$

Using the more convenient functional notation, the formula can be written as follows:

Trapezoidal Rule:

$$\int_a^b f(x)\,dx \approx h\left[\frac{1}{2}f(x_0) + f(x_1) + \cdots + f(x_{n-1}) + \frac{1}{2}f(x_n)\right] \qquad (22.12)$$

where $h = (b - a)/n$.

E X A M P L E **1** Use the trapezoidal rule with $n = 4$ to find the approximate value of

$$\int_0^1 x^2\, dx$$

Solution. For $n = 4$ we have $h = (b - a)/4 = \frac{1}{4}$. Hence $x_0 = 0$, $x_1 = \frac{1}{4}$, $x_2 = \frac{1}{2}$, $x_3 = \frac{3}{4}$, and $x_4 = 1$. So by formula (22.12)

$$\int_0^1 x^2\, dx \approx \frac{1}{4}\left[\frac{1}{2}(0)^2 + \left(\frac{1}{4}\right)^2 + \left(\frac{1}{2}\right)^2 + \left(\frac{3}{4}\right)^2 + \frac{1}{2}(1)^2\right] = 0.344$$

compared to the exact value of $\frac{1}{3}$ obtained by direct integration.
    The calculator sequence is

$$.25\ \boxed{x^2}\ \boxed{+}\ .5\ \boxed{x^2}\ \boxed{+}\ .75\ \boxed{x^2}\ \boxed{+}\ .5\ \boxed{=}\ \boxed{\div}\ 4\ \boxed{=}\ \rightarrow 0.34375 \qquad\blacktriangleleft$$

Simpson's Rule

For our next approximation technique consider the graph of $y = f(x)$ in Figure 22.27. The interval $[a, b]$ is divided into n equal subdivisions, where n is an **even** number. Looking at the first two subdivisions, we can obtain an approximation to the curve by passing a parabola through (x_0, y_0), (x_1, y_1), and (x_2, y_2). Then the area under the curve between $x = x_0$ and $x = x_2$ is approximately equal to the area under the parabola. It can be shown that the area under the parabola determined by (x_0, y_0), (x_1, y_1), and (x_2, y_2) is

$$\frac{h}{3}(y_0 + 4y_1 + y_2) \qquad \text{where } h = \frac{b - a}{n}$$

This is the approximate area under the curve from x_0 to x_2.

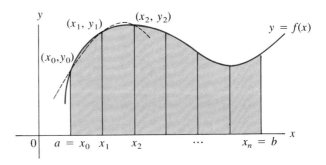

Figure 22.27

Similarly, the area under the parabola determined by (x_2, y_2), (x_3, y_3), and (x_4, y_4) is

$$\frac{h}{3}(y_2 + 4y_3 + y_4)$$

This is the approximate area under the curve from x_2 to x_4. Continuing this process, the total area under the curve from a to b is approximately

$$\frac{h}{3}[(y_0 + 4y_1 + y_2) + (y_2 + 4y_3 + y_4)$$

$$+ (y_4 + 4y_5 + y_6) + \cdots + (y_{n-2} + 4y_{n-1} + y_n)]$$

$$= \frac{h}{3}(y_0 + 4y_1 + 2y_2 + 4y_3 + 2y_4 + 4y_5 + 2y_6$$

$$+ \cdots + 2y_{n-2} + 4y_{n-1} + y_n)$$

This approximation procedure is known as **Simpson's rule** and leads to a better approximation than the trapezoidal rule for the same n (although in the trapezoidal rule n can be even or odd).

Simpson's Rule:

$$\int_a^b f(x)\, dx \approx \frac{h}{3}[f(x_0) + 4f(x_1) + 2f(x_2) + 4f(x_3) +$$

$$\cdots + 2f(x_{n-2}) + 4f(x_{n-1}) + f(x_n)] \qquad (22.13)$$

where $h = (b - a)/n$ and n is an **even** integer.

E X A M P L E **2**   Use Simpson's rule with $n = 4$ to approximate the value of the integral in Example 1.

Solution. Since the x values and h are the same, from formula (22.13) we get

$$\int_0^1 x^2\, dx \approx \frac{\frac{1}{4}}{3}\left[0^2 + 4\left(\frac{1}{4}\right)^2 + 2\left(\frac{1}{2}\right)^2 + 4\left(\frac{3}{4}\right)^2 + 1^2\right] = 0.333$$

which is better than the approximation obtained using the trapezoidal rule.   ◀

The calculations involved in these techniques can be carried out conveniently with a calculator.

E X A M P L E **3**   Find the approximate value of the following definite integral by Simpson's rule with $n = 6$:

$$\int_{-1}^{2} \frac{dx}{x + 2}$$

Solution. For $n = 6$ we have $h = (b - a)/6 = \frac{3}{6} = \frac{1}{2}$. Hence $x_0 = -1$, $x_1 = -\frac{1}{2}$, $x_2 = 0$, $x_3 = \frac{1}{2}$, $x_4 = 1$, $x_5 = \frac{3}{2}$, and $x_6 = 2$. So by (22.13)

$$\int_{-1}^{2} \frac{dx}{x + 2} \approx \frac{\frac{1}{2}}{3} \left[\frac{1}{-1 + 2} + \frac{4}{-\frac{1}{2} + 2} + \frac{2}{0 + 2} \right.$$

$$\left. + \frac{4}{\frac{1}{2} + 2} + \frac{2}{1 + 2} + \frac{4}{\frac{3}{2} + 2} + \frac{1}{2 + 2} \right]$$

$$= \frac{1}{6} \left(1 + \frac{8}{3} + 1 + \frac{8}{5} + \frac{2}{3} + \frac{8}{7} + \frac{1}{4} \right) = 1.388$$

The sequence is

$$1 \boxed{+} 8 \boxed{\div} 3 \boxed{+} 1 \boxed{+} 8 \boxed{\div} 5 \boxed{+} 2 \boxed{\div} 3 \boxed{+} 8 \boxed{\div} 7$$
$$\boxed{+} 1 \boxed{\div} 4 \boxed{=} \boxed{\div} 6 \boxed{=} \to 1.3876984 \qquad \blacktriangleleft$$

Simpson's rule can also be used to integrate a function known only at a discrete set of points, as in the case of experimental data. (See Exercises 13 and 14.)

E X E R C I S E S / S E C T I O N   **22.8**

In Exercises 1–6, find the approximate value of each of the given integrals with the specified value of n by **(a)** the trapezoidal rule; **(b)** Simpson's rule. Use a calculator and round off to three decimal places.

1. $\int_{1}^{3} x^2 \, dx$, $n = 6$ (Check by direct integration.)

2. $\int_{0}^{4} x^3 \, dx$, $n = 6$ (Check by direct integration.)

3. $\int_{0}^{1} \frac{dx}{1 + x^2}$, $n = 4$ **4.** $\int_{1}^{4} \frac{dx}{x}$, $n = 4$

5. $\int_{0}^{2} \sqrt{1 + x} \, dx$, $n = 4$ **6.** $\int_{2}^{6} \frac{dx}{1 + x}$, $n = 8$

7. Use the trapezoidal rule with $n = 5$ to approximate the value of $\int_{0}^{3} \sqrt{x} \, dx$. Check by direct integration.

In Exercises 8–12, find the approximate value of the given integrals by using Simpson's rule with the specified value of n. (Use a calculator and round off to three decimal places.)

8. $\int_{-1}^{1} \sqrt{x + 1} \, dx$, $n = 6$ **9.** $\int_{1}^{4} \sqrt{1 + x^2} \, dx$, $n = 6$

10. $\int_{1}^{3} \sqrt{x^2 - 1} \, dx$, $n = 4$

11. $\int_{0}^{2} \sqrt{1 + x^4} \, dx$, $n = 6$

12. $\int_{1}^{4} \frac{\sqrt{x} \, dx}{1 + x}$, $n = 6$

In Exercises 13 and 14, find the approximate area under the curve defined by each of the following sets of experimental data by using Simpson's rule.

13.

x:	3	4	5	6	7	8	9	10	11
y:	1.3	1.9	3.2	3.8	4.7	6.8	10.2	15.6	20.3

14.

x:	0.3	0.6	0.9	1.2	1.5	1.8	2.1
y:	7.3	8.2	9.8	12.3	10.2	8.7	6.5

REVIEW EXERCISES/CHAPTER 22

1. Evaluate the integral $\int_0^3 3x^2 \, dx$ as a limit of a sum. Check the result by integration.

In Exercises 2–13, perform the integrations.

2. $\int_1^4 2x\sqrt{x} \, dx$

3. $\int (3\sqrt{x} - x^{-4} + 1) \, dx$

4. $\int (5x^2 + 4)^3 (10x) \, dx$

5. $\int (1 - x^2)^5 x \, dx$

6. $\int \dfrac{dx}{\sqrt{x-4}}$

7. $\int \dfrac{3x \, dx}{\sqrt{x^2 - 2}}$

8. $\int (x^3 + 1)^2 (4x^2) \, dx$

9. $\int (x^3 + 1)^2 (3x) \, dx$

10. $\int \dfrac{(\sqrt{x} - 1)^2}{\sqrt{x}} \, dx$

11. $\int (\sqrt{x} - 1)^2 \, dx$

12. $\int x^3 \sqrt{3 - x^4} \, dx$

13. $\int (x - 2)\sqrt{x^2 - 4x} \, dx$

In Exercises 14–24, find the area of the region bounded by the given curves.

14. $y = x^2$, $y = -x^2$, $x = 2$

15. $y = \sqrt{x}$, $y = -x$, $x = 4$

16. $4x + y^2 = 0$, $y^2 - 4x - 8 = 0$

17. $x = 4 - y^2$, y-axis

18. $y = x(x - 1)(x - 3)$, x-axis

19. $x = 4 - y$, $x = y^2 - 4y + 4$

20. $y = x^2 + 4$, $x + y = 6$

21. $y = 1/x^3$, $x = 2$, x-axis, to the right of $x = 2$

22. $y = 1/\sqrt{x}$, $x = 1$, x-axis, to the right of $x = 1$

23. $y = 1/\sqrt{x}$, $x = 1$, coordinate axes

24. $x = 1/(y - 3)^{3/2}$, $y = 4$, y-axis, above the line $y = 4$

25. If the current in a circuit is given by $i = 3.08t^{1/2}$, how many coulombs of charge will pass a point in the first 1.75 s if $q = 0$ when $t = 0$?

26. The voltage across a 4.00-microfarad (μF) capacitor is 10.0 V initially. If the current to the capacitor is $i = 0.310\sqrt{t} - 1.23t$, what is the voltage after 10.0 milliseconds? (1 μF $= 10^{-6}$ F; 1 millisecond $= 10^{-3}$ s.)

27. A woman standing on the roof of a building 121 ft high tosses a stone upward with a velocity of 29 ft/s. How long does it take for the stone to reach the ground? ($g = 32$ ft/s^2)

28. A man hurls an object downward with an initial velocity of 20 m/s from a height of 63 m. How long does the object take to reach the ground and with what speed does it strike? ($g = 10$ m/s^2)

29. Use a calculator to find the approximate value of the integral $\int_{-1}^{2} dx/(x + 3)$ by Simpson's rule with $n = 6$.

CHAPTER 23

Applications of the Integral

23.1 Introduction

The purpose of this chapter is to discuss the basic applications of the integral. Perhaps a more natural place for some of these topics would be after Chapter 25—by then other integration techniques will be available to us. Fortunately, many applications do not require any integration techniques beyond those already considered. Additional problems dealing with applications of other techniques are included in the exercises in Chapter 25.

23.2 Means and Root Mean Squares

Everyone is familiar with the idea of arithmetic **average** or **mean**. We hear about batting averages, average per capita income, average salaries, and even average students. Less obvious is the meaning of the average value of a continuous function. Yet we use this idea all the time: if it takes you half an hour to drive a distance of 25 mi, then you must be averaging 50 mi/h.

To obtain the average value of a continuous function $f(x)$ on $[a, b]$, we divide the interval $[a, b]$ into n *equal* subintervals of length Δx_i (Figure 23.1). Next we take the average of the n values $f(x_1), f(x_2), \ldots, f(x_n)$ by adding the values and dividing by n:

$$\frac{\sum_{i=1}^{n} f(x_i)}{n}$$

Multiplying numerator and denominator by Δx_i, we get

$$\frac{\sum_{i=1}^{n} f(x_i)\,\Delta x_i}{n\,\Delta x_i} = \frac{\sum_{i=1}^{n} f(x_i)\,\Delta x_i}{b - a}$$

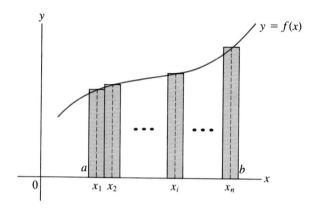

Figure 23.1

since

$$n \, \Delta x_i = \Delta x_i + \Delta x_i + \cdots + \Delta x_i \qquad (n \text{ times})$$

which is the length of the interval $b - a$. The mean value of $f(x)$, denoted by f_{av}, is then defined to be the limit of this average as $n \to \infty$. Since the numerator is a Riemann sum, we get the following formula:

Mean Value of f(x) on [a, b]:

$$f_{av} = \frac{\lim\limits_{n \to \infty} \sum\limits_{i=1}^{n} f(x_i) \, \Delta x_i}{b - a} = \frac{\int_a^b f(x) \, dx}{b - a} \qquad (23.1)$$

Graphically, f_{av} is the ordinate of a point on a curve such that the area of the rectangle of height f_{av} and base $b - a$ is equal to the area beneath the curve on $[a, b]$ (Figure 23.2).

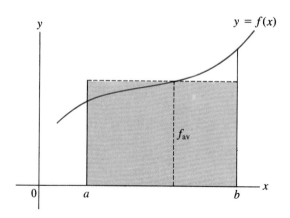

Figure 23.2

E   X   A   M   P   L   E   **1**   Find the mean value of the function $y = \sqrt[3]{x}$ on $[0, 8]$.

Solution. By formula (23.1)

$$f_{av} = \frac{\int_0^8 \sqrt[3]{x}\, dx}{8 - 0} = \frac{1}{8} \int_0^8 x^{1/3}\, dx = \frac{3}{2}$$

◀

E   X   A   M   P   L   E   **2**   If the current in a circuit is given by $i = \sqrt[3]{t}$ A, then the mean current from $t = 0.0$ s to $t = 8.0$ s is 1.5 A by Example 1.   ◀

If the current in a circuit is alternating, then it changes directions periodically. Hence the mean current of any complete cycle is always zero. A more useful measure of the "average" in this case is the **root mean square** (rms), defined as follows:

Root Mean Square: If f is continuous on $[a, b]$, then

$$f_{rms} = \left\{ \frac{1}{b - a} \int_a^b [f(x)]^2\, dx \right\}^{1/2} \qquad\qquad (23.2)$$

Effective current

If i is the current, then i_{rms}, called the **effective current**, is the value of the direct current that generates the same amount of heat. Moreover, if the resistance is constant, then the mean power produced by the current is

$$P = i_{rms}^2 \cdot R \text{ (in watts)}$$

E   X   A   M   P   L   E   **3**   If a current $i = t - t^2$ A flows through a resistor of 30 Ω from $t = 0$ s to $t = 2$ s, find i_{rms} and the power generated.

Solution. From the definition of root mean square we get

$$i_{rms} = \left[\frac{1}{2 - 0} \int_0^2 (t - t^2)^2\, dt \right]^{1/2}$$

$$= \left[\frac{1}{2} \int_0^2 (t^2 - 2t^3 + t^4)\, dt \right]^{1/2}$$

$$= \left[\frac{1}{2} \left(\frac{1}{3} t^3 - \frac{1}{2} t^4 + \frac{1}{5} t^5 \right) \Big|_0^2 \right]^{1/2}$$

$$= \left[\frac{1}{2} \left(\frac{8}{3} - 8 + \frac{32}{5} \right) \right]^{1/2} = \left(\frac{8}{15} \right)^{1/2} \text{ A}$$

Hence

$$P = i^2_{\text{rms}} \cdot R = \left[\left(\frac{8}{15}\right)^{1/2}\right]^2 \cdot 30 = \frac{8}{15} \cdot 30 = 16 \text{ W}$$

◄

E X E R C I S E S / S E C T I O N   23.2

In Exercises 1–4, find the mean value of each function on the given interval.

1. $y = \sqrt{x}$; $[1, 16]$

2. $y = 1 - x$; $[1, 4]$

3. $y = x\sqrt{x^2 + 1}$; $[-2, 2]$

4. $y = \dfrac{1}{\sqrt{x + 2}}$; $[-1, 3]$

In Exercises 5–8, find the root mean square in each case.

5. $y = \dfrac{1}{x}$; $[1, 2]$

6. $y = x^{2/3}$; $[-1, 1]$

7. $y = \sqrt{x}(x^2 + 1)$; $[0, 1]$

8. $y = \sqrt[4]{x}$; $[0, 4]$

9. An object is dropped from a height of 180 m. Find its mean velocity. $(g = 10 \text{ m/s}^2)$

10. A rocket travels a distance $s = 3t^{5/2}$ meters from a launching pad during the first 10 s. Find its mean velocity during that time.

11. Find the mean current from $t = 0.0$ s to $t = 4.0$ s if $i = 1.0t + 1.0\sqrt{t}$ A.

12. Find the root mean square current for the current in Exercise 11.

13. A current $i = 1.0 - 1.0t^2$ A is flowing through a 5.0-Ω resistor. Find the mean power from $t = 0.0$ s to $t = 3.0$ s.

14. The voltage across a resistor of 10 Ω is given by $v = 2\sqrt{t} - t$. Find the mean power generated during the time interval $[0, 8]$.

23.3   Volumes of Revolution: Disk and Washer Methods

We noted in Chapter 22 that finding areas is only one of many applications of the definite integral. In this section we will learn how to find volumes of certain solids, called *solids of revolution*. Examples of such solids are bottles, axles, and funnels.

Let f be a function continuous on $[a, b]$ and consider the region under the curve in Figure 23.3. If the entire region is rotated about the x-axis, we obtain a **solid of revolution**. If the region is approximated by rectangles, then the approximate volume is obtained when each rectangle is rotated about the base.

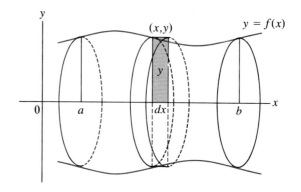

Figure 23.3

Employing the shortcut method ("sloppy Riemann sum"), we draw a typical element and rotate this element about the base, forming a cylindrical disk. The volume of this cylinder is

$$\pi(\text{radius})^2 \cdot \text{height} = \pi[f(x)]^2 \, dx$$

Summing from $x = a$ to $x = b$, the volume is given by

$$V = \pi \int_a^b [f(x)]^2 \, dx = \pi \int_a^b y^2 \, dx \qquad (23.3)$$

Although formula (23.3) can be employed directly to find the volume of a solid of revolution, it is better to draw a typical element and obtain the integral from the figure. The reason is that in this chapter we will be dealing with many different integrals. It is poor practice to rely entirely on memorizing set formulas, and, above all, *a given formula may not be applicable to every situation encountered.*

> **The Disk Method:** To find the volume of a solid of revolution by the **disk method,** draw a typical element, construct a disk, and obtain the desired integral from this disk.
>
> **Volume of disk:** $\pi(\text{radius})^2 \cdot (\text{thickness})$

E X A M P L E **1**   Derive the formula for the volume of a sphere of radius r.

Solution. A sphere is a solid of revolution that can be conveniently generated by the semicircle $y = \sqrt{r^2 - x^2}$ as in Figure 23.4. Taking advantage of the symmetry, we may rotate the quarter-circle in the first quadrant and double the result. Following the plan given above, we draw a typical rectangle in Figure 23.4 and obtain the volume of each cylindrical disk:

$$\pi(\text{radius})^2 \cdot (\text{thickness}) = \pi(\sqrt{r^2 - x^2})^2 \, dx$$

Summing from $x = 0$ to $x = r$ by the operation $\int_0^r$ and multiplying by 2, we get

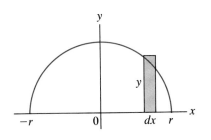

Figure 23.4

$$V = 2 \int_0^r \pi(\sqrt{r^2 - x^2})^2 \, dx$$

$$= 2\pi \int_0^r (r^2 - x^2) \, dx \qquad \text{Note that } r \text{ is a constant.}$$

$$= 2\pi \left(r^2 x - \frac{1}{3}x^3 \right)\Big|_0^r = 2\pi \left(r^3 - \frac{1}{3}r^3 \right) = \frac{4}{3}\pi r^3 \qquad \blacktriangleleft$$

If the region to be rotated is bounded by two curves (Figure 23.5), then the typical rectangle extends between the curves. Upon rotation the typical element

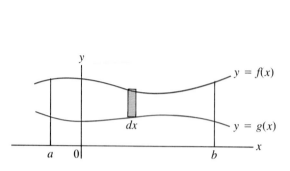

Figure 23.5

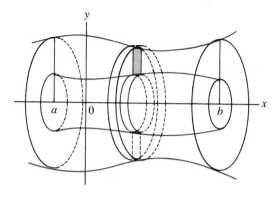

Figure 23.6

generates a *washer* (Figure 23.6) with outer radius $f(x)$ and inner radius $g(x)$. Hence the volume of the washer is

$$\pi[f(x)]^2\, dx - \pi[g(x)]^2\, dx = \pi\{[f(x)]^2 - [g(x)]^2\}\, dx$$

Volume of Washer:

$$\pi\{[f(x)]^2 - [g(x)]^2\}\, dx \qquad\qquad\qquad (23.4)$$

E X A M P L E **2**   The line $y = x + 3$ and the parabola $y = x^2 + 1$ form a bounded region. Find the volume of the solid of revolution formed by revolving the region about the x-axis (Figure 23.7).

Solution. Solving the equations simultaneously, we find that $x = -1$ and $x = 2$ are the limits of integration. The volume of the typical washer is

$$\pi[(x + 3)^2 - (x^2 + 1)^2]\, dx$$

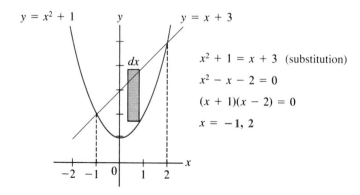

Figure 23.7

Summing from -1 to 2, we have

$$V = \int_{-1}^{2} \pi[(x+3)^2 - (x^2+1)^2]\,dx$$

$$= \pi \int_{-1}^{2} (-x^4 - x^2 + 6x + 8)\,dx$$

$$= \pi \left(-\frac{1}{5}x^5 - \frac{1}{3}x^3 + 3x^2 + 8x \right)\Big|_{-1}^{2}$$

$$= \pi \left[\left(-\frac{32}{5} - \frac{8}{3} + 12 + 16 \right) - \left(\frac{1}{5} + \frac{1}{3} + 3 - 8 \right) \right]$$

$$= \frac{117\pi}{5} \approx 73.5$$ ◄

In the next example the given region is rotated about the y-axis.

E X A M P L E **3** Find the volume of the solid obtained by rotating the region bounded by $x = \sqrt{y}$, $y = 4$, and the y-axis about the y-axis.

Solution. The region is shown in Figure 23.8.

Since the region is to be rotated about the y-axis, we need to interchange the roles of x and y. To this end, we *draw the typical element in the horizontal position* with height x and thickness dy, as shown in Figure 23.8. We now get

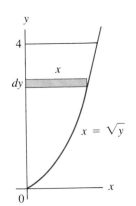

$$\text{Volume of disk} = \pi(\text{radius})^2 \cdot (\text{thickness})$$
$$= \pi x^2\,dy = \pi(\sqrt{y})^2\,dy \qquad x = \sqrt{y}$$

Summing from $y = 0$ to $y = 4$ by the operation $\int_0^4$, we get

$$V = \int_0^4 \pi(\sqrt{y})^2\,dy = \pi \int_0^4 y\,dy = \pi\frac{y^2}{2}\Big|_0^4 = 8\pi \approx 25.1$$ ◄

Figure 23.8

E X A M P L E **4** Revolve the region bounded by $y = x^{2/3}$, $x = 8$, and the x-axis about the y-axis (Figure 23.9).

Solution. Since the region is to be revolved about the y-axis, we need to draw the typical rectangle horizontally. Solving the equation $y = x^{2/3}$ for x, we get $x = y^{3/2}$.

The volume of the typical washer is given by

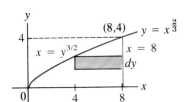

$$\pi(8^2 - x^2)\,dy = \pi[64 - (y^{3/2})^2]\,dy \qquad \text{washer}$$

Figure 23.9   (See Figure 23.9.)

Since $y = 4$ when $x = 8$, we sum from $y = 0$ to $y = 4$:

$$V = \int_0^4 \pi[64 - (y^{3/2})^2]\,dy$$

$$= \pi \int_0^4 (64 - y^3)\,dy = \pi(64y - \tfrac{1}{4}y^4)\big|_0^4$$

$$= \pi(64 \cdot 4 - \tfrac{1}{4} \cdot 4^4) = \pi(4^4 - 4^3) = 4^3\pi(4 - 1) = 192\pi \qquad \blacktriangleleft$$

E X E R C I S E S / S E C T I O N   23.3

In Exercises 1–11, a region R is bounded by the given curves. Find the volume of the solid of revolution obtained by rotating the region R about the x-axis.

1. $y = 2x$, $x = 1$, $x = 4$, x-axis

2. $y = x^3$, $x = 0$, $x = 1$, x-axis

3. $y = x^3$, $y = x$, the part of the region in the first quadrant

4. $y = x^4$, $x = -2$, $x = 2$, x-axis

5. $y = x^{3/2}$, $x = 0$, $x = 2$, x-axis

6. $y^2 = x$, $x = a$, the part of the region in the first quadrant. (The solid is called a *paraboloid*.)

7. $y = x^2 + 1$, $x = 2$, coordinate axes

8. $y = x^{3/2}$, $y = x^2$

9. $y = \sqrt{x^2 + 1}$, $x = 1$, $x = 3$, x-axis

10. $y = \dfrac{1}{x}$, $y = x$, $x = 3$

11. $y = \dfrac{1}{x + 2}$, $x = -1$, $x = 1$, x-axis

In Exercises 12–25 find the volume of the solid of revolution obtained by rotating R, defined by the bounds given in each exercise, about the y-axis.

12. $y = x^2$, $y = 4$ (first quadrant)

13. $y = \tfrac{1}{2}x^2$, $y = 2$ (first quadrant)

14. $x = \sqrt{y + 1}$, coordinate axes

15. $x = \tfrac{1}{2}y$, $y = 4$, y-axis

16. $x = \sqrt{y}$, $y = 6$, y-axis

17. $x = \tfrac{1}{2}y$, $x = 2$, x-axis

18. $x = \sqrt{y}$, $x = 2$, x-axis

19. $y = \tfrac{1}{2}x$, $y = 2$, y-axis

20. $y = \tfrac{1}{3}x$, $y = 2$, y-axis

21. $y = \tfrac{1}{2}x$, $x = 4$, x-axis

22. $y = \tfrac{1}{3}x$, $x = 6$, x-axis

23. $y = x$, $x + y = 2$, x-axis

24. $y^2 = 4x$, $4x - 3y - 4 = 0$

25. $x = y^2$, $x = y + 2$

26. Derive the formula for the volume of a cone of radius r and height h.

27. The headlight on a car has a parabolic mirror with a depth of 12.0 cm and a diameter of 20.0 cm. Determine the volume.

28. A whispering gallery has a flat floor and a ceiling with elliptical cross-sections. The room is 50 ft long and 15 ft high in the center. Find the volume of the room.

29. A hemispherical tank has a radius of 12.0 ft. The water is 3.0 ft deep in the center. Find the volume of the water.

30. A wheel for a certain machine is made by cutting out the center portion of a sphere of radius 9.74 inches. (See Figure 23.10.) Find the volume of the wheel.

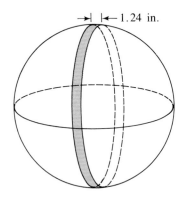

Figure 23.10

In Exercises 31–34, the region R is bounded by the given curves and rotated about the x-axis. Find the volume of the solid in each case, noting that *the resulting integrals are improper.*

31. $y = 1/x$, $x = 1$, x-axis to the right of $x = 1$

32. $x^2y = 2$, $x = 2$, x-axis to the right of $x = 2$

33. $y = 1/x^{3/4}$, $x = 4$, x-axis to the right of $x = 4$ (Show that the area of the region does not exist.)

34. $y = 1/(x + 1)$, coordinate axes (first quadrant)

23.4 Volumes of Revolution: Shell Method

In this section we are going to find the volume of a solid of revolution by a different method: **the shell method**. Consider the region in Figure 23.11 between a and b and the typical element of height $f(x)$ and thickness dx. If the typical element is revolved about the y-axis, we obtain a **cylindrical shell**. To estimate its volume, we cut the shell vertically and lay it out flat to form a box (Figure 23.12). Note that the radius of the shell is x and that the resulting box is extremely thin; its length is approximately equal to the circumference $2\pi x$ of the shell. So the volume of the box is given by

$$\text{Length} \times \text{height} \times \text{width} = 2\pi x\, f(x)\, dx = 2\pi xy\, dx \qquad (23.5)$$

If the region is bounded by two functions f and g (Figure 23.13), then the volume of the typical shell becomes

$$2\pi x[f(x) - g(x)]\, dx \qquad (23.6)$$

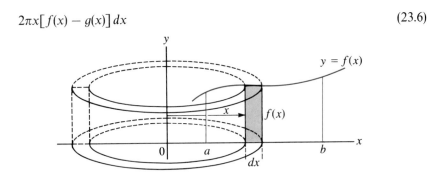

Figure 23.11

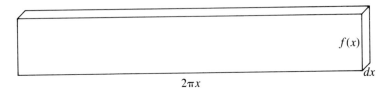

Figure 23.12

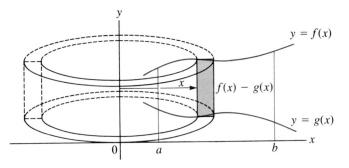

Figure 23.13

Inasmuch as the height of the region may vary at different values of $f(x)$ and $g(x)$, to have a precise formula we must sum from $x = a$ to $x = b$ to find

$$V = 2\pi \int_a^b x[f(x) - g(x)]\,dx \tag{23.7}$$

As in the case of the disk method, *it is best to remember how to construct the shell rather than to memorize the formula* (23.7).

The Shell Method: To find the volume of a solid of revolution by the **shell method**, draw a typical element, construct a shell, and obtain the desired integral from this shell.

Volume of shell: 2π(radius) · (height) · (thickness)

E X A M P L E   **1**   Find the volume of the solid of revolution in Example 4 of Section 23.3 by the shell method.

Solution. We draw the graph of the given function over again (Figure 23.14). This time, however, the typical rectangle has to be drawn vertically to generate shells.

Since $y = x^{2/3}$, the volume of the typical shell is

$$2\pi(\text{radius}) \cdot (\text{height}) \cdot (\text{thickness}) = 2\pi xy\,dx = \boldsymbol{2\pi x \cdot x^{2/3}\,dx}$$

Now sum from $x = 0$ to $x = 8$:

$$V = \int_0^8 2\pi x \cdot x^{2/3}\,dx$$

$$= 2\pi \int_0^8 x^{5/3}\,dx = 192\pi$$

In this particular case the method of shells turns out to be somewhat simpler. It is to be expected, though, that in other problems the disk method will be more convenient than the shell method.

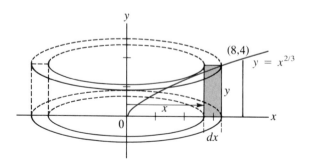

Figure 23.14

E X A M P L E **2**

Use the method of shells to find the volume of the solid obtained by revolving the region bounded by $y = \frac{1}{2}x$, $y = 2$, and the y-axis about the x-axis.

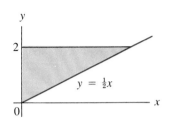

$y = \frac{1}{2}x$

Figure 23.15

Solution. The region is shown in Figure 23.15.

The region is to be rotated about the x-axis. To generate shells, we must draw the typical element in the horizontal position—and interchange the roles of x and y. (See Figure 23.16.)

By Figure 23.16, the volume of the typical shell is

$$2\pi(\text{radius}) \cdot (\text{height}) \cdot (\text{thickness}) = 2\pi \cdot y \cdot x \, dy = 2\pi y(2y) \, dy \qquad x = 2y$$

Summing from $y = 0$ to $y = 2$, we get

$$V = \int_0^2 2\pi y(2y) \, dy = 4\pi \int_0^2 y^2 \, dy$$

$$= 4\pi \left. \frac{y^3}{3} \right|_0^2 = \frac{32\pi}{3}$$

◀

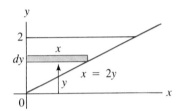

$x = 2y$

Figure 23.16

E X A M P L E **3**

Using the method of cylindrical shells, find the volume of the solid of revolution obtained by revolving the region bounded by

$$x = -y^2 + 4y - 2 \qquad \text{and} \qquad x = y^2 - 4y + 4$$

about the x-axis.

Solution. Solving the equations simultaneously, we find that the curves intersect at $(1, 1)$ and $(1, 3)$. The region is pictured in Figure 23.17.

Since the region is to be revolved about the x-axis, we need to draw the typical element in a horizontal position. Note that the thickness of the typical element is dy and the circumference of the typical shell $2\pi y$. The height of the

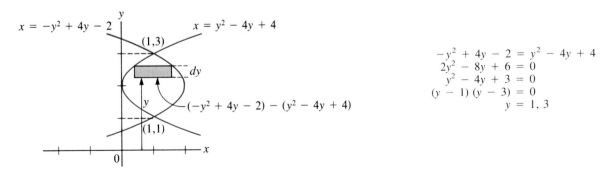

$$-y^2 + 4y - 2 = y^2 - 4y + 4$$
$$2y^2 - 8y + 6 = 0$$
$$y^2 - 4y + 3 = 0$$
$$(y - 1)(y - 3) = 0$$
$$y = 1, 3$$

Figure 23.17

shell is

$$(-y^2 + 4y - 2) - (y^2 - 4y + 4)$$

so that the volume of the typical shell becomes

$$2\pi(\text{radius}) \cdot (\text{height}) \cdot (\text{thickness})$$
$$= 2\pi y[(-y^2 + 4y - 2) - (y^2 - 4y + 4)]dy$$
$$= 2\pi y(-2y^2 + 8y - 6) \, dy$$

Summing from $y = 1$ to $y = 3$, we get

$$V = \int_1^3 2\pi y(-2y^2 + 8y - 6) \, dy = \frac{32\pi}{3} \qquad \blacktriangleleft$$

It is possible to revolve a region about a line other than a coordinate axis. Returning to basic principles is especially important here, since formula (23.7) fails us.

E X A M P L E **4**   Find the volume of the solid obtained by revolving the region bounded by $y = 2x$, $x = 1$, and the x-axis about the line $x = -1$.

Solution. The region is shown in Figure 23.18.

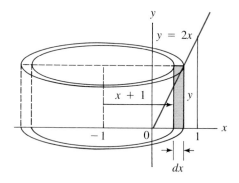

Figure 23.18

The height of the typical element is y and the thickness is dx. Since we are rotating about the line $x = -1$, the radius of the typical shell is the distance from the typical element to the line $x = -1$; that is,

$$\text{Radius} = x - (-1) = x + 1$$

The volume of the typical shell now becomes

$$2\pi(\text{radius}) \cdot (\text{height}) \cdot (\text{thickness}) = 2\pi(x + 1)y \, dx$$
$$= 2\pi(x + 1)(2x) \, dx$$

Summing from $x = 0$ to $x = 1$, we get

$$V = \int_0^1 2\pi(x + 1)(2x)\,dx$$

$$= 4\pi \int_0^1 (x^2 + x)\,dx = \frac{10\pi}{3}$$

◀

E X A M P L E **5**   Find the volume of the solid generated by revolving the region bounded by $y = 8 + 4x - x^2$, $y = 2x$, and the y-axis (first quadrant) about the line $x = 4$ (Figure 23.19).

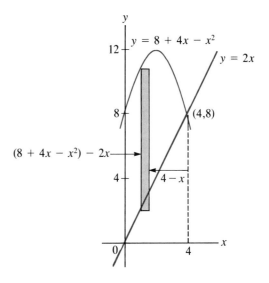

Figure 23.19

Solution. Since the region is to be revolved about the line $x = 4$, the radius of the typical shell is $4 - x$. The height of each shell is

$$(8 + 4x - x^2) - 2x$$

so that the volume of the shell becomes

$$2\pi(\text{radius}) \cdot (\text{height}) \cdot (\text{thickness})$$
$$= 2\pi(4 - x)[(8 + 4x - x^2) - 2x]\,dx$$
$$= 2\pi(4 - x)(8 + 2x - x^2)\,dx$$

Summing from $x = 0$ to $x = 4$, we obtain

$$V = 2\pi \int_0^4 (4 - x)(8 + 2x - x^2)\,dx = 128\pi$$

◀

> **Summary of Disk and Shell Methods:**
>
> In finding a volume of revolution by the **disk method**, place the typical element **perpendicular** to the axis of rotation.
>   In finding a volume of revolution by the **shell method**, place the typical element **parallel** to the axis of rotation.
>   In both methods, if the width of the element is dx, then the variable of integration is x. If the width of the element is dy, then the variable of integration is y.

E X E R C I S E S / S E C T I O N   **23.4**

In Exercises 1–8, use the method of shells to find the volume of the solid obtained by revolving the region bounded by the given curves about the y-axis.

1. $y = x$, $x = 3$, x-axis

2. $y = 4x^2$, $x = 1$, $x = 4$, x-axis

3. $y = x^2$, $y = x$

4. $y = 2\sqrt{x}$, $x = 4$, x-axis

5. $y = x^2$, $y^2 = 8x$

6. $y = \sqrt{x^2 - 1}$, $x = 1$, $x = 4$, x-axis

7. $y = x + 2$, $y = x^2$ (first quadrant)

8. $y = 2x$, $y = 3 - x$, y-axis

In Exercises 9–19, rotate the region bounded by the given curves about the line indicated. Obtain the volume of the solid by the method of shells.

9. $y = x^2$, $x = 1$, x-axis; about the line $x = 1$

10. $y = x^2$, $y = x$; about the line $x = -1$

11. $y = x^2$, $y = \sqrt{x}$; about the x-axis

12. $x = \sqrt{4 - y^2}$, y-axis; about the y-axis

13. $y = x - x^2$, x-axis; about the line $x = 2$

14. $x^2 - y^2 = 1$, $x = \sqrt{5}$ (first quadrant); about the y-axis

15. $x = 2y - y^2$, y-axis; about the x-axis

16. $y = x^2$, $y = 4$; about the x-axis

17. $y = 4x^2$, $x = 4$, x-axis; about the line $x = -1$

18. $y^2 = 8x$, $x = 2$; about the line $x = 4$

19. $y = 2x - 2x^2$, x-axis; about the line $x = -2$

20. Derive the formula for the volume of a sphere by the shell method. *Hint:* Rotate the first-quadrant region bounded by $y = \sqrt{r^2 - x^2}$ and the coordinate axes about the y-axis.

21. Derive the formula for the volume of a cone by the shell method.

22. Use the shell method to find the volume generated by revolving the region bounded by
$$y = x^{3/2}, \ x = 0, \ x = 1, \ x\text{-axis}$$
a. about the x-axis   **b.** about the y-axis

23. Find the volume generated by rotating the region bounded by
$$y = 2\sqrt{x}, \ x = 4, \ x\text{-axis}$$
about the x-axis
a. by the disk method   **b.** by the shell method

24. Repeat Exercise 23 for the region bounded by $y = \sqrt{x - 1}$, $x = 10$, and the x-axis.

25. Consider the first-quadrant region bounded by $y^2 = x^3$, $x = 4$, and the x-axis. Find the volume generated by rotating this region about:
a. the x-axis   **b.** the y-axis
c. the line $x = 4$ **d.** the line $y = 8$

26. Find the volume of the solid obtained by revolving the region bounded by $y = \sqrt{x}$, $x = 4$, and the x-axis about:
a. the x-axis   **b.** the y-axis
c. the line $x = 5$ **d.** the line $y = 3$

27. Find the volume of the solid obtained by revolving the region bounded by $x = \sqrt{y}$, $y = 1$, and the y-axis about:
a. the line $y = -1$ **b.** the line $x = 2$

28. Find the volume of the solid in Exercise 1 by the disk or washer method.

29. Find the volume of the solid in Exercise 3 by the disk or washer method.

In Exercises 30–32, find the volume obtained by rotating the given region about the y-axis. The integrals are improper.

30. $y = 1/x^3$, $x = 1$, x-axis to the right of $x = 1$

31. $y = 1/x^4$, $x = 2$, the x-axis to the right of $x = 2$

32. $y = 1/(x^2 + 1)^2$, coordinate axes (first quadrant)

23.5 Centroids

In this section we will use integration to find centroids of plates and solids of revolution.

Suppose a man wants to play seesaw with his three small children. The children climb up on one side of the fulcrum, while he sits on the other side. Where should he place himself to obtain an exact balance? Let w_1, w_2, and w_3 be the weights of the children and d_1, d_2, and d_3 their respective distances from the fulcrum O (Figure 23.20). If the man's weight is denoted by w_4, then he should position himself a distance d_4 from O such that

$$w_4 d_4 = w_1 d_1 + w_2 d_2 + w_3 d_3$$

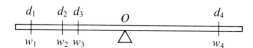

Figure 23.20

Moment

Each distance d_i is the length of the **moment arm**, and each product $w_i d_i$ is the **moment** of w_i with respect to O. Note that moments are additive, so that the moment of a system is equal to the sum of the individual moments. Furthermore, we may pretend that all the weights on the left are concentrated at one point whose distance $\bar{d}$ from O is such that

$$(w_1 + w_2 + w_3)\bar{d} = w_1 d_1 + w_2 d_2 + w_3 d_3$$

or

$$\bar{d} = \frac{w_1 d_1 + w_2 d_2 + w_3 d_3}{w_1 + w_2 + w_3}$$

Centroid

In this formula, $\bar{d}$ is called the **center of mass** or **centroid** of the system on the left and is the point where all the weights appear to be concentrated. For n weights,

$$\bar{d} = \frac{w_1 d_1 + w_2 d_2 + \cdots + w_n d_n}{w_1 + w_2 + \cdots + w_n} = \frac{\displaystyle\sum_{i=1}^{n} w_i d_i}{\displaystyle\sum_{i=1}^{n} w_i} \qquad (23.8)$$

If we let

$$M = \sum_{i=1}^{n} w_i d_i$$

the moment of a system with respect to O, and

$$W = \sum_{i=1}^{n} w_i$$

then we may also write

$$\bar{d} = \frac{M}{W} \tag{23.9}$$

Furthermore, $M = \bar{d}W$.

In words: the moment of a system with respect to O is equal to the total weight of the system multiplied by the distance from O to the center of mass.

These ideas can be extended to systems in which mass is distributed uniformly throughout a plate. The center of mass of a plate is the point at which the entire plate will balance if supported, for instance, by a sharp nail. If the plate consists of a combination of rectangles, finding the center of mass is simple: for each rectangle the center of mass is just the geometric center, provided, of course, that the mass per unit area is constant throughout the plate. We will illustrate the procedure by an example.

E X A M P L E **1**   Find the center of mass of the plate of uniform density ρ pictured in Figure 23.21.

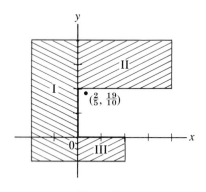

Figure 23.21

Solution. First we divide the plate into rectangles numbered I, II, and III as in Figure 23.21. The coordinate locations of the respective geometric centers are $(-1, \frac{3}{2})$, $(2, 3)$, and $(1, -\frac{1}{2})$. Letting ρ be the density (mass/unit area), the weight per unit area is ρg, while the weight of the plate is proportional to the area. Now, the respective areas are 10, 8, and 2, so the corresponding weights must be $10\rho g$, $8\rho g$, and $2\rho g$. The center of mass has two coordinates $(\bar{x}, \bar{y})$, which have to be found separately. Noting that the x-coordinates of the geometric centers are -1, 2, and 1, respectively, we have, by Equation (23.8)

$$\bar{x} = \frac{[-1(10) + 2(8) + 1(2)]\rho g}{(10 + 8 + 2)\rho g} = \frac{2}{5}$$

Similarly, since the respective y-coordinates of the geometric centers are $\frac{3}{2}$, 3, and $-\frac{1}{2}$, we get

$$\bar{y} = \frac{[\frac{3}{2}(10) + 3(8) + (-\frac{1}{2})(2)]\rho g}{20\rho g} = \frac{19}{10} \qquad \blacktriangleleft$$

Note that the center of mass of the plate in Example 1 lies outside the plate. Also, since the weight ρg cancels, the weight plays no part in locating the center of mass. As a consequence, we assume in this section that the weight per unit area has value 1:

Weight per unit area = 1 unit

Finally, the choice of origin is completely arbitrary; although the values of $\bar{x}$ and $\bar{y}$ depend on the origin, the location of the center of mass will always be the same relative to the plate.

Centroid of an Arbitrary Region

To locate the center of mass of an arbitrary region, it is best to start with Riemann sums rather than the shortcut method. Let the region be bounded by $y = f(x)$, the lines $x = a$ and $x = b$, and the x-axis (Figure 23.22). We divide the interval $[a, b]$ into n subintervals, each of length Δx_i, and let x_i be the center of each subinterval. To calculate the moment M_y about the y-axis, we find the moments of the individual rectangles and add. (Recall that moments are additive.) Since the center of mass of each rectangle is the geometric center $(x_i, \frac{1}{2}f(x_i))$, the moment of the ith rectangle with respect to the y-axis is $x_i f(x_i)\,\Delta x_i$, since x_i is the length of the moment arm and $f(x_i)\,\Delta x_i$ is the area. Hence

$$M_y \approx \sum_{i=1}^{n} x_i f(x_i)\,\Delta x_i.$$

This expression has the form of a Riemann sum, so that by the definition of the definite integral

$$M_y = \lim_{n \to \infty} \sum_{i=1}^{n} x_i f(x_i)\,\Delta x_i = \int_a^b x f(x)\,dx \tag{23.10}$$

Finally, since the area A may be assumed equal to the weight,

$$\bar{x} = \frac{M_y}{A} = \frac{\int_a^b x\,f(x)\,dx}{\int_a^b f(x)\,dx} = \frac{\int_a^b xy\,dx}{\int_a^b y\,dx} \tag{23.11}$$

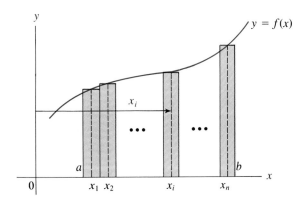

Figure 23.22

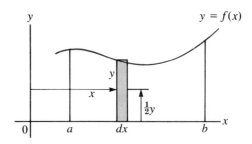

Figure 23.23

Let us now see how this result may be obtained by the shortcut method. Draw the typical element in Figure 23.23. The length of the moment arm is now x and the moment of the typical element with respect to the y-axis is $xy\,dx$. Since moments are additive, we sum the moments over $[a, b]$ to find that

$$M_y = \int_a^b xy\,dx \tag{23.12}$$

To get M_x, the moment with respect to the x-axis, we note that the centroid of the typical rectangle is $(x, \frac{1}{2}y)$. Hence the moment with respect to the x-axis is

Moment arm $\cdot$ area $= \frac{1}{2}y \cdot y\,dx = \frac{1}{2}y^2\,dx$

so that the sum over $[a, b]$ is

$$M_x = \frac{1}{2}\int_a^b y^2\,dx \tag{23.13}$$

Also

$$\bar{y} = \frac{M_x}{A} = \frac{\dfrac{1}{2}\int_a^b y^2\,dx}{\int_a^b y\,dx} \tag{23.14}$$

For many regions we can find $\bar{y}$ by drawing the typical element in the horizontal position, as shown in Figure 23.24. Since the moment arm is y and the area of each typical element is $x\,dy$, we get

Moment arm $\cdot$ area $= y \cdot x\,dy$

Summing from $y = c$ to $y = d$, we have

$$M_x = \int_c^d yx\,dy \tag{23.15}$$

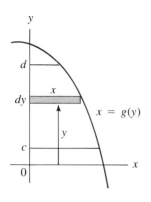

Figure 23.24

Thus,

$$\bar{y} = \frac{M_x}{A} = \frac{\int_c^d yx\,dy}{A} \tag{23.16}$$

E X A M P L E **2** Find the coordinates of the centroid of the region bounded by $y = 4 - 2x$ and the coordinate axes (Figure 23.25).

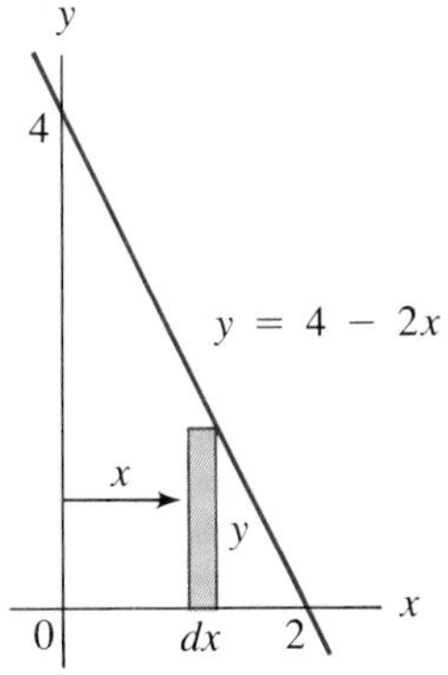

Solution. To find $\bar{x}$, we determine the moment of a typical element about the y-axis:

$$x \cdot y\,dx = x(4 - 2x)\,dx \qquad \text{Moment arm} \cdot \text{area}$$

Summing from $x = 0$ to $x = 2$, we get

$$M_y = \int_0^2 x(4 - 2x)\,dx$$

It follows that

$$\bar{x} = \frac{M_y}{A} = \frac{\int_0^2 x(4 - 2x)\,dx}{\int_0^2 (4 - 2x)\,dx} = \frac{\frac{8}{3}}{4} = \frac{8}{3} \cdot \frac{1}{4} = \frac{2}{3}$$

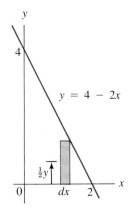

Figure 23.25

To find $\bar{y}$, we determine the moment of a typical element about the x-axis:

$$\tfrac{1}{2}y \cdot y\,dx = \tfrac{1}{2}(4 - 2x)(4 - 2x)\,dx \qquad \text{Moment arm} \cdot \text{area}$$

(See Figure 23.26.)

Summing from $x = 0$ to $x = 2$ we get

$$M_x = \int_0^2 \tfrac{1}{2}(4 - 2x)(4 - 2x)\,dx$$

It follows that

$$\bar{y} = \frac{M_x}{A} = \frac{\int_0^2 \tfrac{1}{2}(4 - 2x)(4 - 2x)\,dx}{4}$$

$$= \frac{\frac{16}{3}}{4} = \frac{16}{3} \cdot \frac{1}{4} = \frac{4}{3}$$

Figure 23.26

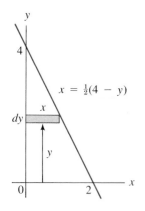

Figure 23.27

We can also find $\bar{y}$ by placing the typical element in the horizontal position (Figure 23.27).

The moment of a typical element about the x-axis can be expressed as

$$y \cdot x \, dy = y \cdot \tfrac{1}{2}(4 - y) \, dy$$

Since the limits of integration along the y-axis are $y = 0$ and $y = 4$, we have

$$\bar{y} = \frac{M_x}{A} = \frac{\displaystyle\int_0^4 y \cdot \tfrac{1}{2}(4 - y) \, dy}{4} = \frac{4}{3} \qquad \blacktriangleleft$$

The next example illustrates the method for finding the centroid of a region bounded by two curves.

E X A M P L E   **3**   Find the centroid of the region bounded by $y = 4x^2$ and $y = 8x$.

Solution. To find $\bar{x}$, we use Figure 23.28:

$$\bar{x} = \frac{M_y}{A} = \frac{\displaystyle\int_0^2 x(8x - 4x^2) \, dx}{\displaystyle\int_0^2 (8x - 4x^2) \, dx} = \frac{\frac{16}{3}}{\frac{16}{3}} = 1$$

To find $\bar{y}$, we use Figure 23.29:

$$\bar{y} = \frac{M_x}{A} = \frac{\displaystyle\int_0^{16} y(\tfrac{1}{2}\sqrt{y} - \tfrac{1}{8}y) \, dy}{A} = \frac{\frac{512}{15}}{\frac{16}{3}} = \frac{32}{5}$$

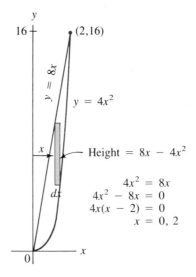

Figure 23.28

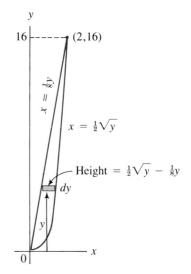

Figure 23.29 $\blacktriangleleft$

The region in the next example is also bounded by two curves, but interchanging the roles of x and y to find $\bar{y}$ is not convenient.

EXAMPLE 4 Find the centroid of the first-quadrant region bounded by $x^2 = 2y$ and $y = x + 4$ (Figure 23.30).

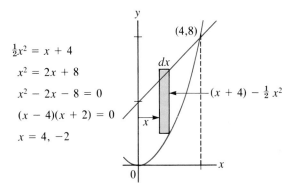

$\frac{1}{2}x^2 = x + 4$

$x^2 = 2x + 8$

$x^2 - 2x - 8 = 0$

$(x - 4)(x + 2) = 0$

$x = 4, -2$

Figure 23.30

Solution. Solving the given equations simultaneously shows that $(4, 8)$ is the point of intersection in the first quadrant. The typical element of thickness dx has a length of $(x + 4) - \frac{1}{2}x^2$. As before, the length of the moment arm is x. We add the moments over the interval $[0, 4]$ to obtain

$$M_y = \int_0^4 x[(x + 4) - \tfrac{1}{2}x^2]\,dx$$

$$= \int_0^4 x(-\tfrac{1}{2}x^2 + x + 4)\,dx = \int_0^4 (-\tfrac{1}{2}x^3 + x^2 + 4x)\,dx = \frac{64}{3}$$

Computing M_x requires a little more care: for a region bounded below by the x-axis, the centroid of the typical element is $(x, \frac{1}{2}y)$. Since the region in this problem is bounded by two functions, the y-coordinate of the centroid of the typical rectangle is the arithmetic average of the upper and lower extremities of the rectangle; that is,

$$\frac{(x + 4) + \frac{1}{2}x^2}{2}$$

Consequently, the moment of the typical rectangle with respect to the x-axis is

$$\tfrac{1}{2}[(x + 4) + \tfrac{1}{2}x^2][(x + 4) - \tfrac{1}{2}x^2]\,dx = \tfrac{1}{2}[(x + 4)^2 - (\tfrac{1}{2}x^2)^2]\,dx$$

$$= -\tfrac{1}{8}x^4 + \tfrac{1}{2}x^2 + 4x + 8$$

Summing, we have

$$M_x = \int_0^4 \left(-\tfrac{1}{8}x^4 + \tfrac{1}{2}x^2 + 4x + 8\right) dx = \frac{736}{15}$$

The area A is easily found to be

$$A = \int_0^4 \left(-\tfrac{1}{2}x^2 + x + 4\right) dx = \frac{40}{3}$$

Hence

$$\bar{x} = \frac{M_y}{A} = \frac{64}{3} \cdot \frac{3}{40} = \frac{8}{5}$$

and

$$\bar{y} = \frac{M_x}{A} = \frac{736}{15} \cdot \frac{3}{40} = \frac{92}{25} \qquad\qquad \blacktriangleleft$$

Centroid of a Solid of Revolution

The method for finding the centroid of a plate can be extended to find the centroid of a solid of revolution. Suppose that the region in Figure 23.31 is rotated about the x-axis. Then the volume of the typical disk is $\pi y^2\, dx$, again assumed to be equal to the weight. Since the length of the moment arm is x, the moment of the typical disk with respect to the y-axis is $x \cdot \pi y^2\, dx = \pi x y^2\, dx$. Summing, we get

$$\bar{x} = \frac{\pi \displaystyle\int_a^b x y^2 \, dx}{\pi \displaystyle\int_a^b y^2 \, dx} \tag{23.17}$$

By symmetry, it follows that $\bar{y} = 0$.

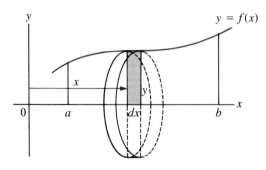

Figure 23.31

E X A M P L E **5**   Find the centroid of the solid of revolution obtained by rotating the first quadrant region bounded by $y = 4 - x^2$ and the coordinate axes about the x-axis
(Figure 23.32).

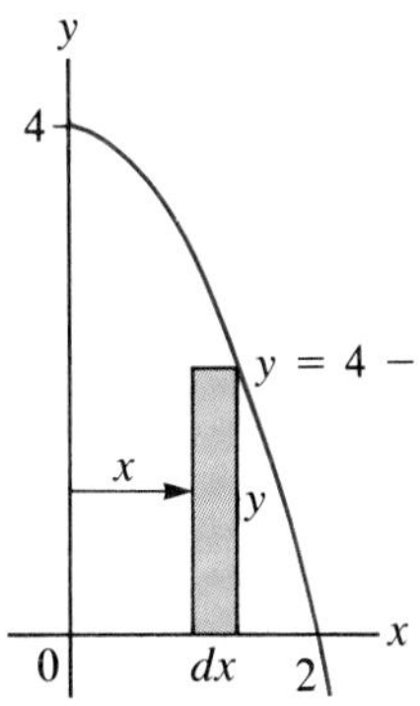

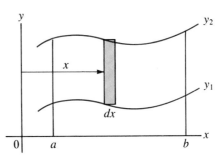

Solution. The volume of the typical cylinder is

$$\pi(\text{radius})^2 \cdot (\text{thickness}) = \pi(4 - x^2)^2 \, dx$$

and the moment M_y about the y-axis* is

$$x \cdot \pi(4 - x^2)^2 \, dx = \pi x(4 - x^2)^2 \, dx$$

We now sum from $x = 0$ to $x = 2$ to get

$$M_y = \pi \int_0^2 x(4 - x^2)^2 \, dx = \frac{32\pi}{3}$$

Since

$$V = \pi \int_0^2 (4 - x^2)^2 \, dx = \frac{256\pi}{15}$$

we have

$$\bar{x} = \frac{M_y}{V} = \frac{32\pi}{3} \cdot \frac{15}{256\pi} = \frac{5}{8}$$

By symmetry, $\bar{y} = 0$. ◀

If the region is bounded by two functions y_1 and y_2 (Figure 23.33), then
the washer method yields

$$\pi x(y_2^2 - y_1^2) \, dx \qquad\qquad (23.18)$$

for the moment of the typical washer with respect to the y-axis.

---

* Strictly speaking, this should be the moment about a plane perpendicular to the plane of the
paper and containing the y-axis. For convenience, we will retain the notation M_y.

Figure 23.32

Figure 23.33

E X E R C I S E S / S E C T I O N   **23.5**

In Exercises 1–4 (Figures 23.34–23.37), find the coordinates of the centroid of each given plate. (See Example 1.)

1.

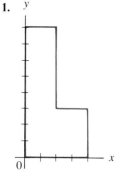

Figure 23.34

2.
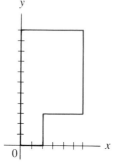

Figure 23.35

3.
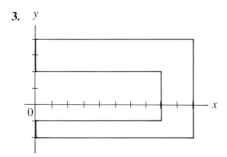

Figure 23.36

4.
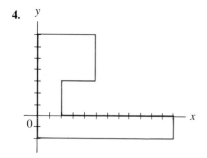

Figure 23.37

In Exercises 5–26, find the centroid of each region bounded by the given curves.

5. $x + y = 1$, coordinate axes

6. $x + 2y = 2$, coordinate axes

7. $2x + y = 2$, coordinate axes

8. $2x + 3y = 12$, coordinate axes

9. $y = x$, $x = 1$, x-axis

10. $y = 2x$, $x = 4$, x-axis

11. $y = x$, $y = 2$, y-axis

12. $y = \frac{1}{2}x$, $y = 1$, y-axis

13. $y = 4 - x^2$, x-axis

14. $y = x^3$, $x = 2$, x-axis

15. $y = x^2$, $x = 2$, x-axis

16. $y = x - x^3$, x-axis (first quadrant)

17. $y = \dfrac{1}{\sqrt[3]{x}}$, $x = 1$, $x = 8$, x-axis

18. $y^2 = x$, $x = 1$

19. $x = y - y^2$, y-axis   **20.** $x = 1 - y^2$, y-axis

21. $y = \sqrt{a^2 - x^2}$, coordinate axes (first quadrant). Note that $A = \frac{1}{4}\pi a^2$ (sector of circle).

22. $y = x^2$, $y^2 = x$ **23.** $y = 2x$, $y = x^2$

24. $y = x^{3/2}$, $y = x$ **25.** $y^2 = 4x$, $y = 2x$

26. $y = 6x - x^2$, $y = x$

27. Find the centroid of a semicircle of radius r.

28. Show that the centroid of the region bounded by $\sqrt{x} + \sqrt{y} = \sqrt{a}$ and the coordinate axes is $(a/5, a/5)$.

29. Find the centroid of a right triangle with sides a and b.

In Exercises 30–34, find the centroid of the solid obtained by rotating the region about the axis indicated.

30. $y = \sqrt{x}$, $x = 1$, $x = 4$, x-axis; about the x-axis

31. a. $y = x$, $x = 1$, x-axis; about the x-axis
   b. $y = x^2$, $y = 2$, y-axis (first quadrant); about the y-axis

32. $y = x^3$, $x = 0$, $x = 2$, x-axis; about the x-axis

33. $y = \sqrt{r^2 - x^2}$ and the coordinate axes (first quadrant); about the x-axis (centroid of a hemisphere)

34. $x + y = 4$, $y = 2$, coordinate axes; about the y-axis

35. Find the centroid of the solid of revolution obtained by rotating the region bounded by $y = x$ and $y = x^2$
   a. about the x-axis   **b.** about the y-axis

36. Find the centroid of the solid obtained by rotating the region in Exercise 32 about the y-axis.

37. Find the centroid of a right circular cone.

38. Find the centroid of the solid obtained by rotating the region bounded by $x = 4/y^2$, $y = 1$, $y = 2$, and the y-axis about the y-axis.

39. Find the centroid of the solid obtained by revolving the region bounded by $y^2 = 4px$, $y = b$, and the y-axis about the y-axis.

40. Show that the centroid of a paraboloid of revolution obtained by rotation of the first-quadrant region bounded by $y^2 = 4px$ and $x = h$ about the x-axis is $(\frac{2}{3}h, 0)$.

23.6 Moments of Inertia

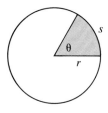

Figure 23.38

In this section we are going to define *moment of inertia* and then use integration to find moments of inertia of plates and solids of revolution.

If a particle of mass m is moving in a straight line with velocity v, then its kinetic energy is given by $K = \frac{1}{2}mv^2$. Now suppose that the particle is moving along a circle of radius r with a constant angular velocity $\omega = d\theta/dt$. From Figure 23.38, $s = r\theta$, so that $ds/dt = r(d\theta/dt) = r\omega$, the speed of the particle along the circle. Hence

$$K = \frac{1}{2}m\left(\frac{ds}{dt}\right)^2 = \frac{1}{2}m(r\omega)^2 = \frac{1}{2}(mr^2)\omega^2$$

If we let $I = mr^2$, then

$$K = \frac{1}{2}I\omega^2$$

which has the same form as the formula $K = \frac{1}{2}mv^2$. This observation suggests the definition

$$I = mr^2 \qquad (23.19)$$

Moment of inertia

which is called the **moment of inertia** of the particle about the axis of rotation.

The moment of inertia is the rotational analog of the mass, as the preceding example suggests. In another analogy, a particle rotating about a fixed axis has *angular momentum* $L = I\omega$, just as a particle of mass m and velocity v moving in a straight line has momentum mv. Also, the analog of Newton's second law, $F = ma$, is $T = I\alpha$, where α is the *rotational acceleration* and T is called the *torque*.

If n particles of mass $m_1, m_2, \ldots, m_n$ revolve about an axis through O perpendicular to the plane of the paper at respective distances $d_1, d_2, \ldots, d_n$ from O (Figure 23.39), then the moment of inertia is defined to be

$$I = \sum_{i=1}^{n} m_i d_i^2$$

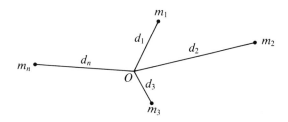

Figure 23.39

We can find a number R such that

$$(m_1 + m_2 + \cdots + m_n)R^2 = m_1 d_1^{\,2} + m_2 d_2^{\,2} + \cdots + m_n d_n^{\,2}$$

That is,

$$R = \left[\frac{\displaystyle\sum_{i=1}^{n} m_i d_i^{\,2}}{\displaystyle\sum_{i=1}^{n} m_i} \right]^{1/2} = \left(\frac{I}{M} \right)^{1/2} \tag{23.20}$$

Radius of gyration

where M is the total mass of the system. R is called the **radius of gyration**. Since the moment of inertia of the system remains the same if all the masses are a distance R from O, the radius of gyration is the rotational analog of the center of mass.

E X A M P L E 1   Refer to Figure 23.39. Suppose $m_1 = 5$ kg, $m_2 = 10$ kg, $d_1 = 5$ m, and $d_2 = 8$ m. If the system rotates about the axis through O with an angular velocity of 4 rad/s, find the kinetic energy of the system.

Solution. The moment of inertia is found to be

$$I = 5 \cdot 5^2 + 10 \cdot 8^2 = 765 \text{ kg} \cdot \text{m}^2$$

Hence

$$K = \tfrac{1}{2}I\omega^2 = \tfrac{1}{2}(765)(4)^2 = 6120 \text{ J} \qquad \blacktriangleleft$$

Moment of Inertia of a Region

If the object is a plate of uniform density ρ, then the moment of inertia can be found by a method very similar to the one used for finding moments. Consider the region in Figure 23.40 and let I_y be the moment of inertia about the y-axis. The distance from the y-axis is x, and the mass of the typical rectangle is $\rho A = \rho[f(x) - g(x)]\,dx$. Thus by formula (23.19) the moment of inertia of the

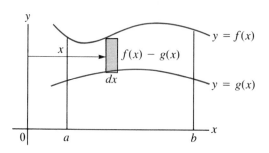

Figure 23.40

typical element is $x^2 \cdot$ mass, or

$$x^2 \cdot \rho[f(x) - g(x)]\,dx = \rho x^2[f(x) - g(x)]\,dx$$

Finally, since moments of inertia are additive, we get

$$I_y = \rho \int_a^b x^2[f(x) - g(x)]\,dx \tag{23.21}$$

E X A M P L E **2** Find I_y of the region bounded by $y = x^3$, $x = 1$, and the x-axis (Figure 23.41).

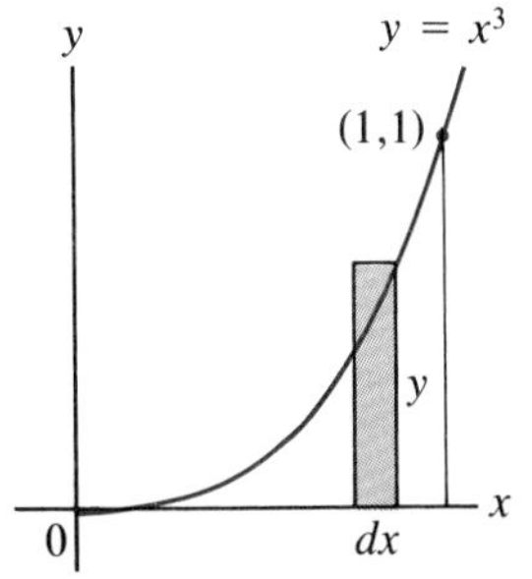

Solution. Since the height of the rectangle is y, we see that the moment of inertia of the typical rectangle about the y-axis is

$$x^2 \cdot \text{mass} = x^2(\rho y\,dx)$$

Summing from $x = 0$ to $x = 1$, we get (since $y = x^3$)

$$I_y = \rho \int_0^1 x^2 \cdot x^3\,dx = \rho \left(\frac{1}{6}x^6\right)\Bigg|_0^1 = \frac{\rho}{6}$$

◄

Figure 23.41

E X A M P L E **3** Find R_x, the radius of gyration about the x-axis, of the region bounded by $y = x^2$ and $y = 3x$ (Figure 23.42).

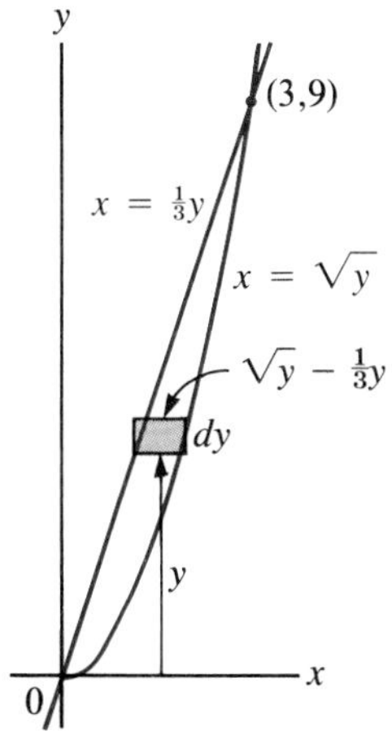

Solution. Since we are seeking the moment of inertia about the x-axis, we are forced to draw the typical element horizontally. Solving the given functions for x, we get $x = \sqrt{y}$ and $x = \frac{1}{3}y$. So the mass of the typical rectangle is $\rho(\sqrt{y} - \frac{1}{3}y)\,dy$ and its moment of inertia about the x-axis is $y^2 \cdot \rho(\sqrt{y} - \frac{1}{3}y)\,dy = \rho y^2(\sqrt{y} - \frac{1}{3}y)\,dy$. Summing from $y = 0$ to $y = 9$, we get

$$I_x = \rho \int_0^9 y^2\left(\sqrt{y} - \frac{1}{3}y\right)dy = \rho \int_0^9 \left(y^{5/2} - \frac{1}{3}y^3\right)dy$$

$$= \rho\left(\frac{2}{7}y^{7/2} - \frac{1}{12}y^4\right)\Bigg|_0^9 = \rho\left[\left(\frac{2}{7}\right)9^{7/2} - \left(\frac{1}{12}\right)9^4\right]$$

$$= \rho\left[\left(\frac{2}{7}\right)(3^2)^{7/2} - \left(\frac{1}{12}\right)(3^2)^4\right] = \rho\left[\left(\frac{2}{7}\right)3^7 - \left(\frac{1}{12}\right)3^8\right]$$

$$= \rho(3^7)\left(\frac{2}{7} - \frac{1}{4}\right) = \frac{2187\rho}{28}$$

Figure 23.42

The area is found to be

$$A = \int_0^3 (3x - x^2)\,dx = \frac{9}{2}$$

so that the mass is $9\rho/2$. Hence

$$R_x = \left(\frac{2187\rho}{28} \cdot \frac{2}{9\rho}\right)^{1/2} = \frac{9\sqrt{42}}{14} \approx 4.17 \qquad R_x = \sqrt{\frac{I_x}{M}}$$

◀

Moment of Inertia of a Solid of Revolution

It is possible to find the moment of inertia of a solid of revolution with respect to its axis. For this case the method of shells turns out to be by far the most convenient.

E X A M P L E **4**   Find the moment of inertia of a right circular cone of radius r and height h about its axis.

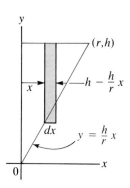

Figure 23.43

Solution. The cone can be generated by rotating the region bounded by $y = (h/r)x$, $y = h$, and the y-axis about the y-axis (Figure 23.43). The mass of the typical shell is

$$\rho \cdot 2\pi(\text{radius}) \cdot (\text{height}) \cdot (\text{thickness}) = \rho \cdot 2\pi x\left(h - \frac{hx}{r}\right)dx$$

by our earlier work. Since the distance from the y-axis is x, we multiply the last expression by x^2 to get the moment of inertia of the typical element:

$$x^2 \cdot 2\pi\rho x\left(h - \frac{hx}{r}\right)dx$$

Summing from $x = 0$ to $x = r$, we have

$$I_y = 2\pi\rho \int_0^r x^2 \cdot x\left(h - \frac{hx}{r}\right)dx$$

$$= 2\pi\rho \int_0^r x^3\left(h - \frac{hx}{r}\right)dx = \frac{\rho}{10}\pi hr^4$$

We also note that since $V = \frac{1}{3}\pi r^2 h$, the mass of the cone is $\rho \cdot \frac{1}{3}\pi r^2 h$. So

$$R_y = \left[\frac{\rho \cdot \frac{1}{10}\pi hr^4}{\rho \cdot \frac{1}{3}\pi r^2 h}\right]^{1/2} = \left[\frac{\frac{1}{10}\pi hr^4}{\frac{1}{3}\pi r^2 h}\right]^{1/2} = \frac{r\sqrt{30}}{10}$$

◀

For certain standard shapes of uniform density, such as the cone in Example 4, the moment of inertia is often expressed in terms of mass. Since the mass of the cone is $\rho(\frac{1}{3}\pi r^2 h)$, we have

$$I_y = \frac{\rho}{10}\pi hr^4 = \frac{3}{10}\left(\frac{1}{3}\rho\pi hr^2\right)r^2 = \frac{3}{10}mr^2$$

E X A M P L E 5   If the cone in Example 4 has a mass of 1.0 kg, a radius of 0.20 m, and rotates about its axis at 120 rev/min, find its kinetic energy K and angular momentum L.

Solution. First we need to convert the rotational velocity to radians per second:

$$120 \frac{\text{rev}}{\text{min}} \times \frac{2\pi \text{ rad}}{1 \text{ rev}} \times \frac{1 \text{ min}}{60 \text{ s}} = 4\pi \frac{\text{rad}}{\text{s}}$$

We now have

$$K = \tfrac{1}{2}I\omega^2 = \tfrac{1}{2}(\tfrac{3}{10}mr^2)\omega^2$$
$$= \tfrac{1}{2}(\tfrac{3}{10} \cdot 1.0 \cdot 0.20^2)(4\pi)^2 = 0.95 \text{ J}$$

and

$$L = I\omega = (\tfrac{3}{10}mr^2)\omega$$
$$= (\tfrac{3}{10} \cdot 1.0 \cdot 0.20^2)(4\pi) = 0.15 \text{ kg·m}^2/\text{s}$$ ◀

As in the case of linear momentum, the angular momentum $L = I\omega$ is conserved. Figure skaters performing a pirouette on one skate take advantage of this principle: with arms and legs extended, the angular velocity may be fairly small. When the limbs are suddenly pulled in, the moment of inertia decreases, but since $I\omega$ remains constant, the angular velocity ω increases.

E X E R C I S E S / S E C T I O N   **23.6**

In Exercises 1–16, R is the region bounded by the given curves.

1. R: $y = x$, $x = 1$, x-axis; find I_y, R_y
2. R: $y = \tfrac{1}{2}x$, $y = 1$, y-axis; find I_x, R_x
3. R: $x = \sqrt{y}$, $y = 1$, y-axis; find R_x
4. R: $x + y = 1$, coordinate axes; find R_y
5. R: $2x + y = 2$, coordinate axes; find I_y, R_y, I_x, R_x
6. R: $y = x^2$, $x = 0$, $x = 1$, x-axis; find I_y, R_y, I_x, R_x
7. R: $y = 4 - x^2$, coordinate axes (first quadrant); find I_y, R_y
8. R: $y = 2x$, $x = 1$, $x = 2$, x-axis; find R_y
9. R: $y^2 = x$, $y = \tfrac{1}{2}x$; find I_y, R_y
10. R: $y = 2x^2$, $y = x^3$; find I_y, R_y, R_x
11. R: $y = 9 - 3x$, $y = 9 - x^2$; find I_y, R_y
12. R: $y = x^3$, $y = \sqrt{x}$; find R_y
13. R: $x = y^2 + 2$, $x = y + 2$; find I_x, R_x
14. R: $y = x^3$, $y = 4x$ (first quadrant); find I_y, R_y, I_x, R_x

15. R: $y = 2x^2$, $y = 4x + 6$; find I_y
16. R: $y^2 = x$, $x = 2$ (first quadrant); find I_y

17. Find the moment of inertia and radius of gyration of a cylinder of radius r and height h about its axis. (*Hint:* Rotate the first-quadrant region bounded by $y = h$, $x = r$, and the coordinate axes about the y-axis.)

18. Find I_y of the solid generated by rotating the first-quadrant region bounded by $y = ax^2$, $y = b$, and the y-axis about the y-axis.

19. Find I_x and R_x of the solid generated by revolving the first-quadrant region bounded by $y = x^2$, $y = 4$, and the y-axis about the x-axis.

20. Find R_x of the following solid: the region bounded by $y^2 = x^3$, $x = 4$, and x-axis (first quadrant) rotated about the x-axis.

21. Find the moment of inertia with respect to its axis of the solid obtained by rotating the region bounded by $y = \tfrac{1}{2}x$, $y = 1$, and the y-axis about the y-axis.

22. Find the radius of gyration with respect to its axis of the solid obtained by revolving the region bounded by $y = x^3$, $x = 2$, and the x-axis about the y-axis.

23. Find the radius of gyration with respect to its axis of the solid obtained by revolving the region bounded by $y = \sqrt{4 - x}$ and the coordinate axes about the x-axis.

24. Find the radius of gyration with respect to its axis of the solid obtained by revolving the region bounded by $y = x$ and $y = \sqrt{x}$ about the y-axis.

25. Repeat Exercise 24 for the region bounded by $y = x^2$ and $x = y^2$.

26. Suppose a certain *paraboloid* is obtained by rotating the first-quadrant region bounded by $y = x^2$ and $y = 1$ about the y-axis (x and y are measured in meters). Assuming that $\rho = 2$ kg/m^3, determine the kinetic energy and angular momentum if the paraboloid rotates at the rate of 20 rev/s.

27. A cylinder has a mass of 2.0 kg, a radius of 0.10 m, and rotates at the rate of 360 rev/min about its axis. Find its kinetic energy and angular momentum. Refer to Exercise 17.

23.7   Work and Fluid Pressure

Work

The problems in this chapter require a lot of work, but in physics the term **work** is used in a different sense: if an object is pushed a distance s along a line by a force acting in the direction of motion, then

$$\text{Work} = \text{force} \times \text{distance}$$

The concept of work is important in technology for determining the energy required to perform certain tasks. If the force is in newtons (or pounds) and s is in meters (or feet), then the work is measured in joules (or foot-pounds). (One joule = 1 newton·meter.)

What happens if the force is not constant? In that case we use the type of approximation procedure that has become familiar. Suppose a body is pushed to the right along the x-axis by a force $f(x)$ from a to b. We divide $[a, b]$ into n subintervals and consider the point x_i in the ith subinterval. As usual, the ith subinterval has length Δx_i. Then the force is approximately constant in the ith subinterval and equal to $f(x_i)$. Consequently, the work done in moving the body across the ith subinterval is approximately $f(x_i) \Delta x_i$ and the total work W is given by the following approximation:

$$W \approx \sum_{i=1}^{n} f(x_i) \Delta x_i$$

This expression has the form of a Riemann sum. Hence

$$W = \lim_{n \to \infty} \sum_{i=1}^{n} f(x_i) \Delta x_i = \int_{a}^{b} f(x) \, dx \qquad (23.22)$$

These principles can be applied to finding the work done in stretching a **spring**. If an ideal spring is stretched x units beyond its natural length, then according to Hooke's law the spring pulls back with a force $f(x) = kx$. (The law also holds if the spring is compressed.) The proportionality constant k depends on the stiffness of the spring and can be determined experimentally.

E X A M P L E **1**   A spring has a natural length of 5 ft. A force of 4 lb stretches the spring $\frac{1}{4}$ ft. Determine how much work is done in stretching the spring:

a. From its natural length to a length of 7 ft

b. From a length of 6 ft to 8 ft

Solution. We first determine the constant k. Since the force is **4 lb** when $x = \frac{1}{4}$ ft, from Hooke's law we have

$$F = kx$$

$$\textbf{4 lb} = k(\tfrac{1}{4}\ \text{ft}) \qquad \text{or} \qquad k = 16 \ \text{lb/ft}$$

Thus $f(x) = kx = \textbf{16x}$

a. Here the spring is stretched from $x = 0$ (no extension) to $x = 2$ (natural length of 5 ft to 7 ft). So by formula (23.22),

$$W = \int_0^2 16x\, dx = 8x^2\Big|_0^2 = 32 \ \text{ft-lb}$$

b. Here the spring is stretched from $x = 1$ to $x = 3$:

$$W = \int_1^3 16x\, dx = 8x^2\Big|_1^3 = 8(9 - 1) = 64 \ \text{ft-lb} \qquad \blacktriangleleft$$

E X A M P L E **2**   A cable is 50 m long and has a density of 3 kg/m. If the cable is hanging from a winch, how much work is done in winding it up?

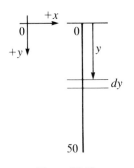

Figure 23.44

Solution. In this problem we are going to return to our shortcut method. To visualize the problem we would need to draw a vertical line representing the cable (Figure 23.44) and subdivide the interval $[0, 50]$. However, it will be easier if, instead of the actual subdivisions, we draw a typical element of length dy that is a distance y units from the top. It will symbolize one subdivision. Each element will have a mass of $3\,dy$ kg, so that $3\,dy$ kg $\times$ 10 m/s^2 = **30 dy** newtons. The work done in moving the element from its initial position to the winch at the top is $y \cdot \textbf{30}\,\textbf{dy} = 30y\,dy$ joules. Then, summing from $y = 0$ to $y = \textbf{50}$, we get

$$W = \int_0^{50} 30y\, dy = 37{,}500 \ \text{J} \qquad \blacktriangleleft$$

E X A M P L E **3**   A cylindrical tank full of water is 10.0 m high and has a radius of 6.0 m. Find the work done in pumping the water to a level 5.0 m above the tank.

Solution. Since the motion is vertical, we subdivide the side of the tank (Figure 23.45) and let the typical element have a thickness dy. The tank itself is subdivided into cylindrical slabs, where the typical slab has a volume of $\pi(6)^2\, dy = 36\pi\, dy$ (with final zeros omitted). Taking the mass of water to be 1000 kg/m^3, each cubic meter of water weighs 10,000 N. Hence the weight of each slab is

$$\textbf{10,000}(36\pi \, dy)$$

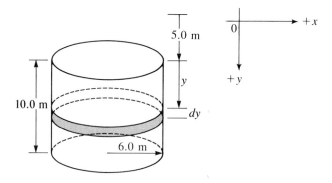

Figure 23.45

Each slab is moved a distance $y + 5$; so the work done is $\mathbf{10{,}000(36\pi\, dy)(y + 5)}$. To get the total work done, we sum from $y = 0$ to $y = 10$:

$$W = \int_0^{10} 10{,}000(36\pi)(y + 5)\,dy$$

$$= 360{,}000\pi \int_0^{10} (y + 5)\,dy = 3.6 \times 10^7\pi = 1.1 \times 10^8 \text{ J} \qquad \blacktriangleleft$$

Fluid Pressure

A well-known law in hydrostatics states that the *pressure of a fluid in an open container is directly proportional to the depth,* or

Pressure

$$p = ky$$

where p is the pressure, y the distance below the surface, and k the weight per unit volume. If ρ is the density of the fluid (mass/unit volume), then by Newton's second law, $F = mg$, we have

$$p = \rho g y$$

which is the force per unit area expressed in N/m^2 or $lb/in.^2$ Moreover, the pressure is independent of the shape and size of the container, and at any point in the container the pressure is the same in all directions.

If a flat plate is submerged in a fluid and placed in a horizontal position, then the pressure on the plate is the same at all points. Since p is the force per unit area, the total force on one side of the plate is pA, where A is the area. If the plate is y units below the surface, then the force becomes

Force

$$F = \rho g y A \qquad \textbf{pressure} \times \textbf{area}$$

If the plate is in a vertical position, the calculation of the total force is more complicated since the pressure now varies as we move downward. But suppose the plate is subdivided into n strips; then the pressure is approximately constant on each strip if n is large.

E X A M P L E 4   Suppose that a rectangular plate 4 m × 8 m is placed in water with the long side parallel to and 2 m below the surface. What is the total force against the plate?

Solution. We divide the plate into horizontal strips and let the typical strip have a thickness dy (Figure 23.46). We have seen that water weighs approximately 10,000 N/m³, to be denoted by w. Since each strip is a distance $y + 2$ m below the surface, the pressure on the strip is $(y + 2)w$. The area is $8\,dy$, so the force against the strip is

$$\text{Pressure} \cdot \text{area} = (y + 2)w \cdot 8\,dy$$
$$= 8w(y + 2)\,dy$$

Adding the forces from $y = 0$ to $y = 4$, we get

$$F = \int_0^4 8w(y + 2)\,dy = 128w = 1{,}280{,}000 \text{ N}$$

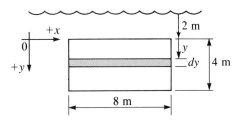

Figure 23.46

E X A M P L E 5   A semicircular gate on a dam has a diameter of 6.0 m and is positioned with the straight portion on top. If the surface of the water is level with the top, find the total force against the gate.

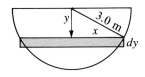

Figure 23.47

Solution. We subdivide the gate into horizontal strips (Figure 23.47). Letting w be the weight per unit volume of water, the force against the typical strip is wy times the area. The real problem is to find the area of the strip. Using the letters in the diagram, the area is given by $2x\,dy$. Since the radius of the gate is 3, from the Pythagorean theorem we get

$$x^2 + y^2 = 9 \qquad \text{or} \qquad 2x = 2\sqrt{9 - y^2}$$

Hence the force against the strip is

$$\text{Pressure} \cdot \text{area} = (wy)(2x\,dy)$$
$$= wy(2x)\,dy = 2wy\sqrt{9 - y^2}\,dy$$

We now sum from $y = 0$ to $y = 3$:

$$F = 2w \int_0^3 y\sqrt{9 - y^2}\,dy$$

To integrate

$$\int y \sqrt{9 - y^2} \, dy = \int y(9 - y^2)^{1/2} \, dy$$

we let $u = 9 - y^2$ and find that $du = -2y \, dy$. Hence

$$\int y \sqrt{9 - y^2} \, dy = -\frac{1}{2} \int (9 - y)^{1/2}(-2y) \, dy$$

$$= -\frac{1}{2} \int u^{1/2} \, du = -\frac{1}{2} \cdot \frac{2}{3} u^{3/2} = -\frac{1}{3} u^{3/2}$$

and (since $w = 10{,}000$ N/m^3),

$$F = 2w \left(-\frac{1}{3} \right)(9 - y^2)^{3/2} \Big|_0^3 = 18w = 180{,}000 \text{ N} \qquad \blacktriangleleft$$

E X E R C I S E S / S E C T I O N   23.7

In the following exercises, let 10,000 N/m^3 be designated by w.

1. A force of 6 lb stretches a spring $\frac{1}{8}$ ft. How much work is done in stretching the spring 2 ft?

2. A spring has a natural length of 6 ft, and a force of 4 lb stretches it 1 ft. Find the work done in stretching the spring **(a)** from 6 ft to 9 ft; **(b)** from 6.5 ft to 10 ft.

3. A force of 12 lb stretches a spring 2 ft. If the spring has a natural length of 8 ft, find the work done in **(a)** compressing it from 8 ft to 6 ft; **(b)** stretching it from 10 ft to 13 ft.

4. A spring has a natural length of 7 ft. A force of 4 lb stretches it $\frac{1}{3}$ ft. Find the work done in compressing the spring from 6 ft to 4 ft.

5. A tank full of water is in the shape of a box 3 m deep, 4 m long, and 3 m wide. Find the work done in pumping out all the water over one of its sides.

6. Repeat Exercise 5, assuming now that the tank is filled to a level of 2 m.

7. A chain 20 m long and weighing 3 N/m is hanging from a winch. Find the work done in winding it up.

8. If only half the chain in Exercise 7 is pulled up, show that the work done is 450 J.

9. A cable weighing 10 lb/ft is 10 ft long. If a 20-lb weight is attached to the end, find the work done in bringing up the weight.

10. A cylindrical tank has a radius of 5 m and a depth of 8 m. If the tank is filled with water to a level of 6 m, how much work is done in pumping the water out over the top?

11. A trough is 12 m long and has a cross-sectional area in the shape of an isosceles triangle 4 m across the top and 3 m high. If the trough is filled with water, how much work is done in pumping the water to a level 2 m above the trough? (Use similar triangles.)

12. A hemispherical tank full of water has a radius of 3 m. Find the work done in pumping the water out over the top.

13. A conical tank (vertex downward) has a radius of 3 m and a height of 5 m. If the tank is filled with water, find the work done in pumping the water out over the top.

14. A trough 6 m long has a semicircular cross-section of radius 1 m. If the trough is full of oil with density 800 kg/m^3, find the work done in pumping the oil out over the top.

15. If two charged particles have opposite charges, they attract each other with a force that is inversely proportional to the square of the distance between them. (If F is the force, then $F = k/s^2$, where s is the distance between the particles.) If the force of attraction is 20.0 dynes when the particles are 1.0 cm apart, find the work done (in ergs) in separating them from **(a)** a distance of 1.0 cm to 10.0 cm; **(b)** a distance of 1.0 cm to an infinite separation.

16. If two particles have like charges, they repel each other with a force that is inversely proportional to the square of the distance between them. If the force of separation is 15 dynes at a distance of 0.50 cm, find the work done (in ergs) in bringing the particles together from an infinite separation to a separation of 1.0 cm.

17. The vertical gate on a dam is a rectangle 5 m long and 2 m high. If the top of the gate is horizontal and 3 m below the surface, find the force against the gate.

18. The vertical gate on a dam is a rectangle 10 m long and 4 m high. If the top of the gate is horizontal and 1 m below the surface, find the force against the gate.

19. A rectangular tank is 3 m wide across its top and is 2 m deep. If the tank is filled with water, find the force against the lower half of one end.

20. A half-filled cylindrical tank of water is lying on its side. If the diameter of the tank is 4 m, find the force against the end.

21. A trough full of water has vertical ends in the shape of an isosceles triangle 4 m across the top and 2 m deep. Find the force against the end.

22. Find the force against the end of the trough in Exercise 21, assuming now that the water is only 1 m deep in the center.

23. A thin plate has the shape of an isosceles triangle 2 m wide at the bottom and 3 m high. If the plate is submerged vertically in water with the top vertex 5 m below the surface and the bottom edge horizontal, find the total force against the (two sides of) the plate.

24. A gate on a vertical dam is an isosceles trapezoid with upper base 6 m, lower base 8 m, and height 3 m. Find the force against the gate if the upper base is 5 m below the surface.

25. A swimming pool is 9 m long. Its bottom is flat but not horizontal. If the depth ranges from 0 m to 3 m, find the total force against one side, assumed vertical.

26. The deep end of a swimming pool is vertical and has the shape of an isosceles trapezoid 6 m across the top, 4 m across the bottom, and 3 m deep. Find the force against the end.

27. Find the force on a vertical dam if the dam is in the shape of a parabola 8 m across the top and 10 m deep.

28. Suppose the face of a dam is inclined at an angle of $30°$ from the vertical. If the face of the dam is a rectangle 40 m wide and slant height 30 m, what is the force against the dam?

29. Suppose the end of the trough in Exercise 21 is tilted so that it makes an angle of $30°$ with the vertical. What is the force against the end?

REVIEW EXERCISES / CHAPTER　23

---

1. A stone is dropped from a height of 256 ft. Find its mean velocity. ($g = 32$ ft/s^2)

2. For a short time interval the current in a certain circuit is $i = 2.0t^{1/3}$ A. Find the mean current from $t = 1.0$ s to $t = 8.0$ s.

3. A current $i = 2.1 - 0.18t^{5/2}$ A is flowing through a 10.0-Ω resistor. Find the mean power from $t = 0.0$ s to $t = 4.0$ s.

4. Derive the formula for the volume of a sphere by the disk and shell methods.

5. Find the volume obtained by revolving the region bounded by $y = x^{3/2}$, $x = 4$, and the x-axis about:
　a. the x-axis　**b.** $x = 4$　**c.** $y = 8$　**d.** the y-axis

6. Find the volume of the solid generated by rotating the region bounded by $y = x^3$, $y = 1$, and $x = 0$ about the line $y = 1$.

7. Find the volume obtained by revolving the following region about the y-axis: bounded by $x + y = 2$, $x = \sqrt{y}$, and the y-axis.

8. Find the volume obtained by revolving the region in Exercise 7 about the x-axis.

9. Find the volume of the solid generated by revolving the region bounded by $y = 2x^2$, $x = 1$, and the x-axis about the line $x = 2$.

10. Find the volume of the solid generated by revolving the region bounded by $y = \sqrt{x - 2}$, $x = 6$, and the x-axis about the x-axis.

11. Find the centroid of the region bounded by $y = 4 - x^2$ and the coordinate axes (first quadrant).

12. Find I_y for the region in Exercise 11.

13. Find the centroid of the region bounded by $y = x^2 - x^3$ and the x-axis.

14. Find the centroid of the region bounded by $y = x$ and $y = 2 - x^2$.

15. Find the centroid of the solid obtained by rotating the region bounded by $x^2 - y^2 = 1$ and $x = 2$ about the x-axis.

16. Find the centroid of the solid obtained by revolving the region bounded by $y = x^2$, $x + y = 2$, and $y = 0$ about the y-axis.

17. Find I_x and R_x of the region bounded by $x = y - y^2$ and the y-axis.

18. Find R_y and R_x of the region bounded by $y = \sqrt{x}$, $x = 4$, and the x-axis.

19. Find the radius of gyration with respect to its axis of the solid obtained by revolving the region bounded by $y = \sqrt{x}$, $x = 4$, and the x-axis about the y-axis.

20. Find the moment of inertia with respect to its axis of the solid obtained by revolving the region bounded by $y = x^3$, $x = 1$, and the x-axis about the x-axis.

21. A spring has a natural length of 4 ft. A force of 2 lb stretches the spring $\frac{1}{2}$ ft. Find the work done in stretching the spring from:

a. its natural length to a length of 6 ft   b. a length of 5 ft to 7 ft

22. A cable 10 m long has a density of 2 kg/m. If the cable hangs from a winch, find the work done in winding it up.

23. A cylindrical reservoir half full of water is 30 m high and has a radius of 10 m. Find the work done in pumping the water out over the top.

24. A conical tank with vertex down has a radius of 5 m and a height of 8 m. If the tank is filled with water, find the work done in pumping the water out over the top.

25. A square plate 3 m on the side is submerged vertically in water with one side parallel to and 5 m below the surface. Find the total force against one side of the plate.

26. A cylindrical tank half full of water is lying on its side. If the radius of the tank is 5 m, find the force against one end.

Derivatives of Transcendental Functions

24.1   Review of Trigonometry

This section consists of a brief review of trigonometry. Many of the ideas discussed here will be needed in the remainder of this chapter.

Standard Position of an Angle

We recall that an angle θ is said to be in **standard position** if its vertex is at the origin and its initial side on the positive x-axis (Figure 24.1). Such an angle θ is considered positive if measured in the counterclockwise direction and negative if measured in the clockwise direction.

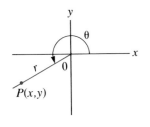

Figure 24.1

The Trigonometric Functions

We define the six trigonometric functions by selecting an arbitrary point $P(x, y)$ on the terminal side of θ, θ being in standard position (Figure 24.1). Let r be the distance from P to the origin. Then the following relationships hold:

$$\sin \theta = \frac{y}{r} \qquad \csc \theta = \frac{r}{y}$$

$$\cos \theta = \frac{x}{r} \qquad \sec \theta = \frac{r}{x}$$

$$\tan \theta = \frac{y}{x} \qquad \cot \theta = \frac{x}{y}$$

We will illustrate these functions by making use of some special angles ($0°$, $30°$, $45°$, $60°$, and so on). These angles occur so often in examples and problems that you should be thoroughly familiar with them. From Figure 24.2 we see that $\sin 30° = \frac{1}{2}$, from Figure 24.3 that $\tan 45° = 1$, and so on. The terminal

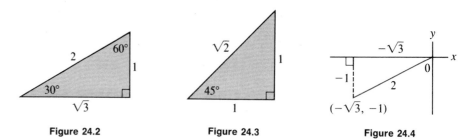

Figure 24.2   Figure 24.3   Figure 24.4

side may not be in the first quadrant. For example, to evaluate sin 210°, we draw the angle as shown in Figure 24.4 and let P be the point $(-\sqrt{3}, -1)$. From the definition, $\sin 210° = -\frac{1}{2}$. Similarly, $\cot 210° = -\sqrt{3}/(-1) = \sqrt{3}$, $\cos 210° = -\sqrt{3}/2$, and so forth.

E X A M P L E 1   Find cos 300°, csc 300°, and tan 300°.

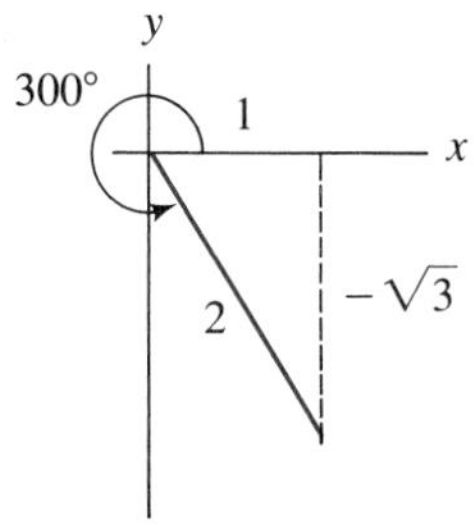

Solution. From Figure 24.5, we have

$$\cos 300° = \frac{x}{r} = \frac{1}{2}$$

$$\csc 300° = \frac{r}{y} = \frac{2}{-\sqrt{3}} = -\frac{2\sqrt{3}}{3}$$

$$\tan 300° = \frac{y}{x} = \frac{-\sqrt{3}}{1} = -\sqrt{3}$$ ◀

Figure 24.5

Radian Measure

Recall the following relationship between radian measure and degree measure:

$$\pi \text{ rad} = 180°$$

As a result, $\pi/2 = 90°$, $\pi/3 = 60°$, $\pi/4 = 45°$, and $\pi/6 = 30°$. Furthermore,

$$1 \text{ rad} = \frac{180°}{\pi} \approx 57.3°$$

and

$$1° = \frac{\pi}{180} \approx 0.01745 \text{ rad}$$

From the relationship π rad = 180°, we obtain the following rules for converting from one system of measurement to the other:

To convert from degree measure to radian measure, multiply by $\pi/180°$.
To convert from radian measure to degree measure, multiply by $180°/\pi$.

E X A M P L E **2**   Convert:

a. $72°$ to radians   **b.** $\dfrac{5\pi}{36}$ to degrees

Solution.

a. $72° = 72° \cdot \dfrac{\pi}{180°} = \dfrac{2\pi}{5}$

b. $\dfrac{5\pi}{36} = \dfrac{5\pi}{36} \cdot \dfrac{180°}{\pi} = 25°$ ◀

It also follows from the definition of radian measure that the arc length s of a circular sector is given by $s = r\theta$.

Formula for the Arc Length of a Circle:

$$s = r\theta \qquad \text{where } \theta \text{ is in radians} \qquad\qquad (24.1)$$

Now we will obtain a useful formula for the area of a circular sector. We know from elementary geometry that the area of a circular sector is always proportional to its central angle. In the special case where the central angle is the whole angle (2π radians), the area is πr^2. Denote the area of the sector by A and the central angle by θ. Since the ratios of the areas to their central angles are equal, we get

$$\frac{A}{\theta} = \frac{\pi r^2}{2\pi} \qquad \text{or} \qquad A = \tfrac{1}{2}r^2\theta$$

Formula for the Area of a Circular Sector:

$$A = \tfrac{1}{2}r^2\theta \qquad \text{where } \theta \text{ is in radians} \qquad\qquad (24.2)$$

Graphs

The relationship between an angle and its sine or cosine can be studied conveniently from the graphs of these functions. The functions $y = \sin x$ (Figure 24.6) and $y = \cos x$ (Figure 24.7) are periodic (repeating) with **period** 2π; they have a maximum value, or **amplitude**, of 1 unit. In general, the functions

$$y = a \sin bx \qquad \text{and} \qquad y = a \cos bx$$

have amplitude a and period

$$P = \frac{2\pi}{b}$$

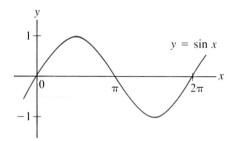

Figure 24.6

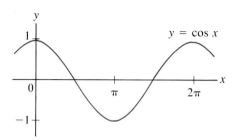

Figure 24.7

E X A M P L E **3**   Sketch the graph of $y = 3 \sin 2x$.

Solution. The amplitude is 3 and the period is $2\pi/2 = \pi$. Noting the basic shape of the sine function, we obtain the sketch directly (Figure 24.8).   ◄

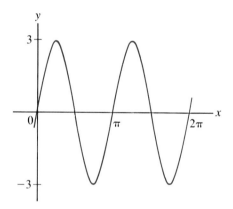

Figure 24.8

Trigonometric Identities

Let us now consider some trigonometric identities of the type that frequently occur in calculus, starting with the basic identities.

Basic Identities:

$$\sec \theta = \frac{1}{\cos \theta} \qquad\qquad \csc \theta = \frac{1}{\sin \theta}$$

$$\cot \theta = \frac{1}{\tan \theta} \qquad\qquad \tan \theta = \frac{\sin \theta}{\cos \theta}$$

$$\sin^2 \theta + \cos^2 \theta = 1 \qquad \cot \theta = \frac{\cos \theta}{\sin \theta}$$

$$1 + \tan^2 \theta = \sec^2 \theta \qquad 1 + \cot^2 \theta = \csc^2 \theta$$

Note especially that all trigonometric functions can be expressed in terms of sines and cosines.

E X A M P L E 4 Change $\cot \theta + \tan \theta$ to sines and cosines and simplify.

Solution.

$$\cot \theta + \tan \theta = \frac{\cos \theta}{\sin \theta} + \frac{\sin \theta}{\cos \theta}$$

$$= \frac{\cos \theta}{\sin \theta} \cdot \frac{\cos \theta}{\cos \theta} + \frac{\sin \theta}{\cos \theta} \cdot \frac{\sin \theta}{\sin \theta}$$

$$= \frac{\cos^2 \theta + \sin^2 \theta}{\sin \theta \cos \theta}$$

$$= \frac{1}{\sin \theta \cos \theta} \qquad \blacktriangleleft$$

Identities are frequently used to convert one form to another.

E X A M P L E 5 Convert:

a. $\cos^2 3x$ to a sine function
b. $1 + \cot^2 4x$ to a cosecant function

Solution.

a. Since $\sin^2 \theta + \cos^2 \theta = 1$, we have

$$\cos^2 3x = 1 - \sin^2 3x$$

b. Since $1 + \cot^2 \theta = \csc^2 \theta$, we get

$$1 + \cot^2 4x = \csc^2 4x \qquad \blacktriangleleft$$

Sum and Difference Formulas:

$$\sin(\theta \pm \phi) = \sin \theta \cos \phi \pm \cos \theta \sin \phi$$
$$\cos(\theta \pm \phi) = \cos \theta \cos \phi \mp \sin \theta \sin \phi$$

E X A M P L E 6 Show that $\cos\left(\frac{\pi}{2} - x\right) = \sin x$.

Solution. From the second identity,

$$\cos\left(\frac{\pi}{2} - x\right) = \cos\frac{\pi}{2}\cos x + \sin\frac{\pi}{2}\sin x$$

$$= 0 \cdot \cos x + 1 \cdot \sin x$$

$$= \sin x \qquad \blacktriangleleft$$

> **Double-Angle Formulas:**
>
> $$\sin 2\theta = 2 \sin \theta \cos \theta$$
>
> $$\cos 2\theta = \cos^2 \theta - \sin^2 \theta = 1 - 2 \sin^2 \theta = 2 \cos^2 \theta - 1$$

E X A M P L E **7**   Write each expression as a single trigonometric function:

a. $\sin 2x \cos 2x$ b. $2 \cos^2 4x - 1$

Solution.

a. From the identity $\sin 2\theta = 2 \sin \theta \cos \theta$, we get

$$\sin 2x \cos 2x = \tfrac{1}{2}(2 \sin 2x \cos 2x)$$
$$= \tfrac{1}{2} \sin 4x$$

b. From the identity $\cos 2\theta = 2 \cos^2 \theta - 1$, we get

$$2 \cos^2 4x - 1 = \cos 8x$$ ◀

> **Half-Angle Formulas:**   **Alternate Half-Angle Formulas:**
>
> $$\sin \frac{\theta}{2} = \pm \sqrt{\frac{1 - \cos \theta}{2}} \qquad \sin^2 \theta = \tfrac{1}{2}(1 - \cos 2\theta)$$
>
> $$\cos \frac{\theta}{2} = \pm \sqrt{\frac{1 + \cos \theta}{2}} \qquad \cos^2 \theta = \tfrac{1}{2}(1 + \cos 2\theta)$$

E X A M P L E **8**   Write $\sin^2 6x$ without the square.

Solution. By the first alternate half-angle formula,

$$\sin^2 6x = \tfrac{1}{2}(1 - \cos 12x)$$ ◀

E X E R C I S E S / S E C T I O N   **24.1**

In Exercises 1–24, evaluate the given expressions without tables or calculators.

1. $\sin 30°$
2. $\cos 120°$
3. $\tan(-45°)$
4. $\csc 240°$
5. $\sec 150°$
6. $\cot 300°$
7. $\sin(-150°)$
8. $\cos 210°$
9. $\csc(-30°)$
10. $\sec 225°$
11. $\tan 390°$
12. $\sin 0°$
13. $\tan 90°$
14. $\sec 180°$
15. $\cot 270°$
16. $\csc 180°$
17. $\sin 135°$
18. $\cos 225°$
19. $\sec 330°$
20. $\csc 240°$
21. $\cot 120°$
22. $\sin 315°$
23. $\cos 0°$
24. $\sin 270°$

In Exercises 25–33, convert the degree measures to radian measures.

25. $60°$ **26.** $45°$ **27.** $150°$

28. $-30°$ **29.** $135°$ **30.** $32°$

31. $144°$ **32.** $15°$ **33.** $20°$

In Exercises 34–43, express the radian measures in degree measures.

34. 0 **35.** $\dfrac{\pi}{4}$ **36.** $\dfrac{5\pi}{4}$ **37.** $\dfrac{\pi}{6}$

38. $\dfrac{\pi}{15}$ **39.** $\dfrac{5\pi}{3}$ **40.** $-\dfrac{5\pi}{4}$ **41.** $\dfrac{11\pi}{10}$

42. $\dfrac{17\pi}{18}$ **43.** $-\dfrac{5\pi}{9}$

In Exercises 44–48, find the period and amplitude, and sketch.

44. $y = 2 \sin x$ **45.** $y = \frac{1}{3} \sin 2x$

46. $y = \frac{1}{2} \sin 3x$ **47.** $y = 3 \cos \frac{1}{2}x$

48. $y = \frac{1}{2} \cos 4x$

In Exercises 49–62, change each expression to an expression involving sines and cosines, and simplify. (See Example 4.)

49. $\cos \theta \tan \theta$ **50.** $1 - \cot \theta \sin \theta$

51. $\tan \theta + \sec \theta$ **52.** $\dfrac{\cot \theta}{\csc \theta}$

53. $\dfrac{\tan^2 \theta - \sec^2 \theta}{\sec \theta}$ **54.** $\tan^2 \theta - \dfrac{\sec^2 \theta}{\csc^2 \theta}$

55. $\csc^2 \theta - \cot^2 \theta$ **56.** $\dfrac{\sec \theta}{\tan \theta + \cot \theta}$

57. $\dfrac{1}{\sin^2 \theta + \tan^2 \theta + \cos^2 \theta}$ **58.** $(1 + \tan^2 \theta) \cos^2 \theta$

59. $\sec^2 \theta + \csc^2 \theta$ **60.** $\dfrac{1 - \tan^2 \theta}{1 + \tan^2 \theta}$

61. $\cos \theta \cot \theta + \sin \theta$

62. $\sin^2 \theta \sec^2 \theta + \sin^2 \theta \csc^2 \theta$

In Exercises 63–70, convert sines to cosines and cosines to sines. (See Example 5.)

63. $1 - \cos^2 4x$ **64.** $1 - \cos^2 3x$

65. $1 - \sin^2 2x$ **66.** $1 - \sin^2 7x$

67. $\cos^2 5x$ **68.** $\sin^2 3x$

69. $\sin^2 6x$ **70.** $\cos^2 8x$

In Exercises 71–78, convert tangents to secants and secants to tangents.

71. $1 + \tan^2 6x$ **72.** $1 + \tan^2 2x$

73. $\sec^2 2x - 1$ **74.** $1 - \sec^2 4x$

75. $\tan^2 5x$ **76.** $\tan^2 9x$

77. $\sec^2 7x$ **78.** $\sec^2 8x$

79. Change $\cot^2 3x$ to a cosecant function.

80. Change $\csc^2 2x - 1$ to a cotangent function.

In Exercises 81–84, verify the given identities. (See Example 6.)

81. $\cos\left(x - \dfrac{\pi}{2}\right) = \sin x$ **82.** $\cos(\pi + x) = -\cos x$

83. $\sin(x + \pi) = -\sin x$ **84.** $\sin\left(x - \dfrac{\pi}{2}\right) = -\cos x$

In Exercises 85–92, write each expression as a single trigonometric function. (See Example 7.)

85. $2 \sin 5x \cos 5x$ **86.** $2 \sin 3x \cos 3x$

87. $\sin \dfrac{x}{2} \cos \dfrac{x}{2}$ **88.** $\sin 4x \cos 4x$

89. $\cos^2 3x - \sin^2 3x$ **90.** $\cos^2 4x - \sin^2 4x$

91. $2 \cos^2 8x - 1$ **92.** $1 - 2 \sin^2 5x$

In Exercises 93–98, write each expression without the square (See Example 8.)

93. $\sin^2 3x$ **94.** $\cos^2 4x$

95. $\cos^2 2x$ **96.** $\sin^2 5x$

97. $\sin^2 \frac{1}{2}x$ **98.** $\cos^2 \frac{1}{2}x$

24.2 Derivatives of Sine and Cosine Functions

In Chapter 20 we developed a number of formulas for differentiating algebraic functions, thereby avoiding the tedious four-step process. In this chapter we will learn how to differentiate **transcendental functions**—functions that are not algebraic. For such functions our earlier formulas offer no help, and we need to return to the definition of derivative.

Two Special Limits

To obtain the derivative of $y = \sin x$, we need to develop two special limits. The first is

$$\lim_{\theta \to 0} \frac{\sin \theta}{\theta} \qquad (\theta \text{ in radians})$$

Since the substitution of zero for θ yields the undefined expression $\frac{0}{0}$, we need to fall back on a geometric argument.

Let θ be the positive acute angle in Figure 24.9. In the figure, BC is an arc of a circle of radius r and AD is an arc of a circle of radius OA; AC is assumed to be perpendicular to OB at A. We can get an inequality between the various areas:

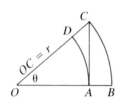

Figure 24.9

$$\text{area of sector } OAD < \text{area of triangle } OAC < \text{area of sector } OBC$$

Since the area of a triangle is $\frac{1}{2}bh$ and that of a circular sector is $\frac{1}{2}r^2\theta$ (formula 24.2), we get

$$\tfrac{1}{2}(OA)^2\theta < \tfrac{1}{2}(OA)(AC) < \tfrac{1}{2}(OC)^2\theta$$

Now, since $OC = r$, we have $OA = r \cos \theta$ and $AC = r \sin \theta$. Hence

$$\tfrac{1}{2}\theta(r \cos \theta)^2 < \tfrac{1}{2}(r \cos \theta)(r \sin \theta) < \tfrac{1}{2}r^2\theta$$

or

$$\theta \cos^2 \theta < \cos \theta \sin \theta < \theta \qquad \textbf{dividing by } \tfrac{1}{2} \, r^2$$

(Because $\frac{1}{2}$ and r^2 are positive, we may divide all the terms without changing the sense of the inequalities.) To obtain an expression for $(\sin \theta)/\theta$, we now divide through by $\theta \cos \theta$:

$$\cos \theta < \frac{\sin \theta}{\theta} < \frac{1}{\cos \theta} \qquad \textbf{dividing by } \theta \textbf{ cos } \theta$$

Finally, since $\cos \theta$ is continuous at zero,

$$\lim_{\theta \to 0} \cos \theta = \lim_{\theta \to 0} \frac{1}{\cos \theta} = 1$$

It follows that the middle expression $(\sin \theta)/\theta$ also approaches 1:

$$\lim_{\theta \to 0} \frac{\sin \theta}{\theta} = 1 \qquad\qquad (24.3)$$

If θ is negative, let $h = -\theta$, so that h is positive. Then

$$\frac{\sin \theta}{\theta} = \frac{\sin(-h)}{-h} = \frac{-\sin h}{-h} = \frac{\sin h}{h}$$

Hence formula (24.3) is valid if $\theta \to 0$ through negative values.
   Formula (24.3) can be used to obtain the other limit:

$$\lim_{\theta \to 0} \frac{\cos \theta - 1}{\theta} = 0 \qquad\qquad (24.4)$$

Note that

$$\lim_{\theta \to 0} \frac{\cos \theta - 1}{\theta} = \lim_{\theta \to 0} \frac{(\cos \theta - 1)(\cos \theta + 1)}{\theta(\cos \theta + 1)}$$

$$= \lim_{\theta \to 0} \frac{\cos^2 \theta - 1}{\theta(\cos \theta + 1)} = \lim_{\theta \to 0} \frac{-\sin^2 \theta}{\theta(\cos \theta + 1)} \qquad \textbf{sin}^2\ \theta + \textbf{cos}^2\ \theta = \textbf{1}$$

$$= \lim_{\theta \to 0} \frac{\sin \theta}{\theta} \cdot \frac{-\sin \theta}{\cos \theta + 1}$$

$$= \lim_{\theta \to 0} \frac{\sin \theta}{\theta} \lim_{\theta \to 0} \frac{-\sin \theta}{\cos \theta + 1} \qquad \textbf{Theorem B on limits}$$

$$= 1 \cdot \frac{0}{1 + 1} = 0 \qquad\qquad \textbf{by formula (24.3)}$$

The Derivative of $y = \sin u$ and $y = \cos u$

The limits (24.3) and (24.4) can now be used to find the derivative of $y = \sin x$ by the four-step process:

Step 1. $f(x + \Delta x) = \sin(x + \Delta x)$

Step 2. $\Delta y = \sin(x + \Delta x) - \sin x$

By the sum formula for the sine function,

$$\Delta y = (\sin x \cos \Delta x + \cos x \sin \Delta x) - \sin x$$
$$= \sin x \cos \Delta x - \sin x + \cos x \sin \Delta x \qquad \textbf{rearranging}$$
$$= \sin x(\cos \Delta x - 1) + \cos x \sin \Delta x \qquad \textbf{factoring sin x}$$

Step 3. $\dfrac{\Delta y}{\Delta x} = \dfrac{\sin x(\cos \Delta x - 1) + \cos x \sin \Delta x}{\Delta x}$

$$= \frac{\sin x(\cos \Delta x - 1)}{\Delta x} + \frac{\cos x \sin \Delta x}{\Delta x}$$

$$= \sin x \frac{\cos \Delta x - 1}{\Delta x} + \cos x \frac{\sin \Delta x}{\Delta x}$$

Step 4. $\displaystyle\lim_{\Delta x \to 0} \frac{\Delta y}{\Delta x} = \sin x \cdot 0 + \cos x \cdot 1$ **by (24.3) and (24.4)**

$$= \cos x$$

We conclude that

$$\frac{d}{dx}(\sin x) = \cos x$$

Remark. Note that we made use of formula (24.3), which depends on formula (24.2), but (24.2) is valid only for angles in radians. Since the derivatives of the other trigonometric functions will be obtained from the derivative of $y = \sin x$, we are going to be using mostly radian measure from now on.

A general form of the derivative of the sine function can be obtained from the chain rule (formula 20.10):

$$\frac{dy}{dx} = \frac{dy}{du}\frac{du}{dx}$$

So if u is a function of x, then

$$\frac{d}{dx}(\sin u) = \frac{d}{du}\sin u \frac{du}{dx}$$

or

$$\frac{d}{dx}(\sin u) = \cos u \frac{du}{dx} \tag{24.5}$$

E X A M P L E **1**   Find the derivative of $y = \sin \sqrt{x^2 + 1}$.

Solution. Since $y = \sin(x^2 + 1)^{1/2}$, from formula (24.5) we get

$$\frac{dy}{dx} = \cos(x^2 + 1)^{1/2}\frac{d}{dx}(x^2 + 1)^{1/2} \qquad \frac{d}{dx}\sin u = \cos u \frac{du}{dx}$$

$$= (\cos \sqrt{x^2 + 1})\cdot\frac{1}{2}(x^2 + 1)^{-1/2}\cdot 2x = \frac{x\cos\sqrt{x^2 + 1}}{\sqrt{x^2 + 1}}$$

(Don't forget to multiply by du/dx.)   ◀

E X A M P L E 2   Differentiate $y = \cos x$.

Solution. From the difference formula for the sine function we get

$$\sin\left(\frac{\pi}{2} - x\right) = \sin\frac{\pi}{2}\cos x - \cos\frac{\pi}{2}\sin x$$

$$= (1)(\cos x) - (0)(\sin x) = \cos x$$

Hence

$$\frac{d}{dx}(\cos x) = \frac{d}{dx}\left[\sin\left(\frac{\pi}{2} - x\right)\right]$$

$$= \cos\left(\frac{\pi}{2} - x\right)\frac{d}{dx}\left(\frac{\pi}{2} - x\right) = \cos\left(\frac{\pi}{2} - x\right)(-1)$$

$$= \left(\cos\frac{\pi}{2}\cos x + \sin\frac{\pi}{2}\sin x\right)(-1) = -\sin x \qquad \blacktriangleleft$$

By Example 2 and the chain rule,

$$\frac{d}{dx}(\cos u) = \frac{d}{du}(\cos u)\frac{du}{dx}$$

or

$$\boxed{\frac{d}{dx}(\cos u) = -\sin u\,\frac{du}{dx}} \qquad (24.6)$$

E X A M P L E 3   Find the derivative of $y = x^2 \cos x^3$.

Solution. By the product rule,

$$\frac{d}{dx}(uv) = u\frac{dv}{dx} + v\frac{du}{dx}$$

we get

$$\frac{dy}{dx} = x^2\frac{d}{dx}\cos x^3 + \cos x^3\frac{d}{dx}x^2$$

$$= x^2(-\sin x^3)(3x^2) + (\cos x^3)(2x) = -3x^4 \sin x^3 + 2x \cos x^3 \qquad \blacktriangleleft$$

E X A M P L E **4**
Find the derivative of $y = \dfrac{\sin^2 x}{\sqrt{x}}$.

Solution. Since $\sin^2 x = (\sin x)^2$, we need both the power and quotient rules. From the quotient rule,

$$\frac{d}{dx}\left(\frac{u}{v}\right) = \frac{v\dfrac{du}{dx} - u\dfrac{dv}{dx}}{v^2}$$

we get

$$\frac{dy}{dx} = \frac{x^{1/2}\dfrac{d}{dx}(\sin x)^2 - (\sin x)^2\dfrac{d}{dx}x^{1/2}}{(\sqrt{x})^2} \qquad \textbf{quotient rule}$$

By the power rule,

$$\frac{d}{dx}u^n = nu^{n-1}\frac{du}{dx}$$

we have

$$\frac{d}{dx}(\sin x)^2 = 2(\sin x)\frac{d}{dx}(\sin x) = (2\sin x)(\cos x)$$

It follows that

$$\frac{dy}{dx} = \frac{x^{1/2}(2\sin x)(\cos x) - (\sin x)^2(\tfrac{1}{2})x^{-1/2}}{x} \qquad \textbf{power rule}$$

$$= \frac{x^{1/2}(2\sin x\cos x) - (\sin x)^2(\tfrac{1}{2})x^{-1/2}}{x}\cdot\frac{2x^{1/2}}{2x^{1/2}}$$

$$= \frac{4x\sin x\cos x - \sin^2 x}{2x\sqrt{x}} \qquad\qquad\qquad\qquad\blacktriangleleft$$

E X E R C I S E S / S E C T I O N   **24.2**

In Exercises 1–38, differentiate the given functions.

1. $y = \cos 5x$

2. $y = 2\sin 3x$

3. $y = 2\cos 4x$

4. $y = 3\sin 5x$

5. $y = \sin x^2$

6. $y = \cos x^2$

7. $y = 3\cos x^3$

8. $y = 4\sin x^3$

9. $y = \sin 3x$

10. $y = \cos 4x$

11. $y = x\sin x$

12. $y = x\sin x^2$

13. $y = \dfrac{\sin x}{x}$

14. $y = \cos\sqrt{x}$

15. $y = \cos(x^2 + 3)$

16. $y = \sin(x + 1)$

17. $y = x\cos 2x$

18. $y = x^2\sin 4x$

19. $y = 2x \sin(2x + 2)$ **20.** $y = 3x \cos(4x - 3)$

21. $y = \sin \dfrac{1}{x}$ **22.** $y = \dfrac{\cos \sqrt{x}}{\sqrt{x}}$

23. $y = \dfrac{x}{\cos x}$ **24.** $y = \dfrac{x^2}{\sin x}$

25. $y = \dfrac{x}{\sin 4x}$ **26.** $y = \dfrac{3x}{\cos 5x}$

27. $y = \dfrac{\cos 2x}{3x}$ **28.** $y = \dfrac{\sin 5x}{x^3}$

29. $y = \sqrt{x} \sin x$ **30.** $y = \sin^2 x$

31. $y = \cos^2 x^3$ **32.** $y = x \sin^2 x$

33. $y = \sin x \cos x$ **34.** $y = \cos^2 x \sin x^2$

35. $y = \dfrac{\sin^3 x}{x}$ **36.** $y = \dfrac{\cos^3 x}{x^3}$

37. $y = x \cos^2 3x$ **38.** $y = \dfrac{x}{\sin^2 2x}$

39. Show that $\dfrac{d^2}{dx^2}(\sin x) = -\sin x$.

40. Show that $\dfrac{d^4}{dx^4}(\cos x) = \cos x$.

41. Show that $\dfrac{d^4}{dx^4}(\sin x) = \sin x$.

42. Show that $\dfrac{d^2}{dx^2}(\cos 3x) = -9 \cos 3x$.

43. Find the slope of the tangent line to the curve $y = x \sin 2x$ at $x = \pi/4$.

44. The charge across a capacitor is $q(t) = 0.25 \cos(t - 1.20)$. Find the current to the capacitor at $t = 3.60$ s. (Set your calculator in radian mode.)

24.3 Other Trigonometric Functions

Using the formulas from the last section, we can readily obtain the derivatives of the remaining trigonometric functions. From the identity $\tan u = \sin u / \cos u$ and the quotient rule,

$$\frac{d}{dx} \tan u = \frac{\cos u \dfrac{d}{dx} \sin u - \sin u \dfrac{d}{dx} \cos u}{\cos^2 u}$$

$$= \frac{\cos^2 u + \sin^2 u}{\cos^2 u} \frac{du}{dx} = \frac{1}{\cos^2 u} \frac{du}{dx} \qquad \text{cos}^2 \, u + \text{sin}^2 \, u = 1$$

$$= \sec^2 u \frac{du}{dx} \qquad\qquad\qquad \text{sec } u = 1/\text{cos } u$$

Hence

$$\frac{d}{dx}(\tan u) = \sec^2 u \frac{du}{dx} \tag{24.7}$$

Similarly, from $\csc u = 1/\sin u$ we find that

$$\frac{d}{dx} \csc u = \frac{d}{dx}(\sin u)^{-1} = -(\sin u)^{-2} \frac{d}{dx} \sin u \qquad \text{power rule}$$

$$= -\frac{1}{\sin^2 u} \cos u \frac{du}{dx} = -\frac{1}{\sin u} \frac{\cos u}{\sin u} \frac{du}{dx}$$

Hence

$$\frac{d}{dx}(\csc u) = -\csc u \cot u \frac{du}{dx} \qquad (24.8)$$

The remaining two derivations are similar and will be left as exercises:

$$\frac{d}{dx}(\cot u) = -\csc^2 u \frac{du}{dx} \qquad (24.9)$$

$$\frac{d}{dx}(\sec u) = \sec u \tan u \frac{du}{dx} \qquad (24.10)$$

Let us now summarize the derivative formulas for the trigonometric functions.

Derivatives of the Trigonometric Functions:

$$\frac{d}{dx}(\sin u) = \cos u \frac{du}{dx} \qquad \frac{d}{dx}(\cos u) = -\sin u \frac{du}{dx}$$

$$\frac{d}{dx}(\tan u) = \sec^2 u \frac{du}{dx} \qquad \frac{d}{dx}(\cot u) = -\csc^2 u \frac{du}{dx}$$

$$\frac{d}{dx}(\sec u) = \sec u \tan u \frac{du}{dx} \qquad \frac{d}{dx}(\csc u) = -\csc u \cot u \frac{du}{dx}$$

To help you remember these forms, note that the derivatives of the functions with prefix *co* (cosine, cotangent, and cosecant) are preceded by a negative sign, while the other three are not.

E X A M P L E **1**   Differentiate the function $y = \sqrt{\tan x}$.

Solution. We write $y = (\tan x)^{1/2}$ and use the power rule:

$$\frac{dy}{dx} = \frac{1}{2}(\tan x)^{-1/2}\frac{d}{dx}\tan x \qquad \frac{du^n}{dx} = nu^{n-1}\frac{du}{dx}$$

$$= \frac{1}{2}(\tan x)^{-1/2}(\sec^2 x)$$

by formula (24.7), so that

$$\frac{dy}{dx} = \frac{\sec^2 x}{2\sqrt{\tan x}}$$

◀

E X A M P L E **2**   Differentiate $y = x \sec x^2$.

Solution. By the product rule,

$$\frac{d}{dx}(uv) = u\frac{dv}{dx} + v\frac{du}{dx}$$

$$\frac{dy}{dx} = x\frac{d}{dx}\sec x^2 + \sec x^2 \frac{dx}{dx}$$

$$= x(\sec x^2 \tan x^2)(2x) + \sec x^2$$

by formula (24.10). Hence

$$\frac{dy}{dx} = 2x^2 \sec x^2 \tan x^2 + \sec x^2$$ ◀

E X A M P L E **3**   Differentiate $y = \sin 2x \cot x^2$.

Solution. We use the product rule and formulas (24.5) and (24.9):

$$\frac{dy}{dx} = \sin 2x\frac{d}{dx}\cot x^2 + \cot x^2\frac{d}{dx}\sin 2x$$

$$= \sin 2x(-\csc^2 x^2)(2x) + \cot x^2(\cos 2x)(2)$$
$$= -2x \sin 2x \csc^2 x^2 + 2\cos 2x \cot x^2$$ ◀

E X A M P L E **4**   Differentiate $y = \sqrt{x + \csc x^3}$.

Solution. By the power rule

$$\frac{dy}{dx} = \frac{1}{2}(x + \csc x^3)^{-1/2}\frac{d}{dx}(x + \csc x^3)$$

and by formula (24.8)

$$\frac{dy}{dx} = \frac{1}{2}(x + \csc x^3)^{-1/2}[1 - \csc x^3 \cot x^3(3x^2)]$$

$$= \frac{1 - 3x^2 \csc x^3 \cot x^3}{2\sqrt{x + \csc x^3}}$$ ◀

E X A M P L E **5**   Find dy/dx by implicit differentiation: $y^2 = \tan y + x$.

Solution. By formula (24.7),

$$2y\frac{dy}{dx} = \sec^2 y\frac{dy}{dx} + 1 \qquad \frac{d}{dx}\tan u = \sec^2 u\frac{du}{dx}$$

Thus

$$2y\frac{dy}{dx} - \sec^2 y \frac{dy}{dx} = 1 \quad \text{or} \quad (2y - \sec^2 y)\frac{dy}{dx} = 1$$

and

$$\frac{dy}{dx} = \frac{1}{2y - \sec^2 y} \qquad \blacktriangleleft$$

E X E R C I S E S / S E C T I O N   **24.3**

In Exercises 1–35, differentiate each of the functions.

35. $y = \dfrac{x}{\csc^2 x^2}$

1. $y = \sec 5x$
2. $y = 3 \tan 4x$
3. $y = 2 \csc 3x$
4. $y = 2 \cot 6x$
5. $y = 3 \cot 4x$
6. $y = 2 \csc 3x$
7. $y = 2 \csc x^2$
8. $y = \tan x^2$
9. $y = \tan 2x$
10. $y = \sec 3x$
11. $y = \cot x^3$
12. $y = \csc(1 - x^2)$
13. $y = \sec(x^3 + 1)$
14. $y = \sqrt{\cot x}$
15. $y = \sec \sqrt{x}$
16. $y = \dfrac{\tan x}{x}$
17. $y = \cot \sqrt{3x}$
18. $y = \sec^3 x$
19. $y = \sqrt{\tan 2x}$
20. $y = \cot^{1/3} x$
21. $y = 2 \tan^4 4x$
22. $y = 4 \cot^6 7x$
23. $y = \sqrt{\csc x^2}$
24. $y = \sqrt[4]{\sec 3x}$
25. $y = x^2 \csc x$
26. $y = \sin x \tan x$
27. $y = \cos^2 x \cot x$
28. $y = \sqrt{x} \cos x$
29. $y = \dfrac{1 + \tan x}{\sin x}$
30. $y = \cot^2 \sqrt{x}$
31. $y = x \sin(1 - x)^2$
32. $y = \dfrac{\sec 4x}{x^2}$
33. $y = \dfrac{x^3}{\tan 3x}$
34. $y = \dfrac{x}{\sec^2 2x}$

36. Show that $(d/dx)(\tan x - x) = \tan^2 x$.
37. Show that $(d/dx)(\frac{1}{3}\tan^3 x + \tan x) = \sec^4 x$.

In Exercises 38–42, find the second derivative of each function.

38. $y = \cot 2x$
39. $y = \sec 4x$
40. $y = \cos 5x$
41. $y = x \tan x$
42. $y = \dfrac{\sin x}{x}$

In Exercises 43–51, find dy/dx by implicit differentiation.

43. $y^2 = \tan x$
44. $y^2 = \sec 4x$
45. $y^2 = x \sec x$
46. $y = \cos(x - y)$
47. $y^2 = \sin(x + y^2)$
48. $x^2 y^2 = \csc 2x$
49. $y = x \cot y^2$
50. $y^2 = y + \sec x^2 - 2$
51. $\cos y = x^2 y - 2x$

52. Derive formulas (24.9) and (24.10).
53. Find the slope of the line normal to the curve $y = 2 \cot 2x$ at $x = \pi/8$.
54. A particle is moving along the x-axis so that its distance x (in feet) from the origin is $x = 2.0\sqrt{\tan 4.0t}$. Find the velocity at $t = 0.96$ s.

24.4   **Inverse Trigonometric Functions**

The purpose of this section is to review the inverse trigonometric functions. We can recall from trigonometry that the equality $\sin(\pi/6) = \frac{1}{2}$ can be written $\arcsin \frac{1}{2} = \pi/6$, or "$\pi/6$ is the number (or angle) whose sine is $\frac{1}{2}$." The problem is that if we start with the expression $\arcsin \frac{1}{2} = y$, then there exist many possible choices for y—namely, $\pi/6$, $5\pi/6$, $13\pi/6$, $17\pi/6$, and so on. It would be highly desirable to restrict the values so that y is unique. In other words, we wish

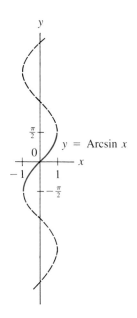

Figure 24.10

$y = \arcsin x$ to represent a **function**, denoted by

$$y = \text{Arcsin } x$$

Notice that the "A" in "Arcsin" is capitalized. (The notation $y = \sin^{-1} x$ is also commonly used.)

We can accomplish such a restriction in several ways. From the graph of $y = \text{Arcsin } x$ (Figure 24.10), we see that if we restrict y to the interval $[-\pi/2, \pi/2]$, then every value of x yields a unique value for y, although many other intervals could have been chosen. The advantage of our choice is that whenever x is positive, the terminal side of angle y will lie in the first quadrant. The function $y = \text{Arcsin } x$ is called the **inverse sine of x**.

Inverses can also be defined for the other trigonometric functions by suitable restriction of the domain of the trigonometric function. The most important cases are listed next.

Inverse Trigonometric Functions:

$$y = \text{Arcsin } x \qquad -\frac{\pi}{2} \le y \le \frac{\pi}{2} \qquad \text{(Figure 24.10)}$$

$$y = \text{Arctan } x \qquad -\frac{\pi}{2} < y < \frac{\pi}{2} \qquad \text{(Figure 24.11)}$$

$$y = \text{Arccos } x \qquad 0 \le y \le \pi \qquad \text{(Figure 24.12)}$$

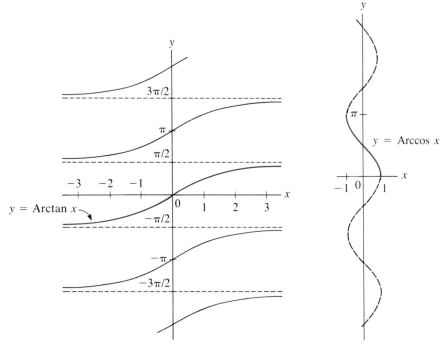

Figure 24.11   **Figure 24.12**

Although the remaining inverse trigonometric functions occur less often, let us list the ranges:

$$y = \text{Arccsc } x \qquad -\frac{\pi}{2} \le y \le \frac{\pi}{2} \qquad (y \ne 0)$$

$$y = \text{Arccot } x \qquad 0 < y < \pi$$

$$y = \text{Arcsec } x \qquad 0 \le y \le \pi \qquad \left(y \ne \frac{\pi}{2}\right)$$

In the following example and the corresponding exercises we are going to rely on our special angles again.

E X A M P L E **1**   Evaluate:

a. $\text{Arcsin } \dfrac{1}{\sqrt{2}}$ **b.** $\text{Arccos}\left(-\dfrac{\sqrt{3}}{2}\right)$ **c.** $\text{Arctan}(-\sqrt{3})$

Solution.

a. We are looking for an angle whose sine is $1/\sqrt{2}$. Inasmuch as y is restricted to the range $[-\pi/2, \pi/2]$, the only admissible value is $\pi/4$.
b. An angle whose cosine is $-\sqrt{3}/2$ cannot lie in the first quadrant. Hence $5\pi/6$ is the only value that lies in the defined range $[0, \pi]$.
c. In the range $(-\pi/2, \pi/2)$ the only admissible value lies in the interval $(-\pi/2, 0]$. Thus

$$\text{Arctan}(-\sqrt{3}) = -\frac{\pi}{3} \qquad \blacktriangleleft$$

E X A M P L E **2**   Solve the equation $y = 1 + \tan 2x$ for x.

Solution. We write the equation in the form $\tan 2x = y - 1$ and use the definition of inverse tangent:

$$2x = \text{Arctan}(y - 1) \qquad \text{or} \qquad x = \tfrac{1}{2} \text{Arctan}(y - 1) \qquad \blacktriangleleft$$

The simplification of certain expressions involving inverse trigonometric functions will be encountered again in Section 25.6. Study the next example.

E X A M P L E **3**   Find an algebraic expression for $\tan(\text{Arcsin } 2x)$.

Solution. Since Arcsin $2x$ in an angle, we let $\theta = \text{Arcsin } 2x$. Then $\sin \theta = 2x = 2x/1$. We now construct the diagram in Figure 24.13. Since $\sin \theta$ is the opposite side over the hypotenuse, we let $2x$ be the length of the opposite side and let

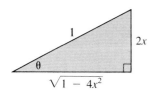

Figure 24.13

1 be the length of the hypotenuse. The radical is obtained from the Pythagorean theorem. The desired expression can now be read from the diagram:

$$\tan(\text{Arcsin } 2x) = \tan\theta = \frac{2x}{\sqrt{1-4x^2}} \qquad \tan\theta = \frac{\text{opposite}}{\text{adjacent}} \qquad \blacktriangleleft$$

Numerical expressions can be evaluated the same way, without tables or calculators.

E X A M P L E **4**   Evaluate $\sin[\text{Arccos}(-\tfrac{3}{4})]$.

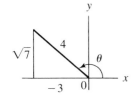

Figure 24.14

Solution. Let $\theta = \text{Arccos}(-\tfrac{3}{4})$, so that $\cos\theta = -\tfrac{3}{4}$. Since $0 \le \text{Arccos } x \le \pi$, θ is in the second quadrant. We now use the definition of the cosine function ($\cos\theta = x/r$) to construct the diagram in Figure 24.14. It follows that

$$\sin\left[\text{Arccos}\left(-\frac{3}{4}\right)\right] = \frac{\sqrt{7}}{4} \qquad \sin\theta = \frac{\text{opposite}}{\text{hypotenuse}} \qquad \blacktriangleleft$$

E X E R C I S E S / S E C T I O N **24.4**

In Exercises 1–24, evaluate each expression.

1. Arctan 1

2. $\text{Arcsin } \dfrac{\sqrt{3}}{2}$

3. $\text{Arcsin}\left(-\dfrac{1}{2}\right)$

4. $\text{Arccos } \dfrac{1}{\sqrt{2}}$

5. Arctan 0

6. $\text{Arccos}\left(-\dfrac{1}{2}\right)$

7. Arcsin(−1)

8. $\text{Arctan } \dfrac{1}{\sqrt{3}}$

9. Arccos(−1)

10. $\text{Arcsin}\left(-\dfrac{\sqrt{3}}{2}\right)$

11. Arctan(−1)

12. $\text{Arctan}\left(-\dfrac{1}{\sqrt{3}}\right)$

13. $\text{Arcsin}\left(-\dfrac{1}{\sqrt{2}}\right)$

14. Arcsin 0

15. sin(Arctan 2)

16. $\cos\left(\text{Arcsin } \dfrac{1}{3}\right)$

17. cos(Arctan 6)

18. $\tan\left(\text{Arccos } \dfrac{2}{3}\right)$

19. $\tan\left(\text{Arcsin } \dfrac{2}{3}\right)$

20. $\sin\left(\text{Arccos } \dfrac{4}{5}\right)$

21. $\sec\left[\text{Arcsin}\left(-\dfrac{1}{3}\right)\right]$

22. cos[Arctan(−2)]

23. $\tan\left[\text{Arccos}\left(-\dfrac{3}{4}\right)\right]$

24. $\cot\left[\text{Arcsin}\left(-\dfrac{2}{5}\right)\right]$

In Exercises 25–33, write each expression as an algebraic expression.

25. sin(Arccos x)

26. sec(Arctan x)

27. csc(Arcsin x)

28. sec(Arcsin x)

29. cos(Arctan $2x$)

30. sin(Arccos $3x$)

31. cot(Arcsin $2x$)

32. cos(Arctan $4x$)

33. sin(Arccos $3x$)

In Exercises 34–43, solve each equation for x.

34. $y = \text{Arcsin } 2x$

35. $y = 2\,\text{Arctan } \dfrac{x}{2}$

36. $y = \text{Arcsec } x^2$

37. $y = 1 + \sin x$

38. $y = 2 - \csc x$

39. $y = 1 + \cos 3x$

40. $y = 1 + 2\sin 2x$

41. $y = 3\tan 4x + 1$

42. $y = 2\cos 3x - 4$

43. $y = \sin \tfrac{1}{2}x - 2$

24.5   Derivatives of Inverse Trigonometric Functions

To find the derivative of $y = \text{Arcsin } u$, where u is a function of x, we write the expression in the form

$$u = \sin y \qquad \left(-\frac{\pi}{2} < y < \frac{\pi}{2}\right) \tag{24.11}$$

Since y is a function of x, formula (24.5) applies:

$$\frac{du}{dx} = \cos y \,\frac{dy}{dx} \qquad \frac{d}{dx}(\sin u) = \cos u \,\frac{du}{dx}$$

or

$$\frac{dy}{dx} = \frac{1}{\cos y}\,\frac{du}{dx} \tag{24.12}$$

We recall that $\sin^2 y + \cos^2 y = 1$, so that

$$\cos y = \pm\sqrt{1 - \sin^2 y}$$

Note, however, that for $-\pi/2 < y < \pi/2$, we must choose the positive square root, since $\cos y > 0$ in this range. Hence formula (24.12) becomes

$$\frac{dy}{dx} = \frac{1}{\sqrt{1 - \sin^2 y}}\,\frac{du}{dx} \qquad \left(|y| < \frac{\pi}{2}\right)$$

By (24.11), we have $u = \sin y$, and the last formula reduces to

$$\frac{d}{dx}(\text{Arcsin } u) = \frac{1}{\sqrt{1 - u^2}}\,\frac{du}{dx} \qquad (|u| < 1) \tag{24.13}$$

The derivation of

$$\frac{d}{dx}(\text{Arccos } u) = -\frac{1}{\sqrt{1 - u^2}}\,\frac{du}{dx} \qquad (|u| < 1) \tag{24.14}$$

is similar and will be left as an exercise.

For $y = \text{Arctan } u$ we have $u = \tan y$, so that

$$\frac{du}{dx} = \sec^2 y \frac{dy}{dx}$$

and, since $\sec^2 y = 1 + \tan^2 y$,

$$\frac{dy}{dx} = \frac{1}{\sec^2 y} \frac{du}{dx} = \frac{1}{1 + \tan^2 y} \frac{du}{dx}$$

Since $u = \tan y$, we get

$$\frac{d}{dx} (\text{Arctan } u) = \frac{1}{1 + u^2} \frac{du}{dx} \qquad\qquad (24.15)$$

E X A M P L E **1** Differentiate $y = \text{Arcsin } 2x^3$.

Solution. By formula (24.13),

$$\frac{dy}{dx} = \frac{1}{\sqrt{1 - (2x^3)^2}} \frac{d}{dx} (2x^3) = \frac{6x^2}{\sqrt{1 - 4x^6}} \qquad u = 2x^3 \qquad\blacktriangleleft$$

E X A M P L E **2** Differentiate $y = (\text{Arctan } x)^2$.

Solution. By the generalized power rule,

$$\frac{dy}{dx} = 2(\text{Arctan } x) \frac{d}{dx} \text{Arctan } x = 2(\text{Arctan } x)\frac{1}{1 + x^2} = \frac{2 \text{ Arctan } x}{1 + x^2}$$

by formula (24.15). $\qquad\blacktriangleleft$

E X A M P L E **3** Differentiate

$$y = \frac{\text{Arccos } 2x}{x}$$

Solution. By the quotient rule,

$$\frac{d}{dx} \left(\frac{u}{v}\right) = \frac{v \dfrac{du}{dx} - u \dfrac{dv}{dx}}{v^2}$$

we have

$$\frac{dy}{dx} = \frac{x\dfrac{d}{dx}\text{Arccos } 2x - (\text{Arccos } 2x)\dfrac{dx}{dx}}{x^2}$$

$$= \frac{x\left(-\dfrac{2}{\sqrt{1-4x^2}}\right) - \text{Arccos } 2x}{x^2}\qquad\text{formula (24.14)}$$

$$= \frac{x\left(-\dfrac{2}{\sqrt{1-4x^2}}\right) - \text{Arccos } 2x}{x^2} \cdot \frac{\sqrt{1-4x^2}}{\sqrt{1-4x^2}}$$

$$= \frac{x\left(-\dfrac{2}{\sqrt{1-4x^2}}\right)\sqrt{1-4x^2} - (\text{Arccos } 2x)\sqrt{1-4x^2}}{x^2\sqrt{1-4x^2}}$$

$$= -\frac{2x + \sqrt{1-4x^2}\,\text{Arccos } 2x}{x^2\sqrt{1-4x^2}}$$

◄

E X E R C I S E S / S E C T I O N 24.5

In Exercises 1–35, differentiate each function.

1. $y = \text{Arctan } 3x$ **2.** $y = \text{Arcsin } 3x$

3. $y = \text{Arccos } 5x$ **4.** $y = \text{Arctan } x^2$

5. $y = \text{Arctan } 2x^2$ **6.** $y = \text{Arctan } \sqrt{x}$

7. $y = \text{Arcsin } 3x^2$ **8.** $y = \text{Arccos } 9x$

9. $y = \text{Arcsin } 2x$ **10.** $y = \text{Arctan } 4x$

11. $y = \text{Arccos } x^2$ **12.** $y = \text{Arcsin } \sqrt{x}$

13. $y = \text{Arcsin } 2x$ **14.** $y = \text{Arctan } 5x$

15. $y = \text{Arctan } 7x$ **16.** $y = \text{Arccos } 2x$

17. $y = x \text{ Arctan } x$ **18.** $y = x^2 \text{ Arcsin } x$

19. $y = x \text{ Arccos } x^2$ **20.** $y = \text{Arccos}(1 - x^2)$

21. $y = x \text{ Arcsin } 3x$ **22.** $y = x \text{ Arccos } x^2$

23. $y = 2x \text{ Arctan } 3x$ **24.** $y = x \text{ Arctan } \sqrt{x}$

25. $y = \dfrac{\text{Arcsin } x}{x}$ **26.** $y = \text{Arctan } \dfrac{1}{x}$

27. $y = \text{Arccos } \sqrt{1 - x}$ **28.** $y = \text{Arctan } \sqrt{1 + x^2}$

29. $y = \sqrt{x} \text{ Arcsin } x$ **30.** $y = \dfrac{x}{\text{Arccos } x}$

31. $y = (\text{Arcsin } x)^2$ **32.** $y = x(\text{Arcsin } x)^2$

33. $y = \sqrt{\text{Arccos } x}$ **34.** $y = \sqrt[3]{\text{Arccos } x}$

35. $y = \dfrac{(\text{Arcsin } x)^2}{x}$

In Exercises 36 and 37, derive the given formulas.

36. $\dfrac{d}{dx}(\text{Arccos } u) = -\dfrac{1}{\sqrt{1 - u^2}}\dfrac{du}{dx}$

37. $\dfrac{d}{dx}(\text{Arccot } u) = -\dfrac{1}{1 + u^2}\dfrac{du}{dx}$

38. A balloon is rising from the ground at the rate of 5.0 ft/s from a point 190 ft from an observer. Determine the rate of increase of the angle of inclination of the observer's line of sight when the balloon is 65 ft high. (*Hint:* If θ is the angle and h the altitude of the balloon, then $\tan \theta = h/190$ and $\theta = \text{Arctan}(h/190)$.)

39. A tractor is moving at 12.4 ft/s toward a building 80.4 ft high. Determine the rate at which the angle of elevation of the top is increasing at the instant when the tractor is 20.6 ft from the base of the building.

24.6 Exponential and Logarithmic Functions

In Chapter 20 we introduced powers with rational exponents. Since discussion of irrational powers is beyond the scope of this book, we will assume that if x is an irrational number and $a > 0$, then a^x is defined and the usual laws of exponents hold.

Exponential Function

E X A M P L E 1   Discuss and sketch the graph of $y = 2^x$.

Solution. Since $2^x > 0$, there is no x-intercept; in fact the graph lies entirely above the x-axis. The y-intercept is $(0, 1)$. If $x > 0$ and increasing, then y increases rapidly, but if $x \to -\infty$, then $y \to 0$. Plotting a few points, we obtain the graph in Figure 24.15.

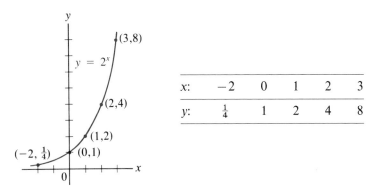

x:	-2	0	1	2	3
y:	$\frac{1}{4}$	1	2	4	8

Figure 24.15 ◀

The function in Example 1 is typical of $y = a^x$, $a > 1$, in that $(0, 1)$ is the y-intercept and

$$\lim_{x \to -\infty} a^x = 0$$

Logarithmic Function

The function $y = a^x$, $a > 0$, $a \neq 1$, can be expressed in a different form by means of **logarithms**. We define x to be the logarithm of y to the base a and write

$$x = \log_a y \qquad (a > 0, a \neq 1)$$

Let us be very clear about this: the expressions $y = a^x$ and $x = \log_a y$ mean exactly the same thing; only the notation is different. Since $x = \log_a y$, we conclude that logarithms are exponents. For example, in the equality $2^3 = 8$ the exponent 3 is the logarithm and we would now say "3 is the logarithm of 8 to the base 2," written as $\log_2 8 = 3$.

Definition of $\log_a y$: If $a > 0$ and $a \neq 1$, then $\log_a y = x$ if, and only if, $a^x = y$.

E X A M P L E **2**   **a.** Change $3^{-2} = \frac{1}{9}$ to logarithmic form and $\log_{32} 2 = \frac{1}{5}$ to exponential form.
b. Show that $\log_a 1 = 0$ and $\log_a a = 1$, $a > 0$.

Solution.

a. By definition, since -2 is the exponent and 3 the base, $3^{-2} = \frac{1}{9}$ is equivalent to $\log_3 \frac{1}{9} = -2$.
 In the expression $\log_{32} 2 = \frac{1}{5}$, the value $\frac{1}{5}$, being the logarithm, is an exponent. Since 32 is the base, we get $(32)^{1/5} = 2$.

b. For these expressions we simply note that $a^0 = 1$ and $a^1 = a$. ◀

Returning now to the logarithmic function $x = \log_a y$, suppose we employ the usual functional form with x as the independent variable. Then the function becomes $y = \log_a x$.

E X A M P L E **3**   Sketch the graph of $y = \log_2 x$.

Solution. The simplest way to sketch this graph is to write the function in its equivalent exponential form: since 2 is the base and y the exponent, the function becomes $x = 2^y$. We plot a few points by assigning values to y and calculating the corresponding values for x. The graph is shown in Figure 24.16. (Note that the equation is identical to the equation in Example 1 with x and y interchanged.)

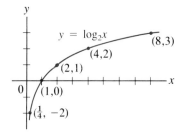

y:	-2	0	1	2	3
x:	$\frac{1}{4}$	1	2	4	8

Figure 24.16 ◀

Figure 24.16 is a typical graph of $y = \log_a x$, $a > 1$, in the sense that the x-intercept is always $(1, 0)$, the negative y-axis the asymptote, and the domain the open interval $(0, \infty)$.

Properties of Logarithms

Let us now recall the three basic laws of logarithms:

Properties of Logarithms:

$$\log_a MN = \log_a M + \log_a N \tag{24.16}$$

$$\log_a \frac{M}{N} = \log_a M - \log_a N \tag{24.17}$$

$$\log_a M^k = k \log_a M \tag{24.18}$$

We will check only formula (24.16); the proofs of the other two are similar. Let $u = \log_a M$ and $v = \log_a N$; the equivalent exponential forms are $M = a^u$ and $N = a^v$. Now

$$MN = a^u a^v = a^{u+v}$$

and the equivalent logarithmic form is

$$\log_a MN = u + v$$

Substituting back, we get

$$\log_a MN = \log_a M + \log_a N$$

E X A M P L E **4**   Write

$$2 \log_a x + 3 \log_a y - \log_a z$$

as a single logarithm.

Solution. We have

$$2 \log_a x + 3 \log_a y - \log_a z = \log_a x^2 + \log_a y^3 - \log_a z \qquad \text{by (24.18)}$$
$$= \log_a x^2 y^3 - \log_a z \qquad \text{by (24.16)}$$
$$= \log_a \frac{x^2 y^3}{z} \qquad \text{by (24.17)} \qquad \blacktriangleleft$$

In the next section we are going to use the laws of logarithms to change certain logarithmic expressions into simpler expressions. The technique is illustrated in the next example.

E X A M P L E **5**   Break up the following logarithms into sums, differences, or multiples of logarithms. (Use the relationships $\log_a 1 = 0$ and $\log_a a = 1$.)

a. $\log_4 4x^2$ **b.** $\log_2 \dfrac{1}{4\sqrt{x}}$

Solution.

a. $\log_4 4x^2 = \log_4 4 + \log_4 x^2$ property (24.16)

$\qquad\qquad = 1 + \log_4 x^2$ $\log_a a = 1$

$\qquad\qquad = 1 + \log_4 x^2$

$\qquad\qquad = 1 + 2\log_4 x$ property (24.18)

b. $\log_2 \dfrac{1}{4\sqrt{x}} = \log_2 1 - \log_2 4\sqrt{x}$ property (24.17)

$\qquad\qquad = 0 - \log_2 4\sqrt{x}$ $\log_a 1 = 0$

$\qquad\qquad = -(\log_2 4 + \log_2 \sqrt{x})$ property (24.16)

$\qquad\qquad = -\log_2 2^2 - \log_2 x^{1/2}$

$\qquad\qquad = -2\log_2 2 - \tfrac{1}{2}\log_2 x$ property (24.18)

$\qquad\qquad = -2 - \tfrac{1}{2}\log_2 x$ $\log_a a = 1$ ◀

Logarithms can be used to solve equations in which the unknown is an exponent.

E X A M P L E **6**   Solve the equation

$$(2.79)^x = 4.68$$

Solution. We take the common logarithm (base 10) of both sides and use property (24.18):

$\qquad (2.79)^x = 4.68$ given equation

$\qquad \log(2.79)^x = \log 4.68$ taking logarithms

$\qquad x \log 2.79 = \log 4.68$ $\log_a M^k = k \log_a M$

$\qquad x = \dfrac{\log 4.68}{\log 2.79}$ solving for x

Using a calculator, we get $x = 1.50$. The sequence is

$\qquad 4.68 \boxed{\text{LOG}} \boxed{\div} 2.79 \boxed{\text{LOG}} \boxed{=} \rightarrow 1.5041282$ ◀

As a final comment, note that the logarithmic function is continuous; that is,

$$\lim_{x \to x_0} \log_a x = \log_a(\lim_{x \to x_0} x) = \log_a x_0 \qquad\qquad (24.19)$$

E X E R C I S E S / S E C T I O N   **24.6**

In Exercises 1–6, write each of the expressions in logarithmic form.

1. $3^3 = 27$ **2.** $2^{-4} = \dfrac{1}{16}$ **3.** $9^0 = 1$

4. $(27)^{1/3} = 3$ **5.** $(32)^{-1/5} = \dfrac{1}{2}$ **6.** $5^{-2} = \dfrac{1}{25}$

In Exercises 7–12, write each of the expressions in exponential form.

7. $\log_3 243 = 5$ **8.** $\log_{10} 1000 = 3$ **9.** $\log_{1/4} \dfrac{1}{16} = 2$

10. $\log_{1/25} \dfrac{1}{5} = \dfrac{1}{2}$ **11.** $\log_9 \dfrac{1}{3} = -\dfrac{1}{2}$ **12.** $\log_5 \dfrac{1}{125} = -3$

In Exercises 13–20, determine the value of x.

13. $\log_{1/3} x = 2$ **14.** $\log_x 4 = \dfrac{1}{2}$ **15.** $\log_3 81 = x$

16. $\log_{16} \dfrac{1}{4} = x$ **17.** $\log_x \dfrac{1}{3} = -\dfrac{1}{3}$ **18.** $\log_x \dfrac{1}{7} = -\dfrac{1}{2}$

19. $\log_3 x = -2$ **20.** $\log_5 x = -1$

In Exercises 21–26, sketch the curves.

21. $y = 3^x$ **22.** $y = 4^x$ **23.** $y = 2^{-x}$
24. $y = 3^{-x}$ **25.** $y = \log_3 x$ **26.** $y = \log_4 x$

In Exercises 27–34, write each of the expressions as a single logarithm.

27. $\log_3 4 + \log_3 6$
28. $\log_5 7 - \log_5 6$
29. $5 \log_5 2 - 3 \log_5 2$

30. $5 \log_a x + 3 \log_a y - \log_a y^2$

31. $\dfrac{1}{2} \log_b 3 - \dfrac{1}{2} \log_b 9$

32. $\dfrac{1}{2} \log_2 4 - 2 \log_2 x$

33. $2 \log_3 y + \dfrac{1}{3} \log_3 8 - 2 \log_3 5$

34. $5 \log_b x + \dfrac{1}{3} \log_b y - 2 \log_b 4$

In Exercises 35–48, write each expression as a sum, difference, or multiple of logarithms. Whenever possible, use $\log_a 1 = 0$ and $\log_a a = 1$. (See Example 5.)

35. $\log_3 27$ **36.** $\log_2 16$
37. $\log_6 \sqrt{6x}$ **38.** $\log_2 4x^2$

39. $\log_5 \dfrac{1}{25x^2}$ **40.** $\log_3 \dfrac{1}{\sqrt{3x}}$

41. $\log_3 \dfrac{1}{\sqrt[3]{3x}}$ **42.** $\log_a \dfrac{1}{\sqrt{a}}$

43. $\log_5 \dfrac{1}{\sqrt{y-2}}$ **44.** $\log_{10} \dfrac{1}{\sqrt{x+1}}$

45. $\log_{10} \dfrac{x}{\sqrt{x+2}}$ **46.** $\log_{10} x\sqrt{x^2+1}$

47. $\log_{10} \dfrac{\sqrt{x}}{x+1}$ **48.** $\log_{10} \dfrac{\sqrt[3]{x}}{1-x}$

In Exercises 49–52, solve each equation for x. (See Example 6.)

49. $(3.62)^x = 12.4$ **50.** $(15.3)^x = 2.30$
51. $(8.04)^x = 2.85$ **52.** $(2.37)^x = 14.4$

24.7   Derivative of the Logarithmic Function

Some time before studying calculus you probably studied logarithms to base 10, which are highly useful for computational work. Logarithms were first used in this way by the Scottish mathematician John Napier (1550–1617), who published his first treatment on the subject in 1614. His work was enthusiastically received since it did much to simplify tedious computations.

Given the usefulness of the base 10, there seems to be little reason to consider any other base. Yet in obtaining the derivative of the logarithmic function, a different base suggests itself quite naturally (hence "natural logarithm").

To see how natural logarithms arise, let us find the derivative of $y = \log_a x$ by the four-step process.

Step 1. $f(x + \Delta x) = \log_a(x + \Delta x)$

Step 2. $\Delta y = \log_a(x + \Delta x) - \log_a x$

$$= \log_a \frac{x + \Delta x}{x} = \log_a\left(1 + \frac{\Delta x}{x}\right) \qquad \log_a \frac{M}{N} = \log_a M - \log_a N$$

Step 3. $\dfrac{\Delta y}{\Delta x} = \dfrac{1}{\Delta x} \log_a\left(1 + \dfrac{\Delta x}{x}\right)$

Step 4. $\displaystyle\lim_{\Delta x \to 0} \frac{\Delta y}{\Delta x} = \lim_{\Delta x \to 0} \frac{1}{\Delta x} \log_a\left(1 + \frac{\Delta x}{x}\right)$

Since the last expression cannot be simplified, it seems as though we might as well give up. Instead, we will take a closer look at the limit on the right:

$$\lim_{\Delta x \to 0} \frac{1}{\Delta x} \log_a\left(1 + \frac{\Delta x}{x}\right) = \lim_{\Delta x \to 0} \frac{x}{\Delta x} \frac{\log_a\left(1 + \frac{\Delta x}{x}\right)}{x} \qquad \text{inserting } x$$

$$= \lim_{\Delta x \to 0} \frac{\log_a\left(1 + \frac{\Delta x}{x}\right)^{x/\Delta x}}{x} \qquad k \log_a M = \log_a M^k$$

$$= \frac{1}{x} \lim_{\Delta x \to 0} \log_a\left(1 + \frac{\Delta x}{x}\right)^{x/\Delta x}$$

$$= \frac{1}{x} \log_a \lim_{\Delta x \to 0} \left(1 + \frac{\Delta x}{x}\right)^{x/\Delta x} \qquad \text{continuity}$$

by formula (24.19). Now we have a chance. The last limit has the form

$$\lim_{h \to 0} (1 + h)^{1/h}$$

whose value can be estimated by using a calculator. When $h = 0$, the function $f(h) = (1 + h)^{1/h}$ is undefined, but the function is continuous at all other points. The following point chart gives some of the values near $h = 0$:

h:	-0.1	-0.01	-0.001	-0.0001	-0.00001
$f(h)$:	2.8680	2.7320	2.7196	2.71844	2.71828

h:	0.00001	0.0001	0.001	0.01	0.1
$f(h)$:	2.71828	2.71815	2.7169	2.7048	2.5937

Based on our calculations, the limit appears to exist and to be approximately equal to 2.71828. (We will see in Chapter 26 that the value is indeed correct to the number of decimal places given.) It is customary to denote the limit by the letter e.

Definition of e

$$e = \lim_{h \to 0} (1 + h)^{1/h} \tag{24.20}$$

It turns out that e is a *transcendental number*, which means that it is not the root of a polynomial equation. (Proving that a number is transcendental can be extremely difficult. For e this was first shown by the French mathematician C. Hermite in 1873.)

Returning now to Step 4, we see that

$$\frac{d}{dx} \log_a x = \frac{1}{x} \log_a \lim_{\Delta x \to 0} \left(1 + \frac{\Delta x}{x} \right)^{x/\Delta x} = \frac{1}{x} \log_a e$$

Finally, since the number e cannot be avoided in the expression for the derivative, it was decided in the early days of calculus to adopt e for the base. Since $\log_e e = 1$, we get the striking formula

$$\frac{d}{dx} \log_e x = \frac{1}{x}$$

Natural logarithm

Logarithms to base e are referred to as **natural logarithms**. To avoid having to write the base each time, we adopt the notation

$$\ln x = \log_e x$$

read "natural log of x" or, if there is no ambiguity, just "log x"; $\ln x$ should not be read "l.n. of x." The last derivative formula is now written

$$\frac{d}{dx} (\ln x) = \frac{1}{x}$$

Furthermore, by the chain rule,

$$\frac{d}{dx} (\log_a u) = \frac{1}{u} (\log_a e) \frac{du}{dx}$$

and

$$\frac{d}{dx} (\ln u) = \frac{1}{u} \frac{du}{dx}$$

These ideas will be summarized next.

Logarithms to base e are called **natural logarithms** and are denoted by

$$\log_e x = \ln x$$

The derivative of the **logarithmic function** is given by

$$\frac{d}{dx}(\log_a u) = \frac{1}{u}(\log_a e)\frac{du}{dx} \qquad (24.21)$$

The derivative of the **natural logarithmic function** is given by

$$\frac{d}{dx}(\ln u) = \frac{1}{u}\frac{du}{dx} \qquad (24.22)$$

In particular,

$$\frac{d}{dx}(\ln x) = \frac{1}{x} \qquad (24.23)$$

E X A M P L E **1**   Differentiate $y = \log_2 x^2$.

Solution. By formula (24.21) we get directly

$$\frac{dy}{dx} = \frac{1}{x^2}\log_2 e\left(\frac{d}{dx}x^2\right) = \frac{2}{x}\log_2 e$$

◀

E X A M P L E **2**   Differentiate $y = \log_{10}(x^2 + x)$.

Solution.

$$\frac{dy}{dx} = \frac{1}{x^2 + x}\log_{10} e(2x + 1)$$

$$= \frac{2x + 1}{x^2 + x}\log_{10} e \qquad (\log_{10} e \approx 0.43429)$$

◀

E X A M P L E **3**   Differentiate $y = \ln \sec x$.

Solution. By formula (24.22) with $u = \sec x$,

$$\frac{dy}{dx} = \frac{1}{\sec x}\frac{d}{dx}(\sec x) = \frac{1}{\sec x}(\sec x \tan x) = \tan x$$

◀

In more complicated cases it is advisable to take advantage of the laws of logarithms.

E X A M P L E **4**   Find dy/dx if $y = \ln \sqrt[3]{x^2 + 1}$.

Solution. Since $y = \ln(x^2 + 1)^{1/3}$, we have, by formula (24.18),

$$y = \frac{1}{3} \ln(x^2 + 1) \qquad \log_a M^k = k \log_a M$$

Hence

$$\frac{dy}{dx} = \frac{1}{3} \frac{1}{x^2 + 1} \frac{d}{dx} (x^2 + 1) = \frac{1}{3} \frac{1}{x^2 + 1} (2x) = \frac{2x}{3(x^2 + 1)}$$

by formula (24.22). ◀

E X A M P L E **5**   Find the derivative of $y = \ln\left(\dfrac{\sin^2 x}{x}\right)$.

Solution. The function can be written in the form

$$y = \ln(\sin x)^2 - \ln x \qquad \log_a \frac{M}{N} = \log_a M - \log_a N$$

$$= 2 \ln \sin x - \ln x \qquad \log_a M^k = k \log_a M$$

It follows that

$$\frac{dy}{dx} = \frac{2}{\sin x} \cos x - \frac{1}{x}$$

$$= 2 \cot x - \frac{1}{x} \qquad \cot x = \frac{\cos x}{\sin x}$$

◀

E X E R C I S E S / S E C T I O N   **24.7**

Differentiate each of the following functions. (Whenever possible, break down the function as in Examples 4 and 5.)

1. $y = \ln 2x$

2. $y = \ln 3x$

3. $y = 4 \ln 3x$

4. $y = 3 \ln 4x$

5. $y = \ln x^2$

6. $y = \ln x^3$

7. $y = 2 \ln x^3$

8. $y = 3 \ln x^4$

9. $y = \log_{10} x^3$

10. $y = \log_5 \sqrt{x}$

11. $y = \ln \sin x$

12. $y = \ln \cos x$

13. $y = x \ln x$

14. $y = \ln(x^2 - 1)^2$

15. $y = \ln \sqrt{x - 1}$

16. $y = \ln \dfrac{1}{\sqrt{x}}$

17. $y = \ln \dfrac{1}{\sqrt{x + 2}}$

18. $y = \ln \sqrt[3]{1 - x}$

19. $y = 3 \ln \sqrt[3]{x^2 + 2}$

20. $y = \ln \dfrac{1}{\sqrt{1 - x^2}}$

21. $y = \ln \dfrac{x^2}{x + 1}$

22. $y = \ln \dfrac{x}{2x - 1}$

23. $y = \ln \dfrac{2x}{x^2 + 1}$

24. $y = \ln \dfrac{\sqrt{x}}{1 - x}$

25. $y = \ln \dfrac{\sqrt{x}}{2 - x^2}$

26. $y = \ln \dfrac{\sqrt[3]{x}}{x + 2}$

27. $y = \ln^2 x$

28. $y = \sqrt{\ln x}$

29. $y = \sqrt{x + 1} \, \ln x$

30. $y = \tan \ln x$

31. $y = (\sec x) \ln x$

32. $y = (\ln x^2) \csc x$

33. $y = \ln(1 + \sqrt{x^2 - 1})$

34. $y = \ln \sqrt{\sec x}$

35. $y = (\ln^2 x) \ln x^2$

36. $y = \tan(\ln \tan x)$

37. $y = \ln(\ln x)$

38. $y = \ln^3 x$

39. $y = \dfrac{\ln x}{x}$

40. $y = \dfrac{x}{\ln x}$

24.8   Derivative of the Exponential Function

Using logarithms to the base e, we can examine the function $y = e^x$. This function has the surprising property of being equal to its derivative. To check this statement, we will obtain a general formula, the derivative of $y = a^u$, $a > 0$, $a \neq 1$. From $y = a^u$ we have $u = \log_a y$. Hence

$$\frac{du}{dx} = \frac{1}{y} \log_a e \frac{dy}{dx}$$

by formula (24.21). Thus

$$\frac{dy}{dx} = \frac{1}{\log_a e} y \frac{du}{dx}$$

and, since $y = a^u$,

$$\frac{d}{dx}(a^u) = \frac{1}{\log_a e} a^u \frac{du}{dx} \tag{24.24}$$

It is readily shown that $1/\log_a e = \ln a$. Let $N = \log_a e$. Then $a^N = e$, and, taking natural logarithms of both sides,

$$\log_e a^N = \log_e e$$
$$N \log_e a = 1 \qquad\qquad \text{log}_a\, M^k = k \log_a M;\ \log_e e = 1$$
$$N = \frac{1}{\log_e a} = \frac{1}{\ln a} \qquad \text{notation for natural logarithm}$$

But $N = \log_a e$, so $\log_a e = 1/\ln a$ and $1/\log_a e = \ln a$. Consequently, formula (24.24) can also be written

$$\frac{d}{dx}(a^u) = a^u(\ln a) \frac{du}{dx}$$

In the special case where $a = e$, this formula reduces to (since $\log_e e = 1$)

$$\frac{d}{dx}(e^u) = e^u \frac{du}{dx}$$

Finally, if $u = x$,

$$\frac{d}{dx}(e^x) = e^x$$

as claimed. These rules are summarized next.

Derivatives of Exponential Functions:

$$\frac{d}{dx}(a^u) = a^u(\ln a)\frac{du}{dx} = \frac{1}{\log_a e}a^u\frac{du}{dx} \qquad (24.25)$$

$$\frac{d}{dx}(e^u) = e^u\frac{du}{dx} \qquad (24.26)$$

$$\frac{d}{dx}(e^x) = e^x \qquad (24.27)$$

E X A M P L E **1** Differentiate $y = 2^{x^2}$.

Solution. By formula (24.25),

$$\frac{d}{dx}(a^u) = a^u(\ln a)\frac{du}{dx}$$

we get

$$\frac{dy}{dx} = 2^{x^2}(\ln 2)\frac{d}{dx}(x^2) = 2^{x^2}(\ln 2)(2x) = 2x(\ln 2)2^{x^2} \qquad \blacktriangleleft$$

E X A M P L E **2** Find dy/dx if $y = e^{\sin^2 x}$.

Solution. By formula (24.26), we have

$$\frac{dy}{dx} = e^{\sin^2 x}\frac{d}{dx}(\sin x)^2 \qquad \frac{d}{dx}e^u = e^u\frac{du}{dx}$$

$$= e^{\sin^2 x}(2\sin x \cos x) \qquad \text{power rule}$$

$$= e^{\sin^2 x}\sin 2x \qquad \text{double-angle formula} \qquad \blacktriangleleft$$

E X A M P L E **3**   Differentiate $y = \ln(\tan e^{3x})$.

Solution. From the derivative of the logarithm,

$$\frac{dy}{dx} = \frac{1}{\tan e^{3x}} \frac{d}{dx} \tan e^{3x} \qquad\qquad \frac{d}{dx} \ln u = \frac{1}{u} \frac{du}{dx}$$

$$= \frac{1}{\tan e^{3x}} \sec^2 e^{3x} \frac{d}{dx} e^{3x} \qquad\qquad \frac{d}{dx} \tan u = \sec^2 u \frac{du}{dx}$$

$$= \frac{1}{\tan e^{3x}} \sec^2 e^{3x}(e^{3x} \cdot 3) \qquad\qquad \frac{d}{dx} e^u = e^u \frac{du}{dx}$$

$$= \frac{3e^{3x} \sec^2 e^{3x}}{\tan e^{3x}} \qquad\qquad\qquad\qquad\qquad\qquad\blacktriangleleft$$

Our methods can be adapted to differentiate functions of the form $y = u^v$, where u and v are both functions of x. The simplest way is to take the natural logarithm of both sides and differentiate.

E X A M P L E **4**   Differentiate $y = x^{\tan x}$.

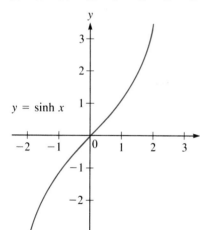

$y = \sinh x$

Figure 24.17

Solution. Since $\ln y = \tan x(\ln x)$, we get

$$\frac{1}{y} \frac{dy}{dx} = \tan x \cdot \frac{1}{x} + \sec^2 x(\ln x) \qquad \textbf{product rule}$$

so that

$$\frac{dy}{dx} = y\left[\frac{1}{x} \tan x + \sec^2 x(\ln x)\right]$$

Substituting back,

$$\frac{dy}{dx} = x^{\tan x}\left[\frac{1}{x} \tan x + \sec^2 x(\ln x)\right] \qquad \textbf{\textit{y}} = \textbf{\textit{x}}^{\tan x}$$

$$= x^{(\tan x - 1)}[\tan x + x \sec^2 x(\ln x)] \qquad \textbf{factoring 1/x} \qquad \blacktriangleleft$$

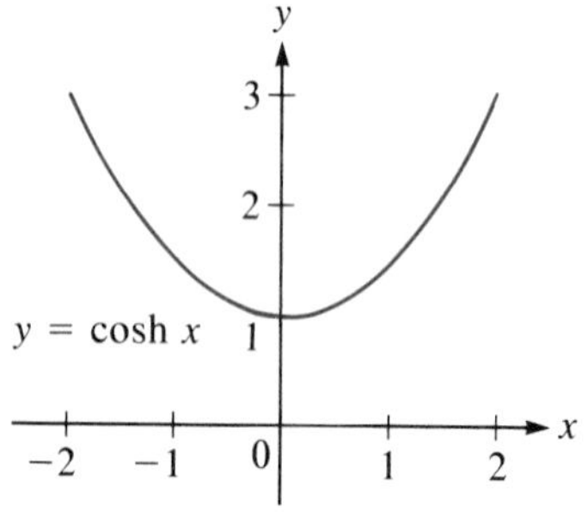

Certain combinations of exponential functions occur so often that they have been given special names; these are **hyperbolic functions**. Two of these functions are referred to in Exercise 29:

$$\sinh x = \tfrac{1}{2}(e^x - e^{-x}) \qquad \text{called the } \textit{hyperbolic sine of } x \text{ (Figure 24.17)}$$

and

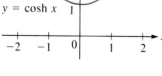

$y = \cosh x$

Figure 24.18

$$\cosh x = \tfrac{1}{2}(e^x + e^{-x}) \qquad \text{called the } \textit{hyperbolic cosine of } x \text{ (Figure 24.18)}$$

EXERCISES / SECTION 24.8

In Exercises 1–26, differentiate the given functions.

1. $y = e^{4x}$ **2.** $y = e^{-2x}$ **3.** $y = e^{x^2}$

4. $y = e^{-x^3}$ **5.** $y = 2e^{-x^2}$ **6.** $y = e^{2x^2}$

7. $y = e^{\tan x}$ **8.** $y = e^{\cot x}$ **9.** $y = 3^{x^3}$

10. $y = 2^{6x}$ **11.** $y = 4^{x^2}$ **12.** $y = e^{\sqrt{x}}$

13. $y = 2xe^x$ **14.** $y = 2^{\sqrt{x}}$ **15.** $y = e^{\sin x}$

16. $y = e^{(1-\cos x)}$ **17.** $y = \sin e^x$ **18.** $y = \dfrac{e^x}{x^2}$

19. $y = e^{2x} \sin x$ **20.** $y = e^{x^2} \cos x$ **21.** $y = \dfrac{\tan x}{e^{x^2}}$

22. $y = \text{Arcsin } e^x$ **23.** $y = (\ln x)e^{\sec x}$

24. $y = e^{\sin x} \ln \sqrt{x}$ **25.** $y = \ln(\sin e^{2x})$

26. $y = e^{\text{Arctan } x}$

27. Find dy/dx implicitly: $e^{y^2} + y = x$.

28. Find dy/dx implicitly: $\ln y - e^{-y} - x^2 = 0$.

29. Two of the hyperbolic functions defined earlier are

$$\sinh x = \tfrac{1}{2}(e^x - e^{-x}) \quad \text{and} \quad \cosh x = \tfrac{1}{2}(e^x + e^{-x})$$

Show that $d(\sinh x)/dx = \cosh x$ and $d(\cosh x)/dx = \sinh x$.

In Exercises 30–35, find the derivative of each of the given functions.

30. $y = x^x$ **31.** $y = x^{\sin x}$ **32.** $y = (\sin x)^x$

33. $y = (\ln x)^x$ **34.** $y = (\sin x)^{\cos x}$ **35.** $y = (\tan x)^x$

36. Show that $\dfrac{d}{dx} u^v = vu^{v-1}\dfrac{du}{dx} + u^v(\ln u)\dfrac{dv}{dx}$.

37. The charge across a capacitor is $q(t) = 1.5e^{-t}\cos 2.0t$. Find the current to the capacitor when $t = 0.34$ s.

38. A radioactive substance decays according to $N = 2.0e^{-0.10t}$, where N is in kilograms and t in minutes. Find the time rate of change of N when $t = 3.4$ min.

24.9 L'Hospital's Rule

In this section we are going to use our differentiation techniques to obtain certain difficult limits in a simple way. Such limits can be found by means of L'Hospital's rule.

L'Hospital's Rule: If $f(x)$ and $g(x)$ are differentiable for every x other than a in some interval, and if $\lim_{x \to a} f(x) = \lim_{x \to a} g(x) = 0$ or if $\lim_{x \to a} f(x) = \lim_{x \to a} g(x) = \infty$, then

$$\lim_{x \to a} \frac{f(x)}{g(x)} = \lim_{x \to a} \frac{f'(x)}{g'(x)} \tag{24.28}$$

provided that the latter limit exists.

EXAMPLE 1   Evaluate

$$\lim_{x \to 0} \frac{1 - \cos x}{x}$$

Solution. Since both numerator and denominator are equal to zero when $x = 0$, we say that $(1 - \cos x)/x$ tends to the indeterminate form $\tfrac{0}{0}$. So, by L'Hospital's rule,

$$\lim_{x \to 0} \frac{1 - \cos x}{x} = \lim_{x \to 0} \frac{\dfrac{d}{dx}(1 - \cos x)}{\dfrac{d}{dx}(x)}$$

$$= \lim_{x \to 0} \frac{\sin x}{1} = 0$$

◀

E X A M P L E **2** Evaluate

$$\lim_{x \to \infty} \frac{e^x}{x^3}$$

Solution. Here we say that the expression tends to the indeterminate form ∞/∞. Again by L'Hospital's rule,

$$\lim_{x \to \infty} \frac{e^x}{x^3} = \lim_{x \to \infty} \frac{e^x}{3x^2}$$

The last limit also tends to the form ∞/∞. So we apply the rule two more times to get

$$\lim_{x \to \infty} \frac{e^x}{6x} = \lim_{x \to \infty} \frac{e^x}{6} = \infty$$

That is, the limit does not exist.   ◀

L'Hospital's rule can be applied to the indeterminate form $0 \cdot \infty$, as illustrated in the next example.

E X A M P L E **3** Evaluate

$$\lim_{x \to \pi/2} (\tan x) \ln \sin x$$

Solution. Since $\tan \pi/2$ is undefined, the function takes on the indeterminate form $0 \cdot \infty$. L'Hospital's rule applies if the expression is written as a quotient:

$$\lim_{x \to \pi/2} (\tan x) \ln \sin x = \lim_{x \to \pi/2} \frac{\ln \sin x}{\cot x} \qquad \tan x = 1/\cot x$$

which tends to $\frac{0}{0}$. Thus

$$\lim_{x \to \pi/2} \frac{\ln \sin x}{\cot x} = \lim_{x \to \pi/2} \frac{\cos x/\sin x}{-\csc^2 x} = \lim_{x \to \pi/2} (-\cos x \sin x) = 0$$

◀

L'Hospital's rule is named after the French marquis G. F. A. de L'Hospital (1661–1704), who was a pupil of the great Swiss mathematician Johann Bernoulli (1667–1748). The rule was discovered by Bernoulli in 1694.

E X E R C I S E S / S E C T I O N   **24.9**

Evaluate each of the following limits.

1. $\lim\limits_{x \to -2} \dfrac{x^2 - 4}{x + 2}$

2. $\lim\limits_{x \to \infty} \dfrac{x^2 + 2x + 1}{2x^2 + 3}$

3. $\lim\limits_{x \to \infty} \dfrac{3x^2 - 4x}{2x^2 + 1}$

4. $\lim\limits_{x \to \infty} \dfrac{4x^2 - 5x + 4}{3x^2 - 6x + 1}$

5. $\lim\limits_{x \to 3} \dfrac{x^2 + x - 12}{x - 3}$

6. $\lim\limits_{x \to -1} \dfrac{x^2 - 3x - 4}{x + 1}$

7. $\lim\limits_{x \to 0} \dfrac{\sin 6x}{x}$

8. $\lim\limits_{x \to 0} \dfrac{\tan 2x}{\sin x}$

9. $\lim\limits_{x \to 0} \dfrac{\tan 3x}{1 - \cos x}$

10. $\lim\limits_{x \to \pi/2} \dfrac{1 - \sin x}{\cos x}$

11. $\lim\limits_{x \to 0} \dfrac{e^x - e^{-x}}{\sin x}$

12. $\lim\limits_{x \to \pi/2} \dfrac{\ln \sin x}{1 - \sin x}$

13. $\lim\limits_{x \to 0} \dfrac{1 - e^x}{2x}$

14. $\lim\limits_{x \to 0} \dfrac{x - \sin x}{x}$

15. $\lim\limits_{x \to \infty} \dfrac{x + \ln x}{x \ln x}$

16. $\lim\limits_{x \to \pi/2} \dfrac{\cos x}{\sin 2x}$

17. $\lim\limits_{x \to 0} \dfrac{x + \sin 2x}{x - \sin 2x}$

18. $\lim\limits_{x \to 0} \dfrac{\ln \sec x}{x^2}$

19. $\lim\limits_{x \to \pi/2} \dfrac{\cos x}{\pi - 2x}$

20. $\lim\limits_{x \to +\infty} \dfrac{\ln^2 x}{x}$

21. $\lim\limits_{x \to 0+} (\sin x) \ln \sin x$

22. $\lim\limits_{x \to 0} x \cot x$

23. $\lim\limits_{x \to 0+} x \ln x$

24. $\lim\limits_{x \to \infty} x \tan \dfrac{1}{x}$

24.10   Applications

Transcendental functions arise in many situations in science and technology. Some of these applications are considered in the following examples and exercises.

E X A M P L E   **1**   If air resistance is neglected, then the range R (in meters) of a projectile fired from a gun that is inclined at an angle θ from the ground is known to be

$$R = \frac{v_0{}^2}{5} \sin \theta \cos \theta$$

where v_0 is the initial velocity. Find the angle for which the range will be as large as possible.

Solution. To find the critical value, we differentiate R with respect to θ. By the product rule,

$$\frac{dR}{d\theta} = \frac{v_0{}^2}{5} \left[\sin \theta (-\sin \theta) + \cos \theta (\cos \theta) \right]$$

$$= \frac{v_0{}^2}{5} (-\sin^2 \theta + \cos^2 \theta) = \frac{v_0{}^2}{5} (-\sin^2 \theta + 1 - \sin^2 \theta)$$

$$= \frac{v_0{}^2}{5} (1 - 2 \sin^2 \theta) = 0$$

Solving for θ, we get

$$2 \sin^2 \theta = 1 \quad \text{or} \quad \sin \theta = \frac{1}{\sqrt{2}}$$

whence $\theta = 45° = \pi/4$. To check this value by the second-derivative test, we need to find

$$\frac{d^2 R}{d\theta^2} = \frac{v_0^2}{5}(-4 \sin \theta \cos \theta)$$

When $\theta = \pi/4$,

$$\frac{d^2 R}{d\theta^2} = -\frac{v_0^2}{5}\left(4 \sin \frac{\pi}{4} \cos \frac{\pi}{4}\right) = -\frac{v_0^2}{5}(4)\left(\frac{1}{\sqrt{2}}\right)\left(\frac{1}{\sqrt{2}}\right)$$

$$= -\frac{2v_0^2}{5} < 0$$

Consequently, the range is maximal when the angle is $\pi/4$.   ◄

E X A M P L E **2**   One of the pioneers in the study of polarized light was the French scientist E. L. Malus, who discovered the following law, called *Malus' law*, experimentally in 1809:

$$I = I_{max} \cos^2 \theta$$

where I_{max} is the maximum amount of light transmitted and I is the amount transmitted at angle θ. Suppose $I = 20$ lumens when $\theta = \pi/4$. If θ changes at the rate of 0.50 rad/min, how fast is I changing at that instant?

Solution. First we need to calculate I_{max} from the given information:

$$20 = I_{max} \cos^2 \frac{\pi}{4} = I_{max} \cdot \frac{1}{2} \qquad I = 20, \theta = \pi/4$$

or $I_{max} = 40$ lumens. Thus

$$I = 40 \cos^2 \theta$$

We are given that $d\theta/dt = 0.50$ and wish to find dI/dt when $\theta = \pi/4$. Recall that both I and θ are treated as functions of time in a related-rates problem. Differentiating with respect to t, we get

$$\frac{dI}{dt} = 40(2 \cos \theta)(-\sin \theta)\frac{d\theta}{dt} \qquad \textbf{power rule}$$

Now substitute the given values:

$$\frac{dI}{dt} = 40\left(2\cos\frac{\pi}{4}\right)\left(-\sin\frac{\pi}{4}\right)(0.50)$$

$$= -20 \text{ lumens/min} \qquad \blacktriangleleft$$

E X A M P L E **3**   The number of bacteria in a culture as a function of time is given by

$$N = Ae^{kt} \qquad \text{(where } A \text{ and } k \text{ are positive constants)}$$

under ideal conditions. Find the rate of change of N with respect to time.

Solution. Differentiation yields directly

$$\frac{dN}{dt} = Ae^{kt} \cdot k = kN \qquad \text{since } N = Ae^{kt}$$

This equation says effectively that "the bigger it is, the faster it grows." $\blacktriangleleft$

E X A M P L E **4**   The retarding force on a falling body owing to air resistance is proportional to the velocity; that is, the force is kv. If air resistance is neglected, then $k = 0$. We will see in Chapter 27 that the motion of a falling body is given by

$$v = \frac{mg}{k}(1 - e^{-kt/m})$$

Show that for $k = 0$, we get $v = gt$, the usual formula without air resistance.

Solution. This problem is an application of L'Hospital's rule. Thus

$$\lim_{k \to 0} v = \lim_{k \to 0} mg\left[\frac{1 - e^{-kt/m}}{k}\right] = \lim_{k \to 0} mg\left[\frac{-e^{-kt/m}\left(-\dfrac{t}{m}\right)}{1}\right] = gt$$

(Note that the derivatives were taken with respect to k.) $\blacktriangleleft$

E X A M P L E **5**   Consider the motion of particle A moving counterclockwise at a constant rate around a circle of radius a in Figure 24.19. We wish to study the motion of point $P(x, 0)$, the projection of A on the x-axis. To this end we denote angle AOP by $\theta = \omega t$, ω a constant, so that the angular velocity is $d\theta/dt = \omega$. Then

$$x = a\cos\theta = a\cos\omega t$$

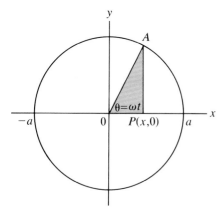

Figure 24.19

As θ ranges from 0 to 2π, t will range from 0 to $2\pi/\omega$. The following table gives some of the other values:

$\theta = \omega t$:	0	$\dfrac{\pi}{2}$	π	$\dfrac{3\pi}{2}$	2π
x:	a	0	$-a$	0	a

We conclude that the point moves back and forth along the x-axis between $x = a$ and $x = -a$.

The motion of point P, given by $x = a \cos \omega t$, is called **simple harmonic motion** and will be studied in greater detail in Chapter 28. In particular, the weight on a spring moves in simple harmonic motion; a condition similar to this mechanical motion can be given for alternating electric current.

Returning to the motion of P, the velocity and acceleration are readily found to be

$$\frac{dx}{dt} = -a\omega \sin \omega t \qquad \text{and} \qquad \frac{d^2x}{dt^2} = -a\omega^2 \cos \omega t$$

respectively. Note especially that

$$\frac{d^2x}{dt^2} = -\omega^2 x \tag{24.29}$$

This *differential equation* is sometimes used to define simple harmonic motion. The equation states that the acceleration is proportional to the displacement with direction opposite to the direction of motion. (See also Exercise 19.) ◀

E X A M P L E 6 Find the minima and maxima and points of inflection of the function $y = x^2 e^{-x}$; sketch the graph.

Solution. The derivatives are

$$f'(x) = -x^2 e^{-x} + 2xe^{-x} = xe^{-x}(2 - x)$$

and

$$f''(x) = x^2 e^{-x} - 4xe^{-x} + 2e^{-x} = e^{-x}(x^2 - 4x + 2)$$

a.   Critical points:

$$f'(x) = xe^{-x}(2 - x) = 0$$

whence $x = 0, 2$. So the critical points are $(0, 0)$ and $(2, 4/e^2)$.

b. Test of critical points:

$$f''(0) = 2; (0, 0) \text{ is a minimum}$$

$$f''(2) = -2e^{-2} < 0; (2, 4/e^2) \text{ is a maximum}$$

c. Concavity and points of inflection:

$$f''(x) = 0 \text{ whenever } x^2 - 4x + 2 = 0$$

so that

$$x = 2 \pm \sqrt{2}$$

Since $f''(x) \neq 0$ at all other points, we substitute arbitrary values such as $x = 0, 2,$ and 4. For example, since $f''(0) > 0, f''(x) > 0$ on the interval $(-\infty, 2 - \sqrt{2}]$. So f is concave up on this interval. Similarly, f is concave down on $[2 - \sqrt{2}, 2 + \sqrt{2}]$ and concave up on $[2 + \sqrt{2}, \infty)$. Hence $(2 - \sqrt{2}, 0.19)$ and $(2 + \sqrt{2}, 0.38)$ are points of inflection.

d. Other: since

$$\lim_{x \to +\infty} x^2 e^{-x} = \lim_{x \to +\infty} \frac{x^2}{e^x}$$

takes on the indeterminate form ∞/∞, we may apply L'Hospital's rule:

$$\lim_{x \to +\infty} \frac{x^2}{e^x} = \lim_{x \to +\infty} \frac{2x}{e^x} = \lim_{x \to +\infty} \frac{2}{e^x} = 0$$

We conclude that the positive x-axis is an asymptote. The graph is shown in Figure 24.20.

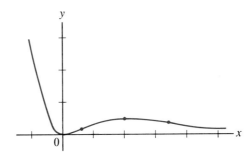

Figure 24.20   ◄

E X A M P L E **7**   A narrow hallway 1 m wide runs perpendicularly into another hallway. A thin pole 8 m long is carried horizontally along the narrow hallway and around the corner. How wide must the other hallway be to permit this?

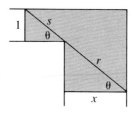

Figure 24.21

Solution. First we draw Figure 24.21. Referring to the figure, the problem is to find the largest distance x, given that the pole must touch the corner and both walls. Put another way, we find θ for which x is a maximum.

To be able to do so, we write x as a function of θ, making use of r and s. From the right triangles in the figure,

$$x = r \cos \theta \quad \text{and} \quad s = \csc \theta$$

Since $r + s = 8$, the length of the pole, we now have

$$r = 8 - s \quad \text{or} \quad r = 8 - \csc \theta$$

and since $x = r \cos \theta$, it follows that

$$x = (8 - \csc \theta) \cos \theta = 8 \cos \theta - \csc \theta \cos \theta = 8 \cos \theta - \cot \theta$$

To maximize x, we find the critical value:

$$x = 8 \cos \theta - \cot \theta$$

$$\frac{dx}{d\theta} = -8 \sin \theta + \csc^2 \theta = 0$$

$$-8 \sin \theta + \frac{1}{\sin^2 \theta} = 0 \qquad \csc \theta = \frac{1}{\sin \theta}$$

$$8 \sin^3 \theta = 1 \qquad \qquad \textbf{multiplying by } \sin^2 \theta$$

and

$$\sin^3 \theta = \tfrac{1}{8} \quad \text{or} \quad \sin \theta = \tfrac{1}{2}$$

Hence $\theta = 30°$ and from $x = 8 \cos \theta - \cot \theta$, we get

$$x = 8 \cos 30° - \cot 30° = 3\sqrt{3} \text{ m}$$

Since

$$\frac{d^2 x}{d\theta^2} = -8 \cos \theta - 2 \csc^2 \theta \cot \theta \bigg|_{\theta = \pi/6} < 0$$

it follows that $x = 3\sqrt{3}$ is indeed a maximum.   ◀

EXERCISES / SECTION 24.10

In Exercises 1–5, find the minima and maxima, the points of inflection, and sketch the graph.

1. $y = xe^{-x}$

2. $y = \dfrac{\ln x}{x} \ (x > 0)$

3. $y = x \ln x \ (x > 0)$

4. $y = \sinh x$ (see Figure 24.17)

5. $y = \cosh x$ (see Figure 24.18)

6. Find the minimum point on the graph of $y = x^x, x > 0$, and sketch.

7. If the current in an AC circuit is given by $i = \cos t + \sin t$, find the first maximum current after $t = 0$.

8. Repeat Exercise 7 for $i = \cos t + 2 \sin t$.

9. Find the voltage across a 0.0030-H inductor at $t = 0$ if the current is $i = 3.0 \sin 200t + 2.0 \cos 200t$.

10. The distance in meters from the origin of a certain object is given by $x = 2 \cos 2t + \sin 4t$. Find the velocity at $t = \pi/4$ s.

11. A certain radioactive substance decays according to the law $N = 12e^{-2.0t}$, where N (in kilograms) is the amount present and t is the time in years. Find an expression for the time rate of decrease of N as a function of time. Show that $dN/dt = -2.0N$.

12. A certain radioactive substance decays according to the law $N = 20(0.80)^{2t}$ where N (in grams) is the amount present. Find the rate of change of N with respect to t when $t = 5.0$ years.

13. A certain radioactive isotope decays according to the law $N = 6.0e^{-0.25t}$, where N is measured in grams. Find the time rate of change of N when $t = 8.5$ min.

14. At $t = 0$, the number of bacteria in a culture is P_0. For $t \geq 0$, the number is $P = P_0e^{kt}$. Show that the time rate of change of the number of bacteria is proportional to the number present.

15. The number N of bacteria in a culture as a function of time is $N = 200e^{(1/2)t}$, where t is measured in hours. Find the rate of increase after 3 h.

16. A warm object is placed in a refrigerator to cool. If the temperature in degrees Celsius at any time is given by $T = 10 + 20e^{-0.0933t}$, find the rate of change of the temperature when $t = 10$ min.

17. An underwater telegraph cable consists of a circular core and a layer of insulation. The speed S of a signal is given by $S = kR^2 \ln(1/R)$, where R is the ratio of the radius of the core to the thickness of the covering and k is a constant. For what value of R is the speed the greatest?

18. Since the human ear is sensitive to a large range of intensities, a logarithmic scale is most convenient. The *intensity level* ß (in decibels) of a sound is given by the equation ß $= 10 \log(I/I_0)$, where $I_0 = 10^{-6}$ W/cm², the intensity corresponding roughly to the faintest sound that can be heard. In testing the noise made by a quiet automobile, the intensity is measured to be $1.1 \times 10^{-11.2}$ W/cm², with a possible error of 4.0×10^{-12} W/cm². Use differentials to determine the approximate error in ß.

19. Consider the simple harmonic motion $x = 5 \cos 2t$. Show that the magnitude of the velocity is a maximum whenever the particle passes the origin.

20. Show that for the straight-line motion described by $x = a \sin 2\pi kt + b \cos 2\pi kt$, where x gives the position as a function of time, the acceleration is proportional to the displacement from the origin and oppositely directed.

21. A balloon is rising from the ground at the rate of 6.0 m/s from a point 100 m from an observer, also on the ground. Use inverse trigonometric functions to determine how fast the angle of inclination of the observer's line of sight is increasing when the balloon is at an altitude of 150 m.

22. A fugitive is running along a wall at 4.0 m/s. A searchlight 20 m from the wall is trained on him. How fast is the searchlight rotating at the instant when he is 10 m from the point on the wall nearest the searchlight?

23. A man is walking toward a building 80.0 m high at the rate of 1.5 m/s. How fast is the angle of elevation of the top increasing when he is 40.0 m away from the building?

24. The largest weight W that can be pulled up a plane inclined at an angle θ with the horizontal by a force F is $W = (F/\mu)(\cos \theta + \mu \sin \theta)$, where μ is the coefficient of friction. Find θ so that W is a maximum.

25. A gutter is to be made from a long piece of metal 24 cm wide by turning up strips 8 cm wide along each side in such a way that they make equal angles θ with the vertical. For what value of θ will the cross-sectional area be the greatest?

26. A wall is 3 m high and 2 m from a building. Find the length of the shortest ladder that touches the ground and building and just clears the wall. (*Hint:* If x is the length of the ladder and θ is the angle made with the ground, then $x = 2 \sec \theta + 3 \csc \theta$.)

27. A hallway is 2 m wide and runs perpendicularly into another hallway 5 m wide. What is the length of the longest thin pole that can be moved horizontally around the corner?

28. An object of weight W is dragged along a horizontal plane by a force F whose line of action makes an angle θ with the plane. The force is given by $F = (\mu W)/(\mu \sin \theta + \cos \theta)$, where μ is the coefficient of friction. Show that the pull is minimal when $\mu = \tan \theta$.

29. The range R of a projectile fired from a gun along an inclined plane making an angle α with the horizontal is

$$R = \frac{2v_0^2 \cos \theta \sin(\theta - \alpha)}{g \cos^2 \alpha}$$

where θ is the angle of elevation of the gun and v_0 is the initial velocity. For what value of θ is the range up the plane a maximum? (*Suggestion:* Write the derivative in the form $K \cos(A + B)$.)

30. The profit P on a certain commodity as a function of the number x of units sold is $P = Axe^{-x/n}$, where A and n are constants. What value of x yields the maximum profit?

REVIEW EXERCISES/CHAPTER 24

In Exercises 1–6, evaluate each expression.

1. Arcsin(-1) **2.** Arctan(-1) **3.** Arccos$(-\frac{1}{2})$

4. Arcsin 0   **5.** cos(Arctan 3)   **6.** sin(Arccos $\frac{1}{3}$)

In Exercises 7 and 8, write each expression as an algebraic expression.

7. sin(Arctan x)   **8.** tan(Arccos x)

In Exercise 9–26, differentiate each of the given functions.

9. $y = x^2 \tan 3x$

10. $y = \dfrac{\sin 4x}{x}$

11. $y = \dfrac{e^{2x}}{x}$

12. $y = \cos x \ln x$

13. $y = \ln \dfrac{1}{\sqrt{x+3}}$ [Use the properties of logarithms.]

14. $y = \ln \dfrac{\sqrt{x}}{1-x}$

15. $y = \ln \dfrac{\sqrt{2x^2+1}}{x}$

16. $y = \sin e^{-x}$

17. $y = \sqrt{\ln 2x}$

18. $y = e^{2 \tan x}$

19. $y = \cos(\ln x)$

20. $y = 5^{x^2}$

21. $y = e^{2x} \cot x$

22. $y = x$ Arcsin x

23. $y = e^{\text{Arccos } 3x}$

24. $y = \ln \sec^2 x$

25. $y = (\cot x)^x$

26. $y = (\sec x)^x$

27. Find dy/dx implicitly: $e^{\sin y} + \csc x = 1$.

28. Find dy/dx implicitly: $\ln y + \tan y + x = 3$.

In Exercises 29–33, evaluate the given limits.

29. $\lim\limits_{x \to 0} \dfrac{\sin 4x}{x}$

30. $\lim\limits_{x \to \infty} \dfrac{\ln x}{x}$

31. $\lim\limits_{x \to \pi/4} (1 - \tan x) \sec 2x$

32. $\lim\limits_{x \to 0} \dfrac{1 - \cos x}{x^2}$

33. $\lim\limits_{x \to 0^+} \dfrac{\sin x - x}{x \sin x}$

34. Find the equation of the line normal to the curve $y = x \ln x$ at $(1, 0)$. (Recall that a normal line is perpendicular to a tangent line.)

35. Find the dimensions of the largest rectangle in the fourth quadrant whose sides lie along the axes with one vertex on the curve $y = \ln x$.

36. The velocity in meters per second of a certain falling object was determined to be $v = 20(1 - e^{-0.1t})$. Find the limiting velocity.

37. A deposit of S dollars that earns $100r\%$ annual interest compounded continuously leaves a balance of $P = Se^{rt}$ dollars after t years. **(a)** What will an amount of $5000 grow to after 15 years at 10% annual interest compounded continuously? **(b)** Determine the rate at which P is growing.

38. The atmospheric pressure at an altitude of h meters above sea level is $P = 10^4 e^{-0.00012h}$ kg/m². If the altitude increases at the rate of 25 m/min, what is the rate of change of pressure at the instant when $h = 1000.0$ m?

39. The vapor pressure p, in millimeters of mercury, of carbon tetrachloride is given by

$$\log p = -\frac{1706.4}{T} - 7.7760$$

where T is in kelvin (K). If T increases at the rate of 2.00 K/min, find the rate at which p increases when $T = 300.0$ K.

40. An inductance L and a resistance R are in series with a battery of E volts. We will see in Chapter 27 that the current is given by

$$I = \frac{E}{R}(1 - e^{-Rt/L})$$

Find an expression for the time-rate of change of the current.

41. A light is hung above the center of a round table of unit radius. Let d be the length of the line segment joining the light to a point on the edge and θ the angle that this line makes with the table. Then the illumination at the edge is given by $I = k(\sin \theta)/d^2$, where k is a constant. Find θ so that the illumination at the edge is a maximum.

42. A tracking device 1 mi from a launching pad is kept trained on a missile traveling along the parabolic path $y^2 = x$ (x and y are measured in miles). By placing the launching pad at the origin and the tracking device at $(-1, 0)$, determine the largest angle of elevation that the device must be capable of.

In Exercises 1–5, perform the indicated integrations.

1. $\int (1 - x\sqrt{x}) \, dx$

2. $\int \sqrt{1 - 3x} \, dx$

3. $\int \frac{x - 1}{\sqrt{x^2 - 2x}} \, dx$

4. $\int (x^3 + 1)^2 x^2 \, dx$

5. $\int (x^3 + 1)^2 x \, dx$

In Exercises 6 and 7, find the area bounded by the given curves.

6. $y = x$, $y = x^2$

7. $y = x^2 - x - 2$, $y = x - 2$

8. Find the approximate value of the integral

$$\int_0^1 \frac{dx}{x^2 + 2}$$

by Simpson's rule with $n = 4$.

9. Find the volume of the solid obtained by rotating the region bounded by $y = \sqrt{x} + 1$ and the coordinate axes about the x-axis.

10. Find the volume of the solid obtained by revolving the region bounded by $y = x - 3$ and $x = y^2 + 1$ about the y-axis.

11. Find the volume of the solid obtained by revolving the region bounded by $y = \sqrt{x}$, $x = 4$, and $y = 0$ about the line $x = 4$.

12. Find the centroid of the region bounded by $y = \frac{1}{4}x^2$ and $y = \frac{1}{2}x$.

13. a. Find I_x of the region bounded by $y = \frac{1}{3}x$, $x = 3$, and $y = 0$.

b. Find the radius of gyration with respect to its axis of the solid obtained by revolving the region in the first quadrant bounded by $y = x^2$, $y = 4$, and the y-axis about the y-axis.

14. The vertical gate on a dam is a rectangle 5 m long and 3 m high. If the top is horizontal and 1 m below the surface, find the force against the gate.

In Exercises 15–19, differentiate the given functions.

15. $y = x \sin 3x$

16. $y = \frac{\tan x}{x}$

17. $y = e^{-2x} \cos x$

18. $y = \ln \sqrt{x + 2}$

19. $y = \text{Arctan } 3x^2$

20. Evaluate $\lim_{x \to 0+} x^2 \ln x$.

21. A stone is hurled downward with an initial velocity of 16 ft/s from a height of 112 ft. How long will it take to reach the ground and with what speed will it strike? (*Note:* $g = 32 \text{ ft/s}^2$.)

22. A certain radioactive substance decays according to the law $N = 2.00e^{-0.0123t}$, where N (in kilograms) is the amount present and t the time in years. Find the rate of decay of N when $t = 4.00$ years.

23. A cable 12 m long and weighing 2.5 N/m is hanging from a winch. Find the work done in winding it up.

Integration Techniques

25.1 The Power Formula Again

In the last chapter we obtained the derivatives of various transcendental functions. As a result we will now be able to integrate a much larger class of functions, both algebraic and transcendental. For our first integration formula, recall the general power formula from Chapter 22.

General Power Formula:

$$\int u^n \, du = \frac{u^{n+1}}{n+1} + C \qquad (n \neq -1) \tag{25.1}$$

Since we are no longer confined to algebraic functions, we may apply formula (25.1) to integrals involving transcendental functions; thus integrals of this form may vary greatly in appearance. Consequently, though the power formula has a very simple structure, integrals of this form are among the hardest to recognize. You must be thoroughly familiar with the differentiation formulas of the last chapter to be able to pick out the function u and its differential du. *Recognizing the type of integral is an important part of the integration process.*

E X A M P L E **1** Integrate $\int \sqrt{\cos x} \, \sin x \, dx$.

Solution. Written as

$$\int (\cos x)^{1/2} \sin x \, dx$$

the integral seems to fit formula (25.1). Indeed, if we let $u = \cos x$, then $du = -\sin x \, dx$. As a consequence, we must introduce -1 to complete the differential and place -1 in front of the integral. Thus

$$-\int (\cos x)^{1/2}(-\sin x)\,dx = -\int u^{1/2}\,du \qquad \begin{bmatrix} u = \cos x \\ du = -\sin x \, dx \end{bmatrix}$$

$$= -\frac{2}{3}u^{3/2} + C = -\frac{2}{3}\cos^{3/2}x + C \qquad \blacktriangleleft$$

E X A M P L E **2** Find

$$\int \frac{e^x \, dx}{(1 + 2e^x)^3}$$

Solution. We try $u = 1 + 2e^x$; then $du = 2e^x \, dx$. For the integral to fit the proper form, we need to introduce a 2 and place $\frac{1}{2}$ in front of the integral. Thus

$$\frac{1}{2}\int \frac{2e^x \, dx}{(1 + 2e^x)^3} = \frac{1}{2}\int \frac{du}{u^3} = \frac{1}{2}\int u^{-3}\,du \qquad \begin{bmatrix} u = 1 + 2e^x \\ du = 2e^x \, dx \end{bmatrix}$$

$$= \frac{1}{2}\frac{u^{-2}}{-2} + C = -\frac{1}{4}\frac{1}{u^2} + C$$

$$= -\frac{1}{4(1 + 2e^x)^2} + C \qquad \blacktriangleleft$$

E X A M P L E **3** Integrate

$$\int \frac{\ln^2 x \, dx}{x}$$

Solution. At first glance this integral appears to be quite different from those above. However, recalling that $d \ln x = (1/x)\,dx$, we see that the substitution $u = \ln x$ leads quite simply to

$$\int (\ln x)^2 \frac{dx}{x} = \int u^2 \, du = \frac{1}{3}u^3 + C = \frac{1}{3}\ln^3 x + C \qquad \begin{bmatrix} u = \ln x \\ du = \frac{1}{x}dx \end{bmatrix} \qquad \blacktriangleleft$$

E X A M P L E **4** Evaluate

$$\int_0^1 \frac{\text{Arctan } x \, dx}{1 + x^2}$$

Solution. Let $u = \text{Arctan } x$; then $du = dx/(1 + x^2)$ and

$$\int_0^1 \frac{\text{Arctan } x \, dx}{1 + x^2} = \frac{1}{2} (\text{Arctan } x)^2 \Big|_0^1 \qquad \int u \, du = \frac{1}{2} u^2 + C$$

$$= \frac{1}{2} [(\text{Arctan } 1)^2 - (\text{Arctan } 0)^2] = \frac{1}{2} \left(\frac{\pi}{4}\right)^2 = \frac{\pi^2}{32} \qquad \blacktriangleleft$$

EXERCISES / SECTION 25.1

Perform the following integrations.

1. $\int x\sqrt{x^2 + 1} \, dx$

2. $\int x^2 \sqrt{1 - x^3} \, dx$

3. $\int \frac{dx}{\sqrt{1 - x}}$

4. $\int \frac{dx}{\sqrt{x + 2}}$

5. $\int \sin^2 x \cos x \, dx$

6. $\int \cos^3 x \sin x \, dx$

7. $\int \tan^2 x \sec^2 x \, dx$

8. $\int \cot^3 x \csc^2 x \, dx$

9. $\int (1 + \tan x)^3 \sec^2 x \, dx$

10. $\int \frac{x \, dx}{\sqrt{1 - 3x^2}}$

11. $\int (1 + e^x)^3 e^x \, dx$

12. $\int e^x \sqrt{1 + 3e^x} \, dx$

13. $\int \frac{\ln x \, dx}{x}$

14. $\int \frac{1 + 2 \ln x}{x} \, dx$

15. $\int \frac{dx}{x\sqrt{\ln x}}$

16. $\int \frac{\text{Arcsin } x \, dx}{\sqrt{1 - x^2}}$

17. $\int \frac{\text{Arccos } x \, dx}{\sqrt{1 - x^2}}$

18. $\int \frac{1 + \cos x}{(x + \sin x)^2} \, dx$

19. $\int_{\pi/6}^{\pi/2} \cos^2 x \sin x \, dx$

20. $\int_0^1 \frac{x \, dx}{\sqrt{x^2 + 1}}$

21. $\int_1^e \frac{\sqrt{\ln x}}{x} \, dx$

22. $\int \frac{e^x \, dx}{\sqrt{2 - e^x}}$

23. $\int \frac{\cot x \, dx}{\sin^2 x}$

24. $\int \frac{dx}{\cot^3 x \cos^2 x}$

25. $\int \sqrt{\tan x} \sec^2 x \, dx$

26. $\int \frac{\sqrt[3]{\ln x}}{x} \, dx$

27. $\int \frac{\text{Arctan } 4x}{1 + 16x^2} \, dx$

28. $\int (1 + \sin x)^3 \cos x \, dx$

29. $\int \frac{(1 + \ln x)^2}{x} \, dx$

30. $\int_1^e \frac{\ln^3 x}{x} \, dx$

25.2   The Logarithmic and Exponential Forms

The general power formula (25.1) is valid for all n except $n = -1$. We can fill this gap by making use of the derivative of the logarithmic function. We know from Chapter 24 that for $u > 0$,

$$\frac{d}{dx} \ln u = \frac{1}{u} \frac{du}{dx}$$

or, in differential form,

$$d \ln u = \frac{du}{u}$$

It follows that

$$\int \frac{du}{u} = \ln u + C \qquad (u > 0)$$

This formula can be extended. First recall the definition of absolute value:

$$|x| = \begin{cases} x & x \geq 0 \\ -x & x < 0 \end{cases}$$

For example,

$$|2| = |+2| = 2 \qquad \text{since x is positive}$$

and

$$|-2| = -(-2) = 2 \qquad \text{since x is negative}$$

Thus

$$\ln|x| = \begin{cases} \ln x & x > 0 \\ \ln(-x) & x < 0 \end{cases}$$

Differentiating, we get

$$\frac{d}{dx}\ln|x| = \begin{cases} \dfrac{d}{dx}\ln x & x > 0 \\ \dfrac{d}{dx}\ln(-x) & x < 0 \end{cases} = \begin{cases} \dfrac{1}{x} & x > 0 \\ \dfrac{-1}{-x} = \dfrac{1}{x} & x < 0 \end{cases}$$

In either case,

$$\frac{d}{dx}\ln|x| = \frac{1}{x} \qquad (x \neq 0)$$

It now follows from the chain rule that

$$\frac{d}{dx}\ln|u| = \frac{1}{u}\frac{du}{dx} \qquad (u \neq 0)$$

By reversing this formula, we get the following integration form:

Logarithmic Form:

$$\int \frac{du}{u} = \ln|u| + C \tag{25.2}$$

This formula fills the gap left by the power formula (25.1).

From $de^u/dx = e^u\, du/dx$, we get the differential form

$$de^u = e^u\, du$$

By reversing this formula, we get the corresponding integration form:

Exponential Form:

$$\int e^u\, du = e^u + C \qquad\qquad\qquad (25.3)$$

For bases other than e we have, from formula (24.25),

$$\int a^u\, du = a^u \log_a e + C = \frac{a^u}{\ln a} + C$$

E X A M P L E **1**   Integrate

$$\int \frac{x\, dx}{x^2 + 1}$$

Solution. If we let $u = x^2 + 1$, then $du = 2x\, dx$; since $n = -1$, formula (25.1) does not apply. But by formula (25.2) we get

$$\frac{1}{2} \int \frac{2x\, dx}{x^2 + 1} = \frac{1}{2} \int \frac{du}{u} = \frac{1}{2} \ln|u| + C \qquad \begin{bmatrix} u = x^2 + 1 \\ du = 2x\, dx \end{bmatrix}$$

$$= \frac{1}{2} \ln|x^2 + 1| + C = \frac{1}{2} \ln(x^2 + 1) + C$$

since $x^2 + 1 > 0$.

Remark. If the quantity $x^2 + 1$ were raised to *any* power besides 1, we would have to use the power formula. For example, to integrate

$$\int \frac{x\, dx}{(x^2 + 1)^2}$$

we let $u = x^2 + 1$, so that $du = 2x\, dx$. The integral then becomes

$$\frac{1}{2} \int \frac{2x\, dx}{(x^2 + 1)^2} = \frac{1}{2} \int \frac{du}{u^2} = \frac{1}{2} \int u^{-2}\, du = \frac{1}{2} \frac{u^{-1}}{-1} + C$$

$$= -\frac{u^{-1}}{2} + C = -\frac{1}{2u} + C = -\frac{1}{2(x^2 + 1)} + C \qquad \blacktriangleleft$$

E X A M P L E 2 Find

$$\int \frac{dx}{x(\ln x + 1)}$$

Solution. Since the derivative of $\ln x$ is $1/x$, the derivative of $\ln x + 1$ is also $1/x$. So we try the substitution $u = \ln x + 1$. Then $du = dx/x$, and we obtain

$$\int \frac{1}{\ln x + 1} \frac{dx}{x} = \int \frac{du}{u} = \ln|u| + C = \ln|\ln x + 1| + C \qquad \left[\begin{array}{l} u = \ln x + 1 \\ du = \dfrac{1}{x} dx \end{array} \right] \blacktriangleleft$$

E X A M P L E 3 Integrate $\int e^{\cos x} \sin x \, dx$.

Solution. This integral seems to fit formula (25.3). Let $u = \cos x$; then $du = -\sin x \, dx$ and

$$\int e^{\cos x} \sin x \, dx = -\int e^{\cos x}(-\sin x) \, dx$$
$$= -\int e^u \, du = -e^u + C = -e^{\cos x} + C \qquad \blacktriangleleft$$

E X A M P L E 4 Integrate

$$\int \frac{e^{\text{Arcsin } 2x}}{\sqrt{1 - 4x^2}} \, dx$$

Solution. The form fits formula (25.3):

$$\int \frac{e^{\text{Arcsin } 2x}}{\sqrt{1 - 4x^2}} \, dx = \frac{1}{2} \int e^{\text{Arcsin } 2x} \frac{2 \, dx}{\sqrt{1 - 4x^2}} \qquad \left[\begin{array}{l} u = \text{Arcsin } 2x \\ du = \dfrac{2 \, dx}{\sqrt{1 - 4x^2}} \end{array} \right]$$
$$= \frac{1}{2} \int e^u \, du = \frac{1}{2} e^u + C$$
$$= \frac{1}{2} e^{\text{Arcsin } 2x} + C \qquad \blacktriangleleft$$

E X A M P L E 5 Evaluate

$$\int_0^1 \frac{e^x \, dx}{1 + e^x}$$

Solution. Even though the integrand contains e^x, the integral is not an exponential form. But since $d(1 + e^x) = e^x\, dx$, we see that formula (25.2) applies. Hence

$$\int_0^1 \frac{e^x\, dx}{1 + e^x} = \ln(1 + e^x)\Big|_0^1 = \ln(1 + e) - \ln 2 = \ln\frac{1 + e}{2}$$ ◀

E X E R C I S E S / S E C T I O N   **25.2**

In Exercises 1–43, perform the designated integrations.

1. $\displaystyle\int \frac{dx}{x - 1}$

2. $\displaystyle\int \frac{dx}{1 - 4x}$

3. $\displaystyle\int \frac{dx}{2 + 3x}$

4. $\displaystyle\int \frac{x\, dx}{1 - 2x^2}$

5. $\displaystyle\int \frac{dx}{1 - 3x}$

6. $\displaystyle\int \frac{dx}{7x + 6}$

7. $\displaystyle\int_0^1 \frac{x\, dx}{x^2 + 1}$

8. $\displaystyle\int \frac{x\, dx}{2x^2 + 4}$

9. $\displaystyle\int e^{-x}\, dx$

10. $\displaystyle\int e^{6x}\, dx$

11. $\displaystyle\int_0^2 2e^{3x}\, dx$

12. $\displaystyle\int 2e^{-4x}\, dx$

13. $\displaystyle\int e^{4x}\, dx$

14. $\displaystyle\int_{-1}^0 3e^{-3x}\, dx$

15. $\displaystyle\int xe^{x^2}\, dx$

16. $\displaystyle\int x^2 e^{x^3}\, dx$

17. $\displaystyle\int e^{\sin x} \cos x\, dx$

18. $\displaystyle\int \frac{\cos x\, dx}{1 + \sin x}$

19. $\displaystyle\int \frac{\sec^2 x\, dx}{1 + \tan x}$

20. $\displaystyle\int e^{\tan x} \sec^2 x\, dx$

21. $\displaystyle\int \frac{e^{\text{Arctan } x}}{1 + x^2}\, dx$

22. $\displaystyle\int \frac{e^{\sqrt{x}}\, dx}{\sqrt{x}}$

23. $\displaystyle\int \frac{dx}{x \ln x}$ (See Example 2.)

24. $\displaystyle\int \frac{dx}{x(1 - \ln x)}$

25. $\displaystyle\int \frac{e^x + 1}{e^x}\, dx$

26. $\displaystyle\int \frac{\sin x\, dx}{(1 + \cos x)^2}$

27. $\displaystyle\int \frac{e^{-x}\, dx}{1 - e^{-x}}$

28. $\displaystyle\int xe^{(1 + x^2)}\, dx$

29. $\displaystyle\int \frac{2x}{(1 + x^2)^2}\, dx$

30. $\displaystyle\int (1 + e^x)^2 e^x\, dx$

31. $\displaystyle\int (1 + e^x)^2\, dx$

32. $\displaystyle\int \frac{\sec x \tan x\, dx}{1 + \sec x}$

33. $\displaystyle\int \frac{x + 1}{x + 2}\, dx$ (*Hint:* Divide numerator by denominator.)

34. $\displaystyle\int \frac{x^2 + x - 1}{x + 1}\, dx$

35. $\displaystyle\int \frac{\cos x}{\sin^2 x}\, dx$

36. $\displaystyle\int \frac{\sec^2 x\, dx}{1 - 2 \tan x}$

37. $\displaystyle\int e^{\sin^2 x} \sin 2x\, dx$

38. $\displaystyle\int_0^1 xe^{-x^2}\, dx$

39. $\displaystyle\int_0^{\pi/3} \frac{\sin x\, dx}{\cos x}$

40. $\displaystyle\int 2^x\, dx$

41. $\displaystyle\int 5^{3x}\, dx$

42. $\displaystyle\int x\, 3^{x^2}\, dx$

43. $\displaystyle\int 2^{\cot x} \csc^2 x\, dx$

44. Find the area of the region bounded by $y = 1/(x + 1)$, $x = 1$, and the coordinate axes.

45. Find I_y, the moment of inertia about the y-axis, of the region bounded by $y = 1/(1 + x^3)$, $x = 2$, and the coordinate axes.

46. Find the first-quadrant area of the region bounded by $y = e^{-x}$ and the coordinate axes.

47. Find the average value of the current $i = e^{(-1/3)t}$ in a certain circuit from $t = 0$ to $t = 2$.

48. Show that the root mean square of the function $(\ln x)/\sqrt{x}$ from $x = 1$ to $x = e$ is $1/\sqrt{3(e - 1)}$.

49. Find the volume generated by rotating the region bounded by $y = e^{-x^2}$, $x = 1$, and the coordinate axes about the y-axis.

50. Find the coordinates of the centroid of the region bounded by $y = 1/\sqrt{x}$, $x = 1$, $x = 4$, and the x-axis.

51. Find the volume of the solid generated by rotating the region bounded by $y = 4/x^2$, $x = 1$, $x = 2$, and the x-axis about the y-axis.

52. The number N ($N > 0$) of bacteria in a culture at any time satisfies the relation $dN/N = k\, dt$, $k > 0$. Show that $N = N_0 e^{kt}$ if $N_0 > 0$ denotes the number of bacteria when $t = 0$.

53. The velocity in meters per second of a falling object of mass 1 kg experiencing a retarding force due to air resistance satisfies the relation $dv/(10 - kv) = dt, k > 0$, $10 - kv > 0$. Find v as a function of time if the body is dropped from rest.

54. Use a calculator to find the approximate value of

$$\int_0^2 e^{-x^2} \, dx$$

by Simpson's rule with $n = 6$.

25.3   Trigonometric Forms

The purpose of this section is to discuss the basic trigonometric integrals. These forms include the integrals of the six trigonometric functions, as well as the forms obtained by reversing the derivative formulas. Other trigonometric forms will be taken up in the next section.

Since integration is the inverse of differentiation, integrating the derivatives of the six trigonometric functions yields the following formulas:

$$\int \cos u \, du = \sin u + C \tag{25.4}$$

$$\int \sin u \, du = -\cos u + C \tag{25.5}$$

$$\int \sec^2 u \, du = \tan u + C \tag{25.6}$$

$$\int \csc^2 u \, du = -\cot u + C \tag{25.7}$$

$$\int \sec u \tan u \, du = \sec u + C \tag{25.8}$$

$$\int \csc u \cot u \, du = -\csc u + C \tag{25.9}$$

The integrals of the remaining trigonometric functions are:

$$\int \tan u \, du = -\ln|\cos u| + C = \ln|\sec u| + C \tag{25.10}$$

$$\int \cot u \, du = \ln|\sin u| + C = -\ln|\csc u| + C \tag{25.11}$$

$$\int \sec u \, du = \ln|\sec u + \tan u| + C \qquad\qquad (25.12)$$

$$\int \csc u \, du = \ln|\csc u - \cot u| + C \qquad\qquad (25.13)$$

Formula (25.10) may be obtained by integrating $\int[(\sin u)/(\cos u)] \, du$ and is left as an exercise. Formula (25.11) is obtained similarly. The derivation of formula (25.13) requires a trick:

$$\int \csc u \, du = \int \frac{\csc u(\csc u - \cot u)}{\csc u - \cot u} \, du$$

$$= \int \frac{\csc^2 u - \csc u \cot u}{\csc u - \cot u} \, du$$

This is recognized to be a logarithmic form; formula (25.13) follows. The derivation of formula (25.12) is similar and will also be left as an exercise.

E X A M P L E **1**   Integrate $\int \sec 4x \tan 4x \, dx$.

Solution. Let $u = 4x$; then $du = 4 \, dx$, and we get

$$\frac{1}{4} \int \sec 4x \tan 4x(4 \, dx) = \frac{1}{4} \int \sec u \tan u \, du$$

$$= \frac{1}{4} \sec u + C \qquad \text{by (25.8)}$$

$$= \frac{1}{4} \sec 4x + C$$

◀

E X A M P L E **2**   Integrate $\int x^2 \cot x^3 \, dx$.

Solution. Let $u = x^3$, so that $du = 3x^2 \, dx$. The integral now reduces to

$$\frac{1}{3} \int \cot x^3 (3x^2 \, dx) = \frac{1}{3} \int \cot u \, du = \frac{1}{3} \ln|\sin u| + C \qquad \text{by (25.11)}$$

$$= \frac{1}{3} \ln|\sin x^3| + C$$

◀

E X A M P L E **3** Find $\int e^{2x} \sec e^{2x} \, dx$.

Solution. Try $u = e^{2x}$; then $du = 2e^{2x} \, dx$. The integral becomes

$$\frac{1}{2} \int (\sec e^{2x})(2e^{2x}) \, dx = \frac{1}{2} \int \sec u \, du \qquad \begin{bmatrix} u = e^{2x} \\ du = 2e^{2x} \, dx \end{bmatrix}$$

$$= \frac{1}{2} \ln|\sec u + \tan u| + C \qquad \text{by (25.12)}$$

$$= \frac{1}{2} \ln|\sec e^{2x} + \tan e^{2x}| + C \qquad \blacktriangleleft$$

E X E R C I S E S / S E C T I O N **25.3**

In Exercises 1–32, perform the designated integrations.

1. $\int \sec^2 2x \, dx$

2. $\int \sin 3x \, dx$

3. $\int \sec 3x \tan 3x \, dx$

4. $\int \csc 2x \, dx$

5. $\int \csc 4x \cot 4x \, dx$

6. $\int \sec 2x \, dx$

7. $\int x \csc x^2 \, dx$

8. $\int x \tan x^2 \, dx$

9. $\int \cos 2x \, dx$

10. $\int \sec^2 5x \, dx$

11. $\int \tan \frac{1}{2}x \, dx$

12. $\int \csc \frac{1}{2}x \cot \frac{1}{2}x \, dx$

13. $\int x \sin x^2 \, dx$

14. $\int x^2 \sec x^3 \tan x^3 \, dx$

15. $\int x \cot \frac{1}{2}x^2 \, dx$

16. $\int \cos^4 x \sin x \, dx$

17. $\int \frac{\cos \sqrt{x}}{\sqrt{x}} \, dx$

18. $\int e^x \tan e^x \, dx$

19. $\int \tan^3 4x \sec^2 4x \, dx$

20. $\int x^3 \sec x^4 \, dx$

21. $\int e^{3x} \csc e^{3x} \, dx$

22. $\int \frac{\sin \ln x}{x} \, dx$

23. $\int \frac{\csc^2 \ln x}{x} \, dx$

24. $\int \frac{\cos x \, dx}{\sin^3 x}$

25. $\int \frac{\sec^2 x \, dx}{\tan^2 x}$

26. $\int (1 + \sec x)^2 \, dx$

27. $\int \frac{\sin x \, dx}{1 + 2 \cos x}$

28. $\int_{\pi/4}^{\pi/2} \frac{dx}{\sin^2 x}$

29. $\int_0^{\sqrt{\pi/2}} x \cos x^2 \, dx$

30. $\int_0^{\pi/16} \sec^2 4x \, dx$

31. $\int_0^{\pi/6} \frac{\cos x}{1 - \sin x} \, dx$

32. $\int \tan^2 2x \sec^2 2x \, dx$

33. The current in a circuit is $i = 2.00 \cos 100t$. Find the voltage across a 100-μF capacitor after 0.200 s, if the initial voltage is zero (one microfarad (μF) $= 10^{-6}$ F).

34. Find the volume of the solid of revolution obtained by rotating the region bounded by $y = \sqrt{\sin x}$, $x = 0$, $x = \pi$, and the x-axis about the x-axis.

35. Find the volume of the solid of revolution obtained by rotating the region bounded by $y = \cos x^2$, $x = 0$, $x = \sqrt{\pi/2}$, and $y = 0$ about the y-axis.

36. Derive formulas (25.10) and (25.12).

37. Find M_y, the moment about the y-axis, of the region bounded by $y = \tan x^2$, $x = 0$, $x = \sqrt{\pi/2}$, and the x-axis. (Assume the weight per unit area to be equal to 1.)

38. A particle is moving along the x-axis in simple harmonic motion. If the velocity is $v = 10 \cos 2t$, find the position of the particle after 4 s if $x = 0$ when $t = 0$.

39. The voltage in a certain circuit is given by $e(t) = E \sin \omega t$. Find the mean voltage over half a period.

40. A certain circuit consists of an inductor connected to a generator with electromotive force $e(t) = E \sin \omega t$. Recall that if i is the current, then $L(di/dt) = E \sin \omega t$. Show that $i = [E/(L\omega)](1 - \cos \omega t)$ if $i = 0$ when $t = 0$.

41. (Set calculator in radian mode.) Use Simpson's rule with $n = 4$ to find the approximate value of $\int_1^2 (\sin x/x) \, dx$.

42. Repeat Exercise 41 for $\int_0^\pi \sqrt{x} \sin x \, dx$ ($n = 6$).

43. The length of arc s of the curve $y = f(x)$ from $x = a$ to $x = b$ is given by $s = \int_a^b \sqrt{1 + (dy/dx)^2} \, dx$. Find the length of the curve $y = \ln \cos x$ from $x = 0$ to $x = \pi/4$.

44. Find the arc length of the curve $y = \cosh x$ from $x = 0$ to $x = 1$. (See Exercise 29 in Section 24.8.)

25.4   **Further Trigonometric Forms**

The techniques of the last section can be extended to more complex trigono-metric integrals by proper use of certain trigonometric identities in Section 24.1. For convenience these will now be repeated.

Trigonometric Identities Needed in This Section:

$$\sin^2 x + \cos^2 x = 1 \tag{25.14}$$

$$1 + \tan^2 x = \sec^2 x \tag{25.15}$$

$$1 + \cot^2 x = \csc^2 x \tag{25.16}$$

$$\cos^2 x = \tfrac{1}{2}(1 + \cos 2x) \tag{25.17}$$

$$\sin^2 x = \tfrac{1}{2}(1 - \cos 2x) \tag{25.18}$$

The integrals discussed in this section are essentially of three types:

Type of Integral:

Type 1: $\displaystyle\int \sin^n u \cos^m u \, du$

Type 2: $\displaystyle\int \tan^n u \sec^m u \, du$

Type 3: $\displaystyle\int \cot^n u \csc^m u \, du$

For each type of integral the method of integration depends on whether the exponents are even or odd. In each case one of the exponents may be zero.

Identity (25.14) can be used to integrate products of powers of sines and cosines, provided that one of the powers is odd. The integral is transformed by means of identity (25.14), so that it consists of powers of sines with a cosine term set aside for the differential du (or powers of cosines with a sine term set aside for du). Although the procedure can best be seen from an example, let us state the general case first.

Type 1—n odd or m odd: Use identity (25.14) to integrate

$$\int \sin^n u \cos^m u \, du$$

E X A M P L E **1**   Integrate $\int \sin^2 x \cos^3 x\, dx$.

Solution. The trick in this integration is to write $\cos^3 x\, dx = \cos^2 x(\cos x\, dx)$ and then change $\cos^2 x$ to $1 - \sin^2 x$. Thus

$$\int \sin^2 x \cos^3 x\, dx = \int \sin^2 x \cos^2 x(\cos x\, dx)$$

$$= \int \sin^2 x(1 - \sin^2 x)(\cos x\, dx)$$

The reason for this step is now clear: if we let $u = \sin x$, then $du = \cos x\, dx$, and the integral becomes

$$\int u^2(1 - u^2)\, du = \int (u^2 - u^4)\, du \qquad \left[\begin{array}{l} u = \sin x \\ du = \cos x\, dx \end{array}\right]$$

$$= \frac{1}{3} u^3 - \frac{1}{5} u^5 + C = \frac{1}{3} \sin^3 x - \frac{1}{5} \sin^5 x + C \qquad \blacktriangleleft$$

Note that the integration cannot be performed by the method of Example 1 if the power of the cosine is even (while the power of the sine is also even). If the power of the sine is odd, then the procedure is similar, as can be seen from the next example.

E X A M P L E **2**   Integrate $\int \sin^5 x \cos^4 x\, dx$.

Solution. We save $\sin x\, dx$ for the differential du and change the remaining sines to cosines:

$$\int \sin^5 x \cos^4 x\, dx = \int \sin^4 x \cos^4 x(\sin x\, dx)$$

$$= \int (\sin^2 x)^2 \cos^4 x(\sin x\, dx)$$

$$= \int (1 - \cos^2 x)^2 \cos^4 x(\sin x\, dx)$$

Now let $u = \cos x$; then $du = -\sin x\, dx$, and the integral can be written as

$$-\int (1 - \cos^2 x)^2 \cos^4 x(-\sin x\, dx)$$

$$= -\int (1 - u^2)^2 u^4\, du \qquad \left[\begin{array}{l} u = \cos x \\ du = -\sin x\, dx \end{array}\right]$$

$$= -\int (u^4 - 2u^6 + u^8)\, du$$

$$= -\frac{1}{5} u^5 + \frac{2}{7} u^7 - \frac{1}{9} u^9 + C$$

$$= -\frac{1}{5} \cos^5 x + \frac{2}{7} \cos^7 x - \frac{1}{9} \cos^9 x + C \qquad \blacktriangleleft$$

The method of Examples 1 and 2 can be used even if one of the exponents is zero.

E X A M P L E 3   Integrate $\int \cos^3 7x \, dx$.

Solution.

$$\int \cos^3 7x \, dx = \int \cos^2 7x (\cos 7x \, dx)$$

$$= \frac{1}{7} \int (1 - \sin^2 7x)(7 \cos 7x \, dx)$$

$$= \frac{1}{7} \int (1 - u^2) du \qquad \left[\begin{array}{l} u = \sin 7x \\ du = 7 \cos 7x \, dx \end{array} \right]$$

$$= \frac{1}{7} \left(u - \frac{1}{3} u^3 \right) + C$$

$$= \frac{1}{7} \sin 7x - \frac{1}{21} \sin^3 7x + C \qquad \blacktriangleleft$$

To integrate even powers of sines and cosines we may use the half-angle identities (25.17) and (25.18) to reduce the powers, as illustrated in Example 4.

Type 1—*n* and *m* even: Use identities (25.17) and (25.18) to integrate

$$\int \sin^n u \, du \qquad \int \cos^m u \, du \qquad \int \sin^n u \cos^m u \, du$$

E X A M P L E 4   Integrate $\int \sin^4 2x \, dx$.

Solution. By identity (25.18),

$$\int \sin^4 2x \, dx = \int (\sin^2 2x)^2 \, dx = \int \left(\frac{1 - \cos 4x}{2} \right)^2 dx$$

$$= \frac{1}{4} \int (1 - 2 \cos 4x + \cos^2 4x) \, dx$$

We now use identity (25.17), to obtain

$$\frac{1}{4} \left(\int dx - 2 \int \cos 4x \, dx + \int \frac{1 + \cos 8x}{2} \, dx \right)$$

The remaining integrations can be performed readily:

$$\int \sin^4 2x \, dx = \frac{1}{4} \int dx - \frac{1}{2} \int \cos 4x \, dx + \frac{1}{8} \int dx + \frac{1}{8} \int \cos 8x \, dx$$

$$= \frac{1}{4} \int dx - \frac{1}{2} \cdot \frac{1}{4} \int \cos 4x(4 \, dx) + \frac{1}{8} \int dx + \frac{1}{8} \cdot \frac{1}{8} \int \cos 8x(8 \, dx)$$

Thus

$$\int \sin^4 2x \, dx = \frac{1}{4} x - \frac{1}{8} \sin 4x + \frac{1}{8} x + \frac{1}{64} \sin 8x + C$$

$$= \frac{3}{8} x - \frac{1}{8} \sin 4x + \frac{1}{64} \sin 8x + C$$

◀

The procedure for integrating powers of sines and cosines is summarized next:

1. If the power of the sine function is odd and positive, save one sine function for the differential and change the remaining sines to cosines:

$$\int \overset{\text{odd}}{\sin^n} u \cos^m u \, du = \int \sin^{n-1} u \cos^m u \overset{\text{save}}{(\sin u)} \, du$$

change to cosines

2. If the power of the cosine function is odd and positive, save one cosine function for the differential and change the remaining cosines to sines:

$$\int \sin^n u \overset{\text{odd}}{\cos^m} u \, du = \int \sin^n u \cos^{m-1} u \overset{\text{save}}{(\cos u)} \, du$$

change to sines

3. If both powers are even, make repeated use of the half-angle identities to eliminate the even powers.

Many integrals of type 2 can be broken down in analogous fashion by means of formula (25.15).

Type 2—*n* odd or *m* even: Use formula (25.15) to integrate

$$\int \tan^n u \sec^m u \, du$$

E X A M P L E **5**   Integrate $\int \tan^4 x \sec^4 x \, dx$.

Solution. The even power of the secant suggests saving $\sec^2 x \, dx$ for the term du and using formula (25.15) to change the remaining secants to tangents:

$$\int \tan^4 x \sec^4 x \, dx = \int \tan^4 x \sec^2 x (\sec^2 x) \, dx$$

$$= \int \tan^4 x (1 + \tan^2 x)(\sec^2 x \, dx)$$

Let $u = \tan x$, so that $du = \sec^2 x \, dx$. We now have

$$\int u^4 (1 + u^2) \, du = \int (u^4 + u^6) \, du = \frac{1}{5} u^5 + \frac{1}{7} u^7 + C \qquad \left[\begin{array}{l} u = \tan x \\ du = \sec^2 x \, dx \end{array} \right]$$

$$= \frac{1}{5} \tan^5 x + \frac{1}{7} \tan^7 x + C \qquad \blacktriangleleft$$

E X A M P L E **6**   Integrate $\int \tan^3 3x \sec 3x \, dx$.

Solution. Since the power of the secant is odd, the method of Example 5 does not work. But since $d(\sec 3x) = 3 \sec 3x \tan 3x \, dx$, the odd power of the tangent suggests the following breakdown:

$$\int \tan^3 3x \sec 3x \, dx = \int \tan^2 3x (\sec 3x \tan 3x \, dx)$$

$$= \int (\sec^2 3x - 1)(\sec 3x \tan 3x) \, dx$$

by formula (25.15). Now let $u = \sec 3x$, so that $du = 3 \sec 3x \tan 3x \, dx$; we get

$$\frac{1}{3} \int (\sec^2 3x - 1)(3 \sec 3x \tan 3x \, dx) = \frac{1}{3} \int (u^2 - 1) \, du$$

$$= \frac{1}{3} \left(\frac{1}{3} u^3 - u \right) + C$$

$$= \frac{1}{9} \sec^3 3x - \frac{1}{3} \sec 3x + C \qquad \blacktriangleleft$$

The procedure for integrating powers of tangents and secants is summarized next:

1. If the power of the secant function is even and positive, save $\sec^2 u$ for the differential and change the remaining secants to tangents:

$$\int \tan^n u \sec^m u \, du = \int \tan^n u \overset{\text{even}}{\sec^{m-2} u} \overset{\text{save}}{(\sec^2 u)} \, du$$

change to tangents

2. If the power of the tangent function is odd and positive, save $\sec u \tan u$ for the differential and change the remaining tangents to secants:

$$\int \underset{\text{odd}}{\tan^n u} \sec^m u\, du = \int \underset{\text{change to secants}}{\tan^{n-1} u} \sec^{m-1} u\, \overset{\text{save}}{(\sec u \tan u)}\, du$$

Powers of tangents (type 2 with $m = 0$) and even powers of secants (type 2 with m even and $n = 0$) can be integrated with the help of formula (25.15).

Type 2—$m = 0$, n even or odd: Use formula (25.15) to integrate

$$\int \tan^n u\, du$$

Type 2—$n = 0$, m even: Use formula (25.15) to integrate

$$\int \sec^m u\, du$$

Odd powers of secants cannot be integrated by present techniques. We will return to this case in Section 25.7.

E X A M P L E **7**   Find $\int \tan^4 x\, dx$.

Solution. By formula (25.15),

$$\int \tan^4 x\, dx = \int \tan^2 x\, \mathbf{tan^2}\, x\, dx$$

$$= \int \tan^2 x(\mathbf{sec^2}\, x - 1)\, dx$$

$$= \int \tan^2 x \sec^2 x\, dx - \int \tan^2 x\, dx$$

$$= \int \tan^2 x \sec^2 x\, dx - \int (\sec^2 x - 1)\, dx$$

$$= \int \tan^2 x \sec^2 x\, dx - \tan x + x + C$$

$$= \frac{1}{3} \tan^3 x - \tan x + x + C \qquad \begin{bmatrix} u = \tan x \\ du = \sec^2 x\, dx \end{bmatrix} \qquad \blacktriangleleft$$

Integrals of type 3 can be integrated with the help of formula (25.16). Otherwise the cases are identical to those for integrals of type 2. For example, if m is even, we set aside $\csc^2 u\, du$ for the differential.

E X A M P L E **8**   Integrate $\int \cot x \csc^4 x \, dx$.

Solution.

$$\int \cot x \csc^2 x \cdot \csc^2 x \, dx = \int \cot x (1 + \cot^2 x)(\csc^2 x \, dx)$$

$$= -\int u(1 + u^2) \, du \qquad \begin{bmatrix} u = \cot x \\ du = -\csc^2 x \, dx \end{bmatrix}$$

$$= -\frac{1}{2} u^2 - \frac{1}{4} u^4 + C$$

$$= -\frac{1}{2} \cot^2 x - \frac{1}{4} \cot^4 x + C \qquad \blacktriangleleft$$

E X E R C I S E S / S E C T I O N   **25.4**

In Exercises 1–36, perform the designated integrations.

1. $\int \sin^2 2x \cos^3 2x \, dx$

2. $\int \sin^3 x \cos^2 x \, dx$

3. $\int \sin^3 x \, dx$

4. $\int \cos^3 2x \, dx$

5. $\int \sin^3 x \cos^4 x \, dx$

6. $\int \sin^4 3x \cos^3 3x \, dx$

7. $\int \sin^3 x \cos^3 x \, dx$

8. $\int \sin^5 x \cos^6 x \, dx$

9. $\int \cos^2 4x \, dx$

10. $\int \sin^2 x \, dx$

11. $\int \sin^2 x \cos^2 x \, dx$

12. $\int \cos^4 x \, dx$

13. $\int \sin^3 2x \cos^2 2x \, dx$

14. $\int \sin^3 3x \, dx$

15. $\int \sin^4 4x \cos^3 4x \, dx$

16. $\int \sin^3 3x \cos^4 3x \, dx$

17. $\int \tan^3 x \, dx$

18. $\int \sec^4 2x \, dx$

19. $\int \tan^2 x \sec^4 x \, dx$

20. $\int \tan^4 3x \sec^4 3x \, dx$

21. $\int \tan x \sec^3 x \, dx$

22. $\int \tan^5 x \sec^4 x \, dx$

23. $\int \tan^3 x \sec^3 x \, dx$

24. $\int \csc^4 x \, dx$

25. $\int \cot^6 2x \csc^4 2x \, dx$

26. $\int \cot x \csc^3 x \, dx$

27. $\int \csc^6 x \, dx$

28. $\int \cot^4 2x \, dx$

29. $\int_0^{\pi/4} \sqrt{\tan x} \sec^4 x \, dx$

30. $\int_0^{\pi/2} \cos^2 x \, dx$

31. $\int_0^{\pi} (1 + \sin x)^2 \, dx$

32. $\int_0^{\pi/3} \tan^3 x \sec x \, dx$

33. $\int_0^{\pi/4} \frac{\sec^2 x}{1 + \tan x} \, dx$

34. $\int \tan^3 2x \sec^3 2x \, dx$

35. $\int \tan^5 4x \sec^4 4x \, dx$

36. $\int \sqrt{\sin x} \cos^3 x \, dx$

37. Find the volume of the solid of revolution obtained by rotating the region bounded by $y = \sin x$, $y = 0$, $x = 0$, and $x = \pi$ about the x-axis.

38. Find the effective current (root mean square value) for $i = 3 \sin 2t$ for one period.

39. Repeat Exercise 38 for $i = 20 \cos 100\pi t$.

40. The mean power supplied to an electrical device in which there is an alternating current is given by

$$P = \frac{1}{T} \int_0^T ei \, dt$$

where T is the common period of e and i. If $e = 5 \sin \omega t$ and $i = 3 \sin(\omega t - \pi/2) = -\cos \omega t$, show that the average power over one period is zero. (During part of the cycle the device returns energy to the circuit.)

41. (Refer to Exercise 40.) Determine the mean power delivered to an electrical device if $e = 3 \sin 2t$ and $i = 5 \sin(2t - \pi/3)$. Note that $T = 2\pi/2 = \pi$.

42. Integrate $\int \sin x \cos x \, dx$ in two ways: first let $u = \sin x$, and then let $u = \cos x$. Explain the result.

43. (Set calculator in radian mode.) Use Simpson's rule with $n = 8$ to find the approximate value of

$$\int_0^{\pi/2} \frac{dx}{\sqrt{1 + \cos x}}$$

44. Repeat Exercise 43 for

$$\int_0^{\pi/2} \sqrt{1 + \sin x} \, dx \qquad (n = 8)$$

25.5 Inverse Trigonometric Forms

In this section we are going to discuss the integrals of certain algebraic functions that lead to inverse trigonometric forms. The simplest of these are

$$\int \frac{du}{\sqrt{a^2 - u^2}} \quad \text{and} \quad \int \frac{du}{a^2 + u^2}$$

Both integrals may be obtained from the derivatives of inverse trigonometric functions. In particular,

$$\frac{d}{dx} \operatorname{Arcsin} \frac{u}{a} = \frac{1/a}{\sqrt{1 - u^2/a^2}} \frac{du}{dx} = \frac{1/a}{\sqrt{\frac{a^2 - u^2}{a^2}}} \frac{du}{dx} = \frac{1/a}{\frac{\sqrt{a^2 - u^2}}{a}} \frac{du}{dx}$$

$$= \frac{1}{\sqrt{a^2 - u^2}} \frac{du}{dx}$$

Hence

$$\int \frac{du}{\sqrt{a^2 - u^2}} = \operatorname{Arcsin} \frac{u}{a} + C \qquad (25.19)$$

Starting with Arctan(u/a), we obtain in similar manner the integration formula

$$\int \frac{du}{a^2 + u^2} = \frac{1}{a} \operatorname{Arctan} \frac{u}{a} + C \qquad (25.20)$$

E X A M P L E **1**   Integrate

$$\int \frac{x \, dx}{\sqrt{4 - x^4}}$$

Solution. Since the method of substitution is one of trial and error, we might try $u = 1 - x^4$, based on earlier experience. Unfortunately, $du = -4x^3 \, dx$, which does not match $x \, dx$. Recognizing the proper form is more than half the battle! Instead, we write

$$\int \frac{x \, dx}{\sqrt{4 - x^4}} = \int \frac{x \, dx}{\sqrt{2^2 - (x^2)^2}}$$

which may be reduced to formula (25.19) by the substitution $u = x^2$, so that $du = 2x\,dx$. Thus

$$\frac{1}{2}\int \frac{2x\,dx}{\sqrt{2^2 - (x^2)^2}} = \frac{1}{2}\int \frac{du}{\sqrt{2^2 - u^2}} = \frac{1}{2}\,\text{Arcsin}\,\frac{u}{2} + C = \frac{1}{2}\,\text{Arcsin}\,\frac{x^2}{2} + C$$

by formula (25.19) with $a = 2$. ◀

E X A M P L E **2**   Integrate

$$\int \frac{x^2\,dx}{9 + 4x^6}$$

Solution. The integral can be written

$$\int \frac{x^2\,dx}{3^2 + (2x^3)^2}$$

Let $u = 2x^3$; then $du = 6x^2\,dx$, and we get

$$\frac{1}{6}\int \frac{6x^2\,dx}{3^2 + (2x^3)^2} = \frac{1}{6}\int \frac{du}{3^2 + u^2} = \frac{1}{6}\cdot\frac{1}{3}\,\text{Arctan}\,\frac{u}{3} + C \qquad \begin{bmatrix} u = 2x^3 \\ du = 6x^2\,dx \end{bmatrix}$$

$$= \frac{1}{18}\,\text{Arctan}\,\frac{2x^3}{3} + C$$

by formula (25.20) with $a = 3$. ◀

More general quadratic forms can often be handled by first completing the square, as shown in the following example.

E X A M P L E **3**   Find

$$\int \frac{dx}{x^2 - 4x + 5}$$

Solution. To complete the square, we add and subtract the square of one-half the coefficient of x: $[\frac{1}{2}(-4)]^2 = 4$. Thus

$$x^2 - 4x + 4 - 4 + 5 = (x^2 - 4x + 4) + 1 = (x - 2)^2 + 1$$

This form suggests the substitution $u = x - 2$, so that $du = dx$:

$$\int \frac{dx}{x^2 - 4x + 5} = \int \frac{dx}{(x - 2)^2 + 1} = \int \frac{du}{u^2 + 1} \qquad \begin{bmatrix} u = x - 2 \\ du = dx \end{bmatrix}$$

$$= \text{Arctan}\,u + C = \text{Arctan}(x - 2) + C$$ ◀

E X A M P L E **4** Find

$$\int \frac{dx}{\sqrt{2 - 2x - x^2}}$$

Solution. The quadratic expression can be written as follows:

$$2 - x^2 - 2x = 2 - (x^2 + 2x) = 2 - (x^2 + 2x + 1 - 1)$$
$$= 3 - (x^2 + 2x + 1) = 3 - (x + 1)^2$$

Hence

$$\int \frac{dx}{\sqrt{2 - 2x - x^2}} = \int \frac{dx}{\sqrt{3 - (x + 1)^2}}$$

Let $u = x + 1$; then $du = dx$ and

$$\int \frac{dx}{\sqrt{3 - (x + 1)^2}} = \int \frac{du}{\sqrt{3 - u^2}} = \text{Arcsin} \frac{u}{\sqrt{3}} + C = \text{Arcsin} \frac{x + 1}{\sqrt{3}} + C$$

by formula (25.19) with $a = \sqrt{3}$.   ◀

E X E R C I S E S / S E C T I O N   **25.5**

In Exercises 1–34, perform the designated integrations.

1. $\int \dfrac{dx}{\sqrt{1 - x^2}}$

2. $\int \dfrac{dx}{\sqrt{4 - 4x^2}}$

3. $\int \dfrac{dx}{9 + 4x^2}$

4. $\int \dfrac{dx}{4 + 3x^2}$

5. $\int \dfrac{x\,dx}{\sqrt{1 - x^2}}$

6. $\int \dfrac{x\,dx}{\sqrt{1 - 4x^2}}$

7. $\int \dfrac{x\,dx}{16 + 9x^2}$

8. $\int \dfrac{dx}{16 + 9x^2}$

9. $\int \dfrac{x\,dx}{4 + x^4}$

10. $\int \dfrac{x\,dx}{4 + x^2}$

11. $\int \dfrac{\csc^2 x\,dx}{\sqrt{4 - \cot^2 x}}$

12. $\int \dfrac{\sec^2 x\,dx}{\sqrt{9 - \tan^2 x}}$

13. $\int \dfrac{dx}{\sqrt{5 - 3x^2}}$

14. $\int \dfrac{x\,dx}{\sqrt{1 - 5x^4}}$

15. $\int \dfrac{\cos x\,dx}{2 + \sin x}$

16. $\int \dfrac{x^2\,dx}{\sqrt{1 - x^6}}$

17. $\int_3^{3\sqrt{3}} \dfrac{3\,dx}{9 + x^2}$

18. $\int_1^2 \dfrac{dx}{\sqrt{4 - x^2}}$

19. $\int \dfrac{dx}{\sqrt{4 - 3x^2}}$

20. $\int \dfrac{dx}{9 + 7x^2}$

21. $\int \dfrac{dx}{x^2 - 6x + 10}$

22. $\int \dfrac{dx}{\sqrt{4x - x^2}}$

23. $\int \dfrac{dx}{\sqrt{1 - x^2 - 4x}}$

24. $\int \dfrac{dx}{x^2 + 2x + 3}$

25. $\int \dfrac{dx}{x^2 + 3x + 3}$

26. $\int \dfrac{1 - x}{\sqrt{1 - x^2}}\,dx$

27. $\int \dfrac{x + 4}{x^2 + 16}\,dx$

28. $\int \dfrac{x\,dx}{\sqrt{9 - x^2}}$

29. $\int_0^{\pi/2} \dfrac{\cos x}{1 + \sin^2 x}\,dx$

30. $\int \dfrac{e^x}{1 + e^{2x}}\,dx$

31. $\int \dfrac{e^{2x}}{1 + e^{2x}}\,dx$

32. $\int \dfrac{x^3}{4 + x^4}\,dx$

33. $\int \dfrac{e^x}{\sqrt{1 - e^{2x}}}\, dx$ 34. $\int_0^{\pi/2} \dfrac{\sin x}{1 + \cos^2 x}\, dx$

35. Find the area "bounded" by $y = 1/(x^2 + 1)$, the y-axis, and the positive x-axis.

36. Find M_y, the moment about the y-axis, of the region bounded by $y = 1/(1 + x^4)$, $y = 0$, $x = 0$, and $x = 1$. (Assume the weight per unit area to be 1.)

37. Find the volume of the solid of revolution obtained by rotating the first-quadrant region "bounded" by $y = 1/(4 + x^4)$ about the y-axis.

38. Find the volume of the solid generated by revolving about the x-axis the region bounded by $y = 2/\sqrt{x^2 + 9}$, $y = 0$, $x = 0$, and $x = 3$.

25.6   Integration by Trigonometric Substitution

So far all of our substitutions in performing integration have been of a certain type: we let u be some function appearing in the integrand. More specifically, if the integral has the form

$$\int f(g(x))g'(x)\, dx$$

then we let $u = g(x)$, so that $du = g'(x)\, dx$. Then, if the resulting integral

$$\int f(u)\, du$$

is in recognizable form, we perform the integration and substitute back.

It is also possible in some cases to make a substitution of a totally different kind: let x be some function of a new variable θ. For example, x can be replaced by some trigonometric function, although other possibilities exist. Suppose we look at the familiar integral

$$\int \frac{dx}{\sqrt{1 - x^2}}$$

If we let $x = \sin \theta$, then $dx = \cos \theta\, d\theta$, and we get

$$\int \frac{dx}{\sqrt{1 - x^2}} = \int \frac{\cos \theta\, d\theta}{\sqrt{1 - \sin^2 \theta}} = \int \frac{\cos \theta\, d\theta}{\sqrt{\cos^2 \theta}} = \int \frac{\cos \theta}{\cos \theta}\, d\theta$$

$$= \int d\theta = \theta + C$$

Since $\sin \theta = x$, it follows that $\theta = \text{Arcsin } x$, so that

$$\int \frac{dx}{\sqrt{1 - x^2}} = \text{Arcsin } x + C$$

which we know to be correct. (See formula (25.19).)

This example reveals the reason why such a substitution works: certain trigonometric forms will collapse in a way that the corresponding algebraic

Table 25.1 *Rules for trigonometric substitution*

For integrals containing . . .	Use . . .	To obtain . . .	
$\sqrt{a^2 - x^2}$	$x = a \sin \theta$	$\sqrt{a^2 - x^2} = a \cos \theta$	$(a > 0)$
$\sqrt{x^2 + a^2}$	$x = a \tan \theta$	$\sqrt{x^2 + a^2} = a \sec \theta$	$(a > 0)$
$\sqrt{x^2 - a^2}$	$x = a \sec \theta$	$\sqrt{x^2 - a^2} = a \tan \theta$	$(a > 0)$

forms will not. As we saw, the radicand $1 - \sin^2 \theta$ can be combined into a single squared term, so that the troublesome radical disappears. The identities (25.14) and (25.15) in Section 25.4 will suffice to eliminate radicals containing $a^2 - x^2$, $a^2 + x^2$, and $x^2 - a^2$. The procedure is summarized in Table 25.1.

One could also employ the cofunctions—that is, $\cos \theta$, $\cot \theta$, or $\csc \theta$—in the trigonometric substitution. For example, letting $x = a \cos \theta$,

$$\sqrt{a^2 - x^2} = \sqrt{a^2 - a^2 \cos^2 \theta} = a \sin \theta$$

Avoiding the cofunctions altogether has the advantage of giving us a one-to-one correspondence between the radicals and the functions to be substituted. Some students claim that they never know what to substitute; according to our scheme, there is really nothing to decide!

E X A M P L E **1**   Integrate

$$\int \frac{\sqrt{x^2 - 4}}{x} \, dx$$

Solution. To eliminate the radical, we must let $x = 2 \sec \theta$, so that $dx = 2 \sec \theta \tan \theta \, d\theta$. (A common error is forgetting to substitute the expression for dx.) Then

$$\sqrt{x^2 - 4} = \sqrt{4 \sec^2 \theta - 4} = \sqrt{4(\sec^2 \theta - 1)} = 2\sqrt{\sec^2 \theta - 1}$$
$$= 2\sqrt{\tan^2 \theta} = 2 \tan \theta$$

and

$$\int \frac{\sqrt{x^2 - 4}}{x} \, dx = \int \frac{(2 \tan \theta)(2 \sec \theta \tan \theta) \, d\theta}{2 \sec \theta}$$

$$= 2 \int \tan^2 \theta \, d\theta = 2 \int (\sec^2 \theta - 1) \, d\theta$$

$$= 2(\tan \theta - \theta) + C$$

It remains to rewrite the last expression in terms of x. As long as θ stands alone, the substituted expression $x/2 = \sec \theta$ gives us

$$\theta = \text{Arcsec} \, \frac{x}{2}$$

directly. For a function of θ we need to go a step further and appeal to the definition of the trigonometric function. Since

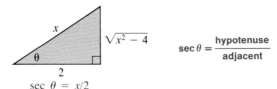

$$\sec\theta = \frac{\text{hypotenuse}}{\text{adjacent}}$$

$$\sec\theta = x/2$$

Figure 25.1

$$\sec\theta = \frac{x}{2}$$

we may construct the triangle in Figure 25.1 with x the length of the hypotenuse and 2 the length of the adjacent side. The length of the opposite side, $\sqrt{x^2 - 4}$, is computed by the Pythagorean theorem. Now we can read off any trigonometric function required. In particular, $\tan\theta = \sqrt{x^2 - 4}/2 = \frac{1}{2}\sqrt{x^2 - 4}$. So

$$\int \frac{\sqrt{x^2 - 4}}{x}\, dx = 2(\tan\theta - \theta) + C$$

$$= 2\left(\frac{1}{2}\sqrt{x^2 - 4} - \text{Arcsec}\,\frac{x}{2}\right) + C$$

$$= \sqrt{x^2 - 4} - 2\,\text{Arcsec}\,\frac{x}{2} + C$$

Remark. Given that the range of $y = \text{Arcsec}\,x$ is $0 \le y \le \pi$, $y \ne \pi/2$ (Section 24.4), it can be shown that the answer should really be written

$$\sqrt{x^2 - 4} - 2\,\text{Arcsec}\,\frac{|x|}{2} + C \qquad\blacktriangleleft$$

E X A M P L E **2**   Integrate $\int x^3\sqrt{x^2 + 4}\, dx$.

Solution. In this case we need to substitute $x = 2\tan\theta$, for then

$$\sqrt{x^2 + 4} = \sqrt{4\tan^2\theta + 4} = \sqrt{4(\tan^2\theta + 1)} = 2\sqrt{\tan^2\theta + 1}$$
$$= 2\sqrt{\sec^2\theta} = 2\sec\theta$$

Also, $dx = 2\sec^2\theta\, d\theta$, and $x^3 = 8\tan^3\theta$. Substitution yields

$$\int x^3\sqrt{x^2 + 4}\, dx = \int (8\tan^3\theta)(2\sec\theta)(2\sec^2\theta)\, d\theta$$

$$= 32\int \tan^3\theta\sec^3\theta\, d\theta$$

$$= 32\int \tan^2\theta\sec^2\theta(\sec\theta\tan\theta)\, d\theta$$

$$= 32\int (\sec^2\theta - 1)(\sec^2\theta)(\sec\theta\tan\theta)\, d\theta$$

$$= 32\int (\sec^4\theta - \sec^2\theta)(\sec\theta\tan\theta\, d\theta)$$

Since $d(\sec \theta) = \sec \theta \tan \theta \, d\theta$, we get

$$32\left(\frac{1}{5}\sec^5 \theta - \frac{1}{3}\sec^3 \theta\right) + C = \frac{32}{5}\sec^5 \theta - \frac{32}{3}\sec^3 \theta + C$$

We now use the substituted expression $\tan \theta = x/2$ to construct the triangle in Figure 25.2. Note that

$$\sec \theta = \frac{\sqrt{x^2 + 4}}{2}$$

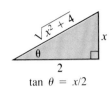

$\tan \theta = x/2$

Figure 25.2

so that

$$\int x^3 \sqrt{x^2 + 4}\, dx = \frac{32}{5}\left(\frac{\sqrt{x^2 + 4}}{2}\right)^5 - \frac{32}{3}\left(\frac{\sqrt{x^2 + 4}}{2}\right)^3 + C$$

$$= \frac{1}{5}(x^2 + 4)^{5/2} - \frac{4}{3}(x^2 + 4)^{3/2} + C$$

◀

E X A M P L E 3 Derive the following formula:

$$\int \frac{du}{\sqrt{u^2 + a^2}} = \ln\left|u + \sqrt{u^2 + a^2}\right| + C \qquad (25.21)$$

Solution. Let $u = a \tan \theta$; then $du = a \sec^2 \theta \, d\theta$. Thus

$$\int \frac{du}{\sqrt{u^2 + a^2}} = \int \frac{a \sec^2 \theta \, d\theta}{\sqrt{a^2 \tan^2 \theta + a^2}} = \int \frac{a \sec^2 \theta \, d\theta}{a \sec \theta}$$

$$= \int \sec \theta \, d\theta = \ln|\sec \theta + \tan \theta| + C'$$

From the substituted expression $\tan \theta = u/a$ and the corresponding triangle (Figure 25.3), we get

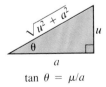

$\tan \theta = \mu/a$

Figure 25.3

$$\int \frac{du}{\sqrt{u^2 + a^2}} = \ln\left|\frac{\sqrt{u^2 + a^2}}{a} + \frac{u}{a}\right| + C'$$

$$= \ln\left|\frac{\sqrt{u^2 + a^2} + u}{a}\right| + C'$$

$$= \ln\left|u + \sqrt{u^2 + a^2}\right| - \ln a + C'$$

Now let $C = C' - \ln a$, and formula (25.21) follows. (If C' is arbitrary, so is $C' - \ln a$.)

◀

The derivation of

$$\int \frac{du}{\sqrt{u^2 - a^2}} = \ln\left|u + \sqrt{u^2 - a^2}\right| + C \tag{25.22}$$

is left as an exercise.

E X E R C I S E S / S E C T I O N   25.6

In Exercises 1–20, perform the designated integrations.

1. $\int \dfrac{\sqrt{4 - x^2}}{x^2}\, dx$

2. $\int \dfrac{\sqrt{x^2 - 9}}{x}\, dx$

3. $\int x\sqrt{x^2 + 9}\, dx$

4. $\int \dfrac{dx}{x\sqrt{x^2 - 1}}$

5. $\int \dfrac{dx}{(x^2 + 25)^{3/2}}$

6. $\int \dfrac{dx}{x\sqrt{9 - x^2}}$

7. $\int \dfrac{dx}{x\sqrt{x^2 - 4}}$

8. $\int \dfrac{x^3\, dx}{\sqrt{16 - x^2}}$

9. $\int \dfrac{dx}{x^2\sqrt{x^2 + 16}}$

10. $\int \dfrac{dx}{(x^2 + 5)^{3/2}}$

11. $\int \dfrac{x\, dx}{(x^2 - 2)^{3/2}}$

12. $\int \dfrac{x\, dx}{(x^2 + 7)^{3/2}}$

13. $\int \dfrac{x^3\, dx}{\sqrt{x^2 - 3}}$

14. $\int x^3\sqrt{x^2 - 6}\, dx$

15. $\int \dfrac{\sqrt{x^2 + 1}}{x^2}\, dx$

16. $\int \dfrac{x\, dx}{\sqrt{x^2 + 1}}$

17. $\int_0^4 \dfrac{dx}{(16 + x^2)^{3/2}}$

18. $\int_1^2 \dfrac{dx}{x\sqrt{16 - x^2}}$

19. $\int_0^3 \sqrt{9 - x^2}\, dx$

20. $\int \dfrac{dx}{(x + 4)\sqrt{1 - (x + 4)^2}}$

21. A circular gate on a vertical dam has a diameter of 4 m, while its center is 20 m below the surface of the water. Find the force against the bottom half of the gate.

22. Derive the formula for the volume of a torus. (A torus is a doughnut-shaped solid formed by revolving the circular disk $(x - c)^2 + y^2 = r^2$ about the y-axis.)

23. Find the moment of inertia of a sphere with respect to its axis.

24. Derive formula (25.22).

25. A cylindrical tank half full of water is lying on its side. If the tank has a radius of 3 m and a length of 8 m, find the work done in pumping the water out of the tank through an opening in the top.

26. Suppose the tank in Exercise 25 is full of water. Find the work done in pumping out half of the water through the opening in the top.

25.7   Integration by Parts

The product rule from differentiation,

$$\frac{d}{dx}\, uv = u\,\frac{dv}{dx} + v\,\frac{du}{dx}$$

can also be written in differential form:

$$d(uv) = u\, dv + v\, du$$

or

$$u\, dv = d(uv) - v\, du$$

Integrating both sides, we get a formula known as the formula for **integration by parts**.

Formula for Integration by Parts:

$$\int u\,dv = uv - \int v\,du \qquad\qquad (25.23)$$

This formula can be extremely useful. If the integral on the right in formula (25.23) is easy to calculate, then the integral on the left can be obtained as well. Unfortunately, the formula does not tell us what part of the given integral $\int u\,dv$ to regard as the function u and which as dv. All we can do is try it out.

E X A M P L E **1** Integrate $\int x \sin x\,dx$.

Solution. Note that this integral does not fit any of our earlier forms. Suppose we make some arbitrary assignment in the integral to the functions u and dv, say $u = x$ and $dv = \sin x\,dx$, and then arrange our work as follows:

$$u = x \qquad\qquad dv = \sin x\,dx \qquad \textbf{Assign } u \textbf{ and } dv.$$
$$du = dx \qquad\qquad v = -\cos x \qquad \textbf{Find } du \textbf{ and } v.$$

The expressions for du and v are obtained from u and dv, respectively. (The constant of integration will be included in the final result.) By formula (25.23)

$$\int x \sin x\,dx = uv - \int v\,du$$

$$= x(-\cos x) - \int (-\cos x)\,dx = -x\cos x + \sin x + C$$

We conclude that

$$\int x \sin x\,dx = -x\cos x + \sin x + C$$

The procedure worked just fine. Suppose we now reverse the choices assigned to u and dv, as follows:

$$u = \sin x \qquad\qquad dv = x\,dx$$
$$du = \cos x\,dx \qquad\qquad v = \tfrac{1}{2}x^2$$

Now, by formula (25.23) we have

$$\int x \sin x\,dx = \frac{1}{2}x^2 \sin x - \frac{1}{2}\int x^2 \cos x\,dx$$

This result is perfectly correct but useless, since the integral on the right is even more complicated than the given integral. ◀

E X A M P L E **2**   Integrate $\int x^2 \ln x \, dx$.

Solution. In this problem we actually have little choice since assigning $dv = \ln x \, dx$ leads nowhere. (We will see in Exercise 6 that $\int \ln x \, dx$ is itself integrated by parts.) Hence we choose $u = \ln x$ and $dv = x^2 \, dx$:

$$
\begin{array}{ll}
u = \ln x & dv = x^2 \, dx \qquad \textbf{Assign } u \textbf{ and } dv. \\[2mm]
du = \dfrac{1}{x} \, dx & v = \tfrac{1}{3}x^3 \qquad \textbf{Find } du \textbf{ and } v.
\end{array}
$$

We now obtain

$$
\int x^2 \ln x \, dx = (\ln x)\left(\frac{1}{3}x^3\right) - \int \frac{1}{3}x^3 \cdot \frac{1}{x} \, dx \qquad \int u \, dv = uv - \int v \, du
$$

$$
= \frac{1}{3}x^3 \ln x - \frac{1}{3}\int x^2 \, dx
$$

$$
= \frac{1}{3}x^3 \ln x - \frac{1}{9}x^3 + C
$$

◀

Integration by parts can sometimes be used on integrals containing a single factor, such as $\int \ln x \, dx$ or $\int \text{Arcsin } x \, dx$. In such a case, we use $dv = dx$, as shown in the next example.

E X A M P L E **3**   Work out the integral $\int \text{Arctan } x \, dx$.

Solution. There is only one possible choice:

$$
\begin{array}{ll}
u = \text{Arctan } x & dv = dx \\[2mm]
du = \dfrac{dx}{1 + x^2} & v = x
\end{array}
$$

Thus

$$
\int \text{Arctan } x \, dx = x \, \text{Arctan } x - \int \frac{x \, dx}{1 + x^2}
$$

$$
= x \, \text{Arctan } x - \frac{1}{2}\ln(1 + x^2) + C
$$

◀

It is sometimes necessary to use integration by parts repeatedly, as shown in the remaining examples.

E X A M P L E **4** Find $\int x^2 e^x \, dx$.

Solution. This integral is similar to the one in Example 1. Let

$u = x^2$	$dv = e^x \, dx$
$du = 2x \, dx$	$v = e^x$

(By now the composition of the right-hand side of formula (25.23) should be familiar.) Hence

$$\int x^2 e^x \, dx = x^2 e^x - 2 \int x e^x \, dx = x^2 e^x - 2 \left[\int x e^x \, dx \right]$$

The integral on the right can be obtained by using integration by parts again:

$u = x$	$dv = e^x \, dx$
$du = dx$	$v = e^x$

Hence

$$\int x^2 e^x \, dx = x^2 e^x - 2 \left[x e^x - \int e^x \, dx \right]$$
$$= x^2 e^x - 2x e^x + 2 \int e^x \, dx$$
$$= x^2 e^x - 2x e^x + 2 e^x + C \qquad \blacktriangleleft$$

E X A M P L E **5** Integrate $\int e^x \cos x \, dx$.

Solution. In this case one choice appears to be as good (or as bad) as another. Suppose we try

$u = e^x$	$dv = \cos x \, dx$
$du = e^x \, dx$	$v = \sin x$

Then

$$\int e^x \cos x \, dx = e^x \sin x - \int e^x \sin x \, dx$$

The resulting integral is of the same type as the given integral, so that no progress appears to have been made. But suppose we repeat this procedure on the resulting integral by letting

$u = e^x$	$dv = \sin x \, dx$
$du = e^x \, dx$	$v = -\cos x$

We now get

$$\int e^x \cos x\, dx = e^x \sin x - \left[-e^x \cos x + \int e^x \cos x\, dx \right]$$

$$= e^x \sin x + e^x \cos x - \int e^x \cos x\, dx$$

At this point we are right back where we started. Or are we? If we just transpose the last term, we end up with

$$2 \int e^x \cos x\, dx = e^x \sin x + e^x \cos x$$

and we obtain the integral after all! Thus

$$\int e^x \cos x\, dx = \tfrac{1}{2}(e^x \sin x + e^x \cos x) + C \qquad \blacktriangleleft$$

E X E R C I S E S / S E C T I O N   **25.7**

In Exercises 1–22, perform the designated integrations.

1. $\int xe^x\, dx$

2. $\int x \cos x\, dx$

3. $\int x \sin 2x\, dx$

4. $\int xe^{-x}\, dx$

5. $\int x \sec^2 x\, dx$

6. $\int \ln x\, dx$

7. $\int x \ln x\, dx$

8. $\int xe^{x^2}\, dx$

9. $\int \text{Arcsin}\, x\, dx$

10. $\int \dfrac{\ln x}{x^2}\, dx$

11. $\int x \cos 3x\, dx$

12. $\int x^2 \sin x\, dx$

13. $\int x^2 e^{-x}\, dx$

14. $\int x^3 \ln x\, dx$

15. $\int \text{Arccot}\, x\, dx$

16. $\int \text{Arcsin}\, 4x\, dx$

17. $\int x \sin x^2\, dx$

18. $\int x \,\text{Arctan}\, x\, dx$

19. $\int e^x \sin x\, dx$

20. $\int e^{-x} \sin x\, dx$

21. $\int e^{-x} \cos \pi x\, dx$

22. $\int \sec^3 x\, dx$ (*Hint:* Let $u = \sec x$ and $dv = \sec^2 x\, dx$, and apply the method of Example 5.)

23. Evaluate $\int_0^1 x\, 2^x\, dx$.   **24.** Evaluate $\int_1^2 x^4 \ln x\, dx$.

25. Find the area of the region bounded by $y = x \sin x$, $x = 0$, $x = \pi$, and $y = 0$.

26. Find the volume of the solid of revolution obtained by rotating the area bounded by $y = \cos x$, $x = 0$, $x = \pi/2$, and $y = 0$ about the y-axis.

27. Find I_y, the moment of inertia about the y-axis, of the region in Exercise 26.

28. Find the average value of the current $i = e^{-2t} \cos 2t$ in a certain circuit on the time interval $0 \le t \le \pi/2$.

29. The current in a certain circuit is $i = e^{-t} \sin 4t$. Find the voltage across a 10-μF capacitor after 50 milliseconds if the initial voltage is zero. (One microfarad (μF) $= 10^{-6}$ F.)

25.8   Integration of Rational Functions*

In this section we study integration of rational functions by means of partial fraction expansions. Partial fractions are taken up again in Chapter 29 in order to keep the presentation of Laplace transforms independent of this section. As a result, a study of the algebraic techniques in this section may be postponed.

* This section may be omitted without loss of continuity.

A *rational function* has the form of a fraction in which both numerator and denominator are polynomials. Although the integrals of some special rational functions have already been considered, a general discussion is more involved. For example, the integral

$$\int \frac{x^3 + x}{x^2 - 1} \, dx$$

has an unfamiliar form. But suppose we divide numerator by denominator:

$$
\begin{array}{r}
x \\
x^2 - 1 \overline{)\, x^3 + x} \\
\underline{x^3 - x} \\
2x
\end{array}
$$

Since the remainder is $2x$, we get

$$\int \frac{x^3 + x}{x^2 - 1} \, dx = \int \left(x + \frac{2x}{x^2 - 1} \right) dx$$
$$= \tfrac{1}{2}x^2 + \ln|x^2 - 1| + C$$

Since the remainder is always of lower degree than the divisor, long division reduces a rational function to either (1) a polynomial or (2) a polynomial plus a *proper* fraction (degree of numerator strictly less than degree of denominator).

Even proper fractions cannot always be integrated by the methods considered so far, as can be seen from the integral

$$\int \frac{2 \, dx}{x^2 - 1}$$

We can readily check, however, that the integrand can be written

$$\frac{2}{(x - 1)(x + 1)} = \frac{1}{x - 1} - \frac{1}{x + 1}$$

so that each term can be integrated separately. The fractions on the right are called *partial fractions*. Thus to integrate a rational function, we factor the denominator and split the whole fraction into a sum of simpler partial fractions.

To obtain the different cases, we are going to rely on the following fact from advanced algebra: a polynomial with real coefficients can be factored into *linear* and irreducible *quadratic* factors with real coefficients. Consequently, we do not have to consider factors of degree greater than 2.

If the fraction is proper, it can be split into a sum of partial fractions according to the following rules.

Rule I. If a linear factor $ax + b$ occurs n times in the denominator, then there exist n partial fractions

$$\frac{A_1}{ax + b} + \frac{A_2}{(ax + b)^2} + \cdots + \frac{A_n}{(ax + b)^n}$$

where $A_1, A_2, \ldots, A_n$ are constants.

Rule II. If a quadratic factor $ax^2 + bx + c$ occurs n times in the denominator, then there exist n partial fractions

$$\frac{A_1 x + B_1}{ax^2 + bx + c} + \frac{A_2 x + B_2}{(ax^2 + bx + c)^2} + \cdots + \frac{A_n x + B_n}{(ax^2 + bx + c)^n}$$

where the A's and B's are constants. In all cases, n may be equal to 1.

Rules I and II will serve only as a general guide. How they are put to use will be illustrated in the examples. To make our job easier, we will classify the integrals according to whether the denominators have:

1. Distinct linear factors
2. Repeating linear factors
3. Distinct quadratic factors
4. Repeating quadratic factors

Case 1: Distinct Linear Factors

E X A M P L E **1**   Integrate

$$\int \frac{x + 8}{(x - 1)(x + 2)}\, dx$$

Solution. Note that the numerator is of first degree and the denominator of second degree, so that the integrand is a proper fraction. Furthermore, the factors are all distinct (each occurs only once). So by Rule I with $n = 1$, we have the following decomposition:

$$\frac{x + 8}{(x - 1)(x + 2)} = \frac{A}{x - 1} + \frac{B}{x + 2} \tag{25.24}$$

(Since there are only two constants, it is better to use A and B rather than subscripts.) The main task is to determine the constants A and B. To this end, we add the fractions on the right to obtain

$$\frac{A}{x - 1} + \frac{B}{x + 2} = \frac{A}{x - 1} \cdot \frac{x + 2}{x + 2} + \frac{B}{x + 2} \cdot \frac{x - 1}{x - 1}$$

$$= \frac{A(x + 2) + B(x - 1)}{(x - 1)(x + 2)}$$

It follows that the numerator of this fraction must be equal to the numerator of the left side of (25.24). In other words,

$$A(x + 2) + B(x - 1) = x + 8 \qquad (25.25)$$

The constants A and B can be determined in two ways: (1) equating coefficients and (2) substituting convenient values for x.

1. *Equating coefficients.* If we multiply the expressions on the left side of (25.25) and combine similar terms, we get

$$A(x + 2) + B(x - 1) = x + 8$$
$$Ax + 2A + Bx - B = x + 8$$
$$(A + B)x + (2A - B) = x + 8$$

Now we equate corresponding coefficients to obtain the following system of equations:

$$
\begin{array}{ll}
A + B = 1 & \text{coefficients of } x \\
\underline{2A - B = 8} & \text{constants} \\
3A \phantom{{}- B} = 9 & \text{adding} \\
A = 3 &
\end{array}
$$

From $A + B = 1$, we get $3 + B = 1$, or $B = -2$.
Since $A = 3$ and $B = -2$, it follows from (25.24) that

$$\frac{x + 8}{(x - 1)(x + 2)} = \frac{3}{x - 1} - \frac{2}{x + 2}$$

2. *Substitution.* To see how A and B may be obtained by substitution, first note that (25.25),

$$A(x + 2) + B(x - 1) = x + 8$$

is an identity and is therefore valid for all values of x. If we let $x = -2$, for example, we obtain B at once:

$$x = -2: \qquad A(-2 + 2) + B(-2 - 1) = -2 + 8$$
$$-3B = 6$$
$$B = -2$$

Similarly, we may substitute 1 for x:

$$x = 1: \qquad A(1 + 2) + B(1 - 1) = 1 + 8$$
$$3A = 9$$
$$A = 3$$

We have shown again that

$$\frac{x + 8}{(x - 1)(x + 2)} = \frac{3}{x - 1} - \frac{2}{x + 2}$$

Finally, rewriting the given integral, we find that

$$\int \frac{x + 8}{(x - 1)(x + 2)}\, dx = \int \left(\frac{3}{x - 1} - \frac{2}{x + 2}\right) dx$$

$$= 3\ln|x - 1| - 2\ln|x + 2| + C$$

$$= \ln|x - 1|^3 - \ln|x + 2|^2 + C$$

$$= \ln\left|\frac{(x - 1)^3}{(x + 2)^2}\right| + C \qquad \blacktriangleleft$$

We can see from Example 1 that the method of substitution is the more efficient, at least for linear factors.

E X A M P L E **2**   Integrate

$$\int \frac{5x + 10}{(x + 1)(x - 2)(x + 3)}\, dx$$

Solution. As in Example 1, the factors are all distinct and linear. By Rule I with $n = 1$, the form is given by

$$\frac{5x + 10}{(x + 1)(x - 2)(x + 3)} = \frac{A}{x + 1} + \frac{B}{x - 2} + \frac{C}{x + 3}$$

$$= \frac{A(x - 2)(x + 3) + B(x + 1)(x + 3) + C(x + 1)(x - 2)}{(x + 1)(x - 2)(x + 3)}$$

Equating numerators, we get

$$A(x - 2)(x + 3) + B(x + 1)(x + 3) + C(x + 1)(x - 2) = 5x + 10$$

We now substitute 2, -3, and -1, respectively:

$$x = 2: \qquad 0 + B(2 + 1)(2 + 3) + 0 = 5(2) + 10$$

$$15B = 20$$

$$B = \tfrac{4}{3}$$

$$x = -3: \qquad 0 + 0 + C(-3 + 1)(-3 - 2) = 5(-3) + 10$$

$$10C = -5$$

$$C = -\tfrac{1}{2}$$

$$x = -1: \qquad A(-1 - 2)(-1 + 3) + 0 + 0 = 5(-1) + 10$$

$$-6A = 5$$

$$A = -\tfrac{5}{6}$$

Substituting the values of A, B, and C, we get

$$\int \frac{5x + 10}{(x + 1)(x - 2)(x + 3)}\, dx = \int \left(-\frac{5}{6}\frac{1}{x + 1} + \frac{4}{3}\frac{1}{x - 2} - \frac{1}{2}\frac{1}{x + 3} \right) dx$$

$$= -\frac{5}{6}\ln|x + 1| + \frac{4}{3}\ln|x - 2| - \frac{1}{2}\ln|x + 3| + C$$

This result can also be written

$$\frac{1}{6}(-5\ln|x + 1| + 8\ln|x - 2| - 3\ln|x + 3|) + C$$

$$= \frac{1}{6}(-\ln|x + 1|^5 + \ln|x - 2|^8 - \ln|x + 3|^3) + C$$

$$= \frac{1}{6}\ln\left|\frac{(x - 2)^8}{(x + 1)^5(x + 3)^3}\right| + C \qquad\blacktriangleleft$$

Case 2: Repeating Linear Factors

E X A M P L E **3**   Integrate

$$\int \frac{x - \frac{1}{2}}{(x - 3)^2}\, dx$$

Solution. By Rule I with $n = 2$, the integrand has the form

$$\frac{x - \frac{1}{2}}{(x - 3)^2} = \frac{A}{x - 3} + \frac{B}{(x - 3)^2} \qquad \text{repeating linear factors}$$

Adding the fractions, we get

$$\frac{A}{x - 3}\frac{x - 3}{x - 3} + \frac{B}{(x - 3)^2} = \frac{A(x - 3) + B}{(x - 3)^2}$$

Equating numerators, we obtain

$$A(x - 3) + B = x - \frac{1}{2}$$

$$x = 3: \qquad A(0) + B = 3 - \frac{1}{2} \qquad \text{substituting } x = 3$$

$$B = \frac{5}{2}$$

Because of the repeating factor, $x = 3$ is the only convenient value we can substitute. If we use the value of B already obtained, however, we can let x be equal to an arbitrary value (such as $x = 0$):

$$A(x - 3) + B = x - \frac{1}{2}$$

$$A(x - 3) + \frac{5}{2} = x - \frac{1}{2} \qquad B = \frac{5}{2}$$

$$x = 0: \qquad A(0 - 3) + \frac{5}{2} = 0 - \frac{1}{2}$$

$$-3A = -\frac{6}{2}$$

$$A = 1$$

Substituting the values for A and B, we get

$$\int \frac{x - \frac{1}{2}}{(x - 3)^2}\, dx = \int \left(\frac{1}{x - 3} + \frac{5}{2} \frac{1}{(x - 3)^2} \right) dx$$

$$= \int \frac{dx}{x - 3} + \frac{5}{2} \int (x - 3)^{-2}\, dx \qquad \begin{bmatrix} u = x - 3 \\ du = dx \end{bmatrix}$$

$$= \ln|x - 3| + \frac{5}{2} \frac{(x - 3)^{-1}}{-1} + C$$

$$= \ln|x - 3| - \frac{5}{2} \frac{1}{x - 3} + C$$

◀

E X A M P L E **4**   Integrate

$$\int \frac{x^5 - x^4 - 8x^3 + 14x^2 - 5x - 8}{(x - 2)^2(x + 3)}\, dx$$

Solution. Since the numerator is of higher degree than the denominator, we first perform the long division. (To be able to do so, we need to multiply out the denominator.)

$$
\begin{array}{r}
x^2 \\
x^3 - x^2 - 8x + 12 \overline{)\, x^5 - x^4 - 8x^3 + 14x^2 - 5x - 8} \\
x^5 - x^4 - 8x^3 + 12x^2 \\
\hline
2x^2 - 5x - 8
\end{array}
$$

The integrand may now be written as

$$x^2 + \frac{2x^2 - 5x - 8}{(x - 2)^2(x + 3)}$$

We need to split up the fraction. Now note that the denominator contains only linear factors, the repeating factor $x - 2$ and the single factor $x + 3$. So by Rule I:

$$\frac{2x^2 - 5x - 8}{(x - 2)^2(x + 3)} = \frac{A}{x - 2} + \frac{B}{(x - 2)^2} + \frac{C}{x + 3}$$

As before, we combine the fractions on the right to obtain

$$\frac{A(x - 2)(x + 3) + B(x + 3) + C(x - 2)^2}{(x - 2)^2(x + 3)}$$

Equating numerators, we get

$$A(x - 2)(x + 3) + B(x + 3) + C(x - 2)^2 = 2x^2 - 5x - 8$$

$$x = 2: \qquad 0 + 5B + 0 = -10 \qquad \text{and} \qquad B = -2$$

$$x = -3: \qquad 0 + 0 + 25C = 25 \qquad \text{and} \qquad C = 1$$

We now substitute the values already obtained for B and C:

$$A(x - 2)(x + 3) - 2(x + 3) + 1(x - 2)^2 = 2x^2 - 5x - 8$$

To find A, we can let x be equal to any value, say $x = 0$:

$$x = 0: \qquad A(-2)(3) - 2(3) + 1(-2)^2 = -8$$

It follows that $A = 1$.

The integral now becomes

$$\int \left(x^2 + \frac{1}{x - 2} - \frac{2}{(x - 2)^2} + \frac{1}{x + 3} \right) dx$$

$$= \tfrac{1}{3}x^3 + \ln|x - 2| + \frac{2}{x - 2} + \ln|x + 3| + C \qquad \blacktriangleleft$$

Case 3: Distinct Quadratic Factors

E X A M P L E **5** Integrate

$$\int \frac{3x^2 + 2x + 4}{(x^2 + 4)(x + 1)} \, dx$$

Solution. Since one of the factors is quadratic, Rule II applies. (The linear factor $x + 1$ leads to the usual form by Rule I.) Thus

$$\frac{3x^2 + 2x + 4}{(x^2 + 4)(x + 1)} = \frac{Ax + B}{x^2 + 4} + \frac{C}{x + 1}$$

After adding the fractions on the right and equating numerators, we get

$$(Ax + B)(x + 1) + C(x^2 + 4) = 3x^2 + 2x + 4$$

Proceeding as we did before, we let $x = -1$.

$$x = -1: \qquad 0 + C(1 + 4) = 3 - 2 + 4$$
$$5C = 5$$
$$C = 1$$

At this point we seem to have run out of values to substitute. By using the value of C already obtained, however, we can let $x = 0$ and solve for B:

$$(Ax + B)(x + 1) + 1(x^2 + 4) = 3x^2 + 2x + 4 \qquad \text{C = 1}$$
$$x = 0: \qquad (0 + B)(1) + 1(4) = 4$$
$$B = 0$$

We now have

$$(Ax)(x + 1) + 1(x^2 + 4) = 3x^2 + 2x + 4 \qquad \text{B = 0}$$

Finally, we let x be any value (say $x = 1$) and solve for A:

$$x = 1: \qquad A(2) + 5 = 9$$
$$A = 2$$

After substituting the values of A, B, and C, the integral becomes

$$\int \left(\frac{2x}{x^2 + 4} + \frac{1}{x + 1} \right) dx = \ln|x^2 + 4| + \ln|x + 1| + C$$
$$= \ln|(x^2 + 4)(x + 1)| + C \qquad \blacktriangleleft$$

In some cases involving quadratic factors, it is best to return to the method of comparing coefficients, as in Example 1.

E X A M P L E **6**   (*Trinomial factor.*) Integrate

$$\int \frac{2x^2 + 10x + 10}{x(x^2 + 4x + 5)} \, dx$$

Solution. By Rules I and II,

$$\frac{2x^2 + 10x + 10}{x(x^2 + 4x + 5)} = \frac{A}{x} + \frac{Bx + C}{x^2 + 4x + 5}$$

It follows that

$$A(x^2 + 4x + 5) + (Bx + C)x = 2x^2 + 10x + 10$$

or

$$(A + B)x^2 + (4A + C)x + 5A = 2x^2 + 10x + 10$$

Equating coefficients, we get the following system of equations:

$$A + B = 2 \qquad \text{coefficients of } x^2$$

$$4A + C = 10 \qquad \text{coefficients of } x$$

$$5A = 10 \qquad \text{constants}$$

From the last equation, $A = 2$. Substituting in the first equation, we get $2 + B = 2$, or $B = 0$. From the second equation, $4(2) + C = 10$, or $C = 2$. Consequently,

$$\int \frac{2x^2 + 10x + 10}{x(x^2 + 4x + 5)}\, dx = \int \left(\frac{2}{x} + \frac{2}{x^2 + 4x + 5} \right) dx$$

$$= \int \frac{2}{x}\, dx + \int \frac{2}{(x + 2)^2 + 1}\, dx$$

$$= 2 \ln|x| + 2 \operatorname{Arctan}(x + 2) + C \qquad \blacktriangleleft$$

Case 4: Repeating Quadratic Factors

E X A M P L E **7**   Integrate

$$\int \frac{x^3\, dx}{(x^2 + 1)^2}$$

Solution. By Rule II,

$$\frac{x^3}{(x^2 + 1)^2} = \frac{Ax + B}{x^2 + 1} + \frac{Cx + D}{(x^2 + 1)^2}$$

Adding fractions again and equating numerators, we get

$$(Ax + B)(x^2 + 1) + (Cx + D) = x^3$$

or, collecting similar terms,

$$Ax^3 + Bx^2 + (A + C)x + (B + D) = x^3$$

The system of equations resulting from comparing coefficients is particularly simple in this case. Note that

$$A = 1 \qquad B = 0 \qquad A + C = 0 \qquad B + D = 0$$

whence $C = -1$ and $D = 0$. Thus

$$\int \frac{x^3 \, dx}{(x^2 + 1)^2} = \int \left(\frac{x}{x^2 + 1} - \frac{x}{(x^2 + 1)^2} \right) dx$$

Now let $u = x^2 + 1$, so that $du = 2x \, dx$. Then we get

$$\frac{1}{2} \int \left(\frac{1}{x^2 + 1} - \frac{1}{(x^2 + 1)^2} \right) 2x \, dx = \frac{1}{2} \int (u^{-1} - u^{-2}) \, du$$

$$= \frac{1}{2} \left(\ln|u| + \frac{1}{u} \right) + C$$

$$= \frac{1}{2} \ln(x^2 + 1) + \frac{1}{2(x^2 + 1)} + C \qquad \blacktriangleleft$$

The partial fraction technique is due to German mathematician Carl Jacobi (1804–1851) and was developed in his Berlin dissertation of 1825. In spite of this modest start, Jacobi soon established himself as a mathematician of first rank through his work on elliptic functions, a difficult branch of the theory of functions of complex variables. The theory of determinants, in the form now taught in algebra, is also due to Jacobi.

E X E R C I S E S / S E C T I O N   **25.8**

Perform the following integrations.

1. $\int \dfrac{dx}{x^2 - 4}$

2. $\int \dfrac{x - 5}{x + 2} \, dx$

3. $\int \dfrac{dx}{x - x^2}$

4. $\int \dfrac{2x + 1}{x^2 + x} \, dx$

5. $\int \dfrac{5x - 4}{(x - 2)(x + 1)} \, dx$

6. $\int \dfrac{x^3 - 3x^2 + 5x - 4}{x^2 - 3x + 2} \, dx$

7. $\int \dfrac{x^3 \, dx}{x^2 - 2x - 3}$

8. $\int \dfrac{3x^2 + 2x - 6}{x(x - 2)(x + 3)} \, dx$

9. $\int \dfrac{x^2 + 10x - 20}{(x - 4)(x - 1)(x + 2)} \, dx$

10. $\int \dfrac{5 - 2x}{(x - 2)^2} \, dx$

11. $\int \dfrac{3x + 5}{(x + 1)^2} \, dx$

12. $\int \dfrac{4 \, dx}{(x + 2)^2}$

13. $\int \dfrac{-2x^2 + 9x - 7}{(x - 2)^2(x + 1)} \, dx$

14. $\int \dfrac{4x^2 - x - 7}{(x - 1)^2(x + 3)} \, dx$

15. $\int \dfrac{2x^2 + 1}{(x - 2)^3} \, dx \left[Hint: \text{ By Rule I, } \dfrac{2x^2 + 1}{(x - 2)^3} = \dfrac{A}{x - 2} + \dfrac{B}{(x - 2)^2} + \dfrac{C}{(x - 2)^3} \right]$

16. $\int \dfrac{x \, dx}{(x - 1)^3}$

17. $\int \dfrac{x^2 - 3x - 2}{(x + 2)(x^2 + 4)} \, dx$ (See Example 5.)

18. $\displaystyle\int \frac{2x^2 + 3x + 9}{(x - 3)(x^2 + 9)}\,dx$

19. $\displaystyle\int \frac{3x^2 + 4x + 3}{(x + 1)(x^2 + 1)}\,dx$

20. $\displaystyle\int \frac{x^3 + 6x^2 + 2x + 3}{(x - 1)(x + 2)(x^2 + 1)}\,dx$

21. $\displaystyle\int \frac{x^5\,dx}{(x^2 + 4)^2}$

22. $\displaystyle\int \frac{x^3 + 3x}{(x^2 + 1)^2}\,dx$

23. $\displaystyle\int \frac{x^2 - 3x + 5}{x(x^2 - 2x + 5)}\,dx$

24. $\displaystyle\int \frac{3x^2 - 4x - 3}{(x - 4)(x^2 + 2x + 5)}\,dx$

25. $\displaystyle\int \frac{dx}{x(x^2 + 2x + 2)}$

25.9 Integration by Use of Tables

It should not be surprising that many basic integrals have been tabulated over the years and collected in tables. (Table 4, Appendix E, contains a short list of such integrals.) Although integration tables may be highly useful, they do have certain natural limitations. To see why, pretend that you have forgotten most of the integration formulas and decided to rely on the table. You are faced with the integral

$$\int \frac{\sin e^{-x}}{e^x}\,dx$$

Neither Table 4 nor any other table contains this particular case. What can be done? Based on our earlier experience, it is not difficult to see that by letting $u = e^{-x}[du = -e^{-x}\,dx = (-1/e^x)\,dx]$, the integral assumes the form

$$-\int \sin u\,du = \cos u + C = \cos e^{-x} + C$$

In other words, the table lists only the general form, which in this case ($\int \sin u\,du$) you would probably have remembered anyway. Consequently, *you have to be familiar with the integration techniques in order to convert a given integral to a recognizable form listed in the table.*

Fortunately, many integrals can be obtained from a table without any special tricks. All that may be required is a little care in identifying the constants.

E X A M P L E **1** Use Table 4 to integrate

$$\int \frac{dx}{x\sqrt{5 + 2x}}$$

Solution. We hunt up the form in the second section of the table ("Forms Containing $\sqrt{a + bu}$"). Note that in formula 12, $a = 5$ and $b = 2$. We now get, by direct substitution,

$$\int \frac{dx}{x\sqrt{5 + 2x}} = \frac{1}{\sqrt{5}} \ln \left| \frac{\sqrt{5 + 2x} - \sqrt{5}}{\sqrt{5 + 2x} + \sqrt{5}} \right| + C$$

◄

E X A M P L E **2**   Integrate $\int \sin 4x \cos 2x\, dx$.

Solution. We look in the table under trigonometric forms and find that formula 63 fits our case. Since $m = 4$ and $n = 2$, we get

$$\int \sin 4x \cos 2x\, dx = -\frac{\cos(4 + 2)x}{2(4 + 2)} - \frac{\cos(4 - 2)x}{2(4 - 2)} + C$$

$$= -\frac{1}{12} \cos 6x - \frac{1}{4} \cos 2x + C$$

◄

E X A M P L E **3**   Integrate $\int \sec^4 x\, dx$.

Solution. Note that formula 71 fits our form. Since the right-hand side contains an integral, such a form is referred to as a *reduction formula*. With $n = 4$, we now get directly

$$\int \sec^4 x\, dx = \frac{\sec^2 x \tan x}{4 - 1} + \frac{4 - 2}{4 - 1} \int \sec^2 x\, dx$$

$$= \frac{1}{3} \sec^2 x \tan x + \frac{2}{3} \tan x + C$$

by formula 53 in Table 4 or directly from memory.   ◄

E X A M P L E **4**   Integrate

$$\int \frac{dx}{x\sqrt{2x^2 + 4}}$$

Solution. As indicated earlier, the use of tables depends on the proper recognition of the form. With $u^2 = 2x^2$, the form can be made to fit formula 34 by adjusting the constants. Since $u = \sqrt{2}x$, we get $du = \sqrt{2}\, dx$ and write the integral as

$$\int \frac{\sqrt{2}\, dx}{\sqrt{2}x\sqrt{2x^2 + 4}}$$

Now formula 34 in the table fits precisely, so that

$$\int \frac{\sqrt{2}\,dx}{\sqrt{2x}\sqrt{2x^2+4}} = \int \frac{du}{u\sqrt{u^2+a^2}} = \frac{1}{a}\ln\left|\frac{u}{a+\sqrt{u^2+a^2}}\right| + C$$

$$= \frac{1}{2}\ln\left|\frac{\sqrt{2}x}{2+\sqrt{2x^2+4}}\right| + C$$ ◄

E X E R C I S E S / S E C T I O N 25.9

Integrate each of the given functions by using Table 4 (Appendix E).

1. $\displaystyle\int \frac{dx}{x(2+x)}$

2. $\displaystyle\int x\sqrt{1+2x}\,dx$

3. $\displaystyle\int \sqrt{x^2-7}\,dx$

4. $\displaystyle\int \frac{dx}{x^2-9}$

5. $\displaystyle\int \frac{dx}{5-x^2}$

6. $\displaystyle\int \frac{dx}{(x^2+2)^{3/2}}$

7. $\displaystyle\int \frac{dx}{x^2\sqrt{5x^2+4}}$

8. $\displaystyle\int xe^x\,dx$

9. $\displaystyle\int \sin 2x \sin x\,dx$

10. $\displaystyle\int e^{-x}\sin 2x\,dx$

11. $\displaystyle\int \frac{dx}{\sqrt{3x^2+5}}$

12. $\displaystyle\int \frac{\sqrt{x^2+8}}{x}\,dx$

13. $\displaystyle\int x^2 e^{2x}\,dx$

14. $\displaystyle\int \sin^4 x\,dx$

15. $\displaystyle\int \tan^6 x\,dx$

16. $\displaystyle\int x\,\text{Arcsin}\,x^2\,dx$

17. $\displaystyle\int \frac{dx}{4x^2-9}$

18. $\displaystyle\int \cos 3x \cos x\,dx$

19. $\displaystyle\int \frac{dx}{x\sqrt{3+x}}$

20. $\displaystyle\int \frac{dx}{3x^2+5}$

21. $\displaystyle\int \frac{dx}{3x^2-5}$

22. $\displaystyle\int x\sin x\,dx$

23. $\displaystyle\int \frac{\sqrt{x^2-10}}{x}\,dx$

24. $\displaystyle\int \frac{dx}{\sqrt{4x^2-9}}$

25. $\displaystyle\int \sin 3x \cos 2x\,dx$

26. $\displaystyle\int \sqrt{5-x^2}\,dx$

27. $\displaystyle\int \frac{dx}{(4x^2+5)^{3/2}}$

28. $\displaystyle\int \sin^2 2x\,dx$

25.10 Additional Remarks

Elementary function

After studying the various integration techniques, you could easily get the impression that any function can be integrated by finding its antiderivative. Such is not the case in the following sense. Suppose we define a function to be **elementary** if it is an algebraic, trigonometric, exponential, logarithmic, inverse trigonometric, or hyperbolic function. It turns out that even apparently simple integrals such as

$$\int e^{x^3}\,dx \qquad \int \sqrt{x}\,\sin x\,dx \qquad \int \sqrt{a^2\cos^2 x + b^2\sin^2 x}\,dx$$

do not lead to elementary antiderivatives. (For example, there is no elementary function whose derivative is e^{x^3}.) Certain definite integrals will be evaluated in Chapter 26 by means of infinite series, which yield nonelementary forms. Other well-known definite integrals such as

$$\int_0^\infty \frac{\sin x}{x}\,dx = \frac{\pi}{2} \qquad \text{and} \qquad \frac{2}{\sqrt{\pi}}\int_0^\infty e^{-x^2}\,dx = 1$$

can only be handled by more advanced methods. Fortunately, numerical techniques have been developed for evaluating many definite integrals. Two such methods were discussed in Chapter 22 (Section 22.8).

REVIEW EXERCISES/CHAPTER   25

One of the problems in integration is recognizing the type of integral. In the previous sections the type was usually known and the idea was to apply the technique correctly. In the following exercises the main task is to decide what technique to apply. *DO NOT USE TABLE 4 OF APPENDIX E.*

1. $\int \dfrac{x\,dx}{x^2 + 1}$

2. $\int \dfrac{x\,dx}{(x^2 + 1)^2}$

3. $\int \dfrac{2\,dx}{x^2 + 1}$

4. $\int \dfrac{e^{\text{Arctan } x}}{x^2 + 1}\,dx$

5. $\int \dfrac{x\,dx}{\sqrt{9 - x^2}}$

6. $\int \dfrac{dx}{\sqrt{9 - x^2}}$

7. $\int x \cos 2x^2 \, dx$

8. $\int x \cos 2x \, dx$

9. $\int \dfrac{e^x\,dx}{4 + e^{2x}}$

10. $\int \dfrac{e^x\,dx}{4 + e^x}$

11. $\int \dfrac{e^x\,dx}{(4 + e^x)^2}$

12. $\int x \ln x \, dx$

13. $\int \dfrac{\ln x}{x}\,dx$

14. $\int \dfrac{\cos(\ln x)}{x}\,dx$

15. $\int \dfrac{x + 2}{x^2 + 4x + 5}\,dx$

16. $\int \dfrac{dx}{x^2 + 4x + 5}$

17. $\int \dfrac{dx}{x^2 + 4x + 4}$

18. $\int \dfrac{dx}{x(1 + \ln^2 x)}$

19. $\int \ln^2 x \, dx$

20. $\int \dfrac{dx}{\sqrt{4 - 5x^2}}$

21. $\int \dfrac{dx}{x\sqrt{4 - x^2}}$

22. $\int x\sqrt{4 - x^2} \, dx$

23. $\int \dfrac{\sin^2 2x \cos 2x\,dx}{1 + \sin^3 2x}$

24. $\int \sin^3 2x \cos^2 2x \, dx$

25. $\int \sin^2 2x \, dx$

26. $\int x e^{2x^2} \, dx$

27. $\int x e^{2x} \, dx$

28. $\int \dfrac{\text{Arctan } 2x\,dx}{1 + 4x^2}$

29. $\int \text{Arctan } 2x \, dx$

30. $\int \dfrac{\sin 2x\,dx}{e^{\sin^2 x}}$

31. $\int \dfrac{e^{\tan x}}{\cos^2 x}\,dx$

32. $\int \dfrac{\sin (1/x)\,dx}{x^2}$

33. $\int \dfrac{x^2 - 1}{x^2 + 3}\,dx$

34. $\int \dfrac{\sin x}{1 - \cos x}\,dx$

35. $\int \dfrac{1 - \cos x}{\sin x}\,dx$

36. $\int \dfrac{dx}{x \sqrt[3]{\ln x}}$

37. $\int \cot^4 x \csc^4 x \, dx$

38. $\int \dfrac{dx}{\sqrt{4x - 2 - x^2}}$

39. $\int \sec^3 2x \tan^3 2x \, dx$

40. $\int e^{\tan x} \sec^2 x \tan x \, dx$

41. $\int \tan^2 x \sec^2 x \, dx$

42. $\int \dfrac{dx}{x(2 - \ln x)}$

43. $\int \sec^4 3x \, dx$

44. $\int \csc^3 4x \cot^3 4x \, dx$

45. $\int e^x \cos 4x \, dx$

46. $\int e^{\cos 4x} \sin 4x \, dx$

47. $\int \dfrac{dx}{5x^2 + 4}$

48. $\int \dfrac{3x\,dx}{5x^2 + 4}$

49. $\int \dfrac{x + 1}{x^2 + 2x - 8}\,dx$

50. $\int \dfrac{x + 2}{x^2 + 2x - 8}\,dx$

51. $\int \dfrac{3x^2 - 4x + 9}{(x - 2)(x^2 + 9)}\,dx$

Infinite Series

26.1 Introduction to Infinite Series

In this section we will see how the idea of an ordinary sum can be extended to the sum of an infinite number of terms, called an *infinite series*.

Suppose that a man standing 1 m away from a wall steps $\frac{1}{2}$ m forward so that the distance to the wall is cut in half. Then he takes another step forward, again cutting the remaining distance in half (Figure 26.1). Imagine that this operation is continued indefinitely. Does it make sense to talk about the total distance covered? In one way it doesn't, for the total distance is given by

$$\frac{1}{2} + \frac{1}{4} + \frac{1}{8} + \frac{1}{16} + \cdots$$

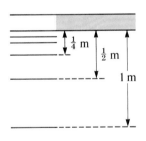

Figure 26.1

which involves an infinite number of terms. Even if you were immortal, you could never complete the addition. Yet according to Figure 26.1 the sum *ought* to be 1 m, even if you cannot physically (or even conceptually) add up infinitely many numbers.

The problem is one of definition: we first have to agree on the meaning of a sum of an infinite number of terms. To this end let a and r be two numbers and consider the expression

$$S = a + ar + ar^2 + ar^3 + \cdots \tag{26.1}$$

Infinite series

(The three dots indicate that the summation is to be continued indefinitely.) To distinguish this case from an ordinary sum, Equation (26.1) is called an **infinite series**. Suppose we consider only the first n terms:

$$S_n = a + ar + ar^2 + ar^3 + \cdots + ar^{n-1} \tag{26.2}$$

called the *nth partial sum.* The sum can be obtained by multiplying S_n by r and subtracting the result from S_n. Thus

$$
\begin{array}{ll}
S_n = a + ar + ar^2 + ar^3 + \cdots + ar^{n-1} & \\
rS_n = \quad\quad ar + ar^2 + ar^3 + \cdots + ar^{n-1} + ar^n & \\
\hline
S_n - rS_n = a \quad\quad\quad\quad\quad\quad\quad\quad\quad\quad\quad - ar^n & \textbf{subtracting}
\end{array}
$$

or

$$S_n(1 - r) = a - ar^n \quad\quad \textbf{factoring } S_n$$

and

$$S_n = \frac{a - ar^n}{1 - r} = \frac{a(1 - r^n)}{1 - r} \quad\quad \textbf{dividing by } 1 - r$$

The only legitimate way to pass from a finite to an infinite number of terms is through the limit process. We define the sum S as the limit of an infinite sequence of partial sums:

$$S = \lim_{n \to \infty} S_n = \lim_{n \to \infty} \frac{a(1 - r^n)}{1 - r} \tag{26.3}$$

If $|r| < 1$, then $r^n \to 0$ as $n \to \infty$. Hence the limit is

$$S = \frac{a}{1 - r}$$

(If $|r| > 1$, the limit does not exist.)

The series in Equation (26.1) is called a *geometric series.*

Geometric Series: A geometric series has the form

$$S = a + ar + ar^2 + \cdots + ar^{n-1} + \cdots$$

If $|r| < 1$, the **sum** is given by

$$S = \frac{a}{1 - r} \tag{26.4}$$

E X A M P L E **1**   Find the sum d of the individual distances in Figure 26.1:

$$\frac{1}{2} + \frac{1}{4} + \frac{1}{8} + \frac{1}{16} + \cdots$$

Solution. To find the sum we only need to observe that a is always the first term and ar the second. It follows that $a = \frac{1}{2}$, $ar = \frac{1}{2}r = \frac{1}{4}$. So $r = \frac{1}{2}$. By formula (26.4),

$$d = \frac{\frac{1}{2}}{1 - \frac{1}{2}} = 1$$

as expected. ◄

Formula (26.3) now provides the motivation for a general definition. Let

$$S = b_1 + b_2 + b_3 + \cdots + b_n + \cdots \qquad (26.5)$$

be an infinite series and let

$$S_n = b_1 + b_2 + b_3 + \cdots + b_n \qquad (26.6)$$

denote the nth partial sum. Then we define the sum S to be

$$(26.7)$$
$$S = \lim_{n \to \infty} S_n$$

whenever the limit exists. This limit of partial sums leads to the definition of *convergence* and *divergence*.

Definition of Convergence and Divergence

If

$$S = b_1 + b_2 + \cdots + b_n + \cdots$$

is an **infinite series**, then

$$S_n = b_1 + b_2 + \cdots + b_n$$

is called the **nth partial sum**.

1. If $\lim_{n \to \infty} S_n$ exists, then the series is said to **converge** or be convergent.
2. If $\lim_{n \to \infty} S_n$ does not exist, then the series is said to **diverge** or be divergent.

The series

$$\frac{1}{2} + \frac{1}{4} + \frac{1}{8} + \frac{1}{16} + \cdots$$

considered earlier is an example of a convergent series. An example of a divergent series is

$$1 + 2 + 3 + \cdots + n + \cdots$$

since

$$S_n = 1 + 2 + 3 + \cdots + n = \frac{n(n + 1)}{2}$$

(Section 22.2) and

$$\lim_{n \to \infty} S_n = \infty$$

On the other hand, the nth partial sum of the series

$$1 - 1 + 1 - 1 + 1 - \cdots$$

does not get very large, but since it oscillates between 0 and 1, it never settles down to a definite limit. As a consequence, the series has to be called divergent.

At first glance the definition of convergence appears to yield a method for finding the sum. Such is not the case: there does not exist a general method for finding the sum of an infinite series, and even (apparently) simple cases require considerable ingenuity. For example, it is known that

$$\frac{1}{1^2} + \frac{1}{2^2} + \frac{1}{3^2} + \frac{1}{4^2} + \cdots + \frac{1}{n^2} + \cdots = \frac{\pi^2}{6}$$

while the exact sum of

$$\frac{1}{1^3} + \frac{1}{2^3} + \frac{1}{3^3} + \frac{1}{4^3} + \cdots + \frac{1}{n^3} + \cdots$$

has never been found.

In the exercises, as well as in later sections, we are going to make regular use of the sigma notation in writing series. Thus

$$\sum_{n=1}^{\infty} b_n = b_1 + b_2 + b_3 + \cdots + b_n + \cdots$$

For example,

$$\sum_{n=1}^{\infty} \frac{1}{n} = \frac{1}{1} + \frac{1}{2} + \frac{1}{3} + \cdots + \frac{1}{n} + \cdots$$

E X E R C I S E S / S E C T I O N **26.1**

In Exercises 1–10, find the sum of each of the geometric series.

1. $\sum_{n=0}^{\infty} \dfrac{1}{3^n} = 1 + \dfrac{1}{3} + \dfrac{1}{3^2} + \dfrac{1}{3^3} + \cdots + \dfrac{1}{3^n} + \cdots$

2. $\sum_{n=1}^{\infty} \left(\dfrac{2}{3}\right)^{n-1} = 1 + \dfrac{2}{3} + \left(\dfrac{2}{3}\right)^2 + \cdots + \left(\dfrac{2}{3}\right)^{n-1} + \cdots$

3. $\sum_{n=1}^{\infty} \dfrac{2}{3^n} = \dfrac{2}{3} + \dfrac{2}{3^2} + \dfrac{2}{3^3} + \cdots + \dfrac{2}{3^n} + \cdots$

4. $\sum_{n=1}^{\infty} \left(\dfrac{5}{6}\right)^n = \dfrac{5}{6} + \left(\dfrac{5}{6}\right)^2 + \left(\dfrac{5}{6}\right)^3 + \cdots + \left(\dfrac{5}{6}\right)^n + \cdots$

5. $\sum_{n=1}^{\infty} \left(\dfrac{3}{4}\right)^{n-1} = 1 + \dfrac{3}{4} + \dfrac{9}{16} + \cdots + \left(\dfrac{3}{4}\right)^{n-1} + \cdots$

6. $\sum_{n=1}^{\infty} \dfrac{3}{4^n} = \dfrac{3}{4} + \dfrac{3}{16} + \dfrac{3}{64} + \cdots + \dfrac{3}{4^n} + \cdots$

7. $\sum_{n=1}^{\infty} \dfrac{4}{9^n} = \dfrac{4}{9} + \dfrac{4}{9^2} + \dfrac{4}{9^3} + \cdots + \dfrac{4}{9^n} + \cdots$

8. $\sum_{n=1}^{\infty} \dfrac{(-1)^{n-1}}{2^{n-1}} = 1 - \dfrac{1}{2} + \dfrac{1}{4} - \dfrac{1}{8} + \cdots + (-1)^{n-1}\dfrac{1}{2^{n-1}}$
$+ \cdots$ (*Hint:* $r = -\frac{1}{2}$.)

9. $\sum_{n=1}^{\infty} (-1)^{n+1}\left(\dfrac{2}{3}\right)^n = \dfrac{2}{3} - \dfrac{4}{9} + \dfrac{8}{27} - \cdots + (-1)^{n+1}\left(\dfrac{2}{3}\right)^n$
$+ \cdots$

10. $\sum_{n=1}^{\infty} \left(\dfrac{4}{3}\right)^n = \dfrac{4}{3} + \left(\dfrac{4}{3}\right)^2 + \left(\dfrac{4}{3}\right)^3 + \cdots + \left(\dfrac{4}{3}\right)^n + \cdots$

In Exercises 11–14, convert the repeating decimals to common fractions.

11. $0.212121 \ldots$ $\left(Hint: 0.212121 \ldots = \dfrac{21}{10^2} + \dfrac{21}{10^4} + \cdots\right)$

12. $0.015015015 \ldots$

13. $0.5070707 \ldots$

14. $0.636363 \ldots$

26.2 Tests for Convergence*

The purpose of this section is to discuss some of the standard tests for determining whether a given infinite series converges or diverges.

For many series it may be difficult to establish convergence. In some cases, however, it is possible to see at a glance that the series could not possibly converge. Let

$$S = a_1 + a_2 + a_3 + \cdots + a_n + \cdots$$

be an infinite series of positive terms and consider the sequence of partial sums

$$S_1 = a_1$$
$$S_2 = a_1 + a_2$$
$$S_3 = a_1 + a_2 + a_3$$
$$\vdots$$
$$S_n = a_1 + a_2 + a_3 + \cdots + a_n$$

If the infinite series converges, then $S = \lim_{n \to \infty} S_n$ exists by definition. Since n is just a dummy subscript, $\lim_{k \to \infty} S_k$ also exists and is equal to the first limit.

---

* This section may be omitted without loss of continuity.

In particular,

$$\lim_{n \to \infty} S_n = \lim_{n \to \infty} S_{n-1}$$

Now, it can be seen from this sequence of partial sums that $a_n = S_n - S_{n-1}$. Consequently,

$$\lim_{n \to \infty} a_n = \lim_{n \to \infty} (S_n - S_{n-1}) = \lim_{n \to \infty} S_n - \lim_{n \to \infty} S_{n-1} = 0$$

We conclude that if $\sum_{n=1}^{\infty} a_n$ converges, then $\lim_{n \to \infty} a_n = 0$.

Necessary Condition for Convergence: If the series

$$S = a_1 + a_2 + \cdots + a_n + \cdots$$

is **convergent,** then

$$\lim_{n \to \infty} a_n = 0 \qquad\qquad\qquad (26.8)$$

Consequently, if $\lim_{n \to \infty} a_n$ is not equal to zero, then the series necessarily *diverges.* The converse is not true, however. If $\lim_{n \to \infty} a_n = 0$, then we may **not** conclude that the series converges—it may diverge anyway.

E X A M P L E **1**   For the series

$$\sum_{n=1}^{\infty} \frac{n}{n+1} = \frac{1}{2} + \frac{2}{3} + \frac{3}{4} + \cdots + \frac{n}{n+1} + \cdots$$

we have

$$\lim_{n \to \infty} a_n = \lim_{n \to \infty} \frac{n}{n+1} = 1 \qquad \textbf{by L'Hospital's rule}$$

so that the series diverges by criterion (26.8).
On the other hand, for the series

$$\sum_{n=1}^{\infty} \frac{1}{n} = 1 + \frac{1}{2} + \frac{1}{3} + \cdots + \frac{1}{n} + \cdots$$

we have

$$\lim_{n \to \infty} a_n = \lim_{n \to \infty} \frac{1}{n} = 0$$

but the series *also diverges*, as we will see shortly.   ◀

Note that in Example 1, $a_n = n/(n+1)$ is a function of n. Using the notation $a_n = f(n)$, we can obtain the *integral test*, a simple and powerful test for convergence.

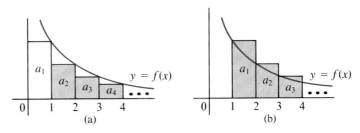

Figure 26.2

Integral Test: Suppose that $\sum_{n=1}^{\infty} a_n$ is a series of positive terms, with $a_n = f(n)$. Then $\sum_{n=1}^{\infty} a_n$ converges if the improper integral $\int_1^{\infty} f(x)\,dx$ exists and diverges if the integral does not exist. (The function $f(x)$ has to be continuous and decreasing.)

The validity of this test can be seen intuitively from a diagram (Figure 26.2). Note that the rectangles in each figure are constructed by plotting the ordinates $f(1), f(2), \ldots$, corresponding to $x = 1, 2, \ldots$.

Suppose the integral exists. Then Figure 26.2(a) shows that the sum of the areas of the rectangles

$$f(2) + f(3) + \cdots + f(n) + \cdots = a_2 + a_3 + \cdots + a_n + \cdots$$

is less than $\int_1^{\infty} f(x)\,dx$, the area under the curve. Since the latter is finite, so is the sum of the rectangles, and the series converges. (The omission of the first term $a_1 = f(1)$ has no bearing on the result.) If $\int_1^{\infty} f(x)\,dx = \infty$, we see from Figure 26.2(b) that the sum of the rectangles

$$f(1) + f(2) + \cdots + f(n) + \cdots = a_1 + a_2 + \cdots + a_n + \cdots$$

cannot be finite either, so that the series diverges.

E X A M P L E **2**   Test the series

$$\sum_{n=2}^{\infty} \frac{1}{e^n} = \frac{1}{e^2} + \frac{1}{e^3} + \cdots + \frac{1}{e^n} + \cdots$$

for convergence by the integral test.

Solution. Since

$$\int_2^{\infty} \frac{1}{e^x}\,dx = \int_2^{\infty} e^{-x}\,dx = \lim_{b \to \infty} \int_2^b e^{-x}\,dx$$

$$= \lim_{b \to \infty} (-e^{-x})\Big|_2^b = \lim_{b \to \infty} (-e^{-b} + e^{-2}) = \frac{1}{e^2}$$

the series converges. (Observe that the lower limit need not be $n = 1$.) ◀

E X A M P L E **3**   Find the values of p for which the series

$$\sum_{n=1}^{\infty} \frac{1}{n^p}$$

converges and diverges.

Solution.

$$\int_1^\infty \frac{dx}{x^p} = \lim_{b \to \infty} \int_1^b x^{-p}\, dx = \lim_{b \to \infty} \frac{x^{-p+1}}{-p+1}\bigg|_1^b$$

$$= \lim_{b \to \infty} \left(\frac{b^{-p+1}}{-p+1} - \frac{1}{-p+1} \right) \qquad (p \neq 1)$$

$$= \begin{cases} \dfrac{1}{p-1} & p > 1 \\[2mm] \infty & p < 1 \end{cases}$$

Consequently, the series converges for $p > 1$ and diverges for $p < 1$ by the integral test. If $p = 1$, we have

$$\int_1^\infty \frac{dx}{x} = \lim_{b \to \infty} \ln x \bigg|_1^b = \lim_{b \to \infty} \ln b = \infty$$

so that the series

$$\sum_{n=1}^{\infty} \frac{1}{n}$$

is divergent.   ◀

A series of the form

$$\sum_{n=1}^{\infty} \frac{1}{n^p} = \frac{1}{1^p} + \frac{1}{2^p} + \frac{1}{3^p} + \cdots + \frac{1}{n^p} + \cdots \tag{26.9}$$

p-series

is called a **p-series**. It was shown in Example 3 that a p-series converges if $p > 1$ and diverges if $p < 1$. The special case

$$\sum_{n=1}^{\infty} \frac{1}{n} = 1 + \frac{1}{2} + \frac{1}{3} + \cdots + \frac{1}{n} + \cdots \tag{26.10}$$

Harmonic series

$(p = 1)$ is called the **harmonic series**. The harmonic series is divergent.

     In the process of discussing convergence and divergence we have encountered three special series: the geometric series, the p-series, and the harmonic series. There exist certain variations on the forms of these series, many of which can be tested by the *comparison test*.

> **Comparison Test:** Let $\sum_{n=1}^{\infty} a_n$ be a series of positive terms.
>
> **1.** If $\sum_{n=1}^{\infty} b_n$ is a convergent series and if $a_n \leq b_n$ for all n, then $\sum_{n=1}^{\infty} a_n$ converges.
>
> **2.** If $\sum_{n=1}^{\infty} c_n$ is a divergent series of positive terms and if $a_n \geq c_n$, then $\sum_{n=1}^{\infty} a_n$ diverges.

To check part (1), let

$$S_n = a_1 + a_2 + \cdots + a_n \qquad \text{and} \qquad T_n = b_1 + b_2 + \cdots + b_n$$

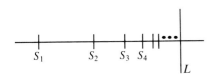

Figure 26.3

Since $\sum_{n=1}^{\infty} b_n$ converges, $\lim_{n \to \infty} T_n = L$ exists. Now $S_n \leq T_n \leq L$; that is, the sequence $S_1, S_2, S_3, \ldots$ is increasing (since each a_n is positive) and bounded above by L. The diagram in Figure 26.3 suggests that $\lim_{n \to \infty} S_n$ also exists. (It is true in general that an increasing sequence bounded above is convergent.) Part (2) is proved similarly.

E X A M P L E **4** Test the series

$$\sum_{n=1}^{\infty} \frac{1}{n(n+2)}$$

for convergence.

Solution. For each n,

$$\frac{1}{n(n+2)} = \frac{1}{n^2 + 2n} < \frac{1}{n^2}$$

Since $\sum_{n=1}^{\infty} (1/n^2)$ is a p-series with $p = 2$, it converges. Consequently, the given series also converges by the comparison test, part (1). ◀

E X A M P L E **5** Test the series

$$\sum_{n=2}^{\infty} \frac{1}{n-1}$$

for convergence.

Solution. For each $n \geq 2$

$$\frac{1}{n-1} > \frac{1}{n}$$

Since $\sum_{n=1}^{\infty} (1/n)$ is the harmonic series (26.10), which diverges, the given series also diverges by the comparison test, part (2). ◀

For completeness we will state without proof the *ratio test*.

Ratio Test

Let $a_1 + a_2 + a_3 + \cdots + a_n + \cdots$ be a series of positive terms and

$$L = \lim_{n \to \infty} \frac{a_{n+1}}{a_n}$$

Then if:

1. $L < 1$, the series converges.
2. $L > 1$, the series diverges.
3. $L = 1$, the test fails. (Series may converge or diverge.)

E X A M P L E **6**   Test the series

$$\sum_{n=1}^{\infty} \frac{2^n}{n!}$$

for convergence. (Recall that $n! = n(n-1)(n-2) \cdots 2 \cdot 1$.)

Solution. Let us first note that the integral test cannot be applied to this series; for the comparison test we have no series with which to make a comparison. By the ratio test,

$$a_n = \frac{2^n}{n!} \quad \text{and} \quad a_{n+1} = \frac{2^{n+1}}{(n+1)!} = \frac{2^n \cdot 2}{(n+1)n!}$$

(To see that last step more clearly, observe that $10! = 10 \cdot 9!$ and $6! = 6 \cdot 5!$) So we have

$$\lim_{n \to \infty} \frac{a_{n+1}}{a_n} = \lim_{n \to \infty} \frac{2^n \cdot 2}{(n+1)n!} \cdot \frac{n!}{2^n} = \lim_{n \to \infty} \frac{2}{n+1} = 0 = L$$

Since $L < 1$, the series converges.   ◀

E X E R C I S E S / S E C T I O N   **26.2**

In Exercises 1–4, show that the series are divergent. (See Example 1.)

1. $\displaystyle\sum_{n=1}^{\infty} \frac{n}{2n+2}$

2. $\displaystyle\sum_{n=1}^{\infty} \frac{2n}{4n+1}$

3. $\displaystyle\sum_{n=2}^{\infty} \frac{5n^2}{2n^2-2}$

4. $\displaystyle\sum_{n=1}^{\infty} \frac{n!}{n^3+1}$

In Exercises 5–20, test the series for convergence by the integral test.

5. $\displaystyle\sum_{n=1}^{\infty} \frac{1}{(n+1)^2} = \frac{1}{2^2} + \frac{1}{3^2} + \frac{1}{4^2} + \cdots + \frac{1}{(n+1)^2} + \cdots$

6. $\displaystyle\sum_{n=2}^{\infty} \frac{1}{2n-2} = \frac{1}{2} + \frac{1}{4} + \frac{1}{6} + \cdots + \frac{1}{2n-2} + \cdots$

7. $\displaystyle\sum_{n=1}^{\infty} \frac{n}{n^2 + 1} = \frac{1}{2} + \frac{2}{5} + \frac{3}{10} + \cdots + \frac{n}{n^2 + 1} + \cdots$

8. $\displaystyle\sum_{n=1}^{\infty} \frac{1}{n^2 + 1} = \frac{1}{2} + \frac{1}{5} + \frac{1}{10} + \cdots + \frac{1}{n^2 + 1} + \cdots$

9. $\displaystyle\sum_{n=0}^{\infty} \frac{1}{(2n + 2)^2}$

10. $\displaystyle\sum_{n=1}^{\infty} \frac{n}{2n^2 - 1}$

11. $\displaystyle\sum_{n=2}^{\infty} \frac{n^2}{n^3 - 2}$

12. $\displaystyle\sum_{n=3}^{\infty} \frac{1}{3n + 2}$

13. $\displaystyle\sum_{n=1}^{\infty} \frac{n}{e^n}$

14. $\displaystyle\sum_{n=2}^{\infty} \frac{1}{n \ln n}$

15. $\displaystyle\sum_{n=1}^{\infty} \frac{n}{(n^2 + 1)^{3/2}}$

16. $\displaystyle\sum_{n=2}^{\infty} \frac{1}{n^2 - 2n + 1}$

17. $\displaystyle\sum_{n=0}^{\infty} \frac{n}{(n^2 + 2)^2}$

18. $\displaystyle\sum_{n=3}^{\infty} \frac{1}{\sqrt{n - 2}}$

19. $\displaystyle\sum_{n=2}^{\infty} \frac{n}{\sqrt{n^2 + 2}}$

20. $\displaystyle\sum_{n=0}^{\infty} \frac{n^2 + 1}{n^2 + 2}$

In Exercises 21–40, test the series for convergence or divergence by the comparison test.

21. $\displaystyle\sum_{n=1}^{\infty} \frac{1}{n^2 + 1}$

22. $\displaystyle\sum_{n=1}^{\infty} \frac{1}{n(n + 4)}$

23. $\displaystyle\sum_{n=6}^{\infty} \frac{1}{n - 5}$

24. $\displaystyle\sum_{n=1}^{\infty} \frac{1}{(n + 2)(n + 3)}$

25. $\displaystyle\sum_{n=0}^{\infty} \frac{1}{n^3 + 2}$

26. $\displaystyle\sum_{n=4}^{\infty} \frac{1}{n - 3}$

27. $\displaystyle\sum_{n=2}^{\infty} \frac{1}{n^2 + n}$

28. $\displaystyle\sum_{n=3}^{\infty} \frac{1}{n^3 + 4}$

29. $\displaystyle\sum_{n=0}^{\infty} \frac{1}{3^n + 1}$

30. $\displaystyle\sum_{n=1}^{\infty} \frac{1}{4^n + 2}$

31. $\displaystyle\sum_{n=2}^{\infty} \frac{1}{3^n - 1}$

32. $\displaystyle\sum_{n=1}^{\infty} \frac{1}{4^n + n}$

33. $\displaystyle\sum_{n=4}^{\infty} \frac{1}{\sqrt{n - 1}}$

34. $\displaystyle\sum_{n=2}^{\infty} \frac{1}{5^n - 1}$

35. $\displaystyle\sum_{n=2}^{\infty} \frac{1}{n^3 - 1}$

36. $\displaystyle\sum_{n=4}^{\infty} \frac{1}{\sqrt[3]{n - 1}}$

37. $\displaystyle\sum_{n=1}^{\infty} \frac{1 + \sin n}{n^3}$

38. $\displaystyle\sum_{n=2}^{\infty} \frac{1}{n^4 - 1}$

39. $\displaystyle\sum_{n=2}^{\infty} \frac{1}{\ln n}$

40. $\displaystyle\sum_{n=1}^{\infty} \frac{1}{n^{5/4} + 1}$

In Exercises 41–56, test the series for convergence by the ratio test; if the test fails, use another test.

41. $\displaystyle\sum_{n=1}^{\infty} \frac{1}{3^n}$

42. $\displaystyle\sum_{n=1}^{\infty} \frac{2}{4^n}$

43. $\displaystyle\sum_{n=1}^{\infty} \frac{1}{n!}$

44. $\displaystyle\sum_{n=1}^{\infty} \frac{n}{n!}$

45. $\displaystyle\sum_{n=1}^{\infty} \frac{4^n}{n!}$

46. $\displaystyle\sum_{n=0}^{\infty} \frac{3^n}{(n + 1)!}$

47. $\displaystyle\sum_{n=1}^{\infty} \frac{n^2}{2^n}$

48. $\displaystyle\sum_{n=1}^{\infty} \frac{n^3 + 1}{n!}$

49. $\displaystyle\sum_{n=1}^{\infty} \frac{n!}{7^n}$

50. $\displaystyle\sum_{n=1}^{\infty} \frac{e^n}{2n^2 + 1}$

51. $\displaystyle\sum_{n=1}^{\infty} \frac{n!}{1 \cdot 3 \cdot 5 \cdots (2n - 1)}$

52. $\displaystyle\sum_{n=1}^{\infty} \frac{1 \cdot 3 \cdot 5 \cdots (2n - 1)}{1 \cdot 4 \cdot 7 \cdots (3n - 2)}$

53. $\displaystyle\sum_{n=1}^{\infty} \frac{1 \cdot 4 \cdot 7 \cdots (3n - 2)}{2 \cdot 4 \cdot 6 \cdots (2n)}$

54. $\displaystyle\sum_{n=1}^{\infty} \frac{n - 1}{n^3}$

55. $\displaystyle\sum_{n=2}^{\infty} \frac{n^2 - 1}{n^3}$

56. $\displaystyle\sum_{n=2}^{\infty} \frac{1}{n \ln^2 n}$

26.3 Maclaurin Series

Power series

While the infinite series considered so far have contained only constant terms, many useful series consist of variable terms. The most important of these are series representing known functions. The main purpose of this section is to study a method by which a function $f(x)$ can be written as a **power series**:

$$f(x) = a_0 + a_1 x + a_2 x^2 + a_3 x^3 + \cdots + a_n x^n + \cdots \tag{26.11}$$

(Series expansions other than power series will be taken up in Section 26.6.)

To express a function as a power series, we need to determine the coefficients in form (26.11). This can be done by means of a simple trick: we differentiate both sides of (26.11) repeatedly, as if it were a regular polynomial, and substitute zero for x. Hence $f(x)$ must be differentiable near $x = 0$ to start with.

Moreover, it is shown in many books on advanced calculus that a power series may be differentiated term by term, provided that it converges for all x in some interval. We now get

$$f(x) \ = \ a_0 + a_1 x + a_2 x^2 + a_3 x^3 \qquad + a_4 x^4 \qquad + a_5 x^5 + \cdots$$

$$f'(x) \ = \ \qquad a_1 + 2a_2 x + 3a_3 x^2 \quad + 4a_4 x^3 \qquad + 5a_5 x^4 + \cdots$$

$$f''(x) \ = \ \qquad\qquad 2 \cdot 1 a_2 \ + 3 \cdot 2a_3 x \quad + 4 \cdot 3a_4 x^2 \ + 5 \cdot 4a_5 x^3 + \cdots$$

$$f'''(x) \ = \ \qquad\qquad\qquad\qquad 3 \cdot 2 \cdot 1 a_3 + 4 \cdot 3 \cdot 2a_4 x + 5 \cdot 4 \cdot 3a_5 x^2 + \cdots$$

$$f^{(4)}(x) \ = \ \qquad\qquad\qquad\qquad\qquad\qquad 4 \cdot 3 \cdot 2 \cdot 1 a_4 \ + 5 \cdot 4 \cdot 3 \cdot 2a_5 x + \cdots$$

$$f^{(5)}(x) \ = \ \qquad\qquad\qquad\qquad\qquad\qquad\qquad\qquad 5 \cdot 4 \cdot 3 \cdot 2 \cdot 1 a_5 + \cdots$$

If we let $x = 0$, all the terms on the right collapse to zero, except for the first in each row. Thus $f(0) = a_0$, $f'(0) = a_1$, $f''(0) = 2 \cdot 1 a_2$, $f'''(0) = 3 \cdot 2 \cdot 1 a_3$, $f^{(4)}(0) = 4 \cdot 3 \cdot 2 \cdot 1 a_4$, and $f^{(5)}(0) = 5 \cdot 4 \cdot 3 \cdot 2 \cdot 1 a_5$. Solving for the constants and recalling that

$$n! = n(n-1)(n-2) \cdots 2 \cdot 1$$

we have

$$a_0 = f(0), \ a_1 = f'(0), \ a_2 = \frac{f''(0)}{2!}, \ldots, \ a_5 = \frac{f^{(5)}(0)}{5!}$$

The pattern is now clear:

$$a_n = \frac{f^{(n)}(0)}{n!}$$

Finally, after substituting in series (26.11) we get the desired form of the *Maclaurin series* of $f(x)$.

Maclaurin Series of f(x)

$$f(x) = f(0) + f'(0)x + \frac{f''(0)}{2!} x^2 + \frac{f'''(0)}{3!} x^3$$

$$+ \cdots + \frac{f^{(n)}(0)}{n!} x^n + \cdots \qquad\qquad (26.12)$$

The Maclaurin series is named after Colin Maclaurin (Scottish mathematician, 1698–1746). Maclaurin made many contributions to geometry, particularly to the development of higher algebraic curves. It is ironic that his name is now attached to a series which is only a special case of the *Taylor series* (Section 26.5). The latter series was published by Brook Taylor (English mathematician, 1685–1731) in 1715 (long before Maclaurin's work) but was known earlier to Johann Bernoulli.

The Maclaurin series can be written in particularly elegant form if we define $0! = 1$.

Maclaurin Series (Sigma Form)

$$f(x) = \sum_{n=0}^{\infty} \frac{f^{(n)}(0)}{n!} x^n \qquad\qquad (26.13)$$

where $f^{(0)}(x) = f(x)$ and $0! = 1$.

(The Maclaurin series of a function is always unique.)

E X A M P L E **1** Expand $f(x) = e^x$ in a Maclaurin series.

Solution. We differentiate first and let $x = 0$:

$$
\begin{aligned}
f(x) &= e^x & f(0) &= 1 \\
f'(x) &= e^x & f'(0) &= 1 \\
f''(x) &= e^x & f''(0) &= 1 \\
f'''(x) &= e^x & f'''(0) &= 1 \\
&\vdots & &\vdots \\
\text{and so on} & & \text{and so on} &
\end{aligned}
$$

Direct substitution in series (26.12) yields

$$e^x = 1 + 1 \cdot x + \frac{1}{2!} x^2 + \frac{1}{3!} x^3 + \cdots + \frac{1}{n!} x^n + \cdots$$

or

$$e^x = 1 + x + \frac{x^2}{2!} + \frac{x^3}{3!} + \cdots + \frac{x^n}{n!} + \cdots$$

Suppose we take a peak ahead to Section 26.5 and replace x by 1; then

$$e = 1 + 1 + \frac{1}{2!} + \frac{1}{3!} + \cdots + \frac{1}{n!} + \cdots$$

Using the convention $0! = 1$, we now get the following elegant representation of the number e:

$$e = \sum_{n=0}^{\infty} \frac{1}{n!} \qquad\qquad \blacktriangleleft$$

E X A M P L E **2**   (Optional) Show that the series

$$e^x = 1 + x + \frac{x^2}{2!} + \frac{x^3}{3!} + \cdots + \frac{x^n}{n!} + \cdots$$

is convergent for all x.

Solution. Convergence may be proved by the ratio test. Since

$$a_n = \frac{x^n}{n!} \quad \text{and} \quad a_{n+1} = \frac{x^{n+1}}{(n+1)!} = \frac{x^n \cdot x}{(n+1)n!}$$

we have

$$\lim_{n \to \infty} \left| \frac{a_{n+1}}{a_n} \right| = \lim_{n \to \infty} \left| \frac{x^n \cdot x}{(n+1)n!} \cdot \frac{n!}{x^n} \right|$$

$$= |x| \lim_{n \to \infty} \frac{1}{n+1} = 0 = L < 1$$

Since $L < 1$ no matter what value we choose for x, the series is convergent for all x by the ratio test.   ◀

E X A M P L E **3**   Find the Maclaurin expansion of the function $f(x) = \cos 2x$.

Solution. As before, we make a list of derivatives and let $x = 0$:

$$f(x) = \cos 2x \qquad\qquad f(0) = 1$$
$$f'(x) = -2 \sin 2x \qquad\quad f'(0) = 0$$
$$f''(x) = -2^2 \cos 2x \qquad\; f''(0) = -2^2$$
$$f'''(x) = 2^3 \sin 2x \qquad\quad f'''(0) = 0$$
$$f^{(4)}(x) = 2^4 \cos 2x \qquad\; f^{(4)}(0) = 2^4$$
$$f^{(5)}(x) = -2^5 \sin 2x \qquad f^{(5)}(0) = 0$$
$$f^{(6)}(x) = -2^6 \cos 2x \qquad f^{(6)}(0) = -2^6$$
$$\vdots \qquad\qquad\qquad\qquad \vdots$$
$$\text{and so on} \qquad\qquad \text{and so on}$$

Substitution in (26.12) yields the desired series:

$$\cos 2x = 1 + 0x + \frac{-2^2}{2!}x^2 + \frac{0}{3!}x^3 + \frac{2^4}{4!}x^4 + \frac{0}{5!}x^5 + \frac{-2^6}{6!}x^6 + \cdots$$

or

$$\cos 2x = 1 - \frac{2^2}{2!}x^2 + \frac{2^4}{4!}x^4 - \frac{2^6}{6!}x^6 + \cdots$$

◀

The following expansions are particularly important and are listed for later reference. (The first has already been obtained and the rest are left as exercises.)

$$e^x = 1 + x + \frac{x^2}{2!} + \frac{x^3}{3!} + \frac{x^4}{4!} + \cdots \qquad \text{(for all } x\text{)} \qquad (26.14)$$

$$\sin x = x - \frac{x^3}{3!} + \frac{x^5}{5!} - \cdots \qquad \text{(for all } x\text{)} \qquad (26.15)$$

$$\cos x = 1 - \frac{x^2}{2!} + \frac{x^4}{4!} - \cdots \qquad \text{(for all } x\text{)} \qquad (26.16)$$

$$\ln(1 + x) = x - \frac{x^2}{2} + \frac{x^3}{3} - \frac{x^4}{4} + \cdots \qquad (-1 < x \le 1) \qquad (26.17)$$

E X E R C I S E S / S E C T I O N 26.3

In Exercises 1–13, verify the Maclaurin series expansions.

1. $\sin x = x - \dfrac{x^3}{3!} + \dfrac{x^5}{5!} - \cdots$

2. $\cos x = 1 - \dfrac{x^2}{2!} + \dfrac{x^4}{4!} - \cdots$

3. $\sin 2x = 2x - \dfrac{2^3}{3!}x^3 + \dfrac{2^5}{5!}x^5 - \cdots$

4. $\cos 3x = 1 - \dfrac{3^2}{2!}x^2 + \dfrac{3^4}{4!}x^4 - \cdots$

5. $e^{-x} = 1 - x + \dfrac{x^2}{2!} - \dfrac{x^3}{3!} + \dfrac{x^4}{4!} - \cdots$

6. $e^{2x} = 1 + 2x + \dfrac{2^2}{2!}x^2 + \dfrac{2^3}{3!}x^3 + \dfrac{2^4}{4!}x^4 + \cdots$

7. $\ln(1 + x) = x - \dfrac{x^2}{2} + \dfrac{x^3}{3} - \dfrac{x^4}{4} + \cdots$

8. $\ln(1 - x) = -x - \dfrac{x^2}{2} - \dfrac{x^3}{3} - \dfrac{x^4}{4} - \cdots$

9. $\sinh x = \dfrac{1}{2}(e^x - e^{-x}) = x + \dfrac{x^3}{3!} + \dfrac{x^5}{5!} + \cdots$

10. $\cosh x = \dfrac{1}{2}(e^x + e^{-x}) = 1 + \dfrac{x^2}{2!} + \dfrac{x^4}{4!} + \cdots$

11. $\text{Arctan } x = x - \dfrac{x^3}{3} + \dfrac{x^5}{5} - \cdots$

12. $\tan x = x + \dfrac{x^3}{3} + \dfrac{2x^5}{15} + \cdots$

13. (Optional) $\text{Arcsin } x = x + \dfrac{1 \cdot x^3}{2 \cdot 3} + \dfrac{1 \cdot 3 \cdot x^5}{2 \cdot 4 \cdot 5} + \dfrac{1 \cdot 3 \cdot 5 x^7}{2 \cdot 4 \cdot 6 \cdot 7}$
$$+ \dfrac{1 \cdot 3 \cdot 5 \cdot 7 x^9}{2 \cdot 4 \cdot 6 \cdot 8 \cdot 9} + \cdots$$

14. (Optional) Show that the series (26.15) and (26.16) converge for all x.

15. Verify the series expansion

$$(1 - x)^{-2} = \sum_{n=0}^{\infty} (n + 1)x^n$$

by **(a)** using the binomial theorem; **(b)** finding the Maclaurin series expansion; **(c)** dividing out $1/(1 - x)^2$.

26.4 Operations with Series

In this section we will study a number of operations that yield new series from series already known.

E X A M P L E **1** Find the Maclaurin series for $\sin x^2$.

Solution. Consider the series

$$\sin x = x - \frac{x^3}{3!} + \frac{x^5}{5!} - \cdots$$

from the last section. If we replace x by x^2, we obtain

$$\sin x^2 = x^2 - \frac{(x^2)^3}{3!} + \frac{(x^2)^5}{5!} - \cdots$$

or

$$\sin x^2 = x^2 - \frac{x^6}{3!} + \frac{x^{10}}{5!} - \cdots$$

Since this series is a power series, it must be the Maclaurin series of $\sin x^2$, since such expansions are unique. ◄

E X A M P L E **2** Find the Maclaurin series for xe^x.

Solution. From

$$e^x = 1 + x + \frac{x^2}{2!} + \frac{x^3}{3!} + \cdots$$

we obtain by direct multiplication

$$xe^x = x\left(1 + x + \frac{x^2}{2!} + \frac{x^3}{3!} + \cdots\right)$$

$$= x + x^2 + \frac{x^3}{2!} + \frac{x^4}{3!} + \cdots$$

◄

It has already been noted that convergent power series can be differentiated termwise; the same is true of integration.

E X A M P L E **3**   Show that $(d/dx)e^x = e^x$ by the use of Maclaurin series.

Solution.

$$\frac{d}{dx} e^x = \frac{d}{dx} \left(1 + x + \frac{x^2}{2!} + \frac{x^3}{3!} + \frac{x^4}{4!} + \cdots \right)$$

$$= 0 + 1 + \frac{2x}{2!} + \frac{3x^2}{3!} + \frac{4x^3}{4!} + \cdots$$

$$= 1 + x + \frac{x^2}{2!} + \frac{x^3}{3!} + \cdots = e^x$$

◄

E X A M P L E **4**   Find the Maclaurin series of Arctan x by integrating (d/dx) Arctan x termwise.

Solution. We recall that

$$\frac{d}{dx} \text{Arctan } x = \frac{1}{1 + x^2}$$

This expression can be written as a geometric series. Let $r = -x^2$ and $a = 1$; then

$$1 - x^2 + x^4 - x^6 + \cdots = \frac{1}{1 - (-x^2)} = \frac{1}{1 + x^2} \qquad S = \frac{a}{1 - r}$$

Consequently,

$$\text{Arctan } x = \int_0^x \frac{dx}{1 + x^2} = \int_0^x (1 - x^2 + x^4 - x^6 + \cdots) \, dx$$

$$= x - \frac{x^3}{3} + \frac{x^5}{5} - \frac{x^7}{7} + \cdots \Big|_0^x = x - \frac{x^3}{3} + \frac{x^5}{5} - \frac{x^7}{7} + \cdots$$

(It is actually poor practice to use x both for the variable of integration and for the upper limit, but the steps are much easier to see this way.)

   Remark: Our main application of the integration of series will be discussed in the next section.   ◄

   The four fundamental operations—addition, subtraction, multiplication, and division—can theoretically be carried out with series. Two of these operations are demonstrated in the following examples.

E X A M P L E **5**   Find the power-series expansion of $e^x \sin x$ by multiplying the series for e^x and $\sin x$.

Solution. We first recall that

$$\sin x = x - \frac{x^3}{6} + \frac{x^5}{120} - \cdots$$

and

$$e^x = 1 + x + \frac{x^2}{2} + \frac{x^3}{6} + \frac{x^4}{24} + \frac{x^5}{120} + \cdots$$

We may now multiply each term in the second series by each term in the first series in exactly the same way that we multiply polynomials. If we decide to carry only powers up to the fifth power, we obtain

$$e^x = 1 + x + \frac{x^2}{2} + \frac{x^3}{6} + \frac{x^4}{24} + \frac{x^5}{120} + \text{higher powers}$$

$$\sin x = x - \frac{x^3}{6} + \frac{x^5}{120} + \text{higher powers}$$

$$\overline{}$$

$$x + x^2 + \frac{x^3}{2} + \frac{x^4}{6} + \frac{x^5}{24} + \cdots \qquad \textbf{multiplying}$$

$$- \frac{x^3}{6} - \frac{x^4}{6} - \frac{x^5}{12} + \cdots$$

$$+ \frac{x^5}{120} + \cdots$$

$$\overline{}$$

$$x + x^2 + \frac{x^3}{3} \qquad - \frac{x^5}{30} + \cdots$$

We now conclude that, up to the fifth power,

$$e^x \sin x = x + x^2 + \frac{1}{3}x^3 - \frac{1}{30}x^5 + \cdots \qquad \blacktriangleleft$$

E   X   A   M   P   L   E   **6**   Use the series in Exercises 7 and 8 of the last section to expand

$$\ln \frac{1+x}{1-x}$$

Solution. We have

$$\ln(1 + x) = x - \frac{x^2}{2} + \frac{x^3}{3} - \frac{x^4}{4} + \frac{x^5}{5} - \cdots$$

and

$$\ln(1 - x) = -x - \frac{x^2}{2} - \frac{x^3}{3} - \frac{x^4}{4} - \frac{x^5}{5} - \cdots$$

Hence

$$\ln\frac{1 + x}{1 - x} = \ln(1 + x) - \ln(1 - x)$$

$$= \left(x - \frac{x^2}{2} + \frac{x^3}{3} - \frac{x^4}{4} + \frac{x^5}{5} - \cdots\right)$$

$$- \left(-x - \frac{x^2}{2} - \frac{x^3}{3} - \frac{x^4}{4} - \frac{x^5}{5} - \cdots\right)$$

$$= 2\left(x + \frac{x^3}{3} + \frac{x^5}{5} + \cdots\right) \qquad \blacktriangleleft$$

As a final exercise we are going to uncover a relationship between three of our transcendental functions by making use of the basic imaginary unit $j = \sqrt{-1}$. As a starting point, notice that the expansion of the sine function has only odd powers and that of the cosine function only even powers. However, all the powers occur in the expansion of e^x; so e^x comes very close to being the sum of the other two—if only the signs matched! Now, by introducing j formally, we find that

$$e^{jx} = 1 + jx + \frac{j^2 x^2}{2!} + \frac{j^3 x^3}{3!} + \frac{j^4 x^4}{4!} + \frac{j^5 x^5}{5!} + \cdots$$

$$= 1 + jx - \frac{x^2}{2!} - \frac{jx^3}{3!} + \frac{x^4}{4!} + \frac{jx^5}{5!} - \cdots \qquad \begin{array}{l} j = j, \ j^2 = -1 \\ j^3 = j^2 j = -j, \ j^4 = j^3 j = 1 \end{array}$$

$$= 1 - \frac{x^2}{2!} + \frac{x^4}{4!} - \cdots + j\left(x - \frac{x^3}{3!} + \frac{x^5}{5!} - \cdots\right)$$

$$= \cos x + j \sin x$$

The resulting formula is known as *Euler's identity* after the Swiss mathematician Leonhard Euler (1707–1783).

Euler's Identity

$$e^{jx} = \cos x + j \sin x \qquad\qquad (26.18)$$

Euler's identity arises in the study of differential equations, as we will see in Chapter 28.

Although there is some room for opinion, it can be argued that the most interesting numbers in mathematics are 0, 1, j, e, and π. By Euler's identity,

$$e^{j\pi} = \cos \pi + j \sin \pi = -1$$

or

$$e^{j\pi} + 1 = 0 \tag{26.19}$$

which involves all five of these numbers. This astounding relationship has been called the eutectic point of mathematics, for no matter how you try to analyze it, it seems to retain an air of mystery not easily explained away.

E X E R C I S E S / S E C T I O N   **26.4**

In Exercises 1–10, use the method of Examples 1 and 2 to find the Maclaurin series of the functions.

1. $\sin 3x$

2. $\cos 2x$

3. e^{-x}

4. e^{-x^2}

5. $\cos \sqrt{x}$

6. $\dfrac{\sin x^2}{x}$

7. $x \cos x$

8. $x^2 e^x$

9. $\ln(1 + x^2)$

10. $\ln(1 - x)$

11. Show that $(d/dx) \sin x = \cos x$ by use of the Maclaurin series. (See Example 3.)

12. Show that $(d/dx) \cos x = -\sin x$ by use of the Maclaurin series.

13. Use the method of Example 4 to find the Maclaurin series of $\ln(1 + x)$.

14. Expand $(\sin x - x)/x^2$ in a Maclaurin series.

15. Use the method of Example 5 to find the Maclaurin series of $e^{-x} \cos x$.

16. Use the method of Example 6 to find the Maclaurin series of $\ln(1 + x)^2$.

17. Expand the function $\ln(1 + x^2)^3$ in a Maclaurin series.

18. Find the Maclaurin series of $\frac{1}{2}(e^x + e^{-x})$ by addition of series.

19. Find the Maclaurin series of $\ln(1 + x) + \text{Arctan } x$.

A complex number $a + bj$ can be written in polar form $r(\cos \theta + j \sin \theta)$, which by Euler's identity becomes $re^{j\theta}$, known as the *exponential form*. In Exercises 20–25, change the complex numbers to exponential form.

20. $1 + j$

21. $-\sqrt{3} + j$

22. $1 - \sqrt{3}j$

23. $3j$

24. $-4j$

25. $-2 + 2j$

26.5   Computations with Series

In this section we are going to do numerical computations by means of power series. By using a sufficiently large number of terms, we can obtain the value of some transcendental functions to any desired degree of accuracy. A particularly important application of these numerical techniques is the evaluation of certain definite integrals.

Before we consider computations involving series, we need to make a few additional observations about series of constants. Suppose that $a_1, a_2, a_3, \ldots,$ $a_n, \ldots$ is a sequence of positive numbers such that each number is less than the preceding one—that is, $a_{n+1} < a_n$ for all n—and consider the series

$$\sum_{n=1}^{\infty} (-1)^{n+1} a_n = a_1 - a_2 + a_3 - a_4 + \cdots + (-1)^{n+1} a_n + \cdots \tag{26.20}$$

Alternating series

called an **alternating series** since the signs alternate. If the series converges, then the sum may be obtained to any desired degree of accuracy by adding the first n terms and estimating the error from the $(n + 1)$st term. To check this statement, suppose we add the first four terms of series (26.20) and estimate the error by writing the series as follows:

$$(a_1 - a_2 + a_3 - a_4) + a_5 - (a_6 - a_7) - (a_8 - a_9) - \cdots$$

Since the numbers a_n are decreasing,

$$(a_6 - a_7) > 0, (a_8 - a_9) > 0, \text{ and so forth}$$

Hence

$$a_5 - (a_6 - a_7) - (a_8 - a_9) - \cdots < a_5$$

So, by adding $a_1 - a_2 + a_3 - a_4$, the error made is less than a_5.

If we wish to add the first five terms, then the error is estimated from

$$(a_1 - a_2 + a_3 - a_4 + a_5) - a_6 + (a_7 - a_8) + (a_9 - a_{10}) + \cdots$$

Again

$$(a_7 - a_8) > 0, (a_9 - a_{10}) > 0, \text{ and so on}$$

so that the error is no worse than $-a_6$.

The **error** made by adding the first n terms of a **convergent alternating series**

$$a_1 - a_2 + a_3 - a_4 + \cdots \qquad (a_n > 0, a_{n+1} < a_n)$$

is numerically less than the first term omitted.

(We state without proof that an alternating series converges if $\lim_{n \to \infty} a_n = 0$.)

E X A M P L E **1**   Compute the value of $e^{-0.1}$ by using the first four terms of the expansion of e^x. Find the maximum possible error and determine the accuracy of the result.

Solution. We let $x = -0.1$ in the series

$$e^x = 1 + x + \frac{x^2}{2!} + \frac{x^3}{3!} + \frac{x^4}{4!} + \frac{x^5}{5!} + \cdots$$

and find the sum of the first four terms:

$$e^{-0.1} = 1 - 0.1 + \frac{(0.1)^2}{2!} - \frac{(0.1)^3}{3!} = 0.904833$$

The sequence is

$$1 \;\boxed{-}\; .1 \;\boxed{+}\; .1 \;\boxed{y^x}\; 2 \;\boxed{\div}\; 2 \;\boxed{!}\; \boxed{-}\; .1 \;\boxed{y^x}\; 3 \;\boxed{\div}\; 3 \;\boxed{!}\; \boxed{=}$$

The error made is no worse than the fifth term:

$$+\frac{(0.1)^4}{4!} = +0.000004 \qquad \begin{array}{r} 0.904833 \\ +\ 0.000004 \\ \hline 0.904837 \end{array}$$

Based on these calculations, the correct value to six decimal places could be any one of the following: 0.904833, 0.904834, ..., or 0.904837. Consequently,

$$e^{-0.1} = 0.9048 \qquad \text{(accurate to four decimal places)} \qquad \blacktriangleleft$$

To see the relationship between e^x and its Maclaurin expansion, let us examine the graphs of the following approximations:

$$y = 1 + x \qquad y = 1 + x + \frac{x^2}{2} \qquad y = 1 + x + \frac{x^2}{2} + \frac{x^3}{6}$$

(See Figure 26.4.)

According to Figure 26.4, the approximation improves as we include more and more terms in the sum, producing a particularly good fit near the origin. As a result, for values near zero the Maclaurin series leads to a good approximation with very few terms, as Example 1 confirms. For values away from the

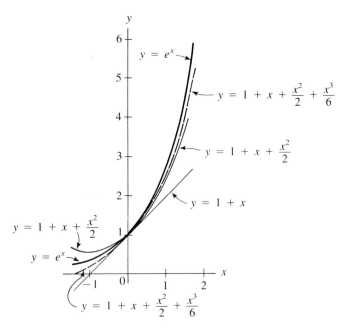

Figure 26.4

origin, a good approximation can be obtained with few terms by means of the Taylor series discussed at the end of the section.

E X A M P L E **2** Find an infinite-series representation of **(a)** e; **(b)** π.

Solution.

a. The representation of e was already obtained in Section 26.3:

$$e = \sum_{n=0}^{\infty} \frac{1}{n!}$$

b. Since

$$\text{Arctan } x = x - \frac{x^3}{3} + \frac{x^5}{5} - \cdots$$

we let $x = 1$ and obtain

$$\text{Arctan } 1 = \frac{\pi}{4} = 1 - \frac{1}{3} + \frac{1}{5} - \frac{1}{7} + \frac{1}{9} - \cdots$$

or

$$\pi = 4\left(1 - \frac{1}{3} + \frac{1}{5} - \frac{1}{7} + \frac{1}{9} - \cdots\right)$$

Although it is a striking relationship, series (b) does not provide us with a practical method of computing π, since the series converges too slowly. A better way is by means of the series for $f(x) = \text{Arcsin } x$ (Exercise 11). ◄

As noted in Chapter 25 (Section 25.10), many functions do not possess elementary antiderivatives. Since a power series can be integrated termwise, many such integrals can be worked out by means of Maclaurin series, leading to *nonelementary functions*. Study the next example and remark.

E X A M P L E **3** Find the approximate value of

$$\int_0^1 \frac{\sin x^2 \, dx}{x}$$

Solution. At $x = 0$ the integrand takes on the indeterminate form $\frac{0}{0}$. Now, by L'Hospital's rule,

$$\lim_{x \to 0} \frac{\sin x^2}{x} = \lim_{x \to 0} \frac{2x \cos x^2}{1} = 0$$

so that the function is bounded on $(0, 1)$. (Otherwise we would be dealing with an improper integral.) From

$$\sin x = x - \frac{x^3}{3!} + \frac{x^5}{5!} - \frac{x^7}{7!} + \cdots$$

we have

$$\sin x^2 = x^2 - \frac{x^6}{3!} + \frac{x^{10}}{5!} - \frac{x^{14}}{7!} + \cdots \qquad \text{replacing } x \text{ by } x^2$$

and

$$\frac{\sin x^2}{x} = x - \frac{x^5}{3!} + \frac{x^9}{5!} - \frac{x^{13}}{7!} + \cdots \qquad \text{dividing by } x$$

Hence

$$\int_0^1 \frac{\sin x^2 \, dx}{x} = \int_0^1 \left(x - \frac{x^5}{3!} + \frac{x^9}{5!} - \frac{x^{13}}{7!} + \cdots \right) dx$$

$$= \frac{x^2}{2} - \frac{x^6}{6 \cdot 3!} + \frac{x^{10}}{10 \cdot 5!} - \frac{x^{14}}{14 \cdot 7!} + \cdots \Big|_0^1 \qquad \text{integrating}$$

$$= \frac{1}{2} - \frac{1}{6 \cdot 3!} + \frac{1}{10 \cdot 5!} - \frac{1}{14 \cdot 7!} + \cdots$$

Adding the first three terms with a calculator, we get

$$\int_0^1 \frac{\sin x^2 \, dx}{x} = 0.473056$$

The sequence is

$$.5 \boxed{-} 1 \boxed{\div} 6 \boxed{\div} 3 \boxed{!} \boxed{+} 1 \boxed{\div} 10 \boxed{\div} 5 \boxed{!} \boxed{=}$$

The error is numerically less than the fourth term:

$$-\frac{1}{14 \cdot 7!} = -0.000014 \qquad \begin{array}{r} 0.473056 \\ -\ 0.000014 \\ \hline 0.473042 \end{array}$$

So the value to five decimal places is 0.47304, 0.47305, or 0.47306. Consequently,

$$\int_0^1 \frac{\sin x^2 \, dx}{x} = 0.473 \qquad \text{(accurate to three decimal places)}$$

◀

Remark. The integral

$$\int \frac{\sin x^2}{x}\,dx$$

in Example 3 is nonelementary—that is, there does not exist an elementary function whose derivative is $(\sin x^2)/x$. The infinite-series form in Example 3,

$$\int \frac{\sin x^2\,dx}{x} = \frac{x^2}{2} - \frac{x^6}{6\cdot 3!} + \frac{x^{10}}{10\cdot 5!} - \frac{x^{14}}{14\cdot 7!} + \cdots$$

is an example of a *nonelementary function*. The extension of the function concept to include nonelementary functions is an important application of our study of infinite series.

Taylor Series

We observed in Example 1 that the Maclaurin series yields a good approximation with only a few terms, provided that x is close to zero. If x is numerically large, the number of terms needed for a good approximation may become prohibitive. We can get around this problem by generalizing the form of the series expansion as follows:

$$f(x) = a_0 + a_1(x - c) + a_2(x - c)^2 + \cdots + a_n(x - c)^n + \cdots$$

To compute the coefficients, we proceed as we did with the Maclaurin series: we find the derivatives of $f(x)$ and let $x = c$. Then the nth coefficient turns out to be

$$a_n = \frac{f^{(n)}(c)}{n!}$$

and the series becomes

$$f(x) = f(c) + f'(c)(x - c) + \frac{f''(c)}{2!}(x - c)^2 + \cdots + \frac{f^{(n)}(c)}{n!}(x - c)^n + \cdots$$

which is called the *Taylor series* of $f(x)$ in powers of $x - c$. We also say that $f(x)$ has been expanded about $x - c$.

Taylor Series of f(x)

$$f(x) = f(c) + f'(c)(x - c) + \frac{f''(c)}{2!}(x - c)^2 + \cdots$$

$$+ \frac{f^{(n)}(c)}{n!}(x - c)^n + \cdots \qquad (26.21)$$

(If $c = 0$, then the Taylor series reduces to the Maclaurin series.)

Although c may have any value, in practice one selects a value that is particularly convenient, as we will see in the next example.

E X A M P L E **4**   Calculate $\sin 61°$ by means of the Taylor series.

Solution. The closest convenient value that we may select for c is $60° = \pi/3$:

$$f(x) = \sin x \qquad\qquad f(\pi/3) = \sqrt{3}/2$$

$$f'(x) = \cos x \qquad\qquad f'(\pi/3) = \frac{1}{2}$$

$$f''(x) = -\sin x \qquad\qquad f''(\pi/3) = -\sqrt{3}/2$$

$$f'''(x) = -\cos x \qquad\qquad f'''(\pi/3) = -\frac{1}{2}$$

$$f^{(4)}(x) = \sin x \qquad\qquad f^{(4)}(\pi/3) = \sqrt{3}/2$$

$$\vdots \qquad\qquad\qquad \vdots$$

and so on   and so on

Substituting in series (26.21), we get

$$\sin x = \frac{\sqrt{3}}{2} + \frac{1}{2}\left(x - \frac{\pi}{3}\right) - \frac{\sqrt{3}}{2}\frac{1}{2!}\left(x - \frac{\pi}{3}\right)^2 - \frac{1}{2}\frac{1}{3!}\left(x - \frac{\pi}{3}\right)^3$$
$$+ \frac{\sqrt{3}}{2}\frac{1}{4!}\left(x - \frac{\pi}{3}\right)^4 + \cdots$$

The reason for choosing $c = \pi/3$ now becomes clear: since $61° = 60° + 1° = \pi/3 + \pi/180$, we get

$$x - c = \left(\frac{\pi}{3} + \frac{\pi}{180}\right) - \frac{\pi}{3} = \frac{\pi}{180}$$

Consequently, the numerical value of $x - c$ is small and so the terms in the series become small as well. Using four terms, we get

$$\sin 61° = \frac{\sqrt{3}}{2} + \frac{1}{2}\left(\frac{\pi}{180}\right) - \frac{\sqrt{3}}{2}\frac{1}{2!}\left(\frac{\pi}{180}\right)^2 - \frac{1}{2}\frac{1}{3!}\left(\frac{\pi}{180}\right)^3 = 0.874620 \qquad \blacktriangleleft$$

E X E R C I S E S / S E C T I O N **26.5**

In Exercises 1–10, find the indicated function values by means of Maclaurin series, using the number of terms indicated. Find the maximum possible error, and determine the accuracy of the result. (See Example 1.)

1. $\sin 0.7$ (3 terms)

2. $\ln 1.5$ (10 terms)

3. $\cos 10°$ (2 terms)

4. $\cos 20°$ (3 terms)

5. $e^{-0.2}$ (4 terms)

6. $\sin 35°$ (3 terms)

7. $\cos 1.2$ (4 terms)

8. $e^{-0.6}$ (6 terms)

9. $\ln 1.1$ (3 terms)

10. Arctan 0.2 (2 terms)

11. Use the expansion of Arcsin x (Exercise 13, Section 26.3) to compute π. (Let $x = \frac{1}{2}$ and use 3 terms.)

12. Compute e by using 10 terms.

In Exercises 13–18, evaluate the integrals to an accuracy of five decimal places. (See Example 3.)

13. $\int_0^{1/2} \dfrac{1 - \cos x}{x}\, dx$

14. $\int_0^1 \dfrac{\sin x - x}{x^2}\, dx$

15. $\int_0^1 \cos\sqrt{x}\, dx$

16. $\int_0^{0.1} \sqrt{x}\, \sin x\, dx$

17. $\int_0^{0.3} e^{-x^2}\, dx$

18. $\int_0^{0.6} \sin x^2\, dx$

19. Evaluate $\sin 29°$ by expanding $f(x) = \sin x$ about $c = \pi/6$, using 3 terms. (See Example 4.)

20. Evaluate $\cos 50°$ by expanding $f(x) = \cos x$ about $c = \pi/4$, using 4 terms.

21. Evaluate $\cos 31°$ by expanding $f(x) = \cos x$ about $c = \pi/6$, using 4 terms.

22. Evaluate $\sin 64°$ by expanding $f(x) = \sin x$ about $c = \pi/3$, using 3 terms.

23. Evaluate $\cos 58°$ by expanding $f(x) = \cos x$ about $c = \pi/3$, using 3 terms.

24. Evaluate $\sin 44°$ by expanding $f(x) = \sin x$ about $c = \pi/4$, using 4 terms.

25. Show that

$$\ln x = (x - 1) - \frac{1}{2}(x - 1)^2 + \frac{1}{3}(x - 1)^3 - \cdots$$

26. Prove that for $x \ge 0$

$$e^x \ge 1 + x + \frac{1}{2}x^2$$

27. Show that the polynomial $5x^5 + 10x^4 - 2x^3 + x^2 + 5$ can be written

$$5(x - 1)^5 + 35(x - 1)^4 + 88(x - 1)^3 + 105(x - 1)^2$$
$$+ 61(x - 1) + 19$$

by using the Taylor series.

28. Find $\ln 3$ by using the series expansion of

$$\ln \frac{1 + x}{1 - x}$$

(Example 6, Section 26.4) with $x = \frac{1}{2}$.

29. a. Use the Maclaurin expansion of $\sin \theta$ to show that $\sin \theta \approx \theta$ for small θ.

b. A simple pendulum consists of a mass m hanging on a light string of length L. If θ is the angle made by the string with the vertical, then θ satisfies the relationship

$$mL\frac{d^2\theta}{dt^2} = -mg \sin \theta$$

where g is the acceleration due to gravity. Use part (a) to show that if θ is sufficiently small, this equation represents a system oscillating with simple harmonic motion. (See Equation (24.29) in Section 24.10.)

c. Show that for small oscillations the period P of a pendulum is

$$P = 2\pi \sqrt{\frac{L}{g}}$$

30. The following sum was first obtained by the Swiss mathematician Leonhard Euler:

$$\sum_{n=1}^{\infty} \frac{1}{n^2} = \frac{\pi^2}{6}$$

Use this sum and Exercise 25 to show that

$$\int_0^1 \frac{\ln x}{x - 1}\, dx = \frac{\pi^2}{6}$$

(This integral is nonelementary and improper.)

31. The function

$$f(x) = \frac{1}{\sqrt{2\pi}}\, e^{(-1/2)x^2}$$

which is called the *standard normal distribution*, was first studied in the eighteenth century when scientists observed that the distribution of errors of measurement is closely approximated by this curve. Calculate to four decimal places the integral

$$P(0 \le x \le 1) = \frac{1}{\sqrt{2\pi}} \int_0^1 e^{(-1/2)x^2}\, dx$$

which is the probability that a measurement x having the standard normal distribution will take on a value between 0 and 1.

32. The charge q on a certain capacitor is

$$q = \int_0^{0.50} \left(1.00 - e^{-0.10t^2}\right) dt$$

Evaluate this integral.

26.6   Fourier Series

In this section we will consider a different kind of series expansion, the **Fourier series**, which expresses a given function in terms of sines and cosines (instead of x^n as in the Maclaurin series). Such expansions are highly useful in the study of periodic phenomena. Moreover, the Fourier series is much less demanding: for example, it is not required that the function to be expanded be differentiable or even continuous.

Let f be the given periodic function with period $2p$ and consider the following expansion:

$$f(t) = \frac{1}{2} a_0 + a_1 \cos \frac{\pi t}{p} + a_2 \cos \frac{2\pi t}{p} + \cdots + a_n \cos \frac{n\pi t}{p} + \cdots$$

$$+ b_1 \sin \frac{\pi t}{p} + b_2 \sin \frac{2\pi t}{p} + \cdots + b_n \sin \frac{n\pi t}{p} + \cdots \qquad (26.22)$$

(Bowing to the usual convention, we will use t for the independent variable.) As in the case of the Maclaurin series, our main task is to find the coefficients in expansion (26.22). To do so, we need the following definite integrals:

$$\int_{-p}^{p} \cos \frac{n\pi t}{p} \, dt = 0 \qquad (n \neq 0) \qquad\qquad\qquad (26.23)$$

$$\int_{-p}^{p} \sin \frac{n\pi t}{p} \, dt = 0 \qquad\qquad\qquad\qquad\qquad (26.24)$$

$$\int_{-p}^{p} \cos \frac{m\pi t}{p} \cos \frac{n\pi t}{p} \, dt = 0 \qquad (m \neq n) \qquad\qquad (26.25)$$

$$\int_{-p}^{p} \cos^2 \frac{n\pi t}{p} \, dt = p \qquad (n \neq 0) \qquad\qquad\qquad (26.26)$$

$$\int_{-p}^{p} \cos \frac{m\pi t}{p} \sin \frac{n\pi t}{p} \, dt = 0 \qquad\qquad\qquad\qquad (26.27)$$

$$\int_{-p}^{p} \sin \frac{m\pi t}{p} \sin \frac{n\pi t}{p} \, dt = 0 \qquad (m \neq n) \qquad\qquad (26.28)$$

$$\int_{-p}^{p} \sin^2 \frac{n\pi t}{p} \, dt = p \qquad (n \neq 0) \qquad\qquad\qquad (26.29)$$

Integrals (26.23), (26.24), (26.26), and (26.29) can be evaluated routinely by the methods of Chapter 25. The remaining integrals can be obtained by using the table of integrals in Appendix E (Table 4, forms 61–63).

The coefficients in expansion (26.22) can now be calculated by multiplying both sides by $\cos(n\pi t/p)$ and $\sin(n\pi t/p)$ and integrating the result from $t = -p$ to $t = p$. Suppose we try this procedure for $\cos(n\pi t/p)$:

$$\int_{-p}^{p} f(t) \cos \frac{n\pi t}{p} \, dt = \int_{-p}^{p} \frac{a_0}{2} \cos \frac{n\pi t}{p} \, dt$$

$$+ a_1 \int_{-p}^{p} \cos \frac{\pi t}{p} \cos \frac{n\pi t}{p} \, dt$$

$$+ a_2 \int_{-p}^{p} \cos \frac{2\pi t}{p} \cos \frac{n\pi t}{p} \, dt + \cdots$$

$$+ a_n \int_{-p}^{p} \cos \frac{n\pi t}{p} \cos \frac{n\pi t}{p} \, dt + \cdots$$

$$+ b_1 \int_{-p}^{p} \sin \frac{\pi t}{p} \cos \frac{n\pi t}{p} \, dt$$

$$+ b_2 \int_{-p}^{p} \sin \frac{2\pi t}{p} \cos \frac{n\pi t}{p} \, dt + \cdots$$

$$+ b_n \int_{-p}^{p} \sin \frac{n\pi t}{p} \cos \frac{n\pi t}{p} \, dt + \cdots$$

Carefully inspecting all the terms in this series with an eye on the special integrals, we see that only one term survives (the fourth one). Thus

$$\int_{-p}^{p} f(t) \cos \frac{n\pi t}{p} \, dt = a_n p$$

or

$$a_n = \frac{1}{p} \int_{-p}^{p} f(t) \cos \frac{n\pi t}{p} \, dt \qquad (n \neq 0) \tag{26.30}$$

By multiplying both sides of expansion (26.22) by $\sin(n\pi t/p)$ and integrating, we obtain

$$b_n = \frac{1}{p} \int_{-p}^{p} f(t) \sin \frac{n\pi t}{p} \, dt \tag{26.31}$$

To get the remaining term a_0, we simply integrate both sides of expansion (26.22) to find that

$$a_0 = \frac{1}{p} \int_{-p}^{p} f(t) \, dt \tag{26.32}$$

(The coefficient $\frac{1}{2}$ of a_0 in expansion (26.22) was introduced so that formulas (26.32) and (26.30) would have the same form.)

The series (26.22), together with formulas (26.30) through (26.32), is called a **Fourier series**, after J. B. J. Fourier (1768–1830), a French mathematician and physicist who was a pioneer in the study of heat conduction. A confidant of Napoleon, Fourier participated in Napoleon's expedition to Egypt. (A child who met Fourier and viewed his collection of Egyptian antiquities—J. F. Champollion—was inspired by this experience to study Egyptian hieroglyphics and twenty years later succeeded in decoding them.)

Fourier Series: The Fourier series for a function $f(t)$ is

$$f(t) = \frac{a_0}{2} + a_1 \cos \frac{\pi t}{p} + a_2 \cos \frac{2\pi t}{p} + \cdots + a_n \cos \frac{n\pi t}{p} + \cdots$$

$$+ b_1 \sin \frac{\pi t}{p} + b_2 \sin \frac{2\pi t}{p} + \cdots + b_n \sin \frac{n\pi t}{p} + \cdots \qquad (26.33)$$

where the coefficients are given by

$$a_0 = \frac{1}{p} \int_{-p}^{p} f(t)\, dt \qquad\qquad (26.34)$$

$$a_n = \frac{1}{p} \int_{-p}^{p} f(t) \cos \frac{n\pi t}{p}\, dt \qquad (n \neq 0) \qquad (26.35)$$

$$b_n = \frac{1}{p} \int_{-p}^{p} f(t) \sin \frac{n\pi t}{p}\, dt \qquad\qquad (26.36)$$

Since $\sin bx$ and $\cos bx$ are periodic functions with period $2\pi/b$, each term in the series is periodic with period

$$\frac{2\pi}{\dfrac{n\pi}{p}} = 2\pi \cdot \frac{p}{n\pi} = \frac{2p}{n} \qquad (n = 1, 2, \ldots)$$

Since a periodic function with period $2p/n$ is also periodic with period $2p$, **the entire series is periodic and represents a periodic function with period $2p$.** Formulas (26.34) through (26.36) yield the Fourier coefficients only for the interval $[-p, p]$. By the periodicity, however, the representation automatically extends over the entire real line.

Since currents and voltages are often periodic, Fourier series are particularly useful in electronics. As examples, we are going to calculate the series of two functions frequently encountered in this field.

E X A M P L E **1** Obtain the Fourier series of the "square wave" in Figure 26.5.

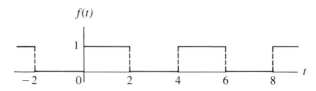

Figure 26.5

Solution. When finding a Fourier series, we use only one period in the calculation. In this problem we use the period from $t = -2$ to $t = +2$, given by the

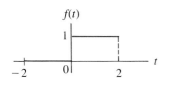

Figure 26.6

function

$$f(t) = \begin{cases} 0 & -2 < t < 0 \\ 1 & 0 < t < 2 \end{cases}$$

(See Figure 26.6.)

Since the period is 4, we have

$$2p = 4 \quad \text{and} \quad p = 2$$

It is important to keep in mind that the first term of the series, $a_0/2$, is always computed separately. We now proceed by direct calculation:

$$a_0 = \frac{1}{p} \int_{-p}^{p} f(t)\,dt \qquad \text{formula (26.34)}$$

Since the function is described differently over different intervals, we need to integrate over the intervals $[-2, 0]$ and $[0, 2]$ separately. Thus

$$a_0 = \frac{1}{2} \int_{-2}^{2} f(t)\,dt \qquad\qquad p = 2$$

$$= \frac{1}{2} \int_{-2}^{0} 0\,dt \qquad\qquad f(t) = 0 \text{ on } (-2, 0)$$

$$+ \frac{1}{2} \int_{0}^{2} 1\,dt \qquad\qquad f(t) = 1 \text{ on } (0, 2)$$

$$= 0 + \frac{1}{2} \int_{0}^{2} dt = \frac{1}{2} t \Big|_{0}^{2} = 1$$

Since $a_0 = 1$, we get for our first term

$$\frac{a_0}{2} = \frac{1}{2}$$

To find a_n, let us evaluate a_1 and a_2 first and then generalize to a_n:

$$a_1 = \frac{1}{p} \int_{-p}^{p} f(t) \cos \frac{1\pi t}{p}\,dt \qquad n = 1 \text{ in formula (26.35)}$$

$$= \frac{1}{2} \int_{-2}^{0} 0\,dt + \frac{1}{2} \int_{0}^{2} 1 \cdot \cos \frac{\pi t}{2}\,dt \qquad p = 2$$

$$= \frac{1}{2} \int_{0}^{2} \cos \frac{\pi t}{2}\,dt$$

$$= \frac{1}{2} \frac{2}{\pi} \sin \frac{\pi t}{2} \Big|_{0}^{2} \qquad\qquad \begin{bmatrix} u = \dfrac{\pi t}{2} \\ du = \dfrac{\pi}{2}\,dt \end{bmatrix}$$

$$= \frac{1}{\pi} \sin \pi = 0$$

Next we evaluate a_2:

$$a_2 = \frac{1}{p} \int_{-p}^{p} f(t) \cos \frac{2\pi t}{p} \, dt \qquad n = 2 \text{ in formula (26.35)}$$

$$= \frac{1}{2} \int_0^2 \cos \pi t \, dt \qquad p = 2$$

$$= \frac{1}{2} \frac{1}{\pi} \sin \pi t \Big|_0^2 \qquad \begin{bmatrix} u = \pi t \\ du = \pi \, dt \end{bmatrix}$$

$$= \frac{1}{2\pi} \sin 2\pi = 0$$

To find the general coefficient a_n, recall the following relationships:

$$\sin n\pi = 0 \qquad \text{(for all } n\text{)}$$

$$\cos n\pi = \begin{cases} 1 & \text{for } n \text{ even} \\ -1 & \text{for } n \text{ odd} \end{cases}$$

$$\cos(-n\pi) = \cos n\pi$$

Thus we have

$$a_n = \frac{1}{p} \int_{-p}^{p} f(t) \cos \frac{n\pi t}{p} \, dt \qquad \text{formula (26.35)}$$

$$= \frac{1}{2} \int_{-2}^0 0 \, dt + \frac{1}{2} \int_0^2 1 \cdot \cos \frac{n\pi t}{2} \, dt \qquad p = 2$$

$$= \frac{1}{2} \int_0^2 \cos \frac{n\pi t}{2} \, dt$$

$$= \frac{1}{2} \frac{2}{n\pi} \sin \frac{n\pi t}{2} \Big|_0^2 \qquad \begin{bmatrix} u = \dfrac{n\pi t}{2} \\ du = \dfrac{n\pi}{2} \, dt \end{bmatrix}$$

$$= \frac{1}{n\pi} \sin n\pi = 0$$

since $\sin n\pi = 0$ for all n.

For our next calculation, we find b_1 and b_2 and then generalize to b_n:

$$b_1 = \frac{1}{p} \int_{-p}^{p} f(t) \sin \frac{1\pi t}{p} \, dt \qquad n = 1 \text{ in formula (26.36)}$$

$$= \frac{1}{2} \int_{-2}^0 0 \, dt + \frac{1}{2} \int_0^2 1 \cdot \sin \frac{\pi t}{2} \, dt \qquad p = 2$$

$$= \frac{1}{2} \int_0^2 \sin \frac{\pi t}{2} \, dt$$

$$= \frac{1}{2} \left(-\frac{2}{\pi} \right) \cos \frac{\pi t}{2} \Big|_0^2 \qquad \qquad \left[\begin{array}{l} u = \dfrac{\pi t}{2} \\[2mm] du = \dfrac{\pi}{2} \, dt \end{array} \right]$$

$$= -\frac{1}{\pi} (\cos \pi - \cos 0)$$

$$= -\frac{1}{\pi} (-1 - 1) = \frac{2}{\pi}$$

$$b_2 = \frac{1}{p} \int_{-p}^{p} f(t) \sin \frac{2\pi t}{p} \, dt \qquad \qquad n = 2 \text{ in formula (26.36)}$$

$$= \frac{1}{2} \int_0^2 \sin \pi t \, dt \qquad \qquad p = 2$$

$$= \frac{1}{2} \left(-\frac{1}{\pi} \right) \cos \pi t \Big|_0^2 \qquad \qquad \left[\begin{array}{l} u = \pi t \\ du = \pi \, dt \end{array} \right]$$

$$= -\frac{1}{2\pi} (\cos 2\pi - \cos 0)$$

$$= -\frac{1}{2\pi} (1 - 1) = 0$$

$$b_n = \frac{1}{p} \int_{-p}^{p} f(t) \sin \frac{n\pi t}{p} \, dt \qquad \qquad \text{formula (26.36)}$$

$$= \frac{1}{2} \int_{-2}^0 0 \, dt + \frac{1}{2} \int_0^2 1 \cdot \sin \frac{n\pi t}{2} \, dt \qquad p = 2$$

$$= \frac{1}{2} \int_0^2 \sin \frac{n\pi t}{2} \, dt$$

$$= \frac{1}{2} \left(-\frac{2}{n\pi} \right) \cos \frac{n\pi t}{2} \Big|_0^2 \qquad \qquad \left[\begin{array}{l} u = \dfrac{n\pi t}{2} \\[2mm] du = \dfrac{n\pi}{2} \, dt \end{array} \right]$$

$$= -\frac{1}{n\pi} (\cos n\pi - \cos 0)$$

$$= -\frac{1}{n\pi} (\cos n\pi - 1)$$

$$= \begin{cases} 0 & \text{for } n \text{ even} \\[2mm] \dfrac{2}{n\pi} & \text{for } n \text{ odd} \end{cases}$$

since $\cos n\pi = 1$ for n even and $\cos n\pi = -1$ for n odd.

We have obtained the following coefficients:

$$\frac{a_0}{2} = \frac{1}{2}, \; a_n = 0, \; b_1 = \frac{2}{\pi}, \; b_2 = 0, \; b_3 = \frac{2}{3\pi}, \; b_4 = 0, \; b_5 = \frac{2}{5\pi}, \text{ and so on}$$

Substituting these values in series (26.33), we get (since $p = 2$)

$$f(t) = \frac{1}{2} + \frac{2}{\pi}\sin\frac{1\pi t}{2} + 0 + \frac{2}{3\pi}\sin\frac{3\pi t}{2} + 0 + \frac{2}{5\pi}\sin\frac{5\pi t}{2} + \cdots$$

or

$$f(t) = \frac{1}{2} + \frac{2}{\pi}\left(\sin\frac{\pi t}{2} + \frac{1}{3}\sin\frac{3\pi t}{2} + \frac{1}{5}\sin\frac{5\pi t}{2} + \cdots\right) \qquad \blacktriangleleft$$

It may seem strange that a series of sines and cosines may approximate a graph consisting of straight lines, especially if such a graph is not even continuous. Suppose we graph the sum of the first two terms of the series in Example 1:

$$\frac{1}{2} + \frac{2}{\pi}\sin\frac{\pi t}{2}$$

This graph (dashed curve in Figure 26.7) is only a crude approximation. If we graph the sum of the first three terms,

$$\frac{1}{2} + \frac{2}{\pi}\sin\frac{\pi t}{2} + \frac{2}{3\pi}\sin\frac{3\pi t}{2}$$

the approximation improves (solid curve in Figure 26.7). If we use more terms of the series, the graph will contain more small oscillations and approximate the given function more and more closely as the number of terms increases. (See Figure 26.8.) At a point of discontinuity, the graph passes through the point midway between the ordinates 0 and 1.

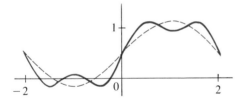

Figure 26.7

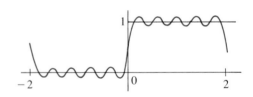

Figure 26.8

E X A M P L E **2**   Obtain the Fourier series of the "sawtooth function" in Figure 26.9, given over
one period by

$$f(t) = t \qquad (-1 < t < 1)$$

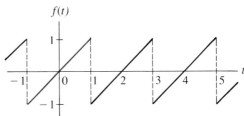

Figure 26.9

Solution. In this problem we use the period $(-1, 1)$ to obtain the coefficients.
Since the period is 2, we have

$$2p = 2 \qquad \text{or} \qquad p = 1$$

As always, we evaluate the first term separately:

$$a_0 = \frac{1}{p} \int_{-p}^{p} f(t)\, dt = \frac{1}{1} \int_{-1}^{1} t\, dt = \frac{1}{2} t^2 \Big|_{-1}^{1} = 0 \qquad p = 1$$

Hence $a_0 = 0$.
    Next we evaluate a_n for $n \geq 1$:

$$a_n = \frac{1}{p} \int_{-p}^{p} f(t) \cos \frac{n\pi t}{p}\, dt = \frac{1}{1} \int_{-1}^{1} t \cos \frac{n\pi t}{1}\, dt \qquad p = 1$$

$u = t$	$dv = \cos n\pi t\, dt$
$du = dt$	$v = \dfrac{1}{n\pi} \sin n\pi t$

integration by parts

$$a_n = \frac{t}{n\pi} \sin n\pi t \Big|_{-1}^{1} - \frac{1}{n\pi} \int_{-1}^{1} \sin n\pi t\, dt$$

$$= \frac{t}{n\pi} \sin n\pi t \Big|_{-1}^{1} - \left(\frac{1}{n\pi}\right)\left(-\frac{1}{n\pi} \cos n\pi t\right)\Big|_{-1}^{1} \qquad \begin{bmatrix} u = n\pi t \\ du = n\pi\, dt \end{bmatrix}$$

$$= \left[\frac{1}{n\pi} \sin n\pi - \frac{-1}{n\pi} \sin(-n\pi)\right] + \frac{1}{n^2\pi^2}[\cos n\pi - \cos(-n\pi)]$$

Since $\sin n\pi = 0$ and $\cos(-n\pi) = \cos n\pi$, we get

$$a_n = 0 + \frac{1}{n^2\pi^2}(\cos n\pi - \cos n\pi) = 0$$

Hence $a_n = 0$.

Finally, we evaluate b_n:

$$b_n = \frac{1}{p}\int_{-p}^{p} f(t)\sin\frac{n\pi t}{p}\,dt = \frac{1}{1}\int_{-1}^{1} t\sin\frac{n\pi t}{1}\,dt \qquad p = 1$$

$u = t$	$dv = \sin n\pi t\,dt$
$du = dt$	$v = -\dfrac{1}{n\pi}\cos n\pi t$

integration by parts

$$b_n = -\frac{t}{n\pi}\cos n\pi t\Big|_{-1}^{1} + \frac{1}{n\pi}\int_{-1}^{1}\cos n\pi t\,dt$$

$$= -\frac{t}{n\pi}\cos n\pi t\Big|_{-1}^{1} + \left(\frac{1}{n\pi}\right)\left(\frac{1}{n\pi}\sin n\pi t\right)\Big|_{-1}^{1} \qquad \begin{bmatrix} u = n\pi t \\ du = n\pi\,dt \end{bmatrix}$$

$$= -\frac{1}{n\pi}\cos n\pi + \frac{-1}{n\pi}\cos(-n\pi) + \frac{1}{n^2\pi^2}\left[\sin n\pi - \sin(-n\pi)\right]$$

Since $\sin n\pi = 0$ for all n and $\cos(-n\pi) = \cos n\pi$, we get

$$b_n = -\frac{1}{n\pi}\cos n\pi - \frac{1}{n\pi}\cos n\pi + 0$$

$$= -\frac{2}{n\pi}\cos n\pi$$

$$= \begin{cases} \dfrac{2}{n\pi} & \text{for } n \text{ odd} \\[2mm] -\dfrac{2}{n\pi} & \text{for } n \text{ even} \end{cases}$$

since $\cos n\pi = 1$ for n even and $\cos n\pi = -1$ for n odd.

We have obtained the following coefficients:

$$a_0 = 0,\; a_n = 0,\; b_1 = \frac{2}{1\pi},\; b_2 = -\frac{2}{2\pi},\; b_3 = \frac{2}{3\pi},\; b_4 = -\frac{2}{4\pi},\text{ and so on}$$

Substituting in series (26.33), we get (since **p = 1**)

$$f(t) = \frac{2}{1\pi}\sin\frac{1\pi t}{1} - \frac{2}{2\pi}\sin\frac{2\pi t}{1} + \frac{2}{3\pi}\sin\frac{3\pi t}{1} - \frac{2}{4\pi}\sin\frac{4\pi t}{1} + \cdots$$

or

$$f(t) = \frac{2}{\pi}\left(\sin \pi t - \frac{1}{2}\sin 2\pi t + \frac{1}{3}\sin 3\pi t - \frac{1}{4}\sin 4\pi t + \cdots\right)$$

◀

EXERCISES / SECTION 26.6

In Exercises 1–8, determine the Fourier series for the periodic functions. (In each case the function is given over one period only.)

1. $f(t) = \begin{cases} 0 & -1 < t < 0 \\ 1 & 0 < t < 1 \end{cases}$

2. $f(t) = \begin{cases} 0 & -\pi < t < 0 \\ 1 & 0 < t < \pi \end{cases}$

3. $f(t) = \begin{cases} 0 & -5 < t < 0 \\ 1 & 0 < t < 5 \end{cases}$

4. $f(t) = \begin{cases} -3 & -2 < t < 0 \\ 3 & 0 < t < 2 \end{cases}$

5. $f(t) = t \qquad -a < t < a$

6. $f(t) = \begin{cases} -t & -2 < t < 0 \\ t & 0 < t < 2 \end{cases}$

7. $f(t) = \begin{cases} 0 & -a < t < 0 \\ t & 0 < t < a \end{cases}$

8. $f(t) = \begin{cases} \pi + t & -\pi < t < 0 \\ \pi - t & 0 < t < \pi \end{cases}$

In Exercises 9 and 10, use the table of integrals in Appendix E (Table 4, forms 61 and 63) to obtain the coefficients for the series.

9. Obtain the Fourier series of the "half-wave rectification" of the sine (Figure 26.10), whose definition in one period

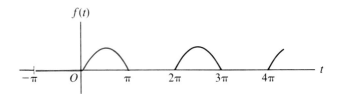

Figure 26.10

is given by

$$\begin{cases} 0 & -\pi < t < 0 \\ \sin t & 0 < t < \pi \end{cases}$$

10. Show that the Fourier series of the "full-wave rectification" of the sine (Figure 26.11) is given by

$$\frac{2}{\pi} - \frac{4}{\pi}\left[\frac{\cos 2t}{2^2 - 1} + \frac{\cos 4t}{4^2 - 1} + \frac{\cos 6t}{6^2 - 1} + \cdots \right]$$

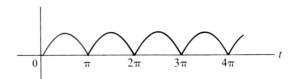

Figure 26.11

REVIEW EXERCISES / CHAPTER 26

1. Find the sum of the geometric series

$$\sum_{n=1}^{\infty} \frac{(-1)^{n-1}}{3^{n-1}} = 1 - \frac{1}{3} + \frac{1}{9} - \frac{1}{27} + \cdots$$

$$+ (-1)^{n-1}\frac{1}{3^{n-1}} + \cdots$$

2. Convert $0.13232\ldots$ to a common fraction.

In Exercises 3–10 (optional), test the series for convergence or divergence.

3. $\sum_{n=1}^{\infty} \frac{2n}{4n+3}$

4. $\sum_{n=1}^{\infty} \frac{1}{n^2+4}$

5. $\sum_{n=2}^{\infty} \frac{1}{n \ln^2 n}$

6. $\sum_{n=1}^{\infty} \frac{1}{n(n+3)}$

7. $\sum_{n=2}^{\infty} \frac{1}{\ln n}$

8. $\sum_{n=0}^{\infty} \frac{n^3}{n!}$

9. $\sum_{n=1}^{\infty} \frac{6^n}{n!}$

10. $\sum_{n=2}^{\infty} \frac{n^2}{n^3 - 3}$

11. Use the formula for the Maclaurin series to verify that

$$e^{-x} = 1 - x + \frac{x^2}{2!} - \frac{x^3}{3!} + \cdots$$

12. Repeat Exercise 11 for $\ln(1 + x) = x - (x^2/2) + (x^3/3) - \cdots$.

13. Use expansion (26.15) to find the Maclaurin series of $\sin x^2$.

14. Use expansion (26.16) to find the Maclaurin series of $\cos\sqrt{x}$.

15. Expand $f(x) = (1 - e^x)/x$ by using (26.14).

16. Expand $f(x) = \sinh x = \frac{1}{2}(e^x - e^{-x})$ by subtracting the appropriate series.

17. Evaluate $\cos(0.5)$ using three terms. Find the maximum possible error and determine the accuracy of the result.

18. Evaluate $\int_0^1 [(\sin x)/x]\,dx$ accurate to five decimal places.

19. Evaluate $\int_1^2 \cos x^3\,dx$ accurate to four decimal places.

20. Evaluate $\int_0^{0.1} e^{-x^3}\,dx$ accurate to three decimal places.

21. The charge on a certain capacitor is given by

$$q = \int_0^{0.10} e^{-0.20t^2}\,dt$$

Find the value of q.

22. Evaluate $\cos 33°$ by expanding $f(x) = \cos x$ about $c = \pi/6$. (Use three terms.)

23. Evaluate $\sin 44°$ by expanding $f(x) = \sin x$ about $c = \pi/4$. (Use four terms.)

24. Show that $\lim_{x \to 0} (\sin x)/x = 1$ by means of Maclaurin series.

25. Show that $\lim_{x \to 0} (1 - \cos x)/x = 0$ by means of Maclaurin series.

26. Show that the Fourier series of the periodic function whose definition in one period is $f(t) = t$, $-2 < t < 2$, is given by

$$\frac{4}{\pi}\left(\sin\frac{\pi t}{2} - \frac{1}{2}\sin\frac{2\pi t}{2} + \frac{1}{3}\sin\frac{3\pi t}{2} - \cdots\right)$$

C H A P T E R **27**

First-Order Differential Equations

27.1　Introduction

Most of the material in this and the next chapter owes its development to the Swiss scientist and mathematician Leonhard Euler (1707–1783). Considered the greatest mathematician of the eighteenth century and easily the most prolific in the history of mathematics, Euler contributed greatly to the development of the calculus, particularly to the study of infinite series. He was also actively engaged in analytic number theory (which he founded); physics, particularly celestial mechanics; the mathematics of finance (which he created), probability; geometry, particularly three-dimensional analytic geometry (which he practically created); parametric equations; and polar coordinates (which he did not create but developed fully for the first time). Perhaps most important of all, he systematized and unified much of the mathematics of his day. In the process he introduced many of the notations that are now standard; for example, the symbols $f(x)$, $\ln x$, $\sum$ (for sum), e, π, and i (for $\sqrt{-1}$) are all due to Euler. Even total blindness during the last seventeen years of his life could not retard the output of his research, and his collected works amount to seventy-five large volumes.

Euler was born in Basel but spent a large part of his life in St. Petersburg, Russia (1727–1741 and 1766–1783) at the invitation of the authorities at the St. Petersburg Academy of Science. Between his stays there he spent twenty-five years at the court of King Frederick the Great of Prussia. (Part of his time there coincided with Voltaire's shorter stay, from 1750 to 1753.) Euler died of a stroke on September 18, 1783.

27.2　What Is a Differential Equation?

Everyone who has studied algebra is familiar with the word *equation*. Equations always seem to involve an unknown, usually denoted by x, and your job is to find its value. In this chapter we are going to look at a completely different kind of equation, one for which the solution is not just a number, but a function. Such equations are therefore called **functional equations**. For example, can you guess the solution of the functional equation $f(x + y) = f(x)f(y)$? A little reflection shows that any exponential function $f(x) = a^x$ will do, since

$$a^{x+y} = a^x a^y$$

Now try $f'(x) - f(x) = 0$. The solution is easily found by inspection once the equation is written in the form

$$f'(x) = f(x)$$

Since $de^x/dx = e^x$, it follows that $f(x) = e^x$. We now define a **differential equation** to be a functional equation that contains derivatives or differentials. For example, the equation $f'(x) - f(x) = 0$ is now seen to be a differential equation; $f(x) = e^x$ is called a **solution**. We may also write $y' - y = 0$ or $dy/dx - y = 0$ for the equation and $y = e^x$ for the solution.

These ideas are summarized next.

A **differential equation** is a functional equation of two variables containing derivatives or differentials.

A **solution** of a differential equation is a relation between the variables that satisfies the equation.

One point should now be emphasized. In the equation under discussion, $y = e^x$ is *a* solution, not *the* solution, since $y = ce^x$, for any arbitrary constant c, satisfies the equation equally well.

It is not hard to see how differential equations arise, since we have already encountered a number of examples. For instance, if v is the velocity of an object moving with constant acceleration a, then $dv/dt = a$, which is indeed a differential equation. It can be readily solved by direct integration:

$$v = \int a\,dt = at + c$$

—again suggesting why solutions are not unique.

Not all differential equations can be solved in this simple manner. Recall that the voltages across an inductor and resistor are $L(di/dt)$ and Ri, respectively. If connected in series with a generator of E volts, then

$$L\frac{di}{dt} + Ri = E$$

by Kirchhoff's voltage law. Here we have an example of a differential equation that cannot be solved by any technique considered so far.

We are going to consider many more of these practical uses of differential equations in Sections 27.5 and 28.4, but first we need to learn some methods for solving these equations.

E X A M P L E **1**   Find the function y if $dy/dx = 2x^3$ and the curve passes through the point $(1, 2)$. (See Section 22.7.)

Solution. The equation $dy/dx = 2x^3$ can be solved directly by integration—that is,

$$y = \frac{1}{2}x^4 + c \tag{27.1}$$

Since $(1, 2)$ lies on the curve, we let $x = 1$ and $y = 2$ and solve for c:

$$2 = \frac{1}{2}(1)^4 + c$$

so that $c = \frac{3}{2}$. So the function is

$$y = \frac{1}{2}x^4 + \frac{3}{2} \tag{27.2}$$

◀

This example may serve to introduce some of our basic terms.

> Solution (27.1) satisfies the equation no matter what value of c is chosen and is called the **general solution**. In solution (27.2) the arbitrary constant was evaluated from the extra condition. Solution (27.2) is therefore called a **particular solution**.

E X A M P L E **2**   Show that the function

$$y = 3xe^{2x} + ce^{2x}$$

is the general solution of the differential equation

$$\frac{dy}{dx} - 2y = 3e^{2x}$$

Solution. To show that the given function solves the equation, we find dy/dx and substitute. By the product rule,

$$\frac{dy}{dx} = \frac{d}{dx}(3xe^{2x} + ce^{2x})$$

$$= 3xe^{2x}(2) + 3e^{2x} + 2ce^{2x}$$

$$= 6xe^{2x} + 3e^{2x} + 2ce^{2x}$$

Substituting in the left side of the given equation

$$\frac{dy}{dx} - 2y = 3e^{2x}$$

we get

$$(6xe^{2x} + 3e^{2x} + 2ce^{2x}) - 2(3xe^{2x} + ce^{2x})$$
$$= 6xe^{2x} + 3e^{2x} + 2ce^{2x} - 6xe^{2x} - 2ce^{2x}$$
$$= 3e^{2x} \quad \text{(the right side)}$$

The equation is therefore satisfied.   ◄

Order

The **order** of a differential equation is the order of the highest derivative. The equation in Example 2 is of first order, while

$$\frac{d^3y}{dx^3} + x\frac{dy}{dx} + y = 0$$

is of third order. In this chapter we will study various types of first-order equations and their applications.

E X E R C I S E S / S E C T I O N   27.2

In Exercises 1–16, show that the given function is a solution of the differential equation. (See Example 2.)

1. $\frac{dy}{dx} - 3y = 0$, $y = 2e^{3x}$

2. $y' - 3y = e^{4x}$, $y = e^{4x}$

3. $\frac{dy}{dx} - y = e^x$, $y = xe^x + 2e^x$

4. $\frac{dy}{dx} + 2y = e^{-2x}$, $y = xe^{-2x} + 3e^{-2x}$

5. $y'' + 4y = 0$, $y = 2\cos 2x + 3\sin 2x$

6. $y'' + y = 0$, $y = c\cos x$

7. $y'' + y' - 6y = 0$, $y = c_1 e^{-3x} + c_2 e^{2x}$ (c_1 and c_2 are arbitrary constants)

8. $\frac{d^2y}{dx^2} - \frac{dy}{dx} - 2y = 0$, $y = c_1 e^{2x} + c_2 e^{-x}$

9. $x\frac{dy}{dx} - y = x^2 + 4$, $y = x^2 - 4$

10. $x\frac{dy}{dx} + y = x + 3$, $y = \frac{1}{2}x + \frac{1}{x} + 3$

11. $x^2\frac{dy}{dx} - xy = x^2 + x$, $y = x + x\ln x - 1$

12. $x^2\frac{dy}{dx} + xy = x + 1$, $y = 1 + \frac{\ln x}{x}$

13. $y'' + 9y = 6\sin 3x$, $y = c\cos 3x - x\cos 3x$

14. $\frac{d^2y}{dx^2} + 16y = 8\cos 4x$, $y = x\sin 4x + c\sin 4x$

15. $\frac{d^2y}{dx^2} + 5\frac{dy}{dx} - 6y = 7e^x$, $y = c_1 e^x + c_2 e^{-6x} + xe^x$

16. $y'' + 2y' - 8y = 6e^{2x}$, $y = c_1 e^{-4x} + c_2 e^{2x} + xe^{2x}$

In Exercises 17–22, solve the differential equations subject to the given conditions. (See Example 1.)

17. $\frac{dy}{dx} = 3x^2$, $y = 5$ when $x = 2$

18. $\frac{dy}{dx} = 2x^3$, $y = 6$ when $x = 2$

19. $\frac{dy}{dx} = \sec^2 x$, $y = 1$ when $x = \frac{\pi}{4}$

20. $\dfrac{dy}{dx} = \cos x$, $y = 2$ when $x = \dfrac{\pi}{2}$

21. $\dfrac{d^2 y}{dx^2} = e^x$, $y = 0$ when $x = 0$ and $y = 1$ when $x = 1$ (The general solution has two arbitrary constants.)

22. $\dfrac{d^2 y}{dx^2} = \sin x$, $y = 0$ when $x = 0$ and $\dfrac{dy}{dx} = 1$ when $x = 0$

27.3 Separation of Variables

In this section we discuss a method for solving a certain type of first-order equation. First recall that if $dy/dx = f'(x)$, then $dy = f'(x)\,dx$; the converse is also true. In other words, a derivative can always be expressed in differential form, and in this section it turns out to be convenient to do so. To see why, consider the following equation of first order:

$$\frac{dy}{dx} = F(x, y) \tag{27.3}$$

Using the differential notation, the equation can be written in the following form:

General First-Order Differential Equation

$$M(x, y)\,dx + N(x, y)\,dy = 0 \tag{27.4}$$

For some equations it is possible to *separate variables* by writing Equation (27.4) in the form $M(x)\,dx + N(y)\,dy = 0$. In other words, M is a function of x alone and N is a function of y alone. In this form, each of the terms can be integrated separately. This procedure is known as the method of **separation of variables**.

Equation with Variables Separated

$$M(x)\,dx + N(y)\,dy = 0 \tag{27.5}$$

The next example illustrates the technique.

E X A M P L E **1**   Use separation of variables to solve the equation

$$\csc 2x\, \frac{dy}{dx} + e^{\cos 2x} = 0$$

Solution. Writing the equation in form (27.4), we get

$$e^{\cos 2x}\, dx + \csc 2x\, dy = 0 \qquad \textbf{formally multiplying by } dx$$

Observe next that if we divide both sides of the equation by csc $2x$, then the equation becomes

$$\frac{e^{\cos 2x}\,dx}{\csc 2x} + dy = 0$$

$$e^{\cos 2x}\sin 2x\,dx + dy = 0 \qquad 1/\csc\theta = \sin\theta$$

In this form all the terms are readily integrated since $M(x)$ is a function of x alone and $N(y)$ is a function of y alone. Thus

$$\int e^{\cos 2x}\sin 2x\,dx + \int dy = c_1$$

or

$$-\frac{1}{2}\int e^{\cos 2x}(-2\sin 2x)\,dx + \int dy = c_1 \qquad \begin{bmatrix} u = \cos 2x \\ du = -2\sin 2x\,dx \end{bmatrix}$$

and

$$-\frac{1}{2}e^{\cos 2x} + c_2 + y + c_3 = c_1 \qquad\qquad (27.6)$$

◀

The constants in the solution of the equation in Example 1 can be combined. Since Equation (27.6) can be written

$$-\frac{1}{2}e^{\cos 2x} + y = c_1 - c_2 - c_3$$

we may let $k = c_1 - c_2 - c_3$ to obtain the simpler form

$$-\frac{1}{2}e^{\cos 2x} + y = k$$

Moreover, multiplying by -2, we get

$$e^{\cos 2x} - 2y = -2k$$

which suggests letting $c = -2k$ to obtain

$$e^{\cos 2x} - 2y = c$$

(If k is arbitrary, so is $-2k$.)

From now on, whenever we integrate the terms in an equation, we simply *place a single arbitrary constant on the right side.*

E X A M P L E **2** Find the general solution of the differential equation

$$2x\,dx + xy\,dx + \cos^2 x^2\,dy = 0$$

Solution. We need to rewrite the equation in the form $M(x)\,dx + N(y)\,dy = 0$, where $M(x)$ is a function of x alone and $N(y)$ is a function of y alone:

$$2x\,dx + xy\,dx + \cos^2 x^2\,dy = 0 \qquad \text{given equation}$$

$$(2 + y)x\,dx + \cos^2 x^2\,dy = 0 \qquad \text{factoring } x\,dx$$

$$\frac{x\,dx}{\cos^2 x^2} + \frac{dy}{2 + y} = 0 \qquad \text{dividing by } (2 + y)\cos^2 x^2$$

$$(\sec^2 x^2)x\,dx + \frac{dy}{2 + y} = 0 \qquad \sec\theta = \frac{1}{\cos\theta}$$

The equation now has the proper form, and we may integrate each term separately and introduce the arbitrary constant k on the right side:

$$\frac{1}{2}\int(\sec^2 x^2)(2x)\,dx + \int \frac{dy}{2 + y} = k \qquad \begin{bmatrix} u = x^2 \\ du = 2x\,dx \end{bmatrix}$$

$$\frac{1}{2}\tan x^2 + \ln|2 + y| = k$$

$$\tan x^2 + 2\ln|2 + y| = 2k \qquad \text{multiplying by 2}$$

$$\tan x^2 + 2\ln|2 + y| = c \qquad \text{letting } c = 2k \qquad \blacktriangleleft$$

E X A M P L E **3** Obtain the general solution of the equation $(y - 2)\,dx + (1 - x)\,dy = 0$.

Solution. Separating variables, we get

$$\frac{dx}{1 - x} + \frac{dy}{y - 2} = 0 \qquad \text{dividing by } (y - 2)(1 - x)$$

If we now integrate each term and introduce the arbitrary constant k on the right side, we get

$$-\int \frac{-dx}{1 - x} + \int \frac{dy}{y - 2} = k \qquad \begin{bmatrix} u = 1 - x \\ du = -dx \end{bmatrix}$$

$$-\ln|1 - x| + \ln|y - 2| = k$$

or

$$\ln\left|\frac{y - 2}{1 - x}\right| = k \qquad \ln A - \ln B = \ln\frac{A}{B}$$

Using the properties of logarithms, we can write the solution much more compactly as follows: for every k, there exists a number $c' > 0$ such that $\ln c' = k$; hence

$$\ln\left|\frac{y-2}{1-x}\right| = \ln c'$$

and, since logarithms are unique, it follows that

$$\left|\frac{y-2}{1-x}\right| = c' \qquad (c' > 0)$$

By the definition of absolute value,

$$\frac{y-2}{1-x} = \pm c'$$

Finally, letting $c = \pm c'$, we get

$$y - 2 = c(1 - x) \qquad (c \neq 0)$$

This type of reduction is possible whenever the solution contains only logarithmic functions. ◀

If the equation satisfies an additional condition, then we get a unique particular solution. In that case the form of the general solution is irrelevant.

E X A M P L E **4**   Find the particular solution of the equation $dx + y^3 \cot x \, dy = 0$, subject to the condition $y = 2$ when $x = 0$.

Solution. Separating variables, we get

$$\frac{dx}{\cot x} + y^3 \, dy = 0 \qquad \text{dividing by cot x}$$

$$\tan x \, dx + y^3 \, dy = 0 \qquad \frac{1}{\cot x} = \tan x$$

and, after performing the integration,

$$\ln|\sec x| + \frac{1}{4} y^4 = c$$

We now let $x = 0$ and $y = 2$, so that $0 + 4 = c$, and the solution becomes $\ln|\sec x| + \frac{1}{4}y^4 = 4$ or

$$4 \ln|\sec x| + y^4 = 16$$

◀

Note that the solution (as well as the conditions) can always be checked. In Example 4 we see that the condition $y = 2$ when $x = 0$ is obviously satisfied. To check the solution, we differentiate implicitly to obtain

$$4 \frac{\sec x \tan x}{\sec x} + 4y^3 \frac{dy}{dx} = 0$$

which reduces to the given equation.

E X A M P L E **5** Show that $\tan x + e^y = c$ satisfies the equation

$$e^{-y} dx + \cos^2 x \, dy = 0$$

Solution. Differentiating $\tan x + e^y = c$ implicitly, we get

$$\sec^2 x + e^y \frac{dy}{dx} = 0 \qquad \text{since } \frac{d}{dx} e^u = e^u \frac{du}{dx}$$

$$\sec^2 x \, dx + e^y \, dy = 0$$

$$\frac{dx}{e^y} + \frac{dy}{\sec^2 x} = 0$$

and

$$e^{-y} dx + \cos^2 x \, dy = 0 \qquad \frac{1}{\sec \theta} = \cos \theta$$

which is the given equation. ◀

E X E R C I S E S / S E C T I O N **27.3**

In Exercises 1–32, find the general solution of each of the equations.

1. $x^2 \, dx + y \, dy = 0$

2. $x^5 \, dx + y^2 \, dy = 0$

3. $(1 + x^2) \, dx = 3y \, dy$

4. $(1 - x) \, dx + 4y^3 \, dy = 0$

5. $2x + (1 + x^2) \dfrac{dy}{dx} = 0$

6. $x + 2 + 2(x + 1)y \dfrac{dy}{dx} = 0$

7. $(1 + x^2) \dfrac{dy}{dx} + y = 0$

8. $y \, dx + \sec x \, dy = 0$

9. $2y \, dx + 3x \, dy = 0$ (See Example 3.)

10. $3y \dfrac{dy}{dx} + \tan x = 0$

11. $1 + (x^2 y - x^2) \dfrac{dy}{dx} = 0$

12. $(1 + x)^2 \dfrac{dy}{dx} = 1$

13. $dx - y \, dx + x \, dy = 0$

14. $xy \, dx + dy + x^2 \, dy = 0$

15. $\dfrac{dV}{dP} = -\dfrac{V}{P}$

16. $\dfrac{di}{dR} = -\dfrac{i}{2R}$

17. $dx + (2 \cos^2 x - y \cos^2 x) \, dy = 0$

18. $xe^{x^2 + y} \, dx + dy = 0$

19. $\cos^2 t + y \csc t \dfrac{dy}{dt} = 0$

20. $(\tan t + y \tan t) \, dt + dy = 0$

21. $\sqrt{v^2 + 1}\, dt + vt^2\, dv = 0$

22. $r\dfrac{dr}{dt} = \dfrac{\sin t \sec t}{\ln r}$

23. $T_1\, dT_1 + (\csc T_1 + T_2 \csc T_1)\, dT_2 = 0$

24. $2s_1 \dfrac{ds_2}{ds_1} - s_2{}^3 \ln s_1 = 0$

25. $(y^2 - 1) \cos x\, dx + 2y \sin x\, dy = 0$

26. $(x^3 - x^3 y^2)\, dx = y\, dy$

27. $xe^y\, dx + e^{-x}\, dy = 0$

28. $xy\, dx + \sqrt{1 - x^2}\, dy = 0$

29. $(e^x \tan y + \tan y)\dfrac{dy}{dx} + e^x = 0$

30. $\dfrac{dy}{dx} = \dfrac{y}{x^2 - 4x + 4}$

31. $y\dfrac{dy}{dx} + 2x \sec y = 0$

32. $3xy\, dx + (1 + x^2)\, dy = 0$

In Exercises 33–37, find the particular solution of the equations satisfying the given conditions.

33. $x\, dy - y\, dx = 0$, $y = 2$ when $x = 1$

34. $\sin x\, dx + \sec x\, dy = 0$, $y = -\dfrac{1}{4}$ when $x = \dfrac{\pi}{4}$

35. $(y + 2)\, dx + (x - 3)\, dy = 0$, $y = 5$ when $x = 2$

36. $(1 + x)\, dy - y^2\, dx = 0$, $y = -1$ when $x = 0$

37. $dx + x \tan y\, dy = 0$, $y = 0$ when $x = 1$

27.4   First-Order Linear Differential Equations

So far we have considered only differential equations that can be solved by separation of variables. In this section we are going to consider another type of equation called *linear*.

> **First-Order Linear Differential Equation**
>
> $$\frac{dy}{dx} + P(x)y = Q(x) \tag{27.7}$$

The method of solution of a first-order linear equation depends on a special device: we multiply both sides of the given equation by a certain function called an **integrating factor**, designed to facilitate the integration. The integrating factor is the function

$$e^{\int P(x)\, dx}$$

where $P(x)$ is the coefficient of y in Equation (27.7). After multiplying both sides of the equation by the integrating factor, we can write the left side in the form

$$\frac{d}{dx}\left(ye^{\int P(x)\, dx}\right) \tag{27.8}$$

as we will see. After removing the differentiation symbol (by integrating both sides), we can solve for y algebraically. Before we justify this procedure, let us consider an example.

E X A M P L E **1**   Solve the linear equation $dy/dx - 2xy = xe^{x^2}$.

Solution. By (27.8), the integrating factor (IF) is the exponential function

$$e^{\int(-2x)\,dx}$$

where $-2x$ is the coefficient of y. Thus

$$\text{IF} = e^{-x^2}$$

Multiplying both sides of the given equation by this function, we get

$$e^{-x^2}\left(\frac{dy}{dx} - 2xy\right) = xe^{x^2}e^{-x^2} = x$$

Now observe that the left side can be written

$$\frac{d}{dx}(ye^{-x^2})$$

as can be readily checked by the product rule:

$$\frac{d}{dx}(ye^{-x^2}) = e^{-x^2}\frac{dy}{dx} + y\frac{d}{dx}(e^{-x^2})$$

$$= e^{-x^2}\frac{dy}{dx} - 2xye^{-x^2} = e^{-x^2}\left(\frac{dy}{dx} - 2xy\right)$$

The resulting equation is therefore

$$\frac{d}{dx}(ye^{-x^2}) = x$$

In this form the equation can be readily solved by integrating both sides:

$$ye^{-x^2} = \frac{1}{2}x^2 + c \quad \text{or} \quad y = \frac{1}{2}x^2e^{x^2} + ce^{x^2} \qquad \blacktriangleleft$$

The method of Example 1 works with all linear equations. To see why, refer to the general form (27.7) and let

$$\text{IF} = e^{\int P(x)\,dx}$$

Multiplying both sides of the equation, we get

$$e^{\int P(x)\,dx}\left[\frac{dy}{dx} + P(x)y\right] = Q(x)e^{\int P(x)\,dx}$$

which can be written

$$\frac{d}{dx}(ye^{\int P(x)\,dx}) = Q(x)e^{\int P(x)\,dx} \tag{27.9}$$

In other words, the left side becomes the derivative of the product of y and IF. To check this statement, we use the product rule:

$$\frac{d}{dx}(ye^{\int P(x)\,dx}) = y\,\frac{d}{dx}e^{\int P(x)\,dx} + e^{\int P(x)\,dx}\frac{dy}{dx} \qquad \textbf{product rule}$$

$$= ye^{\int P(x)\,dx}\,\frac{d}{dx}\int P(x)\,dx + e^{\int P(x)\,dx}\frac{dy}{dx} \qquad \frac{d}{dx}e^{u} = e^{u}\frac{du}{dx}$$

$$= ye^{\int P(x)\,dx} \cdot P(x) + e^{\int P(x)\,dx}\frac{dy}{dx} \qquad \frac{d}{dx}\int P(x)\,dx = P(x)$$

$$= e^{\int P(x)\,dx}\left[\frac{dy}{dx} + P(x)y\right] \qquad \textbf{factoring}$$

If we now integrate both sides of Equation (27.9), namely,

$$\frac{d}{dx}(ye^{\int P(x)\,dx}) = Q(x)e^{\int P(x)\,dx}$$

we obtain the solution

$$ye^{\int P(x)\,dx} = \int Q(x)e^{\int P(x)\,dx}\,dx + c$$

Integrating Factor (IF): An **integrating factor** of the linear equation

$$\frac{dy}{dx} + P(x)y = Q(x) \tag{27.10}$$

is given by

$$\text{IF} = e^{\int P(x)\,dx} \tag{27.11}$$

The procedure for solving first-order linear equations may be summarized as follows:

Procedure for Solving First-Order Linear Differential Equations

1. Write the given equation in the form (27.10).
2. Multiply both sides of the equation by IF.
3. Write the left side as the derivative of the product of y and IF.
4. Integrate both sides.

E X A M P L E **2**   Solve the linear equation

$$\frac{dy}{dx} + 3y = 4$$

Solution. By (27.11),

$$\text{IF} = e^{\int 3\, dx} = e^{3x}$$

Multiplying both sides of the equation by IF, we get

$$e^{3x}\left(\frac{dy}{dx} + 3y\right) = 4e^{3x}$$

By (27.9), the left side can be written as the derivative of the product of y and IF:

$$\frac{d}{dx}(ye^{3x}) = 4e^{3x}$$

Integrating both sides, we get

$$ye^{3x} = \int 4e^{3x}\, dx = \frac{4}{3}e^{3x} + c \qquad \begin{bmatrix} u = 3x \\ du = 3\,dx \end{bmatrix}$$

or

$$y = \frac{4}{3} + ce^{-3x} \qquad\qquad \blacktriangleleft$$

Before continuing with our next example, we need to examine a special form that frequently arises in the integrating factor. If we write $\ln N = \ln N$, or $\log_e N = \log_e N$, it follows from the definition of logarithm that

$$e^{\log_e N} = N$$

The reason is that

$$\log_e A = B \qquad \text{means} \qquad e^B = A$$

$$N = e^{\ln N} \tag{27.12}$$

E X A M P L E **3**   Solve the linear equation

$$x\, dy = (3y + 5x^5)\, dx$$

Solution. First we need to write the equation in form (27.10):

$$x \, dy = (3y + 5x^5) \, dx \qquad \textbf{given equation}$$

$$x \frac{dy}{dx} = 3y + 5x^5 \qquad \textbf{derivative form}$$

$$x \frac{dy}{dx} - 3y = 5x^5 \qquad \textbf{transposing 3y}$$

$$\frac{dy}{dx} - \frac{3}{x} y = 5x^4 \qquad \textbf{dividing by x}$$

The equation now has the proper form (27.10). The integrating factor is

$$\text{IF} = e^{\int(-3/x)\,dx} = e^{-3 \ln x} \qquad \textbf{by (27.11)}$$

$$= e^{\ln x^{-3}} \qquad\qquad \textbf{a ln x = ln x}^a$$

$$= x^{-3} \qquad\qquad\quad \textbf{N = e}^{\ln N}$$

Multiplying both sides of the equation

$$\frac{dy}{dx} - \frac{3}{x} y = 5x^4$$

by x^{-3}, we get

$$x^{-3} \left(\frac{dy}{dx} - \frac{3}{x} y \right) = 5x$$

or, by (27.9),

$$\frac{d}{dx} (x^{-3} y) = 5x \qquad \textbf{derivative of the product of y and IF}$$

Integrating both sides, we have

$$x^{-3} y = 5 \frac{x^2}{2} + c$$

or

$$y = \frac{5}{2} x^5 + cx^3$$

◀

E X A M P L E **4**   Solve the linear equation

$$(\tan x) \frac{dy}{dx} + y = \sec^3 x$$

Solution. The equation can be written in form (27.10) by dividing both sides by tan x:

$$\frac{dy}{dx} + y \cot x = \frac{\sec^3 x}{\tan x} = \frac{\sec^3 x}{\sin x/\cos x} = \frac{\sec^2 x}{\sin x}$$

Hence

$$IF = e^{\int \cot x\, dx} = e^{\ln \sin x} = \sin x \qquad e^{\ln N} = N$$

We now multiply both sides of the equation by IF to get

$$\sin x \left(\frac{dy}{dx} + y \cot x\right) = \sin x \left(\frac{\sec^2 x}{\sin x}\right) = \sec^2 x$$

or, by (27.9),

$$\frac{d}{dx}(y \sin x) = \sec^2 x \qquad \text{derivative of the product of } y \text{ and IF}$$

Integrating both sides, we get

$$y \sin x = \tan x + c$$

$$y = \frac{\tan x}{\sin x} + \frac{c}{\sin x} \qquad \text{or} \qquad y = \sec x + c \csc x \qquad \blacktriangleleft$$

E X E R C I S E S / S E C T I O N   **27.4**

Solve the following linear equations.

1. $\dfrac{dy}{dx} + y = 1$

2. $\dfrac{dy}{dx} - y = 2$

3. $\dfrac{dy}{dx} - 2y = e^{3x}$

4. $\dfrac{dy}{dx} - 2xy = e^{x^2}$

5. $x\, dy + (y - x)\, dx = 0$

6. $x\, dy + (y - 2x)\, dx = 0$

7. $y' = e^x - \dfrac{y}{x}$

8. $y' + \dfrac{2y}{x} = \dfrac{e^x}{x^2}$

9. $\dfrac{dy}{dx} - \dfrac{2y}{x} - x^2 \sec^2 x = 0$

10. $x\left(\dfrac{dy}{dx} - 1\right) + 2y = x^2$

11. $xy' = 3y + x^5 \sin x$

12. $xy' = y + x^3 e^x$

13. $xy' - 2y = x^3 e^x$

14. $x^3 \dfrac{dy}{dx} + 3x^2 y = \csc^2 x$

15. $(y - 1) \sin x\, dx + dy = 0$

16. $y' - y \sec^2 x = \sec^2 x$

17. $y' - y \tan x - \cos x = 0$

18. $y' + y \cot x = x$

19. $(x + 1)y' + y = \dfrac{x + 1}{x - 1}$

20. $x^2 \dfrac{dy}{dx} + xy = 1$

21. $y' - \dfrac{1}{x}y = x^2 \sin x^2$

22. $y' - \dfrac{1}{t}y = t^2 \cos t$

23. $y' + y \tan t = \sec t$

24. $v' + v \cot t = \cos t$

25. $t\dfrac{dr}{dt} + r = t \ln t$

26. $t\dfrac{ds}{dt} - 2s = t^4 \cos 2t$

27. $s\dfrac{dr}{ds} - r = s^3 e^{3s}$

In each of the remaining exercises, find the particular solution.

28. $\dfrac{dy}{dx} - 2y = 4$, $y = 3$ when $x = 0$

29. $\dfrac{dy}{dx} + y = 6e^{-x}$, $y = 2$ when $x = 0$

30. $x\,dy - (2y + x^3 \cos x)\,dx = 0$, $y = \dfrac{\pi^2}{4}$ when $x = \dfrac{\pi}{2}$

31. $dy + x^2 y\,dx = 2x^2\,dx$, $y = 3$ when $x = 0$

32. $\dfrac{dy}{dx} + \dfrac{y}{x} = \cos x + \dfrac{\sin x}{x}$, $y = 1$ when $x = \dfrac{\pi}{2}$

33. $L\dfrac{di}{dt} + Ri = E$, $i = 0$ when $t = 0$

34. $m\dfrac{dv}{dt} + kv = mg$, $v = 0$ when $t = 0$

27.5   Applications of First-Order Differential Equations

In this section we will study various applications of first-order equations. Our first application is concerned with **growth and decay**. Suppose $N = N(t)$ is an amount (or number) that changes at a rate proportional to the amount present. This statement may be expressed as $dN/dt = kN$, where k is the constant of proportionality.

Law of Growth and Decay

$$\frac{dN}{dt} = kN \qquad\qquad (27.13)$$

where N, the amount present, is a function of time and k is a constant.

E X A M P L E **1**

A certain radioactive substance decays at a rate proportional to the amount present. Experimenters observe that after one year 10% of the original amount has decayed. Find the half-life of the substance.

Solution. By (27.13) the differential equation is $dN/dt = kN$. If we now let N_0 be the amount present when $t = 0$ (called the **initial condition**), we have the information needed to compute the arbitrary constant c. The equation is linear, but it may also be solved by separation of variables:

Initial condition

$$N = ce^{kt}$$

At $t = 0$, $N = N_0$; thus $N_0 = ce^0 = c$ and

$$N = N_0 e^{kt}$$

To find k, note that at $t = 1$ we have $N = 0.90N_0$ (90% of the original amount N_0). So

$$0.90N_0 = N_0 e^{k \cdot 1} \qquad \text{or} \qquad 0.90 = e^k \qquad \text{dividing by } N_0$$

We take the natural logarithm of both sides to get

$$\ln(0.90) = \ln e^k$$

$$\ln(0.90) = k \ln e = k \qquad \text{ln } e = 1$$

and from .9 $\boxed{\text{LN}}$ $\boxed{\text{STO}}$ we have

$$k = -0.1054$$

It follows that

$$N = N_0 e^{-0.1054t}$$

The last expression gives us the amount present as a function of time. In particular (to answer the question posed in the problem), we can find the so-called half-life—that is, the time taken for half of the original amount to decay. Let $N = 0.5N_0$, so that

$$0.5N_0 = N_0 e^{-0.1054t} \qquad \text{or} \qquad 0.5 = e^{-0.1054t}$$

Taking logarithms again, we have

$$\ln 0.5 = \ln e^{-0.1054t} = -0.1054t \ln e = -0.1054t$$

and

$$t = \frac{\ln 0.5}{-0.1054} = 6.58 \text{ years}$$

The sequence is .5 $\boxed{\text{LN}}$ $\boxed{\div}$ $\boxed{\text{MR}}$ $\boxed{=}$ ◀

Corresponding problems involving growth lead to a positive constant k; otherwise the technique is the same.

A similar type of problem arises in **Newton's law of cooling**: *the time rate of change of the temperature of a body is proportional to the temperature difference between the body and its surrounding medium.* If M_T denotes the temperature of the medium and T is the temperature of the body at any time, then Newton's law says that

$$\frac{dT}{dt} = -k(T - M_T) \qquad (k > 0)$$

Newton's Law of Cooling

$$\frac{dT}{dt} = -k(T - M_T) \qquad (k > 0) \tag{27.14}$$

where T is the temperature of the body at any time, M_T is the temperature of the surrounding medium, and k is a constant.

E X A M P L E **2**   Suppose a chef baking a cake places the dough at temperature 21°C in an oven kept at a constant temperature of 175°C. It is observed that the temperature of the dough has risen to 50°C after 10 min. How long will it take for the temperature of the dough to reach 100°C?

Solution. Since $M_T = 175$, we get by Equation (27.14)

$$\frac{dT}{dt} = -k(T - 175)$$

—a linear equation with initial condition $T = 21$ when $t = 0$. From

$$\frac{dT}{dt} + kT = 175k$$

we see that $\text{IF} = e^{kt}$, so that the equation becomes

$$\frac{d}{dt}(Te^{kt}) = 175ke^{kt} \qquad \left[\begin{array}{l} u = e^{kt} \\ du = ke^{kt}\,dt \end{array}\right]$$

Integrating, we have $Te^{kt} = 175e^{kt} + c$ or

$$T = 175 + ce^{-kt}$$

Substituting $t = 0$ and $T = 21$, we get $21 = 175 + c$, or $c = -154$. The solution is now written

$$T = 175 - 154e^{-kt}$$

To find k we make use of the fact that $T = 50$ when $t = 10$:

$$50 = 175 - 154e^{-10k}$$

Solving for k, we get $k = 0.0209$, so that

$$T = 175 - 154e^{-0.0209t}$$

Finally, letting $T = 100$ and solving for t, we get $t = 34.4$ min—that is, in 34.4 min the dough will be at 100°C. The rising temperature is illustrated graphically in Figure 27.1. ◀

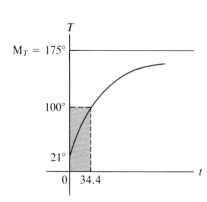

$M_T = 175°$

$100°$

$21°$

$0 \quad 34.4$

Figure 27.1

The force acting on a freely falling object is given by mg, where m is the mass of the object. If the object moves through a medium such as air, it is subjected to a resistance force that depends on the velocity and the size and shape of the object. At relatively low velocities the resistance force appears to be directly proportional to the velocity, or equal to kv, $k > 0$. Consequently, the net force F on the body is given by $F = mg - kv$. So, by Newton's second law, we have

$$m\frac{dv}{dt} = mg - kv$$

> **Motion Through a Resisting Medium**
>
> The velocity v of a body of mass m falling through a resisting medium is found from the equation
>
> $$m\frac{dv}{dt} = mg - kv \qquad (27.15)$$

Since Equation (27.15) is linear and since k has to be given explicitly, the solution is quite straightforward. (However, for some objects the force due to air resistance is approximately kv^2. For this case see Exercise 19.)

E X A M P L E **3** An object of mass 15 kg (weight = 150 N) is dropped from rest. The retarding force due to air resistance is numerically equal to 0.6 of the velocity (that is, $k = 0.6$). Find the velocity after 20 s, as well as the limiting velocity.

Solution. As noted earlier, the force due to air resistance is directly proportional to the velocity. In our problem, the constant of proportionality is $k = 0.6$, so that the force due to air resistance is $0.6v$. Also,

$$m = 15\text{ kg} \qquad \text{and} \qquad mg = 150\text{ N} \qquad \textbf{(15 kg)(10 m/s}^2\textbf{) = 150 N}$$

So, by Equation (27.15),

$$15\frac{dv}{dt} = 150 - 0.6v$$

which is a linear equation with initial condition $v = 0$ when $t = 0$.
It is a straightforward exercise to show that

$$v = 250(1 - e^{-0.04t})$$

At $t = 20$ s,

$$v = 250(1 - e^{-0.04(20)}) = 140\text{ m/s}$$

To find the limiting velocity, we let $t \to \infty$; thus

$$\lim_{t\to\infty} v = \lim_{t\to\infty} 250(1 - e^{-0.04t}) = 250$$

In other words, the velocity of the object approaches 250 m/s. ◄

Our next application is concerned with **orthogonal trajectories**: *given a family of curves, find another family such that every member of the new family intersects every member of the old family at right angles.*

E X A M P L E **4**   Find the equation of the family of orthogonal trajectories of $y = ce^x$ (Figure 27.2).

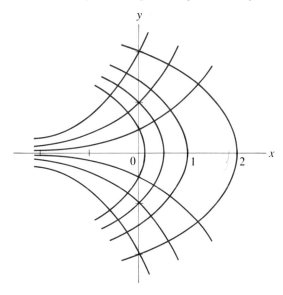

Figure 27.2

Solution. For any point (x, y) not on the x-axis, there exists a member of the given family passing through this point. (For example, the member passing through $(0, -2)$ is $y = -2e^x$.) The slope of the tangent line at any point is $dy/dx = ce^x$. Eliminating c between

$$\frac{dy}{dx} = ce^x \quad \text{and the given equation} \quad y = ce^x$$

yields $dy/dx = y$. (At $(0, -2)$ we get $dy/dx = y = -2$, which is the slope of $y = -2e^x$ at that point.) At any point the slope of the orthogonal trajectory is the negative reciprocal:

$$\frac{dy}{dx} = -\frac{1}{y} \qquad \text{negative reciprocal of } y$$

Hence this condition must be satisfied by the family of orthogonal trajectories. Separating variables, we get

$$y \, dy = -dx$$

and

$$\frac{y^2}{2} = -x + k \qquad \text{or} \qquad y^2 + 2x = k$$

—a family of parabolas. (See Figure 27.2.)   ◄

Orthogonal trajectories play an important role in many technical areas. For example, in a gravitational, magnetic, or electrostatic field the lines of force are orthogonal to the lines of equal potential. On a map the curves of steepest descent are orthogonal to the contour lines. Similar properties hold for heat conduction and fluid flow. (See Exercises 29 and 30.)

E X E R C I S E S / S E C T I O N 27.5

1. An RL circuit with $R = 5\,\Omega$ and $L = 0.2$ H has an applied voltage $e(t) = 5$ volts. Find the current as a function of time if $i = 0$ when $t = 0$. (Recall that $L\,di/dt + Ri = e(t)$.)

2. Repeat Exercise 1 if $e(t) = \sin t$.

3. A radioactive substance decays at a rate proportional to the amount present. If there are 100 g initially and 80 g after 10 days, find an expression for the mass at any time t.

4. A radioactive substance has a half-life of 1000 years. Find how long it takes for 5% of the substance to decay.

5. A radioactive isotope has a half-life of 100 h. Find how long it takes for 10% of the substance to decay.

6. Experimenters have determined that 5% of a certain radioactive substance decays after 28 h. Find the half-life of the substance.

7. It is found that 1% of a certain quantity of some isotope of radium decays after 20 years. Determine the half-life of this isotope.

8. A bacteria culture is known to increase at a rate proportional to the number of bacteria present. It is observed that the size of the culture triples in 3 h. After how many hours should it be 10 times as large?

9. The age of a fossil can be determined by a procedure called *carbon dating*. The procedure is based on the fact that carbon-14 is found in all organisms in a fixed percentage. When the organisms dies, the carbon-14 decays with a half-life of 5600 years. Suppose a fossil contains only 25% of the original amount. How long has the organism been dead?

10. A fossil contains 75% of the original amount of carbon-14. How long ago did the organism die? (Refer to Exercise 9.)

11. If S is an investment earning $100r\%$ annual (compound) interest and ΔS is the interest earned in the time interval Δt, then $\Delta S \approx Sr\,\Delta t$. (That is, ΔS is approximately equal to principal × rate × time, the simple interest earned in this time interval.) Thus $\Delta S/\Delta t \approx Sr$. As $\Delta t \to 0$, $dS/dt = Sr$. Show that $S = S_0 e^{rt}$, where S_0 is the amount invested initially. (The interest is said to be compounded continuously since the length Δt of each interest period approaches zero.)

a. To what sum will $1000 accumulate in 10 years if invested at 7.75% per year compounded continuously?

b. If it takes 10 years for $100 to double, what is the interest rate per year compounded continuously?

c. If a sum is to be invested at 8% per year compounded continuously, how large would the investment have to be in order to accumulate to a sum of $500 at the end of 5 years? The amount sought is called the "present value."

12. A body whose temperature is 25°C is placed outside where the temperature is 0°C. One minute later the temperature of the body has dropped to 22°C. What will the temperature be after 10 min?

13. A thermometer reading 20°F is placed in a room kept at 70°F. Two minutes later the thermometer reads 35°F. Find the temperature as a function of t and the time it takes for the thermometer reading to rise to within one degree of the room temperature.

14. A body whose temperature is 20°C is placed in a freezer kept at -10°C. One minute later the temperature of the body has dropped to 16°C. How long will it take for the temperature to fall to 0°C?

15. A body taken out of a freezer kept at -15°F is placed in a room whose temperature is 60°F. After 3 min the temperature of the body has risen to -5°F. Find the time it takes for the temperature of the body to rise to 50°F.

16. The temperature of a room is 21°C. At 5 P.M. the thermometer is taken outside where the temperature is 10°C. At 5:01 P.M. the thermometer reads 15°C when it is taken back inside. What is the reading at 5:02 P.M.?

17. A body of mass 10 kg (weight = 100 N) is dropped from rest. If the retarding force due to air resistance is numerically equal to 0.2 of the velocity, find the velocity after 10 s. What is the limiting velocity, that is, $\lim_{t \to \infty} v$?

18. A body of mass 5 kg (weight = 50 N) is dropped with an initial velocity of 1 m/s. If the retarding force is directly proportional to the velocity ($k = 1$), find an expression for the velocity as a function of time, as well as its limiting velocity.

19. A 10-kg body (weight = 100 N) attached to a parachute is dropped from rest and encounters a force due to air resistance numerically equal to $4v^2$. Find the velocity as a function of time and the limiting velocity. (Use partial fractions or Table 4.)

20. A body of weight 64 lb (2 slugs) is dropped from rest and encounters a retarding force numerically equal to 0.3 of the velocity. Determine the velocity as a function of time, as well as the limiting velocity.

21. A law similar to the law of growth and decay comes to us from chemistry: in certain chemical reactions in which a substance is converted into another substance, the time rate of change of the amount x of unconverted substance is proportional to x. If only a fourth of the substance has been converted (three-fourths unconverted) at the end of 10 s, find when nine-tenths of the substance will have been converted.

22. Find the orthogonal trajectories of the family of curves $y = cx^2$. Draw the graphs.

23. For a point mass at the origin, the curves of equal gravitational potential are $x^2 + y^2 = c^2$. Find the equations of the lines of force.

In Exercises 24–28, find the orthogonal trajectories of the families of curves.

24. $x^2 - y^2 = c$

25. $y^2 = 4px$

26. $y = cx^3$

27. $y = ce^{-2x}$

28. $x = ce^{y^2}$

29. Suppose the *streamlines* of a certain fluid motion, defined as curves such that the tangent at a point gives the direction of motion, are the hyperbolas $xy = c$. Find the *velocity equipotential curves*, which are the orthogonal trajectories.

30. Suppose in a problem in heat flow the family of isothermal curves, defined as curves joining points at the same temperature, are given by the ellipses $x^2 + \frac{1}{2}y^2 = c^2$. Find the curves along which the heat flows (that is, the orthogonal trajectories).

REVIEW EXERCISES/CHAPTER 27

In Exercises 1–14, solve the differential equations.

1. $y' = x - y$

2. $(x^2 + 1)dx + x^2y^2\,dy = 0$

3. $y' = x - 2xy$ (Solve by two methods.)

4. $dy - (2x + 2xy)dx = 0$ (Solve by two methods.)

5. $(1 + y^2)dx + (x^2y + y)dy = 0$

6. $\sin x \dfrac{dy}{dx} = 1 - 2y \cos x$ **7.** $2y\,dx + x\,dy = 0$

8. $(3x^2y - 3x^2 \tan y)dx - \tan^2 y\,dy = 0$

9. $(x^4 + 2y)dx - x\,dy = 0$

10. $(y + \cos^2 x)dx + \cos x\,dy = 0$

11. $x \sin^2 y\,dx - \cot y\,dy = 0$ **12.** $e^{2x+y}dx + y\,dy = 0$

13. $\dfrac{dy}{dx} + y \sec x = 0$, $y = 2$ when $x = \dfrac{\pi}{4}$ (Solve by two methods.)

14. $x(\ln x) \ln y\,dy + dx = 0$

15. A bacteria culture grows at a rate proportional to the number of bacteria present. If the size of the culture doubles in 2 h, how long will it take for the size to triple?

16. Radium decays at a rate proportional to the amount present. Experimenters have determined that 1.3% of a certain quantity has decayed after 30 years. Determine the half-life of radium.

17. A body is taken out of a freezer of unknown temperature and placed in a room kept at 65°F. After 15 min the temperature of the body is 0°F and after 30 min the temperature of the body is 20°F. Determine the temperature of the freezer.

18. A thermometer reading 65°F is taken outdoors where the temperature is 40°F. One minute later the thermometer reads 60°F. How long will it take for the reading to drop within one degree of the temperature outside?

19. A 6-kg object (weight = 60 N) dropped from rest experiences a retarding force directly proportional to the velocity ($k = 1$). Find its velocity as a function of time.

20. For a point mass at $(-3, 4)$ the curves of equal gravitational potential are given by $(x + 3)^2 + (y - 4)^2 = c^2$. Find the equations of the lines of force.

21. The isothermal curves of a metal plate are given by $x^2 - 2y^2 = c$. Find the curves along which the heat flows (the orthogonal trajectories).

In Exercises 1–11, perform the indicated integrations.

1. $\displaystyle\int \frac{dx}{\sqrt{4 - x^2}}$

2. $\displaystyle\int \frac{dx}{x\sqrt{4 - x^2}}$

3. $\displaystyle\int \frac{x\,dx}{\sqrt{4 - x^2}}$

4. $\displaystyle\int \sin^2 2x \cos^3 2x\,dx$

5. $\displaystyle\int \tan^3 3x\,dx$

6. $\displaystyle\int \sin^2 4x\,dx$

7. $\displaystyle\int x \cos x\,dx$

8. $\displaystyle\int x \cos x^2\,dx$

9. $\displaystyle\int \frac{2x + 1}{x^2 + x - 2}\,dx$

10. $\displaystyle\int \frac{3\,dx}{x^2 + x - 2}$

11. $\displaystyle\int \frac{e^{\cos x}}{\csc x}\,dx$

12. Find the Maclaurin expansion of $f(x) = \sin x$.

13. Use the result of Exercise 12 to evaluate the definite integral

$$\int_0^1 \frac{\sin x}{x}\,dx$$

to five decimal places.

14. Find the Fourier series of the periodic function whose definition over one period is

$$f(t) = \begin{cases} -t, & -3 < t < 0 \\ t, & 0 < t < 3 \end{cases}$$

In Exercises 15–18, solve each of the differential equations.

15. $dx - 2(\cos y\,dx + x \sin y\,dy) = 0$

16. $\dfrac{dy}{dx} = \cos x \tan y$

17. $\dfrac{dy}{dx} - 2xy + 2x = 0$

18. $\dfrac{dy}{dx} + y \cot x = \csc x$

19. The steady-state current in a certain circuit is $i = \cos 2t$. Find the effective current over one period. (Recall that the effective current is the root mean square current.)

20. A radioactive isotope decays at a rate proportional to the amount present. Physicists determined that after 15 days approximately 2.5% of a given amount had decayed. Determine the half-life of this isotope.

21. A falling body weighing 64 lb (2 slugs) is subject to a retarding force numerically equal to 0.42 of the velocity. Find the limiting velocity in feet per second.

Higher-Order Linear Differential Equations

28.1 Higher-Order Homogeneous Differential Equations

In the last chapter we considered only differential equations of first order. In this chapter we will solve linear equations of second and higher order, defined next.

An **nth-order linear differential equation** has the form

$$b_0(x)\frac{d^n y}{dx^n} + b_1(x)\frac{d^{n-1}y}{dx^{n-1}} + \cdots + b_{n-1}(x)\frac{dy}{dx} + b_n(x)y = f(x) \qquad (28.1)$$

If $f(x) = 0$, the equation is called **homogeneous**. If $f(x) \neq 0$, the equation is called **nonhomogeneous**.

In this section we will restrict our attention to the solution of homogeneous equations.

Linear Combinations

The method of solving linear equations depends on the following property: if y_1 and y_2 are two distinct solutions of a homogeneous linear equation, then the *linear combination*

Linear combination

$$y = c_1 y_1 + c_2 y_2$$

is also a solution. To see why, consider the second-order linear equation

$$\frac{d^2 y}{dx^2} - \frac{dy}{dx} - 6y = 0 \qquad (28.2)$$

If y_1 and y_2 are solutions, then

$$\frac{d^2 y_1}{dx^2} - \frac{dy_1}{dx} - 6y_1 = 0 \quad \text{and} \quad \frac{d^2 y_2}{dx^2} - \frac{dy_2}{dx} - 6y_2 = 0 \tag{28.3}$$

Consider next the linear combination $y = c_1 y_1 + c_2 y_2$. Substituting into the left side of Equation (28.2), we get

$$\frac{d^2}{dx^2}(c_1 y_1 + c_2 y_2) - \frac{d}{dx}(c_1 y_1 + c_2 y_2) - 6(c_1 y_1 + c_2 y_2) \tag{28.4}$$

Now observe that

$$\frac{d}{dx}(c_1 y_1 + c_2 y_2) = c_1 \frac{dy_1}{dx} + c_2 \frac{dy_2}{dx}$$

and

$$\frac{d^2}{dx^2}(c_1 y_1 + c_2 y_2) = c_1 \frac{d^2 y_1}{dx^2} + c_2 \frac{d^2 y_2}{dx^2}$$

(Because of this property, the derivative is said to be *linear*.) As a result, expression (28.4) can be rewritten

$$c_1 \frac{d^2 y_1}{dx^2} - c_1 \frac{dy_1}{dx} - 6c_1 y_1 + c_2 \frac{d^2 y_2}{dx^2} - c_2 \frac{dy_2}{dx} - 6c_2 y_2$$

$$= c_1 \left(\frac{d^2 y_1}{dx^2} - \frac{dy_1}{dx} - 6y_1 \right) + c_2 \left(\frac{d^2 y_2}{dx^2} - \frac{dy_2}{dx} - 6y_2 \right)$$

$$= c_1(0) + c_2(0) = 0 \qquad \textbf{by (28.3)}$$

We have therefore shown that the linear combination $y = c_1 y_1 + c_2 y_2$ is also a solution of (28.2). (We will see shortly how y_1 and y_2 can actually be obtained.)

Solution of Linear Equations

Before we turn to the solution of linear equations, we need to make some observations about the left side of (28.1). If the coefficients $b_0, b_1, b_2, \ldots, b_n$ are constants and if we use the symbol D^n for the nth derivative with respect to x, then the left side of (28.1) has the form of a polynomial. For example, a second-order linear equation is now written

$$b_0 D^2 y + b_1 D y + b_2 y = 0$$

or, more commonly,

$$(b_0 D^2 + b_1 D + b_2)y = 0$$

Operator

Now the "coefficient" of y, called an **operator**, not only looks like a polynomial but turns out to have similar algebraic properties. Consider, for example, the equation

$$(D^2 - D - 6)y = 0 \tag{28.5}$$

If we "factor" the polynomial, we get

$$(D - 3)(D + 2)y = 0$$

to be understood in the following sense. First apply the operator $D + 2$ to y and then apply $D - 3$ to the result. As a check, note that

$$(D + 2)y = Dy + 2y$$

Then

$$\begin{aligned}
(D - 3)(D + 2)y &= (D - 3)(Dy + 2y) \\
&= D(Dy + 2y) - 3(Dy + 2y) \\
&= D^2y + 2Dy - 3Dy - 6y \\
&= D^2y - Dy - 6y = (D^2 - D - 6)y
\end{aligned}$$

in agreement with Equation (28.5). Moreover,

$$(D - 3)(D + 2)y = (D + 2)(D - 3)y$$

as can be readily shown.

As a consequence, solving a homogeneous equation with constant coefficients can be reduced essentially to finding roots of polynomial equations. Suppose we split Equation (28.5) into two equations,

$$(D - 3)y_1 = 0 \quad \text{and} \quad (D + 2)y_2 = 0$$

each a first-order linear equation. We proceed to solve each one separately:

$$Dy_1 - 3y_1 = 0 \qquad\qquad Dy_2 + 2y_2 = 0$$

$$\text{IF} = e^{-3x} \qquad\qquad\quad \text{IF} = e^{2x}$$

$$\frac{d}{dx}(y_1 e^{-3x}) = 0 \qquad\quad \frac{d}{dx}(y_2 e^{2x}) = 0$$

$$y_1 e^{-3x} = c_1 \qquad\qquad\quad y_2 e^{2x} = c_2$$

$$y_1 = c_1 e^{3x} \qquad\qquad\quad y_2 = c_2 e^{-2x}$$

Since Equation (28.5) can be written in factored form, it is easily seen that both y_1 and y_2 satisfy the equation. In particular, both e^{3x} and e^{-2x} are solutions. Consequently, the *linear combination*

$$y = c_1 e^{3x} + c_2 e^{-2x}$$

is the general solution of the equation, as we saw earlier. In retrospect the solution could have been obtained directly from

$$(D - 3)(D + 2)y = 0$$

since the coefficients of x are the roots of the polynomial equation

$$(m - 3)(m + 2) = 0 \tag{28.6}$$

Auxiliary equation

—namely, $m = 3$ and $m = -2$. Equation (28.6) is called the **auxiliary equation**. We conclude, then, that the solution of the auxiliary equation may be used to find the general solution of the corresponding differential equation.

Definition of Auxiliary Equation: Let

$$(b_0 D^2 + b_1 D + b_2)y = 0$$

be a homogeneous second-order linear differential equation with constant coefficients. Then

$$b_0 m^2 + b_1 m + b_2 = 0$$

is called the **auxiliary equation**.

The solution of the auxiliary equation determines the solution of the differential equation.

General Solution of a Homogeneous Second-Order Linear Differential Equation: If m_1 and m_2 are distinct real roots of the auxiliary equation, then

$$y = c_1 e^{m_1 x} + c_2 e^{m_2 x}$$

is the **general solution** of the differential equation.

E   X   A   M   P   L   E   **1**   Find the general solution of the equation

$$2\frac{d^2 y}{dx^2} + 5\frac{dy}{dx} - 12y = 0$$

Solution. In terms of the D-operator the equation can be written

$$(2D^2 + 5D - 12)y = 0$$

Hence the auxiliary equation is

$$2m^2 + 5m - 12 = 0 \quad \text{or} \quad (2m - 3)(m + 4) = 0$$

Since the roots are $m = \frac{3}{2}$ and $m = -4$, the solution of the differential equation is given by

$$y = c_1 e^{(3/2)x} + c_2 e^{-4x}$$

or

$$y = c_1 e^{3x/2} + c_2 e^{-4x} \qquad \blacktriangleleft$$

E X A M P L E **2**   Solve the equation

$$\frac{d^2 y}{dx^2} + 2\frac{dy}{dx} - 2y = 0$$

Solution. Even without the D-operator the auxiliary equation is seen to be

$$m^2 + 2m - 2 = 0$$

By the quadratic formula we get

$$m = \frac{-2 \pm \sqrt{2^2 - 4(-2)}}{2}$$

$$= \frac{-2 \pm \sqrt{12}}{2} = \frac{-2 \pm 2\sqrt{3}}{2} = -1 \pm \sqrt{3}$$

Hence

$$y = c_1 e^{(-1+\sqrt{3})x} + c_2 e^{(-1-\sqrt{3})x}$$
$$= c_1 e^{-x} e^{\sqrt{3}x} + c_2 e^{-x} e^{-\sqrt{3}x}$$
$$= e^{-x}(c_1 e^{\sqrt{3}x} + c_2 e^{-\sqrt{3}x}) \qquad \text{factoring } e^{-x} \qquad \blacktriangleleft$$

E X A M P L E **3**   Solve the differential equation $(D^2 - 4)y = 0$, subject to the following conditions:

$$y = 1 \text{ and } Dy = 2 \text{ when } x = 0$$

Solution.

Auxiliary equation: $m^2 - 4 = 0$

Roots of auxiliary equation: $m = \pm 2$

(1) General solution: $y = c_1 e^{2x} + c_2 e^{-2x}$

To evaluate the constants c_1 and c_2, we substitute the given conditions and solve the resulting system of equations for c_1 and c_2:

(2) $Dy = y' = 2c_1 e^{2x} - 2c_2 e^{-2x}$ derivative of y

$$1 = c_1 + c_2$$ equation 1: $x = 0, y = 1$
$$2 = 2c_1 - 2c_2$$ equation 2: $x = 0, Dy = 2$
$$\overline{1 = c_1 + c_2}$$
$$1 = c_1 - c_2$$ dividing by 2
$$\overline{2 = 2c_1 + 0}$$ adding

$$c_1 = 1, c_2 = 0$$

Substituting the values of c_1 and c_2 in the general solution (1), we get the particular solution

$$y = e^{2x} \qquad \mathbf{c_1 = 1, c_2 = 0}$$ ◀

The technique for solving second-order linear equations can be extended to linear equations of higher order. If

$$b_0 \frac{d^n y}{dx^n} + b_1 \frac{d^{n-1} y}{dx^{n-1}} + \cdots + b_{n-1} \frac{dy}{dx} + b_n y = 0$$

is a homogeneous linear equation with constant coefficients, then

$$b_0 m^n + b_1 m^{n-1} + \cdots + b_{n-1} m + b_n = 0$$

is the **auxiliary equation**.

General Solution of a Homogeneous Linear Equation: If $m_1, m_2, \ldots, m_n$ are distinct real roots of the auxiliary equation, then

$$y = c_1 e^{m_1 x} + c_2 e^{m_2 x} + \cdots + c_n e^{m_n x}$$

is the **general solution** of the differential equation.

E X A M P L E **4**   Solve the equation $(D^3 - 3D^2 - D + 3)y = 0$.

Solution. Since the auxiliary equation

$$m^3 - 3m^2 - m + 3 = 0$$

is a cubic equation, we try to find one root by inspection and apply the factor theorem from algebra. Note first that the only possible rational roots are ± 1 and ± 3. It is easily checked that $m = 1$ is a root, so $m - 1$ must be a factor.

To divide $m - 1$ into the polynomial, we may use synthetic division:

$$\begin{array}{r} 1 - 3 - 1 + 3 \,)\, 1 \\ \underline{1 - 2 - 3} \\ 1 - 2 - 3 + 0 \end{array}$$

We now have

$$(m - 1)(m^2 - 2m - 3) = (m - 1)(m + 1)(m - 3) = 0$$

or $m = 1, -1,$ and 3. Thus

$$y = c_1 e^x + c_2 e^{-x} + c_3 e^{3x}$$ ◀

Note that the solution of the equation in Example 4 has three arbitrary constants. In general, an nth-order equation has n arbitrary constants in its general solution.

Remark. We saw in this section that forming linear combinations of solutions to obtain new solutions is essential to the technique for solving linear equations. This technique does not work with nonlinear equations, however, and as a consequence, *there does not exist a general method for solving nonlinear equations.* Some nonlinear equations can be solved by special techniques, as we have seen in the case of separable equations.

E X E R C I S E S / S E C T I O N 28.1

Solve each differential equation.

1. $(D^2 - 13D + 42)y = 0$

2. $\dfrac{d^2y}{dx^2} + \dfrac{dy}{dx} - 20y = 0$

3. $6\dfrac{d^2y}{dx^2} - \dfrac{dy}{dx} - 2y = 0$

4. $(D^2 - 4)y = 0$

5. $4D^2y + 7Dy - 2y = 0$

6. $(D^2 - 3)y = 0$

7. $\dfrac{d^2y}{dx^2} - \dfrac{dy}{dx} - y = 0$

8. $(D^2 + D - 3)y = 0$

9. $2D^2y - 3Dy + y = 0$

10. $(D^2 + 3D - 3)y = 0$

11. $(D^2 - 9)y = 0$; $y = 0$ and $Dy = 6$ when $x = 0$

12. $(D^2 - 3D + 2)y = 0$; if $x = 0$, then $y = 0$ and $Dy = 1$

13. $(D^2 - D - 2)y = 0$; $y = 0$ when $x = 0$, and $y = 1$ when $x = 1$

14. $(D^2 + 4D - 5)y = 0$; $y = 1$ and $Dy = 3$ when $x = 0$

15. $(D^3 - 7D + 6)y = 0$

16. $(D^3 - 2D^2 - 2D + 3)y = 0$

17. $(D^3 - D^2 - 4D - 2)y = 0$

18. $(4D^3 - 4D^2 - 11D + 6)y = 0$

19. $(D^2 + D - 1)y = 0$

20. $(D^2 + 2D - 4)y = 0$

21. $(D^2 - 2D - 2)y = 0$

22. $(D^2 + 2D - 1)y = 0$

23. $(D^2 - 4D - 2)y = 0$

24. $(D^2 - 6D - 2)y = 0$

25. $(D^2 + 6D - 6)y = 0$

26. $(2D^2 - 2D - 1)y = 0$

27. $(2D^2 + 4D + 1)y = 0$

28. $(3D^2 - 7D + 2)y = 0$

29. $(3D^2 - D - 2)y = 0$

30. The equation in Exercise 29, subject to the following conditions: $y = 1$ and $Dy = 0$ when $x = 0$.

28.2 Auxiliary Equations with Repeating or Complex Roots

In the preceding section all the roots of the auxiliary equations were distinct and real. In this section we will consider the cases of **repeating** and **complex roots**.

Real Repeating Roots

Consider the **linear** differential equation $(D^2 - 2D + 1)y = 0$. The auxiliary equation is

$$m^2 - 2m + 1 = (m - 1)^2 = 0$$

leading to the repeating root $m = 1, 1$. The method of the previous section now gives

$$y = c_1 e^x + c_2 e^x = (c_1 + c_2)e^x = ce^x$$

containing only one arbitrary constant. It is easily checked, however, that $y = xe^x$ is also a solution, so that the general solution is now given by the linear combination

$$y = c_1 e^x + c_2 xe^x$$

Real Repeating Roots: If the auxiliary equation has real repeating roots

$$m = a, a$$

then the **general solution** of the differential equation is

$$y = c_1 e^{ax} + c_2 xe^{ax}$$

E X A M P L E **1**   The auxiliary equation of $(4D^2 + 12D + 9)y = 0$ is

$$4m^2 + 12m + 9 = (2m + 3)^2 = 0$$

whence

$$m = -\frac{3}{2}, -\frac{3}{2}$$

The general solution of the differential equation is therefore given by

$$y = c_1 e^{-(3/2)x} + c_2 xe^{-(3/2)x}$$

◀

It is true in general that

$$(D - a)^n(x^k e^{ax}) = 0 \qquad \text{(for } k = 0, 1, 2, \ldots, n - 1)$$

Thus if

$$m = a, a, \ldots, a \qquad (n \text{ times})$$

then the general solution of the differential equation is

$$y = c_1 e^{ax} + c_2 x e^{ax} + c_3 x^2 e^{ax} + \cdots + c_n x^{n-1} e^{ax}$$

Combinations of distinct and repeating roots may also occur. Study the next example.

E X A M P L E **2**   Solve the equation $D^3(D - 2)y = 0$.

Solution. From the auxiliary equation

$$m^3(m - 2) = 0$$

we get $m = 0, 0, 0, 2$. Since zero is a repeating root but 2 is a single root, we get

$$y = c_1 e^{0x} + c_2 x e^{0x} + c_3 x^2 e^{0x} + c_4 e^{2x}$$
$$= c_1 + c_2 x + c_3 x^2 + c_4 e^{2x} \qquad \blacktriangleleft$$

Complex Roots

If the auxiliary equation with real coefficients

$$b_0 m^2 + b_1 m + b_2 = 0$$

has complex roots, then we know from the quadratic formula that the roots have the form

$$m = a \pm bj$$

—a pair of complex conjugates. (If the coefficients b_0, b_1, and b_2 are not real, this is not true.) The solution of the corresponding differential equation then becomes

$$y = c_3 e^{(a+bj)x} + c_4 e^{(a-bj)x}$$

which can be put into a more convenient form by applying Euler's identity from Section 26.4:

$$y = e^{ax}(c_3 e^{bxj} + c_4 e^{-bxj})$$
$$= e^{ax}\{c_3(\cos bx + j \sin bx) + c_4[\cos(-bx) + j \sin(-bx)]\}$$
$$= e^{ax}[c_3(\cos bx + j \sin bx) + c_4(\cos bx - j \sin bx)]$$
$$= e^{ax}[(c_3 + c_4) \cos bx + j(c_3 - c_4) \sin bx]$$
$$= e^{ax}(c_1 \cos bx + c_2 \sin bx)$$

where c_1 and c_2 are new arbitrary (complex) constants.

> **Complex Roots:** If the auxiliary equation has complex roots
>
> $$m = a \pm bj$$
>
> then the **general solution** of the differential equation is
>
> $$y = e^{ax}(c_1 \cos bx + c_2 \sin bx) \tag{28.7}$$

E X A M P L E **3**   Find the general solution of the equation $(D^2 - 4D + 6)y = 0$.

Solution. Solving the auxiliary equation

$$m^2 - 4m + 6 = 0$$

we get by the quadratic formula

$$m = \frac{4 \pm \sqrt{16 - 24}}{2} = \frac{4 \pm \sqrt{-8}}{2} = \frac{4 \pm 2\sqrt{2}j}{2}$$

$$m = 2 \pm \sqrt{2}j$$

By solution (28.7),

$$y = e^{2x}(c_1 \cos \sqrt{2}x + c_2 \sin \sqrt{2}x) \qquad \blacktriangleleft$$

Combinations of real and complex roots may also occur. For example, if

$$m = 2, 3 \pm 4j$$

then

$$y = c_1 e^{2x} + e^{3x}(c_2 \cos 4x + c_3 \sin 4x)$$

(For the case of repeating complex roots, see Exercise 49.)

E X E R C I S E S / S E C T I O N   **28.2**

In Exercises 1–47, solve the differential equations.

1. $(D^2 + 6D + 9)y = 0$

2. $(D^2 - 8D + 16)y = 0$

3. $(4D^2 - 4D + 1)y = 0$

4. $(4D^2 + 4D + 1)y = 0$

5. $9\dfrac{d^2y}{dx^2} + 12\dfrac{dy}{dx} + 4y = 0$

6. $9\dfrac{d^2y}{dx^2} + 30\dfrac{dy}{dx} + 25y = 0$

7. $(4D^2 - 20D + 25)y = 0$

8. $(16D^2 - 8D + 1)y = 0$

9. $(D^2 - 4D + 5)y = 0$

10. $(D^2 - D + 1)y = 0$

11. $\dfrac{d^2y}{dx^2} + 4\dfrac{dy}{dx} + 8y = 0$

12. $\dfrac{d^2y}{dx^2} - 2\dfrac{dy}{dx} + 5y = 0$

13. $2\dfrac{d^2y}{dx^2} - 2\dfrac{dy}{dx} + y = 0$

14. $\dfrac{d^2y}{dx^2} - 2\dfrac{dy}{dx} + 6y = 0$

15. $(D^2 + 25)y = 0$

16. $(4D^2 + 9)y = 0$

17. $(D^2 - 6D + 9)y = 0$

18. $\dfrac{d^2y}{dx^2} + 8\dfrac{dy}{dx} + 16y = 0$

19. $(D^4 + 2D^3)y = 0$

20. $D^5y = 0$

21. $\dfrac{d^2y}{dx^2} + \dfrac{dy}{dx} + 2y = 0$

22. $\dfrac{d^2y}{dx^2} - 3\dfrac{dy}{dx} + 2y = 0$

23. $(D^2 - 3D + 5)y = 0$

24. $(2D^2 - 2D + 1)y = 0$

25. $(2D^2 - 4D + 5)y = 0$

26. $(3D^2 - 2D + 2)y = 0$

27. $(2D^2 + 4D - 1)y = 0$

28. $3\dfrac{d^2y}{dx^2} + 6\dfrac{dy}{dx} - 2y = 0$

29. $\dfrac{d^2y}{dx^2} - 100y = 0$

30. $(D^2 - 256)y = 0$

31. $\dfrac{d^2y}{dx^2} + 100y = 0$

32. $(D^2 + 256)y = 0$

33. $(3D^3 - 2D^2 + D)y = 0$

34. $(D^2 - 5D + 7)y = 0$

35. $(D^2 - 4D + 2)y = 0$

36. $(2D^2 - 4D + 3)y = 0$

37. $(D^2 + 4)y = 0$

38. $(D^3 + 2D)y = 0$

39. $(D^3 - 4D^2 + 4D)y = 0$ **40.** $(D^3 + 9D)y = 0$

41. $(D^3 - 2D^2 + 2D - 1)y = 0$

42. $(D^3 + 3D^2 + 3D + 1)y = 0$

43. $(D - 2)^4y = 0$ **44.** $(D - 1)^2(D + 2)^3y = 0$

45. $(D^2 + 1)y = 0$; $y = 0$ when $x = 0$, and $y = 1$ when $x = \pi/2$

46. $(D^2 + 4D + 4)y = 0$; if $x = 0$, $y = 0$ and if $x = -1$, $y = 2$

47. $(D^2 - 2D + 2)y = 0$; if $x = 0$, then $y = 0$ and $Dy = -1$

48. Repeat Exercise 47 for the following conditions: $y = 0$ when $x = 0$ and $y = 2$ when $x = \pi/4$.

49. Solve the equation

$$(D^4 + 18D^2 + 81)y = 0$$

by writing the solution in the form

$$y = c_1e^{Ax} + c_2xe^{Ax} + c_3e^{Bx} + c_4xe^{Bx}$$

and using Euler's identity.

28.3   Nonhomogeneous Equations

So far in this chapter all the equations have been homogeneous. We are now going to turn our attention to **nonhomogeneous equations**, equations for which the right side is not zero.

To get an overview of the problem, let us examine a typical nonhomogeneous equation of second order:

$$b_0D^2y + b_1Dy + b_2y = f(x) \tag{28.8}$$

If the equation *were* homogeneous, then we would know how to solve it. Let us denote by y_c the solution of the corresponding homogeneous equation, so that

$$b_0D^2y_c + b_1Dy_c + b_2y_c = 0 \tag{28.9}$$

However, being able to find y_c does not seem to accomplish much. So let us pretend that we have somehow managed to get some particular solution y_p (no arbitrary constants) to Equation (28.8). In other words, y_p is such that

$$b_0D^2y_p + b_1Dy_p + b_2y_p = f(x) \tag{28.10}$$

Now we draw a surprising conclusion: the function

$$y = y_c + y_p$$

is the general solution of Equation (28.8). This statement can be checked by direct substitution:

$$b_0 D^2(y_c + y_p) + b_1 D(y_c + y_p) + b_2(y_c + y_p)$$
$$= (b_0 D^2 y_c + b_1 D y_c + b_2 y_c) + (b_0 D^2 y_p + b_1 D y_p + b_2 y_p)$$
$$= 0 + f(x) = f(x)$$

by (28.9) and (28.10). Since $y = y_c + y_p$ contains the required number of arbitrary constants, it must be the general solution. The physical significance of y_c and y_p will be considered in the next section.

General Solution of a Nonhomogeneous Equation: The **general solution of a nonhomogeneous equation** is

$$y = y_c + y_p$$

where y_p is some particular solution of the equation and y_c is the general solution of the corresponding homogeneous equation.

To solve a nonhomogeneous equation, we find y_c, the solution of the corresponding homogeneous equation, by the methods studied in the last two sections; y_c is called the **complementary solution**. The real task is to find a particular solution y_p. Several techniques for finding y_p are available. We will confine ourselves to the **method of undetermined coefficients**, which is entirely adequate for most physical problems. This method has two subheadings, the **method of inspection**, which we try to use whenever possible, and the **annihilator method**, to be used whenever mere inspection fails. We will illustrate the different cases by several examples.

Inspection Method

E X A M P L E **1** Solve the equation $(D^2 + 2)y = 2e^x$.

Solution. The first step is to find y_c, the solution of

$$(D^2 + 2)y = 0$$

From the auxiliary equation $m^2 + 2 = 0$, we find that $m = \pm\sqrt{2}j$, so that

$$y_c = c_1 \cos\sqrt{2}x + c_2 \sin\sqrt{2}x$$

That was easy. Since we really have no idea how to find y_p, we first ask ourselves: what would a particular solution look like? Could it have the form $y_p = Ax$ or $y_p = A \sin x$? Not likely. A reasonable assumption is $y_p = Ae^x$ for some constant A, as we can readily see from the equation itself. This is what is meant by "inspection." To find the constant A, we simply substitute $y_p = Ae^x$

in the given equation, noting that $y_p'' = Ae^x$. We get

$$(D^2 + 2)y = 2e^x \qquad \text{given equation}$$

$$y'' + 2y = 2e^x \qquad \text{same equation}$$

$$Ae^x + 2Ae^x = 2e^x \qquad y_p = Ae^x,\ y_p'' = Ae^x$$

or

$$3Ae^x = 2e^x$$

Equality holds only if $3A = 2$ or $A = \frac{2}{3}$. (Hence the name *undetermined coefficients*.) So $y_p = \frac{2}{3}e^x$, and the solution of the equation is therefore given by

$$y = y_c + y_p = c_1 \cos \sqrt{2}x + c_2 \sin \sqrt{2}x + \tfrac{2}{3}e^x$$

The same method can be applied to equations of higher order.   ◄

Inspection Method: To see how the form y_p can be obtained by inspection, observe that the function $f(x)$ on the right side must be a linear combination of y_p and its derivatives. Consequently, y_p has to contain $f(x)$ as well as its derivatives up to the order of the equation. For example, if $f(x) = \sin x$ or $f(x) = \cos x$, we would choose $y_p = A \cos x + B \sin x$. If $f(x) = 2x^2$, then $y_p = Ax^2 + Bx + C$. If $f(x) = 2 \sin 3x - 3e^{-x}$, we let $y_p = A \sin 3x + B \cos 3x + Ce^{-x}$, and so on.

E X A M P L E **2**   **a.** If the right side of the equation is $f(x) = 5x^2$, then y_p has the form $y_p = Ax^2 + Bx + C$. Here Ax^2 is included to account for the term $5x^2$, Bx to account for any derivative of the Ax^2 term, and C to account for any second derivative of Ax^2.

b. If the right side is $2x - e^{3x}$, then y_p has the form $y_p = Ax + B + Ce^{3x}$. Here Ax is included to account for the term $2x$ and B to account for any derivative of the Ax term. To account for $-e^{3x}$, we include Ce^{3x}. No other terms are needed, however, since the derivatives of Ce^{3x} all have this same form.

c. If the right side is $2e^{-x} + 4 \cos 2x$, then $y_p = Ae^{-x} + B \cos 2x + C \sin 2x$. As in part (b), Ae^{-x} is included to account for the term $2e^{-x}$, but no other terms are required. To account for $4 \cos 2x$, we include $B \cos 2x$; $C \sin 2x$ is needed to account for any derivative of $B \cos 2x$. Since the second derivative of $B \cos 2x$ is another cosine function, no other terms are required.

d. If $f(x) = 3 + 6 \cos 4x - 7 \sin 4x$, then $y_p = A + B \cos 4x + C \sin 4x$.

e. If $f(x) = 6x^2 - 4$, then $y_p = Ax^2 + Bx + C$.   ◄

E X A M P L E **3**   Solve the differential equation $(D^2 - 3D + 2)y = 3x$.

Solution. To find y_c we note that the auxiliary equation $m^2 - 3m + 2$ has roots $m = 1$ and 2. Hence

$$y_c = c_1 e^x + c_2 e^{2x}$$

The right side, $3x$, must be a linear combination of y_p and its derivatives. Since $d(3x)/dx = 3$, a constant, we choose

$$y_p = Ax + B$$

Now we compute $y'_p = A$ and $y''_p = 0$ and substitute into the given equation:

$$(D^2 - 3D + 2)y = 3x$$
$$D^2y - 3Dy + 2y = 3x$$
$$0 - 3A + 2(Ax + B) = 3x \qquad y''_p = 0,\ y'_p = A,\ y_p = Ax + B$$

or

$$2Ax + (-3A + 2B) = 3x + 0$$

Comparing the corresponding coefficients, we get the following system of equations:

$$\left. \begin{array}{l} 2A = 3 \\ -3A + 2B = 0 \end{array} \right\} \qquad \begin{array}{l} \text{coefficients of } x \\ \text{constants} \end{array}$$

The solution is $A = \frac{3}{2}$ and $B = \frac{9}{4}$. Hence

$$y_p = \frac{3}{2}x + \frac{9}{4}$$

Since $y = y_c + y_p$, the general solution is

$$y = c_1 e^x + c_2 e^{2x} + \frac{3}{2}x + \frac{9}{4} \qquad \blacktriangleleft$$

Caution. Lack of experience often results in the wrong choice for y_p. Suppose in Example 3 we had used $y_p = Ax$, ignoring the derivative of $3x$; then $y'_p = A$ and, after substituting,

$$-3A + 2Ax = 3x$$

This implies that $A = 0$ and $A = \frac{3}{2}$, which is impossible. We have simply "overworked" the A.

E X A M P L E **4**   Solve the equation $(D^2 + 1)y = 4e^x - \sin 2x$.

Solution. The complementary solution is

$$y_c = c_1 \cos x + c_2 \sin x$$

Since $d(4e^x)/dx = 4e^x$ and $d(-\sin 2x)/dx = -2\cos 2x$, we use

$$y_p = Ae^x + B\cos 2x + C\sin 2x$$

Also,

$$y_p' = Ae^x - 2B\sin 2x + 2C\cos 2x$$

and

$$y_p'' = Ae^x - 4B\cos 2x - 4C\sin 2x$$

Substituting in $(D^2 + 1)y = 4e^x - \sin 2x$, we get

$$D^2 y_p + y_p = (Ae^x - 4B\cos 2x - 4C\sin 2x) + (Ae^x + B\cos 2x + C\sin 2x)$$
$$= 4e^x - \sin 2x$$

or

$$2Ae^x - 3B\cos 2x - 3C\sin 2x = 4e^x - \sin 2x$$

Comparing coefficients,

$$2A = 4 \qquad -3B = 0 \qquad -3C = -1$$

so that $A = 2$, $B = 0$, and $C = \frac{1}{3}$. So $y_p = 2e^x + \frac{1}{3}\sin 2x$, and the general solution is

$$y = c_1\cos x + c_2\sin x + 2e^x + \frac{1}{3}\sin 2x$$

◀

Annihilator Method (Optional)

Remark. The remainder of this section is devoted to a brief discussion of the "annihilator method." If desired, this topic may be omitted. The exercises in which this method is needed are so designated.

E X A M P L E **5**   Solve the equation

$$(D^2 - 4D + 3)y = e^x \tag{28.11}$$

Solution. From the auxiliary equation

$$m^2 - 4m + 3 = (m - 1)(m - 3) = 0$$

we obtain $m = 1$ and $m = 3$. Hence

$$y_c = c_1 e^{3x} + c_2 e^x$$

Proceeding exactly as we did before, we choose

$$y_p = Ae^x$$

Substituting, we get

$$Ae^x - 4Ae^x + 3Ae^x = 0 \neq e^x$$

The failure of the inspection method could have been predicted from the outset since Ae^x is one of the terms in y_c. Consequently, Ae^x satisfies the corresponding homogeneous equation automatically and so cannot be a solution to the given equation.

To find the correct form of y_p, we start with the right side e^x and work backward: if e^x is a solution to an equation, this equation can be constructed by noting that the root of its auxiliary equation is $m' = 1$ (to yield ce^x). Hence $m' - 1 = 0$ is the auxiliary equation and

$$(D - 1)y = 0$$

the corresponding differential equation. Hence

$$(D - 1)e^x = 0$$

Annihilator

The operator $D - 1$ is called the **annihilator** since applying $D - 1$ to both sides of the given equation "annihilates" the right side:

$$(D - 1)(D^2 - 4D + 3)y = (D - 1)e^x = 0 \tag{28.12}$$

Moreover, any solution of Equation (28.11) must be a solution of Equation (28.12). Solving the homogeneous equation (28.12), we get from

$$(D - 1)^2(D - 3)y = 0$$

the solution

$$y = c_1 e^{3x} + c_2 e^x + c_3 x e^x$$

Since the first two terms on the right side coincide with y_c, the last term must be the form of y_p. Hence

$$y_p = Axe^x$$

The constant A is found in the usual way and turns out to be $-\frac{1}{2}$. The solution is therefore given by

$$y = c_1 e^{3x} + c_2 e^x - \frac{1}{2} x e^x$$

◀

The annihilator method can be employed to solve any nonhomogeneous equation in this section but is needlessly complicated if y_p can be determined directly by inspection.

E X A M P L E **6** Determine the form of y_p in the solution of the equation

$$(D^2 - 4D + 3)y = 4xe^x$$

Solution. The left side is identical to that of the equation in Example 5. To construct the annihilator, note that if $4xe^x$ is the solution of an equation, then $m' = 1$ and 1, a repeating root. Hence the auxiliary equation is $(m' - 1)^2 = 0$ and the annihilator $(D - 1)^2$. Applying this operator to both sides of the given equation, we get

$$(D - 1)^2(D^2 - 4D + 3)y = (D - 1)^2(4xe^x) = 0$$

or

$$(D - 1)^3(D - 3)y = 0$$

Hence

$$y = c_1e^{3x} + c_2e^x + c_3xe^x + c_4x^2e^x$$

Since

$$y_c = c_1e^{3x} + c_2e^x$$

we conclude that

$$y_p = Axe^x + Bx^2e^x$$

(Note that $m = 1$ is also one of the roots of the auxiliary equation $m^2 - 4m + 3 = 0$. In fact, *the annihilator method is ordinarily used whenever a root of the auxiliary equation coincides with a root associated with the annihilator.*) ◀

E X E R C I S E S / S E C T I O N   **28.3**

In Exercises 1–26, solve the differential equations.

1. $(D^2 - 6D + 9)y = e^x$
2. $(D^2 - 6D + 9)y = 2$
3. $(D^2 - 6D + 9)y = 9x$
4. $(D^2 - 6D + 9)y = 18x^2$
5. $(D^2 - D - 2)y = 2x^2$
6. $(D^2 - 3D - 4)y = -4x$
7. $(D^2 - D + 1)y = 1 - x^2$
8. $(D^2 + 4)y = 2x^2 - x$
9. $(D^2 + 4D + 3)y = 6 + e^x$
10. $(D^2 - 4)y = -5\cos x$
11. $(D^2 + 1)y = 6\sin 2x$
12. $(D^2 - 3D - 4)y = \cos x$
13. $(D^2 + 5D + 6)y = 4\cos x + 6\sin x$

14. $(D^2 - 4)y = 8x + 12$
15. $(D^2 - 2D + 1)y = 10$
16. $(D^2 - 1)y = e^{2x} + 4$
17. $(D^3 - 2D^2 - D + 2)y = 8e^{3x}$
18. $(D^2 - 2D + 1)y = x^2 - 1$
19. $(D^2 - 2D + 5)y = 4xe^x$ (Since $(d/dx)(4xe^x) = 4xe^x + 4e^x$,
$y_p = Axe^x + Be^x$.)
20. $(D^2 + 3)y = 8xe^x + 3$

21. $(D^2 + 9)y = 9e^{3x}$; if $x = 0$, then $y = 1$ and $Dy = \frac{3}{2}$

22. $(D^2 + 1)y = x$; if $x = 0$, then $y = 1$ and $Dy = 1$

23. $(D^2 + 1)y = 6 \cos 2x$; if $x = 0$, then $y = 3$ and $Dy = 1$

24. $(D^2 - 1)y = -16 \sin x$; if $x = 0$, then $y = 0$ and $Dy = 1$

25. $(D^2 + 2D + 5)y = 10 \cos x$; if $x = 0$, then $y = 5$ and $Dy = 6$

26. $(D^2 - 2D + 5)y = 5 \sin x$; if $x = 0$, then $y = -2$ and $Dy = 0$

In Exercises 27–40, find y_p.

27. $(D^2 + 4)y = 10 \sin 3x$

28. $(D^2 - 6)y = 5 \cos 2x$

29. $(D^2 - D + 2)y = \cos x$

30. $(D^2 - 2D + 2)y = 5 \sin 2x$

31. $(D^2 + D - 2)y = 2x^2 + 1$

32. $(D^2 + 5D + 1)y = x$

33. $(D^2 + 2)y = 2e^{2x} + 2$

34. $(D^2 + 5)y = 3xe^x$

35. $(D^2 - D - 4)y = x + 2e^{3x}$

36. $(D^2 - 6)y = e^x - \sin x$

37. $(D^2 + 3)y = 2x + \cos x$

38. $(D^2 - 2)y = x^2 + 2e^{-x}$

39. $(D^2 - D - 3)y = 6xe^x$

40. $(D^2 - 2D + 3)y = 4xe^x$

In Exercises 41–51, use the annihilator method.

41. $(D^2 - 4)y = 8e^{2x}$

42. $(D^2 - 9)y = 10e^{-3x}$

43. $(D^2 - D - 6)y = 10e^{-2x}$

44. $(D^2 + 2D - 15)y = 12e^{3x}$

45. $(D^2 + 3D - 28)y = 11e^{4x}$

46. $(D^2 + 4D - 12)y = 4e^{2x}$

47. $(D^2 - 1)y = 2e^x$

48. $(D^2 + D - 2)y = 3e^x$

49. $(D^2 - 4)y = 4 + e^{2x}$

50. $(D^3 - 3D^2 + 3D - 1)y = e^x - 3$

51. $(D^2 + 1)y = 2 \sin x$ (*Hint:* Since $m' = \pm j$, the annihilator is $D^2 + 1$.)

28.4 Applications of Second-Order Equations

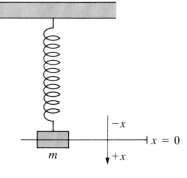

Figure 28.1

The main applications of second-order linear equations involve a weight oscillating on a spring and the electrical analog.

Recall that a spring stretched x units beyond its natural length pulls back with a force kx by Hooke's law; k is called the **spring constant**. If the spring is compressed, then it pushes back with the same force kx. Suppose a mass hanging on a spring is allowed to come to rest (Figure 28.1). Let $x = 0$ be the equilibrium position; the x-coordinate in the downward direction is considered positive and the upward, negative. Then by Newton's second law, the force exerted by the mass on the spring is given by

$$m \frac{d^2x}{dt^2} \qquad \text{mass} \times \text{acceleration}$$

Since the spring exerts a force kx by Hooke's law, it follows that

$$m \frac{d^2x}{dt^2} = -kx \tag{28.13}$$

The negative sign indicates that the forces act in opposite directions. If, as in the case of a falling body, we assume that the damping force due to the resistance of the surrounding medium is proportional to the velocity dx/dt, then formula (28.13) must be modified as follows:

$$m \frac{d^2x}{dt^2} = -kx - b \frac{dx}{dt} \qquad \text{or} \qquad m \frac{d^2x}{dt^2} + b \frac{dx}{dt} + kx = 0$$

Fundamental Equation of Damped Oscillatory Motion

$$m \frac{d^2x}{dt^2} + b \frac{dx}{dt} + kx = 0 \tag{28.14}$$

E X A M P L E **1**   A 2-lb weight stretches a spring 6 inches. The weight is attached to the spring and allowed to come to rest (equilibrium position). The weight is then pulled 4 inches below the equilibrium position and released. Determine the motion of the weight as a function of time, assuming that the damping force is negligible.

Solution. To determine the spring constant, we use the information in the first sentence: **2** lb stretches the spring 6 inches $= \frac{1}{2}$ ft. So by Hooke's law,

$$F = kx$$

$$2 = k \left(\frac{1}{2} \right) \qquad \text{or} \qquad k = 4$$

Since $g = 32$ ft/s^2, we have 32 lb $= 1$ slug, so that the mass of the 2-lb weight is

$$\frac{2}{32} = \frac{1}{16} \text{ slug}$$

Finally, since the damping force is negligible, $b = 0$. So by Equation (28.14),

$$m \frac{d^2x}{dt^2} + b \frac{dx}{dt} + kx = 0$$

$$\frac{1}{16} \frac{d^2x}{dt^2} + 4x = 0 \qquad m = \frac{1}{16}, b = 0, k = 4$$

or

$$\frac{d^2x}{dt^2} + 64x = 0 \qquad \text{multiplying by 16}$$

Since the weight is initially at rest 4 inches below the equilibrium position, we have the following initial conditions:

1. When $t = 0$, $x = 4$ in. $= \frac{1}{3}$ ft.   initial position
2. When $t = 0$, $dx/dt = 0$.   initial velocity

We now solve the equation, making use of the initial conditions:

$$\frac{d^2x}{dt^2} + 64x = 0$$

$$m^2 + 64 = 0 \qquad \textbf{auxiliary equation}$$

$$m = \pm\sqrt{-64} = \pm 8j$$

It follows that

$$x(t) = c_1 \cos 8t + c_2 \sin 8t$$

Substituting $t = 0$ and $x = \frac{1}{3}$, we get

$$\frac{1}{3} = c_1 \cos 0 + c_2 \sin 0 \qquad \text{or} \qquad c_1 = \frac{1}{3}$$

Thus

$$x(t) = \frac{1}{3}\cos 8t + c_2 \sin 8t$$

Now

$$\frac{dx}{dt} = x'(t) = -\frac{8}{3}\sin 8t + 8c_2 \cos 8t$$

From the condition $dx/dt = 0$ when $t = 0$, we have

$$0 = 0 + 8c_2 \qquad \text{or} \qquad c_2 = 0$$

Hence from

$$x(t) = \frac{1}{3}\cos 8t + c_2 \sin 8t$$

the final solution is (since $c_2 = 0$)

$$x(t) = \frac{1}{3}\cos 8t \qquad \blacktriangleleft$$

So far we have assumed that $g = 10 \text{ m/s}^2$ for the acceleration due to gravity. In this section we will use the more accurate value $g = 9.8 \text{ m/s}^2$.

E X A M P L E **2**   A weight of mass 0.50 kg (weight = 4.9 N) stretches a spring 0.70 m. The weight is pushed 0.40 m above the equilibrium position and released. Find the position of the weight as a function of time, if a damping force numerically equal to twice the velocity is present.

Solution. First we determine the spring constant. Since $F = 4.9$ N when $x = 0.70$ m, we get by Hooke's law

$$4.9 = k(0.70) \quad \text{or} \quad k = 7.0$$

The initial conditions are:

1. When $t = 0$, $x = -0.40$. upward negative
2. When $t = 0$, $dx/dt = 0$. initial velocity

Since the damping force is numerically equal to twice the velocity, this force must be $2\, dx/dt$, so that $b = 2$. The resulting equation is

$$m\frac{d^2x}{dt^2} + b\frac{dx}{dt} + kx = 0$$

$$0.50\frac{d^2x}{dt^2} + 2\frac{dx}{dt} + 7.0x = 0 \qquad m = 0.50,\ b = 2,\ k = 7.0$$

or

$$\frac{d^2x}{dt^2} + 4\frac{dx}{dt} + 14x = 0 \qquad \text{multiplying by 2}$$

$$m^2 + 4m + 14 = 0 \qquad \text{auxiliary equation}$$

$$m = \frac{-4 \pm \sqrt{16 - 4(14)}}{2} = \frac{-4 \pm \sqrt{-40}}{2}$$

$$= -2 \pm \sqrt{10}j$$

The general solution is

$$x(t) = e^{-2t}(c_1 \cos \sqrt{10}t + c_2 \sin \sqrt{10}t)$$

From the condition $x = -0.40$ when $t = 0$, we get

$$-0.40 = e^0(c_1 \cos 0 + c_2 \sin 0)$$

or

$$c_1 = -0.40$$

So

$$x(t) = e^{-2t}(-0.40 \cos \sqrt{10}t + c_2 \sin \sqrt{10}t)$$

Now

$$\frac{dx}{dt} = e^{-2t}(0.40\sqrt{10} \sin \sqrt{10}t + c_2\sqrt{10} \cos \sqrt{10}t)$$

$$+ (-2e^{-2t})(-0.40 \cos \sqrt{10}t + c_2 \sin \sqrt{10}t)$$

Since $dx/dt = 0$ when $t = 0$, we have

$$0 = (0 + c_2\sqrt{10}) - 2(-0.40 + 0)$$
$$0 = c_2\sqrt{10} + 0.80$$
$$c_2 = -\frac{0.80}{\sqrt{10}}$$

The solution is therefore given by

$$x(t) = e^{-2t}\left(-0.40 \cos\sqrt{10}t - \frac{0.80}{\sqrt{10}} \sin\sqrt{10}t\right)$$

Using two significant digits, we get

$$x(t) = e^{-2.0t}(-0.40 \cos 3.2t - 0.25 \sin 3.2t)$$

The graph of the solution is shown in Figure 28.2.

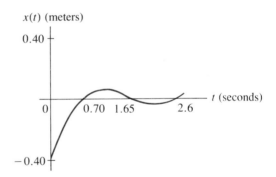

Figure 28.2   ◀

Note that the motion of the weight in Example 2 is oscillatory, but the oscillations gradually damp out due to the factor $e^{-2.0t}$. (See Figure 28.2.) If the damping force is neglected (as in Example 1), the equation becomes

$$0.50\frac{d^2x}{dt^2} + 7.0x = 0$$

$$\frac{d^2x}{dt^2} + 14x = 0$$

and

$$x(t) = c_1 \cos\sqrt{14}t + c_2 \sin\sqrt{14}t$$

or

$$x(t) = c_1 \cos 3.7t + c_2 \sin 3.7t$$

From the initial conditions we now obtain

$$x(t) = -0.40 \cos 3.7t$$

This last equation describes **simple harmonic motion**.

In the general case (28.14) the roots of the auxiliary equation are given by

$$\frac{-b \pm \sqrt{b^2 - 4mk}}{2m}$$

Consequently, the solution is sinusoidal (containing sines and cosines) and the motion oscillatory whenever these roots are complex numbers—that is, whenever

$$b^2 - 4mk < 0 \qquad \text{or} \qquad b < 2\sqrt{mk}$$

Such a system is said to be **underdamped**. If

$$b > 2\sqrt{mk}$$

then the motion is not oscillatory and the system is said to be **overdamped**. If

$$b = 2\sqrt{mk}$$

the system is said to be **critically damped**.

An additional vertical force $f(t)$ may act on the system. (For example, $f(t)$ may be due to the motion of the support or the presence of a magnetic field.) Then Equation (28.14) becomes

$$m\frac{d^2x}{dt^2} + b\frac{dx}{dt} + kx = f(t) \qquad\qquad (28.15)$$

Such cases are called **forced oscillations**.

E X A M P L E **3**   Suppose the system in Example 1 is acted on by an external force $f(t) = \frac{1}{8}\sin 4t$. Determine the motion of the weight.

Solution. The equation in Example 1,

$$\frac{1}{16}\frac{d^2x}{dt^2} + 4x = 0$$

must be modified to take into account the external force:

$$\frac{1}{16}\frac{d^2x}{dt^2} + 4x = \frac{1}{8}\sin 4t$$

or

$$\frac{d^2x}{dt^2} + 64x = 2 \sin 4t$$

The initial conditions are still the same:

1. When $t = 0$, $x = \frac{1}{3}$ ft.
2. When $t = 0$, $dx/dt = 0$.

The solution obtained in Example 1 now serves as the complementary solution:

$$x_c = c_1 \cos 8t + c_2 \sin 8t$$

By the method of Section 28.3,

$$x_p = \frac{1}{24} \sin 4t$$

so that the general solution is

$$x(t) = c_1 \cos 8t + c_2 \sin 8t + \frac{1}{24} \sin 4t$$

The constants are now evaluated as in Examples 1 and 2. The final solution is

$$x(t) = \frac{1}{3} \cos 8t - \frac{1}{48} \sin 8t + \frac{1}{24} \sin 4t \qquad \blacktriangleleft$$

The physical situation we have dealt with so far has an electrical analog (Figure 28.3). Recall that the voltage across an inductor, resistor, and capacitor is given by

$$L\frac{di}{dt}, \quad Ri, \text{ and } \frac{q}{C}$$

respectively. Since $i = dq/dt$, these expressions can be written

$$L\frac{d^2q}{dt^2}, \quad R\frac{dq}{dt}, \text{ and } \frac{q}{C}$$

If the components are connected in series with a generator, then the impressed voltage $e(t)$ is equal to the sum of the voltages across the components, known as **Kirchhoff's voltage law**.

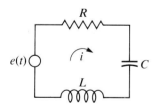

Figure 28.3

Differential Equation of the Circuit in Figure 28.3:

$$L\frac{d^2q}{dt^2} + R\frac{dq}{dt} + \frac{q}{C} = e(t) \qquad (28.16)$$

As before, if $e(t) = 0$, then the roots of the auxiliary equation tell us whether or not the solution is sinusoidal (Exercise 20). In the oscillatory case, the damping factor will cause the current to die out quickly unless a voltage source is present.

E X A M P L E **4**   Find an expression for the charge on the capacitor in Figure 28.3 as a function of time if $L = 1$ H, $R = 15 \ \Omega$, $C = 10^{-2}$ F, and $e(t) = 100 \sin 60t$.

Solution. By Equation (28.16)

$$\frac{d^2q}{dt^2} + 15 \frac{dq}{dt} + 100q = 100 \sin 60t$$

To obtain the complementary solution q_c, we note that

$$m^2 + 15m + 100 = 0$$

and

$$m = \frac{-15 \pm \sqrt{225 - 400}}{2} = -\frac{15}{2} \pm \frac{5}{2} \sqrt{7}j$$

Hence

$$q_c = e^{-15t/2} \left(c_1 \cos \frac{5}{2} \sqrt{7}t + c_2 \sin \frac{5}{2} \sqrt{7}t \right)$$

For the particular solution we choose

$$q_p = A \cos 60t + B \sin 60t$$

so that

$$\frac{dq_p}{dt} = -60A \sin 60t + 60B \cos 60t$$

$$\frac{d^2q_p}{dt^2} = -3600A \cos 60t - 3600B \sin 60t$$

Substituting, we get

$$-3600A \cos 60t - 3600B \sin 60t - 900A \sin 60t$$
$$+ 900B \cos 60t + 100A \cos 60t + 100B \sin 60t = 100 \sin 60t$$

Collecting terms,

$$(-3500A + 900B) \cos 60t + (-900A - 3500B) \sin 60t = 100 \sin 60t$$

Comparing coefficients, we get the system of equations

$$\left.\begin{array}{r} -3500A + 900B = 0 \\ -900A - 3500B = 100 \end{array}\right\}$$

whose solution set is $A = -0.0069$ and $B = -0.027$. Finally,

$$q(t) = q_c + q_p = e^{-15t/2}\left(c_1 \cos \frac{5}{2}\sqrt{7}t + c_2 \sin \frac{5}{2}\sqrt{7}t\right)$$
$$-0.0069 \cos 60t - 0.027 \sin 60t \qquad \blacktriangleleft$$

Suppose we look at the solution in Example 4 more closely. The complementary solution q_c contains the exponential decaying factor $e^{-15t/2}$, but q_p does not. Consequently, q_c will die out quickly, leaving only q_p. For this reason q_c is called the **transient** part of the solution, while q_p is called the **steady-state solution**. In other words, after a certain time period the solution is essentially given by q_p, hence the name *steady state*. Moreover, in many problems only the steady-state solution may actually be of interest.

Transient
Steady state

E X A M P L E **5** Find the steady-state current in the circuit in Example 4.

Solution. As noted above, the steady-state solution is

$$q_p = -0.0069 \cos 60t - 0.027 \sin 60t$$

Consequently, the steady-state current is

$$i = \frac{dq_p}{dt} = 0.41 \sin 60t - 1.62 \cos 60t \qquad \blacktriangleleft$$

E X E R C I S E S / S E C T I O N **28.4**

1. A spring is such that a 4-lb weight stretches it 6 inches. The weight is attached to the spring and allowed to reach the equilibrium position. The weight is then pulled 3 inches below the equilibrium position and released. Find the motion of the weight as a function of time, assuming no damping.

2. A 2-lb weight stretches a spring 6 inches. The weight is pushed 7 inches above the equilibrium position and released. Find the motion of the weight as a function of time, assuming no damping.

3. Find the motion of the weight in Exercise 1 if an external force $f(t) = \frac{1}{4}\cos 6t$ acts on the system.

4. Find the motion of the weight in Exercise 2 if the external force is $\frac{1}{4}\sin 4t$.

5. A 12-lb weight stretches a spring 2 ft. The weight is pulled 8 inches below the equilibrium position and given an initial downward velocity of 3 ft/s. (Recall that the down-

ward direction is positive.) Find the motion of the weight as a function of time, assuming that the damping force may be neglected.

6. A 5-lb weight stretches a spring 6 inches. The weight is pulled 4 inches below the equilibrium position and given an initial downward velocity of 4 ft/s. Find the motion of the weight as a function of time. (Assume that the damping force may be neglected.)

7. Find the motion of the weight in Exercise 5 if the initial velocity is 4 ft/s in the upward direction.

8. Find the motion of the weight in Exercise 6 if the initial velocity is 6 ft/s in the upward direction.

9. A 4-lb weight stretches a spring 2 ft. The weight is pulled 6 inches below the equilibrium position and released. Find the motion of the weight as a function of time, given that a damping force numerically equal to one-half the velocity is present.

10. A 2-lb weight stretches a spring 6 inches. The weight is pulled 3 inches below the equilibrium position and released. Find $x(t)$, given that a damping force numerically equal to the velocity is present.

11. Repeat Exercise 10, given that an external force $2 \sin 8t$ acts on the system.

12. Repeat Exercise 10, given that the system is acted on by an external force $\frac{1}{4} \cos 8t$.

13. A weight of mass 2.0 kg (19.6 N) hanging on a spring stretches the spring 0.098 m. If this weight is pulled 0.25 m below the equilibrium position and released, find the motion of the weight as a function of time. (We are assuming that the damping force may be neglected.)

14. Find the motion of the weight in Exercise 13 if the system is acted on by an external force $20 \sin 5t$.

15. Find the motion of the weight in Exercise 13 if a damping force equal to four times the velocity is present.

16. In Exercise 15, for what values of b will the motion be oscillatory?

17. A mass of 2.0 kg (19.6 N) stretches a certain spring 0.098 m. Suppose that the damping force is numerically equal to four times the velocity and that the external force is $20 \sin 5t$. If the weight is attached to the spring and pulled 0.25 m below the equilibrium position and released, find the transient and steady-state solutions. (Refer to Exercise 15.)

18. A weight of 0.50 kg (4.9 N) stretches a spring 0.49 m. If a damping force equal to 2.8 times the velocity and an external force of $3.0 \cos 2t$ are present, find the steady-state motion of the attached weight.

19. For a given LC circuit, $L = 1$ H and $C = 1.0 \times 10^{-4}$ F. Find the charge and current as functions of time if $i = 10$ and $q = 0$ when $t = 0$. (Assume that $e(t) \equiv 0$.)

20. If $L = 1$ H and $C = 1.0 \times 10^{-4}$ F in an LRC circuit, find the range on R for which the current is oscillatory. (Assume that $e(t) \equiv 0$.)

21. For a given LC circuit, $L = 0.5$ H, $C = 8 \times 10^{-4}$ F, and $e(t) = 50 \sin 100t$. Find the charge as a function of time if $i(0) = q(0) = 0$.

22. Find the steady-state current of the following LRC circuit: $L = 1$ H, $C = 10^{-2}$ F, $R = 50$ Ω, and $e(t) = 100 \sin 50t$.

23. A weight of mass 1.2 kg is attached to a spring with spring constant $k = 80$ N/m. If the damping force is numerically equal to $1.5 \, dx/dt$ and the external force is $10 \sin 5t$, find the steady-state motion of the weight.

24. A weight of mass 2.4 kg is attached to a spring with spring constant $k = 150$ N/m. If the damping force is numerically equal to $1.25 \, dx/dt$ and an external force of $20 \sin 3t$ is present, find the steady-state motion of the attached weight.

25. Find the steady-state current of the following LRC circuit: $L = 1$ H, $R = 10$ Ω, $C = \frac{1}{100}$ F, and $e(t) = 50 \cos 10t$.

REVIEW EXERCISES/CHAPTER 28

In Exercises 1–22, solve the differential equations.

1. $D^4 y = 0$

2. $(D^2 - 4)y = 0$

3. $(D^2 + 4)y = 0$

4. $(D - 4)^2 y = 0$

5. $(D^2 - 2D - 2)y = 0$

6. $(D^2 - 2D + 2)y = 0$

7. $(3D^2 - D + 1)y = 0$

8. $(2D^2 + D + 2)y = 0$

9. $(D - 2)^2(D^2 + 1)y = 0$

10. $(D^5 + 9D^3)y = 0$

11. $2\dfrac{d^2 y}{dx^2} - \dfrac{dy}{dx} + y = 0$

12. $2\dfrac{d^2 y}{dx^2} - 2\dfrac{dy}{dx} - y = 0$

13. $3\dfrac{d^2 y}{dx^2} - 2\dfrac{dy}{dx} - 2y = 0$

14. $\dfrac{d^2 y}{dx^2} - 2\dfrac{dy}{dx} + 2y = 0$

15. $(D^2 - 3D - 4)y = 6e^x$

16. $(D^2 - 3D - 4)y = 8x$

17. $(D^2 - 3D - 4)y = 2 \sin x$

18. $(D^2 - 3D - 4)y = 20e^{-x}$ (annihilator)

19. $(D^2 - 3D - 4)y = 10xe^{-x}$ (annihilator)

20. $(D^2 - 3D - 4)y = e^{4x}$ (annihilator)

21. $(D^2 - 2D - 3)y = 0$; if $x = 0$, then $y = 0$ and $Dy = -4$

22. $(D^2 + 1)y = 0$; if $x = 0$, then $y = 2$ and $Dy = 0$

23. Find the steady-state current of the following LRC circuit: $L = 0.100$ H, $C = 2.00 \times 10^{-3}$ F, $R = 40.0$ Ω, and $e(t) = 100.0 \cos 20.0t$.

24. A weight of 2.45 N (0.25 kg) stretches a spring 4.9 cm. If the weight is attached to the spring, show that the motion is oscillatory if $b < 5\sqrt{2} \approx 7.1$.

25. A 5-lb weight stretches a spring 6 inches. The weight is pulled 3 inches below the equilibrium position and given an initial upward velocity of 5 ft/s. Assume that an external force $\frac{1}{8} \cos 4t$ acts on the system and that the damping force may be neglected. Find the motion of the weight as a function of time.

The Laplace Transform

29.1 Introduction and Basic Properties

With the **Laplace transform** we are finally leaving the age of Euler for a brief glimpse into the twentieth century. Because of our sudden jump in time, it is difficult to motivate the definition of Laplace transform. Also, this concept was developed gradually over a period spanning several decades. But it is safe to say that in its modern form the Laplace transform can be traced to the English electrical engineer Oliver Heaviside (1850–1925), who discovered a unique method for solving differential equations arising in electrical circuit theory. Later attempts to justify Heaviside's methods led to the following definition:

> **Definition of the Laplace Transform:** If $f(t)$ is defined for $0 \leq t < \infty$, then the **Laplace transform** of f is defined to be the integral
>
> $$F(s) = L\{f(t)\} = \int_0^\infty e^{-st} f(t)\, dt \qquad (29.1)$$

Similar transforms had been studied earlier by the French mathematician Pierre Simon de Laplace (1749–1827) and even by Euler.

It is not difficult to see how Laplace transforms may be employed to solve differential equations. Since that is the topic of Section 29.4, we will first try to become acquainted with the transform and its basic properties.

Our main task in this section is to find the Laplace transforms of some special functions. Consider the following examples.

Find the Laplace transform of the function $f(t) = t$, $t \geq 0$.

Solution. By definition (29.1)

$$L\{t\} = \int_0^\infty e^{-st} \cdot t\, dt = \lim_{b \to \infty} \int_0^b t e^{-st}\, dt \qquad \text{improper integral}$$

We integrate by parts, letting $u = t$ and $dv = e^{-st} dt$. Since the variable s is a constant as far as the integration is concerned, we obtain $du = dt$ and $v = (-1/s)e^{-st}$. Hence

$$L\{t\} = \lim_{b \to \infty} \left(-\frac{t}{s} e^{-st} \Big|_0^b + \frac{1}{s} \int_0^b e^{-st} dt \right)$$

$$= \lim_{b \to \infty} \left[-\frac{t}{s} e^{-st} \Big|_0^b + \frac{1}{s} \left(-\frac{1}{s} \right) e^{-st} \Big|_0^b \right] \qquad d(-st) = -s\,dt$$

$$= \lim_{b \to \infty} \left(-\frac{b}{s} e^{-sb} - \frac{1}{s^2} e^{-sb} + \frac{1}{s^2} \right)$$

$$= \lim_{b \to \infty} \left(-\frac{b}{se^{sb}} \right) - \frac{1}{s^2} \lim_{b \to \infty} \frac{1}{e^{sb}} + \lim_{b \to \infty} \frac{1}{s^2}$$

$$= \lim_{b \to \infty} \left(-\frac{b}{se^{sb}} \right) - 0 + \frac{1}{s^2}$$

provided that $s > 0$. The remaining limit may be evaluated using L'Hospital's rule:

$$L\{t\} = \lim_{b \to \infty} \left(-\frac{1}{s^2 e^{sb}} \right) + \frac{1}{s^2} = \frac{1}{s^2}$$

We conclude that

$$L\{t\} = \frac{1}{s^2} \qquad (s > 0) \tag{29.2}$$

(Observe that s has to be positive to ensure the existence of the improper integral. Otherwise the variable s will play no role in our work, as we will see.) ◀

E X A M P L E **2**   Find the Laplace transform of the function $f(t) = e^{at}$.

Solution. Again, by formula (29.1),

$$L\{e^{at}\} = \int_0^\infty e^{-st} e^{at} \, dt$$

$$= \lim_{b \to \infty} \int_0^b e^{-(s-a)t} \, dt$$

$$= \lim_{b \to \infty} \frac{1}{-(s-a)} e^{-(s-a)t} \Big|_0^b \qquad \left[\begin{array}{l} u = -(s-a)t \\ du = -(s-a)\,dt \end{array} \right]$$

$$= \lim_{b \to \infty} \left(\frac{1}{-(s-a)} e^{-(s-a)b} + \frac{1}{s-a} \right) = \frac{1}{s-a}$$

provided that $s > a$. Hence

$$L\{e^{at}\} = \frac{1}{s - a} \qquad (s > a) \tag{29.3}$$

◀

We can see from these examples that a function $f(t)$ has a Laplace transform whenever the improper integral

$$\int_0^\infty e^{-st} f(t)\, dt$$

exists. For example, the functions $f(t) = \tan t$ and $f(t) = e^{t^2}$ do not possess transforms.

Table of Transforms

Rather than continuing with these calculations, we refer you to the accompanying table of common transforms, leaving a few additional cases as exercises.

Short Table of Laplace Transforms

$f(t)$	$F(s)$	$f(t)$	$F(s)$
1. 1	$\dfrac{1}{s}$	**9.** $t^n e^{at}$	$\dfrac{n!}{(s - a)^{n+1}}$
2. t	$\dfrac{1}{s^2}$	**10.** $1 - \cos at$	$\dfrac{a^2}{s(s^2 + a^2)}$
3. t^n	$\dfrac{n!}{s^{n+1}}$	**11.** $at - \sin at$	$\dfrac{a^3}{s^2(s^2 + a^2)}$
4. e^{at}	$\dfrac{1}{s - a}$	**12.** $\sin at - at \cos at$	$\dfrac{2a^3}{(s^2 + a^2)^2}$
5. $\sin at$	$\dfrac{a}{s^2 + a^2}$	**13.** $t \sin at$	$\dfrac{2as}{(s^2 + a^2)^2}$
6. $\cos at$	$\dfrac{s}{s^2 + a^2}$	**14.** $\sin at + at \cos at$	$\dfrac{2as^2}{(s^2 + a^2)^2}$
7. $e^{at} \sin bt$	$\dfrac{b}{(s - a)^2 + b^2}$	**15.** $t \cos at$	$\dfrac{s^2 - a^2}{(s^2 + a^2)^2}$
8. $e^{at} \cos bt$	$\dfrac{s - a}{(s - a)^2 + b^2}$		

Observe next that the definition of the Laplace transform implies that

$$L\{af(t) + bg(t)\} = aL\{f(t)\} + bL\{g(t)\} \tag{29.4}$$

Since it possesses property (29.4), the Laplace transform is said to be *linear*.

E X A M P L E **3**   Use the table and the linearity property to obtain

$$L\{4t^3 + 2 \sin 3t\}$$

Solution. By property (29.4) and transforms 3 and 5 in the table,

$$L\{4t^3 + 2 \sin 3t\} = 4L\{t^3\} + 2L\{\sin 3t\}$$

$$= 4\frac{3!}{s^4} + 2\frac{3}{s^2 + 9} = \frac{24}{s^4} + \frac{6}{s^2 + 9}$$

◀

E X A M P L E **4**   Find $L\{e^{-3t} + e^t \cos 4t\}$.

Solution. By transforms 4 and 8 in the table we get

$$\frac{1}{s + 3} + \frac{s - 1}{(s - 1)^2 + 16}$$

◀

29.2   Inverse Laplace Transforms

The procedure in the last section can be reversed: we can look up $f(t)$, given $F(s)$. For example, if $F(s) = 1/(s + 4)$, then $f(t) = e^{-4t}$ by transform 4. Here $f(t)$ is called the **inverse transform**, denoted by L^{-1}. Thus we may write

$$L^{-1}\left\{\frac{1}{s + 4}\right\} = e^{-4t}$$

Unfortunately, the forms in the table do not always fit, in which case an adjustment is required. For example,

$$L^{-1}\left\{\frac{3}{s + 4}\right\} = 3L^{-1}\left\{\frac{1}{s + 4}\right\} = 3e^{-4t}$$

which can be readily checked by reversing the steps. In general, then,

$$L^{-1}\{aF(s) + bG(s)\} = aL^{-1}\{F(s)\} + bL^{-1}\{G(s)\}$$

E X A M P L E **1**   Find

$$L^{-1}\left\{\frac{1}{s(s^2 + 4)}\right\}$$

Solution. By transform 10 for $a = 2$, we get

$$L^{-1}\left\{\frac{1}{s(s^2 + 4)}\right\} = L^{-1}\left\{\frac{\frac{1}{4} \cdot 4}{s(s^2 + 4)}\right\}$$

$$= \frac{1}{4} L^{-1}\left\{\frac{4}{s(s^2 + 4)}\right\} = \frac{1}{4}(1 - \cos 2t)$$ ◀

E X A M P L E **2** Find

$$L^{-1}\left\{\frac{s + 1}{s^2 + 4s + 8}\right\}$$

Solution. For trinomial denominators that are not factorable we may use transforms 7 and 8 after completing the square:

$$L^{-1}\left\{\frac{s + 1}{s^2 + 4s + 8}\right\} = L^{-1}\left\{\frac{s + 1}{(s + 2)^2 + 4}\right\}$$

Even now the form does not quite fit transform 8, but noting that $a = -2$, we may proceed as follows:

$$L^{-1}\left\{\frac{s + 1}{(s + 2)^2 + 4}\right\} = L^{-1}\left\{\frac{s + 2 - 2 + 1}{(s + 2)^2 + 4}\right\}$$

$$= L^{-1}\left\{\frac{(s + 2) - 1}{(s + 2)^2 + 4}\right\}$$

$$= L^{-1}\left\{\frac{s + 2}{(s + 2)^2 + 4}\right\} - L^{-1}\left\{\frac{1}{(s + 2)^2 + 4}\right\}$$

$$= L^{-1}\left\{\frac{s + 2}{(s + 2)^2 + 4}\right\} - \frac{1}{2} L^{-1}\left\{\frac{2}{(s + 2)^2 + 4}\right\}$$

$$= e^{-2t} \cos 2t - \frac{1}{2} e^{-2t} \sin 2t \qquad a = -2, b = 2$$

by transforms 8 and 7, respectively.   ◀

29.3 Partial Fractions

Transforms more complex than those considered in the previous section can often be broken up to fit the forms in the table. For example,

$$L^{-1}\left\{\frac{5}{(s - 1)(s + 4)}\right\} = L^{-1}\left\{\frac{1}{s - 1} - \frac{1}{s + 4}\right\} \qquad (29.5)$$

$$= e^t - e^{-4t}$$

by transform 4. The fractions on the right in Equation (29.5) are called **partial fractions**. Certain proper fractions (degree of the numerator less than the degree of the denominator) can be written as a sum of partial fractions according to the following rules:

I. If a *linear* factor $as + b$ occurs n times in the denominator, then there exist n partial fractions

$$\frac{A_1}{as + b} + \frac{A_2}{(as + b)^2} + \cdots + \frac{A_n}{(as + b)^n}$$

where $A_1, A_2, \ldots, A_n$ are constants.

II. If a *quadratic* factor $as^2 + bs + c$ occurs n times in the denominator, then there exist n partial fractions

$$\frac{A_1 s + B_1}{as^2 + bs + c} + \frac{A_2 s + B_2}{(as^2 + bs + c)^2} + \cdots + \frac{A_n s + B_n}{(as^2 + bs + c)^n}$$

where the A's and B's are constants. In all cases n may be equal to 1.

Rules I and II are only intended to be a general guide. How they are put to use will be illustrated in the examples.

E X A M P L E **1**　(*Distinct linear factors.*) Find

$$L^{-1}\left\{\frac{6s^2 + 12s - 6}{(s - 1)(s + 2)(s + 3)}\right\}$$

Solution. Since the factors are all distinct, we get by Rule I

$$\frac{6s^2 + 12s - 6}{(s - 1)(s + 2)(s + 3)} = \frac{A}{s - 1} + \frac{B}{s + 2} + \frac{C}{s + 3} \qquad (29.6)$$

(Since there are only three constants, it is more convenient to use A, B, and C, rather than subscripts.) The main task is to determine the constants. To this end we add the fractions on the right of Equation (29.6) to obtain

$$\frac{A(s + 2)(s + 3) + B(s - 1)(s + 3) + C(s - 1)(s + 2)}{(s - 1)(s + 2)(s + 3)}$$

The numerator of this fraction must be equal to the numerator of the left side of Equation (29.6). Thus

$$A(s + 2)(s + 3) + B(s - 1)(s + 3) + C(s - 1)(s + 2) = 6s^2 + 12s - 6$$

$$(29.7)$$

To find the constants, we let s be equal to certain convenient values. For example, if $s = -2$, then Equation (29.7) collapses to

$$0 + B(-3)(1) + 0 = -6$$

so that $B = 2$. Similarly, if $s = -3$, we get

$$0 + 0 + C(-4)(-1) = 12$$

or $C = 3$. Finally, if $s = 1$, then

$$A(3)(4) + 0 + 0 = 12$$

or $A = 1$. We now have (since $A = 1$, $B = 2$, and $C = 3$)

$$L^{-1}\left\{\frac{6s^2 + 12s - 6}{(s - 1)(s + 2)(s + 3)}\right\} = L^{-1}\left\{\frac{1}{s - 1} + \frac{2}{s + 2} + \frac{3}{s + 3}\right\}$$

$$= e^t + 2e^{-2t} + 3e^{-3t}$$

by transform 4. ◀

E X A M P L E **2** (*Repeating linear factors.*) Find

$$L^{-1}\left\{\frac{s}{(s - 2)^2(s + 1)}\right\}$$

Solution. This form contains only linear factors, but one of them is repeating. Even with this repetition Rule I applies:

$$\frac{s}{(s - 2)^2(s + 1)} = \frac{A}{s - 2} + \frac{B}{(s - 2)^2} + \frac{C}{s + 1}$$

(Note that the factor $s + 1$ does not repeat and hence occurs only once on the right side.) As before, we combine the fractions on the right side to obtain

$$\frac{s}{(s - 2)^2(s + 1)} = \frac{A(s - 2)(s + 1) + B(s + 1) + C(s - 2)^2}{(s - 2)^2(s + 1)}$$

Equating numerators, we get

$$A(s - 2)(s + 1) + B(s + 1) + C(s - 2)^2 = s.$$

To determine the constants, we make the appropriate substitutions:

$$s = 2: \qquad 0 + B(3) + 0 = 2 \qquad \text{or} \qquad B = \frac{2}{3}$$

$$s = -1: \qquad 0 + 0 + C(-3)^2 = -1 \qquad \text{or} \qquad C = -\frac{1}{9}$$

At this point we seem to have run out of values to substitute. However, if we **use the values already obtained for B and C**, we can let s be equal to any number, say $s = 0$.

$$A(s - 2)(s + 1) + \frac{2}{3}(s + 1) + \left(-\frac{1}{9}\right)(s - 2)^2 = s \qquad B = \frac{2}{3}, C = -\frac{1}{9}$$

$$s = 0: \qquad A(-2)(1) + \frac{2}{3}(1) + \left(-\frac{1}{9}\right)(-2)^2 = 0 \qquad \text{or} \qquad A = \frac{1}{9}$$

It follows that

$$L^{-1}\left\{\frac{s}{(s - 2)^2(s + 1)}\right\} = L^{-1}\left\{\frac{1}{9}\frac{1}{s - 2} + \frac{2}{3}\frac{1}{(s - 2)^2} - \frac{1}{9}\frac{1}{s + 1}\right\}$$

$$= \frac{1}{9}e^{2t} + \frac{2}{3}te^{2t} - \frac{1}{9}e^{-t}$$

by transforms 4 and 9, respectively. ◀

E X A M P L E **3** (*Distinct quadratic factors.*) Find

$$L^{-1}\left\{\frac{4}{(s - 1)(s + 1)(s^2 + 1)}\right\}$$

Solution. Since one of the factors is quadratic, Rule II applies. (The linear factors lead to the usual form by Rule I.) Thus

$$\frac{4}{(s - 1)(s + 1)(s^2 + 1)} = \frac{A}{s - 1} + \frac{B}{s + 1} + \frac{Cs + D}{s^2 + 1}$$

$$= \frac{A(s + 1)(s^2 + 1) + B(s - 1)(s^2 + 1) + (Cs + D)(s - 1)(s + 1)}{(s - 1)(s + 1)(s^2 + 1)}$$

Equating numerators, we get

$$A(s + 1)(s^2 + 1) + B(s - 1)(s^2 + 1) + (Cs + D)(s - 1)(s + 1) = 4$$

Once again we substitute convenient values for s:

$$s = 1: \qquad A(2)(2) + 0 + 0 = 4 \qquad A = 1$$

$$s = -1: \qquad 0 + B(-2)(2) + 0 = 4 \qquad B = -1$$

To get the remaining coefficients, we use the values already found for A and B:

$$1(s + 1)(s^2 + 1) + (-1)(s - 1)(s^2 + 1) + (Cs + D)(s - 1)(s + 1) = 4$$

Now, because of the factor $Cs + D$, we let $s = 0$ and solve for D:

$$s = 0: \qquad 1(1)(1) + (-1)(-1)(1) + D(-1)(1) = 4$$

whence $D = -2$. Finally, to get C we let $s =$ any value, say $s = 2$:

$$s = 2: \qquad 1(3)(5) + (-1)(1)(5) + (2C - 2)(1)(3) = 4 \qquad A = 1, B = -1, D = -2$$

and $C = 0$. Thus

$$L^{-1}\left\{\frac{4}{(s-1)(s+1)(s^2+1)}\right\} = L^{-1}\left\{\frac{1}{s-1} - \frac{1}{s+1} + \frac{-2}{s^2+1}\right\}$$

$$= e^t - e^{-t} - 2\sin t$$

by transforms 4 and 5, respectively. ◀

E X A M P L E **4**   (*Trinomial factor.*) Find

$$L^{-1}\left\{\frac{3s - 7}{(s^2 - 2s + 5)(s + 2)}\right\}$$

Solution. Since one factor is quadratic and one linear, we get

$$\frac{3s - 7}{(s^2 - 2s + 5)(s + 2)} = \frac{As + B}{s^2 - 2s + 5} + \frac{C}{s + 2}$$

$$= \frac{(As + B)(s + 2) + C(s^2 - 2s + 5)}{(s^2 - 2s + 5)(s + 2)}$$

Equating numerators, we get

$$(As + B)(s + 2) + C(s^2 - 2s + 5) = 3s - 7$$

$$s = -2: \qquad C(4 + 4 + 5) = -6 - 7 \qquad \text{or} \qquad C = -1$$

$$(As + B)(s + 2) - 1(s^2 - 2s + 5) = 3s - 7$$

$$s = 0: \qquad B(2) - 1(5) = -7 \qquad \text{or} \qquad B = -1$$

$$s = 1: \qquad (A - 1)(3) - 1(4) = -4 \qquad \text{or} \qquad A = 1 \qquad C = -1, B = -1$$

Hence

$$L^{-1}\left\{\frac{3s - 7}{(s^2 - 2s + 5)(s + 2)}\right\} = L^{-1}\left\{\frac{s - 1}{s^2 - 2s + 5} - \frac{1}{s + 2}\right\}$$

$$= L^{-1}\left\{\frac{s - 1}{(s - 1)^2 + 4} - \frac{1}{s + 2}\right\}$$

$$= e^t \cos 2t - e^{-2t}$$

by transforms 8 and 4, respectively. ◀

EXERCISES/SECTIONS   29.1–29.3

1. Use definition (29.1) to verify transforms 1, 5, 3 (for $n = 2$), and 9 (for $n = 1$).

In Exercises 2–13, use the table to find the transforms of the functions.

2. $f(t) = 5e^{2t}$

3. $f(t) = 2 + 3e^{-t}$

4. $f(t) = 1 - \sin t$

5. $f(t) = t + \cos 2t$

6. $f(t) = 2t^2 - 3 \cos t$

7. $f(t) = e^{2t} \sin 5t$

8. $f(t) = e^{-2t}\cos 3t$

9. $f(t) = t^3 e^{-4t}$

10. $f(t) = 2t^2 e^{3t}$

11. $f(t) = 2t^4 e^{-t}$

12. $f(t) = 5t + 3e^{2t}$

13. $f(t) = 4 - 5 \sin 2t$

In Exercises 14–29, use the table to find the inverse transforms of the functions.

14. $F(s) = \dfrac{3}{s - 5}$

15. $F(s) = \dfrac{10}{s^2 + 4}$

16. $F(s) = \dfrac{s}{s^2 + 7}$

17. $F(s) = \dfrac{1}{(s + 2)^3}$

18. $F(s) = \dfrac{1}{s^2(s^2 + 9)}$

19. $F(s) = \dfrac{4}{(s^2 + 4)^2}$

20. $F(s) = \dfrac{s^2 - 4}{(s^2 + 4)^2}$

21. $F(s) = \dfrac{4s + 2}{(s^2 + 4)^2}$

22. $F(s) = \dfrac{1}{(s - 4)^3}$

23. $F(s) = \dfrac{5s}{s^2 + 6}$

24. $F(s) = \dfrac{5}{s^2 + 9}$

25. $F(s) = \dfrac{2}{s^2(s^2 + 4)}$

26. $F(s) = \dfrac{s}{(s^2 + 16)^2}$

27. $F(s) = \dfrac{2s + 4}{(s + 2)^2 + 4}$

28. $F(s) = \dfrac{3s - 3}{(s - 1)^2 + 16}$

29. $F(s) = \dfrac{1}{(s + 3)^2 + 5}$

In Exercises 30–36, find the inverse transforms of the functions by the method of Example 2 (Section 29.2).

30. $\dfrac{s + 1}{s^2 + 2s + 5}$

31. $\dfrac{\sqrt{10}}{s^2 - 2s + 11}$

32. $\dfrac{6}{s^2 + 2s + 5}$

33. $\dfrac{s}{s^2 - 6s + 10}$

34. $\dfrac{1}{s^2 + 6s + 12}$

35. $\dfrac{s}{s^2 - 2s + 6}$

36. $\dfrac{2s + 1}{s^2 + 4s + 9}$

In Exercises 37–55, use the method of partial fractions to find the inverse transforms of the functions.

37. $\dfrac{1}{s(s + 1)}$

38. $\dfrac{2}{s(s - 1)(s + 1)}$

39. $\dfrac{2s + 1}{(s - 2)(s + 3)}$

40. $\dfrac{s}{(s - 1)(s + 3)}$

41. $\dfrac{s^2}{(s - 2)(s + 2)(s - 4)}$

42. $\dfrac{1}{s^2(s - 2)}$

43. $\dfrac{3s^2}{(s + 2)^2(s - 1)}$

44. $\dfrac{1}{(s + 2)(s^2 + 4)}$

45. $\dfrac{1}{(s + 1)(s^2 + 1)}$

46. $\dfrac{5}{(s - 1)(s^2 + 4)}$

47. $\dfrac{s}{(s + 1)(s^2 + 1)}$

48. $\dfrac{1}{(s + 1)^2(s + 2)}$

49. $\dfrac{1}{(s^2 + 1)(s - 1)^2}$

50. $\dfrac{2a^2 s}{s^4 - a^4}$

51. $\dfrac{s}{(s + 1)^2}$

52. $\dfrac{s^2}{(s - 1)^3}$

53. $\dfrac{2s^2 + 2s + 1}{(s^2 + 2s + 2)(s - 1)}$

54. $\dfrac{5}{(s^2 + 2s + 5)(s + 2)}$

55. $\dfrac{1}{(s^2 + 4s + 7)(s + 4)}$

29.4   Solution of Linear Equations by Laplace Transforms

In this section we are finally going to return to differential equations. To see how differential equations may be solved by using Laplace transforms, let us find the transform of $f'(t)$ in terms of the transform of $f(t)$. Assume that

$F(s) = L\{f(t)\}$ exists. Then

$$L\{f'(t)\} = \int_0^\infty e^{-st}f'(t)\,dt$$

$u = e^{-st}$	$dv = f'(t)\,dt$	**integration by parts**
$du = -se^{-st}\,dt$	$v = f(t)$	

It follows that

$$L\{f'(t)\} = \lim_{b\to\infty}\left[e^{-st}f(t)\Big|_0^b + s\int_0^b e^{-st}f(t)\,dt\right]$$
$$= \lim_{b\to\infty}\left[e^{-sb}f(b) - f(0)\right] + s\int_0^\infty e^{-st}f(t)\,dt$$

So, if

$$\lim_{t\to\infty} e^{-st}f(t) = 0$$

then

$$L\{f'(t)\} = -f(0) + s\int_0^\infty e^{-st}f(t)\,dt$$

or

$$L\{f'(t)\} = sL\{f(t)\} - f(0) \tag{29.8}$$

We can see, then, that the transform of $f'(t)$ may be expressed in terms of the transform of $f(t)$ itself. Since we are now interested in differential equations, let us adopt the following notation: if $y = f(t)$, denote the transform of y by $Y(s)$ and $f(0)$ by $y(0)$. Formula (29.8) then becomes

$$L\{y'\} = sY(s) - y(0) \tag{29.9}$$

Now consider the differential equation

$$y' - 2y = e^t \qquad y(0) = 0$$

To solve this equation, we take the Laplace transform of both sides, making use of formula (29.9):

$$L\{y'\} - 2L\{y\} = L\{e^t\}$$

or

$$sY(s) - y(0) - 2Y(s) = \frac{1}{s-1} \qquad \textbf{by (29.9)}$$

Since $y(0) = 0$, the initial condition, we have

$$sY(s) - 2Y(s) = \frac{1}{s-1} \qquad y(0) = 0$$

Note that formula (29.9) has "destroyed" the derivative, so that we are left with a simple algebraic equation. Solving this equation for $Y(s)$, we get

$$Y(s) = \frac{1}{(s-2)(s-1)} = \frac{1}{s-2} - \frac{1}{s-1}$$

Hence the inverse transform—namely,

$$y = e^{2t} - e^t$$

must be the solution. This example points out another critical feature of the transform method: since the condition $y(0) = 0$ was used in the third step, no arbitrary constants appear in the solution.

The method for solving equations by Laplace transforms will now be summarized.

Solution of Differential Equations by Laplace Transforms:

1. Find the Laplace transform of both sides of the differential equation and substitute the initial values.
2. Solve the resulting algebraic equation for $Y(s)$.
3. Find the inverse transform $y = L^{-1}\{Y(s)\}$.

Repeated use of formula (29.9) yields derivative formulas for higher derivatives:

$$L\{y''\} = L\{(y')'\} = sL\{y'\} - y'(0) = s[sY(s) - y(0)] - y'(0)$$

or

$$L\{y''\} = s^2 Y(s) - sy(0) - y'(0)$$

Similarly,

$$L\{y'''\} = s^3 Y(s) - s^2 y(0) - sy'(0) - y''(0) \qquad (29.10)$$

and so on.

The first two formulas will now be restated for easy reference.

Laplace Transforms of Derivatives:

$$L\{y'\} = sY(s) - y(0) \qquad (29.11)$$

$$L\{y''\} = s^2 Y(s) - sy(0) - y'(0) \qquad (29.12)$$

E X A M P L E **1**   Solve the equation $y'' - 4y' + 8y = 0$, with $y(0) = 1$, $y'(0) = 0$.

Solution. Taking the transform of both sides, we get

$$L\{y''\} - 4L\{y'\} + 8L\{y\} = L\{0\} = 0$$

or, using formulas (29.11) and (29.12),

$$s^2 Y(s) - sy(0) - y'(0) - 4[sY(s) - y(0)] + 8Y(s) = 0$$

Making use of the initial conditions, the last equation reduces to

$$s^2 Y(s) - s - 4sY(s) + 4 + 8Y(s) = 0 \qquad y(0) = 1, \; y'(0) = 0$$

We now solve for $Y(s)$:

$$s^2 Y(s) \quad 4sY(s) + 8Y(s) = s - 4$$

Thus

$$(s^2 - 4s + 8)Y(s) = s - 4 \qquad \text{factoring Y(s)}$$

or

$$Y(s) = \frac{s - 4}{s^2 - 4s + 8} \qquad \text{dividing by (s}^2 - 4\text{s} + 8\text{)}$$

To find the inverse transform, we complete the square in the denominator. Thus

$$Y(s) = \frac{s - 4}{(s - 2)^2 + 4} = \frac{s - 2 - 2}{(s - 2)^2 + 4}$$

$$= \frac{s - 2}{(s - 2)^2 + 4} - \frac{2}{(s - 2)^2 + 4}$$

From the table

$$y = e^{2t} \cos 2t - e^{2t} \sin 2t \qquad \blacktriangleleft$$

E X A M P L E **2**   Solve the differential equation $y'' + 2y' + y = te^{-t}$, with $y(0) = 0$, $y'(0) = -2$.

Solution. Transforming, we get by transform 9 in the table

$$s^2 Y(s) - sy(0) - y'(0) + 2[sY(s) - y(0)] + Y(s) = \frac{1}{(s + 1)^2}$$

Using the initial conditions, we have

$$s^2 Y(s) - (-2) + 2s Y(s) + Y(s) = \frac{1}{(s+1)^2} \qquad y(0) = 0, \ y'(0) = -2$$

$$s^2 Y(s) + 2s Y(s) + Y(s) = \frac{1}{(s+1)^2} - 2$$

$$(s^2 + 2s + 1) Y(s) = \frac{1}{(s+1)^2} - 2$$

$$(s+1)^2 Y(s) = \frac{1}{(s+1)^2} - 2$$

$$Y(s) = \frac{1}{(s+1)^4} - \frac{2}{(s+1)^2}$$

Finally,

$$y = \frac{1}{6} t^3 e^{-t} - 2te^{-t}$$

by transform 9 in the table.   ◀

E X E R C I S E S / S E C T I O N   29.4

In Exercises 1–27, solve the differential equations by the method of Laplace transforms.

1. $y' - y = 0$, $y(0) = 1$

2. $y' + 3y = 0$, $y(0) = 2$

3. $y' - 2y = 4$, $y(0) = 0$

4. $y' + 2y = 1$, $y(0) = 1$

5. $y' - 2y = e^{2t}$, $y(0) = 0$

6. $y' - 3y = e^{3t}$, $y(0) = -2$

7. $y' + 4y = te^{-4t}$, $y(0) = 3$

8. $y'' + 4y = 0$, $y(0) = 1$, $y'(0) = 0$

9. $y'' + 9y = 0$, $y(0) = 1$, $y'(0) = -2$

10. $y' - y = 4e^{-3t}$, $y(0) = 0$

11. $y'' + y = 2 \sin t$, $y(0) = y'(0) = 0$

12. $y'' + 4y = \sin 2t$, $y(0) = 0$, $y'(0) = 1$

13. $y' - y = \cos 2t$, $y(0) = 0$

14. $y'' + y = \sin t$, $y(0) = 0$, $y'(0) = 1$

15. $y'' + 4y = 4t$, $y(0) = 1$, $y'(0) = 0$

16. $y'' - 6y' + 9y = 12t^2 e^{3t}$, $y(0) = y'(0) = 0$

17. $y'' + 6y' + 13y = 0$, $y(0) = 1$, $y'(0) = -2$

18. $y'' - 4y' + 6y = 0$, $y(0) = 2$, $y'(0) = 0$

19. $y'' + y = 4e^t$, $y(0) = y'(0) = 0$

20. $y'' - 4y = 4e^{3t}$, $y(0) = y'(0) = 0$

21. $y'' - 4y = 3 \cos t$, $y(0) = y'(0) = 0$

22. $y'' - y' - 6y = 50 \sin t$, $y(0) = y'(0) = 0$

23. $y'' + 2y' + 5y = 8e^t$, $y(0) = y'(0) = 0$

24. $y'' - 4y' + 5y = 4e^t$, $y(0) = 1$, $y'(0) = 0$

25. In an LRC circuit, $L = 0.1$ H, $R = 6.0$ Ω, $C = 0.02$ F, and $e(t) = 6.0$ V. Find q as a function of time if $q(0) = 0$ and $i(0) = q'(0) = 0$.

26. A 64-lb weight (2 slugs) stretches a certain spring 2 ft. With this weight attached, the spring is stretched 2 ft below its equilibrium position and released. Find the resulting motion, assuming no damping but an external force $f(t) = 2 \sin 4t$.

27. Exercises 1 and 9, Section 28.4

In Exercises 1–8, solve each of the differential equations.

1. $(D^2 - D - 20)y = 0$

2. $(2D^2 - 2D - 3)y = 0$

3. $(D^5 - D^4)y = 0$

4. $(D^2 - 8D + 16)y = 0$

5. $(D^2 - 4D + 7)y = 0$

6. $(D^2 + 1)y = 0$; if $x = 0$, then $y = 3$ and $Dy = 0$

7. $(D^2 + 9)y = 3 \cos 2x$

8. $(D^2 - 2D - 8)y = 4x - e^{2x}$

In Exercises 9–12, solve the differential equations by the method of Laplace transforms.

9. $y' + 2y = 1$, $y(0) = 3$

10. $y' - y = 2e^{3t}$, $y(0) = -2$

11. $y'' - 6y' + 18y = 0$, $y(0) = 2$, $y'(0) = 1$

12. In an LRC circuit $L = 0.20$ H, $R = 7.0\ \Omega$, $C = 0.020$ F, and $e(t) = 12$ V. Find q as a function of time if $q(0) = 0$ and $i(0) = q'(0) = 0$.

13. Solve the equation $(D^2 - 2D - 8)y = 12e^{-2x}$.

REVIEW EXERCISES/CHAPTER   29

In Exercises 1–6, use the table to find the transform of each of the given functions.   •

1. $f(t) = 2e^{-3t}$

2. $f(t) = 1 - 2 \cos 2t$

3. $f(t) = 2t^3 + \sin 3t$

4. $f(t) = e^{-3t} \cos 6t$

5. $f(t) = 2t - \sin 2t$

6. $f(t) = 5t^4 e^{3t}$

In Exercises 7–14, find the inverse transform in each case.

7. $F(s) = \dfrac{s - 2}{(s - 2)^2 + 5}$

8. $F(s) = \dfrac{1}{(s + 3)^4}$

9. $F(s) = \dfrac{s}{s^2 - 2s + 5}$

10. $F(s) = \dfrac{1}{(s^2 + 4)^2}$

11. $F(s) = \dfrac{s}{(s + 1)(s - 2)}$

12. $F(s) = \dfrac{s}{(s + 1)^2(s + 4)}$

13. $F(s) = \dfrac{1}{(s + 2)(s - 3)(s - 4)}$

14. $F(s) = \dfrac{5}{(s + 2)(s^2 + 1)}$

In Exercises 15–24, use the method of Laplace transforms to solve the differential equations.

15. $y' + 2y = 0,\ y(0) = 1$

16. $y' - y = e^t,\ y(0) = 0$

17. $y' + 2y = te^{-2t},\ y(0) = -1$

18. $y'' + 2y = 0,\ y(0) = 0,\ y'(0) = 2$

19. $y'' - 2y' - 3y = 0,\ y(0) = 0,\ y'(0) = -4$

20. $y'' + y = e^{-t},\ y(0) = y'(0) = 0$

21. $y'' + 2y' + 5y = 0,\ y(0) = 1,\ y'(0) = 0$

22. $y'' - 2y' + 5y = 0,\ y(0) = 0,\ y'(0) = 1$

23. $y'' + 2y' + 5y = 3e^{-2t},\ y(0) = 1,\ y'(0) = 1$

24. $y'' + 4y' + 4y = 2 \sin t,\ y(0) = y'(0) = 0$

Newton's Method

Newton's method is an application of calculus to the solution of equations. The method is useful and practical, especially if carried out with the aid of a calculator.

To solve the equation $f(x) = 0$ for x, we let $y = f(x)$ and use Newton's method to find the x-intercept. (See Figure 30.1.) Let x_0 be a rough estimate of the intercept on the right and consider the tangent line to the graph at $(x_0, f(x_0))$. The point $(x_1, 0)$, where the tangent line crosses the x-axis, is an approximation of the root. We can find the value of x_1 by noting that the slope of the tangent line is

$$\frac{f(x_0) - 0}{x_0 - x_1} = f'(x_0)$$

It follows that

$$f(x_0) = f'(x_0)(x_0 - x_1)$$

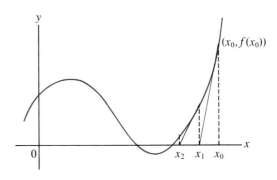

Figure 30.1

Solving for x_1, we get the following formula:

Newton's Method

$$x_1 = x_0 - \frac{f(x_0)}{f'(x_0)}$$

The procedure is now repeated: we use x_1 for our new estimate and find

$$x_2 = x_1 - \frac{f(x_1)}{f'(x_1)}$$

and so on. (See Figure 30.1.)

E X A M P L E   Find the root of the cubic equation $x^3 - 3x - 4 = 0$.

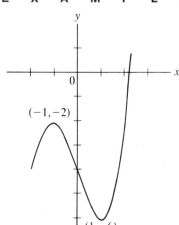

Figure 30.2

Solution. To obtain the first estimate x_0 of the root, we need to graph the equation by the methods of Chapter 21. (The graph is shown in Figure 30.2.) As a rough first guess, **$x = 2$** seems to do. Using a calculator, we quickly find x_1, x_2, and so on. If we decide to carry out the solution to six decimal places, we continue the iteration until the first six decimal places no longer change:

$$x_1 = x_0 - \frac{f(x_0)}{f'(x_0)} = x_0 - \frac{x_0^3 - 3x_0 - 4}{3x_0^2 - 3} \qquad \mathbf{f(x) = x^3 - 3x - 4}$$

or

$$x_1 = 2 - \frac{2^3 - 3(2) - 4}{3(2^2) - 3} = 2.222222 \qquad \mathbf{x_0 = 2}$$

A possible sequence is

2 STO y^x 3 $-$ 3 $\times$ MR $-$ 4 $=$ $\div$ (3 $\times$ MR x^2 $-$ 3) $=$
$+/-$ $+$ MR $=$ STO

(Using the stored value, we can use the same sequence (after 2 STO) to obtain x_2.)

$$x_2 = x_1 - \frac{x_1^3 - 3x_1 - 4}{3x_1^2 - 3}$$

$$= 2.222222 - \frac{(2.222222)^3 - 3(2.222222) - 4}{3(2.222222)^2 - 3}$$

$$= 2.196215$$

For the next iteration, we get

$$x_3 = x_2 - \frac{x_2^3 - 3x_2 - 4}{3x_2^2 - 3}$$

$$= 2.196215 - \frac{(2.196215)^3 - 3(2.196215) - 4}{3(2.196215)^2 - 3}$$

$$= 2.195823$$

From this point on the values no longer change, so that $x = 2.195823$ must be the root accurate to six decimal places.   ◄

E X E R C I S E S / C H A P T E R   **30**

Use Newton's method to find the root (or roots) of each of the following equations to six decimal places. In some of the exercises the calculator must be set in the **radian mode**.

1. $\cos x + x = 0$

2. $e^x + 2x - \dfrac{3}{2} = 0$

3. $4 \sin x - x = 0$; find the positive root

4. $\sin x = 1 - x$

5. $\sin x = x^2$; find the positive root

6. $e^x = 2 \cos x$; find the positive root

7. $x^2 - 2x - 12 = 0$; find all the roots

8. $x^2 + 3x - 11 = 0$; find all the roots

9. $x^3 - 6x^2 - 4x + 6 = 0$; find all the roots

10. $x^3 + 3x - 6 = 0$

Approximation and Measurement

Approximate and Exact Numbers

In mathematics we generally assume that numbers are **exact**. Thus "5" means "exactly 5." In technology numbers are often obtained by measurement, and measured quantities are necessarily **approximate**. For example, if the length of a beam is given as 32.6 ft, it is understood that 32.6 is only approximate.

On the other hand, 1 ft = 12 in. by definition. There is no measurement involved. So the number 12 is exact. Similarly, there are exactly 60 minutes in an hour. A number is ordinarily exact if it is obtained

1. by counting,
2. from a definition,
3. as a result of a computation with other exact numbers.

Accuracy

The **accuracy** of a measurement refers to the number of significant digits used to express the measurement. Suppose that the measured length of a beam has been rounded off to 32.6 ft. We conclude that the correct length lies between 32.55 ft and 32.65 ft. Thus all three digits contain significant information, and we say that 32.6 has three **significant digits**. If the length were rounded off to 30 ft, then we would know only that the length is closer to 30 ft than it is to 20 ft or 40 ft. So the final zero serves only to place the decimal point and is therefore not significant. Put another way, we can "believe" the 6 in 32.6, but we cannot "believe" the 0 in 30.

In general, the digits 1 through 9 are always significant. The digit 0 may or may not be significant: If its only function is to place the decimal point, then it is not significant. For example, 0.0032 has two significant digits, since the zeros are only placeholders. On the other hand, 1,031.5 has five significant digits, since the digit 0 is not just a placeholder in this case.

A length of 920 ft appears to have only two significant digits, since the true length is only known to lie between 910 ft and 930 ft, but can we always be sure? For example, if the length were really rounded off to the nearest foot, then the zero would indeed be significant; in other words, we could then "believe" the zero. However, unless this information is provided, we must presume that 920 has only two significant digits.

On the other hand, final zeros are significant if they occur to the right of the decimal point. For example, 1.3200 has five significant digits, since the zeros do more than just place the decimal point. The rules for determining significant digits are listed next.

Significant Digits:

1. The digits 1 through 9 are always significant.
2. Zeros are not always significant:
   a. Zeros are significant if they lie between two significant digits.
   b. Unless otherwise specified, final zeros are significant only if they lie to the right of the decimal point.
   c. Initial (beginning) zeros are not significant.

Remark. Why initial zeros are not considered significant can be seen from the following: Given that 10 mm = 1 cm, it follows that

$$0.1 \text{ mm} = 0.01 \text{ cm}$$

Here each number has one significant digit. This is reasonable, since 0.1 mm and 0.01 cm describe exactly the same measurement, but with different units of measure.

If you carefully follow the above rules, you should have no trouble determining the number of significant digits. Study the following example.

E X A M P L E **1**
a. 12.4 has three significant digits. (Nonzero digits are always significant.)
b. 102 has three significant digits, and 102.1 has four significant digits, since the zero lies between two significant digits in each case.
c. 302.00 has five significant digits, since the final zeros lie to the right of the decimal point.
d. 39,200 has only three significant digits. (The final zeros do not lie to the right of the decimal point and are therefore not significant.)
e. 0.013 has two significant digits, since the initial zeros are not significant.
f. 0.0130 has three significant digits. (Note that the final zero is significant since it lies to the right of the decimal point.)
g. 1.0020 has five significant digits.   ◀

Numbers must frequently be rounded off to a certain level of accuracy— that is, to a certain number of significant digits. In such cases zeros may be needed as placeholders.

E X A M P L E **2**   **a.** 9,043 rounded to three significant digits is 9,040.
b. 9,046 rounded to three significant digits is 9,050.
c. 999 rounded to one significant digit is 1,000.
d. 31,512 rounded to two significant digits is 32,000.
e. 1.35 rounded to two significant digits is 1.4.
f. 1.256 rounded to three significant digits is 1.26.
g. 1.012 rounded to two significant digits is 1.0.
h. 21.97 rounded to three significant digits is 22.0.   ◄

Calculations, especially if done with a calculator, often produce answers that contain more digits than the input data warrant. For example, if the dimensions of a room are 10.1 ft by 15.9 ft, then the calculated area of

$$(10.1 \text{ ft}) \times (15.9 \text{ ft}) = 160.59 \text{ ft}^2$$

is more accurate than is warranted by the given information and therefore should be rounded off. Since both dimensions are given to three significant digits, the answer should also be rounded off to three significant digits, namely, 161 ft². Any more accuracy cannot be justified. For example, the measurement of 160.6 ft² may seem acceptable, since both dimensions are expressed to the nearest tenth of a foot. However, 10.1 is only known to lie between 10.05 and 10.15, and 15.9 between 15.85 and 15.95. So the area must lie between

$$(10.05) \times (15.85) = 159.3 \text{ ft}^2$$

and

$$(10.15) \times (15.95) = 161.9 \text{ ft}^2$$

So 160.6 ft² is not correct.

> **1.** When approximate numbers are multiplied or divided, the result is no more accurate than the least accurate number.
> **2.** The square root of an approximate number is as accurate as the number.

E X A M P L E **3**   The dimensions of a box are 3.2 ft × 0.98 ft × 1.98 ft. Find the volume.

Solution. Using a calculator, we have the following sequence:

$$3.2 \boxed{\times} 0.98 \boxed{\times} 1.98 \boxed{=} \rightarrow 6.20928$$

Since the least accurate number is 3.2 (two significant digits), we round off 6.20928 to two significant digits and obtain 6.2 ft³.   ◄

E X A M P L E **4**   Evaluate $\sqrt{4.786}$.

Solution.   4.786 $\boxed{\sqrt{}}$ → 2.1876928

Since the square root is as accurate as the number, we have $\sqrt{4.786} = 2.188$, rounded off to four significant digits. ◄

Precision

The **precision** of a number refers to the right-most significant digit. For example, 12.43 ft is more precise than 12.4 ft since 12.43 was measured to the nearest hundredth of a foot, but 12.4 ft only to the nearest tenth of a foot.

E X A M P L E **5**   0.0014 ft is more precise than 1.4 ft. However, the accuracy is the same (two significant digits). ◄

The rule for addition and subtraction of approximate numbers is given in terms of precision.

> When approximate numbers are added or subtracted, the result is only as precise as the least precise number.

E X A M P L E **6**   Add the following weights:

10.132 lb,   1.95 lb,   3.7 lb,   and   0.13 lb

Solution. The sequence is

10.132 $\boxed{+}$ 1.95 $\boxed{+}$ 3.7 $\boxed{+}$ 0.13 $\boxed{=}$ → 15.912

Since 3.7 lb has only one decimal place, the answer is rounded off to 15.9 lb. ◄

E X E R C I S E S / A P P E N D I X **A**

In Exercises 1–12, round off to the number of significant digits indicated.

1. 3.1416 (four)
3. 1.039 (three)
5. 0.000155 (two)
7. 985,732 (three)

2. 0.039 (one)
4. 7.173 (three)
6. 1.000155 (six)
8. 1,325 (three)

9. 876,512 (two)
11. 0.0201 (two)

10. 13,705 (four)
12. 1.396 (three)

In Exercises 13–36, carry out the indicated operations using a calculator. Round off the answers according to the rules given in this section.

13. $1.35 + 7.94 - 1.987 + 0.128$ **14.** $0.31 + 4.9 - 1.782$

15. $8.965 + 2.1314$

16. $35.86 + 4.037 - 10.1176$

17. $10.98 - 1.71 + 11.3$

18. $9.99 - 2.651 + 0.3$

19. $(2.1)(5.6)(1.11)$

20. $(0.28)(0.165)(1.1)$

21. $(15.3)(14.6)(0.12)$

22. $\dfrac{100.7}{3.55}$

23. $\dfrac{87.51}{2.71}$

24. $\dfrac{(8.34)(0.73)}{19.3}$

25. $\dfrac{(68.1)(0.3)}{(0.5)}$

26. $\dfrac{(100.3)(500.4)}{(10.11)(3.4456)}$

27. $(606.6)(0.26)$

28. $(98.34)(0.92)$

29. $(22.9)(3.80)$

30. $(0.110)(0.091)$

31. $(2.600)(0.250)$

32. $(3.20)(0.70)$

33. $\sqrt{7.8}$

34. $\sqrt{0.361}$

35. $\sqrt{28.367}$

36. $\sqrt{1496.2}$

37. The respective resistances of three resistors connected in series are $15.3 \ \Omega$, $19.21 \ \Omega$, and $19.0 \ \Omega$. Find the resistance of the combination, which is the sum of the individual resistances.

38. Two forces of 9.732 lb and 6.60 lb are acting on an object in opposite directions. Determine the net force, which is $(9.732 \ \text{lb}) - (6.60 \ \text{lb})$.

39. The power (in watts) developed in a circuit is given by $P = RI^2$. If $I = 0.132 \ \text{A}$ and $R = 9.6 \ \Omega$, find P.

40. A circular metal plate has a diameter of 10.6 cm. Find the area $(A = \pi r^2)$.

Units of Measurement and Conversions

Two basic systems of measurement are in use today, the **metric system** and the **British system**. In addition, there is a standard system derived from metric units, the **International System of Units (SI)**. A table of SI units is given on the inside front cover of this book.

Basic Units

The base unit of length is the *foot* in the British system and the *meter* in the metric system. (Converting from one system of measurement to the other will be discussed shortly.)

Weight
Two sets of very important units are those of mass and weight. The **weight** of an object is the measure of the gravitational force on the object, which varies with altitude. The **mass** of an object is the measure of its inertia, which does

Mass
not vary with altitude. Although mass and weight are different quantities, they are closely related: The weight of an object is equal to its mass times the acceleration due to gravity, or $w = mg$. This is a special case of Newton's second law, $F = ma$.

In the SI system, the base unit of mass is the *kilogram*. In the metric system, mass can also be expressed in *grams*. In the British system, mass is expressed in *slugs* (since mass is a measure of "sluggishness"), where

$$1 \text{ slug} = 1 \frac{\text{lb} \cdot \text{s}^2}{\text{ft}}$$

(The base unit of weight is the pound.)

To see why 1 slug = 1 (lb·s²)/ft, consider Newton's second law, $F = ma$. Given that the acceleration due to gravity g near the surface of the earth is

$$32 \frac{\text{ft/s}}{\text{s}} = 32 \frac{\text{ft}}{\text{s}^2}$$

and that $mg = F$, then with $m = 1$ slug,

$$1\left(\frac{lb \cdot s^2}{ft}\right)\left(32\frac{ft}{s^2}\right) = 32 \text{ lb}$$

As already noted, the base unit of weight in the British system is the pound. In the SI system, the unit of weight is the newton (N), where

$$1 \text{ N} = 1\frac{kg \cdot m}{s^2}$$

As a check, given that the acceleration due to gravity is 9.8 m/s² and that $mg = F$, then with $m = 1$ kg,

$$(1 \text{ kg})\left(9.8\frac{m}{s^2}\right) = 9.8\frac{kg \cdot m}{s^2} = 9.8 \text{ N}$$

(For other units, see the table on the inside front cover of this book.)

A special feature of the metric system is the use of prefixes to denote different orders of magnitude. These are given in the following table:

Table B.1

Prefix	Factor	Symbol	Prefix	Factor	Symbol
exa-	10^{18}	E	deci-	10^{-1}	d
peta-	10^{15}	P	centi-	10^{-2}	c
tera-	10^{12}	T	milli-	10^{-3}	m
giga-	10^{9}	G	micro-	10^{-6}	μ
mega-	10^{6}	M	nano-	10^{-9}	n
kilo-	10^{3}	k	pico-	10^{-12}	p
hecto-	10^{2}	h	femto-	10^{-15}	f
deca-	10^{1}	da	atto-	10^{-18}	a

The following equivalencies illustrate the use of these prefixes:

$$1 \text{ millimeter (mm)} = 10^{-3} \text{ m}$$

$$1 \text{ kilometer (km)} = 10^{3} \text{ m}$$

$$1 \text{ milliohm (m}\Omega) = 10^{-3} \, \Omega$$

$$1 \text{ microfarad } (\mu F) = 10^{-6} \text{ F}$$

$$1 \text{ hectoliter (hL)} = 10^{2} \text{ L}$$

Reduction of Units

Reduction
Conversion

Changing units of measurement within a system is called a **reduction**. A change from one system to another is called a **conversion**. Conversions will be taken up in the last part of this section.

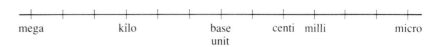

Figure B.1

We have seen that reduction within the SI system can be accomplished by multiplying or dividing a given measurement by a multiple of 10. Since multiplication and division by a multiple of 10 moves the decimal point, a reduction can be performed readily with the help of the diagram in Figure B.1.

Reduction in the SI System: To change from one unit to another, move the decimal point as many places as you move along the scale.

1. To change to a unit farther to the right, move the decimal point to the right.
2. To change to a unit farther to the left, move the decimal point to the left.

E X A M P L E **1**   Change 6.8 μs to milliseconds.

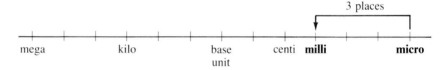

Figure B.2

Solution. We see from Figure B.2 that the decimal point is moved 3 units to the left. To do so, we need to insert zeros:

6.8 μs = 0.0068 ms   ◀

E X A M P L E **2**   Change 0.3047 L to milliliters.

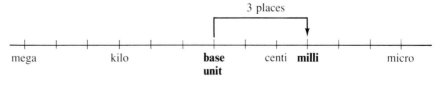

Figure B.3

Solution. Figure B.3 shows that the decimal point must be moved 3 places to the right:

0.3047 L = 304.7 mL   ◀

Conversion of Units

Converting between the British and SI or metric systems is fairly complicated. To illustrate the technique, we need the basic conversion units given in the box below.

1 in. = 2.54 cm (exact)	1 kg = 2.205 lb
1 km = 0.6214 mi	1 lb = 4.448 N
1 lb = 454 g	1 ft^3 = 28.32 L

To change the units of measure of a given number, we use two basic principles:

1. *Perform algebraic operations with the units of measure as you would perform them with any algebraic symbols.*

For example, the operation

$$1\frac{\text{ft}}{\text{s}} = 1\frac{\text{ft}}{\cancel{s}} \times 60\frac{\cancel{s}}{\text{min}} = 60\frac{\text{ft}}{\text{min}}$$

converts feet per second to feet per minute. Note that the units are canceled as if they were algebraic symbols.

2. *To convert units, multiply by quantities each of which has quotient 1.*

To see how this rule is used, note that the values in the box above are given as equal pairs. Thus

$$\frac{1\text{ in.}}{2.54\text{ cm}} = 1, \qquad \frac{1\text{ km}}{0.6214\text{ mi}} = 1, \qquad \frac{454\text{ g}}{1\text{ lb}} = 1$$

Multiplying a given quantity by any of these fractions does not change the value. For example,

$$2.1\text{ mi} = 2.1\,\cancel{\text{mi}}\left(\frac{1\text{ km}}{0.6214\,\cancel{\text{mi}}}\right) = \frac{2.1}{0.6214}\text{ km} = 3.4\text{ km}$$

E X A M P L E **3**   Express 20.0 mi/h in kilometers per minute.

Solution. Since 1 km = 0.6214 mi, we multiply the given quantity by

$$\frac{1\text{ km}}{0.6214\text{ mi}}$$

in order to cancel mi and introduce km. Similarly, we multiply by

$$\frac{1\text{ h}}{60\text{ min}}$$

to cancel h and introduce 1/min. Successive multiplication yields

$$20.0\,\frac{\cancel{mi}}{\cancel{h}}\left(\frac{1\ km}{0.6214\,\cancel{mi}}\right)\left(\frac{1\,\cancel{h}}{60\ min}\right)=\frac{20.0}{(0.6214)(60)}\,\frac{km}{min}=0.536\,\frac{km}{min} \quad \blacktriangleleft$$

E X A M P L E **4**   Convert 28.6 oz to newtons.

Solution. From the table, 1 lb = 4.448 N, the weight units. To use this relationship, we must first change ounces to pounds by multiplying the given quantity by

$$\frac{1\ lb}{16\ oz}$$

Next, we multiply by

$$\frac{4.448\ N}{1\ lb}$$

to cancel lb and introduce N. Successive multiplication yields

$$\frac{28.6\,\cancel{oz}}{1}\times\frac{1\,\cancel{lb}}{16\,\cancel{oz}}\times\frac{4.448\ N}{1\,\cancel{lb}}=7.95\ N \quad \blacktriangleleft$$

E X A M P L E **5**   Express 53.7 lb/in.^{2} in newtons per square meter.

Solution. Since 1 in. = 2.54 cm, we multiply the given quantity by

$$\left(\frac{1\ in.}{2.54\ cm}\right)^2$$

to cancel in.^{2} The multiplication introduces 1/cm^2. To cancel cm^2, we multiply by

$$\left(\frac{100\ cm}{1\ m}\right)^2$$

Thus

$$53.7\,\frac{lb}{in.^2}=\left(53.7\,\frac{lb}{in.^2}\right)\left(\frac{1\ in.}{2.54\ cm}\right)^2\left(\frac{100\ cm}{1\ m}\right)^2\left(\frac{4.448\ N}{1\ lb}\right)$$

$$=53.7\,\frac{\cancel{lb}}{\cancel{in.^2}}\times\frac{1^2\,\cancel{in.^2}}{(2.54)^2\,\cancel{cm^2}}\times\frac{(100)^2\,\cancel{cm^2}}{1^2\ m^2}\times\frac{4.448\ N}{1\,\cancel{lb}}$$

$$=\frac{(53.7)(100)^2(4.448)}{(2.54)^2}\,\frac{N}{m^2}=3.70\times10^5\,\frac{N}{m^2} \quad \blacktriangleleft$$

E X A M P L E **6** Express 0.124 in.^{3} in centiliters.

Solution. $0.124 \; \cancel{\text{in.}^3} \left(\dfrac{1^3 \cancel{\text{ft}^3}}{(12)^3 \cancel{\text{in.}^3}} \right) \cdot \dfrac{28.32 \cancel{L}}{1 \cancel{\text{ft}^3}} \cdot \dfrac{100 \; \text{cL}}{1 \cancel{L}} = 0.203 \; \text{cL}$ ◄

E X E R C I S E S / A P P E N D I X   **B**

In Exercises 1–20, use the diagram in Figure B.4 to reduce the given units to the units indicated.

1. 0.65 m to millimeters
2. 3.48 cm to millimeters
3. 0.736 L to milliliters
4. 5.609 Ω to milliohms
5. 46 mg to grams
6. 2.72 kg to grams
7. 2.08 kW to watts
8. 48 ms to seconds
9. 206 ms to seconds
10. 25 ms to microseconds
11. 0.76 ms to microseconds
12. 62 μs to milliseconds
13. 26 μΩ to milliohms
14. 1.4 mΩ to microohms
15. 47.4 μF to millifarads
16. 34.1 μF to millifarads
17. 0.064 MΩ to milliohms
18. 0.00028 MΩ to milliohms
19. 0.000815 kW to centiwatts
20. 0.0000050 kW to milliwatts

In Exercises 21–46, carry out the indicated conversions.

21. 1.10 m to inches
22. 5.30 lb to newtons
23. 7.710 in.^{2} to square centimeters
24. 10.5 cm^2 to square feet
25. $50.0 \dfrac{\text{mi}}{\text{h}}$ to kilometers per minute
26. $3.8 \dfrac{\text{mi}}{\text{s}}$ to meters per minute
27. $223 \dfrac{\text{lb}}{\text{ft}}$ to newtons per centimeter
28. $50.3 \dfrac{\text{km}}{\text{h}}$ to miles per second
29. $251 \dfrac{\text{ft}}{\text{min}}$ to meters per second

30. $101.1 \dfrac{\text{km}}{\text{s}}$ to miles per minute
31. $3.7 \dfrac{\text{N}}{\text{cm}^2}$ to pounds per square inch
32. $10.3 \dfrac{\text{lb}}{\text{ft}^2}$ to newtons per square meter
33. $1.50 \dfrac{\text{lb}}{\text{ft}^3}$ to newtons per cubic centimeter
34. $2.78 \dfrac{\text{N}}{\text{m}^3}$ to pounds per cubic inch
35. $3.200 \dfrac{\text{N}}{\text{cm}^3}$ to pounds per cubic foot
36. 2.30 ft^3 to liters
37. 50.1 L to cubic feet
38. 10.12 mL to cubic inches
39. 142.3 mL to cubic inches
40. 48 in.^{3} to liters
41. $2.76 \dfrac{\text{mi}}{\text{h}}$ to kilometers per minute
42. $73,000 \dfrac{\text{km}}{\text{s}}$ to miles per minute
43. $3.92 \dfrac{\text{lb}}{\text{in.}^2}$ to newtons per square centimeter
44. $10.12 \dfrac{\text{lb}}{\text{ft}^2}$ to newtons per square meter
45. $72.40 \dfrac{\text{N}}{\text{m}^2}$ to pounds per square inch
46. $10.34 \dfrac{\text{lb}}{\text{in.}^3}$ to newtons per cubic meter

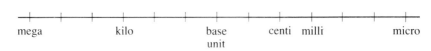

mega   kilo   base   centi   milli   micro
                        unit

Figure B.4

Introduction to BASIC

The language of BASIC (*B*eginners *A*ll-Purpose *S*ymbolic *I*nstruction *C*ode) is the most widely used computer language today. The purpose of this appendix is to give a brief introduction to BASIC.

Algebraic Expressions, Operations, and Functions

Algebraic expressions and operations are discussed in Section 1.12 and in Section 2.4. Functions are discussed in Section 2.8. For trigonometric functions, see Section 7.4, and for logarithmic and exponential functions, see Section 12.6.

Format

A computer program is a list of step-by-step instructions that a computer can follow. These instructions consist of a list of numbered statements. Although the statements may be numbered 1, 2, 3, 4, . . . , it is more practical to use a sequence such as 10, 20, 30, 40, . . . This numbering scheme allows you to insert additional lines, if necessary, without retyping the entire program. A typical program, then, has the following form:

     1∅ Statement 1
     2∅ Statement 2
     3∅ Statement 3
     and so on

(Note that the symbol ∅ is used for zero.) The lines do not have to be typed in order. The computer automatically orders the statements when entered.

To correct a line, it is sufficient to enter the corrected line—the last typed line replaces the previous one.

E X A M P L E **1**   Suppose the following lines are part of a program:

> 1∅ LET A = 1
> 2∅ LET B = 2
> 3∅ LET C = 6
> 2∅ LET B = 4

Since the second line 20 replaces the first line 20, the computer reads these lines as if they had been typed as follows:

> 1∅ LET A = 1
> 2∅ LET B = 4
> 3∅ LET C = 6   ◀

System Commands

A **system command** is a general command to a computer. It is not part of a program. The most important system commands are LIST, DELETE, and RUN.

> LIST: Lists an entered program. (This command is usually used to check or correct an entered program.)
> DELETE: Erases a designated line.
> RUN: Executes the program.

LET Statement

The LET statement is a program statement that assigns a value to a variable.

E X A M P L E **2**   The program statement

> 3∅ LET X = 2.71

assigns the value 2.71 to the memory location named X. It also assigns 2.71 to X in all statements that follow.   ◀

E X A M P L E **3**   The statement

> 5∅ LET A = A + 1

replaces the value of A stored in the memory by A + 1. (Note that in BASIC, the symbol = means "replaces.")   ◀

INPUT Statement

The INPUT statement causes the computer to pause until a set of data is entered. (See Example 6.)

PRINT Statement

The PRINT statement instructs the computer to print or display an expression or the value of a variable or expression. For example, the statement

$5\emptyset$ PRINT X

causes the computer to print the *value* of the variable X. On the other hand, the statement

$5\emptyset$ PRINT "X"

(X in quotation marks) causes the *letter* X to be printed, not the value of X.

It is possible to print or display more than one variable. For example, depending on the system, the statement

$6\emptyset$ PRINT X, X + 1

causes the values of X and X + 1 to be printed

1. On two separate lines, or
2. On the same line, separated by a wide space.

If you want the results printed closer together on the same line, separate the variables by a semicolon:

$6\emptyset$ PRINT X; X + 1

(See Example 4.)

END Statement

The END statement tells the computer that the program is finished.

Examples of Simple Programs

Given the information we have so far, it is possible to write some simple programs.

E   X   A   M   P   L   E   **4**   Consider the program:

$1\emptyset$ LET X = 4
$2\emptyset$ LET Y = X + 1
$3\emptyset$ PRINT X; Y
$4\emptyset$ END

Note that the last line consists of an END statement.

The command RUN will result in the following display:

4   5

(The numbers are printed close together on the same line because a semicolon is used in line 30.)   ◄

E   X   A   M   P   L   E   **5**   A circuit contains a resistor and inductor. If R is the resistance and X_L the inductive reactance, then the impedance Z is given by $Z = \sqrt{R^2 + (X_L)^2}$. Write a BASIC program for computing Z, if $R = 32.4 \ \Omega$ and $X_L = 27.6 \ \Omega$.

Solution. In the program, X1 represents X_L:

```
10 PRINT "IMPEDANCE"
20 LET R = 32.4
30 LET X1 = 27.6
40 LET Z = SQR(R↑2 + X1↑2)
50 PRINT "Z="; Z
60 END
```

(Line 10 merely identifies the program.)
The RUN command will result in the following display:

```
IMPEDANCE
Z = 42.56195484
```
◀

In the next program, the descriptive labels are used to identify not only the program but also the data to be entered.

E   X   A   M   P   L   E   **6**   Write a program for the solution of the equation

$$ax + b = c$$

in terms of a, b, and c.

Solution. Note that the solution is

$$x = \frac{c - b}{a}, \qquad a \neq 0$$

Program:

```
10 PRINT "SOLUTION OF LINEAR EQUATION"
20 PRINT "A="
30 INPUT A
40 PRINT "B="
50 INPUT B
60 PRINT "C="
70 INPUT C
80 LET X = (C − B)/A
90 PRINT "X="; X
100 END
```

On some systems the INPUT statement may contain a reminder in quotation marks. On such a system, the program can be written as follows:

```
10 PRINT "SOLUTION OF LINEAR EQUATION"
20 INPUT "A ="; A
30 INPUT "B ="; B
40 INPUT "C ="; C
50 LET X = (C − B)/A
60 PRINT "X ="; X
70 END
```

The values of A, B, and C are entered while the program is running. The computer prints or displays a question mark on lines 20, 30, and 40, allowing the constants to be "input."

When solving the equation $2x + 3 = 12$, the RUN command will result in the following display:

SOLUTION OF LINEAR EQUATION
A = ?   **2**   **Input 2**
B = ?   **3**   **Input 3**
C = ?   **12**   **Input 12**
X = 4.5   ◄

GO TO and IF-THEN Statements

The GO TO statement directs the computer to go to a specific line.

The IF-THEN statement tells the computer to go to a specific line if a certain condition is met. Consider, for example, the following programming steps:

```
100 IF N = 8 THEN 120
110 GO TO 80
```

When the computer reads line 100, it jumps to line 120 **if N = 8**. Otherwise it reads line 110 and follows the instruction to go to line 80.

The next program uses an inequality. In BASIC, $\leq$ is expressed as $<=$ and $\geq$ as $>=$.

E X A M P L E **7**   Write a program to print the squares of the first 12 positive integers.

Solution.   10 LET N = 1
                      20 PRINT N↑2
                      30 LET N = N + 1
                      40 IF N < = 12 THEN 20
                      50 END

Line 20 instructs the computer to print N^2 (starting with $N = 1$). On line 30 the index is increased by 1. Line 40 says that if N is *less than or equal to 12*, the computer is to jump to line 20 to repeat the cycle. When N *exceeds* 12, the condition in line 40 is not met, so the computer goes to line 50, thereby terminating the program.   ◄

In the next program, the sum of the squares of the first 12 positive integers is found. Note how the equality symbol is used to perform the addition.

E X A M P L E **8**   Write a program for finding the sum of the squares of the first 12 positive integers.

Solution.   1∅ LET M = ∅
2∅ LET N = 1
3∅ LET M = M + N↑2
4∅ LET N = N + 1
5∅ IF N < = 12 THEN 3∅
6∅ PRINT "SUM OF SQUARES="; M
7∅ END

Note that in line 30, M is replaced by the original value plus N^2, the next term in the sum.

The RUN command will cause the following result to be printed or displayed:

SUM OF SQUARES = 650   ◄

FOR-NEXT Statement

The FOR-NEXT statement provides a convenient method for programming simple loops of the type in Example 8. The FOR-NEXT statement has the following form:

FOR variable = *n* TO *m* STEP *k*

initial value ⟋ increment

final value

NEXT variable

For example,

FOR I = 1 TO 1∅ STEP 1
   . . .
NEXT I

repeatedly executes a specified command while the value of I changes from 1 to 10 in increments of 1 unit. (If the increment is 1, as in this case, STEP 1 can be omitted.)

E X A M P L E **9**   Write a program for constructing a table of values for the function $y = x^2 + 1$ from $x = -2$ to $x = 2$ in increments of 0.1. (See also Sections 2.8 and 2.9.)

Solution.   10 FOR X = −2 TO 2 STEP 0.1
20 LET Y = X↑2 + 1
30 PRINT X; Y
40 NEXT X
50 END   ◀

E X E R C I S E S / A P P E N D I X   **C**

All the programs for these exercises are given in the answer section.

1. Write a program entitled "Pythagorean theorem" for finding the hypotenuse of a right triangle, given the legs.

2. Write a program for solving a pure quadratic equation of the form $ax^2 - b = 0$ $(a, b > 0)$.

3. Write a program to print the sum of the cubes of the first eight positive integers, using the IF-THEN statement. (See Example 8.)

4. Write the program in Exercise 3 using the FOR-NEXT statement. (See Example 9.)

5. Write a program for constructing a table of values for the function $f(x) = \sin 2x$ for $x = 0$ to $x = 2$ (radians) in increments of 0.05.

6. Repeat Exercise 5 for $f(x) = \ln x^2$ for $x = 1$ to $x = 2$ in increments of 0.1.

7. Write a program to print the volume and surface area of a sphere of radius 2.47 in.

8. Write a program for finding the length of a side of a triangle, given the lengths of the other two sides and the angle between them. Use the law of cosines.

9. Write a program for solving the quadratic equation

$$ax^2 + bx + c = 0, \qquad a \neq 0$$

valid for real roots.

10. Write a program for finding the area of a circular sector of radius r and central angle θ, measured in degrees.

Review of Geometry

The purpose of this appendix is to recall some of the basic concepts from geometry.

Angles

It is stated in Section 4.2 that an **angle** is generated by rotating a ray (half-line) about its endpoint, called the **vertex** of the angle (Figure D.1). The original position is called the **initial side**, and the final position is called the **terminal side**.

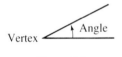

Vertex — Angle

Figure D.1

One complete rotation about the endpoint, called a **whole angle**, is defined to be a measure of 360°. (See Figure D.2.) One-half a complete rotation is called a **straight angle** and measures 180°. (See Figure D.2.) One-fourth of a complete rotation is a **right angle** and measures 90° (Figure D.2).

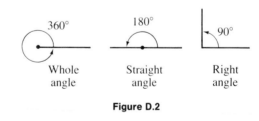

| 360° | 180° | 90° |
| Whole angle | Straight angle | Right angle |

Figure D.2

Acute angle   Obtuse angle

Figure D.3

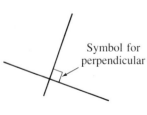

Symbol for perpendicular

Figure D.4

An angle between 0° and 90° is called an **acute angle**, and an angle between 90° and 180° is called an **obtuse angle**. (See Figure D.3.)

If two lines meet so that they form a right angle, the lines are said to be **perpendicular**. (Note the symbol for perpendicular in Figure D.4.) The point of intersection is called the **foot** of the perpendicular.

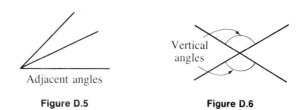

Figure D.5   **Figure D.6**

Two angles may be given special names depending on their sizes or relative positions. Two angles are **adjacent angles** if they have a common vertex and one common side (Figure D.5). A pair of angles formed by two intersecting lines and lying on opposite sides of the point of intersection are called **vertical angles** (Figure D.6). (Two vertical angles are always equal.) Two angles are **complementary** if their sum is 90°. Two angles are **supplementary** if their sum is 180°.

Parallel Lines

Two lines in a plane are **parallel** if they do not intersect (Figure D.7).

If two lines (not necessarily parallel) are crossed by a third line, this third line is called a **transversal**. In Figure D.8, the line EF is a transversal.

Whenever two parallel lines are cut by a transversal, then any pair of alternate interior and any pair of corresponding angles are equal. In Figure D.8, the **alternate interior angles** are: $\angle 3$ and $\angle 5$; $\angle 4$ and $\angle 6$. The **corresponding angles** are: $\angle 1$ and $\angle 5$; $\angle 2$ and $\angle 6$; $\angle 4$ and $\angle 8$; $\angle 3$ and $\angle 7$.

Polygons

A **polygon** is formed by three or more line segments that enclose a portion of a plane. Polygons are named according to the number of sides, as shown in Figure D.9.

A polygon is called a **regular polygon** if all sides have the same length.

Parallel lines

Figure D.7

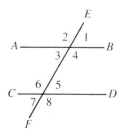

Figure D.8

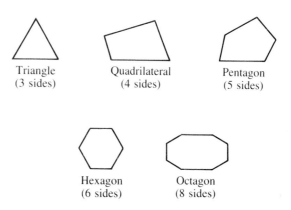

Triangle   Quadrilateral   Pentagon
(3 sides)   (4 sides)   (5 sides)

Hexagon   Octagon
(6 sides)   (8 sides)

Figure D.9

Triangles

A triangle is said to be an **equilateral triangle** if all three sides have the same length and all three angles are equal (60°). (See Figure D.10.) A triangle is

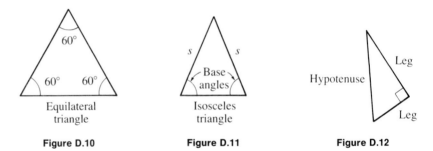

Figure D.10   **Figure D.11**   **Figure D.12**

isosceles if two sides have the same length (Figure D.11). The angles opposite the equal sides are called the **base angles**, which are always equal (Figure D.11). In a **scalene triangle** no two sides or angles are equal. A **right triangle** has one right angle. The side opposite the right angle is called the **hypotenuse**. The other two sides are called the **legs**. (See Figure D.12.)

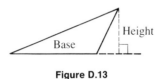

Figure D.13

The **altitude** or **height** of a triangle is the perpendicular from a vertex to the base (side opposite). (See Figure D.13.)

The three altitudes of a triangle meet at a common point (Figure D.14). A **median** of a triangle is the line segment from a vertex to the midpoint of the opposite side. The three medians of a triangle meet at a common point called the *centroid* (Figure D.15).

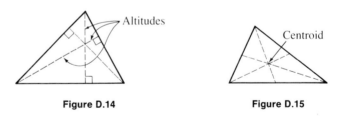

Figure D.14   **Figure D.15**

An important special property of a triangle is the following:

The sum of the angles of a triangle is 180°.

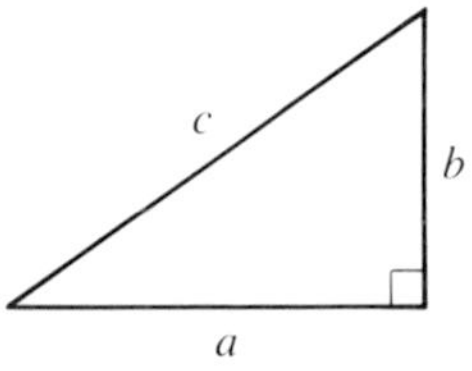

A particularly useful relationship between the sides of a right triangle is the **Pythagorean theorem**. (Refer to Figure D.16.)

Figure D.16

$$a^2 + b^2 = c^2 \qquad \text{(Pythagorean theorem)}$$

Two triangles are **congruent** if the corresponding angles and sides are equal. Two triangles are **similar** (denoted by the symbol $\sim$) if the corresponding angles are equal and the corresponding sides are proportional. Referring to Figure D.17:

If $\triangle ABC \sim \triangle A'B'C'$, then

1. $\angle A = \angle A'$, $\angle B = \angle B'$, $\angle C = \angle C'$

2. $\dfrac{a}{a'} = \dfrac{b}{b'} = \dfrac{c}{c'}$

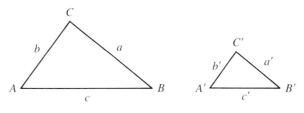

Figure D.17

Quadrilaterals

A **parallelogram** is a quadrilateral having two pairs of parallel sides (Figure D.18). In a parallelogram the parallel sides are equal.

There are several special parallelograms. A **rhombus** is a parallelogram with four equal sides (Figure D.19). A **rectangle** is a parallelogram in which the intersecting sides are perpendicular (Figure D.20). A **square** is a rectangle with four equal sides (Figure D.21). A **trapezoid** is a quadrilateral with exactly one pair of parallel sides (Figure D.22).

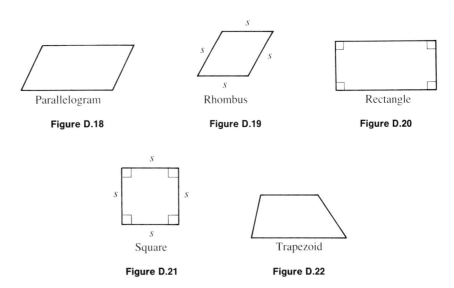

Parallelogram

Figure D.18

Rhombus

Figure D.19

Rectangle

Figure D.20

Square

Figure D.21

Trapezoid

Figure D.22

Circles

A **circle** is a figure in the plane all points of which are the same distance from the fixed point called the **center**. (See Figure D.23.) The distance from the center is called the **radius** (Figure D.23). The distance between two points on a circle and on a line through the center is called the **diameter** (Figure D.24). The diameter is twice the radius.

Other lines or line segments associated with circles are shown in Figure D.24. Note especially that a tangent line (a line that touches a circle at one point) is always perpendicular to the radius.

A **central angle** is an angle formed by two radii (Figure D.25). An **inscribed angle** of a circle is an angle whose vertex is on the circle (Figure D.25).

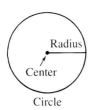

Circle

Figure D.23

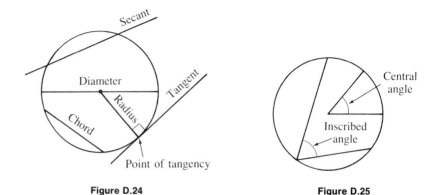

Figure D.24   **Figure D.25**

Perimeter and Area

Two of the most common measures associated with plane figures are **perimeter** and **area**. The perimeter of a figure is the distance around the figure. The perimeter of a circle is called the **circumference**. Various formulas pertaining to these measures are given at the end of this appendix.

Solid Geometric Figures

A **polyhedron** is a solid figure bounded by planes. A **prism** is a polyhedron whose **bases** (top and bottom) are parallel congruent polygons and whose sides are parallelograms. (See Figure D.26.) Each plane surface (including the bases) is called a **face**. The intersection of two faces is called an **edge**. The intersection of edges is called a **vertex**. (See Figure D.26.)

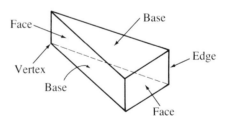

Figure D.26

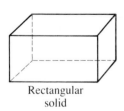

Rectangular
solid

Figure D.27

A **rectangular solid** is a prism whose faces are rectangles (Figure D.27). A **pyramid** is a solid whose base is a polygon and whose **lateral faces** are triangles that meet at a common point called the **vertex**. The **altitude** or **height** is the perpendicular distance from the vertex to the base (Figure D.28).

A **right circular cylinder** is formed by revolving a rectangle about one of its sides (Figure D.29). A **right circular cone** is generated by revolving a right triangle about one of its legs (Figure D.30). The **altitude** or **height** of a right circular cone is perpendicular to the circular base. The **slant height** s is drawn from the vertex to the base, as shown in Figure D.30.

A **sphere** is a curved surface formed by rotating a circle about a diameter (Figure D.31).

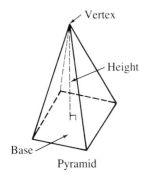

Vertex

Height

Base

Pyramid

Figure D.28

Cylinder

Figure D.29

Cone

Figure D.30

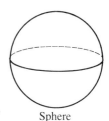

Sphere

Figure D.31

Geometric Formulas

In the following formulas, A = area, B = area of base, C = circumference, S = lateral surface area, V = volume, b = base, h = height, r = radius, and s = slant height.

1. Triangle: $A = \frac{1}{2}bh$ (Figure D.32)
2. Parallelogram: $A = bh$ (Figure D.33)
3. Trapezoid: $A = \frac{1}{2}h(b_1 + b_2)$ (Figure D.34)
4. Circle: $A = \pi r^2$, $C = 2\pi r$ (Figure D.35)
5. Rectangular solid: $V = lwh$ (Figure D.36)
6. Prism $V = Bh$ (Figure D.37)
7. Right circular cylinder: $V = \pi r^2 h$, $S = 2\pi rh$ (Figure D.38)
8. Pyramid: $V = \frac{1}{3}Bh$ (Figure D.39)
9. Right circular cone: $V = \frac{1}{3}\pi r^2 h$, $S = \pi rs$ (Figure D.40)
10. Sphere: $V = \frac{4}{3}\pi r^3$, $A = 4\pi r^2$ (Figure D.41)

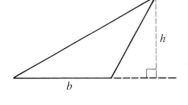

Figure D.32

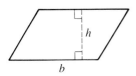

Figure D.33

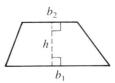

Figure D.34

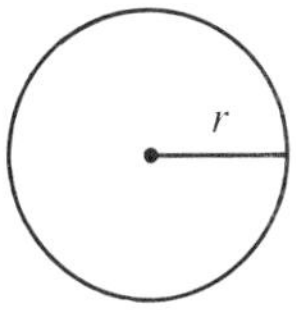

Figure D.35

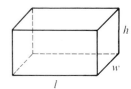

Figure D.36

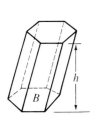

Figure D.37

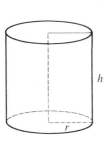

Figure D.38

Figure D.39

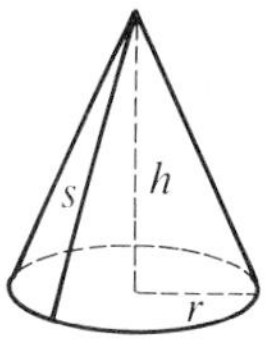

Figure D.40

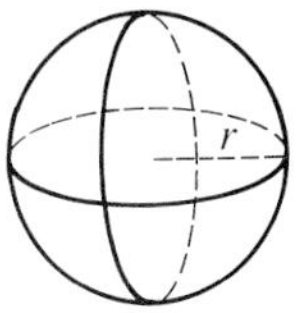

Figure D.41

EXERCISES/APPENDIX D

All answers to these exercises are given in the answer section.

1. In Figure D.42, find $\angle DBE$ and $\angle ABD$.

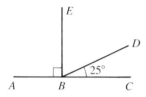

Figure D.42

2. In Figure D.43, find $\angle ABD$ and $\angle A$.

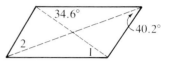

Figure D.43

3. In Figure D.44, DE is parallel to AB. Find $\angle A$, $\angle B$, and $\angle ACB$.

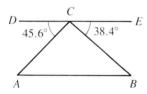

Figure D.44

4. In Figure D.45, L_1 is parallel to L_2. Find $\angle 1$, $\angle 2$, and $\angle 3$.

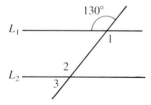

Figure D.45

5. In the parallelogram in Figure D.46, find $\angle 1$ and $\angle 2$.

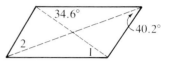

Figure D.46

6. Find c in Figure D.47.

Figure D.47

7. In Figure D.48, AB is tangent to the circle at A. Find the length of AB.

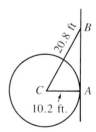

Figure D.48

8. Find x in Figure D.49.

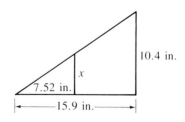

Figure D.49

9. Find x in Figure D.50.

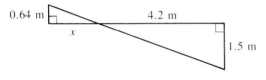

Figure D.50

10. Find the distance from first to third base. (A baseball diamond is a 90 foot square.)

In Exercises 11–16, find the perimeter or circumference of each figure.

11. A triangle whose sides are 8.0 ft, 6.5 ft, and 10.4 ft, respectively

12. A rectangle 7 cm long and 5 cm wide

13. A square whose side is 3.17 mm

14. An isosceles triangle whose equal sides are 7.0 in. long and whose remaining side is 4.6 in. long

15. A circle of radius 3.570 ft

16. A circle of diameter 8.06 in.

In Exercises 17–23, find the area of each figure.

17. A rectangle 10.00 yd long and 8.00 yd wide

18. A parallelogram whose base is 3.652 cm and whose height is 5.400 cm

19. A triangle of base 11.40 mm and height 6.39 mm

20. A right triangle whose legs are 0.917 in. and 0.546 in., respectively

21. A trapezoid whose bases are 3.8 ft and 7.5 ft, respectively, and whose height is 1.6 ft

22. A circle of radius 23.4 yd

23. A circle of diameter 7.6 cm

In Exercises 24–32, find the volume of each solid.

24. A rectangular solid 8.367 ft long, 6.902 ft wide, and 10.300 ft high

25. A cube whose edge is 29.4 ft long

26. Prism—area of base: 4.83 cm^2; height: 5.07 cm

27. Cylinder—radius of base: 11.020 cm; height: 5.368 cm

28. Cone—radius of base: 3.2 yd; height: 4.6 yd

29. Cone—diameter of base: 8.30 m; height: 4.70 m

30. Pyramid—area of base: 25.1 in.^{2}; height: 18.0 in.

31. Sphere of radius 4.8 cm

32. Sphere of diameter 16.90 ft

In Exercises 33–38, find the total surface area of each figure.

33. Cube whose edge is 12.0 in. long

34. Sphere of radius 6.41 yd

35. Cylinder—radius of base: 12.40 cm; height: 7.37 cm

36. Cylinder—radius of base: 1.0090 mm; height: 0.4716 mm

37. Cone—radius of base: 8.30 in.; slant height: 3.75 in.

38. Cone—radius of base: 4.36 m; height: 3.94 m

Tables

Table 1 *Trigonometric functions*

Degrees	Sin θ	Cos θ	Tan θ	Cot θ	Sec θ	Csc θ	
0°00′	0.0000	1.0000	0.0000	—	1.000	—	90°00′
10	0.0029	1.0000	0.0029	343.8	1.000	343.8	50
20	0.0058	1.0000	0.0058	171.9	1.000	171.9	40
30	0.0087	1.0000	0.0087	114.6	1.000	114.6	30
40	0.0116	0.9999	0.0116	85.94	1.000	85.95	20
50	0.0145	0.9999	0.0145	68.75	1.000	68.76	10
1°00′	0.0175	0.9998	0.0175	57.29	1.000	57.30	89°00′
10	0.0204	0.9998	0.0204	49.10	1.000	49.11	50
20	0.0233	0.9997	0.0233	42.96	1.000	42.98	40
30	0.0262	0.9997	0.0262	38.19	1.000	38.20	30
40	0.0291	0.9996	0.0291	34.37	1.000	34.38	20
50	0.0320	0.9995	0.0320	31.24	1.001	31.26	10
2°00′	0.0349	0.9994	0.0349	28.64	1.001	28.65	88°00′
10	0.0378	0.9993	0.0378	26.43	1.001	26.45	50
20	0.0407	0.9992	0.0407	24.54	1.001	24.56	40
30	0.0436	0.9990	0.0437	22.90	1.001	22.93	30
40	0.0465	0.9989	0.0466	21.47	1.001	21.49	20
50	0.0494	0.9988	0.0495	20.21	1.001	20.23	10
3°00′	0.0523	0.9986	0.0524	19.08	1.001	19.11	87°00′
10	0.0552	0.9985	0.0553	18.07	1.002	18.10	50
20	0.0581	0.9983	0.0582	17.17	1.002	17.20	40
30	0.0610	0.9981	0.0612	16.35	1.002	16.38	30
40	0.0640	0.9980	0.0641	15.60	1.002	15.64	20
50	0.0669	0.9978	0.0670	14.92	1.002	14.96	10
4°00′	0.0698	0.9976	0.0699	14.30	1.002	14.34	86°00′
10	0.0727	0.9974	0.0729	13.73	1.003	13.76	50
20	0.0756	0.9971	0.0758	13.20	1.003	13.23	40
30	0.0785	0.9969	0.0787	12.71	1.003	12.75	30
40	0.0814	0.9967	0.0816	12.25	1.003	12.29	20
50	0.0843	0.9964	0.0846	11.83	1.004	11.87	10
5°00′	0.0872	0.9962	0.0875	11.43	1.004	11.47	85°00′
	Cos θ	Sin θ	Cot θ	Tan θ	Csc θ	Sec θ	Degrees

Table 1 *Trigonometric functions* (continued)

Degrees	Sin θ	Cos θ	Tan θ	Cot θ	Sec θ	Csc θ	
5°00′	0.0872	0.9962	0.0875	11.43	1.004	11.47	85°00′
10	0.0901	0.9959	0.0904	11.06	1.004	11.10	50
20	0.0929	0.9957	0.0934	10.71	1.004	10.76	40
30	0.0958	0.9954	0.0963	10.39	1.005	10.43	30
40	0.0987	0.9951	0.0992	10.08	1.005	10.13	20
50	0.1016	0.9948	0.1022	9.788	1.005	9.839	10
6°00′	0.1045	0.9945	0.1051	9.514	1.006	9.567	84°00′
10	0.1074	0.9942	0.1080	9.255	1.006	9.309	50
20	0.1103	0.9939	0.1110	9.010	1.006	9.065	40
30	0.1132	0.9936	0.1139	8.777	1.006	8.834	30
40	0.1161	0.9932	0.1169	8.556	1.007	8.614	20
50	0.1190	0.9929	0.1198	8.345	1.007	8.405	10
7°00′	0.1219	0.9925	0.1228	8.144	1.008	8.206	83°00′
10	0.1248	0.9922	0.1257	7.953	1.008	8.016	50
20	0.1276	0.9918	0.1287	7.770	1.008	7.834	40
30	0.1305	0.9914	0.1317	7.596	1.009	7.661	30
40	0.1334	0.9911	0.1346	7.429	1.009	7.496	20
50	0.1363	0.9907	0.1376	7.269	1.009	7.337	10
8°00′	0.1392	0.9903	0.1405	7.115	1.010	7.185	82°00′
10	0.1421	0.9899	0.1435	6.968	1.010	7.040	50
20	0.1449	0.9894	0.1465	6.827	1.011	6.900	40
30	0.1478	0.9890	0.1495	6.691	1.011	6.765	30
40	0.1507	0.9886	0.1524	6.561	1.012	6.636	20
50	0.1536	0.9881	0.1554	6.435	1.012	6.512	10
9°00′	0.1564	0.9877	0.1584	6.314	1.012	6.392	81°00′
10	0.1593	0.9872	0.1614	6.197	1.013	6.277	50
20	0.1622	0.9868	0.1644	6.084	1.013	6.166	40
30	0.1650	0.9863	0.1673	5.976	1.014	6.059	30
40	0.1679	0.9858	0.1703	5.871	1.014	5.955	20
50	0.1708	0.9853	0.1733	5.769	1.015	5.855	10
10°00′	0.1736	0.9848	0.1763	5.671	1.015	5.759	80°00′
10	0.1765	0.9843	0.1793	5.576	1.016	5.665	50
20	0.1794	0.9838	0.1823	5.485	1.016	5.575	40
30	0.1822	0.9833	0.1853	5.396	1.017	5.487	30
40	0.1851	0.9827	0.1883	5.309	1.018	5.403	20
50	0.1880	0.9822	0.1914	5.226	1.018	5.320	10
11°00′	0.1908	0.9816	0.1944	5.145	1.019	5.241	79°00′
10	0.1937	0.9811	0.1974	5.066	1.019	5.164	50
20	0.1965	0.9805	0.2004	4.989	1.020	5.089	40
30	0.1994	0.9799	0.2035	4.915	1.020	5.016	30
40	0.2022	0.9793	0.2065	4.843	1.021	4.945	20
50	0.2051	0.9787	0.2095	4.773	1.022	4.876	10
12°00′	0.2079	0.9781	0.2126	4.705	1.022	4.810	78°00′
10	0.2108	0.9775	0.2156	4.638	1.023	4.745	50
20	0.2136	0.9769	0.2186	4.574	1.024	4.682	40
30	0.2164	0.9763	0.2217	4.511	1.024	4.620	30
40	0.2193	0.9757	0.2247	4.449	1.025	4.560	20
50	0.2221	0.9750	0.2278	4.390	1.026	4.502	10
13°00′	0.2250	0.9744	0.2309	4.331	1.026	4.445	77°00′
	Cos θ	Sin θ	Cot θ	Tan θ	Csc θ	Sec θ	Degrees

Table 1 *Trigonometric functions* (continued)

Degrees	Sin θ	Cos θ	Tan θ	Cot θ	Sec θ	Csc θ	
13°00′	0.2250	0.9744	0.2309	4.331	1.026	4.445	77°00′
10	0.2278	0.9737	0.2339	4.275	1.027	4.390	50
20	0.2306	0.9730	0.2370	4.219	1.028	4.336	40
30	0.2334	0.9724	0.2401	4.165	1.028	4.284	30
40	0.2363	0.9717	0.2432	4.113	1.029	4.232	20
50	0.2391	0.9710	0.2462	4.061	1.030	4.182	10
14°00′	0.2419	0.9703	0.2493	4.011	1.031	4.134	76°00′
10	0.2447	0.9696	0.2524	3.962	1.031	4.086	50
20	0.2476	0.9689	0.2555	3.914	1.032	4.039	40
30	0.2504	0.9681	0.2586	3.867	1.033	3.994	30
40	0.2532	0.9674	0.2617	3.821	1.034	3.950	20
50	0.2560	0.9667	0.2648	3.776	1.034	3.906	10
15°00′	0.2588	0.9659	0.2679	3.732	1.035	3.864	75°00′
10	0.2616	0.9652	0.2711	3.689	1.036	3.822	50
20	0.2644	0.9644	0.2742	3.647	1.037	3.782	40
30	0.2672	0.9636	0.2773	3.606	1.038	3.742	30
40	0.2700	0.9628	0.2805	3.566	1.039	3.703	20
50	0.2728	0.9621	0.2836	3.526	1.039	3.665	10
16°00′	0.2756	0.9613	0.2867	3.487	1.040	3.628	74°00′
10	0.2784	0.9605	0.2899	3.450	1.041	3.592	50
20	0.2812	0.9596	0.2931	3.412	1.042	3.556	40
30	0.2840	0.9588	0.2962	3.376	1.043	3.521	30
40	0.2868	0.9580	0.2994	3.340	1.044	3.487	20
50	0.2896	0.9572	0.3026	3.305	1.045	3.453	10
17°00′	0.2924	0.9563	0.3057	3.271	1.046	3.420	73°00′
10	0.2952	0.9555	0.3089	3.237	1.047	3.388	50
20	0.2979	0.9546	0.3121	3.204	1.048	3.356	40
30	0.3007	0.9537	0.3153	3.172	1.049	3.326	30
40	0.3035	0.9528	0.3185	3.140	1.049	3.295	20
50	0.3062	0.9520	0.3217	3.108	1.050	3.265	10
18°00′	0.3090	0.9511	0.3249	3.078	1.051	3.236	72°00′
10	0.3118	0.9502	0.3281	3.047	1.052	3.207	50
20	0.3145	0.9492	0.3314	3.018	1.053	3.179	40
30	0.3173	0.9483	0.3346	2.989	1.054	3.152	30
40	0.3201	0.9474	0.3378	2.960	1.056	3.124	20
50	0.3228	0.9465	0.3411	2.932	1.057	3.098	10
19°00′	0.3256	0.9455	0.3443	2.904	1.058	3.072	71°00′
10	0.3283	0.9446	0.3476	2.877	1.059	3.046	50
20	0.3311	0.9436	0.3508	2.850	1.060	3.021	40
30	0.3338	0.9426	0.3541	2.824	1.061	2.996	30
40	0.3365	0.9417	0.3574	2.798	1.062	2.971	20
50	0.3393	0.9407	0.3607	2.773	1.063	2.947	10
20°00′	0.3420	0.9397	0.3640	2.747	1.064	2.924	70°00′
10	0.3448	0.9387	0.3673	2.723	1.065	2.901	50
20	0.3475	0.9377	0.3706	2.699	1.066	2.878	40
30	0.3502	0.9367	0.3739	2.675	1.068	2.855	30
40	0.3529	0.9356	0.3772	2.651	1.069	2.833	20
50	0.3557	0.9346	0.3805	2.628	1.070	2.812	10
21°00′	0.3584	0.9336	0.3839	2.605	1.071	2.790	69°00′
	Cos θ	Sin θ	Cot θ	Tan θ	Csc θ	Sec θ	Degrees

Table 1 *Trigonometric functions* (continued)

Degrees	Sin θ	Cos θ	Tan θ	Cot θ	Sec θ	Csc θ	
21°00′	0.3584	0.9336	0.3839	2.605	1.071	2.790	69°00′
10	0.3611	0.9325	0.3872	2.583	1.072	2.769	50
20	0.3638	0.9315	0.3906	2.560	1.074	2.749	40
30	0.3665	0.9304	0.3939	2.539	1.075	2.729	30
40	0.3692	0.9293	0.3973	2.517	1.076	2.709	20
50	0.3719	0.9283	0.4006	2.496	1.077	2.689	10
22°00′	0.3746	0.9272	0.4040	2.475	1.079	2.669	68°00′
10	0.3773	0.9261	0.4074	2.455	1.080	2.650	50
20	0.3800	0.9250	0.4108	2.434	1.081	2.632	40
30	0.3827	0.9239	0.4142	2.414	1.082	2.613	30
40	0.3854	0.9228	0.4176	2.394	1.084	2.595	20
50	0.3881	0.9216	0.4210	2.375	1.085	2.577	10
23°00′	0.3907	0.9205	0.4245	2.356	1.086	2.559	67°00′
10	0.3934	0.9194	0.4279	2.337	1.088	2.542	50
20	0.3961	0.9182	0.4314	2.318	1.089	2.525	40
30	0.3987	0.9171	0.4348	2.300	1.090	2.508	30
40	0.4014	0.9159	0.4383	2.282	1.092	2.491	20
50	0.4041	0.9147	0.4417	2.264	1.093	2.475	10
24°00′	0.4067	0.9135	0.4452	2.246	1.095	2.459	66°00′
10	0.4094	0.9124	0.4487	2.229	1.096	2.443	50
20	0.4120	0.9112	0.4522	2.211	1.097	2.427	40
30	0.4147	0.9100	0.4557	2.194	1.099	2.411	30
40	0.4173	0.9088	0.4592	2.177	1.100	2.396	20
50	0.4200	0.9075	0.4628	2.161	1.102	2.381	10
25°00′	0.4226	0.9063	0.4663	2.145	1.103	2.366	65°00′
10	0.4253	0.9051	0.4699	2.128	1.105	2.352	50
20	0.4279	0.9038	0.4734	2.112	1.106	2.337	40
30	0.4305	0.9026	0.4770	2.097	1.108	2.323	30
40	0.4331	0.9013	0.4806	2.081	1.109	2.309	20
50	0.4358	0.9001	0.4841	2.066	1.111	2.295	10
26°00′	0.4384	0.8988	0.4877	2.050	1.113	2.281	64°00′
10	0.4410	0.8975	0.4913	2.035	1.114	2.268	50
20	0.4436	0.8962	0.4950	2.020	1.116	2.254	40
30	0.4462	0.8949	0.4986	2.006	1.117	2.241	30
40	0.4488	0.8936	0.5022	1.991	1.119	2.228	20
50	0.4514	0.8923	0.5059	1.977	1.121	2.215	10
27°00′	0.4540	0.8910	0.5095	1.963	1.122	2.203	63°00′
10	0.4566	0.8897	0.5132	1.949	1.124	2.190	50
20	0.4592	0.8884	0.5169	1.935	1.126	2.178	40
30	0.4617	0.8870	0.5206	1.921	1.127	2.166	30
40	0.4643	0.8857	0.5243	1.907	1.129	2.154	20
50	0.4669	0.8843	0.5280	1.894	1.131	2.142	10
28°00′	0.4695	0.8829	0.5317	1.881	1.133	2.130	62°00′
10	0.4720	0.8816	0.5354	1.868	1.134	2.118	50
20	0.4746	0.8802	0.5392	1.855	1.136	2.107	40
30	0.4772	0.8788	0.5430	1.842	1.138	2.096	30
40	0.4797	0.8774	0.5467	1.829	1.140	2.085	20
50	0.4823	0.8760	0.5505	1.816	1.142	2.074	10
29°00′	0.4848	0.8746	0.5543	1.804	1.143	2.063	61°00′
	Cos θ	Sin θ	Cot θ	Tan θ	Csc θ	Sec θ	Degrees

Table 1 *Trigonometric functions* (continued)

Degrees	Sin θ	Cos θ	Tan θ	Cot θ	Sec θ	Csc θ	
29°00′	0.4848	0.8746	0.5543	1.804	1.143	2.063	61°00′
10	0.4874	0.8732	0.5581	1.792	1.145	2.052	50
20	0.4899	0.8718	0.5619	1.780	1.147	2.041	40
30	0.4924	0.8704	0.5658	1.767	1.149	2.031	30
40	0.4950	0.8689	0.5696	1.756	1.151	2.020	20
50	0.4975	0.8675	0.5735	1.744	1.153	2.010	10
30°00′	0.5000	0.8660	0.5774	1.732	1.155	2.000	60°00′
10	0.5025	0.8646	0.5812	1.720	1.157	1.990	50
20	0.5050	0.8631	0.5851	1.709	1.159	1.980	40
30	0.5075	0.8616	0.5890	1.698	1.161	1.970	30
40	0.5100	0.8601	0.5930	1.686	1.163	1.961	20
50	0.5125	0.8587	0.5969	1.675	1.165	1.951	10
31°00′	0.5150	0.8572	0.6009	1.664	1.167	1.942	59°00′
10	0.5175	0.8557	0.6048	1.653	1.169	1.932	50
20	0.5200	0.8542	0.6088	1.643	1.171	1.923	40
30	0.5225	0.8526	0.6128	1.632	1.173	1.914	30
40	0.5250	0.8511	0.6168	1.621	1.175	1.905	20
50	0.5275	0.8496	0.6208	1.611	1.177	1.896	10
32°00′	0.5299	0.8480	0.6249	1.600	1.179	1.887	58°00′
10	0.5324	0.8465	0.6289	1.590	1.181	1.878	50
20	0.5348	0.8450	0.6330	1.580	1.184	1.870	40
30	0.5373	0.8434	0.6371	1.570	1.186	1.861	30
40	0.5398	0.8418	0.6412	1.560	1.188	1.853	20
50	0.5422	0.8403	0.6453	1.550	1.190	1.844	10
33°00′	0.5446	0.8387	0.6494	1.540	1.192	1.836	57°00′
10	0.5471	0.8371	0.6536	1.530	1.195	1.828	50
20	0.5495	0.8355	0.6577	1.520	1.197	1.820	40
30	0.5519	0.8339	0.6619	1.511	1.199	1.812	30
40	0.5544	0.8323	0.6661	1.501	1.202	1.804	20
50	0.5568	0.8307	0.6703	1.492	1.204	1.796	10
34°00′	0.5592	0.8290	0.6745	1.483	1.206	1.788	56°00′
10	0.5616	0.8274	0.6787	1.473	1.209	1.781	50
20	0.5640	0.8258	0.6830	1.464	1.211	1.773	40
30	0.5664	0.8241	0.6873	1.455	1.213	1.766	30
40	0.5688	0.8225	0.6916	1.446	1.216	1.758	20
50	0.5712	0.8208	0.6959	1.437	1.218	1.751	10
35°00′	0.5736	0.8192	0.7002	1.428	1.221	1.743	55°00′
10	0.5760	0.8175	0.7046	1.419	1.223	1.736	50
20	0.5783	0.8158	0.7089	1.411	1.226	1.729	40
30	0.5807	0.8141	0.7133	1.402	1.228	1.722	30
40	0.5831	0.8124	0.7177	1.393	1.231	1.715	20
50	0.5854	0.8107	0.7221	1.385	1.233	1.708	10
36°00′	0.5878	0.8090	0.7265	1.376	1.236	1.701	54°00′
10	0.5901	0.8073	0.7310	1.368	1.239	1.695	50
20	0.5925	0.8056	0.7355	1.360	1.241	1.688	40
30	0.5948	0.8039	0.7400	1.351	1.244	1.681	30
40	0.5972	0.8021	0.7445	1.343	1.247	1.675	20
50	0.5995	0.8004	0.7490	1.335	1.249	1.668	10
37°00′	0.6018	0.7986	0.7536	1.327	1.252	1.662	53°00′
	Cos θ	Sin θ	Cot θ	Tan θ	Csc θ	Sec θ	Degrees

Table 1 *Trigonometric functions* (continued)

Degrees	Sin θ	Cos θ	Tan θ	Cot θ	Sec θ	Csc θ	
37°00'	0.6018	0.7986	0.7536	1.327	1.252	1.662	53°00'
10	0.6041	0.7969	0.7581	1.319	1.255	1.655	50
20	0.6065	0.7951	0.7627	1.311	1.258	1.649	40
30	0.6088	0.7934	0.7673	1.303	1.260	1.643	30
40	0.6111	0.7916	0.7720	1.295	1.263	1.636	20
50	0.6134	0.7898	0.7766	1.288	1.266	1.630	10
38°00'	0.6157	0.7880	0.7813	1.280	1.269	1.624	52°00'
10	0.6180	0.7862	0.7860	1.272	1.272	1.618	50
20	0.6202	0.7844	0.7907	1.265	1.275	1.612	40
30	0.6225	0.7826	0.7954	1.257	1.278	1.606	30
40	0.6248	0.7808	0.8002	1.250	1.281	1.601	20
50	0.6271	0.7790	0.8050	1.242	1.284	1.595	10
39°00'	0.6293	0.7771	0.8098	1.235	1.287	1.589	51°00'
10	0.6316	0.7753	0.8146	1.228	1.290	1.583	50
20	0.6338	0.7735	0.8195	1.220	1.293	1.578	40
30	0.6361	0.7716	0.8243	1.213	1.296	1.572	30
40	0.6383	0.7698	0.8292	1.206	1.299	1.567	20
50	0.6406	0.7679	0.8342	1.199	1.302	1.561	10
40°00'	0.6428	0.7660	0.8391	1.192	1.305	1.556	50°00'
10	0.6450	0.7642	0.8441	1.185	1.309	1.550	50
20	0.6472	0.7623	0.8491	1.178	1.312	1.545	40
30	0.6494	0.7604	0.8541	1.171	1.315	1.540	30
40	0.6517	0.7585	0.8591	1.164	1.318	1.535	20
50	0.6539	0.7566	0.8642	1.157	1.322	1.529	10
41°00'	0.6561	0.7547	0.8693	1.150	1.325	1.524	49°00'
10	0.6583	0.7528	0.8744	1.144	1.328	1.519	50
20	0.6604	0.7509	0.8796	1.137	1.332	1.514	40
30	0.6626	0.7490	0.8847	1.130	1.335	1.509	30
40	0.6648	0.7470	0.8899	1.124	1.339	1.504	20
50	0.6670	0.7451	0.8952	1.117	1.342	1.499	10
42°00'	0.6691	0.7431	0.9004	1.111	1.346	1.494	48°00'
10	0.6713	0.7412	0.9057	1.104	1.349	1.490	50
20	0.6734	0.7392	0.9110	1.098	1.353	1.485	40
30	0.6756	0.7373	0.9163	1.091	1.356	1.480	30
40	0.6777	0.7353	0.9217	1.085	1.360	1.476	20
50	0.6799	0.7333	0.9271	1.079	1.364	1.471	10
43°00'	0.6820	0.7314	0.9325	1.072	1.367	1.466	47°00'
10	0.6841	0.7294	0.9380	1.066	1.371	1.462	50
20	0.6862	0.7274	0.9435	1.060	1.375	1.457	40
30	0.6884	0.7254	0.9490	1.054	1.379	1.453	30
40	0.6905	0.7234	0.9545	1.048	1.382	1.448	20
50	0.6926	0.7214	0.9601	1.042	1.386	1.444	10
44°00'	0.6947	0.7193	0.9657	1.036	1.390	1.440	46°00'
10	0.6967	0.7173	0.9713	1.030	1.394	1.435	50
20	0.6988	0.7153	0.9770	1.024	1.398	1.431	40
30	0.7009	0.7133	0.9827	1.018	1.402	1.427	30
40	0.7030	0.7112	0.9884	1.012	1.406	1.423	20
50	0.7050	0.7092	0.9942	1.006	1.410	1.418	10
45°00'	0.7071	0.7071	1.000	1.000	1.414	1.414	45°00'
	Cos θ	Sin θ	Cot θ	Tan θ	Csc θ	Sec θ	Degrees

Table 2 *Common logarithms*

n	0	1	2	3	4	5	6	7	8	9
1.0	.0000	.0043	.0086	.0128	.0170	.0212	.0253	.0294	.0334	.0374
1.1	.0414	.0453	.0492	.0531	.0569	.0607	.0645	.0682	.0719	.0755
1.2	.0792	.0828	.0864	.0899	.0934	.0969	.1004	.1038	.1072	.1106
1.3	.1139	.1173	.1206	.1239	.1271	.1303	.1335	.1367	.1399	.1430
1.4	.1461	.1492	.1523	.1553	.1584	.1614	.1644	.1673	.1703	.1732
1.5	.1761	.1790	.1818	.1847	.1875	.1903	.1931	.1959	.1987	.2014
1.6	.2041	.2068	.2095	.2122	.2148	.2175	.2201	.2227	.2253	.2279
1.7	.2304	.2330	.2355	.2380	.2405	.2430	.2455	.2480	.2504	.2529
1.8	.2553	.2577	.2601	.2625	.2648	.2672	.2695	.2718	.2742	.2765
1.9	.2788	.2810	.2833	.2856	.2878	.2900	.2923	.2945	.2967	.2989
2.0	.3010	.3032	.3054	.3075	.3096	.3118	.3139	.3160	.3181	.3201
2.1	.3222	.3243	.3263	.3284	.3304	.3324	.3345	.3365	.3385	.3404
2.2	.3424	.3444	.3464	.3483	.3502	.3522	.3541	.3560	.3579	.3598
2.3	.3617	.3636	.3655	.3674	.3692	.3711	.3729	.3747	.3766	.3784
2.4	.3802	.3820	.3838	.3856	.3874	.3892	.3909	.3927	.3945	.3962
2.5	.3979	.3997	.4014	.4031	.4048	.4065	.4082	.4099	.4116	.4133
2.6	.4150	.4166	.4183	.4200	.4216	.4232	.4249	.4265	.4281	.4298
2.7	.4314	.4330	.4346	.4362	.4378	.4393	.4409	.4425	.4440	.4456
2.8	.4472	.4487	.4502	.4518	.4533	.4548	.4564	.4579	.4594	.4609
2.9	.4624	.4639	.4654	.4669	.4683	.4698	.4713	.4728	.4742	.4757
3.0	.4771	.4786	.4800	.4814	.4829	.4843	.4857	.4871	.4886	.4900
3.1	.4914	.4928	.4942	.4955	.4969	.4983	.4997	.5011	.5024	.5038
3.2	.5051	.5065	.5079	.5092	.5105	.5119	.5132	.5145	.5159	.5172
3.3	.5185	.5198	.5211	.5224	.5237	.5250	.5263	.5276	.5289	.5302
3.4	.5315	.5328	.5340	.5353	.5366	.5378	.5391	.5403	.5416	.5428
3.5	.5441	.5453	.5465	.5478	.5490	.5502	.5514	.5527	.5539	.5551
3.6	.5563	.5575	.5587	.5599	.5611	.5623	.5635	.5647	.5658	.5670
3.7	.5682	.5694	.5705	.5717	.5729	.5740	.5752	.5763	.5775	.5786
3.8	.5798	.5809	.5821	.5832	.5843	.5855	.5866	.5877	.5888	.5899
3.9	.5911	.5922	.5933	.5944	.5955	.5966	.5977	.5988	.5999	.6010
4.0	.6021	.6031	.6042	.6053	.6064	.6075	.6085	.6096	.6107	.6117
4.1	.6128	.6138	.6149	.6160	.6170	.6180	.6191	.6201	.6212	.6222
4.2	.6232	.6243	.6253	.6263	.6274	.6284	.6294	.6304	.6314	.6325
4.3	.6335	.6345	.6355	.6365	.6375	.6385	.6395	.6405	.6415	.6425
4.4	.6435	.6444	.6454	.6464	.6474	.6484	.6493	.6503	.6513	.6522
4.5	.6532	.6542	.6551	.6561	.6571	.6580	.6590	.6599	.6609	.6618
4.6	.6628	.6637	.6646	.6656	.6665	.6675	.6684	.6693	.6702	.6712
4.7	.6721	.6730	.6739	.6749	.6758	.6767	.6776	.6785	.6794	.6803
4.8	.6812	.6821	.6830	.6839	.6848	.6857	.6866	.6875	.6884	.6893
4.9	.6902	.6911	.6920	.6928	.6937	.6946	.6955	.6964	.6972	.6981
5.0	.6990	.6998	.7007	.7016	.7024	.7033	.7042	.7050	.7059	.7067
5.1	.7076	.7084	.7093	.7101	.7110	.7118	.7126	.7135	.7143	.7152
5.2	.7160	.7168	.7177	.7185	.7193	.7202	.7210	.7218	.7226	.7235
5.3	.7243	.7251	.7259	.7267	.7275	.7284	.7292	.7300	.7308	.7316
5.4	.7324	.7332	.7340	.7348	.7356	.7364	.7372	.7380	.7388	.7396
5.5	.7404	.7412	.7419	.7427	.7435	.7443	.7451	.7459	.7466	.7474
5.6	.7482	.7490	.7497	.7505	.7513	.7520	.7528	.7536	.7543	.7551
5.7	.7559	.7566	.7574	.7582	.7589	.7597	.7604	.7612	.7619	.7627
5.8	.7634	.7642	.7649	.7657	.7664	.7672	.7679	.7686	.7694	.7701
5.9	.7709	.7716	.7723	.7731	.7738	.7745	.7752	.7760	.7767	.7774

Table 2 *Common logarithms* (continued)

n	0	1	2	3	4	5	6	7	8	9
6.0	.7782	.7789	.7796	.7803	.7810	.7818	.7825	.7832	.7839	.7846
6.1	.7853	.7860	.7868	.7875	.7882	.7889	.7896	.7903	.7910	.7917
6.2	.7924	.7931	.7938	.7945	.7952	.7959	.7966	.7973	.7980	.7987
6.3	.7993	.8000	.8007	.8014	.8021	.8028	.8035	.8041	.8048	.8055
6.4	.8062	.8069	.8075	.8082	.8089	.8096	.8102	.8109	.8116	.8122
6.5	.8129	.8136	.8142	.8149	.8156	.8162	.8169	.8176	.8182	.8189
6.6	.8195	.8202	.8209	.8215	.8222	.8228	.8235	.8241	.8248	.8254
6.7	.8261	.8267	.8274	.8280	.8287	.8293	.8299	.8306	.8312	.8319
6.8	.8325	.8331	.8338	.8344	.8351	.8357	.8363	.8370	.8376	.8382
6.9	.8388	.8395	.8401	.8407	.8414	.8420	.8426	.8432	.8439	.8445
7.0	.8451	.8457	.8463	.8470	.8476	.8482	.8488	.8494	.8500	.8506
7.1	.8513	.8519	.8525	.8531	.8537	.8543	.8549	.8555	.8561	.8567
7.2	.8573	.8579	.8585	.8591	.8597	.8603	.8609	.8615	.8621	.8627
7.3	.8633	.8639	.8645	.8651	.8657	.8663	.8669	.8675	.8681	.8686
7.4	.8692	.8698	.8704	.8710	.8716	.8722	.8727	.8733	.8739	.8745
7.5	.8751	.8756	.8762	.8768	.8774	.8779	.8785	.8791	.8797	.8802
7.6	.8808	.8814	.8820	.8825	.8831	.8837	.8842	.8848	.8854	.8859
7.7	.8865	.8871	.8876	.8882	.8887	.8893	.8899	.8904	.8910	.8915
7.8	.8921	.8927	.8932	.8938	.8943	.8949	.8954	.8960	.8965	.8971
7.9	.8976	.8982	.8987	.8993	.8998	.9004	.9009	.9015	.9020	.9025
8.0	.9031	.9036	.9042	.9047	.9053	.9058	.9063	.9069	.9074	.9079
8.1	.9085	.9090	.9096	.9101	.9106	.9112	.9117	.9122	.9128	.9133
8.2	.9138	.9143	.9149	.9154	.9159	.9165	.9170	.9175	.9180	.9186
8.3	.9191	.9196	.9201	.9206	.9212	.9217	.9222	.9227	.9232	.9238
8.4	.9243	.9248	.9253	.9258	.9263	.9269	.9274	.9279	.9284	.9289
8.5	.9294	.9299	.9304	.9309	.9315	.9320	.9325	.9330	.9335	.9340
8.6	.9345	.9350	.9355	.9360	.9365	.9370	.9375	.9380	.9385	.9390
8.7	.9395	.9400	.9405	.9410	.9415	.9420	.9425	.9430	.9435	.9440
8.8	.9445	.9450	.9455	.9460	.9465	.9469	.9474	.9479	.9484	.9489
8.9	.9494	.9499	.9504	.9509	.9513	.9518	.9523	.9528	.9533	.9538
9.0	.9542	.9547	.9552	.9557	.9562	.9566	.9571	.9576	.9581	.9586
9.1	.9590	.9595	.9600	.9605	.9609	.9614	.9619	.9624	.9628	.9633
9.2	.9638	.9643	.9647	.9652	.9657	.9661	.9666	.9671	.9675	.9680
9.3	.9685	.9689	.9694	.9699	.9703	.9708	.9713	.9717	.9722	.9727
9.4	.9731	.9736	.9741	.9745	.9750	.9754	.9759	.9763	.9768	.9773
9.5	.9777	.9782	.9786	.9791	.9795	.9800	.9805	.9809	.9814	.9818
9.6	.9823	.9827	.9832	.9836	.9841	.9845	.9850	.9854	.9859	.9863
9.7	.9868	.9872	.9877	.9881	.9886	.9890	.9894	.9899	.9903	.9908
9.8	.9912	.9917	.9921	.9926	.9930	.9934	.9939	.9943	.9948	.9952
9.9	.9956	.9961	.9965	.9969	.9974	.9978	.9983	.9987	.9991	.9996

Table 3 *Natural logarithms*

n	$\log_e n$	n	$\log_e n$	n	$\log_e n$	n	$\log_e n$
0.0	—	3.5	1.2528	7.0	1.9459	15	2.7081
0.1	−2.3026	3.6	1.2809	7.1	1.9601	16	2.7726
0.2	−1.6094	3.7	1.3083	7.2	1.9741	17	2.8332
0.3	−1.2040	3.8	1.3350	7.3	1.9879	18	2.8904
0.4	−0.9163	3.9	1.3610	7.4	2.0015	19	2.9444
0.5	−0.6931	4.0	1.3863	7.5	2.0149	20	2.9957
0.6	−0.5108	4.1	1.4110	7.6	2.0281	25	3.2189
0.7	−0.3567	4.2	1.4351	7.7	2.0412	30	3.4012
0.8	−0.2231	4.3	1.4586	7.8	2.0541	35	3.5553
0.9	−0.1054	4.4	1.4816	7.9	2.0669	40	3.6889
1.0	0.0000	4.5	1.5041	8.0	2.0794	45	3.8067
1.1	0.0953	4.6	1.5261	8.1	2.0919	50	3.9120
1.2	0.1823	4.7	1.5476	8.2	2.1041	55	4.0073
1.3	0.2624	4.8	1.5686	8.3	2.1163	60	4.0943
1.4	0.3365	4.9	1.5892	8.4	2.1282	65	4.1744
1.5	0.4055	5.0	1.6094	8.5	2.1401	70	4.2485
1.6	0.4700	5.1	1.6292	8.6	2.1518	75	4.3175
1.7	0.5306	5.2	1.6487	8.7	2.1633	80	4.3820
1.8	0.5878	5.3	1.6677	8.8	2.1748	85	4.4427
1.9	0.6419	5.4	1.6864	8.9	2.1861	90	4.4998
2.0	0.6931	5.5	1.7047	9.0	2.1972	95	4.5539
2.1	0.7419	5.6	1.7228	9.1	2.2083	100	4.6052
2.2	0.7885	5.7	1.7405	9.2	2.2192	200	5.2983
2.3	0.8329	5.8	1.7579	9.3	2.2300	300	5.7038
2.4	0.8755	5.9	1.7750	9.4	2.2407	400	5.9915
2.5	0.9163	6.0	1.7918	9.5	2.2513	500	6.2146
2.6	0.9555	6.1	1.8083	9.6	2.2618	600	6.3969
2.7	0.9933	6.2	1.8245	9.7	2.2721	700	6.5511
2.8	1.0296	6.3	1.8405	9.8	2.2824	800	6.6846
2.9	1.0647	6.4	1.8563	9.9	2.2925	900	6.8024
3.0	1.0986	6.5	1.8718	10	2.3026		
3.1	1.1314	6.6	1.8871	11	2.3979		
3.2	1.1632	6.7	1.9021	12	2.4849		
3.3	1.1939	6.8	1.9169	13	2.5649		
3.4	1.2238	6.9	1.9315	14	2.6391		

Table 4 *A Short Table of Integrals*

Forms Containing $a + bu$

1. $\displaystyle\int (a + bu)^n \, du = \frac{(a + bu)^{n+1}}{b(n + 1)} + C, \; n \neq -1$

2. $\displaystyle\int \frac{du}{a + bu} = \frac{1}{b} \ln |a + bu| + C$

3. $\displaystyle\int \frac{u \, du}{a + bu} = \frac{1}{b^2} [(a + bu) - a \ln |a + bu|] + C$

4. $\displaystyle\int \frac{u^2 \, du}{a + bu} = \frac{1}{b^3} \left[\frac{1}{2} (a + bu)^2 - 2a(a + bu) + a^2 \ln |a + bu| \right] + C$

5. $\displaystyle\int \frac{du}{u(a + bu)} = \frac{1}{a} \ln \left| \frac{u}{a + bu} \right| + C$

6. $\displaystyle\int \frac{du}{u^2(a + bu)} = -\frac{1}{au} + \frac{b}{a^2} \ln \left| \frac{a + bu}{u} \right| + C$

7. $\displaystyle\int \frac{u \, du}{(a + bu)^2} = \frac{1}{b^2} \left(\ln |a + bu| + \frac{a}{a + bu} \right) + C$

Forms Containing $\sqrt{a + bu}$

8. $\displaystyle\int u \sqrt{a + bu} \, du = -\frac{2(2a - 3bu)(a + bu)^{3/2}}{15b^2} + C$

9. $\displaystyle\int u^2 \sqrt{a + bu} \, du = \frac{2(8a^2 - 12abu + 15b^2u^2)(a + bu)^{3/2}}{105b^3} + C$

10. $\displaystyle\int \frac{u \, du}{\sqrt{a + bu}} = -\frac{2(2a - bu)\sqrt{a + bu}}{3b^2} + C$

11. $\displaystyle\int \frac{u^2 \, du}{\sqrt{a + bu}} = \frac{2(3b^2u^2 - 4abu + 8a^2)\sqrt{a + bu}}{15b^3} + C$

12. $\displaystyle\int \frac{du}{u \sqrt{a + bu}} = \frac{1}{\sqrt{a}} \ln \left| \frac{\sqrt{a + bu} - \sqrt{a}}{\sqrt{a + bu} + \sqrt{a}} \right| + C, \quad a > 0$

13. $\displaystyle\int \frac{du}{u \sqrt{a + bu}} = \frac{2}{\sqrt{-a}} \text{Arctan} \sqrt{\frac{a + bu}{-a}} + C, \quad a < 0$

14. $\displaystyle\int \frac{\sqrt{a + bu} \, du}{u} = 2 \sqrt{a + bu} + a \int \frac{du}{u \sqrt{a + bu}}$

Forms Containing $a^2 \pm u^2$ and $u^2 \pm a^2$

15. $\displaystyle\int \frac{du}{a^2 + u^2} = \frac{1}{a} \text{Arctan} \frac{u}{a} + C$

16. $\displaystyle\int \frac{du}{a^2 - u^2} = \frac{1}{2a} \ln \left| \frac{a + u}{a - u} \right| + C, \quad a^2 > u^2$

17. $\displaystyle\int \frac{du}{u^2 - a^2} = \frac{1}{2a} \ln \left| \frac{u - a}{u + a} \right| + C, \quad a^2 < u^2$

Table 4 *A Short Table of Integrals* (continued)

Forms Containing $\sqrt{a^2 - u^2}$

18. $\int \sqrt{a^2 - u^2} \, du = \dfrac{u}{2} \sqrt{a^2 - u^2} + \dfrac{a^2}{2} \operatorname{Arcsin} \dfrac{u}{a} + C$

19. $\int \dfrac{du}{\sqrt{a^2 - u^2}} = \operatorname{Arcsin} \dfrac{u}{a} + C$

20. $\int \dfrac{du}{(a^2 - u^2)^{3/2}} = \dfrac{u}{a^2 \sqrt{a^2 - u^2}} + C$

21. $\int \dfrac{u^2 \, du}{\sqrt{a^2 - u^2}} = -\dfrac{u}{2} \sqrt{a^2 - u^2} + \dfrac{a^2}{2} \operatorname{Arcsin} \dfrac{u}{a} + C$

22. $\int \dfrac{u^2 \, du}{(a^2 - u^2)^{3/2}} = \dfrac{u}{\sqrt{a^2 - u^2}} - \operatorname{Arcsin} \dfrac{u}{a} + C$

23. $\int \dfrac{du}{u \sqrt{a^2 - u^2}} = -\dfrac{1}{a} \ln \left| \dfrac{a + \sqrt{a^2 - u^2}}{u} \right| + C$

24. $\int \dfrac{du}{u^2 \sqrt{a^2 - u^2}} = -\dfrac{\sqrt{a^2 - u^2}}{a^2 u} + C$

25. $\int \dfrac{\sqrt{a^2 - u^2} \, du}{u^2} = -\dfrac{\sqrt{a^2 - u^2}}{u} - \operatorname{Arcsin} \dfrac{u}{a} + C$

26. $\int \dfrac{\sqrt{a^2 - u^2} \, du}{u} = \sqrt{a^2 - u^2} - a \ln \left| \dfrac{a + \sqrt{a^2 - u^2}}{u} \right| + C$

Forms Containing $\sqrt{u^2 \pm a^2}$

27. $\int \sqrt{u^2 \pm a^2} \, du = \dfrac{1}{2} \left[u \sqrt{u^2 \pm a^2} \pm a^2 \ln \left| u + \sqrt{u^2 \pm a^2} \right| \right] + C$

28. $\int u^2 \sqrt{u^2 \pm a^2} \, du = \dfrac{1}{8} u (2u^2 \pm a^2) \sqrt{u^2 \pm a^2} - \dfrac{1}{8} a^4 \ln \left| u + \sqrt{u^2 \pm a^2} \right| + C$

29. $\int \dfrac{\sqrt{u^2 - a^2}}{u} \, du = \sqrt{u^2 - a^2} - a \operatorname{Arccos} \dfrac{a}{|u|} + C$

30. $\int \dfrac{\sqrt{u^2 + a^2}}{u} \, du = \sqrt{u^2 + a^2} - a \ln \left| \dfrac{a + \sqrt{u^2 + a^2}}{u} \right| + C$

31. $\int \dfrac{\sqrt{u^2 \pm a^2}}{u^2} \, du = -\dfrac{\sqrt{u^2 \pm a^2}}{u} + \ln \left| u + \sqrt{u^2 \pm a^2} \right| + C$

32. $\int \dfrac{du}{\sqrt{u^2 \pm a^2}} = \ln \left| u + \sqrt{u^2 \pm a^2} \right| + C$

33. $\int \dfrac{du}{u \sqrt{u^2 - a^2}} = \dfrac{1}{a} \operatorname{Arccos} \dfrac{a}{|u|} + C$

34. $\int \dfrac{du}{u \sqrt{u^2 + a^2}} = \dfrac{1}{a} \ln \left| \dfrac{u}{a + \sqrt{u^2 + a^2}} \right| + C$

35. $\int \dfrac{du}{u^2 \sqrt{u^2 \pm a^2}} = -\dfrac{(\pm \sqrt{u^2 \pm a^2})}{a^2 u} + C$

36. $\int \dfrac{u^2 \, du}{\sqrt{u^2 \pm a^2}} = \dfrac{1}{2} \left(u \sqrt{u^2 \pm a^2} \mp a^2 \ln \left| u + \sqrt{u^2 \pm a^2} \right| \right) + C$

Table 4 *A Short Table of Integrals* (continued)

37. $\displaystyle\int \frac{du}{(u^2 \pm a^2)^{3/2}} = \frac{\pm u}{a^2 \sqrt{u^2 \pm a^2}} + C$

38. $\displaystyle\int \frac{u^2\, du}{(u^2 \pm a^2)^{3/2}} = \frac{-u}{\sqrt{u^2 \pm a^2}} + \ln \left| u + \sqrt{u^2 \pm a^2} \right| + C$

Exponential and Logarithmic Forms

---

39. $\displaystyle\int e^u\, du = e^u + C$

40. $\displaystyle\int a^u\, du = \frac{a^u}{\ln a} + C$

41. $\displaystyle\int u e^{au}\, du = \frac{e^{au}}{a^2} (au - 1) + C$

42. $\displaystyle\int u^n e^{au}\, du = \frac{u^n e^{au}}{a} - \frac{n}{a} \int u^{n-1} e^{au}\, du$

43. $\displaystyle\int \frac{e^{au}}{u^n}\, du = -\frac{e^{au}}{(n-1)u^{n-1}} + \frac{a}{n-1} \int \frac{e^{au}\, du}{u^{n-1}}$

44. $\displaystyle\int \ln u\, du = u \ln u - u + C$

45. $\displaystyle\int u^n \ln u\, du = \frac{u^{n+1} \ln u}{n+1} - \frac{u^{n+1}}{(n+1)^2} + C$

46. $\displaystyle\int \frac{du}{u \ln u} = \ln \left| \ln u \right| + C$

Trigonometric Forms

---

47. $\displaystyle\int \sin u\, du = -\cos u + C$

48. $\displaystyle\int \cos u\, du = \sin u + C$

49. $\displaystyle\int \tan u\, du = -\ln \left| \cos u \right| + C = \ln \left| \sec u \right| + C$

50. $\displaystyle\int \cot u\, du = \ln \left| \sin u \right| + C = -\ln \left| \csc u \right| + C$

51. $\displaystyle\int \sec u\, du = \ln \left| \sec u + \tan u \right| + C$

52. $\displaystyle\int \csc u\, du = \ln \left| \csc u - \cot u \right| + C$

53. $\displaystyle\int \sec^2 u\, du = \tan u + C$

54. $\displaystyle\int \csc^2 u\, du = -\cot u + C$

55. $\displaystyle\int \sec u \tan u\, du = \sec u + C$

56. $\displaystyle\int \csc u \cot u\, du = -\csc u + C$

Table 4 *A Short Table of Integrals* (continued)

---

57. $\int \sin^2 u \, du = \dfrac{1}{2} u - \dfrac{1}{4} \sin 2u + C$

58. $\int \cos^2 u \, du = \dfrac{1}{2} u + \dfrac{1}{4} \sin 2u + C$

59. $\int \sin^n u \cos u \, du = \dfrac{\sin^{n+1} u}{n + 1} + C$

60. $\int \cos^n u \sin u \, du = -\dfrac{\cos^{n+1} u}{n + 1} + C$

61. $\int \sin mu \sin nu \, du = -\dfrac{\sin (m + n)u}{2(m + n)} + \dfrac{\sin (m - n)u}{2(m - n)} + C$

62. $\int \cos mu \cos nu \, du = \dfrac{\sin (m + n)u}{2(m + n)} + \dfrac{\sin (m - n)u}{2(m - n)} + C$

63. $\int \sin mu \cos nu \, du = -\dfrac{\cos (m + n)u}{2(m + n)} - \dfrac{\cos (m - n)u}{2(m - n)} + C$

64. $\int u \sin u \, du = \sin u - u \cos u + C$

65. $\int u \cos u \, du = \cos u + u \sin u + C$

66. $\int \sin^n u \cos^m u \, du = \dfrac{\sin^{n+1} u \cos^{m-1} u}{n + m} + \dfrac{m - 1}{n + m} \int \sin^n u \cos^{m-2} u \, du$

67. $\int \sin^n u \, du = -\dfrac{1}{n} \sin^{n-1} u \cos u + \dfrac{n - 1}{n} \int \sin^{n-2} u \, du$

68. $\int \cos^n u \, du = \dfrac{1}{n} \cos^{n-1} u \sin u + \dfrac{n - 1}{n} \int \cos^{n-2} u \, du$

69. $\int \tan^n u \, du = \dfrac{\tan^{n-1} u}{n - 1} - \int \tan^{n-2} u \, du$

70. $\int \cot^n u \, du = -\dfrac{\cot^{n-1} u}{n - 1} - \int \cot^{n-2} u \, du$

71. $\int \sec^n u \, du = \dfrac{\sec^{n-2} u \tan u}{n - 1} + \dfrac{n - 2}{n - 1} \int \sec^{n-2} u \, du$

72. $\int \csc^n u \, du = -\dfrac{\csc^{n-2} u \cot u}{n - 1} + \dfrac{n - 2}{n - 1} \int \csc^{n-2} u \, du$

Other Forms

---

73. $\int e^{au} \sin bu \, du = \dfrac{e^{au}(a \sin bu - b \cos bu)}{a^2 + b^2} + C$

74. $\int e^{au} \cos bu \, du = \dfrac{e^{au}(a \cos bu + b \sin bu)}{a^2 + b^2} + C$

75. $\int \text{Arcsin } u \, du = u \text{ Arcsin } u + \sqrt{1 - u^2} + C$

76. $\int \text{Arctan } u \, du = u \text{ Arctan } u - \dfrac{1}{2} \ln (1 + u^2) + C$

Answers to Selected Exercises

Chapter 1

Section 1.3 (page 10)

1. 3   **3.** 7   **5.** π **7.** $\sqrt{3}$ **9.** 3   **11.** r   **13.** r   **15.** i   **17.** r   **19.** i   **21.** -3 **23.** 8   **25.** 11   **27.** -17

29. -5 **31.** -11 **33.** 16   **35.** 8.4   **37.** $-\dfrac{1}{2}$ **39.** -78 **41.** -96 **43.** 96   **45.** 0.91   **47.** 28.014   **49.** $-\dfrac{5}{3}$

51. undefined   **53.** $-\dfrac{8}{225}$ **55.** 0   **57.** 1   **59.** -20 **61.** -3 **63.** 3   **65.** 20   **67.** -4 **69.** 21   **71.** -64 **73.** 5

75. $-\dfrac{1}{25}$ **77.** $-\dfrac{1}{8}$ **79.** $-7°C$ **81.** 2.7 volts   **83.** 20,602 ft   **85.** $-4°F$ **87.** 2.0 ft/s

Section 1.4 (page 19)

1. 1,052   **3.** 456   **5.** -896 **7.** -324 **9.** -442 **11.** 4   **13.** 676   **15.** 361   **17.** 27   **19.** 35   **21.** 9.7168
23. 4.6088   **25.** 19   **27.** -3.50 **29.** -3.709 **31.** 1.98   **33.** -1.5 **35.** -12.6 **37.** 57   **39.** -7.7 **41.** -3.14
43. 1.15

Section 1.5 (page 22)

1. numerical coefficient: 4; factors: x and y **3.** numerical coefficient: -2; factors: a and b **5.** numerical coefficient: 12; factors: s and t
7. monomial   **9.** binomial   **11.** binomial   **13.** monomial   **15.** trinomial   **17.** binomial   **19.** trinomial

Section 1.6 (page 25)

1. $-6x - 5y$ **3.** $-2x - 4y$ **5.** $-6a^2 - 2a$ **7.** $xy^2 - 3x^2y$ **9.** $-x - 5y + z$ **11.** $4x^2 - 6y^2 - 3z^2$ **13.** a
15. $-xy + 2y^2$ **17.** $-6x - 5y + z$ **19.** $-2a - 2b - 7c$ **21.** $11a - b + 7c$ **23.** $a + 6b + c$ **25.** $-7a^2 + 10c^2 - d^2 + e$
27. $-3a + 6b^2 + c$ **29.** $-b - 4c$ **31.** $41a^2 + 12b^2$ **33.** $2x + 2y$ **35.** total weight of contents   **37.** $5I_1 - 4I_2$
39. $(2C + 4a)$ dollars

Section 1.7 (page 29)

1. $-12 + 3x$ **3.** $-x + 10y$ **5.** $x - 2y$ **7.** $-2x^2 - 4xy$ **9.** $3a - 4b$ **11.** $3p - q$ **13.** $3V - C$ **15.** $3s - 7t$ **17.** $7 - 8z$
19. $-8x + 2y$ **21.** $-3a + b$ **23.** $2xy + y - 3$ **25.** $-2x$ **27.** $8a - 60$ **29.** $a + b + (-c + 2d), a + b - (c - 2d)$
31. $x - 2y + (-3x^2 - 2y^2), x - 2y - (3x^2 + 2y^2)$ **33.** $7 + (2a - 5b), 7 - (-2a + 5b)$ **35.** $6w + 5z + (-3x + 5y), 6w + 5z - (3x - 5y)$
37. $7x^2y + 2xy^2 + (-6xz + 5yz), 7x^2y + 2xy^2 - (6xz - 5yz)$ **39.** $-3x^2 + 2y^2 + (-7z^2 - 8w^2), -3x^2 + 2y^2 - (7z^2 + 8w^2)$
41. $-5x + (-7a - 7b), -5x - (7a + 7b)$ **43.** $2R_1 + 4R_2$ **45.** $-(R_1I_1 - R_2I_2 + R_3I_3)$ **47.** $v = \sqrt{19.6y_2 - 19.6y_1}$

Section 1.8 (page 35)

1. -8 **3.** 2   **5.** x^8 **7.** $-2x^3y^3$ **9.** x^5 **11.** $-\dfrac{z^4}{2}$ **13.** 1   **15.** $\dfrac{1}{2}$ **17.** $-\dfrac{y}{2x}$ **19.** $\dfrac{1}{x}$ **21.** $\dfrac{2y^2}{x^2}$ **23.** $\dfrac{2}{a^3b^2}$

25. $2\sqrt{2}$ **27.** $3\sqrt{6}$ **29.** $5\sqrt{3}$ **31.** $2\sqrt[3]{2}$ **33.** $2\sqrt{2}$ **35.** $4\sqrt{2}$ **37.** $\dfrac{1}{5\alpha}$ **39.** $4A_2{}^2$ **41.** v_1 **43.** $8a^{12}b^{11}$ **45.** $2xy^3$

47. $-2v_1v_2$ **49.** $\dfrac{\sqrt{5}}{5}$ **51.** $\sqrt{3}$ **53.** $\dfrac{a\sqrt{b}}{b}$ **55.** $\dfrac{2\sqrt{5}}{15}$ **57.** $\dfrac{\sqrt{2}}{4}$ **59.** $\dfrac{\sqrt{3}}{2}$ **61.** 9 cm   **63.** $\dfrac{2\sqrt{\pi ad}}{ad}$ **65.** $14\sqrt{5}$ cm^2

67. a^6 **69.** a^4 **71.** a **73.** x^6 **75.** $4x^3$ **77.** $6a^2$ **79.** $7p^4q^2$ **81.** $9V_2{}^3$ **83.** $\dfrac{C_1{}^4 C_2{}^8}{2}$

Section 1.9 (page 41)

1. 2.6×10^4 **3.** 3.792×10^8 **5.** 1.3×10^{-4} **7.** 8.927×10^{-5} **9.** 1,200,000   **11.** 0.006273   **13.** 0.00000027   **15.** 956,000
17. 2.4×10^5 mi   **19.** 0.00000000006670 N·m^2/kg^2 **21.** 3,600,000 J   **23.** 101,300 N/m^2 **25.** 3.92×10^{12} **27.** 1.9×10^{-9}
29. 2.34×10^{-8} **31.** 6.94×10^{10} **33.** 1.2×10^{-10} **35.** 3.0×10^{-12} **37.** 2.0×10^{15} **39.** 4.07×10^9 **41.** 7.4×10^{-7}
43. 9.17×10^{-6} 7.09×10^{-6}

Section 1.10 (page 44)

1. $12a^3b^4$ **3.** $24a^4b^5x$ **5.** $3x^2y + 2x^2y^2$ **7.** $-30x^3y^4 + 20x^4y^3$ **9.** $2a^3b^2 + 3a^2b^2 - 4a^2b^3$ **11.** $-12a^4bx + 10a^5x^2 - 4a^3bx$
13. $16x^2y^2 - 60x^3y^3$ **15.** $7a^5b^2 - 5a^3b^3$ **17.** $x^2 - 3x + 2$ **19.** $2x^2 - 7x + 3$ **21.** $10x^2 - 4x - 14$ **23.** $x^2 - y^2$
25. $4x^2 - 1$ **27.** $9x^2 - 4y^2$ **29.** $9x^2 - 6x + 1$ **31.** $4x^2 - 8xy + 4y^2$ **33.** $4x^2 + 4xy + y^2$ **35.** $x^4 - y^4$
37. $5x^2 - 11bx + 2b^2$ **39.** $x^3 - 3x^2 - x + 3$ **41.** $a^3 - 4a^2b + 4ab^2 - 3b^3$ **43.** $25x^2 - 30xy + 9y^2$
45. $5x^2 - 7xy + 2xz + 2y^2 + yz - 3z^2$ **47.** $8x^8 - 8x^6 + 8x^4 - 3x^2 - 5$ **49.** $2x^5y^2 - 13x^3y^4 + 6x^2y^5 + 2xy^6$ **51.** $x^5 + x^2 - x - 1$
53. $x^2 + 2xy + 2xz + y^2 + 2yz + z^2$ **55.** $4a^2 - 12ab + 9b^2 + 8ac - 12bc + 4c^2$ **57.** $3l^2 + 3lw - 18w^2$ **59.** $6R_1{}^2 + 18R_1R_2 + 12R_2{}^2$
61. $(x^2 + 4x + 4)$ square meters   **63.** $(140a + 35b)$ dollars   **65.** $65x^2 - 14x$ **67.** $0.04t^2 + 0.1t - 18$ **69.** $E = kT^4 - kT_0{}^4$
71. $\frac{1}{2}(y_2{}^2 - y_1{}^2)$

Section 1.11 (page 49)

1. a^2 **3.** $-2x^2y$ **5.** $-4x^4 - 2x^2$ **7.** $4c - 3ac^2$ **9.** $1 - 2xy^2 + 3x^3y^2$ **11.** $4x^5y^3a^2 + 2x^4y - 3x^3a^4$ **13.** $3x - 2$
15. $2x + 1$ **17.** $2x + 4$ **19.** $4x - 4$ **21.** $3x + 4$ **23.** $3x + 6y$ **25.** $a + b$ **27.** $3w - 4z$ **29.** $x^2 - 2xy + y^2$
31. $a^2 + 2ab - 5b^2$ **33.** $v + 4$ **35.** $2V - 3$ **37.** $3s^2 + 3st + t^2$ **39.** $2p^2 - 3pq - q^2$ **41.** $2x^2 - xy + y^2$ **43.** $x^2 - xy + y^2$

45. $2x^2 - 3x + 1$ **47.** $2x - 3y$ **49.** $x^2 - 2xy + y^2 - \dfrac{y^3}{x + 2y}$ **51.** $x^3 + x^2 + x - 2 - \dfrac{5}{x - 4}$ **53.** $\dfrac{1}{C} = \dfrac{1}{C_1} + \dfrac{1}{C_2}$

55. $(x^2 + 2x + 2)\dfrac{\text{mi}}{\text{gal}}$

Section 1.12 (page 51)

1. $-2 * A\uparrow3$ **3.** $3 * X\uparrow3 - 4 * W$ **5.** $(A - 2 * B)/C$ **7.** $\text{SQR}(2 - 6 * P)$ **9.** $\text{SQR}(S1 - 2 * S2)$ **11.** $(A - B)/\text{SQR}(T1)$
13. $(I\uparrow2 + K\uparrow2)/\text{SQR}(M)$ **15.** $\text{SQR}(C1 - C2)/(5 * T1)$

Review Exercises for Chapter 1 (page 51)

1. 0   **2.** -2 **3.** -16 **4.** -16 **5.** -10 **6.** 8   **7.** $\dfrac{6}{5}$ **8.** undefined   **9.** -20 **10.** 1   **11.** 4   **12.** $-\dfrac{7}{4}$

13. $7x - 8y$ **14.** $-3x - 7y$ **15.** $5x^2 - 8xy$ **16.** $-3a - b$ **17.** b **18.** $-3x$
19. a. $2x + a + (-x^2 + y - 4w)$ **b.** $2x + a - (x^2 - y + 4w)$ **20. a.** $a - 3b + (-2a^2 - b^2 + c^2)$ **b.** $a - 3b - (2a^2 + b^2 - c^2)$

21. $\dfrac{3}{5s^5}$ **22.** $-\dfrac{R_3}{R_2}$ **23.** $5\sqrt{2}$ **24.** $4\sqrt{3}$ **25.** $3\sqrt[3]{3}$ **26.** $\dfrac{2\sqrt{A}}{A}$ **27.** $\dfrac{\sqrt{5}}{5}$ **28.** $\dfrac{\sqrt{2}}{2}$ **29.** binomial   **30.** monomial

31. trinomial   **32.** binomial   **33.** $x^3 - 2xy$ **34.** $2v - 4$ **35.** $-2x^2 + 2xy + 12y^2$ **36.** $3a^3 - 4a^2 - 3a - 2$

37. $4s^3 + 10s^2t - 7st^2 - 3t^3$ **38.** $3x^2 - 6xy - ax + 2ay$ **39.** $x - 2y$ **40.** $x + 3y$ **41.** $5a - 3b + \dfrac{2b^2}{a + b}$ **42.** $2a^2 - 3ab + b^2$

43. $a^2 + 2ab + 4b^2$ **44.** $8x^3 + y^3$ **45.** $32x^2$ **46.** $10 + 6t - 15t^2 - 9t^3$ **47.** $2L^2 - 2Lx - 2Lx^2 + 2x^3$ **48.** $\pi(r_2{}^2 - r_1{}^2)$
49. 1.55   **50.** 10.68   **51.** 0.39   **52.** 25.2   **53.** 0.00951   **54.** 17,000   **55.** 1.0   **56.** 2,800
57. 5,980,000,000,000,000,000,000,000 kg   **58.** 8.64×10^5 mi   **59.** 1.164×10^{-23} g   **60.** 0.00000003 cm
61. 0.000000000000000000160 C   **62.** 1.86×10^5 mi/s   **63.** 2.8×10^{-14} **64.** 6.5×10^{-13}

Chapter 2

Section 2.1 (page 56)

1. $x = 7$ **3.** $x = -6$ **5.** $x = -2$ **7.** $x = -4$ **9.** $x = \dfrac{5}{4}$ **11.** $x = 1$ **13.** $x = 5$ **15.** $x = -\dfrac{7}{4}$

Section 2.2 (page 61)

1. $x = 9$ **3.** $x = 4$ **5.** $x = 4$ **7.** $x = 3$ **9.** $x = -4$ **11.** $x = -3$ **13.** $x = -\dfrac{4}{7}$ **15.** $x = \dfrac{1}{13}$ **17.** no solution

19. no solution   **21.** $x = -4$ **23.** $x = -3$ **25.** $x = -5$ **27.** no solution   **29.** $x = \dfrac{b}{a}$ **31.** $x = -\dfrac{3}{a}$ **33.** $x = -\dfrac{b}{a}$

35. $x = 0$ **37.** $x = \dfrac{4}{7}$ **39.** $x = \dfrac{5}{b}$ **41.** $x = -1$ **43.** no solution   **45.** $x = -12$ **47.** $x = -2$

Section 2.3 (page 64)

1. $a = \dfrac{b}{2}$ **3.** $q = \dfrac{s-3}{p}$ **5.** $T_2 = \dfrac{T_1}{ab}$ **7.** $R = \dfrac{Q+S}{P}$ **9.** $a = \dfrac{4-2c}{b}$ **11.** $b = \dfrac{d-2a}{2}$ **13.** $b = a - PS$ **15.** $n = \dfrac{1}{3}(Lc - 4)$

17. $T_2 = \dfrac{6T_1 - bT_0}{2}$ **19.** $D = \dfrac{B - S_0 a + S_0 D_0 d}{S_0 d}$ **21.** $b = \dfrac{aM - s + 3}{M}$ **23.** $C_1 = \dfrac{C_0 C_2 - t C_2}{t}$ **25.** $V = \dfrac{aP + ab}{r}$

27. $P_2 = \dfrac{P_1 V_1 T_2}{T_1 V_2}$ **29.** $b_2 = \dfrac{2A - h b_1}{h}$ **31.** $t = \dfrac{A-P}{Pr}$ **33.** $K = \dfrac{E_d I - H^2}{H^2}$ **35.** $k_2 = AL - k_1 A^2$

Section 2.4 (page 65)

1. $3 * (X - 5) = X$ **3.** $4 * (2 * X - 3) = 0$ **5.** $V = I * R$ **7.** $P = K/V$ **9.** $1/R = 1/R1 + 1/R2$
11. $S = (1/2) * A * T\uparrow 2 + V * T$ **13.** $L = K1 * A + K2/A$

Section 2.5 (page 70)

1. $13\dfrac{1}{2}$, $16\dfrac{1}{2}$ **3.** \$10.50, \$13.10   **5.** 1.2 A, 3.1 A   **7.** 12.6 cm, 9.4 cm   **9.** $\dfrac{19}{3}$ in. $\times \dfrac{38}{3}$ in.   **11.** 8.4 Ω, 14.0 Ω

13. 21.3 Ω, 42.6 Ω, 56.9 Ω   **15.** 8 L   **17.** 60 lb of 10% alloy, 40 lb of 15% alloy   **19.** 7.5 gal   **21.** $\dfrac{1}{2}$ L   **23.** 5 mi/h   **25.** 20 mi/h

27. 6 mi/h   **29.** 10 mi/h   **31.** 100 h   **33.** 8.47 min   **35.** 45.0 min   **37.** 13   **39.** 22 days
41. 90 mi for slower car, 110 mi for faster car

Section 2.6 (page 75)

1. 2.3   **3.** 624 lb/ft^2 **5.** 126   **7.** 82 mi/h   **9.** 80.67 ft/s   **11.** 3.94 in.   **13.** 0.16 Ω   **15.** 9 lb   **17.** \$84.80   **19.** \$1,035
21. 50/7 lb   **23.** 30 ft   **25.** 70 Ω

Section 2.7 (page 79)

1. $y = kx$ **3.** $w = ki^2$ **5.** $y = \dfrac{k}{x^2}$ **7.** $y = kxw$ **9.** $A = k\left(\dfrac{a}{b}\right)$ **11.** $F = k\left(\dfrac{mM}{r^2}\right)$ **13.** $y = 4x$ **15.** $P = \dfrac{2}{V}$ **17.** 6.9 lb

19. $S = kwd^2$ **21.** 100 units   **23.** $f = k\dfrac{q_1 q_2}{r^2}$ **25.** 11.6 lb/ft^2 **27.** \$120   **29.** 4.43 Ω

Section 2.8 (page 84)

1. independent variable: x; dependent variable: y **3.** independent variable: R; dependent variable: E
5. independent variable: r; dependent variable: A **7.** independent variable: F; dependent variable: C **9.** $f(x) = x^2 - 3$

11. $f(w) = \sqrt{6w + 2}$ **13.** $f(v) = 1 - 3v^2$ **15.** $f(n) = \sqrt{4n + 6}$ **17.** 1, 1, 1, 61   **19.** 0, 3   **21.** $\dfrac{1}{2}$, 1   **23.** 1, 3   **25.** 0, -28

27. $a^2 + 2$, $a^4 + 2$ **29.** 3.42 Ω, 25.2 Ω   **31.** $A = s^2$ **33.** $V = \dfrac{4}{3}\pi r^3$ **35.** $V = 4\pi h$ **37.** $C = 3A$ **39.** $C = 50 + 25t$

Section 2.9 (page 88)

1. all x **3.** $x \neq 0$ **5.** $x \neq 1, -1$ **7.** $x \geq -2$ **9.** $x \leq 2$

11.
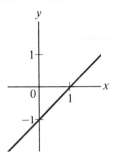

13.

15.
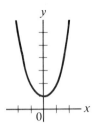

17.

19.
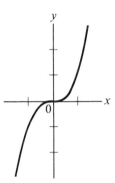

21.
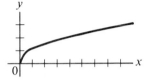

23.
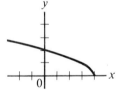

25.
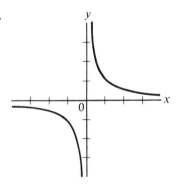

27.
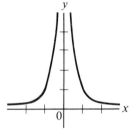

29.
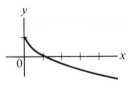

31.
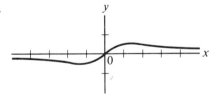

33. $R = 1 \, \Omega$
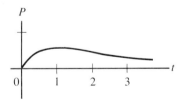

Review Exercises for Chapter 2 (page 89)

1. $x = 9$ **2.** $x = -1$ **3.** $x = 1$ **4.** $x = -3$ **5.** $x = 3$ **6.** $x = -\dfrac{1}{2}$ **7.** $x = -2$ **8.** $x = \dfrac{1}{2}$ **9.** $x = \dfrac{b}{a}$ **10.** $x = \dfrac{7}{a}$

11. $x = \dfrac{3b + 3}{2a}$ **12.** $x = \dfrac{2b}{a}$ **13.** $R = \dfrac{V}{I}$ **14.** $a = \dfrac{A - bn}{n}$ **15.** $b = \dfrac{3a}{Q - C}$ **16.** $t = \dfrac{A - p}{pr}$ **17.** 2.5 A, 0.7 A   **18.** 2 h

19. 40 lb   **20.** 3, 6, 8   **21.** $22\dfrac{2}{9}$ min   **22.** 4 ft   **23.** 60%   **24.** 0.0404   **25.** 29.7 cm   **26.** 48 ft   **27.** $y = k\left(\dfrac{st}{r}\right)$

28. $y = x^2$ **29.** $w = \dfrac{k}{d^2}$ **30.** $C = \dfrac{k}{d}$ **31.** $R = k\left(\dfrac{L}{D^2}\right)$ **32.** $P = k\sqrt{L}$

33. 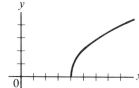   **34.** 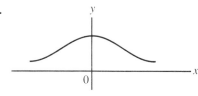   **35.** $x \geq 4$ **36.** 1, 0

Chapter 3

Section 3.1 (page 94)

1. 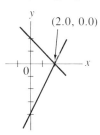   **3.** 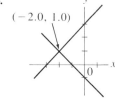   **5.**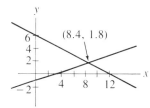

7. 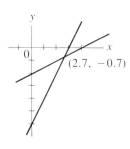   **9.** 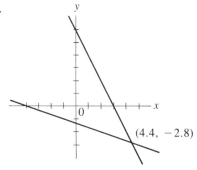   **11.**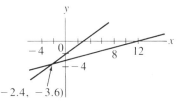

Section 3.2 (page 101)

1. (3, 1)   **3.** (−5, −8)   **5.** (3, −6)   **7.** $\left(-\dfrac{1}{2}, 3\right)$ **9.** $\left(\dfrac{7}{20}, \dfrac{3}{20}\right)$ **11.** (1, −1)   **13.** (3, 3)   **15.** (−1, 3)   **17.** $\left(-1, \dfrac{1}{2}\right)$

19. $\left(\dfrac{5}{3}, -\dfrac{8}{3}\right)$ **21.** (0, −7)   **23.** inconsistent   **25.** (1, 1)   **27.** $\left(\dfrac{5}{4}, 5\right)$ **29.** $\left(-2, -\dfrac{1}{2}\right)$ **31.** $\left(-\dfrac{1}{3}, 3\right)$ **33.** (0, 1)

35. dependent   **37.** inconsistent   **39.** (1, 2)

Section 3.3 (page 106)

1. 0   **3.** 0   **5.** 1   **7.** −10   **9.** 2   **11.** −69   **13.** 869   **15.** 450   **17.** (−13, 10)   **19.** (22, 31)   **21.** (−83, −68)

23. $\left(-3, \dfrac{17}{3}\right)$ **25.** $\left(\dfrac{13}{44}, -13\right)$ **27.** $\left(-\dfrac{19}{12}, -\dfrac{19}{68}\right)$ **29.** $(20, 10)$ **31.** $\left(\dfrac{7}{3}, \dfrac{4}{3}\right)$ **33.** $\left(2, \dfrac{7}{2}\right)$ **35.** inconsistent   **37.** $(2.43, 1.07)$

39. $(-2.32, -0.974)$ **41.** $(0.669, -1.28)$

Section 3.4 (page 110)

1. $v_1 = 3$ ft/s, $v_2 = -2$ ft/s   **3.** $S = 566 - 0.00200T$ **5.** $w_1 = 0.5$ N, $w_2 = 3.0$ N   **7.** 18, 26   **9.** 4.3, 8.6   **11.** 14 ft, 40 ft
13. 16, 35   **15.** 70 Ω, 80 Ω   **17.** 30.1 V, 25.0 V   **19.** 8.9 in. by 3.3 in.   **21.** \$4,110 at 12%, \$4,390 at 11%

23. 40 mL of 10% solution, 60 mL of 20% solution   **25.** $33\dfrac{1}{3}$ lb of 12% alloy, $16\dfrac{2}{3}$ lb of 6% alloy   **27.** 20 at \$1.95, 60 at \$2.50

29. 21 dimes, 16 quarters   **31.** 8 smaller offices, 12 larger offices

Section 3.5 (page 114)

1. $(1, 2, 3)$ **3.** $(1, 1, -2)$ **5.** $(1, -1, 2)$ **7.** $\left(2, -\dfrac{3}{2}, \dfrac{7}{2}\right)$ **9.** $\left(\dfrac{2}{3}, -\dfrac{5}{3}, 4\right)$ **11.** $\left(\dfrac{1}{6}, \dfrac{7}{20}, -\dfrac{7}{25}\right)$ **13.** $(1, 2, -2, 3)$

15. $\left(-\dfrac{1}{2}, 2, 3, -1\right)$

Section 3.6 (page 122)

1. 4   **3.** -55 **5.** 115   **7.** -234 **9.** 940   **11.** -129 **13.** $(1, 2, 3)$ **15.** $(1, 1, -2)$ **17.** $(1, -1, 2)$ **19.** $\left(2, -\dfrac{3}{2}, \dfrac{7}{2}\right)$

21. $\left(-\dfrac{8}{19}, -\dfrac{12}{19}, -\dfrac{1}{19}\right)$ **23.** $w_1 = 1$ lb, $w_2 = 1$ lb, $w_3 = 2$ lb   **25.** \$1,200 at 8%, \$1,550 at 10%, \$3,200 at 12%   **27.** \$20, \$6, \$14

29. $I_1 = \dfrac{24}{11}$ A, $I_2 = -\dfrac{10}{11}$ A, $I_3 = \dfrac{14}{11}$ A   **31.** $I_1 = 4$ A, $I_2 = 3$ A, $I_3 = -1$ A   **33.** $I_1 = \dfrac{13}{17}$ A, $I_2 = \dfrac{9}{17}$ A, $I_3 = -\dfrac{14}{17}$ A, $I_4 = \dfrac{8}{17}$ A

Review Exercises for Chapter 3 (page 124)

1. -7 **2.** 47   **3.** 160   **4.** -69 **5.** 3   **6.** 288   **7.** 238   **8.** -270

9. $(-1.5, 2.0)$

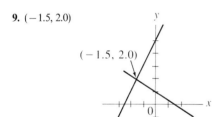

10. $(1.0, 0.3)$

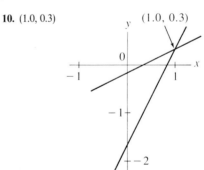

11. $(2.5, 2.0)$

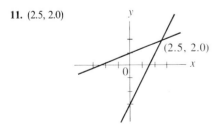

12. $(2.0, -3.0)$

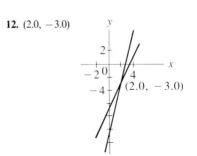

13. $\left(\dfrac{14}{11}, \dfrac{1}{11}\right)$ **14.** $(5, 1)$ **15.** $\left(-\dfrac{5}{2}, -\dfrac{3}{2}\right)$ **16.** $\left(-\dfrac{1}{3}, -\dfrac{5}{3}\right)$ **17.** $\left(\dfrac{9}{5}, -\dfrac{1}{5}\right)$ **18.** $(3, 1)$ **19.** $\left(2, \dfrac{1}{3}\right)$ **20.** $\left(-\dfrac{1}{2}, 3\right)$

21. $\left(\dfrac{37}{38}, -\dfrac{17}{38}\right)$ **22.** $\left(\dfrac{1}{10}, \dfrac{7}{10}\right)$ **23.** $\left(-\dfrac{9}{10}, \dfrac{13}{5}\right)$ **24.** $\left(\dfrac{16}{11}, -\dfrac{13}{11}\right)$ **25.** $(2, -1)$ **26.** $\left(\dfrac{5}{11}, -\dfrac{5}{3}\right)$ **27.** $\left(-\dfrac{1}{4}, \dfrac{11}{8}\right)$

28. $\left(1, \dfrac{1}{2}\right)$ **29.** $\left(\dfrac{3}{35}, \dfrac{3}{37}\right)$ **30.** $\left(\dfrac{13}{11}, \dfrac{13}{3}\right)$ **31.** $(2, 4)$ **32.** $(1, 2)$ **33.** $(2, 3)$ **34.** $\left(\dfrac{1}{2}, 4\right)$ **35.** $\left(-\dfrac{1}{2}, 3\right)$ **36.** $\left(-\dfrac{1}{3}, \dfrac{1}{6}\right)$

37. $(1, -1, 3)$ **38.** $\left(-\dfrac{1}{2}, 2, 3\right)$ **39.** $\left(-\dfrac{1}{2}, 1, \dfrac{1}{3}\right)$ **40.** $\left(\dfrac{1}{2}, -2, 1\right)$ **41.** $(-1, 3, -2)$ **42.** $\left(\dfrac{7}{5}, 0, -\dfrac{2}{5}\right)$ **43.** $\left(\dfrac{2}{3}, -2, -2\right)$

44. $\left(\dfrac{9}{2}, -\dfrac{3}{2}, -7\right)$ **45.** $a = -1$ **47.** \$172 at 10%, \$430 at 8%, \$412.80 at $12\frac{1}{2}$%   **48.** $\left(\dfrac{5}{13}, \dfrac{5}{4}\right)$

49. 60 gal of 1% milk, 30 gal of 4% milk   **50.** 3 in., 5 in., 6 in.   **51.** $I_1 = 1$ A, $I_2 = 1$ A, $I_3 = 0$ A
52. $I_1 = 0$ A, $I_2 = -1$ A, $I_3 = 0$ A, $I_4 = -1$ A   **53.** $w_1 = 1$ lb, $w_2 = 2$ lb, $w_3 = 2$ lb   **54.** 20.8 Ω, 36.4 Ω

Cumulative Review Exercises for Chapters 1–3 (page 126)

1. $-7T_a - 10T_b + 2T_c$ **2.** $2L - 2C$ **3.** $5\sqrt{6}$ **4.** $L\sqrt{\pi}/(2\pi)$ **5.** $3p^4q^3$ **6.** $4A^2$ **7.** $a^3 - 5a^2b + 4ab^2 + 4b^3$ **8** $x^2 + 2x + 3$

9. 6.4×10^{-14} **10.** $x = \dfrac{29}{11}$ **11.** $x = 18$ **12.** $t_1 = \dfrac{L_1 + L_1\beta t_2 - L}{L_1\beta}$ **13.** $x \le 4$ **14.** $\sqrt{2}, 2$ **15.** $(2, 0)$ **16.** $(26, 15)$

17. $\left(\dfrac{7}{2}, 2, -\dfrac{3}{2}\right)$ **18.** $6.00I_1 - 5.00I_2$ **19.** $3F_1 + 2F_2$ **20.** 2.4×10^5 mi   **21.** 1.2 mm by 3.6 mm   **22.** 6.56 lb

Chapter 4

Section 4.2 (page 129)

1. a.
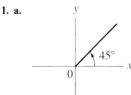
b.
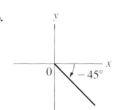
c.
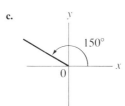
d.
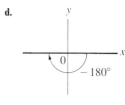

e.
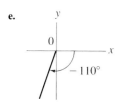
f.
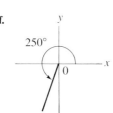
g.
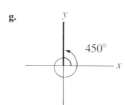
h.
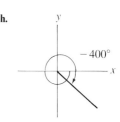

3. $205°$ **5.** $287°20'$ **7.** $180°3'$ **9.** $213°55'$ **11.** $178°6'$ **13.** $34°19'$ **15.** $355°$
17. a. $-333°$ **b.** $-209°$ **c.** $-143°20'$ **d.** $-47°35'$

Section 4.3 (page 135)

For Exercises 1–27, the trigonometric functions are given in the following order: $\sin \theta$, $\cos \theta$, $\tan \theta$, $\csc \theta$, $\sec \theta$, and $\cot \theta$.

1. $\dfrac{2}{\sqrt{5}} = \dfrac{2\sqrt{5}}{5}, \dfrac{1}{\sqrt{5}} = \dfrac{\sqrt{5}}{5}, 2, \dfrac{\sqrt{5}}{2}, \sqrt{5}, \dfrac{1}{2}$ **3.** $\dfrac{2}{\sqrt{29}} = \dfrac{2\sqrt{29}}{29}, \dfrac{5}{\sqrt{29}} = \dfrac{5\sqrt{29}}{29}, \dfrac{2}{5}, \dfrac{\sqrt{29}}{2}, \dfrac{\sqrt{29}}{5}, \dfrac{5}{2}$ **5.** $\dfrac{1}{\sqrt{5}} = \dfrac{\sqrt{5}}{5}, \dfrac{2}{\sqrt{5}} = \dfrac{2\sqrt{5}}{5}, \dfrac{1}{2}, \sqrt{5}, \dfrac{\sqrt{5}}{2}, 2$

7. $\dfrac{4}{3\sqrt{2}} = \dfrac{2\sqrt{2}}{3}, \dfrac{1}{3}, \dfrac{4}{\sqrt{2}} = 2\sqrt{2}, \dfrac{3\sqrt{2}}{4}, 3, \dfrac{\sqrt{2}}{4}$ **9.** $\dfrac{\sqrt{6}}{\sqrt{42}} = \dfrac{1}{\sqrt{7}} = \dfrac{\sqrt{7}}{7}, \dfrac{6}{\sqrt{42}} = \dfrac{\sqrt{42}}{7}, \dfrac{\sqrt{6}}{6}, \sqrt{7}, \dfrac{\sqrt{42}}{6}, \sqrt{6}$

11. $\dfrac{\sqrt{10}}{\sqrt{59}} = \dfrac{\sqrt{590}}{59}, \dfrac{7\sqrt{59}}{59}, \dfrac{\sqrt{10}}{7}, \dfrac{\sqrt{590}}{10}, \dfrac{\sqrt{59}}{7}, \dfrac{7\sqrt{10}}{10}$ **13.** $\dfrac{\sqrt{3}}{6}, \dfrac{\sqrt{33}}{6}, \dfrac{\sqrt{11}}{11}, 2\sqrt{3}, \dfrac{2\sqrt{33}}{11}, \sqrt{11}$ **15.** $\dfrac{\sqrt{10}}{4}, \dfrac{\sqrt{6}}{4}, \dfrac{\sqrt{15}}{3}, \dfrac{2\sqrt{10}}{5}, \dfrac{2\sqrt{6}}{3}, \dfrac{\sqrt{15}}{5}$

17. $\dfrac{\sqrt{7}}{5}, \dfrac{3\sqrt{2}}{5}, \dfrac{\sqrt{14}}{6}, \dfrac{5\sqrt{7}}{7}, \dfrac{5\sqrt{2}}{6}, \dfrac{3\sqrt{14}}{7}$ **19.** $\dfrac{2\sqrt{142}}{71}, \dfrac{3\sqrt{497}}{71}, \dfrac{2\sqrt{14}}{21}, \dfrac{\sqrt{142}}{4}, \dfrac{\sqrt{497}}{21}, \dfrac{3\sqrt{14}}{4}$ **21.** 0.866, 0.500, 1.73, 1.15, 2.00, 0.577

23. 0.421, 0.907, 0.464, 2.38, 1.10, 2.15   **25.** 0.792, 0.610, 1.30, 1.26, 1.64, 0.770   **27.** 0.24, 0.97, 0.25, 4.1, 1.0, 4.0   **29.** $\dfrac{3}{4}$ **31.** $\dfrac{\sqrt{3}}{2}$

33. $\dfrac{\sqrt{34}}{5}$ **35.** $\dfrac{4}{9}$ **37.** $\dfrac{6}{7}$ **39.** $\dfrac{4\sqrt{19}}{19}$ **41.** $\dfrac{\sqrt{6}}{3}$ **43.** $\dfrac{\sqrt{33}}{6}$ **45.** $\sqrt{10}$ **47.** $\sqrt{10}$ **53. a.** $\dfrac{2}{\sqrt{5}} = \dfrac{2\sqrt{5}}{5}$ **b.** 2   **c.** $\dfrac{\sqrt{5}}{2}$

Section 4.4 (page 144)

1. $\dfrac{\sqrt{3}}{3}$ **3.** $\dfrac{\sqrt{3}}{2}$ **5.** 1   **7.** 2   **9.** 1   **11.** $\dfrac{2\sqrt{3}}{3}$ **13.** 1   **15.** $\sqrt{2}$ **17.** $\dfrac{1}{2}$ **19.** $\dfrac{\sqrt{2}}{2}$ **21.** 0.3692   **23.** 1.695   **25.** 0.3357

27. 0.4906   **29.** $17°20'$ **31.** $26°25'$ **33.** 0.2147   **35.** 0.6794   **37.** 3.801   **39.** 1.010   **41.** 1.000   **43.** 0.4574   **45.** 5.059
47. 0.7701   **49.** 0.1851   **51.** 1.319   **53.** 11.10   **55.** 1.217   **57.** 1.047   **59.** 3.030   **61.** 0.3169   **63.** 0.9960   **65.** 0.7950
67. 4.843   **69.** $46.84°$ **71.** $51.17°$ **73.** $85.64°$ **75.** $81.42°$ **77.** $6.08°$ **79.** $38°10'$ **81.** $18°38'$ **83.** $46°15'$ **85.** $46°58'$
87. $18°11'$ **89. a.** $W = 16.1$ g   **b.** $W = 17.1$ g   **c.** $W = 19.3$ g   **d.** $W = 23.4$ g   **e.** $W = 25.1$ g   **91.** 46.1 V

Section 4.5 (page 150)

1. 1.67   **3.** 73.5   **5.** $B = 59.48°$, $a = 2.075$, $c = 4.086$ **7.** $A = 33.8°$, $B = 56.2°$, $b = 783$ **9.** $B = 67°20'$, $a = 0.771$, $b = 1.85$
11. $A = 82°5'$, $b = 0.01252$, $c = 0.09087$ **13.** 29.6 ft   **15.** 71.7 m   **17.** 65.0 ft   **19.** 7.0 ft   **21.** 3.72 ft   **23.** $31.3°$ **25.** $43.0°$
27. 0.160 in.   **29.** 1,250 ft   **31.** 134 ft   **33.** 345.4 ft   **35.** 27.60 m   **37.** 471 ft   **39.** 932 ft^2 **41.** 8.16 m

Review Exercises for Chapter 4 (page 152)

In Exercises 1–4, the trigonometric functions are given in the following order: $\sin \theta$, $\cos \theta$, $\tan \theta$, $\csc \theta$, $\sec \theta$, and $\cot \theta$.

1. 0.46, 0.89, 0.52, 2.2, 1.1, 1.9   **2.** $\dfrac{\sqrt{14}}{4}, \dfrac{\sqrt{2}}{4}, \sqrt{7}, \dfrac{2\sqrt{14}}{7}, 2\sqrt{2}, \dfrac{\sqrt{7}}{7}$ **3.** $\dfrac{2}{3}, \dfrac{\sqrt{5}}{3}, \dfrac{2\sqrt{5}}{5}, \dfrac{3}{2}, \dfrac{3\sqrt{5}}{5}, \dfrac{\sqrt{5}}{2}$

4. $\dfrac{2\sqrt{2}}{3}, \dfrac{1}{3}, 2\sqrt{2}, \dfrac{3\sqrt{2}}{4}, 3, \dfrac{\sqrt{2}}{4}$ **5.** $\dfrac{2\sqrt{6}}{5}$ **6.** $\dfrac{\sqrt{21}}{3}$ **7.** $\dfrac{5\sqrt{7}}{14}$ **8.** $\dfrac{\sqrt{10}}{10}$ **9.** $\sqrt{3}$ **10.** $\dfrac{\sqrt{5}}{2}$ **11.** $\dfrac{\sqrt{38}}{6}$ **12.** $\dfrac{2\sqrt{7}}{7}$

13. $\dfrac{1}{2}\sqrt{4 - a^2}$ **14.** $\sqrt{b^2 - 1}$ **17.** $\dfrac{\sqrt{3}}{2}$ **18.** 1   **19.** 1   **20.** $\dfrac{2\sqrt{3}}{3}$ **21.** 0   **22.** $\sqrt{2}$ **23.** 0   **24.** $\dfrac{\sqrt{3}}{3}$ **25.** 0.5336

26. 1.918   **27.** 1.757   **28.** 0.9849   **29.** 7.069   **30.** 0.2583   **31.** $10°45'$ **32.** $78°58'$ **33.** $10°59'$ **34.** $71°32'$ **35.** $30°25'$
36. $75°50'$ **37.** 0.8059   **38.** 0.2553   **39.** 1.075   **40.** 1.461   **41.** 6.581   **42.** 6.021   **43.** 0.9582   **44.** 2.393   **45.** $47.01°$
46. $16.06°$ **47.** $61.52°$ **48.** $61.17°$ **49.** $63.26°$ **50.** $42.11°$ **51.** $36°3'$ **52.** $68°20'$ **53.** $69.5°$ **54.** 2.50 cm
55. 11.3 in.^{2} **56.** 6.30 cm   **57.** 1.28 in.   **58.** 130.4 ft   **59.** 65.4 ft   **60.** 73.66 ft   **61.** 1,100 ft   **62.** 52 cm^2

Chapter 5

Section 5.1 (page 156)

1. $2x - 6y$ **3.** $2a^4 - 2a^2b^2 - 6a^2c$ **5.** $VR_1^2 - VR_2^2$ **7.** $2PQ^2 - 2P^3 + 2P$ **9.** $2ab^2c - 4b^3c - 2b^2c$ **11.** $x^2 - 4y^2$
13. $36A^2 - B^2$ **15.** $a^4 - b^2$ **17.** $4x^4 - y^4$ **19.** $S^2 - 4T^4$ **21.** $x^2y^2 - 1$ **23.** $9R^2S^4 - 1$

Section 5.2 (page 161)

1. $2(a + b)$ **3.** $5(x - 1)$ **5.** $3x(x - 2)$ **7.** $2xy(2xy - 3x + 6y)$ **9.** $14c^2d^2(c + 2cd + d)$ **11.** $11R_1R_2(R_1 - 2R_2 + 1)$
13. $(x - y)(x + y)$ **15.** $4(x - y)(x + y)$ **17.** $16(2m - n)(2m + n)$ **19.** $2(2f_1 - f_2)(2f_1 + f_2)$ **21.** $2(5x - 4z)(5x + 4z)$

23. $\left(\dfrac{1}{2}x - y\right)\left(\dfrac{1}{2}x + y\right)$ **25.** $\left(\dfrac{1}{3}x - \dfrac{1}{4}y\right)\left(\dfrac{1}{3}x + \dfrac{1}{4}y\right)$ **27.** $\left(\dfrac{x}{a} - B\right)\left(\dfrac{x}{a} + B\right)$ **29.** $\left(1 - \dfrac{b}{a}\right)\left(1 + \dfrac{b}{a}\right)$

31. $(t_1^2 + t_2^2)(t_1 - t_2)(t_1 + t_2)$ **33.** $3(x - 3a)(x + 3a)$ **35.** $(8a^2 - 1)(8a^2 + 1)$ **37.** $(x - y)(x^2 + xy + y^2)$
39. $2(x + a)(x^2 - ax + a^2)$ **41.** $(3a - y)(9a^2 + 3ay + y^2)$ **43.** $(5m - 3a)(25m^2 + 15am + 9a^2)$ **45.** $(ab - 1)(a^2b^2 + ab + 1)$
47. $(1 + xy^2)(1 - xy^2 + x^2y^4)$ **49.** $(a + b - 1)(a + b + 1)$ **51.** $(1 - x - y)(1 + x + y)$ **53.** $(x - y + z)(x^2 - 2xy + y^2 - xz + yz + z^2)$

55. $(v_1 - v_2 - v_3)(v_1^2 - 2v_1v_2 + v_2^2 + v_1v_3 - v_2v_3 + v_3^2)$ **57.** $(F_1 + F_2 - 2)(F_1^2 + 2F_1F_2 + F_2^2 + 2F_1 + 2F_2 + 4)$
59. $(2 + 3r + 2s)(4 - 6r - 4s + 9r^2 + 12rs + 4s^2)$ **61.** $16(x - 2a)(x^2 + 2ax + 4a^2)$ **63.** $E = mg(y_2 - y_1)$ **65.** $A = \pi(r_2 - r_1)(r_2 + r_1)$
67. $E_d = a(T_1^2 + T_2^2)(T_1 + T_2)(T_1 - T_2)$ **69.** $\dfrac{1}{2}m(v_2 - v_1)(v_2 + v_1)$ **71.** $\dfrac{4}{3}\pi(r_2 - r_1)(r_2^2 + r_2r_1 + r_1^2)$

Section 5.3 (page 166)

1. $x^2 + 4x + 4$ **3.** $C_1^2 + 2C_1C_2 + C_2^2$ **5.** $x^2 - x - 2$ **7.** $15x^2 - 7x - 2$ **9.** $12x^2 - 11xy + 2y^2$ **11.** $s^2 - 5st + 6t^2$
13. $49i_1^2 + 28i_1i_2 + 4i_2^2$ **15.** $9x^2 - 12xy + 4y^2$ **17.** $2i^2 + 11ik + 15k^2$ **19.** $2x^2 - 8xy - 10y^2$ **21.** $6x^2 - 31xy + 40y^2$
23. $3v_1^2 - 12v_2^2$ **25.** $81x^2 - 36y^2$ **27.** $7x^2 - 28y^2$ **29.** $9x^3 - xy^2$ **31.** $2x^2 - 10xy + 12y^2$ **33.** $12ax^2 + 17axy - 5ay^2$
35. $x^2 + 2xy + y^2 + 4xz + 4yz + 4z^2$ **37.** $a^2 + 4ab + 4b^2 + 2ac + 4bc + c^2$ **39.** $x^2 + 2xy + y^2 + xz + yz - 6z^2$
41. $p^2 + 2pq + q^2 - 7pr - 7qr + 10r^2$ **43.** $2x^2 + 3ax + 3bx + a^2 + 2ab + b^2$

Section 5.4 (page 171)

1. $(x + 2y)^2$ **3.** $(3x - 4y)^2$ **5.** $(x - 1)(x - 3)$ **7.** $(x - 4)(x + 3)$ **9.** $2(a - 2b)^2$ **11.** $2(x + 6)(x + 1)$ **13.** $(x - 6y)(x + y)$
15. $(D + 7)(D - 2)$ **17.** $(2x - y)(x - y)$ **19.** $(5x - y)(x - 2y)$ **21.** $(4x + y)(x + 3y)$ **23.** $(2x - 3y)(3x + 4y)$
25. $(5w_1 - 2w_2)(w_1 - 4w_2)$ **27.** $8(L - 3C)(L + 2C)$ **29.** $2(3f - 4g)^2$ **31.** $x^2(x - 2)^2$ **33.** not factorable
35. $(a + b - 3)(a + b + 2)$ **37.** $(n + m - 2)(n + m - 1)$ **39.** $(2a + 2b - 1)(a + b - 4)$ **41.** $(f_1 + 2f_2)^2(f_1 + 2f_2 - 1)$
43. $(1 - x + y)(1 + x - y + x^2 - 2xy + y^2)$ **45.** $(7a - 2b)(4a + b)$ **47.** $(5x - y)(8x + 3y)$ **49.** $(3\alpha - 2\beta)(4\alpha - 5\beta)$
51. $t = 3$ s   **53.** $t = 1.33$ s

Section 5.5 (page 174)

1. $(x - y)(a + b)$ **3.** $(x + 3y)(2x + 1)$ **5.** $(a - b)(4c + 1)$ **7.** $(x - y)(5b - 1)$ **9.** $2(x + y)(a - c)$ **11.** $3(R - r)(a - 2b)$
13. $(x + y)(x - y - z)$ **15.** $(x + y)(a - x + y)$ **17.** $(x - y)(x + y - 2z)$ **19.** $(x - y)(x + 4y + 1)$ **21.** $(2x - y - z)(2x - y + z)$
23. $(x + 2y - z)(x + 2y + z)$ **25.** $(3a - 2b - c)(3a + 2b + c)$ **27.** $(x - y)(3 - x - 4y)$ **29.** $(x + 2y)(a - x + 3y)$
31. $(x - y)(z + x + 3y)$ **33.** $(V_1 - V_2 - V_3)(V_1 - V_2 + V_3)$ **35.** $a(A - 1)(Aa - 4)$

Section 5.6 (page 178)

1. $\dfrac{x}{2}$ **3.** $\dfrac{a}{2x}$ **5.** x **7.** $\dfrac{2}{x}$ **9.** $x + 4$ **11.** $x^2 - xy + y^2$ **13.** $\dfrac{x - y}{3}$ **15.** $R^2 + 2R + 4$ **17.** $x - 4$ **19.** $i + 5$

21. $\dfrac{3x - 4}{x + 3}$ **23.** $\dfrac{7x + 2}{2x + 1}$ **25.** $\dfrac{m_1 + m_2}{m_1 - 2m_2}$ **27.** -1 **29.** $\dfrac{2L + C}{3L - 4C}$ **31.** $\dfrac{x - 4}{x - 6}$ **33.** $\dfrac{a - 4b}{2a - b}$ **35.** $\dfrac{1}{2x + 1}$ **37.** $\dfrac{1}{2l + 1}$

39. $\dfrac{1}{3t + 1}$ **41.** $\dfrac{1}{3E - 2}$ **43.** $P - Q$ **45.** $\dfrac{M^2 + m^2}{(M + m)^2}$ **47.** $2t - 6$ (amperes)

Section 5.7 (page 182)

1. $\dfrac{3xy}{2zw^2}$ **3.** $\dfrac{8x^2y}{3a^2b}$ **5.** $\dfrac{6}{5}bcxy$ **7.** $\dfrac{dx}{ac^2}$ **9.** $2x(x - y)$ **11.** $\dfrac{x + y}{x - y}$ **13.** $-\dfrac{a + 2b}{3}$ **15.** $-(1 + 3a)(x^2 - 3xy + 9y^2)$

17. $\dfrac{4(2a + b)}{(3x - 1)(a + b)}$ **19.** $\dfrac{(3v_0 - 4)(v_1 + 6)}{(2v_0 - 3)(2v_1 + 1)}$ **21.** $\dfrac{(3T + J)(5K - 6)}{(4T + J)(K - 7)}$ **23.** $\dfrac{x - 3y}{(x + y)(x^2 + 2xy + 4y^2)}$ **25.** $\dfrac{x - y}{x + 2y}$

27. $\dfrac{1}{(R + 2r)(R + r - 1)}$ **29.** $\dfrac{2(2L - 7)}{(L + 4)(C - 5)}$ **31.** $\dfrac{2}{c - d}$ **33.** 2 **35.** $i = \dfrac{t + 4}{2t + 1}$ **37.** $\dfrac{2w_1w_2}{w_1 - w_2}$ **39.** $\dfrac{1}{8}w(s - 2H)$

Section 5.8 (page 187)

1. 1 **3.** $\dfrac{5x + 1}{9}$ **5.** $\dfrac{3x - 3}{4x}$ **7.** $\dfrac{4a - 1}{12b}$ **9.** $\dfrac{8xy}{(x - y)(x + y)}$ **11.** $\dfrac{1}{y - x}$ **13.** $\dfrac{2x^2 + 2xy - y^2}{(x + 2y)(x - y)}$ **15.** $\dfrac{1}{x - 3}$ **17.** $\dfrac{1}{x - y}$

19. $\dfrac{2(2x^2 - y^2)}{(3x - y)(2x - 3y)}$ **21.** $-\dfrac{x}{y(x + y)}$ **23.** $\dfrac{2(2a + b)}{a + 2b}$ **25.** $\dfrac{A + B}{A}$ **27.** 1 **29.** $\dfrac{1}{(x - y)(x + y)}$ **31.** $\dfrac{4}{(x - 2)(x + 2)}$

33. $\dfrac{c(c - d)}{(c + 2d)(2c + d)}$ **35.** $\dfrac{3y}{(x + y)(2x + y)(x + 3y)}$ **37.** $\dfrac{k(2np - p^2)}{n^2(n - p)^2}$ **39.** $\dfrac{k^2}{k^2 + L^2}$ **41.** $R = \dfrac{R_1R_2 + R_1R_3 + R_2R_3}{R_1 + R_2}$

Section 5.9 (page 192)

1. $\dfrac{1}{2}$ **3.** $\dfrac{8}{11}$ **5.** $\dfrac{x}{3x+1}$ **7.** $\dfrac{x-4}{x}$ **9.** $C_1 + C_2$ **11.** $\dfrac{x+5}{x-2}$ **13.** $\dfrac{h-5}{h+4}$ **15.** $\dfrac{w}{w+1}$ **17.** $\dfrac{1}{\beta+1}$ **19.** $\dfrac{2E-3}{E^2-2E+1}$

21. $\dfrac{k+3}{k+4}$ **23.** $\dfrac{x+1}{x-4}$ **25.** $\dfrac{(t+1)(t-4)}{(t-3)(2t+1)}$ **27.** $\dfrac{R_1 R_2}{R_1 + R_2}$ **29.** $\dfrac{pf}{p-f}$ **31.** $\dfrac{R_1(R_2 + R_3)}{R_1 + R_2 + R_3}$

Section 5.10 (page 197)

1. 3   **3.** -2 **5.** 10   **7.** 1   **9.** -8 **11.** no solution   **13.** -2 **15.** 12   **17.** no solution   **19.** $\dfrac{3}{2}$ **21.** 0   **23.** 10

25. -5 **27.** 9   **29.** -2 **31.** 1   **33.** $d = \dfrac{c}{3c-1}$ **35.** $R_1 = \dfrac{3R}{3-R}$ **37.** $f = \dfrac{pq}{p+q}$ **39.** $R = \dfrac{R_1 R_2}{R_1 + R_2}$ **41.** $10\,\Omega$

43. $k = \dfrac{k_1 k_2}{k_1 + k_2}$

Review Exercises for Chapter 5 (page 198)

1. $8s^3 t^3 - 12s^3 t^5 + 20s^4 t^4$ **2.** $F_1^2 + 8F_1 F_2 + 16F_2^2$ **3.** $4i_1^2 - 12i_1 i_2 + 9i_2^2$ **4.** $x^2 - 9z^2$ **5.** $4s^2 - 25t^2$
6. $25w_1^2 - 110w_1 w_2 + 121w_2^2$ **7.** $2c^2 + cd - 15d^2$ **8.** $9a^2 - 18ay - 7y^2$ **9.** $x^3 y - 9xy^3$ **10.** $6a^2 - 36ab + 48b^2$
11. $v^2 + 4vw + 4w^2 + 2v + 4w + 1$ **12.** $a^2 - 6ab + 9b^2 + 2a - 6b + 1$ **13.** $4x(a-b)$ **14.** $4R^2 C^2(2R + 3C - 1)$
15. $pq(3p^2 q^3 + 1)$ **16.** $(A - W)(A + W)$ **17.** $(2\beta - \gamma)(2\beta + \gamma)$ **18.** $2(a - 3b)(a + 3b)$ **19.** $16(L - 2C)(L + 2C)$
20. $(b^2 + 1)(b - 1)(b + 1)$ **21.** $(3h + g)(9h^2 - 3hg + g^2)$ **22.** $(2F - 1)(4F^2 + 2F + 1)$ **23.** $(a + 2b - 1)(a^2 + 4ab + 4b^2 + a + 2b + 1)$
24. $(1 - x + y)(1 + x - y + x^2 - 2xy + y^2)$ **25.** $(a - 1)(a + 1)(a^4 + a^2 + 1)$ or $(a - 1)(a + 1)(a^2 - a + 1)(a^2 + a + 1)$
26. $(x + 4)(x - 2)$ **27.** $(v_0 - 4)(v_0 + 3)$ **28.** $(2a - b)(a + 4b)$ **29.** $(3v_1 - v_2)(v_1 + v_2)$ **30.** $(PS^2 + 1)(P^2 S^4 - PS^2 + 1)$
31. $(C_1 - 5C_2)(C_1 + 3C_2)$ **32.** $(4x - w)(x - 2w)$ **33.** not factorable   **34.** $2(a - 4c)(a + 3c)$ **35.** $(x - y - 1)(x - y + 1)$
36. $(a + d - 2)(a + d + 2)$ **37.** $(s + t)(s - 2t + 1)$ **38.** $(x + b - 3)(x + b + 2)$ **39.** $(a + b)(x - y)$ **40.** $(v - 2w)(s + 2t)$

41. $(2x + y - a)(2x + y + a)$ **42.** $(1 - a + 2b)(1 + a - 2b)$ **43.** $x + 3y$ **44.** $a - 3c$ **45.** $\dfrac{q + 7r}{q + 9r}$ **46.** $\dfrac{3}{a + 2t}$ **47.** $\dfrac{2(a + d)}{a^2 d}$

48. $\dfrac{x + 4w}{x^2 + 2xw + 4w^2}$ **49.** $x + y$ **50.** $\dfrac{2}{3}(2f - g + 1)$ **51.** $\dfrac{x + 4y}{x + 2y}$ **52.** $\dfrac{9c - 5a}{a - 3c}$ **53.** $\dfrac{w^2}{(y - w)(y + w)}$ **54.** $\dfrac{y}{x + y}$ **55.** 1

56. $\dfrac{1}{(n + m)(n + 2m)}$ **57.** $\dfrac{a}{(a - b)(a + b)(a + 2b)}$ **58.** $\dfrac{2x^2 - 3y^2}{(x - 3y)(2x - y)}$ **59.** $\dfrac{w(2v - w)}{(v - 2w)(v + w)}$ **60.** $\dfrac{3(\theta - 1)}{(\theta - 3)(\theta + 3)}$ **61.** $\dfrac{1}{\omega - 2}$

62. $\dfrac{y + 3}{2y + 5}$ **63.** $\dfrac{i}{i - 2}$ **64.** $\dfrac{v + 2}{v + 5}$ **65.** $\dfrac{1}{r_1}$ **66.** $\dfrac{G + 2}{G + 1}$ **67.** 1   **68.** $\dfrac{C}{2(A - 2C)}$ **69.** 3   **70.** 11   **71.** $-\dfrac{12}{5}$ **72.** -14

73. no solution   **74.** 2   **75.** 6   **76.** 5   **77.** 1   **78.** $R = \dfrac{R_1 R_2 R_3}{R_1 R_2 + R_1 R_3 + R_2 R_3}$ **79.** $a = \dfrac{S(1 - r)}{1 - r^n}$ **80.** $m_1 = \dfrac{m_2 w}{m_2 - w}$

81. $t = 4.5$ s, $t = 5$ s   **82.** $r = \dfrac{r_1 r_2}{r_1 + r_2}$ **83.** $q = 12$ cm   **84.** $v = (t^2 - 4)^3(9t^2 - 4)$ **85.** $\dfrac{m_2}{m_1 + m_2}$ **86.** $f = \dfrac{f_1 f_2}{f_1 + f_2}$

Chapter 6

Section 6.1 (page 204)

1. $1, -1$ **3.** $6, -6$ **5.** $3, -3$ **7.** $\sqrt{10}, -\sqrt{10}$ **9.** $4, -4$ **11.** $\dfrac{5}{6}, -\dfrac{5}{6}$ **13.** $1, -2$ **15.** $4, -6$ **17.** $3, -\dfrac{1}{2}$ **19.** $-2, -\dfrac{1}{3}$

21. $\dfrac{3}{4}, -2$ **23.** $-3, \dfrac{7}{5}$ **25.** $-\dfrac{5}{2}, \dfrac{3}{2}$ **27.** $-\dfrac{3}{5}, \dfrac{5}{6}$ **29.** $-\dfrac{9}{4}, \dfrac{5}{2}$ **31.** $-\dfrac{11}{7}, 1$ **33.** $-\dfrac{3}{2}, \dfrac{5}{9}$ **35.** $-\dfrac{1}{11}, 7$ **37.** $-\dfrac{4}{3}, -\dfrac{4}{3}$

39. $\dfrac{1}{4}, \dfrac{1}{4}$ **41.** $\dfrac{v_0^2}{32}$ ft   **43.** $x = L, \dfrac{1}{3}L, 0$

Section 6.2 (page 209)

1. 2, 4   **3.** −6, 2   **5.** −4, 3   **7.** −5, −2   **9.** $\frac{1}{2}(-5 \pm \sqrt{17})$ **11.** $-3 \pm \sqrt{3}$ **13.** $\frac{1}{2}(3 \pm \sqrt{7})$ **15.** $\frac{1}{4}(-3 \pm \sqrt{33})$

17. $-1, \frac{1}{3}$ **19.** $\frac{1}{3}(2 \pm \sqrt{19})$ **21.** $-\frac{3}{4}, 1$ **23.** $\frac{1}{12}(-1 \pm \sqrt{47}j)$ **25.** $2 \pm j$ **27.** $\frac{5}{8} \pm \frac{\sqrt{23}}{8}j$ **29.** $\frac{1}{7}(-1 \pm 2\sqrt{2})$

31. $\frac{1}{12}(5 \pm \sqrt{73})$ **33.** $-\frac{10}{3}, \frac{5}{2}$ **35.** $\frac{1}{10} \pm \frac{\sqrt{19}}{10}j$ **37.** $\frac{1}{2}(b \pm \sqrt{b^2 - 8})$ **39.** $\frac{1}{2a}(-5 \pm \sqrt{4a + 25})$

Section 6.3 (page 213)

1. −3, 2   **3.** 4, 5   **5.** $\frac{1}{2}, 2$ **7.** $-\frac{1}{2}, \frac{2}{3}$ **9.** $\frac{3 \pm \sqrt{17}}{4}$ **11.** $\frac{-1 \pm \sqrt{7}}{3}$ **13.** $1 \pm j$ **15.** $-1 \pm \sqrt{3}j$ **17.** $\frac{-3 \pm \sqrt{3}j}{6}$

19. $\frac{-1 \pm \sqrt{3}j}{4}$ **21.** $\frac{-5 \pm \sqrt{73}}{8}$ **23.** $\pm \frac{\sqrt{5}}{5}j$ **25.** $-\frac{3}{2}, 0$ **27.** $\frac{3c \pm \sqrt{9c^2 - 8}}{4}$ **29.** $\frac{-3 \pm \sqrt{9 - 4b}}{2b}$ **31.** $\frac{3}{2}, \frac{3}{2}$ **33.** $\frac{5}{2}, \frac{5}{2}$

35. 0.70, −2.26   **37.** 5.77, −0.18   **39.** 0.66, −0.49   **41.** −5, 4   **43.** $-\frac{6}{5}, 4$ **45.** $\frac{4}{3}, 8$ **47.** $\frac{7 \pm \sqrt{33}}{2}$ **49.** $1 \pm y$

Section 6.4 (page 218)

1. $t = 4$ s   **3.** 0.75 s, 3.4 s   **5.** 0.919 A or 0.486 A   **7.** 7, 15   **9.** 9.59, 6.61   **11.** 7.21 and 9.21, or −7.21 and −9.21
13. 31 mm × 62 mm   **15.** 9.21 cm × 7.21 cm   **17.** 18.19 in. for base, 8.19 in. for height   **19.** 10 Ω, 15 Ω   **21.** 12 μF, 18 μF

23. $2\frac{2}{3}$ h, 8 h   **25.** 60 min, 40 min   **27.** 25 mi/h, 10 mi/h   **29.** 5.00 in. × 17.0 in.   **31.** 30 ft × 40 ft or $\frac{80}{3}$ ft × 45 ft

33. 10 ft × 15 ft   **35.** 80 shares

Review Exercises for Chapter 6 (page 220)

1. −2, 5   **2.** $-\frac{1}{2}, 4$ **3.** $-\frac{3}{5}, 2$ **4.** $\frac{3}{2}, -9$ **5.** $-\frac{2}{3}, -\frac{1}{2}$ **6.** $\frac{5}{4}, 3$ **7.** $\frac{5}{6}, 1$ **8.** $-\frac{3}{2}, \frac{6}{5}$ **9.** −2, 4   **10.** $-\frac{5}{2}, 1$

11. $\frac{-5 \pm \sqrt{13}}{2}$ **12.** $\frac{2 \pm \sqrt{2}}{2}$ **13.** $\frac{5}{4} \pm \frac{\sqrt{7}}{4}j$ **14.** $\frac{1}{3} \pm \frac{2\sqrt{2}}{3}j$ **15.** $\frac{1}{8} \pm \frac{\sqrt{47}}{8}j$ **16.** $-\frac{1}{2}, -2$ **17.** 2, 4   **18.** $-\frac{3}{2}, \frac{1}{3}$

19. $\frac{1}{2} \pm \frac{\sqrt{3}}{6}j$ **20.** $4 \pm \sqrt{10}$ **21.** $\frac{2 \pm \sqrt{10}}{2}$ **22.** $1 \pm \frac{1}{2}\sqrt{2}j$ **23.** $\frac{4 \pm \sqrt{34}}{6}$ **24.** $-\frac{1}{10} \pm \frac{1}{10}\sqrt{19}j$ **25.** 0.806, −1.90

26. 0.27, −0.39   **27.** $x = 2 + y, x = 2 - y$ **28.** $x = 2 \pm \sqrt{4 + a^2}$ **29.** $-\frac{5}{9}, 4$ **30.** $\frac{1 \pm \sqrt{13}}{2}$ **31.** 3 Ω, 6 Ω

32. 10 ft × 20 ft   **33.** 6 h, 12 h   **34.** 12 ft × 16 ft   **35.** 20 mi/h, 35 mi/h

Cumulative Review Exercises for Chapters 4–6 (page 221)

1. 122°19′   **2.** $\sin \theta = \frac{\sqrt{3}}{4}$, $\cos \theta = \frac{\sqrt{13}}{4}$, $\tan \theta = \frac{\sqrt{39}}{13}$, $\csc \theta = \frac{4\sqrt{3}}{3}$, $\sec \theta = \frac{4\sqrt{13}}{13}$, $\cot \theta = \frac{\sqrt{39}}{3}$ **3.** $\frac{3\sqrt{7}}{7}$ **4.** $\frac{\sqrt{3}}{3}$ **5.** 1.688

6. 56.90°   **7.** 20°31′   **8.** $(V_a - V_b)(s - 1)$ **9.** $\frac{1}{2}(L^2 + LC + C^2)$ **10.** $\frac{x + y}{2(x + 3y)}$ **11.** $\frac{2a - 1}{a(2x + y)}$ **12.** $\frac{3st - t^2 + 1}{(s - t)(s + t)}$ **13.** $\frac{L}{L + 4}$

14. $x = 4, -2$ **15.** $x = \frac{1}{4}, -3$ **16.** $x = 1, 3$ **17.** $x = 1 \pm j$ **18.** $x = -0.975, 0.427$ **19.** 0.433 A   **20.** 1.55 in.

21. $\frac{R_1 R_2 R_3}{R_2 R_3 + R_1 R_3 + R_1 R_2}$ **22.** 3.4 in. by 5.4 in.

Chapter 7

Section 7.1 (page 228)

In Exercises 1–19, the trigonometric functions are given in the following order: $\sin\theta$, $\cos\theta$, $\tan\theta$, $\csc\theta$, $\sec\theta$, and $\cot\theta$.

1. $\dfrac{3}{5}$, $-\dfrac{4}{5}$, $-\dfrac{3}{4}$, $\dfrac{5}{3}$, $-\dfrac{5}{4}$, $-\dfrac{4}{3}$　　**3.** $-\dfrac{12}{13}$, $\dfrac{5}{13}$, $-\dfrac{12}{5}$, $-\dfrac{13}{12}$, $\dfrac{13}{5}$, $-\dfrac{5}{12}$　　**5.** $\dfrac{2\sqrt{2}}{3}$, $\dfrac{1}{3}$, $2\sqrt{2}$, $\dfrac{3\sqrt{2}}{4}$, 3, $\dfrac{\sqrt{2}}{4}$　　**7.** $-\dfrac{3\sqrt{10}}{10}$, $-\dfrac{\sqrt{10}}{10}$, 3, $-\dfrac{\sqrt{10}}{3}$, $-\sqrt{10}$, $\dfrac{1}{3}$

9. $\dfrac{\sqrt{15}}{4}$, $-\dfrac{1}{4}$, $-\sqrt{15}$, $\dfrac{4\sqrt{15}}{15}$, -4, $-\dfrac{\sqrt{15}}{15}$　　**11.** $\dfrac{2\sqrt{5}}{5}$, $-\dfrac{\sqrt{5}}{5}$, -2, $\dfrac{\sqrt{5}}{2}$, $-\sqrt{5}$, $-\dfrac{1}{2}$　　**13.** -0.93, 0.37, -2.5, -1.1, 2.7, -0.40

15. 0.78, -0.63, -1.2, 1.3, -1.6, -0.81　　**17.** -0.462, -0.887, 0.521, -2.16, -1.13, 1.92　　**19.** -0.770, 0.638, -1.21, -1.30, 1.57, -0.829

21. $\dfrac{5\sqrt{29}}{29}$　　**23.** $-\dfrac{\sqrt{30}}{5}$　　**25.** $-\dfrac{\sqrt{7}}{3}$　　**27.** $\dfrac{\sqrt{6}}{2}$　　**29.** $-\dfrac{\sqrt{13}}{4}$　　**31.** $-\dfrac{\sqrt{10}}{10}$　　**33.** $\sqrt{3}$　　**35.** $\dfrac{\sqrt{6}}{2}$　　**37.** -2.206　　**39.** -1.178

41. 0.3014　　**43.** -4.789　　**45.** -1.170

Section 7.2 (page 232)

1. $\dfrac{1}{2}$　　**3.** $-\dfrac{\sqrt{3}}{3}$　　**5.** $-\dfrac{\sqrt{2}}{2}$　　**7.** 2　　**9.** $-\dfrac{1}{2}$　　**11.** 1　　**13.** 2　　**15.** 0　　**17.** -1　　**19.** $-\dfrac{\sqrt{3}}{3}$　　**21.** undefined　　**23.** $-\dfrac{\sqrt{3}}{2}$

25. $-\dfrac{2\sqrt{3}}{3}$　　**27.** $-\dfrac{\sqrt{3}}{3}$　　**29.** $-\dfrac{2\sqrt{3}}{3}$　　**31.** 0　　**33.** $-\dfrac{\sqrt{2}}{2}$　　**35.** $-\dfrac{\sqrt{3}}{3}$　　**37.** $\dfrac{2\sqrt{3}}{3}$　　**39.** $\dfrac{1}{2}$　　**45.** $30°$, $150°$　　**47.** $30°$, $150°$

49. $240°$, $300°$　　**51.** $180°$　　**53.** $60°$, $240°$　　**55.** $60°$, $300°$　　**57.** $225°$, $315°$　　**59.** $90°$, $270°$　　**61.** $135°$, $315°$　　**63.** $135°$, $225°$
65. $60°$, $120°$　　**67.** $120°$, $300°$　　**69.** $90°$, $270°$　　**71.** $120°$, $300°$　　**73.** $45°$, $135°$　　**75.** $30°$, $330°$

Section 7.3 (page 235)

1. 0.8678　　**3.** -5.593　　**5.** -1.502　　**7.** 1.381　　**9.** 0.6357　　**11.** 1.304　　**13.** 0.3569　　**15.** 0.8996　　**17.** 0.8675　　**19.** -5.575
21. -1.502　　**23.** 1.381　　**25.** -0.8234　　**27.** 1.180　　**29.** 2.8006　　**31.** $17.00°$, $163.00°$　　**33.** $79.77°$, $280.23°$　　**35.** $191.15°$, $348.85°$
37. $242.25°$, $297.75°$　　**39.** $165.47°$, $345.47°$　　**41.** $21.29°$, $158.71°$　　**43.** $137°2'$, $222°58'$　　**45.** $65°5'$, $294°55'$　　**47.** $49°42'$, $229°42'$
49. $130°5'$, $229°55'$　　**51.** $241.89°$　　**53.** $236.59°$　　**55.** $140.30°$　　**57.** $221.25°$　　**59.** 53.8 m　　**61.** 91.6 lb　　**63.** $90°$

Section 7.4 (page 241)

1. $\dfrac{\pi}{6}$　　**3.** $\dfrac{\pi}{3}$　　**5.** $\dfrac{\pi}{9}$　　**7.** $\dfrac{2\pi}{5}$　　**9.** $-\dfrac{\pi}{3}$　　**11.** $\dfrac{7\pi}{6}$　　**13.** $\dfrac{11\pi}{20}$　　**15.** $\dfrac{11\pi}{9}$　　**17.** $\dfrac{3\pi}{5}$　　**19.** $\dfrac{19\pi}{90}$　　**21.** $-\dfrac{28\pi}{15}$　　**23.** $\dfrac{13\pi}{20}$　　**25.** $45°$

27. $-210°$　　**29.** $25°$　　**31.** $198°$　　**33.** $320°$　　**35.** $378°$　　**37.** $3°$　　**39.** $140°$　　**41.** 0.3068　　**43.** 2.3178　　**45.** 5.5920
47. -1.7482　　**49.** $23.8°$　　**51.** $153.6°$　　**53.** $186.6°$　　**55.** $-51.7°$　　**57.** 0.7606　　**59.** 1.6450　　**61.** 5.8819　　**63.** -0.4321
65. 0.8660　　**67.** 9.5668　　**69.** -0.227 A

Section 7.5 (page 246)

1. 19.8 cm　　**3.** 2.78 m　　**5.** 1.52 ft　　**7.** 0.9661 m　　**9.** 0.52 ft　　**11.** 16.3 cm^2　　**13.** 60.5 m^2　　**15.** 20.3 ft^2　　**17.** 2.812 m^2
19. 0.260 ft^2　　**21.** 13.6 ft　　**23.** $62°$　　**25.** $2{,}160$ mi　　**27.** 67.88 cm^2　　**29.** 39.5 cm^2　　**31.** 5.0 cm/s　　**33.** 161 ft/s　　**35.** 10.4 ft/s
37. 249 m/min　　**39.** $1{,}050$ mi/h　　**41.** 55.0 mi/h　　**43.** 20.2 mi/h　　**45.** 6.0 rev/s　　**47.** 17.8 rev/min　　**49.** 18.5 mi/s

Review Exercises for Chapter 7 (page 248)

In Exercises 1–4, the trigonometric functions are given in the following order: $\sin\theta$, $\cos\theta$, $\tan\theta$, $\csc\theta$, $\sec\theta$, and $\cot\theta$.

1. $\dfrac{\sqrt{5}}{3}$, $-\dfrac{2}{3}$, $-\dfrac{\sqrt{5}}{2}$, $\dfrac{3\sqrt{5}}{5}$, $-\dfrac{3}{2}$, $-\dfrac{2\sqrt{5}}{5}$　　**2.** $-\dfrac{\sqrt{10}}{10}$, $-\dfrac{3\sqrt{10}}{10}$, $\dfrac{1}{3}$, $-\sqrt{10}$, $-\dfrac{\sqrt{10}}{3}$, 3　　**3.** $-\dfrac{2\sqrt{5}}{5}$, $\dfrac{\sqrt{5}}{5}$, -2, $-\dfrac{\sqrt{5}}{2}$, $\sqrt{5}$, $-\dfrac{1}{2}$

4. 0.913, -0.408, -2.24, 1.10, -2.45, -0.447　　**5.** $-\dfrac{2\sqrt{2}}{3}$　　**6.** $\dfrac{\sqrt{2}}{2}$　　**7.** $-\dfrac{\sqrt{13}}{4}$　　**8.** -4.02　　**9.** $\dfrac{\sqrt{2}}{2}$　　**10.** 0　　**11.** $-\dfrac{1}{2}$　　**12.** $\sqrt{2}$

13. 1　　**14.** 2　　**15.** $-\dfrac{2\sqrt{3}}{3}$　　**16.** $-\sqrt{3}$　　**17.** 1　　**18.** undefined　　**19.** $-\dfrac{\sqrt{3}}{2}$　　**20.** $-\dfrac{2\sqrt{3}}{3}$　　**21.** $60°$, $300°$　　**22.** $210°$, $330°$

23. $120°$, $300°$　　**24.** $30°$, $210°$　　**25.** $90°$, $270°$　　**26.** $180°$　　**27.** $150°$, $210°$　　**28.** $225°$, $315°$　　**29.** -1.525　　**30.** -1.062
31. 0.7528　　**32.** -3.404　　**33.** $144.16°$, $215.84°$　　**34.** $45.71°$, $134.29°$　　**35.** $157.36°$, $337.36°$　　**36.** $15.74°$, $164.26°$

37. $29°15', 209°15'$ **38.** $108°26', 251°34'$ **39.** $\dfrac{2\pi}{9}$ **40.** $\dfrac{8\pi}{9}$ **41.** $-\dfrac{5\pi}{9}$ **42.** $-\dfrac{\pi}{5}$ **43.** $\dfrac{3\pi}{10}$ **44.** $\dfrac{59\pi}{45}$ **45.** $150°$ **46.** $35°$

47. $234°$ **48.** $75°$ **49.** $340°$ **50.** $230°$ **51.** 0.4416 **52.** 1.3905 **53.** 2.5150 **54.** -0.4163 **55.** -4.4884 **56.** 5.9417
57. 6.8068 **58.** 6.9813 **59.** $41.0°$ **60.** $82.4°$ **61.** $149.4°$ **62.** $221.2°$ **63.** -1.493 **64.** -0.3598 **65.** -0.2106
66. -1.014 **67.** 1.002 **68.** -1.7013 **69.** 3.35 in.   **70.** 10.1 in.^{2} **71.** 48 ft   **72.** $97°40'$ **73.** 12.6 mi/h   **74.** $18,000$ mi/h
75. $6,900$ mi/h

Chapter 8

Section 8.1 (page 257)

1. $A = 2,\ P = 2\pi$
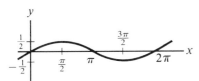

3. $A = 1,\ P = 2\pi$
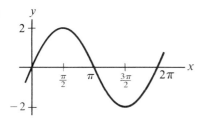

5. $A = \dfrac{1}{2},\ P = 2\pi$
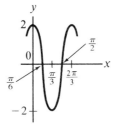

7. $A = 2,\ P = \dfrac{2\pi}{3}$
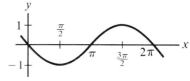

9. $A = 5,\ P = \pi$
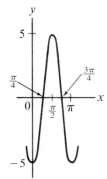

11. $A = \dfrac{1}{2},\ P = \dfrac{2\pi}{3}$
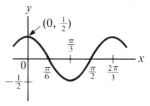

13. $A = 4,\ P = 3\pi$
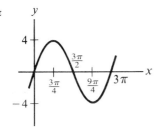

15. $A = 10,\ P = 8\pi$
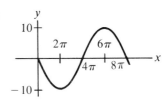

17. $A = 5$, $P = 10\pi$

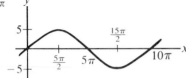

19. $A = 4$, $P = \dfrac{6\pi}{7}$

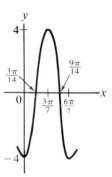

21. $A = 1$, $P = 2$

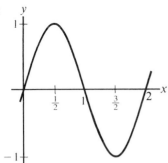

23. $A = \dfrac{1}{3}$, $P = \dfrac{2}{3}$

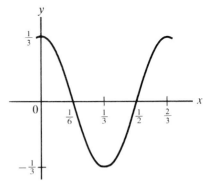

Section 8.2 (page 262)

1. $A = 2$, $P = 2\pi$, shift: $\dfrac{\pi}{4}$ to the right

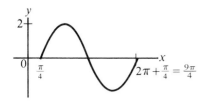

3. $A = \dfrac{1}{2}$, $P = 2\pi$, shift: $\dfrac{\pi}{8}$ to the left

5. $A = \dfrac{1}{2}$, $P = 4\pi$, shift: $\dfrac{\pi}{4}$ to the right

7. $A = 3$, $P = 2\pi$, shift: $\dfrac{3\pi}{2}$ to the left

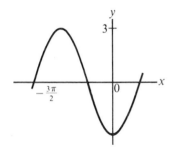

9. $A = 2$, $P = 2\pi$, shift: $\dfrac{\pi}{3}$ to the right

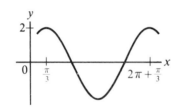

11. $A = 3$, $P = 2\pi$, shift: 2 to the right

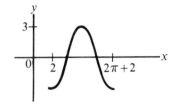

13. $A = 3$, $P = \pi$, shift: $\dfrac{\pi}{4}$ to the left

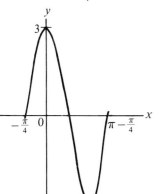

15. $A = 10$, $P = 6\pi$, shift: $\dfrac{12\pi}{5}$ to the right

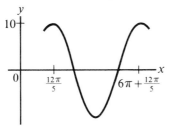

17. $A = 2$, $P = 2$, shift: 1 to the left

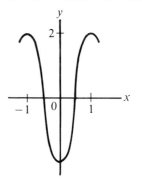

19. $A = 3$, $P = 4$, shift: $\dfrac{2}{\pi}$ to the right

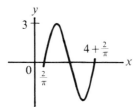

Section 8.3 (page 267)

1. $y = 3.0 \sin 2t$, $A = 3.0$ in., $P = \pi$ s

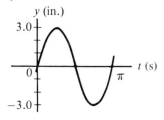

3. $y = 5 \cos 8\pi t$, $A = 5$ cm, $P = \dfrac{1}{4}$ min

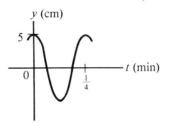

5. $A = 6.00$ V, $P = \dfrac{1}{40}$ s, shift: $\dfrac{1}{320}$ s to the left

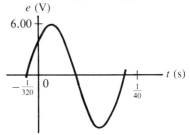

7. $A = 6.0$ A, $P = \dfrac{1}{60}$ s, shift: $\dfrac{1}{480}$ s to the left

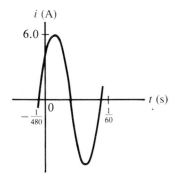

9. $A = 5$, $P = \pi$

11. $A = 0.002$ in., $P = \dfrac{1}{224}$ s = 0.0045 s

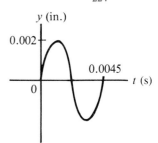

13. $y = 2.1 \sin \dfrac{\pi}{3.0}(t - 0.65)$, $A = 2.1$ ft, $P = 6.0$ s, shift: 0.65 s to the right

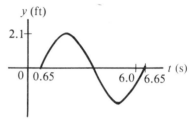

Section 8.4 (page 271)

1. $P = \pi$

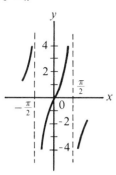

3. $P = \pi$

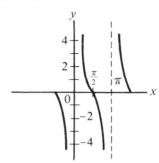

5. $P = \pi$

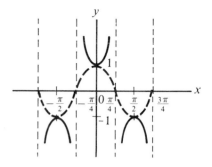

7. $P = \pi$

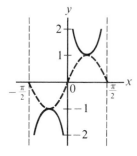

9. $P = \dfrac{\pi}{3}$

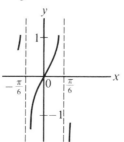

11. $P = \pi$

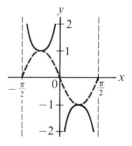

13. $P = 4\pi$

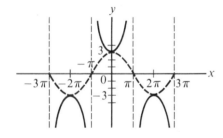

15. $P = 3\pi$

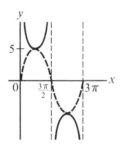

17. $P = \pi$, shift: $\dfrac{\pi}{6}$ to the left

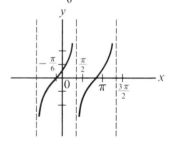

19. $P = \dfrac{1}{3}$

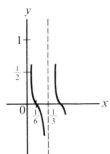

21. $P = 2\pi$

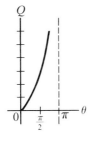

Section 8.5 (page 274)

1.

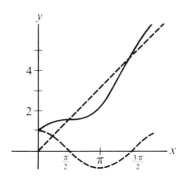

3.

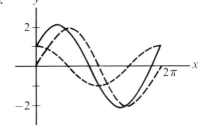

5.

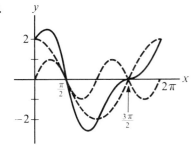

7.

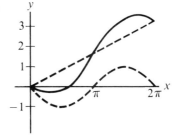

9.

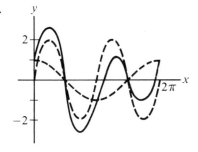

11.

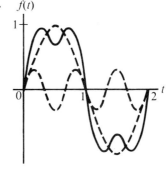

13. i (A)

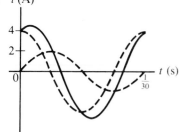

Review Exercises for Chapter 8 (page 274)

1. $A = 2,\ P = 2\pi$

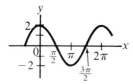

2. $A = \dfrac{1}{2},\ P = \pi$

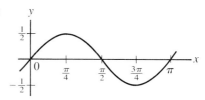

3. $A = 4,\ P = \dfrac{\pi}{2}$

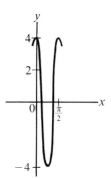

4. $A = 2,\ P = 2\pi$

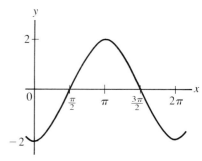

5. $A = \dfrac{1}{2},\ P = \pi$

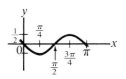

6. $A = 3,\ P = 4\pi$

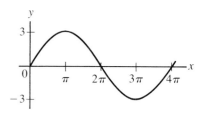

7. $A = 4,\ P = 4\pi$

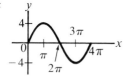

8. $A = 8,\ P = 8\pi$

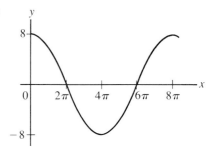

9. $A = 2$, $P = 2\pi$, shift: $\dfrac{\pi}{4}$ to the left

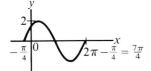

10. $A = 2$, $P = 2\pi$, shift: $\dfrac{\pi}{4}$ to the right

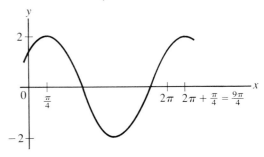

11. $A = \dfrac{1}{2}$, $P = 2\pi$, shift: $\dfrac{\pi}{8}$ to the left

12. $A = \dfrac{4}{3}$, $P = 2\pi$, shift: $\dfrac{\pi}{4}$ to the right

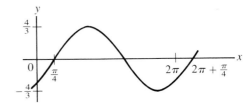

13. $A = 4$, $P = 4\pi$, shift: $\dfrac{\pi}{4}$ to the right

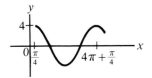

14. $A = 2$, $P = 4\pi$, shift: $\dfrac{\pi}{4}$ to the left

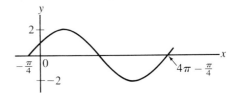

15. $A = \dfrac{1}{2}$, $P = \pi$, shift: $\dfrac{\pi}{8}$ to the right

16. $A = 3$, $P = \pi$, shift: 1 to the right

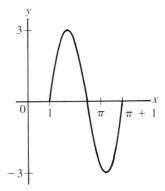

17. $P = \pi$

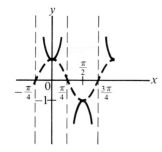

18. $P = \dfrac{\pi}{2}$

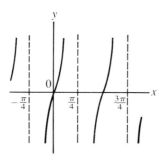

19. $P = 2\pi$

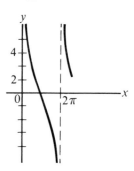

20. $P = \dfrac{2\pi}{3}$

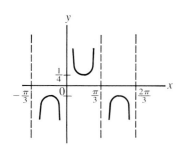

21. $y = 4 \sin 2t$, $A = 4$ cm, $P = \pi$ s

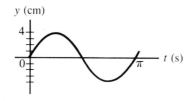

22. $y = 4 \cos 2t$, $A = 4$ cm, $P = \pi$ s

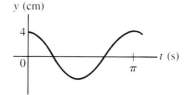

23. $y = 6 \cos 10\pi t$, $A = 6$ cm, $P = \dfrac{1}{5}$ min

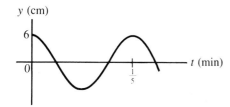

24. $y = 6 \sin 10\pi t$, $A = 6$ cm, $P = \dfrac{1}{5}$ min

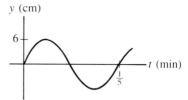

25. $A = 5.00$ V, $P = \dfrac{1}{30}$ s, shift: $\dfrac{1}{180}$ s to the left

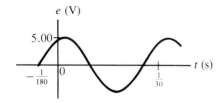

26. $A = 4.00$ A, $P = \dfrac{1}{45}$ s, shift: $\dfrac{1}{540}$ s to the left

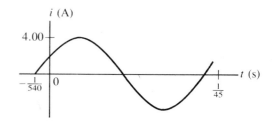

27. $y = 0.003 \sin 2\pi(236)t,\ A = 0.003$ in., $P = \dfrac{1}{236}$ s

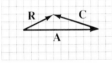

28.

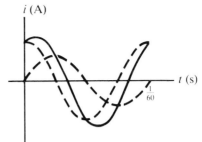

29.

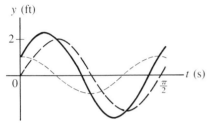

30.

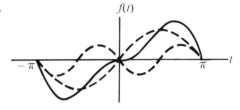

Chapter 9

Section 9.1 (page 279)

1.

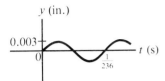

3.

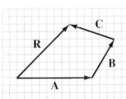

5.

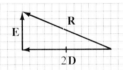

7.

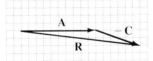

9.

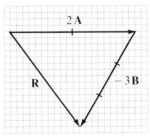

11.

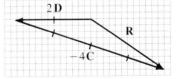

13.

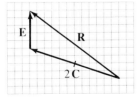

15.

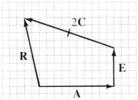

17.

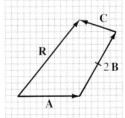

Section 9.2 (page 284)

1. $\sqrt{2}, 45°$ **3.** $3\sqrt{2}, 225°$ **5.** $\sqrt{13}, 56.3°$ **7.** 3, 234.7°   **9.** 4, 228.6°   **11.** 4, 154.3°   **13.** $\sqrt{5}, 296.6°$ **15.** $\sqrt{17}, 256.0°$
17. (0.52, 2.95)   **19.** (−2.01, 1.40)   **21.** (−0.63, −3.95)   **23.** (2.95, −0.52)   **25.** (−0.53, 1.31)   **27.** 4.2, 40.6°   **29.** 5.6, 141.7°
31. 7.4, 261.6°   **33.** 5.5, 7.2°   **35.** 5.0, 214.0°

Section 9.3 (page 288)

1. 11.7 lb, 29.2° from F_1 **3.** 20.15 N, 73.27° from F_1 **5.** 16.2 mi/h, 14°40′ with line across   **7.** 55 lb, 35° with larger force
9. 83.2 mi/h   **11.** 9.03 mi/h, 25.5° upstream with line across   **13.** 401 km/h, 4° east of north   **15.** 223 km/h, 7.7° east of south
17. 623.9 lb   **19.** 1,450 lb   **21.** 101 lb, 14°   **23.** 188.6 lb   **25.** 152 mi/h, 2.2° north of east   **27.** 437 km/h, 10.9° south of west
29. 55.4 lb, 18.2° with larger force   **31.** 95.9 lb, 32.1° with horizontal   **33.** 2,150 lb, 1.8° north of east

Section 9.4 (page 295)

1. $B = 109.1°, a = 1.46, c = 3.28$ **3.** $A = 23.9°, a = 8.02, b = 8.62$ **5.** $B = 8°50′, C = 151°20′, c = 144$ **7.** no solution
9. $C = 68.4°, B = 82.3°, b = 145$ **11.** $A = 80.0°, b = 172, c = 194$ **13.** $A = 73.2°, B = 43.2°, b = 8.87$
15. $A = 13°7′, C = 133°8′, c = 3.684$ **17.** $A = 39°33′, C = 106°42′, c = 1.724$ **19.** $B = 43.10°, b = 115.9, c = 30.63$
21. 510 mi/h   **23.** 66.2 ft   **25.** 953 ft

Section 9.5 (page 302)

1. $B = 82.2°, A = 40.9°, C = 56.9°$ **3.** $A = 153°31′, B = 15°29′, C = 11°0′$ **5.** $a = 1.76, C = 109.8°$ (by cosine law), $B = 23.9°$
7. $c = 280, A = 31°20′, B = 23°30′$ **9.** $c = 81.8, A = 102.2°$ (by cosine law), $B = 38.4°$ **11.** $a = 76.8, C = 93.0°$ (by cosine law), $B = 24.0°$
13. 349 ft   **15.** 27.1 km   **17.** 17.9° north of east, 158 mi/h   **19.** 22.0 in.   **21.** 269.4 N, 12.42° with larger force

Review Exercises for Chapter 9 (page 302)

1. $\sqrt{2}, 135°$ **2.** 2, 240°   **3.** $\sqrt{5}, 63.4°$ **4.** $\sqrt{10}, 288.4°$ **5.** 4, 14.5°   **6.** 4, 138.6°   **7.** $\sqrt{6}, 234.7°$ **8.** $\sqrt{11}, 64.8°$
9. (1.29, 4.83)   **10.** (−1.27, 2.72)   **11.** (−2.03, −1.70)   **12.** (1.47, −0.92)   **13.** 2.0, 116.8°   **14.** 2.5, 325.6°
15. $B = 80.0°, a = 3.86, c = 4.30$ **16.** $C = 35.4°, a = 53.8, b = 99.2$ **17.** $A = 39°0′, a = 195, b = 57.3$ **18.** $C = 87°, a = 18, c = 28$
19. $B = 14°, C = 140°, b = 14$ **20.** $c = 35.5, B = 114.0°$ (by cosine law), $A = 34.4°$ **21.** $a = 19.80, C = 149°46′$ (by cosine law), $B = 10°51′$
22. $b = 1.5, A = 82°$ (by cosine law), $C = 47°$ (by sine law: $A = 80°$)   **23.** $B = 76.1°, A = 38.3°, C = 65.6°$
24. $A = 75°26′, B = 56°16′, C = 48°18′$ **25.** 1.2 tons   **26.** 313 lb   **27.** 55.5 ft   **28.** 27.7 in.   **29.** 12.4 lb   **30.** 244 lb
31. 44.7°   **32.** 67.5 ft   **33.** 11.0 mi/h   **34.** 330 mi/h, 18°50′ south of west

Cumulative Review Exercises for Chapters 7–9 (page 304)

1. $\dfrac{\sqrt{5}}{2}$ **2. a.** $-\dfrac{2\sqrt{3}}{3};$ **b.** $-\dfrac{\sqrt{3}}{3}$ **3. a.** 45°, 315°;   **b.** 150°, 330°   **4.** 1.388   **5.** 228°13′, 311°47′   **6.** $\dfrac{28\pi}{45}$

7. 500°   **8.** 1.11 cm   **9.** 98.0 in.^{2} **10.** 85.0 m/s   **11.** 19 m/s   **12.**

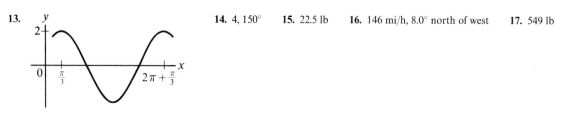

13.   **14.** 4, 150°   **15.** 22.5 lb   **16.** 146 mi/h, 8.0° north of west   **17.** 549 lb

18. 23.6 knots, 17.9° east of south

Chapter 10

Section 10.1 (page 307)

1. 72 **3.** $\dfrac{27}{64}$ **5.** $6x^5y^9$ **7.** $2a^4b^3$ **9.** R^4S^8 **11.** $-512C_1{}^9C_2{}^{12}$ **13.** $\dfrac{3sv}{7t}$ **15.** $\dfrac{1}{3}ns$ **17.** $3V^3$ **19.** $\dfrac{x^2}{2y^3z^6}$ **21.** $3abc$

23. $\dfrac{7v_0}{s_2{}^5}$ **25.** $\dfrac{8x^3}{27y^3}$ **27.** $\dfrac{16q^{12}}{81p^8r^4}$ **29.** $5{,}625x^8y^{16}z^{10}$ **31.** $\dfrac{3bxy^4}{2a^4}$ **33.** $\dfrac{R^2V^2}{32C}$ **35.** x **37.** y **39.** xy^a **41.** a

43. x^{5b} **45.** x^{6b}

Section 10.2 (page 313)

1. $\dfrac{1}{16}$ **3.** 16 **5.** 36 **7.** 1 **9.** x^5 **11.** $\dfrac{b^6}{a^3}$ **13.** $\dfrac{1}{4z}$ **15.** $\dfrac{4f_1f_2{}^5}{f_3{}^5}$ **17.** $\dfrac{t^2}{su^7}$ **19.** $\dfrac{16q^7}{3r^2s^7}$ **21.** $\dfrac{b}{a}$ **23.** $\dfrac{x^6}{8y^6}$ **25.** $\dfrac{b^8}{256x^4}$

27. 1 **29.** $\dfrac{64\omega^6}{\pi^6}$ **31.** $\dfrac{1}{16a^8b^8}$ **33.** $\dfrac{8L^{12}C^{15}}{R^{12}}$ **35.** $\dfrac{216a^{12}c_1{}^3}{c_2{}^3}$ **37.** $s^4r^{22}t^6$ **39.** $\dfrac{1}{81a^8b^2y^2}$ **41.** π^{2e} **43.** $\dfrac{b^y}{3a^x}$

45. $a^{b^2}x^{bc}$ **47.** $2x^2$ **49.** $\dfrac{x+y}{xy}$ **51.** $1+ab$ **53.** $\dfrac{b}{b+1}$ **55.** $\dfrac{ab}{a+b}$ **57.** $\dfrac{z^4}{(z^2-1)^2}$ **59.** $\dfrac{1}{b-a}$ **61.** $\dfrac{b^3}{b^2+1}$

63. $\dfrac{1}{(b-a)^2}$ **65.** b^2-a^2 **67.** 0.000000038 cm **71.** $\dfrac{n}{n-1}$ **73.** \$4,038.83

Section 10.3 (page 318)

1. 9 **3.** 9 **5.** $\dfrac{1}{7}$ **7.** 0.3 **9.** 3 **11.** -4 **13.** -8 **15.** $\dfrac{1}{343}$ **17.** $\dfrac{1}{5}$ **19.** $\dfrac{1}{25}$ **21.** 1 **23.** $m^{3/7}$ **25.** $x^{25/12}$

27. $\omega^{1/2}$ **29.** $\dfrac{10}{x^{1/4}}$ **31.** $\dfrac{a^{1/2}}{2b^{1/20}}$ **33.** $\dfrac{3}{5R_1{}^{7/12}R_2{}^{1/12}}$ **35.** $\dfrac{2a^{7/30}}{w^{7/8}}$ **37.** $2x^3y^{16}$ **39.** $\dfrac{y^3}{3a^2}$ **41.** $\dfrac{x}{2y^2z^3}$ **43.** $\dfrac{c^{11/4}}{2a^{5/3}}$

45. $\dfrac{2}{m^{17/10}d^2}$ **47.** $\dfrac{1}{6}m^3v^{6/5}$ **49.** $\dfrac{8L^3}{T^5}$ **51.** $\dfrac{2x+2}{(x+2)^{1/2}}$ **53.** $\dfrac{2x-1}{(x-3)^{1/2}}$ **55.** $\dfrac{2x+1}{(x+1)^{2/3}}$ **57.** $\dfrac{7}{(x-1)^{2/3}}$ **59.** $\dfrac{-x-1}{(x-1)^{1/3}}$

61. $\dfrac{-2x-6}{(x+2)^{4/5}}$ **63.** $\dfrac{-3}{(x+2)^{6/7}}$ **65.** x^{a+1} **67.** $\dfrac{1}{x}$ **69.** 1.69 **71.** 10.29 **73.** 3.18 **75.** 2.83 **77.** $p=\dfrac{k}{(\sqrt[5]{v})^7}$ **79.** $\dfrac{32}{7}$ cm

83. $\dfrac{14t^2+12t}{3(1+t)^{2/3}}$

Section 10.4 (page 326)

1. $\sqrt[4]{x}$ **3.** $\sqrt[12]{b}$ **5.** $\sqrt{abc^3}$ **7.** $\sqrt[4]{u^2v^3w^4}$ **9.** $\sqrt[3]{v^2m^2g^3}$ **11.** $\sqrt[4]{5p^2q^4r^5}$ **13.** $2x\sqrt{x}$ **15.** $3ac^2\sqrt[3]{2ac^2}$ **17.** $2u^3v^4\sqrt[4]{2uv^2}$

19. $3mns^3t^5\sqrt{2nst}$ **21.** $\dfrac{\sqrt{5}}{5}$ **23.** $\dfrac{\sqrt{3x}}{x}$ **25.** $\dfrac{\sqrt{abc}}{bc}$ **27.** $\dfrac{\sqrt{3pqv}}{3pq}$ **29.** $\dfrac{\sqrt[3]{18}}{3}$ **31.** $\dfrac{\sqrt[3]{2\pi^2V}}{2\pi}$ **33.** $\dfrac{\sqrt[4]{6L^3V^2}}{2L^2V}$ **35.** $\dfrac{\sqrt{21xy}}{3x^2y^3}$

37. $\dfrac{\sqrt[3]{3aRC^2}}{3C}$ **39.** $\dfrac{\sqrt{2pq}}{2pqr}$ **41.** $\dfrac{2a^2\sqrt{3b}}{3b^2}$ **43.** $\dfrac{3\sqrt{wz}}{w^3z}$ **45.** $\dfrac{2x\sqrt{2z}}{z^2}$ **47.** $\dfrac{v\sqrt{3uvw}}{3u^2w^4}$ **49.** $\dfrac{\sqrt[3]{4c^2}}{2ac}$ **51.** $\dfrac{\sqrt[3]{6uv^2}}{2uv}$ **53.** $\dfrac{\sqrt[4]{2xyz}}{2xy^2}$

55. $\dfrac{\sqrt{7stu}}{49s^2t^3u^3}$ **57.** $5\sqrt{6}$ **59.** $3\sqrt{5}$ **61.** $-\sqrt[3]{3}$ **63.** $7\sqrt{5}-5\sqrt[3]{2}$ **65.** $-ab\sqrt{ab}$ **67.** $(5mn-m^2n+2m^2n^2)\sqrt{mn}$

69. $\dfrac{2\sqrt{2\pi kmT}}{\pi m}$ **71.** $1.215\sqrt{273(T+273)}$ **73.** $x=\pm\dfrac{\sqrt{16\pi^2-k^2}}{4\pi}$ **75.** $t=C\sqrt[3]{Lm^2i}/mi$ **77.** 2.45 **79.** 1.709

Section 10.5 (page 329)

1. $2y\sqrt{3}$ **3.** $ab\sqrt{6}$ **5.** $2x\sqrt[3]{y^2}$ **7.** $2mc\sqrt[3]{2c}$ **9.** $2st\sqrt{5}$ **11.** $3\pi\omega\sqrt[4]{\pi^3\omega}$ **13.** $\sqrt[8]{243}$ **15.** $\sqrt[4]{x^2y}$ **17.** $\sqrt[10]{c^5d^2}$

19. $\sqrt[12]{\pi^3e^2}$ **21.** -4 **23.** 5 **25.** $5+2\sqrt{6}$ **27.** $17+4\sqrt{13}$ **29.** -68 **31.** 44 **33.** x^2-y **35.** $1-k$

37. $4+4\sqrt{q}+q$ **39.** $2a+\sqrt{ab}-6b$ **41.** $5\sqrt{35}-74$ **43.** $a\sqrt{a}+b\sqrt{b}$ **45.** $a\sqrt{a}-2b\sqrt{a}+b\sqrt{b}$

Section 10.6 (page 332)

1. $\dfrac{\sqrt{2ab}}{b}$ **3.** $\dfrac{2\sqrt{cd}}{d}$ **5.** $\dfrac{\sqrt{6y}}{3}$ **7.** $\dfrac{\sqrt[3]{6v_2v_3}}{3v_3}$ **9.** $\dfrac{\sqrt[4]{6\pi r}}{2}$ **11.** $\sqrt[6]{a}$ **13.** $-1-\sqrt{2}$ **15.** $4(\sqrt{5}-2)$ **17.** $\sqrt{2}$ **19.** $\dfrac{1}{7}(5+4\sqrt{2})$

21. $5-2\sqrt{6}$ **23.** $\sqrt{2}$ **25.** $3\sqrt{3}+2\sqrt{6}$ **27.** $\dfrac{2(1+\sqrt{a})}{1-a}$ **29.** $\dfrac{3(\sqrt{b}+2)}{b-4}$ **31.** $\dfrac{a+2\sqrt{ab}+b}{a-b}$ **33.** $\dfrac{\sqrt{a}+\sqrt{a-b}}{b}$

35. $\dfrac{a^{2/3}+a^{1/3}b^{1/3}+b^{2/3}}{a-b}$ **39.** $\dfrac{\mu_1-2\sqrt{\mu_1\mu_2}+\mu_2}{\mu_1-\mu_2}$ **41.** $\dfrac{a^{1/3}(a^{2/3}-a^{1/3}b^{1/3}+b^{2/3})}{a+b}$

Review Exercises for Chapter 10 (page 333)

1. $\dfrac{1}{25}$ **2.** 9   **3.** 64   **4.** $\dfrac{1}{4}$ **5.** 0.2   **6.** $-\dfrac{1}{3}$ **7.** $\dfrac{2V^3}{9W^{10}}$ **8.** $\dfrac{r^3}{16\pi^2}$ **9.** $\dfrac{C_2}{2C_1}$ **10.** $27p^{18}q^9c^{12}$ **11.** $\dfrac{w^8}{256v^{12}z^4}$ **12.** $\dfrac{16}{9f^2g^{12}}$

13. $\dfrac{r^{18}}{9R^{12}}$ **14.** $\dfrac{v_1v_3^{\,2}}{98v_2^{\,4}}$ **15.** x^a **16.** $1+VI$ **17.** $\dfrac{C}{C-1}$ **18.** $(a+b)^2$ **19.** $\dfrac{2v}{w^{1/2}}$ **20.** $\dfrac{8w^6}{v^{3/2}}$ **21.** $10x^{3/4}$ **22.** $\dfrac{A^{5/12}}{2B^{7/15}}$

23. $\dfrac{3^{1/3}F_2^{\,1/12}}{F_1^{\,5/7}}$ **24.** $\dfrac{c^4}{4b^2}$ **25.** $\dfrac{2n^{1/6}}{m^{17/8}}$ **26.** $\dfrac{2x-3}{(x-3)^{1/2}}$ **27.** $\dfrac{2x}{(x-1)^{2/3}}$ **28.** $\dfrac{3x+4}{(x+2)^{3/4}}$ **29.** $\dfrac{9-3x}{(x-2)^{4/5}}$ **30.** x^{a-1} **31.** $\sqrt[6]{x}$

32. $\sqrt[15]{b}$ **33.** $\sqrt{4a^2b^3c^5}$ **34.** $\sqrt[4]{4x^2y^3z^6}$ **35.** $2Rr^3\sqrt{3Rr}$ **36.** $2\pi^3q^2\sqrt[3]{3\pi q^2}$ **37.** $2ab^2\sqrt[4]{ab^3}$ **38.** $2xy^2z^3\sqrt[4]{2x^2z}$ **39.** 0

40. $4\sqrt{5}$ **41.** $8\sqrt{5}-5\sqrt[3]{3}$ **42.** $(6xy-x^2y^2+4y)\sqrt{xy}$ **43.** $\dfrac{2\sqrt{5}}{5ab^2}$ **44.** $\dfrac{r^2\sqrt{R}}{3R^2}$ **45.** $\dfrac{\sqrt{6st}}{3st}$ **46.** $\dfrac{a\sqrt{15ab}}{5}$ **47.** $\dfrac{\sqrt[3]{9s^2}}{3Rs}$

48. $\dfrac{\sqrt[3]{2x^2y^2}}{y}$ **49.** $\dfrac{\sqrt[3]{3a^2b}}{3a^2b}$ **50.** $\dfrac{\sqrt[4]{6u^3v^3}}{2uv}$ **51.** $\dfrac{\sqrt[3]{3\pi^2\omega^3}}{3\pi\omega}$ **52.** $\dfrac{\sqrt[3]{27ab}}{3ab}$ **53.** $3x\sqrt[3]{y^2}$ **54.** $2R\sqrt[5]{V^3}$ **55.** $\sqrt[12]{a^7}$ **56.** $\sqrt[12]{a^3b^4}$

57. 4   **58.** $8-2\sqrt{15}$ **59.** $1-9x$ **60.** $3a-\sqrt{ab}-10b$ **61.** $\dfrac{4}{3}(3+\sqrt{6})$ **62.** $3-\sqrt{5}$ **63.** $\dfrac{1}{3}(4\sqrt{2}+\sqrt{5})$ **64.** $\dfrac{\sqrt{R}-\sqrt{C}}{R-C}$

65. $\sqrt{\pi}-\sqrt{\pi-1}$ **66.** $-\dfrac{1}{2}(\sqrt{x-2}+\sqrt{x})$ **67.** \$6,805.83   **68.** $r=\dfrac{\sqrt[3]{6\pi^2V}}{2\pi}$ **69.** $\dfrac{2\pi}{5gL}\sqrt{5gL(5L^2+2r^2)}$

Chapter 11

Section 11.1 (page 340)

1. $3j$ **3.** $2\sqrt{2}j$ **5.** $2\sqrt{5}j$ **7.** $5\sqrt{3}j$ **9.**   **11.**   **13.** $5-j$

15. $2+3j$ **17.** $5+6j$ **19.** $-4+6j$

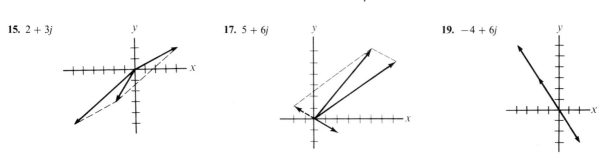

21. $7j$

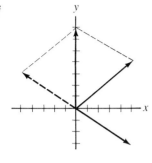

23. 10

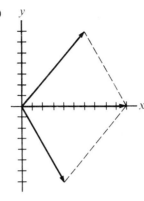

25. $-7 - j$

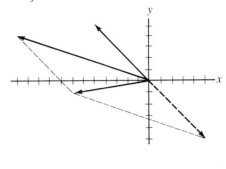

27. $-2 - 14j$

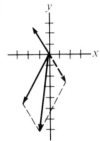

29. $x = 2, y = 3$ **31.** $x = 2, y = -3$ **33.** $x = 3, y = -1$

35. $x = 6, y = 1$ **37.** $x = \dfrac{5}{2}, y = \dfrac{1}{2}$

Section 11.2 (page 343)

1. $9 + 10j$ **3.** $3\sqrt{5} - 4j$ **5.** $6 - 6j$ **7.** $4 - 6j$ **9.** $-9 + 13j$ **11.** $-12 + j$ **13.** $7 + j$ **15.** $4 - 7j$ **17.** 25

19. $-11 - 17j$ **21.** $2j$ **23.** $-1 - 4\sqrt{3}j$ **25.** $1 - 2\sqrt{2}j$ **27.** $38 + 6j$ **29.** $\dfrac{1}{10} + \dfrac{3}{10}j$ **31.** $\dfrac{4}{5} - \dfrac{2}{5}j$ **33.** $-\dfrac{3}{5} - \dfrac{4}{5}j$

35. $-\dfrac{7}{25} - \dfrac{24}{25}j$ **37.** $-\dfrac{10}{17} + \dfrac{11}{17}j$ **39.** $-\dfrac{1}{68} - \dfrac{21}{68}j$ **41.** $\dfrac{7}{5} - \dfrac{1}{5}j$ **43.** $-2 - \dfrac{7}{2}j$ **45.** $\dfrac{7}{10} + \dfrac{1}{10}j$ **47.** $\dfrac{62}{85} - \dfrac{41}{85}j$ **49.** $-j$

51. j **53.** -1 **55.** $-j$ **57.** -1 **59.** 1

Section 11.3 (page 348)

1. $1\underline{/0^\circ}$ **3.** $4\underline{/180^\circ}$ **5.** $2\underline{/90^\circ}$ **7.** $5\underline{/270^\circ}$ **9.** $\sqrt{2}\underline{/45^\circ}$ **11.** $2\underline{/300^\circ}$ **13.** $3\sqrt{2}\underline{/225^\circ}$ **15.** $6\underline{/150^\circ}$ **17.** $2e^{(\pi/2)j}$

19. $\sqrt{2}e^{(7\pi/4)j}$ **21.** $2e^{(11\pi/6)j}$ **23.** $2\sqrt{2}e^{(3\pi/4)j}$ **25.** $\sqrt{2} + \sqrt{2}j$ **27.** $-1 + \sqrt{3}j$ **29.** $-\dfrac{3}{2} - \dfrac{3\sqrt{3}}{2}j$ **31.** $-\dfrac{5\sqrt{3}}{2} + \dfrac{5}{2}j$ **33.** 3

35. $5j$ **37.** $\sqrt{29}\underline{/68.2^\circ}$ **39.** $2\sqrt{5}\underline{/243.4^\circ}$ **41.** $3\sqrt{2}\underline{/331.9^\circ}$ **43.** $160 + 1.21j$ **45.** $-0.271 - 0.238j$ **47.** $0.508 - 6.32j$

49. $2.73 + 1.31j$ **51.** $9.7\underline{/325^\circ}$

Section 11.4 (page 350)

1. $6\underline{/93^\circ}$ **3.** $18\underline{/350^\circ}$ **5.** $30\underline{/170^\circ}$ **7.** $54\underline{/185^\circ}$ **9.** $4\underline{/60^\circ}$ **11.** $3\underline{/260^\circ}$ **13.** $\dfrac{3}{2}\underline{/211^\circ}$ **15.** $\dfrac{5}{3}\underline{/153^\circ}$ **17.** $2\underline{/226^\circ}$

19. $\dfrac{1}{14}\underline{/181^\circ}$

Section 11.5 (page 355)

1. $-4 - 4j$ **3.** 4,096   **5.** $-16 - 16\sqrt{3}j$ **7.** -64 **9.** $-128 + 128\sqrt{3}j$ **11.** 1   **13.** $-7 - 24j$ **15.** $-1121 + 404j$

17. $3^{10}/\underline{33°}$ **19.** $3^{15}/\underline{261°}$ **21.** $1, -\dfrac{1}{2}+\dfrac{\sqrt{3}}{2}j, -\dfrac{1}{2}-\dfrac{\sqrt{3}}{2}j$ **23.** $-\dfrac{\sqrt{2}}{2}+\dfrac{\sqrt{2}}{2}j, \dfrac{\sqrt{2}}{2}-\dfrac{\sqrt{2}}{2}j$

25. $\dfrac{3\sqrt{2}}{2}+\dfrac{3\sqrt{2}}{2}j, -\dfrac{3\sqrt{2}}{2}+\dfrac{3\sqrt{2}}{2}j, -\dfrac{3\sqrt{2}}{2}-\dfrac{3\sqrt{2}}{2}j, \dfrac{3\sqrt{2}}{2}-\dfrac{3\sqrt{2}}{2}j$ **27.** $2\sqrt{2}+2\sqrt{2}j, -2\sqrt{2}-2\sqrt{2}j$

29. $2/\underline{18°+k\cdot 72°}, k=0,1,2,3,4; 1.90+0.618j, 2j, -1.90+0.618j, -1.18-1.62j, 1.18-1.62j$

31. $2^{1/6}/\underline{10°+k\cdot 60°}, k=0,1,\ldots,5; 1.11+0.195j, 0.384+1.05j, -0.722+0.860j, -1.11-0.195j, -0.384-1.05j, 0.722-0.860j$

33. $2/\underline{54°+k\cdot 72°}, k=0,1,\ldots,4$ **35.** $2^{1/16}/\underline{\dfrac{1}{8}(225°+k\cdot 360°)}, k=0,1,\ldots,7$ **37.** $2^{1/3}/\underline{35°+k\cdot 60°}, k=0,1,\ldots,5$

39. $6^{1/8}/\underline{37.5°+k\cdot 45°}, k=0,1,\ldots,7$ **41.** $3^{1/9}/\underline{30°+k\cdot 40°}, k=0,1,\ldots,8$ **43.** $2^{1/10}/\underline{24°+k\cdot 36°}, k=0,1,\ldots,9$

Section 11.6 (page 357)

1. $Z=3.0+4.0j, |Z|=5.0\ \Omega, \theta=53.1°$ **3.** $Z=12.0+9.0j, |Z|=15.0\ \Omega, \theta=36.9°$ **5.** $Z=20.0-17.4j, |Z|=26.5\ \Omega, \theta=-41.0°$
7. 29.5 V   **9.** 55.3 V

Review Exercises for Chapter 11 (page 358)

1. $3+j$ 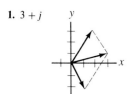   **2.** $-3+5j$ 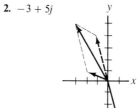   **3.** $1-3j$ **4.** $3+j$ **5.** $x=4, y=2$ **6.** $x=-1, y=-1$

7. $x=3, y=-1$ **8.** $x=6, y=2$ **9.** $3\sqrt{2}$ **10.** $2\sqrt{5}+3j$ **11.** 25   **12.** 50   **13.** $-8-6j$ **14.** $26+13j$

15. $\dfrac{12}{25}+\dfrac{9}{25}j$ **16.** $\dfrac{8}{17}+\dfrac{2}{17}j$ **17.** $-j$ **18.** $-\dfrac{4}{5}-\dfrac{7}{5}j$ **19.** $2/\underline{0°}$ **20.** $2/\underline{90°}$ **21.** $3/\underline{180°}$ **22.** $4/\underline{270°}$ **23.** $2\sqrt{2}/\underline{45°}$

24. $2\sqrt{2}/\underline{315°}$ **25.** $2/\underline{120°}$ **26.** $4/\underline{240°}$ **27.** $3e^{(\pi/2)j}$ **28.** $2e^{\pi j}$ **29.** $3\sqrt{2}e^{(3\pi/4)j}$ **30.** $4e^{(5\pi/6)j}$ **31.** $6e^{(5\pi/3)j}$ **32.** $2e^{(7\pi/6)j}$

33. $-\sqrt{2}+\sqrt{2}j$ **34.** $-1-\sqrt{3}j$ **35.** $-2\sqrt{2}-2\sqrt{2}j$ **36.** $3\sqrt{3}-3j$ **37.** $-\dfrac{3\sqrt{3}}{2}-\dfrac{3}{2}j$ **38.** $2-2\sqrt{3}j$ **39.** $-2.30+2.22j$

40. $-0.449-1.24j$ **41.** $5.46+2.56j$ **42.** $3.63-3.38j$ **43.** $\sqrt{34}/\underline{301.0°}$ **44.** $\sqrt{53}/\underline{254.1°}$ **45.** $\sqrt{10}/\underline{161.6°}$ **46.** $5/\underline{323.1°}$

47. $6/\underline{142°}$ **48.** $12/\underline{183°}$ **49.** $2\sqrt{5}/\underline{12°}$ **50.** $\dfrac{7}{2}/\underline{138°}$ **51.** $7/\underline{236°}$ **52.** $3/\underline{196°}$ **53.** $\dfrac{1}{2}/\underline{211°}$ **54.** $6/\underline{205°}$ **55.** $8-8j$

56. 4,096   **57.** -64 **58.** $-64+64\sqrt{3}j$ **59.** $-38+41j$ **60.** $597-122j$ **61.** $28+96j$ **62.** $28-96j$

63. $\sqrt{3}+j, -\sqrt{3}+j, -2j$ **64.** $\dfrac{\sqrt{2}}{2}+\dfrac{\sqrt{2}}{2}j, -\dfrac{\sqrt{2}}{2}+\dfrac{\sqrt{2}}{2}j, -\dfrac{\sqrt{2}}{2}-\dfrac{\sqrt{2}}{2}j, \dfrac{\sqrt{2}}{2}-\dfrac{\sqrt{2}}{2}j$ **65.** $1/\underline{k\cdot 72°}, k=0,1,2,3,4$

66. $2^{1/4}/\underline{22.5°+k\cdot 180°}, k=0,1$ **67.** $2^{1/6}/\underline{5°+k\cdot 60°}, k=0,1,\ldots,5$

Chapter 12

Section 12.1 (page 361)

1. $\log_3 81=4$ **3.** $\log_{10} 1{,}000=3$ **5.** $\log_4 256=4$ **7.** $\log_2 \dfrac{1}{16}=-4$ **9.** $\log_3 1=0$ **11.** $\log_{3/4}\dfrac{9}{16}=2$ **13.** $\log_{3/2}\dfrac{8}{27}=-3$

15. $\log_3 \dfrac{1}{81}=-4$ **17.** $5^3=125$ **19.** $5^0=1$ **21.** $2^1=2$ **23.** $3^{-3}=\dfrac{1}{27}$ **25.** $\left(\dfrac{1}{2}\right)^2=\dfrac{1}{4}$ **27.** $10^{-3}=\dfrac{1}{1{,}000}$

29. $6^{-2} = \dfrac{1}{36}$ **31.** $25^{-1} = \dfrac{1}{25}$ **33.** $x = 81$ **35.** $a = 3$ **37.** $x = 10{,}000$ **39.** $b = 2$ **41.** $x = 5$ **43.** $x = \dfrac{1}{64}$

45. $x = \dfrac{16}{9}$ **47.** $b = \dfrac{2}{3}$ **49.** $a = -\dfrac{1}{2}$

Section 12.2 (page 365)

1.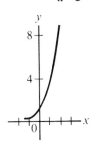

3.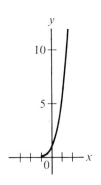

5.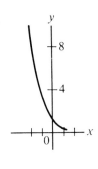

7.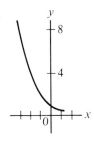

9.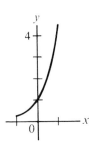

11.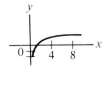

13.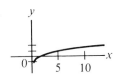

15.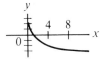

17.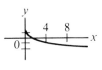

19.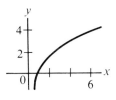

21.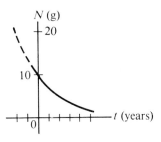

23.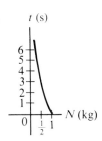

25. $T = 10^{0.5\theta}$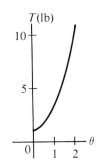

Section 12.3 (page 369)

1. $4 \log_{10} 3$ **3.** 3 **5.** $2 \log_7 6$ **7.** 2 **9.** 4 **11.** $\dfrac{3}{2}$ **13.** $2 + 2 \log_5 x$ **15.** $1 + \log_3 5 + 2 \log_3 x$ **17.** $1 + \log_4 5 + \dfrac{1}{2} \log_4 a$

19. $-2 - 2 \log_3 a$ **21.** $-\frac{1}{2}(1 + \log_3 C)$ **23.** $-\left(1 + \log_5 6 + \frac{1}{2} \log_5 u\right)$ **25.** -2 **27.** $-\frac{1}{3}$ **29.** $2 \log_e 5 - 3$

31. $-\frac{1}{2}(\log_e \pi + 1)$ **33.** $\log_e 2 + \frac{1}{3}$ **35.** $-2(\log_{10} 2 + 1)$ **37.** $2 \log_{10} 3 - 3$ **39.** $\log_{10} x - \log_{10} 3 - 2$ **41.** $\log_2 15$

43. $\log_4 10$ **45.** $\log_2 \frac{35}{3}$ **47.** $\log_2 \frac{25}{3}$ **49.** $\log_{10} 4$ **51.** $\log_2 \frac{\sqrt{x}}{3}$ **53.** $\log_e \sqrt{mv}$ **55.** $\log_7 \frac{\sqrt{st}}{2}$ **57.** $p = p_0 \, 10^{-0.0149k/T}$

59. $(5.97 \times 10^7)p = 10^{-1706.4/T}$

Section 12.4 (page 373)

1. 0.7474   **3.** 4.9212   **5.** $7.7193 - 10$ **7.** 3.3941   **9.** 5.9310   **11.** 1.2601   **13.** -2.1927 **15.** 4.4721   **17.** 0.4542
19. 1.8042   **21.** 73.1   **23.** 5,370   **25.** 0.856   **27.** 4,438   **29.** 0.008528   **31.** 7.8741   **33.** 42.9141   **35.** 110.6369
37. 0.0181   **39.** 0.1434   **41.** 7   **43.** 0.380 W

Section 12.5 (page 375)

1. 81.4   **3.** 3.192   **5.** 20.52   **7.** 0.211   **9.** 2.721   **11.** 219.0   **13.** 37.56   **15.** 61.73

Section 12.6 (page 379)

1. 1.9906   **3.** -0.6591 **5.** -1.7510 **7.** 1.3432   **9.** 0.8399   **11.** -3.5659 **13.** 2.7026   **15.** 20.0855   **17.** 0.1353
19. 0.3985   **21.** 1.0131   **23.** 0.9704   **25.** 16 years   **27.** \$5,422.65   **29.** 0.16 atmosphere   **31.** 214 ft   **33.** 17.8 W

Section 12.7 (page 383)

1. 2.3219   **3.** 2.1610   **5.** -0.6309 **7.** 1.4307   **9.** 1.3123   **11.** -1.1338 **13.** 9   **15.** 1.5647   **17.** 4   **19.** $\frac{e^2}{4}$

21. 2   **23.** no solution   **25.** $t = -RC \ln \frac{Ri}{E}$ **27.** $V_2 = V_1 \left[10^{Q/(P_1 V_1)} \right]$ **29. a.** 3.25   **b.** 5   **31.** 10 decibels   **33.** 0.0032 W/m^2

35. $10^{5.25} \approx 180{,}000$ times   **37. a.** 11,000   **b.** 4.3 h   **39.** 11,200 years   **41.** 6.5 million years   **43.** 1,750 m

Section 12.8 (page 389)

1.

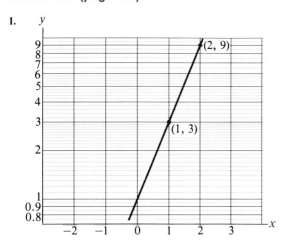

3.

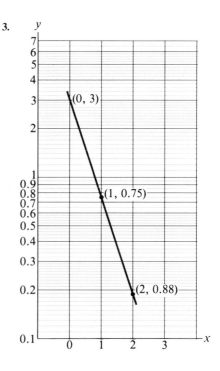

5.

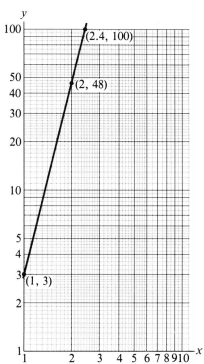

7.

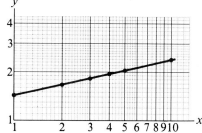

9.

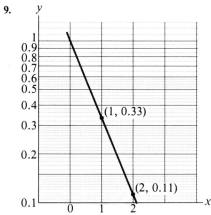

11.

13.

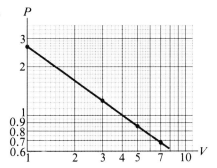

Review Exercises for Chapter 12 (page 389)

1. $\log_3 \dfrac{1}{27} = -3$ **2.** $\log_{1/2} \dfrac{1}{4} = 2$ **3.** $\log_7 \dfrac{1}{49} = -2$ **4.** $\log_a b = 3$ **5.** $4^1 = 4$ **6.** $3^{-2} = \dfrac{1}{9}$ **7.** $\left(\dfrac{1}{4}\right)^2 = \dfrac{1}{16}$ **8.** $\left(\dfrac{1}{2}\right)^{-1} = 2$

9. $x = \dfrac{1}{25}$ **10.** $x = 8$ **11.** $a = -4$ **12.** $b = 6$ **13.** $x = \dfrac{27}{8}$ **14.** $b = \dfrac{1}{3}$ **15.** $b = 2$ **16.** $a = -\dfrac{1}{3}$

17. 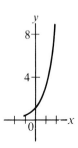   **18.** 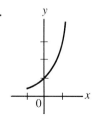   **19.**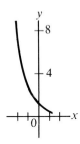

20. 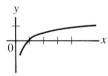   **21.** 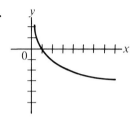   **22.**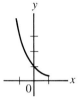

23. 5   **24.** $1 + \log_5 7$ **25.** 3   **26.** 4   **27.** $3 + 3\log_5 x$ **28.** 3   **29.** $\dfrac{1}{2}(1 + \log_6 x)$ **30.** $-1 - \log_3 b$ **31.** $-4 - 2\log_2 a$

32. $-\dfrac{1}{2}(1 + \log_{10} a)$ **33.** $4\ln 2 - 2$ **34.** $\dfrac{1}{2}(\ln \pi + 1)$ **35.** $-\dfrac{1}{3}$ **36.** $2\log_{10} 2 - 3$ **37.** $\log_5 2$ **38.** $\ln a\sqrt{b}$ **39.** $\log_6 \dfrac{\sqrt{x}}{b^2}$

40. $\ln 3$ **41.** 0.4116   **42.** 3.4116   **43.** 3.9610   **44.** 1.8650   **45.** $7.7608 - 10 = -2.2392$
46. $6.6548 - 10 = -3.3452$ (table), -3.3451 (calculator)   **47.** 448.4   **48.** 6,798   **49.** 0.3409 by interpolation   **50.** 0.002745
51. 0.4216   **52.** 0.9708   **53.** 3.6222   **54.** 1.5731   **55.** 2.3740   **56.** 2.0832   **57.** 0.3012   **58.** 0.1003   **59.** 14.8797
60. 1.4405   **61.** 0.2019   **62.** 0.7103   **63.** 0.8488   **64.** 5.1538   **65.** -0.1631 **66.** -3.2203 **67.** 2.5850   **68.** 0.3691

69. -0.8340 **70.** 3   **71.** $x = \dfrac{1}{2}(-1 + \sqrt{13})$ **73.** 100 times   **74.** 3.69   **75.** 31°C   **76.**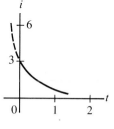

77. $t = -\dfrac{1}{2}\ln \dfrac{i}{3.0}$ **78.** \$2,406.53   **79.** 7.0 years   **80.** 2.0 min   **81.** 2.9 h   **82.** 9,690 years   **83.** 7.1×10^{-4} mol/L

84. 2,500 m

Cumulative Review Exercises for Chapters 10–12 (page 392)

1. $\dfrac{x^4}{y^2}$ **2.** $\dfrac{g_2{}^5}{18g_1{}^6 g_3{}^2}$ **3.** $\dfrac{b - a}{b}$ **4.** $\dfrac{R^3}{4L^4}$ **5.** $\dfrac{3y^{4/15}}{x^{5/6}}$ **6.** $-\dfrac{5}{(a + 3)^{1/2}}$ **7.** $\dfrac{\sqrt[4]{a^3 b}}{ab}$ **8.** $\dfrac{\sqrt[3]{2G^2 H}}{2G^2 H^2}$ **9.** $\dfrac{1}{3}(2 + \sqrt{10} - \sqrt{5} - \sqrt{2})$

10. $2 + 2j$ **11.** $1 + 5j$ **12.** $\frac{1}{13}(-1 - 5j)$ **13. a.** $4\underline{/210°}$;   **b.** $4e^{(7\pi/6)j}$ **14.** $\frac{1}{2}\underline{/220°}$ **15.** $2^{1/4}\underline{/30° + k \cdot 90°}$, $k = 0, 1, 2, 3$

16. $\log_6 \dfrac{Q\sqrt{P}}{R^2}$ **17.** $x = 1.771$ **18.** $t = \dfrac{1}{k}\ln\dfrac{S_2}{S_1 - S}$ **19.** $p = \dfrac{fq}{q - f}$ **20.** \$1,838.84   **21.** $N = 10^{-0.4t}$ **22.** 203 days

Chapter 13

Section 13.1 (page 398)

1. $(2.0, 2.0), (-2.0, -2.0)$ **3.** $(0.7, 1.0)$ **5.** $(2.8, 2.4), (-1.4, 0.3)$ **7.** $(3.3, 1.8)$

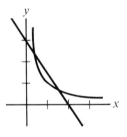

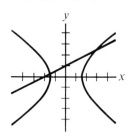

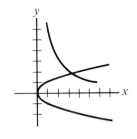

9. $(1.6, 0.6), (0.4, 2.4)$ **11.** $(2.6, 1.5), (2.6, -1.5), (-2.6, 1.5), (-2.6, -1.5)$

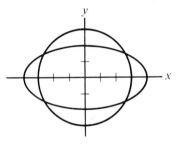

13. $(2.9, -4.3), (-2.9, -4.3)$ **15.** $(0.4, 0.7)$ **17.** $(1.5, 0.4)$

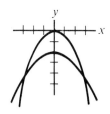

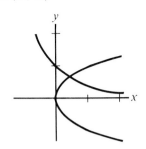

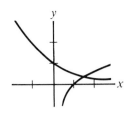

Section 13.2 (page 402)

1. $(2, 4), (-1, 1)$ **3.** $(2, 5), (-1, 2)$ **5.** $(-2, 0), (1, 3)$ **7.** $(0, 0), (2, -2)$ **9.** $(2, 1), (-1, -2)$ **11.** $(-2, 0), \left(\dfrac{5}{2}, \dfrac{9}{4}\right)$

13. $(0, 2), (-2, 0)$ **15.** $\left(1 - \dfrac{\sqrt{6}}{3}j, 1 + \dfrac{\sqrt{6}}{6}j\right), \left(1 + \dfrac{\sqrt{6}}{3}j, 1 - \dfrac{\sqrt{6}}{6}j\right)$ **17.** $(-1, -1)$ **19.** $\left(\dfrac{11}{4}, -\dfrac{9}{8}\right), (1, -2)$

21. $(1, 2), (1, -2), (-1, 2), (-1, -2)$ **23.** $\left(\sqrt{2}, \dfrac{2\sqrt{3}}{3}j\right), \left(\sqrt{2}, -\dfrac{2\sqrt{3}}{3}j\right), \left(-\sqrt{2}, \dfrac{2\sqrt{3}}{3}j\right), \left(-\sqrt{2}, -\dfrac{2\sqrt{3}}{3}j\right)$

25. $(\sqrt{5}, \sqrt{11}), (\sqrt{5}, -\sqrt{11}), (-\sqrt{5}, \sqrt{11}), (-\sqrt{5}, -\sqrt{11})$ **27.** $(\sqrt{2}, \sqrt{14}), (\sqrt{2}, -\sqrt{14}), (-\sqrt{2}, \sqrt{14}), (-\sqrt{2}, -\sqrt{14})$ **29.** 2.5 cm × 10 cm
31. 3.8 s **33.** 6.47 Ω, 2.53 Ω **35.** 1 m/s, 2 m/s

Section 13.3 (page 407)

1. $\pm 1, \pm 2$ **3.** $\pm 2, \pm j$ **5.** $2, -1 \pm \sqrt{3}j, -1, \dfrac{1}{2} \pm \dfrac{1}{2}\sqrt{3}j$ **7.** $\dfrac{1}{3}, -\dfrac{1}{2}$ **9.** $\pm\dfrac{1}{2}, \pm\dfrac{\sqrt{6}}{6}j$ **11.** $\pm 2, \pm\dfrac{\sqrt{2}}{2}j$ **13.** 1, 25 **15.** $\dfrac{16}{9}$

17. 5, 26 **19.** $8, -27$ **21.** $2, 2, 2 \pm \sqrt{6}$ **23.** $\pm\dfrac{\sqrt{3}}{2}, \pm\dfrac{\sqrt{42}}{6}j$ **25.** 1, 16 **27.** $6\sqrt{10}$ in. × $2\sqrt{10}$ in. **29.** 0°, 90°, 180°

31. 0°, 120°, 240°

Section 13.4 (page 411)

1. 8 **3.** no solution **5.** 0 **7.** 2 **9.** 9 **11.** 3 **13.** 6 **15.** 1 **17.** no solution **19.** 6 **21.** 5 **23.** 5, 13

25. 2, 6 **27.** -3 **29.** $v = \left(\dfrac{L^2 + 1}{2L}\right)^2$ **31.** 10 in., 24 in. **33.** $x = 3$ cm, $x = \dfrac{153}{26}$ cm **35.** $C = \dfrac{Z}{2\pi f \sqrt{1 - RZ^2}}$

Review Exercises for Chapter 13 (page 412)

1. (1.5, 0.5) **2.** (0.4, 0.9) **3.** $(1, 3), \left(-\dfrac{1}{2}, \dfrac{3}{2}\right)$

4. $\left(\dfrac{1 + \sqrt{5}}{2}, \dfrac{-1 + \sqrt{5}}{2}\right), \left(\dfrac{1 - \sqrt{5}}{2}, \dfrac{-1 - \sqrt{5}}{2}\right)$ **5.** $(2, 0), (1, -3)$ **6.** $(-2, -1), (-1, -2)$

7. $\left(\dfrac{\sqrt{2}}{2}j, \dfrac{\sqrt{6}}{2}\right), \left(\dfrac{\sqrt{2}}{2}j, -\dfrac{\sqrt{6}}{2}\right), \left(-\dfrac{\sqrt{2}}{2}j, \dfrac{\sqrt{6}}{2}\right), \left(-\dfrac{\sqrt{2}}{2}j, -\dfrac{\sqrt{6}}{2}\right)$ **8.** $\left(\dfrac{\sqrt{70}}{5}, \dfrac{\sqrt{15}}{5}j\right), \left(\dfrac{\sqrt{70}}{5}, -\dfrac{\sqrt{15}}{5}j\right), \left(-\dfrac{\sqrt{70}}{5}, \dfrac{\sqrt{15}}{5}j\right), \left(-\dfrac{\sqrt{70}}{5}, -\dfrac{\sqrt{15}}{5}j\right)$

9. $(\sqrt{2}, \sqrt{2}), (-\sqrt{2}, -\sqrt{2})$ **10.** $(2, 1), (-2, -1), (j, -2j), (-j, 2j)$ **11.** $\pm 2, \pm 4$ **12.** $\pm\dfrac{3}{2}, \pm j$ **13.** $\pm\dfrac{\sqrt{3}}{2}, \pm 2j$

14. $\dfrac{2}{3}, -5$ **15.** $\dfrac{2}{5}, -\dfrac{1}{3}$ **16.** $\pm\dfrac{1}{2}, \pm\dfrac{1}{4}j$ **17.** $\pm 2, \pm\dfrac{1}{2}$ **18.** $\dfrac{9}{4}$ **19.** $\dfrac{16}{9}$ **20.** $\dfrac{4}{9}$ **21.** 3, 15 **22.** $8, -64$ **23.** $-1, -1, -1 \pm j$

24. no solution **25.** -2 **26.** $\dfrac{9}{4}$ **27.** -5 **28.** 25 **29.** 7 **30.** 3 **31.** 3.19 Ω, 6.81 Ω

32. $v_1 = 6$ cm/s, $v_2 = 10$ cm/s; or $v_1 = 9.56$ cm/s, $v_2 = 5.56$ cm/s **33.** 36 in. × 15 in. **34.** base = 16 in., height = 12 in.

35. $v = \dfrac{c}{m}\sqrt{m^2 - m_0^2}$

Chapter 14

Section 14.2 (page 416)

1. 3 **3.** -36 **5.** -10 **7.** 40 **9.** 3 **11.** 36 **13.** 40 **15.** 60 **17.** 12 **19.** 0 **21.** yes **23.** no **25.** yes
27. yes **29.** no

Section 14.3 (page 421)

1. $x^2 + 3x + 2$, $R = 3$ **3.** $x^2 - x - 4$, $R = -10$ **5.** $x^3 - x^2 + 2x - 7$, $R = 13$ **7.** $4x^2 - 18$, $R = 60$ **9.** $4x^4 + 8x - 5$, $R = 0$
11. $2x^3 + 3x + 3$, $R = -4$ **13.** $4x^3 + 4x^2 - 2x$, $R = 2$ **15.** $x^4 + 2x^3 + 4x^2 - 4$, $R = -2$ **17.** $9x^4 + 3x^3 + 3x^2 - 3x - 6$, $R = 0$
19. $x^3 + 2x^2 + 4x + 3$, $R = -4$ **21.** yes   **23.** no   **25.** yes   **27.** yes   **29.** yes   **31.** yes   **33.** yes   **35.** no   **37.** yes

39. 19   **41.** $\dfrac{5}{2}$ **43.** 430   **45.** -7

Section 14.4 (page 425)

1. degree 5, roots: -2 (multiplicity 2), 3 (multiplicity 2), -1 (single)

3. degree 7, roots: 0 (multiplicity 2), -4 (multiplicity 3), 5 (multiplicity 2)   **5.** degree 9, roots: $-\dfrac{1}{2}$ (multiplicity 4), 10 (multiplicity 5)

7. degree 9, roots: 0 (multiplicity 2), $\dfrac{3}{4}$ (multiplicity 3), -7 (multiplicity 4)   **9.** $1, -3$ **11.** $4, -1$ **13.** $\dfrac{1}{2}(-1 \pm \sqrt{13})$

15. $-1 \pm \dfrac{1}{2}\sqrt{2}j$ **17.** $2, 4$ **19.** $-\dfrac{1}{2} \pm \dfrac{1}{2}\sqrt{3}j$ **21.** $-j, 2, -1$ **23.** $2j, -3, 2$ **25.** $-j, -1 \pm \sqrt{3}j$ **27.** $1 - j, 1, 2$ **29.** $1, 2$

31. $\dfrac{1}{2}(1 \pm \sqrt{5})$

Section 14.5 (page 432)

1. $1, -1, 2$ **3.** $-3, -2, 2$ **5.** $-1, -3, 1, 2$ **7.** $-1, -3, -3, 2$ **9.** $3, 3, \dfrac{1}{2}(-3 \pm \sqrt{5})$ **11.** $4, 4, 1 \pm \sqrt{2}j$ **13.** $\dfrac{1}{3}, -\dfrac{1}{2} \pm \dfrac{1}{2}\sqrt{23}j$

15. $\dfrac{1}{2}, -1 \pm \sqrt{11}j$ **17.** $-\dfrac{1}{2}, -\dfrac{1}{4} \pm \dfrac{1}{4}\sqrt{15}j$ **19.** $\dfrac{1}{2}, \dfrac{1}{2}, -1, -2$ **21.** $-2, -\dfrac{1}{2}, -\dfrac{1}{2}, 1$ **23.** $1, 1, \dfrac{4}{3}, -\dfrac{3}{2}$ **25.** $-2, -1, \dfrac{1}{2}, \dfrac{5}{3}$

27. $\dfrac{3}{2}, \dfrac{3}{2}, 1, -2$ **29.** $\dfrac{1}{3}, \dfrac{1}{3}, 2, -4$ **31.** 3 in.   **33.** 4 ft × 4 ft × 2 ft   **35.** 3 μF, 6 μF, 12 μF

Section 14.6 (page 437)

1. 2.21   **3.** 1.10   **5.** 0.75   **7.** 3.27   **9.** -1.87 **11.** 0.90   **13.** 2.672   **15.** -0.414 **17.** 3.24   **19.** 1.44 in. or 1.90 in.

Review Exercises for Chapter 14 (page 438)

1. yes   **2.** yes   **3.** no   **4.** yes   **5.** $x^2 - x + 2$, $R = 3$ **6.** $x^3 + 3x^2 + 6x + 20$, $R = 50$
7. $2x^4 + 2x^3 - 10x^2 + 20x - 20$, $R = 20$ **8.** $3x^3 + x - 2$, $R = 2$ **9.** $4x^4 + 4x^3 + 2x^2 + 3$, $R = -1$
10. $3x^4 - x^3 + 3x^2 + x + 2$, $R = -2$ **11.** yes   **12.** no   **13.** no   **14.** yes   **15.** no   **16.** yes   **17.** -2 **18.** 2   **19.** 4

20. -6 **21.** $2, 2$ **22.** $-\dfrac{1}{2} \pm \dfrac{1}{2}\sqrt{7}j$ **23.** $1 \pm \sqrt{2}$ **24.** $j, -4$ **25.** $-j, 3, -5$ **26.** $1 - j, -3$ **27.** $1, -2, -6$

28. $4, \dfrac{1}{2}(1 \pm \sqrt{7}j)$ **29.** $-2, -2, 1, 3$ **30.** $3, 3, -1 \pm \sqrt{3}j$ **31.** $\dfrac{1}{2}, \dfrac{1}{2}, -3$ **32.** $2, -\dfrac{2}{3}, -\dfrac{2}{3}$ **33.** $\dfrac{1}{3}, \dfrac{1}{3}, \dfrac{1}{2}(1 \pm \sqrt{13})$

34. $-\dfrac{1}{2}, -\dfrac{1}{2}, 3, 4$ **35.** 0.59   **36.** 1.63   **37.** -0.62 **38.** 0.94   **39.** 2.196   **40.** 6.47, 0.75, -1.23

Chapter 15

Section 15.2 (page 441)

1. 12   **3.** 20   **5.** 5   **7.** 21   **9.** 189   **11.** 440   **13.** 366

Section 15.3 (page 448)

1. 0   **3.** -11 **5.** 85   **7.** -935 **9.** 14   **11.** -104 **13.** 394   **15.** -50 **17.** 450   **19.** 406   **21.** 124   **23.** -27

Section 15.4 (page 452)

1. $(1, 2, 3)$ **3.** $(1, 1, -2)$ **5.** $(1, -1, 2)$ **7.** $\left(2, -\dfrac{3}{2}, \dfrac{7}{2}\right)$ **9.** $\left(\dfrac{2}{3}, -\dfrac{5}{3}, 4\right)$ **11.** $\left(6, \dfrac{20}{7}, -\dfrac{25}{7}\right)$ **13.** $(1, 2, -2, 3)$

15. $\left(-\dfrac{1}{2}, 2, 3, -1\right)$ **17.** $\left(\dfrac{3}{2}, \dfrac{1}{3}, -\dfrac{7}{6}, \dfrac{2}{3}\right)$ **19.** $(7, -2, -4, -6, 2)$ **21.** $I_1 = -\dfrac{3}{171}A = -\dfrac{1}{57}A, I_2 = \dfrac{39}{171}A, I_3 = \dfrac{109}{171}A, I_4 = \dfrac{145}{171}A$

23. $\$25, \$15, \$10, \$5$

Section 15.5 (page 461)

1. $x = -3, y = 2$ **3.** $a = 2, b = -7, x = 6, y = 3$ **5.** $\begin{bmatrix} 3 & -9 & 3 \\ 9 & 3 & 6 \end{bmatrix}$ **7.** $\begin{bmatrix} -8 & -1 & -5 \\ -2 & -1 & 1 \end{bmatrix}$ **9.** $\begin{bmatrix} -3 & 9 & 6 \\ -6 & 0 & -18 \end{bmatrix}$

11. $\begin{bmatrix} -2 & -19 & 10 \\ 19 & 8 & 1 \end{bmatrix}$ **13.** $\begin{bmatrix} 5 \\ 4 \end{bmatrix}$ **15.** $\begin{bmatrix} -15 & 4 \end{bmatrix}$ **17.** $\begin{bmatrix} 62 & 80 \\ -5 & 0 \end{bmatrix}$ **19.** $\begin{bmatrix} -2 & -4 & -3 \\ -13 & -47 & -24 \\ 8 & 58 & 21 \end{bmatrix}$ **21.** $\begin{bmatrix} -94 & 13 \\ -60 & -22 \end{bmatrix}$

23. $\begin{bmatrix} 87 & 46 \\ 27 & 47 \end{bmatrix}$ **25.** $AB = \begin{bmatrix} -10 & 3 \\ 0 & 6 \end{bmatrix}, BA = \begin{bmatrix} 4 & 7 \\ 4 & -8 \end{bmatrix}$ **27.** $AB = \begin{bmatrix} -40 \\ -20 \end{bmatrix}, BA$ undefined

Section 15.6 (page 469)

1. $\dfrac{1}{2}\begin{bmatrix} -4 & 3 \\ 2 & -1 \end{bmatrix}$ **3.** $\dfrac{1}{8}\begin{bmatrix} -2 & 4 \\ 3 & -2 \end{bmatrix}$ **5.** $\begin{bmatrix} 1 & 0 & 0 \\ 0 & 1 & 0 \\ -1 & 0 & 1 \end{bmatrix}$ **7.** $\begin{bmatrix} 1 & -4 & 2 \\ \frac{1}{3} & 1 & -1 & -1 \\ 1 & 2 & -1 \end{bmatrix}$ **9.** $\dfrac{1}{6}\begin{bmatrix} -4 & 3 & -4 \\ 2 & 0 & -4 \\ 2 & 0 & 2 \end{bmatrix}$

11. $\begin{bmatrix} -3 & 5 & -1 \\ 2 & -3 & 1 \\ 2 & -3 & 0 \end{bmatrix}$ **13.** $\dfrac{1}{20}\begin{bmatrix} 5 & 0 & 5 \\ 15 & -4 & 3 \\ -10 & 8 & -6 \end{bmatrix}$ **15.** $\dfrac{1}{16}\begin{bmatrix} -5 & -4 & 7 \\ 6 & 8 & -2 \\ 3 & -4 & -1 \end{bmatrix}$ **17.** $\begin{bmatrix} -3 & -2 & 0 & 2 \\ 2 & 1 & 0 & -1 \\ -4 & -2 & 2 & 3 \\ 2 & 1 & -1 & -1 \end{bmatrix}$

19. $\dfrac{1}{24}\begin{bmatrix} 22 & -3 & -16 & 1 \\ -14 & 3 & 8 & 7 \\ 6 & 9 & 0 & -3 \\ -2 & -3 & 8 & 1 \end{bmatrix}$ **21.** $\dfrac{1}{6}\begin{bmatrix} 17 & -4 & -5 & 24 \\ -10 & 2 & 4 & -12 \\ -8 & 4 & 2 & -12 \\ -10 & 2 & 4 & -18 \end{bmatrix}$ **23.** $\dfrac{1}{4}\begin{bmatrix} 14 & -4 & 2 & -6 \\ 10 & -4 & 2 & -4 \\ -4 & 2 & 0 & 1 \\ -14 & 4 & -2 & 8 \end{bmatrix}$

Section 15.7 (page 475)

1. $X = \begin{bmatrix} -\frac{5}{2} \\ \frac{3}{2} \end{bmatrix}$ **3.** $X = \begin{bmatrix} -2 \\ 5 \\ 2 \end{bmatrix}$ **5.** $X = \begin{bmatrix} 2 \\ 3 \\ 2 \end{bmatrix}$ **7.** $X = \begin{bmatrix} 7 \\ -1 \\ -2 \end{bmatrix}$ **9.** $X = \begin{bmatrix} 3 \\ -1 \\ 0 \end{bmatrix}$ **11.** $X = \begin{bmatrix} 4 \\ -1 \\ -4 \end{bmatrix}$ **13.** $X = \begin{bmatrix} 1 \\ 2 \\ -2 \end{bmatrix}$

15. $X = \begin{bmatrix} -1 \\ 1 \\ 0 \end{bmatrix}$ **17.** $X = \begin{bmatrix} -9 \\ 5 \\ -9 \\ 3 \end{bmatrix}$ **19.** $X = \begin{bmatrix} 1 \\ -1 \\ 1 \\ 0 \end{bmatrix}$ **21.** $X = \begin{bmatrix} 2 \\ -1 \\ -1 \\ -1 \end{bmatrix}$ **23.** $X = \begin{bmatrix} \frac{5}{2} \\ 2 \\ -\frac{5}{4} \\ -1 \end{bmatrix}$

25. $A^{-1} = \begin{bmatrix} 4 & -2 & -1 \\ -3 & 2 & 1 \\ 2 & -1 & -1 \end{bmatrix}, X = \begin{bmatrix} -1 \\ 2 \\ -2 \end{bmatrix}$ **27.** $A^{-1} = \dfrac{1}{4}\begin{bmatrix} -4 & 1 & 2 \\ -4 & 2 & 0 \\ 8 & -1 & -2 \end{bmatrix}, X = \begin{bmatrix} \frac{3}{4} \\ \frac{1}{2} \\ -\frac{3}{4} \end{bmatrix}$

29.
$$A^{-1} = \frac{1}{3}\begin{bmatrix} -6 & 3 & 6 & 0 \\ -2 & -1 & 2 & 1 \\ -8 & 2 & 5 & 1 \\ 1 & -1 & -1 & 1 \end{bmatrix}, \quad X = \begin{bmatrix} 2 \\ 1 \\ 3 \\ 0 \end{bmatrix}$$

31.
$$A^{-1} = \frac{1}{5}\begin{bmatrix} -3 & 14 & -7 & -10 \\ -6 & 3 & 1 & 0 \\ 2 & 4 & -2 & -5 \\ 4 & -2 & 1 & 0 \end{bmatrix}, \quad X = \begin{bmatrix} -5 \\ 2 \\ -3 \\ -1 \end{bmatrix}$$

33. $\dfrac{13}{17}, \dfrac{9}{17}, -\dfrac{14}{17}, \dfrac{8}{17}$ (in amps)   **35.** $2\,\Omega, 2\,\Omega, 4\,\Omega$

Review Exercises for Chapter 15 (page 476)

1. -4 **2.** -22 **3.** 0 **4.** 490 **5.** -6 **6.** -24 **7.** $\left(\dfrac{1}{3}, -\dfrac{11}{15}, -\dfrac{1}{15}\right)$ **8.** $\left(\dfrac{23}{16}, \dfrac{1}{16}, -\dfrac{3}{16}\right)$ **9.** $(-3, 2, 0, 1)$

10. $\left(\dfrac{7}{3}, \dfrac{2}{3}, 1, -1\right)$ **11.** $x = 2, y = 1, z = -3$ **12.** $a = 4, b = 2, c = -1, d = 3$ **13.** $\begin{bmatrix} -1 & 4 \\ -2 & 3 \\ 3 & 7 \end{bmatrix}$ **14.** $\begin{bmatrix} 2 & -5 \\ 2 & -1 \\ 1 & -2 \end{bmatrix}$

15. $\begin{bmatrix} -2 & 3 \\ 2 & -8 \\ -5 & -5 \end{bmatrix}$ **16.** $\begin{bmatrix} 1 & -4 \\ -6 & 14 \\ -5 & -10 \end{bmatrix}$ **17.** $\begin{bmatrix} -9 \\ -20 \end{bmatrix}$ **18.** $\begin{bmatrix} -5 & 21 \end{bmatrix}$ **19.** $\begin{bmatrix} 1 & -20 & 13 \\ 0 & -48 & 30 \\ -4 & 24 & -17 \end{bmatrix}$ **20.** $\begin{bmatrix} 12 & -2 \\ 2 & -4 \end{bmatrix}$

21. $\dfrac{1}{4}\begin{bmatrix} 11 & -6 \\ -3 & 2 \end{bmatrix}$ **22.** $\dfrac{1}{27}\begin{bmatrix} 9 & -6 \\ -3 & 5 \end{bmatrix}$ **23.** $\dfrac{1}{3}\begin{bmatrix} -11 & -15 & 5 \\ 6 & 9 & -3 \\ -14 & -21 & 8 \end{bmatrix}$ **24.** $\begin{bmatrix} -2 & 4 & -1 \\ 1 & -2 & 1 \\ 4 & -7 & 2 \end{bmatrix}$ **25.** $\dfrac{1}{25}\begin{bmatrix} 10 & 5 & -15 \\ 1 & -2 & -14 \\ 2 & -4 & -3 \end{bmatrix}$

26. $\begin{bmatrix} 1 & 0 & 0 & 0 \\ 2 & 0 & -1 & 0 \\ 4 & -1 & -2 & -1 \\ 4 & -1 & -2 & 0 \end{bmatrix}$ **27.** $\dfrac{1}{3}\begin{bmatrix} 5 & -6 & 11 & -3 \\ 4 & -6 & 13 & -3 \\ -7 & 9 & -19 & 6 \\ -4 & 6 & -10 & 3 \end{bmatrix}$ **29.** $\begin{bmatrix} -\dfrac{2}{3} \\ 0 \\ \dfrac{1}{3} \end{bmatrix}$ **30.** $\begin{bmatrix} -4 \\ 4 \\ 9 \end{bmatrix}$ **31.** $\begin{bmatrix} \dfrac{7}{5} \\ \dfrac{11}{25} \\ -\dfrac{3}{25} \end{bmatrix}$ **32.** $\begin{bmatrix} 1 \\ 7 \\ 13 \\ 11 \end{bmatrix}$

33. $\begin{bmatrix} \dfrac{1}{3} \\ -\dfrac{1}{3} \\ -\dfrac{2}{3} \\ \dfrac{1}{3} \end{bmatrix}$ **34.** $2, 2, 3, 4$ (in pounds)   **35.** $3, -\dfrac{4}{3}, -\dfrac{7}{3}, 4$ (in amps)

Cumulative Review Exercises for Chapters 13–15 (page 478)

1. $(0.6, 0.4), (-1.6, 2.6)$

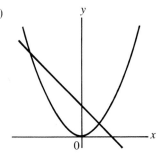

2. $(2, 4), (4, 2)$ **3.** $(\sqrt{14}, \sqrt{2}), (\sqrt{14}, -\sqrt{2}), (-\sqrt{14}, \sqrt{2}), (-\sqrt{14}, -\sqrt{2})$

4. $\pm\dfrac{1}{2}, \pm j$ **5.** $x = 9$ **6.** no   **7.** yes   **8.** 5 **9.** $1, 1, 2, 4$ **10.** $1, 1, \dfrac{4}{3}, -\dfrac{3}{2}$ **11.** 0.75 **12.** 394 **13.** $(1, 2, -2, 3)$

14. $\begin{bmatrix} 2 & -3 & -2 \\ -1 & 7 & 5 \\ -6 & 6 & 11 \end{bmatrix}$ **15.** $\begin{bmatrix} 0 & 8 \\ 10 & -17 \end{bmatrix}$ **16.** $\begin{bmatrix} -3 & 2 & 2 \\ 5 & -3 & -3 \\ -1 & 1 & 0 \end{bmatrix}$ **17.** $(-1, 1, 0)$ **18.** $3.10\ \Omega,\ 6.90\ \Omega$

Chapter 16

Section 16.1 (page 483)

1. $\dfrac{\cos \beta}{\sin \beta}$ **3.** $\sin \theta$ **5.** 1 **7.** $\dfrac{\cos x + 1}{\sin x}$ **9.** $\cos \theta$ **11.** $\dfrac{1}{\sin^2 s}$ **13.** $-\dfrac{\sin^2 \theta}{\cos^2 \theta}$ **15.** $\csc x$ **17.** $\csc^2 \theta$ **19.** $\csc \theta$ **21.** 1

23. $\csc^2 t$ **25.** $\cos \theta$ **27.** 1 **29.** $\sec^3 x$ **31.** $\cot \theta$

Section 16.2 (page 488)

41. $a\omega$ **43.** $y = \dfrac{2v_0^2 x \sin \alpha \cos \alpha - gx^2}{2v_0^2 \cos^2 \alpha}$

Section 16.3 (page 494)

1. $\dfrac{\sqrt{6}+\sqrt{2}}{4}$ **3.** $\dfrac{\sqrt{6}+\sqrt{2}}{4}$ **5.** $-\dfrac{\sqrt{6}+\sqrt{2}}{4}$ **7.** $\dfrac{\sqrt{2}-\sqrt{6}}{4}$ **9.** $\dfrac{\sqrt{3}}{2}$ **11.** $\dfrac{\sqrt{2}}{2}$ **13.** $\dfrac{\sqrt{2}}{2}$ **15.** $\sin 2x$ **17.** $\sin 3x$ **19.** $\cos 2x$

21. $\cos 9x$ **23.** $\sin x$ **25.** $\dfrac{1}{2}(\sqrt{3}\cos x - \sin x)$ **27.** $-\sin 2x$ **29.** $-\cos x$ **31.** $\sin 2x$ **33.** $\cos 2x$ **35.** $\dfrac{1}{2}(\sin x - \sqrt{3}\cos x)$

37. $\dfrac{1}{2}(\sqrt{3}\cos x - \sin x)$ **39.** $\dfrac{\sqrt{2}}{2}(\sin x + \cos x)$ **41.** $\dfrac{1 + \tan x}{1 - \tan x}$ **61.** $y = 2A\cos\left(\dfrac{2\pi t}{T}\right)\cos\left(\dfrac{2\pi x}{\lambda}\right)$

Section 16.4 (page 499)

1. $\dfrac{24}{25}$ **3.** $\dfrac{7}{25}$ **5.** $-\dfrac{120}{169}$ **7.** $-\dfrac{\sqrt{3}}{2}$ **9.** $\dfrac{17}{25}$ **11.** $-\dfrac{31}{49}$ **13.** $\cos 6y$ **15.** $\sin 6\theta$ **17.** $\cos 4\beta$ **19.** $-\cos 8y$ **21.** $\dfrac{1}{2}\sin 8\omega$

23. $2\sin 4x$ **37.** $v = 4\sin 2t$

Section 16.5 (page 504)

1. $\dfrac{\sqrt{2-\sqrt{3}}}{2} = 0.2588$ **3.** $\dfrac{\sqrt{2+\sqrt{2}}}{2} = 0.9239$ **5.** $\dfrac{\sqrt{2+\sqrt{2}}}{2} = 0.9239$ **7.** $\dfrac{7\sqrt{2}}{10}$ **9.** $\dfrac{2\sqrt{13}}{13}$ **11.** $-\dfrac{5\sqrt{26}}{26}$ **13.** $-\dfrac{\sqrt{26}}{26}$

15. $\sin 2\theta$ **17.** $\sqrt{2}\cos 3\theta$ **19.** $\sqrt{10}\sin 2\theta$ **21.** $2\cos 3\theta$ **23.** $\sqrt{14}\sin\dfrac{\theta}{2}$ **25.** $\dfrac{1}{2}(1 - \cos 8x)$ **27.** $\dfrac{1}{2}(1 + \cos 4x)$

29. $1 - \cos 6x$ **31.** $6(1 - \cos 2x)$ **37.** $\dfrac{\sqrt{2}}{2}\csc\dfrac{x}{2}$ **39.** $2\sin\dfrac{\theta}{2}$

Section 16.6 (page 508)

1. $\dfrac{\pi}{6}, \dfrac{5\pi}{6}$ **3.** $\dfrac{\pi}{4}, \dfrac{5\pi}{4}$ **5.** $0, \pi$ **7.** $0, \dfrac{\pi}{2}, \pi$ **9.** $\dfrac{\pi}{6}, \dfrac{5\pi}{6}, \dfrac{7\pi}{6}, \dfrac{11\pi}{6}$ **11.** $\dfrac{\pi}{4}, \pi, \dfrac{5\pi}{4}$ **13.** $\dfrac{\pi}{2}, \dfrac{7\pi}{6}, \dfrac{11\pi}{6}$ **15.** 0 **17.** $\dfrac{\pi}{4}, \dfrac{5\pi}{4}$ **19.** π

21. $\dfrac{\pi}{4}, \dfrac{3\pi}{4}, \dfrac{5\pi}{4}, \dfrac{7\pi}{4}$ **23.** $\dfrac{\pi}{4}, \dfrac{3\pi}{4}, \dfrac{5\pi}{4}, \dfrac{7\pi}{4}$ **25.** $\dfrac{\pi}{2}, \dfrac{3\pi}{2}$ **27.** $\dfrac{\pi}{3}, \pi, \dfrac{5\pi}{3}$ **29.** $\dfrac{\pi}{2}$ **31.** $0, \dfrac{\pi}{2}, \pi, \dfrac{3\pi}{2}$ **33.** $54.7°, 125.3°, 234.7°, 305.3°$

35. $90°, 194.5°, 270°, 345.5°$ **37.** $0°, 41.4°, 180°, 318.6°$ **39.** $54.7°, 125.3°, 234.7°, 305.3°$ **41.** $22.5°, 112.5°$ **43.** $t = \dfrac{\pi}{24} = 0.13$ s

45. $t = \dfrac{\pi}{6\omega}$ s

Section 16.7 (page 511)

1. $\dfrac{\pi}{3}, \dfrac{2\pi}{3}$ **3.** $\dfrac{\pi}{4}, \dfrac{5\pi}{4}$ **5.** $\dfrac{\pi}{2}$ **7.** $0, \pi$ **9.** $\dfrac{\pi}{3}, \dfrac{5\pi}{3}$ **11.** $\dfrac{\pi}{2}, \dfrac{3\pi}{2}$ **13.** $\dfrac{7\pi}{6}, \dfrac{11\pi}{6}$ **15.** $\dfrac{\pi}{3}, \dfrac{4\pi}{3}$ **17.** $x = \dfrac{1}{3}\arcsin y$

19. $x = \dfrac{1}{2}\arctan(1 - y)$ **21.** $x = \text{arccsc}(3y - 6)$ **23.** $x = 2 + \arccos\dfrac{y}{4}$ **25.** $x = \tan y$ **27.** $x = \sin(1 - y)$ **29.** $x = \csc y - 1$

Section 16.8 (page 516)

1. $\dfrac{\pi}{3}$ **3.** $\dfrac{\pi}{4}$ **5.** 0   **7.** $\dfrac{\pi}{2}$ **9.** 0   **11.** $-\dfrac{\pi}{4}$ **13.** $-\dfrac{\pi}{3}$ **15.** $\dfrac{\pi}{3}$ **17.** $\dfrac{5\pi}{6}$ **19.** $-2\sqrt{2}$ **21.** $-\dfrac{4}{3}$ **23.** $\dfrac{\sqrt{5}}{5}$ **25.** $\dfrac{\sqrt{5}}{3}$

27. $-\dfrac{5}{3}$ **29.** $-\dfrac{5}{12}$ **31.** 4   **33.** $-\sqrt{15}$ **35.** $\dfrac{\sqrt{21}}{5}$ **37.** $\dfrac{1}{5}$ **39.** $\dfrac{1}{4}$ **41.** $\dfrac{x}{\sqrt{1 - x^2}}$ **43.** $\sqrt{1 + x^2}$ **45.** $\dfrac{\sqrt{1 - 4x^2}}{2x}$

47. $\dfrac{\sqrt{9x^2 + 1}}{3x}$ **49.** $\sqrt{1 - 4x^2}$ **51.** 1.1071, 63.4°   **53.** $-0.3398, -19.5°$ **55.** $-0.9203, -52.7°$ **57.** 2.0846, 119.4°

59. 0.5201, 29.8°   **63.** $R = X \cot\theta$ **65.** $t = \sqrt{\dfrac{m}{k}}\,\text{Arccos}\,\dfrac{x}{A}$

Review Exercises for Chapter 16 (page 517)

1. $\dfrac{\sqrt{2 - \sqrt{2}}}{2}$ **2.** $-\dfrac{\sqrt{2 - \sqrt{2}}}{2}$ **3.** $\dfrac{\sqrt{3}}{2}$ **4.** 1   **5.** $\sin 6x$ **6.** $\cos 5x$ **7.** $\sin 3x$ **8.** $\cos x$ **9.** $-\cos 2x$ **10.** $-\cos x$

11. $-\sin 2x$ **12.** $\cos x$ **13.** $\dfrac{1}{2}(\sqrt{3}\sin x - \cos x)$ **14.** $\dfrac{\sqrt{2}}{2}(\cos x + \sin x)$ **15.** $-\dfrac{24}{25}$ **16.** $\dfrac{4\sqrt{2}}{9}$ **17.** $\dfrac{119}{169}$ **18.** $\dfrac{3\sqrt{10}}{10}$

19. $-\dfrac{3}{5}$ **20.** $-\dfrac{5\sqrt{26}}{26}$ **21.** $\cos 6x$ **22.** $-\cos 4x$ **23.** $\cos 8x$ **24.** $\cos 6\beta$ **25.** $\sin 6x$ **26.** $\dfrac{1}{2}\sin 8x$ **27.** $\sin 2\theta$

28. $\cos 2\theta$ **29.** $\sqrt{2}\cos 2\theta$ **30.** $2\sin 4\theta$ **31.** $\dfrac{1}{2}(1 - \cos 6x)$ **32.** $\dfrac{1}{2}(1 + \cos 8x)$ **33.** $1 + \cos 6x$ **34.** $2(1 - \cos 8x)$ **57.** $0, \pi$

58. $0, \pi$ **59.** $\dfrac{\pi}{3}, \pi, \dfrac{5\pi}{3}$ **60.** $0, \dfrac{\pi}{4}, \pi, \dfrac{5\pi}{4}$ **61.** $\pi, \dfrac{4\pi}{3}$ **62.** $\text{Arccos}\dfrac{2}{3} = 0.84, 2\pi - \text{Arccos}\dfrac{2}{3} = 5.44$ **63.** $\theta = \dfrac{\pi}{4} + \dfrac{\alpha}{2}$ **65.** $\dfrac{\pi}{3}, \dfrac{5\pi}{3}$

66. $\dfrac{\pi}{2}$ **67.** $\dfrac{5\pi}{6}, \dfrac{11\pi}{6}$ **68.** $\dfrac{\pi}{6}, \dfrac{7\pi}{6}$ **69.** $\dfrac{2\pi}{3}$ **70.** $-\dfrac{\pi}{6}$ **71.** $-\sqrt{35}$ **72.** $-\dfrac{2\sqrt{5}}{5}$ **73.** $\sqrt{1 - 4x^2}$ **74.** $x = \dfrac{1}{4}\arcsin\dfrac{1}{3}(2 - y)$

75. $x = \tan\dfrac{y}{2} - 2$ **77. a.** $5\sin(\theta + \alpha)$, where $\alpha = \text{Arctan}\dfrac{4}{3}$ **b.** $2\sqrt{13}\sin(\theta + \alpha)$, where $\alpha = \text{Arctan}\dfrac{3}{2}$

Chapter 17

Section 17.1 (page 524)

1. $x < 3$ **3.** $x \le 2$ **5.** $x < -6$ **7.** $x \ge -5$ **9.** $x \le \dfrac{4}{3}$ **11.** $x > -\dfrac{3}{2}$ **13.** $x \ge \dfrac{7}{3}$ **15.** $x > \dfrac{5}{2}$ **17.**

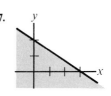

19. 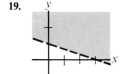   **21.** 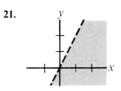   **23.** 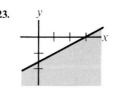   **25.**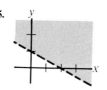

27. 7.6 h   **29.** $0.0 \le F \le 80$ (pounds)

Section 17.2 (page 529)

1. $x > -3$ **3.** $-2 < x < 2$ **5.** $x \le -3, x \ge -2$ **7.** $x < -2, 0 < x < 3$ **9.** $x \le -5, -2 \le x \le 3$ **11.** $-8 < x < 6, x > 7$
13. $1 < x < 2$ **15.** $x \le 6, x > 7$ **17.** $3 < x < 4$ **19.** $x \le -4, 1 < x \le 3$ **21.** $-3 < x < 3, x > 6$ **23.** $x \le -6, 1 \le x < 2, x > 3$

25. $x > 4$ **27.** all x **29.** $x < -1$ **31.** $x < -4, 1 < x \le 4$ **33.** $0 \le t \le \dfrac{4}{3}$ (seconds)   **35.** $x \le 5, x \ge 15$ (feet)   **37.** $1 \le x < 20$

Section 17.3 (page 532)

1. $-1 < x < 3$ **3.** $-4 \le x \le -2$ **5.** $-1 < x < 0$ **7.** $x \le \dfrac{1}{2}, x \ge \dfrac{7}{2}$ **9.** $-\dfrac{13}{2} < x < -\dfrac{1}{2}$ **11.** $x < -1, x > 5$

13. $x < -3, -2 < x < 3, x > 4$ **15.** $x < 0, x > 4$ **17.** $-2 < x < 1 - \sqrt{7}, 1 + \sqrt{7} < x < 4$
19. $-3 - \sqrt{10} < x < -5, -1 < x < -3 + \sqrt{10}$ **21.** $|d - 2.5550| \le 0.0001$

Section 17.4 (page 536)

1.
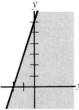

3.

5.

7.

(1, 1)
(2, 0)

9.
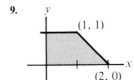
(1, 1)
(2, 0)

11.
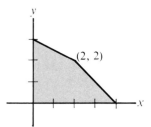
(2, 2)

13.
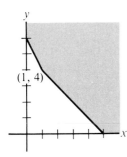
(1, 4)

15.
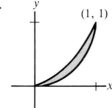
(1, 1)

17.
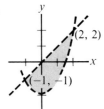
(2, 2)
(−1, −1)

Section 17.5 (page 541)

1. $P = 4$ at $(2, 0)$ **3.** $P = 3$ at $(1, 0)$ **5.** $P = 13$ at $(3, 1)$ **7.** $P = 17$ at $(3, 2)$ **9.** $P = 25$ at $(3, 8)$
11. 90 acres of soybeans, no corn; $8,100   **13.** 20 smaller trees, 10 larger trees; $9,600   **15.** 4 of Type I, 11 of Type II; $1,720/month
17. 2 lb of Product I, 4 lb of Product II; $2.10 per bag

Review Exercises for Chapter 17 (page 542)

1. $x < -5$ **2.** $x \ge \dfrac{5}{4}$ **3.** $x < 3$ **4.** $x \le -\dfrac{2}{5}$ **5.** $-3 \le x \le 3$ **6.** $-2 \le x \le 3$ **7.** $-4 < x < -1, x > 3$ **8.** $0 < x < 1$

9. $-2 < x \le 1$ **10.** $-4 < x \le 2$ **11.** $x < 1, x > 6$ **12.** $x < 2, x > 5$ **13.** $-4 < x \le 1, x \ge 2$ **14.** $x \le -5, -3 \le x < 1$

15. $x \le -2, x = 1$ **16.** $x \ge -3$ **17.** $2 \le x \le 6$ **18.** $x < -2, x > 4$ **19.** $-\frac{5}{2} < x < \frac{1}{2}$ **20.** $-1 < x < 3$

21. $-1 < x < 2 - \sqrt{7}, 2 + \sqrt{7} < x < 5$ **22.** $x < -1, 1 - \sqrt{2} < x < 1 + \sqrt{2}, x > 3$

23.
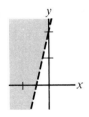

24.
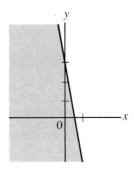

25.
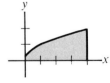

26.
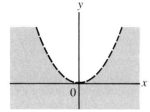

27.
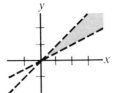

28.
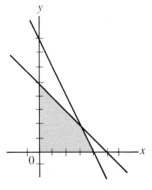

29. 6.5 h   **30.** $t \ge 5\frac{1}{3}$ h   **31.** $t \ge 6$ s   **32.** $125 \le S \le 175$ **33.** $P = 48$ at $(8, 6)$ **34.** $P = 7$ at $(4, 3)$

35. 60 small appliances, 20 large appliances; $8,600

Chapter 18

Section 18.1 (page 546)

1. 28   **3.** 24   **5.** 66   **7.** 5   **9.** 3   **11.** 272   **13.** 343   **15.** 845   **17.** $s = \dfrac{n(n + 1)}{2}$ **19.** $2,700

Section 18.2 (page 550)

1. $\dfrac{128}{243}$ **3.** $-\dfrac{5}{512}$ **5.** 162   **7.** 63   **9.** 242   **11.** $\dfrac{127}{64}$ **13.** $\dfrac{20}{27}$ **15.** $\dfrac{422}{81}$ **17.** $\dfrac{129}{64}$

19. a. $540   **b.** $540.80   **c.** $541.22   **d.** $541.50   **21.** $977.34   **23.** $671.56   **25. a.** $676.45   **b.** $679.02

27. $2,484.04   **29.** $1,024N$ **31.** $\dfrac{N}{4,096}$ **33.** $\dfrac{155}{4}$ ft   **35.** $\dfrac{36,447}{512}$ ft

Section 18.3 (page 553)

1. $\dfrac{4}{9}$ **3.** $\dfrac{13}{99}$ **5.** $\dfrac{104}{333}$ **7.** $\dfrac{133}{990}$ **9.** 1   **11.** $\dfrac{3}{4}$ **13.** $\dfrac{4}{5}$ **15.** $\dfrac{27}{4}$ **17.** $\dfrac{3}{8}$ **19.** $\dfrac{16}{7}$ **21.** 100 in.   **23.** 48 ft   **25.** 8 mm

Section 18.4 (page 557)

1. $x^4 - 4x^3y + 6x^2y^2 - 4xy^3 + y^4$ **3.** $s^4 + 8s^3v + 24s^2v^2 + 32sv^3 + 16v^4$ **5.** $x^3 - 3x^2y + 3xy^2 - y^3$
7. $s^4 + 8s^3t + 24s^2t^2 + 32st^3 + 16t^4$ **9.** $a^5 - 5a^4b^2 + 10a^3b^4 - 10a^2b^6 + 5ab^8 - b^{10}$
11. $S^6 - 12S^5W + 60S^4W^2 - 160S^3W^3 + 240S^2W^4 - 192SW^5 + 64W^6$ **13.** $R^6 + 12R^5 + 60R^4 + 160R^3 + 240R^2 + 192R + 64$
15. $R^6 - 18R^5 + 135R^4 - 540R^3 + 1215R^2 - 1458R + 729$ **17.** $x^{10} + 20x^9 + 180x^8 + 960x^7 + \cdots$
19. $k^{12} - 12k^{11}m^2 + 66k^{10}m^4 - 220k^9m^6 + \cdots$ **21.** $r^8 + 16\pi r^7 + 112\pi^2 r^6 + 448\pi^3 r^5 + \cdots$ **23.** $1 - 2x + 3x^2 - 4x^3 + \cdots$

25. $1 + \dfrac{1}{2}s - \dfrac{1}{8}s^2 + \dfrac{1}{16}s^3 - \cdots$ **27.** $1 + x + x^2 + x^3 + \cdots$ **29.** 0.9698 **31.** 1.096 **35.** $1 + \dfrac{1}{2}x^2 - \dfrac{1}{8}x^4 + \dfrac{1}{16}x^6 - \cdots$

Review Exercises for Chapter 18 (page 558)

1. 47 **2.** 42 **3.** 279 **4.** 72.8 **5.** $\dfrac{2}{243}$ **6.** $\dfrac{243}{256}$ **7.** $\dfrac{43}{64}$ **8.** $\dfrac{61}{81}$ **9.** $\dfrac{1{,}055}{81}$ **10.** $\dfrac{406}{125}$ **11.** $\dfrac{266}{243}$ **12.** $\dfrac{312}{125}$ **13.** $\dfrac{3}{4}$

14. $\dfrac{3}{5}$ **15.** $\dfrac{10}{3}$ **16.** 4 **17.** 12 **18.** 5 **19.** $\dfrac{7}{45}$ **20.** $\dfrac{5}{18}$ **21.** $\dfrac{38}{99}$ **22.** $\dfrac{533}{990}$ **23.** $a^3 - 3a^2b + 3ab^2 - b^3$

24. $a^4 - 4a^3b + 6a^2b^2 - 4ab^3 + b^4$ **25.** $M^5 - 10M^4T + 40M^3T^2 - 80M^2T^3 + 80MT^4 - 32T^5$ **26.** $\pi^4 + 8\pi^3 + 24\pi^2 + 32\pi + 16$
27. $e^4 - 8e^3 + 24e^2 - 32e + 16$ **28.** $s^4 + 4s^3t^2 + 6s^2t^4 + 4st^6 + t^8$ **29.** $m^5 - 5m^4v^2 + 10m^3v^4 - 10m^2v^6 + 5mv^8 - v^{10}$
30. $a^6 + 12a^5 + 60a^4 + 160a^3 + 240a^2 + 192a + 64$ **31.** $x^{10} - 20x^9 + 180x^8 - 960x^7 + \cdots$ **32.** $R^9 - 9R^8 + 36R^7 - 84R^6 + \cdots$

33. $1 + 2t + 3t^2 + 4t^3 + \cdots$ **34.** $1 + \dfrac{1}{2}v - \dfrac{1}{8}v^2 + \dfrac{1}{16}v^3 - \cdots$ **35.** $1 + \dfrac{1}{3}c + \dfrac{2}{9}c^2 + \dfrac{14}{81}c^3 + \cdots$ **37.** 0.9680 **38.** 1.0223

39. $\$1{,}591.98$ **40.** $\$2{,}575.08$ **41.** $\$1{,}030.32$ **42.** $\$638.11$ **43.** $6{,}561N$ **44.** $\dfrac{N}{1{,}024}$ **45.** $\dfrac{4{,}850}{243}$ ft   **46.** 56.95 in.

47. 100 in.   **48.** 30 cm   **49.** $M = \dfrac{E}{2}(1 - s + s^2 - s^3 + \cdots)$

Cumulative Review Exercises for Chapters 16–18 (page 560)

8. $\dfrac{\sqrt{2}}{2}$ **9.** $\dfrac{\sqrt{2+\sqrt{3}}}{2}$ **10.** $0°, 180°, 63.4°, 243.4°$ **11. a.** $-\dfrac{\pi}{3}$;   **b.** $\dfrac{2\pi}{3}$ **12.** $-4 < x \le -3, 1 < x \le 2$ **13.** $x < 0, x > 4$

14. $\dfrac{1}{3}$ **15.** $x^5 - 10x^4y + 40x^3y^2 - 80x^2y^3 + 80xy^4 - 32y^5$ **16.** $R = \dfrac{v^2}{g}\sin 2\theta$ **17.** 0.084 s   **18.** $\theta = \text{Arccos}\left(\dfrac{I}{I_{\text{max}}}\right)$

19. $\dfrac{315}{8}$ ft   **20.** 100 ft   **21.** 84.0 m/s

Chapter 19

Section 19.2 (page 562)

1. $\sqrt{13}$ **3.** $4\sqrt{5}$ **5.** $\sqrt{7}$ **7.** $\sqrt{6}$ **9.** $2\sqrt{2}$ **11.** positive in quadrants I and III   **13. (a)** y-axis   **(b)** x-axis
21. $y^2 - 4x + 4 = 0$

Section 19.3 (page 567)

1. -1 **3.** $\dfrac{3}{2}$ **5.** $\dfrac{6}{5}$ **7.** 2 **9.** undefined   **11.** 0 **13.** $\dfrac{3}{4}$

15. a.

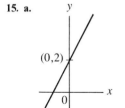

b.

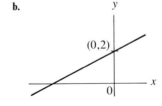

c.

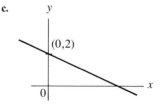

d.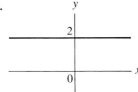

e.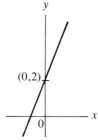

f.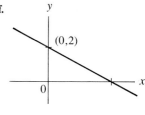

17. $\dfrac{13}{6}$ **23.** $(0, 1)$ **25.** $(7, 2)$ **27.** $\left(-\dfrac{9}{2}, 5\right)$ **29.** $\dfrac{7}{2}$ **31.** $-\dfrac{1}{5}, -\dfrac{8}{7}, -\dfrac{7}{2}$ **33.** $x = -4$

Section 19.4 (page 571)

1. $x - 2y + 11 = 0$ **3.** $3x - y - 13 = 0$ **5.** $x + 3y = 0$ **7.** $y = 0$ (x-axis)   **9.** $5x + 3y + 3 = 0$
11. $x + y - 5 = 0$ **13.** $x + 8y - 26 = 0$ **15.** $x - y + 10 = 0$ **17.** $x + 3y + 6 = 0$

19. $y = -3x + \dfrac{5}{2}$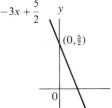

21. $y = \dfrac{2}{3}x + 0$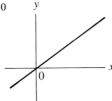

23. $y = 0x + \dfrac{7}{2} = \dfrac{7}{2}$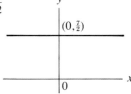

25. parallel   **27.** neither   **29.** perpendicular   **31.** $3x - 4y + 7 = 0$ **33.** $20x + 28y - 27 = 0$

35.

37. $F = 6x$

39. $F = \dfrac{9}{5}C + 32$ **41.** $R = 0.01T + 50$

Section 19.5 (page 580)

In the following answers the intercepts are given first, followed by symmetry, asymptotes, and extent.

1. $y = -1, x = \dfrac{1}{2}$; none; none; all x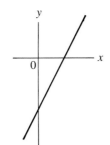

3. $y = 1, x = \pm 1$; y-axis; none; all x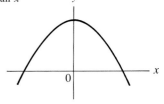

5. origin; x-axis; none; $x \geq 0$

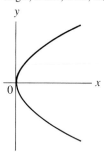

7. $y = \pm 1$, $x = -1$; x-axis; none; $x \geq -1$

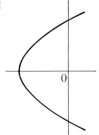

9. none; origin; both axes; $x \neq 0$

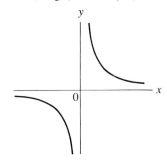

11. $y = -15$, $x = -5$, 3; none; none; all x

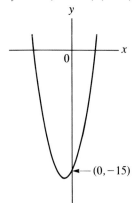

13. $y = 0$, $x = 0$, 1, 2; none; none; all x

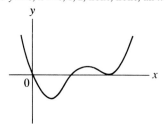

15. $y = -6$, $x = -2$, 3; none; none; all x

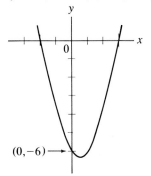

17. $y = 0$, $x = 0$, ± 1; origin; none; $-1 \leq x \leq 1$

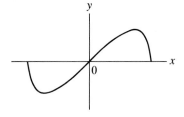

19. $y = 1$; none; $x = -2$, $y = 0$; $x \neq -2$

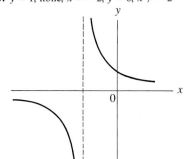

21. $y = 2$; none; $x = 1$, $y = 0$; $x \neq 1$

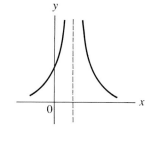

23. $y = 0$, $x = 0$; none; $x = 1$; $x \neq 1$

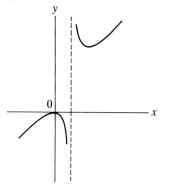

25. $y = -\dfrac{1}{2}$, $x = -1$; none; $x = -2$, 1, $y = 0$; $x \neq -2$, 1

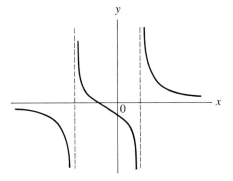

27. $y = 4$, $x = \pm 2$; y-axis; $x = \pm 1$, $y = 1$; $x \neq \pm 1$

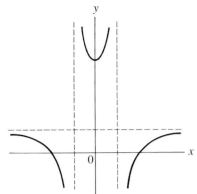

29. $x = -5$, 3; x-axis; none; $x \leq -5$, $x \geq 3$

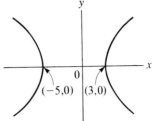

$(-5,0)$ $(3,0)$

31. $y = 0$, $x = 0$; x-axis; $x = 2$, 3, $y = 0$; $0 \leq x < 2$, $x > 3$

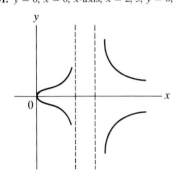

33. $y = \pm 2$, $x = \pm 2$; both axes; $x = \pm 1$, $y = \pm 1$; $x \leq -2$, $-1 < x < 1$, $x \geq 2$

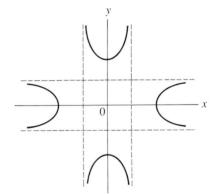

35. $C = 0$, $C_1 = 0$; none; $C = 10^{-2}$; $C_1 \geq 0$

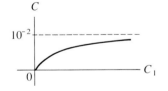

37.

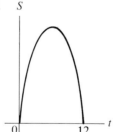

39.

Section 19.7 (page 584)

1. $x^2 + y^2 = 25$ **3.** $x^2 + y^2 = 100$ **5.** $(x + 2)^2 + (y - 5)^2 = 1$ or $x^2 + y^2 + 4x - 10y + 28 = 0$

7. $(x + 1)^2 + (y + 4)^2 = 17$ or $x^2 + y^2 + 2x + 8y = 0$ **9.** $\left(x + \frac{1}{2}\right)^2 + \left(y + \frac{1}{2}\right)^2 = \frac{65}{2}$ or $x^2 + y^2 + x + y - 32 = 0$

11. $(x - 1)^2 + (y - 1)^2 = 4$, $(1, 1)$, $r = 2$ **13.** $(x + 2)^2 + (y - 4)^2 = 16$, $(-2, 4)$, $r = 4$ **15.** $(x + 2)^2 + (y + 1)^2 = 3$, $(-2, -1)$, $r = \sqrt{3}$

17. $(x + 2)^2 + (y - 1)^2 = 9$, $(-2, 1)$, $r = 3$ **19.** $\left(x - \frac{1}{2}\right)^2 + (y - 1)^2 = 1$, $\left(\frac{1}{2}, 1\right)$, $r = 1$ **21.** $(x - 2)^2 + \left(y + \frac{1}{2}\right)^2 = 2$, $\left(2, -\frac{1}{2}\right)$, $r = \sqrt{2}$

23. $\left(x + \frac{3}{2}\right)^2 + (y + 2)^2 = 5$, $\left(-\frac{3}{2}, -2\right)$, $r = \sqrt{5}$ **25.** point circle, $\left(\frac{5}{2}, \frac{1}{2}\right)$ **27.** point circle, $(3, -4)$ **29.** imaginary circle

31. $x^2 + y^2 = 4.00$, $x^2 + y^2 = 11.6$

Section 19.8 (page 590)

1. $y^2 = 12x$ **3.** $x^2 = -20y$ **5.** $y^2 = -16x$ **7.** $y^2 = 4x$ **9.** $y^2 = -8x$ **11.** $x^2 = -12y$ **13.** $y^2 = -8x$
15. $(0, 2)$, $y + 2 = 0$ **17.** $(0, -3)$, $y - 3 = 0$ **19.** $(4, 0)$, $x + 4 = 0$ **21.** $(-1, 0)$, $x - 1 = 0$ **23.** $(0, 1)$, $y + 1 = 0$

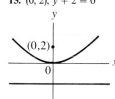 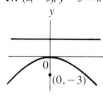 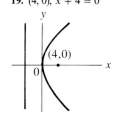 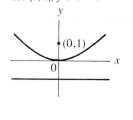

25. $\left(\frac{9}{4}, 0\right)$, $x + \frac{9}{4} = 0$ **27.** $\left(-\frac{1}{4}, 0\right)$, $x - \frac{1}{4} = 0$ **29.** $\left(-\frac{1}{6}, 0\right)$, $x - \frac{1}{6} = 0$ **31.** $x^2 + y^2 - 6y - 27 = 0$ **33.** $y^2 - 2y - 8x + 17 = 0$

35. $x^2 = -y$, $y^2 = -8x$ **37.** 26.3 m   **39.** $16\sqrt{2} = 22.6$ m   **41.** $2\sqrt{3}$ m

Section 19.9 (page 596)

1. $(\pm 5, 0)$, $(\pm 3, 0)$, 4   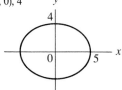   **3.** $(\pm 3, 0)$, $(\pm\sqrt{5}, 0)$, 2   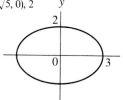   **5.** $(\pm 4, 0)$, $(\pm\sqrt{15}, 0)$, 1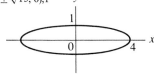

7. $(0, \pm 4)$, $(0, \pm\sqrt{7})$, 3   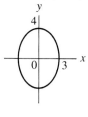 **9.** $(0, \pm\sqrt{10})$, $(0, \pm\sqrt{6})$, 2   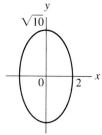   **11.** $(0, \pm\sqrt{5})$, $(0, \pm 2)$, 1   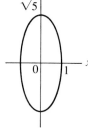   **13.** $(\pm\sqrt{6}, 0)$, $(\pm\sqrt{3}, 0)$, $\sqrt{3}$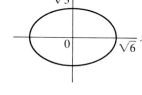

15. $(0, \pm\sqrt{15})$, $(0, \pm 2\sqrt{2})$, $\sqrt{7}$ **17.** $(\pm 5, 0)$, $(\pm\sqrt{15}, 0)$, $\sqrt{10}$ **19.** $\dfrac{x^2}{16} + \dfrac{y^2}{7} = 1$ **21.** $4x^2 + 3y^2 = 48$ **23.** $2x^2 + y^2 = 18$

25. $\dfrac{x^2}{39} + \dfrac{y^2}{64} = 1$ **27.** $x^2 + 4y^2 = 16$ **29.** $\dfrac{x^2}{64} + \dfrac{y^2}{28} = 1$ **31.** $3x^2 + 4y^2 = 48$ **33.** circle; $x^2 + y^2 - 8x + 12 = 0$

35. $9x^2 + 16y^2 = 36$ **37.** $\dfrac{12\sqrt{5}}{5} = 5.4$ m   **39.** $\dfrac{x^2}{16{,}810{,}000} + \dfrac{y^2}{16{,}809{,}600} = 1$

Section 19.10 (page 602)

1. $(\pm 4, 0)$, $(\pm 5, 0)$ 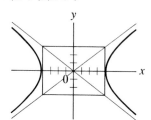   **3.** $(\pm 3, 0)$, $(\pm 5, 0)$ 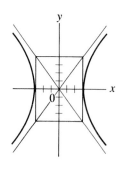   **5.** $(0, \pm 2)$, $(0, \pm 2\sqrt{2})$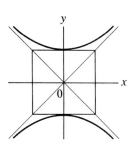

7. $(\pm 1, 0), (\pm \sqrt{6}, 0)$

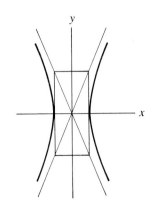

9. $(0, \pm 2\sqrt{3}), (0, \pm 2\sqrt{5})$

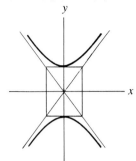

11. $(0, \pm \sqrt{2}), (0, \pm \sqrt{5})$

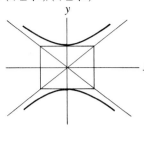

13. $4x^2 - 9y^2 = 36$ **15.** $16y^2 - 9x^2 = 144$ **17.** $\dfrac{y^2}{36} - \dfrac{x^2}{28} = 1$ **19.** $\dfrac{x^2}{16} - \dfrac{y^2}{16} = 1$ **21.** $3x^2 - y^2 = 27$ **23.** $4x^2 - y^2 = 4$

25. $16y^2 - 9x^2 = 144$ **27.** $3x^2 - y^2 + 6x + 4y - 4 = 0$ **29.** $y^2 - 25x^2 = 144$ **31.** $pV = 36$

Section 19.11 (page 607)

1. circle, center at $(1, 2)$, radius $\sqrt{3}$ **3.** parabola, vertex at $(2, -3)$, focus at $(4, -3)$

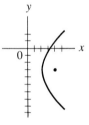

5. $\dfrac{(y + 2)^2}{6} - \dfrac{(x + 2)^2}{9} = 1$; hyperbola, center at $(-2, -2)$ **7.** $(x + 2)^2 + \dfrac{(y - \frac{3}{2})^2}{4} = 1$; ellipse, center at $\left(-2, \dfrac{3}{2}\right)$

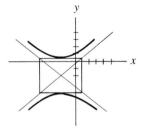

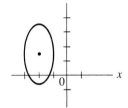

9. $(x + 1)^2 + (y - 1)^2 = 0$; point: $(-1, 1)$ **11.** $\left(y - \dfrac{5}{2}\right)^2 - \dfrac{x^2}{6} = 1$; hyperbola   **13.** $\left(x - \dfrac{1}{8}\right)^2 + \left(y - \dfrac{3}{4}\right)^2 = 1$; circle, center at $\left(\dfrac{1}{8}, \dfrac{3}{4}\right)$

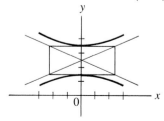

15. $3(x - 3)^2 + (y + 1)^2 = -1$; imaginary locus   **17.** $(y - 2)^2 = 16(x + 1)$ **19.** $\dfrac{(x + 3)^2}{9} + \dfrac{y^2}{4} = 1$ **21.** $\dfrac{(x + 3)^2}{16} - \dfrac{(y - 1)^2}{20} = 1$

23. $\dfrac{(x-2)^2}{25} + \dfrac{(y-3)^2}{4} = 1$ **25.** $\dfrac{(x-1)^2}{4} - y^2 = 1$ **27.** $\dfrac{(x+3)^2}{9} + \dfrac{(y-1)^2}{25} = 1$ **29.** $(y+2)^2 = -12(x-4)$

31. $(x+2)^2 = -16(y+4)$ **33.** $\dfrac{(y-1)^2}{4} - \dfrac{(x+1)^2}{5} = 1$

Section 19.12 (page 612)

3. $(-1, \sqrt{3})$ **5.** $(-3, -3\sqrt{3})$ **7.** $(1.97, -0.35)$ **9.** $\left(\sqrt{2}, \dfrac{7\pi}{4}\right)$ **11.** $(5, 0.93)$ **13.** $r = 2 \sec \theta$ **15.** $r = \sqrt{2}$

17. $r^2 = \sec 2\theta$ **19.** $r^2 = \dfrac{1}{2(1 + \sin^2 \theta)}$ **21.** $y = 2$ **23.** $y = x$ **25.** $x^2 + y^2 = x$ **27.** $(x^2 + y^2 - x)^2 = x^2 + y^2$

29. $(x^2 + y^2 + x)^2 = x^2 + y^2$ **31.** $(x^2 + y^2)^2 = 2xy$ **33.** $y^2 = 4(x + 1)$ **35.** $x^2 = 8(y + 2)$ **37.** $(x^2 + y^2)^2 = 3(x^2 - y^2)$

39. $(x^2 + y^2)^2 = a(3x^2 y - y^3)$ **41.** $(x^2 + y^2 + 2y)^2 = x^2 + y^2$

Section 19.13 (page 616)

1.
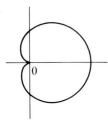

3.
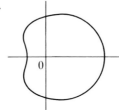

5.
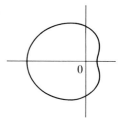

7.

9.

11.

13.
15.
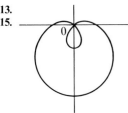

17.
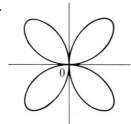

19.
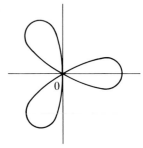

21.
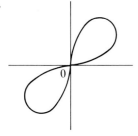

23.

25.
27.
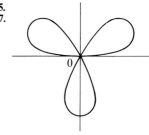

29.

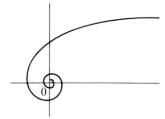

31.

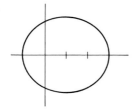

33.

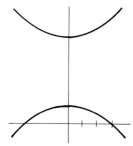

Review Exercises for Chapter 19 (page 616)

2. $x + y = 3$ **3.** -40 **6.** $2x - y - 7 = 0$ **7.** $3x + y - 2 = 0$ **9.** $x^2 + y^2 - 2x + 4y = 0$ **10.** $x^2 + y^2 - 4x - 10y + 4 = 0$
11. $(-1, -1), \sqrt{2}$
13. ellipse, $(0, \pm 4), (0, \pm \sqrt{7})$

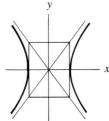

14. ellipse, $(\pm 1, 0), (\pm \sqrt{3}/2, 0)$ **15.** hyperbola, $(0, \pm 2), (0, \pm \sqrt{11})$

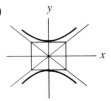

16. hyperbola, $(\pm 3, 0), (\pm 5, 0)$

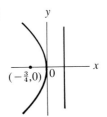

17. parabola, $(0, 0), \left(-\dfrac{3}{4}, 0\right)$

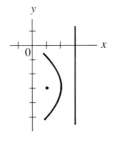

18. parabola, $(0, 0), \left(0, \dfrac{9}{4}\right)$

19. $(y + 3)^2 = -4(x - 2), (2, -3)$, parabola   **20.** $(x - 4)^2 + (y + 5)^2 = 45$, circle, center at $(4, -5), r = 3\sqrt{5}$

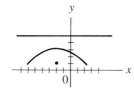

21. $\dfrac{(x - 2)^2}{9} + \dfrac{(y + 1)^2}{16} = 1, (2, -1)$, ellipse   **22.** $(x + 2)^2 = -8(y - 3), (-2, 3)$, parabola   **23.** $\dfrac{(x - 2)^2}{9} - \dfrac{(y - 4)^2}{9} = 1, (2, 4)$, hyperbola

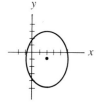

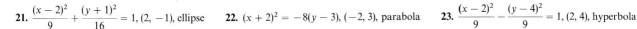

24. $y^2 = 8x$ **25.** $(x - 1)^2 = 12(y - 3)$ **26.** $\dfrac{(x + 2)^2}{12} + \dfrac{(y + 4)^2}{16} = 1$ **27.** $\dfrac{x^2}{7} + \dfrac{y^2}{16} = 1$ **28.** $9x^2 - 16y^2 = 81$

29. $\dfrac{(y - 2)^2}{9} - \dfrac{x^2}{7} = 1$ **30.** $(y - 2)^2 = -16(x + 4)$ **31.** $x^2 = 4(y - 2)$ **32.** $\dfrac{(y + 1)^2}{4} - \dfrac{(x - 2)^2}{5} = 1$ **33.** $\dfrac{(x - 4)^2}{12} + \dfrac{(y + 1)^2}{16} = 1$

34. intercepts: $x = -1$, $y = 1$; symmetry: none   **35.** intercepts: $x = -1$, $y = 1$; symmetry: none

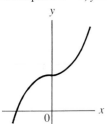

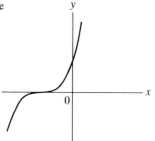

36. intercepts: $x = 0$, $\pm\sqrt{2}$, $y = 0$; symmetry: origin   **37.** intercepts: $x = 0, 4$, $y = 0$; symmetry: x-axis; extent: $x \leq 0$, $x \geq 4$

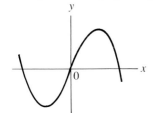

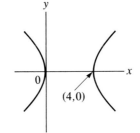

38. intercept: $(0, 0)$; symmetry: y-axis; asymptote: $y = 1$; extent: all x **39.** intercept: $(0, 0)$; symmetry: origin; asymptotes: $x = \pm2$, $y = 0$; extent: $x \neq 2, -2$

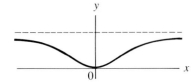

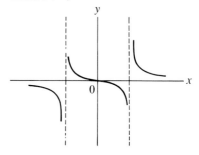

40. intercepts: $x = 0, -1$, $y = 0$; symmetry: none; extent: $x \geq -1$ **41.** 0.10 ft from vertex   **42.** 12.0 ft

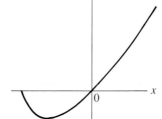

43. $V = x(6 - 2x)^2$, $x = 1$ in.   **44.** 144 ft   **45.** 15 million kilometers

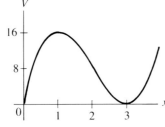

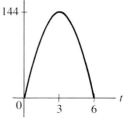

46. $\theta = \dfrac{\pi}{4}$ **47.** $r = 3 \cot \theta \csc \theta$ **48.** $y = 4$ **49.** $x(x^2 + y^2) = 4y$ **50.** $(x^2 + y^2 + 3x)^2 = 4(x^2 + y^2)$

51. circle   **52.** cardioid   **53.** four-leaf rose

                    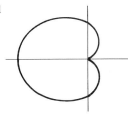   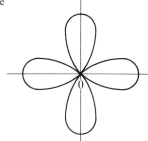

54. limaçon with loop   **55.** lemniscate   **56.** three-leaf rose

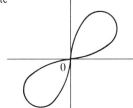      

Chapter 20

Section 20.2 (page 622)

1. $A = x^2$ **3.** $V = \frac{2}{3}\pi r^2$ **5.** $R_T = 10 + R$ **7.** $d = 50t$ **9.** $(-\infty, \infty); (-\infty, \infty)$ **11.** $[2, \infty); [0, \infty)$ **13.** $[-2, 2]; [0, 2]$
15. $x = 0$ and $x \ge 3;\ [0, \infty)$ **17.** $x = 1$ and $x \ge 2;\ [0, \infty)$ **19.** all x except $x = 1;\ y \neq 0$ **21.** $0, 12$ **23.** $3, 15$

25. $1, 2, -7$ **27.** $\dfrac{1}{3}, \dfrac{1}{a}$ **29.** $\sqrt{a^4 - 1}, \sqrt{x^2 - 2x}$ **31.** $1 - x^2 - 2x\,\Delta x - (\Delta x)^2,\ 1 - x^2 + 2x\,\Delta x - (\Delta x)^2$

33. (a) $x^2 + 2x + 1$ **(b)** $x^2 + 1$ **(c)** x^4 **35.** $-1, 1, 1, 1$

Section 20.3 (page 629)

1. 1   **3.** 1   **5.** 0   **7.** 0   **9.** 0   **11.** 4   **13.** -1 **15.** -8 **17.** 0   **19.** 5   **21.** $\dfrac{3}{2}$ **23.** 0   **25.** -6 **27.** 10

29. 2   **31.** 2   **33.** $\dfrac{4}{5}$ **35.** $\dfrac{3}{4}$ **37.** 0   **39.** $\dfrac{1}{3}$ **41.** 2   **43.** 1   **45.** 0   **47. (a)** ∞ **(b)** 0

Section 20.5 (page 638)

1. $y' = 2$ **3.** $y' = -3$ **5.** $y' = 2x$ **7.** $y' = 4x - 1$ **9.** $y' = 3x^2 - 6x$ **11.** $y' = -\dfrac{1}{(x+1)^2}$ **13.** $y' = -\dfrac{2}{x^3}$

15. $y' = \dfrac{2x}{(1 - x^2)^2}$ **17.** $y' = \dfrac{1}{2\sqrt{x}}$ **19.** $y' = -\dfrac{1}{2\sqrt{1 - x}}$ **21.** $-2, 4, 6$ **23.** $-\dfrac{1}{16}, -\dfrac{1}{54}$

Section 20.6 (page 641)

1. $y' = 1$ **3.** $y' = 2x$ **5.** $y' = 2x + 1$ **7.** $y' = 6x + 4$ **9.** $y' = 15x^2 - 14x$ **11.** $y' = 21x^2 - 2x - 1$ **13.** $y' = x^2 + x + 1$

15. $y' = x - x^2$ **17.** $y' = 200x^9 - 144x^5 + 6x^2$ **19.** $y' = \dfrac{7}{5}x^6 - \dfrac{5}{\sqrt{2}}x^4$ **21.** $y' = x^5 + \dfrac{4}{5}x^3$

Section 20.7 (page 644)

1. $v = 4t,\ a = 4,\ t = 0$ **3.** $v = 2t - 2,\ a = 2,\ t = 1$ **5.** $v = 2 - 2t,\ a = -2,\ t = 1$ **7.** $v = 12 - 4t,\ a = -4,\ t = 3$

9. $v = 6t - 6,\ a = 6,\ t = 1$ **11.** $v = 9t^2 + 4t,\ a = 18t + 4,\ t = 0$ and $t = -\dfrac{4}{9}$ **13.** 1000 m/s   **15.** 4 cm² per cm   **17.** 2.95 Ω/°C

19. 63 W/A   **23.** $\omega = 6$ rad/s, $\alpha = 6$ rad/s²   **25.** $i = 4t + 1$ **27.** 0.20 A   **29.** 22 J/s = 22 W (watts)

Section 20.8 (page 651)

Group A   **1.** $16x^3 - 8x$ **3.** $-\dfrac{1}{x^2}$ **5.** $5x^4 + 9x^{-4} - 4x^{-3}$ **7.** $\dfrac{5}{2}x\sqrt{x} - \dfrac{1}{2x\sqrt{x}}$ **9.** $16x(2x^2 - 3)^3$ **11.** $100x^9(x^{10} + 1)^9$

13. $-\dfrac{4x}{(2x^2 + 3)^{3/2}}$ **15.** $\dfrac{1}{(6 - x)^{4/3}}$ **17.** $\dfrac{3x^2 - 3}{2(x^3 - 3x)^{1/2}}$ **19.** $\dfrac{x^2}{(x^3 - 3)^{2/3}}$ **21.** $x^2(x + 1)(5x + 3)$

23. $4x^3(x + 2)(3x + 4)$ **25.** $2x(x^2 - 5)(3x^2 - 5)$ **27.** $\dfrac{-1}{(x - 1)^2}$ **29.** $(x^2 - 2x)(5x^2 - 2x - 4)$ **31.** $\dfrac{5x^2 + 4x}{2\sqrt{x + 1}}$

Group B   **1.** $-6x^{-3} + 6x^{-4}$ **3.** $4x^{1/3} + x^{-3/4}$ **5.** $(x^3 - 3)(8x^3 - 40x^4) + 3x^2(2x^4 - 8x^5)$ **7.** $(168x^3 - 36x)(7x^4 - 3x^2)^5$

9. $-\dfrac{1}{2\sqrt{1 - x}}$ **11.** $\dfrac{x}{\sqrt{x^2 + 2}}$ **13.** $\dfrac{1 - 2x}{(x - x^2)^{3/4}}$ **15.** $\dfrac{x^2 - 2x}{(x - 1)^2}$ **17.** $\dfrac{x^2 - 4x + 3}{(x - 2)^2}$ **19.** $\dfrac{x^4 - 26x^2 - 16}{(x^2 - 8)^2}$ **21.** $\dfrac{2x^2 - 1}{\sqrt{x^2 - 1}}$

23. $-\dfrac{x + 4}{2(x - 4)^2\sqrt{x}}$ **25.** $\dfrac{3x^2 + 4x}{2(x + 1)\sqrt{x + 1}}$ **27.** $\dfrac{2 - x^2}{x^3\sqrt{x^2 - 1}}$ **29.** $\dfrac{x\sqrt{x}(x^2 + 15)}{2(x^2 + 3)^2}$ **31.** $\dfrac{3x - 5}{2\sqrt{x - 2}}$ **33.** $\dfrac{2x^2 + 9x - 6}{2(2x + 3)^2\sqrt{x - 1}}$

35. $\dfrac{2x^3 - 30x}{(x + 3)^{3/2}(x - 5)^{1/2}}$ **37.** 18   **39.** $\dfrac{dZ}{dX} = \dfrac{X}{\sqrt{16 + X^2}}$ **41.** $t = 2$ s

Section 20.9 (page 656)

1. $\dfrac{dy}{dx} = -\dfrac{2}{3}$ **3.** $\dfrac{dy}{dx} = \dfrac{x}{y}$ **5.** $\dfrac{dy}{dx} = \dfrac{2x}{3y}$ **7.** $\dfrac{dy}{dx} = -\dfrac{x}{3y^2}$ **9.** $\dfrac{dy}{dx} = \dfrac{9x^2}{16y^3}$ **11.** $\dfrac{dy}{dx} = \dfrac{1 - 10x}{18y^2}$ **13.** $\dfrac{dy}{dx} = -\dfrac{b^2x}{a^2y}$ **15.** $\dfrac{dy}{dx} = -\dfrac{y}{x}$

17. $\dfrac{dy}{dx} = -\dfrac{2y}{x}$ **19.** $\dfrac{dy}{dx} = -\dfrac{2x + 2xy^2 + 1}{2x^2y}$ **21.** $\dfrac{dy}{dx} = \dfrac{8xy^2 - 3x^2}{2y - 8x^2y}$ **23.** $\dfrac{dy}{dx} = \dfrac{6x^2 - 10xy^3}{15x^2y^2 - 4y^3}$ **25.** $\dfrac{dy}{dx} = -\dfrac{4x^3y^4 + 5}{4x^4y^3 - 6y}$ **27.** $\dfrac{1}{4}$

Section 20.10 (page 658)

1. $y'' = 60x^2 + 30x$ **3.** $y'' = \dfrac{-1}{4(x - 1)^{3/2}}$ **5.** $\dfrac{d^3y}{dx^3} = 120x^3 - 120x^2 - 24x$ **7.** $f'''(x) = \dfrac{3}{8(5 + x)^{5/2}}$ **9.** $\dfrac{d^2y}{dx^2} = \dfrac{48}{(3 - 2x)^3}$

Review Exercises for Chapter 20 (page 658)

1. $-1, 0, 1$ **2.** $\sqrt[3]{x^2 + 2},\ x^{2/3} + 2$ **3.** $0, 0, 2,$ undefined   **4.** $[0, 1), (1, \infty);\ 0$ and 2 **5. (a)** $x \geq 1;\ y \geq 0$ **(b)** all x and y **6.** 18

7. 8   **8.** 8   **9.** 3   **10.** 2   **11. (a)** -3 **(b)** -3 **12.** 7   **13.** $\dfrac{1}{2}$ **14.** 2   **15.** 2   **16.** 0   **17.** 0   **18. (a)** 2   **(b)** 1

19. $1 - 6x$ **20. (a)** $3x^2$ **(b)** $-\dfrac{2}{x^2}$ **21. (a)** $\dfrac{1}{(4 - x)^2}$ **(b)** $\dfrac{1}{2\sqrt{x}}$ **22.** $-\dfrac{1}{2\sqrt{3 - x}}$ **23.** $12x^2(x^3 - 2)^3$ **24.** $-\dfrac{4x^3}{(x^4 + 3)^2}$

25. $\dfrac{5}{(x + 1)^2}$ **26.** $\dfrac{5x^4}{2(7 - x^5)^{3/2}}$ **27.** $\dfrac{8x - x^3}{(4 - x^2)^{3/2}}$ **28.** $(x^2 + 1)(5x^2 - 12x + 1)$ **29.** $\dfrac{4 - 2x^2}{\sqrt{4 - x^2}}$

30. $\dfrac{dy}{dx} = -\dfrac{2x + 3}{2y}$ **31.** $\dfrac{dy}{dx} = -\dfrac{2xy + y^2}{x^2 + 2xy + 3y^2}$ **32.** $\dfrac{dy}{dx} = \dfrac{1 - 4xy^2 + 4y}{4x^2y - 4x}$

33. $\dfrac{dy}{dx} = -\dfrac{3x^2y + y^3}{x^3 + 3xy^2}$ **34.** $-\dfrac{9}{2}$ **35.** $-\dfrac{14}{5}$ **36.** $\dfrac{3}{8}(x + 3)^{-5/2}$ **37. (a)** 2   **(b)** 1   **38.** 0   **39.** $-\dfrac{2k}{r^3}$

41. 0.21 V/s   **42.** 0.13   **43.** 30¢ per widget   **44.** $t = 50$ s   **45.** 0.0158 dyne/cm

Chapter 21

Section 21.1 (page 662)

1. $y = 2x, x + 2y - 5 = 0$ **3.** $4x - y - 1 = 0, x + 4y - 13 = 0$ **5.** $2x - y - 1 = 0, x + 2y - 3 = 0$
7. $4x - y + 2 = 0, x + 4y + 9 = 0$ **9.** $x + 16y - 8 = 0, 64x - 4y - 255 = 0$ **11.** $x + 48y - 32 = 0, 96x - 2y - 767 = 0$
13. $3x + 4y + 25 = 0$ **15.** $5x - 8y - 9 = 0$ **17.** $5x + y - 6 = 0, x - 5y + 4 = 0$ **19.** $y = -\frac{3}{4}x$

Section 21.2 (page 667)

1. min. at $x = 1$ **3.** max. at $x = -1$ **5.** min. at $x = 1$, max. at $x = -2$

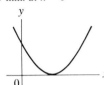

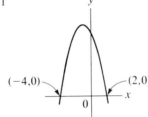

7. min. at $x = 1$, max. at $x = 3$ **9.** min. at $x = \pm 1$, max. at $x = 0$ **11.** critical values: $x = 0, -1$; min. at $x = -1$

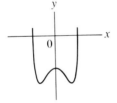

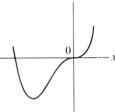

13. critical values: $x = 0, -1$; max. at $x = -1$ **15.** no min. or max., vertical tangent at $(0, 0)$ **17.** 320 m   **19.** 2 Ω

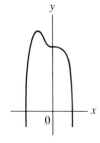

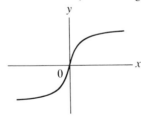

Section 21.3 (page 673)

1. min.: $(1, -2)$; concave up everywhere   **3.** max.: $(\frac{1}{2}, \frac{3}{2})$; concave down everywhere

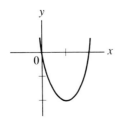

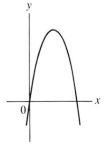

5. max.: $(-3, \frac{1}{2})$; concave down everywhere

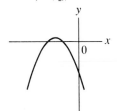

7. max.: $(-1, 5)$, min.: $(1, -3)$; concave up on $[0, \infty)$, concave down on $(-\infty, 0]$; $(0, 1)$ is an inflection point

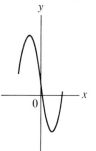

9. max.: $(1, 1)$, min.: $(3, -3)$; concave up on $[2, \infty)$, concave down on $(-\infty, 2]$; $(2, -1)$ is an inflection point

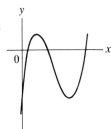

11. max.: $(-1, 16)$, min.: $(3, -16)$; concave up on $[1, \infty)$, concave down on $(-\infty, 1]$; $(1, 0)$ is an inflection point

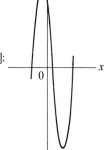

13. min. at $x = 3$; concave up everywhere.

15. min. at $x = 1$; concave up on $(-\infty, 0]$ and $[\frac{2}{3}, \infty)$, concave down on $[0, \frac{2}{3}]$; inflection points at $x = 0, \frac{2}{3}$

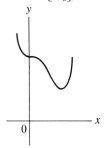

17. min. at $x = -\frac{3}{4}$; concave up on $(-\infty, -\frac{1}{2}]$ and $[0, \infty)$, concave down on $[-\frac{1}{2}, 0]$; inflection points at $x = -\frac{1}{2}, 0$

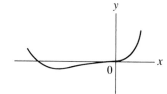

19. min. at $x = 3$; concave up everywhere

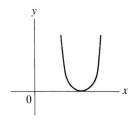

21. intercept $(0, 0)$; vertical asymptote: $x = 3$; horizontal: $y = 1$; no min. or max.; concave down, $x < 3$, concave up, $x > 3$

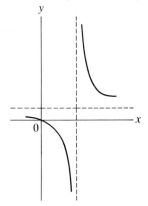

23. vertical asymptote: y-axis; asymptotic curve: $y = x^2$; min. at $x = \sqrt[3]{4}$; concave up on $(-\infty, -2]$, concave down on $[-2, 0)$, concave up on $(0, \infty)$; inflection point at $x = -2$

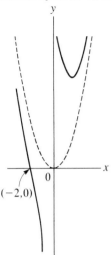

$(-2, 0)$

25. vertical asymptote: $x = 2$; horizontal asymptote: $y = 1$; no min. or max.; decreasing everywhere; concave down, $x < 2$, concave up, $x > 2$

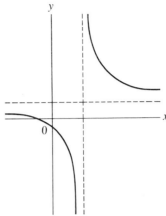

27. domain: $x \geq 0$; vertical tangent at $(0, 0)$; min. at $x = 0$; concave down

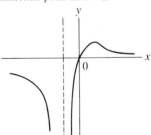

29. no min. or max.; vertical tangent at $(3, 0)$; concave down, $x \geq 3$, concave up, $x \leq 3$; inflection point: $(3, 0)$

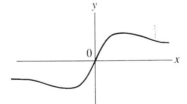

31. vertical asymptote: $x = -1$; max. at $x = 1$; concave down on $(-\infty, -1)$, concave down on $(-1, 2]$, concave up on $[2, \infty)$; inflection point at $x = 2$

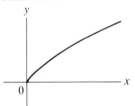

33. max. at $x = \sqrt{3}$, min. at $x = -\sqrt{3}$; inflection points at $x = 0, \pm 3$

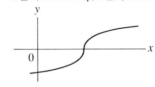

35. min. at $x = -1$, max. at $x = 1$; concave up on $(-\infty, -1/\sqrt{2}]$ and $[0, 1/\sqrt{2}]$; concave down on $[-1/\sqrt{2}, 0]$ and $[1/\sqrt{2}, \infty)$; inflection points at $x = 0, \pm 1/\sqrt{2}$

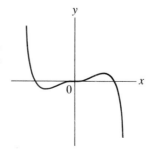

37. min. at $x = 0$; concave up everywhere

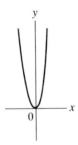

Section 21.4 (page 678)

3. $L/2$ **5.** 30, 30   **7.** 20 m × 10 m   **9.** 2 cm   **11.** $(\sqrt{2}, \sqrt{2})$ **13.** square with side $10\sqrt{2}$ cm   **15.** $\frac{1}{2}$ C

17. base: 4 in. × 4 in.; height: 2 in.   **19.** 35   **21.** width $= \sqrt{3}$ ft, depth $= \sqrt{6}$ ft   **23.** $h = r$ **25.** $r = \dfrac{5}{4 + \pi}$ m   **27.** $\frac{1}{3}$ height of cone

29. $\frac{1}{2}$ **31.** 8 km from nearest point   **33.** 20 ft, 60 ft

Section 21.5 (page 684)

1. $x - 3y + 3 = 0$ **3.** $y^2 = x - 1$ **5.** $y = 2 + e^x$ **7.** $\mathbf{v} = (-2, 2), 2\sqrt{2}, 135°$ **9.** $\mathbf{v} = \left(\dfrac{1}{4}, -\dfrac{1}{8}\right), \dfrac{\sqrt{5}}{8}$ m/s, $-26.6°$

11. $\mathbf{v} = (-5, 1), \sqrt{26}$ m/s, $168.7°$ **13.** $\mathbf{v} = (11, -12), 16.28$ m/s, $-47.5°$; $\mathbf{a} = (6, -12), 6\sqrt{5}$ m/s², $-63.4°$

15. $\mathbf{v} = (-1, 3), \sqrt{10}$ m/s, $108.4°$; $\mathbf{a} = \left(\dfrac{3}{2}, -2\right), \dfrac{5}{2}$ m/s², $-53.1°$ **17.** $(10\sqrt{13}$ m/s, $16.1°), (10$ m/s², $-90°)$

Section 21.6 (page 690)

1. 4 ft/s   **3.** $-\dfrac{9}{4}$ ft/s   **5.** -2 A/s   **7.** 0.16 Ω/s   **9.** Rate of change $= -\dfrac{8}{3}$ m/min   **11.** $\dfrac{1}{20\pi} = 0.0159$ cm/min   **13.** 53 Pa/min

15. 2.2 ft/s   **17.** 210 km/h   **19.** 25 km/h   **21.** $-\frac{1}{4}$ unit/min   **23.** $\dfrac{2\sqrt{13}}{3} = 2.4$ m/min   **25.** $\dfrac{3}{16\pi} = 0.060$ ft/s   **27.** 14 ft/s

29. 2.6 m/s   **31.** $\dfrac{4}{\pi} = 1.3$ ft/min   **33.** $\frac{4}{75} = 0.053$ m/min

Section 21.7 (page 694)

1. $dy = (3x^2 - 1)\,dx$ **3.** $dy = -\dfrac{dx}{(x - 1)^2}$ **5.** $\Delta y = 0.31, dy = 0.3$ **7.** 0.24 cm², 0.67%   **9.** ±0.75 in.³, 0.6%   **11.** 3%, 2%

13. ±0.07 s, 2.5%   **15.** 4.2 W   **17.** $2\pi r\,dr$

Review Exercises for Chapter 21 (page 694)

1. $x - 2y - 1 = 0$; $2x + y - 7 = 0$ **2.** $2x + 9y + 23 = 0$; $9x - 2y - 24 = 0$

3. min. at $x = 2$; concave up everywhere

4. min. at $x = 3$, max. at $x = 1$; concave down on $(-\infty, 2]$, concave up on $[2, \infty)$; inflection point at $x = 2$

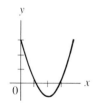

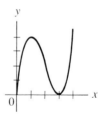

5. min. at $x = -2$, max. at $x = 2$; concave up on $(-\infty, 0]$, concave down on $[0, \infty)$; inflection point at $x = 0$

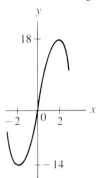

6. min. at $x = 2$; concave up on $(-\infty, 0]$ and $[\frac{4}{3}, \infty)$, concave down on $[0, \frac{4}{3}]$; inflection points at $x = 0$ and $x = \frac{4}{3}$

7. min. at $x = 1$; concave up on $(-\infty, 0]$ and $[\frac{2}{3}, \infty)$, concave down on $[0, \frac{2}{3}]$; inflection points at $x = 0$ and $x = \frac{2}{3}$

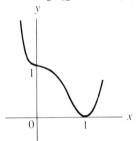

8. max.: $(0, 0)$; concave down on $(-1, 1)$, concave up on $(-\infty, -1)$ and $(1, \infty)$; no inflection points; vertical asymptotes: $x = \pm 1$, horizontal: $y = 1$

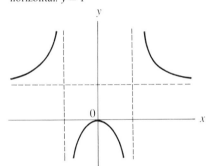

9. min. at $x = -1/\sqrt[3]{2}$; concave up on $(-\infty, 0)$ and $[1, \infty)$, concave down on $(0, 1]$; inflection point: $(1, 0)$; vertical asymptote: $x = 0$, asymptotic curve: $y = x^2$

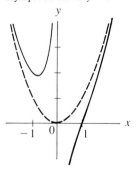

10. max. at $x = \sqrt{3}$, min. at $x = -\sqrt{3}$; concave up on $[-\sqrt{6}, 0)$ and $[\sqrt{6}, \infty)$, concave down on $(-\infty, -\sqrt{6}]$ and $(0, \sqrt{6}]$; inflection points at $x = \pm\sqrt{6}$; vertical asymptote: $x = 0$, horizontal: $y = 0$

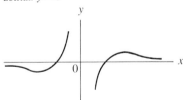

11. $\dfrac{2a}{3}$ **14.** 24 ft × 60 ft   **17.** on land $(50 - 8\sqrt{5})$ m   **18.** $2\sqrt{6}$ mi from nearest point   **19.** 0.095 A/s

20. 3.75 cm/min toward lens   **21.** $\dfrac{1}{\pi}$ m³/min   **22.** $6\sqrt{3}$ m/min   **23.** 840 (lb/in.²)/s   **24.** $\Delta y = 0.0150062$, $dy = 0.015$

25. ± 6 cm³, 0.6%   **26.** 15%   **27.** ± 0.63 in.², 0.8%   **28.** $\mathbf{v} = (1, -2)$, $\sqrt{5}$ m/s, $-63.4°$; $\mathbf{a} = \left(-\dfrac{1}{8}, -2\right)$, 2.00 m/s², 266.4°

29. $\mathbf{v} = (-1, 4)$, $\sqrt{17}$ m/s, 104.0°; $\mathbf{a} = \left(\dfrac{2}{3}, 4\right)$, $\dfrac{2}{3}\sqrt{37}$ m/s², 80.5°

Cumulative Review Exercises for Chapters 19–21 (page 696)

1. $5x - 3y + 13 = 0$ **3.** $\sqrt{26}$ **4.** $(0, \pm 3), (0, \pm \sqrt{6})$ **5.** $y^2 = -16(x - 1)$ **6.** hyperbola, $(1, -1)$

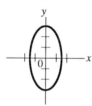

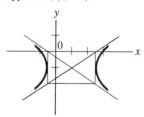

7. $r^2 = 4 \sec 2\theta$ **8.** 6   **9.** 2/3   **10.** $y' = -\dfrac{2x}{(x^2 - 4)^2}$ **11.** $\dfrac{-6x}{(1 - 3x^2)^{1/2}}$ **12.** $\dfrac{3x + 4}{2\sqrt{x + 2}}$ **13.** $\dfrac{3 - x}{2\sqrt{x}(x + 3)^2}$

14a. $\dfrac{dy}{dx} = -\dfrac{2xy - 3y^2}{x^2 - 6xy}$ **b.** $5x - y - 4 = 0$ **15.** $(-1, 8), (1, -4)$

16. min. at $x = 1$; concave up on $(-\infty, 0]$ and $\left[\dfrac{2}{3}, \infty\right)$, concave down on $\left[0, \dfrac{2}{3}\right]$; inflection points at $x = 0, \dfrac{2}{3}$

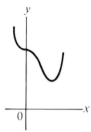

17. 3 in. from vertex   **18.** $x^2 + y^2 = (0.206x + 3.442 \times 10^7)^2$ **19.** Cardioid   **20.** 1.80×10^{-3} V   **21.** 6 ft $\times$ 6 ft $\times$ 3 ft

22. 9.1 in.^{3}/min   **23.** $\mathbf{v} = \left(-\dfrac{1}{2}, 2\right), \dfrac{\sqrt{17}}{2}$ m/s, 104.0°   **24.** 8%

Chapter 22

Section 22.1 (page 699)

1. $3x + C$ **3.** $x - x^3 + C$ **5.** $\dfrac{1}{2}x^4 - x^3 + \dfrac{1}{2}x^2 + C$ **7.** $\dfrac{1}{4}x^4 - x^3 + C$ **9.** $\dfrac{1}{6}x^6 - \dfrac{6}{5}x^5 + \dfrac{1}{2}x^4 + 3x + C$ **11.** $-\dfrac{1}{x} - 2x + C$

13. $-\dfrac{3}{x} + 3x^{2/3} + C$ **15.** $-\dfrac{2}{3x} - \dfrac{5}{8x^2} + \dfrac{2}{3}x^{3/2} + C$

Section 22.2 (page 703)

1. $\dfrac{1}{2}$ **3.** $\dfrac{1}{3}$ **5.** 8

Section 22.3 (page 706)

3. 1   **5.** $\dfrac{5}{4}$ **7.** $\dfrac{32}{3}$ **9.** $\dfrac{4}{9}$

Section 22.5 (page 712)

1. $\frac{2}{3}x\sqrt{x} + C$ **3.** $-\frac{1}{2x^2} + \frac{3}{x} + C$ **5.** $\frac{4}{3}x^{3/2} - x^3 + x + C$ **7.** $-\frac{1}{3x^3} + 2\sqrt{x} - 4x + C$ **9.** $\frac{1}{4}(2x^2 - 3)^4 + C$

11. $-\frac{1}{10}(2 - x^2)^5 + C$ **13. (a)** $x - \frac{1}{2}x^2 + C$ **(b)** $-\frac{1}{5}(1 - x)^5 + C$ **15.** $-\frac{1}{2(x^2 - 1)} + C$ **17.** $\frac{1}{4}(2x^2 + x)^4 + C$

19. $-2\sqrt{1 - x} + C$ **21.** $-\sqrt{1 - x^2} + C$ **23.** $-\frac{1}{12}(1 - 3x^3)^{4/3} + C$ **25.** $4\sqrt[4]{x} + C$ **27.** $\frac{5}{12}(x^4 - 2x)^{6/5} + C$

29. $\frac{1}{5}x^5 + \frac{2}{3}x^3 + x + C$ **31.** $5(x^2 - 3)^{1/2} + C$ **33.** $x + \frac{4}{3}x^{3/2} + \frac{1}{2}x^2 + C$ **35.** $-\frac{3}{35}(1 - 5x)^{7/3} + C$ **37.** $\frac{8}{3}(x - x^2)^{3/4} + C$

39. $\frac{3}{8}x^8 + \frac{6}{5}x^5 + \frac{3}{2}x^2 + C$ **41.** $\frac{5}{12}(x^3 + 1)^4 + C$ **43.** $24x^8 - \frac{96}{5}x^5 + 6x^2 + C$ **45.** $\frac{1}{2}(4x - 2x^2)^{3/2} + C$ **47.** $\frac{2}{3}(1 + \sqrt{x})^{3/2} + C$

49. $3x^8 + \frac{24}{5}x^5 + 3x^2 + C$ **51.** $\frac{1}{3}(x^4 + 2)^3 + C$ **53.** $\frac{1}{8}x^8 + \frac{2}{5}x^5 + \frac{1}{2}x^2 + C$ **55.** $-\frac{1}{12}(3 - x^4)^3 + C$ **57.** $\frac{1}{2}$ **59.** $\frac{45}{4}$

61. $\frac{2}{3}$ **63.** $\frac{8}{15}$ **65.** 2   **67.** $\frac{26}{3}$ **69.** 7   **71.** $\sqrt{3}$

Section 22.6 (page 718)

1. 1   **3.** 1   **5.** 1   **7.** $\frac{9}{2}$ **9.** $\frac{1}{2}$ **11.** $\frac{1}{4}$ **13.** $\frac{32}{3}$ **15.** 12   **17.** $\frac{1}{3}$ **19.** $\frac{1}{12}$ **21.** $\frac{32}{3}$ **23.** $\frac{32}{3}$ **25.** $\frac{9}{2}$ **27.** $\frac{16}{3}$

29. $\frac{2}{3}$ **31.** 1   **33.** 1   **35.** 2   **37.** 4   **39.** 8   **41.** does not exist

Section 22.7 (page 724)

1. $y = \frac{3}{2}x^2 + 1$ **3.** $y = 2x^3 + x + 4$ **5.** $y = x^3 + 2x - 3$ **7.** 3 s   **9.** 45 m   **11.** 5 s   **13.** $t = 2.3$ s, $v = 33$ m/s

15. 5 s, 35 m/s   **17.** no; he stops in 56 m   **19.** $10\sqrt{3}$ m/s   **21.** $v = \frac{1}{2}\left(21 - \frac{1}{t^2 + 1}\right)$ **23.** 4.7 C   **25.** 0.38 C   **27.** 56.7 V

Section 22.8 (page 729)

1. 8.704, 8.667 $(\frac{26}{3})$ **3.** 0.783, 0.785   **5.** 2.793, 2.797   **7.** 3.376   **9.** 8.146   **11.** 3.6535   **13.** 56.7

Review Exercises for Chapter 22 (page 730)

1. 27   **2.** $\frac{124}{5}$ **3.** $2x\sqrt{x} + \frac{1}{3}x^{-3} + x + C$ **4.** $\frac{1}{4}(5x^2 + 4)^4 + C$ **5.** $-\frac{1}{12}(1 - x^2)^6 + C$ **6.** $2\sqrt{x - 4} + C$ **7.** $3\sqrt{x^2 - 2} + C$

8. $\frac{4}{9}(x^3 + 1)^3 + C$ **9.** $\frac{3}{8}x^8 + \frac{6}{5}x^5 + \frac{3}{2}x^2 + C$ **10.** $\frac{2}{3}(\sqrt{x} - 1)^3 + C$ **11.** $\frac{1}{2}x^2 - \frac{4}{3}x\sqrt{x} + x + C$ **12.** $-\frac{1}{6}(3 - x^4)^{3/2} + C$

13. $\frac{1}{3}(x^2 - 4x)^{3/2} + C$ **14.** $\frac{16}{3}$ **15.** $\frac{40}{3}$ **16.** $\frac{16}{3}$ **17.** $\frac{32}{3}$ **18.** $\frac{37}{12}$ **19.** $\frac{9}{2}$ **20.** $\frac{9}{2}$ **21.** $\frac{1}{8}$ **22.** does not exist   **23.** 2

24. 2   **25.** 4.75 C   **26.** 46.3 V   **27.** 3.8 s   **28.** 2.1 s, $v = 41$ m/s   **29.** 0.916

Chapter 23

Section 23.2 (page 734)

1. $\frac{14}{5}$ **3.** 0   **5.** $\frac{\sqrt{2}}{2}$ **7.** $\frac{\sqrt{42}}{6}$ **9.** 30 m/s   **11.** 3.3 A   **13.** 56 W

Section 23.3 (page 738)

1. 84π **3.** $\dfrac{4\pi}{21}$ **5.** 4π **7.** $\dfrac{206\pi}{15}$ **9.** $\dfrac{32\pi}{3}$ **11.** $\dfrac{2\pi}{3}$ **13.** 4π **15.** $\dfrac{16\pi}{3}$ **17.** $\dfrac{32\pi}{3}$ **19.** $\dfrac{32\pi}{3}$ **21.** $\dfrac{64\pi}{3}$ **23.** 2π

25. $\dfrac{72\pi}{5}$ **27.** 1880 cm^3 **29.** $99\pi = 310 \text{ ft}^3$ **31.** π **33.** π

Section 23.4 (page 744)

1. 18π **3.** $\dfrac{\pi}{6}$ **5.** $\dfrac{24\pi}{5}$ **7.** $\dfrac{16\pi}{3}$ **9.** $\dfrac{\pi}{6}$ **11.** $\dfrac{3\pi}{10}$ **13.** $\dfrac{\pi}{2}$ **15.** $\dfrac{8\pi}{3}$ **17.** $\dfrac{2048\pi}{3}$ **19.** $\dfrac{5\pi}{3}$ **23.** 32π

25. (a) 64π (b) $\dfrac{512\pi}{7}$ (c) $\dfrac{1024\pi}{35}$ (d) $\dfrac{704\pi}{5}$ **27.** (a) $\dfrac{32\pi}{15}$ (b) $\dfrac{13\pi}{6}$ **31.** $\dfrac{\pi}{4}$

Section 23.5 (page 754)

1. $\left(\dfrac{17}{11}, \dfrac{73}{22}\right)$ **3.** $\left(\dfrac{17}{3}, \dfrac{4}{3}\right)$ **5.** $\left(\dfrac{1}{3}, \dfrac{1}{3}\right)$ **7.** $\left(\dfrac{1}{3}, \dfrac{2}{3}\right)$ **9.** $\left(\dfrac{2}{3}, \dfrac{1}{3}\right)$ **11.** $\left(\dfrac{2}{3}, \dfrac{4}{3}\right)$ **13.** $\left(0, \dfrac{8}{5}\right)$ **15.** $\left(\dfrac{3}{2}, \dfrac{6}{5}\right)$ **17.** $\left(\dfrac{62}{15}, \dfrac{1}{3}\right)$

19. $\left(\dfrac{1}{10}, \dfrac{1}{2}\right)$ **21.** $\left(\dfrac{4a}{3\pi}, \dfrac{4a}{3\pi}\right)$ **23.** $\left(1, \dfrac{8}{5}\right)$ **25.** $\left(\dfrac{2}{5}, 1\right)$ **27.** along the axis, $\dfrac{4r}{3\pi}$ units from center

29. $\left(\dfrac{a}{3}, \dfrac{b}{3}\right)$, if legs are placed along positive axes   **31.** (a) $\bar{x} = \dfrac{3}{4}$ (b) $\bar{y} = \dfrac{4}{3}$ **33.** $\bar{x} = \dfrac{3r}{8}$ **35.** (a) $\bar{x} = \dfrac{5}{8}$ (b) $\bar{y} = \dfrac{1}{2}$

37. along the axis, one-fourth of the way from base   **39.** $\left(0, \dfrac{5}{6}b\right)$

Section 23.6 (page 759)

1. $\dfrac{\rho}{4}, \dfrac{\sqrt{2}}{2}$ **3.** $\dfrac{\sqrt{21}}{7}$ **5.** $I_y = \dfrac{\rho}{6}, R_y = \dfrac{\sqrt{6}}{6}, I_x = \dfrac{2\rho}{3}, R_x = \dfrac{\sqrt{6}}{3}$ **7.** $I_y = \dfrac{64\rho}{15}, R_y = \dfrac{2\sqrt{5}}{5}$ **9.** $I_y = \dfrac{32\rho}{7}, R_y = \dfrac{2\sqrt{42}}{7}$

11. $I_y = \dfrac{243\rho}{20}, R_y = \dfrac{3\sqrt{30}}{10}$ **13.** $I_x = \dfrac{\rho}{20}, R_x = \dfrac{\sqrt{30}}{10}$ **15.** $I_y = \dfrac{192\rho}{5}$ **17.** $I_y = \dfrac{1}{2}\pi r^4 h\rho = \dfrac{1}{2}mr^2; R_y = \dfrac{r\sqrt{2}}{2}$

19. $I_x = \dfrac{2^{11}\pi}{9} \approx 715, R_x = \dfrac{4\sqrt{5}}{3}$ **21.** $\dfrac{8\pi\rho}{5}$ **23.** $\dfrac{2\sqrt{3}}{3}$ **25.** $\dfrac{\sqrt{30}}{9}$ **27.** $7.1 \text{ J}, 0.38 \text{ kg}\cdot\text{m}^2/\text{s}$

Section 23.7 (page 764)

(Recall that $w = 10,000 \text{ N/m}^3$.)   **1.** 96 ft-lb **3.** (a) 12 ft-lb (b) 63 ft-lb **5.** $54w \text{ J}$ **7.** 600 J **9.** 700 ft-lb **11.** $216w \text{ J}$

13. $\dfrac{75\pi w}{4} \text{ J}$ **15.** (a) 18 ergs (b) 20 ergs **17.** $40w \text{ N}$ **19.** $\left(\dfrac{9}{2}\right)w \text{ N}$ **21.** $\dfrac{8w}{3} \text{ N}$ **23.** $42w \text{ N}$ **25.** $\dfrac{27w}{2} \text{ N}$ **27.** $\dfrac{640w}{3} \text{ N}$

29. $\dfrac{4w\sqrt{3}}{3} \text{ N}$

Review Exercises for Chapter 23 (page 765)

1. 64 ft/s **2.** 3.2 A **3.** 30 W **4.** $\dfrac{4}{3}\pi r^3$ **5.** (a) 64π (b) $\dfrac{1024\pi}{35}$ (c) $\dfrac{704\pi}{5}$ (d) $\dfrac{512\pi}{7}$ **6.** $\dfrac{9\pi}{14}$ **7.** $\dfrac{5\pi}{6}$

8. $\dfrac{32\pi}{15}$ **9.** $\dfrac{5\pi}{3}$ **10.** 8π **11.** $\left(\dfrac{3}{4}, \dfrac{8}{5}\right)$ **12.** $\dfrac{64\rho}{15}$ **13.** $\left(\dfrac{3}{5}, \dfrac{2}{35}\right)$ **14.** $\left(-\dfrac{1}{2}, \dfrac{2}{5}\right)$ **15.** $\left(\dfrac{27}{16}, 0\right)$ **16.** $\left(0, \dfrac{7}{22}\right)$ **17.** $\dfrac{\rho}{20}, \dfrac{\sqrt{30}}{10}$

18. $\dfrac{4\sqrt{21}}{7}, \dfrac{2\sqrt{5}}{5}$ **19.** $\dfrac{4\sqrt{5}}{3}$ **20.** $\dfrac{\pi\rho}{26}$ **21.** (a) 8 ft-lb (b) 16 ft-lb **22.** 1000 J **23.** $1.06 \times 10^9 \text{ J}$ **24.** $4.2 \times 10^6 \text{ J}$

25. $585,000 \text{ N}$ **26.** $8.3 \times 10^5 \text{ N}$

Chapter 24

Section 24.1 (page 772)

1. $\dfrac{1}{2}$ **3.** -1 **5.** $-\dfrac{2\sqrt{3}}{3}$ **7.** $-\dfrac{1}{2}$ **9.** -2 **11.** $\dfrac{\sqrt{3}}{3}$ **13.** undefined   **15.** 0   **17.** $\dfrac{\sqrt{2}}{2}$ **19.** $\dfrac{2\sqrt{3}}{3}$ **21.** $-\dfrac{\sqrt{3}}{3}$ **23.** 1

25. $\dfrac{\pi}{3}$ **27.** $\dfrac{5\pi}{6}$ **29.** $\dfrac{3\pi}{4}$ **31.** $\dfrac{4\pi}{5}$ **33.** $\dfrac{\pi}{9}$ **35.** $45°$ **37.** $30°$ **39.** $300°$ **41.** $198°$ **43.** $-100°$

45. $P = \pi,\ A = \dfrac{1}{3}$

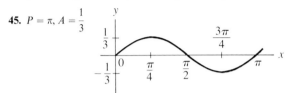

47. $P = 4\pi,\ A = 3$

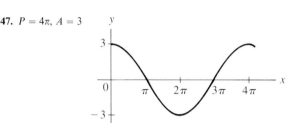

49. $\sin\theta$ **51.** $\dfrac{\sin\theta + 1}{\cos\theta}$ **53.** $-\cos\theta$ **55.** 1   **57.** $\cos^2\theta$ **59.** $\dfrac{1}{\cos^2\theta\,\sin^2\theta}$ **61.** $\dfrac{1}{\sin\theta}$ **63.** $\sin^2 4x$ **65.** $\cos^2 2x$

67. $1 - \sin^2 5x$ **69.** $1 - \cos^2 6x$ **71.** $\sec^2 6x$ **73.** $\tan^2 2x$ **75.** $\sec^2 5x - 1$ **77.** $1 + \tan^2 7x$ **79.** $\csc^2 3x - 1$

85. $\sin 10x$ **87.** $\dfrac{1}{2}\sin x$ **89.** $\cos 6x$ **91.** $\cos 16x$ **93.** $\dfrac{1}{2}(1 - \cos 6x)$ **95.** $\dfrac{1}{2}(1 + \cos 4x)$ **97.** $\dfrac{1}{2}(1 - \cos x)$

Section 24.2 (page 778)

1. $-5\sin 5x$ **3.** $-8\sin 4x$ **5.** $2x\cos x^2$ **7.** $-9x^2\sin x^3$ **9.** $3\cos 3x$ **11.** $x\cos x + \sin x$ **13.** $\dfrac{x\cos x - \sin x}{x^2}$

15. $-2x\sin(x^2 + 3)$ **17.** $\cos 2x - 2x\sin 2x$ **19.** $4x\cos(2x + 2) + 2\sin(2x + 2)$ **21.** $-\dfrac{\cos(1/x)}{x^2}$ **23.** $\dfrac{\cos x + x\sin x}{\cos^2 x}$

25. $\dfrac{\sin 4x - 4x\cos 4x}{\sin^2 4x}$ **27.** $-\dfrac{2x\sin 2x + \cos 2x}{3x^2}$ **29.** $\sqrt{x}\cos x + \dfrac{\sin x}{2\sqrt{x}}$ **31.** $-6x^2\cos x^3\sin x^3$

33. $\cos^2 x - \sin^2 x = \cos 2x$ **35.** $\dfrac{\sin^2 x(3x\cos x - \sin x)}{x^2}$ **37.** $\cos 3x(\cos 3x - 6x\sin 3x)$ **43.** slope $= 1$

Section 24.3 (page 782)

1. $5\sec 5x\tan 5x$ **3.** $-6\csc 3x\cot 3x$ **5.** $-12\csc^2 4x$ **7.** $-4x\csc x^2\cot x^2$ **9.** $2\sec^2 2x$ **11.** $-3x^2\csc^2 x^3$

13. $3x^2\sec(x^3 + 1)\tan(x^3 + 1)$ **15.** $\dfrac{\sec\sqrt{x}\tan\sqrt{x}}{2\sqrt{x}}$ **17.** $-\dfrac{3\csc^2\sqrt{3x}}{2\sqrt{3x}}$ **19.** $\dfrac{\sec^2 2x}{\sqrt{\tan 2x}}$ **21.** $32\tan^3 4x\sec^2 4x$

23. $-x\sqrt{\csc x^2}\cot x^2$ **25.** $x\csc x(2 - x\cot x)$ **27.** $-\cot^2 x - 2\cos^2 x$ **29.** $\dfrac{\sin x\sec^2 x - \cos x - \sin x}{\sin^2 x}$

31. $2x(x - 1)\cos(1 - x)^2 + \sin(1 - x)^2$ **33.** $\dfrac{3x^2(\tan 3x - x\sec^2 3x)}{\tan^2 3x}$ **35.** $\dfrac{1 + 4x^2\cot x^2}{\csc^2 x^2}$ **39.** $16\sec 4x(\sec^2 4x + \tan^2 4x)$

41. $2\sec^2 x(x\tan x + 1)$ **43.** $\dfrac{\sec^2 x}{2y}$ **45.** $\dfrac{\sec x(x\tan x + 1)}{2y}$ **47.** $\dfrac{\cos(x + y^2)}{2y - 2y\cos(x + y^2)}$ **49.** $\dfrac{\cot y^2}{1 + 2xy\csc^2 y^2}$ **51.** $\dfrac{2 - 2xy}{x^2 + \sin y}$

53. $\dfrac{1}{8}$

Section 24.4 (page 785)

1. $\dfrac{\pi}{4}$ **3.** $-\dfrac{\pi}{6}$ **5.** 0   **7.** $-\dfrac{\pi}{2}$ **9.** π **11.** $-\dfrac{\pi}{4}$ **13.** $-\dfrac{\pi}{4}$ **15.** $\dfrac{2\sqrt{5}}{5}$ **17.** $\dfrac{\sqrt{37}}{37}$ **19.** $\dfrac{2\sqrt{5}}{5}$ **21.** $\dfrac{3\sqrt{2}}{4}$ **23.** $-\dfrac{\sqrt{7}}{3}$

25. $\sqrt{1-x^2}$ **27.** $\dfrac{1}{x}$ **29.** $\dfrac{1}{\sqrt{1+4x^2}}$ **31.** $\dfrac{\sqrt{1-4x^2}}{2x}$ **33.** $\sqrt{1-9x^2}$ **35.** $2\tan\dfrac{y}{2}$ **37.** $\text{Arcsin}(y-1)$ **39.** $\dfrac{1}{3}\text{Arccos}(y-1)$

41. $x=\dfrac{1}{4}\text{Arctan}\dfrac{y-1}{3}$ **43.** $x=2\text{Arcsin}(y+2)$

Section 24.5 (page 788)

1. $\dfrac{3}{1+9x^2}$ **3.** $\dfrac{-5}{\sqrt{1-25x^2}}$ **5.** $\dfrac{4x}{1+4x^4}$ **7.** $\dfrac{6x}{\sqrt{1-9x^4}}$ **9.** $\dfrac{2}{\sqrt{1-4x^2}}$ **11.** $-\dfrac{2x}{\sqrt{1-x^4}}$ **13.** $\dfrac{4x}{\sqrt{1-4x^4}}$ **15.** $\dfrac{7}{1+49x^2}$

17. $\dfrac{x}{1+x^2}+\text{Arctan }x$ **19.** $-\dfrac{2x^2}{\sqrt{1-x^4}}+\text{Arccos }x^2$ **21.** $\dfrac{3x}{\sqrt{1-9x^2}}+\text{Arcsin }3x$ **23.** $\dfrac{6x}{1+9x^2}+2\text{Arctan }3x$

25. $\dfrac{x-\sqrt{1-x^2}\,\text{Arcsin }x}{x^2\sqrt{1-x^2}}$ **27.** $\dfrac{1}{2\sqrt{x-x^2}}$ **29.** $\dfrac{\sqrt{x}}{\sqrt{1-x^2}}+\dfrac{\text{Arcsin }x}{2\sqrt{x}}$ **31.** $\dfrac{2\,\text{Arcsin }x}{\sqrt{1-x^2}}$ **33.** $-\dfrac{1}{2}\left[(1-x^2)\text{Arccos }x\right]^{-1/2}$

35. $\dfrac{2x\,\text{Arcsin }x-\sqrt{1-x^2}\,(\text{Arcsin }x)^2}{x^2\sqrt{1-x^2}}$ **39.** 0.145 rad/s

Section 24.6 (page 793)

1. $\log_3 27=3$ **3.** $\log_9 1=0$ **5.** $\log_{32}\dfrac{1}{2}=-\dfrac{1}{5}$ **7.** $3^5=243$ **9.** $\left(\dfrac{1}{4}\right)^2=\dfrac{1}{16}$ **11.** $9^{-1/2}=\dfrac{1}{3}$ **13.** $\dfrac{1}{9}$ **15.** 4   **17.** 27

19. $\dfrac{1}{9}$ **21.** 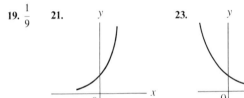   **23.** 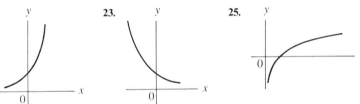   **25.**

27. $\log_3 24$ **29.** $\log_5 4$ **31.** $\log_b\dfrac{\sqrt{3}}{3}=-\dfrac{1}{2}\log_b 3$ **33.** $\log_3\dfrac{2y^2}{25}$ **35.** 3   **37.** $\dfrac{1}{2}(1+\log_6 x)$ **39.** $-2(1+\log_5 x)$

41. $-\dfrac{1}{3}(1+\log_3 x)$ **43.** $-\dfrac{1}{2}\log_5(y-2)$ **45.** $\log_{10} x-\dfrac{1}{2}\log_{10}(x+2)$ **47.** $\dfrac{1}{2}\log_{10} x-\log_{10}(x+1)$ **49.** 1.96   **51.** 0.502

Section 24.7 (page 797)

1. $\dfrac{1}{x}$ **3.** $\dfrac{4}{x}$ **5.** $\dfrac{2}{x}$ **7.** $\dfrac{6}{x}$ **9.** $\dfrac{3}{x}\log_{10} e$ **11.** $\cot x$ **13.** $1+\ln x$ **15.** $\dfrac{1}{2(x-1)}$ **17.** $-\dfrac{1}{2(x+2)}$ **19.** $\dfrac{2x}{x^2+2}$

21. $\dfrac{x+2}{x(x+1)}$ **23.** $\dfrac{1-x^2}{x(x^2+1)}$ **25.** $\dfrac{2+3x^2}{2x(2-x^2)}$ **27.** $\dfrac{2\ln x}{x}$ **29.** $\dfrac{\sqrt{x+1}}{x}+\dfrac{\ln x}{2\sqrt{x+1}}$ **31.** $\dfrac{\sec x}{x}+(\sec x\tan x)\ln x$

33. $\dfrac{x}{x^2-1+\sqrt{x^2-1}}$ **35.** $\dfrac{6\ln^2 x}{x}$ **37.** $\dfrac{1}{x\ln x}$ **39.** $\dfrac{1-\ln x}{x^2}$

Section 24.8 (page 801)

1. $4e^{4x}$ **3.** $2xe^{x^2}$ **5.** $-4xe^{-x^2}$ **7.** $e^{\tan x}\sec^2 x$ **9.** $3x^2(\ln 3)3^{x^3}$ **11.** $\dfrac{2x}{\log_4 e}4^{x^2}=2x(\ln 4)4^{x^2}$ **13.** $2(x+1)e^x$ **15.** $e^{\sin x}\cos x$

17. $e^x\cos e^x$ **19.** $e^{2x}(\cos x+2\sin x)$ **21.** $\dfrac{\sec^2 x-2x\tan x}{e^{x^2}}$ **23.** $e^{\sec x}\left(\sec x\tan x\ln x+\dfrac{1}{x}\right)$ **25.** $2e^{2x}\cot e^{2x}$ **27.** $\dfrac{1}{2ye^{y^2}+1}$

31. $x^{\sin x}[(\sin x)/x+\cos x\ln x]=x^{(\sin x-1)}(\sin x+x\cos x\ln x)$ **33.** $(\ln x)^x[\ln(\ln x)+(1/\ln x)]$ **35.** $(\tan x)^{x-1}(x\sec^2 x+\tan x\ln\tan x)$

37. $i=-2.2$ A

Section 24.9 (page 803)

1. -4 **3.** $\dfrac{3}{2}$ **5.** 7   **7.** 6   **9.** ∞ **11.** 2   **13.** $-\dfrac{1}{2}$ **15.** 0   **17.** -3 **19.** $\dfrac{1}{2}$ **21.** 0   **23.** 0

Section 24.10 (page 808)

1. max. at $x = 1$; inflection point at $x = 2$ **3.** min. at $x = \dfrac{1}{e}$; concave up everywhere   **5.** min. at $x = 0$; concave up everywhere

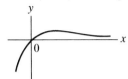

7. $i = \sqrt{2}$ A   **9.** 1.8 V   **11.** $\dfrac{dN}{dt} = -24e^{-2.0t}$ **13.** -0.18 g/min   **15.** 448 bacteria per hour   **17.** $\dfrac{1}{\sqrt{e}}$ **21.** 0.018 rad/s

23. 0.015 rad/s   **25.** $\dfrac{\pi}{6}$ **27.** 9.58 m   **29.** $\dfrac{\pi}{4} + \dfrac{\alpha}{2}$

Review Exercises for Chapter 24 (page 810)

1. $-\dfrac{\pi}{2}$ **2.** $-\dfrac{\pi}{4}$ **3.** $\dfrac{2\pi}{3}$ **4.** 0   **5.** $\dfrac{\sqrt{10}}{10}$ **6.** $\dfrac{2\sqrt{2}}{3}$ **7.** $\dfrac{x}{\sqrt{1+x^2}}$ **8.** $\dfrac{\sqrt{1-x^2}}{x}$ **9.** $3x^2 \sec^2 3x + 2x \tan 3x$

10. $\dfrac{4x \cos 4x - \sin 4x}{x^2}$ **11.** $\dfrac{e^{2x}(2x-1)}{x^2}$ **12.** $\dfrac{1}{x}\cos x - \sin x \ln x$ **13.** $-\dfrac{1}{2(x+3)}$ **14.** $\dfrac{1+x}{2x(1-x)}$ **15.** $-\dfrac{1}{x(2x^2+1)}$

16. $-e^{-x}\cos e^{-x}$ **17.** $\dfrac{1}{2x\sqrt{\ln 2x}}$ **18.** $2e^{2\tan x}\sec^2 x$ **19.** $-\dfrac{\sin(\ln x)}{x}$ **20.** $(2x)(\ln 5)5^{x^2}$ **21.** $e^{2x}(2\cot x - \csc^2 x)$

22. $\dfrac{x}{\sqrt{1-x^2}} + \text{Arcsin } x$ **23.** $-\dfrac{3e^{\text{Arccos } 3x}}{\sqrt{1-9x^2}}$ **24.** $2\tan x$ **25.** $(\cot x)^x(\ln \cot x - x \csc x \sec x)$ **26.** $(\sec x)^x(x \tan x + \ln \sec x)$

27. $\dfrac{dy}{dx} = e^{-\sin y}\sec y \csc x \cot x$ **28.** $\dfrac{dy}{dx} = -\dfrac{y}{1 + y\sec^2 y}$ **29.** 4   **30.** 0   **31.** 1   **32.** $\dfrac{1}{2}$ **33.** 0   **34.** $x + y - 1 = 0$

35. $\dfrac{1}{e} \times 1$ **36.** 20 m/s   **37. (a)** \$22,408.45   **(b)** rP **38.** decreases at 27 kg/m^2 per min   **39.** 3.00×10^{-15} mm of mercury per min

40. $\dfrac{dI}{dt} = \dfrac{E}{L}e^{-(Rt)/L}$ **41.** $\theta = \text{Arcsin } \dfrac{1}{\sqrt{3}} \approx 35.3°$ **42.** $\text{Arctan } \dfrac{1}{2} \approx 26.6°$

Cumulative Review Exercises for Chapters 22–24 (page 811)

1. $x - \dfrac{2}{5}x^{5/2} + C$ **2.** $-\dfrac{2}{9}(1 - 3x)^{3/2} + C$ **3.** $\sqrt{x^2 - 2x} + C$ **4.** $\dfrac{1}{9}(x^3 + 1)^3 + C$ **5.** $\dfrac{1}{8}x^8 + \dfrac{2}{5}x^5 + \dfrac{1}{2}x^2 + C$ **6.** $\dfrac{1}{6}$ **7.** $\dfrac{4}{3}$

8. 0.435   **9.** $\dfrac{\pi}{2}$ **10.** $\dfrac{117\pi}{5}$ **11.** $\dfrac{256\pi}{15}$ **12.** $\left(1, \dfrac{2}{5}\right)$ **13. a.** $\dfrac{\rho}{4}$ **b.** $\dfrac{2\sqrt{3}}{3}$ **14.** $\left(\dfrac{75}{2}\right)w$ N   **15.** $\sin 3x + 3x \cos 3x$

16. $\dfrac{x \sec^2 x - \tan x}{x^2}$ **17.** $-e^{-2x}\sin x - 2 e^{-2x}\cos x$ **18.** $\dfrac{1}{2(x+2)}$ **19.** $\dfrac{6x}{1 + 9x^4}$ **20.** 0   **21.** 2.2 s, 86 ft/s

22. $\dfrac{dN}{dt} = -0.0234$ kg/year   **23.** 180 J

Chapter 25

Section 25.1 (page 814)

1. $\frac{1}{3}(x^2+1)^{3/2}+C$ **3.** $-2\sqrt{1-x}+C$ **5.** $\frac{1}{3}\sin^3 x+C$ **7.** $\frac{1}{3}\tan^3 x+C$ **9.** $\frac{1}{4}(1+\tan x)^4+C$ **11.** $\frac{1}{4}(1+e^x)^4+C$

13. $\frac{1}{2}\ln^2 x+C$ **15.** $2\sqrt{\ln x}+C$ **17.** $-\frac{1}{2}(\text{Arccos } x)^2+C$ **19.** $\frac{\sqrt{3}}{8}$ **21.** $\frac{2}{3}$ **23.** $-\frac{1}{2}\cot^2 x+C$ **25.** $\frac{2}{3}(\tan x)^{3/2}+C$

27. $\frac{1}{8}(\text{Arctan } 4x)^2+C$ **29.** $\frac{1}{3}(1+\ln x)^3+C$

Section 25.2 (page 818)

1. $\ln|x-1|+C$ **3.** $\frac{1}{3}\ln|2+3x|+C$ **5.** $-\frac{1}{3}\ln|1-3x|+C$ **7.** $\ln\sqrt{2}$ **9.** $-e^{-x}+C$ **11.** $\frac{2}{3}(e^6-1)$ **13.** $\frac{1}{4}e^{4x}+C$

15. $\frac{1}{2}e^{x^2}+C$ **17.** $e^{\sin x}+C$ **19.** $\ln|1+\tan x|+C$ **21.** $e^{\text{Arctan } x}+C$ **23.** $\ln|\ln x|+C$ **25.** $x-e^{-x}+C$ **27.** $\ln|1-e^{-x}|+C$

29. $-\frac{1}{1+x^2}+C$ **31.** $x+2e^x+\frac{1}{2}e^{2x}+C$ **33.** $x-\ln|x+2|+C$ **35.** $-\csc x+C$ **37.** $e^{\sin^2 x}+C$ **39.** $\ln 2$

41. $\frac{1}{3}(5^{3x})\log_5 e+C=\frac{5^{3x}}{3\ln 5}+C$ **43.** $-2^{\cot x}\log_2 e+C=\frac{-2^{\cot x}}{\ln 2}+C$ **45.** $\frac{2}{3}\rho\ln 3$ **47.** $\frac{3}{2}(1-e^{-2/3})$ A   **49.** $\pi\left(1-\frac{1}{e}\right)$

51. $8\pi\ln 2$ **53.** $v=\frac{10}{k}(1-e^{-kt})$ m/s

Section 25.3 (page 821)

1. $\frac{1}{2}\tan 2x+C$ **3.** $\frac{1}{3}\sec 3x+C$ **5.** $-\frac{1}{4}\csc 4x+C$ **7.** $\frac{1}{2}\ln|\csc x^2-\cot x^2|+C$ **9.** $\frac{1}{2}\sin 2x+C$ **11.** $2\ln\left|\sec\frac{1}{2}x\right|+C$

13. $-\frac{1}{2}\cos x^2+C$ **15.** $\ln\left|\sin\frac{1}{2}x^2\right|+C$ **17.** $2\sin\sqrt{x}+C$ **19.** $\frac{1}{16}\tan^4 4x+C$ **21.** $\frac{1}{3}\ln|\csc e^{3x}-\cot e^{3x}|+C$

23. $-\cot(\ln x)+C$ **25.** $-\cot x+C$ **27.** $-\frac{1}{2}\ln|1+2\cos x|+C$ **29.** $\frac{1}{2}$ **31.** $\ln 2$ **33.** 183 V   **35.** π **37.** $\frac{1}{4}\ln 2$

39. $\frac{2E}{\pi}$ V   **41.** 0.66   **43.** $\ln(1+\sqrt{2})$

Section 25.4 (page 828)

1. $\frac{1}{6}\sin^3 2x-\frac{1}{10}\sin^5 2x+C$ **3.** $\frac{1}{3}\cos^3 x-\cos x+C$ **5.** $\frac{1}{7}\cos^7 x-\frac{1}{5}\cos^5 x+C$ **7.** $\frac{1}{4}\sin^4 x-\frac{1}{6}\sin^6 x+C$

9. $\frac{1}{2}x+\frac{1}{16}\sin 8x+C$ **11.** $\frac{1}{8}x-\frac{1}{32}\sin 4x+C$ **13.** $\frac{1}{10}\cos^5 2x-\frac{1}{6}\cos^3 2x+C$ **15.** $\frac{1}{20}\sin^5 4x-\frac{1}{28}\sin^7 4x+C$

17. $\frac{1}{2}\tan^2 x+\ln|\cos x|+C$ **19.** $\frac{1}{3}\tan^3 x+\frac{1}{5}\tan^5 x+C$ **21.** $\frac{1}{3}\sec^3 x+C$ **23.** $\frac{1}{5}\sec^5 x-\frac{1}{3}\sec^3 x+C$

25. $-\frac{1}{14}\cot^7 2x-\frac{1}{18}\cot^9 2x+C$ **27.** $-\cot x-\frac{2}{3}\cot^3 x-\frac{1}{5}\cot^5 x+C$ **29.** $\frac{20}{21}$ **31.** $4+\frac{3}{2}\pi$ **33.** $\ln 2$

35. $\frac{1}{24}\tan^6 4x+\frac{1}{32}\tan^8 4x+C$ **37.** $\frac{\pi^2}{2}$ **39.** $10\sqrt{2}$ A   **41.** $\frac{15}{4}$ W   **43.** 1.25

Section 25.5 (page 831)

1. $\text{Arcsin } x+C$ **3.** $\frac{1}{6}\text{Arctan}\frac{2}{3}x+C$ **5.** $-\sqrt{1-x^2}+C$ **7.** $\frac{1}{18}\ln(16+9x^2)+C$ **9.** $\frac{1}{4}\text{Arctan}\frac{1}{2}x^2+C$

11. $-\text{Arcsin}\left(\frac{1}{2}\cot x\right) + C$ **13.** $\frac{\sqrt{3}}{3}\text{Arcsin}\frac{\sqrt{15}}{5}x + C$ **15.** $\ln(2 + \sin x) + C$ **17.** $\frac{\pi}{12}$ **19.** $\frac{1}{\sqrt{3}}\text{Arcsin}\frac{\sqrt{3}}{2}x + C$

21. $\text{Arctan}(x - 3) + C$ **23.** $\text{Arcsin}\frac{x+2}{\sqrt{5}} + C$ **25.** $\frac{2}{\sqrt{3}}\text{Arctan}\frac{2x+3}{\sqrt{3}} + C$ **27.** $\frac{1}{2}\ln(x^2 + 16) + \text{Arctan}\frac{1}{4}x + C$ **29.** $\frac{\pi}{4}$

31. $\frac{1}{2}\ln(1 + e^{2x}) + C$ **33.** $\text{Arcsin }e^x + C$ **35.** $\frac{\pi}{2}$ **37.** $\frac{\pi^2}{4}$

Section 25.6 (page 836)

1. $-\frac{\sqrt{4-x^2}}{x} - \text{Arcsin}\frac{x}{2} + C$ **3.** $\frac{1}{3}(x^2 + 9)^{3/2} + C$ **5.** $\frac{x}{25\sqrt{x^2+25}} + C$ **7.** $\frac{1}{2}\text{Arcsec}\frac{x}{2} + C$ **9.** $-\frac{\sqrt{x^2+16}}{16x} + C$

11. $-\frac{1}{\sqrt{x^2-2}} + C$ **13.** $\frac{1}{3}(x^2 + 6)\sqrt{x^2-3} + C$ **15.** $-\frac{\sqrt{x^2+1}}{x} + \ln\left|x + \sqrt{x^2+1}\right| + C$ **17.** $\frac{\sqrt{2}}{32}$ **19.** $\frac{9\pi}{4}$

21. $(40\pi + \frac{16}{3})$ w newtons, where $w = 10{,}000$ N/m³   **23.** $\frac{8a^5\pi\rho}{15} = \frac{2}{5}ma^2$ **25.** $(108\pi + 144)w$ J $(w = 10{,}000$ N/m³$)$

Section 25.7 (page 840)

1. $xe^x - e^x + C$ **3.** $\frac{1}{4}\sin 2x - \frac{1}{2}x\cos 2x + C$ **5.** $x\tan x + \ln|\cos x| + C$ **7.** $\frac{1}{4}x^2(2\ln x - 1) + C$ **9.** $x\text{ Arcsin }x + \sqrt{1-x^2} + C$

11. $\frac{1}{3}x\sin 3x + \frac{1}{9}\cos 3x + C$ **13.** $-e^{-x}(x^2 + 2x + 2) + C$ **15.** $x\text{ Arccot }x + \frac{1}{2}\ln(1 + x^2) + C$ **17.** $-\frac{1}{2}\cos x^2 + C$

19. $\frac{1}{2}e^x(\sin x - \cos x) + C$ **21.** $\frac{1}{\pi^2+1}\left[e^{-x}(\pi\sin\pi x - \cos\pi x)\right] + C$ **23.** $\frac{2\ln 2 - 1}{(\ln 2)^2}$ **25.** π **27.** $\left(\frac{\pi^2}{4} - 2\right)\rho$ **29.** 482 V

Section 25.8 (page 850)

1. $\frac{1}{4}\ln\left|\frac{x-2}{x+2}\right| + C$ **3.** $\ln\left|\frac{x}{1-x}\right| + C$ **5.** $\ln\left|(x-2)^2(x+1)^3\right| + C$ **7.** $\frac{1}{2}x^2 + 2x + \frac{27}{4}\ln|x-3| + \frac{1}{4}\ln|x+1| + C$

9. $\ln\left|\frac{(x-4)^2(x-1)}{(x+2)^2}\right| + C$ **11.** $-\frac{2}{x+1} + 3\ln|x+1| + C$ **13.** $-\frac{1}{x-2} - 2\ln|x+1| + C$ **15.** $2\ln|x-2| - \frac{8}{x-2} - \frac{9}{2(x-2)^2} + C$

17. $\ln|x+2| - \frac{3}{2}\text{Arctan}\frac{x}{2} + C$ **19.** $\ln\left|(x+1)(x^2+1)\right| + 2\text{ Arctan }x + C$ **21.** $\frac{1}{2}x^2 - 4\ln(x^2 + 4) - \frac{8}{x^2+4} + C$

23. $\ln|x| - \frac{1}{2}\text{Arctan}\frac{1}{2}(x - 1) + C$ **25.** $\frac{1}{2}\ln|x| - \frac{1}{4}\ln|x^2 + 2x + 2| - \frac{1}{2}\text{Arctan}(x + 1) + C$

Section 25.9 (page 853)

1. $\frac{1}{2}\ln\left|\frac{x}{2+x}\right| + C$ **3.** $\frac{1}{2}(x\sqrt{x^2-7} - 7\ln|x + \sqrt{x^2-7}|) + C$ **5.** $\frac{1}{2\sqrt{5}}\ln\left|\frac{\sqrt{5}+x}{\sqrt{5}-x}\right| + C$ **7.** $-\frac{\sqrt{5x^2+4}}{4x} + C$

9. $-\frac{1}{6}\sin 3x + \frac{1}{2}\sin x + C$ **11.** $\frac{1}{\sqrt{3}}\ln|\sqrt{3}x + \sqrt{3x^2+5}| + C$ **13.** $\frac{1}{2}x^2e^{2x} - \frac{1}{2}xe^{2x} + \frac{1}{4}e^{2x} + C$

15. $\frac{1}{5}\tan^5 x - \frac{1}{3}\tan^3 x + \tan x - x + C$ **17.** $\frac{1}{12}\ln\left|\frac{2x-3}{2x+3}\right| + C$ **19.** $\frac{1}{\sqrt{3}}\ln\left|\frac{\sqrt{3+x}-\sqrt{3}}{\sqrt{3+x}+\sqrt{3}}\right| + C$

21. $\frac{1}{2\sqrt{15}}\ln\left|\frac{\sqrt{3x}-\sqrt{5}}{\sqrt{3x}+\sqrt{5}}\right| + C$ **23.** $\sqrt{x^2-10} - \sqrt{10}\text{ Arccos}\frac{\sqrt{10}}{x} + C$ **25.** $-\frac{1}{10}\cos 5x - \frac{1}{2}\cos x + C$ **27.** $\frac{x}{5\sqrt{4x^2+5}} + C$

Review Exercises for Chapter 25 (page 854)

1. $\frac{1}{2}\ln(x^2 + 1) + C$ **2.** $-\frac{1}{2(x^2 + 1)} + C$ **3.** $2\,\text{Arctan }x + C$ **4.** $e^{\text{Arctan }x} + C$ **5.** $-\sqrt{9 - x^2} + C$ **6.** $\text{Arcsin }\frac{x}{3} + C$

7. $\frac{1}{4}\sin 2x^2 + C$ **8.** $\frac{1}{4}\cos 2x + \frac{1}{2}x \sin 2x + C$ **9.** $\frac{1}{2}\text{Arctan }\frac{1}{2}e^x + C$ **10.** $\ln(4 + e^x) + C$ **11.** $\frac{-1}{4 + e^x} + C$

12. $\frac{x^2}{4}(2\ln x - 1) + C$ **13.** $\frac{1}{2}\ln^2 x + C$ **14.** $\sin(\ln x) + C$ **15.** $\frac{1}{2}\ln|x^2 + 4x + 5| + C$ **16.** $\text{Arctan}(x + 2) + C$

17. $-\frac{1}{x + 2} + C$ **18.** $\text{Arctan}(\ln x) + C$ **19.** $x\ln^2 x - 2x \ln x + 2x + C$ **20.** $\frac{1}{\sqrt{5}}\text{Arcsin }\frac{\sqrt{5}}{2}x + C$

21. $\frac{1}{2}\ln\left|\frac{2 - \sqrt{4 - x^2}}{x}\right| + C$ **22.** $-\frac{1}{3}(4 - x^2)^{3/2} + C$ **23.** $\frac{1}{6}\ln|1 + \sin^3 2x| + C$ **24.** $\frac{1}{10}\cos^5 2x - \frac{1}{6}\cos^3 2x + C$

25. $\frac{1}{2}x - \frac{1}{8}\sin 4x + C$ **26.** $\frac{1}{4}e^{2x^2} + C$ **27.** $\frac{1}{4}e^{2x}(2x - 1) + C$ **28.** $\frac{1}{4}(\text{Arctan }2x)^2 + C$ **29.** $x\,\text{Arctan }2x - \frac{1}{4}\ln(1 + 4x^2) + C$

30. $-e^{-\sin^2 x} + C$ **31.** $e^{\tan x} + C$ **32.** $\cos\frac{1}{x} + C$ **33.** $x - \frac{4}{\sqrt{3}}\text{Arctan }\frac{x}{\sqrt{3}} + C$ **34.** $\ln|1 - \cos x| + C$

35. $\ln|\csc x\,(\csc x - \cot x)| + C = -\ln|1 + \cos x| + C$ **36.** $\frac{3}{2}(\ln x)^{2/3} + C$ **37.** $-\frac{1}{5}\cot^5 x - \frac{1}{7}\cot^7 x + C$ **38.** $\text{Arcsin }\frac{x - 2}{\sqrt{2}} + C$

39. $\frac{1}{10}\sec^5 2x - \frac{1}{6}\sec^3 2x + C$ **40.** $e^{\tan x}(\tan x - 1) + C$ **41.** $\frac{1}{3}\tan^3 x + C$ **42.** $-\ln|2 - \ln x| + C$ **43.** $\frac{1}{3}\tan 3x + \frac{1}{9}\tan^3 3x + C$

44. $\frac{1}{12}\csc^3 4x - \frac{1}{20}\csc^5 4x + C$ **45.** $\frac{1}{17}e^x(4\sin 4x + \cos 4x) + C$ **46.** $-\frac{1}{4}e^{\cos 4x} + C$ **47.** $\frac{1}{2\sqrt{5}}\text{Arctan }\frac{\sqrt{5}}{2}x + C$

48. $\frac{3}{10}\ln(5x^2 + 4) + C$ **49.** $\frac{1}{2}\ln|x^2 + 2x - 8| + C$ **50.** $\frac{1}{3}\ln|(x + 4)(x - 2)^2| + C$ **51.** $\ln|(x - 2)(x^2 + 9)| + C$

Chapter 26

Section 26.1 (page 859)

1. $\frac{3}{2}$ **3.** 1 **5.** 4 **7.** $\frac{1}{2}$ **9.** $\frac{2}{5}$ **11.** $\frac{7}{33}$ **13.** $\frac{251}{495}$

Section 26.2 (page 864)

5. conv. **7.** div. **9.** conv. **11.** div. **13.** conv. **15.** conv. **17.** conv. **19.** div. **21.** conv. **23.** div. **25.** conv.
27. conv. **29.** conv. **31.** conv. **33.** div. **35.** conv. **37.** conv. **39.** div. **41.** conv. **43.** conv. **45.** conv. **47.** conv.
49. div. **51.** conv. **53.** div. **55.** div.

Section 26.4 (page 874)

1. $3x - \frac{3^3 x^3}{3!} + \frac{3^5 x^5}{5!} - \cdots$ **3.** $1 - x + \frac{x^2}{2!} - \frac{x^3}{3!} + \frac{x^4}{4!} - \cdots$ **5.** $1 - \frac{x}{2!} + \frac{x^2}{4!} - \cdots$ **7.** $x - \frac{x^3}{2!} + \frac{x^5}{4!} - \frac{x^7}{6!} + \cdots$

9. $x^2 - \frac{x^4}{2} + \frac{x^6}{3} - \frac{x^8}{4} + \cdots$ **15.** $1 - x + \frac{1}{3}x^3 - \frac{1}{6}x^4 + \cdots$ **17.** $3\left(x^2 - \frac{x^4}{2} + \frac{x^6}{3} - \frac{x^8}{4} + \cdots\right)$

19. $2x - \frac{x^2}{2} - \frac{x^4}{4} + \frac{2x^5}{5} - \frac{x^6}{6} - \frac{x^8}{8} + \frac{2x^9}{9} - \cdots$ **21.** $2e^{5\pi j/6}$ **23.** $3e^{\pi j/2}$ **25.** $2\sqrt{2}e^{3\pi j/4}$

Section 26.5 (page 880)

1. 0.644234, max. error: -0.000016, 0.6442 **3.** 0.98477, max. error: 0.00004, 0.9848 **5.** 0.818667, max. error: 0.000067, 0.8187
7. 0.36225, max. error: 0.0001, 0.362 **9.** 0.095333, max. error: -0.000025, 0.0953 **11.** 3.14 **13.** 0.06185 **15.** 0.76355
17. 0.29124 **19.** 0.4848 **21.** 0.85717 **23.** 0.5299 **31.** 0.3413

Section 26.6 (page 891)

1. $\frac{1}{2} + \frac{2}{\pi}\left(\sin \pi t + \frac{1}{3}\sin 3\pi t + \frac{1}{5}\sin 5\pi t + \cdots\right)$ **3.** $\frac{1}{2} + \frac{2}{\pi}\left(\sin \frac{\pi t}{5} + \frac{1}{3}\sin \frac{3\pi t}{5} + \frac{1}{5}\sin \frac{5\pi t}{5} + \cdots\right)$

5. $\frac{2a}{\pi}\left(\sin \frac{\pi t}{a} - \frac{1}{2}\sin \frac{2\pi t}{a} + \frac{1}{3}\sin \frac{3\pi t}{a} - \cdots\right)$ **7.** $\frac{a}{4} - \frac{2a}{\pi^2}\left(\cos \frac{\pi t}{a} + \frac{1}{3^2}\cos \frac{3\pi t}{a} + \frac{1}{5^2}\cos \frac{5\pi t}{a} + \cdots\right) + \frac{a}{\pi}\left(\sin \frac{\pi t}{a} - \frac{1}{2}\sin \frac{2\pi t}{a} + \frac{1}{3}\sin \frac{3\pi t}{a} - \cdots\right)$

9. $\frac{1}{\pi} + \frac{1}{2}\sin t - \frac{2}{\pi}\left(\frac{1}{3}\cos 2t + \frac{1}{15}\cos 4t + \frac{1}{35}\cos 6t + \frac{1}{63}\cos 8t + \cdots\right)$

Review Exercises for Chapter 26 (page 891)

1. $\frac{3}{4}$ **2.** $\frac{131}{990}$ **3.** div.   **4.** conv.   **5.** conv.   **6.** conv.   **7.** div.   **8.** conv.   **9.** conv.   **10.** div.

13. $\sin x^2 = x^2 - \frac{x^6}{3!} + \frac{x^{10}}{5!} - \cdots$ **14.** $\cos \sqrt{x} = 1 - \frac{x}{2!} + \frac{x^2}{4!} - \cdots$ **15.** $-1 - \frac{x}{2!} - \frac{x^2}{3!} - \cdots$ **16.** $\sinh x = x + \frac{x^3}{3!} + \frac{x^5}{5!} + \cdots$

17. 0.877604, max. error: -0.000022, 0.8776   **18.** 0.94609   **19.** 0.4994   **20.** 0.100   **21.** 0.10 C   **22.** 0.8387   **23.** 0.69466

Chapter 27

Section 27.2 (page 896)

17. $y = x^3 - 3$ **19.** $y = \tan x$ **21.** $y = e^x + (2 - e)x - 1$

Section 27.3 (page 901)

1. $2x^3 + 3y^2 = c$ **3.** $6x + 2x^3 = 9y^2 + c$ **5.** $\ln(1 + x^2) + y = c$ **7.** $\ln|y| + \text{Arctan } x = c$ **9.** $x^2 y^3 = c$ **11.** $xy^2 - 2xy - 2 = cx$
13. $x = c(1 - y)$ **15.** $PV = c$ **17.** $2 \tan x + 4y - y^2 = c$ **19.** $3y^2 - 2\cos^3 t = c$ **21.** $t\sqrt{v^2 + 1} - 1 = ct$
23. $2\sin T_1 - 2T_1 \cos T_1 + 2T_2 + T_2{}^2 = c$ **25.** $(y^2 - 1)\sin x = c$ **27.** $xe^x - e^x - e^{-y} = c$ **29.** $(e^x + 1)\sec y = c$
31. $y \sin y + \cos y + x^2 = c$ **33.** $y = 2x$ **35.** $(y + 2)(x - 3) = -7$ **37.** $x = \cos y$

Section 27.4 (page 907)

1. $y = 1 + ce^{-x}$ **3.** $y = e^{3x} + ce^{2x}$ **5.** $y = \frac{1}{2}x + \frac{c}{x}$ **7.** $xy = xe^x - e^x + c$ **9.** $y = x^2 \tan x + cx^2$

11. $y = x^3 \sin x - x^4 \cos x + cx^3$ **13.** $y = x^2 e^x + cx^2$ **15.** $y = 1 + ce^{\cos x}$ **17.** $4y \cos x = 2x + \sin 2x + c$

19. $y(x + 1) = x + 2\ln|x - 1| + c$ **21.** $y = -\frac{1}{2}x \cos x^2 + cx$ **23.** $y \sec t = \tan t + c$ **25.** $r = \frac{1}{2}t \ln t - \frac{1}{4}t + \frac{c}{t}$

27. $r = \frac{1}{3}s^2 e^{3s} - \frac{1}{9}se^{3s} + cs$ **29.** $y = 2e^{-x}(3x + 1)$ **31.** $y = 2 + e^{(-1/3)x^3}$ **33.** $i = \frac{E}{R}(1 - e^{-Rt/L})$

Section 27.5 (page 913)

1. $i = 1 - e^{-25t}$ **3.** $N = 100e^{-0.0223t}$ **5.** 15.2 h   **7.** 1380 years   **9.** 11,200 years   **11. (a)** \$2170.59   **(b)** 6.9%   **(c)** \$335.16
13. $T = 70 - 50e^{-0.1783t}$; 21.9 min   **15.** 42.2 min   **17.** 91 m/s; 500 m/s
19. $\ln[(5 + v)/(5 - v)] = 4t$ or $v = 5(1 - e^{-4t})/(1 + e^{-4t})$; $v = 5$ m/s   **21.** 80 s   **23.** $y = kx$ **25.** $2x^2 + y^2 = k$
27. $y^2 = x + k$ **29.** $y^2 - x^2 = k$

Review Exercises for Chapter 27 (page 914)

1. $y = x - 1 + ce^{-x}$ **2.** $3x - \frac{3}{x} + y^3 = c$ **3.** $y = \frac{1}{2} + ce^{-x^2}$ **4.** $y = -1 + ce^{x^2}$ **5.** $2\text{ Arctan } x + \ln(1 + y^2) = c$

6. $y\sin^2 x = -\cos x + c$ **7.** $x^2 y = c$ **8.** $x^3 + \ln|y - \tan y| = c$ **9.** $y = \frac{1}{2}x^4 + cx^2$ **10.** $y(\sec x + \tan x) = \cos x - x + c$

11. $x^2 + \csc^2 y = c$ **12.** $e^{2x} - 2ye^{-y} - 2e^{-y} = c$ **13.** $y(\sec x + \tan x) = 2(\sqrt{2} + 1)$ **14.** $y \ln y - y + \ln|\ln x| = c$ **15.** 3.2 h

16. 1590 years   **17.** $-29°F$ **18.** 14.4 min   **19.** $v = 60(1 - e^{-(1/6)t})$ **20.** $y - 4 = k(x + 3)$ **21.** $y = \frac{k}{x^2}$

Cumulative Review Exercises for Chapters 25–27 (page 915)

1. $\text{Arcsin} \dfrac{x}{2} + C$ **2.** $\dfrac{1}{2} \ln \left| \dfrac{2 - \sqrt{4 - x^2}}{x} \right| + C$ **3.** $-\sqrt{4 - x^2} + C$ **4.** $\dfrac{1}{6} \sin^3 2x - \dfrac{1}{10} \sin^5 2x + C$ **5.** $\dfrac{1}{6} \tan^2 3x + \dfrac{1}{3} \ln|\cos 3x| + C$

6. $\dfrac{1}{2} x - \dfrac{1}{16} \sin 8x + C$ **7.** $x \sin x + \cos x + C$ **8.** $\dfrac{1}{2} \sin x^2 + C$ **9.** $\ln|x^2 + x - 2| + C$ **10.** $\ln \left| \dfrac{x - 1}{x + 2} \right| + C$ **11.** $-e^{\cos x} + C$

12. $\sin x = x - \dfrac{x^3}{3!} + \dfrac{x^5}{5!} - \cdots$ **13.** 0.94608 **14.** $\dfrac{3}{2} - \dfrac{12}{\pi^2} \left(\cos \dfrac{\pi t}{3} + \dfrac{1}{3^2} \cos \dfrac{3\pi t}{3} + \dfrac{1}{5^2} \cos \dfrac{5\pi t}{3} + \cdots \right)$ **15.** $x = c(1 - 2\cos y)$

16. $\ln|\sin y| = \sin x + C$ **17.** $y = 1 + ce^{x^2}$ **18.** $y = (x + c) \csc x$ **19.** $\dfrac{\sqrt{2}}{2}$ A   **20.** 410 days (two significant digits)

21. 150 ft/s (two significant digits)

Chapter 28

Section 28.1 (page 922)

1. $y = c_1 e^{6x} + c_2 e^{7x}$ **3.** $y = c_1 e^{2x/3} + c_2 e^{-x/2}$ **5.** $y = c_1 e^{x/4} + c_2 e^{-2x}$ **7.** $y = e^{x/2}(c_1 e^{\sqrt{5}x/2} + c_2 e^{-\sqrt{5}x/2})$ **9.** $y = c_1 e^x + c_2 e^{x/2}$

11. $y = e^{3x} - e^{-3x}$ **13.** $y = \dfrac{e}{1 - e^3}(e^{-x} - e^{2x})$ **15.** $y = c_1 e^x + c_2 e^{2x} + c_3 e^{-3x}$ **17.** $y = c_1 e^{-x} + c_2 e^{(1 + \sqrt{3})x} + c_3 e^{(1 - \sqrt{3})x}$

19. $y = e^{-x/2}(c_1 e^{\sqrt{5}x/2} + c_2 e^{-\sqrt{5}x/2})$ **21.** $y = e^x(c_1 e^{\sqrt{3}x} + c_2 e^{-\sqrt{3}x})$ **23.** $y = e^{2x}(c_1 e^{\sqrt{6}x} + c_2 e^{-\sqrt{6}x})$ **25.** $y = e^{-3x}(c_1 e^{\sqrt{15}x} + c_2 e^{-\sqrt{15}x})$
27. $y = e^{-x}(c_1 e^{(\sqrt{2}/2)x} + c_2 e^{-(\sqrt{2}/2)x}$ **29.** $y = c_1 e^x + c_2 e^{-(2/3)x}$

Section 28.2 (page 925)

1. $y = c_1 e^{-3x} + c_2 x e^{-3x}$ **3.** $y = c_1 e^{(1/2)x} + c_2 x e^{(1/2)x}$ **5.** $y = c_1 e^{-(2/3)x} + c_2 x e^{-(2/3)x}$ **7.** $y = c_1 e^{(5/2)x} + c_2 x e^{(5/2)x}$

9. $y = e^{2x}(c_1 \cos x + c_2 \sin x)$ **11.** $y = e^{-2x}(c_1 \cos 2x + c_2 \sin 2x)$ **13.** $y = e^{(1/2)x}\left(c_1 \cos \dfrac{1}{2}x + c_2 \sin \dfrac{1}{2}x \right)$

15. $y = c_1 \cos 5x + c_2 \sin 5x$ **17.** $y = c_1 e^{3x} + c_2 x e^{3x}$ **19.** $y = c_1 + c_2 x + c_3 x^2 + c_4 e^{-2x}$ **21.** $y = e^{-x/2}\left(c_1 \cos \dfrac{\sqrt{7}}{2}x + c_2 \sin \dfrac{\sqrt{7}}{2}x \right)$

23. $y = e^{3x/2}\left(c_1 \cos \dfrac{\sqrt{11}}{2}x + c_2 \sin \dfrac{\sqrt{11}}{2}x \right)$ **25.** $y = e^x\left(c_1 \cos \dfrac{\sqrt{6}}{2}x + c_2 \sin \dfrac{\sqrt{6}}{2}x \right)$ **27.** $y = e^{-x}(c_1 e^{(\sqrt{6}/2)x} + c_2 e^{-(\sqrt{6}/2)x})$

29. $y = c_1 e^{10x} + c_2 e^{-10x}$ **31.** $y = c_1 \cos 10x + c_2 \sin 10x$ **33.** $y = c_1 + e^{x/3}\left(c_2 \cos \dfrac{\sqrt{2}}{3}x + c_3 \sin \dfrac{\sqrt{2}}{3}x \right)$

35. $y = e^{2x}(c_1 e^{\sqrt{2}x} + c_2 e^{-\sqrt{2}x})$ **37.** $y = c_1 \cos 2x + c_2 \sin 2x$ **39.** $y = c_1 + c_2 e^{2x} + c_3 x e^{2x}$

41. $y = c_1 e^x + e^{x/2}\left(c_2 \cos \dfrac{\sqrt{3}}{2}x + c_3 \sin \dfrac{\sqrt{3}}{2}x \right)$ **43.** $y = c_1 e^{2x} + c_2 x e^{2x} + c_3 x^2 e^{2x} + c_4 x^3 e^{2x}$ **45.** $y = \sin x$ **47.** $y = -e^x \sin x$

49. $y = c_1 \cos 3x + c_2 \sin 3x + c_3 x \cos 3x + c_4 x \sin 3x$

Section 28.3 (page 932)

1. $y = c_1 e^{3x} + c_2 x e^{3x} + \dfrac{1}{4} e^x$ **3.** $y = c_1 e^{3x} + c_2 x e^{3x} + x + \dfrac{2}{3}$ **5.** $y = c_1 e^{2x} + c_2 e^{-x} - x^2 + x - \dfrac{3}{2}$

7. $y = e^{(1/2)x}\left(c_1 \cos \dfrac{\sqrt{3}}{2}x + c_2 \sin \dfrac{\sqrt{3}}{2}x \right) - x^2 - 2x + 1$ **9.** $y = c_1 e^{-x} + c_2 e^{-3x} + \dfrac{1}{8} e^x + 2$ **11.** $y = c_1 \cos x + c_2 \sin x - 2 \sin 2x$

13. $y = c_1 e^{-2x} + c_2 e^{-3x} - \dfrac{1}{5} \cos x + \sin x$ **15.** $y = c_1 e^x + c_2 x e^x + 10$ **17.** $y = c_1 e^x + c_2 e^{-x} + c_3 e^{2x} + e^{3x}$

19. $y = e^x(c_1 \cos 2x + c_2 \sin 2x) + x e^x$ **21.** $y = \dfrac{1}{2}(\cos 3x + e^{3x})$ **23.** $y = 5 \cos x + \sin x - 2 \cos 2x$

25. $y = e^{-x}(3 \cos 2x + 4 \sin 2x) + 2 \cos x + \sin x$ **27.** $y_p = -2 \sin 3x$ **29.** $y_p = \dfrac{1}{2} \cos x - \dfrac{1}{2} \sin x$ **31.** $y_p = -x^2 - x - 2$

33. $y_p = \dfrac{1}{3} e^{2x} + 1$ **35.** $y_p = -\dfrac{1}{4} x + \dfrac{1}{16} + e^{3x}$ **37.** $y_p = \dfrac{2}{3} x + \dfrac{1}{2} \cos x$ **39.** $y_p = -2x e^x - \dfrac{2}{3} e^x$ **41.** $y = c_1 e^{-2x} + c_2 e^{2x} + 2x e^{2x}$

43. $y = c_1 e^{3x} + c_2 e^{-2x} - 2xe^{-2x}$ **45.** $y = c_1 e^{-7x} + c_2 e^{4x} + xe^{4x}$ **47.** $y = c_1 e^{-x} + (c_2 + x)e^x$ **49.** $y = c_1 e^{-2x} + c_2 e^{2x} + \dfrac{1}{4} xe^{2x} - 1$

51. $y = c_1 \cos x + c_2 \sin x - x \cos x$

Section 28.4 (page 941)

1. $x(t) = \dfrac{1}{4} \cos 8t$ **3.** $x(t) = \dfrac{5}{28} \cos 8t + \dfrac{1}{14} \cos 6t$ **5.** $x(t) = \dfrac{2}{3} \cos 4t + \dfrac{3}{4} \sin 4t$ **7.** $x(t) = \dfrac{2}{3} \cos 4t - \sin 4t$

9. $x(t) = e^{-2t}\left(\dfrac{1}{2} \cos 2\sqrt{3}t + \dfrac{1}{2\sqrt{3}} \sin 2\sqrt{3}t \right)$ **11.** $x(t) = \dfrac{1}{2} e^{-8t} + 4te^{-8t} - \dfrac{1}{4} \cos 8t$ **13.** $x(t) = 0.25 \cos 10t$

15. $x(t) = e^{-t}\left(0.25 \cos \sqrt{99}t + \dfrac{0.25}{\sqrt{99}} \sin \sqrt{99}t \right)$ or $x(t) = e^{-1.0t}(0.25 \cos 9.9t + 0.025 \sin 9.9t)$ **17.** $x_c = e^{-1.0t}(0.27 \cos 9.9t - 0.039 \sin 9.9t)$
$x_p = -0.017 \cos 5t + 0.13 \sin 5t$

19. $q = 0.1 \sin 100t, \; i = 10 \cos 100t$ **21.** $q = \dfrac{2}{75} \sin 50t - \dfrac{1}{75} \sin 100t$ **23.** $x_p = 0.20 \sin 5t - 0.029 \cos 5t$

25. $q_p = \dfrac{1}{2} \sin 10t, \; i_p = 5 \cos 10t$

Review Exercises for Chapter 28 (page 942)

1. $y = c_1 + c_2 x + c_3 x^2 + c_4 x^3$ **2.** $y = c_1 e^{2x} + c_2 e^{-2x}$ **3.** $y = c_1 \cos 2x + c_2 \sin 2x$ **4.** $y = c_1 e^{4x} + c_2 xe^{4x}$

5. $y = e^x(c_1 e^{\sqrt{3}x} + c_2 e^{-\sqrt{3}x})$ **6.** $y = e^x(c_1 \cos x + c_2 \sin x)$ **7.** $y = e^{(1/6)x}\left(c_1 \cos \dfrac{\sqrt{11}}{6} x + c_2 \sin \dfrac{\sqrt{11}}{6} x \right)$

8. $y = e^{-x/4}\left(c_1 \cos \dfrac{1}{4}\sqrt{15}x + c_2 \sin \dfrac{1}{4}\sqrt{15}x \right)$ **9.** $y = c_1 e^{2x} + c_2 xe^{2x} + c_3 \cos x + c_4 \sin x$

10. $y = c_1 + c_2 x + c_3 x^2 + c_4 \cos 3x + c_5 \sin 3x$ **11.** $y = e^{(1/4)x}\left(c_1 \cos \dfrac{\sqrt{7}}{4} x + c_2 \sin \dfrac{\sqrt{7}}{4} x \right)$ **12.** $y = e^{(1/2)x}(c_1 e^{(\sqrt{3}/2)x} + c_2 e^{-(\sqrt{3}/2)x})$

13. $y = e^{(1/3)x}(c_1 e^{(\sqrt{7}/3)x} + c_2 e^{-(\sqrt{7}/3)x})$ **14.** $y = e^x(c_1 \cos x + c_2 \sin x)$

In Exercises 15–20, $y_c = c_1 e^{4x} + c_2 e^{-x}$

15. $y = y_c - e^x$ **16.** $y = y_c - 2x + \dfrac{3}{2}$ **17.** $y = y_c + \dfrac{3}{17} \cos x - \dfrac{5}{17} \sin x$ **18.** $y = y_c - 4xe^{-x}$ **19.** $y = y_c - \dfrac{2}{5} xe^{-x} - x^2 e^{-x}$

20. $y = y_c + \dfrac{1}{5} xe^{4x}$ **21.** $y = e^{-x} - e^{3x}$ **22.** $y = 2 \cos x$ **23.** $i(t) = -1.08 \sin 20.0t + 1.88 \cos 20.0t$

25. $x(t) = \dfrac{7}{30} \cos 8t - \dfrac{5}{8} \sin 8t + \dfrac{1}{60} \cos 4t$

Chapter 29

Section 29.1–29.3 (page 952)

3. $\dfrac{5s + 2}{s(s + 1)}$ **5.** $\dfrac{s^3 + s^2 + 4}{s^2(s^2 + 4)}$ **7.** $\dfrac{5}{(s - 2)^2 + 25}$ **9.** $\dfrac{6}{(s + 4)^4}$ **11.** $\dfrac{48}{(s + 1)^5}$ **13.** $\dfrac{4}{s} - \dfrac{10}{s^2 + 4}$ **15.** $5 \sin 2t$ **17.** $\dfrac{1}{2} t^2 e^{-2t}$

19. $\dfrac{1}{4} (\sin 2t - 2t \cos 2t)$ **21.** $\left(t + \dfrac{1}{8} \right) \sin 2t - \dfrac{1}{4} t \cos 2t$ **23.** $5 \cos \sqrt{6}t$ **25.** $\dfrac{1}{2} t - \dfrac{1}{4} \sin 2t$ **27.** $2e^{-2t} \cos 2t$ **29.** $\dfrac{1}{\sqrt{5}} e^{-3t} \sin \sqrt{5}t$

31. $e^t \sin \sqrt{10}t$ **33.** $e^{3t}(\cos t + 3 \sin t)$ **35.** $e^t(\cos \sqrt{5}t + \dfrac{1}{\sqrt{5}} \sin \sqrt{5}t)$ **37.** $1 - e^{-t}$ **39.** $e^{2t} + e^{-3t}$ **41.** $\dfrac{1}{6} e^{-2t} - \dfrac{1}{2} e^{2t} + \dfrac{4}{3} e^{4t}$

43. $\dfrac{8}{3} e^{-2t} - 4te^{-2t} + \dfrac{1}{3} e^t$ **45.** $\dfrac{1}{2} e^{-t} - \dfrac{1}{2} \cos t + \dfrac{1}{2} \sin t$ **47.** $-\dfrac{1}{2} e^{-t} + \dfrac{1}{2} \cos t + \dfrac{1}{2} \sin t$ **49.** $\dfrac{1}{2} \cos t - \dfrac{1}{2} e^t + \dfrac{1}{2} te^t$ **51.** $e^{-t} - te^{-t}$

53. $e^{-t} \cos t + e^t$ **55.** $\dfrac{1}{7} e^{-4t} - \dfrac{1}{7} e^{-2t} \cos \sqrt{3}t + \dfrac{2\sqrt{3}}{21} e^{-2t} \sin \sqrt{3}t$

Section 29.4 (page 956)

1. e^t **3.** $2e^{2t} - 2$ **5.** te^{2t} **7.** $\frac{1}{2}t^2 e^{-4t} + 3e^{-4t}$ **9.** $\cos 3t - \frac{2}{3}\sin 3t$ **11.** $\sin t - t\cos t$ **13.** $\frac{1}{5}e^t - \frac{1}{5}\cos 2t + \frac{2}{5}\sin 2t$

15. $t - \frac{1}{2}\sin 2t + \cos 2t$ **17.** $e^{-3t}\left(\cos 2t + \frac{1}{2}\sin 2t\right)$ **19.** $2(e^t - \cos t - \sin t)$ **21.** $\frac{3}{10}e^{2t} + \frac{3}{10}e^{-2t} - \frac{3}{5}\cos t$

23. $e^t - e^{-t}\cos 2t - e^{-t}\sin 2t$ **25.** $q(t) = 0.03e^{-50t} - 0.15e^{-10t} + 0.12$

Review Exercises for Chapter 29 (page 957)

1. $\frac{2}{s+3}$ **2.** $\frac{4-s^2}{s(s^2+4)}$ **3.** $\frac{12}{s^4} + \frac{3}{s^2+9}$ **4.** $\frac{s+3}{(s+3)^2+36}$ **5.** $\frac{8}{s^2(s^2+4)}$ **6.** $\frac{120}{(s-3)^5}$ **7.** $e^{2t}\cos\sqrt{5}t$ **8.** $\frac{1}{6}t^3 e^{-3t}$

9. $e^t\left(\cos 2t + \frac{1}{2}\sin 2t\right)$ **10.** $\frac{1}{16}(\sin 2t - 2t\cos 2t)$ **11.** $\frac{1}{3}e^{-t} + \frac{2}{3}e^{2t}$ **12.** $\frac{4}{9}e^{-t} - \frac{1}{3}te^{-t} - \frac{4}{9}e^{-4t}$ **13.** $\frac{1}{30}e^{-2t} - \frac{1}{5}e^{3t} + \frac{1}{6}e^{4t}$

14. $e^{-2t} + 2\sin t - \cos t$ **15.** e^{-2t} **16.** te^t **17.** $e^{-2t}\left(\frac{1}{2}t^2 - 1\right)$ **18.** $\sqrt{2}\sin\sqrt{2}t$ **19.** $e^{-t} - e^{3t}$ **20.** $\frac{1}{2}(e^{-t} - \cos t + \sin t)$

21. $e^{-t}\left(\cos 2t + \frac{1}{2}\sin 2t\right)$ **22.** $\frac{1}{2}e^t \sin 2t$ **23.** $\frac{3}{5}e^{-2t} + \frac{2}{5}e^{-t}\cos 2t + \frac{13}{10}e^{-t}\sin 2t$ **24.** $\frac{8}{25}e^{-2t} + \frac{2}{5}te^{-2t} - \frac{8}{25}\cos t + \frac{6}{25}\sin t$

Cumulative Review Exercises for Chapters 28–29 (page 958)

1. $y = c_1 e^{5x} + c_2 e^{-4x}$ **2.** $y = e^{x/2}(c_1 e^{\sqrt{7}x/2} + c_2 e^{-\sqrt{7}x/2})$ **3.** $y = c_1 + c_2 x + c_3 x^2 + c_4 x^3 + c_5 e^x$ **4.** $y = c_1 e^{4x} + c_2 x e^{4x}$

5. $y = e^{2x}(c_1 \cos\sqrt{3}x + c_2 \sin\sqrt{3}x)$ **6.** $y = 3\cos x$ **7.** $y = c_1 \cos 3x + c_2 \sin 3x + \frac{3}{5}\cos 2x$ **8.** $y = c_1 e^{4x} + c_2 e^{-2x} - \frac{1}{2}x + \frac{1}{8} + \frac{1}{8}e^{2x}$

9. $y = \frac{1}{2} + \frac{5}{2}e^{-2t}$ **10.** $y = e^{3t} - 3e^t$ **11.** $y = 2e^{3t}\cos 3t - \frac{5}{3}e^{3t}\sin 3t$ **12.** $q(t) = 0..24 - 0.40e^{-10t} + 0.16e^{-25t}$

13. $y = c_1 e^{4x} + c_2 e^{-2x} - 2xe^{-2x}$

Chapter 30 (page 961)

1. -0.739085 **3.** 2.474577 **5.** 0.876726 **7.** $4.605551, -2.605551$ **9.** $6.474667, 0.754138, -1.228805$

Appendix A (page 965)

1. 3.142   **3.** 1.04   **5.** 0.00016   **7.** 986,000   **9.** 880,000   **11.** 0.020   **13.** 7.43   **15.** 11.096   **17.** 20.6   **19.** 13   **21.** 27
23. 32.3   **25.** 40   **27.** 160   **29.** 87.0   **31.** 0.650   **33.** 2.8   **35.** 5.3261   **37.** 53.5 Ω **39.** 0.17 W

Appendix B (page 972)

1. 650 mm   **3.** 736 mL   **5.** 0.046 g   **7.** 2,080 W   **9.** 0.206 s   **11.** 760 μs   **13.** 0.026 mΩ **15.** 0.0474 mF

17. 64,000,000 mΩ **19.** 81.5 cW   **21.** 43.3 in.   **23.** 49.74 cm^2 **25.** 1.34 $\frac{\text{km}}{\text{min}}$ **27.** 32.5 $\frac{\text{N}}{\text{cm}}$ **29.** 1.28 $\frac{\text{m}}{\text{s}}$ **31.** 5.4 $\frac{\text{lb}}{\text{in.}^2}$

33. $2.36 \times 10^{-4}\ \frac{\text{N}}{\text{cm}^3}$ **35.** 20,370 $\frac{\text{lb}}{\text{ft}^3}$ **37.** 1.77 ft^3 **39.** 8.683 in.^{3} **41.** 0.0740 $\frac{\text{km}}{\text{min}}$ **43.** 2.70 $\frac{\text{N}}{\text{cm}^2}$ **45.** 0.01050 $\frac{\text{lb}}{\text{in.}^2}$

Appendix C (page 979)

1.
```
1Ø PRINT "PYTHAGOREAN THEOREM"
2Ø INPUT "X="; X
3Ø INPUT "Y="; Y
4Ø LET Z = SQR(X↑2 + Y↑2)
5Ø PRINT "Z="; Z
6Ø END
```

2.
```
1Ø INPUT "A="; A
2Ø INPUT "B="; B
3Ø LET X = SQR(B/A)
4Ø PRINT X, −X
5Ø END
```
or
```
1Ø PRINT "ENTER THE COEFFICIENT OF THE SQUARE OF X"
2Ø INPUT A
3Ø PRINT "ENTER THE CONSTANT"
4Ø INPUT B
5Ø LET X = SQR(B/A)
6Ø PRINT X, −X
7Ø END
```

3. 1∅ LET M = ∅
   2∅ LET N = 1
   3∅ LET M = M + N↑3
   4∅ LET N = N + 1
   5∅ IF N < =8 THEN 3∅
   6∅ PRINT M
   7∅ END

4. 1∅ LET M = ∅
   2∅ FOR N = 1 TO 8
   3∅ LET M = M + N↑3
   4∅ NEXT N
   5∅ PRINT M
   6∅ END

5. 1∅ FOR X = ∅ TO 2 STEP ∅.∅5
   2∅ LET Y = SIN(2 ∗ X)
   3∅ PRINT X; Y
   4∅ NEXT X
   5∅ END

6. 1∅ FOR X = 1 TO 2 STEP ∅.1
   2∅ LET Y = LOG(X↑2)
   3∅ PRINT X; Y
   4∅ NEXT X
   5∅ END

7. 1∅ LET V = (4/3) ∗ 3.14159265 ∗ 2.47↑3
   2∅ LET S = 4 ∗ 3.14159265 ∗ 2.47↑2
   3∅ PRINT "V ="; V
   4∅ PRINT "S ="; S
   5∅ END

8. 1∅ INPUT "A ="; A
   2∅ INPUT "B ="; B
   3∅ INPUT "ANGLE ="; T
   4∅ LET C = SQR(A↑2 + B↑2 − 2 ∗ A ∗ B ∗ COS(T))
   5∅ PRINT "C ="; C
   6∅ END

9. 1∅ INPUT "A ="; A
   2∅ INPUT "B ="; B
   3∅ INPUT "C ="; C
   4∅ LET R = SQR(B↑2 − 4 ∗ A ∗ C)
   5∅ LET X1 = (−B + R)/(2 ∗ A)
   6∅ LET X2 = (−B − R)/(2 ∗ A)
   7∅ PRINT X1, X2
   8∅ END

10. 1∅ INPUT "R ="; R
   2∅ INPUT "ANGLE ="; T
   3∅ LET B = T ∗ 3.14159265/18∅
   4∅ LET A = ∅.5 ∗ R↑2 ∗ B
   5∅ PRINT "A ="; A
   6∅ END

Appendix D (page 986)

1. $65°, 155°$ **2.** $110°, 30°$ **3.** $45.6°, 38.4°, 96.0°$ **4.** $130°, 130°, 50°$ **5.** $34.6°, 40.2°$ **6.** 7.9 cm   **7.** 18.1 ft   **8.** $x = 4.92$ in.
9. $x = 1.8$ m   **10.** 127 ft   **11.** 24.9 ft   **12.** 24 cm   **13.** 12.7 mm   **14.** 18.6 in.   **15.** 22.43 ft   **16.** 25.3 in.   **17.** 80.0 yd^2
18. 19.72 cm^2 **19.** 36.4 mm^2 **20.** 0.250 in.^{2} **21.** 9.0 ft^2 **22.** 1,720 yd^2 **23.** 45 cm^2 **24.** 594.8 ft^3 **25.** 25,400 ft^3
26. 24.5 cm^3 **27.** 2,048 cm^3 **28.** 49 yd^3 **29.** 84.8 m^3 **30.** 151 in.^{3} **31.** 460 cm^3 **32.** 2,527 ft^3 **33.** 864 in.^{2}
34. 516 yd^2 **35.** 1,540 cm^2 **36.** 9.387 mm^2 **37.** 314 in.^{2} **38.** 140 m^2

I N D E X

Geometric Formulas

Areas

Square: $A = s^2$

Rectangle: $A = bh$

Triangle: $A = \dfrac{1}{2} bh$

Circle: $A = \pi r^2$

Sphere: $A = 4\pi r^2$

Volumes

Cube: $V = s^3$

Rectangular solid: $V = lwh$

Sphere: $V = \dfrac{4}{3} \pi r^3$

Cylinder: $V = \pi r^2 h$

Cone: $V = \dfrac{1}{3} \pi r^2 h$

Properties of Logarithms

$\log_b xy = \log_b x + \log_b y$

$\log_b \dfrac{x}{y} = \log_b x - \log_b y$

$\log_b x^r = r \log_b x$

$\log_b 1 = 0 \qquad \log_b b = 1$

$\log_b x = \dfrac{\log_a x}{\log_a b} \qquad$ (change of base)

Special Triangles

Angle Measurement

$1° = \dfrac{\pi}{180}$ radians

$1 \text{ radian} = \left(\dfrac{180}{\pi}\right)^{\circ}$

Integration Formulas

$\int [f(x) \pm g(x)]\, dx = \int f(x)\, dx \pm \int g(x)\, dx$

$\int cf(x)\, dx = c\int f(x)\, dx \quad$ if c is a constant

$\int u\, dv = uv - \int v\, du \quad$ (integration by parts)

$\displaystyle\int \dfrac{du}{u} = \ln |u| + C$

$\int e^u\, du = e^u + C$

$\int \cos u\, du = \sin u + C$

$\int \cot u\, du = \ln |\sin u| + C$

$\int \csc u\, du = \ln |\csc u - \cot u| + C$

$\int \csc^2 u\, du = -\cot u + C$

$\int \csc u \cot u\, du = -\csc u + C$

$\displaystyle\int \dfrac{du}{u^2 + a^2} = \dfrac{1}{a} \text{Arctan} \dfrac{u}{a} + C$

$\displaystyle\int u^n\, du = \dfrac{u^{n+1}}{n + 1} + C, \quad n \neq -1$

$\int a^u\, du = \dfrac{a^u}{\ln a} + C$

$\int \sin u\, du = -\cos u + C$

$\int \tan u\, du = \ln |\sec u| + C$

$\int \sec u\, du = \ln |\sec u + \tan u| + C$

$\int \sec^2 u\, du = \tan u + C$

$\int \sec u \tan u\, du = \sec u + C$

$\displaystyle\int \dfrac{du}{\sqrt{a^2 - u^2}} = \text{Arcsin} \dfrac{u}{a} + C$

$\displaystyle\int \dfrac{du}{u^2 - a^2} = \dfrac{1}{2a} \ln \left|\dfrac{u - a}{u + a}\right| + C$